Occupational Health

Recognizing and Preventing
Work-Related Disease and Injury

Fourth Edition

D1472917

Occupational Health

Recognizing and Preventing
Work-Related Disease and Injury

Fourth Edition

Editors

Barry S. Levy, M.D., M.P.H.
Director, Barry S. Levy Associates
Sherborn, Massachusetts
Adjunct Professor of Community Health
Tufts University School of Medicine
Boston, Massachusetts

David H. Wegman, M.D., M.Sc.
Professor and Chair
Department of Work Environment
University of Massachusetts Lowell
Lowell, Massachusetts

LIPPINCOTT WILLIAMS & WILKINS
A **Wolters Kluwer** Company
Philadelphia • Baltimore • New York • London
Buenos Aires • Hong Kong • Sydney • Tokyo

Acquisitions Editor: Joyce-Rachel John
Developmental Editor: Sonya L. Seigafuse
Production Editor: Steven Martin
Manufacturing Manager: Tim Reynolds
Cover Designer: Mark Lerner
Compositor: Bi-Comp
Printer: Maple Press

© 2000 by LIPPINCOTT WILLIAMS & WILKINS
227 East Washington Square
Philadelphia, PA 19106-3780 USA
LWW.com

Photographs copyrighted by Earl Dotter: Figures 1-1, 1-2, 1-3, 1-4, 2-2, 2-4, 2-5, 2-7, 3-3, 4-2, 4-3, 4-5, 5-1, 5-2, 5-4, 5-5, 8-1, 8-6(A), 9-2, 9-6, 9-7, 10-1, 12-1, 18-3, 19-2, 19-3, 21-2, 21-3, 21-4, 26-2, 29-3, 33-1, 36-2, 37-1, 37-2, 38-4, 39-3, 39-6, 39-7, and 40-2.

Photographs copyrighted by Ken Light: Figures 8-7, 37-3, 41-1, 41-2, and 42-5.

Printed in the USA

Library of Congress Cataloging-in-Publication Data

Occupational health : recognizing and preventing work-related disease and injury /
    edited by Barry S. Levy, David H. Wegman. — 4th ed.
        p. cm.
    Includes bibliographical references and index.
    ISBN 0-7817-1954-2 (pbk.)
    1. Medicine, Industrial.   I. Levy, Barry S.   II. Wegman, David H.
    [DNLM: 1. Occupational Diseases—prevention & control.   2. Occupational
Health.   WA 440 0149 2000]
RC963.022   2000
616.9'803—dc21
DNLM/DLC
for Library of Congress   99-35194
                                                                              CIP

10 9 8 7 6 5 4 3 2 1

To our families
Nancy, Laura, and Ben Levy
Bernice and Jerome Levy
Peggy, Marya, and Jesse Wegman
Myron Wegman
for their never-ending
encouragement and support
and to the memory of two wonderful people
Betty Petersen and Isabel Wegman

Contents

III. Hazardous Exposures

V. Selected Groups of Workers

Appendices

Contributors

Torbjörn Åkerstedt, Ph.D.
Professor
Institutet för Psykosocial Medicin (IPM) and Department
of Stress Research
Karolinska Institute
Box 230
17177 Stockholm, Sweden

Devendra Amre, M.D., M.Sc.
Joint Departments of Epidemiology and
Biostatistics, and Occupational Health
McGill University
Faculty of Medicine
1020 Pine Avenue West
Montreal, Quebec, Canada H3A 1A2

Gunnar B. J. Andersson, M.D., Ph.D.
Professor of Orthopedic Surgery
Rush Medical College
Chairman
Department of Orthopedic Surgery
Rush-Presbyterian-St. Luke's Medical Center
1653 West Congress Parkway
1471 Jelke
Chicago, Illinois 60612

Kenneth A. Arndt, M.D.
Professor of Dermatology
Harvard Medical School
Chief of Dermatology
Beth Israel Deaconess Medical Center
330 Brookline Avenue
Boston, Massachusetts 02215

Nicholas A. Ashford, Ph.D., J.D.
Professor of Technology and Policy
Center for Technology, Policy and Industrial
Development
Massachusetts Institute of Technology
77 Massachusetts Avenue, Room E40-239
Cambridge, Massachusetts 02139

Dean B. Baker, M.D., M.P.H.
Professor and Director
Center for Occupational and Environmental Health
University of California at Irvine
19722 MacArthur Boulevard
Irvine, California 92612

Edward L. Baker, Jr., M.D., M.P.H.
Assistant Surgeon General and Director
Public Health Practice Program Office
Centers for Disease Control and Prevention
1600 Clifton Road
Atlanta, Georgia 30333

Michael E. Bigby, M.D.
Assistant Professor of Dermatology
Harvard Medical School
Department of Dermatology
Beth Israel Deaconess Medical Center
330 Brookline Avenue
Boston, Massachusetts 02215

Leslie I. Boden, Ph.D.
Professor of Environmental Health
Boston University School of Public Health
715 Albany Steet T2E
Boston, Massachusetts 02118

Bill Borwegen, M.P.H.
Director, Occupational Health and Safety
Service Employees International Union, AFL-CIO
1313 L Street, NW
Washington, DC 20005

Marianne Parker Brown, M.P.H.
Director
UCLA-Labor Occupational Safety and Health (LOSH)
Program
School of Public Policy and Social Research
6350 B Public Policy Building
Los Angeles, California 90095-1478

Elise Caccappolo, Ph.D.
Post-doctoral Fellow
Department of Environmental and Community Medicine
Robert Wood Johnson Medical School
170 Frelinghuysen Road
Piscataway, New Jersey 08854

Dawn N. Castillo, Ph.D.
Supervisory Research Epidemiologist
Division of Safety Research
National Institute for Occupational Safety and Health
1095 Willowdale Road, MS P180
Morgantown, West Virginia 26505

David C. Christiani, M.D., M.P.H., M.S.
Professor of Occupational Medicine and Epidemiology
Harvard School of Public Health
665 Huntington Avenue
Boston, Massachusetts 02115
Associate Physician
Pulmonary and Critical Care Unit
Massachusetts General Hospital

James E. Cone, M.D., M.P.H.
Assistant Clinical Professor of Medicine
University of California, San Francisco
Chief, Occupational Health Branch
California Department of Health Services
1515 Clay Street, #1901
Oakland, California

Serge A. Coopman, M.D.
Head of Dermatology
Eeuwfeest Clinic
Belgielei 56
2018 Antwerpen, Belgium

Mark R. Cullen, M.D.
Professor of Medicine and Public Health
Director, Occupational Health Program
Yale University School of Medicine
135 College Street, 3rd Floor
New Haven, Connecticut 06510

Letitia Davis, Ph.D.
Director, Occupational Health Surveillance Program
Massachusetts Department of Public Health
250 Washington Street
Boston, Massachusetts 02108

Scott D. Deitchman, M.D., M.P.H.
Medical Officer, Office of the Deputy Director
National Institute for Occupational Safety and Health
Centers for Disease Control and Prevention
1600 Clifton Road, NE D32
Atlanta, Georgia 30333

Earl Dotter
Photographer
1714 Luzerne Avenue
Silver Spring, Maryland 20910

Derek E. Dunn, Ph.D.
Director, Division of Biomedical and Behavioral Science
National Institute for Occupational Safety and Health
Chief Scientific Officer
Office of the Surgeon General
Department of Health and Human Services
200 Independence Avenue SW
Room 7166, Humphrey Building
Washington, DC 20201

Ellen A. Eisen, Sc.D.
Professor
Department of Work Environment
University of Massachusetts Lowell
One University Avenue
Lowell, Massachusetts 01854

Anders Englund, M.D.
Medical Director
Swedish National Board of Occupational Safety and
 Health
Arbetarskyddsstyrelsen
Eklundsvagen 16
S-17184 Solna, Sweden

Richard A. Fenske, Ph.D., M.P.H.
Professor
Department of Environmental Health
Director, Pacific Northwest
Agricultural Safety and Health Center
Box 357234
Health Sciences Building, F-233
University of Washington
Seattle, Washington 98195

Nancy Fiedler, Ph.D.
Associate Professor
Department of Environmental and Community Medicine
University of Medicine and Dentistry of New Jersey
Robert Wood Johnson Medical School
170 Frelinghuysen Road, Room 210
Piscataway, New Jersey 08855

Lawrence J. Fine, M.D., Dr.P.H.
Director
Division of Surveillance, Hazard Evaluations, and Field
 Studies
National Institute for Occupational Safety and Health
 (NIOSH)
4676 Columbia Parkway
Cincinnati, Ohio 45226-1998

Will Forest, M.P.H.
Associate Toxicologist
Hazard Evaluation System and Information Service
Occupational Health Branch
California Department of Health Services
1515 Clay Street, Suite 1900
Oakland, California 94602

Robert Forrant, Ph.D.
Associate Professor
Department of Regional Economic and Social
 Development
University of Massachusetts Lowell
One University Avenue
Lowell, Massachusetts 01854

Linda Frazier, M.D., M.P.H.
Associate Professor
Department of Preventive Medicine
University of Kansas School of Medicine - Wichita
1010 North Kansas Avenue
Wichita, Kansas 67214

Howard Frumkin, M.D., Ph.D.
Professor and Chairman
Department of Occupational and Environmental Health
Rollins School of Public Health
Emory University
1518 Clifton Road, NE
Atlanta, Georgia 30322

Nelson M. Gantz, M.D.
Clinical Professor of Medicine
MCP Hahnemann School of Medicine
Chairman, Department of Medicine
Chief, Division of Infectious Diseases
Pinnacle Health Hospitals
2601 North Third Street
Harrisburg, Pennsylvania 17110

Lynn R. Goldman, M.D.
Principal Investigator
Pew Environmental Health Commission
Visiting Scholar
Health Policy and Management
Johns Hopkins University School of Hygiene and Public
* Health*
111 Market Place, Suite 850
Baltimore, Maryland 21202

Bernard D. Goldstein, M.D.
Professor and Chairman
Department of Environmental and Community Medicine
Robert Wood Johnson Medical School
Director
Environmental and Occupational Health Sciences
* Institute*
University of Medicine and Dentistry of New Jersey-
* Robert Wood Johnson Medical School*
170 Frelinghuysen Road
Piscataway, New Jersey 08854

John D. Groopman, Ph.D.
Professor and Chair
Department of Environmental Health Sciences
Johns Hopkins University School of Hygiene and Public
* Health*
615 North Wolfe Street
Baltimore, Maryland 21205

William E. Halperin, M.D., M.P.H.
Physician and Epidemiologist
Division of Surveillance, Hazard Evaluations, and Field
* Studies*
National Institute for Occupational Safety and Health
4676 Columbia Parkway
Cincinnati, Ohio 45226

Jay S. Himmelstein, M.D., M.P.H.
Professor
Department of Family and Community Medicine
University of Massachusetts Medical School
Director
Center for Health Policy and Health Services Research
University of Massachusetts Medical School
222 Maple Avenue/Higgins Building
Shrewsbury, Massachusetts 01545

Howard Hu, M.D., M.P.H., Sc.D.
Associate Professor of Environmental Health and
* Medicine*
Harvard School of Public Health
Associate Physician
Department of Medicine
Brigham and Women's Hospital
665 Huntington Avenue
Boston, Massachusetts 02115

Robert A. Karasek, Ph.D.
Professor
Department of Work Environment
University of Massachusetts Lowell
One University Avenue
Lowell, Massachusetts 01854

W. Monroe Keyserling, Ph.D.
Professor, Industrial and Operations Engineering
Director, Center for Occupational Health and Safety
* Engineering*
Department of Industrial and Operations Engineering
The University of Michigan
1205 Beal Avenue
Ann Arbor, Michigan 48109-2117

Howard M. Kipen, M.D., M.P.H.
Professor and Director
Division of Occupational Health
Robert Wood Johnson Medical School
Director
Division of Occupational Health
Environmental and Occupational Health Science Institute
170 Frelinghuysen Road
Piscataway, New Jersey 08854

Anders Knutsson, M.D., Ph.D.
Associate Professor
Department of Public Health and Clinical Medicine
University of Umeå
Chief
Department of Occupational Medicine
Umeå University Hospital
90185 Umeå, Sweden

Kathleen Kreiss, M.D.
Chief, Field Studies Branch
Division of Respiratory Disease Studies
National Institute for Occupational Safety and Health
1095 Willowdale Road, MS 234
Morgantown, West Virginia 26505

Charleen Kubota (APPA)
Acting Director
Occupational and Environmental Health Library
University of California Berkeley
Public Health Library No. 7360
Berkeley, California 9472-7360

Christopher J. Kuczynski, J.D., L.L.M.
Assistant Legal Counsel
Director, Americans with Disabilities Act Policy Division
U.S. Equal Employment Opportunity Commission
1801 L Street
Washington, DC 20507

Anthony D. LaMontagne, Ph.D.
Instructor in Occupational Health
Occupational Health Program
Harvard School of Public Health
Center for Community-Based Research
Dana-Farber Cancer Institute
44 Binney Street
Boston, Massachusetts 02115

Christopher T. Leffler, M.D., M.P.H.
Department of Environmental Health and Medicine
Harvard School of Public Health
Medical Director
NeuroMetrix, Inc.
One Memorial Drive
Cambridge, Massachusetts 02142

Richard A. Lemen, Ph.D.
Former Deputy Director and Former Acting Director
National Institute for Occupational Safety and Health
3495 Highgate Hiss Drive
Duluth, Georgia 30155

Nancy Lessin
Health and Safety Coordinator
Massachusetts AFL-CIO
8 Beacon Street
Boston, Massachusetts 02108

Charles Levenstein, Ph.D., M.S.
Professor
Department of Work Environment
University of Massachusetts Lowell
One University Avenue
Lowell, Massachusetts 01854

Barry S. Levy, M.D., M.P.H.
Director, Barry S. Levy Associates
20 North Main Street, No. 125
P.O. Box 1230
Sherborn, Massachusetts 01770
Adjunct Professor of Community Health
Department of Family Medicine and Community Health
Tufts University School of Medicine
Boston, Massachusetts 02111

Ken Light, B.G.S.
Photographer
55 Via Farallon
Orinda, California 94563

Jane Lipscomb, R.N., Ph.D.
Associate Professor
University of Maryland School of Nursing
655 W. Lombard Street
Baltimore, Maryland 20201

Michael McCann, Ph.D., CIH
Director of Ergonomics and Safety
The Center to Protect Workers' Rights
111 Massachusetts Ave., NW, 5th Floor
Washington, DC 20001

James Melius, M.D., Dr.P.H.
Director
NYS Laborers' Health and Safety Trust Fund
18 Corporate Woods Boulevard
Albany, New York 12211

David Michaels, Ph.D., M.P.H.
Professor of Epidemiology
Department of Community Health and Social Medicine
The City University of New York Medical School
138th Street and Convent Avenue
New York, New York 10031

Franklin E. Mirer, Ph.D., CIH
Director
Health and Safety Department
United Autoworkers International Union
8000 East Jefferson Avenue
Detroit, Michigan 48214

Rafael Moure-Eraso, Ph.D., CIH
Associate Professor
Department of Work Environment
University of Massachusetts Lowell
One University Avenue
Lowell, Massachusetts 01854

Linda Rae Murray, M.D., M.P.H.
Medical Director
Woodlawn Health Center
6337 South Woodlawn Street
Chicago, Illinois 60637

David L. Parker, M.D., M.P.H.
Director, Occupational Health Programs
Physician
Park Nicollet Clinic
Minnesota Department of Health
717 Delaware Street, SE
Minneapolis, Minnesota 55440-9441

Maureen Paul, M.D., M.P.H.
Medical Director
Planned Parenthood
Boston, Massachusetts 02215

John M. Peters, M.D., Sc.D.
Professor of Preventive Medicine
University of Southern California School of Medicine
1540 Alcazar Street
Los Angeles, California 90033

Timothy J. Pizatella, M.S.
Deputy Director
Division of Safety Research
National Institute for Occupational Safety and Health
1095 Willowdale Road, MS P180
Morgantown, West Virginia 26505

Glenn S. Pransky, M.D., M.Occ.H.
Associate Professor
Department of Family Medicine and Community Health
University of Massachusetts Medical School
Director
Center for Disability Research
Liberty Mutual Research Center for Safety and Health
71 Frankland Road
Hopkinton, Massachusetts 01748

Laura Punnett, Sc.D.
Professor
Department of Work Environment
University of Massachusetts Lowell
One University Avenue
Lowell, Massachusetts 01854

Margaret M. Quinn, Sc.D., CIH
Associate Professor
Department of Work Environment
University of Massachusetts Lowell
One University Avenue
Lowell, Massachusetts 01854

Kathleen M. Rest, Ph.D., M.P.A.
Associate Professor
Occupational and Environmental Health Program
Department of Family and Community Medicine
University of Massachusetts Medical School
55 Lake Avenue North
Worcester, Massachusetts 01655

Knut Ringen, Dr.P.H.
Public Health Researcher
Former Director
Center to Protect Workers' Rights
2610 SW 151 Place
Seattle, Washington 98166

Beth J. Rosenberg, Sc.D., M.P.H.
Assistant Professor
Department of Family Medicine and Community Health
Tufts University School of Medicine
136 Harrison Avenue
Boston, Massachusetts 02111

Thomas Schneider, M.Sc.
Director of Research
Department of Indoor Climate
National Institute for Occupational Health
105 Lerso Parkalle
2100 Copenhagen, Denmark

Jane L. Seegal, M.S.
Communications Director
The Center to Protect Workers' Rights
111 Massachusetts Avenue, NW, 5th Floor
Washington, DC 20001

Barbara Silverstein, Ph.D., M.P.H.
Research Director
Safety and Health Assessment and Research for
  Prevention (SHARP)
Washington State Department of Labor and Industries
P.O. Box 44330
Olympia, Washington 98504-4330

Michael Silverstein, M.D., M.P.H.
Assistant Director
Washington State Department of Labor and Industries
PO Box 44600
Olympia, Washington 98504-4600

Nancy J. Simcox, M.S.
Research Industrial Hygienist
Department of Environmental Health
University of Washington
Box 357234
Seattle, Washington 98195-4695

David H. Sliney, Ph.D.
Director, Laser Program
Center for Health Promotion and Preventive Medicine
Building E 1950
Aberdeen Proving Ground, Maryland 21010-5422

Thomas J. Smith, Ph.D.
Professor of Industrial Hygiene
Harvard University School of Public Health
665 Huntington Avenue
Boston, Massachusetts 02115

Emily A. Spieler, J.D.
Professor of Law
West Virginia University College of Law
P. O. Box 6130
Morgantown, West Virginia 26506

Nancy Stout, Ed.D.
Director
Division of Safety Research
National Institute for Occupational Safety and Health
1095 Willowdale Road, MS P180
Morgantown, West Virginia 26505

Charles P. Sweet, M.D., M.P.H.
Associate Professor
Department of Family Medicine and Community Health
University of Massachusetts Medical School
55 Lake Avenue North
Worcester, Massachusetts 01655-0309

Andrea Kidd Taylor, Dr.P.H., M.S.P.H.
Board Member
US Chemical Safety and Hazard Investigation Board
2175 K Street NW, Suite 400
Washington, DC 20037

Gilles Thériault, M.D., Dr.P.H.
Professor and Chair
Joint Departments of Epidemiology and Biostatistics,
  and Occupational Health
McGill University
1020 Pine Avenue West
Montreal, Quebec, Canada H3A 1A2

Nick Thorkelson
Graphic Artist
237A Holland Street
Somerville, Massachusetts 02144

Michael J. Thun, M.D., M.S.
Vice President for Epidemiology and Surveillance
  Research
American Cancer Society
1599 Clifton Road, NE
Atlanta, Georgia 30329-4251

Patricia Hyland Travers, Sc.M., M.S.,
  COHN-S
Corporate Health Strategies Manager
Compaq Computer Corporation
200 Forest Street, MR01-1/K28
Marlboro, Massachusetts 01752

Arthur C. Upton, M.D.
Clinical Professor of Environmental and Community
  Medicine
Department of Environmental and Community Medicine
Robert Wood Johnson Medical School
675 Hoes Lane
Piscataway, New Jersey 08854-5635

Terrence A. Valen
Research Associate
UCLA-Labor Occupational Safety and Health (LOSH)
  Program
School of Public Policy and Social Research
Institute of Industrial Relations
6350 B Public Policy Building
Box 951478
Los Angeles, California 90095-1478

Paul F. Vinger, M.D.
Clinical Professor of Ophthalmology
Tufts University School of Medicine
Ophthalmologist
Lexington Eye Associates
297 Heath's Bridge Road
Concord, Massachusetts 01742

Jia-Sheng Wang, M.D., Ph.D.
Research Assistant Professor
Department of Environmental Health Sciences
Johns Hopkins School of Hygiene and Public Health
615 North Wolfe Street, Room 1102
Baltimore, Maryland 21205

Richard P. Wedeen, M.D.
Associate Chief of Staff for Research and Development
New Jersey Department of Veterans Affairs Healthcare
  System
Professor
Departments of Medicine and Preventive Medicine and
  Community Health
UMDNJ-New Jersey Medical School
East Orange Campus
385 Tremont Avenue
East Orange, New Jersey 07018-1095

James L. Weeks, Ph.D.
Associate Research Professor
George Washington University School of Public Health
2300 K Street, NW, Suite 201
Washington, DC 20037

David H. Wegman, M.D., M.Sc.
Professor and Chair
Department of Work Environment
University of Massachusetts Lowell
One University Avenue
Lowell, Massachusetts 01854

John Wooding, Ph.D.
Professor
Department of Regional Economic and Social
  Development
University of Massachusetts Lowell
One University Avenue
Lowell, Massachusetts 01854

Susan R. Woskie, Ph.D.
Professor
Department of Work Environment
University of Massachusetts Lowell
One University Avenue
Lowell, Massachusetts 01854

Stephen Zoloth, Ph.D., M.P.H.
Associate Provost
Hunter College
695 Park Avenue
New York, New York 10021

Preface

The publication of this book on the cusp of the new century and the new millennium has been designed to focus attention both on the great achievements in occupational health in the past and the great challenges that still remain.

All occupational illnesses and injuries are preventable. As a result, each occupational illness and injury represents a failure of the system to have prevented it. This book is intended to provide information to health and safety professionals in many different disciplines and students in these disciplines, and to motivate and inspire them to recognize and prevent work related disease and injury.

The Chinese word for "crisis" has two symbols: one symbol stands for danger; the other, for opportunity. There are many dangers and opportunities that face health and safety professionals, workers, employers, and others as we begin the new century and the new millennium. These dangers and opportunities relate to many important changes that are taking place in the United States and other nations: changes in the nature of work and the evolution of various types of occupational hazards, changes in the workforce, advances in science and technology, changes in the organization and financing of health care and other health and human services, the information and communications revolution, the biotechnology and genetics revolution, the aging and the increasing diversity of populations, the globalization of the economy, the evolution of government structures and functions, the growth and roles of nongovernmental organizations, and major societal changes.

With this edition, we and our contributing authors have updated chapters that were present in the third edition of this book, and have added six new chapters on: environmental health (and its relationship to occupational health), occupational health services, injuries, young (child and adolescent) workers, older workers, and health care workers — reflecting increasing recognition of the importance of these subjects. In order to recognize appropriately the importance that both acute and chronic occupational injuries have for workers and for occupational health and safety professionals, we have also added the words "and injury" to the subtitle of the book.

Although this book focuses on many occupational health problems in the United States, it includes many examples from other countries — ranging from workers' compensation systems to child labor policies, and is designed for use by health and safety professionals and students in health and safety professions throughout the world. Occupational health and safety problems remain great in the United States and other more developed nations. However, the magnitude and severity of these problems are substantially greater in the "newly industrializing" and less developed nations. We all bear a responsibility to address these problems, especially in nations where the needs are greatest and the resources to deal with them are often meager. We should ensure that unsafe equipment and hazardous substances, such as pesticides, that are banned or restricted for use in more developed nations are not exported to nations that are economically disadvantaged.

The book is divided into five parts. Part I provides an overview of occupational health, including from a social perspective, and also extensive information on environmental health and its relationship to occupational health. Part II focuses on the recognition of occupational disease and injury, at both the individual and the group, or population, level, as well as various approaches to its prevention. In addition, this section includes chapters on workers' compensation, disability evaluation, ethics, and occupational health services. Part III presents information on various types of chemical, physical, biological, and psychosocial hazards in the workplace and how these hazards can be reduced or eliminated. Part IV considers injuries and disorders by organ system, with emphasis on clinical features and prevention of these disorders. Part V focuses on selected groups of workers who are especially vulnerable to occupational illnesses and injuries as a result of employment patterns, and the nature of work hazards for agricultural, construction, and health care workers. Appendices provide guidance on how to obtain additional information on workplace hazards and on organizations working in occupational health and related fields.

Information alone will not prevent occupational diseases and injuries. Prevention also depends, in part, on developing the popular and political will, to support it and to implement specific preventive measures. Our society woefully undervalues the importance of prevention. Informed health and safety professionals and students, through their values, vision, and leadership, can help develop the popular and political will to ensure that occupational diseases and injuries are indeed prevented.

B.S.L. and D.H.W.

Acknowledgments

We greatly appreciate the assistance and support of many people in the development of the fourth edition of *Occupational Health*. We thank the many chapter authors, whose work is appropriately credited within the text. In addition, there have been many other people working behind the scenes, to whom we are deeply grateful and appreciative.

Several people at Lippincott, Williams & Wilkins deserve credit for their outstanding work. We are especially grateful for the assistance of Joyce-Rachel John, acquisitions editor; Sonya Seigafuse, developmental editor; and Steven Martin, production editor.

The illustrative materials throughout the book are included to offer understanding and insights not easily gained from the text. We call special attention to the work of Earl Dotter, who provided many outstanding photographs of workers and workplaces. We are also grateful for the photographic contributions of Marvin Lewiton, Ken Light, David Parker, Elise Morse, Nick Kaufman, Joe LaDou, and others. We also thank graphic artist Nick Thorkelson for sharing his talent and perspective in a series of creative drawings that convey concepts and perspectives that are difficult to capture in words or photographs.

We are grateful to the following individuals, who critically reviewed draft chapters and concepts for this edition and made many helpful comments and suggestions that were incorporated into the final text: Gregory W. Wagner, L. Christine Oliver, John Peters, Robert Feldman, Bruce Bernard, Nina Wallerstein, Mark Cullen, Tom Hodous, Christer Hogstedt, and Eva Vingård.

We thank Lisa Campbell for her excellent secretarial assistance and support, and Nell Switzer.

Finally, we express our appreciation to students and colleagues who, over the years, have broadened—and continue to broaden—our understanding of occupational health.

B.S.L. and D.H.W.

PART 1

Work and Health

1 Occupational Health
An Overview

Barry S. Levy and David H. Wegman

Occupational health is the multidisciplinary approach to the recognition, diagnosis, treatment, and prevention and control of work-related diseases, injuries, and other conditions. It is part of public health—what we, as a society, do collectively to assure the conditions in which people can be healthy—and it is an integral part of many clinical disciplines, as illustrated by the following examples:

A pregnant woman who works as a laboratory technician asks her obstetrician if she should change her job or stop working because of the chemicals to which she and her fetus are exposed.

A middle-aged man sees an orthopedic surgeon and states that he is totally disabled from chronic back pain, which he attributes to lifting heavy objects for many years as a construction worker.

A long-distance truck driver asks a cardiologist how soon after his recent myocardial infarction will he be able to return to work, and what kinds of tasks he will be able to perform.

A chemical manufacturer, aware that a pesticide that they produce is carcinogenic and has recently been banned from sale in the United States, makes arrangements for the export of the pesticide for sale and use in developing countries

A former asbestos worker with lung cancer asks his surgeon if he can submit a claim for workers' compensation for his disease.

An oncologist observes an unusual cluster of bladder cancer cases among middle-aged women in a small town.

The vice president of a small tool and die company asks his family physician to advise his company regarding prevention of occupational disease among his employees.

A pediatric nurse practitioner diagnoses lead poisoning in a young child and wonders if the source of the lead may be due to dust brought home on the workclothes of the father, who works in a battery plant

A United States–based physician epidemiologist visits a developing country to advise its occupational health program on identifying and controlling workplace hazards.

Three women who work for a plastics company, all complaining of severe rashes on their hands and forearms, consult an internist, who believes their problems may be work related.

These are but a few examples of the numerous occupational health challenges facing health professionals. Virtually all health professionals need to be able to deal effectively with the challenges of occupational health.

THE UNITED STATES WORKFORCE

Work and the work ethic are basic to life throughout the world. Most adults in the

B. S. Levy: Barry S. Levy Associates, Sherborn, Massachusetts 01770.

D. H. Wegman: Department of Work Environment, University of Massachusetts Lowell, Lowell, Massachusetts 01854.

FIG. 1-1. Workers who perform repetitive tasks at work are at risk for cumulative trauma disorders. This woman, a data entry clerk in Virginia, contracted carpal tunnel syndrome as a result of work at a computer keyboard. (Photograph by Earl Dotter.)

FIG. 1-2. Health care workers, including these laundry workers in New York, face a number of occupational hazards, including human immunodeficiency virus, hepatitis B, hepatitis C, and other infections associated with needlestick injuries. These laundry workers found these sharps in soiled bed linens over the course of a year. (Photograph by Earl Dotter.)

TABLE 1-1. *Employed civilians in the United States, by industry (1998)*

Industry	Size of workforce (in millions)
Services	47.2
Wholesale and retail trade	27.2
Manufacturing	20.7
Finance, insurance, real estate	8.6
Transportation and public utilities	9.3
Construction	8.5
Agriculture, forestry, and fishing	3.5
Mining	0.6
Public administration	5.9
Total	131.5

From the Bureau of Labor Statistics, U.S. Department of Labor, 1999. Available: http://www.bls.gov.

United States spend almost one-fourth of their time at work, and despite the high degree of automation and computerization of American industry, many workers are exposed to safety and ergonomic hazards (Fig. 1-1; see Chapters 8, 9, 24, and 26); chemical hazards, such as lead and other heavy metals, benzene and other organic solvents, dusts, and pesticides (see Chapters 15 and 16); physical hazards, such as radiation and noise

TABLE 1-2. *Employed civilians in the United States, by occupational category (1998)*

Occupational category	Number of workers (in millions)
Administrative support workers, including clerical workers	18.4
Operators, fabricators, and laborers	18.3
Professional workers	19.9
Service workers	17.8
Executive, administrative, and managerial workers	19.0
Sales workers	15.9
Precision production, craft, and repair workers	14.4
Technicians and related support workers	4.2
Farming, forestry, and fishing industry workers	3.5
Total	131.5[a]

[a] Because of rounding, the sum of the components does not add up to the total.

From the Bureau of Labor Statistics, U.S. Department of Labor, 1999. Available: http://www.bls.gov.

(see Chapters 17 through 19); biologic hazards, including hepatitis B virus, human immunodeficiency virus, and the tubercle bacillus (Fig. 1-2; see Chapter 20); and psychosocial hazards, such as stress and the hazards of shift work (see Chapters 21 and 22). Exposure to such hazards can cause disorders of organ systems throughout the body (see Chapters 25 through 35).

The U.S. civilian workforce was approximately 130 million in 1997. Table 1-1 categorizes workers in the United States by industry and Table 1-2 by occupation. They work in a wide range of jobs, from shop floors in traditional heavy manufacturing plants (Fig. 1-3), to mines (Fig. 1-4), to clean rooms in the semiconductor industry (Fig. 1-5). An

FIG. 1-3. Although a declining percentage of workers in the United States work in heavy manufacturing, such as this worker at a wheel stamping plant in Michigan, manufacturing still represents a major part of the economy and a source of many occupational health and safety hazards. (Photograph by Earl Dotter.)

FIG. 1-4. Roof bolting in coal mines is essential to prevent roofs from collapsing. Miners, like this man testing the mine roof support bolts in a Pennsylvania coal mine, face many other injury risks as well as exposure to hazardous dusts, gases, and other substances. (Photograph by Earl Dotter.)

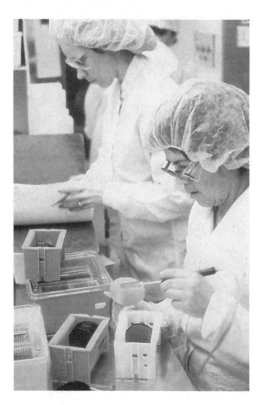

FIG. 1-5. Semiconductor industry workers in clean rooms may still be exposed to many hazardous substances, such as glycol ethers, which may be associated with adverse reproductive outcomes. (Photograph by Joe LaDou.)

Box 1-1. Workers Caught in a Squeeze

Robert Forrant

For several years, new forms of work organization have been introduced by U.S. management to cut labor costs, improve productivity, and increase their shop floor control. Once seemingly secure jobs in diverse industries like aviation, jet engines, machine tools, and computer chips are no longer so stable. Rather than embrace true partnerships on the shop floor, managers move only as far as they are forced in order to capture workers' production knowledge. It is not a virtuous turn on the part of managers, but the imperative of global competition that has forced many corporations to reconsider their approach to shop floor management to make some space for participation schemes. However, the outcomes for workers are neither job enrichment nor even a modicum of workplace security.

Scholars of Japanese manufacturing contend that much of the success enjoyed by firms there in the 1970s and 1980s was derived from the full engagement of the hands and heads of front-line employees. This requires a workforce using all of its faculties and a management orientation that considers worker intellect to be an asset, not a liability. But in the drive to maximize production and add to shareholder value, empowerment and team building easily fall victim to output demands and "line speed-up."

Yet, the forces of globalization have placed workers and their unions squarely between a rock and a hard place. They are damned as backward thinkers if they refuse to consider work changes that might give their plant a chance to prosper, and they are just as damned if they "go along" only to have managers "pick their brains" and pull the work, placing it where wages are cheaper.

A review of recent U.S. economic history helps us to understand the pressures unions and workers contend with. One in five American workers saw his or her job disappear during the recession of the early 1980s. The rules of the game have changed significantly since Congress passed three pieces of important legislation in the 1930s that buttressed the economy and worker protections. The National Labor Relations Act (1935) protected workers' rights to organize and bargain for contracts. The Fair Labor Standards Act (1938) set national standards for minimum pay and child labor, and The Social Security Act (1935) established a national retirement system (this section draws heavily on Herzenberg, Alic, and Wial, 1998). American industry, undamaged by the war, accounted for close to half of global manufacturing output in the mid-1950s. Productivity rose and workers enjoyed rising standards of living and the protections of the liberal state and Keynesian fiscal and monetary policy.

However, this choice position in the global economy was not sustainable as competition from Japan and Germany challenged U.S. preeminence in autos, steel, major appliances, and electronics. Between 1979 and 1983, employment in highly unionized durable goods declined by slightly over 2 million jobs, or 15.9%. For the Fortune 500 largest manufacturers between 1980 and 1990, employment dropped to 12.4 million from 15.9 million. Management threats to relocate work were often interjected during contract negotiations. Indeed, a precipitating factor in the momentous summer of 1998 United Auto Workers–General Motors strike was the removal of stamping dies from a factory that unionists rightly interpreted as an attack on any modicum of job security they had.

A profound consequence of downsizing has been the rapid increase in new employment relationships that place tremendous burdens on workers. Variously called contingent, contract, or part-time work, these arrangements are now spread across the manufacturing and service sector. In all cases, workers have a tenuous relationship to their job and lack any benefits. According to the U.S. Bureau of Labor Statistics, on-call and day laborers, temporary agency employees, contract workers, and independent contractors comprise 10% of service and 5% of manufacturing workers. Part-time work (less than 35 hours a week) constitutes approximately 25% of the jobs in the service sector and 5% of employment in manufacturing in the United States. Virtually all of these workers lack basic job security

Box 1-1   (*continued*)

and the range of benefits that most American workers once took for granted—and there is heightened anxiety among this growing number of people as they continuously scramble to secure more stable jobs.

Herzenberg and colleagues report that in 1996 "about three-quarters of all employed Americans worked in service industries, up from two-thirds in 1979" (p. 21). Giant retailer Wal-Mart is the fourth-largest U.S. corporation measured by revenues. It has created over a half a million new jobs since the late 1980s. Today, General Electric generates more than half its revenues from financial and other services. General Motors, Ford, Boeing, and General Electric collectively eliminated close to 250,000 jobs from 1990 and 1998. Income inequality has grown as well-paying jobs decline; adjusted for inflation, the median income of American employees in the mid-1990s was roughly 5% lower than it was in the late 1970s. With a global capacity glut in steel, automobile, computer chip, and aircraft production, the opening of Eastern Europe, and the downturn in Asia, in all likelihood these downward trends will continue.

In the context of these global changes, new, more flexible forms of work organization have been introduced by management in the United States and Europe to cut direct and indirect labor costs, improve productivity and quality, and achieve greater production flexibility in an effort to remain price competitive. "High performance practices," "employee involvement," "mutual gains enterprises," and "flexible work organizations" strive for what economist Bennett Harrison calls *functional flexibility*. Functional flexibility refers to the efforts of managers to redefine work tasks, redeploy resources, and reconfigure relationships with suppliers to achieve rapid product development and faster changeovers from one product to another. Corporations also availed themselves of computer-controlled machinery to eliminate large numbers of their blue-collar workers and to decrease their reliance on the tacit knowledge of these workers. Absent a concerted and collective global labor voice, global producers will maintain their substantial bargaining leverage, and worsening wages and increasingly stressful working conditions for workers will likely be the result.

increasing proportion of U.S. workers are in the service sector, including health care. Over 40% of U.S. workers are women—not accounting for full-time homemakers, who usually are not included in data on the U.S. workforce. Approximately 14% of American workers belong to unions, a lower percentage than in most industrialized nations; according to the Bureau of Labor Statistics, approximately 10% of wage and salary workers in private industry are affiliated with a union, compared with approximately 37% of government workers. Most U.S. workers are employed by small or moderate-sized firms that do not employ physicians or other health professionals to provide health and safety programs. Although, as of this writing, the U.S. economy is strong, with unemployment

officially reported below 5%, many workers feel caught in a squeeze (Box 1-1).

Approximately 2,400 U.S. physicians now in practice have been certified in Occupational Medicine by the American Board of Preventive Medicine. Therefore, most workers with work-related injuries and illnesses are treated by physicians or other clinicians who have little or no training in occupational health. Similar shortages exist for qualified occupational health nurses and other personnel in the field.

Although the workforce in Canada, Western Europe, and developed countries elsewhere has many similarities to the workforce in the United States, the same is not true for the workforce of developing countries—in the aggregate potentially much larger than the workforce of developed countries. In developing countries, the workforce has several distinct characteristics:

1. Most people who are employed work in the informal sector of the economy, mainly in agriculture (Fig. 1-6), or in small-scale industries, in which they are often self-employed or working in a small workplace.
2. There are high rates of unemployment, often reaching 25% or higher, and underemployment rates in many developing countries are increasing each year.
3. In general, workers are at greater risk of occupational hazards for a variety of reasons, including lower rates of education and literacy; unfamiliarity with work processes and exposures, in part due to inadequate training; predisposition not to complain about working conditions or exposures because jobs, even if they are hazardous, are relatively scarce; high prevalence of endemic (mainly infectious) diseases and malnutrition; and inadequate infrastructure and human resources to recognize, diagnose, treat, and prevent and control work-related illnesses and injuries.

4. Wages are low: annual per capita income in many developing countries is $500 (U.S.) or less per year.
5. Groups that comprise vulnerable populations in any country are at even greater risk, including:
   - Women, who make up a large proportion of the workforce in many developing countries and often face significant physical and psychological hazards in their work, in addition to the physical and psychological stresses of being homemakers and mothers. In the rural areas of many developing countries, from which men have left to seek jobs in the cities, women often do most of the physical labor in raising crops on small plots of land, in addition to raising children and maintaining the household (Fig. 1-7).
   - Children, sometimes even very young children, who account for a significant part of the workforce in many developing countries and often undertake some of the most hazardous tasks at

FIG. 1-6. Agricultural workers in developing countries are at high risk of poisoning from pesticides, including those banned or restricted in developed countries. (Photograph by Barry S. Levy.)

FIG. 1-7. Women in rural Kenya, carrying heavy loads of firewood. Rural women in many developing countries bear heavy physical and psychosocial stresses as men leave to seek jobs in the cities. (Photograph by Barry S. Levy.)

FIG. 1-8. Child laborers, such as this girl working as a garbage picker in India, are exploited and are exposed to many serious occupational health and safety hazards. (Photograph by David L. Parker.)

work. In many of these countries, primary education is not required and there are no legal protections against child labor (Fig. 1-8).

• Migrants, both within countries and between countries, who, for a variety of reasons (such as uncertain legal status and unfamiliarity with the culture in and around the workplace), face significant health and safety hazards at work.

MAGNITUDE OF THE PROBLEM

An estimated 10 million work-related injuries and 430,000 new work-related illnesses occur each year in the United States. In developing countries, occupational injury and illness rates are much higher than in the United States. Each day in the United States, an average of 9,000 workers sustain disabling injuries on the job, 16 workers die from a workplace injury, and 137 workers die from work-related diseases. However, the number of occupational diseases and injuries reported is much lower. Table 24-1 analyzes reported injuries by industry. Reported illnesses by category are shown in Table 1-3. In addition, it is estimated that there may be 100,000 or more work-related deaths in the United States each year. According to the National Institute for Occupational Safety and Health (NIOSH) National Traumatic Occupational Fatality database, approximately 5,400 traumatic occupational fatalities occur in the United States each year; the highest rates are in mining (30 per 100,000 workers per year), construction (23 per 100,000; see Chapter 42), and agriculture (20 per 100,000; see Chapter 41). Although these statistics provide some idea of the scope and types of occupational medical problems, they grossly underestimate the role of the workplace in causing new diseases and injuries and exacerbating existing ones. In addition, statistics do not represent the relative distribution of various work-related diseases. For example, because skin disorders are easy to recognize and relate to working conditions, their representation in Table 1-3 exaggerates their relative importance.

A study funded by NIOSH and published

TABLE 1-3. *Distribution of new cases of reported occupational illnesses in the United States, by category of illness, private sector (1997)*

Category of illness	Number[a]	Percentage
Disorders associated with repeated trauma	276,600	64
Skin diseases or disorders	57,900	13
Respiratory conditions due to toxic agents	20,300	5
Disorders due to physical agents	16,600	4
Poisoning	5,100	1
Dust diseases of the lungs	2,900	1
All other occupational illnesses	50,600	12
Total	429,800	100

[a] Excludes farms with fewer than 11 employees.
From the Bureau of Labor Statistics, U.S. Department of Labor, 1999. Available: http://www.bls.gov.

Box 1-2. Twelve Myths about Work-Related Disease

John M. Peters

1. "It can't be a bad place to work. She's been working here 45 years and there's nothing wrong with her."
2. "It's only a statistical relationship. It cannot be of any clinical significance."
3. "Our working population is healthier than the general population."
4. "There is no problem—the exposure level is below the government standard."
5. "It will cost too much to control the problem—it will put us out of business."
6. "It's only a mortality study, and you know how inaccurate death certificates are."
7. "Okay, there is a problem. But you know how much smoking and drinking the workers do."
8. "Exposures are good for workers because it keeps up their defenses. If we stop exposing them, their defenses will break down."
9. "If there is a problem, it is because of all the former exposures that don't exist anymore."
10. "We have this party each year for retired workers, and they all look great."
11. "Don't tell the workers—they will only worry."
12. "If there is a problem, it would show up in their medical records because the workers would tell the nurses and doctors in the clinic about it."

---

in 1997 showed that in 1992, indirect and direct (including administrative) costs of occupational injuries were $145 billion, and of occupational illnesses, $26 billion: a total of $171 billion. (These costs can be compared with the costs of the following diseases for the same year in the United States: $33 billion for acquired immunodeficiency syndrome, $57 billion for Alzheimer's disease, $164 billion for circulatory diseases, and $171 billion for cancer.)

As indicated in Box 1-2, myths about work-related disease impede the identification of occupational health problems.

The difficulty in obtaining accurate estimates of the frequency of work-related diseases is due to several factors, as indicated below (see also Fig. 1-9):

1. Many problems do not come to the attention of health professionals and employers, and therefore are not included in data collection systems. A worker may not recognize a medical problem as being work related, but even when the connection is obvious, a variety of disincentives may deter the worker from reporting such a problem, the greatest being fear of losing a job. Training workers about both occupational hazards and legal rights has been helpful.

2. Many occupational medical problems that do come to the attention of physicians and employers are not recognized as work related. Recognition of work-related disease is often difficult because of the long period between initial exposure and onset of symptoms (or time of diagnosis), making cause–effect relationships difficult to assess. It is also difficult because of the many and varied occupational and nonoccupational hazards to which most workers are exposed. Training of health professionals in occupational health has begun to improve health care providers' knowledge of these factors, resulting in increased recognition of the work-relatedness of diseases and chronic injuries.

3. Some medical problems recognized by health professionals or employers as work related are not reported because the association with work is equivocal and because reporting requirements are not strict. The initiation of occupational disease surveillance activities by federal and state governments has begun to address this problem.

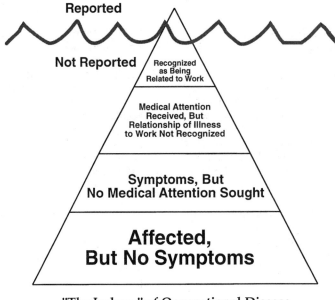

"The Iceberg" of Occupational Disease

FIG. 1-9. The iceberg of occupational disease.

4. Because many occupational medical problems are preventable, their very persistence implies that some individual or group is legally and economically responsible for creating or perpetuating them.

There are few good statistics on the occurrence of occupationally related illnesses and injuries in developing countries, or on the magnitude or degree of health and safety hazards at work. This inadequacy is a function of the inadequate infrastructure and human resources for occupational health and safety in developing countries, as well as the inherent difficulty in diagnosing and obtaining data on work-related health problems.

In those situations for which data are available, there is evidence that the occurrence of occupational illnesses and injuries in developing countries is much higher than in developed countries. The International Labor Organization (ILO), for example, collected data on annual rates of fatal injuries in 24 developing countries from 1971 to 1975 and 1976 to 1980. Although rates of fatal injuries at work decreased in developed countries during the 1970s, these rates stayed the same, or even increased, in developing countries; in almost half of the 24 countries studied, the rates increased between the first and second halves of the decade. In addition, ILO found that in 21 developing countries, nonfatal occupational injury rates rose 5.3% during 1976 to 1980, while they were declining elsewhere.

There are a variety of reasons for the poor occupational *safety* situation in developing countries, including use of outdated machinery, often imported in used condition from more developed countries; poor maintenance and little safety guarding of machinery; inadequate training of workers; poor design of equipment and workstations; and lack of personal protective equipment, which, even when available, may be difficult to wear because of working conditions or workers' health status. In addition, workers' underlying health may be compromised by endemic diseases, poor housing and sanitation, and inadequate nutrition.

The same picture is true for occupational *health* hazards in developing countries, and resultant occupational diseases. Pesticide exposures—often to highly toxic or carcinogenic pesticides that have been banned or restricted in developed countries that produce them and then export them to developing countries—and resultant cases of acute and chronic poisoning are a major problem. This problem is exacerbated by intensive use of pesticides, especially on coffee, cotton, and other "cash crops"; inadequate labeling on pesticide containers; inadequate training of workers on pesticide safety; inadequate personal protective equipment; and inadequate knowledge of health care providers about the diagnosis, treatment, and prevention of pesticide poisoning. In addition, pesticide formulation, mixing, and application are often done by small groups or individual workers, making control of this hazard even more difficult (see Fig. 31-1). A number of studies have confirmed the seriousness and magnitude of pesticide poisoning in developing countries (see also Chapter 41).

The problems of pesticide poisoning in developing countries are compounded by the numerous other chemicals and occupational health hazards to which workers are exposed. Exposure to organic and inorganic dusts, organic solvents, heavy metals, and other chemicals represents a serious problem; in some countries, for example, miners experience rates of pneumoconiosis of 10% to 12% or higher. Physical hazards, such as noise, biologic hazards related to exposure to infectious agents (see Box 20-1, Infectious Diseases in Developing Countries), and psychosocial problems directly related to work (e.g., work stress) and indirectly related to work (e.g., alcoholism) are also usually much more prevalent in developing countries than in developed ones.

In general two important aspects of occupational disease distinguish it from other medical problems:

1. *Recognition of work-related medical problems depends almost totally on ob-*

taining occupational information in the medical history. The health professional must know when to suspect work-related medical problems and what questions to ask in an occupational history to evaluate the work relatedness of that problem (see Chapter 4).

2. *In contrast to many nonoccupational diseases, occupational diseases can almost always be prevented.* Most occupational medical problems can be prevented by a variety of approaches in which professionals in health, safety, and other fields play vital roles (see Chapters 5 through 9).

ILLUSTRATIVE OCCUPATIONAL HEALTH ISSUES

Legislation, social activism, educational activities, and other developments have contributed to increased interest in work-related medical problems in recent years. Some of these developments are summarized in the following sections.

Governmental Role

With the passage of the Federal Coal Mine Safety and Health Act of 1969 and the Occupational Safety and Health Act (OSHAct) of 1970, the federal government began taking a more active role in the creation and enforcement of standards for a safe and healthful workplace (see Chapter 10). In addition, the OSHAct established NIOSH, which (a) has greatly expanded epidemiologic and laboratory research into the causes of occupational diseases and injuries and the methods of preventing them; and (b) is strengthening the training of occupational health and safety professionals.

Occupational Safety and Health Education

A variety of factors have contributed to a recent growth in education and training opportunities for workers, employers, health professionals, and others. Unions have directed more attention to occupational health and safety through collective bargaining agreements, hiring of health professionals, workplace health and safety committees, educational programs, and support of epidemiologic studies. Worker education has been facilitated by right-to-know laws and regulations, independent coalitions for occupational safety and health (COSH groups; see Appendix), and employer-sponsored programs, and academic institutions concerned with occupational health have improved existing professional training opportunities and created new ones. Furthermore, alerted to the critical problems associated with asbestos, lead, pesticides, and ionizing radiation, the mass media have made the public aware of many workplace hazards.

Social and Ethical Questions

Serious social and ethical problems have arisen over such subjects as the allegiance of occupational physicians who are employed by management, workers' "right-to-know" about job hazards, confidentiality of workers' medical records kept by employers, and the restriction of female workers of childbearing age from certain jobs (see Chapters 2, 13, and 36). The controversies surrounding these subjects will eventually be settled by labor–management negotiations and by the deliberations of the courts and legislative and executive bodies of government. As an example, the United States Supreme Court has upheld a worker's right to refuse hazardous work, stating that a worker could not be discharged or discriminated against for exercising a right not to work under conditions reasonably believed to be very dangerous.

Workplace-Related Health and Medical Programs

Health care is increasingly available in or near the workplace, often emphasizing education, screening, and other approaches to prevention. In recent years, there has been substantial growth in occupational health services in academic medical centers, community hospitals, managed care organizations, and free-standing clinics (see Chapter 14). A number of hospital-based programs have been developed, in part, to increase revenues; the impact of these programs on occupational disease and injury prevention is not yet clear.

Liability

An extremely controversial force in occupational health is the product liability suit. Workers, barred from suing their employers under workers' compensation laws and unhappy about deregulation of the workplace, have increasingly turned to "third-party," or product liability, suits as a means of getting redress for occupational disease (see Chapter 11). These lawsuits always require that the worker show that the third party was negligent—the mere fact of the injury is never enough. The fear of liability suits has driven many employers to focus on preventive activities. Such suits play an important role in directing attention to prevention of some diseases, although this approach can be cumbersome and the outcomes may not be equitable. In some jurisdictions, some of the most egregious health and safety offenders have been criminally prosecuted.

Advances in Technology

Advances in technology continue to facilitate identification of workplace hazards and potential hazards, including *in vitro* assays to determine the mutagenicity (and therefore the possible carcinogenicity) of substances, improvements in ways of determining the presence and measuring the levels of workplace hazards, and new methods of monitoring concentrations of hazardous substances in body fluids and the physiologic impairments they cause (see Chapters 7, 15, and 16).

The Environmental Movement

Public concern for a safe and healthful environment—from protecting air, soil, and water from contamination to ensuring safety in consumer products—extends to the workplace. Whether considering asbestos or pesticides, noise or radiation, there is a continuum of exposures that extends from the workplace to the general environment (see Chapter 3).

Globalization of the Economy

The growth of multinational corporations, reduction in trade barriers, and development of regional treaty arrangements like the North American Free Trade Agreement (NAFTA) and global organizations like the World Trade Organization (WTO) are having an increasing impact on occupational and environmental health, some of which may be positive and some of which may be negative (see Chapters 2 and 10).

Additional Challenges in Developing Countries

All of these issues are important in nations throughout the world. In addition, developing countries face the following challenges:

Export of hazards. Developed countries often export their most hazardous industries, as well as hazardous materials and hazardous wastes, to developing countries, where laws and regulations concerning these substances are more lax or nonexistent and workers may be less aware of these hazards (see Chapter 31).

Inadequate infrastructure and human resources. In developing countries, there are far fewer adequately trained personnel to recognize, diagnose, treat, and prevent and control occupational health problems. Governments and often employers have fewer resources to devote to occupational health, and labor unions, facing other challenges such as low wages and high unem-

ployment rates, may give little attention to occupational health.

Transnational problems. Occupational and environmental health problems in developing countries often involve several countries in the same region, requiring transnational or regional approaches to problems, such as development and implementation of transnational standards.

Relationships between work and the environment. In developing countries, where so many people work in or near their homes, the distinction between the workplace and the general environment is blurred. As a result, family members may often be exposed to workplace hazards.

Economic development. Governments of developing countries often give high priority to economic development, sometimes even over the health of their people. In the context of economic development, and accompanying rapid industrialization and urbanization, there is often pressure to overlook occupational and environmental health issues, given limited resources and the fear that attention to these issues may drive away potential investors or employers. Similarly, workers desperate for jobs in economies with high unemployment rates are unlikely to complain about occupational health and safety hazards once they are employed.

Occupational health services and primary health care. Given limited resources and infrastructure, many developing countries are exploring ways to integrate occupational health services with primary medical care and with a broader range of public health services. Although some successes have been achieved with this approach, there remains much untapped potential in fully achieving this kind of integration.

CONCLUSION

Many health professionals will eventually work with management, labor, or government on occupational health and safety issues, and some will become occupational health and safety specialists. But almost all health professionals, in one way or another, will be involved with the recognition, diagnosis, treatment, or prevention and control of work-related illnesses and injuries.

BIBLIOGRAPHY

Selected Books

Ashford NA, Caldart CC. Technology, law, and the working environment. New York: Van Nostrand Reinhold, 1991.
An in-depth analysis of the technical, legal political, and economic problems in occupational health and safety.

Burgess W. Recognition of health hazards in industry: a review of materials and processes, 2nd ed. New York: John Wiley & Sons, 1995
An excellent summary of industrial hazards, updated, made more comprehensive, and well illustrated with photographs, drawings, and graphs in this second edition.

Environmental Health Criteria Series, Environmental Program. Geneva: World Health Organization.
A collection of monographs, each of which provides a succinct and comprehensive review of the relevant literature on the human health effects of exposure to specific chemicals.

Fleming LE, Herzstein JA, Bunn WB III. Issues in international occupational and environmental medicine. Beverly, MA: OEM Press, 1997.
A very good, succinct primer on international issues.

Hamilton A. Exploring the dangerous trades: an autobiography. Boston: Little, Brown, 1943.
A classic historical reference.

Hathaway GJ, Proctor NH, Hughes JP. Proctor and Hughes' chemical hazards of the workplace. 4th ed. New York: Van Nostrand Reinhold, 1996.
Brief summaries of many chemical hazards, including basics about their chemical, physical, and toxicologic characteristics; diagnostic criteria, including special tests; and treatment and medical control measures.

Harrington MM, Gill FS, Aw TC, Gardiner K. Occupational health, 4th ed. Oxford: Blackwell, 1998.

LaDou J. Occupational and environmental medicine, 2nd ed. Stamford, CT: Appleton & Lange, 1997.
Both of these well organized, well written textbooks are valuable resources.

Rom WN, ed. Environmental and occupational medicine, 3rd ed. Philadelphia: Lippincott–Raven, 1998.

Rosenstock L, Cullen M. Textbook of clinical occupational and environmental medicine. Philadelphia: WB Saunders, 1994.
Both of these are excellent, comprehensive in-depth references.

Waldron HA, Edling C, eds. Occupational health practice, 4th ed. Oxford: Butterworth-Heinemann, 1998
A general overview of occupational disease and health services with a British orientation.

Weeks JL, Levy BS, Wagner GR, eds. Preventing occupational disease and injury. Washington, DC: American Public Health Association, 1991.

A systematically organized handbook designed for the primary care clinician and public health worker.

Zenz C, ed. Occupational medicine, 3rd ed. Chicago: Year Book, 1993.
Broad, detailed review designed for occupational health physicians.

Selected Periodical Publications

Occupational Health and Occupational Medicine

American Journal of Industrial Medicine, published monthly by Wiley-Liss, 605 Third Avenue, New York, NY 10158-0012.

American Journal of Public Health, published monthly by the American Public Health Association, 800 I Street, NW, Washington, DC 20001.

International Journal of Occupational and Environmental Health, published quarterly by Hanley & Belfus, Inc., 210 S. 13th Street, Philadelphia, PA 19107.

Journal of Occupational and Environmental Medicine, published monthly by the American College of Occupational and Environmental Medicine, Lippincott Williams & Wilkins, 351 West Camden Street, Baltimore, MD 21201-2436.

New Solutions: A Journal of Occupational and Environmental Health Policy, published by the Baywood Publishing Company, Inc., 26 Austin Avenue, Amityville, NY 11701.

Occupational and Environmental Medicine, published monthly by the BMJ Publishing Group, BMA House, Tavistock Square, London WC1H 9JR, United Kingdom.

Scandinavian Journal of Work, Environment & Health, published bimonthly by occupational health agencies and boards in Finland, Sweden, Norway, and Denmark; Topeliuksenkatu 41A, FIN-00250, Helsinki 29, Finland.

State of the Art Reviews: Occupational Medicine, published quarterly by Hanley & Belfus, Inc., 210 S. 13th Street, Philadelphia, PA 19107.

Occupational Health Nursing

American Association of Occupational Health Nurses Journal, published monthly, 50 Lenox Pointe, Atlanta, GA 30324.

Occupational Hygiene (Industrial Hygiene)

American Industrial Hygiene Association Journal, published monthly by the American Industrial Hygiene Association, 2700 Prosperity Ave., Suite 250, Fairfax, VA 22031.

The Annals of Occupational Hygiene, published bimonthly by Elsevier Science, 660 White Plains Road, Tarrytown, NY 10591-5153 (alternate address: The Boulevard, Langford Lane, Kidlington, Oxford OX5 1GB, United Kingdom).

Applied Industrial and Environmental Hygiene, published monthly by Applied Industrial Hygiene, Inc., a subsidiary of the American Conference of Governmental Industrial Hygienists, 1330 Kemper Meadow Drive, Cincinnati, OH 45240.

Occupational Safety

Professional Safety, published monthly by the American Society of Safety Engineers, 1800 E. Oakton Street, Des Plaines, IL 60018-2187.

Safety and Health, published monthly by the National Safety Council, 1121 Spring Lake Drive, Itasca, IL 60143-3201.

Occupational Ergonomics

Applied Ergonomics, published bimonthly by Butterworth-Heinemann, Ltd., Linacre House, Jordan Hill, Oxford OX2 8DP, United Kingdom.

Ergonomics, published monthly by Taylor & Francis Headquarters, 11 New Fetter Lane, London EC4P 4EE, United Kingdom.

Human Factors, published quarterly by The Human Factors Society, Inc., Box 1369, Santa Monica, CA 90406.

International Journal of Industrial Ergonomics, published monthly by Elsevier Science Publishers, B.V., P.O. Box 521, Amsterdam, 1000 AM, The Netherlands.

General News and Scientific Update Publications

BNA Occupational Safety and Health Reporter, published weekly by the Bureau of National Affairs, 1231 25th Street N.W., Washington, DC 20037.

The Occupational and Environmental Medicine Report, published monthly by OEM Health Information, 8 West Street, Beverly Farms, MA 01915.

BIBLIOGRAPHY FOR BOX 1-1

Cappelli P, Bassi L, Katz H, Knoke D, Osterman P, Useem M. Change at work. New York: Oxford University Press, 1997.
Analyzes the changes in the nature of the employment relationship caused by corporate downsizing, heightened global competition, globalization, and the increased deployment of a contingent workforce as corporations seek ways to lower their labor costs as much as practicable.

Forrant R. Restructuring for flexibility and survival: a comparison of two metal engineering plants in Massachusetts. Geneva: International Labor Organization, 1998.
Case studies and discussion of two U.S. metal-working firms that restructured into team-based production.

Herzenberg SA, Alic JA, Wial H. New rules for a new economy: employment and opportunity in postindustrial America. Ithaca, NY: Cornell University Press, 1998.
Presents a detailed analysis of the growth of the service economy and its implications for jobs, long-term careers, and shared prosperity in the United States.

Graham L. On the line at Subaru-Isuzu: the Japanese model and the American worker. Ithaca, NY: ILR Press, 1995.

Rinehart J, Huxley C, Robertson J. Just another car factory? Lean production and its discontents. Ithaca, NY: Cornell University Press, 1997.
Both books discuss work life under a lean production framework.

APPENDIX
Training and Career Opportunities

Howard Frumkin, Howard Hu, and Patricia Hyland Travers

EDUCATION AND TRAINING

Few medical schools and training programs teach occupational health in anything more than a cursory fashion. The same is true for most nursing schools and other clinically oriented health professional schools.

Some medical and nursing schools have developed formal courses, and others include occupational health in their basic science and clinical curricula, often as the result of student efforts. Many students find it easier to incorporate occupational health into existing courses rather than add extra courses to already crowded schedules.

Students can also participate in full-time structured programs that include occupational health field work. In such programs, students learn the fundamentals of recognizing occupational disease, occupational history taking, and work process analysis. They visit various worksites and conduct health evaluations, sometimes designing and implementing specific projects. Students can also participate in a wide variety of extracurricular activities, some of which are acceptable for elective credit by medical schools. Examples include work with public interest organizations, labor unions, companies, and academic research groups.

A variety of opportunities are available to those who seek training at the postgraduate level. A good source for information on current training programs is the Division of Training at NIOSH (800-356-4674). The most complete programs are located at the NIOSH Educational Resource Centers (ERCs). As of 1999, there were 15 ERCs in the United States aiming to provide full- and part-time academic career training, cross-training of occupational health and safety practitioners, mid-career training in the field of occupational safety and health, and access

---

Box 1-3. Board Certification in Occupational Health for Professionals

Physicians may pursue board certification in occupational medicine. Eligibility is based on academic training, clinical training, and practical experience. A current list of approved academic and field-training programs and criteria for board certification may be obtained from the American Board of Preventive Medicine, 9950 West Lawrence Avenue, Suite 106, Schiller Park, IL 60176 (telephone: 847-671-1750).

Certification for industrial hygienists is based on academic preparation, experience, and written examination. Further information can be obtained from the American Board of Industrial Hygiene, 6015 West St. Joseph, Suite 102, 4600 West Saginaw, Suite 101, Lansing, MI 48917 (telephone: 517-321-2638).

Certification for industrial safety professionals is also based on academic prepara-tion, experience, and written examinations. Further information, including a list of training programs, can be obtained from the Board of Certified Safety Professionals, 208 Burwash Avenue, Savoy, IL 61874 (telephone: 217-359-9263).

The designation of Occupational Health and Safety Technologist is a joint certification offered by the American Board of Industrial Hygiene and the Board of Certified Safety Professional (BCSP). Further information can be obtained from the Board of Certified Safety Professionals.

The American Board of Occupational Health Nurses offers two credentials: a COHN, which focuses on the nurse's role as a clinician; and a COHN-S, which reflects the nurse's role in direct care, management, education, and consulting. Information on board certification can be obtained by contacting the American Board for Occupational Health Nurses, 201 East Ogden, Suite 114, Hinsdale, IL 60521 (telephone: 630-789-5799).

---

to relevant courses for students pursuing various degrees. The ERCs are listed in Appendix B.

Educational Resource Centers have developed core areas of instruction for graduate and postgraduate students and continuing education programs for professionals. Areas of study include epidemiology and biostatistics, toxicology, industrial hygiene, safety and ergonomics, policy issues, administration, and clinical occupational medicine. Many ERCs are located at schools of public health, where they draw on the strength of departments in related disciplines. Emphasis is placed on training occupational physicians, industrial hygienists, occupational health nurses, and other professionals to work as a team in preventing occupational disease and injury.

The schools of public health that house ERCs, and many others as well, offer masters and doctoral programs that focus on occupa-tional health. Some medical students earn Master of Public Health (M.P.H.) or equivalent degrees during medical school elective time or during time off from medical school. Interested professionals may enroll in public health programs after completing other professional study; some have done so after many years of practice.

Both occupational health specialists and other physicians can seek continuing education credits in occupational health by attending appropriate seminars and short courses. These are frequently offered by ERCs. Another source is the American College of Occupational and Environmental Medicine, through both its national office and its regional affiliates. Announcements appear in its *Journal of Occupational and Environmental Medicine* and other journals in the field.

Finally, physicians—and a few nonphysicians—who seek full-time, on-the-job training may join a 2-year program administered by NIOSH in conjunction with the Epidemic Intelligence Service of the Centers for Disease Control and Prevention. Members of this program learn occupational disease epidemiology through rigorous field studies, data analyses, and seminars. Information may be obtained from NIOSH, Centers for Disease Control and Prevention, Atlanta, GA 30333.

Board certification is available for several different types of occupational health professions (Box 1-3).

CAREER OPPORTUNITIES

The following is an overview of career opportunities available to professionals in occupational health, with personal examples of career pathways.

Physicians Employed by Companies

Industry has traditionally provided most employment in occupational medicine. Those employed by companies have responsibilities in three general areas: prevention and early detection of occupational disease and injury; diagnosis and treatment of occupational disease and injury (emphasizing return of workers to their jobs); and diagnosis and treatment of nonoccupational disease or injury in emergency situations or when community resources are unavailable.

Physicians who work at the upper management level are more involved in questions of policy, whereas those at the plant level are more involved in clinical duties. A clinical toxicologist for a petrochemical corporation writes:

> I provide guidance and technical assistance to the company on issues pertaining to the health and safety of our processes and products. One of my responsibilities is to review the company's existing medical surveillance programs. I am called upon to develop medical department procedures to ensure compliance with company standards for workplace and product safety and the medical requirements of the Occupational Safety

and Health Administration (OSHA) and other federal and state regulations. I respond to customer and employee concerns regarding the health and safety of our operations, such as the safety of video display terminals and the use of contact lenses in our various work environments.

> My office works very closely with environmental affairs, toxicology, industrial hygiene, and product safety professionals in performing risk assessments, setting internal exposure standards, planning our toxicology testing programs, providing recommendations concerning the medical information to be included on labels and material safety data sheets, and evaluating adverse effect allegations received by the company. Recent areas of review have included the potential toxicity of ceramic fiber insulation and thermal degradation products of plastics, commonly referred to as "blue haze." I participate as a member of the company's Product Safety Emergency Response Team and in disaster preparedness activities of the company.

> My group has also had a role in developing a process for evaluating the reproductive risks of our operations.

> Providing assistance to our attorneys has become another important responsibility. I am frequently called upon to review medical records related to lawsuits against the company and to provide recommendations and opinions.

> I also contribute to policy development, program planning, and the administration of the Medical Department. For example, my office has helped in developing the Medical Department's role in the company's Alcohol and Drug Control Program.

> Finally, I represent the company in outside professional organizations and teaching activities.

Physicians Employed by Labor Unions

Physicians employed by labor unions work closely with other members of their union's health and safety departments. Their job responsibilities are in general less well defined and more flexible than those of company physicians and nurses. Responsibilities may include worker education, health hazard evaluation, participation in contract negotiations, research, and maintenance of disease registries.

Health care providers employed by labor unions may function as worker advocates. This perspective is described by a former United Auto Workers (UAW) union physician:

> The basic thrust of the union's health and safety program is to identify hazardous plant conditions and eliminate them before damage has been done to workers' health. In situations of uncertainty, the benefit of the doubt goes to the worker.
>
> In other words, it is assumed that an exposure or condition is hazardous until and unless we can demonstrate otherwise.
>
> I spent my time responding to requests for service and assistance from local unions; providing education and training as well as developing educational materials on health and safety issues; participating in collective bargaining concerning specific issues in which we have particular expertise, such as guarantees to members of certain rights to health and safety protection of various types; and helping in defining and implementing the union's public posture and activity in the area of health and safety, including work on health and safety legislation, OSHA standard setting, and workers' compensation reform.
>
> Within this framework, I have found myself able to apply a substantial portion of my medical skills and resources to a diversified set of problems in a manner that has been professionally challenging and politically satisfying. I fall among those who approach occupational medicine as a public health discipline rather than an internal medicine specialty.

Federal Government Physicians

Physicians work with NIOSH and have worked with OSHA, assisting in the functions of these two agencies (see Chapters 5 and 10). Because NIOSH is primarily a research agency, its medical staff is devoted primarily to research. This research may take the form of health hazard evaluations or field investigations, or it may involve basic scientific or epidemiologic investigations.

A physician epidemiologist who has worked in the Hazard Evaluation Program and in the Surveillance Branch writes:

> During my time at Hazard Evaluation, I worked on outbreaks of phytophotodermatitis in grocery workers; cumulative trauma disorders in machine operators in the bookbinding and wire die industries and among police transcribers; breast cancer and spontaneous abortion clusters among women exposed to radiofrequency heat sealers in the manufacture of loose-leaf binders; lead and arsenic exposures in a copper smelter; health effects among firefighters exposed to polychlorinated biphenyls (PCBs) and PCB pyrolysis products during a dumpsite fire; skin sensitization and restrictive lung disease in workers exposed to hard metals in the manufacture of hardened blade tips; control of ethylene oxide (EtO) exposures in hospitals; acute neuropathy and cataracts in EtO-exposed workers in the manufacture of hospital supplies; an assessment of carboxyhemoglobin levels in workers exposed to methylene chloride in the electronics industry; health effects from exposure to chlordane following a misapplication; and eye irritation associated with use of soft contact lenses among university biochemists.
>
> I have also responded to numerous telephone calls and letters from workers or their families requesting information and advice concerning workplace exposures and their potential health effects, and to requests for information from the media, other governmental agencies, and Congress.
>
> Later, I worked on surveillance and studied workers' compensation claims in Ohio as a source for describing the epidemiologic characteristics of occupational lead poisoning, skin disease, cumulative trauma disorders, and work-related violent crime injuries. Work in surveillance has offered me the opportunity to learn how to use data sets to create surveillance systems for the identification and follow-back of companies with lead exposure, skin disease, or cumulative trauma problems, and to generate hypotheses relating disease outcomes with industries at high risk. This work is very satisfying in that it allows me to work in the area where policy planning, epidemiology, and health overlap.

Among the advantages of his job, he counts the NIOSH mission to promote worker safety and health, the access to workplaces for study purposes, the variety of problems he confronts, the freedom and responsibility it entails, the opportunity for interdisciplinary collaboration, and the opportunity to publish. He notes that work at NIOSH offers little opportunity for clinical practice.

Other federal agencies that employ physicians may also become involved in occupational health issues. For example, a medical epidemiologist at the National Cancer Institute might study mortality patterns among certain occupational groups, or a staff physician at the OSHA might review the health effects of certain occupational exposures.

State and Provincial Government Physicians

Some states and provinces that are active in occupational health and safety regulation employ physicians. In some cases, physicians work in environmental health or cancer prevention programs and naturally become involved with issues of occupational safety and health.

An occupational hygiene physician for Massachusetts describes her work as follows:

The Division of Occupational Hygiene inspects workplaces and assists employers and employees in correcting health hazards and improving working conditions. It also publishes recommended exposure limits and safe practice bulletins. I participate in worksite health hazard evaluations, conduct small-scale epidemiologic projects, perform educational activities, answer telephone inquiries, and supervise a full-time occupational hygiene nurse who surveys health programs in industrial plants.

For workplace evaluations, I work with an industrial hygienist to assess health hazards and evaluate workers. We usually make recommendations to eliminate or diminish the hazard; sometimes we recommend a program of ongoing medical surveillance. For example, in evaluating workers at a sewage treatment plant with possible lead and chromate exposure, I administered brief questionnaires, performed physical

examinations, and obtained blood analyses for chromium and lead. We were able to determine that the current control measures usually limited exposure and protected workers, but that in a few parts of the work process additional control measures were warranted.

The opportunity for plant access provides the potential to perform epidemiologic surveys, such as a recent survey of automobile radiator repair shops and workers for lead poisoning. Previous state occupational hygiene physicians in Massachusetts have performed important studies of the health effects of beryllium, cadmium, toluene diisocyanate, lead, and talc, which have often led to important preventive measures.

I also educate employers and employees daily on hazardous exposures and control measures, update and write the medical aspects of safe practice guides and material safety data sheets, supervise medical residents and industrial hygiene students learning about worksite evaluations, and participate in preparing state regulations and proposed statues. I am also a member of a state interagency task force that is developing methods for surveillance of and intervention for occupational health problems.

Although a strong national program is integral to controlling workplace hazards, state agencies perform additional important functions: they provide services more readily to local small plants not covered by OSHA, respond quickly to emergencies, provide enforcement on problems not covered by OSHA, help employers comply with OSHA standards, and respond to complaints.

Occupational Medicine in the Community Setting

A growing number of opportunities to practice occupational medicine can be found in clinical settings in the community. In some cases, occupational medical clinics exist as independent contractors that sell to client companies such services as preplacement and return-to-work evaluations, drug screening, and trauma care. Alternatively, some clinical occupational medicine units exist as parts of hospital staffs, usually within departments of medicine or family practice. A chief of occupational health at a community hospital in Massachusetts reports that both community hospitals and companies have an interest in such arrangements, the former, in part, to increase revenues and the latter to cope with regulatory demands and workers' compensation costs. He indicates that a wide range of options may be available in the community hospital setting, including extension of emergency services that triage the injured worker, extension of employee health services, development of free-standing occupational health departments, and development of satellite sites.

He reports that his clinical practice is a combination of general internal medicine and office orthopedics. He performs "complex injury management," evaluates patients for insurance companies, manages an active executive health and wellness program, and establishes medical surveillance programs on a consultative basis, subcontracting with industrial hygiene and clinical laboratory facilities when appropriate. He notes that one potential controversy is that some community physicians might feel threatened by an occupational medicine service if they have been previously performing such duties on an informal basis.

Nurses in Occupational Health

Occupational health nursing has undergone a metamorphosis since its beginnings in the late 1800s, when nurses were employed by industries to care for ailing workers and their families. Attention in the field of occupational health nursing has shifted from a narrow focus on communicable disease, maternal and child health issues, and emergency treatment of injured workers to a much broader focus today. The occupational health nurse (OHN) applies public health principles to meet the needs of workers in an ever-changing work environment. The focus of the OHN has thus expanded to include integration of many areas, including

epidemiology, industrial hygiene, environmental health, toxicology, safety, management, health education, early disease detection, disease prevention, health promotion, and health and environmental surveillance.

The American Association of Occupational Health Nurses (AAOHN) defines occupational health nursing as:

> . . . the application of nursing principles in conserving the health of workers in all occupations. It emphasizes prevention, recognition, and treatment of illnesses and injuries and requires special skills and knowledge in the fields of health education and counseling, environmental health, and human relations.

Most OHNs work for company medical departments. The OHN, whether employed as a single health care provider at a small plant or as a member of a multidisciplinary health unit, must balance ethical and clinical responsibilities to employees with ethical and administrative responsibilities to management. This balance requires the OHN to assist management in providing a safe and healthful work environment through disease prevention and health promotion activities. Some of the responsibilities include the daily operation of a comprehensive health care program; development of treatment and surveillance protocols; keeping informed about health and safety legislation; maintenance of a toxic substance list; identification of high-risk areas; clinical intervention, including delivery of health care and counseling services; record keeping; liaison with managers, workers, and health and safety colleagues; and implementation of health-related programs on a primary, secondary, and tertiary level. (See P. Travers. A Comprehensive Guide for Establishing an Occupational Health Service. New York: AAOHN, 1987. Available from AAOHN, 2120 Brandywine Road, Atlanta, GA 30341; 1-800-241-8014.)

A manager of nursing and health services for a U.S.-based multinational corporation writes:

> I am responsible for overseeing nursing functions and health services for approxi-

mately 150 locations throughout the United States, Canada, Puerto Rico, Haiti, and Central America. My responsibilities include performing audits of health services throughout the company, developing policy, arranging for continuing education, and apprising the nurses who report to me of current events in medicine, legislation, health care delivery, and other fields as they relate to employee health and safety. I act as a resource, consultant, and liaison among health services and other areas of the company, and as an advocate for the nurses.

I find occupational health very exciting. In no other area of nursing does one need such a broad base of nursing knowledge or have such a varied role. In no other area of nursing does one have such a captive audience in which to promote health and wellness, or such an opportunity to express individuality and creativity within a practice.

Opportunities for OHNs also exist in other sectors, such as organized labor and government.

Industrial Hygiene

The practice of industrial hygiene includes the recognition, evaluation, and control of occupational hazards. Most industrial hygienists are employed directly by companies; however, many are also employed by government agencies concerned with regulation, independent consulting groups, and academic institutions. A few are also employed by labor unions. The following is an account by an industrial hygienist that reflects the breadth of opportunities in this field. An industrial hygienist with a labor occupational health program at a university in California writes:

My job as an industrial hygienist involves the development and implementation of education and training programs on all aspects of industrial hygiene for workers, their representatives, managers, and members of labor–management health and safety committees. This work has included training programs that address basic hazard recognition procedures, hearing conservation, dust exposures (silica and asbestos), chemical exposures, hazardous waste, OSHA standards, ventilation, control measures used in the workplace, ergonomics, and manual materials handling. I also teach in university courses for occupational and environmental health sciences students and in continuing education courses for occupational health professionals at a NIOSH educational resource center. I have trained contractors and workers in asbestos abatement techniques at an EPA asbestos training center.

The creation of written materials has been a key part of my career. I have written a number of slide and videotape training programs and a series of guidebooks on various industrial hygiene topics. I am the editor of the third edition of *Fundamentals of Industrial Hygiene,* a basic textbook in the field.

Providing technical assistance to workers, managers, and the general public has also been a part of my work. Acting as a technical resource is a large part of industrial hygienists' work, whether they work for industry, labor, academic institutions, or the government.

My career as an industrial hygienist reflects a heavy emphasis on writing and teaching. Many industrial hygienists may find themselves also performing a training and education function, but as a smaller part of their jobs. They may be in the field, performing sampling and evaluations much of the time. These evaluations may cover a wide range of chemical and physical workplace health hazards. Some industrial hygienists specialize in one particular hazard. For example, health physicists specialize in ionizing radiation, whereas other industrial hygienists may specialize in hearing conservation and noise control technologies or in asbestos abatement work.

Government jobs for industrial hygienists include program development, training and education, and compliance inspections and evaluations of worksites.

Industrial hygienists who work for private industry may work mainly as field hygienists, performing daily sampling and evaluation; as trainers and program developers; or as program administrators. The typical job incorporates all of these functions, depending on the size of the company.

Industrial hygienists in academic organizations typically divide their time between teaching and research functions. Some are program administrators.

The field of industrial hygiene is exciting

and challenging. It draws on many other disciplines (chemistry, biology, toxicology, epidemiology, occupational medicine, health physics, engineering, and health education) to meet its goal of protecting the health of workers.

The coordinator of a graduate occupational hygiene program in Venezuela, writes:

Our graduate program, which was established in 1970, is the first one in Venezuela. Now, there are two other programs in other universities more focused on occupational medicine. Our program is a multidisciplinary program that is oriented principally toward prevention. It is integrated, with engineers, physicians, epidemiologists, and other professionals.

In general, industrial hygienists in Venezuela are limited in their activity in factories. Only in the larger ones are there occupational health policies on which industrial hygienists can base preventive activities. Another problem is that access to technical resources is too limited, leading to inadequate quantity and quality of equipment in laboratories. In small- and medium-sized facilities, occupational health and safety problems are greater. Government policies are not structured to improve the work environment. Venezuela has a deficit of over 1,000 industrial hygienists.

My job involves the development and implementation of preventive programs in different areas of industrial hygiene. The most important activity developed is recognition and control programs. Evaluation is limited because we do not have sufficient equipment, laboratories, and other resources. Education and training programs are common activities in my professional practice. We have developed introductory courses in occupational health and industrial hygiene for managers, technical employees, and workers in several industries in our state.

Providing technical assistance to workers, managers, and governmental agencies has also been part of my work. Now, the unions and the manager organizations have begun to take interest in this area, and day to day, more requests for services and technical assistance are arising.

My career as an industrial hygienist reflects a heavy emphasis on teaching and researching, but this is not the most common activity of industrial hygienists in Venezuela. In our program, we develop different educational course levels—

introductory, intermediate, and intensive courses in the occupational health—with special emphasis on workers' training programs. We have agreements with unions in our state to provide technical assistance, a priority activity for our work.

Researching is another area of my activity. The possibility of creating a multidisciplinary team in this area is very important, and now we have one. The development of new research models that are adequate for our conditions and that promote worker participation is a policy of our team. We are also developing different research programs—for example, on neurotoxicity and occupational exposures to solvents, mercury, and lead, on new technologies and worker health effects.

Academic Occupational Health

Many occupational health specialists have faculty appointments at schools of medicine, nursing, public health, or other disciplines. Often these are part-time appointments that complement clinical appointments or employment in other settings.

Academic positions in occupational health entail the same types of duties as do other faculty posts, including research and publication, classroom and clinical teaching, outside consulting, and patient care. Faculty members may specialize further, depending on their interests and training. Many collaborate with statisticians, epidemiologists, toxicologists, and others in their research and with clinicians from other fields in their clinical duties. Specific job responsibilities are variable and are often largely defined by the individual.

A former director of a university-based occupational health program writes:

Academic medicine is traditionally described as demanding productivity in three areas: clinical service, teaching, and research. Increasingly, administrative activities are added responsibilities even for junior faculty. Although these expectations may be viewed as overly demanding in some disciplines, they add to the variety and challenge of working in the emerging academic field of occupational medicine.

My colleagues and I have responsibility

for teaching medical students in the basic science years and providing elective opportunities to them in the clinical years. Residents in internal medicine, family medicine, and occupational medicine also rotate through our program, participating in clinical and research activities and interacting with a multidisciplinary staff of physicians, industrial hygienists, and nurses. Other trainees are also involved, including students in industrial hygiene and occupational nursing.

Our main objective, determining whether an individual's health problem is work related, is largely carried out in a consultative clinic setting. Here, patients are referred for evaluation by themselves or by physicians, unions, companies, and workers' compensation and other agencies. After our review of medical records and available information about exposures, each patient undergoes a comprehensive interview eliciting information about work and exposure history, a physical examination, and appropriate laboratory tests. It is against this background that a determination about the individual's medical condition and its relation to workplace factors is made. Often this task is relatively straightforward, but sometimes in the process we recognize unexpected or new associations between the workplace and health.

Probably in no other field is there this same potential for discovering and detecting new etiologies and syndromes. But, regardless of the complexity of evaluating disease and its cause in an individual, we are always thinking about the implications of our findings in terms of prevention—for the individual in terms of returning to the workplace and for others who may be similarly affected or at risk.

This process—from initial evaluation to diagnosis and follow-through—is, I think, the most rewarding part of clinical occupational medicine. In our setting, all aspects of a patient's situation are discussed by the entire staff in case conference; these discussions are open and sometime heated, recognizing that occupational diseases have not only medical, but also social, economic, legal, and political components.

Scholarly investigation and research are also fundamental to our activities and often brought about by problems encountered in the clinic. Occasionally, this leads to basic "bench" research, but the objective of such inquiry—to provide new knowledge—can also be achieved by describing individual cases or clusters of cases, or by undertaking population-based studies to answer predetermined questions systematically.

We bring a collaborative spirit to all these activities. They are largely enjoyable (and often fun), and rewarding. It is an opportunity and challenge to work in an evolving field, to be humble in recognizing what we do not know, to be vigilant about opportunities to broaden knowledge, and to recognize that in occupational health we are dealing with the social context of disease, a context that is central to the recognition, determination of causation, and prevention of occupational disease.

A chairman of a department of preventive medicine and environmental health at a university also describes teaching, research, patient care, and administration as his four duties. He writes:

As an academic occupational health physician, I see to it that our medical students get a reasonable amount of occupational health in their curriculum (with all the usual constraints and competing educational needs). I have started a masters program, on which to base a new residency program in occupational medicine. Patient care activities are in many ways similar to traditional occupational medicine, and patients are seen in the university occupational medicine clinic. In addition, the department operates occupational health programs for several companies in such fields as coal mining and electronic repair, and I have served as medical consultant to an automobile manufacturing company. My busy program of work at the university is supplemented by regional and national activities; I serve as a board member of several national organizations and as a consultant to NIOSH.

Occupational Health: Recognizing and Preventing Work-Related Disease and Injury, 4th ed. Edited by Barry S. Levy and David H. Wegman, Lippincott Williams & Wilkins, Philadelphia © 2000.

2

Occupational Health
A Social Perspective

Charles Levenstein, John Wooding, and Beth Rosenberg

Why should scientists and health professionals concern themselves with the social and political context of occupational health and safety problems? Is it not sufficient to learn about the characteristics of risk factors, the diagnosis of occupational disease, and technical approaches to prevention?

- There has been evidence for centuries about the health hazards of lead. Why are workers and children poisoned by lead exposures?
- Pesticides are designed to kill pests, but they are also toxic for other living things. Why do we know so little about the human health effects of most pesticides in use today?
- The textile industry has been the leader of the Industrial Revolution throughout the world. Why did byssinosis, a respiratory disease of cotton mill workers go unrecognized in the United States until the late 1960s?
- Asphalt fume has been identified as a carcinogen in Denmark. Why is it regulated in the United States not as a carcinogen, but only as an "air contaminant?"

C. Levenstein: Department of Work Environment, University of Massachusetts Lowell, Lowell, Massachusetts 01854.

J. Wooding: Department of Regional Economic and Social Development, University of Massachusetts Lowell, Lowell, Massachusetts 01854.

B. Rosenberg: Department of Family Medicine and Community Health, Tufts University School of Medicine, Boston, Massachusetts 02111.

- A major transnational automobile company has clear internal guidelines for reviewing possible equipment purchases to prevent hearing loss among its employees. Why are these guidelines ignored by plant managers?
- There is less full-time work, more temporary work, more work speed-up, and more shift work, all of which increase physical and psychological health problems. Although there are scientific and technical aspects of this situation that are worth studying, are the solutions solely scientific and technical?
- When hazardous technologies and hazardous substances find their way to developing countries, after being prohibited from use in the developed countries where they originated, what kind of solutions are available?
- What can economically challenged workers or countries do when confronted with the choice between jobs or health?

The effective understanding of workplace injury and disease requires a full comprehension of the nature of work and the social, political, and economic context of the workplace. Work is a necessary human activity.

People work to survive, yet work is more than a way to gain an income. Work provides a host of rewards and problems: it can be laborious and numbing, stimulating and satisfying, frustrating and demeaning. All too often it is dangerous and unhealthful.

Work occupies a central place in most people's lives, but the overall context of work is often ignored or poorly understood. To recognize and prevent work-related disease and injury requires that health care providers and other health and safety professionals appreciate the full context of work and workplaces in the world today. This chapter focuses primarily on the U.S. experience, although parallels to other countries are drawn when possible. It is believed that many of the underlying issues being addressed here cross national boundaries. Before we discuss the current situation, we review some history.

A SOCIAL AND POLITICAL PERSPECTIVE ON THE HISTORY OF OCCUPATIONAL HEALTH AND SAFETY

Occupational health has rarely received much attention in most societies. Historically, our commitment to economic advance through technology has made us blind to its toll on workers' health. Workers have been engaged in the more pressing task of making a living for their families to pay too much attention to widespread occupational safety and health problems. The labor movement in the United States has not been strong enough to force public attention to these issues on a continual basis. As a result of a number of interrelated historical and ideological factors, relatively little attention has been paid until recently to the problem of occupational illness and injury in the United States.

In most countries, the process of industrialization that resulted in the creation of the factory system radically changed people's experience of work. Forced by economic necessity into the newly created factories of the machine age, workers found themselves controlled by bosses whose sole concern was the maximization of profit. Working in large-scale plants and using the new technology of modern industry, workers confronted a whole new set of conditions; powerless and

tied to the speed of the machine they served, facing the ever-present dangers of physical injury from conveyor belts and speeding looms, and exposed to a range of dyes, bleaches, and gases—for workers, the workplace had become a source of injury, disease, disability, and death.

With the help of social reformers and professionals, workers, newly organized into unions, fought back against these conditions in countries such as Britain and Germany in the middle and end of the 19th century, and they were somewhat successful in improving conditions through government regulation. Laws restricting working hours and the employment of women and children, and promoting protection against safety hazards and some hazardous chemical exposures, increased. A system of factory inspection was established in Britain by the mid-19th century, and Germany moved to control working conditions by the beginning of the 20th century. In Europe, these efforts built on an earlier tradition of occupational medicine, an acceptance of government intervention and paternalism, and a relatively powerful workers' movement. By the 20th century, workers and unions had achieved political representation in the form of labor, socialist, or social democratic parties. This gave workers powers to demand reform and was a major factor in establishing laws to improve working conditions.

In the 19th century, the Industrial Revolution brought to the United States, as it had to Europe, many safety problems and a level of public concern about these problems. Massachusetts created the first factory inspection department in the United States in 1867 and in subsequent years enacted the first job safety laws in the textile industry. The Knights of Labor, one of the earliest labor unions, agitated for safety laws in the 1870s and 1880s. Social reformers and growing union power did gain, by 1900, minimal legislation to improve workplace health and safety in the most heavily industrialized states. The regulations and the system of inspection were, however, inadequate. Those states that had some legislated protections rarely enforced them and focused largely on safety issues; little was done to protect workers from exposure to the growing number of chemicals in the workplace.

After 1900, the rising tide of industrial accidents resulted in passage of state workers' compensation laws, so that by 1920 virtually all states had adopted these no-fault insurance programs. Britain had passed its Workmen's Compensation Act in 1897 for occupational injuries, and occupational diseases were added in 1906. Germany, too, had a system of compensation in place by the turn of the century.

Throughout the 1920s in the United States, the rise of company paternalism was accompanied by the development of occupational medicine programs. Much attention was paid to preemployment physical examinations rather than to industrial hygiene and accident prevention. Occasional scandals reached the public eye, like cancer in young radium watch dial painters. However, it was not until the resurgence of the labor movement in the 1930s that there was important national legislation: the Walsh-Healey Public Contracts Acts of 1936 required federal contractors to comply with health and safety standards, and the Social Security Act of 1935 provided funds for state industrial hygiene programs. During this period, the Bureau of Mines was authorized to inspect mines; this helped to a minimal extent to improve working conditions in the mining industries.

The mobilization for World War II required that the U.S. government become involved in the organization of production. Concern for the health of workers increased during this period because a healthy workforce was considered indispensable to the war effort. However, after the war, health and safety receded from public attention. An exception to the general neglect of the field was passage of the Atomic Energy Act in 1954 which included provision for radiation safety standards.

Not until the 1960s, when labor regained some political clout under the Democratic administrations of Presidents Kennedy and Johnson, did the issue reemerge as significant. Injury rates rose 29% during the 1960s, prompting union concern, but it was a major mine disaster in 1968 in Farmington, West Virginia, when 78 miners were killed, that captured public sympathy. In 1969, the Coal Mine Health and Safety Act was passed and, finally, the first comprehensive federal legislation to protect workers was created when the Occupational Safety and Health Act (OSHAct) became law In 1970.

This brief history illustrates just some of the dimensions of the struggle to provide a safe and healthful workplace. Although many countries provide regulatory protection for workers and unions often demand safe working conditions through collective bargaining agreements, the problems facing workers have increased. New chemicals in the workplace, limits in regulatory enforcement, and the demands of an increasingly competitive global economy exacerbate the need to maintain and improve working conditions.

These problems are global in scope. The globalization of production, trade, and consumption has resulted in occupational and environmental safety and health problems becoming ubiquitous. Workers in developing and newly industrialized countries now face a range of workplace hazards. Stricter environmental regulations in the industrialized countries make it attractive for companies to use countries in Latin America, Asia, and Africa as dumping grounds for toxic waste and as places to export highly toxic substances and hazardous industries.

Perhaps the most pressing problems in occupational health stem from the increasing integration of the world economy. In North America, the development of continental free trade may threaten the more advanced work environment standards of Canada and the United States, while bringing many new hazards to Mexico. In Europe, integration has made the movement of capital and labor across borders much easier; industries can

move to countries with less strict occupational and environmental standards. In some
cases, this intrusion has led to threats to
worker and environmental health; in others,
the more advanced standards of some countries are being imposed on the less advanced,
improving working and living conditions. In
both situations, conflict over standards has
arisen. The export of hazardous technologies, hazardous products, and hazardous
wastes poses increasing challenges to public
health worldwide. On the one hand, our understanding of the nature of health hazards
to workers has been improving; on the other
hand, however, the restructuring of the world
economy may undercut the political will to
control these hazards.

THE GLOBAL CONTEXT OF
OCCUPATIONAL HEALTH

The magnitude and pattern of occupational
disease and injury in a particular society are
strongly affected by the level of economic
and technological development, by the societal distribution of power, and by the dominant ideology of a particular social and political system. These factors bear on the way in
which diseases and injuries are "produced,"
on the recognition and prevention of these
problems, and on the extent to which workers receive compensation for them. Fully understanding occupational injury and disease
requires, therefore, an understanding of the
broad context in which production takes
place. This context includes the economic
and technological basis of production, ideological and cultural factors driving the design
and organization of work and the workplace,
and the main social actors in decisions that
affect the work environment.

The Social Actors in Occupational
Health and the Role of Ideology

The medical/scientific model focuses on disease and injury causation, using scientific
methods to discover, explain, and solve problems in the work environment; it rarely ad-

dresses the critical economic, social, and
political context of work organization. Occupational diseases and injuries are distinct
from other health issues because they are the
direct, although unintended, result of economic activity. The analysis presented here
provides a different perspective (Fig. 2-1),
one that places management's control of the
workplace, technological decisions, and the
labor process at the center of occupational
health.

The structure in Fig. 2-1 suggests that the
key relationship for understanding the work
environment is a "triangle" of control: workers, any potential hazards, and management—and its dominance of the workplace.
This relationship exists in a historical and
ideological context, influenced by a number
of other institutions and individuals. These
actors include professional consultants, universities, and research institutes that typically provide scientific information about
workplace hazards and how to control them.
These research institutes may or may not
work in collaboration with government; government, however, plays a key role in provid-

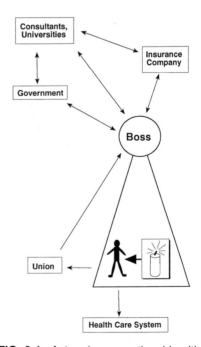

FIG. 2-1. Actors in occupational health.

ing and initiating research about work environment hazards in most countries. More specifically, government typically sets and enforces occupational safety and health standards.

Insurance companies are also key actors. They provide the economic context in which firms obtain workers' compensation insurance and may, by "experience rating" premiums, encourage firms to improve health and safety conditions. Unions project the collective strength of organized workers. They negotiate working conditions and may provide a counterweight to management's prerogatives. Many unions in the United States have their own health and safety staff to provide information and services to workers and workers' representatives. They also push government to act on workplace hazards by lobbying for establishing and enforcing regulations.

The organization of work and the roles played by key actors are deeply influenced by ideology—a set of beliefs, norms, and values. Ideologies of workers, managers, government officials, scientists, and others reflect what they think about society and about themselves. Ideologies also reflect what they expect from work and from employers, government, and each other.

A capitalist, free-market economic system incorporates presumptions about human behavior that most people have come to accept: notions about individual "choice" and "rights" and a belief in the primacy of private property and the efficiency of markets. Americans, in particular, are deeply suspicious of government. It is, therefore, necessary to examine the role of ideology to identify the assumptions that determine power relations in the workplace and how they are reflected in the problems of occupational health and safety.

The typical workplace in the United States is organized hierarchically. In large workplaces, the model is owner or owners at top, followed by leading managers, supervisors, and then the workers. Smaller workplaces compress this structure. The hierarchy re-flects the distribution of power; owners and managers have complete control over investment decisions, the budget, the structure of production, what is produced, how and when production occurs, and hiring and firing of workers, and ultimately control the conditions of work.

Labor unions, considered to be a counterweight to this power, have had some success in gaining better wages and working conditions. They have usually been constrained, however, by a number of factors: the strength of the general economy, the level of unemployment, their own economic and political strength, and an ideology that supports the rights of property. Labor's achievements have also depended importantly on the level of government support for protecting and promoting the rights of workers.

In Europe, although the rights of private property remain relatively sacrosanct, the power of unions and workers' parties, as well as the acceptance and expectation of government regulation of working conditions, has led to a greater ability by government to regulate private industry and working conditions than is found in the United States.

The culture of most liberal democracies, including the United States, has supported belief in the rationality and apolitical nature of science and technology—a belief that social and public health problems (indeed, most societal problems) are amenable to technical solutions. Remarkably enduring has been the ideology of the "technical fix," and the notion that science can be separated from politics and from issues of power and control.

Economic and Technological Development

Changes in the national and international economic order—growth of new markets and the disappearance of old ones, new technologies, new competitors, demographic shifts, and shifts in investment—all directly affect the structure of production and work.

In contrast to the 20 years after World

War II, when American economic power was at its peak, the United States now faces fierce competition in heavy manufacturing, in the service sector, and in high-technology production. By 1970, the United States found itself confronting a new and highly competitive world economy in which American goods and American companies no longer dominated. In addition, multinational corporations based in Europe and the United States began to spread their activities across the globe, setting up production facilities in many developed and developing countries. These multinational corporations invested heavily abroad, seeking new markets and new places of production with lower wages, less regulation, and less taxation. Aided by new communications systems and new opportunities for investment, industry and investment capital have become increasingly mobile. This situation undercuts the ability of advanced industrial countries to regulate domestic industry for fear that industry might flee regulation. At the same time, it spreads hazards, some of which are associated with advanced technologies, to countries without the social or scientific infrastructures to protect their citizens.

Some particular economic developments have led to this situation, such as the major increase in oil prices by the Oil Producing and Exporting Countries (OPEC) in the 1970s, which led many developing countries to borrow heavily to buy oil. This resulted in a vicious cycle for these countries, involving the siphoning off of domestic savings to service the debt, domestic austerity programs imposed by institutions such as the World Bank and the International Monetary Fund (IMF), and a shift to export-oriented production. As a consequence, developing countries increasingly had to accept foreign investment and foreign technologies to survive. At the same time, the dire economic situation in these countries forced many of their most productive and mobile citizens to migrate to other countries in search of work, often at substandard conditions, in Europe and elsewhere.

By the end of the 1980s, the world economy had undergone a fundamental realignment, with four major effects on the United States and developed countries of Europe:

1. Their economies shifted from heavy manufacturing (of chemicals and steel) toward the service sector (banking, insurance, food service, and clerical work). American businesses lost approximately 38 million manufacturing jobs during the 1970s and 1980s (1).
2. Their economies became dominated by extremely mobile and mostly large international corporations.
3. In the United States, ownership of industry became concentrated in a smaller number of very large firms. The frequent buying and selling of companies during the 1980s and 1990s led to the U.S. economy coming increasingly under the control of the banking and finance sector.
4. With decreasing profitability, management in the United States could not afford, and was not willing to accept, the "social contract" with labor—a commitment to maintaining decent wages and working conditions in return for some job security and rising standards of living for most workers, a contract that it had maintained for most of the period since World War II. Companies tried to cut the costs of production by demanding reductions in wages or benefits, and they fought health, safety, and environmental regulation. In Europe, similar economic changes ushered in a period of political conservatism, resulting in the deregulation of the market and reduction in government control over private industry.

All this had an impact on workers; for example, in the United States, real average wages were $9 per hour in 1973 and $8 per hour in 1998 (2). Housing, education, and medical costs have all increased at a rate of approximately 9% faster than inflation over this period. Despite more two-earner families, American workers are much worse off than they had been in 1970.

As the 1990s drew to a close, globalization continued to be the major factor in the social and economic life of all countries. Despite the return of left-leaning parties in much of Western Europe, and one of the longest economic boom periods in the United States in recent history, workers everywhere still struggle to maintain their standards of living. The collapse of many of the Southeast Asian economies in the second half of the decade, the chronic high levels of unemployment in Europe, and the continued economic slump in Japan are in stark contrast to the economic growth of the United States. But the consequent shrinking of world markets poses an enormous threat to U.S. economic prosperity and can only exacerbate the record levels of social and economic inequality still painfully evident in the United States. In addition, pressure to adopt neoliberal policies (reduced public spending, weakening of government regulation, privatization of state industries and services, and the virtual eradication of social welfare spending), imposed by the domination of American-led free-market ideology and orchestrated through the activities of international organization such as the IMF and the World Trade Organization, place a heavy burden on workers everywhere. In short, global economic and technological change continue to transform the workplace.

Management Theory and the Structure of Work

Although under attack and reconsideration in recent years, the general tendency in management theory from the time of Adam Smith, the father of economic liberalism, to the present has been to divide work into ever more discrete units to increase productivity, cheapen the cost of labor, and increase management's control over the labor process (Fig. 2-2). This quest for "efficiency" became more self-conscious and explicit in the early 20th century with the work of such promoters

FIG. 2-2. Automobile assembly line worker. (Photograph by Earl Dotter.)

of scientific management as Frederick Winslow Taylor (3,4). In Taylor's view, the worker should be treated not as a whole person but rather as a collection of machine-like movements: walk, bend, grasp, sit, depress typewriter key. Such motions can be analyzed, timed, and reassembled into a program for maximum productivity. This "scientific" approach to management was widely accepted, both in capitalist and noncapitalist economies. Taylorism's impact is well illustrated by the following comment by an automobile assembly line worker (5):

> My father worked in auto for 35 years and he never talked about the job. What's there to say? A car comes, I weld it; a car comes, I weld it; a car comes, I weld it. One hundred and one times an hour . . . . There is a lot of variety in the paint shop . . . you clip on the color hose, bleed out the old color, and squirt. Clip, bleed, squirt, think; clip, bleed, squirt, yawn; clip, bleed, squirt, scratch your nose. Only now the [company has] taken away the time to scratch your nose.

Taylorism had a wide-ranging impact on the quality of work life. It meant the separation of conception from performance and the division of performance into multiple repetitive tasks. The intrinsic satisfaction of "work," craftsmanship, and the ability to take pride in the whole finished product necessarily diminished. Employers increasingly relied on supervisory hierarchies and monetary rewards and punishment, such as piece rates and bonuses, to motivate workers in a carrot-and-stick fashion (5):

> You're too busy to talk. Can't hear. They got these little guys coming around in white shirts and if they see you running your mouth, "This guy needs more work." A lot of guys who've been in jail they say you don't work as hard in jail. They say, "Man, jail ain't never been this bad."

Another profound influence on modern production and the workplace has been the rapid increase in the use of chemicals, especially since World War II. There are currently 70,000 chemicals in use in the United States, with 1,000 new chemicals introduced each year (6) (Fig. 2-3). A similar number of chemicals and chemical processes exist in most of the industrialized world, and increasingly so in developing countries as production is shifted to them. Most of these chemicals are unregulated and their human health effects unknown. They are used in a variety of production settings to produce a wide range of products, but they are also encoun-

FIG. 2-3. Hazardous waste worker sealing abandoned drums of waste chemicals.

tered in a range of occupations not traditionally considered dangerous. From typists and stockroom workers to janitors and artists, workers confront some potentially toxic chemicals on a daily basis.

Technology has increased the speed of production enormously, putting greater pressure on workers to perform rapid and repetitive motions that are damaging to mental and physical health. Stress and related psychological and physiologic illnesses are increasing in industrialized countries, including the United States, as the pace of work and life increases, as well as pressures to work longer hours to compensate for falling wage rates and a declining standard of living (7). In some countries, however, such as those in Europe, because of historical and cultural reasons and pressure from powerful trade unions, a shorter work week with reduced working hours has been adopted since World War II. More recently, unemployment pressures have furthered the call for a shorter work week (8).

With speed-up has come automation. Apart from obvious physical hazards associated with use of robots, robotic systems, and highly automated machinery, automation also eliminates jobs and de-skills others, leaving fewer workers responsible for complex systems. With the help of automation, one worker can do a job that may have required 10 workers before. This advance, however, has been accompanied by greater stress and, in general, more overtime work. Under these circumstances, rather than achieving its promise replacing grueling, mindless labor, automation has resulted in more stress, longer hours, and overwhelming responsibility at work (9).

Economic and technological changes go together. The spread of new technologies, the globalization of the world economy, and vast changes in the international division of labor both directly and indirectly affect not only the work environment, but general power relations in society. Class, race, and gender are key dimensions in the power relationships in the United States that shape substantial aspects of the work environment.

The Distribution of Power

All societies are composed of classes, of interest groups, of sects and sectarians, of minorities and majorities, with varying degrees of power and influence. The distribution of such political and economic power and influence is another essential factor shaping the work environment. In the most simple formulation, there are "workers" and "owners." In advanced industrial societies, such a formulation cannot capture the complex features of the contemporary class system. In such societies, a middle stratum has developed that is composed of independent professionals, an enduring class of small business owners, and a growing group of government employees with a wide range of social functions and with their own roles, interests, and power. The varying degrees of political power among lower, middle, and upper classes set limits on what can happen in a particular workplace or a particular industry. Class and the distribution of power in society, therefore, shape the work environment.

In many liberal democracies, although the stated goal is equality, power in society at large is unevenly distributed along the lines of class, race, and gender. Because the workplace is a microcosm of society, the power relationships in society are reproduced at work. Inequities in the distribution of power have a profound influence on work and health because power determines who does what work and under what conditions, who gets exposed to risks, and what is considered acceptable risk. Furthermore, the people affected are not the people deciding the acceptability of workplace hazards (10):

> All the cliches and pleasant notions of how the old class divisions . . . have disappeared are exposed as hollow phrases by the simple fact that American workers must accept serious injury and even death as a part of their daily reality while the middle class does not. Imagine . . . the universal outcry that would occur if every year several corporate headquarters routinely collapsed like mines, crushing sixty or seventy executives. Or suppose that all the banks were filled with an invisible noxious dust that constantly produced cancer in the managers, clerk and tellers. Finally, try to imagine the horror . . . if thousands of university professors were deafened every year or lost fingers, hands, sometimes eyes, while on their jobs.

Social class and class-based assumptions have been widely discussed from a variety of perspectives: sociologic, economic, and political. Class is clearly related to family background, level of education, occupation, and a variety of cultural factors. The lower a person's social class, the less likely he or she will have a range of educational and employment options. Class determines levels of material well-being and health. Because class influences employment options, it affects the probability of becoming ill or injured at work.

Impact of Racism

In the workplace and in society as a whole, racism plays a role in determining who does what job, how much he or she will be paid for it, and what alternatives are open. For most of its history, the United States has depended on minorities to do the least desirable and dangerous work. Immigrant and minority communities have been the major sources of labor to build the railways, pick cotton and weave it in the mills, work in the foundries in the automobile industry, run coke oven operations in the steel industry, sew in the sweatshops, and provide migrant agricultural labor (Fig. 2-4). Minorities are still overrepresented in the most hazardous and least desirable occupations (Fig. 2-5). Minority workers may leave a hazardous work environment only to arrive home to a hazardous community environment. Since the early 1980s, in the United States, scientific evidence has increasingly pointed to discriminatory environmental practices of certain industries, of state and local govern-

FIG. 2-4. Migrant worker picking cotton. (Photograph by Earl Dotter.)

ments, and in some instances, of the federal government. One well documented example is that minority communities experience a disproportionate number of toxic threats to health (see Chapter 3) (11).

A social system with strongly racist elements bars members of minority groups from significant positions of power and, consequently, elevates the concerns of dominant

FIG. 2-5. Worker in a commercial laundry. (Photograph by Earl Botter.)

racial or national groups. For example, one of the essential reasons for the lack of attention to hazards faced by farm workers in the United States (most of whom are African-American or Latino) is their relative lack of power in the American political system.

Impact of Sexism

Any discussion of power relations must include the situation of women, whose experience of work is in general different from that of men. Most obviously, this is reflected in the wage differentials paid to women for comparable work. Despite a political and legal commitment to equality in the United States, as of 1998, women were earning 74 cents to every dollar earned by a man, and the gap widens as one goes up the career ladder (12,13). African-American women and Latinas earn only 50% of white men's pay (see Chapters 36 and 37) (13).

Even though women frequently work outside the home for as many hours as their spouses, domestic duties are rarely shared equally. Working mothers sleep less, get sick more, and have less leisure time than their husbands. One study finds that women who are employed full time outside the home and whose youngest child is less than 5 years of age spend an average of 47 hours per week on household work, whereas their male counterparts spend a mere 10 hours (14). Although the situation may have improved somewhat over the last 10 to 20 years, the stress and fatigue from balancing work life and home life remain a serious problem. The average working woman puts in an estimated 80 hours a week in both job and household work, and up to 105 hours if she has sole responsibility for children.

Women are also the main targets of sexual harassment at work. Any unwanted verbal or physical sexual advance constitutes harassment, and this can range from sexual comments and suggestions, to pressure for sexual favors accompanied by threats concerning one's job, to physical assault, including rape. Studies indicate that 40% to 60%

of women have experienced some form of sexual harassment at work (15). An estimated one-third of the largest 500 companies in the United States spend approximately $6.7 million in dealing with sexual harassment (15).

Gender relations have political, and hence work environment implications. Cultural assumptions about gender can have a strong impact on the distribution of power in society. A strongly patriarchal society that bars women from positions of power is also likely to have a profoundly sex-segregated labor market. As a result, sexual harassment and occupational health in female-dominated retail trade jobs may not be considered important.

Thus, in addition to the development of the market, the level of technology, ideological considerations, and changes in the global economy, power distribution related to class, race, and gender constitutes the framework in which the actors in an industrial system attempt to create a "web of rules" governing the work environment. Management, labor, and government are constrained in their behavior by these broad social-environmental factors.

THE MICROCONTEXT OF OCCUPATIONAL HEALTH: LABOR–MANAGEMENT RELATIONS

The First Key Actors: Workers

A hundred years ago, when a cobbler woke up in the morning, the decision to make boots or shoes, to buy hides, or to take some of his wares to the neighboring town was under his control. If the cobbler acquired an allergy to a certain polish or was told that it caused cancer, he could choose not to use it. If he found that carving heels bothered his elbow, he could do a few every other day instead of spending a long stretch of time on a bothersome or painful task, or he could try redesigning the tools or using alternative carving methods that might be better for him. He was his own manager. He set the pace and conditions of his work.

Contrast the cobbler's situation with the working lives of most people today. These options are not open to modern-day shoemakers, or to nurses, auto workers, bank tellers, or employees in countless other occupations. Management controls the work environment; the hours of work, the pace, the tasks, the tools, and the technologies are all determined by someone other than the worker.

In addition to the detrimental effects of lack of control, which by itself causes stress (see Chapter 21), workers' interests conflict with those of management. Management's goal is to maximize profit; labor's goal is a fair wage for a fair day's work. Expenditures on health and safety are often seen by management as limiting profit. As a business school textbook advises (16):

> In making decisions about their workplace, managers have two choices. They can remedy health and safety problems or they can provide risk compensation to workers. If reducing risk is less costly than the additional compensation, then working conditions will be improved. However, if the marginal cost of worker compensation is less than the marginal cost of safety improvements, then the firm will choose the compensation alternative. This outcome represents an efficient allocation of resources in that the firm minimizes its total costs.

Although one would hope that the conscience of managers will go against their training and the incentive system in their businesses, history has shown that it is unwise for workers to depend on the benevolence of management. The sociopolitical structure provides only weak motivation for management to construct a safe workplace. Government regulations exist but they are not always enforced, which is not surprising given that current Occupational Safety and Health Administration (OSHA) resources would allow the federal government to inspect each of 6 million workplaces once every 84 years (17).

Labor–management relations may be particularly problematic when jobs are pitted against improving occupational or environmental conditions. The most frequent example of this contradiction occurs under conditions of "job blackmail," a colloquial term for the problem created when workers are forced to choose between remaining in a hazardous job or finding employment elsewhere (18). Examples include employers who threaten to fire workers or relocate the plant if workers or regulatory agencies try to impose controls over hazardous production. Job blackmail is found more often in those workplaces where workers have little or no power of control over their jobs as well as in workplaces that are not unionized. Although not unique to minority workers, job blackmail takes a heavy toll on them, because they are more likely than nonminority workers to hold hazardous jobs. Although job blackmail may occur in a variety of direct and indirect ways, the end result is to force workers to choose between being employed or not.

In job blackmail the choices are seldom, if ever, favorable to the worker. The worker who chooses to remain on a hazardous job may, in the short term, avoid unemployment, but may seriously jeopardize his or her future health and safety. The worker who chooses not to question "unfair" compensation will continue to receive a paycheck but will still earn less than she or he is worth. The worker who chooses not to unionize may remain employed, but will likely remain employed in an unjust, unsafe, and unhealthful workplace. Even in those situations in which a worker remains on the job, she or he may be labeled a troublemaker and ostracized to the point of quitting the job anyway (Box 2-1).

Although unions are a force in spurring companies to attend to health and safety problems, typically through collective bargaining agreements or, where they exist, union-controlled health and safety committees, and government regulation provides a further stimulus, there are two other motivational sources for improving health and safety: (a) corporate reputation ("public relations"), which functions to press management not to appear negligent in its provisions for workplace safety (although this tends to function more effectively for pollution prob-

lems and environmental concerns and, more often, in large corporations); and (b) the cost of replacing labor. If a company has invested in developing a skilled and loyal workforce, it is unlikely to want to damage that investment by exposing workers to dangerous conditions. This factor helps to explain why low-skilled, easily replaced workers, such as migrant laborers or poultry workers, are so vulnerable (Fig. 2-6).

The Changing Structure of Work

The economy of the United States and many other developed countries is changing rapidly. The shift from heavy manufacturing toward the service sector affects the structure of work and the work experience for many Americans. In general, in service industries, the most rapidly growing sector of the econ-

FIG. 2-6. Nonunion demolition worker in East Africa. (Photograph by Barry S. Levy.)

Box 2-1. History of a Secondary Lead Smelter in an Urban Environment

A small smelting company made news in the early 1980s after OSHA charged the company with administering chelating drugs to employees to lower the level of lead in their blood. The small scrap recycling company, located in Massachusetts, was charged with illegally providing these lead-purging drugs to employees while they continued to work in a lead-contaminated environment. After the administration of these drugs, two employees became severely ill. One had kidney failure and was ultimately diagnosed with kidney cancer. The other employee died; lead poisoning was listed as a significant contributing factor on his death certificate.

Although OSHA cited the company for many serious violations of standards, the fines and some of the charges were significantly reduced after negotiation. The company agreed to clean up the plant and reduce employee lead exposures.

Ten years later, however, not much had changed. One of the first reports of multiple poisonings in a single workplace listed in the state's new adult lead poisoning registry came from employees of the smelting company. Every "shop floor" employee was reported to the registry as having an elevated blood lead level, the average being 40 μg/dL, a level associated with adverse health effects in adults.

Shortly before the registry was established, the employees decided they had been poisoned long enough. A complaint was filed with OSHA and an inspection was conducted. The inspector was told by company officials that they no longer used lead and that there was no need to perform industrial hygiene sampling. The inspector took them at their word.

Frustrated, the employees approached the Massachusetts Coalition for Occupational Safety and Health (MassCOSH), a worker health and safety advocacy group (see COSH groups in Appendix B). There they told their story to staff members of the Latino Workers Project of MassCOSH. They explained that conditions had not changed in years and that employees were routinely sick from processes that were kept secret from regulatory agencies. They described conditions that were not unlike those of smelters from another century.

The workers were mostly non–English-speaking immigrants from Central America, many of whom had entered the United States illegally. They worked among family and friends at the plant and were unlikely to find other jobs. But they were not comfortable with their failing health and the daily compromises they were expected to make for a meager living.

The most vocal of the workers was fired after being accused of reporting working conditions to OSHA. The Immigrants' Rights, Advocacy, Training and Education Project (IRATE) became involved in the case and worked with MassCOSH and the employees to establish strategies to improve working conditions. A union-organizing drive was initiated with the International Ladies' Garment Workers Union (ILGWU).

By the time the state's Division of Occupational Hygiene investigated the reports of lead poisoning at the plant, the employees' organized struggle was well under way. The inspection revealed years of accumulated lead dust and debris, as well as very high levels of airborne lead from incineration of insulated wire and a process of sifting scrap metal dust and grit. Workers were not given clean washing facilities, and were forced to eat their lunches in a filthy washroom where a microwave oven was set up in a toilet stall. There was no soap, hot water, or towels. Many serious safety hazards were also observed.

OSHA was called back in, and this time it found many violations. It cited the company with 48 serious violations and 3 willful violations, accompanied by a fine of more than $200,000.

A review of the Division of Occupational Hygiene files later revealed that the company had a record that extended back to the 1930s, with nearly 50 site inspections. Each report was almost identical, with the company saying the process that had once produced lead was no longer practiced and that cleanup was in progress.

Why was the company able to elude the full authority of the regulatory agencies

Box 2-1   (*continued*)
for so many years? What happened that
brought about the ultimate rigor of these
agencies in this case? The answer appears
to be that for the first time in 50 years, the
workers were organized. Although they finally lost their union election, they did understand their rights and recognized the consequences of their working conditions.

Through the efforts of the advocacy
groups, the workers were able to communicate critical information to the health and
safety inspectors regarding lead-generating
processes, so that inspections could be conducted under typical working conditions. All
of the workers were interviewed during inspections in the presence of interpreters.
The workers had been empowered to hold
the inspectors accountable for workplace
health and safety.

In addition to the issues of worker health
and safety, for many years community residents had expressed concerns about their
exposures to the environmental pollution
produced by the plant. Inspections were
conducted by the state's Department of Environmental Protection and the company
was ordered to halt certain processes and
reduce emissions from others. Local oversight by the city's health department was
critical in informing the Department of Environmental Protection when conditions
worsened, so that unannounced inspections
could be conducted.

The conditions at the smelting company
were reported to the state attorney general's
Environmental Strike Force. After months
of scrutiny by that office, both civil and criminal charges were sought against the company. Under a civil consent order, the company removed all of the lead debris and
cleaned the facility. Most of the lead-generating processes were eliminated, although
the company still conducts a brisk business
in metal scrap recycling.

Under the company's settlement with
OSHA, workers' lead exposure is regularly
monitored and they are protected against
airborne contamination through a combination of engineering controls, personal protective equipment, and safe work practices.
Blood lead levels have been significantly reduced.

Although the fines from OSHA and the
threat of additional fines from the attorney
general motivated the cleanup, publicizing
the outcome of the criminal investigation is
likely to be an effective deterrent to other
companies that are exposing their employees to unsafe working conditions.

omy, wages are low, benefits scanty, job
security limited, and unions virtually nonexistent. Much of this work is part time or temporary.

In response to the shrinking economic pie
of the 1980s, employers are increasingly using part-time and temporary workers to cut
costs. The average part-time worker earns
only 60% of a full-time worker on an hourly
basis. Fewer than 25% of part-time workers
have employer-paid health insurance, compared with nearly 80% of full-time workers.
Sixty percent of full-time workers have pensions provided by employers, whereas only
20% of part-time workers have this coverage
(13). In 1990, in the United States, there were
5 million involuntary part-time workers—

that is, workers who would prefer to be working full time but were unable to do so.

In addition to lower pay and fewer benefits, there are other negative aspects to this
trend toward temporary and part-time work.
Temporary workers live with the stress of
not knowing when and for how long they
will work. They have little or no job security.
Neither part-time nor temporary workers receive equal protection under government
laws, including occupational safety and
health regulations, unemployment insurance, and pension regulations. Few are represented by unions (13). A case study commissioned by OSHA of contract labor in the
petrochemical industry (usually small contractors of nonunion workers, brought into

a plant to do maintenance and other work) showed that contract workers get less health and safety training and have higher injury rates than do noncontract workers (19). The consequence for occupational health and safety of an increasingly unorganized, temporary, and part-time workforce should not be underestimated.

Another increasingly common characteristic of changes in the structure of work in the United States is the rise in home-based industry. In 1949, the U.S. Congress passed a law making industrial home work illegal, largely because it was almost impossible to enforce workplace regulations and labor standards (such as the minimum wage) for home work. Under the Reagan administration, Congress reversed this policy and legalized home work. The consequence was a rapid growth of home-based manufacturing and service work throughout the 1980s (13). A large number of home workers are women, typically garment and clerical workers who are paid on a piece-rate system. Piece-rate payments encourage speed, increase the risks of accidents, and exacerbate repetitive motion injuries, resulting in numerous ergonomic problems in workplaces not designed for the type of work being undertaken. Chemical exposures also pose a problem. Semiconductor manufacture undertaken at home, for example, not only exposes workers and their families to toxic agents used in the manufacturing process but also may contaminate local sewage systems.

Work and Labor

Of the 131 million American workers, approximately 92% work for other people. Approximately one-fourth have professional, managerial, or supervisory employment, with varying degrees of partial autonomy and control over their own jobs. These people both work and labor. Most workers, however, only labor. They find what jobs they can and, by and large, do what they need to do to keep them. They do not choose what

they will make, under what conditions they will make it, or what will happen to it afterward.

These choices are made for them by their employers, the sales and labor markets, and the working of the economy as a whole. Whatever control most workers have over how much they receive in return for their labor, how long they labor, how hard they labor, and the quality of the workplace environment is acquired in a contractual situation in which the workers' desire for comfort, income, safety, and leisure is continually counterbalanced by the employers' need for profit.

Many workers have profound ambivalence about their jobs. Although labor provides an income, workers also seek less tangible satisfactions from their work. For example, an unemployed miner reflected (20):

> Some no doubt will find this a sad thing, the fact of not having any work, I mean. Others simply won't notice, while still others, with a more fundamental way of looking at things and sadly lacking a working-class consciousness, will utter some such expression as "Lucky bastard!". . . Frankly, I hate work. Of course I could also say with equal truth that I love work; that it is a supremely interesting activity; that it is often fascinating; that I wish I did not have to do it; that I wish I had a job at which I could earn a decent wage . . . .

The contradictions in this statement cannot be dismissed as the contrariness of human nature; they correspond to contradictions in the real situation. What workers love about work is the opportunity to guide their own lives and to do meaningful things. On the other hand, workers oppose their work being made meaningless by the ways it is organized by others and bleached of integrity, autonomy, and creativity for reasons of efficiency, productivity, and profit.

Modern production and market competition lead employers to seek the highest possible rates of productivity. The normal social interactions among workers that, in a less mechanized and fragmented work process, appear as part of the rhythm of work itself, are seen as disruptive to production. Attempts on the part of workers to establish some level of control and sociability in the workplace are often misconstrued. Employers and managers who see such acts as threats to productivity and efficiency consider them to be indications of laziness. Workers, even in nonunion settings, may view them as efforts to protect themselves against the requirements of a fragmented division of labor that treats them as tools rather than as people. These attempts to take greater control actually represent, consciously or unconsciously, the individual's desire to replace labor with work. The structure of the contemporary workplace undercuts such acts of rebellion and self-assertion, however.

Innovations such as word-processing technology, computerized record keeping, electronic mail, and computer and video monitoring have turned large offices into assembly lines. New forms of work organization have broken the close personal tie that frequently existed between secretaries and their employers, and new technology has downgraded the skills required. With these changes, clerical work becomes subject to the same kind of machine-like analysis and control as factory work.

Similar situations are often found in service, retail, distributive, and other types of work. What is true for the auto worker, the word processor, and the keypunch operator is increasingly the case for the short-order cook, the checkout clerk, and the telephone operator. One young woman describes her sense of powerlessness and alienation as a grocery store cashier (21):

It was extremely repetitive work. Pushing numbers all day sort of got to me. I used to have dreams, or should I say nightmares, all night long of ringing up customers' orders when it was after closing time. I have even woken up and found myself sitting up in bed talking to customers. That job ended when the whole building exploded one night because of some faulty electrical work. The summer of my senior year in high school I got another job as a cashier in a discount department store, doing the same thing, pushing numbers again. My nightmare of talking to customers in my sleep began again. . . . This was a job that was an extremely strict one. There was no leeway about anything. They had cameras above the registers watching us to see if we were polite, if we checked inside of containers for any hidden merchandise, checked the tags to see if they were switched, etc. If we failed to do something we were given a written warning . . . .

Everyone who worked there, with the exception of the management, was part-time. The schedules were made so that no one had exactly 40 hours. I worked for 3 months, 35 to 38 hours per week. By not giving us those few extra hours, they saved themselves a lot of money by not having to give their employees benefits, insurance, etc. Of course, their hiring, firing, quitting went on week after week. There weren't too many loyal employees.

A fractionated division of labor and "scientific" work discipline are ways of exerting managerial control in the interests of efficiency and profit (Fig. 2-7). The experience

FIG. 2-7. Long-distance telephone operators. Monotony characterizes many jobs. (Photograph by Earl Dotter.)

of alienation and powerlessness on the part of the workers, however, is not limited to workplaces where this type of organization is imposed. Many jobs in small shops—particularly in the service and retail sectors, which employ the largest number of women and youth—are equally unattractive despite their lack of specialization.

The characteristic jobs of a service sector economy tend, therefore, to replicate quite often the alienating, repetitive work once associated with assembly-line production and the monotony of the modern factory. Today, in developed countries, however, improved technology and the ubiquity of computers have enormously increased the potential pace of work as well as the ability of the work rate to be monitored. Technology combines with pressures for increased productivity in an increasingly competitive world economy. "Competitiveness" and the drive for productivity incur enormous costs in terms of worker health and well-being.

The constant demand to do work faster and more efficiently, to "produce" under the whip of being fired or laid-off, takes a huge toll on the mental and physical health of workers. The dignity of work is not evident in the voices quoted previously. As American international competitiveness declines further and goods and services enter the United States from developing countries where workers get wages little above subsistence and often work in horrendous conditions, there is even greater pressure on domestic manufacturing to compete. The reality of that competition for most American workers has been demands for wage give-backs, compulsory overtime, work speed-up, and increasingly less attention to workplace health and safety.

Organized Labor

Unions are a way to counteract the disempowering, disenfranchising effects of class, race, and gender (see Chapter 40). They provide workers a voice in determining the rules and conditions of work, wage rates, and benefits. They are the collective strength that provides a counterweight to management power and prerogative. Some unions have been deeply involved in health and safety issues but, for most unions, such issues are only a few among many. In the United States, given the weakness of unions and the historic antagonism to organized labor, unions have not always been able to give the necessary resources to protect their members from workplace hazards. In Europe, organized labor has been more successful in combating the prerogatives of management and, in a number of European countries, social democratic political parties supported by labor movements have frequently been in power. Even in the United States, with its relatively weak labor movement and the absence of social democratic or labor parties, unions do offer some protection against arbitrary exercise of power.

Formally, unionized workers try to regain some control over the labor process through collective bargaining—the negotiation of work rules and grievance mechanisms, the institutionalized process for adjudicating individual complaints. However, only approximately 15% of workers in the United States are unionized, and even where grievance mechanisms exist, they are not always respected. Informally, workers seek what escapes they can find or fabricate. They sneak a surreptitious cigarette, they fantasize, they horse around, and they fight. "Anything so that you don't feel like a machine" is a common refrain.

Organized labor in the United States is now weaker numerically and politically than at any time since World War II. This decline began in the 1970s and continues through the late 1990s. Over the decade from 1985 to 1995, unionization rates declined 21% in the United States (22). The decline is evident across the whole range of union activity: loss of negotiating strength, decrease in membership, decline in strike activity, and a vast increase in "concessionary" collective bargaining agreements between unions and industry.

In contrast to the United States, in Great Britain, 55% of workers are in unions and labor governments have ruled the country. In Sweden, more than 95% of blue-collar workers are organized, and approximately 75% of white-collar employees are in unions. Over the past decade, unlike in the United States, Swedish unionization rates increased by 8.7% (22). For most of the past 45 years, Sweden has had a labor government, and the labor laws reflect that power (23). In Germany, France, and many other countries, the existence of a labor party (or a social democratic party) has enabled workers to push for and defend significant legislation to control workplace hazards and provide extensive schemes of social insurance and welfare.

The strength of a labor movement determines a host of issues that directly influence worker health, including what information is generated about workplace hazards, who has access to it, what workplace standards are set and by whom they are enforced, the options open to workers encountering a hazard, and the effectiveness of workers' compensation (23).

Unionized workers are more likely to be informed about the presence of health and safety hazards than are nonunion members in the same jobs (24). In addition to union-sponsored education programs, the union provides a shield against employer discrimination. This shield is extremely important for health and safety because employers may fire a worker for raising concerns about health and safety problems.

Unions in the United States and elsewhere have fought to create legislation requiring employers to clean up the workplace, to control the employment of women and children, to limit the hours of work, and to set and enforce industrial hygiene standards. In the United States, where OSHA requires that workers be informed about the hazards associated with the chemicals with which they work, unions have pushed to make sure that employers comply with these "right-to-know" regulations. When there was no federal right-to-know law, some unions negoti-

ated this right, as well as the right to refuse unusually hazardous work (see Chapter 10).

Unemployment

It is striking that, even though unsatisfying jobs produce hostility in many workers, almost all workers would rather have a job than no job at all. One worker says of unemployment: "Lovely life if you happen to be a turnip. . . . One does not willingly opt for near-the-bone life on the dole. My personal problem is easily solved. All I need is work" (25).

Unemployment is more destructive to physical and mental health than all but the most dangerous jobs. Studies have even suggested a correlation between unemployment and mortality from heart disease, liver disease, suicide, and other stress-related ailments (26). Changing levels of unemployment have an impact not only on unemployed workers but also on their families. For example, households in which the husband is unemployed or underemployed show rates of domestic violence two to three times greater than in households of fully employed men (26). Many studies have shown that workers internalize the experience of joblessness as personal lack of worth. This sense of worthlessness appears completely unrelated to a worker's actual degree of responsibility in losing his or her job (27).

In the 1980s, the unemployment rate in the United States fluctuated between 6% and 11%. Some economists have proposed that a 5% unemployment rate be considered "full employment." The late 1990s have brought the unemployment rate in the United States to just below 5%, but the memories of unemployment, along with feelings of expendability, remain fresh to many Americans.

Unemployment has also become a regional and international problem, with the official average unemployment rate among the 12 member states of the European Community almost 13% in 1993. Furthermore, these unemployment rates are based only on those actively seeking work; by excluding those jobless who, through discouragement, have stopped looking or never began to look for work, the official data understate the magnitude of the problem. Unemployment figures rarely include the underemployed, those working part-time who seek full-time work, and those women who would be working if good, well paying jobs were available. In the developing world, the percentages of workers who are unemployed or underemployed is often much higher than in the United States or Europe.

Unemployment has significant economic effects. The existence of many unemployed people keeps wages down as more people compete for jobs and are willing to take lower wages in the struggle to earn a living. It also makes union organizing extremely difficult. As workers lose jobs in manufacturing—where unions had strength—and as management campaigns against unions, the barriers to encouraging people to join unions or to mount significant organizing drives become larger. All these factors weaken the movements to protect workers from occupational hazards.

The Second Key Actor: Management

Unquestionably, there are firms that seek to maintain safe and healthful work environments. These are frequently large, profitable companies that have relatively secure markets for their products and that have decided that their continued success depends on a well motivated, high-quality, and healthy workforce. Frequently these are firms that have a deep commitment to collective bargaining and to negotiating industrial peace. Some firms have decided that the only way that they can attract and keep highly skilled workers is to ensure the quality of working life. Other firms, concerned about product safety because of consumer concerns or the inherent risks of their technology, have attended to worker health and safety virtually as a spillover from their other essential activities.

The remarkable success of Japanese industry in reducing its injury rate, probably as a consequence of its attention to quality in general and its abhorrence of waste, may have beneficial consequences in American and European firms pursuing Japanese-style manufacturing success. Sometimes these company efforts may miss the problems associated with low-level chemical exposures, because they focus primarily on the more obvious safety hazards. Nevertheless, such efforts are to be applauded.

Some small firms pay serious attention to safety and health hazards because the owner or manager came up from the ranks, knows the processes well, and maintains close social contact with the employees. The economic pressures on small companies, however, may undercut even the most decent employer. For small or large firms, the pressures of the market are hard to resist. In these cases, the role of government in enforcing work environment standards is particularly important.

The Third Key Actor: Government

A third key actor in the complex of workplace health and safety is "government" in the form of regulatory intervention (see Chapter 10). The impact of state intervention in health and safety is defined by that set of institutions—legislature, executive, judiciary, and civil service—that responds to needs and initiates policy, establishes laws and regulations, and implements them. In this century, social policies created by government have embraced such measures as unemployment benefits, pensions, and medical insurance, and have protected consumers (such as by control of food additives and laws on advertising and product liability), preserved the environment, and promoted public health and safety.

Why should the state in free-market economies interfere in the operation of that market to ensure the achievement of public welfare goals? What prompts the state to ascribe to itself a regulatory role? The effort to protect the health and safety of workers provides an excellent example of the contradictory forces operating on the state.

On the one hand, such regulation helps ensure the continued existence of a healthy workforce capable of continual productivity, resulting in a positive effect for the economy as a whole. By creating national rules and regulations, it equalizes the responsibilities, as well as the penalties, among industries by requiring certain minimum standards in the workplace, thus giving stability and continuity to all forms of production. Further, by establishing the apparently neutral and regulatory role of the state, such intervention increases the legitimacy of the existing political order. The state must respond to public pressures and to demands that it intervene to prevent illness, injury, and death on the job. It must appear to be responsible and responsive to the concerns of trade unions, workers, and public opinion.

On the other hand, such regulation may have enormous costs for capitalism as a whole and for individual firms. By controlling activities at the point of production, such direct state intervention challenges control of the workplace and, by requiring certain levels of safety and minimum health measures, it imposes costs on industry that may affect profitability. In addition, such intervention gives specific rights to workers (such as the right to refuse unsafe work), which, again, threaten managerial and corporate control of the production process. Of course, the particular ways in which government develops and implements policy are constrained by the constitutional and governing structures of a given country and by the ideological and cultural mix arising from history and traditions.

As countries industrialized and factory production became centralized, the issue of working conditions emerged as a serious cause for concern. In England, the state began to develop laws to prevent the worst abuses. In most countries, as modern industrial production became established, the state was forced to take action to improve working conditions. In nearly all cases, legal protections grew in piecemeal fashion, reflecting class pressures as well as moral outrage at working conditions.

Because most countries faced similar problems, the solutions have taken largely the same form. In Britain, the body of laws (until the passage of the Health and Safety at Work Act in 1974) reflected the incremental progress of legislative action. In Germany, France, Sweden, and some other European countries, specific problems in the field of health and safety were dealt with as information about a given issue became available or as pressures built up to demand legal or administrative action.

In the United States, the history of health and safety legislation has taken a somewhat different course. Influenced by the federal structure of the country, it was not until the passage of OSHAct in 1970 that the United States had a comprehensive federal law to control workplace conditions. The creation of OSHA resulted in extensive debate about the role of the state in the American polity, and the agency has often been stigmatized as a vivid example of too much government (see Chapter 10).

The United States case is especially interesting in that, alone among the developed capitalist economies of the West, the commitment to welfarism has remained embryonic. Although the characteristic structures and programs of the welfare state are not entirely absent, they remain relatively undeveloped in the United States compared with Europe.

Since the passage of the OSHAct and the creation of OSHA, the struggle for healthful and safe working conditions has raised issues of control of the workplace and organization of production. Throughout the 1970s, intense debates occurred over the role and actions of OSHA, the validity of the scientific evidence on the dangers posed by chemicals, the enforcement of standards, the extent and legitimacy of government regulation, and workers' rights to a hazard-free working environment. From the time the first draft of the OSHAct appeared before Congress, industry associations, chambers of commerce, trade unions, and government agencies all became deeply embroiled in highly politicized and emotionally charged issues.

In Europe, the existence of government regulatory agencies and the idea of government inspection of workplaces has, on the whole, been less controversial. The United States, on the other hand, is unique in the openness of its government institutions and the extent of the potential citizen participation in policy making. Most Western European and former Eastern Bloc countries allow much less public input into policies and provide far fewer channels for public supervision. In principle, unions, representatives of workers, serve as partners in policy making by participating at the highest governmental levels on an equal footing with business and industry. The efficacy of their participation, however, may be diluted in Europe if unions are numerically and economically weaker than they once were. Community groups and nonunionized workers have also questioned whether union partnerships and involvement with government and business representative bodies may exclude others from having their interests served.

Science Professionals and the Work Environment

What about the other actors? What can be said about the importance of understanding the social context of occupational health? First, ideology shapes the way we think about problems and how to solve them. Second, the changing structure of the economy, and, hence, of work, has presented new problems for workers and for people involved with occupational health. Third, the global sociotechnical environment has an enormous impact on the work environment. Fourth, the direction of technological development may create new hazards and may set limits on our ability to remedy them. Finally, the distribution of social power can have a profound impact on the attention given to worker health and safety.

The relationships among major social actors—labor, management, and government—define the rules of the work environment, including health and safety standards and practices as well as boundaries within which health care providers, occupational health specialists, and health and safety advocates operate. Although the web of rules sets real limits on reform at the point of production, changes in global factors can open up new possibilities to provide a safe and healthful work environment.

HEALTH AND SAFETY SPECIALISTS

What is the significance, then, of this analysis for the actual work of health care providers and, in particular, occupational health specialists?

In the United States, the largest group of people working in occupational health are occupational health nurses. Other professionals include occupational health physicians, industrial hygienists and industrial hygiene technicians, safety engineers, ergonomists, health and safety educators, and program administrators. Ideological assumptions determine aspects of scientific investigation and research. Scientific disciplines focus attention on the technical aspects of occupational hazards and underestimate the importance of the macrosocial and microsocial, economic, and political context. In this regard, some workplaces have worker or union safety stewards and, increasingly, joint labor–management occupational safety and health committees comprising nonprofessionals who are involved in hazard surveillance as well as injury and illness prevention. These people tend not to be imbued with the scientific model of research and hazard control, a tendency which, in some circumstances, may be advantageous in dealing with workplace hazards.

Where do these different types of people responsible for occupational health work? Some are blue-collar workers in factories with special assignments on health and

safety. Most occupational health professionals work for companies as staff. In most companies, they are part of human resources or
labor relations departments or, much less
commonly, part of a safety and health department that is directly responsible to top
management. With surprising frequency,
health is separated from safety, with professionals from the different fields reporting
through different hierarchies. Rarely are
work environment professionals given direct
responsibility and authority over production;
they are advisory staff and can be influential,
but basic decisions are made typically by
"production" managers, even in service industries. In the private market, profit making
is the prime commitment of the enterprise.

Small companies—where most people in
the United States and the rest of the world
work—rarely have professionals in health
and safety on their payrolls. If they do, the
professional most often is an occupational
health nurse. Usually, such firms rely on *ad
hoc* consultations with independent professionals or simply on emergency medical services. Occasionally, there may be a relationship with specialist occupational health
clinics or services. Such consulting operations must sell their services and are sometimes confronted with ethical difficulties because their clients are companies, not sick or
injured workers (or workers at risk because
of workplace hazards). Large and small companies buy the services of a wide range of
consultants, often without understanding the
degree of specialist knowledge and training
necessary for effective management—and
prevention—of health and safety problems.

In addition, many small firms, and some
large ones, rely on professionals employed
by workers' compensation insurance carriers. "Loss prevention" departments of the
insurance companies, however, may be as
concerned with reducing short-term financial
losses as reducing injury rates, and they may
focus on case management rather than prevention of disease and injury. A new type of
consulting firm has emerged to reduce workers' compensation costs through "managed

care" for injured workers. It is probably too
soon to tell whether these firms, working on
contract with employers, will attempt to reform the work environment or seek to reduce
company expenditures in other ways.

Some professionals in this field work for
labor unions (see Chapter 40). Although
groups such as the United Mine Workers of
America have had health and safety staff
members for many years, the real growth
of occupational health professionals in labor
unions has happened since the 1970 passage
of the OSHAct. Nevertheless, the number
of physicians, industrial hygienists, and other
work environment professionals employed
by the labor movement remains quite small
and is usually at the national or international
level. These people usually provide policy
assistance rather than direct services to
workers. An exception is the government-
subsidized growth in the number of health
educators working for unions, some providing or facilitating general health and safety
education, others working on targeted programs, such as hazardous materials training
for emergency responders and other hazardous waste–related workers.

Finally, and perhaps most important,
many practitioners, including those in the full
range of work environment professions, are
employed by government agencies, usually
as inspectors but sometimes as technical advisers to government or industry, or as educators. In the United States, practitioners are
employed by such institutions as OSHA, the
Mine Safety and Health Administration, the
Department of Energy, the Environmental
Protection Agency, the National Institute for
Occupational Safety and Health, and state
departments of labor and of health.

CONCLUSION

The significance, then, of a social analysis
of the fundamental, but often unrecognized,
problem facing health care providers and
others working in occupational health is that
they frequently are in the difficult situation
of having responsibilities for worker health

while working in organizations with other priorities. Management and government organizations are influenced by economic responsibilities that may compromise worker health and safety. Even labor organizations with their key responsibility to rank-and-file workers may find health and safety low down on a list of concerns and demands. In Fig. 2-1, professionals in occupational health are not separately identified because they fall either under or between the listed categories.

Health professionals can be successful in improving the working environment, especially if they understand the social and economic context of their efforts and work toward "win-win" situations. For example, where workers' compensation costs to a company are high, it may be possible to improve the economic performance of the company and improve worker health through preventive measures. In cotton textile manufacturing, new equipment increased productivity and reduced cotton dust exposure of mill workers. When OSHA mandated reductions in vinyl chloride exposure, the controls introduced by the companies resulted in increased profits. Some have argued that health and safety regulation may stimulate companies to technological innovation they might not otherwise have considered (28). Health and safety practitioners need to master economic as well as humanist arguments for change.

Sometimes, however, the economic arguments alone are not sufficiently convincing to sway management. Many industrial hygienists are members of regional and local professional groups that exchange technical information. These groups have codes of ethics (see Chapter 13) that can inspire and strengthen efforts to improve the work environment. An important source of support, as well technical and strategic ideas, for professionals in occupational health and professional education are the professional societies, such as the American Association of Occupational Health Nurses, the American College of Occupational and Environmental Medicine, the American Industrial Hygiene Association, the American Conference of Governmental Industrial Hygienists, the Human Factors Society, the American Public Health Association, and the American Society of Safety Engineers. Professionals in occupational health, however, need to think in broader terms than usual when confronting difficult situations and recalcitrant employers. In many states, occupational health professionals have played important roles in new coalitions of labor activists and environmentalists. Committees or coalitions for occupational safety and health (COSH groups) have engaged in worker education and advocacy since the early 1970s and have been instrumental in establishing right-to-know laws in some states, in improving workers' compensation in others, and in focusing the attention of labor unions and the general public on health and safety issues (see Appendix B). These groups represent a grassroots movement that links professionals and concerned citizens in a new way to improve the work environment.

REFERENCES

1. Kuhn S, Wooding J. The changing structure of work in the U.S. Part 1: the impact on income and benefits. New Solutions 1994;4(2):43–56.
2. Global Sustainable Development Resolution of 1999. To be introduced to the House of Representatives by Rep. Bernie Sanders, I-VT.
3. Braverman H. Labor and monopoly capital. New York: Monthly Review Press, 1974.
4. Buroway M. Manufacturing consent: changes in the labor process under monopoly capitalism. Chicago: University of Chicago Press, 1979.
5. Garson B. All the livelong day: the meaning and demeaning of work. New York: Penguin, 1977.
6. Rizer-Roberts E. Bioremediation of petroleum contaminated sites. Boca Raton, FL: CRC Press, 1992:4.
7. Karasek R, Theorell T. Healthy work. New York: Basic Books, 1990.
8. Schor J. The overworked American. New York: Basic Books, 1991.
9. Zuboff S. In the age of the smart machine. New York: Basic Books, 1988.
10. Levison A. The working class majority. New York: Coward; McGann Geohegan, 1974.
11. Bullard RD. Reviewing the EPA's Draft Environmental Equity Report. New Solutions 1993; 3(3):78–86.
12. U.S. Bureau of Labor Statistics. BLS Bulletin 2340. Washington, DC: Government Printing Offices, 1998.
13. Amott T. Caught in the crisis: women and the U.S.

economy today. New York: Monthly Review Press, 1993:56.

14. Bryant WK, Zick CD, Kim H. The dollar value of household work. Ithaca, NY: Cornell University Press, 1992:3.

15. Spangler E. Sexual harassment: labor relations by other means. New Solutions 1992;3(1):23–30.

16. Peterson HC. Business and government, 3rd ed. New York: Harper & Row, 1989:429–430.

17. AFL-CIO. Death on the job: the toll of neglect. AFL-CIO Department of Occupational Safety and Health, Washington, DC, April 1992.

18. Kazis R, Grossman RL. Fear at work: job blackmail, labor and the environment. New York: Pilgrim Press, 1982.

19. Wells JC, Kochan T, Smith M. Managing workplace safety and health. Beaumont, TX: John Gray Institute, Lamar University, July 1991.

20. Keenan J. On the dole. In: Fraser R, ed. Work: twenty personal accounts. Harmondsworth, United Kingdom: Penguin, 1968:271–279.

21. Miller L (interviewer). (Anonymous interviewee.) Unpublished interview, Southeastern Massachusetts University, Dartmouth, MA, 1980.

22. International Labor Organization. World Labor Report. Geneva: ILO, 1998:97–98.

23. Elling R. The struggle for workers' health. Farmingdale, NY: Baywood, 1986.

24. Weil D. Reforming OSHA: modest proposals for major change. New Solutions 1992;2(4):26–36.

25. Whiteside N. Unemployment and health: an historical perspective. Social Policy 1988;17:177–194.

26. Leeflang RL, Klein-Hesselink DJ, Spruit IP. Health effects of unemployment. Soc Sci Med 1992;34:351–63.

27. Lerner M. Surplus powerlessness: the psychodynamics of everyday life. Oakland, CA: Institute for Labor and Mental Health, 1989.

28. Ashford NA, Heaton GR Jr. Regulation and technological innovation in the chemical industry. Law and Contemporary Problems, 1983;46:109–57.

BIBLIOGRAPHY

Ashford N, Caldart C. Technology, law and the working environment. Washington, DC: Island Press, 1996.
A detailed compendium of laws and important court cases that govern the workplace in the United States.

Brodeur P. Expendable Americans. New York: Viking, 1974.
A detailed account of the sociopolitical forces that have worked against the dissemination of research information on occupational hazards and effective enforcement of occupational health and safety laws. Focus on asbestos.

Wooding J, Levenstein C. The point of production. New York: Guilford Press, 1999.
Analyzes the political economy of the work environment by creating a theoretical framework for understanding occupational safety and health based on a return to contradictions arising at the point of production.

Wooding J, Levenstein C, eds. Work, health and the environment: old problems, new solutions. New York: Guilford Press, 1997.
A compendium of selected articles from New Solutions: A Journal of Occupational and Environmental Health Policy.

3 Environmental Health and Its Relationship to Occupational Health

Lynn R. Goldman

Environmental health can best be understood within the overall context of health. In 1945, the World Health Organization (WHO) defined health as "a state of complete physical, mental and social well-being and not merely the absence of disease or infirmity" (1). In 1993, the WHO stated, "Environmental health comprises those aspects of human health, including quality of life, that are determined by physical, chemical, biological, social and psychosocial processes in the environment. It also refers to the theory and practice of assessing, correcting, controlling, and preventing those factors in the environment that can potentially affect adversely the health of present and future generations" (2). Thus, as defined, environmental health encompasses a wide array of determinants that can affect a person's health, including occupational exposures—that is, in a sense, the workplace can be viewed as a subset of the environment, an environment unique to those who are employed at that workplace and very much dependent on their specific tasks and protections provided therein.

This chapter focuses on the nonoccupational environment, which is actually a series of different environments a person may inhabit in the course of a day. The household environment refers to the home and the mixture of indoor determinants unique to an individual and family. There are other indoor environments that are not occupational in nature, such as the classroom environment for school children. The ambient environment refers to the shared environment in public areas; determinants of health are factors like air pollution. Ambient environments vary by geography and can vary markedly even within a single city.

Many factors modify the relationship between environment and health. There is much individual variability in response to the environment. Differences in age, sex, and genetic makeup influence both exposure and susceptibility to environmental agents. A challenge in environmental health is to consider all age groups, as well as the very ill and the very healthy, in evaluation of hazards. In addition, social differences can affect exposure. For example, diets vary greatly across different cultural groups. People who live in poverty may experience multiple environmental threats, dietary inadequacies, and other factors that contribute to increased risk from environmental exposures.

Most broadly, environmental health encompasses a wide array of agents that may cause acute or chronic health effects in the population. For example:

- Lead-based paint in housing
- Ambient air pollutants (e.g., ozone, particulate matter, and toxic chemicals)
- Indoor air pollutants (e.g., molds, formaldehyde, carbon monoxide, and tobacco smoke)
- Pathogens in food and drinking water (e.g., cryptosporidia and *Escherichia coli* O157:H7)

L. R. Goldman: Johns Hopkins University School of Hygiene and Public Health, Baltimore, Maryland 21202.

- Pesticide residues in food (e.g., organophosphates and pesticides that are suspected carcinogens)
- Disinfection byproducts in drinking water
- Stressors that cause injury (e.g., automobiles and firearms in the home)
- Hazards in the work environment.

The practice of environmental health also encompasses the examination of rates and causes of diseases of often unknown etiology, such as childhood cancers, asthma, and birth defects. This chapter focuses on a somewhat narrower definition of *environment*—factors that affect health and are more traditionally referred to as "environmental," that is, chemical and physical stressors in the ambient and household environment, but not other environmental factors, like diet and physical activity.

This chapter primarily is concerned with environmental health issues in the United States and, by extension, in other developed countries. There are a number of serious nonoccupational environmental health problems that are much more important in developing countries. For example, drinking water contamination with microorganisms and toxic substances is much more prevalent, and consequent morbidity and mortality more serious. Indoor and outdoor air pollution are considerably aggravated by the burning of coal, wood, and other biomass fuel sources for cooking and heating homes. Air is much more polluted because many of the controls and technological changes that have been required in developed countries have not yet been applied. Injuries, on the road and in the home, occur with a much higher incidence and severity because environmental controls have yet to be applied. Chemical spills and plant accidents are more common and there are fewer means to protect nearby communities and passers-by. Not mentioned in this chapter, but very important worldwide, is disaster prevention and management. Worldwide, there are large numbers of unnecessary deaths and injuries due to earthquakes, storms, and floods that would

be completely preventable with appropriate environmental measures like construction standards for homes and buildings.

Rene Dubos in 1965 noted that indices of environmental health are "expressions of the success or failure experienced by the [human] organism in its efforts to respond adaptively to environmental challenges" (3). This effort to respond adaptively to environmental challenges becomes ever more complex as the environment is changed by humans at a rapid pace. Despite the difficulty of adapting to an environment that has been changed dramatically within just a few generations, there is evidence of remarkable success in this century. The sanitation movement of the 1800s resulted in enormous reductions in mortality due to infectious diseases and marked increases in life expectancy. This has been responsible for much of the increase in life expectancy in the United States, from 47 years in 1900 to almost 77 years in 1997. Since the mid-1970s in the United States, stronger environmental laws have resulted in cleaner air, safer drinking water, and recovery of some water bodies that in 1970 were unsafe for fishing and swimming. Even pesticide usage patterns have changed dramatically; we no longer allow dichlorodiphenyltrichloroethane (DDT) and a number of other persistent organochlorine pesticides to be manufactured in the United States. Today, air lead levels are 98% lower than they were 20 years ago, and from 1976 to 1993, the percentage of children 1 to 5 years of age with elevated blood lead levels (BLLs) decreased from 88% to 4%. Hazardous wastes are handled and disposed of more safely, and the largest sources of pollution discharges to water and air have been controlled. Reports to the nation's Toxic Release Inventory (TRI) indicate that emissions of toxic wastes from U.S. manufacturers decreased by nearly 50% between 1988 and 1996.

Despite these successes, there are numerous challenges that remain. To a great extent, the easiest problems have been addressed, leaving environmental threats that are much more difficult to control and require more participation from a broader range of society. Often, the problems that must be faced today involve multiple small sources of pollutants rather than a few large and visible ones. Many of these small sources are from sectors like agriculture and small business, which are less familiar with environmental regulations and may be very resistant to change. Clearly they will need to be involved, yet they do not have the resources of large industries to address environmental issues. Today there are many efforts at all levels of government to work with these sectors and to provide the needed technical information and assistance to promote better understanding and compliance with environmental standards, as well as to put in place new regulatory efforts to address the emissions. As automobile emissions become a larger component of air pollution, land use, transportation planning, and urban sprawl are becoming greater concerns. Further, problems like non–point-source pollution engage everyone in society, from the farmer to the weekend car mechanic who needs to know how to properly dispose of used motor oil. In most of the world, population is exerting an enormous pressure on resources and contributing to pollution. Even in the United States, there are shortages of potable drinking water in many parts of the country. All of this means that new tools for assessing and managing environmental hazards will be needed to continue to achieve gains in environmental health.

mentally induced diseases and injuries are completely preventable through pollution prevention, product design, engineering controls, personal protection, and education. So much of environmental medicine falls outside the realm of traditional medicine because the focus is usually on primary prevention, preventing exposures before the development of disease. Yet, many of the interventions flow directly from a physician encounter that diagnoses the health problem and forms a connection between that problem and an environmental exposure. As with occupational disease, single or small numbers of diagnosed cases can be sentinels for more widespread population exposures and disease. Environmental medicine is different from occupational medicine in several important respects. The patients include not only the relatively healthy working population, but those who are too young, too old, or too ill to work.

It is important that those with specialized knowledge be involved. For example, a child with a suspected environmental cause of a neurodevelopmental problem needs an assessment by practitioners trained in the assessment of child development and behavior. An elderly person may need a practitioner with not only environmental but also geriatric expertise. In addition, environmental exposures may be more complex, more difficult to identify, and further removed from control than occupational exposures.

ENVIRONMENTAL MEDICINE

As members of the public have become more concerned about the impacts of the environment on their health and that of their children, physicians are increasingly called on to address this issue. As with occupational medicine, environmental medicine is prevention oriented (4). In fact, many of the most important environmental exposures are also occupational exposures, and there is a strong link between environmental and occupational medicine. For the most part, environ-

Taking an Environmental Exposure History

All patient encounters should include a very brief environmental history (see Chapter 4). Basic information to elicit from all patients includes occupations of patient or parents and smoking behaviors of patient and members of the household. Physicians also should have some sense of the community where patients live and what environmental hazards are likely to be present. For children younger than 6 years of age, specific questions need to be asked regarding lead poisoning risks.

Patients with a possible environmentally related disease under evaluation, such as asthma, or those concerned about an environmental hazard, require a more thorough environmental history. A number of specific areas for focus in taking an environmental exposure history have been identified. Some of these areas overlap with the occupational health history, and others go well beyond the history that would be taken for an occupational health examination (Table 3-1) (5). This chapter covers issues unique to the environmental history.

Children and Other Susceptible Populations

Children are not just small adults. Children develop very rapidly in the first few years of life, their diets vary from those of adults, and they require more caloric intake, oxygen, and water for their body weight than adults. Children's metabolism changes over the first few years of life, affecting how their systems handle pharmaceuticals and toxic substances. Normal childhood behavior includes intense exploration of the environment and hand-to-mouth activities that can lead to increased exposures to contaminants in soil and around the home. Children lack judgment and thus cannot avoid exposures unless adults ensure that their environments are safe.

These differences between children and adults influence toxicity and exposure assessments for children, as well as options for risk management. A National Research Council (NRC) committee in its 1993 report, *Pesticides in the Diets of Infants and Children*, concluded that the toxicity of, and exposures to pesticides are frequently different for children and adults. It found that despite a wealth of scientific information to warrant addressing risks to children, the U.S. Environmental Protection Agency (EPA) rarely did so in making regulatory decisions about pesticides. The committee advised the EPA to incorporate information about dietary exposures to children in risk assessments, and augment pesticide testing with new assessments of neurotoxicity, developmental toxicity, endocrine effects, immunotoxicity, and developmental neurotoxicity. It recommended that the EPA include cumulative risks from pesticides that act through a common mechanism of action and aggregate risks from nonfood exposures when developing a tolerance for a pesticide. Since then, there has been a major undertaking by government to incorporate these recommendations into federal management of the use of pesticides (6).

In 1997, President Clinton signed an Executive Order requiring that all federal agencies ensure that their policies and rules address these and other disproportionate environmental health and safety risks to infants and children (7). The U.S. Department of Health and Human Services and the EPA recently moved to establish the first federal research centers dedicated solely to studying children's environmental health hazards. Grants of between $1.2 and $1.6 million were awarded to establish eight federal research centers. Five of the centers will study the links between the rise in asthma rates in children and environmental factors, such as second-hand smoke, smog, and other pollutants; the other three will conduct research on children's special vulnerability to pesticides (8).

There are many other vulnerable populations as well, many of whom are not in the workplace. Those who live in poverty are vulnerable because of the potential for multiple exposures, poorer diets, and lack of access to medical care. For example, children who are relatively deficient in iron or calcium absorb more lead per gram of intake than children who have adequate nutrition. The elderly population may be particularly susceptible to some environmental exposures; for example, they may have slower elimination of many toxicants. Those who have chronic illnesses are often more susceptible as well. For example, people who have human immunodeficiency virus infection or are

TABLE 3-1. *The environmental history*

Components of the environmental history	Specific questions and issues
Present and prior home locations	Neighborhood is strongly associated with housing age and condition as well as proximity to sources of air pollution and other ambient hazards.
Jobs of household members	Poor industrial hygiene practices at work can bring contaminants into the home, such as lead. Children may inappropriately be brought to the worksite, such as occurs frequently for farm work.
Environmental tobacco smoke	How many smokers are in the home and other environments? What about the automobile and other environments?
Lead exposure history	Centers for Disease Control and Prevention questions: Does the child live in or visit a house built before 1950? Does the child live in or visit a house built before 1978 with recent or ongoing renovation or remodeling? Does the child have a sibling or playmate who did have or now has lead poisoning? Other exposure issues are presence of lead in pipes solder in household plumbing, use of imported pottery or ethnic folk remedies, and parents' occupations or hobbies in lead-related industries.
Home insulating, heating and cooking system	Poorly vented stoves can result in high CO levels in the home.
Household building materials	Formaldehyde-containing materials may cause irritative and respiratory symptoms. A study indicated that vinyl wall and floor coverings may be a risk factor for obstructive bronchitis in children.
Home cleaning agents and other household products	Many household products contain toxic, allergenic, or irritant chemicals. For example, fragrance in detergent is a known cause of atopic dermatitis in infants.
Presence of pests, mold, pets, dust in the home	Asthmatic and other atopic people may be allergic to cockroaches, molds, animal danders, or dust mites. Is there a damp basement or recent flooding that might be conducive to mold growth? Are there pillows and stuffed animals that can be a reservoir for house dust mites? Shag carpets, which also serve as a reservoir for allergens?
Pesticide usage	Indoor, outdoor, and pet uses may result in exposure to household members.
Water supply	People on small water systems or with private wells are especially at risk. Bottled water may not be "safer" than municipal water. If a private well, when was the last time it was tested?
Diet	If a foodborne illness, what food was eaten and what was its source during the time of likely exposure? Was lead-glazed pottery used for food preparation?
Recent renovation/remodeling	Lead dust hazards can be created during renovation and remodeling activities. Newly installed carpets, wall coverings, and the like may off-gas irritating or toxic fumes.
Air pollution, indoor and outdoor	For asthmatics, what is the timing of asthma attacks vis-à-vis change in environments, air pollution alert days, and so forth? Where does the patient spend time, and what is the air quality in these locations? Include home, school, and other locations, including the automobile.
Hobbies	Hobbies such as painting, sculpting, welding, woodworking, piloting, autos, firearms, stained glass, ceramics, and gardening may bring chemicals and heavy metals into the home environment.
Ethnic/folk remedies or dietary supplements	Many ethnic or folk remedies contain lead and other heavy metals. So-called "dietary supplements" are not required by law to be carefully regulated by the Food and Drug Administration and may be of uncertain potency, purity, and composition.
Hazardous wastes/spills exposure	What is the proximity and what is known about the waste site? For spills, it is important to get a complete history of the episode and what the patient was doing at the time of and hours after the episode.
Recreational history	Obtain history of recent travel or fishing or swimming in polluted beaches, lakes, rivers, or streams, and history of foreign travel and food eaten and drinking water consumed.

immunosuppressed as a result of cancer therapy are much more at risk for serious infections from cryptosporidia in drinking water. Pregnant women are at risk not only from the perspective of exposure to the developing child but also because of altered physiology and metabolism of many toxic agents. For women, menopause may be another time of vulnerability; for example, there is strong evidence that at the time of menopause, BLLs increase because of liberation of stored lead from bones.

FIG. 3-1. Petrochemical plant in the Czech Republic, that accounts for much ambient air pollution in the surrounding area. (Photograph by Barry S. Levy.)

Specific Problems

The following is a review of some of the major issues in environmental medicine. This is by no means intended to be an exhaustive list of issues, nor necessarily a list of the most important issues; rather it includes a number of problems that are illustrative of the important challenges in environmental medicine. Some of these are organized by medium or vector for disease (i.e., ambient air pollution, indoor air pollution, water pollution, and contaminated food); some are by hazardous substance (i.e., pesticides, chemicals, lead, other heavy metals, and radiation), and some by adverse health effects (i.e., injuries, asthma, birth defects, and cancer).

Ambient Air Pollution

With the onset of the Industrial Revolution, air pollution became a major public health problem (Fig. 3-1) (9). In October 1948, in Donora, Pennsylvania, an air pollution episode resulted in the deaths of 19 people in a community of 14,000. In 1952, a similar episode in London, England, with a severe air inversion, resulted in the buildup of sulfur dioxide and particulate pollution and 4,000 excess deaths. Today, air pollution is still the cause of extensive rates of morbidity and mortality and is especially risky for children.

Air pollution is a complex mixture of substances discharged into the air in a myriad of ways. Incinerators and combustion sources, including vehicles, emit large quantities of carbon dioxide, carbon monoxide, nitrous oxides, and sulfur dioxide; more complex combustion byproducts, such as polycyclic aromatic hydrocarbons, dioxins, furans, and benzo[a]pyrene; and feedstock materials that do not burn, such as cadmium, lead, chromium, mercury, and other metals. The constituents of the emissions from any one source depend on such factors as the feedstock or fuel, the temperature of combustion, the oxygen availability for combustion, the chemical reactions between feedstock chemicals and byproducts, and the use of pollution controls such as after-burners and bag houses. Toxic air contaminants may be emitted as products of incomplete combustion or as a result of their manufacture, processing, use, or disposal. Examples are benzene emissions from gasoline tank nozzles, solvent emissions from automobile spray painting operations, and trichlorethylene emissions from waste disposal activities. For 1996, the EPA reported that manufacturing facilities alone released more than 1.4 billion pounds of TRI-listed contaminants into the air in the United States. Although this is a large amount of chemicals, the 1996 report also showed a 50% reduction in air releases since the baseline year of 1988. The most significant contributions to air emissions came from the following sectors: chemicals, paper, primary metals, plastics, and transportation/equipment. These five sectors accounted for two-thirds of the total emissions to air reported TRI.

So-called "priority" air pollutants are very important in causation and prevention of asthma and other chronic lung diseases. The six priority air pollutants regulated under the Clean Air Act are ozone, sulfur dioxide, respirable particulate matter, nitrogen dioxide, carbon monoxide, and lead. Lead is discussed later in this chapter. Priority air pollutants are to be regulated by the EPA on strictly a public health basis, with an adequate margin of safety to protect the population and special attention given to protection of vulnerable populations. Air pollutants, especially those that are the product of uncontrolled coal combustion, are responsible for increased levels of morbidity and mortality in children (10). They are also especially hazardous for people who have asthma, bronchitis, or other chronic respiratory diseases. Regulations for priority air pollutants are called National Ambient Air Quality Standards.

Ozone is the molecule O_3. Ozone naturally occurs at only very low levels in the troposphere, the part of the atmosphere where humans live. [This ozone is distinct from the stratospheric ozone, at very high altitudes, that helps protect the earth from ultraviolet (UV) radiation and is discussed later in this chapter.] Ozone pollution is the primary constituent of the smog and haze that can often be observed on days with higher pollution levels. Ozone formation is driven by photochemical reactions triggered by the action of UV radiation on precursor chemicals, nitrogen oxides (NO_x), and volatile organic compounds. The sources of these pollutants are mobile sources (automobiles and trucks), stationary sources (power plants and boilers), manufacturing plants, and personal-use products (e.g., lighter fluid). Because ozone formation is driven by UV light, levels are highest from late spring to early fall. In the course of a day, ozone formation begins in the mid-morning and peaks late in the afternoon. Ozone levels become most hazardous on days with strong UV irradiation, high lev-

els of automobile and other emissions of ozone-forming chemicals, and air inversion that causes poor mixing and dilution of air. Ideal conditions for ozone formation occur most often in cities like Los Angeles and Mexico City, where there are often low inversion layers in summer months, heavy automobile use, and strong sunlight. Ozone is a potent oxidant, and with acute exposure many people have substernal pain on deep inspiration, irritative cough, and reduced vital capacity and 1-second forced expiratory volume. Ozone levels have been associated with emergency department visits for acute respiratory illness, inflammatory lung changes, and, in clinical studies, with increased airway resistance and more severe symptoms in patients with asthma. Chronic ozone exposures seem to produce both obstructive and restrictive changes in the lungs.

Acid aerosols are also health threats. Combustion sources release sulfur dioxide, nitrogen dioxide, and particles; in the atmosphere, these are transformed to acid particles. Sulfate particles are more prevalent in the eastern United States, probably because they are produced from burning high-sulfur coal and sulfur-containing petroleum fuels for energy. In the western United States, nitrate species are more common, because automobiles comprise most of the acid source and because there is less use of sulfate-containing fuels. Acid aerosols cause chronic bronchitis and premature mortality from respiratory diseases.

Particulate pollution is actually a mixture of numerous pollutants. Fine particles (soot) are mostly the result of combustion. Larger particles can be generated by natural or human activities that generate soil disturbances and dusty areas. Particulate pollution causes the decreased visibility that can often be observed on a polluted day. Fine particles, so-called PM 2.5 (i.e., particulate matter of 2.5 μm or less in diameter), are of most concern because they are inspirable and therefore have more serious pulmonary effects. Fine particulate pollution exacerbates asthma, is associated with increased rates of hospitalization for respiratory diseases, and may increase the rates of lower respiratory infections in children. Fine particulates are also known to increase death rates among the elderly and those with chronic lung disease.

A major source of ozone and particulate pollution in the United States today is the internal combustion engine that powers automobiles and other vehicles (Fig. 3-2). From 1990 to 1995, the average car's emission of smog-causing compounds was cut nearly in half. Unfortunately, this trend was more than matched by two trends in the opposite direction. First, more miles are driven because of growth in numbers of automobiles per person and annual vehicle-miles driven per person. Second, automobile manufacturers, taking advantage of an exception built into the Clean Air Act for trucks for farmers, are marketing popular "sport utility vehicles" that are less efficient than other cars and thus pollute more per mile driven.

Hazardous air pollutants (HAPs) came into the public eye in 1985, with the catastrophic release of methyl isocyanate at a pesticide production facility in Bhopal, India. In the 1990 Clean Air Act, Congress directed the EPA to establish standards for nearly 200 HAPs. Unlike the standards for priority air pollutants, the HAP standards aim at the maximum achievable control technology (MACT). The EPA is moving rapidly to establish these MACT standards for various categories of industry. In a later phase, Congress directed the EPA to conduct "residual risk" analyses and to tighten the MACT standards if they provide inadequate public health protection.

Indoor Air

A 6-year-old child with asthma comes to the emergency department. He has been ill for the past 2 weeks and the family appears to be complying well with the medication that was prescribed. Both parents are smokers, but the mother smokes only outside the home. The family owns a dog, which lives outdoors. The parents own an automobile body shop that is located next door to the family's home. Both parents work in the shop during the day; the father supervises repair work and the mother is the bookkeeper and receptionist for the shop. Each day, the mother picks the child up from school and brings him home, where he does homework in the office in the shop while his parents are working. The child was referred for allergy testing and did not have allergies to dogs or other tested allergens.

In this case, a very brief environmental history has identified a number of factors that may be exacerbating the child's asthma. At home, in the automobile to and from school, and at the car shop, there are smokers and the child has exposures to environmental tobacco smoke. There is a pet; if the child were allergic to dogs, this could be a factor even though the dog lives outdoors. The car shop may be a hazardous environment for this child. If the ventilation is poor and engines are often running in the garage, he may be exposed to high levels of a number of potentially hazardous combustion products. Although he is receiving good medical care, the lack of effective environmental management is probably complicating the treatment of this child.

Indoor air exposures are as important in the general environment as they are in the workplace (11). Indoor air contains the pollutants found outdoors plus any substances added by the home environment (12,13). Chapter 23 describes the occupational disor-

FIG. 3-2. Traffic jams like this one account for much ambient air pollution. (Photograph by Stephen Delaney, U.S. Environmental Protection Agency.)

ders related to indoor air environments. All of these are also relevant to school and home environments. For example, there have been examples of "sick building syndrome" occurring in students in schools with indoor air problems and even concerns about development of multiple chemical sensitivity.

Environmental tobacco smoke is one of the most common hazardous pollutants found indoors and is associated with a number of risks. Children with higher levels of exposure to environmental tobacco smoke show decreased growth, increased incidence of ear infections, increased incidence and severity of respiratory infections, and exacerbation of asthma. In addition, they are much more likely to become smokers and to begin smoking earlier than other children. Epidemiologic studies have shown that nonsmoking adults who are married to smokers have a higher incidence of lung cancer as well. Environmental tobacco smoke is not just a problem in the home, but in other indoor spaces, such as vehicles (14).

Homes may have sources of carbon monoxide, such as poorly ventilated gas stoves or coal or wood burners for heating or cooking. Homes should be equipped with carbon monoxide monitors. Schools may use temporary buildings that are inadequately ventilated for many children in a classroom and may off-gas organic solvents.

Molds are also important indoor air contaminants. They occur because of excessive moisture or poor building maintenance. Many molds cause sensitization, allergic rhinitis, and asthma exacerbation. These molds may exist in basements, in ventilation ducts, and in automobile air conditioning systems. They can be difficult to eliminate once established in ventilation equipment. Some molds also have toxic properties. The *Stachybotrys atra* fungus was associated with a cluster of pulmonary hemosiderosis cases in infants in Cleveland, Ohio (15,16). The environmental study found a strong association between the outbreak, flooding of homes, and inadequate remediation after the flood. Primary prevention, prevention and treatment of moisture

in homes and other buildings, and maintenance of ventilation systems is the best way to address this problem.

Drinking Water

Water pollution by microorganisms and toxins (Fig. 3-3) is a major problem worldwide. Historically and in most countries in the world today, contaminated drinking water and food are responsible for large outbreaks of infectious disease and concomitant morbidity and mortality. In the United States, the sanitation movement of the 1800s resulted in the imposition of engineering controls to divert and treat wastewater and to dispose of garbage.

Today in the United States, most consumers obtain drinking water from a public water system regulated under federal standards and receive annual "consumer confidence" letters informing them of which contaminants were found in the prior year, and their health significance. Clinicians may be asked to help interpret these letters, and it is probably helpful for consumers to bring these letters if they have medical questions about them. Consumers living in areas served by small water systems or who drink water from their own wells may not have access to such information.

Herbicides are the most rapidly growing usage category of pesticides on food. Herbicide use has gone up partly in an effort to conserve soil by reducing tillage, but the result has been growing levels of herbicides in agricultural runoff and in groundwater in agricultural areas. Examples are the triazine pesticides, atrazine, simizine, and propazine, and a number of chloraniline herbicides, alachlor, metolochlor, and acetochlor. Many of these herbicides have known carcinogenic and other chronic health risks. Occurrence in surface water tend to show seasonal peaks, with highest levels during the active growing season in spring and summer. This is of particular concern in growing areas where large acreage is planted with a single crop,

FIG. 3-3. Although non-point sources account for increasing amounts of water pollution in the United States, stationary point sources still account for a substantial amount of water pollution, such as with dioxin, a byproduct of the manufacture of bleached white paper at this Mississippi plant. (Photograph by Earl Dotter.)

such as the corn, soy, wheat, sorghum, and canola growing areas in the Midwest. When groundwater contamination occurs, it is difficult to remediate, and these pesticides break down very slowly in that environment. Pollution prevention through pesticide use reduction and substitution with safer methods of weed control, including safer pesticides and measures that prevent runoff and movement into groundwater, are the strategies of choice. Also available are charcoal filtration systems that water purveyors can use to remove the pesticides from finished drinking water. These provide the additional benefit of removal of chlorination byproducts as well.

In April of 1993, Milwaukee reported a major outbreak of cryptosporidiosis, an infectious diarrheal illness caused by the protozoan *Cryptosporidium* (17). More than 400,000 people were symptomatic, with more than 4,000 requiring hospitalization. A total of 104 deaths were attributed to this epidemic. A dramatic increase in turbidity was noticed at the Milwaukee water treatment plants in association with the outbreak. A historical analysis of patterns of water turbidity and outpatient, emergency department, and inpatient medical care for gastroenteritis in Milwaukee found that increases in water turbidity over time were strongly associated with increases in gastroenteritis, especially emergency department and inpatient admissions. In this case, the epidemic was halted by advising people either to boil water for 5 minutes to kill spores or to drink bottled water. Milwaukee and many other cities have moved to increase treatment and filtration of drinking water, and the EPA is developing more sophisticated tests to detect cryptosporidia and has proposed the first federal standard for this pathogen. The source of pathogens was probably animal wastes upstream of the drinking water collection point, and efforts are underway under the Safe Drinking Water Act of 1996 to put in place more source protection efforts as well as water treatment facilities.

Worldwide, the most important health problem associated with drinking water is pathogens, including cryptosporidia. In developing countries, infectious diarrhea is still the largest cause of infant mortality. The pathogens often originate in human wastes or agricultural runoff. A variety of pathogens are found in contaminated drinking water, including protozoa (Cryptosporidia, Cyclospora, *Giardia*, and amebas), bacteria (*Vibrio* and *E. coli*), and viruses (hepatitis A and enterovirus). The primary line of protection is proper treatment of human wastes and preventing runoff of animal wastes into drinking water sources. This is easier said than done; sewage treatment plants are very costly to build and maintain. Also important is preventing cross-connections between systems that provide drinking water from systems that handle wastewater. Because millions of people have access to the drinking water infrastructure in a large urban area, this maintenance can be a very large job in itself. Filtering can remove many organisms. Disinfection is a third alternative. In the United States, disinfection is usually done through chlorination; another option is ozonation. The advantage of chlorination is that it leaves a residual level in the water that prevents contamination downstream of the drinking water treatment plant. The disadvantage is that chlorination byproducts are formed, as described later. Although in the United States many communities are taking steps to protect their source waters or to increase treatment, clinicians should be vigilant about identifying and reporting possible cases of infection with cryptosporidia and other reportable diseases. Some commu-

nities are advising immunocompromised people to boil water for 5 minutes before consumption when there is an increased risk of cryptosporidial contamination.

Chlorination byproducts (trihalomethanes), which result from treating drinking water to kill pathogens, are weak carcinogens. The evidence is disputed, but many epidemiologists believe that there are good human data to support the contention that increased exposure to trihalomethanes in drinking water is associated with increased risk of cancer (18). The EPA's Science Advisory Board published a report on future directions for safe drinking water that pointed to uncertainties in both the cancer risk assessment and our knowledge of the risks of microbial hazards. It suggested that new technologies are under development that may reduce the levels of chlorination required for the microbiologic safety of drinking water in the future (19).

A less frequent concern is well water contaminated with nitrates, a hazard for newborn infants. If nitrate-contaminated drinking water is used to mix infant formula, the child is at risk for development of blue baby syndrome, or methemoglobinemia. This is because newborn infants lack sufficient levels of methemoglobin reductase to reverse the oxidation of hemoglobin caused by nitrates.

Fluoridation of drinking water is a public health measure that adds fluoride to finished drinking water to supplement fluoride intake for the population. Some U.S. scientists are concerned that fluoride may be carcinogenic. The question of fluoride's carcinogenicity was reviewed by the National Toxicology Program (NTP) in the 1980s, and the conclusion by the Department of Health and Human Services was that there was no concern (20). In areas of the country that naturally have very high levels of fluoride in drinking water, fluorosis has occurred. At levels of ambient exposure, this has resulted in tooth mottling in children. Although dental health experts consider this to be a cosmetic defect, many of these communities are making efforts to reduce the fluoride levels in the their water supplies. Even in other areas, fluorosis is a theoretic possibility because of multiple sources of fluoride and use of fluoride-containing toothpaste and mouthwashes. It is important that children rinse and spit out toothpaste, especially fluoride rinses, to avoid excessive fluoride intake.

Food Safety

The earlier part of the 20th century was also an important time for progress in food safety. After the publication of Upton Sinclair's *The Jungle*, the U.S. Department of Agriculture (USDA) and the Food and Drug Administration (FDA) were given authority over the food industry to put in place a food inspection and surveillance system largely focused on meat-packing houses. These actions resulted in a dramatic decrease in morbidity and mortality from communicable diseases in the United States. Although the U.S. food supply is among the safest in the world, there are still serious hazards, especially for infants and children. Today, with the globalization of food production and the industrialization of both production and distribution of foods, identifying and tracking outbreaks of foodborne communicable disease has become quite complex (21).

In 1997, the U.S. Centers for Disease Control and Prevention (CDC) identified a multistate outbreak of a protozoan infection called Cyclospora. An elaborate investigation identified raspberries as the common source of exposure among the cases. The raspberries were traced to a producer in Guatemala. Although it was initially not possible to isolate Cyclospora from the raspberries, it was found that the producer was irrigating crops with contaminated water. The FDA embargoed shipments of Guatemalan raspberries until measures were taken to eliminate the practice, and the federal government began to develop methods for rapid detection of Cyclospora in foods.

A number of contaminants are found in food; toxic chemicals in food fall into three broad categories:

- Residues of pesticides deliberately applied to food crops or stored or processed foods
- Colorings, flavorings, and other chemicals deliberately added to food during processing
- Chemicals that inadvertently enter the food supply, such as aflatoxins, polychlorinated biphenyls (PCBs), mercury, and persistent pesticide residues like DDT.

Pathogens in food include:

- Viruses such as hepatitis A virus and Norwalk virus
- Bacteria like *Salmonella*, *Campylobacter*, *E. coli* O157:H7, and *Vibrio*
- Protozoa, such as *Toxoplasma gondii*, *Cryptosporidium*, and Cyclospora
- Aquatic microorganisms that elaborate toxins, like *Pfiesteria* and dinoflagellates responsible for producing red tides.

The problem of pesticides on food is discussed elsewhere in this chapter. Deliberate food additives are regulated by the FDA. The FDA requires stringent approval before allowing food additives to go to market. However, some types of food additives receive much less scrutiny and are potentially greater sources of concern. First are those that were grandfathered in by legislation—that is, Congress set less stringent requirements for existing food additives than for new ones. That decision created the potential to miss chronic and low-level hazards. There is some concern about additives that enter food through packaging materials, especially endocrine-disrupting chemicals like phthalates and bisphenol-A, and antimicrobial chemicals. The regulation of so-called "food supplements" is not very stringent. Inadvertent contaminants in food continue to be a public health problem. Chemicals like DDT and PCBs, long banned in the United States, continue to show up in the food supply. The major exposure route for dioxins is the food chain.

In 1996, the EPA and FDA conducted a

routine survey of dioxin levels in chickens in the United States to carry out a national exposure assessment. Two of 80 chickens sampled had elevated levels of dioxins, in the range of 25 parts per trillion in fat (compared with less than a fraction of a part per trillion in other chickens.) Believing a possible laboratory quality assurance problem was the cause, the chemist picked up two additional chickens to run as controls and found, to his surprise, that those too had elevated levels. On investigation, all four chickens with higher dioxin levels were traced back to farms using chicken feed with a ball clay additive. The ball clay was analyzed and contained high levels of dioxins. A trace forward found that the ball clay had been added to feeds for at least 30% of the catfish farms in the United States and a significant percentage of chicken and egg farms, as well as a number of more minor crops such as emus, alligators, and tilapia. The ball clay feed additive was banned on an emergency basis and all catfish, chicken, and egg products recalled by the FDA and the USDA. Animals fed the ball-clay–adulterated feed were kept out of the marketplace until dioxin levels in eggs or in meat were at or below 1 part per trillion fresh weight. Dioxin levels in food animals declined rapidly as they grew (by dilution) or laid eggs (by depuration) (22).

The U.S. food supply is among the safest in the world, but there are still millions stricken by foodborne illness every year and an estimated 9,000 per year, mostly the elderly and very young, die as a result. Infectious organisms are ubiquitous in the environment and can enter the food supply in a multitude of ways. Chickens infected with *Salmonella* can excrete them into eggs or in feces on the eggs. Shellfish and other seafood can become contaminated by pathogens like hepatitis A virus in manure runoff and sewage overflows. Nitrogen pollution can stimulate the growth of toxin-producing organisms like dinoflagellates and *Pfiesteria*. In both cases, the concern is overgrowth of microorganisms that elaborate toxic agents. When these toxins are present at high levels, fish

and shellfish can become contaminated, causing potentially serious neurologic illnesses in those who eat them. Animal feces can contaminate foods through polluted irrigation water, poor handling of manure, and unsanitary production and processing activities. Food can become contaminated in retail facilities, institutional settings, and homes from poor food-handling activities.

Many pathogens are particularly risky for children. *Salmonella*, *Listeria*, cyclospora, cryptosporidia, *E. coli* O157:H7, *Toxoplasma*, and *Campylobacter* are among many foodborne pathogens that pose particular risks to young children. The food production system is becoming more centralized and global, adding to the complexity of the picture of foodborne pathogen exposures. Outbreaks of hemolytic uremic syndrome and death due to *E. coli* O157:H7–contaminated hamburger meat have involved the transport and blending of meats from various parts of the country, often with distribution to regions remote from the source of the contamination. There have been a number of major outbreaks of *Listeria monocytogenes* infection due to contamination of lunch meats and hot dogs. *Listeria* is a cause of serious infections in the elderly, stillbirths in pregnancy, and meningitis in newborns. It is thought that longer shelf-lives for these products are providing *Listeria*, which can continue to grow in cold and even freezing temperatures, with the opportunity to increase to infectious levels in these delicatessen meats.

The USDA sets standards for pathogens on meat, poultry, and other foods and performs inspections of food items (Fig. 3-4). The CDC monitors incidence of foodborne illness, along with state and local public health agencies. The EPA regulates the discharge of pollutants into waters that may later contaminate food. Given the complexity of the food distribution system, it is now more important than ever that physicians recognize and report foodborne illnesses to public health agencies.

Food contamination is best prevented by application of good agricultural and manu-

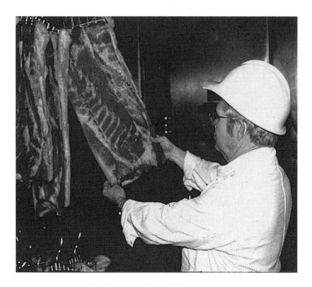

FIG. 3-4. Inspector with the U.S. Department of Agriculture examines sides of beef. Governmental meat and poultry inspections do not adequately prevent distribution of contaminated food products. (Photograph courtesy of the U.S. Department of Agriculture.)

facturing practices, as well as careful food preparation and storage. In the case of pesticides, this is accomplished primarily by reduction of unnecessary uses of pesticides on foods at all stages of production. Integrated pest management, using information about pest biology to control pests, is a means to reduce the risks and use of pesticides by using only what is necessary. Advances in farming technology include precision pesticide application to prevent spillage and drift, use of biologic control systems to avoid the need for chemical pesticides, and surveillance techniques to use pesticides only when needed rather than for preventive purposes. "Organic" farming does not yet have a single definition, but loosely means farming without the use of chemical pesticides and with a minimum of other pesticides as well. Federal standards for organic food have been proposed but are not yet finalized. However, it has been demonstrated that organic food has lower levels of pesticides, and consumers can reduce levels of pesticides in food and risks to the environment of pesticide usage by purchasing organic foods.

In the case of pathogens, care must be taken in every step of food production, storage, and preparation to prevent their introduction into the food supply. In the manufacturing process, these methods are referred to as hazard analysis and critical control point (HACCP) systems. HACCP systems require that food processors, farmers, and others identify points where contamination is likely to occur and implement control processes to prevent it. Also important are education for everyone on proper preparation and storage of foods, pasteurization, protection of animal health, prevention of discharge of pathogens and nitrogen into water bodies, and numerous other efforts to ensure that at every stage of production pathogens are not introduced into the food supply.

Antibiotic use in agriculture plays a role in the development of antibiotic resistance in foodborne pathogens. Although the magnitude of the problem is controversial, there is little doubt that the emergence of antibiotic-resistant organisms is related to selection of resistant strains through antibiotic use. Bacteria are capable of sharing resistance genes across strains and species. Therefore, on a theoretic basis alone, massive antibiotic use in agriculture is certainly contributing to the more rapid development of antibiotic-resistant organisms. The FDA is, at the time of this writing, considering limiting the use of therapeutic antibiotics in food animals. In Europe, such use is not allowed. Moreover, in Europe, the practice of subtherapeutic use of antibiotics for food animals has been banned. However, this practice continues in the United States because it is thought that it would not be economically feasible to abandon it.

Enforcement of the food safety laws plays an important role in prevention at all levels. This job is shared by two federal agencies, the FDA and the USDA, state agriculture and public health agencies, and local public health, environmental, and agricultural agencies. The FDA is responsible for ensuring the safety of all foods except meat and poultry, which are regulated by the USDA. In addition, the USDA is responsible for the federal school lunch program. Thus, regulation and enforcement involve a very complex patchwork of federal, state, and local laws and regulations. Some enforcement efforts involve routine monitoring and surveillance of the food supply; other efforts are in response to reports of problems and incidents. Health care providers play an important role in reporting food poisoning incidents to state and local public health agencies. For example, physician reports of outbreaks of hemolytic uremic syndrome caused by *E. coli* O157:H7 have led to stronger enforcement efforts to ensure that products as diverse as hamburger meat and apple juice are safe for children.

Food safety also depends on the efforts of those who prepare and consume food. There are a number of measures that should be taken by consumers:

- Thorough washing of fruits and vegetables with water removes some pathogens as well as many pesticide residues. Wash before peeling. It is not necessary to use soap or chemicals when washing food.
- Raw milk products (including cheese) and raw eggs, fish, and meat should be avoided in small children, the elderly, and the immunocompromised.
- Thoroughly cook meat, poultry, and eggs to ensure that pathogens are killed.
- When preparing poultry, be careful to wash hands, cutting boards, and any implements that were used to prepare the raw poultry

with soap and hot water before touching any other foods. Cook stuffing for poultry separately rather than inside the birds.
- Store food appropriately. Refrigeration of prepared food prevents growth of many microorganisms responsible for food poisoning.
- Incorporation of chemical agents into kitchen sponges, high chair trays, and cutting boards and similar practices do not have a role in preventing foodborne infections. It is also unnecessary to use chemical disinfectants for washing hands in the home; soap and water are effective.

Pesticides

A pesticide is a chemical or biologic agent used to control (or kill) a nonhuman organism considered by humans to be a "pest"— that is, inimical to human interest. Thus, the term *pesticide* encompasses insecticides, fungicides, herbicides, rodenticides, antimicrobial disinfectants, and biocides. Pesticides are applied extensively to food crops in nations around the world. More than 400 different pesticide active ingredients, formulated into thousands of products, are registered for use on food in the United States. Pesticides are used at all stages of food production, to protect against pests in the field, and in shipping and storage.

The sheer volume of pesticide use necessitates a pollution prevention approach. In 1993, for instance, an estimated 4.23 billion pounds of pesticides were used in the United States; this total is based on the amount of active ingredients only. The figure includes conventional pesticides used in agriculture, wood preservatives, disinfectants, and water treatment, such as swimming pool chemicals. Table 3-2 gives some examples of pesticide active ingredients. Chapter 41 presents additional information on pesticides.

There are numerous pathways for exposure to pesticides in the ambient environment. Of concern to communities near farming areas is pesticide spray drift that can

TABLE 3-2. *Examples of major classes of pesticides*

Insecticides
Organophosphates and carbamates
 Malathion, chlorpyrifos, parathion, carbofuran, aldicarb
Organochlorines
 DDT/DDE, aldrin/dieldrin, kepone, mirex, HCB, chlordane, heptachlor, methoxychlor, lindane
Others
 Biologics, pheromones, pyrethrins and pyrethroids, chitin inhibitors, Bt products
Herbicides
Triazines
 Atrazine, simizine, cyanazine
Chloranilines
 Alachlor, acetochlor, metolochlor
Others
 Glyphosate, sulfonylureas, 2,4,-D
Fungicides
Benylate, chlorthalanil, vinclozolin

and cancer. Because these effects may occur after years of exposure, with a long latency period, they may be difficult to attribute to exposure to residues in food. The EPA sets standards for allowable levels of pesticides on food, called *tolerances*. The FDA and USDA monitor the food supply for pesticide residues. It is estimated that approximately 1% of food sold in the United States has pesticide residues above the legal limit. Further, the EPA has only begun to implement a new standard that accounts for the cumulative effects of multiple pesticides with the same mechanism of action. Clearly, it is prudent for consumers to avoid pesticide exposures by, for example, carefully washing fruits and vegetables.

result in significant exposures to children residing near farms (Fig. 3-5).

Agricultural pesticides also find their way to people through foods. Both acute and chronic hazards are associated with pesticides. The most important acute effects related to food contamination are cholinesterase inhibition (from high levels of organophosphates or carbamates) and developmental toxicity (due to exposure to teratogens during a sensitive period *in utero*). Important chronic health risks include developmental neurotoxicity, endocrine effects,

In 1994 the FDA, in a routine market-basket study of pesticides in the diet, identified an unapproved pesticide, chlorpyrifos, on ready-to-eat oat cereals. On further investigation, it was found that an applicator had applied chlorpyrifos, rather than a similar approved pesticide, chlorpyrifos methyl, to oat grain in storage silos. The FDA and the industry conducted extensive sampling and found that the illegal pesticide contaminated millions of pounds of grain as well as millions of packages of oat cereal in commerce, factories, and warehouses. EPA scientists conducted a risk assessment that concluded that the residues were well below any levels of health concern. However, be-

A   B

FIG. 3-5. Much environmental contamination and human exposure to pesticides still occurs as a result of **(A)** aerial spraying, and **(B)** runoff from fields of pesticides applied by rigs, such as this one. (Photographs by Stephen Delaney, U.S. Environmental Protection Agency.)

cause the food contained an illegally applied pesticide and was adulterated, the FDA ordered a recall of all affected cereal products upstream of the retail market. Millions of boxes of cereal were destroyed. The FDA allowed adulterated oat grain to be used for animal feed (because that would not have resulted in illegal residues), but not for human consumption. Consumers were not advised to send back or discard product on their shelves, but nonetheless the cereal company reported receiving thousands of telephone calls and a decrease in retail sales over a several-month period as a result of this incident. The pesticide applicator was prosecuted by the state and federal government and served time in jail.

On July 4, 1985 in northern California (23,24), an emergency department physician on duty at 4 a.m. admitted a 62-year-old woman on digoxin with hypotension, severe bradycardia (heart rate of 31 per minute), atrial fibrillation, diaphoresis, vomiting, diarrhea, lacrimation, salivation, and muscle twitching. He treated her with atropine with resolution of symptoms. The patient had been driving from her home to a fishing resort. She and two other family members reporting consuming watermelon 30 minutes before coming to the emergency department. The other adults were also ill and had similar but milder symptoms. The physician alerted a regional poison center that, in turn, alerted the state health department of the possible association of watermelon and symptoms of cholinesterase inhibitor poisoning.

Later that day, the State of Oregon notified California that it had traced back to California watermelons a similar cluster of illnesses related to watermelon ingestion. Laboratory evaluation of the watermelon found aldicarb sulfoxide, the first (and toxic) metabolite of the pesticide aldicarb (Temik). Because contaminated watermelons could not be traced to specific store chains, warehouses, or suppliers, it was necessary for three states—California, Washington, and Oregon—to order an immediate embargo and ultimately a destruction of all watermelons, as well as alerting consumers, on July 4(!), to avoid eating watermelons. Despite these precautions 1,376 illnesses were reported to public health authorities, of which 77% were classified as being probably or possibly due to aldicarb ingestion.

This episode illustrates the importance of an alert clinician, a thorough exposure history, and rapid response to what can be a major poisoning episode even with just a few index cases. Aldicarb, a systemic pesticide, was not approved for use on watermelons and other cucurbits because of the potential for hazard, as seen in this episode. It was probably applied illegally but also could have drifted from neighboring fields, where it was used legally. The symptoms experienced by those who ate the aldicarb-contaminated watermelons were those of cholinesterase inhibition. Both organophosphates and carbamates cause these symptoms, but organophosphates have a longer time of onset and duration of symptoms than carbamates, which are rapidly metabolized and detoxified. This symptom complex is sometimes referred to by its mnemonic, MUDDLES, where:

M = miosis (contraction of pupils)
U = urination
D = diarrhea and diaphoresis
D = disorientation
L = lacrimation
E = excitation
S = salivation

In 1993, the NRC found that the EPA was not adequately accounting for children's diets and risks in setting standards for pesticides in food (6). In response, the EPA changed its methodology for dietary exposure assessment so that it could incorporate available information about children's diets. The EPA also updated a number of test guidelines to generate more complete information about developmental, neurologic, and endocrine effects from pesticides. In 1996, Congress enacted the Food Quality Protection Act (FQPA), which codified a number of these new changes (Table 3-3). New in FQPA were requirements for cumulative and aggregate risk assessment. *Aggregate risk* means considering all routes of ex-

TABLE 3-3. *Food Quality Protection Act of 1996: New provisions related to protection of infants and children*

Health-based standard: A new standard of a "reasonable certainty of no harm" that prohibits taking into account economic considerations when children are at risk.

Additional margin of safety: Requires that the EPA use an additional 10-fold margin of safety to protect children. Less than a 10-fold margin of safety may be used when there are adequate data to assess prenatal and postnatal developmental risks.

Account for children's diets: Requires the use of age-appropriate estimates of dietary consumption in establishing allowable levels of pesticides on food to account for children's unique dietary patterns.

Account for all exposures: In establishing acceptable levels of a pesticide on food, the EPA must account for exposures that may occur through other routes, such as drinking water and residential application of the pesticide.

Cumulative impacts: The EPA must consider the cumulative impacts of all pesticides that may share a common mechanism of action.

Tolerance reassessments: All existing pesticide food standards must be reassessed over a 10-year period to ensure that they meet the new standard to protect children.

Endocrine disruptor testing: The EPA must screen and test all pesticides and pesticide ingredients for estrogen effects and other endocrine disruptor activity.

Registration renewal: Establishes a 15-year renewal process for all pesticides to ensure that they have up-to-date science evaluations over time.

EPA, Environmental Protection Agency.

posure and uses of a pesticide rather than approving uses one at a time. The EPA already aggregated food uses, but the 1996 statute required a review of drinking water and household exposures as well. *Cumulative risk* means considering all pesticides that may share a common mechanism of action. An early decision made by the EPA is that all 40 organophosphate pesticides need to be assessed cumulatively. Another challenging new provision requires "an additional tenfold margin of safety to protect children." Congress went on to say that "the Administrator may use a different margin of safety . . . only if, on the basis of reliable data, such margin will be safe for infants and children." This provision, which reflects the views of the NRC, is requiring the development of

new policies, which in turn are resulting in new approaches to risk assessments for pesticides. The 10-fold safety factor and accompanying effort to define adequate data to assess children's exposures and risks are likely to have far-reaching effects on public policy. At the time of this writing, there is still much debate about application of the 10-fold FQPA factor. At the heart of the debate are two questions: first, what is the scientific (as opposed to the legal) justification for applying the 10-fold factor? Second, what constitutes reliable and complete scientific data on which to base that decision? Both of these questions have implications for establishing child protective standards across the board, not just for pesticides.

A substantial number of preschool children are exposed to or poisoned by pesticides each year. Based on calls to poison control centers, there were an estimated 125,000 such exposure-related cases in 1989. Based on a 1% statistical sample of the nation's emergency departments, there were 11,600 pesticide exposures in 1985 in this age group. Ten percent of these cases were admitted for hospitalization and 25% reported symptoms. Insecticides and rodenticides accounted for 87% of these cases, and ingestion had occurred in 76% of the cases. From 1977 to 1982, there were an estimated 900 childhood hospitalizations per year due to pesticides (25,26).

A number of factors account for these ingestions. It is only since the early 1990s that the EPA has required child-resistant packaging and it is expected that, as was the experience with consumer products, the rates of poisonings will decline. In addition, since the late 1980s, a number of pesticides have been taken out of home use by being either cancelled or restricted, meaning that they can be applied only by certified applicators. However, a survey conducted by the EPA indicated that there are additional preventive measures needed in the home. Almost half of all households (47%) with children younger than 5 years of age had at least one pesticide stored in an unlocked cabinet that was less than 4 feet off the ground or in easy reach of children; approximately 75% of households with no children younger than 5 years of age store pesticides in an unlocked cabinet. This is significant because 13% of children's pesticide poisoning incidents occur in homes other than their own. Pesticides need to be kept out of reach of children, not only in their homes but in other homes where they are likely to spend time.

Pesticides are also applied directly to residential areas. Household applications can lead to exposure, especially if pesticides spill or leak into ventilation systems or cracks in slabs in homes. Homeowners may overuse pesticides like flea bombs or fail to follow precautionary labeling, such as wearing gloves, or reentry instructions. Misuses can occur. In 1996, there was widespread misuse of the agricultural pesticide methyl parathion for household pest control, resulting in millions of dollars of cleanups to make the homes habitable. In other instances, pest control operators may fail to clean up spills or leaks properly. It is prudent for all consumers, but especially parents of young children, to be cautious about the use of household pesticides (27).

> A 12-week-old girl was brought in for evaluation by her parents because they noticed increased arm and leg tone. Her physician gave her a diagnosis of mild cerebral palsy and arranged for pediatric child development intervention. Several weeks later, on a follow-up visit, the physician obtained a history of diazinon (an organophosphate insecticide) spraying in her home in the days before symptom onset. A local university pesticide toxicologist was contacted and an environmental investigation was conducted. The investigator found diazinon residues in the home. At age 8 months, the child's urine was examined and it contained high levels of alkyl phosphate, a metabolite of diazinon. At that point, the child was removed from the environment. Her symptoms resolved in 6 weeks (28).

This case, which occurred in Oregon in 1993, illustrates how difficult even serious chronic toxicity is to detect. Rarely would an environmental history be taken in the case of an infant with a diagnosis of cerebral palsy, and very few clinicians have access to investigative resources to monitor homes and children for residues of diazinon and other pesticides.

Industrial Chemicals

There are some 85,000 chemicals on the EPA's inventory of industrial chemicals, but most of these have never been in commerce. However, there are approximately 15,000 chemicals produced or imported in the United States at amounts of at least 10,000 pounds per year, and approximately 3,000 chemicals at at least a million pounds per year. These are called high–production-volume chemicals. Studies indicate that we know very little about these chemicals (29). In 1998, only 7% had a complete set of basic screening information, called the Screening Inventory Data Set (SIDS) battery. Approximately 40% had no SIDS data at all. For chemicals with Occupational Safety and Health Administration permissible exposure limits, only approximately one-fourth had complete SIDS data. A voluntary effort currently underway by the chemical industry should result in availability of screening level data by the year 2003 for all high–production-volume chemicals. Acute toxicity from chemicals is rarely seen in environmental medicine, although cases do occur in association with use and misuse of consumer products. However, most concerns involve low-level exposures and long-term effects like cancer, reproductive toxicity, and neurotoxicity. Because of the great gaps in testing information, very little is known about the risks of most of the chemicals in the human environment.

Dioxins and PCBs are persistent toxic chemicals that are classified by the NTP as probable carcinogens and are considered by the EPA to be reproductive toxicants. PCBs were partially phased out of use in the United States in the 1980s, but there is still a significant amount of environmental PCB exposure, especially through consumption of fish from contaminated waters. Low-level PCB exposure has been found to cause developmental neurotoxicity in children, and thus there are PCB fish advisories for the Great Lakes and in several other parts of the United States. Dioxins have a very long half-life in the environment and today are mostly produced as a result of combustion processes.

An *endocrine disruptor* is defined by the EPA as "an exogenous agent which interferes with the synthesis, secretion, transport, binding action, or elimination of natural hormones in the body which are responsible for homeostasis, reproduction, development, or behavior" (see Chapter 36). The EPA has begun a program to set priorities among the 15,000 chemicals produced at rates of least 10,000 pounds per year for endocrine disruptor screening and testing (30).

Consumer products contain many of these chemicals. Poison control centers are an excellent resource if exposures are suspected, and they play a useful role in contributing to surveillance efforts. Serious adverse effects should be reported to the product maker and to the EPA Office of Pollution Prevention and Toxics.

Lead

Lead poisoning has been recognized since antiquity. In the second century B.C., Dioscorides, a Greek physician, said "lead makes the mind give way." Lead poisoning in adults continues to occur today, mostly as a consequence of occupational exposures. However, adults can be exposed through some of the same sources as children. So, although the discussion in this chapter focuses on childhood lead poisoning, many of the same exposures for children are also important for adults. (Also see Chapters 15, 29, 31, and 35.)

Childhood lead poisoning from lead-based paint was first described in Brisbane, Australia, in 1897. The cause of this endemic illness was identified as painted porch railings. In 1920, the city of Brisbane passed the first lead paint poisoning prevention act. In the

United States, plumbism from lead-based paint was described in the first decade of the 20th century. It was initially believed that if a child recovered from the acute illness, there were no sequelae. Byers and Lord refuted this in 1943 in their report of 20 children who had recovered from acute lead intoxication; 19 had obvious behavioral disorders or mental retardation. Better-designed and more sophisticated studies have been performed since then, and there is a general consensus of opinion about the relationship between lead and cognitive function (31).

The most recent data from the CDC document that the average (geometric mean) children's BLL fell from 12.8 to 2.8 μg/dL between 1976 and 1980, and 1988 and 1991. Most significantly, during the same period the percentage of U.S. children between 1 and 5 years of age with BLLs above 10 μg/dL (the current CDC definition of toxicity) fell from 88% to 9%. Similar drops in BLLs were seen for all age groups and income levels, and for inner-city and rural residents alike. This is surely one of the most remarkable public health achievements of the decade. However, whereas 4% of the U.S. population had BLLs of 10 μg/dL or greater in 1988 to 1991, almost 12% of children 1 to 2 years of age had BLLs in that range. Exposure is greatest to minority children, with 10% of Mexican-American and 22% of black children having BLLs in this range, and to low-income children, children living in larger urban areas, and children living in the Northeast. Thus, there is exposure in all strata of society, yet there are children at much higher risk who deserve the most attention. Certainly any child living in a house containing lead-based paint may be at risk. Such housing and other sources of lead are found throughout the United States. A national survey of housing conducted by the Department of Housing and Urban Development (HUD) found that the age of housing, not geographic location, is the best predictor for presence of lead-based paint (32,33).

Lead is absorbed by ingestion or inhalation. The relationship between exposure and

BLL is a dynamic process in which BLL represents a product of recent exposures, excretion, and equilibration with other tissues. Children deficient in iron, protein, calcium, or zinc absorb lead more readily. Most retained lead is stored in the bones. At high BLLs ($>70 \mu g/dL$), lead may cause encephalopathy and death in children. Survivors of encephalopathy almost always have lifelong severe disabilities, such as seizures and mental retardation. Lead toxicity affects almost every organ system, but most importantly the central nervous system, peripheral nervous system, kidneys, and blood. Lead interferes with enzymes that catalyze the formation of heme, inhibits prenatal and postnatal growth, and impairs hearing acuity. Lead is a carcinogen in laboratory animals, and there is some evidence for carcinogenicity in workers, but not in children.

Although the impairment of cognition in young children at the level $10 \mu g/dL$ has been established, no threshold has been identified. At lower BLL values, the impact on an individual child may be undetectable. In contrast, there may be a significant impact on a population of children with such BLLs. This body of literature has been examined by meta-analysis, which is a way of synthesizing data from multiple studies. The relationship between lead and IQ deficits was found to be remarkably consistent. A number of studies have found that for every BLL increase of 10 to 15 $\mu g/dL$, within the range of 5 to 35 $\mu g/dL$, there is a lowering of the mean IQ in children by 2 to 4 points (34).

Evidence suggests that the effects of early lead exposure can persist. A follow-up study of a group of subjects classified by dentin lead levels in the first and second grade into adulthood showed that those with high tooth leads as children were seven times more likely not to graduate from high school and six times more likely to have reading scores at least two grades below expected, after adjustment for a number of factors, including socioeconomic status and parental IQ (35). They also had higher absenteeism in the final year of school, lower class rank, poorer vocabulary and grammatical reasoning scores, longer reaction times, and poorer hand–eye coordination.

Lead paint is the major source of lead exposure for children in the United States. As lead paint deteriorates or is removed, house dust and soil become contaminated, and lead enters the body through normal hand-to-mouth activity (Fig. 3-6A). Children may also ingest paint chips. Before 1955, most house paint was so-called "white lead," 50:50 lead and linseed oil. In 1955, manufacturers adopted a voluntary house paint lead standard of 1%, but house paint with higher levels continued to be manufactured (36). In the United States, the amount of lead allowable in paint was lowered by law in two steps, to 1% in 1971 and then to 0.06% in 1977. Occasionally, lead paint manufactured for nonresidential purposes continues to be used to paint houses. It is estimated that 5 million tons of lead have been applied to houses in the United States. Seventy percent of the homes built before 1960 are estimated to have lead paint. Most dangerous are the 3.8 million homes with decaying or deteriorating lead paint, in which 2 million children younger than 6 years of age live.

Uncontaminated soil contains lead concentrations of less than 50 parts per million (ppm), but soil lead levels in many urban areas exceed 200 ppm. Areas near lead mines, industries, and smelters may have high levels of soil contamination (up to 60,000 ppm). In the United States, the use of leaded gasoline released an estimated 30 million tons of lead into the air. Lead in house dust is an important source of exposure, and comes from paint, soil, and other sources. Acidic water of low mineral content can leach large amounts of lead from lead pipes or solder; this is particularly apt to occur when water has been standing in pipes for extended periods. Hot water may be of particular concern. Lead solder and fittings can also be found in older drinking water coolers and coffee urns. Brass fixtures may also be lead contaminated. Lead-contaminated water has been linked to lead poison-

FIG. 3-6. Lead-based paint in many older homes still represents a serious health hazard to many young children. **(A)** Potential child exposure to peeling lead-based paint on a windowsill, a common site for such exposure. Although the most important pathway of exposure to lead-based paint is through house dust and hand-to-mouth activity by young children, paint chips may be directly ingested, and toddlers often stand at a windowsill while chewing or sucking on the paint. **(B)** Lead abatement workers, with personal protective equipment, perform postabatement cleanup. Workers performing lead abatement must be trained and certified, and they must carefully adhere to safe practice standards. (Photographs by the California Department of Health Services.)

ing in children given reconstituted infant formula.

Lead may also contaminate food. Soil lead is taken up by root vegetables and atmospheric lead may fall on to leafy vegetables. Lead may also be added to food during processing. Cans with soldered seams can add lead to foods. In the United States, soldered cans have largely been replaced by seamless aluminum containers, but imported and large, commercial-sized cans still have lead-soldered seams. Other modes of food contamination include some ceramic tableware (especially imported), certain "natural" calcium supplements, and bright red and yellow paints on bread bags.

Other lead sources include ethnic folk remedies (*azarcon* and *greta* used by Hispanics and *pay-loo-ah* used by Southeast Asians), eye cosmetics (kohl used by Moslems and surma by Hindus), hobbies (e.g., stained glass, artist paints, and shooting ranges), household fixtures (e.g., plastic miniblinds manufactured in lead molds), and accidental ingestion of small lead objects (e.g., fishing weights or curtain weights). Parents employed in the lead industry may bring lead dust home on clothing or expose children by allowing them to visit worksites.

Identification of the source of lead requires a careful history and usually investigation of the household. There are several strategies for preventing childhood lead poisoning.

Primary Prevention

Remove Lead from Children's Environment. This is the most effective preventive measure. For past contamination problems, source reduction involves removal of lead or modification of the environment to prevent children's contact with lead (see Fig. 3-6B). Screening of lead-poisoned children is useful in identifying areas in most need of environmental cleanup and preventing other cases of lead exposure.

Management in Place. Stringent dust control measures like frequent wet mopping and vacuuming of homes with a high-efficiency particle vacuum cleaner (hepavac) lower children's exposures to lead. However, this approach is subject to breakdown when families move and properties change hands.

Secondary Prevention

Screen Children for Lead. Lead poisoning is one of the causes of mental retardation and neurologic impairment that can be prevented with routine screening. In 1991, the CDC called for universal screening of children for lead exposure (37). However, a national survey in 1994 showed that only one-fourth of children were screened and that only one-third of low-income children, those most at risk, were screened. Therefore, in 1997, the CDC revised its recommendation to recommend that, on a state-by-state basis, plans be developed to ensure that children who need to be screened are tested for lead exposure. In the absence of a state plan, children should be screened if they live in targeted zip codes (with older housing), are on public assistance, or if the parents answer "yes" to the CDC exposure questions (see Table 3-1).

Tertiary Prevention

Low-Level Exposures. If a child has a BLL between 10 and 14 μg/dL, surveillance and anticipatory guidance should be initiated. The toxicity of lead is a function of both dose and duration of exposure. It is the role of the practitioner to give realistic reassurance that early detection and source control can minimize intellectual and behavioral consequences for the individual child. Urgency of timing for follow-up of an initial or screening BLL with a diagnostic BLL depends on the initial BLL (Table 3-4).

Further treatment and evaluation depends on the follow-up BLL. If the diagnostic BLL is 10 to 14 μg/dL, the recommendation is to retest the child within 3 months. If it is be-

tween 15 and 19 μg/dL, retest in 2 months. If the BLL is 20 μg/dL or higher, or if the child has two venous BLLs between 15 and 19 μg/dL at least 3 months apart, then clinical management is needed. Clinical management includes clinical evaluation of consequences of lead exposure to the child, family lead education and referrals, chelation therapy (as appropriate), and follow-up lead testing.

Patient Education. Education of parents about nutritional sources of calcium, iron, zinc, and ascorbate is important for all children, but especially children with BLLs of 15 μg/dL and higher. Parents' attention should be directed to the following steps to avoid lead exposure in their children: (a) having paint removal, renovation, and remodeling in the home done by trained, experienced personnel with the family out of the home and proper cleanup done before they return; (b) controlling dust and paint chip debris; (c) preventing the children from eating dirt or other foreign substances; (d) changing workclothes and cleaning up before going home from a lead-related job; (e) avoiding the use of lead around the home for hobbies and other purposes; (f) handwashing; and (c) using cold tap water for drinking and especially for mixing infant formula. Any child in the same home with a child with a BLL of 20 μg/dL or higher, where lead exposure is believed to have occurred, should also be tested. If other locations, such as day care centers, schools, playgrounds, or baby sitters' homes, are identified as being lead contaminated, children in those environments should be tested as well.

TABLE 3-4. *Urgency and extent of follow-up depend on the initial blood lead level*

If the result of lead screening (μg/dL) is:	Perform diagnostic test on venous blood within:
10–19	3 mo
20–44	1 mo–1 wk[a]
45–59	48 h
60–69	24 h
70 or higher	Immediately as an emergency laboratory test

[a] The higher the blood lead level, the more urgent the need for a follow-up diagnostic test.

A 3-year-old Hispanic boy was transferred by air to a major California medical center with an encephalopathy of unknown etiology. He had been admitted to a regional hospital 3 days earlier because of deteriorating mental status. The child had an unremarkable medical history except for having been treated for iron deficiency anemia for the last year. On admission, blood and urine samples were sent for toxicology screens, along with a battery of other medical tests.

His BLL was 65 μg/dL. The child was treated with chelation therapy and support and a referral was made to the local environmental health agency for an environmental investigation to identify the source of lead poisoning. It identified lead-based paint on the windowsills of the home, and abatement of the lead-painted surfaces was carried out. The child was discharged to home with a BLL of 15 μg/dL and instructions for follow-up care. On return for follow-up 1 month later, the BLL was 25 μg/dL.

"Rebound" is often seen after chelation, but 2 weeks later the BLL was 30 . Again, the local agency was called in, and on further investigation it was found that the child's father worked in a radiator repair shop. Every day, he came home wearing his workclothes and he often held the child in his lap. Instructions were given for changing and showering before having contact with the boy, but nonetheless on return visit 2 weeks later the BLL was 40 μg/dL. The child was again admitted for chelation. On further questioning, family members admitted to having given the child doses of an ethnic folk remedy for *empacho*, or stomachache. On analysis, this remedy was found to contain a high concentration of lead.

This episode illustrates the multiple potential pathways of lead exposure in the environment. All three sources of lead in this child's environment probably contributed to his exposure. Investigators need to do a thorough exposure investigation, including an occupational assessment, in the investigation of a lead-exposed child. Family members are often reluctant to admit to physicians the use of ethnic folk remedies, and careful and sensitive questioning is required to elicit this history.

Other Heavy Metals

Mercury exposures are also common and serious health hazards in the environment. Elemental mercury is present in many products. A unique syndrome called *acrodynia* can develop in children exposed to elemental mercury. Biologic monitoring of elemental mercury levels in children is done using urinary levels, and at high levels of exposure chelation therapy may be warranted. The clinician should refer to or consult with a pediatric toxicologist if possible. Organic mercury (methylmercury) is neurotoxic to the fetus and young child. It is found in fish caught in contaminated waters, and there are local fish advisories in place to limit consumption by children and pregnant women. At most risk are sports and subsistence fishers who habitually catch fish from the same contaminated area, and family members who consume the fish.

Radiation

Radiation exposures include ionizing and nonionizing radiation, and occur against a background of natural exposures. The nature, measurement, sources, environmental levels, and effects of ionizing radiation are reviewed in Chapter 17 Radon is probably the most important environmental radiation hazard in terms of risk and exposure. Exposures in the home occur when radon-containing gas enters the home through the foundation or cracks in slabs. Homes that are "tight" (i.e., poorly ventilated) are at high risk because the gas can accumulate. The occupational experience with lung cancer in hard-rock and uranium mining indicates that radon, once inhaled, can emit harmful alpha particles that cause tissue damage and, ultimately, lung cancer. Radon levels in homes vary according to the geology of the area, and this problem therefore varies in importance by geographic area. Homes that are tight and in radon-rich areas may have high exposures. A radon test is recommended to evaluate the risk, and there are mitigation options available for homeowners (38).

Ultraviolet light is also an exposure of concern. Excessive exposure to UVB rays causes not only sunburn but increased rates of cataracts and skin cancers, particularly melanoma. The deterioration of the stratospheric ozone layer (the "good" ozone), which was brought about by human use of chlorofluorocarbons and other ozone-destructive chemicals, has resulted in increased exposure to UVBs to humans and the ecosystem. Although steps have been taken to phase out the use of these chemicals, it will be decades before the ozone layer completely recovers because chlorofluorocarbons have a long half-life in the stratosphere. Therefore, physicians need to continue to recommend personal protective measures, such as limiting sun exposure where possible (especially at mid-day) and using eye protection, sunscreens, and hats. Young children in particular should be protected; their skin is most sensitive to exposure to UVB rays.

Ionizing radiation exposures are often of concern to communities near nuclear facilities. There is little opportunity for direct exposure, but indirect pathways, such as movement of radioactive chemicals in groundwater or food chain uptake of radioactive cesium, are of concern and can be monitored by health authorities.

Nonionizing radiation (see Chapter 19) includes electromagnetic fields, which have been listed by the NTP as a possible human carcinogen. They are implicated in a number of relatively small studies that show a small increase in relative risk for childhood leukemia among children living close to power lines. Many believe that a strategy of prudent avoidance, avoiding exposure to electromagnetic fields when the costs are low, is a sensible approach to the uncertainties about the risks (39).

Injuries

Unintentional injuries result in nearly 70,000 deaths and millions of nonfatal injuries each year. Unintentional injuries are the leading cause of death in the United States for people aged 1 to 44 years. The leading causes of death from unintentional injury are motor vehicle accidents, fires, falls, drownings, and poisonings (40,41). Injuries have a large impact on the health of infants and children. Each year, between 20% and 25% of children have injuries severe enough to require medical care, missed school days, or bed rest. The leading causes of injury for all children are motor vehicles, fires/burns, drowning, falls,

and poisoning. For older adolescents (15 to 19 years of age), firearm deaths move up into second place (also see Chapters 8, 24, and 38).

Motor vehicle injuries are by far the greatest cause of death. The two factors most commonly involved in motor vehicle deaths and injuries are inadequate passenger restraints and elevated blood alcohol levels while driving. Most states have laws requiring seat belt use by adults, and all states have laws requiring use of appropriate restraints for young children. Children weighing less than 20 lbs. and younger than 1 year of age should be in a rear-facing car seat. Children weighing 20 to 40 lbs. should be in a front-facing car seat, and children above 40 lbs. in weight should be in a booster seat until the lap and shoulder belts fit properly. Children 12 years of age and younger should always sit in the back seat of the car, wearing seat belts or other appropriate restraints, to avoid front-seat air bag injuries. Alcohol is often involved in motor vehicle accidents, especially for teenage drivers. Thus, prevention efforts must focus on eliminating this deadly combination of drinking and driving. In 1996, a total of 21 states had child-endangerment laws that make it a separate violation to drive under the influence of alcohol with a child in the car.

Residential fires are another important cause of injuries in the home environment. In 1997 in the United States, 3,360 deaths were reported due to residential fires. Fatality rates are highest for children younger than age 5 and for seniors over age 65 years. The most common causes are cooking and heating equipment, smoking, and children playing with matches and other sources of fire. Alcohol intoxication plays a significant role as a risk factor as well.

Drowning in young children (<5 years of age) is primarily an environmental problem and completely preventable. A swimming pool or water body is an irresistible lure to a curious young toddler. Swimming lessons do not help; in fact, they can give the child and parent a false sense of security. Many fences can be scaled by adventuresome tots. Toddlers can slip under pool covers and drown, delaying by vital minutes the time they are found because parents assume they are not in the pool. Infants and toddlers can also drown in very shallow ponds and pools. Probably the safest course of action is for young children not to have access to a pool. If there is a pool or nearby water body, it should be appropriately fenced and secured and toddlers must be carefully watched at all times.

Falls are also preventable. Among people older than 65 years of age, falls account for approximately 7,390 deaths a year. For children, one of the most common causes of fatal and severe falls is falling out of windows in homes. Any child who can pull to stand should be assumed capable of climbing out of an open window. Most of these falls can be prevented by using window guards, or opening the top half rather than the bottom half of double-hung windows.

Poisoning is discussed earlier in the Pesticides section.

Firearm deaths are epidemic among adolescents and young adults in the United States today, especially among black men. This is increasingly understood as having a large environmental component, with concentrations of gun stores and firearm advertisements in low-income communities contributing to the ready availability of firearms. Prevention efforts need to focus on reducing the availability of firearms to teenagers as well as underlying problems like drug trafficking and poverty.

Asthma

Since 1980, the CDC has observed rising rates of asthma morbidity and mortality for children in the United States. The greatest increase has been observed for children younger than 5 years of age, in whom rates rose 160% between 1980 and 1995 (42). Asthma is a multifactorial disease. There is a genetic component; people with asthma are more likely to have family members who either have asthma or another atopic disease like eczema or allergic rhinitis. However, genetics cannot explain the increases in incidence and severity over time. Little is known about the cause of asthma, but there is much evidence about risk factors for asthma severity and death. Numerous pollutants in indoor and outdoor air can aggravate or provoke attacks of asthma. Environmental tobacco smoke has been implicated. In addition, allergens play an important role. Here, biologic agents are likely to be the culprits, especially dust mites, molds, pollens, cockroach antigens, and pet danders. Certain pesticides, like the organophosphates, also aggravate asthma. Many asthma deaths are completely preventable with sound medical management. Fragmented and inadequate medical care can result in undertreatment of asthma or in overdose through self-administered medications (see Chapter 25).

Asthma management requires a comprehensive approach that ties together medical, environmental, and social assessment so that the patient is protected from environmental insults and has optimal medical management. For society, preventive measures involve addressing preventable causes of asthma. This means abatement of air pollution, but it also means providing safe homes for families. Cockroach-infested tenements should be a thing of the past, but all too frequently are still the best housing available for children in the United States today. For addressing cockroach problems, there are nonorganophosphate alternatives, including integrated pest management and safer chemical pesticides. This is a difficult social issue in multiunit dwellings because the only effective way to reduce or eliminate the cockroaches is if concerted action is taken for an entire building.

Birth Defects

Between 3% and 5% of all pregnancies result in a serious birth defect. Birth defects are the leading cause of death in the first year of life; more than 8,000 infants die each year in the United States from birth defects (43). Birth defects comprise an array of congenital malformations, most of which are thought to be due to events in the first trimester of pregnancy. Approximately 25% of birth defects have identified etiologic factors; for the rest, the etiology is unknown, and environmental factors are strongly suspected. In the 1990s, it has been recognized that a large percentage of neural tube defects can be prevented by beginning folic acid supplementation before conception. The neural tube defects are anencephaly (which is inevitably fatal) and spina bifida (which can cause varying degrees of lifetime disability, depending on the severity of the defect). Today, approximately 30 regions of the country have birth defects registries, which enable variations in birth defects to be studied on a geographic basis.

Cancer

Chapter 16 describes the contribution of environmental and occupational exposures to cancer, as well as occupational carcinogens and classifications that are risky for workers. In the ambient environment, people are exposed to a complex mixture of many carcinogens, usually at very low levels. Thus, using the tools of epidemiology, it is in general very difficult to study cancer in the environment, and much of what we know is derived from occupational studies or from laboratory testing. The attributable risk is the fraction of a disease that can be attributed to a specific etiology. For environmental cancer, it is difficult to estimate an attributable risk because of uncertainties about the degree of hazard of most chemicals and a paucity of information about population exposures. Because of long latency periods for cancer induction in adults and the difficulty in measuring low-level exposures, there are very few epidemiology data. Therefore, the usual approach is to extrapolate risks from occupational studies and laboratory testing of animals, as described later in this chapter. For childhood cancer, the latency periods are much shorter and therefore exposure assessment is less complicated.

In 1982, a rural California county reported to the state an unusual occurrence of brain cancers over a 10-year period in a small town in the Mojave Desert (44). In all, in a town of 10,000 people, there had been 6 cases of childhood brain cancer. On investigation, an additional case was identified. Hospital records and tumor pathology slides were hunted down and examined by a single neuropathologist, who confirmed that in all cases the child had been diagnosed with a single tumor type, medulloblastoma. A statistical analysis showed that this was a very unusual occurrence, expected by chance alone in less than 1 in 10,000 communities. A questionnaire was administered to families; the only common thread was residence in this rural desert town before the child's diagnosis. Parents did not share the same occupational history; families were not related; all of the families and children were not acquainted with each other. Community members complained of air pollution and nighttime industrial flares with strange blue and orange flames.

An environmental investigation was conducted to identify any unusual environmental exposures. The only unusual factors that turned up were a large, and not well regulated, waste disposal industry. The community was just north of the border of a large urban county. That county was known for having very stringent waste disposal regulations. Within the community, there were several poorly controlled incineration facilities: one that burned circuit boards, transformers, and other electrical equipment; a second that incinerated used railroad cars in a crude earthen kiln; and a third that burned wires and other metal wastes to carry out metal recovery operations. There was also a large carbon black facility.

The state and the EPA ordered cleanups of a number of facilities, and in one case carried out an emergency removal action. The state health department carried out a dioxin investigation and found very high levels of dioxins in incineration ash from the metal recovery site and several other areas. Elevated levels of dioxins were also seen in pigs and chickens raised downstream of some of the incinerators.

The cause of the brain cancers could never be proved, and the increased rate did not continue through the 1990s. However, the environmental investigation turned up a number of serious environmental hazards in the community, with evidence of increased exposures to dioxins through food chain contamination.

Even when a cancer cluster consists of cases of the same disease, it is almost always impossible to determine the etiology. Unfortunately, most cancer clusters consist of multiple diagnoses that may be completely unrelated diseases. In this case, the occurrence may well have been from chance alone. However, the investigation did result in a major environmental cleanup in a community that had been neglected because of its rural desert location and the financial advantage for illegal waste haulers in the neighboring, more affluent county, to haul wastes over the county line and use poor disposal practices.

Childhood cancer has been a particular concern despite the gains in cancer therapy and resultant decreased mortality. Although mortality is decreasing, the National Cancer Institute detects an increase in the incidence of childhood cancer: between 1973 and 1991, the rate of cancers for children younger than 15 years of age increased by approximately 1% per year. Childhood cancer is rare, occurring at the rate of 14 cases per 100,000 children per year. Half of childhood cancers occur before the age of 5 years. Leukemia and central nervous system cancers each account for approximately 25% of the incidence. Infants (<12 months of age) are at relatively high risk (22 cancers per 100,000 per year). The annual increase in cancer incidence for infants between 1973 and 1992 was an average of 3% per year (45–47).

There are several known genetic associations with childhood cancer. A number of genetic defects are linked to central nervous system tumors, non-Hodgkin's lymphoma,

and Wilms' tumor. Children with trisomy 21 (Down's syndrome) have an increased risk of leukemia. Despite these associations, genetics explains only a small fraction of all cases of childhood cancer, and cannot explain the observed increases over time. Ionizing radiation can cause childhood leukemia; other environmental factors have yet to be elucidated. Epidemiologic studies have indicated that other environmental risk factors, such as pesticides in the home, electromagnetic fields, and nitrosamines in food, may play a role as well (48).

EVALUATION AND MANAGEMENT OF ENVIRONMENTAL RISK

In the United States, there are a number of governmental entities involved with environmental health. The EPA is the federal regulatory agency assigned most of the national environmental regulatory responsibilities, with a few exceptions. The FDA sets standards for food additives, other than for pesticides. Both the FDA and USDA set standards related to pathogens in foods and they, not the EPA, jointly enforce the food laws. The Consumer Product Safety Commission is responsible for safety of consumer products, and the National Highway Transportation and Safety Administration for motor vehicle and highway safety. The Department of Energy and the Nuclear Regulatory Commission set standards for radiation risks, whereas the EPA performs risk assessments to support those standards. Finally, most environmental regulation is delegated to state and local government levels; for example, municipal waste disposal is primarily a state and local function in the United States.

There are also numerous agencies involved in environmental health assessment and research. The EPA has a research component, the Office of Research and Development, that has several laboratories throughout the country and funds extramural research as well. The CDC National Center for Environmental Health (NCEH) carries out surveillance, epidemiology, and laboratory research in environmental hazards, asthma, air pollution, childhood lead exposure, birth defects, developmental disabilities, and environmental radiation exposure. The Agency for Toxic Substances and Disease Registry, a sister agency of CDC, conducts public health assessments for hazardous waste sites and by petition for communities. It also develops toxicologic profiles for chemicals involved with hazardous waste sites, and conducts health education programs. The National Institute for Environmental Health Sciences is part of the National Institutes of Health. It conducts and funds environmental health research, and also houses the NTP. The FDA also has a research program, the National Center for Toxicologic Research. There are also significant environmental health research efforts in the EPA, the NCEH, the National Science Foundation, the Department of Defense, and the Department of Energy.

Evaluation of Risk

There are a number of tools used for assessment and management of environmental hazards in the community. Environmental health relies extensively on the use of *risk assessment* to evaluate environmental stressors. Risk assessment allows investigators to extrapolate either between human populations or from laboratory animals to humans. It involves weighing all of the evidence to develop estimates of the risks to populations who may be exposed. The current practice of risk assessment in environmental health is largely based on a set of principles developed by the National Academy of Sciences in 1983 (Fig. 3-7). *Hazard* is a measure of the intrinsic ability of the stressor to cause harm. *Dose* is the amount of the stressor delivered to the person, organism, or ecosystem. *Risk* is a function of hazard and dose. Four steps in risk assessment have been delineated: hazard identification, dose–response evaluation, exposure assessment, and risk characterization (49).

The weight-of-evidence approach in environmental health inevitably involves a multitude of disciplines, including toxicology, epidemiology, statistics, modeling, environmental fate and transport, engineering, and chemistry. The principles are those used in the evaluation of epidemiology, the nine Bradford Hill principles: strength of association, consistency, specificity, temporality, biologic gradient, biologic plausibility, coherence of evidence, experimental evidence, and reasoning by analogy.

Fundamental to the current practice of environmental risk assessment are several important assumptions. Although all of these are being challenged today, they form an underpinning for risk assessment as it is practiced in most countries in the world.

1. Finding a dose–response relationship in the laboratory or in human studies is critical to the inference of causality.
2. A hazard to laboratory animals is a hazard to humans unless proven otherwise.
3. Tenfold uncertainty factors are sufficient to account for variability between species (to extrapolate from, say, a mouse to a human) and the variability within a species (to account for the much greater variability within the human population than within, say, a single strain of mouse).
4. In the absence of measurements of exposures, we can use population-based exposure estimates, such as that an adult man drinks 2 liters of water a day.
5. Noncancerous effects have a threshold below which there is no risk.
6. There is no threshold for cancer effects (50).

The challenges to these practices come from all sides, but clearly, as the scientific literature grows, there are many findings that call into question many of these assumptions. With regard to the first assumption, for example, there are many who are concerned that current practices may miss effects at very low dose levels and who are calling for new approaches, especially in the area of endocrine disruptors (see Chapter 36). Although

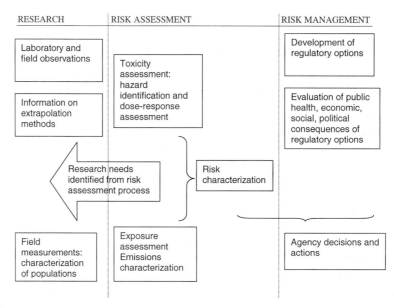

FIG. 3-7. Risk assessment/risk assessment paradigm. (Adapted from National Academy of Sciences/National Resource Council, 1983.)

this is an issue that needs research, so far no substances have been identified for which testing at higher doses would fail to identify the substance as hazardous. However, this phenomenon, if significant for a number of chemicals in the environment today, could have a profound effect on risk assessment. For the second assumption, as described later, there are examples of animal studies that are not relevant to humans. However, what we know about genetics suggests that there is enough similarity between species to continue to rely on animal testing data unless proven otherwise. In the case of the third assumption, the NRC recently concluded that past practices were unable to encompass age-related as well as other sources of variability (6). There is much ongoing research to try to understand better the full range of variability in the human population based on genetics, age, and other differences. Thus, this assumption may be replaced by better information in the future. The fourth assumption has also been questioned. The EPA generally uses a "reasonable high-end" exposure estimate for risk assessment. The default of 2 liters of water consumed a day

is based on data showing that level of consumption of fluids in people who work hard in a warm environment. Certainly with drinking water, for example, most people drink water other than from the tap, such as in milk, juices, and sodas. Most also drink water away from home and perhaps from a different water supply. This type of assumption therefore is a protective one; it ensures that those with the most exposure to, for example, water from a given tap will be protected. For the fifth assumption, there are notable exceptions. For example, the neurotoxicity of lead to children may not have a threshold or may have a threshold that cannot yet be detected in epidemiologic studies (34). In this instance, the assumption tend to be underprotective. At the same time, scientists are identifying new mechanisms of chemical carcinogenesis, which in some cases may lead to use of threshold models for carcinogens. Here again, ongoing research is likely to lead to more chemical-specific approaches to dose–response modeling. Thus, the science and practice of risk assessment is often very complex and difficult and in a continuous state of flux as science advances.

Identification of Hazards

Hazard identification relies in general on two types of information, data from epidemiologic studies and data from animal testing and other scientific studies of animals. There are many sources of toxicity data (Table 3-5). *Epidemiology* is the study of the distribution and determinants of health-related status or events in specific populations, and the application of this study to the control of health problems. *Environmental epidemiology* is defined as "the study of the effect on human health of physical, biologic, and chemical factors in the external environment, broadly conceived. By examining specific populations or communities exposed to different ambient environments, it seeks to clarify the relationship between physical, biologic or chemical factors and human health" (51). Many of the best-understood environmental hazards, such as lead hazards to children, asthma exacerbation by air pollutants, and risks of cryptosporidia in drinking water, have been described and their risks quantified using environmental epidemiology studies. (See also Chapter 6.)

Occupational studies have proven to be important in assessing health risks from some exposures. A good example of this is radon, where studies of underground miners of uranium and other ores demonstrated that radon decay products could damage cells and cause lung cancer. The EPA used the quantitative dose–response relationship between radon levels and cancer to estimate that ap-proximately 15,000 lung cancers every year are attributable to indoor radon. For other risks, direct study of the general population has been important. For example, studies of fish eaters in the Great Lakes area demonstrated a dose–response relationship between consumption of PCB-contaminated fish and neurologic outcomes of children who were exposed *in utero*. The subtle behavioral and intellectual decrements measured in the most exposed children would have been difficult to demonstrate in any other fashion. Both the radon and PCB studies were strengthened by the ability to measure directly individual exposures and disease outcomes. This is often difficult in environmental health, but a major focus today is the development of biomarkers, or measurable biochemical indicators of exposure and disease to enhance the ability to perform environmental epidemiology studies.

Environmental health surveillance, an important tool for community environmental health, is defined as the systematic collection, analysis, and dissemination and use of data concerning disorders possibly due to environmental factors and environmental exposure levels. Examples of environmental health surveillance include air pollution monitoring, childhood lead poisoning monitoring, and maintaining birth defects registries. All of these are tools for monitoring trends and identifying opportunities to prevent and control environmental disease. In some of these areas there have been remarkable successes. The CDC surveillance of lead levels in children in the United States demonstrated the benefits of the EPA's phase-out of lead in gasoline at a time when this was in doubt and there were efforts to overturn the decision. Despite this and other successes, the capacity for environmental surveillance at the federal, state, and local level is quite limited. Another form of surveillance is postmarket monitoring for adverse effects. There are provisions under both the pesticide and chemical laws for reporting to the EPA adverse health (as well as environmental) effects of toxic chemicals. This can be an

TABLE 3-5. *Sources of toxicity data*

Observational studies	Controlled studies
Humans	**Humans**
Case reports	*Clinical trials*
Epidemiology	*Dosing studies*
Geographic studies	
Temporal studies	
Ecologic	**Animals/plants**
Incident reports	*General toxicity*
Field studies	*Specialized toxicity*
	In vitro studies
	Microbiologic
	Mammalian

important safety mechanism for chemicals approved as a result of animal testing alone, because such limited testing cannot detect effects that are expected to occur in a small percentage of the population, especially idiosyncratic effects that are not completely dose dependent.

Environmental epidemiology suffers from some limitations. For one, it cannot detect risks of concern when there is little variation in exposure across the population. For example, dioxin exposures are difficult to evaluate in the general population because most people have dioxin body burdens within a narrow range. Second, epidemiology cannot be applied before approving the introduction into commerce of a chemical, product, or technology. Third, studies of environmental exposures often rely on measurements for the ambient environment as a whole rather than measurements of individual exposures. Such studies are known as *ecologic studies*, and they are often the only feasible way to study exposures; air pollution is often studied this way. In general, the larger the area over which exposures are averaged, the greater the study's methodologic limitations.

The major limitation of ecologic studies is the *ecologic fallacy*, which in some circumstances can result from making causal inferences based on ecologic data. There are two components of ecologic fallacy. One is aggregation bias, due to grouping of individuals. For example, if group A has a higher disease rate than group B, and also a higher *average* exposure to X, aggregation bias occurs when the *individuals* in group A with the disease are not more highly exposed to X. The second component is specification bias. Members of group A not only have more exposure to X, but a confounding factor—for example, a behavior like cigarette smoking that was not measured differs between the two groups and is related to the outcome under study. Problems with ecologic studies can be minimized by using appropriate statistical techniques and by using data grouped in the smallest geographic units possible, but critical evaluation is important (52).

Animal *toxicity testing* allows examination of a wide range of exposures, use of experimental controls to limit the possibility of confounding, and premarket prediction of hazards. Testing performed in a regulatory context is performed according to predefined test protocols or guidelines that have been developed by the EPA. Most test guidelines used in the United States are part of a set of internationally harmonized guidelines developed by the Organization for Economic Cooperation and Development (OECD). Requirements for testing vary with the type of substance and the statute under which the substance is covered. The most highly tested substances are food-use pesticides, for which numerous health tests are required, including tests of acute and chronic toxicity, neurotoxicity tests, cancer bioassays, and multiple-generation studies to assess reproductive and developmental toxicity. In addition, there are new requirements for tests of immunotoxicity, developmental neurotoxicity, and endocrine toxicity that are being implemented by the EPA. The EPA plans to develop new and enhanced assays for endocrine disruption for estrogen, androgen, and thyroid effects (as described later), and a developmental immunotoxicity test.

The least tested substances are new chemicals. These are brought to the EPA as "Premanufacture Notifications," or PMNs, and no testing is required unless, within 90 days of a notification, the EPA denies manufacture of the chemical pending the performance of specific tests. Under law, the EPA must have a justification to require testing, even though manufacturers are allowed to file a PMN with very limited information (usually the molecular structure and some physical properties.) The EPA uses *structure–activity relationships*, with the help of computer models and software, to identify PMNs that are likely to pose environmental health risks to humans. In contrast, in the European Union, a new chemical is reviewed before marketing, not before manufacture, and the European Commission requires a minimum set of toxicity data for the review.

In a side-by-side evaluation between the United States and the European Commission in 1994, it was found that use of structure–activity relationships to evaluate new chemicals was most likely to miss chronic human health hazards.

Toxicity testing is done using *good laboratory practices* that attempt to standardize the practices between laboratories and eliminate factors, such as poor nutrition of animals or unclean environments, that tend to bias or distort the results of laboratory tests. These practices also include record-keeping requirements that allow intensive peer review of studies to ensure their quality. Good laboratory practices are encoded in the regulations and guidelines issued by the EPA and FDA. In addition, there is an internationally agreed-on set of practices for chemicals adopted by the OECD.

Despite efforts to carry out accurate toxicity tests, these tests have limitations. To be cost effective, they are designed with as few animals as statistically possible, and the animals are dosed at high levels. Toxicologists usually first do "range finding" to find a dose that is high enough to begin to produce effects but low enough to be tolerated by the animals. This dose is called the *maximum tolerated dose* and is usually given to the "high-dose" group under study. There usually are also a control group and other dosage groups. Outcome measures have been refined over the years, but may be cruder than the measurements that can be taken in humans; for example, a mouse cannot report a headache. Some phenomena in the high-dose group may not be relevant to human risk assessment. For example, a cancer bioassay of the pesticide fosetyl-a demonstrated an increased rate of bladder cancers in high-dose rats. However, on closer examination, it was noted that because the pesticide dose was so high, phosphate stones were crystallizing out of the urine of the high-dose group. Further, toxicologists had observed that bladder stones were a risk factor for bladder cancer in rats. Thus, the high-dose group exhibited a lesion (bladder stones) not seen at environmental levels of exposure and a mechanism for disease (bladder stones causing cancer) not seen in humans. Unfortunately, this kind of case has led to a perception that animal testing is irrelevant. Fortunately, although these cases occur, they are the exception to the general observation that, when we have both epidemiologic and animal testing data, there is a striking concordance between the two with respect to relevance to risk assessment. Further, most chemicals that have been subjected to high-dose testing do not cause cancer, refuting the often-made assertion that "everything causes cancer if you give a high enough dose." (See also Chapter 16.)

Assessment of Dose–Response

The practice of dose–response assessment differs significantly between a carcinogen and a noncarcinogen. *Cancer assessment* is one of the most established areas of risk assessment. A number of fundamental scientific issues involved with cancer dose–response assessment are covered in detail in Chapter 16. This chapter emphasizes some of the environmental aspects. There are several authoritative bodies, all of which conduct cancer risk assessment in a similar fashion. On the international level, there is the International Agency for Research in Cancer, which publishes monographs on assessments of individual carcinogens. There are many bodies in the United States, but the most important is the NTP, which, in its *Biennial Report on Carcinogens*, reviews the evidence and lists substances likely to be carcinogenic. At the EPA, there are two major foci for cancer assessment, the Integrated Risk Information System and the Office of Pesticide Programs, which conducts cancer risk assessments for pesticides. All of these bodies conduct cancer risk assessment in a similar fashion, so, rather than explain each difference, this section focuses on the state of the practice for two steps, hazard assessment and dose–response assessment.

At the hazard assessment phase, all studies

relevant to the assessment of cancer are reviewed. If there is definitive human evidence of cancer causation, all of these bodies would rate the chemical as a human carcinogen. A substance can also be rated as a human carcinogen when the human evidence alone does not prove a causal relationship, but the weight of the evidence is convincing. (This is a change from earlier days, when only human data could be used to make this judgment.) When there is strong, but not probative evidence of carcinogenicity to humans, the substance is considered to be a "probable" human carcinogen. Most systems then have a category for "possible" carcinogens, those with weaker evidence, and noncarcinogens, chemicals that despite testing show no evidence for carcinogenicity.

At the dose–response assessment phase, the default assumption is that the dose–response curve is linear at low doses and starts at zero. This means that we assume that for every additional exposure there is additional cancer risk. In other words, we generally assume that if 200 of 1,000 people exposed at 1 part per 1,000 in air will get cancer, the risk for an exposure to a much lower level of 1 part per million would be 200 cancers for every 1 million people exposed. This relationship is assumed unless there is compelling evidence for a different dose–response relationship at low doses.

There are many mechanisms for carcinogenicity, and it is believed that not all of these mechanisms have linear dose–response relationships at low doses. However, there are rigorous criteria for accepting arguments to depart from the low-dose linear default model, and most carcinogens are still considered to have linearity at low doses. Whether from human or animal data, the dose–response curve is modeled using statistical techniques that extrapolate the curve from the higher doses in the occupational or laboratory setting to the lower doses that are often of concern in environmental settings. Because of the uncertainties in extrapolating from high to low doses, and to account for the variability in the general human population,

the dose–response curve is plotted with 95th percentile confidence limits and the upper 95th percentile bound is usually used for risk assessment. This estimate is combined with the exposure assessment to give a probabilistic estimate of risk, such as 10^{-3}, 10^{-5}, or 10^{-6}.

An alternative to cancer risk analysis is the Delaney Clause, which previously applied, in part, to the regulation of cancer risks from pesticides in food. On the basis of a hazard identification alone (carcinogenicity in animals), it set a standard of no allowable residues of that pesticide in processed food. In the 1950s, scientists began to identify a number of carcinogenic substances that had been added to the food supply. Appropriately, Congress acted on this to protect the public from cancer. At that time, the science of the day indicated that there might be no safe level for a carcinogen. The Delaney Clause for food additives and food colorants resulted in removal of a number of carcinogens from the food supply. Later science—in the form of increasingly sensitive methods in analytic chemistry, tools for quantitative risk assessment, and new understandings of cancer mechanisms—eventually led Congress to change its policy and remove pesticides from governance by the Delaney Clause. However, the Delaney Clause still applies to food additives and colorants that are intentionally added to the food supply. Although this may seem paradoxical, perhaps from the standpoint of consumers even a negligible risk may be too high in return for the benefit of having an additional food color additive.

Noncancer risk assessment usually involves use of the *reference dose (RfD)* approach. For toxicity end points presumed to have a threshold, the EPA and other U.S. government agencies establish an RfD for a given exposure duration (usually acute or chronic, but sometimes intermediate). A chronic RfD is an estimate of a daily exposure to a population that, over a 70-year life span, is likely to have no significant deleterious effects (53). An acute RfD considers a 1-day exposure only. Other federal agencies, such as the

FDA, refer to the RfD as an *acceptable daily intake*. To calculate an RfD, the risk assessor first chooses the most appropriate, usually the most sensitive, effect from a chronic or acute study. Next, the "no-observable adverse effect level," or NOAEL, is identified for this effect. Finally, uncertainty factors are applied to the NOAEL. Modifying factors between 3- and 10-fold are also applied when critical studies are missing. In general, the RfD for an acute exposure may be much higher than the RfD for a chronic exposure, but this depends strongly on the nature of the chemical and effects under study.

The FDA first developed this approach and has used it for decades in the regulation of food additives under the Federal Food, Drug and Cosmetics Act (FFDCA). As a default, it is assumed that the average human could be 10 times more sensitive than the laboratory test animal—the interspecies uncertainty factor—and that the most sensitive human might be 10 times more sensitive than the average human—the intraspecies uncertainty factor. Although application of these factors is fundamental to risk assessment as we know it today, few are aware of their origins. The FDA based them on evaluation of a modest database, along with a large portion of scientific judgment that supported the view that initially one factor of 100 would cover all sources of variability. Over time, this 100-fold factor was split into two factors of 10 each. For pesticides, a third 10-fold factor is required to account for differences between children and adults in exposure and susceptibility, unless the EPA can demonstrate that a different factor is adequately protective of children.

Assessment of Exposure

The purpose of exposure assessment is to estimate dose. Dose is determined by a number of factors:

$$Dose = \frac{[Stressor] \times Rate \times Duration \times \%Absorbed}{Body\ weight}$$

where *stressor* is the concentration of the agent in the medium, *rate* is the rate of intake, and *duration* is the time over which intake occurs. Toxicologists are often interested in the average daily lifetime exposure as well:

Average daily lifetime exposure

$$= \frac{Total\ dose}{Days\ of\ life}$$

Thus, assessment of exposure involves numerous factors (Table 3-6). In risk assessment, the investigator usually does not have access to precise measurements of all of these exposure attributes, but they are all important in calculating an average daily lifetime exposure. We would like to know the rate and duration of exposure and the amount absorbed, as well as the body weight. In a laboratory experiment, a toxicologist has almost complete control over these factors.

As a practical matter in risk assessment, actual exposure measurements are often replaced by defaults. At the EPA, the policy is to assess a *reasonable high-end exposure*—that is, an exposure at the upper 90th or 95th percentile. However, at times Congress has directed the EPA to be more protective, such as in risk assessments under the Clean Air Act, which in some cases has required looking at the *maximally exposed individual*. Summation of numerous high-end exposures can greatly overestimate exposure, however. Exposure to pesticides in food is a good example of this. If the upper 90th percentile bounds for all foods are added up, the result is a theoretic individual who eats 5,000 calories per day—not exactly a reasonable high-end estimate of exposure. If there are data

TABLE 3-6. *Exposure issues and assumptions*

Extent and frequency of exposure
Number/proportion exposed
Timing of exposure
Duration of exposure
Degree of exposure by various routes
Use of average or typical individual and high-risk
  groups

on distributions of food consumption and on pesticide levels in the food, it is possible to use *probabilistic modeling*, which incorporates those distributions for all foods to compute the distribution of exposure to pesticide residues in the food. Most frequently, this is done using Monte Carlo modeling techniques, not only for pesticide residues on food but for other aggregate exposure situations. Monte Carlo and other probabilistic modeling techniques simulate the distributions of individual combinations of multiple exposures to produce a theoretic distribution of an aggregate exposure to the population.

Characterization of Level of Risk

Risk characterization is the final step of the risk assessment process. No additional scientific information is added during this phase, which involves estimating the magnitude of the public health or environmental problem. Much judgment is needed in appropriate selection of populations and exposure levels for analysis. In addition, relevant statistical and biologic uncertainties must be made clear at this stage. This part of the risk assessment process is the largest nexus between risk assessment and risk managers, and it is important that risk managers receive a complete set of information to guide decisions.

A community was exposed to a mixture of low-level carcinogens; the EPA conducted a risk assessment of toxic air contaminants near a Superfund site in the area and found that there were a number of carcinogenic air contaminants, including vinyl chloride and benzene, in ambient air and homes. The EPA assumed that residents near the site might breathe the air every day during a 70-year lifetime and, adding the risks together, found that the total excess cancer risk over a lifetime was as high as 4.4×10^{-3} or as low as zero. On this basis, the EPA decided to require the owner of the facility to lower the emissions from the surface of the site by installing an extensive landfill gas collection system. In this way,

it was possible to protect the community from being exposed to this hazard over a lifetime.

How do we talk about this kind of risk? First, in plain language, this assessment means that if 1,000 people were exposed to this air over a lifetime, in theory 0 to 4 would be expected to get cancer, in addition to the 400 to 500 who are already expected to get cancer over a lifetime. (Today, approximately 20% of the population dies from cancer.) On a public health basis, this is considered to be a significant exposure. In the United States, we tend to strive for exposures that will have less risk than 10^{-5} or 10^{-6}, and in the Superfund statute (Comprehensive Environmental Response, Compensation and Liability Act), the highest acceptable lifetime cancer risk is 10^{-4}. This has been articulated in food law (FFDCA) as a standard of a "reasonable certainty of no harm"—that is, no one will become ill from acute or chronic exposures and that there is a "negligible risk" of exposure to a carcinogen. What can be difficult to communicate to the public, and especially to a community near such a landfill, is that although an exposure of this magnitude does warrant a swift public health response, such as containment of landfill gases, it may not be a cause for emergency public health action (50).

In 1992, there was an invasion of Mediterranean fruit flies (medflies) in the urban Los Angeles area. These are alien pests that not only damage valuable fruit crops, but cause trade irritants and increased pesticide use for food crops because other countries close their borders to medfly-contaminated agricultural produce. To protect domestic agriculture from invasive pest species, countries establish quarantines of imports from countries that are known to have medfly outbreaks. The medflies unfortunately had become well established in urban fruit trees and orchards, necessitating the use of aerial sprays with malathion to suppress and ultimately eradicate the population. Public concern over spraying an urban area with an organophosphate pesticide was predictable and swift. In response, the California Department of Health Services conducted monitoring and a risk assessment to determine the population risk from aerial malathion spraying. State toxicologists established an intermediate-term oral RfD of 2 μg per kg per day (54,55). It was found that because spraying was done frequently enough to leave residues, both intermediate and chronic exposure scenarios were relevant. There were three possible pathways for exposure: inhalation, ingestion, and skin absorption. Malathion has a major toxic metabolite, malaoxon, which also needed to be included in the risk assessment.

The exposure assessment was complex and involved modeling of air emissions and assumptions about transfer to vegetation, playground surfaces, and skin of a malathion–bait mixture that was applied by helicopters. The state found that there was no significant risk for the average person. However, it also found that children were most likely to have significant exposures to the pesticide by either the oral or dermal route of exposure. Significant oral pathways were eating unwashed back-yard vegetables or pica behavior, which can lead to directly eating bait. Level of ingestion for a reasonable high-end exposure (not a child with pica) was between 10 and 20 times the RfD. A child playing outdoors on contaminated surfaces could have high dermal absorption of the pesticide because of the time spent and relatively more skin contact with the environment. Level of intake for a reasonably high-end exposure was between 10 and 30 times the RfD. These risk calculations indicated that, although exposures were not high enough to anticipate any illnesses as a result of spraying, in theory children with higher exposures could exceed the RfD established by the state health department, and thus these exposures were of public health concern. This was in spite of the fact that there were uncertainties in estimating the concentrations of the pesticides on environmental surfaces because of limited sampling data. Therefore, recommendations were made to the public to wash produce before eating, cover or wash outdoor play surfaces, and stay indoors during bait applications. The California Department of Food and Agriculture adopted a policy that

use of malathion bait spraying should be a last resort, and began a study of alternative methods for medfly control. Further, it decided that when spraying is necessary, it should be done only in the evening when most people (including children) are indoors, and as infrequently as possible.

Risk Management

Command and Control

The major federal environmental statutes are shown in Table 3-7. In general, regulations and *standards* are promulgated by four major regulatory arms of the EPA: the Office of Air and Radiation, the Office of Water, the Office of Solid Waste and Emergency Response, and the Office of Prevention, Pesticides and Toxic Substances. In addition, for chemicals and pesticides there are new chemical notifications and pesticide registrations. Many states also have environmental statutes and states usually can set higher standards than the federal standard, but are not allowed to establish a lower standard. *Permitting* of facilities for air emissions, water discharges, and waste disposal are essential to controlling point sources of pollution, as is a strong environmental *enforcement* presence. The first line of responsibility usually lies with local and state health and environmental agencies. The EPA's regional offices play numerous roles, ranging from funding state and local efforts, to program implementation, to taking on more complex or neglected areas of enforcement. *Environmental monitoring* is also an important tool for evaluating the success of efforts and for targeting future regulatory and enforcement actions.

Risk Management Tools

There are a number of tools used in risk management. *Environmental engineering* has played an important role. Many environmental standards are based in whole or in part on best available technology, such as

TABLE 3-7. *Major federal environmental statutes, United States, 1999*

Statute	Date of last major amendment	Scope
Clean Air Act (CAA)	1990	Standards for toxic air contaminants and priority air pollutants, and means to achieve reductions in pollution
Clean Water Act (CWA)	1976	Standards to protect rivers, lakes, and streams
Safe Drinking Water Act (SDWA)	1996	Standards for contaminants in drinking water and methods to treat and prevent contamination
Resource Conservation and Recovery Act (RCRA)	1976	Management, treatment, and disposal of hazardous wastes
Comprehensive Environmental Response, Compensation and Liability Act (CERCLA)	1988	Hazardous waste cleanup; created the Agency for Toxic Substances and Disease Registry
Emergency Planning and Community Right to Know Act (EPCRA)	1986	Prevention of hazardous releases and emergency planning; Toxic Release Inventory
Toxic Substances Control Act (TSCA)	1976	Management of chemicals in commerce
Pollution Prevention Act (PPA)	1990	Pollution prevention reporting and goals
Federal Insecticide, Fungicide and Rodenticide Act (FIFRA)	1996	Management and registration of pesticides
Federal Food, Drug and Cosmetics Act (FFDCA)	1996	Standards for pesticides on food (also numerous other provisions that apply to the Food and Drug Administration)

the air toxics MACT standards under the Clean Air Act. In some cases, the EPA must consider an *economic analysis* of costs or feasibility in developing standards, such as the MCLs (maximum contaminant levels) under the Safe Drinking Water Act. In others, such as the Toxic Substances Control Act, a cost–benefit analysis is required. In response to the escalating costs of hazardous waste cleanup and other consequences of pollution, Congress in 1990 enacted the Pollution Prevention Act, which set out the *pollution prevention* hierarchy. The rungs of the ladder go from the most preferable strategy, reduction of pollution at the source (source reduction), to waste minimization, reuse, recycling, emissions controls, and, least preferably, cleanup. As a rule, it is cheaper and smarter to reduce pollution at the source (see Box in Chapter 5). For example, in 1998, to address the problem of mercury emissions from medical waste incinerators, the American Hospital Association launched a voluntary effort to phase out all uses of mercury in hospitals.

The *precautionary principle* often plays a role in environmental regulation. As governments agreed in 1992: "In order to protect the environment, the precautionary approach shall be widely applied by States according to their capabilities. Where there are threats of serious or irreversible damage, lack of full scientific certainty shall not be used as a reason for postponing cost-effective measures to prevent environmental degradation" (56). For example, DDT was banned in the United States long before its precise mechanisms of action had been described by scientists. Pollution and its consequences are not distributed equally in society, and thus it is important to consider *environmental justice* issues in assessment of hazard (57). Unfortunately, in the past there was a failure to do so, accounting for concentrations of polluting industries, sources of air pollution, and waste disposal operations in certain low-income and minority communities. The EPA has begun to apply the Civil Rights Act, Title VI, to address issues of environmental inequity in permitting and other regulatory decisions, but it is too early to predict the impact of this.

Community Right to Know

Community right to know is a powerful driver for reducing pollution (see Chapter 10). It was introduced at the federal level in 1986 with passage of the Emergency Preparedness and Community Right to Know Act and establishment of the TRI, which initially required the manufacturing industry to report releases of some 300 chemicals to the public. Like the material safety data sheets in workplaces, community right to know is designed to empower citizens to make informed decisions either as individuals or as a community. Community right to know is a powerful tool to inform not only citizens but workers in plants as well as plant and corporate managers (Fig. 3-8). Before the establishment of the TRI, no one was aware of the large quantities of wastes that were sent up stacks or dumped in water or on land. In many cases, these wastes were actually the starting materials for products. Industry eventually recognized that it was more cost effective to prevent the pollution by better managing the flows of materials into, in, and through facilities. In other cases, the reports on TRI allowed citizens near the fenceline to find out about poor environmental performance in plants. Companies, in turn, made efforts to improve their standing, especially in the relative rankings that were printed in the newspaper each year when the reports were announced. Workers and occupational hygienists found that TRI reports helped to point to chemicals that were likely to expose workers. For example, large amounts of fugitive air emissions may be indicative of even higher air levels inside a facility.

In 1991, the EPA challenged manufacturers with the "33/50 program" to reduce TRI emissions of several toxic air contaminants by 33% by 1992 and 50% by 1995 from the TRI baseline year of 1988. Overall, industry met this target, and those enrolled in 33/50 did so a year early. Since then, the EPA has expanded the TRI several times, first by doubling the number of chemicals on the list to approximately 600, to ensure that the most

FIG. 3-8. Explosion and fire at a pentachlorophenol wood treatment plant in California, which created elevated dioxin levels at the site of the fire. Such an event can cause acute exposures to workers and community residents, who may need follow-up medical evaluation. Such situations highlight the importance of worker and community right-to-know laws (also see Chapter 10). (Photograph by Lynn R. Goldman.)

hazardous chemicals are listed. In 1996, the EPA added a number of new sectors to the TRI: metals mining, coal mining, oil- and coal-fired utilities, solvent recycling, hazardous waste disposal, and bulk petroleum terminals. Most recently, in 1998, the EPA proposed to lower the threshold for reporting for persistent, bioaccumulative, and toxic chemicals on the TRI. In 1996, the EPA reported an overall reduction of 45.6% of all TRI releases for "base" facilities and chemicals for a total of 2.4 billion pounds of chemicals released in the environment or transferred for disposal (58).

The concept of environmental right to know has expanded beyond the TRI in a number of ways, providing a wealth of information for scientists and communities alike. Some examples:

- Proposition 65: In California in 1988, voters established, by initiative, a right-to-know law that requires product labels when use of a product is likely to exceed a 10^{-5} cancer risk or 1/1,000 of the RfD for a reproductive toxicant. The law is enforced by a citizen's suit provision that enables anyone to identify unlabeled products and bring suit, which may be joined by the State Attorney General. To date, suits brought under this law have resulted in changes in manufacture of dinnerware and plumbing fixtures formerly containing lead, and a number of other products, as well as voluntary actions by industry to avoid labels or lawsuits, like the removal of trichloroethylene from certain office products.
- Materials Accounting Reporting: Two states, New Jersey and Massachusetts, have expanded their state release inventory to require reporting of materials flows in and out of manufacturing facilities. Such reporting helps communities understand what materials are transported through their neighborhoods and helps citizens learn about the toxic materials in products. Workers at facilities can achieve a better understanding of the quantities of materials they are handling.
- Drinking Water Consumer Confidence Letters: Under the federal Safe Drinking Water Act, water providers are now required to provide customers with an annual report summarizing occurrences of contaminants found in the drinking water. (This is in addition to a previously required public health advisory when a very high level is detected.)
- Sector Facility Indexing: The EPA's Office of Enforcement now provides on the internet a database that allows, for a number of major sectors in the economy, comparison of facilities by environmental performance and violations of environmental statutes.

- Surf Your Watershed: On the internet, the EPA's Office of Water provides a comprehensive report card for all of the major watersheds in the country. This contains health information ranging from chemical contaminants to fish advisories to beach closures.
- Real-Time Air Monitoring: The EPA provides major metropolitan areas with up-to-date reports and forecasts of air pollution levels. These are especially useful for those with chronic lung disease, who need to avoid exercise during severe air pollution episodes.
- Lead-Based Paint Real Estate Disclosures: The EPA and HUD now require that any known lead hazards in housing be disclosed at the time of sale or lease to prospective buyers or tenants. The regulation requires that prospective buyers have an opportunity to do a lead inspection before final sale of the home.

The current expansion of information available online is likely to continue. A challenge for environmental health professionals will be keeping up with the available information, and helping communities and individuals sift through it to understand what is important and relevant for their communities and how to place it into perspective. Keeping up with and understanding these information sources is a critical part of environmental health practice. Law enforcement agencies have raised concerns about some information posing security threats, such as information about facilities that might allow terrorists to commit sabotage. Finally, industry has long been concerned that provision of information is damaging to competitiveness. Although the need for information should be balanced against other concerns, the right to know has proven to be a useful tool for environmental protection.

Community and International Risk Management Approaches

Another important tool at a community level is an ecosystem-based or a community-based

TABLE 3-8. *UNCED[a] principles of sustainable development most relevant to environmental health*

Principle 1
Human beings are at the center of concerns for sustainable development. They are entitled to a healthy and productive life in harmony with nature.

Principle 2
States have, in accordance with the Charter of the United Nations and the principles of international law, the sovereign right to exploit their own resources pursuant to their own environmental and developmental policies, and the responsibility to ensure that activities within their jurisdiction or control do not cause damage to the environment of other States or of areas beyond the limits of national jurisdiction.

Principle 3
The right to development must be fulfilled so as to equitably meet developmental and environmental needs of present and future generations.

Principle 4
In order to achieve sustainable development, environmental protection shall constitute an integral part of the development process and cannot be considered in isolation from it.

Principle 5
All States and all people shall cooperate in the essential task of eradicating poverty as an indispensable requirement for sustainable development, in order to decrease the disparities in standards of living and better meet the needs of the majority of the people of the world.

Principle 10
Environmental issues are best handled with the participation of all concerned citizens, at the relevant level. At the national level, each individual shall have appropriate access to information concerning the environment that is held by public authorities, including information on hazardous materials and activities in their communities, and the opportunity to participate in decision-making processes. States shall facilitate and encourage public awareness and participation by making information widely available. Effective access to judicial and administrative proceedings, including redress and remedy, shall be provided.

Principle 11
States shall enact effective environmental legislation. Environmental standards, management objectives and priorities should reflect the environmental and developmental context to which they apply. Standards applied by some countries may be inappropriate and of unwarranted economic and social cost to other countries, in particular developing countries.

Principle 13
States shall develop national law regarding liability and compensation for the victims of pollution and other environmental damage. States shall also cooperate in an expeditious and more determined manner to develop further international law regarding liability and compensation for adverse effects of environmental damage caused by activities within their jurisdiction or control to areas beyond their jurisdiction.

Principle 14
States should effectively cooperate to discourage or prevent the relocation and transfer to other States of any activities and substances that cause severe environmental degradation or are found to be harmful to human health.

Principle 15
In order to protect the environment, the precautionary approach shall be widely applied by States according to their capabilities. Where there are threats of serious or irreversible damage, lack of full scientific certainty shall not be used as a reason for postponing cost-effective measures to prevent environmental degradation.

Principle 18
States shall immediately notify other States of any natural disasters or other emergencies that are likely to produce sudden harmful effects on the environment of those States. Every effort shall be made by the international community to help States so afflicted.

Principle 19
States shall provide prior and timely notification and relevant information to potentially affected States on activities that may have a significant adverse transboundary environmental effect and shall consult with those States at an early state and in good faith.

Principle 22
Indigenous people and their communities and other local communities have a vital role in environmental management and development because of their knowledge and traditional practices. States should recognize and duly support their identity, culture and interests and enable their effective participation in the achievement of sustainable development.

Principle 24
Warfare is inherently destructive of sustainable development. States shall therefore respect international law providing protection for the environment in times of armed conflict and cooperate in its further development, as necessary.

Principle 25
Peace, development and environmental protection are interdependent and indivisible.

[a] United Nations Commission for Environment and Development.

approach to environmental protection. Under the Clean Air Act, it has long been recognized that, for many communities, it would not be possible to meet standards unless management is undertaken for an entire airshed. An example is the Los Angeles air basin, in which more than 10 million people reside. Environmental management in such situations needs to connect better with other activities, like land use and transportation planning. This approach is now being adopted for protection of large and complex watersheds under the Clean Water Act. Increasingly, nonpoint sources of pollution to air and water—that is, sources that are diffuse rather than from large industrial stacks and outfalls—are important. Ecosystem-based approaches are more effective than individual permitting activities in controlling such sources.

The global environment has taken on an increased importance over the last several years and was highlighted when more than 100 nations signed the United Nations Commission on Environment and Development treaty in 1992, formally adopting the goal of sustainable development and 27 principles of sustainable development (Table 3-8). The threats of large-scale changes to the earth's environment, such as destruction of the tro-

pospheric ozone layer and global climate change, are encouraging nations to cooperate on environmental management issues. For example, air pollutants can persist and travel long distances, creating environmental damage. Thus, there are global and regional agreements to control emission of ozone-depleting chemicals (Montreal protocol), acid rain, and persistent pollutants (Long Range Transport of Atmospheric Pollutants Persistent Organic Pollutants protocol), and climate change. At the time of this writing, negotiations are underway to create a global agreement on persistent organic pollutants and activity to agree on a voluntary global system for classification and labeling of chemicals in commerce.

The Kyoto climate change treaty was negotiated in 1998, and are efforts are underway for countries to begin to ratify and implement that agreement. In the United States, there continues to be debate about the urgency of addressing global climate change. There is very little doubt that we have observed a gradual and subtle increase in the earth's temperature over the last century, and that the cause is the increased levels of greenhouse gases, especially carbon dioxide, in the atmosphere (Fig. 3-9). These gases act as an insulating blanket, reflecting back

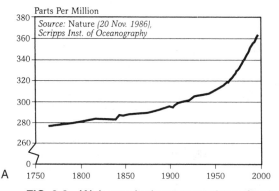
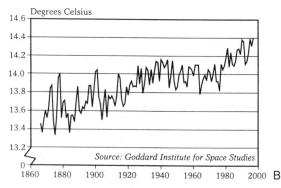

FIG. 3-9. (A) Atmospheric concentrations of carbon dioxide, 1764 to 1997. **(B)** Average temperature at the earth's surface, 1866 to 1997. (From Brown LR, Renner M, Flavin C, Starke L (Ed.). Vital Signs 1997. New York; W. W. Norton & Company, 1997:17.)

TABLE 3-9. *International organizations involved in environmental protection*

Acronym	Organization	Environmental health scope
UNEP	United Nations Environment Program	Environmental agreements, chemical information systems, technical assistance, right to know
WHO	World Health Organization	Toxicology and epidemiology, International Agency for Research on Cancer, technical assistance
UNCED	United Nations Commission for Environment and Development	Implementation of Agenda 21 treaty signed in 1992
UNDP	United Nations Development Program	Sustainable development, growth and population issues
FAO	Food and Agriculture Organization	Pesticides and other agricultural health issues Food safety (Codex Alimentarius)
IMO	International Maritime Organization	Seafood safety and protection of the seas
OECD	Organization for Economic Cooperation and Development	Harmonization of chemicals testing and classification, cooperation on waste disposal, climate, and other issues
IPCC	International Program on Climate Change	Scientific assessment of climate change
IFCS	Intergovernmental Forum on Chemical Safety	Cooperation on global chemical safety issues
ILO	International Labor Organization	Workplace health and safety; chemical labeling in the workplace (material safety data sheets)
UNCTDG	United Nations Commission on Transport of Dangerous Goods	Harmonization of classification and labels for chemicals in transport

to earth the heat of the sun in much the way that the glass of a greenhouse has a warming effect. At the time of this writing, there is no federal legislation either to ratify the Kyoto agreement or to create an alternative U.S. government approach to controlling the emission of greenhouse gases. Warming of the earth is predicted to have a number of adverse consequences. First, many scientists believe that the climate is already more erratic than historically, increasing the likelihood of regional flooding and drought and severe storm episodes. Second, as the polar ice caps melt, low-lying areas will be inundated. There is already some scientific evidence that this process is beginning to occur. The ultimate result would be flooding, with many coastal and low-lying areas becoming uninhabitable or requiring elaborate dikes and drainage systems. Third, the ecosystem will not be able to adapt readily to a rapid shift in climate. This could result in spread of vectors of infectious disease, poor health, death of forests and other ecosystems, disruption of agriculture in many areas, and concomitant effects on the health of other species and humans due to spread of infectious disease and disruption of habitat and food supplies.

Numerous international organizations participate in global environmental health concerns (Table 3-9), but the United Nations Environment Program, the World Health Organization, and the International Labor Organization have the major responsibilities. For addressing chemical hazards, they have joined together to form the International Program on Chemical Safety.

REFERENCES

1. World Health Organization (WHO). Preamble: Constitution. Geneva: WHO, 1948.
2. World Health Organization (WHO). Consultation: Sofia, Bulgaria. Geneva: WHO, 1993.
3. Dubos R. Man adapting. New Haven, CT: Yale University Press, 1965.
4. Pope AM, Rall DP. Environmental medicine: integrating a missing element into medical education. Washington, DC: National Academy Press, 1995.
5. Mayer J, Balk S. A pediatrician's guide to environmental toxins. Contemporary Pediatrics 1988;5:22–40, 63–76.
6. National Research Council. Pesticides in the diets of infants and children. Washington, DC: National Academy Press; 1993.
7. Clinton WJ. Executive Order 13045: Protection of

children from environmental health and safety risks. Washington, DC: 1997.

8. Gore A. Press release. Washington, DC: White House, 1998.

9. Committee of the Environmental and Occupational Health Assembly of the American Thoracic Society. Health effects of outdoor air pollution: state of the art. Parts 1 and 2. Am J Respir Crit Care Med 1996;153:3–50, 477–498.

10. Bates DV. The effects of air pollution on children. Environ Health Perspect 1995;103:49–53.

11. Spengler J, Sexton K. Indoor air pollution: a public health perspective. Science 1983;221:9–17.

12. Samet J, Marbury M, Spengler J. Health effects and sources of indoor air pollution: part I. American Review of Respiratory Disease 1987;136:1486–1508.

13. Samet J, Marbury M, Spengler J. Health effects and sources of indoor air pollution: part II. American Review of Respiratory Disease. 1988;137:221–242.

14. U.S. Department of Health and Human Services. The health consequences of involuntary smoking: a report of the Surgeon General. Washington, DC: U.S. Department of Health and Human Services, 1986.

15. Montana E, Etzel R, Allan T, Horgan TE, Dearborn DG. Environmental risk factors associated with pediatric idiopathic pulmonary hemorrhage and hemosiderosis in a Cleveland community. Pediatrics 1997;99:1–8.

16. Centers for Disease Control and Prevention. Update: pulmonary hemorrhage/hemosiderosis among infants—Cleveland, Ohio, 1993–1996. MMWR Morb Mortal Wkly Rep 1997;46:33–35.

17. MacKenzie W, Hoxie N, Proctor M, et al. A massive outbreak in Milwaukee of *Cryptosporidium* infection transmitted through the public water supply. N Engl J Med 1994;331:161–167.

18. Morris R, Audet A-M, Angelillo I, Chalmers T, Mosteller F. Chlorination, chlorination by-products, and cancer: a meta-analysis. Am J Public Health 1992;82:955–963.

19. U.S. Environmental Protection Agency, Science Advisory Board. An SAB report: safe drinking water: future trends and challenges. Washington, DC: U.S. Environmental Protection Agency, 1995.

20. Committee to Coordinate Environmental Health and Related Programs, Ad Hoc Subcommittee on Fluoride. Review of fluoride: benefits and risks. Washington, DC: Public Health Service, U.S. Department of Health and Human Services, 1991.

21. Institute of Medicine, National Research Council. Ensuring safe food: from production to consumption. Washington, DC: National Academy Press, 1998.

22. Hayward D, Nortrup D, Gardiner A, Clower M. Elevated TCDD in chicken eggs and farm-raised catfish fed a diet containing ball clay from a Southern United States mine. Environ Res 1999;80(10) (in press).

23. Goldman LR, Smith DF, Neutra RR, et al. Pesticide food poisoning from contaminated watermelons in California, 1985. Arch Environ Health 1990;45:229–236.

24. Goldman LR, Beller M, Jackson RJ. Aldicarb food poisonings in California, 1985–1988: toxicity estimates for humans [published erratum appears in Arch Environ Health 1990;45:following 380]. Arch Environ Health 1990;45:141–147.

25. Blondell J. Epidemiology of pesticide poisonings in the United States, with special reference to occupational cases. Occup Med 1997;12:209–220.

26. Litovitz T, Manoguerra A. Comparison of pediatric poisoning hazards: an analysis of 3.8 million exposure incidents. A report from the American Association of Poison Control Centers [see comments]. Pediatrics 1992;89:999–1006.

27. Esteban E, Rubin C, Hill R, Olson D, Pearce K. Association between indoor residential contamination with methyl parathion and urinary para-nitrophenol. Journal of Exposure Analysis and Environmental Epidemiology 1996;6:375–387.

28. Wagner SL, Orwick DL. Chronic organophosphate exposure associated with transient hypertonia in an infant. Pediatrics 1994;94:94–97.

29. U.S. Environmental Protection Agency. Office of Prevention Pesticides and Toxic Substances. Chemical hazard data availability study: what do we really know about the safety of high production volume chemicals? EPA's 1998 baseline of hazard information that is readily available to the public. Washington, DC: U.S. Environmental Protection Agency, 1998.

30. U.S. Environmental Protection Agency. Endocrine disruptor screening program: notice. Federal Register 1998;63:42852.

31. American Academy of Pediatrics, Committee on Environmental Health. Lead poisoning: from screening to primary prevention. Pediatrics 1993;92:176–183.

32. Pirkle J, Brody D, Gunter E, et al. The decline in blood lead levels in the United States: the National Health and Nutrition Examination Surveys (NHANES). JAMA 1994;272:284–291.

33. Brody D, Pirkle J, Kramer R, et al. Blood lead levels in the United States: the National Health and Nutrition Examination Surveys (NHANES III, 1988 to 1991). JAMA 1994;272:277–283.

34. National Research Council. Measuring lead exposure in infants, children and other sensitive populations. Washington, DC: National Academy Press, 1993.

35. Needleman HL, Schell A, Bellinger D, Leviton A, Allred EN. The long-term effects of exposure to low doses of lead in childhood: an 11-year follow-up report. N Engl J Med 1990;322:83–88.

36. Rabin R. Warnings unheeded: a history of child lead poisoning. Am J Public Health 1989;79:1668–1674.

37. Centers for Disease Control and Prevention (CDC). Screening young children for lead poisoning: guidance for state and local public health officials. Atlanta: CDC, 1997.

38. National Research Council, Commission on Life Sciences, Board on Radiation Effects Research, Committee on Health Risks of Exposure to Radon. Health effects of exposure to radon: BEIR 6. Washington, DC: National Academy Press, 1999.

39. National Research Council (U.S.), Committee on the Possible Effects of Electromagnetic Fields on Biologic Systems. Possible health effects of expo-

sure to residential electric and magnetic fields. Washington, DC: National Academy Press, 1997.

40. Centers for Disease Control and Prevention (CDC), National Center for Injury Prevention and Control. Home and leisure injuries in the United States: a compendium of articles from The Morbidity and Mortality Weekly Report 1985–1995. Atlanta: CDC, 1996.

41. Centers for Disease Control and Prevention (CDC), National Center for Injury Prevention and Control. Prevention of motor vehicle-related injuries: a compendium of articles from The Morbidity and Mortality Weekly Report 1985–1996. Atlanta: CDC, 1997.

42. Mannino DM, Homa DM, Pertowski CA, et al. Surveillance for asthma: United States, 1960–1995. MMWR Morb Mortal Wkly Rep 1998;47:1–27.

43. Centers for Disease Control and Prevention. Economic costs of birth defects and cerebral palsy: United States, 1992. MMWR Morb Mortal Wkly Rep 1995;44:695–699.

44. Harnly M, Stephens R, McLaughlin C, Marcotte J, Petreas M, Goldman L. Polychlorinated dibenzo-p-dioxin and dibenzofuran contamination at metal recovery facilities, open burn sites, and a railroad car incineration facility. Environmental Science and Technology 1995;29:677–684.

45. Gurney JG, Ross JA, Wall DA, Bleyer WA, Severson RK, Robison LL. Infant cancer in the U.S.: histology-specific incidence and trends, 1973 to 1992. J Pediatr Hematol Oncol 1997;19:428–432.

46. Gurney JG, Davis S, Severson RK, Fang JY, Ross JA, Robison LL. Trends in cancer incidence among children in the U.S. Cancer 1996;78:532–541.

47. Kenney LB, Miller BA, Ries LA, Nicholson HS, Byrne J, Reaman GH. Increased incidence of cancer in infants in the U.S.: 1980–1990. Cancer 1998; 82:1396–1400.

48. Miller RW. Cancer rates and risks: childhood risk factors. Bethesda, MD: National Cancer Institute, 1995.

49. National Research Council. Risk assessment in the federal government: managing the process. Washington, DC: National Academy Press, 1983.

50. National Research Council. Science and judgment in risk assessment. Washington, DC: National Academy Press, 1994.

51. National Research Council. Environmental epidemiology: public health and hazardous wastes. Washington, DC: National Academy Press, 1991.

52. Morgenstern H. Uses of ecologic analysis in epidemiologic research. Am J Public Health 1982; 72:1336–1344.

53. Barnes D. DM. Reference dose (RfD): description and use in health risk assessments. Regul Toxicol Pharmacol 1988;8:471–486.

54. Bradman MA, Harnly ME, Goldman LR, Marty MA, Dawson SV, Dibartolomeis MJ. Malathion and malaoxon environmental levels used for exposure assessment and risk characterization of aerial applications to residential areas of southern California, 1989–1990. Journal of Exposure Analysis and Environmental Epidemiology 1994;4:49–63.

55. Marty M, Dawson S, Bradman M, Harnly M, DiBartolomeis M. Assessment of exposure to malathion and malaoxon due to aerial application over urban areas of southern California. Journal of Exposure Analysis and Environmental Epidemiology 1994; 4:65–81.

56. The United Nations Conference on Environment and Development. Rio declaration on environment and development, principle 15. Rio de Janeiro, Brazil: United Nations, 1992.

57. Lee C. Unequal protection: the racial divide in environmental law, a special investigation. National Law Journal 1992;9/21/92:S12.

58. U.S. Environmental Protection Agency, Office of Pollution Prevention and Toxics. 1996 Toxics Release Inventory: public data release—ten years of right-to-know. Washington, DC: U.S. Environmental Protection Agency, 1998.

BIBLIOGRAPHY

Agency for Toxic Substances and Disease Registries (ATSDR) site on the World Wide Web. Available: http://www.atsdr.cdc.gov/
Contains links to ATSDR's toxicologic profiles for chemicals as well as ATSDR's public health assessments for individual sites. Also available is a free Environmental Medicine Case Studies course, for which continuing education credit is available.

Casarett LJ, Amdur MO, Klassen CD, eds. Casarett and Doull's toxicology: the basic science of poisons, 5th ed. New York: McGraw-Hill, 1996
Reviews basic toxicologic principles and methods relevant to environmental health.

Centers for Disease Control and Prevention (CDC). Screening young children for lead poisoning: guidance for state and local public health officials. Atlanta: CDC, 1997.
The CDC's current policies on childhood lead poisoning prevention for public health agencies. Guidance is given on screening children for risk of lead poisoning and elevated blood lead levels, follow-up of elevated levels, and public health approaches for assessing lead risks for communities.

Centers for Disease Control and Prevention National Center for Environmental Health (NCEH) site on the World Wide Web. Available: http://www.cdc.gov/nceh
Contains links to key information related to surveillance and epidemiology of environmental disease, including asthma and air pollution; birth defects monitoring and prevention; developmental disabilities; the NCEH environmental toxicology laboratory; and environmental radiation-related research.

Committee of the Environmental and Occupational Health Assembly of the American Thoracic Society. Health effects of outdoor air pollution: state of the art. Parts 1 and 2. Am J Respir Crit Care Med 1996;153:3–50, 477–498.
A complete and clinically relevant review of the state of the science of outdoor air pollution and health effects.

EPA Integrated Risk Information System (IRIS) on the World Wide Web. Available: http://www.epa.gov/ordntrnt/ORD/dbases/iris/index.html
Contains detailed description and EPA's chronic noncancer and cancer assessments for more than 500 chemicals.

EPA Publications on the World Wide Web. Available: www.epa.gov/ocepa111/NNEMS/oeecat
Contains a wealth of publications about community, home, and school environments. Some highlights: EPA's publication "Integrated Pest Management in Schools," a guide to school districts and parents for safer pest control in schools and a video and toolkit called "Tools for Schools," a program to assist schools in preventing and correcting indoor air pollution problems; and "Citizen's Guide to Radon: Guide to Protecting Yourself and Your Family from Radon."

EPA Right-to-Know Web Pages, (a) Center for Environmental Information and Statistics (CEIS), Available: http://www.epa.gov/ceisweb1/ceishome/ceis_home.html; and (b), Envirofacts: http://www.epa.gov/enviro
These are the EPA right-to-know sites on the World Wide Web and provide access to most other EPA information about communities. Both of these have local information about TRI emissions, air pollution, and waste, and the CEIS site gives watershed and drinking water information as well. Both sites have mapping capabilities. Envirofacts can access EPA's "Master Chemical Integrator," which helps to search for information on chemicals. The CEIS site can access a number of documents, including reports on the state of the environment for states in the United States, as well as other nations and the world.

National Institute for Environmental Health Sciences Environmental Health Information Service site on the World Wide Web. Available: http://ehis.niehs.nih.gov/
Contains reams of useful information, including summaries and background information for the National Toxicology Program Report on Carcinogens; NIEHS technical reports; and current contents for the NIEHS

journal, Environmental Health Perspectives.

National Pesticide Telecommunications Network (NPTN). Telephone: 1-800 858-7378; or World Wide Web: http://ace.orst.edu/info/nptn/
A partnership between EPA and Oregon State University, the NPTN provides emergency advice for poison centers, clinicians, and the public as well as a tremendous amount of information about pesticides, ranging from use statistics to toxicity to tips for poisoning prevention, including worker protection. It contains a comprehensive array of useful links to the EPA and other pesticide resources.

National Research Council. Risk assessment in the federal government: managing the process. Washington, DC: National Academy Press, 1983.
A classic in environmental risk assessment, this book describes the process for assessment and management of risks that is used by most U.S. government agencies.

Pope AM, Rall DP, eds. Environmental medicine: integrating a missing element into medical education. Washington, DC: National Academy Press, 1995.
This 1995 Institute of Medicine book presents a core curriculum and a series of case studies that are useful for those who wish to teach the practice of environmental medicine or further their education in this area. Also contains an extensive list of resources and contacts for clinicians in the area of environmental medicine.

Samet J, Marbury M, Spengler J. Health effects and sources of indoor air pollution: parts I and II. American Review of Respiratory Disease 1987;136:1486–1508, and 1988;137:221–242.
This two-part review gives a comprehensive and clear description of the state of the science for indoor air pollution.

Occupational Health: Recognizing and Preventing Work-Related Disease and Injury, 4th ed.
Edited by Barry S. Levy and David H. Wegman, Lippincott Williams & Wilkins, Philadelphia © 2000.

Recognition and Prevention of Occupational Disease and Injury

4 Recognizing Occupational Disease and Injury

Barry S. Levy, David H. Wegman, and William E. Halperin

To effectively prevent occupational disease and injury, health care providers must know how to recognize work-related conditions, not only in workers who present with symptoms but also in those who are presymptomatic and in those for whom individual and group health information is available. A systematic approach facilitates consideration of all aspects of prevention in reducing or eliminating occupational hazards.

This chapter is organized to highlight the three levels of recognition that serve the three levels of prevention. *Primary prevention* is designed to deter or avoid the occurrence of disease or injury. *Secondary prevention* is designed to identify and adequately treat a disease or injury process as soon as possible. *Tertiary prevention* is designed to treat a disorder when it has advanced beyond its early stages so as to avoid complications and limit disability, or, if the condition is too advanced, to address rehabilitative and palliative needs.

The correct diagnosis and approach to treatment of a worker with an occupational illness or injury is essential to maximize opportunities for tertiary prevention and can also promote primary and secondary prevention. The selection and use of screening and monitoring tests that are appropriate to identify workplace risks promotes secondary prevention. A carefully designed occupational surveillance program, using both case- and rate-based approaches, promotes primary prevention.

When properly planned and integrated, these approaches contribute to (a) controlling risks at the source, (b) identifying new risks at the earliest possible time, (c) delivering the best level of therapeutic care and rehabilitation for workers who are ill or injured, (d) preventing recurrence of disease and injury of affected workers and occurrence of disease and injury in other workers who are exposed to similar risks, (e) ensuring that affected workers receive economic compensation legally due them, and (f) discovering new relationships between work exposures and disease.

The remainder of this textbook provides necessary information needed to recognize and prevent occupational disease and injury. This chapter introduces a systematic approach for the health care provider to recognize occupational disease and injury, with an eye toward prevention.

B. S. Levy: Barry S. Levy Associates, Sherborn, Massachusetts 01770.

D. H. Wegman: Department of Work Environment, University of Massachusetts Lowell, Lowell, Massachusetts 01854.

W. E. Halperin: National Institute for Occupational Safety and Health, Cincinnati, Ohio 45226.

DIAGNOSIS OF SYMPTOMATIC WORKERS

Proper diagnosis of illness or injury related to work requires information from a variety of sources. Successful identification of the work association rarely results from a single laboratory test or diagnostic procedure, but

Physicians and other health professionals have a vital role in recognizing occupational disease. Contrary to the drawing above, there is no simple test. The suspicion and the determination of work-relatedness depend primarily on a careful occupational history. (Drawing by Nick Thorkelson.)

rather depends critically on a comprehensive and appropriate patient history that adequately explores the relation of the illness to the occupation.

The more specialized use of the laboratory for biomonitoring and clinical testing, the need for proper environmental exposure assessment, and the important concerns with ethical, legal, and socioeconomic factors are addressed in subsequent chapters. In this section, attention is devoted exclusively to the task of obtaining and interpreting the occupational history.

The Occupational History

Consider the following four cases:

1. A woman who worked in a high-tech manufacturing plant had numbness in her distal arms and legs that her physician attributed to her diabetes.
2. A machinist was noted by his supervisor to have loss of balance on the job and was diagnosed at a nearby emergency department as being acutely intoxicated with alcohol.
3. A garment worker was told by her primary care physician that the numbness and weakness in some of her fingers was caused by her rheumatoid arthritis.
4. A man working at a bottle-making factory was told by his internist that the worsening of his chronic cough was caused by cigarette smoking.

In each of these situations, the physician made a reasonable and considered evaluation and diagnosis. The facts fit together and resulted in a coherent story, leading each physician to recommend a specific therapeutic and preventive regimen. In each of these cases, however, the physician made an incorrect diagnosis because of a common oversight—failure to take an occupational history.

The first patient had a peripheral neuropathy and the second had acute central nervous system (CNS) intoxication, both caused by exposure to solvents at work. The garment worker had carpal tunnel syndrome, possibly caused by some combination of her rheumatoid arthritis and the strenuous repetitive movements she performed with her hands and wrists hundreds of times an hour. The man working in the bottle-making factory had worsening of his chronic cough and other respiratory tract symptoms as a result of exposure to hydrochloric acid fumes at work.

This is not to say that the associations noted by the physicians were unrelated to the conditions diagnosed. They were probably contributory in at least the first, third, and fourth cases, but without the occupational history, proper therapy and prevention could not be planned.

The identification of work-related medical problems depends most importantly on the occupational history. Physical examination findings and laboratory test results may sometimes raise suspicion or help confirm that a medical problem is work related, but ultimately it is information obtained from the occupational history that determines the likelihood that this is the case. A phrase or two in the psychosocial section of the medical history is not enough; the physician should obtain data on the current and the two major past occupations for all patients. The extent of detail depends largely on the physician's level of suspicion that work may have caused or contributed to the patient's illness. The history should be recorded with great care and precision so that the data may be used for legal or research purposes.

What Questions to Ask

The occupational history has five key parts (Box 4-1): (a) a description of all of the patient's pertinent jobs, both past and present; (b) a review of exposures faced by the patient in these jobs; (c) information on the timing of symptoms in relation to work; (d) data on similar problems among coworkers; and (e) information on nonwork factors, such as

---

Box 4-1. Outline of the Occupational History

1. Descriptions of all jobs held
2. Work exposures
3. Timing of symptoms
4. Epidemiology of symptoms or illness among other workers
5. Nonwork exposures and other factors

---

smoking and hobbies, that may cause or contribute to disease or injury.

Some hospitals and clinics have standardized forms for recording the occupational history, which can expedite the taking and recording of this information. Ideally, such forms should include a grid with column headings for job, employer, industry, major job tasks, dates of starting and stopping the job, and major work exposures. It may be helpful to ask questions about whether the patient has had any exposures to hazardous substances or physical factors such as noise or radiation, from a list prepared in advance. On such an occupational history form (Fig. 4-1), the rows of the grid should be completed with information on each job, starting with the current or most recent job.

Further elaboration on each of the key parts of an occupational history may be helpful.

Descriptions of All Jobs Held

The history should include descriptions of all jobs held by the patient; in some cases, it may be important also to obtain information on summer and part-time jobs held while attending school. (Generally, details of these jobs are sought only on second interviews.) Job titles alone are not sufficient: An electrician may work in a plant where lead storage batteries are manufactured, a clerk may work in a pesticide-formulating company, or a physician may perform research with hepatitis B virus. It is important to remember that workers in heavy industry are not the only ones prone to occupational diseases—so are clerks, electronic equipment assembly work-

1. Please provide the following information on your work history.

Job	Employer	Industry	Major job tasks	Dates of starting	stopping	Major work exposures*
CUSTODIAN	City of Boston	Day Care	Repair, cleaning	10/91	→	Flu, kid's infections, asbestos, cleaners
GRINDER	Hudson Engine	Engine Mfg.	metal machining	10/86	10/91	Oil mist,
LATHE OPER.	Nash Engine	Engine Mfg.	metal machining	10/76	10/86	Noise,
BORE MACHINE OPERATOR	Kaiser	Die Making	Cutting metals	10/70	10/76	Lifting/Twisting
VOLUNTARY FIREFIGHTER	Town of Salem	—	Fighting House Fires	10/68	10/79	Fumes, gases
STUDENT	—	—	mechanic Student	6/68	9/70	Noise, oils
Military? YES	US Air Force	Helicopter Mech.	motor Repair	1/67	1/68	Noise, Stress
Part-time work?	Town General Store	Retail Food	Checkout Clerk	1/64	1/67	Repetitive motion

2. Have you had any possibly hazardous exposures outside of work? __Yes__ If yes, complete the following.

Major exposures	Associated activity	Location	Dates of starting	stopping
Wood dust	Cabinet making	Home	~1971	→

3. Have you ever smoked cigarettes? __Yes__ If yes, please answer the following questions.
How old were you when you started smoking? __18__
On average, how many packs have you smoked a day? __1/2__
Do you currently smoke? __No__ If no, how old were you when you stopped smoking? __31__

*Such as chemicals, fumes, dusts, vapors, gases, noise, and radiation.

FIG. 4-1. Sample occupational history form. (From BS Levy, Wegman DH. The occupational history in medical practice: what questions to ask and when to ask them. Postgrad Med 1986;79:301.)

ers, domestic workers, food service employees, and virtually all other types of workers. To learn exactly what the patient does at work, it may be useful to have the patient describe a typical work shift from start to finish and simulate the performance of work tasks by demonstrating the body movements associated with them. A visit to the patient's workplace by the physician may be necessary and is always informative.

The history should describe routine tasks (unless the job title is self-explanatory); unusual and overtime tasks, such as cleaning out tanks, should also be noted, because they may be the most hazardous assignments in which a patient is involved. It is important to ask about second or part-time jobs, the patient's work in the home as a homemaker or parent, and service in the military.

Work Exposures

The patient should be questioned carefully about working conditions and past or present chemical, physical, biologic, and psychological exposures. As in other parts of a medical history, to avoid limiting the responses, open-ended questions are asked initially, "What have you worked with?". Then more specific questions are asked, such as, "Were you ever exposed to lead? Other heavy metals? Solvents? Asbestos? Dyes?" Some knowledge of the most likely exposures in the jobs listed can help focus additional questions, and it is important to remember that tasks performed in adjacent parts of the workplace can also contribute to a worker's exposures. It is often worthwhile to rephrase important questions and

It is crucial to clearly understand working conditions and exposures. (Drawing by Nick Thorkelson.)

ask them at two points in the interview, because patients sometimes recall, on repeat questioning, exposures that they initially overlooked. It is also wise to inquire about unusual accidents or incidents that may be related to the patient's problem (such as spills of hazardous materials), work in confined spaces (Fig. 4-2), and new substances or changed processes at work.

Many workplace chemicals and other substances are referred to only by brand names, slang terms, or code numbers. It should be possible for a physician to obtain a list of the ingredients of most chemicals and to determine the nature of any hazard (see Appendix A). The federal Hazard Communication Standard is also helpful (see Chapter 10). In many states and localities, right-to-know laws facilitate the process whereby workers and their health care providers, with only limited information, can determine the toxic effects of these substances.

It is important to quantify these exposures as accurately as possible. Clinicians can esti-

FIG. 4-2. Many jobs require work in confined spaces. (Photograph by Earl Dotter.)

mate the degree of exposure by determining the duration of exposure and route of entry. Large amounts of volatile substances such as solvents can be inhaled unknowingly, especially if they do not irritate the upper respiratory tract or do not have a strong odor. Large amounts of certain substances—again, solvents are a good example—can be absorbed through the skin without the worker's being aware of the degree of this exposure. The patient should be asked to describe the amount of a potentially hazardous material that contacts skin or clothes or is inhaled on a typical workday. The patient should also provide information on eating, drinking, and smoking in the workplace, because contamination of hands can lead to inadvertent ingestion of toxic materials (Fig. 4-3). Handwashing and showering at work, changing of workclothes, and who cleans the workclothes may also be relevant.

It should be determined whether personal protective equipment (PPE)—such as gloves, workclothes, masks, respirators, and hearing protectors—has been provided, and if, when, and how often the worker has used this equipment. If PPE is not being used, it

is important to determine the reasons. Masks and respirators frequently are not worn because of poor fit, discomfort in hot weather, and difficulty in communicating when the mask is worn. In addition, masks and respirators that are not properly maintained are ineffective. If PPE is being used, it should be determined whether the equipment appears to fit and work properly. The presence of protective engineering systems and devices (e.g., ventilation systems) in the workplace and whether they seem to function adequately should be determined.

Timing of Symptoms

Information on the time course of the patient's symptoms is often vital in determining whether a given disease or syndrome is work related. The following questions are often useful: "Do the symptoms begin shortly after the start of the workday? Do they disappear shortly after leaving work? Are they present during weekends or vacation periods? Are they time-related to certain processes, work tasks, or work exposures? Have you recently begun a new job, worked with a new process, or been exposed to a new chemical in the workplace?"

Questions on recent changes at work are often critical in suspecting or proving that a disease is work related. On the basis of the responses to these and related questions the physician can determine whether the period from the start of exposure to the onset of symptoms and the time course of the patient's symptoms are consistent with those of the suspected illness. For example, certain irritants with low water solubility produce severe pulmonary damage and even fatal pulmonary edema with onset about 12 to 18 hours after work ceases. Symptoms of byssinosis are characteristically worse on returning to work on Monday morning. Nitroglycerin workers, whose blood vessels have dilated because of work exposure to nitrates, may suffer "withdrawal" angina while away from work. Latent periods vary, and occupational causes should not be ruled out simply

FIG. 4-3. Workers eating in the workplace may ingest toxic substances. (Photograph by Earl Dotter.)

because the timing of symptoms does not initially correlate with time spent at work.

Epidemiology of Symptoms or Illness Among Other Workers

The patient's knowledge of other workers at the same workplace or in similar jobs elsewhere who have the same symptoms or illness may be the most important clue to recognizing work-related disease. The physician should inquire further what the affected workers share in common, such as similar job, exposure, physical location in the workplace, age, or gender. Queries should be made regarding birth defects among offspring, fertility problems, cancer incidence, and high turnover or early retirement for health reasons. Workers and then their physicians linked the pesticide dibromochloropropane (DBCP) to male sterility and the catalyst dimethylaminopropionitrile (DMAPN) to bladder neuropathy by recognizing that similarly exposed workers had the same medical problems. However, workers may not always be aware of symptoms present in coworkers.

Nonwork Exposures and Other Factors

Sometimes there is a synergistic relation between occupational and nonoccupational factors in causing disease. The clinician should ask whether the patient smokes cigarettes or drinks alcohol; if so, amount and duration should be quantified. For skin problems, questions should be asked regarding recent exposure to new soaps, cosmetics, or clothes. The clinician should also ask whether the patient has any hobbies (e.g., woodworking, gardening) or other nonwork activities that involve potentially hazardous chemical, physical, biologic, or psychological exposures that may account for the symptoms; whether the patient lives near any factories, toxic waste sites, or contaminated sources of water; and whether the patient lives with someone who brings hazardous workplace substances home on workclothes,

shoes, or hair. The same suggestions noted in the Work Exposures section apply here: repeated questioning, quantification of exposure to the degree possible, and obtaining generic names of substances. Questioning should be aimed at determining both current and past exposures.

Other information that the clinician obtains may supplement the occupational history. It is useful to know whether the patient has had preplacement or periodic physical and laboratory examinations at work. For example, preplacement audiograms or pulmonary function test results may be helpful in determining whether hearing impairment or respiratory symptoms are work related. Because Occupational Safety and Health Administration (OSHA) regulations mandate periodic screening of workers with certain exposures (such as asbestos or coke oven emissions), and because many employers voluntarily provide health screening in the workplace, it is increasingly likely that such information may be available to a physician, if the worker approves its release.

Finally, it is often useful to ask the patient whether there is some reason to suspect that the symptoms may be work related.

When to Take a Complete Occupational History

A work history should always contain information on past and present jobs of the patient to provide a good understanding of how the workday is spent and what potential health hazards may exist. It is impossible to obtain a detailed occupational history on every patient seen, but every medical history should include at least the two major previous jobs and the current job.

In the following situations, the clinician should have a strong suspicion of occupational factors or influences on the development of the problem and take a detailed, complete occupational history. Many symptoms appear to be nonspecific but may have their origin in occupational exposures.

(Drawing by Nick Thorkelson.)

Respiratory Disease

Virtually any respiratory symptoms can be work related. It is all too easy to diagnose acute respiratory symptoms as acute tracheobronchitis or viral infection when the actual diagnosis is occupational asthma, or to attribute chronic respiratory symptoms as chronic obstructive pulmonary disease when the actual diagnosis is asbestosis. Viruses and cigarettes are too often assumed to be the sole agents responsible for respiratory disease. Adult-onset asthma is frequently work related but often not recognized as such. In addition, patients with preexisting asthma may have exacerbations of their otherwise quiescent condition when exposed to workplace sensitizers. Less commonly, pulmonary edema can be caused by workplace chemicals such as phosgene or oxides of nitrogen; a detailed work history should be obtained for anyone with acute pulmonary edema when no likely nonoccupational cause can be identified (see Chapter 25).

Skin Disorders

Many skin disorders are nonspecific in nature, bothersome but not life-threatening, and self-limited. Diagnoses often are nonspecific, and physicians all too often fail to take a brief occupational history that might identify the offending irritant, sensitizer, or other factor. Contact dermatitis, which accounts for about 90% of all work-related skin disease, does not have a characteristic appearance. Determination of the etiologic agent and work-relatedness depends on a carefully obtained work history (see Chapter 27).

Hearing Impairment

Many cases of hearing impairment are falsely attributed to aging (presbycusis) or other nonoccupational causes. Millions of American workers have been exposed to hazardous noise at work; for this reason, a detailed occupational history should be obtained from anyone with hearing impairment. Recommendations for the prevention of future hearing loss should also be made (see Chapter 1).

Back and Joint Symptoms

Most back pain is at least partially work related, but there are no tests or other procedures that can differentiate work-related from non–work-related back problems; the determination of likelihood depends on the occupational history. A surprising number of cases of arthritis and tenosynovitis are caused by unnatural repetitive movements associated with work tasks. Ergonomics, the study of the complex interactions among workers, their workplace environments, job demands, and work methods, can help prevent some of these problems (see Chapters 9 and 26).

Cancer

A significant percentage of cancer cases are caused by work exposures, and, as time goes by, more occupational carcinogens are discovered. Often the initial suspicion that a workplace substance may be carcinogenic comes from individual clinicians' reports. This effort would be facilitated if occupational histories were obtained from all patients with cancer. Of importance in considering occupational cancer is that exposure to the carcinogen may have begun 20 or more years before diagnosis of the disease and that the exposure need not have been continued over the entire time interval (see Chapter 16).

Exacerbation of Coronary Artery Disease Symptoms

Exposure to stress (see Chapter 21) and to carbon monoxide and other chemicals in the workplace (see Chapter 15) may increase the frequency or severity of symptoms of coronary artery disease (see Chapter 32).

Liver Disease

As with respiratory disease, it is all too easy to give liver ailments common diagnoses such as viral hepatitis or alcoholic cirrhosis, rather than the less common diagnoses of work-related toxic problems. It is always important to take a good occupational history from a patient with liver disease. Hepatotoxins encountered in the workplace are discussed in Chapter 34.

Neuropsychiatric Problems

The possible relation of neuropsychiatric problems to the workplace is often overlooked. Peripheral neuropathies are more frequently attributed to diabetes, alcohol abuse, or "unknown etiology"; CNS depression, to substance abuse or psychiatric problems; and behavioral abnormalities (which may be the first sign of work-related stress or, less frequently, a neurotoxic problem), to psychosis or personality disorder. More than 100 chemicals (including virtually all solvents) can cause CNS depression, and several neurotoxins (including arsenic, lead, mercury, and methyl n-butyl ketone) can produce peripheral neuropathy. Carbon disulfide exposure can cause symptoms that mimic a psychosis (see Chapters 29 and 30).

Illnesses of Unknown Cause

A detailed, complete occupational history is essential in all cases in which the cause of illness is unknown or uncertain (such as fever of unknown origin) or the diagnosis is obscure. The need to search carefully for a work-related source in such illnesses results from the increasing awareness of low-level environmental exposures as a cause of symptoms or disease. Although this issue has been raised most forcefully by groups concerned about hazardous waste disposal sites and indoor air quality (see Chapter 23), medical authorities increasingly have found reason to look more closely at this complex topic.

A key principle in toxicology and occupational health is that the biologic response to a chemical or physical agent is primarily a function of exposure dose. Although health effects from high levels of exposure typically are more frequent and more severe than those caused by low levels, more people are subject to low levels of exposure in the workplace and in the ambient environment. It is important for health professionals who are approached by workers with symptoms they think are related to low levels of exposure to chemical substances to develop a caring and careful approach to addressing these concerns.

Symptoms associated with low-level exposures are often difficult to evaluate because of difficulty in documenting the exposure and because the symptom pattern is much less specific than that of a well-established disorder. Health professionals may be skeptical and may wish to dismiss such complaints or to direct these patients to other specialists. This attitude is supported when there is the impression that complaints are being driven by psychosocial aspects of the job or other non–health-related factors.

As the complex nature of human physiology and its response to toxic materials evolves, however, there are compelling reasons why such cases should be examined systematically before they are set aside. Indirect toxic responses (allergic responses) may well not have a clear threshold dose below which effects do not occur. Moreover, human variability is such that even a normal distribution of responses includes a few individuals who respond at the low dose extremes. In addition, increasing experience with confusing problems such as the sick building syndrome makes clear that some patterns of response are environmentally related despite the absence of readily identifiable causal factors.

Although laboratory investigation of the syndromes represented in these workers may predominate, the history is still central in the final determination of how to care for these patients as indicated below:

1. If the problem is related to classic allergy, it may be possible to identify patterns of response of those who are severely atopic that effectively explain the nonspecific stimuli associated with lower symptom severity between actual allergic attacks.

2. Anxiety disorders may be associated with chemical or other environmental stimuli resulting in symptoms interpreted as being caused by the environment. A careful medical history should identify the need to have such patients evaluated by a specialist, especially because the relevant diagnoses may be ones of exclusion.

3. Sick building syndrome, a disorder caused by poor ventilation, is discussed further in Chapter 23. Characteristic symptoms of fatigue and respiratory, dermal, and CNS complaints are reported in association with a specific environment. Here the history should identify similar illnesses in coworkers and relate the symptoms to presence in a specific environment.

4. Attention has been drawn to a syndrome referred to as multiple chemical sensitivities (MCS) (1,2). This syndrome is reported to affect multiple systems and to occur in multiple, unrelated environments. Stimulants are reported to include such seemingly unrelated low-level chemical exposures as perfumes, petroleum derivatives, and smoke. Although there is often a willingness to attribute these symptoms to a primary anxiety disorder, few MCS cases meet the established criteria for this psychiatric diagnosis. Careful and well-documented medical and environmental histories may shed light on MCS as more knowledge develops.

In concluding this section on the diagnosis of symptomatic workers, it is surprising how current the following advice still is (3):

> If the recording intern would only treat the poison from which the man is suffering with as much interest as he gives to the coffee the patient has drunk and tobacco he has smoked, if he would ask as carefully about the length of time he was exposed to the

poison as about the age at which he had measles, the task of the searcher for the truth about industrial poisons would be made so very much easier.

SCREENING FOR OCCUPATIONAL DISEASE

Screening for occupational disease is the search for previously unrecognized diseases or physiologic conditions that are caused or influenced by work-associated factors. It may be part of an individual physician's evaluation of a patient's health or part of a large-scale prevention effort by an employer, union, or other organization. Screening methods can include questionnaires seeking suggestive symptoms or exposures, examinations and laboratory tests, or other procedures. To be widely used, the methods should be simple, noninvasive, safe, rapid, and relatively inexpensive. Screening is one technique in a continuum for the prevention of occupational disease. Other techniques include eliminating hazards from the workplace; containing hazards with engineering controls; protecting workers with PPE such as gloves and respirators; measuring intoxicants in the environment (environmental monitoring) or in biologic samples (biologic monitoring); and detecting, screening, and treating occupational diseases at early stages, when they are reversible or more easily treatable (4). As with screening for nonoccupational diseases, screening for work-related diseases only *presumptively* identifies those individuals who are likely (and those who are unlikely) to have a particular disease. Further diagnostic tests are almost always necessary to confirm the diagnosis and assess the severity of the worker's condition.

Although screening data may eventually lead to more effective primary prevention measures, the purpose of screening is the identification of conditions already in existence at a stage when their progression can be slowed, halted, or even reversed. Screening is therefore a secondary prevention measure. Primary prevention measures that re-

duce workers' exposure to occupational hazards are, in general, more likely to improve health and prevent disease (see Chapters 5, 7, 8, and 9).

The main goal of screening is early detection and treatment of disease; other goals include evaluation of the adequacy of exposure control and other means of primary prevention, detection of previously unrecognized health effects suspected on the basis of toxicologic and other studies, and suitable job placement. Clearly screening data, in addition to their clinical use for the protection of the individual screened, may be useful in a surveillance system in which they analyzed epidemiologically for the protection of the community of similarly exposed workers (5).

The employees at a particular workplace are a logical target for screening for occupational disease because they have some risk factors in common (their workplace exposures) and a clear opportunity for prevention in common (reduction or elimination of those exposures). In addition, a workplace can provide excellent opportunities for screening for treatable nonoccupational diseases such as hypertension. To be effective, screening programs for occupational disease must meet the following five criteria:

1. Screening must be selective, applying only the appropriate tests to the population at risk for development of a specific disease, given exposures, demographic features, and other factors. A "shotgun" approach, involving a battery of tests (e.g., "chemistry profile") applied indiscriminately without regard to the diseases for which the population is at risk, is generally not effective. The natural history of the exposure-disease relationship should be considered in the application of screening tests. For example, screening of workers exposed to asbestos during the first few years after the start of exposure may lead to a false sense of security, because there has not been sufficient time for the disease process to become detectable on screening examination.

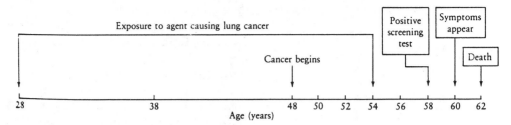

FIG. 4-4. Phases of cancer development. If the course of the disease cannot be positively influenced by early detection and effective treatment, there is no advantage to screening an individual for early detection of the disease. Current screening tests for lung cancer have yet to be proved effective. Screening may detect cancer earlier than would occur without screening, but the eventual time of death is not significantly changed.

2. Identification of the disease in its latent stage, instead of after symptoms appear, must lead to treatment that impedes progression of the disease in a given patient or to measures that prevent additional cases (Fig. 4-4). The major justification for screening for a disease for which there is no therapy is to allow an opportunity to control exposure and prevent disease in others similarly exposed.

3. Adequate follow-up is critical, and further diagnostic tests and effective management of the disease must be available, accessible, and acceptable both to examiner and worker. Lack of follow-up is a frequent deficiency in screening programs for occupational disease. Workers who have been screened should receive test reports along with interpretation of test results and summary data for the entire group tested. (OSHA requires that records of medical surveillance be made available to the affected employee. These records may be transmitted to third parties only with the written consent of the worker.) Follow-up also entails action to reduce or eliminate the hazard. An example is job transfer for the ill worker combined with improvements in the ventilation systems of the plant; job transfer without action to control the underlying problem may result in exposure of another worker to the same hazard.

4. The screening test must have good relia-

bility and validity. Reliability reflects the reproducibility of the test. Validity reflects the ability of the test to identify correctly which individuals have the disease and which do not. Validity is evaluated by examining sensitivity and specificity. *Sensitivity* is the proportion of those with the disease that the test identifies correctly; *specificity* is the proportion of those without the disease that the test identifies correctly. Another measure of a screening test is the *predictive value positive*, which is often more useful clinically than either sensitivity or specificity; it indicates the proportion of those with a positive screening test who actually have the disease (Table 4-1). The prevalence of the disease affects the predictive value. The pre-

TABLE 4-1. *Hypothetical data: Screening of 100,000 workers for colon cancer**

Test outcome	Colon cancer present		
	Yes	No	Total
Positive	150	300	450
Negative	50	99,500	99,550
Total	200	99,800	100,000

Sensitivity = 150/200 = 75%. The test was (correctly) positive for 75% of actual cancer cases, but 25% of the actual cases were not detected.
Specificity = 99,500/99,800 = 99%. The test was (correctly) negative for 99% of those who actually did not have colon cancer.
Predictive value positive = 150/450 = 33%. Of those with a positive test, 33% actually had colon cancer.

dictive value rises as prevalence rises, even if the sensitivity and specificity of the test remain the same.

5. The benefits of the screening program should outweigh the costs. Benefits consist primarily of improved quality and length of life—that is, reduced morbidity and mortality. Costs include both economic costs (the expenses of performing the screening tests and further diagnostic tests and of managing the disease in affected workers) and human costs (the risks, inconvenience, discomfort, and anxiety of screening and of diagnostic workups for those with false-positive results). Screening tests in the community must be inexpensive because they compete with other public health resources, such as immunization. It should not be assumed that effective screening tests for occupational disease must be inexpensive, because they do not compete for the same resources. The cost-benefit equation is often difficult to determine and relies on tenuous assumptions. Such analysis should not be allowed to obscure the primary objective of screening: early identification of work-related disease. Advocates of screening should be cautious, because increased survival in those determined to have the disease by screening, compared with those detected after they become symptomatic, may be a result of lead-time bias or length bias. In lead-time bias, the apparently increased survival time results from adding part of the preclinical detection period to the postdiagnosis survival time, and not from altering the actual duration of survival after the disease is contracted. In length bias, an apparently increased survival time results from the greater probability of detecting indolent, more benign disease than quickly developing disease, which is less likely to be detected because it is present for a shorter period.

There must be mutual trust among the individuals who have requested or authorized the screening program, the health professionals who are administering it, and the workers being screened. Without such trust, workers may be reluctant to be screened. This trust is developed, in part, by management personnel and health professionals assuring that screening data will be kept strictly confidential, will be used only for the stated purpose of the screening program, and will not adversely affect the worker's salary or other benefits. In addition, for any screening program to be effective, it cannot be used as a tool to discriminate—sexually, racially, or otherwise—against a specific group of workers.

Screening Approaches

The following paragraphs review current screening approaches to five major categories of work-related disease: nonmalignant respiratory disease, hearing impairment, toxic effects, cancer, and back problems. It may be noted that few of the current screening approaches for occupational disease meet all five criteria for effective screening.

Nonmalignant Respiratory Disease

Screening for acute work-related respiratory diseases such as irritant pneumonitis generally is not possible. Pathologic changes caused by exposure to irritant or toxic gases and fumes develop so quickly that there is no opportunity to screen for these disorders during a latent stage. However, many chronic work-related respiratory diseases are amenable to screening. The time from initial exposure to first appearance of symptoms in these diseases usually is very long—often years. Early identification of workers with asymptomatic pulmonary disease and reduction or elimination of their hazardous exposure may reverse the disease process or at least halt or slow its progression. Once well-established, however, most of these diseases are not reversible by currently available treatment, and they account for much morbidity and mortality (see Chapter 25).

Screening approaches for occupational respiratory diseases range from simple questions to sophisticated tests of pulmonary function. The four basic approaches are history of respiratory symptoms, physical examination, chest radiographs, and pulmonary function tests. Each has its strengths and weaknesses, and usually two or more approaches are used in combination.

History. By means of direct questioning or use of a standardized questionnaire,[1] information can be elicited on the presence of respiratory symptoms, including cough, sputum production, wheezing, and dyspnea. The worker is questioned about the presence of these symptoms, their time course, and their relation to airborne substance exposure, exertion, and work habits. The worker is also questioned about work history and in detail about cigarette smoking history. Although this approach is simple, inexpensive, without risk, usually acceptable to the worker, and capable of being performed by paramedical personnel or the worker, it has major weaknesses. With some diseases, such as asbestosis, cough, dyspnea on exertion, and other symptoms often do not appear until the disease is moderately advanced. The worker with lung disease may fail to report certain symptoms, such as "smoker's cough," that may be considered acceptable or unimportant. The worker may choose not to report certain symptoms for fear of losing a job or being labelled an unhealthy person. Finally, respiratory symptoms often result from causes other than chronic lung disease. (The first three of these weaknesses have to do with low sensitivity, the last with low specificity.)

Physical Examination. Performing a physical examination is generally a less helpful screening approach than obtaining a history of previously unrecognized respiratory symptoms. It, too, is simple, inexpensive, without risk, acceptable to the worker, and

capable of being done by paramedical personnel; but it, too, has low sensitivity and specificity and is rarely helpful in screening for work-related pulmonary disease. For example, by the time basilar rales are heard in a person with asbestosis, significant fibrosis has already occurred, and such physical signs as clubbing and cyanosis usually are associated with far-advanced disease.

Chest Radiographs. Chest radiographs also have significant limitations in the early detection of chronic respiratory disease. This screening approach is more expensive and requires special equipment. In addition, with periodic radiographs a given worker may face some cumulative radiation hazard. Moreover, chest films usually are not very sensitive or very specific: the presence or absence of abnormalities does not always correlate with the intensity of symptoms, early physiologic abnormalities, or actual pathology. Chest films also fail to reveal early changes of chronic obstructive pulmonary disease, and they are subject to much variation in technique and interpretation. Despite these limitations, chest films can play an important role in the early diagnosis and assessment of work-related restrictive diseases, especially if chest radiographic abnormalities begin to appear relatively early in the course of chronic disease. (Chest radiographs are discussed again later in the context of lung cancer screening.)

Pulmonary Function Testing. Although it requires special equipment and therefore is more costly than performing histories or physical examinations, pulmonary function testing is a reasonably sensitive screening approach for work-related respiratory diseases, and it generally provides more useful screening information than the other approaches (Fig. 4-5). Pulmonary function tests used for screening are relatively easy to perform, and if they are properly done the results are reproducible.

Pulmonary function testing suffers from two general limitations: the range of normal is wide, and if the worker being tested does not cooperate fully artifacts can appear, es-

---

[1] An excellent questionnaire developed by the American Thoracic Society has been published (Ferris BG. Epidemiology standardization project. Am Rev Respir Dis 1978;118:7–54). It is also available from the National Heart, Lung and Blood Institute, Bethesda, MD 20892.

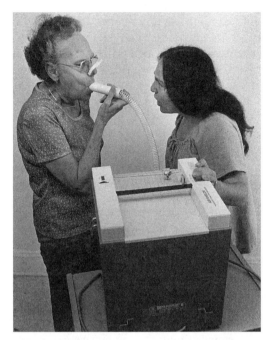

FIG. 4-5. Pulmonary function testing. (Photograph by Earl Dotter.)

pecially in tests requiring maximal effort. The first limitation can be countered by periodic testing of the same individual; test results in a given worker can be monitored over time and abnormalities can be identified when greater than expected decreases in function occur. The second limitation can be addressed by applying standardized rules for acceptable tests (see Chapter 25).

The two most frequently used screening tests for pulmonary function are forced vital capacity (FVC) and forced expiratory volume in the first second of expiration (FEV$_1$). FVC is the maximal volume of air that can be exhaled forcefully after maximal inspiration. For most people it closely approximates the vital capacity without a forced effort. FVC is reduced relatively early in restrictive diseases such as asbestosis. FEV$_1$ is the volume of air that can be forcefully expelled during the first second of expiration with a maximal effort after the lungs have been filled completely. It is reduced in both restrictive and obstructive disease but relatively more so

than FVC in the latter. In the early course of asthma, it returns toward normal as the attack ends spontaneously or with the use of bronchodilators. Advances in assessing pulmonary dysfunction are occurring but generally require validation before they are ready for routine use (6). For further discussion, see Chapter 25.

Evaluation of Nonoccupational Risk Factors. Evaluation of workers for non–work-related risk factors that may predispose them to occupational pulmonary disease is another approach, albeit a controversial one. This is actually a method of primary prevention rather than screening. For example, some employers identify smokers, who may be at increased risk for a variety of occupational pulmonary diseases, and restrict them from certain jobs. This approach is sometimes opposed by workers who believe that it represents unfair discrimination. One large asbestos company has prohibited workers from smoking in the workplace—even at its corporate offices—and has refused to hire new workers who smoke. This approach is controversial also because it can encourage employers to avoid eliminating hazardous conditions and instead find workers with "iron constitutions" who can withstand these conditions. Decisions on this subject obviously involve both scientific assessments and public policy considerations of equity.

Screening for α_1-antitrypsin deficiency is an example of this approach to risk factor identification. People with a severe deficiency of this protein (1 in 5,000 of the general population) are at very high risk for development of emphysema and chronic bronchitis and should not work in a dusty workplace. It has not been established, however, whether those persons with lesser degrees of this deficiency are at increased risk for development of respiratory diseases.

Although there are many screening opportunities, occupational respiratory disease is difficult to detect before significant loss of lung function has occurred. Therefore, more reliance should be placed on methods of primary prevention, such as ventilation systems, changed work practices, and substitution of nonhazardous substances for hazardous ones.

Hearing Impairment

Several million Americans suffer chronic work-related hearing impairment, and several million American workers are exposed to loud noise at work that poses a threat to hearing (see Chapter 18). Even at the current OSHA standard of 90 dBA (decibels of sound pressure) for 8 hours of workplace noise exposure, it is estimated that 10% of those exposed for a lifetime will have significant hearing impairment. By the time a worker notices hearing impairment, irreversible sensorineural damage affecting the sound frequencies of human conversation has usually occurred. Long before a worker notices any hearing impairment, significant changes can be seen in the audiogram. Screening for hearing impairment is therefore important for workers exposed to loud noise.

The first sign of hearing impairment is a dip in the audiogram, usually at 3,000 to 6,000 Hz (cycles per second). If hearing impairment progresses, the audiologic abnormality becomes more severe and covers a broader range of frequencies. Discovery of an abnormal audiogram can indicate the need to prevent hearing impairment by reduction of noise at the source, modification of work procedures creating the noise, use of PPE (earmuffs or earplugs), or removal of the worker from a noisy work environment.

Audiograms should be performed as part of the preplacement examination of workers who will be exposed to loud noise at work, so that baseline findings will be available for comparison with later audiograms. They should be repeated for exposed workers every year. Audiograms should not be performed within 14 hours of any significant noise exposure; if they are preformed sooner, a temporary threshold shift may be mistakenly identified as a permanent one. As with other screening tests, it usually is best to compare test results repeated on an individual over time rather than with "normal limits." Because deterioration of hearing caused by loud noise exposure is fairly rapid for the first years of exposure, the effectiveness of a screening program is maximal during this initial work period.

Audiometry is generally accepted as a useful screening approach and is widely performed in industry in the United States. However, its value can be undermined by poor technique, such as inadequate calibration of equipment, excess noise in the testing room, headphone position variations, headset pressure against the external ear, examiner and tester biases, improved performance of the subject after familiarization with the testing procedure, obstruction of the ear canal, tinnitus, simulation or malingering, and fluctuation of the subject's criterion for threshold identification of the test tone. A study performed by the National Institute of Occupational Safety and Health (NIOSH) indicated that 80% of industries surveyed used inadequate audiometric equipment. However, most of these problems can be minimized with appropriately trained technicians and adequate equipment.

Toxic Effects

Three components are involved in the prevention of the toxic effects of workplace chemicals: evaluating the toxicity of the chemical itself (preferably before it is introduced, as the Toxic Substances Control Act now mandates) by means of animal studies and short-term *in vitro* assays; environmental monitoring of levels of the chemical in workplace air to determine whether it is controlled in accordance with recommended or mandated standards; and biologic monitoring (see also Chapter 15).

Biologic monitoring is the testing of blood, urine, or exhaled air of workers to determine either (a) the body's level of a hazardous chemical and its metabolites, or (b) reasonably specific biochemical changes that are associated with cellular damage. The first of these findings actually provides evidence of exposure; the second is considered screening in the strict sense. For the latter approach to be effective, it must detect early ("sentinel") biologic effects before serious health effects occur. As with other evaluation approaches, once biologic monitoring of any kind indicates that workers have excessive exposure or early toxic effects, measures must be taken to reduce their exposure to the responsible agents. Following are three examples of biologic screening:

1. Several volatile organic compounds, including benzene and toluene, if inhaled or absorbed through the skin, produce metabolites that can be measured in urine.
2. Organophosphate pesticides, before exerting any known health effects, begin to inhibit both plasma and red blood cell cholinesterase. The amount of plasma cholinesterase (pseudocholinesterase) reflects absorption of the organophosphate, and the activity of the red cell cholinesterase correlates well with the degree of adverse effect.
3. Various tests have been used to evaluate lead exposure or body burden and the biologic effects of lead. Biologic monitoring for lead is particularly useful because the early effects of lead poisoning may be reversible and because symptoms are nonspecific (such as headache or fatigue) or may be absent (see Chapters 15, 29, 31, and 35).

Biologic monitoring has a different use from environmental monitoring because it takes into consideration host differences in, for example, susceptibility to toxic effects and absorption, distribution, and biotransformation of the substance. It also considers possible multiple exposures (both occupational and nonoccupational) and multiple routes of absorption. A crucial issue is the relation between environmental monitoring and biologic monitoring. Because environmental monitoring leads to control of exposure before the absorption of the hazardous chemical, it is preferable. However, biologic monitoring should not be considered a substitute for environmental monitoring. Given the potential for multiple routes of exposure not well assessed by environmental monitoring (such as percutaneous absorption), biologic monitoring should be used as a valuable adjunctive, fail-safe technique.

Biologic monitoring is still in its infancy and has several limitations. For many substances, there is no known health effect parameter; as each new parameter is developed, its relation to both the amount of exposure and the disease must be established. The biologic half-lives of many toxins are not known, and screening may be done at the wrong time to identify acute intoxication or transient effects. For many known biologic parameters, the range of normal is wide, so it is necessary to base the interpretation of testing on a series of tests on the same individual over time. This demonstrates the importance of performing baseline biologic screening studies that are specific to known or anticipated hazards during the preplacement examination. Quality control in laboratories varies; it is essential that the laboratories performing the biologic monitoring ensure accurate results.

These potential problems make it crucial to plan biologic monitoring carefully. Workers who may be exposed must be identified. The appropriate parameter to monitor them must be chosen. Baseline measurements made before exposure and measurements made after exposure must be appropriately timed. There is also much room for error in the choice of specimen, the storage and handling of specimens, and the interpretation of results. However, biologic monitoring holds much promise, and with the increase of toxic substances in the workplace and greater recognition of toxic hazards, it can play an important preventive role.

There is no central repository of information for choosing biologic monitoring tests or appropriate laboratory methodology. NIOSH has developed a chart that provides a summary of the recommendations that have been made by a variety of experts for biomonitoring for industrial chemicals. It can be accessed via the World Wide Web at http://www.cdc.gov/niosh/nmed/medstart.html.

The World Wide Web is also a valuable source of policy papers on screening and monitoring and of summaries of recommendations that are being made in this evolving area. For example, see Guidelines for Health Surveillance from the Australian National Occupational Health and Safety Commission, which can be accessed at http://www.worksafe.gov.au/worksafe/fulltext/toc/01997- 01.htm.

Cancer

Screening has a limited role in occupational cancer control. National Cancer Institute data support the concept that early detection of cancer followed by appropriate treatment can increase survival time in some patients with certain cancers. Approaches that have been used to screen for different cancers include examination of exfoliated cells by the Papanicolaou technique (Pap smear); radiographs; proctosigmoidoscopy; identification of a substance in the blood or other body fluid that may be a specific marker for a given malignancy; breast self-examination; measures of organ function; and tests to detect colon cancer by identifying occult blood in the stool. These approaches have widely varying degrees of effectiveness (see Chapter 16).

Few screening approaches of any kind, however, have been proved to reduce mortality from cancer. As with screening for most chronic diseases, discussion of the effectiveness of cancer screening has been greatly confused by studies that do not differentiate between true mortality reduction and mere earlier identification.

A dramatic increase in the lung cancer mortality rate has taken place in the past 50 years. Attempts to detect lung cancer early have focused on periodic chest radiographs and cytologic examinations of sputum, which tend to complement one another: chest radiographs are more useful for detecting peripherally situated cancers, whereas sputum cytology can identify early squamous cell carcinoma involving major airways. Relatively few cases of lung cancer give a positive result on both tests at the same time. Although these tests are often used to screen for lung cancer, neither has convincingly been shown to be effective. Usually, by the time either of these tests presumptively identifies a lung cancer, it has metastasized and is incurable. Well-controlled studies have demonstrated that the addition of sputum cytology to chest radiographic screening does not significantly reduce the mortality rate from all types of lung cancer but have suggested that mortality from squamous cell carcinoma is reduced (7). A report of three randomized trials of screening for early lung cancer indicated that sputum cytology detects 15% to 20% of all lung cancers, mostly squamous cell carcinoma, with a relatively good prognosis, and that chest radiography alone may be a more effective test for early-stage lung cancer than previous reports suggested. However, a randomized clinical trial at the Mayo Clinic showed that performance of both procedures every 4 months resulted in no survival advantage compared with standard medical practices (8). Recent reports raise the possibility that CT scans may have a role in early detection of lung cancer.

The status of attempts to screen for bladder cancer is much the same as for lung cancer. The approaches used most frequently are a search for occult blood and cytologic examination of exfoliated cells in urine. These approaches successfully identified asymptomatic persons with early bladder cancer in a population at high risk because of exposure to aromatic amines (9). However, whether screening in this or similar high-risk groups leads to prolongation of life or decreased morbidity has not been evaluated in a controlled clinical trial. Given the continued high incidence of cancer (and the substantial incidence of occupational cancer), its severity, and the frequent lack of effective treatment, attempts will no doubt continue to develop better screening tests. In the meantime, health professionals should not raise false hopes of workers by using screening tests of unproven effectiveness and should concentrate on measures of primary prevention. These measures include testing workplace substances for carcinogenicity, limiting exposure to proven or suspected carcinogens, and encouraging smoking cessation, since smoking is associated with lung, bladder, and other cancers.

Back Problems

When the term "screening" is used to refer to back problems in the workplace, it usually refers to preplacement identification of pre-existing back problems, both work-related and non–work-related, or of a predilection for back problems. Three methods have traditionally been used to try to identify workers at high risk for work-related back problems: history, physical examination, and radiographs of the lumbosacral spine. None of these methods has been effective in controlling low back injuries. Radiographs were used on the basis of a hypothesis, now shown to be false, that developmental abnormalities of the spine predispose to low back injury. The persistent use of back radiographs to detect such abnormalities not only is without benefit but also discriminates unnecessarily against prospective workers with radiographic abnormalities. Radiographs do not necessarily predict future back injury risk, and they create unnecessary exposure to x-rays. Although the only effective control for back problems today seems to be the ergonomic approach of designing the job to fit the worker, some evidence indicates that measurements of strength and fitness before the start of work can predict back injuries (see Chapters 9 and 26). In addition, strength measurements can be used to match a worker's strength to job requirements.

Possibilities for Improved Screening

Opportunities for effective screening for occupational diseases at present are relatively limited, and most available screening approaches do not meet the criteria outlined earlier in this chapter. Unless screening approaches are improved, much time, effort, and limited resources may be wasted; workers may face unnecessary risks and experience unnecessary anxiety and inconvenience; and workers and employers may become disillusioned with preventive approaches in general.

The general industry standards for specific hazardous exposures, published by OSHA, specify requirements for medical surveillance of exposed workers (10). These may include preplacement and periodic screening histories, examinations, and tests. Table 4-2 illustrates some of the specific screening tests required by OSHA. OSHA also requires employers to keep records of this surveillance and to make these records available to affected employees. The records can also be made available to physicians or other third parties on specific written request.

Suggested principles for screening and biologic monitoring of the effects of exposure in the workplace and many related articles were the subject of an intensive national conference held in 1984 and published as the August and October 1986 issues of the *Journal of Occupational Medicine*. A central theme expressed in these discussions was the following: "Screening and monitoring, in and of themselves, prevent nothing; only the appropriate intervention, in response to results of these tests, can prevent" (11).

OCCUPATIONAL SURVEILLANCE FOR DISEASE CONTROL

Occupational surveillance is the systematic and ongoing collection, analysis, and dissemination of information on disease, injury, or

TABLE 4-2. *Illustrative components of medical surveillance in selected OSHA standards*

Exposure	History	Physical examination	Other tests/procedures
Airborne asbestos	Especially respiratory symptoms	Especially chest examination	Chest x-ray FVC and FEV_1
Vinyl chloride	Especially alcohol use, history of hepatitis, transfusions	Especially liver, spleen, and kidneys	Liver function tests
Inorganic arsenic	Especially respiratory symptoms	Especially nasal and skin examinations	Chest x-ray Sputum cytology
Benzene	Including alcohol use and medications	If respirator used >30 days a year, specific attention to cardiopulmonary exam	Complete blood count Reticulocyte count Serum bilirubin
Cadmium	Including respiratory and renal symptoms, and medications	Especially blood pressure, respiratory and genitourinary system	Urinalysis Blood cadmium
Methylene chloride	Including neurological symptoms and heart, liver, and blood disease	Particular attention to lungs, cardiovascular system, liver, skin, and neurological system	Based on medical and work history

Source: Occupational Safety and Health Administration, U.S. Department of Labor. Code of Federal Regulations (CFR) Title 29: General industry.

hazard for the prevention of morbidity and mortality. Surveillance as it applies to populations should be differentiated from *medical surveillance* of individuals. Medical surveillance, also known as "medical monitoring" and sometimes as "periodic medical screening," is focused on the interview and examination of the individual. *Public health surveillance,* of which occupational surveillance is a subset, is focused on populations. Although the overriding goals of medical surveillance and public health surveillance are the same—that is, prevention, the specific goals are different. There are five goals of public health surveillance as it is applied to occupational disease:

1. To identify illnesses, injuries, and hazards that represent new opportunities for prevention. New opportunities can arise from new problems, such as might occur with the introduction of a new hazardous machine, or from belated identification of a long-standing but ignored problem or the recurrence of a problem previously controlled.

2. To define the magnitude and distribution of the problem in the workforce. Information on magnitude and distribution is use-

ful for planning intervention programs. Although no hazard is acceptable, the more common and severe problems deserve more immediate attention.

3. To track trends in the magnitude of the problem as a rudimentary method of assessing the effectiveness (or lack of effectiveness) of prevention efforts. Epidemics can be tracked on their rise or their decline.

4. To target (identify) categories of occupations, industries, and specific worksites that require attention in the form of consultation, educational efforts, or inspection for compliance with established regulations.

5. To publicly disseminate information so that wise personal and societal decisions can be made.

There is a *continuum of outcomes* that could be monitored. The continuum may range from the presence of an exposure or hazard, to early and subclinical health effects of that hazard, to morbidity and associated medical care and disability, and finally to mortality. The choice of an appropriate exposure or health outcome for surveillance should depend on the goal of the surveillance. Other considerations should include an assessment of whether the proposed reporting entity (such as physician or employer) will report the occurrence; the accuracy of the system in detecting real problems and minimizing false-positive leads; the timeliness of the system in producing useful information; and the cost of the system in relation to other systems that could be supported instead.

There are two kinds of surveillance. One is based on the intensive investigation of cases (*case-based*); the other is more embedded in epidemiologic methods, especially determination of the distribution or rate of disease, injury, or hazard in the population (*rate-based*). An underlying philosophy for case-based surveillance has been called the *sentinel health event (occupational) method,* or SHE(O) (12). A SHE(O) is defined as a case of disease, injury, or exposure that represents a failure of the system for prevention. Although a list of SHE(O)s has been published, this should not inhibit focusing on other adverse entities that are more germane to a local situation.

Rate-based surveillance is embedded in epidemiology in that it seeks to establish the rate of occurrence of the disease, injury, or exposure and to track that rate over time or compare it with the rate in some other population. Surveillance differs from epidemiologic research, however, in that surveillance is an ongoing activity with goals directly related to the functioning of the public health system, whereas epidemiologic research is concerned with assessing the association between effect and etiologic agent. Epidemiologic research also involves intensive collection of data during a limited period, rather than the ongoing collection and assessment of data that is part of surveillance. Although it is valuable to discern the differences between surveillance and research in their pure forms, in reality these distinctions often blur.

Surveillance can be used to monitor either the occurrence of diseases (or physiologic abnormalities) or the presence of hazardous substances and worker exposures to them.

This section focuses on the use of surveillance in monitoring the occurrence of injury and disease. For chronic diseases caused by workplace exposures, monitoring of exposures may be more useful. This possibility exists because a number of exposure-effect relationships are now sufficiently well described so that the long-term exposures that predictably result in chronic illness are known. Furthermore, the long latency period between exposure and onset of chronic work-related disease makes it difficult to associate the exposure with the disease in an individual case. For diseases of shorter latency, direct disease surveillance may be useful.

In contrast to communicable disease surveillance, which is largely based on physician reporting, there are a variety of models for occupational disease surveillance. Some of these are broadly based, and others can be done on a workplace-specific or job-specific basis.

Broad-Based Occupational Surveillance Programs

Death Certificates. The National Occupational Mortality System (NOMS) of NIOSH collects and codes mortality and occupational information from about 500,000 death certificates annually from 23 states in the United States. This allows analysis of differential mortality patterns among occupations and industries and comparison of the distributions of industries and occupations among diseases. It is one of the few systems capable of providing information about women and minority workers in the workforce. NIOSH also conducts surveillance for fatalities from injuries through the National Traumatic Occupational Fatalities (NTOF) system, which collects from all states death certificates in which the cause of death was an injury at work.

Employer Records. An annual survey of a large sample of employers is performed by the Bureau of Labor Statistics (BLS) of the U.S. Department of Labor. Using information from the required "OSHA 200" log of injuries and illnesses, these data provide broad estimates of work-related disease and injury. However, the survey is limited by the absence of specific criteria for determining the work-relatedness of disease, the limited sensitivity of the OSHA 200 log for detecting cases, and the assurance of confidentiality, which limits the usefulness of the survey for identifying cases or workplaces for in-depth follow-up investigations.

Workers' Compensation Records. Although readily available in most states, workers' compensation data are limited because they include only those who file (generally workers with the more severe injuries and illnesses), they exclude most cases of chronic work-related disease, and they are limited by adjudication procedures and diagnostic criteria that vary from state to state (see Chapter 11). However, these data have been very useful in identifying new problems, such as violence toward women workers, and in providing estimates of the magnitude of newly identified problems, such as disability from knee disease in carpet installers. In Ohio, workers' compensation readily identified companies with excess cases of dermatitis, as well as the offending agent (13).

Cancer Registries. Hospital-based, regional, or statewide cancer incidence registers can be useful sources of surveillance data on cancer but often provide only limited, if any, information on occupation.

Physician Reporting. In locations such as Alberta (Canada), Great Britain, Germany, and some states in the United States, the law requires physicians to report all work-related diseases and injuries or certain specified ("scheduled") conditions. Where this is effectively enforced, the scheduled diseases can be tracked and epidemics identified early.

Laboratory-Based Reporting. A state-based national system, the Adult Blood Lead Epidemiology and Surveillance (ABLES), collects information from the 26 U.S. states that require laboratories to report cases of excessive lead levels. This information has

proved useful in making national estimates of lead poisoning, tracking trends, identifying underserved occupations and industries, and targeting specific worksites with excessive cases. The limitations of laboratory-based reporting include the limited number of conditions for which laboratories can be involved; an irony is that those workers with the most inadequate resources for assistance are also the least likely to be monitored for lead.

Sentinel Event Approaches. Examples of sentinel event approaches exist in both Great Britain and the United States. In Great Britain, the SWORD system was developed to identify new and survey known types of occupational respiratory disease, using reports from thoracic and occupational physicians (14). Preliminary success has led to efforts to replicate the model for occupational dermatitis. In the United States, NIOSH is working with 36 states to develop state-based systems for surveillance of occupational disease and injury. A central element of this effort is the Sentinel Event Notification System for Occupational Risks (SENSOR), which has included silicosis, occupational asthma, amputations, cadmium poisoning, carpal tunnel syndrome, child-labor injuries and illnesses, noise-induced hearing loss, pesticide poisoning, spinal cord injuries, and tuberculosis (15). New conditions are also being explored for inclusion in SENSOR. For example, states are now using workers' compensation reports and networks of dermatologists to report occupational dermatitis, and states are testing surveillance of severe occupational burns, using reports from hospital burn units.

Focused Occupational Surveillance Programs

Surveys of Workers. Interviews and examinations of workers represent an effective surveillance approach, especially for estimating the magnitude and distribution of occupational problems in the workforce. In addition to focused efforts at specific work locations, large interview surveys addressing the prevalence of cumulative trauma disorders, dermatitis, and other conditions have been conducted by the National Center for Health Statistics (NCHS). NCHS conducts other large examination surveys that contain limited information relevant to occupation, such as blood lead level.

Union Records. Unions may have morbidity or mortality data, often related to medical or death-benefit programs, that can be used for surveillance. Even without this information, union records can define the exposed, or at-risk, population, information needed in the search for adverse health outcomes in state vital registry records or cancer registries.

Employer Records. Employer records can be helpful for finding morbidity data, although such data are likely to underestimate the actual incidence of disease; such records may also provide valuable information on exposure.

Disability Records. Disability records can be examined as a potentially useful source of surveillance data (see Chapter 12).

Conclusion

With time, it is likely that improved surveillance of occupational disease will yield additional useful information. In evaluating occupational surveillance programs, it is most important to clearly understand the goals of the specific surveillance system and to recognize that not every system will meet every goal.

More information on surveillance of occupational disease and injury can be obtained from (a) NIOSH (see Appendix B); (b) workers' compensation system agencies in most states; (c) the BLS of the U.S. Department of Labor in Washington, DC; and (d) the occupational disease and injury epidemiologists within health or labor departments in most states.

REFERENCES

1. Ashford NA, Miller CS. Chemical exposures: low levels and high stakes. New York: Van Nostrand Reinhold, 1991. (WE LL SOON REPLACE THIS WITH 2nd edition of this book.)
2. Cullen MR. Low level chemical exposure. In: Rosenstock L, Cullen MR, eds. Textbook of clinical occupational and environmental medicine. Philadelphia: Saunders, 1994.
3. Hamilton A. Industrial poisons in the U.S. 1925.
4. Halperin WE, Frazier TM. Surveillance for the effects of workplace exposure. Annu Rev Public Health 1985;6:419–432.
5. Halperin WE, Ratcliffe J, Frazier TM, Wilson L, Becker SP, Schulte PA. Medical screening in the workplace: proposed principles. J Occup Med 1986; 28:547–552.
6. Kreiss K. Approaches to assessing pulmonary dysfunction and susceptibility in workers. J Occup Med 1986;28:664–669.
7. Frost JK, Ball WC Jr, Levin ML, et al. Sputum cytopathology: use and potential in monitoring the workplace environment by screening for biological effects of exposure. J Occup Med 1986;28:692–703.
8. Fontana RS. Lung cancer screening: the Mayo program. J Occup Med 1986;28:746–750.
9. Schulte P, Ringen K, Hemstreet G. Optimal management of asymptomatic workers at high risk of bladder cancer. J Occup Med 1986;28:13–17.
10. Occupational Safety and Health Administration, U.S. Department of Labor. General industry: OSHA safety and health standards (29 CFR 1910). Washington, D.C.: U.S. Government Printing Office, 1978.
11. Millar JD. Screening and monitoring: tools for prevention. J Occup Med 1986;28:544–546.
12. Rutstein D, Mullen R, Frazier T, Halperin W, Melius J, Sestito J. The sentinel health event (occupational): a framework for occupational health surveillance and education. Am J Public Health 1983;73:1054–1062.
13. O'Malley M, Thun M, Morrison J, Mathias T, Halperin W. Surveillance of occupational skin disease using the supplementary data system. Am J Ind Med 1988;13:291–300.
14. Meredith SK, Taylor VM, McDonald JC. Occupational respiratory disease in the United Kingdom 1989: a report to the British Thoracic Society and the Society of Occupational Medicine by the SWORD project group. Br J Ind Med 1991;48:292–298.
15. Baker EL. Sentinel Event Notification System for Occupational Risks (SENSOR): the concept. Am J Public Health 1989;79[Suppl]:18–20.

BIBLIOGRAPHY

Recognition

Bureau of Labor Statistics, U.S. Department of Labor. Towards improved measurement and reporting of occupational illness and disease. (Symposium Proceedings, Albuquerque, NM, 1985.) Washington, DC: U.S. Department of Labor, 1987.

State-of-the-art review of practical issues in occupational disease surveillance and proposals for the future.

Froines JR, Dellenbaugh CA, Wegman DH. Occupational health surveillance: a means to identify work-related risk. Am J Public Health 1986;76:1089–96.
Introduction to the concept of hazard surveillance.

Goldman RH, Peters JM. The occupational and environmental health history. JAMA 1981;246:2831–6.
An excellent article with more detail on the occupational history.

Screening

Halperin WE, Schulte PA, Greathouse DG (eds, Part I) and Mason TJ, Prorok PC, Costlow RD (eds, Part II). Conference on medical screening and biological monitoring for the effects of exposure in the workplace. J Occup Med 1986;28:543–788,901–1126.
An in-depth, comprehensive review on screening in the workplace.

Halperin WE, Ratcliffe J, Frazier TM, Wilson L, Becker SP, Schulte PA. Medical screening in the workplace: proposed principles. J Occup Med 1986;28:547–552.
Questions the adequacy of current recommendations on screening in the workplace and proposes a revised set of principles for such screening.

Hathaway GJ, Proctor NH, Hughes JP. Proctor & Hughes Chemical Hazards of the Workplace, 4th ed. New York: Van Nostrand Reinhold, 1996.
Includes recommended screening examinations and tests for workers exposed to some of the 600 substances covered in this book.

Lauwerys RR. Industrial chemical exposure: guidelines for biological monitoring, 2nd ed. Davis CA: Biomedical Publications, 1993.
Presents concepts of biologic monitoring and reviews current knowledge on numerous specific agents.

Morrison AS. Screening in chronic disease. Monographs in epidemiology and biostatistics, vol 7, 2nd ed. New York: Oxford University Press, 1992.
An excellent text on the epidemiology of screening.

Silverstein M. Analysis of medical screening and surveillance in 21 OSHA standards: support of a generic medical surveillance standard. Am J Ind Med 1994;26:283–295.
An excellent review article.

World Health Organization. Early detection of occupational diseases. Geneva: WHO, 1986.

An excellent guide on the principles of early detection and approaches to early detection and control of various occupational diseases.

Surveillance

Ashford NA, et al. Monitoring the worker for exposure and disease: scientific, legal, and ethical considerations in the use of biomarkers. Baltimore: Johns Hopkins Press, 1990.
The considerations given to both screening and surveillance issues in this monograph raise a number of important questions concerning the objectives of efforts to evaluate biologic materials from workers, how these measurements are used effectively, and how they can be of little or no use for the objectives identified.

Bureau of Labor Statistics, U.S. Department of Labor. Towards improved measurement and reporting of occupational illness and disease. (Symposium Proceedings, Albuquerque, NM, 1985.) U.S. Department of Labor, Washington, DC, 1987.
Review of practical issues in occupational disease surveillance and proposals for the future.

Halperin WE, Frazier TM. Surveillance and the effects of workplace exposure. Annu Rev Public Health 1985;6:419.
A systematic review that provides a careful integration of the range of issues related to surveillance in work settings.

Halperin W, Baker EL, Monson RR, eds. Public health surveillance. New York: Van Nostrand Reinhold, 1992.
This book covers basic principles of public health surveillance and provides discussions of specific subject areas particularly relevant to the occupational setting, including occupational disease, hazard surveillance, AIDS, chronic disease, and injury.

Mullan RJ, Murthy LI. Occupational sentinel health events: an updated list for physician recognition and public health surveillance. Am J Ind Med 1991; 19:775–799.
Adaptation of the general concept of sentinel health events to occupational disease.

Wegman DH. Hazard surveillance. In: Halperin W, Baker EL, Monson RR, eds. Public health surveillance. New York: Van Nostrand Reinhold, 1992.
Provides conceptual framework for the surveillance of hazards. This is a primary prevention approach especially relevant for long latency diseases.

5 Preventing Occupational Disease and Injury

Barry S. Levy and David H. Wegman

Occupational diseases and injuries are, in principle, preventable. Among the approaches to preventing occupational diseases and injuries are developing awareness of occupational health and safety hazards among workers and employers, assessing the nature and extent of hazards, and introducing and maintaining effective control measures (see Chapter 7). Sometimes these approaches are undertaken solely by employers and workers within a specific workplace. Other times there is also the need for external involvement, ranging from encouragement by appropriate individuals or agencies outside the specific workplace to promulgation and rigorous enforcement of occupational health and safety regulations.

Although the specific circumstances necessitate a variety of approaches, it is important for physicians and other health professionals to recognize the vital role they can play in preventing work-related disease and injury. This role requires going beyond treatment of a given patient to help prevent both recurrence in that patient and initial occurrence of disease or injury in other, similarly exposed workers.

The opportunities for prevention of work-related disease or injury vary according to whether the role of the health professional is related solely to an individual patient or extends to one or more groups of workers in the same workplace (or with the same employer), in the same industry, or in the same occupation.

This chapter covers opportunities and responsibilities that present themselves to all health care providers in caring for individual patients with newly-identified work-related diseases and injuries, and summarizes preventive measures that are possible when the health professional also has a formal relationship with, and responsibility to, an employer.

METHODS OF PREVENTION

Measures to prevent occupational disease and injury can be categorized as directed primarily either toward the workplace and work practices or toward the worker. These two approaches are described in this chapter and illustrated in many places in this textbook. Chapters 7 through 9 further address general approaches to prevention.

Measures Directed Primarily Toward the Process or Workplace

Substitution of a Nonhazardous Substance for a Hazardous One. An example of this type of action is the substitution of synthetic vitreous fibers, such as fibrous glass, for asbestos. Substitution carries certain risks, because substitute materials often have not been adequately tested for health effects and may, in fact, be hazardous. For example, years ago fire protection was enhanced by replacing flammable cleaning solvents with

B. S. Levy: Barry S. Levy Associates, Sherborn, Massachusetts 01770.

D. H. Wegman: Department of Work Environment, University of Massachusetts Lowell, Lowell, Massachusetts 01854.

Box 5-1. Avoiding the Transfer of Risk: Pollution Prevention and Occupational Health

Rafael Moure-Eraso

The growing national concern with environmental pollution became acute in the last decade because of the increase in waste-generating activities by industry. The U.S. Congress responded to this concern by enacting the Pollution Prevention Act of 1990. Congress reflected the consensus of the scientific community that waste management and control alone will not resolve environmental problems in the long run and that a change of approach—a "paradigm shift"—from *pollution control* to *pollution prevention* was necessary. Source reduction is the strategy of choice to achieve pollution prevention. Only to the degree that this cannot be achieved is it appropriate to turn to pollution control activities such as treatment, disposal, and remediation.

Pollution prevention has begun to take hold. It provides, for the first time, a coordinated effort of primary prevention, eliminating the possibility of pollution-related health effects and superseding "end-of-pipe" interventions. In 1996, the Environmental Protection Agency reported that more than 7,200 companies in the United States had established pollution prevention programs and predicted that the number would exceed 18,000 by the year 2000.

These developments have critical implications for occupational health. The important conceptual change from *control* of environmental exposures to their *prevention* through source reduction and changes in process methods allows the workplace to be seen as a separate source of pollution when undertaking a comprehensive and systematic pollution source evaluation. When industries that use chemicals as raw materials begin to look at changing materials and processes as an environmental health strategy, the opportunity exists to incorporate workplace pollution exposures into the equation and prevent the choice of substitute materials without consideration for the impacts of any proposed changes on the exposures within the plant. Consequently, the working population and work environment can be given equal footing with the general population when pollution prevention strategies are planned.

The establishment of this understanding as the foundation for pollution prevention activities requires a change in both environmental health and occupational hygiene practice. Just as previous environmental health activities did not consider root causes and their prevention, traditional workplace-based exposure control activities have been end-of-pipe interventions designed to control exposure without systematically examining root causes. Consequently, it was not recognized that a preferred engineering control such as local exhaust ventilation tended to shift the burden from the workplace to the ambient environment in the form of air pollution or solid hazardous waste (via contaminated filters or other pollution collection media). Unless source reduction or process modifications are examined comprehensively, occupational hygienists may be equally responsible for shortsightedness. So work environment scientists must join environmental health scientists in an unified effort to avoid simply shifting risk among different media (Table 5-1).

For this conceptual potential to be realized, the occupational health professional must be at the table during discussion of pollution prevention strategies. The six general pollution prevention (source reduction) strategies that most directly affect occupational health are raw material substitution or reduced use, closed-loop recycling, process or equipment modification, improvement of maintenance, reformulation of products, and improvement of housekeeping and training.

Some examples of pollution prevention interventions that incorporate concern for reduction or elimination of work exposures are the following:

1. In industrial textile dry-cleaning operations, water-based solvents have been successfully substituted for perchloroethylene. This change eliminates exposures to a potential human carcinogen but also leads to improvement in dry-cleaning job organization and reduction in ergonomic risk factors.
2. In the offset lithographic industry, the solvent with the lowest concentration of aliphatic organic chemicals has been successfully substituted for regular organic solvents to clean printing

Box 5-1 (*continued*)

ink from metal surfaces. Products with high organic chemical content were found to perform no better than those with lowest.

3. In painting of small metal parts, the introduction of an electrostatically delivered coating was successful in replacing a resin-based epoxide paint. Not only were the respiratory and skin hazards from epoxide exposures eliminated, but the paint dispenser was made substantially lighter, avoiding an ergonomic hazard.

Occupational hygiene should strive to change its most common practice from secondary to primary prevention, by addressing workplace problems as comprehensive production and materials problems rather than end-of-pipe solutions (see Table 5-1). Pollution inside and outside of the point of production cannot be compartmentalized, nor can worker and community concerns.

TABLE 5-1. *Pollution prevention and occupational health: occupational hygiene as an instrument for primary prevention*

Occupational hygiene model of action	Primary prevention	Secondary prevention
Anticipation	*Hazard surveillance*	—
Identification	*Hazard identification*	*Medical surveillance*
Evaluation	*Exposure assessment*	—
Controls	*Exposure prevention*	*Control of generated exposures*
	Comprehensive	End-of-pipe
	Source reduction	Engineering Controls
	Materials changes	Enclosure
	Substitution	Local exhaust
	Process changes	Wet methods
	Physical conditions	General ventilation
	Machinery	Administrative controls
	Operations	Personal protective equipment
	Work organization	*Early therapeutic intervention*

carbon tetrachloride. Increased use of carbon tetrachloride led to identification of its hepatotoxicity and its subsequent replacement by less toxic chlorinated hydrocarbons. Now there is concern that use of chlorinated hydrocarbons should be reduced to better protect the general environment. The lesson in this evolution is not that substitution is hopeless but that the introduction of a substituted material should be considered only a first step and that the impact of the substitution must always be monitored to determine whether initially unrecognized problems develop after increased use of the new material.

The substitution approach is embodied in the broader concept of pollution prevention, described in Box 5-1.

Installation of Engineering Controls and Devices. These approaches are more often available than substitution and cover a wide range of effective options to reduce both chemical and ergonomic hazards. Some common approaches are the following:

- Installing ventilation exhaust systems that remove hazardous dusts (Fig. 5-1)
- Using jigs or fixtures to reduce static muscle contractions while holding parts or tools
- Applying appropriately designed sound-proofing materials to reduce loud noises that cannot be engineered out of a work process
- Installing tools on overhead balancers to eliminate torque and vibration transmitted to the hand
- Constructing enclosures to isolate hazardous processes

FIG. 5-1. Local exhaust ventilation used to protect a worker from asbestos dust generated in working with clutch plates. (Photograph by Earl Dotter.)

- Installing hoists to eliminate manual lifting of containers or parts
- Carefully maintaining process equipment to reduce or eliminate fugitive emissions from processes designed as closed systems or the development of unwanted vibrations as equipment ages.

Although installation of these engineering controls can involve a substantial initial capital expenditure, they often save money by reducing materials use, reducing toxic and other material wastes, and reducing costs of disease, injury, and absenteeism. Often, such approaches are not considered or implemented because of lack of awareness that such solutions are available.

Job Redesign, Work Organization Changes, and Work Practice Alternatives. A number of changes can be introduced that take advantage of methods that directly reduce or eliminate various types of risks in work processes. These include job redesign, changes in work organization, and alternative work practices. Job redesign, which often combines engineering and administrative aspects, typically seeks several related objectives: to increase job content, to make the physical work less redundant or repetitive, and to improve workers' opportunity to exercise individual or collective autonomy in decision making (see Chapter 22).

Changes in work organization, often closely integrated with individual job redesign, are directed at elimination of undesirable features in the structure of work processes. For example, a change from piece-rate work (with incentive wages) to hourly-rate work removes inappropriate pressure and tension—both physical and mental—on affected workers. Piece-rate work has been associated with higher rates of musculoskeletal problems in a variety of work settings. Another example is the elimination of machine pacing, which tend to enforce repetitive and mind-numbing work.

Work practice alternatives can, through relatively limited changes, lead to important improvements in the work environment. For example, dust exposures in a variety of settings can be significantly reduced by the introduction of vacuum cleaning in place of compressed air to clean dusty surfaces and wet mopping in place of dry sweeping wherever possible.

As a rule, these preventive measures are more effective than methods that primarily affect the worker. The four measures that follow potentially reduce the damage that may result from workplace hazards without actually removing the source of the problem.

Measures Directed Primarily Toward the Worker

Education and Advice. Education and advice concerning specific work hazards are essential. Workers should always be given full information about workplace hazards and means of reducing their risk (Fig. 5-2). Many safety measures necessitate changed behavior by workers, which also requires education or training. Workers who are not aware of job hazards will not take the health and safety precautions necessary to protect themselves and their coworkers (see Box 5-2, Fig. 5-3, and information on the OSHA Hazard Communication Standard in Chapter 10).

Personal Protective Equipment. Use of personal protective equipment (PPE), such as respirators, earplugs, gloves, and protective clothing (Fig. 5-4) will continue to be necessary in a number of settings where it is the only available protective measure. However, this approach to controlling a hazard often has important limitations; for example, workers often resist wearing such protection because it is cumbersome or causes other difficulties. It is important that the effectiveness of PPE be evaluated in actual use where the experimentally determined effectiveness claimed by its manufacturer may not apply. In the United States, the Occupational Safety and Health Administration (OSHA) has developed lists of acceptable PPE that can be helpful in proper selection and use of

FIG. 5-2. Hazardous labeling regulations require that labels such as this one be improved to gain attention readily and in appropriate languages. (Photograph by Earl Dotter.)

these devices. OSHA and other authorities have also emphasized the need for and importance of developing a complete program for PPE, not only a requirement for its use. Adequate programs include requirements for proper fitting of the equipment (especially with respirators), education about proper use, and a plan for maintenance, cleaning, and replacement of equipment or parts. The costs of an effective PPE program are significant, making it particularly important to recognize that use of such equipment should be accepted only when no alternative control is present.

Organization Measures. Organization measures taken by the employer may offer some protection. For example, exposure can be reduced somewhat by implementing work schedules such that workers spend carefully limited amounts of time in areas with potential exposure. Such measures require good environmental monitoring data to design appropriate schedules, and care must be taken that the result is not simply to distribute more widely the exposure to substances that can be controlled by engineering approaches. Another preventive administrative measure is preplacement examination to avoid assigning workers to jobs in which individual risk factors place them at higher risk for specific diseases or injuries. The requirements of the Americans with Disabilities Act in the United States place a special responsibility on those carrying out preplacement examinations (see Box 12-1 in Chapter 12).

Screening and Surveillance. Screening and surveillance for early detection of disease, either separately or together, may lead to the identification of need for control measures to prevent further hazardous exposure to workers (see Chapter 4). Unlike the methods described previously, which are designed to prevent occurrence of occupational disease or injury by primary prevention, screening and surveillance activities are part of secondary prevention. Both screening and surveillance are directed toward identification of health events or documentation of early evidence for adverse health effects that have

Box 5-2. Effectively Educating Workers

Margaret Quinn and Nancy Lessin

A prerequisite to effective health and safety programs is education. The most effective approach to teaching health and safety acknowledges that the worker is the one most familiar with his or her job. Workers can identify hazards, both apparent and hidden, that may be associated with their work. Worker involvement in prioritizing educational needs and in the design and content of training are major determinants of meaningful and useful programs. Workers also should be included in the development and implementation of the solutions to health and safety problems. Education regarding solutions should include a discussion of the traditional industrial hygiene hierarchy of hazard controls that emphasizes hazard elimination and should not be limited to training on the use of personal protective equipment. In addition to worker involvement with needs assessment and program design, additional guidelines for successful educational programs include the following:

1. Develop an educational program in the trainee's literal and technical language. The educator should also understand the social context and psychosocial factors of a workplace that may affect a worker's ability to participate in an educational program or to perform certain work practices in response to a potential hazard.
2. Define specific and clearly stated goals for each session based on a needs assessment that has involved representatives of the workers to be trained. Begin each program with a concise overview, and reinforce the key issues that come up during the session.
3. Build an evaluation mechanism that can easily be adapted to each program. The evaluation process should be designed to judge the effectiveness of the educational program in attaining goals set by both the trainer and trainees.
4. Use participatory teaching methods, which draw on the experience that workers already have, in place of a traditional lecture approach.

Participatory teaching methods, also called *empowerment approaches,* are designed to foster maximum worker participation and interaction. They constitute an approach to labor education that is based on the understanding that adults bring an enormous amount of experience to the classroom and that this experience should be used in the training program. In addition, adults learn more effectively by doing rather than listening passively. Learners' experiences are incorporated into the course material and are used to expand their grasp of new concepts and skills. Basing new knowledge on prior practical experience helps the learner solve problems and develop safe solutions to unforeseen hazards. Instructors offer specialized knowledge; workers have direct experience. It is the combination of these that leads to effective, long-lasting solutions to health and safety problems.

Participatory learning generally requires more trainer-trainee interaction than lecture-style presentations. Groups should be limited to approximately 20 participants, and these may be broken down into groups of three to six for small-group exercises. Participatory teaching methods may include use of the following techniques:

1. *Speakouts* (large-group discussions): The participants share their experiences in relation to a particular hazard or situation.
2. *Brainstorming sessions*: The instructor throws out a particular question or problem; the participants call out their ideas. The ideas are recorded on a flipchart so that they become a collective work. In this activity, the trainer elicits information from the participants rather than presenting it in a didactic manner.
3. *Buzz groups* (a small-group discussion or exercise): Each group of three to six participants discusses a particular problem, situation, or question and records the answers or views of the group.
4. *Case studies* (a small-group exercise): Participants apply new knowledge and skills in the exploration of solutions to a particular problem or situation.
5. *Discovery exercises:* Participants go back into the workplace to obtain certain items such as OSHA 200 logs or perform activities such as interviewing coworkers regarding a particular hazard; this information is brought back into the classroom for discussion.

Box 5-2   (*continued*)

6. *Hands-on training*: The participants practice skills such as testing respirator fit, simulating asbestos removal or hazardous waste cleanup (Fig. 5-3), calculating lost workday injury rates from OSHA 200 logs, or handling and learning the uses and limitations of industrial hygiene monitoring equipment.

7. *Report-back sessions*: After buzz groups, the class reconvenes as a larger group, and a spokesperson for each buzz group reports the group's answers or views; similarities and differences among groups are noted, and patterns may be discovered.

Empowerment or participatory learning techniques are well-established methods practiced in labor education programs, schools of education, labor unions, and Committees for Occupational Safety and Health (COSH groups). These groups have demonstrated that it is possible to use participatory methods even for educational programs that require conveyance of specific, technical knowledge. For example, the OSHA Hazard Communication Standard has worker training requirements for use of material safety data sheets (MSDSs), forms that contain brief information regarding chemical and physical hazards, health effects, proper handling, storage, and personal protection for a particular substance. Training on MSDSs should cover how to obtain the data sheets, how to interpret them, and their uses and limitations and should give the participants practice in each of these areas. Rather than presenting the MSDS in a lecture-style format, the information can be taught more effectively with a participatory exercise such as the one that follows:

In the first part of the exercise, workers go back into their work areas, find a labeled chemical container, and seek an MSDS for that substance. This requires workers to become familiar with where MSDSs are located in their particular workplace and the process required to find one. It also serves to identify problems in the system that can be corrected, such as missing MSDSs or locked file cabinets to which no one on the shift has a key. In the second part of the exercise, the class is divided into small groups; the groups review sample MSDSs and collectively answer questions such as, "Is the substance flammable?" "What are the health effects associated with it?" "Does it require wear

FIG. 5-3. Hands-on field training for hazardous waste workers.

Box 5-2   *(continued)*

ing of gloves?" and "What ventilation is required?" During the report-back session, the instructor asks for the answers from all of the groups and reviews how to read and interpret MSDSs in general. In the final part of the exercise, participants look up the chemicals covered in the sample MSDSs in other sources, such as the *NIOSH Pocket Guide to Chemical Hazards.* In some situations, more hazards, especially health hazards, are discovered when other sources are consulted. In this way students learn about the uses and lim-

itations of MSDSs and get practice in using additional sources.

Participatory or empowerment approaches not only make learning active but give workers the skills and support necessary to recognize hazards and to improve health and safety conditions. These approaches broaden the objectives of worker training curricula to include learning of knowledge, attitudes, skills, and behaviors, as well as problem-solving, critical thinking, and social action skills.

already occurred. However, both screening and surveillance can lead to primary prevention measures: early detection of disease or abnormality can identify inadequate control measures, allowing them to be corrected so that other workers can be protected.

By recognizing potential or existing work-related disease or injury, health professionals can initiate activities leading to one or more of these methods of prevention. They can play an active role in education by informing workers and employers about potentially hazardous workplace exposures and ways of

minimizing them. They can advise appropriate use of respirators or other PPE. They can also screen workers and facilitate screening of coworkers who may be at high risk for certain diseases. Consultation with specialists in occupational medicine, occupational hygiene, occupational safety, or occupational ergonomics may be necessary to facilitate these activities.

INITIATING PREVENTIVE ACTION

Once a health professional has identified a probable case of work-related disease or in-

A   B

FIG. 5-4. A: Spray painter with respiratory protections. **B:** Makeshift PPE. Cotton plugs are not effective as PPE; only adequately fitting earplugs or earmuffs are effective. (Photographs by Earl Dotter.)

jury, it is crucial to take preventive action while also providing appropriate treatment and rehabilitation services. Failure to consider the prevention opportunities along with the necessary therapeutic measures may lead to recurrence or worsening of the disease or injury in the affected worker and the continuation or new occurrence of similar cases among workers in similar jobs, either at the same workplace or other workplaces. A health professional has at least the following five opportunities for preventive action after identifying a case of work-related disease or injury: advise the patient, contact the patient's union or other labor organization, contact the patient's employer, inform the appropriate government authority, and contact an appropriate research or expert group. Often some combination of these approaches is undertaken.

Advise the Patient

The health professional should always advise the worker concerning the nature and prognosis of the condition; the possibility that there may be appropriate engineering controls to remove the hazard; the need, even if only temporarily, for PPE at work; or, in extreme circumstances, the necessity to change jobs. The health professional should alert the patient to the need to file a workers' compensation report to protect the worker's rights to income replacement and both medical and rehabilitation services (see Chapter 11). These reports also trigger the employer to consider listing the health event as a reportable injury or illness and may lead the insurance carrier to provide consultative services to the employer to assess the problem area and consider appropriate control measures.

At times, the health professional may be called on to provide advice to the patient concerning legal remedies should a health problem result in a contested workers' compensation claim or the need for registering a complaint with an appropriate government agency (as discussed later). A worker's options may be limited; the worker may not wish to file a claim or register a complaint, fearing job loss or other punitive action. However, it is essential to inform a worker of potential hazards. It is not appropriate to withhold this information because of the possibility of upsetting the patient. A health professional cannot assume that even a large and relatively sophisticated employer has adequately educated its workers about workplace hazards. It should be noted that once a patient is informed of the work-relatedness of a disease in writing, this may start the time clock on notification procedures and statutes of limitations for workers' compensation (see Chapter 11).

Contact the Patient's Union or Other Labor Organization

If it is agreeable to the affected worker, the health professional should inform the appropriate labor organization of the health hazards suspected to exist in the workplace. The provision of this information may help to alert other workers to a potential workplace hazard, facilitate investigation of the problem, identify additional similar cases, and eventually facilitate implementation of any necessary control measures. (Keep in mind, however, that fewer than 15% of workers in the United States belong to a union.)

Contact the Patient's Employer

The health professional, again only with the patient's consent, may choose to report the problem to the employer. This can be effective in initiating preventive action. Many employers do not have the staff to deal with reported problems adequately, but they can obtain assistance from insurance carriers, government agencies, academic institutions, or private firms. In addition to triggering workplace-based prevention activity, discussions with the employer may lead to obtaining useful information concerning exposures and the possibility of similar cases among other workers. Depending on the circumstance, it can be particularly helpful to the health professional to arrange with an employer to visit a patient's work area. This presents the opportunity to observe the possibly hazardous environment firsthand and to establish the necessary rapport with managers to involve them in prevention.

Although the law prohibits employers from firing workers for making complaints to OSHA, it does not prohibit them from firing workers who have a potentially work-related diagnosis. In the United States, only the federal lead and cotton dust standards mandate removal of workers from jobs that are making them sick. The medical removal protection section of the federal lead standard provides temporary medical removal for workers at risk of health impairment from continued lead exposure, as well as temporary economic protection for workers so removed. (See Chapters 15, 29, 31, 33, and 35 for more information on lead.) It states, "During the period of removal, the employer must maintain the worker's earnings, seniority and other employment rights and benefits as though the worker had not been removed." The cotton dust standard does not offer such protection.

Inform the Appropriate Governmental Regulatory Agency

If a case of occupational disease or injury appears to be serious or may be affecting other workers in the same workplace, company, or industry, it is wise for the worker or the health professional to consider filing a complaint with the appropriate governmental agency (OSHA or the appropriate state occupational safety and health agency). OSHA establishes and enforces standards for hazardous exposures in the workplace and undertakes inspections, both routinely and in response to complaints from workers, physicians, and others (see Chapter 10). In about half of the states in the United States, the program is implemented directly by OSHA, which is part of the U.S. Department of Labor; in the other states, a state agency—

often the state department of labor— implements the program. Both OSHA and the state agency may investigate a workplace in response to a complaint. Most state agencies make recommendations to improve the situation, but only those states with OSHA-delegated authority can order changes to improve health and safety in the workplace and impose fines if these changes are not made.

The health professional should always inform the patient in advance of notifying federal or state governmental agencies. Although regulations of OSHA and the Mine Safety and Health Administration (MSHA) protect U.S. workers who file health and safety complaints against resultant discrimination by the employer (loss of job, earnings, or benefits), this protection is difficult to enforce, and workers' fears are not unfounded. Health professionals should familiarize themselves with pertinent laws and regulations. For example, if the worker does not file an "11(c)" (antidiscrimination) complaint within 30 days of a discriminatory act, the worker's rights are lost. In the United States, health professionals and workers (or their union, if one exists) have the right, guaranteed by the Freedom of Information Act, to obtain the results of an OSHA inspection.

Contact an Appropriate Research or Expert Group

The health professional may choose to refer the patient to an occupational health academic center or to report the identified situation to an agency or other organization conducting research on occupational safety and health. Such agencies include the following:

- The National Institute for Occupational Safety and Health (NIOSH), especially the hazard evaluation group at NIOSH, which responds to complaints of possibly serious occupational health or safety hazards (see Appendix to this chapter)
- An appropriate state agency, usually within the state departments of labor or public health

Advice to employees and employers should be practical. (Drawing by Nick Thorkelson.)

- A medical school or school of public health, several of which have NIOSH-ponsored occupational health and safety Educational Resource Centers (see Appendix B)
- Some other group with expertise, experience, and interest in research concerning work-related diseases and injuries.

Occasionally, the health professional who is reporting a work-related medical problem may undertake or assist in a research investigation of this problem. No matter who conducts the research, investigation of the workplace and identification and analysis of additional cases often lead to new information. Publication of epidemiologic studies or case reports alerts others to newly discovered hazards and ways of controlling them.

The health professional may also assist with research to evaluate the effectiveness of preventive approaches, such as the impact of OSHA regulations (Box 5-3).

ROLE OF THE CONSULTING PHYSICIAN

Many employed people in the United States work in workplaces that do not have formal health and safety programs. Although there are publications and other sources of information available to guide employers to provide such programs, there is little guidance for health care providers who might have the opportunity to offer components of such programs on a consultant basis or from a community-based clinical practice. Increasingly, physicians are being asked to advise, consult with, or provide a full range of services to an industry or union on a part-time basis. Initial invitation for such work may be focused on very limited activity, such as preplacement physical examinations, return-to-work evaluations, or reviews of cases in which there have been workers' compensation claims. However, a physician consulting on a part-time basis to an employer or union can and should insist on being permitted to undertake preventive measures. In any such relationship, the physician should have the authority to obtain data; to share data with those who need to have this information, including the workers; and to take preventive and corrective action.

The following are reasons that can be introduced to encourage an employer to adopt components of a comprehensive program:

- *Government regulations.* With the incentive of government regulations, small employers need to consider a basic in-plant health and safety program as a high priority.
- *Pressure from organized labor.* Where workers are represented by a union, health and safety concerns are often given higher priority and may be dealt with through collective bargaining. Such programs are most effective when health care providers are involved in efforts for which both the union and the employer have developed common objectives.
- *Enlightened self-interest.* Effective workplace health and safety programs should

Box 5-3. Evaluation of OSHA Health Standards

Anthony D. LaMontagne

The 35 agent-specific and generic health standards promulgated by OSHA between 1970 and 1998 have been among the most controversial of the agency's regulatory activities. Yet, there has been relatively little systematic evaluation of these standards (1–4). Standards evaluation is needed to assess progress in reducing occupational hazards, illnesses, and injuries and to support the continuing improvement of OSHA standards.

OSHA standards are based on the best available evidence regarding risks and how to control them. Therefore, they contain implicit or explicit expectations about how requirements will be implemented and the impacts of implementation. Fundamental evaluation questions focus on implementation ("Was the standard implemented as intended?") and effectiveness ("Did implemented measures result in decreases in exposures and health effects of concern?").

Implementation studies are important complements to effectiveness studies and are valuable in their own right. As examples, detailed implementation studies have been conducted on the generic hazard communication and the agent-specific ethylene oxide standards (5,6). The hazard communication study showed that roughly one fourth of responding employers provided no worker training, with small employers being the least likely to provide training. Noncompliance involved one or more requirements in each of the following three areas: training (53%), material safety data sheets (46%), and labeling (41%). The ethylene oxide evaluation showed that most hospitals had implemented the requirements for initial personal exposure monitoring, worker training, and medical surveillance; however, workers were being exposed in accidental releases of ethylene oxide that were not being captured by personal monitoring, training was most commonly video-based, and OSHA's action level trigger for medical surveillance used in many health standards was neither understood nor related to providing surveillance. In summary, process evaluation studies, when modeled on the implied or explicit logic of a standard, provide information needed for action to improve the implementation, enforcement, and impact of the standard.

Effectiveness questions can be asked at three general levels: "Was implementation of the standard associated with decreased exposures to the hazard of interest?" "Was implementation of the standard associated with decreases in health outcomes of interest?" and "Did the OSHA standards cause observed changes in exposures and health outcomes?" Studies at the first two levels are observational, with the usual limitations on causal inference. However, combining qualitative and quantitative approaches can greatly improve the opportunities to interpret such observational studies (7,8). Evidence of positive impacts at the first and second levels include the following: (a) a hazard communication evaluation showing that interactive small-group training methods were associated with positive changes in work practices and working conditions (proxies for decreased exposures) (9); (b) decreases in blood lead level (as a biomarker of both exposure and health outcome) after implementation of the Lead Exposure in Construction Standard (10); and (c) decreases in the incidence of hepatitis B (health outcome) after implementation of the bloodborne pathogens standard (11). Studies at the third level are rarely feasible owing to the practical, ethical, and legal constraints of conducting randomized, controlled experiments in this context. A sensible and economical approach to evaluating occupational interventions has been proposed in which qualitative and quasiexperimental studies would be performed, after which randomized, controlled trials would be conducted where they were both necessary and feasible (7).

Opportunities for future improvements in standards evaluation are many. Traditional etiologic epidemiology perspectives need to be complemented by more eclectic and action-oriented perspectives. Examples include borrowing from the field of program evaluation, adopting alternative paradigms such as participatory action research, and expanding the use of qualitative research methods (7,12). More population-based, rather than worksite-based, studies are needed. In addition, many impact and outcome measures have been underutilized to date. For example, there is a need for economic studies that focus on health costs to affected workers and implementation costs to employers. In addition, greater use of exposures, hazards, and biomarkers would provide more measurable performance pa-

---

Box 5-3 (*continued*)

rameters than health outcomes do, and far greater probability of demonstrating the impacts of OSHA standards (13). Expanded standards evaluation research will foster the development of standards that are minimally burdensome to employers and maximally effective in reducing exposures and health effects. The naming of intervention effectiveness research, including evaluation of OSHA health standards, as 1 of 21 priority areas in the 1996 National Occupational Research Agenda provided new support and impetus for urgently needed work in this area.

---

reduce workers' compensation and medical care costs while increasing productivity.

Opportunities should be actively sought to become involved as a consultant to a workplace-based health and safety program. Although special knowledge is required for the most comprehensive aspects of such a program, a well-educated health professional should be able to establish and maintain the basic components of a program. There are a variety of types of consulting activities that may be requested. Every effort should be made not only to provide services as requested but to follow a prevention-based order of priorities. The following is a recommended set of priorities for program activities:

1. *Programs designed to identify and control both new and old risks.* The identification and control of risks depends on a series of activities designed to build an effective, ongoing program. The critical first action should be to undertake a comprehensive baseline hazard survey documenting major chemical, physical, and biologic hazards along with important sources of psychosocial, safety, and ergonomic risks. The results of this survey should be organized in terms of priority for improvement and frequency of follow-up evaluation. The results should be maintained in an easily accessible form, preferably an electronic database system, facilitating review and additions. The baseline survey of risks should be followed by implementation of a hazard-based surveillance program designed to update the status of risks and to document successful reduction in risks that previously were not appropriately controlled. When possible, objective measurements should be collected and maintained; however, regular observations alone can be an important surveillance tool. Close cooperation with workplace safety managers and other consultants is important, especially when introducing new control strategies. Also important is close cooperation between labor and management, such as that demonstrated by health and safety committees (Box 5-4). The importance of systematic and periodic evaluation of the effectiveness of controls cannot be overemphasized.

2. *Programs designed to properly match jobs to workers.* A basic principle of occupational health and safety is that jobs should be designed to fit the worker. Too often the reverse has been the case. A consultant should set high priority on the need to evaluate every job and determine how the job design can be changed or equipment altered or replaced to best fit human capacities (see Chapters 8, 9, 21, and 26). Although cost may be used as an argument against such alterations, the development of a program to implement changes over time, as opportunities present themselves, is essential.

A consultant should carefully examine any limitations listed for job eligibility

Box 5-4. Labor-Management Health and Safety Committees

The benefits that accrue from seeking the participation of labor unions and workers in the development and implementation of occupational health and safety programs and research can be substantial. As a consequence of their experience and intimate knowledge of the actual work processes, workers and their unions often can add significantly to the understanding of a health or safety problem and determine the best approach to prevention of risks. Their participation also aids in understanding and explaining the nature and importance of programs and research efforts and in interpreting the impact and meaning of such work to individual workers (Fig. 5-5).

One effective means for including workers and their labor unions in the development and improvement of approaches to prevention is joint labor-management health and safety committees in the workplace. These committees consist of representatives of workers and managers. They meet periodically to systematically review workplace health and safety hazards and their control and to respond to specific complaints concerning workplace health and safety. For these committees to function effectively, labor representatives must be truly representative of workers and not simply appointed by management.

Joint labor-management health and safety committees have been legally authorized and are more generally active in some countries, such as Canada. In the United States they are less common and usually are established through collectively bargained agreements. Proposed OSHA reform legislation in the United States would require operation of health and safety committees in many more workplaces than at present.

Studies in Canada, where joint health and safety committees have been mandated, suggest that this particular form of involvement can be unusually effective. Reduction in work injuries and resolution of health and safety problems without the need for governmental intervention have been documented. Effective committees tend to have cochairs and equal representation, readily available training and information, and well-established procedures. An important feature of successful committees is sufficient authority for action, either as a committee or on the part of the management representatives.

Typically, labor-management health and safety committees meet on a monthly basis for 1 to 2 hours. They review, evaluate, and respond to worker and manager complaints and concerns about working conditions and workplace hazards. They periodically walk through the workplace to observe and assess working conditions and possible health and safety hazards. In addition, they systematically evaluate work practices and procedures and materials used in the workplace in regard to their impacts on workplace health and safety.

As effective as labor-management health and safety committees can be, they are most effective when seen as one component of a more general prevention program that also relies on the development and enforcement of government regulations.

A

B

FIG. 5-5. Joint labor-management health and safety committees are increasingly important in ongoing workplace prevention activity. **A:** Medical surveillance, screening programs, and a wide variety of other occupational health issues are discussed by committee. **B:** A worker points out a faulty oil line in a grinder to the union health and safety representative (man in white shirt). (Photograph by Earl Dotter.)

and determine that these are essential, rather than items of convenience or simply common practice. The implementation of the Americans with Disabilities Act (see Chapter 12) is stimulating more attention to designing the job to fit the worker.

Two circumstances exist in which attention should focus on the worker and not the job. The first occurs when an employee returns to work after an absence caused by illness or injury. Too often this reentry is either inappropriately accelerated or characterized by an overly cautious approach. The proper middle course can be achieved if careful estimates are made of both the requirements for the job and the temporary or permanent limitations of the returning worker. Directives that are too nonspecific, such as "light work," are likely to impede successful reentry to active work, unless they are more specific. The second situation occurs when there is an appropriately justified job requirement, such as high-frequency hearing ability or the need to respond to great physical demands on short notice. Only when the consultant is convinced that the worker characteristic is essential should he or she be involved in accepting such limitations on job eligibility or determining that a worker meets the stated conditions.

Health care providers should be asked to play an important role in a comprehensive PPE program in regard to issues of fitting and dispensing equipment. Although care should be taken to determine that the proper equipment has been selected and is being used appropriately, questions should always be asked about why such equipment is required and what alternative engineering or other measures can be implemented to better protect the worker from the hazard—or to entirely eliminate the hazard.

3. *Programs designed to identify early evidence of work-related health effects while they are still reversible.* Medical monitor-ing or screening data are designed to identify early evidence of control failures at the workplace. Generally, results are applied to the individual to guard against future damage. However, all such data should also be used in a prevention context. Collection and analysis of data obtained from screening or monitoring, along with data on disease or injury morbidity, disability, or mortality of workers, must be routinely compared with information concerning hazards experienced in the workplace and levels of exposure to these hazards (see first item in this list) in order to seek unrecognized opportunities for intervention and risk reduction.

4. *Programs designed for treatment and rehabilitation of occupational illnesses and injuries.* Generally, state workers' compensation regulations require employers to provide treatment and appropriate rehabilitative measures for work-related illnesses and injuries. This requirement may cause employers to seek consulting arrangements with health care providers. Although the determination of work-relatedness of an injury or illness is difficult in some cases, it is important that prompt and appropriate treatment and rehabilitative measures be initiated regardless of this determination. Numerous studies have documented the importance of early, effective medical intervention to successful return to work at full capacity. It may be possible to plan to have treatment services provided at a workplace-based medical department; increasingly, however, appropriate arrangements can be made for prompt and efficient care in a nearby community-based clinic, hospital, or other health care facility. (See Chapter 14.) Underlying this activity is the principle that the best possible treatment is most effective and least expensive in the long run.

5. *Programs designed for hazard communication.* All members of a workforce should understand the nature of work-related risks, the actions taken by man-

agement to control these risks, and the importance of engineering controls and substitutions as primary prevention priorities. Those unavoidable circumstances in which individual worker responsibility is an important component of control must be explained in a way that encourages and supports cooperation. The need to use PPE (when it is the only recourse) or the role of administrative rules (such as cleaning spills or maintaining clear aisles) should be understood as components of a control program without promoting the idea that workers are to blame when failures occur.

Any risk communication effort directed at specific hazard incidents needs to begin by ascertaining what the affected workers perceive to be a health or exposure concern. It should explain also what is generally known about these hazards, what is known about the possibility of risk related to the work setting, what efforts (present and future) are being undertaken to determine whether the risk is work-related, and the nature of the control effort at the given workplace. Experts in risk communication should be consulted in planning and implementing this effort.

Most health professionals are reluctant to go beyond treating or advising their patients when they suspect work-related medical problems. Although it is understandable that they might not want to get involved in legal proceedings or other complex situations that may require siding with either employer or employee, such a decision really represents a failure in the responsibilities of a health professional. Although only a few health professionals may have gained the medical and medicolegal expertise to provide the full range of needs for patients with work-related medical problems, participation in the active efforts at prevention related to each occupational disease or injury is essential. Because each event is, in principle, preventable, there

is an obligation to seek to prevent the next occurrence.

Prevention of these problems requires the active participation of all health professionals, not only those who have a particular interest or expertise in the field. All health professionals should be willing to take appropriate action to ensure that workplace hazards are correctly identified, investigated, and controlled. To do this effectively, health professionals must develop appropriate methods for advice, consultation, and referral. They also must recognize that persistence—sometimes extraordinary persistence—is necessary to solve complex problems. Nevertheless, there are few other areas of medicine that offer such an opportunity to practice prevention and to make a difference.

References for Box 5-3

1. Stayner L, Kuempel E, Rice F, Prince M, Althouse R. Approaches for assessing the efficacy of occupational health and safety standards. Am J Ind Med 1996;29:353–357.
2. Boden LI. Policy evaluations: better living through research. Am J Ind Med 1996;29:346–352.
3. Office of Technology Assessment. Gauging control technology and regulatory impacts in occupational safety and health: an appraisal of OSHA's analytical approach. Washington, DC: US Congress, OTA, 1995.
4. McQuiston TH, Zackocs RC, Loomis D. The case for stronger OSHA enforcement: evidence from evaluation research. Am J Public Health 1998; 88:1022–1024.
5. General Accounting Office. OSHA action needed to improve compliance with the Hazard Communication Standard. Washington, DC: US GAO, 1991.
6. LaMontagne AD, Kelsey KT. OSHA's renewed mandate for regulatory flexibility review: in support of the 1984 ethylene oxide standard. Am J Ind Med 1998;34:95–104.
7. Zwerling C, Daltroy LH, Fine LJ, Johnston JJ, Melius J, Silverstein BA. Design and conduct of occupational injury intervention studies: a review of evaluation strategies. Am J Ind Med 1997;32:164–179.
8. LaMontagne AD, Needleman C. Overcoming practical challenges in intervention research in occupational health and safety. Am J Ind Med 1996;29:367–372.
9. Robins TG, Hugentobler MK, Kaminski M, Klitzman S. Implementation of the federal Hazard Communication Standard: does training work? J Occup Med 1990;32:1133–1140.
10. Levin SM, Goldberg M, Doucette JT. The effect of the OSHA Lead Exposure in Construction Stan-

dard on blood levels among iron workers employed in bridge rehabilitation. Am J Ind Med 1997;31: 303–309.

11. Yodaiken RE. Annotation: evaluating OSHA's ethylene oxide standard and evaluating OSHA. Am J Public Health 1997;87:1096–1097.

12. Schulte PA, Goldenhar LM, Connally LB. Intervention research: science, skills, and strategies. Am J Ind Med 1996;29:285–288.

13. Gomez MR. Exposure surveillance tools needed in agency GPRA plans. Am Ind Hyg Assoc J 1998;59:371–374.

BIBLIOGRAPHY FOR BOX 5-1

Ellenbecker MJ. Engineering controls as an intervention to reduce worker exposure. Am J Ind Med 1996;29:303–307.
This paper broadens the definition of substitution to include process changes and presents field examples of interventions. It also describes the general methods of pollution prevention (toxic use reduction).

Goldschmidt G. An analytical approach for reducing workplace health hazards through substitution. Am Ind Hyg Assoc J 1993;54:36–43.
This reference describes a systematic approach involving analysis of health characteristics of raw materials (162 examples from Denmark are summarized) for the purpose of choosing as alternatives more environmentally and occupationally benign materials.

Quinn MM, Kriebel D, Geiser K, Moure-Eraso R. Sustainable production: a proposed strategy for the work environment. Am J Ind Med 1998;34:297–304.
This paper calls for expansion of the role of the occupational health professional to include evaluation and redesign of production processes. It also calls for new research to develop these activities as the scientific and public health policy basis of sustainable production.

BIBLIOGRAPHY FOR BOX 5-2

Auerbach E, Wallerstein N. ESL for action: problem-posing at work: teachers guide and student manual. Reading, MA: Addison-Wesley, 1987.

Briggs D, Cameron B, Johnson H, et al. The evolution of worker training methods at Boeing commercial airplane group. New Solutions 1997;7:31–36.

Cary M, Van Belle G, Morris SI, et al. The role of worker participation in effective training. New Solutions 1997;7:23–30.

Colligan M. Occupational safety and health training. Occup Med 1994;9.

Hecker S. Education and training. In: Hecker S, ed. ILO Encyclopaedia of Occupational Health and Safety. Geneva: International Labor Organization, 1998, pp. 18.1–18.32.

McQuiston T, Coleman P, Wallerstein N, et al. Hazardous waste worker education. J Occup Med 1994;36:1310–1323.

Massachusetts Coalition for Occupational Safety and Health. English as a second language health and safety curriculum for working people. Boston: Mass-COSH, 1993.

National Clearinghouse for Worker Safety and Health

Training for Hazardous Materials, Waste Operations and Emergency Response, National Institute of Environmental Health Sciences (NIEHS), 5107 Benton Avenue, Bethesda, MD 20814 (telephone 301-571-4226). Materials from all NIEHS regional training centers, including *Chemical and Radioactive Hazardous Material Workbook* (Oil, Chemical and Atomic Workers/Labor Institute, New York, NY, 1993) and *Hazardous Waste Workers Health and Safety Training Manual, Version 3.1* (The New England Consortium, University of Massachusetts Lowell, Lowell, MA, 1997).

Slatin C, ed. Health and safety training [Special section]. New Solutions: A Journal of Environmental and Occupational Health Policy 1995;Winter.

The Labor Institute. Sexual harassment at work: a training workbook for working people. New York: The Labor Institute, 1994.

Wallerstein N, Pillar C, Baker R. Labor Educator's health and safety manual. Berkeley, CA: Labor Occupational Health Program, Center for Labor Research and Education, Institute of Industrial Relations, University of California, 1981.

Wallerstein N, Rubenstein H. Teaching about job hazards: a guide for workers and their health providers. Washington, DC: American Public Health Association, 1993.

Wallerstein N, Weinger M, eds. Empowerment education [Special issue]. Am J Ind Med 1992;22(5).

APPENDIX
Summary of Information Concerning NIOSH

Although the National Institute for Occupational Safety and Health (NIOSH) and the Occupational Safety and Health Administration (OSHA) were both created by the Occupational Safety and Health Act of 1970, they are two distinct agencies with separate responsibilities. OSHA is part of the Department of Labor and is responsible for creating and enforcing workplace safety and health regulations. NIOSH, which is part of the Centers for Disease Control and Prevention in the Department of Health and Human Services, is responsible for conducting research, training, and field studies and making recommendations for the prevention of work-related illnesses and injuries.

The main responsibilities of NIOSH include investigating potentially hazardous working conditions, as requested by employers or employees; evaluating hazards in the workplace, ranging from chemicals and repetitive trauma to machinery and radiation;

creating and disseminating methods for preventing disease, injury, and disability; conducting research and providing scientifically valid recommendations for exposure limits to hazardous chemicals, physical agents, and processes; and providing education and training to persons preparing for careers in the field of occupational safety and health.

NIOSH responds to requests for investigations of workplace hazards through the Health Hazard Evaluation (HHE) program. An HHE is a worksite study designed to evaluate potential workplace health hazards. HHEs can be requested by a management official, a current employee (provided that two other current employees sign the request), or any officer of a labor union representing the employee. NIOSH reports the results of the investigation to the workers, the employer, and the Department of Labor and makes recommendations for reduction or removal of the hazard. The HHE program serves as a useful surveillance tool, keeping NIOSH abreast of emerging workplace concerns.

NIOSH conducts a wide range of surveillance activities to determine the number of workers exposed to specific hazards and which industries and occupations are at risk (see Chapter 4). Ongoing surveillance allows NIOSH to address the most critical problems in occupational safety and health.

NIOSH supports research through intramural programs that it conducts, through cooperative agreements that it initiates and in which it participates, and through research grants that extramural investigators initiate and conduct. In 1996, NIOSH, working with a broad array of partners, announced the National Occupational Research Agenda (NORA), a framework to guide occupational safety and health research into the next decade—not only for NIOSH, but for the occupational health and safety community at large. NORA priority research areas are as follows:

Disease and injury
Allergic and irritant dermatitis
Asthma and chronic obstructive pulmonary disease
Fertility and pregnancy abnormalities
Hearing loss
Infectious diseases
Low back disorders
Musculoskeletal disorders of the upper extremities
Traumatic injuries

Work environment and workforce
Engineering technologies
Indoor environment
Mixed exposures
Organization of work
Special populations at risk

Research tools and approaches
Cancer research methods
Control technology and personal protective equipment
Exposure assessment methods
Health services research

Intervention effectiveness research
Risk assessment methods
Social and economic consequences of workplace illness and injury
Surveillance research methods

To disseminate research findings, NIOSH publishes a variety of reports and other materials. NIOSH publications are designed to inform workers, employers, and occupational safety and health professionals of hazards and how to avoid them. Additionally, NIOSH develops documents to forward its scientific recommendations to OSHA and to the Mine Safety and Health Administration (MSHA) for consideration in the standards-setting process.

Finally, to improve and maintain the competence of occupational safety and health professionals, NIOSH provides a wide variety of training programs that focus on occupational medicine, occupational nursing, industrial hygiene, and occupational safety.

NIOSH is headquartered in Washington, DC, with administrative offices in Atlanta and with seven working divisions located in Morgantown and Cincinnati and two laboratories in Pittsburgh and Spokane. The divisions are as follows:

Division of Biomedical and Behavioral Science (DBBS). It conducts research in toxicology, neurologic and behavioral science, and ergonomics. Responsibilities include laboratory and field studies of biomechanical, psychological, neurobehavioral, and physiologic effects of physical, psychological, biomechanical, and selected chemical stressors. It also develops biologic monitoring and diagnostic procedures to improve worker health.

Division of Physical Sciences and Engineering (DPSE). It conducts research to develop procedures and equipment for the measurement of occupational safety and health hazards and for the development of effective engineering controls and work practices. It also maintains a quality control reference program for industrial hygiene laboratories.

Division of Respiratory Disease Studies (DRDS). It conducts epidemiologic, environmental, clinical, and laboratory research focusing on all aspects of occupational respiratory disease. It also has specific responsibilities from the Mine Safety and Health Act (the National Coal Workers X-Ray Surveillance and the National Coal Workers Autopsy Study, certification of x-ray facilities, mine plan approvals, and B-reader examination).

Division of Safety Research (DSR). It conducts research on occupational injury prevention through studies of risk factors and the effectiveness of prevention efforts. It conducts research to provide criteria for improving personal protective equipment and devices.

Education and Information Division (EID). It has responsibility for development of NIOSH policy and recommendations, with special attention to new occupational health and safety standards. It publishes Current Intelligence Bulletins to disseminate new scientific information, and Alerts to identify opportunities for preventative interventions. It also undertakes quantitative risk assessment efforts to prioritize issues for regulatory attention. It coordinates NIOSH testimony for U.S. Department of Labor hearings and maintains databases such as NIOSHTIC (NIOSH Technical Information Clearinghouse) and RTECS (Registry of Toxic Effects of Chemical Substances), library services, technical information services, the Institute Archives, and the toll-free telephone number (see Appendix A and Appendix B).

Division of Surveillance, Hazard Evaluations and Field Studies (DSHEFS). It has responsibility for surveillance of the extent of hazards and occupational illnesses. It conducts legislatively mandated health hazard evaluations at the request of employees or employers. It also conducts a broad range of industry-wide epidemiologic and industrial hygiene research programs, with wide responsibility for occupational illnesses not included in DBBS or DRDS. It is also responsible for energy-related health research related to workers at U.S. Department of Energy facilities.

Health Effects Laboratory Division (HELD). It conducts basic, applied, and preventive laboratory research, develops intervention programs, and designs and implements methods for health communications in the area of occupational injury and disease. HELD collaborates with researchers throughout NIOSH and in other public and private institutions to apply the latest scientific research to workplace health problems

To further assist professionals and the public, NIOSH provides a toll-free information system. It can be accessed by telephone at 1-800-35-NIOSH, or 1-800-356-4674. NIOSH specialists provide technical advice and information on subjects in occupational safety and health.

Occupational Health: Recognizing and Preventing Work-Related Disease and Injury, 4th ed.
Edited by Barry S. Levy and David H. Wegman, Lippincott Williams & Wilkins, Philadelphia © 2000.

6 Epidemiology

Ellen A. Eisen and David H. Wegman

Epidemiologic studies further our understanding about environmental determinants of disease and serve as a basis for developing public health policy. All health professionals rely to some degree on epidemiologic literature. To remain current in their fields they need to be able to assess the quality of studies. This chapter is designed to assist health professionals in understanding how epidemiology is applied to occupational health and in critically interpreting the relevant epidemiologic literature.

The relation between any group of workers and their work environment is dynamic. New workers are hired and others leave the workforce. Exposures vary over time because of job transfers, changes in technology or production processes, and other factors. The workforce ages, and individual workers alter their personal habits, such as cigarette smoking. The epidemiologist uses analytic tools to examine this complex mix of variables in an attempt to understand the effects of workplace conditions on injury, disease, disability, and death.

Epidemiology complements clinical medicine in addressing occupational health problems. The clinical approach focuses on the individual and is concerned with diagnosis, treatment, and education of the worker regarding risk factors and preventive behavior. In contrast, the epidemiologic approach focuses on groups and is concerned with de-scribing the distribution of injuries and disease in work groups, identifying population subgroups at high risk for a particular outcome, providing evidence for causal associations, estimating dose-response relations, and determining the effectiveness of preventive measures.

In epidemiology, the health outcome may be a discrete end point, such as diagnosis of a disease, or measurement of a biologic parameter, such as pulmonary function. The measure of exposure may be crude, such as membership in an occupational group, or more refined, such as the average daily time-weighted average exposure to a particular substance. Epidemiologists collaborate with toxicologists, ergonomists, environmental scientists, statisticians, and others to collect exposure data and to develop more precise methods for estimating exposures to chemical, physical, biomechanical, biologic, and psychosocial factors.

MEASURING EXPOSURE

Exposure is characterized by intensity or concentration, such as parts per million, and the duration over which it occurs. Cumulative or aggregate exposure is the product of intensity and duration and is an approximation of dose (to an organ or tissue). There are several degrees of refinement for approximating dose.

Potential Exposure

The most common available measure of exposure is simply a history of employment in

E. A. Eisen and D. H. Wegman: Department of Work Environment, University of Massachusetts Lowell, Lowell, Massachusetts 01854.

a specific industry or a specific job. Although this is a crude surrogate for exposure, if the relation between exposure and outcome is sufficiently strong a true association can be seen. For example, lung cancer was associated with asbestos in a study of shipyard workers, despite the fact that fewer than half of the shipyard workers had asbestos exposure (1). Nevertheless, the estimate of risk associated with exposure to a specific agent is greatly diluted by such a surrogate measure. A study of diesel exposure among railroad workers was largely negative, but only 7% of workers were found to have had exposure to diesel fumes.

Quantity of Exposure

Measures of exposure should include both intensity and duration. Because data on duration of employment may be more easily and accurately determined than intensity of exposure, duration is frequently used as the dose surrogate. It is often possible to document the number of years employed from payroll records or from union seniority records. Sometimes length of employment is unknown, but data such as pension plan eligibility may provide at least a dichotomous measure of duration (e.g., more than or less than 10 years of employment).

Exposure estimates are improved when industrial hygiene input is available, either as judgments of potential exposure or as measurements of actual exposure. Variation in exposure occurs as a result of changes in work assignments between days and within any given day, differences in work habits, seasonal changes in ventilation patterns, use of personal protective equipment, and other factors. Knowledge about these variations is used to adjust job-specific exposure estimates. Current exposure estimates alone can be used to study acute effects, but for the study of chronic effects such estimates must be integrated with job histories to develop a measure of cumulative exposure (2). A complete work history ideally includes documentation of time spent in specific jobs together with information on gaps in employment, such as prolonged sick leaves, periods of layoff, or military leaves.

Estimations of cumulative exposure ideally rely on compilation of current and historical industrial hygiene data and interviews of plant personnel about the history of changes in the production process and exposure controls. Estimates can be made of past exposures by reconstructing and testing old work environments. For example, in studies of pulmonary function in the Vermont granite industry, there was a need to account for old exposures, but no measurements were available. An old granite shed was reopened and operated without modern exhaust ventilation controls to arrive at appropriate estimates of the old exposures (3).

To compute cumulative exposure, estimated exposure levels are weighted by the number of years in successive jobs and summed over all jobs held by each worker. An implicit assumption in the computation of cumulative exposure is that 1 year of exposure to 20 fibers/cubic centimeter of asbestos is equivalent to 10 years of exposure to 2 fibers/cc. Furthermore, exposure that occurred years ago is assumed to be biologically equivalent to the exposure last year. More complex weighting schemes are possible but should be based on specific biologic hypotheses about the relative importance of different exposure patterns. For example, exposures in the distant past can be weighted more heavily than those in the recent past for diseases such as silicosis, in which irreversible changes are believed to accumulate gradually over years.

Biologic Monitoring

Evaluation of workers for toxic agents (or their metabolites) in blood, urine, or exhaled air sometimes permits improved estimation of real dose. One advantage of a biologic index is that it accounts for exposures from multiple routes of absorption, including inhalation, skin absorption, and ingestion. For example, urinary hippuric acid levels can be used to estimate total recent dose of toluene to an exposed worker via both inhalation and skin absorption. Another advantage of biologic markers is that they may reflect exposure over specific time intervals. For example, although blood lead levels indicate recent lead exposure, x-ray fluorescence of bone provides an estimate of total body burden of lead, reflecting long-term exposure (4). Although no biologic monitoring tests currently exist for a substantial number of hazardous workplace substances, biologic monitoring is receiving more attention today and new measures of the body burden of toxic agents can be expected.

In summary, there are a variety of ways to estimate both current and past exposures. An accurate measurement of exposure is equally as important as an accurate measurement of health outcome in arriving at an unbiased and precise estimate of the exposure-outcome relationship.

COMMON MEASURES OF DISEASE FREQUENCY

If a disease is extremely rare, the occurrence of even a few cases can prompt further investigation of a possible workplace hazard. For example, three cases of angiosarcoma diagnosed during a 3-year period among a group of workers exposed to vinyl chloride was sufficient to make a plant physician suspect that the chemical was a carcinogen (5). In most instances, however, a count of cases cannot be interpreted without knowing the size of the population from which the affected workers came. The problem is illustrated by the example of a study of workers in a coated fabrics plant, 68 of whom were found to have a peripheral neuropathy (6). Even though this end point is uncommon in the general population, it is not sufficiently rare that the expected number of cases can be treated as zero. A case count by itself has little or no meaning without a standard of reference; that is, disease frequency can be interpreted only in relation to the size of the population at risk. The measures of disease frequency that are most commonly used are prevalence and incidence.

Prevalence

The simplest quantity, known as *point prevalence,* is the ratio between the number of cases present and the size of the population at risk *at a single point in time.*

Point prevalence
$$= \text{No. cases} \div \text{Total population}$$

To interpret the public health significance of the 68 cases of peripheral neuropathy in the coated fabrics plant, we first need a denominator. The total plant population was 1,157, so the point prevalence was $68 \div 1157 = 5.9\%$. To determine whether this is excessive, the prevalence in the plant must be compared with the prevalence in the general population or in another, more appropriate comparison group. A limitation of point prevalence is that it does not distinguish between old and new cases.

Incidence Rate

By contrast, *incidence* measures the occurrence of new cases. The incidence rate is based on the number of new cases occurring *during a specified period of time.*

Incidence rate
$$= \text{No. new cases} \div \text{Total population at risk}$$

In the coated fabrics plant, only 50 affected workers had onset of the disease within the past year; 18 of the 68 prevalent cases occurred more than 1 year ago. Therefore, the population at risk for development of a new case within the past year was $1,157 - 18 = 1,139$. Because the number of new cases in that period was 50, the plant-wide annual incidence rate was $50 \div 1,139 = 4.4\%$ for the preceding year.

The incidence rate can also be refined to reflect monitoring of individual workers for varying lengths of time. The appropriate denominator incorporates the concept of *person-time,* usually expressed in units of per-

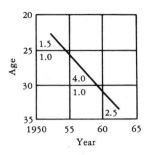

Time Period	Age Group	Person-years
1950 – 1954	20 – 24	1.5
	25 – 29	1.0
1955 – 1959	25 – 29	4.0
	30 – 34	1.0
1960 – 1964	30 – 34	2.5
		——
		10.0

FIG. 6-1. Person-years experienced by a worker entering a follow-up program at age 23 years 6 months in mid-1952 and leaving in mid-1962. (Adapted from Monson RR. Occupational epidemiology, 2nd ed. Boca Raton, FL: CRC Press, 1989.)

son-years. This denominator takes into account not only the number of at-risk persons but also the period during which they were at risk for development of the disease. An example of how to calculate the contribution of a single worker to a person-years denominator is illustrated in Fig. 6-1.

COMPARISONS OF RATES

To understand whether an incidence rate in an exposed population is excessive, it is necessary to compare it with the rate in an unexposed population. The two most common comparisons, or estimates of risk, are *relative risk* (the ratio of rates) and *attributable risk* (the difference between rates).

Relative Risk

The relative risk, or rate ratio, is designed to communicate the relative importance of an exposure by comparing rates from an exposed population with those from a nonexposed or normal population. In its simplest form, it is the ratio of two rates (Table 6-1). In the case of the fabrics plant, the suspect neurotoxin was in the print department, so it was possible to create a within-plant comparison. Of the 1,139 disease-free workers in the plant, 169 worked in the print department and 34 of those had onset of peripheral neuropathy in the past year, resulting in an annual incidence rate of 34 ÷ 169 = 20.1%.

Among the remaining 970 workers, there were 16 new cases, resulting in an annual incidence rate of 16 ÷ 970 = 1.6%. The relative risk (or *incidence rate ratio*) therefore was .201 ÷ .016 = 12.6.

When examining different diseases or the effects of different hazards, relative risks can be compared directly. For example, the relative risk of lung cancer in heavy smokers compared with nonsmokers is very large (32.4), whereas that for cardiovascular disease is small (1.4). This suggests that smoking is more potent as a lung carcinogen than as a cardiotoxic agent.

Attributable Risk

Whereas the relative risk is a measure of the potency of the hazard, attributable risk measures the magnitude of the disease bur-

TABLE 6-1 *Derivation of relative and attributable risk**

	Exposure		
Disease	Present	Absent	Total
Present	a	c	a + c
Absent	b	d	b + d
Total	a + b	c + d	a + b + c + d

* Calculations:
Exposed disease rate = a/(a + b)
Nonexposed disease rate = c/(c + d)
Relative risk = a/(a + b) ÷ c/(c + d)
Attributable risk = a/(a + b) − c/(c + d)

den in the population ascribed to the exposure under study. This concept is particularly useful in occupational disease studies because occupational exposure is generally only one of several possible causes of any specific disease. The attributable risk is calculated by subtracting the rate of the particular disease in the nonexposed population from that in the exposed population (see Table 6-1). This risk difference is attributed to the exposure. In the coated fabrics plant, the incidence rate in the unexposed population (.016) is subtracted from the rate in the exposed population (.20), yielding an attributable risk of .184.

In the example of the impact of cigarette smoking on health, Table 6-2 shows that the smoking-attributable risk for lung cancer (2.20/1,000) is smaller than the smoking-attributable risk for cardiovascular disease (2.61/1,000). The attributable risk takes account of both the potency and the magnitude of the disease in the population. Despite the lower relative risk of cardiovascular disease due to smoking, the larger attributable risk indicates that reduction of smoking has a greater impact on cardiovascular disease than on lung cancer in a population.

Relative risks are commonly presented in epidemiologic studies as a measure of association between an exposure and a disease outcome. In contrast, attributable risks are useful in setting priorities for public health interventions or control.

INTERPRETING RATES

Crude Rates. When rates are calculated without consideration of factors such as age or calendar year, they are referred to as *crude* rates. Crude rates can be misleading. For example, if the exposed group includes a high proportion of elderly persons and disease incidence increases with age, then observed differences in crude rates may only reflect differences in age.

Specific Rates. These are rates estimated for homogeneous subgroups of a population defined by specific levels of a factor, such as age-specific rates. Sometimes an elevated disease risk exists only in one subgroup.

Adjusted Rates. Although specific rates can sometimes provide valuable information, it is cumbersome to compare many specific rates. Methods have been developed for estimating a single summary rate that takes account of differences in the distribution of population characteristics such as age. Such rates are known as *adjusted* or *standardized* rates. Two types of adjustment are commonly used: *direct* adjustment (rates in the study population are weighted by person-time in a reference population) and *indirect* adjustment (rates in a reference population are weighted by person-time in the study population). These methods can be illustrated with examples of adjustment for age (Table 6-3). For a description of these types of adjustment, see the appendix at the end of the chapter.

TABLE 6-2 *Relative and attributable risk of death among British male physicians from selected causes associated with heavy cigarette smoking*

| | Death rate* | | | |
| | | Heavy | Relative | Attributable |
Cause of death	Nonsmokers	smokers†	risk	risk
Lung cancer	0.07	2.27	32.4	2.20
Other cancers	1.91	2.59	1.4	0.68
Chronic bronchitis	0.05	1.06	21.2	1.01
Cardiovascular disease	7.32	9.93	1.4	2.61
All causes	12.06	19.67	1.6	7.61

*Number of deaths per 1,000 per year.
†Smokers of ≥25 cigarettes per day
From Doll R, Hill AB. Mortality in relation to smoking: ten years' observations of British doctors. Br Med J 1964;1:1399.

TABLE 6-3 *Age effect on incidence of myocardial infarction**

Location	Workers <45 yr			Workers ≥45 yr			All workers			
	Cases	Population at risk	Age-specific incidence rate	Cases	Population at risk	Age-specific incidence rate	Cases	Population at risk	Crude incidence rate	Age-adjusted incidence rate†
Factory 1	4	400	10.0	18	600	30.0	22	1,000	22.0	18.0
Factory 2	10	800	12.5	10	200	50.0	20	1,000	20.0	27.5

* The incidence rate is expressed as new myocardial infarctions occurring in a 10-year period of observation per 1,000 population.
† Based on age distribution summed for factory 1 and factory 2.

Drawing by Nick Thorkelson.

TYPES OF EPIDEMIOLOGIC STUDY DESIGNS

Epidemiologic studies can be categorized into three general types: cohort, case-control, and a hybrid called cross-sectional. The population in a cohort study is defined on the basis of exposure status and often represents a complete enumeration of both current employees and past workers. The cohort is monitored over time, and the incidence of symptoms, functional abnormalities, disease, and death are observed. By contrast, the study group in a case-control study is defined on the basis of health status (Fig. 6-2), and exposures are compared between subjects with and without disease. The cross-sectional design typically focuses on active employees at a single point in time, collecting both exposure and health information simultaneously.

Cross-Sectional Studies

The cross-sectional approach is commonly used in field investigations because it is the simplest study design to execute. Either the prevalence of disease is compared between groups defined by exposure status, or the prevalence of exposure is compared between groups defined by disease status. Exposure can be classified dichotomously (e.g., exposed versus nonexposed) or along a gradient (e.g., high, medium, and low). Exposure classification can be based on either current or lifetime exposure.

Example: Cross-sectional Study, Exposure-Based

A pathology resident died of an acute heart attack at the age of 28 years. In discussing this incident, a number of the other pathology residents noted that they had been experiencing abnormal heart rhythms (palpitations). Those with palpitations had all worked with fluorocarbon propellants, which were used to prepare frozen sections of pathology tissue and to clean instruments or specimen slides. This discovery led to a study of all pathology department employees (the exposed group). Employees of a radiology department of similar size and distribution of physicians and nonphysician staff members were selected as a nonexposed comparison group. Each person was asked about occur-

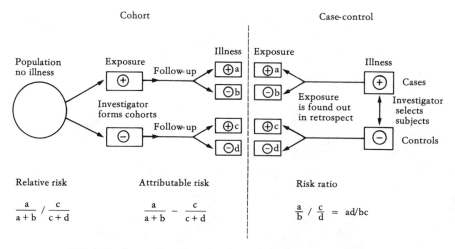

FIG. 6-2. General outline of cohort and case-control studies.

rence of palpitations and current use of fluorocarbon propellants. Those exposed to the propellants had twice the prevalence of palpitations as those not exposed (7).

Example: Cross-sectional Study, Disease-Based

In the study just described, the investigators wanted to further explore their hypothesis that exposure to fluorocarbon propellants accounted for the elevated risk of palpitations among pathology department employees. A follow-up analysis was designed in which groups were defined on the basis of disease prevalence (7). The pathology department staff was divided into those with palpitations (cases) and those without palpitations (controls). Forty percent of the cases but only 20% of the controls had exposure to fluorocarbon propellants, for an odds ratio (OR) of 2.7. The OR is an approximation of the relative risk (see later discussion).

Cohort Studies

In a cohort study design, an exposed group is identified and monitored forward in time to measure the occurrence of adverse health outcomes. The incidence is observed in the study group and compared with that in a nonexposed, reference group. Cohort studies are described as *retrospective* (the cohort is defined at some point in the past and monitored to the present) or *prospective* (the cohort is defined at the present and monitored into the future).

Cohort Mortality Studies. Although the cohort design can be used to examine nonfatal outcomes, most occupational cohort studies examine mortality from specific causes. The most common of type of cohort study is the standardized mortality study, in which the cause-specific mortality rate of the exposed cohort is compared with that of the general population (assumed to be nonexposed). This comparison results in an approximation of relative risk, known as the *standardized mortality ratio* (SMR). If the number of deaths observed in the exposed cohort is equal to the number expected based on death rates in the standard population, the SMR equals 1.0, which indicates neither an excess nor a deficit of risk. If the SMR is greater than 1.0, the data suggest an increased risk in the exposed population.

To conduct an SMR study, the following information must be obtained for each member of the cohort: date of birth, date of entry into cohort, date of leaving cohort, vital status (alive or dead), and cause of death for those who died. With these data, person-years at risk can be determined that take into consideration times when workers entered or left during the study period. This permits a calculation of years at risk, adjusting for length of time since entry into the study and for age. This type of study requires personnel records with accurate employment data; if such data on the total population at risk are lacking, the mortality experience can be evaluated by proportional mortality analysis.

Cause-specific *proportional mortality ratios* (PMRs) are calculated as the proportion of all deaths attributed to each specific cause of death. These ratios in the study population are compared with those from the general population and are adjusted for age, sex, race, and year of death (again, indirect standardization). In contrast to an SMR study, an excess or deficit of deaths from causes other than the one under scrutiny in a PMR study can affect the proportional distributions. Therefore, PMR study results are less reliable than SMR study results.

Example: Retrospective Cohort Mortality Study, Standardized Mortality Ratio

A study of mortality in steelworkers was planned in 1962 (8). The workers were selected for study if they were employed in 1953; they were monitored until the end of 1962 for vital status. More than 59,000 workers participated, and 4,716 deaths recorded. When the numbers of deaths from specific causes were compared with the numbers expected, based on deaths in the study county, there appeared to be no excesses. (SMR for all cancers combined was only .92.) The study population was large enough, however, to examine the SMRs for particular subgroups and compare them in each case with the SMR for the rest of the workers (documented as nonexposed). This in-depth evaluation led to the discovery that lung cancer risk appeared to be higher among the coke plant workers and that the excess was much greater for the nonwhite employees. Further analysis showed that those who worked on

TABLE 6-4 *Standardized mortality ratios for lung cancer among nonwhite male coke oven workers for ≥5 yr*

Work area	Observed	Expected	SMR
Side oven	19	9.5	2.00
Part-time topside	9	2.3	3.91
Full-time topside	23	1.9	12.10
Total	51	19.0	2.68

Adapted from Redmond CK, Wieand HS, Rockette HE, et al. Long-term mortality study of steelworkers. Department of Health and Human Services (NIOSH) Publication No. 81-120, U.S. Government Printing Office, Washington, DC, 1981.

top of the coke ovens (the most heavily exposed job assignment) had the highest risk for lung cancer (Table 6-4). The large size of the study population permitted detailed examination of a number of subgroups. As a result, the very high risk of lung cancer in coke oven workers was extracted from the overall unremarkable results.

Cohort Morbidity Studies. Increasingly, the cohort design is being used to study occupational risks associated with a variety of nonfatal health outcomes. Retrospective studies can be conducted if information on past health status is available (e.g., in medical records or collected in health surveys). More often, morbidity studies require prospective study designs so that the health information can be collected directly by administering medical examinations, physiologic tests, or surveys of current health status. Studies that look at episodic health events, such as recurrent symptoms or changes in pulmonary function, are referred to as *longitudinal* studies, and the change in health status over time becomes the outcome.

Example: Retrospective Cohort Morbidity Study

A cohort of approximately 1,000 hospital nurses was studied to examine possible reproductive effects associated with use of sterilizing agents (9). Questionnaires and medical records were used to collect information retrospectively about both exposure and pregnancy history as far back as 30 years.

The frequency of spontaneous abortion among nurses currently using the sterilizing agents was only slightly higher than that for currently nonexposed nurses. A more striking difference was observed when results were stratified according to whether exposure to sterilizing agents had occurred during a past pregnancy. Among those exposed, the rate of spontaneous abortion was 16%, compared with 6% among the nonexposed. Of the three specific sterilizing agents considered, ethylene oxide showed the strongest association with spontaneous abortion.

Example: Prospective Cohort Morbidity Study

A study was designed to characterize adverse respiratory effects associated with exposure to toluene diisocyanate (TDI). Because TDI was already known to be a cause of asthma, the study was designed to measure other types of acute and chronic respiratory effects. Pulmonary function tests (including measurement of the forced expiratory volume in 1 second, FEV_1) were administered to all workers at a polyurethane-foam manufacturing firm on the first shift on a Monday morning. The workers were then divided into three exposure groups and retested at the end of the workday—a 1-day, prospective study design.

Generally, FEV_1 changes only slightly over the course of a workday. However, on average these workers were losing lung function over the duration of the shift, and the amount of loss increased with exposure (Table 6-5). To examine whether this acute response (presumably caused by bronchospasm) reflected a more persistent chronic effect, those who were still employed were retested 4 years later. The annual decline in FEV_1 was estimated for the same three exposure groups (10). Again, an exposure-related response was observed (see Table 6-5), with the high-exposure group losing lung function at the greatest annual rate. Cigarette smoking habits did not explain the

TABLE 6-5 *Acute and chronic change in FEV_1, by exposure group, in polyurethane foam manufacturing workers**

| | FEV_1 differences from beginning to end of study period | |
Exposure group	Acute (1-day) change (ml)	Chronic change over 4 yr (ml/yr)
Low	−78	0.5
Medium	−108	−33.3
High	−180	−60.5

* Negative change means loss over time.
Adapted from Wegman DH, Musk AW, Main DM, et al. Accelerated loss of FEV-1 in polyurethane production workers: a four year prospective study. Am J Indus Med 1982;3:209–215.

effects noted in either the 1-day or the 4-year prospective study.

Case-Control Studies

In the case-control, or case-referent, study design, the investigator compares the frequency of exposure between groups with and without the disease of interest (see Fig. 6-2). The case-control design is particularly well suited to study diseases that occur with low incidence; a cohort study would have to be prohibitively large to generate enough cases to study.

There are three types of case-control studies: (a) studies nested within occupational cohort studies, (b) population-based case-control studies, and (c) registry-based case-control studies. In nested case-control studies, all cases of the selected disease are identified from the cohort, and controls are sampled from among those without the disease. In a mortality study, disease status may be determined at death from a particular disease; in a morbidity study, disease status may be determined by disease incidence based on diagnosis. In a population-based case-control study, all cases occurring in residents of a defined geographic area are included, and controls are selected from the same defined population. In a registry-based case-control study, cases of disease that are reported to the registry with onset during a defined time period are identified, and controls are selected from the same registry base. Because registries often are not population-based (e.g., a hospital cancer registry), the selection of controls requires identification of patients with other diseases from the same source as the cases (e.g., from the same hospital).

The measure of risk typically calculated in a case-control study is the OR. The OR is a ratio of the odds of exposure (exposed to non-exposed) among the cases compared with the odds of exposure among the controls. From Table 6-1, it can be seen that a/b is the odds of exposure among the cases, and c/d is the odds of exposure among the controls. ORs approximate the incidence rate ratios that are obtained in cohort studies. Their interpretations are similar: OR = 1 means that there is no excess or deficit of risk.

A case-control study need not include all the cases within a defined population. Valid results may still be obtained when the case group includes only a sample of all cases. The major requirement for a valid case-control study is that the controls selected be comparable to the population from which cases were identified and that both cases and controls be selected without prior knowledge of past exposure history.

Example: Population-Based Case-Control Study

Non-Hodgkin's lymphoma has been associated with agricultural pesticide use in men, but little is known about risks in women. To address this lack of knowledge, National Cancer Institute investigators conducted a population-based case-control study (11) in which cases were defined as incident cases of non-Hodgkin's lymphoma among women residing in 66 counties in eastern Nebraska, diagnosed between 1983 and 1986 in all area hospitals. Controls were selected from female residents in the same counties using random digit dialing. No risk was found to be related to living or working on a farm. Small risks were observed for women who

personally handled insecticides (OR = 1.3) or herbicides (OR = 1.2), and women who personally handled organophosphate insecticides had a 4.5-fold increased risk (Table 6-6). Because non-Hodgkin's lymphoma is a rare disease with a long latency, the case-control design was more feasible than a cohort study. Because exposures occur on farms, each of which employs a small number of workers, a community-based study was more practical than a workplace-based study.

Example: Nested Case-Control Study

The carcinogenic risk of pulsed electromagnetic fields was studied in a series of case-control studies nested in a cohort of electric utility workers (12). Case groups were defined as all diagnosed cases of selected cancers that occurred at any time after entry into the cohort until the end of follow-up in 1988. Controls were chosen at random from sets of cohort members matched to each case who had survived to the date of diagnosis of the case. Cumulative exposures were estimated up to the date of diagnosis of the case. Smoking information was obtained from company medical records. No associations were found between exposure to pulsed electromagnetic fields and cancers previously suspected of being associated with magnetic fields. However, the investigators reported a clear association between cumulative exposure to pulsed electromagnetic fields and lung cancer (after adjusting for cigarette smoking history), with an OR of 3.1 in the highest exposure category.

SELECTION OF TYPE OF STUDY

The study designs described have relative strengths and weaknesses. The choice of design is based on a variety of factors.

Cross-Sectional Studies

Cross-sectional studies have several advantages over cohort studies. First, cross-

TABLE 6-6 *Non-Hodgkin's lymphoma according to insecticide use among women in eastern Nebraska*

Insecticide class	Used on farms			Personally handled		
	Cases	OR	95% CI	Cases	OR	95% CI
Any insecticide	56	0.8	0.5–1.3	22	1.3	0.7–2.3
Chlorinated hydrocarbons	20	1.6	0.8–3.1	5	1.7	0.5–5.8
Organophosphates	14	1.2	0.6–2.5	6	4.5	1.1–17.9
Metals	3	1.6	0.3–7.5	0	—	—

OR, odds ratio; 95% CI, 95% confidence interval.
Adapted from Zahm SH, Weisenburger DD, Saal RC, Vaught JB, Babbitt PA, Blair A. The role of agricultural pesticide use in the development of non-Hodgkin's lymphoma in women. Arch Environ Health 1993;48:353–358.

sectional studies permit the examination of disease morbidity or measures of physiologic function. Second, because the subjects are alive at the time of the study, it is often possible to collect information directly on nonoccupational risk factors such as cigarette smoking or diet (potential confounders). Finally, because both disease prevalence and exposure data are collected at one point in time, cross-sectional studies usually require less time to complete than cohort or case-control studies.

These studies also have important limitations. They are regarded as less appropriate for investigating causal relations because they are based on prevalent, rather than incident, cases of disease. A second limitation is that they are based on actively employed workers and do not include employees who retired or terminated their employment before the beginning of the study. In the presence of an occupational hazard, workers whose health has been affected are more likely to leave the workforce; therefore, the absence of such workers may result in an underestimate of the association of interest.

Cohort Studies

These studies focus on exposure and look ahead to outcome or disease incidence. Several outcomes can be studied in the same population. Data collected on exposure retrospectively depends on the quality of past records, in contrast to data collected prospectively according to a specific study plan. If questionnaires or interviews about past ex-

posures are used, selective recall can be a source of bias. Furthermore, because retrospective studies typically rely on outcomes recorded for other purposes, the end point is more likely to be cause of death than an earlier marker of disease. In comparison to cross-sectional studies, of cohort studies have the advantage of including the entire population of interest. However, the difficulty of long-term follow-up means that some subjects are inevitably "lost."

Case-Control Studies

The principal advantage of the case-control study is its relative simplicity and relatively low cost. Case-control studies are valuable when multiple exposures are being explored in the etiology of a disease. If the investigator wishes to examine a spectrum of diseases associated with an exposure, such as lead, a cohort study is desirable; but if the interest is in the causes of a specific disease, such as bladder cancer, then the case-control study is more suitable.

Case-control studies are regarded as slightly more susceptible to biases than cohort studies (see below). For example, exposure information may be recalled differently by subjects with and without disease. Moreover, there is the need to identify a control group with the same general exposure history of the population that generated the cases.

PROBLEMS RELATED TO VALIDITY

Because epidemiologic studies are observational studies rather than randomized experiments, they are prone to biases, some of which are unavoidable. Careful consideration needs to be given to a study's validity (lack of bias). *Bias* is defined as a distortion of the measure of association between exposure and health outcome, such as an SMR or an OR. The degree to which the inferences drawn from a study are warranted is determined largely by the absence of bias. Reports of epidemiologic studies should provide sufficient information for the reader to understand what potential sources of bias were present and how these biases were addressed. There are three sources of bias: selection, misclassification, and confounding.

Selection Bias

Selection bias results from the inappropriate inclusion or exclusion of subjects in the study population. For example, in the past it was customary in studies of pulmonary function to exclude subjects who did not perform reproducible pulmonary function tests. It was subsequently discovered that subjects who had difficulty performing a reproducible forced expiratory maneuver had compromised respiratory health (13). The exclusion of such subjects could result in an overestimation of the respiratory health of a working population and possibly the underestimation of a dose-response association, if one exists.

Most types of selection bias, such as exclusion of short-term workers from the study population, cannot easily be corrected or controlled for in the analysis; they can only be prevented. To prevent selection bias in a cohort study, investigators should be kept unaware ("blinded") of cohort members' outcome status. Similarly, in case-control studies, investigators should be blinded as to the exposures status of cases and controls. Furthermore, selection of subjects should not be influenced by prior knowledge or suspicion of health outcome in a cohort study or of exposure status in a case-control study.

The most common type of selection bias in occupational epidemiologic studies is the "healthy worker effect" (HWE). This bias results from workers' selecting themselves out of the study groups rather than investigator error or oversight. For example, as described earlier, cross-sectional studies may result in an underestimate of the dose-response association if the occupational exposure causes disease which, in turn, causes workers to leave the workforce. Another example of HWE, common to cohort mortality studies, occurs because employed people are healthier than the general population, which includes the aged, the chronically ill, and those who are otherwise unfit to obtain and maintain employment.

As a result of the HWE, studies of illness or death among working populations often show lower rates of chronic diseases (e.g., cardiovascular diseases) than in the general population. In the mortality study of steelworkers described previously, the overall SMR, expected to be 1.00, was only .82 when the mortality rates of the surrounding county were used for comparison.

It is rare that an appropriate alternative comparison group of sufficient size is available. When possible, HWE bias is minimized by using a nonexposed comparison group drawn from within the study population. The HWE is reduced, although not necessarily eliminated, when analyses are based on this sort of "internal" comparison between exposed and nonexposed workers.

Misclassification

Misclassification (information bias) refers to an investigator's inadvertent placement of a worker into an incorrect category or group. Either disease or exposure can be misclassified. However, for purposes of illustration, the focus here is on exposure misclassification, because it is more relevant for occupational epidemiology. There are two types of misclassification: nondifferential and differential.

Exposure misclassification that is *nondifferential* is random misassignment of exposure that occurs regardless of disease status. Nondifferential misclassification is common in occupational studies, in which there is often little information on subjects' exposures and subjects cannot be well classified into exposure categories. The problem is generally worse in retrospective studies, because adequate documentation of historical exposures is more difficult. The net effect is to reduce a study's ability to detect exposure-disease associations when such associations truly exist. Thus, nondifferential misclassification may result in an existing occupational hazard going unrecognized.

Bias of a different sort is presented by *differential* misclassification, in which the likelihood of misassignment of exposure is related to disease status. This type of bias can result in either a stronger or a weaker association than truly exists. In cohort studies, differential misclassification is commonly prevented by keeping the investigators blinded to exposure status during collection of outcome information. In this manner, any errors in collection should be randomly distributed among both exposed and nonexposed groups. In case-control studies, control of differential bias is much more complicated; it is difficult for the investigator, and usually impossible for the subject, to be unaware of the disease status when exposure information is being obtained. Prevention of differential misclassification depends on collecting data as objectively as possible.

Confounding

Confounding is present when two study groups (e.g., exposed and nonexposed) are not comparable with respect to a characteristic that is *also* a risk factor for the disease. For example, in a study comparing stomach cancer in coal miners and iron miners, chewing tobacco was considered to be a potential confounder because (a) it is used more commonly by coal miners who are prohibited from smoking in coal mines and, (b) it may be an independent risk factor for stomach cancer.

Confounding can be controlled either in the design of the study or in the analysis of the data. In case-control studies, matching of study subjects on potential confounders in the design phase can facilitate control of confounding in the analysis. To control confounding in the example of the stomach cancer study, subjects could be matched on tobacco-chewing habits so that the proportion of tobacco chewers is the same among cases and controls.

Stratification is the major approach to control of confounding in the analysis phase of cross-sectional, cohort, and case-control studies. A confounder such as age is used to define strata (e.g., 10-year age groups). The exposure-response association is then estimated in each stratum. Stratification, however, becomes problematic as the number of confounders increases, because the strata become too small to allow stable measures of risk. For example, if age, smoking, race, and gender must be controlled for simultaneously, there may be no nonsmoking 40-to-45-year-old white females in the study population. In this case, stratification becomes an inadequate method of controlling confounding, and mathematical modeling must be used to control confounding statistically.

Multivariate models impose particular mathematical forms on the dose-response relations, such as a linear or exponential form. By restricting the data to a specific structure, one can interpolate between sparse strata. Mathematical modeling generally involves "smoothing" of the data the distributions of confounders and exposure categories.

INTERPRETATION OF EPIDEMIOLOGIC STUDIES

The interpretation of epidemiologic studies depends on the strength of the association, the validity of the observed association, and supporting evidence for causality (Box 6-1). The strength of an association usually is measured by the size of the relative risk in studies of discrete health outcomes such as cancer, or by the magnitude of the difference between groups in studies of physiologic pa-

---

Box 6-1. Guide for Evaluating Epidemiologic Studies

To assist health professionals in reading, understanding, and critically evaluating epidemiologic studies, the following questions, adapted from Monson's *Occupational Epidemiology* (15), should serve as a useful guide.

Collection of Data

1. What were the *objectives* of the study? What was the association of interest?
2. What was the primary *outcome* of interest? Was it accurately measured?
3. What was the primary *exposure* of interest? Was it accurately measured?
4. What *type of study* was conducted?
5. What was the *study base*? Consider the process of subject selection and sample size.
6. *Selection bias:* Was subject selection based on the outcome or the exposure of interest? Could the selection have differed with respect to other factors of interest? Were these likely to have introduced a substantial bias?
7. *Misclassification:* Was subject assignment to exposure or disease categories accurate? Were possible misassignments equally likely for all groups? Were these likely to have introduced a substantial bias?

8. *Confounding:* What provisions, such as study design and subject restrictions, were made to minimize the influence of external factors before analysis of the data?

Analysis of the Data

9. What *methods* were used to control for confounding bias?
10. What *measure of association* was reported in the study? Was this appropriate?
11. How was the *stability* of the measure of association reported in the study?

Interpretation of Data

12. What was the *major result* of the study?
13. How was the interpretation of this result affected by the previously noted *biases*?
14. How was the interpretation affected by any nondifferential *misclassification*?
15. To what larger population may the results of this study be *generalized*?
16. Did the *discussion* section adequately address the limitations of the study? Was the final conclusion of the paper a balanced summary of the study findings?

---

rameters such as FEV[1]. Further evidence of the strength of an effect is provided by dose-response relations, in which the effect estimate rises over increasing categories of exposure.

When an association appears to be present, the validity of the association must be evaluated. This can be done in studies that provide adequate detail on design and results. The internal validity should be evaluated by examining for selection bias, by misclassification, and confounding. All studies suffer to some degree from problems with validity, so a judgment must be made concerning the importance of the biases. The important biases are those that could explain the findings—that is, biases large in magnitude and operating in the direction of the finding (away from the null in positive studies, toward the null in negative studies).

Finally, the consistency of the association—that is, the repeated demonstration of a particular association in different populations and by different investigators—is valuable supporting evidence that the association truly exists. Toxicology data and reasonable consistency with a postulated biologic mechanism may also assist in determining causal associations.

Results of statistical tests of significance, probability values (p-values) or confidence intervals, usually are presented along with estimates of the relative risk. These results contribute to interpretation of studies by providing a measure of stability of the associations reported. Statistical tests evaluate the probability that the observed association could have occurred by chance alone (assuming that no effect is expected *a priori*). For example, a p-value less than .05 indicates that the likelihood of observing an effect at least as large as the one actually observed is less than 5%, given that no association truly exists. Some investigators define significance as a probability value less than .01; others require only that it be less than .10. Confidence intervals provide more information than probability values alone because they provide information on the magnitude of the association as well as the stability of the estimate.

The statistical power of a study to detect a true effect depends on the background prevalence of the disease or exposure, the size of the group studied, the length of follow-up, and the level of statistical significance required. Monitoring of a small cohort for a brief period can yield a falsely negative result. For this reason, it is important, when interpreting a negative study, to examine whether the design itself precluded a positive finding. For example, a retrospective cohort study of formaldehyde exposure had only 80% power to detect a fourfold risk in nasal cancer mortality, despite having 600,000 person-years of observation (14). The power was low because nasal cancer has a very low background prevalence. Formulas for calculating the statistical power associated with a given sample size are available in standard biostatistics and epidemiology texts.

REFERENCES

1. Blot WJ, Harrington JM, Toledo A, et al. Lung cancer after employment in shipyards during World War II. N Engl J Med 1978;299:620–624.
2. Corn M, Esmen NA. Workplace exposure zones for classification of employee exposures to physical and chemical agents. Am Ind Hyg Assoc J 1979;40: 47–60.
3. Ayer HE, Dement JM, Busch KA, et al. A monumental study: reconstruction of a 1920 granite shed. Am Ind Hyg Assoc J 1973;34:206–211.
4. Hu H, Rabinowitz M, Smith D. Bone lead as a biological marker in epidemiologic studies of chronic toxicity: conceptual paradigms. Environ Health Perspect 1998;106:1–8.
5. Creech JL, Johnson MN. Angiosarcoma of liver in the manufacture of polyvinyl chloride. J Occup Med 1974;16:150–151.
6. Billmaier D, Yee HT, Allen N, et al. Peripheral neuropathy in a coated fabrics plant. J Occup Med 1974;16:668–671.
7. Speizer FE, Wegman DH, Ramirez A. Palpitation rates associated with fluorocarbon exposure in a hospital setting. N Engl J Med 1975;292:624–626.
8. Redmond CK, Wieand HS, Rockette HE, et al. Long term mortality experience of steelworkers. Department of Health and Human Services, U.S. Government Printing Office, Washington, DC, (NIOSH) Publication No. 81-120.1981.
9. Hemminki K, Mutanen P, Saloniemi I, Niemi M-L, Vainio H. Spontaneous abortions in hospital staff engaged in sterilizing instruments with chemical agents. Br Med J 1982;285:1461–1463.
10. Wegman DH, Musk AW, Main DM, et al. Accelerated loss of FEV-l in polyurethane production workers: a four-year prospective study. Am J Ind Med 1982;3:209–215.
11. Zahm SH, Weisenburger DD, Saal RC, Vaught JB, Babbitt PA, Blair A. The role of agricultural pesticide use in the development of non-Hodgkin's lymphoma in women. Arch Environ Health 1993;48:353–358.
12. Armstrong B, Theriault G, Guenel P, Deadman J, Goldberg M, Heroux P. Association between exposure to pulsed electromagnetic fields and cancer in electric utility workers in Quebec, Canada and France. Am J Epidemiol 1994;140:805–820.
13. Eisen EA, Robins JM, Greaves IA, Wegman DH. Selection effects of repeatability criteria applied to lung spirometry. Am J Epidemiol 1984;120:734–742.
14. Blair A, Stewart P, O'Berg M, et al. Mortality among industrial workers exposed to formaldehyde. J Natl Cancer Inst 1986;76:1071–1084.
15. Monson RR. Occupational epidemiology, 2nd ed. Boca Raton, FL: CRC Press, 1989.
16. Doll R, Hill AB. Mortality in relation to smoking: ten years' observations of British doctors. Br Med J 1964;1:1399.

BIBLIOGRAPHY

Ahlbom A, Norell S. Introduction to modern epidemiology, 2nd ed. Chestnut Hill, MA: Epidemiology Resources Inc, 1990.

Beaglehole R, Bonita R, Kjellstrøm T. Basic epidemiology. Geneva: World Health Organization, 1993.

Two introductory texts on the core ideas underlying epidemiologic research and useful starting points for more advanced reading. The second book is available worldwide (through WHO), and a teacher's guide can be obtained for use with the text.

Checkoway H, Pearce NE, Crawford-Brown DJ. Research methods in occupational epidemiology. New York: Oxford University Press, 1989.

Very readable full text on epidemiologic approaches specific to occupational studies. Numerous examples are provided to guide the reader in understanding both the simple and the complex issues that must be addressed.

Hernberg S. Introduction to occupational epidemiology. Chelsea, MI: Lewis Publishers, 1992.

An excellent introductory text that is well written and illustrated. Aimed at the reader new to occupational epidemiology but somewhat familiar with principles of epidemiology.

Monson RR. Occupational epidemiology, 2nd ed. Boca Raton, FL: CRC Press, 1989.

A systematic review of methods as applied specifically to occupational settings. A practical textbook for those doing occupational studies.

Olsen J, Merletti R, Snashall D, Vuylsteek K. Searching for causes of work-related diseases: an introduction to epidemiology at the work site. Oxford: Oxford Medical Publications, 1991.

A practical introduction to epidemiology for health professionals with no formal training in the discipline. It is written to assist professionals to better plan and carry out investigation of worksite health problems.

Pagano M, Gauvreau K. Principles of biostatistics. Belmont, CA: Duxbury Press, 1993.

Basic statistics text written in a reasonable fashion with a functional index. Good general reference for statistics.

Rothman KJ, Greenland S. Modern epidemiology, 2nd ed. Philadelphia: Lippincott-Raven, 1998.

Probably the best general text on epidemiologic methods designed both for the novice and the expert. Provides principles of epidemiology in substantial detail as well as the quantitative basis for the research methods. Particularly useful as a reference.

Steenland K. Case studies in occupational epidemiology. New York: Oxford University Press, 1993.

Provides the reader the opportunity to explore further many of the questions discussed in this chapter through practical and detailed presentation of case studies of various types of epidemiologic studies.

APPENDIX
Adjustment of Rates

For purposes of illustration, adjusting for differences in age is examined in detail. Table 6-3 presents a hypothetical problem involving the myocardial infarction experience in two viscose rayon factories. To compare the incidence of myocardial infarction, a summary rate is calculated for each factory. If crude rates were calculated, it would appear that workers in factory 2 have a slightly greater risk. Comparison of these rates, however, ignores the rather striking difference in age distribution of the populations in the two factories. These can be taken into account by adjusting for age differences by either the direct method or the indirect method.

Direct Adjustment

The principle of direct adjustment is to apply the age-specific rates determined in the study groups to a set of common age weights, such as a standard age distribution. The selection of the standard is somewhat arbitrary, but often the sum of the specific age groups for the study groups is chosen. In Table 6-3, the standard population is 1,200 persons younger than 45 years and 800 persons 45 years or older. The specific rates are applied to this set of weights and then added to create an adjusted rate.

$$\text{Factory 1} = \frac{(.010 \times 1{,}200) + (.030 \times 800)}{2{,}000}$$

$$= .018$$

$$\text{Factory 2} = \frac{(.0125 \times 1{,}200) + (.050 \times 800)}{2{,}000}$$

$$= .0275$$

Not only is the magnitude of the rate of myocardial infarction affected by the adjustment procedure, but the rank order is reversed. Note that if another age distribution had been selected as the standard, the standardized rates would change. For example, for 1,500 persons younger than 45 years and 500 age 45 or older, the rate for factory 1 would become .015 and that for factory 2 would become .022. Although the absolute magnitudes of the two adjusted rates have no inherent meaning, the relative magnitudes do. While the size of the ratio will change slightly, it will be closely duplicated regardless of the weights. In these two examples of weighting, the ratios of the adjusted rates are 1.53 and 1.47.

INDIRECT ADJUSTMENT

In indirect adjustment, standard rates are applied to the observed weights or the distribution of specific characteristics (e.g., age, sex or race) in the study populations. This provides a value for the number of cases (events) that would be expected if the standard rates were operating. The expected number of cases can be compared with the number actually observed for each study group in the form of a ratio. In Table 6-3, assume a national standard rate for myocardial infarction of 1 in 1,000 (.001) for those younger than 45 years of age and 2 in 1,000 (.002) for those 45 years or older. The expected number of cases in the two factories would then be as follows:

$$\text{Factory 1} = (.001 \times 400) + (.002 \times 600)$$
$$= 1.6$$
$$\text{Factory 2} = (.001 \times 800) + (.002 \times 200)$$
$$= 1.2$$

These expected values are compared with the observed values to calculate a standardized morbidity ratio, as follows:

$$\text{Factory 1 SMR} = \frac{22}{1.6} = 13.8$$

$$\text{Factory 2 SMR} = \frac{20}{1.2} = 16.7$$

It is tempting to compare the two SMRs and calculate a ratio similar to that calculated for the directly standardized rates. However, a drawback of indirect standardization is that SMRs cannot be compared. Because the age distributions and age-specific rates are significantly different for the two factories, the resulting comparison of the two SMRs would not distinguish differences caused by a different disease incidence rate from differences caused by a different age distribution.

It is reasonable, then, to ask why indirectly standardized rates are used. One reason is that often only one population is being studied, so comparison with the general population experience is convenient and possibly the only reasonable comparison available. Probably of greater importance is the instability of observed rates. In the example presented here, if five rather than two age groups were used and it was also necessary to adjust for both race and sex, then the total number of subdivisions necessary would be $5 \times 2 \times 2 = 20$. With a maximum of 22 cases in either factory, several of the subdivisions would contain no cases and therefore have no reliable rate estimate. Even in the illustration provided, one case more or one case less among the group of younger workers in factory 1 would have changed the age-specific incidence rate to 12.5 or 7.5, respectively, a very large difference.

Occupational Health: Recognizing and Preventing Work-Related Disease and Injury, 4th ed.
Edited by Barry S. Levy and David H. Wegman, Lippincott Williams & Wilkins, Philadelphia © 2000.

7 Occupational Hygiene

Thomas J. Smith and Thomas Schneider

Occupational hygiene (industrial hygiene) is the environmental science of anticipating, recognizing, evaluating, and controlling health hazards in the working environment with the objectives of protecting workers' health and well-being and safeguarding the community at large. It encompasses the study of chronic and acute conditions emanating from hazards posed by physical agents, chemical agents, biologic agents, and stress in the occupational environment as well as concern for the outdoor environment. For example, an occupational hygienist determines the composition and concentrations of air contaminants in a workplace where there have been complaints of eye, nose, and throat irritation and determines whether the contaminant exposures exceed the permissible exposure limits set by the Occupational Safety and Health Administration (OSHA) or other national limits. If the problem is caused by airborne materials, which might be determined in consultation with a physician or epidemiologist, then the hygienist would be responsible for (a) selecting the techniques to be applied to reduce or eliminate the exposure, such as installing exhaust ventilation around the source of the air contaminants and isolating the source from the general work area, and (b) performing follow-up sampling to verify that the controls were effective.

Harvard School of Public Health, Boston, Massachusetts 02115.

T. Schneider: National Institute for Occupational Health, Copenhagen, Denmark.

Most occupational hygienists have earned either a bachelor's degree in science or engineering or a master of science degree in industrial hygiene. Occupational hygienists tend to specialize in specific technical areas because the scope of the field has so greatly expanded. Occupational hygienists must work with physicians to develop comprehensive occupational health programs and with epidemiologists to perform research on health effects. It has been traditional to separate occupational hygiene and occupational safety, but more recently the trend has been to broaden the training for each discipline to include that of the other. This has led to the specialty of risk management for evaluating and controlling all types of workplace hazards. At present, occupational hygienists generally do not deal with mechanical hazards or job activities that can cause physical injuries; these are the responsibility of safety specialists (see Chapter 8). However, in private companies it is not uncommon for a single individual to be responsible for both occupational hygiene and safety and for this person to have no formal training in either area.

Most occupational hygienists work for large companies or governmental agencies. A small but growing number work for labor unions. In any case, occupational hygienists often are located in organizational units where they have little power to bring about necessary changes. Hygienists who work for labor unions may be restricted in their access to the workplace for sampling and exposure measurements, which can limit their ability to assess and control hazards.

The closeness of working relationships between occupational hygienists and occupational physicians varies. Some have close collaborative activities with an extensive exchange of information, while others operate with almost complete independence and have little more than formal contact. A physician who is familiar with the workplace, job activities, and health status of workers in all parts of the process may be very helpful in guiding the occupational hygienist's assessment of environmental hazards, and vice versa. Within a framework of multidisciplinary approaches, occupational hygienists and physicians should collaborate with safety specialists, workers in production units, staff members in personnel departments, worker representatives, and delegates of health and safety committees. When contact among these groups is minimal, many opportunities are lost for improving the effectiveness of health hazard control and the prevention of adverse effects.

The integration of occupational hygiene into an overall program for occupational health is an important issue that may limit the effectiveness of intervention strategies. An effective hygiene program must involve good working relationships among production, personnel, and health and safety departments and strong support from upper management. Core activities of hazard anticipation, recognition, evaluation, and control must be well integrated with the day-to-day activities of the enterprise. There is no single organizational structure that is optimal (see Chapter 14).

ANTICIPATION, RECOGNITION, EVALUATION, AND CONTROL OF HAZARDS

Formal strategies for workplace assessment have not been well developed. The American Hygiene Association (AIHA) Industrial monograph on exposure assessment, *A Strategy for Occupational Exposure Assess-ment*, (1), is focused on sampling and does not address program management issues. European Union regulations now require workplace assessment to identify occupational hazards. However, the formal mechanism has not been established nor validated to demonstrate its consistency. It is expected that European Union regulations when fully implemented will be beneficial to small and medium-sized enterprises because they will create awareness and an expectation of controlling the problems identified.

Anticipation

Anticipation of hazards has become an important responsibility of the occupational hygienist. *Anticipation* refers to the application of and mastery of knowledge that permits the occupational hygienist to foresee the potential for disease and injury. The occupational hygienist should therefore be involved at an early stage in the planning of technology, process development, and workplace design.

> An electronics company was developing a new process for making microcomputer chips. The process involved dissolving a photographic masking agent in toluene and then spraying the mask on a large surface covered with chips. The company's hygienist noted that this would expose the workers to potentially high airborne levels of toluene. She suggested they substitute xylene, which has a lower vapor pressure, and modify the process to use smaller amounts of solvent, which would reduce the amount of hazardous waste generated by the process.

Commonly, process engineers or industrial researchers propose use of hazardous materials or do not consider the interaction of the worker and the process or machine. Hygienists can prevent many problems that would be expensive to fix after installation by reviewing early plans and findings of pilot plant experiments.

An overview of the production process may be most easily obtained by describing the complete flow from raw material to final

TABLE 7-1. *Common unit processes and associated hazards by route of entry**

Unit process	Route of entry and hazard
Abrasive blasting (surface treatment with high-velocity materials, such as sand, steel shot, pecan shells, glass, or aluminum oxide)	Inhalation: silica, metal, and paint dust Noise
Acid/alkali treatments (dipping metal parts in open baths to remove oxides, grease, oil, and dirt)	
Acid pickling (with HCl, HNO$_3$, H$_2$SO$_4$, H$_2$CrO$_4$, HNO$_3$/H$_2$SO$_4$)	Inhalation: acid mist with dissolved metals Skin contact: burns and corrosion, HF toxicity
Acid bright dips (with HNO$_3$/H$_2$SO$_4$)	Inhalation: NO$_2$, acid mists
Molten caustic descaling bath (high temperature)	Inhalation: smoke and vapors Skin contact: burns
Blending and mixing (powders and/or liquids are mixed to form products, or undergo reactions)	Inhalation: dusts and mists of toxic materials Skin contact: toxic materials
Cleaning (application of cleansers, solvents and strong detergents to clean surfaces and articles; operation of devices to aid cleaning such as floor washers, waxers, polishers and vacuums)	Inhalation: dust, vapors Skin contact: defatting agents, solvents, strong bases
Crushing and sizing (mechanically reducing the particle size of solids and sorting larger from smaller with screens or cyclones)	Inhalation: dusts and mists of toxic materials Noise
Degreasing (removing grease, oil, and dirt from metal and plastic with solvents and cleaners)	
Cold solvent washing (cleaning parts with ketones, cellosolves, and aliphatic, aromatic, and stoddard solvents	Inhalation: vapors Skin contact: dermatitis and absorption Fire and explosion (if flammable) Metabolic: carbon monoxide formed from methylene chloride
Vapor degreasers (with trichloroethylene, methyl chloroform, ethylene dichloride, and certain fluorocarbon compounds)	Inhalation: vapors, thermal degradation may form phosgene, hydrogen chloride, and chlorine gases Skin contact: dermatitis and absorption
Electroplating (coating metals, plastics and rubber with thin layers of metals, such as copper, chromium, cadmium, gold, or silver)	Inhalation: acid mists, HCN, alkali mists, chromium, nickel, cadmium mists Skin contact: acids, alkalis Ingestion: cyanide compounds
Forging (deforming hot or cold metal by presses or hammering)	Inhalation: hydrocarbons in smokes (hot processes), including polyaromatic hydrocarbons, SO$_2$, CO, NO$_x$, and other metals sprayed on dies (e.g., lead and molybdenum) Heat stress Noise
Furnace operations (melting and refining metals; boilers for steam generation)	Inhalation: metal fumes, combustion gases (e.g., SO$_2$, CO) Noise from burners Heat stress Infrared radiation, cataracts in eyes
Grinding, polishing, and buffing (an abrasive is used to remove or shape metal or other material)	Inhalation: toxic dusts from both metals and abrasives Vibration from hand tools Noise
Industrial radiography (x-ray or gamma ray sources used to examine parts of equipment)	Radiation exposure
Machining (metals, plastics, or wood are worked or shaped with lathes, drills, planers, or milling machines)	Inhalation: airborne particles, cutting oil mists, toxic metals, nitrosamines formed in some water based cutting oils, solvents Noise
Materials handling and storage (conveyors, forklift trucks are used to move materials to/from storage)	Inhalation: CO, exhaust particulate, dusts from conveyors, emissions from spills or broken containers
Mining (drilling, blasting, mucking to remove loose material, and material transport)	Inhalation: silica dust, NO$_2$ from blasting, gases from the mine Vibration stress Heat stress Noise

(continued)

TABLE 7-1. *Continued*

Unit process	Route of entry and hazard
Painting and spraying (applications of liquids to surfaces, for example, paints, pesticides, coatings)	Inhalation: solvents as mists and vapors, toxic materials
	Skin contact: solvents, toxic materials
Repair and maintenance (servicing malfunctioning equipment; cleaning production equipment and control systems)	Inhalation: dusts, vapors, and gases from the operation
	Skin contact: grease, oil, solvents
Quality control (collection of production samples, performace of test procedures that produce emissions)	Inhalation: dusts, vapors and gases
	Skin contact: solvents
Soldering (joining metals with molten lead or silver alloys)	Inhalation: lead or cadmium particulate (fumes) and flux fumes
Welding and metal cutting (joining or cutting metals by heating them to molten or semi-molten state)	Inhalation: metal fumes, toxic gases and materials, flux particulate, etc.
Arc or resistance welding	Noise: from burner
Flame cutting and welding	Eye and skin damage from infrared and ultraviolet
Brazing	radiation

* The health hazards may also depend on the toxicity and physical form (gas, liquid, solid, powder) of the materials used.

Adapted from WA Burgess. Recognition of health hazards in industry: a review of materials and processes, 2nd ed., New York: Wiley & Sons, 1995.

product. Production can be subdivided into its component unit processes. In this stepwise fashion, the processes involving hazards can be recognized, worker exposures can be evaluated, and the exposures in nearby areas can be assessed. Examples of some common unit processes and their hazards are given in Table 7-1. This general approach and the hazards of a wide range of common industrial processes are discussed in more detail in Burgess' *Recognition of Health Hazards in Industry: A Review of Materials and Processes* (2) (see Bibliography).

This approach can be illustrated by considering a small company that manufactures tool boxes from sheets of steel by a six-step process: (a) sheets of steel are cut into the specified shape; (b) sharp edges and burrs are removed by grinding; (c) sheets are formed into boxes with a sheet metal bender; (d) box joints are spot-welded; (e) boxes are cleaned in a vapor degreaser in preparation for painting; and (f) boxes are painted in a spray booth.

Production steps b, d, e, and f use unit processes with known sources of airborne emissions, and their hazards, which are given in Table 7-1, should be evaluated. It may also

be necessary to evaluate the exposures of workers involved with steps a and c, because they may be located near enough to the operations with hazards to have significant exposures.

The design of job tasks and an individual's work habits can both have an important influence on exposures. For example, a furnace tender's exposure to metal fumes depends on the length of the tools used to scrape slag away from the tapping hole in the furnace and on the instructions for performing the task. Lack of adequate tools or sufficient operating instructions may cause excessive exposure to fumes emitted by molten materials. Similarly, the furnace tender who is positioned close to the slag as it runs out of the furnace may receive a much higher exposure to fumes than a coworker who stands farther away from the slag. Therefore, an important part of an evaluation is the observation of work practices used in hazardous unit processes.

Recognition

Recognition of problems in a new or unfamiliar workplace generally requires that the occupational hygienist engage in the collection of background information on production layout, processes, and raw materials.

Visits to the workplace to become familiar with the production processes and their hazards are crucial for detecting unique aspects of the workplace that may strongly affect exposures. Information is collected on (a) the type, composition, and quantities of substances and materials, including raw materials, intermediate products, and additives; (b) design of work processes and tasks; (c) emission sources; and (d) design and capacity of ventilation systems or other control measures.

Flow visualization with smoke tubes (glass tubes with a packing that produces dense white smoke when air is forced through it) can yield information on the effectiveness of local exhausts or process ventilation.

Work practices, worker positions in relation to sources, and task duration are recorded. Cleaning routines and performance and general tidiness are important determinants of exposure.

> Farm workers were experiencing episodes of depressed blood cholinesterase levels from exposure to organophosphates despite the fact that they were observing the required waiting times before reentry into sprayed fields and were wearing long-sleeved shirts and gloves to prevent skin contact. The pesticide had a very low vapor pressure, so there was no significant inhalation exposure. However, it was known that environmental moisture decomposes this type of pesticide. Because the weather was very dry during these episodes, there was concern that the pesticide was not decomposing as rapidly as expected. Consequently, despite the skin protection, there could still be sufficient skin absorption of the pesticide to affect cholinesterase levels. Skin sampling with patches showed that fine dust was sifting through the cloth of the shirt sleeves and depositing pesticide on the workers' arms in substantial quantities. The problem was solved by extending the standard reentry times.

If the initial appraisal cannot definitely rule out a hazard, a basic survey must be performed to provide quantitative information concerning exposure of workers. Particular account must be taken of tasks with high exposure. Sources of information include (a) earlier measurements, (b) measurements from comparable installations or work processes, (c) reliable calculations or modeling based on relevant quantitative data, and (d) air sampling to determine the range of exposures.

Sampling may show that sensory impressions underestimate or overestimate exposures; for example, the odor threshold for most solvents is well below the level at which they present a toxic exposure hazard.

If this information is insufficient to enable valid comparisons to be made with the limit values, a full-scale survey must be performed. The full-scale survey examines all phases of workplace activities, both normal activities and abnormal or infrequent ones, such as maintenance, reactor cleaning, or simulation of malfunctions. The survey activities can take several weeks or months in a complex manufacturing or chemical plant.

Evaluation

The evaluation of recognized or suspected hazards by the hygienist uses techniques based on the nature of the hazards, the emission sources, and the routes of environmental contact with the worker. For example, air sampling can show the concentrations of toxic particulates, gases, and vapors that workers may inhale; skin wipes can be used to measure the degree of skin contact with toxic materials that may penetrate the skin; biologic samples (blood or urine) can provide data when there are multiple routes of entry; and noise dosimeters record and electronically integrate workplace noise levels to determine total daily exposure. Both acute and chronic exposures should be considered in the evaluation, because they may be associated with different types of adverse health effects.

The workplace is not a static environment: exposures may change by orders of magnitude within short distances from exposure sources (e.g., welding activities) or over short time intervals because of intermittent source output or incomplete mixing of air contaminants. In addition, operations and materials used or produced commonly change, as do job titles and definitions. The nature of these changes and their possible effects must be recognized and taken into consideration by the occupational hygienist.

> In the 1940s, a company that produced cadmium pigment and other products decided to operate a long-term monitoring program for cadmium exposures because of their concern about the chronic effects of cadmium. This program was thoughtfully designed around existing sampling and analysis methodologies. It used routine fixed-location air sampling with chemical analysis of cadmium in the dust. All of the work areas where cadmium dust might have been present were sampled at the same location on a regular schedule so that the time profile of exposure could be determined. Later, when more sophisticated methods for personal sampling were developed, both methods were used for several years; this allowed the personal exposures in each area to be correlated with the fixed location data. As a result, it was possible to develop a clear picture of the inhalation exposures of the workers and to estimate their personal doses of cadmium over a substantial period. These chronic exposure estimates were critical for studies of exposure-response relations for pulmonary and renal effects of the cadmium exposures.

All monitoring programs for both long-term and acute problems should be structured with a clear focus on the individual worker's sources of exposures and the ultimate objective of estimating dose. Monitoring organized solely around compliance with today's standards will probably be unable to answer tomorrow's questions about hazards associated with personal exposures. The effects of environmental controls, such as ventilation and personal protective equipment (PPE), that intervene between the emission source and the worker must also be considered.

The hygienist's decision on whether a hazard is present is based on three sources of information:

1. Scientific literature and various exposure limit guidelines such as the American Conference of Governmental Industrial Hygienists (ACGIH) Threshold Limit Values (TLVs), (3)[1] a set of consensus standards developed by occupational hygienists, toxicologists, and physicians from governmental agencies and academic institutions, or the recommendations of the World Health Organization (WHO)[2]
2. The legal requirements of OSHA (in some cases these are less stringent than the TLVs because the TLVs have been updated) or legal requirements of other countries (4)
3. Interactions with other health professionals who have examined the exposed workers and evaluated their health status.

In cases where health effects are present but exposures do not exceed the TLVs, WHO recommendations, or OSHA or other national requirements, the prudent hygienist nevertheless concludes that there is a relation between adverse health effects and the workplace exposures if such a conclusion is consistent with the facts. Exposure limits are designed to prevent adverse effects in most exposed workers, but they are not absolute levels below which effects cannot occur. The supporting data for many of these exposure limits are sometimes insufficient, out-of-date, or based too much on evidence of acute toxic effects and not enough on recent evidence of carcinogenicity, mutagenicity, or teratogenicity.

Control

Once a hazard has been identified and the extent of the problem evaluated, the hygienist's next step is to design a control strategy or plan to reduce exposure to an acceptable

---

[1] These guidelines may be obtained from the American Conference of Governmental Industrial Hygienists, 6500 Glenway Avenue, Building D-5, Cincinnati, OH 45211.
[2] These recommendations may be obtained from the World Health Organization, Avenue Appia 20, 1211 Geneva 27, Switzerland.

level. Such controls may include the following:

1. Changing the industrial process or the materials used to eliminate the source of the hazard, such as changing to clean technologies
2. Isolating the source and installing engineering controls such as ventilation systems
3. Using administrative directives to limit the amount of exposure a worker receives, or, as a final resort, to require the use of PPE (see Chapter 5).

The controls in the last of these three approaches are less reliable because they depend on enforcement by managers and conscientious application by the workers, either of which can fail. Usually the control strategy includes a combination of these approaches, particularly if there are delays in installing engineering controls.

In designing control strategies, account should also be taken of the environmental impact of emissions, waste, accidents, storage, spills, and leaks. Action can be taken at the process, materials, component, system, or workplace levels. Education of both workers and supervisors is an important part of any control strategy; both must understand the nature of the hazards and support the efforts taken to control or eliminate them. Implementation of control measures should be supervised and their efficiency evaluated.

Automobile manufacturers have been very concerned about the hazards of coolants used in machining and grinding operations. Workers complain of skin and inhalation problems associated with exposures to liquids splashed on their skin and mists in the air. In the recent past controls were installed based on hypotheses about the causal factors but without an investigation of the specific causes for exposures. As is often the case, some hypotheses were later found to be incorrect, and it was determined that incomplete control had been achieved despite substantial expenditures. Inhalation exposures had been only partially controlled by local exhaust ventilation and enclosure of processes; a relation was still found between symptoms and reduced pulmonary function associated with exposures at levels below the current allowable exposure. Analysis of the coolants also revealed that material safety data sheets (MSDSs) were inaccurate and more hazardous materials were being used than were known to the machining department. Investigations are now underway to determine what engineering controls will be needed to further reduce the exposures. Substitution of alternate types of coolant components and better control of microbial contaminants are part of the planned investigations.

This example indicates that controlling hazards in large, complex manufacturing operations is frequently a stepwise process. Control strategies are most effective when they are based on complete knowledge of the nature of the problems.

After hazards are controlled, the hygienist may recommend a routine hazard surveillance program to ensure that controls remain adequate. This type of surveillance is most effective when it is done in close association with a medical surveillance program designed to detect subtle effects that may occur at low levels of exposure.

The following sections indicate how assessment and control techniques are used. The approach for toxic materials is used as a paradigm that can be applied to other environmental hazards, including noise, vibration, ionizing and nonionizing radiation, temperature extremes, poor lighting, and infectious agents.

TOXIC MATERIALS

Exposure Pathways

The hazard of a given exposure to a toxic material depends on the toxicity of the substance and on the duration and intensity of contact with the substance. Adverse effects can result from chronic low-level exposure to a substance or from short-term exposure to a dangerously high concentration of it. However, the pharmacologic mechanisms by which effects are caused differ for acute and chronic effects. Occupational hygienists are concerned with both long-term, low-level exposures and brief, acute exposures, and they strive to characterize both.

In assessing a given hazardous material, the hygienist determines the route of exposure by which workers contact the material and by which it may enter their bodies. There are four major routes of exposure: (a) direct contact with skin or eyes; (b) inhalation, with deposition in the respiratory tract; (c) inhalation, with deposition in the upper respiratory tract, subsequent transport to the throat, and ingestion; and (d) direct ingestion with gastrointestinal uptake from food or drink. In the workplace, several concurrent routes of exposure may occur for a single toxic substance.

Inhalation of airborne particulates, vapors, or gases is by far the most common route of exposure, and this topic is the focus of many of a hygienist's assessment and control activities. Skin absorption may be important if the substance is lipid soluble or the skin's barrier is damaged or otherwise compromised. Ingestion of contaminated food and drink is a problem, especially for particulate and liquid materials, whose degree of risk may depend on the worker's level of awareness of the hazard and personal hygiene habits and on the availability of adequate facilities for washing and eating at the workplace. Contamination of cigarettes with toxic materials and their subsequent inhalation is also a problem for some substances.

For example, workers handling lead ingots are exposed to a low-level hazard from eating contaminated food or inhaling fumes from contaminated cigarettes. However, workers refining lead at temperatures above 800°F are exposed to a serious hazard from inhaling large amounts of lead fumes if they work close to unventilated refining kettles for several hours daily. Workers handling liquid nitric acid are exposed to the hazard of direct contact with the liquid on their skin, but they may also be exposed to a respiratory hazard from inhaling acid mist generated by an electroplating process using the nitric acid. In these examples, the toxic materials cause different types and magnitudes of hazards because their physical forms vary: solid material versus small-diameter airborne particulates, and liquid material versus airborne droplets.

Anticipation and Recognition

The first problem the hygienist faces in evaluating an unfamiliar workplace for toxic hazards is the identification of toxic materials. In many cases, such as a lead smelter or pesticide manufacturing process, the emission sources for toxic materials are clearly evident. But even in these examples some hazards may not be evident without careful examination of an inventory of the chemicals to be used or in use in the facility, including raw materials, byproducts, products, wastes, solvents, cleaners, and special-use materials. Lead smelter workers are also exposed to carbon monoxide and sometimes to arsenic and cadmium; pesticide workers are subject to solvent exposures. Relatively nontoxic chemicals may be contaminated with highly toxic ones; for example, low-toxicity chlorinated hydrocarbons used in weed killers (e.g., 2,4-T) may contain dioxin, which is highly toxic, and technical-grade toluene may contain significant amounts of highly toxic benzene. In some cases, toxic materials may not be hazardous because there are no emissions into the workplace and only small amounts are handled, such as in a chemical laboratory.

MSDSs, which list the compositions of commercial products, are available from manufacturers and can be useful, but they are sometimes too general or out of date. Toxicity data on specific substances can be obtained by literature searches (manual or electronic) or by searches of toxic data indices (see Appendix A).

Because exposure to toxic substances can occur through contamination of food, drink, or cigarettes, the hygienist determines whether eating and drinking facilities are physically well separated from the work area, whether the facilities for washing are close enough to the eating area, and whether sufficient time is permitted for workers to use both of these facilities. Protective clothing and facilities for showering at the end of a work shift should also be provided. Workers' understanding of hazards from toxic materials they are using must also be assessed. Finally, the hygienist determines the existence and enforcement of rules prohibiting eating, drinking, and smoking in areas with toxic substances.

Evaluation

Measurement Techniques. Two types of environmental sampling techniques are available.

Direct-reading instruments have sensors that detect the instantaneous air concentration; they may produce a reading on a dial, or they may store a complete 8-hour time profile in a small datalogger for later retrieval. Some are expensive, and all require careful calibration and maintenance to obtain accurate data. The *detector tube* is another type of direct-reading instrument of considerable use in determining approximate concentrations of air contaminants. This simple device uses a small hand pump to draw air through a bed of reagent in a glass tube that changes color or develops a length of stain that is proportional to the concentration of a given gaseous air contaminant. The conventional tube is suitable for short-term sampling (e.g., 10 minutes), but short-term samples can misrepresent long-term average exposures. Tubes with 8-hour collection times are available that are capable of measuring time-weighted average (TWA) exposure levels. Detector tubes are available that have been manufactured under strict quality control, and their degree of measurement uncertainty is specified. Consideration must always be given to interference from other substances (cross-sensitivity), which usually is specified on the tube's data sheets.

Sample collectors that remove substances from the air for analysis in a laboratory are less expensive than direct-reading instruments. *Personal sampling* is a common approach used by the occupational hygienist to obtain accurate and precise measurements of workers' exposures to particulate and gaseous air contaminants. The worker wears the sampler like a radiation dosimeter. Particulate contaminants are collected with filters, and gases and vapors are collected by solid adsorbents or liquid bubblers. The sampling apparatus is generally quite simple, consisting of a small air pump usually worn on a worker's belt, connected by tubing to the collector, and attached to the worker's lapel. (Some gas and vapor collectors are passive, using diffusion instead of an air pump to move the contaminant into the sampler.) With the appropriate selection of a gas or particulate collector, or both combined in a sampling train, it is possible to measure the average concentration of an air contaminant in the worker's breathing zone during an 8-hour work shift.

Collection devices for toxic particulates may capture either total dust (all particle sizes that can enter the collector) or only the respirable dust (only particles that can penetrate the terminal airways and alveolar spaces—less than 5 μm). Total particulate samples are collected if the toxic substance causes systemic health problems, as lead and pesticides do. Respirable dust samples are collected if the particulate causes chronic pulmonary disease, such as pneumoconiosis. There is some controversy about the size of particles that cause chronic bronchitis and, therefore, which type of sample to collect. However, the type of sampler should be matched to the route of entry, type of effect, and target tissue.

Charcoal and other sorbents packed into tubes have been the most common adsorption collectors for gases and vapors; a small amount of charcoal inside a small glass tube acts as an activated surface that retains nonpolar materials, such as benzene. These collectors are commonly used to measure inhalation exposures to solvents, such as vapor exposures of printers. Other types, such as impingers or bubblers, collect gases and vapors into liquid from air drawn through them. Bubblers are less convenient to use than charcoal tubes but may be required for some compounds, such as sulfur dioxide. New absorbent types and use of chemosorption (e.g., in impregnated filters) have greatly extended the range of substances that can be sampled, and these collectors have largely replaced bubblers for collection of reactive gases. The specific methods are discussed in detail in the *OSHA Manual of Analytical Methods* (5).

Passive or badge-type samplers are much more convenient to use for gas and vapor sampling than the collectors that require air pumps, and they have better worker acceptance because many workers do not like the weight of the pumps. After the sampling period is completed, the cover is replaced on the badge and it is sent to a laboratory for analysis. These samplers are convenient and relatively inexpensive. Several passive samplers have well-documented sampling rates. They may surpass active samplers in accuracy, if contamination from liquid splashes during use can be avoided.

Sampling Strategy. The hygienist must design a sampling strategy that takes into account the types of hazards, variations in exposure, routes of exposure, and the uses for the data (e.g., risk assessment, source evaluation, and control). The approach should enable the most efficient use of resources. Personal measurements are designed to reflect the accumulation of exposure from a variety of sources that a worker may encounter during a workshift. In some cases exposure may occur only during certain operations. *Worst-case sampling* is the approach used when it is clear that high emissions from certain activities or sources occur and it is decided that sampling will be done only during the period of highest exposure. Workers in adjacent areas not directly involved with the air contaminant of interest are frequently found to have significant exposures because the air contaminant drifts into their work areas.

Variability in exposure levels can be large owing to day-to-day variations in work pattern, changes in production rate, and differences in the process. Differences in personal work habits, wind velocity, and direction also cause variation. The exposed populations should be subdivided into smaller, well-defined groups of workers performing identical or similar tasks.

Proper selection of subgroups reduces within-group variability so that resources can be concentrated on the groups with the highest exposures, although these may be difficult to identify *a priori.* Single samples are avoided wherever possible, because it is difficult to know what the sample value represents. Additionally, because workers have different work habits and techniques, the average exposure may vary among workers (6). Several replicate samples on workers may indicate how important these differences are and to what degree the assumption of uniform mean exposure within groups is violated.

In addition to personal sampling, the occupational hygienist also uses *fixed-location sampling* in the sampling strategy. In this approach, the sampler is set at a given location that has some useful relation to a source of exposure. This type of sampling is advantageous because it can enable determination of features of the exposure that would be difficult with personal samples. For example, a large sampler can be used to determine the particle size distribution of airborne dust in a work area or to provide sufficient airborne material for detailed chemical analysis if the composition of the contaminants is not known. These samplers can be very useful for identifying and characterizing sources of exposure and assessing the effectiveness of engineering controls. As with personal sampling, to get the most out of the effort, careful selection of the sampling location and strategy is important. In some cases, a combination of personal and fixed-location sampling is used to describe a given problem completely. For example, personal samples are used to describe the highly variable exposures of steelworkers tending a blast furnace, whereas stationary samples measure exposures to the uniform, well-mixed air levels they experience while waiting in the lunchroom for their next job assignment (2 to 4 hours per work shift). In some cases, such as cotton dust exposures, only fixed-location sampling techniques are available.

In general, the principal applications of fixed-location sampling are to identify and characterize sources within a work area and to evaluate the effectiveness of emission control systems, such as local exhaust ventilation. Many large plants use continuous multipoint sampling of gases and vapors with central analysis. Instant action can be taken if concentrations exceed specified limits. Continuous monitoring at stationary sites should be part of the total quality management process.

Assessment of indoor air pollution requires a special approach (Box 7-1).

In some occupational settings, the most important route of exposure is skin contact. Skin contact is difficult to evaluate with environmental sampling because, even if the amount of skin contamination can be determined, it is not possible to know how much of the contaminant has already entered the body, or would enter given sufficient time.

Two principal sampling approaches are employed. First, *cloth patches* can be used to cover given locations of skin, such as the forehead, back of the neck, back of the hands, and forearms, to measure the amount of contamination per unit area during a given period of exposure. The second approach is that of *wipe sampling,* in which an area of skin is washed with an appropriate, nontoxic solvent to determine the quantity of contamination. Both of these techniques have been used to estimate pesticide exposures of agricultural workers. Addition of a fluorescent whitening agent to the pesticide as a tracer allows visualization of contamination (see Chapter 41). Additionally, wipe sampling on surfaces can be used as a method to detect and control the indiscriminate distribution of toxic materials throughout the workplace

environment. This type of sampling is also useful in estimating the risk for one person or area of the workplace in relation to another. However, it is difficult to know in absolute terms the quantity of contaminant that may actually penetrate the skin and become a health problem. Biologic monitoring is probably the best method for determining the intensity of skin exposures to substances for which such monitoring tests are available (see Chapter 4).

Some nonpolar substances, such as pesticides and solvents, may enter the body via both the respiratory tract and the skin. In these cases, both skin contact and air exposure must be evaluated to completely assess the risk. Biologic sampling that integrates these two routes of intake may be a practical necessity. However, two important theoretical problems are associated with biologic monitoring. Some types of tests may represent detection of adverse effects (e.g., monitoring of red blood cell cholinesterase in pesticide-exposed workers). As a result, they detect excessive exposures only after the effects have occurred. Because exposure is rarely constant, tissue levels of environmental contaminants represent a dynamic interaction. There is a complex relation between exposures and levels of compounds and metabolites in blood, urine, exhaled breath, and other biologic media. This relation is controlled by toxicokinetics of the particular agents (7). Consequently, proper interpretation of findings from biologic monitoring for a given worker requires some knowledge of the temporal variations in the worker's exposure. In many situations, biologic monitoring should be used only to verify that exposures have been controlled. Its use in detecting high exposures should be limited—for example, when absorption is primarily through the skin.

It is almost never possible to evaluate ingestion as a route of exposure with sampling. Occasionally, samples of food and drink may be collected to assess the level of contamination; however, this sort of exposure is likely to be extremely variable and episodic in na-

---

Box 7-1. Assessing Indoor Air Pollution

In many cases indoor problems cannot be characterized by a generally uniform clinical picture and a specific cause (see Chapter 23). These problems must be considered as being a result of complex interactions between several factors, including air contaminants (including their odor), temperature, ventilation, air movement, illumination, noise, ergonomics, and psychological and social factors. Emerging complaints about indoor environmental quality should be dealt with immediately, through assessment and control of the problems, while making it clear that management cares. The latter is important to set a positive social context. Actions taken could follow a cost-effective, stepwise approach such as the following:

1. Check whether operational conditions are normal, such as for the heating, ventilation, and air conditioning (HVAC) system. Instruct workers who complain regarding the possibilities for individual control of the HVAC system.
2. Determine the type and extent of problem by using a standardized sick building questionnaire or, if there are few employees, structured interviews.
3. Perform a technical survey to assess risk factors inherent in the building or its use and operation.
4. Assess the building, its construction materials and furniture, quality of cleaning, moisture damage and mold growth, temperature, air movement (using smoke tubes), and carbon dioxide concentration. Estimate the degree of recirculation of air and the possible contamination of intake air.
5. Measure ventilation efficiencies with tracer gas studies.
6. Make a detailed assessment of contaminant sources and concentrations.
7. Perform clinical examinations of affected persons and additional occupational hygiene investigations, such as detailed chemical analyses of complex mixtures or assessment of individualized work habits to guide training interventions.

This approach is recommended because buildings are now recognized as a possible source of hazardous exposures. Such exposures may be caused by emissions from building materials (e.g., formaldehyde) or by biologic agents or their toxins (see Chapter 20). All of these problems are of interest to the occupational hygienist.

---

ture, so that environmental sampling is usually an ineffective method of assessment. On some occasions, workers may ingest particles that were initially inhaled. Again, biologic monitoring is the method of choice to monitor this route of exposure.

It has already been noted that exposure measurements commonly show substantial variation in mean exposure between workers doing the same job under similar conditions. These differences are the primary limitation to what can be achieved with exposure controls. There are many reasons why differences might occur. First, individual workers have differences in skill, training, and experience that may lead them to perform a particular job with diverse techniques that affect personal exposure. Second, workers may have differ-

ent levels of concern about the hazards of the job and take more or fewer precautions to avoid exposure through the use of PPE or engineering controls. Differences among workers in regard to these factors are generally assigned to the category of "work practices" and dismissed. As a result, there has been little systematic investigation of the nature of these differences, especially the behavioral components. Research is needed to better understand the causes of these differences in exposure and to identify effective methods of intervention to reduce the exposures.

Controls

Substitution. Substances and materials that pose risks to health and safety should not be used if substitutes for them exist. Substitution is part of the concept of toxics use reduction and waste management. Potential benefits to health and safety must be balanced against technological and economic consequences. This balance should include product properties, production process, environment, and reliability of supply. Substitution is the method of choice, whenever it is possible. For example, less-toxic toluene may be an adequate replacement for benzene. Regular auditing of use of substances and materials provides inspiration for substitution and keeps the substitution process active.

Limitation of Release and of Buildup of Contamination. If substitution is not possible, then the next step is to control or limit releases and prevent the buildup of toxic materials in the worker's environment (Fig. 7-1). Local exhaust ventilation combined with source isolation is used to control process emissions. General room ventilation is used to prevent the buildup of hazardous concentrations in the work area caused by contaminants escaping local exhaust, spills, or fugitive emissions (from seals, valves, or pumps). An example of these two ventilation approaches is shown in Fig. 7-2.

Local exhaust systems surround the point of emission with a partial or complete enclosure and attempt to capture and remove the emissions before they are released into the

A   B

FIG. 7-1. Monitoring equipment can be used to collect samples for measurement of personal exposure on the job. **A.** Particulate sampler is connected to portable pump located on worker's right hip. Here, more sophisticated sampling is being accomplished by adding a real-time direct-reading aerosol sampler and logging device (black box on chest and package on left hip). **B.** Vapor sampling tubes on worker's chest usually collect a time-weighted sample to measure chemical exposure. Here, a direct-reading aerosol measurement system has been added (in the backpack) to permit collection of real-time data in a logging device. (Photographs by Susan Woskie.)

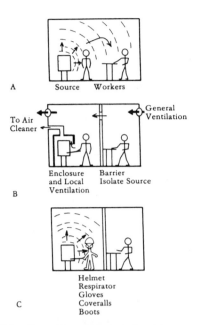

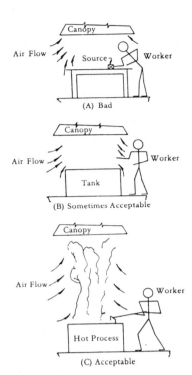

FIG. 7-2. Examples of controls for airborne exposures. **A.** Workers with primary and secondary exposure to source emissions. **B.** Ventilation and source isolation to control exposures. **C.** Personal protection and source isolation to control exposures. (Diagrams prepared by T. J. Smith, Harvard School of Public Health, Boston, Massachusetts.)

FIG. 7-3. The proper use of a canopy hood, which does not allow the air contaminants to be drawn through the worker's breathing zone. The worker's location is crucial. (From National Institute for Occupational Safety and Health. The industrial environment: its evaluation and control. Washington, DC: NIOSH, 1973:599.)

worker's breathing zone. Figures 7-3 through 7-5 show several examples of local ventilation systems; various types, with differing degrees of effectiveness are available. It is not possible before installation to determine precisely the effectiveness of a particular system, although this is an area of active research. As a result, it is important to measure exposures and evaluate how much control has been achieved after a system is installed. Unless contaminant sources are totally enclosed, collection will capture only a percentage of the total emission. Release of smoke from smoke tubes at the point of contaminant generation is a useful technique for visualizing the flow of air toward the exhaust. It may reveal whether the distance to the exhaust is too large, whether there are cross-drafts or strong air disturbances, or whether the worker creates wakes, all of which greatly reduce the collection efficiency. A good sys-

tem may collect 80% to 99% of emissions, but a poor system may capture 50% or less. Careful maintenance must be performed on the system to maintain efficiency. Poor maintenance is probably most responsible for system failures.

The rising cost of energy has made the practice of ventilating work areas with outside fresh air an increasingly expensive process; considerable effort is being directed to the design of systems that can recirculate decontaminated air or use heat exchangers so that the heat value is not lost.

Limitation of Contact. The third important approach to controlling exposures to toxic materials is to limit worker contact by (a) automating processes, (b) isolating processes that use toxic materials from the

FIG. 7-4. Local exhaust ventilation successfully captures dust produced by stone cutting. (From W. A. Burgess, Harvard School of Public Health, Boston, Massachusetts.)

remainder of the work area so that the potential for contact with these materials is limited (Fig. 7-6), or (c) furnishing workers with PPE, such as dust or gas masks (respirators), hoods, or suits with externally supplied air for controlling inhalations of toxic materials. Many people mistakenly think that the use of respirators is a simple and inexpensive way to control exposure to toxic airborne materials. However, these masks are uncomfortable to wear, there is poor worker acceptance, and variable levels of protection are achieved. There are extensive OSHA requirements for an adequate respirator program to ensure that the quality of the devices is maintained and that workers are receiving adequate protection. The annual cost of a good respirator program for lead dust exposures is reported to be approximately $1,000 per worker or more. Fitting of respirators is extremely important but is often neglected; a poorly fitting respirator provides substantially less protection than expected because, even if the filters are highly efficient, air leaks around the edges of the face mask.

The use of rubber gloves and protective

FIG. 7-5. Electroplating workers are protected by local slot exhaust ventilation. (From W. A. Burgess, Harvard School of Public Health, Boston, Massachusetts.)

FIG. 7-6. A glovebox enclosure system prevents solvent exposure to workers gluing shoes. (Photograph by Barry S. Levy.)

clothing does not automatically ensure that workers are protected adequately. Toluene and other aromatic solvents readily penetrate rubber gloves; therefore, glove composition must be matched to the chemical nature of the substance. Similarly, long-sleeved shirts or coveralls may not prevent skin contact with toxic dusts, because small dust particles can sift through the openings between threads in woven cloth. A study of orchard workers showed the effects of pesticide exposures even though the workers had been wearing dust masks to prevent inhalation of the dust and long-sleeved shirts to prevent skin exposure. Special testing indicated that, despite the shirts, the arms of workers were covered with dust that contained pesticide. Tests with fluorescent dusts showed similar findings for impermeable protective suits, which are difficult to seal against migration of dust past cuffs and the neck opening.

An important part of limiting contact with hazardous substances is the requirement that protective clothing be changed each day and not worn outside the work area. This effort is also facilitated by the requirement for showers after the work shift.

Some reduction in exposure can be obtained by administrative controls, such as scheduling workers to spend limited amounts of time in areas with potential exposure. Such efforts may reduce exposures to below recommended guidelines. Although this approach can be effective in certain situations, it requires good exposure data to demonstrate its effectiveness and the careful attention of supervisory personnel. It also may impose an inefficient use of labor. Finally, it may be inappropriate for controlling exposures to carcinogens.

Ideally, all the control approaches described should be used together to develop an overall control strategy that deals with all aspects of toxic material exposure in a particular workplace. Short-term measures, such as extensive use of PPE, may be adopted immediately after a problem is recognized to allow time for development of engineering controls or process modifications that can provide more long-term control. Despite their undesirable aspects (OSHA policy is to use them only as a last resort), respirators may be the only effective control device for some exposures, such as those faced by maintenance or cleanup workers.

NOISE PROBLEMS

Occupational exposure to excessive noise is an important problem that is evaluated and controlled in part by occupational hygienists (see Chapter 18). Hygienists are trained to measure the intensity and quality of noise, assess its potential for producing damage, and devise means to control noise exposures. Two principal types of workplace noise, continuous and impact, have somewhat different techniques of evaluation and control. *Continuous noise* is produced by high-velocity air flow in compressors, fans, gas burners, and motors. Crushing, drilling, and grinding are important sources of continuous noise because a large amount of energy is used in a small space. *Impact noise* results from sharp or explosive inputs of energy into some object or process, such as hammering or pounding on metal or stone, dropping heavy objects, or handling materials.

During the evaluation of a workplace, a hygienist looks for sources of excessive noise, determines which workers are exposed, and then selects an evaluation strategy to clarify the nature and extent of the exposures. If the noise is continuous or almost continuous, a hand-held noise survey meter is used to determine noise levels at the worker's location. If the exposure involves impact noises, an electronic instrument that records and averages high-intensity, short-duration pulses is used to characterize the source and exposures.

Typically, workers spend variable amounts of time exposed to noise sources, and they may work at different distances from the sources, which alters their exposure. Exposures may also vary if the output of noise sources changes over time. Therefore, the TWA may not be easy to estimate even when the sources present clear potential for overexposure. This problem has been solved by the use of small noise dosimeters worn by workers; these devices electronically record sound levels and indicate the average noise level during a work shift. Dosimeters are very useful for describing average exposures. Some store 8-hour time traces in a datalogger, where they can be displayed and linked to records of worker activities. A typical noise evaluation includes both source noise level and dosimeter measurements.

National requirements and TLV guidelines are used by the hygienist to evaluate noise data and decide whether a hazard is present. In addition to the possible damage to hearing, noise also affects verbal communication, which may create a hazard by interfering with warnings and worker detection of safety hazards such as moving equipment. The current OSHA standard for continuous noise for 8 hours is 90 dBA;[3] higher levels are permitted for shorter periods (4). The OSHA standard allows levels of noise exposure that will protect some, but not all, workers from the adverse effects of workplace noise. The TLV for an 8-hour exposure to noise (1993) is 85, which is significantly lower than the 90 dBA OSHA standard for (continuous) noise (3).

Although techniques exist to obtain an overall TWA of noise exposures received in several different work settings, no means are available to assess the hearing risks of combined exposure to both continuous and impulse noise. Many workers are exposed to both types of noise. For example, brass foundry workers are exposed to continuous noise from gas burners and to impulse noise from brass ingots being dropped into metal bins from conveyors.

The strategies for controlling noise are similar to those used for control of toxic materials:

Substitution. Use another process or piece of equipment. For example, electrically heated pots for melting metal can be used instead of gas-heated pots to eliminate burner noise.

Prevent or Reduce Release of Noise. Modify the source to reduce its output, enclose and soundproof the operation, or install mufflers or baffles. For example, noisy air compressors can be fitted with mufflers and placed in soundproofed rooms; impact-absorbing materials can be installed to eliminate impulse noise from ingots dropping into a metal bin.

Prevent Excessive Worker Contact. Provide PPE, such as earplugs or earmuffs, or provide a control booth.

As with toxic materials, the overall strategy to control noise exposures usually involves separate approaches for various aspects of the problem. It may be necessary to consult an acoustical engineer with advanced evaluation and engineering expertise for dealing with complex noise problems. If engineering controls are not completely effective or are impractical, ear protectors may be required; however, the effectiveness of these devices is limited because sound may also reach the ear through bone conduction. A full-shift exposure greater than 120 dBA cannot be controlled adequately by earplugs or muffs (see Chapter 18).

RADIATION PROBLEMS

Radiation hazards are commonly first identified by occupational hygienists, but the responsibility for their evaluation and control overlaps among the occupational hygienist, the health physicist, and the radiation protection officer (see Chapters 17 and 19).

Exposure to *ionizing radiation* may be external (from x-ray machines or radioactive materials) or internal (from radioactive substances in the body). External exposures can be monitored instrumentally by several

---

[3] The unit dBA denotes decibels on the "A" scale, which is related to the human ear's response to sound.

methods; the type of detector system chosen for a given problem depends on the nature of the ionizing radiation. Personal monitoring is commonly performed with badges of photographic emulsions, thermal luminescent materials, or induced-radiation materials that indicate the cumulative dose during the period worn. Data from these measurement systems can be used to construct relatively accurate estimates of tissue exposure. If there are also detailed supporting data on worker activities, sources of exposure and points of intervention can also be identified.

Nonionizing radiation is also an external exposure problem. This type of radiation includes a variety of electromagnetic waves, ranging from short-wavelength ultraviolet, to visible and infrared, to long-wavelength microwaves and radiowaves. Exposures to ultraviolet, visible, and infrared radiation are measured with photometers of various types. Microwaves and radiowaves can also be measured by several standardized techniques, but there is some controversy regarding the exposure intensities required to produce adverse health effects.

Exposures to radioactive materials can be evaluated with methodologies similar to those used for toxic substances. Personal air sampling, surface sampling, and skin contamination measurements can be used to quantify exposures by route of contact or entry into the body. For example, personal air sampling in uranium mines can measure miners' exposure to respirable radioactive particles that will be deposited in their respiratory tracts. Internal levels of some radionuclides can be detected outside the body and measured directly if they emit sufficient penetrating radiation (e.g., gamma rays emitted from radioactive cobalt). However, most cannot be detected externally. The quantities of radioactive substances reaching sensitive tissues usually must be estimated by determining the worker's external exposures and making assumptions about the amount entering the body and being transported to the sites of adverse effects.

The Nuclear Regulatory Commission has set standards for allowable ionizing radiation exposures from both external and internal sources. These exposure limits, based on the work of an earlier governmental agency, the National Council on Radiation Protection and Measurements, can be used, like TLVs, to decide whether a given exposure presents a health risk. They also have many of the same problems as TLVs: they were based on limited data obtained at high dose levels and include many assumptions and extrapolations that have not been verified. They are especially controversial because they contain the inherent assumption of a threshold below which there is no cancer risk. Radiation protection programs have strict requirements regarding techniques for handling radioactive materials and working with radiation sources, and they also require extensive routine exposure monitoring and medical monitoring.

Exposure limits for nonionizing radiation have been set by OSHA, based on published scientific data, the TLVs developed by the ACGIH, and standards developed by the American National Standards Institute. Equivalent limits have been developed by WHO and a number of countries (see Bibliography). The eyes and the skin are critical organs to be protected, and the standards are set to protect the most susceptible areas. There is concern about reproductive hazards for these agents. Standards also have been developed for lasers based on ophthalmoscopic data and irreversible functional changes in visual responses. As with other types of standards, the numeric limits cannot be treated as absolute, and the margin of safety in many cases is uncertain.

Control of external ionizing and nonionizing radiation exposures is achieved by minimizing the amounts of radiation used, isolating the processes. shielding the sources, using warning devices and interlocking door and trigger mechanisms to prevent accidental exposures, educating workers and supervisors about the hazards, and, if necessary, requiring use of PPE.

To control radiation exposures, an industrial x-ray machine used to check castings for flaws was placed in a separate room with extensive lead shielding and was controlled so that the machine could not be triggered when the door was open. The room also had signs warning of the hazard. A red warning light inside the room was designed to be lit for 30 seconds before the x-rays were released so that a worker who was inside the room when the door closed could activate an emergency override switch to prevent operation of the machine. All personnel working around the x-ray operation were required to wear film badges to monitor their accumulated x-ray exposure.

Control of internal radiation exposures from radioactive materials is similar to controls for toxic materials. The objectives are to use minimal amounts of radioactive materials; isolate the work areas; enclose any operations that are likely to produce airborne emissions; use work procedures that prevent or minimize worker contact with contaminated air or materials; have workers wear PPE to prevent skin contact, eye exposure, or inhalation of materials; monitor environmental contamination levels; and educate workers about the hazards. Some or all of these measures are used in typical radiation control situations. Careful supervision of work activities and monitoring of program implementation are required to provide adequate protection.

Traditional work environments, such as factories and other workplaces in heavy industry, have been long-term concerns of hygienists. These are now a growing concern in developing countries that seek to balance the economic benefits of industry against the health costs of insufficient worker protection. In some developed countries there has been a growing concern about office environments and the health effects associated with energy-efficient, tightly sealed buildings.

CONCERNS FOR THE FUTURE

The scientific basis for occupational hygiene practice has been eroding because of limited research funding and the small number of researchers. Examination of the scientific literature in occupational hygiene shows it to be narrowly focused on limited issues. Internal research funded by companies is often not published because it could aid competitors or raise liability concerns. Gaining access to workplaces for academic studies has also become more difficult. There is a reluctance on the part of industries to examine the hazards of their operations because of concerns about the costs of additional government regulation and legal liability for health claims from previously unrecognized hazards. As a result, there has been little development and refinement of exposure assessment methods, control technology, or intervention strategies despite the extensive worldwide development of new materials, production technologies, biomedical and drug manufacturing, and other advances. Occupational hygiene has not kept up with these developments. A number of occupational hygienists in North America and other parts of the world are concerned about this situation and have begun working to strengthen local research and to develop more collaborative international research programs. Joint labor-management research programs supported by company funds have also become increasingly important sources of workplace access and research funding; an example is the joint health and safety research programs established by the United Auto Workers with Daimler Chrysler Corporation and General Motors Corporation.

REFERENCES

1. Hawkins NC, Norwood SK, Rock JC. A strategy for occupational exposure assessment. Fairfax, VA: American Industrial Hygiene Association, 1991.
2. Burgess WA. Recognition of health hazards in industry: a review of materials and processes, 2nd ed. New York: Wiley & Sons, 1995.
3. American Conference of Governmental Industrial Hygienists. Threshold limit values for chemical substances and physical agents: biological exposure indices (TLV/BEI Booklet) Cincinnati, OH: ACGIH, 1999.
4. Occupational Safety and Health Administration, U.S. Department of Labor. General industry: OSHA safety and health standards (29 CFR 1910). Washington, DC: US Government Printing Office, 1984.

5. Occupational Safety and Health Administration. Analytical methods manual: part I (organic substances), vol 1–3, 1990; part II (inorganic substances), vol 1–2, 1991. Washington D.C.: US Department of Labor.
6. Rappaport SM. Assessment of long-term exposures to toxic substances in air [Review]. Ann Occup Hyg 1991;35:61–121.
7. Fiserova-Bergerova V. Modeling of inhalation exposure to vapors: uptake, distribution, and elimination. Boca Raton, FL: CRC Press, 1983.

BIBLIOGRAPHY

American Conference of Governmental Industrial Hygienists. Industrial ventilation, 23th ed. Cincinnati: ACGIH, 1999.

Colton CB, Birkner LR, Brosseau L. Respiratory protection: a manual and guideline. Akron, OH: American Industrial Hygiene Association, 1991.

Schwope AD, Costas PP, Jackson JO. Guidelines for selection of chemical protective clothing. Cincinnati, OH: American Conference of Governmental Industrial Hygienists, 1987.

Manuals containing recommendations for the design and operation of ventilation systems to control air contaminants, along with other protective approaches.

Burgess WA. Recognition of health hazards in industry: a review of materials and processes, 2nd ed. New York: Wiley & Sons, 1995.

The source of much of the data in Table 7-1 and a highly recommended reference for all occupational health professionals.

Clayton GD, Clayton FE, eds. Patty's industrial hygiene and toxicology. New York: Wiley. Vol. I, Parts A and B, 1991; Vol. II, Parts A through F, 1994; and Vol. III, Parts A and B, 1994.

DiNardi S, ed. The industrial environment: its evaluation and control. Fairfax, VA: American Industrial Hygiene Association, 1993.

Two useful and comprehensive general references on industrial hygiene. Patty's works (now edited by the Claytons), although somewhat unwieldy, are the professional's reference works. A new edition is planned for 2000. The AIHA volume is very comprehensive but somewhat uneven in quality.

Considine DM. Chemical and process technology encyclopedia. New York: McGraw-Hill, 1974.

Although out-of-date for newer technologies such as semiconductors or biotechnology, it is a good technical source for gathering background information on basic industries and industrial processes.

Harvey B, Crockford G, Silk S, eds. Handbook of occupational hygiene, vol 1–3. London: Crone Publications Ltd, 1992.

World Health Organization. Occupational hygiene in Europe: development of the profession. European Occupational Health Series No. 3. Copenhagen: WHO Regional Office for Europe, 1992.

These references contain relevant material about exposure standards in Europe and a number of other countries.

Hawkins NC, Norwood SK, Rock JC, eds. A strategy for occupational exposure assessment. Akron, OH: American Industrial Hygiene Association, 1991.

Rappaport SM. Assessment of long-term exposures to toxic substances in air [Review]. Ann Occup Hyg 1991;35:61–121.

Although these two works contain much useful information, they are difficult to follow in places and require a strong background in statistics. They are important because they lay out the rationale for sampling strategies to determine compliance with OSHA's permissible exposure limits.

Occupational Safety and Health Administration: Analytical methods manual: part I (organic substances), vol 1–3, 1990; part II (inorganic substances), vol 1–2, 1991. Washington, DC: US Department of Labor.

Designed for the laboratory chemist and industrial hygienist. New methods are continuously being added and can be found at the OSHA website: www.osha.gov.

Occupational Health: Recognizing and Preventing Work-Related Disease and Injury, 4th ed.
Edited by Barry S. Levy and David H. Wegman, Lippincott Williams & Wilkins, Philadelphia © 2000.

8 Occupational Safety
Preventing Accidents and Overt Trauma

W. Monroe Keyserling

An *accident* is an unanticipated, sudden event that may cause an undesired outcome such as property damage, bodily injury, or death. *Injury* is physical damage to body tissues caused by an accident or by exposure to environmental stressors. Many work injuries result from accidents, such as a heavy object falling and crushing bones in a worker's foot. Some work injuries, however, are not associated with an accident, but are caused by normal work activities, such as tendinitis on a repetitive assembly line or data entry job. The occupational basis of these nonacute injuries may not always be recognized because of the lack of a well defined causative event.

Each year, work accidents produce staggering costs in terms of loss of life, pain and suffering, lost wages for the injured, damage to facilities and equipment, and lost production opportunity. Based on information collected by Bureau of Labor Statistics (BLS) and the National Safety Council, the estimated annual toll of workplace accidents for 1997 in the United States was:

- 5,100 deaths (3.9 per 100,000 workers)
- 3,800,000 disabling injuries
- More than 125 million lost workdays.

The estimated cost of injuries, including lost wages, medical and rehabilitation payments, insurance administrative costs, property losses, and other indirect costs was ap-

proximately $128 billion in 1997 (1). These figures underestimate both the count and cost of all work accidents and injuries, however, because of the underreporting of certain nonacute injuries and the omission of certain indirect costs. For example, carpal tunnel syndrome, tendinitis, and related upper extremity cumulative trauma disorders are classified as occupational *diseases* and do not show up in the aforementioned statistics. Furthermore, these costs do not include governmental transfer payments, such as welfare and Social Security to the families of deceased or permanently disabled workers.

Aggressive implementation of industrial safety programs in the United States after World War II and the passage of the Occupational Safety and Health Act in 1970 has dramatically reduced fatal work accidents. The rate of occupational fatalities per 100,000 workers declined from 33 in 1945 to approximately 4 in the late 1990s. Certain industries, such as mining, agriculture, fishing, and construction, however, have continued to experience high rates of fatal accidents. Death rates for these industries ranged between 13 and 24 per 100,000 workers in 1997 (1). The safest U.S. industries in 1997 were retail trade and service (including such industries as finance, insurance, real estate, and education), where death rates ranged between 1 and 2 per 100,000 workers (1). Although nonfatal accidents also declined during the last four decades, workers in certain industries experience high rates of disabling injuries. In the construction, agri-

W.M. Keyserling: Department of Industrial and Operations Engineering, University of Michigan, Ann Arbor, Michigan 48109-2117.

TABLE 8-1. *Selected examples of occupational injuries and affected workers*

Cause Injury or disorder	Affected occupations
Overt trauma	
Mechanical energy	
Lacerations	Sheet metal workers, press operators, meat/poultry processors, woodworkers, carpenters, forestry workers, fabric cutters, glass workers
Fractures	Construction workers, miners, materials handlers, truck drivers, press operators, roofers
Contusions	Any worker exposed to low-energy impacts
Amputations	Press operators, meat/poultry processors, saw operators, machine operators and repairers
Crushing injuries	Materials handlers, construction workers, press operators, calender operators, maintenance workers, miners
Eye injuries (struck by foreign objects)	Miners, grinders, machine tool operators, carpenters, roofers
Strains/sprains (overt)	Materials handlers, miners, baggage handlers, mail handlers, construction workers, warehouse workers, truck drivers
Thermal extremes	
Burns	Foundry workers, smelter workers, cooks, welders, roofers, glass workers, mechanics
Heat strain	Firefighters, smelter workers, steel workers, hazardous waste cleanup workers, cooks
Cold strain	Utility workers, forestry workers, meat/poultry processors, firefighters
Chemical reactivity	
Burns	Masons and cement workers, hazardous waste workers, chemical plant workers
Toxicity (including asphyxiation)	Firefighters, hazardous waste workers, confined space workers
Electrical energy	
Electrocution, shocks, burns	Utility workers, construction workers, operators of electrical equipment and tools, electricians
Radiation	
Burns, radiation illness	Hospital workers, industrial radiographers, nuclear workers
Cumulative trauma	
Heavy lifting, prolonged sitting, awkward posture	
Back pain	Materials handlers, nurses, truck drivers, sewers, assemblers, mechanics, construction workers, miners, maintenance workers
Repetitive motions, prolonged postures, awkward postures, forceful exertions	
Upper extremity cumulative trauma disorders (e.g., tendinitis and carpal tunnel syndrome)	Assemblers, data entry operators, word processors, clerical workers, journalists, meat/poultry processors, packers, sewing machine operators, press operators, musicians
Vibration	
Raynaud's syndrome	Forestry workers (power-saw operators), grinders, air-hammer operators

culture, manufacturing, transportation, and public utility sectors, disabling injury rates exceeded 7 per 100 worker-years in the mid-1990s.

In certain instances, the consequences of a workplace accident can affect people outside the confines of a plant, sometimes with catastrophic results. Such was the case in the 1984 accident in Bhopal, India, where more than 2,000 people died and over 300,000 people were injured after the release of a toxic gas from a chemical plant.

The principal responsibility of the safety professional is to reduce the risk of injuries and disorders by preventing accidents and controlling hazardous exposures in the work environment. Common examples of these hazards and affected occupational groups are listed in Table 8-1. Traditionally, safety professionals have concentrated their efforts on the prevention of accidents and *overt traumatic injuries* such as amputations, fractures, lacerations, burns, and electrocution. These injuries typically are immediately apparent to the victim and can be directly linked to a well defined event or exposure. In recent years, safety professionals have also become concerned with the prevention of *overexertion injuries and disorders* such as strains and sprains, chronic back pain, tendinitis, and carpal tunnel syndrome. Unlike the general downward trend in the rates of fatalities and traumatic injuries, overexertion cases have increased dramatically during the 1980's, particularly in the manufacturing and service sectors.

The remainder of this chapter emphasizes topics related to the prevention of overt traumatic injuries. For additional discussion of overexertion injuries and disorders, see Chapters 9 and 26.

SAFETY HAZARDS IN THE WORK ENVIRONMENT

Safety hazards and the injuries they cause can be categorized in many different ways. The BLS has adopted a system originally developed by the American National Standards Institute (ANSI; see Appendix B) for classifying the cause of work injuries according to the type of *accident event* or *exposure*. This system, summarized in Table 8-2, considers the conditions and events that caused the injury. The BLS/ANSI classification system is used to organize the following discussion of the causes and prevention of common work injuries.

Contact with Objects and Equipment

Mechanical injuries caused by unintentional contact with objects and equipment in the

TABLE 8-2. *Major causes of work injury classified by accident type*

Struck by
Caught in, under, or between (CIUB)
Struck against
Fall from elevation
Fall on same level
Motor vehicle accident
Overexertion (see Chapter 26)
Contact with electric current
Contact with temperature extremes (see Chapter 19)
Rubbed or abraded
Bodily reaction
Contact with radiation, caustics, or toxic and noxious substances (see Chapters 15–17)
Public transportation accident
Other
Unknown

From American National Standards Institute. Method of recording basic facts relating to the nature and occurrence of work injuries. Standard no. Z16.2. New York: American National Standards Institute.

work environment accounted for 26% of all lost-time work injuries in the United States in 1996. These injuries can be subclassified into three groups based on the nature of the accident: (a) struck by, (b) caught in or between, and (c) struck against.

Struck By

The classification "struck by" applies to a broad variety of cases where a worker is hit by a moving object or particle. Struck-by accidents accounted for 13% of lost-time work accidents in 1996 and 579 fatalities. Typical accident scenarios include a construction worker who has a concussion when hit by a hammer that is accidentally kicked off a scaffold; a warehouse worker who fractures a leg when hit by a moving forklift; and a grinding machine operator who is permanently blinded when struck in the eyes by the fragments of an exploding grinding wheel. Certain struck-by injuries can be self-inflicted, such as when using a knife or cleaver (Fig. 8-1).

Falling objects cause approximately one-third of these accidents. Items dropped from overhead, such as hand tools, construction materials, and equipment being hoisted to

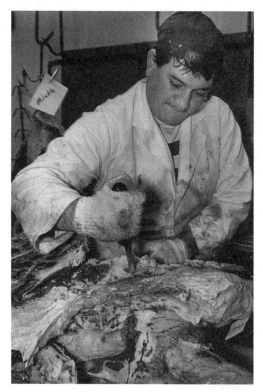

FIG. 8-1. Meat cutting is a high-risk job. Using chain-mail gloves can reduce this risk. (Photograph by Earl Dotter.)

6. Painting safety warnings on floors to indicate work zones where overhead hazards are present.

Objects dropped during materials handling tasks, such as lifting and carrying, can cause severe struck-by injuries to the feet and toes; crush injuries, fractures, and contusions are quite common. Objects that are manually carried should be equipped with handles that allow workers to maintain a firm grasp and should not exceed safe weight-lifting limits. (See discussion of manual materials handling in Chapters 9 and 26.) Finally, foot protection (safety shoes and metatarsal guards) must be provided and worn whenever these hazards exist.

Flying objects are another cause of struck-by injuries. Small airborne particles released during operations such as grinding, chipping, and machining can cause severe eye injuries and blindness. These operations should be fully enclosed whenever possible. The use of compressed air for cleaning dust and chips may also be hazardous because small particles can rapidly accelerate to very high velocities. Occupational Safety and Health Ad-

an upper floor of a building, can cause severe head, neck, and shoulder injuries, such as fractures, concussions, and lacerations. In some instances, these injuries are fatal. Effective methods for preventing these events include:

1. Installing covers or side rails to enclose or contain materials carried on overhead conveyor systems (Fig. 8-2). Nets or gratings can also be placed under conveyors to catch falling objects or material.
2. Placing toe boards on all overhead work platforms to prevent materials or tools from being kicked off.
3. Installing nets or gratings above workers assigned to hazardous areas.
4. Providing and requiring the use of safety helmets.
5. Training workers to follow safe rigging practices when hoisting materials overhead.

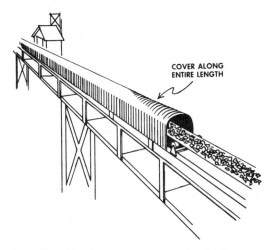

COVER ALONG ENTIRE LENGTH

FIG. 8-2. Overhead conveyors should be enclosed to prevent loose materials from falling on people walking or working below. (From McElroy FE, ed. Accident prevention manual for business and industry: engineering and technology, 10th ed. Chicago: National Safety Council, 1992:279.)

ministration (OSHA) regulations limit the pressure of cleaning air to a maximum of 30 pounds per square inch to minimize this hazard. Eye and face protection is a critical factor in the prevention of eye injuries. Workers exposed to operations that have the potential to release flying particles and fragments must be provided with and *constantly* wear the appropriate personal protective equipment, such as safety glasses, safety goggles, and face shields (see Chapter 28).

The remainder of struck-by injuries are caused by a variety of factors. In-plant vehicles such as fork lifts, automatic guided vehicles, and other industrial trucks can strike workers, causing fractures and severe trauma to internal organs. Construction tools, such as nail guns used by roofers, use high-pressure air or explosive charges to drive nails and other fasteners into wood and concrete. Because of design defects or improper use, these tools are capable of shooting high-velocity projectiles that can cause serious injuries. Programmable robots, now common in many manufacturing and assembly operations, present a special hazard. Robots have very limited sensory capabilities to detect the presence of a worker. If a person enters the work zone of a robot without taking the necessary precautions to deactivate and lock out all sources of energy, he or she may be struck or entrapped during subsequent movements of the system. Research is underway to improve the sensory abilities of robots to prevent this type of accident. In the meantime, however, the safest practice is to enclose fully the robot's work zone to prevent the entry of any person when the robot is operational. Workers who must enter the zone to perform maintenance or programming require special training in robot safety.

Caught In, Under, or Between

Accidents classified as "caught in, under, or between" (frequently abbreviated as CIUB) include those where the injury is caused by the crushing, squeezing, or pinching of a body part between a moving object and a stationary object, or between two moving objects. CIUB events accounted for 4% of lost-time work accidents in 1996 and 283 fatalities. Frequently, these accidents are associated with operations that involve the use of mechanized equipment, such as power presses and calenders. Power presses are used to cut and shape contoured products from sheet metal or plastic composite, such as automobile body panels, enclosures for household appliances, and metal furniture. When operating a power press, unprocessed material is loaded in the machine and positioned on the lower die. The machine is then activated, causing the upper die to close down on the working stock with tremendous force (up to several tons). If any part of the body is between the dies during this action, it can be severely crushed or amputated.

Calenders are large, heavy rolls used to compress raw materials, such as slabs of steel, plastic, or rubber, into sheets of precise thickness. It is possible for a limb to become caught in the "in-running pinch point" between these rollers (top panel of Fig. 8-3) and to be crushed or amputated. When working around large calenders, the worker's entire body may be drawn into the dies or rollers, killing the worker.

When power presses and calenders are in use, the most hazardous operations involve the feeding and removal of stock because these tasks place workers' hands near moving machine parts. Effective systems for preventing accidents during these activities include *barrier guards*, devices that enclose the dangerous zone before the machine can be activated; *two-handed safety buttons*, devices that are positioned in a safe location and require simultaneous activation to start the machine (Fig. 8-4); and *presence-sensing systems* such as electric eyes that prevent the operation of a machine while any part of the body is in the danger zone. In addition, all machines that can catch clothing or body parts in a pinch point should be equipped with emergency stop buttons that can be easily reached by the operator with either hand.

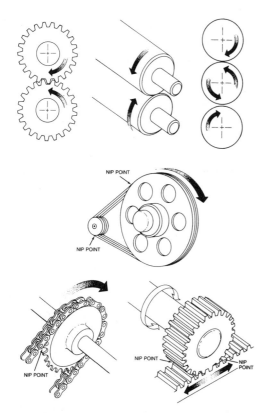

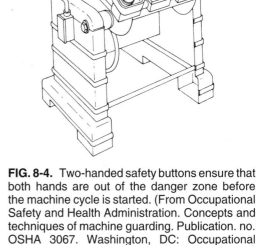

FIG. 8-3. Common examples of in-running pinch points. (From Occupational Safety and Health Administration. Concepts and techniques of machine guarding. Publication no. OSHA 3067. Washington, DC: Occupational Safety and Health Administration, 1980:3–4.)

FIG. 8-4. Two-handed safety buttons ensure that both hands are out of the danger zone before the machine cycle is started. (From Occupational Safety and Health Administration. Concepts and techniques of machine guarding. Publication. no. OSHA 3067. Washington, DC: Occupational Safety and Health Administration, 1980:38.)

Because all of these systems are quite sophisticated, they must be designed and installed by qualified personnel.

Power transmission systems, such as belts and pulleys, chains and sprockets, and intermeshing gears (see Fig. 8-3), have pinch points that can trap and injure body parts. Moving parts of power transmission systems that can either strike or ensnare a person should be fully enclosed within a barrier guard (Fig. 8-5). Access and inspection panels are frequently built into power transmission guards to facilitate maintenance. These panels should be designed so that they cannot be opened or removed while internal components are moving.

Maintenance activities, such as setup,

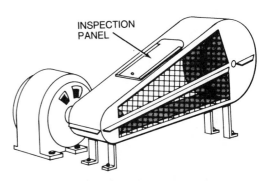

FIG. 8-5. The belt drive in this power transmission system is totally enclosed with a barrier guard. (From Occupational Safety and Health Administration. Concepts and techniques of machine guarding. Publication. no. OSHA 3067. Washington, DC: Occupational Safety and Health Administration, 1980:13.)

cleanup, clearing or unjamming parts, and repair are also associated with CIUB accidents. These operations are particularly hazardous because of the necessity to remove or bypass guards to reach locations that require maintenance. Maintenance workers require special training in safe procedures for working with powered machines and equipment that are temporarily unguarded. Maintenance activities should be performed only when the machine has been put into a nonpowered state. This can be done using lockout procedures that prevent the machine from being started during servicing (Fig. 8-6). Lock-out is required by OSHA and provides a simple procedure for protecting workers during maintenance. Before performing any service on a piece of equipment, the power switch is placed in the "off" position. The switch is then secured in the "off" position using the personal padlock of each maintenance worker assigned to the job. Because each worker carries the key to his or her padlock, it is practically impossible to restart the machine until all workers have left the danger zone and removed their locks. Effective lock-out programs require the support and participation of both management and workers. Standard operating procedures must be developed and followed for deenergizing, servicing, and reactivating powered equipment. Policies that prohibit any maintenance activities before placing machines in a safe, locked-out state must be enforced and rigorously adhered to by all personnel.

Because of the Occupational Safety and Health Act of 1970 and increasing product liability lawsuits settled in favor of injured

 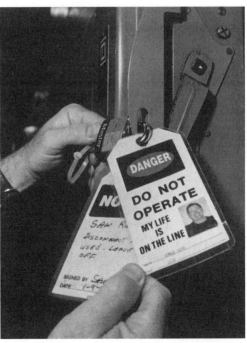

A B

FIG. 8-6. Lock-out is a control measure used to prevent premature activation of a power switch, particularly during maintenance activities. **(A)** Maintenance worker has locked operating switch in the "off" position with a personal key before entering machine. Without lock-out, he could be caught in the machine if another worker starts it. (Photograph by Earl Dotter.) **(B)** Three workers' locks are in place on this machine power switch. All must remove their locks before the machine can be started. (From Occupational Safety and Health Administration. Concepts and techniques of machine guarding. Publication. no. OSHA 3067. Washington, DC: Occupational Safety and Health Administration, 1980:60.)

workers, employers and machine tool manufacturers in the United States have been under increasing legal and economic pressure to upgrade guarding systems on presses, calenders, and other machines. This has resulted in a decrease in the frequency of CIUB and struck-by incidents in the United States since OSHA promulgated the lock-out Standard. Unfortunately, the older machines that have been rendered obsolete because of inadequate guarding and other safety deficiencies are sometimes exported to manufacturers in developing countries. Workers who use these machines are not afforded the protection provided by contemporary guarding technology or legislation. In addition, workers in developing countries often lack the necessary training and personal protective equipment needed to operate these machines safely.

Struck Against

The classification "struck against" is used to describe cases where a worker collides with a stationary object. Although these events accounted for 7% of disabling work accidents in 1996, they were rarely fatal.

Typical injuries include contusions, abrasions, lacerations, and fractures. Head injuries are common and are associated with low ceilings and working in confined spaces. Hand and finger injuries are also common and are frequently caused by forceful exertions with poorly designed or improperly selected tools. The severity of struck-against injuries can sometimes be controlled by using personal protective equipment, such as safety helmets and gloves, and through improved design of workstations and tools.

Fall from Elevation

This classification covers incidents where a worker falls to a lower level and is injured on impact against an object or the ground. Although these events accounted for only 5% of disabling work injuries in 1996, they caused approximately 10% of all work fatalities.

Fall injuries include fractures, sprains, strains, contusions, and severe damage to internal organs. The nature and severity of the injury are primarily determined by the velocity and orientation of the body at the time of impact. Over 40% of these falls occur in the construction industry, where work activities are frequently performed on temporary, elevated structures that are exposed to rain, snow, and ice (Fig. 8-7; see Chapter 42). Maintenance workers in manufacturing plants frequently fall while performing repairs when standing on ladders and other temporary work surfaces. Falls from elevated scaffolds, walkways, and work platforms are usually due to insufficient or nonexistent guardrails. ANSI (see Appendix B) has developed standards for the design and safe use of scaffolds and elevated work platforms. Many of these standards have been adopted by OSHA in the Fall Protection Standard and other rules.

Falls from ladders account for over 20% of work injuries in this category. In a common fall scenario, the accident is initiated by a slippage of the worker's shoe on a ladder

FIG. 8-7. Construction workers have high rates of occupational injuries, many of which are due to falls. (Photograph by Ken Light.)

rung, which is followed by a loss of grip at the hands. For fixed vertical ladders, this type of accident is sometimes caused by insufficient toe space between the ladder and the structure to which it is mounted. In other cases—particularly outdoor locations where the ladder is slippery—the slip may be caused by insufficient friction between the shoe and the ladder rung. In another common scenario, the accident is initiated when the worker attempts an extended reach to the left or right of the ladder. Here, a loss of grip or foothold can initiate a fall. Furthermore, if a portable ladder is used (one that is not tied to the structure), the center of gravity of the worker–ladder system may fall outside the ladder footings, causing the ladder to topple sideways. Falls may also be caused by slippage or breakage of the ladder itself. Many of these accidents can be prevented by worker training and compliance with structural and work practice standards established by OSHA and ANSI for the selection and safe setup of ladders.

Personal protective devices have been developed to reduce the risk of severe injuries and fatalities that result from falls from elevations, including safety belts, harnesses, lanyards, and lifelines. In spite of the protection offered by these devices, they frequently are not used. Furthermore, contusions, fractures, strains, and deceleration injuries to internal organs can still occur when fall protection devices are used.

Fall on the Same Level

Although rarely fatal, serious injuries such as fractures, sprains, strains, and contusions occur when workers lose their footing or balance and fall to the surface supporting them. These events are common, with same-level falls accounting for approximately 12% of disabling injuries in 1996. Slipping, a loss of traction between the shoes and the floor, accounts for about half of all same-level falls in industry. Most slips are caused by an unanticipated reduction in shoe–floor friction, such as when walking from a dry surface to a wet or oily surface, or from a nonskid surface to a highly polished one. These accidents can be controlled by using similar floor surfaces throughout a work area and by preventing and quickly cleaning up liquid leaks and spills. Many slips occur on floor surfaces where friction is relatively uniform but low, such as in slaughterhouses, food processing plants, and on floors where oil spills are common. In these environments, floor surfaces and shoe soles must be carefully selected to provide adequate friction. Sole selection is also important for workers who encounter snow and icy conditions while performing outdoor activities during the winter.

Trips and missteps, usually because of unseen objects on a floor or unexpected changes in floor elevation, are another cause of same-level falls. These accidents can be prevented through good housekeeping (keeping floors and aisles clear), good maintenance (repairing cracks and uneven surfaces), and good lighting. Finally, warning devices, such as caution signs, are useful in situations where a hazardous floor condition is temporary and being corrected, such as during maintenance activities. However, warning signs should never be used on a permanent basis. Instead, the underlying problem must be eliminated.

Motor Vehicle Accidents

Motor vehicle accidents are frequently overlooked as a serious occupational hazard. Although motor vehicle accidents account for less than 5% of occupational injuries, they are responsible for approximately one-third of work-related fatalities. Controlling injuries and deaths due to vehicle accidents is a complex problem. Motor vehicle drivers frequently work nontraditional shifts, encounter a variety challenging environmental conditions (poor lighting, rain, ice, and snow), and are rarely directly supervised while on the road. Driver selection (including screening for substance abuse) and training may help reduce the frequency and seriousness of accidents. Daily limits on driving time may reduce operator fatigue as a cause of accidents. Vehicle inspection programs, routine preventive maintenance, passive safety devices such as air bags, and regular use of seat belts also help. Investigations to determine the causes of motor vehicle accidents can lead to measures that prevent recurrence. Details on these and other loss control programs for motor vehicles are available from the National Safety Council, (see Appendix B) the Insurance Institute for Highway Safety, and the National Highway Traffic Safety Administration.

Overexertion and Repetitive Trauma

Overexertion and repetitive trauma injuries, such as low back pain, strains, sprains, tendinitis, and carpal tunnel syndrome, are caused by jobs that involve excessive physical effort or highly repetitive use of localized muscles and joints. Awkward work posture is frequently a confounding or aggravating factor in these injuries. Overexertion and repetitive trauma cases account for approximately one-third of disabling work injuries/diseases in the United States. In some workplaces, these injuries account for well over one-half of workers' compensation costs. These injuries frequently occur on jobs that involve manual materials handling, such as lifting, pushing, pulling, or carrying; or highly repetitive hand motions, such as working on an assembly line, processing poultry or beef, hand packing operations, or high-speed keyboard operations. For additional information on overexertion injuries, see Chapters 9 and 26.

Other Causes of Injury and Death

Approximately one-fourth of work accidents do not fall into any of the categories discussed thus far. *Contact with electric current* results in shocks and electrical burns, which are sometimes fatal. Approximately 5% of all work-related fatalities are electrocutions. The critical factor determining the severity of electric shock is the amount of current that flows through the victim, particularly through the chest cavity. Studies have shown that an alternating current of less than 100 mA at the commercial frequency of 60 Hz (standard in North America) may cause ventricular fibrillation or cardiac arrest if the current passes through the chest cavity. Power supplies commonly found at construction sites, in manufacturing facilities, and even in households are sufficiently high to drive fatal currents through the body.

Electric utility workers and electricians are the most frequent victims of electrical accidents because their jobs involve many situations where they must work in close proximity to energized wires and equipment. Construction workers are also victims because they use electrically powered tools and equipment in wet outdoor locations. Control programs for construction sites include effective grounding of all electrical equipment, double insulation of all electrical hand tools, or ground fault circuit interrupters. Regular inspections of all equipment and power cords should be performed to ensure that insulation is intact. Personal protective devices, such as insulated gloves, boots, and clothing must be worn. Metal ladders and heavy equipment, such as cranes and "cherry pickers," present a special problem at construction sites because they can accidentally become energized on contact with live overhead wires. Special precautions must be taken when working below or near overhead lines, such as substituting wood ladders and deenergizing power sources.

Contact with temperature extremes refers to incidents where tissue damage (either burning or freezing) results from exposure to hot or cold solids, liquids, or gases. Control measures include insulation, using robots and other automation in extreme thermal environments, and personal protection such as heat-resistant clothing and gloves. (For additional information on thermal extremes, see Chapter 19.)

The category *rubbed, abraded, or scratched* refers to relatively minor tissue damage resulting from prolonged contact with rough or sharp objects. Common sites of these injuries are the hands (using abrasives for cleaning or polishing) and the knees (prolonged kneeling on a rough surface). In most instances, these incidents can be controlled by covering the objects with padded materials or handholds, or by wearing protective clothing.

ELEMENTS OF A SAFETY PROGRAM

Organization and Responsibilities

An effective safety program results from a multidisciplinary effort involving interactions among many groups within a work organization. An effective site-based program is built around a workplace-level safety committee with overall authority and responsibility for safety. This steering committee includes upper-level managers (typically the site manager or designate, and heads of production departments), the physician or nurse, the safety manager, staff managers (from the engineering, purchasing, maintenance, and industrial relations departments), and labor representatives (the president or safety steward in facilities with union representation, the employee representative in facilities without unions). This committee establishes policy, sets goals, and oversees the activities of department-level safety teams that run the day-to-day safety program and solve "floor-level" problems. For the safety program to be effective, the departmental teams must encourage active participation from first-line supervisors and production workers.

Upper-level managers establish a safety policy, develop the policy into a program, and provide the leadership and resources needed to execute the program. Although safety policies vary from one organization to another, most safety policies include the following:

1. A commitment to provide the greatest possible safety to all employees and to ensure that all facilities and processes are designed with this objective. Similarly, purchasing policies must provide that all

equipment, machines, and tools meet the highest safety standards.
2. A requirement that all occupational injuries and accidents be reported and investigated, with corrective actions taken to ensure that similar incidents do not recur.
3. Clear explanations to all employees of their exposures to safety and health hazards, and the establishment of training programs to inform employees of how to minimize their risk of being affected.
4. Regularly scheduled systems safety analyses of all processes and workstations to identify potential safety hazards so that corrective actions can be taken before accidents occur. (Systems safety is discussed later in this chapter.)
5. Disciplinary procedures for employees who engage in unsafe behavior and for supervisors who encourage or permit unsafe activities.

Although the chief executive officer is ultimately responsible for the safety of all employees, this responsibility should be delegated throughout all levels of management.

First-line supervisors play a key role in the execution of safety programs because of their direct contact with employees. Supervisors must ensure that all equipment and tools comply with applicable safety standards and regulations, and that employees use safe work practices. In addition, the supervisor must make certain that all injuries are promptly reported, treated, and investigated. Some organizations use "safety contests," in which supervisors and work teams vie to achieve the best safety record. These contests yield beneficial results when supervisors are encouraged to bring their departments into compliance with applicable safety standards and regulations. Unfortunately, however, such competitions sometimes discourage the accurate reporting of accidents or appropriate medical care for injuries. For this reason, safety competitions may lead to unintended and counterproductive results, and should be undertaken with caution. Rewards should be based on maintaining safe conditions and compliance with safe work practices. Care should also be taken to avoid giving supervisors incompatible goals, such as unreasonably high production standards, when lower rates are necessary to ensure safety.

Larger worksites usually have a full-time *safety director*, a manager responsible for the day-to-day administration of the safety program. Typical responsibilities include developing and presenting safety training, inspecting facilities and operations for unsafe conditions and practices, conducting accident investigations, maintaining accident records and performing analyses to identify causal factors, and developing programs for hazard control. The safety director must work with the *engineering* and *purchasing* departments to ensure that equipment and facilities are designed and purchased in compliance with all applicable safety standards. The safety director also works closely with the medical staff to ensure that all injuries and illnesses are properly recorded and investigated. A full-time safety director should be certified by the Board of Certified Safety Professionals. For smaller worksites that do not employ a full-time safety director, the duties described previously should be assigned to managerial personnel on a part-time basis or obtained from a qualified safety consultant.

Regardless of the size and structure of the worksite medical services, *physicians and nurses* play an essential role in the safety program through primary treatment of injured workers and by helping to identify workplace hazards. Although the causes of overt trauma injuries are usually obvious, causes of cumulative trauma are often subtle and difficult to identify. By evaluating patterns of employee injuries, disorders, and complaints, the health professional can provide early detection of potentially hazardous operations and processes. Whenever disorders or complaints are suspected of being work related, this information should be reported to the plant safety director and the responsible supervisor. The physician or nurse should participate in the subsequent worksite investigations to identify specific hazards or stresses that could be causing the observed injuries, and to plan the subsequent hazard control programs. Health professionals must work closely with supervisors to ensure the prompt reporting and treatment of all work-related health and safety problems. Finally, physicians or nurses are called on to assist managers and supervisors in placing workers with permanent or temporary disabilities. This requires matching job demands to the specific capabilities and limitations of the worker.

The *maintenance department* and *skilled trades* (electricians, millwrights, and others) play a critical role in the success of the safety program by routinely inspecting facilities, equipment, and tools, and servicing them when necessary. Maintenance work presents special hazards because of constantly changing tasks, a variety of locations where work is performed, tight time constraints, and work in locations not intended for full-time human occupancy. People in these groups should receive special training to enhance their knowledge of safety hazards and control technology. Standard operating procedures should be developed for all anticipated tasks to ensure that exposures to risks are minimized.

Finally, *workers* play an essential role in the successful execution of safety program. Before a new assignment, a worker must be educated about specific hazards associated with his or her new job. Training should include both hazard recognition and control techniques. If personal protective equipment is required to ensure safety, training must cover how to inspect, maintain, and wear such equipment. Training must emphasize the responsibility of each worker to maintain a safe workstation and to comply with safe work practices. Part of this responsibility includes the necessity to report unsafe conditions to supervisors and employee safety representatives in order that corrective action can be taken.

Employee participation is an important

component of the total safety program. Each worker is an expert in his or her job, and should be actively involved in inspections and systems safety analyses. If modifications are deemed necessary to reduce hazards, worker acceptance of new equipment, tools, and work methods is an essential ingredient in the successful implementation of change. For this reason, workers should actively participate in the design of equipment and process safety features and the selection of personal protective equipment.

Hazard Discovery and Identification

A completely successful safety program would identify and eliminate all hazards *before* any accidents occur. *Systems safety analysis* is a subdiscipline of safety engineering concerned with the discovery and evaluation of hazards so that preventive actions can be taken to prevent or substantially reduce the likelihood of an accident. Over 30 different systems safety methodologies have been developed for specific applications, such as aircraft design, and consumer product safety. One of these methodologies, *job safety analysis* (JSA), has been found to be particularly useful for identifying hazards in the work environment.

Job safety analysis is performed by an interdisciplinary team composed of the worker, supervisor, and safety/health specialist. If the analyzed job or process is technically complex, the responsible engineers and skilled trades workers should also participate. The first step in JSA is to break the job down into a sequence of work elements. The next step is to scrutinize each element to identify existing or potential hazards. To do this effectively, the worker should simulate or "walk through" each element, explaining the details of the element to the team and describing previous accidents or "near misses" and the associated causes. Team members rely on experience and expertise in their specialty areas to identify any additional hazards. The results of this analysis are used to recommend changes to the workstation, process, or methods to eliminate or effectively control all identified hazards.

Certain operations pose special hazards to both workers and the surrounding community because of the storage, use, or production of materials that are either highly toxic or reactive (having the potential to cause a major explosion or fire). These types of operations are usually associated with the chemical and petroleum industries, but can also be found in the pharmaceutical, food processing, paper, automotive, and electronic industries. Because these operations have the potential for a catastrophic accident with multiple deaths and widespread property damage, OSHA issued a process safety standard in 1992. The standard is performance based and addresses 12 major points, including:

• Developing a written description of potential hazards
• Performing comprehensive analyses to evaluate hazards and establish priorities for implementing changes to enhance safety
• Formalizing and enforcing safe operating procedures to reduce the likelihood of an accident
• Instituting emergency response planning, including developing evacuation procedures and coordinating with community law enforcement and firefighting agencies

Job safety analysis and the OSHA process safety standard are widely regarded as useful techniques for the identification and control of hazards *before* accidents, injuries, or illnesses. To maximize their effectiveness, these techniques must be formally incorporated into the safety program and practiced on a regular basis.

REFERENCES

1. National Safety Council. Accident facts (1998 edition). Itasca, IL: National Safety Council, 1998.

BIBLIOGRAPHY

Brauer RL. Safety and health for engineers, New York: Van Nostrand Reinhold, 1990.

A comprehensive text that is significantly more quantitative and engineering oriented than most contemporary books on safety and health. This book covers a broad range of topics and problems faced by engineers and other safety and health professionals. Numerous illustrations, sample problems, and reference citations enhance the presentation of technical topics.

Clemens PL. A compendium of hazard identification and evaluation techniques for systems safety application. Hazard Prevention March/April 1982:11–18.

Provides a brief summary of 25 frequently used systems safety techniques. An excellent bibliography directs the reader to detailed descriptions of each approach.

Fullman JB. Construction safety, security, and loss prevention. New York: Wiley-Interscience, 1984.

Discusses management and engineering aspects of construction safety. Covers hazards unique to the construction environment, such as scaffolding and excavations.

Goetsch DL. Occupational safety and health for technologists, engineers, and managers. Upper Saddle River, NJ: Prentice-Hall, 1999.

A useful reference or instructional text that covers both engineering and management aspects of occupational safety and industrial hygiene. Engineering chapters are organized by generic hazard types and include coverage of machine guarding, ergonomics, temperature ex-tremes, electrical hazards, fire safety, product safety/liability, and an overview of industrial hygiene. Management chapters provide broad topical coverage of many issues, including employee training, workers' compensation, safety legislation, violence in the workplace, accident investigation, and data management.

Krieger GR, Montgomery JF, eds. Accident prevention manual for business and industry: engineering and technology, 11th ed. Itasca, IL: National Safety Council, 1997.

Reference manual for identification and control of generic safety topics such as machine guarding, fire, electricity, materials handing, and building maintenance. Well illustrated with topical chapters that summarize relevant safety standards and established control technology.

National Safety Council. Power press safety manual, 4th ed. Itasca, IL: National Safety Council, 1989.

Covers basic safety issues related to the design, setup, operation and guarding of mechanical power presses. Illustrations provide good examples for safeguarding a wide variety of stamping situations.

Winburn DC. Practical electrical safety. New York: Marcel Dekker, 1988.

Covers basic issues related to effects of electrical current on the human body and controlling exposures to electrical hazards in residential, occupational, and construction settings. Summarizes major points of the National Electrical Code and relevant federal standards.

9 Occupational Ergonomics Promoting Safety and Health Through Work Design

W. Monroe Keyserling

Ergonomics is the study of humans at work in order to understand the complex interrelationships among people, their work environment (e.g., facilities, equipment, and tools), job demands, and work methods. A basic principle of ergonomics is that all work activities place some level of physical and mental demands on the worker. If these demands are kept within reasonable limits, work performance should be satisfactory and the worker's health and well-being should be maintained. If demands are excessive, however, undesirable outcomes may occur in the form of errors, accidents, injuries, or a decrement in physical or mental health.[1]

Ergonomists evaluate demands associated with work and the corresponding abilities of people to react and cope. The goal of an occupational ergonomics program is to create a safe work environment by designing facilities, furniture, machines, tools, and job demands to be compatible with workers' attributes, such as size, strength, aerobic capacity, information-processing capacity, and expectations. A successful ergonomics program should simultaneously improve health and enhance productivity.

W.M. Keyserling: Department of Industrial and Operations Engineering, University of Michigan, Ann Arbor, Michigan 48109-2117.

[1] An *accident* is defined as an unanticipated, sudden event that results in an undesired outcome such as property damage, bodily injury, or death. An *injury* is defined as physical damage to body tissues caused by an accident or by exposure to environmental stressors. Injuries can be associated with accidents, but can also result from normal stresses in the environment.

The following examples call attention to ergonomic issues that may affect health and safety in the contemporary workplace:

Accident Prevention

- Designing a machine guard that allows a worker to operate equipment with smooth, comfortable, time-efficient motions. This minimizes inconveniences introduced by the guard and decreases the likelihood that it will be bypassed or removed, thus exposing the worker to mechanical hazards. A well designed guard may also eliminate awkward postures that lead to musculoskeletal disorders in vulnerable body parts such as the lower back, shoulder, and upper extremity.

- Evaluating the biomechanics of human gait to determine forces acting between the floor surface and the sole of the shoe. This information is used to determine friction requirements between the floor surface and shoe sole to reduce the risk of a slip or fall. Falls can also be prevented by alerting workers to slip and trip hazards, such as puddles of oil on the floor, uneven floor surfaces, and changes in floor elevation. Good lighting enhances the ability of workers to perceive and react to these hazards.

- Designing warning signs for hazardous equipment and work locations so that workers take appropriate actions to avoid accidents. Warnings are particularly important for inexperienced workers or if the hazards are hidden or subtle.

Preventing Excessive Fatigue and Discomfort

- Designing a computer workstation (equipment and furniture) and associated tasks so that an operator can use a monitor, mouse, and keyboard for an extended period without experiencing visual fatigue or musculoskeletal discomfort. Discomfort in the neck or upper extremities may be a precursor of serious problems, such as tendinitis or carpal tunnel syndrome.
- Evaluating the metabolic demands of a job performed in a hot, humid environment to develop a work–rest regimen that prevents heat stress.
- Establishing maximum work times for transportation workers, such as truck drivers and airline pilots, to reduce the risk of drowsiness, performance errors, and accidents caused by sleep deprivation.

Preventing Musculoskeletal Disorders Caused by Overexertion or Overuse

- Evaluating lifting tasks to determine biomechanical strain on the lower back and designing lifting tasks to prevent back injuries.
- Evaluating workstation layouts to discover causes of postural stress and implementing changes to eliminate awkward work postures (such as torso bending and twisting, and overhead work with the arms and hands) associated with the development of musculoskeletal disorders. Eliminating awkward postures may also reduce fatigue.
- Evaluating highly repetitive manual assembly line jobs and developing alternative hand tools and work methods to reduce the risk of cumulative trauma disorders such as tendinitis, epicondylitis, tenosynovitis, and carpal tunnel syndrome.

The remainder of this chapter describes several subdisciplines of ergonomics concerned with occupational safety and health.

HUMAN FACTORS ENGINEERING

Human factors engineering—also called engineering psychology or cognitive ergonomics—is concerned with perceptual, information-processing, and psychomotor aspects of work. Engineering psychologists design displays, controls, procedures, software, equipment, and the general work environment to improve work performance and to reduce accidents caused by human error. Common causes of work accidents due to human error include:

1. *Failure to perceive or recognize a hazardous condition or situation.* To react to a dangerous situation, a worker must first perceive that danger exists. Many workplace hazards, such as excessive pressure inside a boiler, a forklift or automatic guided vehicle approaching from behind in a noisy factory, uneven flooring in a poorly lit room, or the sudden release of an odorless, colorless toxic gas, are not easily perceived through human sensory channels. These situations require special informational displays. The boiler should be equipped with a gauge that displays internal pressure, coupled with an audible alarm that activates when the pressure exceeds safe limits. Forklifts and automatic guided vehicles should have beepers and flashing lights that operate when the vehicle moves. Good lighting is required near trip hazards, and alarm systems should sound if toxic gases are released. Warning signs at locations with concealed hazards, such as confined space entry points, enhance awareness and help to prevent accidents.
2. *Failure in information-processing or decision-making processes.* Decision making involves combining new information with existing knowledge to provide a basis for action. Errors can occur at this stage if the information-processing load is excessive. For example, during the Three Mile Island nuclear power plant accident, operators were required to react to an overwhelming number of simultaneous alarms. Decision-

making errors can also occur if previous training was incorrect or inappropriate for handling a specific situation.

3. *Failures in motor actions after correct decisions.* After a decision, it is frequently necessary for a worker to perform a motor action such as flipping a switch or adjusting a knob to control the status of a system or machine. Problems can occur if required actions exceed motor abilities. For example, the force required to adjust a control valve in a chemical plant should not exceed a worker's strength. Errors can occur if controls are not clearly labeled or if manipulation of the control causes an unexpected response. Switches that start potentially dangerous machinery or equipment should be guarded to prevent accidental activation. This is accomplished by covering the switch, locking it in the "off" position, or placing it in a location where it cannot be accidentally touched.

Workers with disabilities may require special accommodations to work safely. For example, fire alarms with strobe lights are needed to warn the hearing disabled, whereas computers equipped with voice recognition hardware and software can accommodate people who have lost the use of their hands or where traditional data entry devices (keyboards and mice) cause or aggravate upper extremity musculoskeletal disorders. Older workers may also require accommodations because many human capabilities, such as vision, hearing, balance, aerobic endurance, reaction time, and strength, begin slowly to decline starting at approximately 40 years of age. The aging of the workforce will present increasing accommodation challenges in the 21st century. When designing work, ergonomists must consider the capabilities and needs of older workers. (See Chapter 39.)

WORK PHYSIOLOGY

Work physiology is concerned with stresses that occur during the metabolic conversion of stored biochemical energy sources to mechanical work. If these stresses are excessive, the worker experiences fatigue. Fatigue may be localized to a relatively small number of muscles or may affect the entire body.

Static Work and Local Muscle Fatigue

Static work occurs when a muscle remains in a contracted state for an extended period. High levels of static work can be caused by sustained awkward posture, such as a mechanic who must continuously flex the trunk while repairing an automobile engine; or by high strength demands associated with a specific task, such as using a tire iron to unfreeze a badly rusted wheel nut when changing a tire.

When a muscle contracts, its blood vessels are compressed by the adjacent contractile tissue. Vascular resistance increases with the level of muscle tension and the blood supply to the working muscle decreases. If the muscle cannot relax periodically, the demand for metabolic nutrients exceeds the supply and metabolic wastes accumulate. The short-term effects of this condition include ischemic pain, tremor, or a reduced capacity to produce tension. Any of these effects can severely inhibit work performance [1].

Figure 9-1 shows the relationship between the intensity and duration of a static exertion. A contraction of maximum intensity can be held for only about 6 seconds. At 50% of maximum intensity, the limit is approximately 1 minute. To sustain a static contraction indefinitely, muscle tension must be kept below 15% of maximum strength. The endurance curve shown in Fig. 9-1 reflects time to exhaustion. It is not desirable for workers to exert themselves to the point of exhaustion, and therefore static work demands should stay below the curve. A Swedish study used physiologic measurements, such as electromyography and measurements of blood lactate levels, to document residual fatigue in muscles 24 hours after sustained handgrip exertions at only 10% maximum strength [2]. There is increasing epidemiologic evidence showing an association between forceful

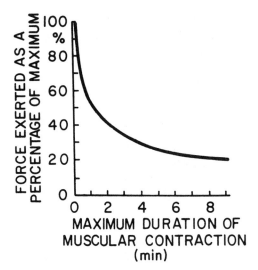

FIG. 9-1. Maximum duration of a static muscle contraction for various levels of muscular contraction. (From Keyserling WM, Armstrong TJ. Ergonomics. In: Last JM, ed. Maxcy-Rosenau public health and preventive medicine, 12th ed. Norwalk, CT: Appleton, Century, Crofts, 1986:734–750.)

static exertions and elevated rates of musculoskeletal disorders (see later, and Chapter 26). Work activities should be designed so that static exertions are of limited duration and adequate recovery time is built into the job. Dynamic activities involving cyclical contraction and relaxation of working muscle are in general preferable to static work.

Static work also causes a temporary increase in the peripheral resistance of the cardiovascular system. Significant increases in heart rate and mean arterial blood pressure have been observed in conjunction with short-duration static contractions (3). Caution should be exercised when placing a person with a history of cardiovascular disease on a job that requires moderate or heavy static exertions. If feasible, the job should be modified to reduce the intensity and duration of these exertions.

Dynamic Work and Whole-Body Fatigue

Whole-body dynamic work occurs when large skeletal muscle groups repeatedly con-

tract and relax while performing a task. Common examples of dynamic work include walking on a level surface, pedaling a bicycle, climbing stairs, shoveling snow, and carrying a load.

The intensity of whole-body dynamic work is limited by the capacity of the pulmonary and cardiovascular systems to deliver adequate supplies of oxygen and glucose to the working muscles and to remove the products of metabolism. Whole-body fatigue occurs when the collective metabolic demands of working muscles throughout the body exceed this capacity. Symptoms of whole-body fatigue include shortness of breath, weakness in working muscles, and a general feeling of fatigue. These symptoms continue and may increase until the work activity is stopped or decreased in intensity.

For extremely short durations of whole-body dynamic activity (typically 4 minutes or less), a person can work at an intensity equal to his or her aerobic capacity before a significant rest break is required. As the duration of the work period increases, the work intensity must decrease. If a task continues for 1 hour, the average energy expenditure for this period should not exceed 50% of the worker's aerobic capacity. For a job that is performed for an 8-hour shift, the average energy expenditure should not exceed 33% of the worker's aerobic capacity.

Aerobic capacity varies considerably within the population. Table 9-1 presents mean aerobic capacities for untrained men and women (nonathletes) of various ages. Aerobic capacity peaks in the third decade

TABLE 9-1. *Average aerobic capacities (kilocalories per minute) of untrained men and women for various ages*

Age (yr)	Men	Women
20	15.0	11.0
30	15.0	9.5
40	13.0	8.5
50	12.0	8.0
60	10.5	7.5
70	9.0	6.5

From Stegemann J. *Exercise physiology: physiologic bases of work and sport.* Chicago: Year Book, 1981.

(20 to 29 years) for both men and women. At 50 years of age, average aerobic capacity decreases to approximately 90% of the peak value; by 65 years of age, it falls to approximately 70% of the peak (4). These are average values for each age–sex category and do not reflect the full range of variability among the adult population. This variability is an important consideration when evaluating ergonomic stress; a job that is relatively easy for a person with high aerobic capacity can be extremely fatiguing for a person with low capacity. In particular, some older workers may have difficulty performing jobs with high energy expenditure requirements.

The prevention of whole-body fatigue is accomplished through good work design. The energy demands of a job should be sufficiently low to accommodate the adult working population, including people with limited aerobic capacity. This can be accomplished by designing the workplace to minimize unnecessary body movements (excessive walking or climbing) and providing mechanical assists, such as hoists or conveyors for handling heavy materials. If these approaches prove infeasible, it may be necessary to provide additional rest allowances to prevent excessive fatigue. This is particularly true in hot, humid work environments because of the metabolic contribution to heat stress (see Chapter 19).

In establishing metabolic criteria for jobs that involve repetitive manual lifting, the National Institute for Occupational Safety and Health (NIOSH) (5,6) recommends that the average energy expenditure during an 8-hour work shift should not exceed 3.5 kcal/minute. Applying the "33%" rule to the values in Table 9-1, the NIOSH rate would be acceptable to most of the adult population. Caution should be practiced when placing people with low levels of physical fitness on metabolically strenuous jobs.[2]

To assess the potential for whole-body fatigue, it is necessary to determine the energy expenditure rate for a specific job. This is usually done in one of three ways:

1. *Table reference:* Extensive tables of the energy costs of various work activities have been developed and can be consulted. (The text by McArdle, Katch, and Katch cited in the Bibliography provides tables describing the energy cost of many work tasks.)
2. *Indirect calorimetry:* Energy expenditure can be estimated for a specific job by measuring a worker's oxygen uptake while performing the job (4,7).
3. *Modeling:* The job can be analyzed and broken down into fundamental tasks such as walking, carrying, and lifting. Parameters describing each task are inserted into equations to predict energy expenditure (8).

There is no "best" method for determining energy expenditure. The selection of a method is often a trade-off between the availability of published tables or prediction equations for the specific work activities of interest versus the time and expenses associated with data collection for indirect calorimetry. Indirect calorimetry is indicated when a precise measure of energy expenditure is required.

BIOMECHANICS

Biomechanics is concerned with the mechanical properties of human tissue and the response of tissue to mechanical stresses. Some injury-causing mechanical stresses in the work environment are associated with overt accidents, such as bones in the feet crushed by the impact of a dropped object. The hazards that produce these injuries can usually be controlled through safety engineering techniques (see Chapter 8). Other injurious mechanical stresses are more subtle and can cause cumulative trauma injuries. These stresses can be external, such as a chain saw that causes vibration white finger (or Raynaud's) syndrome, or internal, such as com-

---

[2] Aerobic capacity can be determined by measuring oxygen uptake and carbon dioxide production during a stress test. For additional information on measuring or estimating aerobic capacity, see the texts by McArdle, Katch, and Katch (4), and by Rodahl (7).

pression on spinal discs during strenuous lifting.

Work-related overexertion disorders (also called *cumulative trauma disorders*) often result from excessive biomechanical stress. These disorders are frequently seen in the lower back, neck, shoulders, or upper extremities and include a variety of injury and disease entities such as sprains, strains, tendinitis, bursitis, and carpal tunnel syndrome (9, 10). Because these disorders impair mobility, strength, tactile capabilities, or motor control, affected workers may be unable to perform their jobs. In many industries, overexertion is the leading cause of workers' compensation expenditures. At companies whose operations include a large amount of manual materials handling or repetitive assembly, overexertion disorders may account for well over one-half of all occupational health and safety expenditures. (For additional information on musculoskeletal disorders and related overexertion syndromes, see Chapter 26.)

Ergonomists and other health professionals are often called on to perform job analyses to identify and control exposures to biomechanical risk factors that cause overexertion injuries and disorders. These risk factors can be grouped into the following categories (9–12):

- Forceful exertions
- Awkward postures
- Localized mechanical contact stresses
- Vibration
- Temperature extremes
- Repetitive exertions
- Sustained or prolonged static exertions or postures

In addition to identifying the presence of these risk factors, job analysis determines specific aspects of the job, such as workstation layout, production standards, incentive systems, work organization, or work methods, that cause or contribute to worker exposures. This information must be obtained to design and implement job modifications effectively.

The following sections present a brief discussion of each risk factor.

Forceful Exertions

"Whole-body" exertions such as strenuous lifting, pushing, and pulling can cause back pain and other injuries and disorders (Fig. 9-2). Because lifting and handling of heavy weights are the most commonly cited activities associated with occupational low back pain, NIOSH (5,13) has issued guidelines for the evaluation and design of jobs that require manual lifting. These guidelines consider task factors such as lift frequency, work duration, workplace geometry, and posture to establish the amount of weight that a person can safely lift. Factors other than object weight play a significant role in the amount of force that workers can safely exert during lifting and other manual transfer tasks. Be-

FIG. 9-2. Lift assists are necessary to prevent the risk of back injury on this job.

cause of the effect of long moment arms, handling relatively light loads can stress muscles in the back and shoulder if the load is held at a long horizontal distance in front or to the side of the body.

One or more of the following approaches may prove useful in reducing the magnitude of forces exerted during whole-body exertions:

1. Reduce the weight of the lifted object by decreasing the size of a unit load, such as by placing fewer parts in a tote bin or purchasing smaller bags of powdered or granular materials.
2. Reduce extended reach postures by removing obstructions, such as rails on storage bins, that prevent the worker from getting close to the lifted object.
3. Use gravity or mechanical aids, such as conveyors, hoists, conveyors, or articulating arms, to assist the worker or eliminate the manual exertion (Fig. 9-3).

Forceful exertions of the hands, such as cutting with knives or scissors, tightening screws, snapping together electrical connectors, and using the hands/fingers to sand or buff parts, can cause upper extremity disorders such as tendinitis or carpal tunnel syndrome. Pinch grips, heavy tools, poorly balanced tools, poorly maintained tools, such as dull knives or scissors, or low friction between the hand and tool increase the forces exerted in the finger flexor muscles and tendons. Gloves may increase force requirements of some jobs because of reduced tactile feedback, reduced friction, or resistance of the glove itself to stretching or compression. Environmental conditions may also increase force requirements because some rubber and plastic materials lose their flexibility when cold and become more difficult to shape or manipulate. One or more of the following approaches may prove useful in reducing the forcefulness of hand exertions:

1. Substitute power tools (electric or pneumatic) for manual tools. If a power tool is infeasible, redesign the manual tool to increase mechanical advantage or otherwise decrease required hand forces.
2. Suspend heavy tools with "zero-gravity" balance devices.
3. Treat slippery handles on tools and other objects with friction-enhancing coatings to minimize slippage in the hands.
4. Move the handle of an off-balance tool closer to the center of gravity or suspend the tool in a way that minimizes off-balance characteristics.
5. Use torque control devices (reaction arms, automatic shut-off) on power tools such as air wrenches, nut runners, or screwdrivers (Fig. 9-4) (6,14,15). Investi-

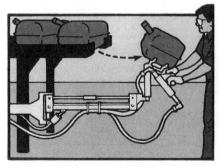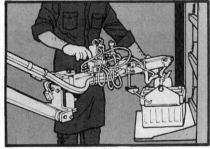

FIG. 9-3. Mechanical assist devices can reduce or eliminate forceful exertions during manual materials handling activities such as lifting or carrying. (Courtesy of The University of Michigan and the UAW-Ford National Joint Committee on Health and Safety. From The University of Michigan Center for Ergonomics. Fitting jobs to people: an ergonomics process. Ann Arbor, MI: The Regents of the University of Michigan, 1991.)

FIG. 9-4. Torque control devices can substantially reduce the amount of force exerted when using air wrenches and similar tools. Note that the weight of the tool is also borne by the device, further reducing the force exerted by the worker. (Courtesy of The University of Michigan and the UAW-Ford National Joint Committee on Health and Safety. From The University of Michigan Center for Ergonomics. Fitting jobs to people: an ergonomics process. Ann Arbor, MI: The Regents of the University of Michigan, 1991.)

gate if torque can be reduced without adversely affecting product quality.

6. If high force is required to assemble poorly fitting parts, consider improving quality control to achieve better fit, or

using a lubricant to facilitate the assembly of tightly fitting parts.

7. Prewarm rubber and plastic components if these become cold and unmalleable during storage.

Awkward Posture

Awkward posture at any joint may cause transient discomfort and fatigue. Prolonged awkward postures may contribute to disabling injuries and disorders of musculoskeletal tissue or peripheral nerves. Awkward trunk postures such as those shown in Fig. 9-5 increase the risk of back injuries (10,16). Raising the elbow above shoulder height or reaching behind the torso can increase the likelihood of musculoskeletal problems in the neck and shoulders. The worker shown in Fig. 9-6 must position his arms in an extended forward reach because of poor workstation layout. The garment worker in Fig. 9-7 is at risk for carpal tunnel syndrome.

Most awkward postures of the trunk and shoulder result from excessive reach distances, such as bending into bins or behind the body to place or retrieve parts, reaching overhead to high shelves and conveyors, or

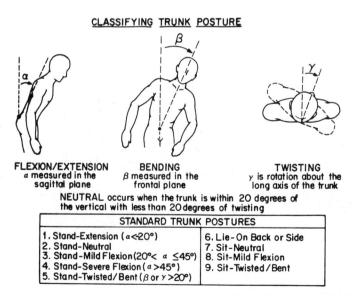

FIG. 9-5. A method for classifying nonneutral trunk postures. (From Keyserling WM. Postural analysis of the trunk and shoulders in simulated real time. Ergonomics 1986;29:569–583.)

FIG. 9-6. This job involves exposure to several risk factors associated with the development of upper extremity cumulative trauma disorders. (Photograph by Earl Dotter.)

reaching overhead or in front of the body to activate machine controls. These postures can be eliminated through improved workstation layout. In general, workers should not reach below knee height or above shoulder

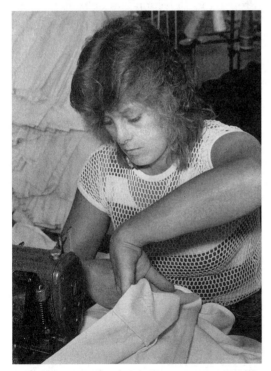

FIG. 9-7. Garment workers, who perform many awkward and repetitive tasks in their jobs, are at high risk for carpal tunnel syndrome. (Photograph by Earl Dotter.)

height for prolonged periods. Routine forward reaches should be performed with the trunk upright and the upper arms nearly parallel to the trunk. Where possible, workstations and equipment should offer adjustability to accommodate workers of different body sizes. Anthropometry is the branch of ergonomics concerned with the study of body size and dimensions. Anthropometry is also concerned with designing facilities and equipment to accommodate work populations of varying body dimensions. A detailed presentation of anthropometry is beyond the scope of this chapter. (For additional information, refer to the textbook by Pheasant in the Bibliography.)

Allowing workers to sit while working reduces fatigue and discomfort in the legs and feet and can increase stability of the upper body. (A high level of body stability is essential for precision manual tasks.) However, prolonged sitting may be a factor in the development of back pain. A well designed workseat, such as one with a good lumbar support and adjustability of the seat pan and backrest, enhances comfort and can reduce the risk of health problems. Layouts that allow workers to alternate between standing and sitting postures are also desirable.

Awkward upper extremity postures can occur at the shoulder (discussed earlier), elbow, or wrist. It is important to avoid frequent or prolonged activities that require a worker to bend the wrist. Hand tool features, such as the shape and orientation of handles, in combination with workstation layout (location and orientation of work surfaces) play an important role in determining wrist postures.

Localized Contact Stresses

Local mechanical stresses result from concentrated pressure during contact between body tissues and an object or tool. "Hand hammering" (using the palm as a striking tool) is used in some manufacturing operations as a method for joining two parts. This activity, which can irritate nerves and other tissues in the palm, can be avoided by using a mallet. Hand tools with hard, sharp, or small-diameter handles, such as knives, pliers, and scissors, can irritate nerves and tendons in the palm or fingers. This problem can be controlled by either padding the handles or increasing the radius of curvature of tool handles. In some bench assembly activities and office jobs, contact stresses result from resting the forearms or wrists against a sharp, unpadded workbench edge. This problem can usually be controlled by either rounding or padding the edge, or by providing a support for the forearm and wrist.

Seated workstations that produce localized pressure on the posterior knee and thigh can impair circulation to the lower extremities, causing swelling and discomfort in the lower legs, ankles, or feet. A common cause of this condition in both factories and offices is a work seat that is too high, allowing the lower legs to dangle. Because the full weight of the lower extremities hangs from the work seat, concentrated compressive forces can squeeze tissues in the area where the thighs contact the front edge of the seatpan. Solutions to this problem include adjustable seats or providing a foot rest to support some of the weight of the lower extremities.

Vibration

Exposure to whole-body vibration while driving or riding in motor vehicles (including forklifts and off-road vehicles) may be a factor in increased risk of back pain (10). Because driving tasks are usually performed in a seated posture, most drivers are exposed to two back pain risk factors. Driving over rough surfaces for prolonged periods while sitting in a poorly suspended seat can increase vibration exposure. Standing on vibrating floors, such as near power presses in a stamping plant or near shakeout equipment in a foundry, may also result in exposure to whole-body vibration.

Localized vibration of the upper extremity (also called segmental vibration) can occur when using powered hand tools such as screwdrivers, nutrunners, grinders, jackhammers, and chippers. Other exposures include holding parts against grinding wheels or prolonged gripping of a vibrating steering wheel. Localized vibration may contribute to the development of hand–arm vibration syndromes such as vibration white finger (10). Proper tool selection can help in reducing exposure. For example, an air wrench that uses an automatic shut-off system produces less vibration exposure than a slip-clutch mechanism (15). Many manufacturers offer a variety of low-vibration hand tools. For additional discussion of vibration, see Chapter 19.

Temperature Extremes

Exposure to unusually hot or cold ambient temperatures can produce a variety of adverse health effects, as discussed in Chapter 19. In addition to considering the general thermal characteristics of the work room (air temperature, air movement, and relative humidity), it is also necessary to look at temperature extremes that affect the hands. For example, handling extremely hot or cold parts may require the use of special gloves that increase the force requirements of the job (see earlier). In jobs that involve the use of pneumatic tools, air from high-pressure lines and tool exhaust ports may be directed onto the hands, causing local chilling and reducing manual dexterity and tactile sensitivity. This exposure can be controlled by eliminating leaks or by directing exhaust air away from the hands.

Repetitive and Prolonged Activities

The biomechanical and physiologic strain experienced by a worker is related to the cumulative exposure to all the risk factors discussed previously. Because ergonomic risk factors are often related to specific work activities, jobs that involve high repetition or prolonged activities, such as driving 5,000 screws a day on an assembly line or continuous word processing in an office, typically involve higher exposures than nonrepetitive jobs, such as inspection tasks in a factory or a supervisory position in an office. Studies have shown that jobs with a basic cycle time of 30 seconds or less (a production rate of two or more parts per minute) have an elevated rate of carpal tunnel syndrome and related disorders. Longer cycle times do not necessarily result in lower risk of injury if basic hand motions are repeated within the cycle or if the intensity of work is so high that the worker is unable to have brief micropauses in hand exertions. Jobs where over 50% of the work cycle involves similar hand motions may have elevated rates of upper extremity disorders (17).

Repetitiveness as a risk factor is not limited to upper extremity problems. Frequent lifting and other manual materials handling activities increase the risk of back pain (5,6,13).

Repetitiveness can often be measured or estimated using industrial engineering records and other work standards. For example, on an assembly line, repetitiveness is a function of the line speed or the time allowed to complete one unit of work. For a clerk in a bank or insurance office, repetitiveness can be a function of the number of forms processed a day. For a supermarket checker, repetitiveness is a function of the number of items scanned over the course of a work shift. Repetitiveness can also be measured using an observational technique where the rapidity and intensity of hand motions are compared against benchmarks or a scale with verbal anchors (12). A scale for describing the repetitiveness of hand-intensive work is presented in Fig. 9-8.

Resolving problems of repetition and prolonged exertions can be a major challenge. Two possible approaches are job enrichment and job rotation. The premise behind these approaches is to increase the overall variety of activities performed by a worker to reduce the repetitiveness of any specific stressful activity. Although good in theory, these approaches may be very difficult to implement. Job enrichment and job rotation are not feasible in work locations where there are no "low-repetition" jobs to combine with the "high-repetition" jobs. Even in situations where a good mix of low- and high-repetition jobs exists there may be other factors, such as increased learning time and seniority practices, that present significant barriers. In these instances, it may be necessary to estab-

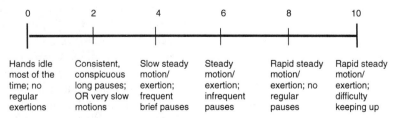

FIG. 9-8. Visual analog scale for rating repetition/hand activity with verbal anchors. (From Latko WA, Armstrong TJ, Foulke JA, Herrin GD, Rabourn RA, Ulin SS. Development and evaluation of an observational method for assessing repetition in hand tasks. Am Ind Hyg Assoc J 1997;58:278–285.)

lish a participative ergonomics program and to educate management and workers before attempting these interventions.

COMPONENTS OF AN ERGONOMICS PROGRAM

An effective program for controlling overexertion injuries and disorders starts with the commitment and involvement of management to provide the organizational resources and motivation to control ergonomic hazards in the workplace. Management must also perform regular reviews and evaluations of the program to ensure program goals are met in a deliberate and timely manner. Because ergonomic programs focus on improving the complex interrelationships among workers and their jobs, employee involvement is essential to ensuring the success of the program (18). The United States and other nations are working toward developing ergonomics standards and guidelines (Box 9-1).

An effective ergonomics program should include the following components:

- Surveillance of health and safety records to identify patterns of overexertion injuries and illnesses
- Job analysis to identify worker exposures to risk factors that cause overexertion injuries and illnesses
- Job design, and redesign if necessary, to reduce or eliminate ergonomic risk factors
- Training of managers, engineers, and workers in the recognition and control of ergonomic risk factors
- Medical management of injured workers to improve the chances for a speedy return to work.

Limited resources must be directed at those jobs with the greatest ergonomic problems. One approach for identifying high-hazard jobs is to analyze available medical, insurance, and safety records, such as the Occupational Safety and Health Administration (OSHA) 200 log, for evidence of high rates of cumulative trauma disorders in certain departments, job classifications, or workstations. This approach is called *passive surveillance* because it relies on previously collected information. Passive surveillance may underestimate the true level of cumulative trauma problems. (For example, at small plants that do not have in-plant medical services, a worker may seek treatment from his or her personal physician. Unless the worker requests coverage under the workers' compensation system, the complaint and associated treatment may not appear in any company records.) *Active surveillance* involves a more aggressive approach to identifying potential problems. Active surveillance may include employee surveys to identify jobs associated with elevated rates of discomfort in the back, neck, shoulders, and upper extremities. Active surveillance may also include interviews with supervisors and personnel managers to identify jobs with high turnover. If other employment opportunities are available, workers often seek relief by leaving jobs with unusually high physical stresses before a cumulative trauma injury develops.

Once the high-risk jobs have been identified, the next step is to determine the specific causes of exposure so that corrective actions can be taken. This activity involves job analysis to identify the various risk factors discussed previously and the development of engineering or administrative controls to reduce or eliminate exposures. The appropriateness of an intervention to reduce ergonomic stress varies among and within facilities. Changes that are practical at one workstation may not be appropriate for others. Alternatives must be evaluated to determine the best strategy for resolving each ergonomic problem. It is also important to recognize that most solutions require some degree of fine tuning to ensure that they are acceptable to workers and accomplish the intended reductions in ergonomic stress. Follow-up job analyses should be performed to ensure that the solution is working effectively and that no new stresses have been introduced. Follow-up health surveillance is also recommended to detect any changes in

Box 9-1. Key Elements for Ergonomics Standards and Guidelines

*Laura Punnett and
Barbara Silverstein*

There are ongoing efforts to develop ergonomics standards and guidelines to prevent musculoskeletal disorders. In the United States, this has included federal and state government regulations as well as voluntary guidelines by professional and trade associations. Standards and guidelines have also been developed and implemented in Western Europe, Brazil, Japan, Australia, New Zealand, and British Columbia (Canada). Those working to design useful and appropriate regulations or guidelines often use a model with most or all of these seven key elements (see also Chapter 26):

1. A system to document job tasks that require repetitive or forceful exertions, awkward postures, and vibration to reduce hazardous exposures (primary prevention). Hazard surveillance may use checklists, review of employee complaints, or formal job analyses. It should be carried out when a control program is first put into place and updated periodically or whenever there are changes in the work process. Jobs with obvious risks need to be modified independent of whether health effects have been reported. Hazard surveillance responsibility should be assigned to management personnel or labor–management health and safety/ergonomic committee members who have been trained to recognize ergonomic problems in the worksite.

2. A system to document health-related events that help target need for intervention and secondary prevention. Morbidity surveillance should use both routine reporting systems and targeted surveillance. At a minimum, there should be timely reviews of workers' compensation records, logs of medical department visits, or OSHA 200 logs of workplace injuries and illnesses. Although these vary in their sensitivity, usually jobs or departments with the greatest problems have the highest rates, using any of these data sources. Targeted surveillance might include routine surveys for worker reports of musculoskeletal aches and pains in sus-

pect areas. Nonspecific indicators of possible ergonomic problems include high absenteeism rates and high job turnover.

3. Implementation of effective controls for tasks that pose a risk of musculoskeletal injury and, once instituted, evaluate these approaches to ensure they reduce or eliminate the problem without transferring stressors from one body area (or worker) to another. The four basic principles include (a) increase task variety; (b) keep loads close to the body; (c) reduce fixed positions (the body is made to move); and (d) keep all joints in relatively neutral postures.

4. Design rules for new work processes and operations that minimize risk factors for musculoskeletal disorders. Use of anthropometric data to design workstation dimensions to fit most workers is only one example of the many ways that ergonomic stresses can be reduced or eliminated before a job is even in the prototype stage. Designing jobs to avoid overly routinized and monotonous work is also far easier before personnel have been hired. Incentives should promote product designs that incorporate manufacturability (ergonomic factors related to the manufacturing process) into design criteria and service delivery systems that enhance workers' skill utilization and informed decision making. It is less costly to build good design and procedures into the workplace than to redesign or retrofit later.

5. Training programs to expand management and worker ability to evaluate potential musculoskeletal problems. All employees and managers should understand the nature of work-related musculoskeletal disorders and their causes. Include workers and supervisors in the process of identifying, evaluating, and controlling ergonomic hazards and evaluating control measures. Training should address the importance of early symptom reporting, so that disorders still in the early stages can be correctly diagnosed, treated, and reversed. It may take a substantial period of time before a program reduces the rate or severity of work-related health problems.

6. Health care management programs to emphasize the importance of early detection and treatment of musculoskeletal disorders to prevent impairment and

Box 9-1　(*continued*)

disability. The health care provider must be made aware of the nature of the work performed and the ergonomic stressors involved, both for diagnostic insights and to select appropriate alternative duty assignments. Health care providers should insist on job descriptions that include the magnitude, duration, and frequency of exposure. The health status and job tasks of an injured or recuperating worker must be reviewed on a regular basis to ensure that the work demands have not changed adversely and that the recovery process is on track. Accommodations for injured workers should be provided as early as possible—to maximize the possibility of recovery—and should include controls such as reducing work hours or implementing engineering modifications to reduce the physical workload.

7. Management commitment to address problems at any level and to encourage worker involvement in problem-solving activities. This can be done by specific allocation of resources and time and by recognition of worker–management efforts. The program should have a budget to purchase engineering controls or sponsor training, and the authority to conduct workplace walk-throughs and, if necessary, to shut down imminently hazardous jobs. Workers should be free to make their problems known without fear of reprisal. Further, their input should be actively encouraged because they often have unique insights into the nature of the hazards and ideas for potential solutions.

the pattern of injuries, illnesses, or employee complaints.

REFERENCES

1. Lieber RL, Frieden J. Skeletal muscle metabolism, fatigue, and injury. In: Gordon SL, Blair SJ, Fine LJ, eds. Repetitive motion disorders of the upper extremity. Rosemont, IL: American Academy of Orthopedic Surgeons, 1995:287–300.
2. Bystrom S, Fransson-Hall C. Acceptability of intermittent handgrip contractions based on physiologic response. Hum Factors 1994;36:158–171.
3. Armstrong TJ, Chaffin DB, Faulkner JA, Herrin GD, Smith, RG. Static work elements and selected circulatory responses. Am Ind Hyg Assoc J 1980; 41:254-260.
4. McArdle WD, Katch FI, Katch VL. Exercise physiology: energy, nutrition, and human performance, 4th ed. Baltimore: Williams & Wilkins, 1996.
5. National Institute for Occupational Safety and Health. Work practices guide for manual lifting. NIOSH publication no. 81-122. Cincinnati, OH: National Institute for Occupational Safety and Health, 1981.
6. National Institute for Occupational Safety and Health. Applications manual for the revised NIOSH lifting equation. NIOSH publication no. 94-110. Cincinnati, OH: National Institute for Occupational Safety and Health, 1994.
7. Rodahl K. The physiology of work. London: Taylor and Francis, 1989.
8. Garg A, Chaffin DB, Herrin GD. Prediction of metabolic rates for manual materials handling jobs. Am Ind Hyg Assoc J 1978;39:661–674.
9. Kuorinka I, Forcier L, Hagberg M, et al, eds. Work related musculoskeletal disorders (WMSDs): a reference book for prevention. London: Taylor and Francis, 1995.
10. National Institute for Occupational Safety and Health. Musculoskeletal disorders and workplace factors: a critical review of epidemiologic evidence for work-related musculoskeletal disorders of the neck, upper extremity, and low back. NIOSH publication no. 97-141. Cincinnati, OH: National Institute for Occupational Safety and Health, 1997.
11. Keyserling WM, Armstrong TJ, Punnett L. Ergonomic job analysis: a structured approach for identifying risk factors associated with overexertion injuries and disorders. Applied Occupational and Environmental Hygiene 1991;6:353–363.
12. Latko WA, Armstrong TJ, Foulke JA, Herrin GD, Rabourn RA, Ulin SS. Development and evaluation of an observational method for assessing repetition in hand tasks. Am Ind Hyg Assoc J 1997;58:278–285.
13. Waters TR, Putz-Anderson V, Garg A, Fine LJ. Revised NIOSH lifting equation for the design and evaluation of manual lifting tasks. Ergonomics 1993;36:749–776.
14. Freivalds A, Eklund J. Reaction torques and operator stress while using powered nutrunners. Appl Ergon 1993;24:158–164.
15. Kihlberg S, Kjellberg A, Lindbeck L. Discomfort from pneumatic tool torque reaction: acceptability limits. International Journal of Industrial Ergonomics 1995;15:417–426.
16. Punnett L, Fine, LJ, Keyserling WM, Herrin GD, Chaffin DB. A case-referent study of back disorders in automobile assembly workers: the health effects of non-neutral trunk postures. Scand J Work Environ Health 1991;17:337–346.
17. Silverstein BA, Fine LJ, Armstrong TJ. Occupational factors and carpal tunnel syndrome. Am J Ind Med 1987;11:343–358.
18. Cohen AL, Gjessing CC, Fine LJ, Bernard BP,

McGlothlin JD. Elements of ergonomics programs: a primer based on workplace evaluations of musculoskeletal disorders. NIOSH publication no. 97-117. Cincinnati, OH: National Institute for Occupational Safety and Health, 1997.

BIBLIOGRAPHY

Chaffin DB, Andersson GBJ, Martin BJ. Occupational ergonomics, 3rd ed. New York: Wiley-Interscience, 1999.
Discusses in detail the biomechanical basis of many occupational injuries and disorders, with special coverage of the lower back and upper extremities. Quantitative methods of job analysis are presented with numerous examples of ergonomic approaches to equipment, tool, and workstation design.

Cohen AL, Gjessing CC, Fine LJ, Bernard BP, McGlothlin JD. Elements of ergonomics programs: a primer based on workplace evaluations of musculoskeletal disorders. NIOSH publication no. 97-117. Cincinnati, OH: National Institute for Occupational Safety and Health, 1997.
Presents general guidance for establishing and managing worksite-based ergonomic programs. Topics include organizational elements of effective ergonomics programs, training for managers and workers, health surveillance, job analysis, engineering interventions to reduce exposure to ergonomic stress, and medical management. Text includes many illustrations, examples, and references.

Keyserling WM, Armstrong TJ. Ergonomics and work-related musculoskeletal disorders. In: Wallace RB, ed. Maxcy-Rosenau-Last public health and preventive medicine, 14th ed. Norwalk, CT: Appleton & Lange, 1998:645–660.
A general review of occupational ergonomics with emphasis on causes and prevention of musculoskeletal injuries and disorders of the lower back and upper extremity, anthropometry, work physiology, and biomechanics. A total of 104 references include many recent journal articles describing laboratory and epidemiologic studies.

Kroemer KHE. Fitting the task to the human: a textbook of occupational ergonomics. London: Taylor and Francis, 1997.
A well written survey text that covers all aspects of ergonomics. Chapters on fatigue, work physiology, anthropometry, biomechanics, and cognitive ergonomics provide an excellent introduction to these topics.

Kuorinka I, Forcier L, Hagberg M, et al, eds. Work related musculoskeletal disorders (WMSDs): a reference book for prevention. London: Taylor and Francis, 1995.
Written by a multidisciplinary team of international experts in occupational health and ergonomics, this book provides comprehensive and sophisticated coverage of musculoskeletal diseases and disorders resulting from repeated trauma. Topics include a conceptual model that describes the development of WMSDs, a review of the clinical and epidemiologic literature, evaluation of workplace risk factors, and medical management of affected workers. Although this is an excellent reference book, it does not cover occupational low back pain.

McArdle WD, Katch FI, Katch VL. Exercise physiology: energy, nutrition, and human performance, 4th ed. Baltimore: Williams & Wilkins, 1996.
This comprehensive textbook covers a wide range of issues in work and exercise physiology. Early chapters cover basic exercise physiology (nutrition, energy conversion during exercise, structure and function of the pulmonary, cardiovascular, and neuromuscular systems), whereas advanced chapters cover applied topics such as measurement of human energy expenditure, training for muscle strength and aerobic power, and rehabilitation training programs. Appendices include comprehensive tables of energy expenditure costs of common household, occupational, and recreational activities.

National Institute for Occupational Safety and Health. Applications manual for the revised NIOSH lifting equation. NIOSH publication no. 94-110. Cincinnati, OH: National Institute for Occupational Safety and Health, 1994.
A "hands-on" users' guide for evaluating work activities that require manual lifting. Numerous examples demonstrate application of the 1991 revised NIOSH lifting equation in a variety of work environments. The guide includes a brief summary of the scientific basis of the 1991 lifting equation with references to biomechanical, physiologic, psychophysical, and epidemiologic research. Many illustrations and examples.

Pheasant S. Bodyspace: anthropometry, ergonomics, and the design of work. London: Taylor and Francis, 1996.
Provides comprehensive coverage of the anthropometric aspects of ergonomics. Introductory chapters describe methodologies for measuring and statistically summarizing human body dimensions. This is followed by an excellent presentation of how anthropometric principles are used to design furniture, equipment, and workstations in the home, office, and factory environment. Numerous examples, illustrations, and anthropometric tables make this text an indispensable reference book for both novice ergonomists and experienced ergonomic designers.

Wickens CD, Gordon SE, Liu Y. An introduction to human factors engineering. New York: Addison Wesley Longman, 1998.
Provides a good introduction to all aspects of ergonomics with special emphasis on human factors engineering. Introductory chapters cover human sensory mechanisms, displays, cognition, decision making, and design of controls. Advanced chapters cover a variety of topics, including human–computer interaction, human factors in transportation, usability testing, stress and work performance, and the role of human error in accidents.

Occupational Health: Recognizing and Preventing Work-Related Disease and Injury, 4th ed.
Edited by Barry S. Levy and David H. Wegman, Lippincott Williams & Wilkins, Philadelphia © 2000.

10 Government Regulation of Occupational Health and Safety

Nicholas A. Ashford

The use of chemicals, materials, tools, machinery, and equipment in industrial, mining, and agricultural workplaces is often accompanied by health and safety hazards or risks. Ergonomic injuries account for 34% of all injuries, whereas slips, trips, and falls account for another 25%. These and other hazards cause occupational disease and injury that place heavy economic and social burdens on both workers and employers. Because voluntary efforts in the free market have not succeeded historically in reducing the incidence of these diseases and injuries, government intervention into the activities of the private sector has been demanded by workers. This intervention takes the form of the regulation of health and safety hazards through standard setting, enforcement, and the transfer of information.

In the United States, toxic substances in the workplace were historically regulated primarily through three federal laws: the Mine Safety and Health Act of 1969 (Box 10-1 and Figs. 10-1 and 10-2), the Occupational Safety and Health Act (OSHAct) of 1970, and the Toxic Substances Control Act (TSCA) of 1976. These federal laws have remained essentially unchanged since their passage, although serious attempts at reform are made from time to time. Sudden and accidental releases of chemicals (chemical accidents), which may affect both workers and commu-

nity residents, are now also regulated under the Clean Air Amendments of 1990.

The OSHAct established the Occupational Safety and Health Administration (OSHA) in the Department of Labor to enforce compliance with the act, the National Institute for Occupational Safety and Health (NIOSH) in the Department of Health and Human Services (under the Centers for Disease Control and Prevention) to perform research and conduct health hazard evaluations, and the independent, quasijudicial Occupational Safety and Health Review Commission to hear employer contests of OSHA citations. The Office of Pollution Prevention and Toxic Substances in the Environmental Protection Agency (EPA) administers TSCA. The Office of Chemical Preparedness and Emergency Response in EPA is responsible for the chemical safety provisions of the Clean Air Amendments.

The evolution of regulatory law under the OSHAct has profoundly influenced other environmental legislation, especially the evolution of TSCA. This chapter addresses federal regulation, focusing on standard setting, enforcement mechanisms, and right-to-know provisions.

The OSHAct requires OSHA to (a) encourage employers and employees to reduce hazards in the workplace and to implement new or improved safety and health programs; (b) develop mandatory job safety and health standards and enforce them effectively; (c) establish "separate but dependent responsibilities and rights" for employers and em-

N. A. Ashford: Center for Technology, Policy and Industrial Development, Massachusetts Institute of Technology, Cambridge, Massachusetts 02139.

FIG. 10-1. Mine hazards such as the increased dust exposure from continuous mining machines are regulated by the Mine Safety and Health Administration (MSHA). (Photograph by Earl Dotter.)

ployees for the achievement of better safety and health conditions; (d) establish reporting and record keeping procedures to monitor job-related injuries and illnesses; and (e) encourage states to assume the fullest responsibility for establishing and administering their own occupational safety and health programs, which must be at least as effective as the federal program.

As a result of these responsibilities, OSHA inspects workplaces for violations of existing health and safety standards; establishes advisory committees; holds hearings; sets new or revised standards for control of specific substances, conditions, or use of equipment; enforces standards by assessing fines or by other legal means; and provides for consultative services for management and for employer and employee training and education. In all of its procedures, from the development of standards through their implementation and enforcement, OSHA guarantees employers and employees the right to be fully informed, to participate actively, and to appeal its decisions (although employees are limited somewhat in the latter activity).

The coverage of the OSHAct initially extended to all employers and their employees, except self-employed people; family-owned and -operated farms; state, county, and municipal workers (Fig. 10-3); and workplaces already protected by other federal agencies or other federal statutes. In 1979, however, Congress exempted from routine OSHA safety inspections approximately 1.5 million businesses with 10 or fewer employees. (Exceptions to this are allowed if workers claim there are safety violations.) Because federal agencies (except the U.S. Postal Service) are not subject to OSHA regulations and enforcement provisions, each agency is required to establish and maintain its own effective and comprehensive job safety and health program. OSHA provisions do not apply to state and local governments in their role as employers. OSHA requires, however, that any state desiring to gain OSHA support or funding for its own occupational safety and health program must provide a program to cover its state and local government workers that is at least as effective as the OSHA program for private employees.

Box 10-1. Essentials of the Mine Safety and Health Administration

James L. Weeks

The Mine Safety and Health Administration (MSHA) in the U.S. Department of Labor has responsibility for writing and enforcing regulations to protect the health and safety of the approximately 200,000 miners in the United States. These miners work in underground and surface mines that produce coal, metal ore, other nonmetal commodities, such as salt and trona, and in sand, stone, and gravel quarries. Mining is one of the most dangerous industries worldwide and in the United States. There are high rates of fatal and nonfatal traumatic injuries, occupational lung disease (coal workers' pneumoconiosis, silicosis, and lung cancer), and noise-induced hearing loss. Underground miners are also exposed to high concentrations of exhaust from diesel engines.

Historically, federal government intervention in mine safety and health was the responsibility of the U.S. Bureau of Mines in the Department of the Interior. The Bureau was organized in 1910 for the purpose of investigating coal mine disasters, and over the next six decades, it acquired increasing authority and responsibility to enter and inspect mines and promote mine safety, but it had limited authority to compel compliance with safety regulations. When Congress passed the Federal Coal Mine Health and Safety Act of 1969, it significantly changed the relationship between the federal government and the mining industry. This Act was passed after a widespread miners strike for compensation for black lung and a spectacular and disastrous explosion that caused 78 deaths in a mine in West Virginia. Among other things, the Act created an agency to perform epidemiologic research (NIOSH), an agency to continue its engineering research and development to develop safe mining practices (Bureau of Mines, since then absorbed into NIOSH), and a federal program to compensate miners totally disabled by pneumoconiosis.

The 1969 Act created the federal black lung program to compensate miners totally disabled by pneumoconiosis. This program has been controversial, in part, because of the many manifestations of disease caused by inhaling coal mine dust. One innovative aspect of the program is that it allowed for decisions about eligibility when etiology was ambiguous by establishing a series of presumptions based on the miner's clinical status and work history. Originally, claims were paid out of the general treasury, but in 1981, claims were paid by the operator who last employed the miner or, if that operator could not be found, by a disability trust fund to which operators contribute based on their tons of coal produced. The 1969 Act also created the Mining Enforcement and Safety Administration. (MESA), which enforced the basic structure and function of regulation as described later.

The 1969 Act was amended in 1977, with passage of the Mine Safety and Health Act. The 1977 Mine Act moved MESA to the Department of Labor, changed its name to MSHA, preserved the basic structure of the 1969 Act, and extended authority beyond coal mining to all other mines and quarries. The 1977 Act also required that miners receive 40 hours training in safety and health when first hired and 8 hours annually thereafter.

MSHA is structurally similar to OSHA but differs in some important ways. Both agencies write and enforce regulations, and disputes are adjudicated by administrative law review commissions with opportunities to appeal decisions to federal district courts. The standards-setting language in both acts is practically identical. Regulations covering toxic substances must be based on the best available evidence, must be designed to prevent material impairment of health for all miners, even if exposed for their entire working life, and standards must be feasible. Consequently, for the purpose of establishing regulations covering exposure to hazardous substances, the legal and scientific requirements of MSHA and OSHA are essentially the same.

But MSHA is significantly different from OSHA in its enforcement capabilities. Under MSHA, underground mines must be inspected four times and surface mines must be inspected twice each year. Most OSHA inspections are discretionary. Under MSHA, an inspector may, on his or her own authority, close all or part of a mine in case of imminent danger; the OSHA inspector does not have this authority and must get a court order. All mines are covered under MSHA, without exception; under OSHA,

Box 10-1   (*continued*)

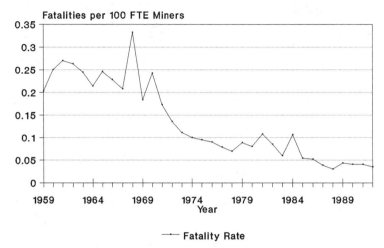

Fatalities per 100 FTE Miners

—•— Fatality Rate

FIG. 10-2. Underground bituminous coal mine fatality rates, 1959 to 1991. FTE, full-time equivalent miners. (From the Mine Safety and Health Administration.)

employers with 10 or fewer employees are exempt from general schedule inspection. Mine operators must submit a mine plan and have it approved before it can produce; only with confined spaces must employers under OSHA's jurisdiction obtain a permit and only then under limited conditions. Some numerical comparisons are informative. OSHA has jurisdiction over approximately 100 million workers and MSHA has jurisdiction over less than a quarter of a million, even though both agencies have approximately the same number of inspectors (including state plans). Thus, the number of inspectors per worker under MSHA is approximately 400 times that under OSHA.

Information about injuries and accidents in mining is more pertinent and more available. Mine-specific data on the number and rates of injuries, hours worked, and (coal) production is reported by mine operators to MSHA every quarter and some of it is available on the internet. Surveillance data on exposure to dust, crystalline silica, other hazardous materials, and noise is also available from MSHA. Under OSHA, estimates of injury rates are available for SIC (Standard Industrial Classification) categories based on an annual survey of a sample of employers conducted by the Bureau of Labor Statistics. Employer-specific data are not available. Employers must post injury data annually, but they are not required to report it to OSHA. OSHA or workers' representatives may request it from each employer, but it is not available from a single

source, as are MSHA's data. The accuracy and reliability of all surveillance data, however, is not guaranteed. Most injury and exposure data are provided by employers and passed on by either MSHA, OSHA, or the Bureau of Labor Statistics with little, if any, validation.

What has this regulatory intervention into the mining industry achieved? Before the passage of the 1969 Coal Mine Act, the fatality rate of U.S. miners was approximately 0.25 fatalities per 100 workers per year, four times that of miners in Western European coal mining countries. For the first 10 years after the Act, it declined each year to a level approximately the same as that in European mines. Since then, it has declined further, so that now, coal mines in the United States are among the safest in the world at an annual fatality rate of approximately 0.03 fatalities per 100 workers (Fig. 10-2). Even so, the fatality rate in mining remains the highest of any major industrial group in the United States.

Trends in nonfatal injury rates are harder to measure because occurrence of these injuries varies significantly by occupation and among different age and experience cohorts. Trends in age- and experience-specific injury rates are not available. The crude rate of nonfatal injuries in coal mining has declined steadily, but this could be because very few new and inexperienced miners have been hired at the same time that the population of working miners is getting older and more experienced. This

Box 10-1 (*continued*)

change in the age and experience distribution alone could account for the steady decline in the overall injury rate. Mine operators also must report certain accidents that do not cause injury but that signal the existence of hazards that could cause serious injury. These accidents include non-planned roof falls, inundations with water, fires, and failure of ventilation.

This regulatory scheme has also significantly reduced miners' exposure to respirable dust and has reduced the prevalence of coal worker's pneumoconiosis (CWP). Respirable coal mine dust was measured at 6 to 8 mg/m^3 before the 1969 Act but, for the same job, declined to less than 3 mg/m^3 within 6 months and to approximately 2 mg/m^3 in another year. For continuous mining operators, the level is now regularly below 1 mg/m^3. This progress was achieved in spite of mine operators claiming, in 1969, that it was impossible to reduce exposure to the statutory limit of 2 mg/m^3. Exposure remains high at some mines and with some mining methods, such as longwall mining. Consistent with this reduction in exposure, the experience-adjusted prevalence of CWP has also been reduced since passage of the 1969 and 1977 Mine Act. Problems persist, however. Noise exposure remains high, exposure to crystalline silica is also elevated, where it is known, and underground miners are exposed to high levels of diesel exhaust.

MSHA's program of surveillance and control of exposure to respirable dust and its enforcement of dust regulations is part of a more comprehensive effort to prevent the occurrence and progression of CWP. Other aspects of this plan include a federal program to compensate underground coal miners totally disabled by CWP, a prospective study of a cohort of miners, engineering research and development on methods of monitoring and controlling exposure to dust, and a program to allow miners to transfer to less dusty jobs in a mine if they have a positive chest radiograph for CWP. All these facets of the prevention effort are and have been controversial, but nevertheless they contain the essential elements for preventing occupational disease: exposure monitoring, enforcement, disease surveillance, right to transfer, epidemiologic research, and engineering research and development.

In sum, MSHA is an intensive intervention in a dangerous industry and, as such, is a laboratory on a number of issues important to worker health and safety generally. One important lesson from MSHA is that a concerted and multifaceted effort at controlling occupational hazards can succeed at reducing rates of traumatic fatalities and of pneumoconiosis. The important aspects of such an effort include sufficient resources, surveillance, exposure monitoring, worker training, epidemiologic research, and engineering research and development—all of which are supported, in one way or another, by regulatory authority.

OSHA can begin standard-setting procedures either on its own or on petitions from other parties, including the Secretary of Health and Human Services, NIOSH, state and local governments, any nationally recognized standards-producing organization, employer or labor representatives, or any other interested person. The standard-setting process involves input from advisory committees and from NIOSH. When OSHA develops plans to propose, amend, or delete a standard, it publishes these intentions in the *Federal Register*. Subsequently, interested parties have opportunities to present arguments and pertinent evidence in writing or at public hearings. Under certain conditions, OSHA is authorized to set emergency temporary standards, which take effect immediately, but which are supposed to be followed up by the establishment of permanent standards within 6 months. OSHA must first determine that workers are in grave danger from exposure to toxic substances or new hazards and are not adequately protected by existing standards. Standards can be appealed through the federal courts, but filing an appeals petition does not delay the enforcement of the standard unless a court of appeals specifically orders it. Employers may make application to OSHA for a variance

FIG. 10-3. The Occupational Safety and Health Administration's (OSHA) positive impact on general industry health and safety in the United States unfortunately does not extend to municipal workers such as firefighters. (Photograph by Marvin Lewiton.)

from a standard or regulation if they lack the means to comply readily with it or if they can prove that their facilities or methods of operation provide employee protection that are at least as effective as that required by OSHA.

OSHA requires employers of more than 10 employees to maintain records of occupational injuries and illnesses as they occur. All occupational injuries and diseases must be recorded if they result in death, one or more lost workdays, restriction of work or motion, loss of consciousness, transfer to another job, or medical treatment (other than first aid).

STANDARD SETTING AND OBLIGATIONS OF THE EMPLOYER AND THE MANUFACTURER OR USER OF TOXIC SUBSTANCES

Legal Background for OSHA Obligations

The OSHAct provides two general means of protection for workers: (a) a statutory general duty to provide a safe and healthful workplace, and (b) adherence to specific standards by employers. The act imposes on virtually every employer in the private sector a general duty "to furnish to each of his employees employment and a place of employment which are free from *recognized hazards* that are causing or are likely to cause death or serious physical harm. . .." (emphasis added). A recognized hazard may be a substance for which the likelihood of harm has been the subject of research, giving rise to reasonable suspicion, or a substance for which an OSHA standard may or may not have been promulgated. The burden of proving that a particular substance is a recognized hazard and that industrial exposure to it results in a significant degree of exposure is placed on OSHA. Because standard setting is a slow process, protection of workers through the employer's general duty obligation is especially important, but it is crucially dependent on the existence of reliable health effects data.

The OSHAct addresses specifically the subject of toxic materials. It states, under Section 6(b)(5) of the act, that the Secretary of Labor (through OSHA), in promulgating standards dealing with toxic materials or harmful physical agents, shall set the standard that "most adequately assures, *to the extent feasible*, on the basis of the *best available evidence* that no employee will suffer material impairment of health or functional capacity, even if such employee has a regular exposure to the hazard dealt with by such standard for the period of his working life" (emphasis added). These words indicate that the issue of exposure to toxic chemicals or carcinogens that have long latency periods, as well as to reproductive hazards, is covered by the act in specific terms.

Under Section 6(a) of the act, without critical review, OSHA initially adopted as standards, called permissible exposure limits (PELs), the 450 threshold limit values (TLVs) recommended by the American Conference of Governmental Industrial Hygienists (ACGIH) as guidelines for protection against the toxic effects of these materials. In the 1970s, under Section 6(b), OSHA set formal standards for asbestos, vinyl chloride, arsenic, dibromochloropropane, coke oven emissions, acrylonitrile, lead, cotton dust, and a group of 14 carcinogens. In the 1980s, OSHA regulated benzene, ethylene oxide, and formaldehyde as carcinogens and asbestos more rigidly as a carcinogen at 0.2 fibers/cm³. In the early 1990s, OSHA regulated cadmium, bloodborne pathogens, glycol ethers, and confined spaces. OSHA also lowered the PEL for formaldehyde from 1 to 0.75 parts per million (ppm; over an 8-hour period) and issued a process safety management (PSM) rule (see discussion, later). More recent rule-making activity by OSHA is discussed later in this chapter.

The burden of proving the hazardous nature of a substance is placed on OSHA, as is the requirement that the proposed controls are technologically feasible. The necessarily slow and arduous task of setting standards substance by substance makes it impossible to deal realistically with 13,000 toxic substances or approximately 250 suspect carcinogens on NIOSH lists. Efforts were made to streamline the process by proposing generic standards for carcinogens, and by proposing a generic standard updating the TLVs (PELs). As discussed later, neither was successful.

The inadequacy of the 450 TLVs adopted under Section 6(a) of the act is widely known. The TLVs originated as guidelines recommended by the ACGIH to protect the average worker from either recognized acute effects or easily recognized chronic effects. The standards were based on animal toxicity data or the limited epidemiologic evidence available at the time of the establishment of the TLVs. They do not address the sensitive populations within the workforce or those with prior exposure or existing disease, nor do they address the issues of carcinogenicity, mutagenicity, and teratogenicity. These standards were adopted *en masse* in 1971 as a part of the consensus standards that OSHA adopted along with those dealing primarily with safety.

As an example of the inadequacy of protection offered by the TLVs, the 1971 TLV for vinyl chloride was set at 250 ppm, whereas the later protective standard (see later) recommended no greater exposure than 1 ppm (as an average over 8 hours)—a level still recognized as unsafe, but the limit that the technology could detect. Another example is the TLV for lead, which was established at 200 μg/m³, whereas the later lead standard was established at 50 μg/m³, also recognizing that that level was not safe for all populations, such as pregnant women or those with prior lead exposure. The ACGIH updates its TLV list every 2 years. Although useful, an updated list would have little legal significance unless formally adopted by OSHA. OSHA did try, unsuccessfully, to adopt an updated and new list of PELs in its Air Contaminants Standard in 1989 (see discussion, later). However, OSHA is intent on revising the list.

Under Section 6(b) of the OSHAct, new health standards dealing with toxic substances were to be established using the mechanism of an open hearing and subject to review by the U.S. Circuit Courts of Appeals. The evolution of case law associated with the handful of standards that OSHA promulgated through this section is worth considering in detail. The courts addressed the difficult issue of what is adequate scientific information necessary to sustain the requirement that the standards be supported by "substantial evidence on the record as a whole." The cases also addressed the extent to which economic factors were permitted or required to be considered in the setting of the standards, the meaning of "feasibility," OSHA's technology-forcing authority, the question of whether a cost–benefit analysis was required or permitted, and, finally, the extent of the jurisdiction of OSHAct in addressing different degrees of risk.

The 14 Carcinogens Standard

In an early case challenging OSHA's authority to regulate 14 carcinogens, the District of Columbia Circuit Court of Appeals first addressed the issue of substantial evidence. For 8 of the 14 carcinogens, there were no human (epidemiologic) data. Industry challenged OSHA's ability to impose controls on employers in the absence of human data. Here the court expressed its view that some facts, such as the establishment of human carcinogenic risk from animal data, were on the "frontiers of scientific knowledge" and that the requirement for standards to be supported by substantial evidence in these kinds of social policy decisions could not be subjected to the rigors of other kinds of factual determinations. Thus, OSHA was permitted to require protective action against substances known to produce cancer in animals but with no evidence of producing cancer in humans. It was not until 1980 that the U.S. Supreme Court in the benzene case (see later) placed limits on the extent of OSHA's policy determination on carcinogenic risk.

The Asbestos Standard

In the challenge to OSHA's original asbestos standard, in which asbestos was regulated as a classic lung toxin and not as a carcinogen, the Industrial Union Department of the American Federation of Labor and Congress of Industrial Organizations (AFL-CIO) unsuccessfully challenged the laxity of the standard, claiming that OSHA improperly weighed economic considerations in its determination of feasibility. OSHA indeed was permitted to consider economic factors in establishing feasibility. The District of Columbia Circuit Court of Appeals went on to state, however, that a standard might be feasible even if some employers were forced out of business, as long as the entire asbestos-using industry was not disrupted. In 1986, OSHA revised the standard from 2.0 to 0.2 fibers/cm^3, thus finally acknowledging it as a carcinogen.

The Vinyl Chloride Standard

In the industry challenge to OSHA's regulation of vinyl chloride at 1 ppm, the Second Circuit Court of Appeals reiterated OSHA's ability to make policy judgments with regard to matters "on the frontiers of scientific knowledge" when it declared that there could be no safe level for a carcinogen. In addition, the court said that because 1 ppm was the lowest feasible level, OSHA was permitted to force employers to comply even though it had performed no formal risk assessment or knew how many tumors would be prevented by the adoption of this protective level. Another noteworthy aspect of the case was the recognition that OSHA could act as a "technology forcer" and require controls not yet fully developed at the time of the setting of the standard.

The Lead Standard

Protection from lead exposure had been provided through the TLV of 200 μg/m^3. This level was long recognized as inadequate for workers who accumulated lead in their body tissues and for women (and possibly men) who intended to have children. As a result, based on the limits of technological feasibility, OSHA promulgated a new standard that permitted no exposure greater than 50 μg/m^3 averaged over an 8-hour period. In addition, because this was still unsafe for many workers, OSHA also provided that workers be removed with pay and employment security if their blood lead levels (BLLs) exceeded 50 μg/dL or if there were grounds to remove them based on risks to their reproductive system. The legality and necessity of this additional provision, known as Medical Removal Protection (MRP), was unsuccessfully challenged by the Lead Industries Association. (MRP has since been required in a limited way in the cotton dust and benzene standards.) OSHA specifically provided that workers in workplaces with air lead levels over an "action level" of 30 μg/m^3 have the benefit of a continuing medical surveillance program, including periodic sampling of BLLs and removal from exposure above the action level after finding a BLL in an individual worker above 50 μg/dL, with job return when the worker's BLL fell below 40 μg/dL.

Removal could also be triggered by other medical conditions deemed especially sensitive to risks associated with lead exposure (e.g., pregnancy). OSHA provided that workers' pay and seniority be maintained by the employer during any periods of medical removal (up to 18 months), even if such removal entailed sending the worker home. In actual practice, many employers have reduced the ambient air lead level well below 50 μg/m^3, which results in the removal of fewer workers.

The Benzene Standard

After the first serious successful industry challenge of an OSHA benzene standard in the Fifth Circuit Court of Appeals, the U.S. Supreme Court, in a controversial and divided majority opinion, chided OSHA for not attempting to evaluate the benefits of changing the PEL for benzene from 10 ppm (the former TLV) to 1 ppm. The Court argued that OSHA is obligated to regulate only "significant risks" and that without a risk assessment of some kind, OSHA could not know whether the proposed control addressed a significant risk. The Court was careful to state that it was not attempting to "statistically straitjacket" the agency, but that at a minimum the benefits of regulation needed to be addressed to meet the substantial evidence test. The Court did not give useful guidance concerning what constituted a significant risk. It stated that a risk of death of 1 in 1,000 was clearly significant, whereas a risk of 1 in 1 billion was clearly not so. This six-orders-of-magnitude range, of course, represents the area on which the arguments have always been centered. The implications of the benzene decision for subsequent standards reflect the political and philosophical leanings of particular OSHA administrations.

There is little question that had OSHA performed a risk assessment for benzene at the time, it could have argued that the risk it was attempting to address was actually significant. The precise requirement and nature of a risk assessment sufficient to meet the substantial evidence test remains unclear. In late 1985, OSHA again proposed to lower the PEL from 10 to 1 ppm, and in 1987, the standard was set at that level. OSHA, however, after intervention by the Office of Management and Budget, declined to establish a short-term exposure limit.

The petroleum industry argued in the benzene case that not only must a risk assessment be performed, but a cost–benefit analysis must be done in which the risks of exposure are balanced against the benefits of the chemical. The question, however, was not decided in the benzene case but was addressed in a later case challenging OSHA's cotton dust standard. The Supreme Court not only acknowledged that cotton dust did represent a significant risk but indicated that a cost–benefit balancing was neither required nor permitted by the OSHAct because Congress had already struck the balance heavily in favor of worker health and safety.

The Generic Carcinogen Standard

In 1980, OSHA promulgated a generic carcinogen standard by which questions of science policy, already settled as law in cases dealing with other standards, were codified in a set of principles. During the process of developing the generic carcinogen standard, OSHA and NIOSH developed lists of chemical substances that would probably be classified as suspect carcinogens. Each agency composed a list of approximately 250 substances. Thus far, OSHA has declined formally to list any substance under the carcinogen standard.

In setting or revising standards for formaldehyde, ethylene oxide, asbestos, and benzene, the agency has proceeded to act as if the generic carcinogen standard did not exist, thus following the historically arduous and slow path to standard setting.

Emergency Temporary Standards

In Section 6(c), the OSHAct authorizes OSHA to set emergency temporary (6-month) standards for toxic exposures constituting a "grave danger" on publication in the *Federal Register* and without recourse to a formal hearing.

Before OSHA lowered its permanent standard for asbestos from 2.0 to 0.2 fibers/cm^3, it attempted to protect workers by promulgating an Emergency Temporary Standard (ETS) at 0.5 fibers/cm^3. In 1984, the Fifth Circuit Court of Appeals denied OSHA the ETS, arguing that the cost involved defeated the requirement that the ETS be "necessary" to protect workers. Attempts by OSHA to establish an ETS for hexavalent chromium likewise failed court review. As a result, OSHA now avoids setting ETSs, and instead proceeds directly to establishing permanent standards for toxic substances under Section 6(b)(5).

Short-Term Exposure Limits

Short-term exposures to higher levels of carcinogens are in general considered more hazardous than longer exposures to lower levels. OSHA issued a new standard for exposure to ethylene oxide in 1984, but excluded a short-term exposure limit (STEL) that had originally been prepared, in deference to objections from the Office of Management and Budget. Ralph Nader's Health Research Group sued the Secretary of Labor in 1986 over OSHA's continuing failure to issue the STEL. In 1987, the District of Columbia Circuit Court of Appeals ordered OSHA to establish a STEL for ethylene oxide by March 1988. OSHA complied by setting a STEL of 5 ppm over a 15-minute period.

The Air Contaminants Standard

It is obvious that the slow, arduous process of promulgating individual health standards under Section 6(b)(5) of the OSHAct could never catch up with advances in scientific knowledge concerning the toxicity of chemicals. The ACGIH has updated its TLV list every 2 years, and although not as protective as workers and their unions would have liked, the recent updated lists did advance protection over the 1969 list that OSHA adopted into law in 1971. In 1989, OSHA decided to update the original list in a single rule-making effort through the 6(b) standard revision route. The agency issued more protective limits for 212 substances and established limits for 164 chemicals that were previously unregulated. Neither industry nor labor was satisfied with the standards. Industry, although giving general support, objected to the stringency of some of the PELs. Labor objected to their laxity, citing NIOSH recommendations not adopted, and generally objected to the rush-it-through process. The Eleventh Circuit Court of Appeals vacated the standard in 1992, ruling that OSHA failed to establish that a significant risk of material health impairment existed for each regulated substance (required by the benzene decision), and that the new exposure limit for each substance was feasible for the affected industry. OSHA decided not to appeal the decision to what it perceived as a conservative Supreme Court. Thus, the original and inadequate TLV list remains in effect and 164 new substances remain unregulated. However, OSHA is intent on updating the list through new rule making. In the meantime, OSHA could argue that those 164 substances are "recognized hazards" and enforceable through OSHA's general duty clause (see later).

The Toxic Substances Control Act

TSCA enables the EPA to require data *from industry* on the production, use, and health and environmental effects of chemicals. The EPA may regulate by requiring labeling, setting tolerances, or banning completely and requiring repurchase or recall. The EPA may also order a specific change in chemical process technology. In addition, TSCA gives aggrieved parties, including consumers and workers, specific rights to sue for damages under the act, with the possibility of awards for attorneys' fees. (This feature was missing in the OSHAct.)

Under TSCA, the EPA must regulate "unreasonable risks of injury to human health or the environment." The EPA has issued a regulation for worker protection from asbestos at the new OSHA limit of 0.2 fibers/cm^3, which applies to state and local government asbestos abatement workers not covered by OSHA. The EPA has declared formaldehyde a "probable carcinogen" but has not taken regulatory action on this substance. Although the potential for regulating workplace chemicals is there, the EPA has not been aggressive in this area. Between 1977 and 1990, of the 22 regulatory actions taken on existing chemicals, 15 addressed polychlorinated biphenyls (PCBs). Only regulations pertaining to asbestos, hexavalent chromium, and metal-working fluids had a strong occupational exposure component.

One strength of TSCA is its ability to shift onto the producer the requirement to prove that a substance is safe to the extent that exposure to it does not present an "unreasonable risk of injury to human health or the environment." Used together, the OSHAct and TSCA provide potentially comprehensive and effective information-generation and standard-setting authority to protect workers. In particular, the information-generation activities under TSCA can provide the necessary data to have a substance qualify as a "recognized hazard" that, even in the absence of specific OSHA standards, must be controlled in some way by the employer to meet the general duty obligation under the OSHAct to provide a safe and healthful workplace.

The potentially powerful role of TSCA regulation was seriously challenged by the Fifth Circuit Court of Appeals in 1991, when it overturned the EPA's omnibus asbestos phase-out rule issued in 1989. The Court, which is generally unfriendly to environmental, occupational health, and consumer product regulation, argued that under TSCA, the EPA should have considered alternatives to a ban that would have been less burdensome to industry. The case was not appealed to the Supreme Court, and the EPA regards regulation of chemicals under TSCA to be a dead letter for now, except under specific statutory directives with regard to PCBs. With an unsympathetic Congress, there are no current attempts to resurrect the regulatory authority of TSCA. However, TSCA continues to be important for its surviving authority to require the testing of chemicals and for its information reporting and retaining requirements (see the discussion on the right to know, later).

The Chemical Safety Provisions of the Clean Air Act

Although the first congressional response to the country's "Bhopal" concerns was the Emergency Planning and Community Right to Know Act of 1986, the chemical safety provisions of that law are focused almost solely on mitigation, and not on accident prevention. A much greater potential for a direct focus on accident prevention can be found in the 1990 amendments to the Clean Air Act, although that potential has yet to be realized by the EPA and OSHA.

As amended in 1990, Section 112 of the Clean Air Act directs the EPA to develop regulations regarding the prevention and detection of accidental chemical releases, and to publish a list of at least 100 chemical substances (with associated threshold quantities) to be covered by the regulations. The regulations must include requirements for the development of risk management plans (RMPs) by facilities using any of the regulated substances in amounts above the relevant threshold. These RMPs must include a hazard assessment, an accident prevention program, and an emergency release program. Similarly, Section 304 of the Clean Air Amendments of 1990 directed OSHA to promulgate a "chemical process safety standard" under the OSHAct.

Section 112 of the revised Clean Air Act

also imposes a "general duty" on *all* "owners and operators of stationary sources," regardless of the particular identity or quantity of the chemicals used on site. These parties have a duty to:

". . . *identify hazards* that may result from [accidental chemical] releases using appropriate hazard assessment techniques,

. . . *design and maintain a safe facility* taking such steps as are necessary to prevent releases, and

. . . *minimize the consequences* of accidental releases which do occur." [emphasis added]

Thus, firms are now under a general duty to *anticipate, prevent, and mitigate* accidental releases.

In defining the nature of this duty, Section 112 specifies that it is "a general duty in the same manner and to the same extent as" that imposed by Section 5 of the OSHAct. Because Section 112 specifically ties its general duty obligation to the general duty clause of the OSHAct, case law interpreting the OSHAct provision should be directly relevant. Specifically, in the *General Dynamics* case discussed previously, the D.C. Circuit Court of Appeals held that standards and the general duty obligation are distinct and independent requirements, and that compliance with a standard does not discharge an employer's duty to comply with the general duty obligation. Similarly, compliance with other Clean Air Act chemical safety requirements should not relieve a firm's duty to comply with the Act's general duty clause.

The Clean Air Act also requires each state to establish programs to provide small business with technical assistance in addressing chemical safety. These programs could provide information on alternative technologies, process changes, products, and methods of operation that help reduce emissions to air. However, these state mandates are unfunded and may not be uniformly implemented. Where they are established, linkage with state offices of technical assistance, especially those that provide guidance on pollution prevention, could be particularly beneficial.

Finally, the 1990 amendments established an independent Chemical Safety and Hazard Investigation Board. The Board is to investigate the causes of accidents, perform research on prevention, and make recommendations for preventive approaches, much like the Air Transportation Safety Board does with regard to airplane safety.

As required by the 1990 Clean Air Amendments, OSHA, in 1992, promulgated a standard requiring chemical PSM in the workplace that became effective later that year. The PSM standard is designed to protect employees working in facilities that use "highly hazardous chemicals," and employees working in facilities with more than 10,000 pounds of flammable liquids or gases present in one location. The list of highly hazardous chemicals in the standard includes acutely toxic, highly flammable, and reactive substances. The PSM standard requires employers to compile safety information (including process flow information) on chemicals and processes used in the workplace, complete a workplace process hazard analysis every 5 years, conduct triennial compliance safety audits, develop and implement written operating procedures, conduct extensive worker training, develop and implement plans to maintain the integrity of process equipment, perform pre-startup reviews for new (and significantly modified) facilities, develop and implement written procedures to manage changes in production methods, establish an emergency action plan, and investigate accidents and near-misses at their facilities. Many aspects of chemical safety are not covered by specific workplace standards. Most that do apply to chemical safety have their origin in the consensus standards adopted under section 6(a) of the OSHAct in 1971, and hence are greatly out of date. Arguably, the general duty obligation of the OSHAct imposes a duty to seek out technological improvements that would improve safety for workers.

In 1996, the EPA promulgated regulations setting forth requirements for the RMPs specified in the Clean Air Act. The RMP rule is modeled after the OSHA PSM standard, and is estimated to affect some 66,000 facilities. The rule requires a hazard assessment (involving an offsite consequence analysis—including worst-case risk scenarios—and compilation of a 5-year accident history), a prevention program to address the hazards identified, and an emergency response program.

ENFORCEMENT ACTIVITIES

Regulations and standard setting, of course, are only the beginning of the regulatory process. For a regulatory system to be effective, there must be a clear commitment to the enforcement of standards. Under OSHA, a worker can request workplace inspection if the request is in writing and signed. Anonymity is preserved on request. When an inspector visits a workplace, a representative of the workers can accompany the inspector on the "walk-around."

If specific requests for inspections are not made, OSHA makes random inspections of those workplaces with worse-than-average safety records. However, the inspection frequency is low. Furthermore, firms with significant exposures to chemicals may not be routinely inspected, simply because their record for *injuries* (which dominate the reported statistics) is good.

Inspections are usually conducted without advance notice, but an employer may insist that OSHA inspectors obtain a court order before entering the workplace. Federal OSHA continues to have approximately 1,000 inspectors, and state agencies have approximately another 2,000. OSHA and OSHA-approved state programs conducted approximately 90,000 annual inspections in both fiscal years 1997 and 1998, focusing inspections on the most hazardous industries, construction and manufacturing. Federal OSHA issued approximately 320,000 citations for fiscal year 1997. Clearly, not all 5 million workplaces covered by the OSHAct could be inspected on anything like a regular basis. With the recent expansion of OSHA authority to cover U.S. post offices, the agency continues to be short of the resources needed to perform its statutory duties. In sharp contrast, the number of inspectors per worker is 10 times larger in British Columbia, Canada, and in many European countries.

OSHA can fine employers up to $7,000 for each violation of the act that is discovered during a workplace inspection, and up to $70,000 or up to 6 months imprisonment if the violation is willful or repeated. The failure to abate hazards can result in a $7,000 fine per day. These penalties are very much less than those for violations of environmental statutes. Management can appeal violations, amounts of fines, methods of correcting hazards, and deadlines for correcting hazards (abatement dates). Workers can appeal only deadlines. All appeals are processed through the Occupational Safety and Health Review Commission, established by the OSHAct.

The act requires OSHA to encourage states to develop and operate their own job safety and health programs. State programs, when "at least as effective" as the federal program, can take over enforcement activities. Once a state plan is approved, OSHA funds half of its operating costs. Approximately 20 state plans, which OSHA monitors, are in effect. State safety and health standards under such approved plans must keep pace with OSHA standards, and state plans must guarantee employer and employee rights, as does OSHA.

During the 1980s, OSHA inspection policy resulted in directives given to the field staff to deemphasize general duty violations. In addition, inspectors were actually evaluated by the managers of the establishments they inspected. Follow-up inspections after violations were often restricted to checks by telephone. Thus, incentives for aggressive inspection activity were not great under the Reagan and Bush administrations.

THE RIGHT TO KNOW

The transfer of information regarding workplace exposure to toxic substances has received considerable public attention. It is clear that workers need an accurate picture of the nature and extent of probable chemical exposures to decide whether to enter or remain in a particular workplace. Workers also need to have knowledge regarding past or current exposures to be alert to the onset of occupational disease. Regulatory agencies must have timely access to such information if they are to devise effective strategies to reduce disease and death from occupational exposures to toxic substances. Accordingly, laws designed to facilitate this flow of information have been promulgated at the federal, state, and local levels. Indeed, the right to know has become a political battleground in many states and communities and has been the subject of intensive organizing efforts by business, labor, and citizen-action groups.

In essence, the right to know embodies a democratization of the workplace. It is the mandatory sharing of information between management and labor. Through a variety of laws, manufacturers and employers are directed to disclose information regarding toxic substance exposure to workers, to unions in their capacity as worker representatives, and to governmental agencies charged with the protection of public health. The underlying rationale for these directives is the assumption that this transfer of information will prompt activity that will improve worker health.

Although the phrase *right to know* is a useful generic designation, it is an inadequate description of the legal rights and obligations that govern the transfer of workplace information on toxic substances. A person cannot have a meaningful *right* to information unless someone else has a corresponding *duty* to provide that information. Thus, a worker's right to know is secured by requiring a manufacturer or employer to disclose. The disclosure requirement can take a variety of forms, and the practical scope of that

requirement may depend on the nature of the form chosen. In particular, a duty to disclose only such information as has been requested may provide a narrower flow of information than a duty to disclose all information, regardless of whether it has been requested. The various rights and obligations in the area of toxics information transfer may be grouped into three categories. Although they share a number of similarities, each category is conceptually distinct:

1. *The duty to generate or retain information* refers to the obligation to compile a record of certain workplace events or activities or to maintain such a record for a specified period of time if it has been compiled. An employer may, for example, be required to monitor its workers regularly for evidence of toxic exposures (biologic monitoring) and to keep written records of the results of such monitoring.

2. *The right of access* (and the corresponding duty to disclose on request) refers to the right of a worker, a union, or an agency to request and secure access to information held by a manufacturer or employer. Such a right of access would provide workers with a means of obtaining copies of biologic monitoring records pertaining to their own exposure to toxic substances.

3. Finally, *the duty to inform* refers to an employer's or manufacturer's obligation to disclose, without request, information pertaining to toxic substance exposures in the workplace. An employer may, for example, have a duty, independent of any worker's exercise of a right to access, to inform workers whenever biologic monitoring reveals that their exposure to a toxic substance has produced bodily concentrations of that substance above a specified level.

In general, the broadest coverage is found in rights and duties emanating from the OSHAct. By its terms, that act is applicable to all *private* employers and thus covers the bulk of workplace exposures to toxic substances. Most private industrial workplaces are also subject to the National Labor Relations Act (NLRA). Farm workers and workers subject to the Railway Labor Act, however, are exempt from NLRA coverage. TSCA provides a generally narrower scope. Although many of the act's provisions apply broadly to both chemical manufacture and use, its information transfer requirements extend only to chemical manufacturers, processors, and importers. On the state level, the relevant coverage of the various rights and duties depends on the specifics of the particular state and local law defining them. In general, common-law rights and duties evidence much less variation than those created by state statute or local ordinance.

Under OSHA's Hazard Communication Standard, employers have a duty to inform workers of the identity of substances with which they work through labeling the product container and disclosing to the purchaser (the employer) using material safety data sheets (MSDSs).

Employers are under no obligation to amend inadequate, insufficient, or incorrect information provided by the manufacturer. Employers must, however, transmit certain information to their employees: (a) information on the standard and its requirements, (b) operations in their work areas where hazardous chemicals are present, and (c) the location and availability of the company's hazard communication program. The standard also requires that workers must be trained in (a) methods to detect the presence or release of the hazardous chemicals; (b) the physical and health hazards of the chemicals; (c) protective measures, such as appropriate work practices, emergency procedures, and personal protective equipment; and (d) the details of the hazard communication program developed by the employer, including an explanation of the labeling system and the MSDSs, and how employees can obtain and use hazard information.

Rights and duties governing toxic information transfer in the workplace can originate from a variety of sources. Some are grounded in state common law, whereas others arise out of specific state statutes or local ordinances. Although the states have been increasingly active in this field, the primary source of regulation is federal law. Most federal regulation in this area emanates from three statutes: the OSHAct, TSCA, and NLRA, the last of which is administered by the National Labor Relations Board (NLRB).

The scope of a particular right or duty depends on many factors. The first, and perhaps most important, is the nature of the information required to be transferred.

Scientific information refers to data concerning the nature and consequences of toxic substance exposures. These data, in turn, can be divided into three subcategories:

1. *Ingredients information* provides the worker with the identity of the substances to which he or she is exposed. Depending on the circumstances, this information may constitute only the generic classifications of the various chemicals involved or may include the specific chemical identities of all chemical exposures and the specific contents of all chemical mixtures.

2. *Exposure information* encompasses all data regarding the amount, frequency, duration, and route of workplace exposures. This information may be of a general nature, such as the results of ambient air monitoring at a central workplace location, or may take individualized form, such as the results of personal environmental or biologic monitoring of a specific worker.

3. *Health effects information* indicates known or potential health effects of workplace exposures. This information may be general data regarding the effects of chemical exposure, usually found in an MSDS or a published or unpublished workplace epidemiologic study, or it may be individualized data, such as worker medical records compiled as a result of medical surveillance.

The federal standard preempts state right-to-know laws in the worker notification area in a minority of jurisdictions; it would appear to be coexistent with state requirements in most jurisdictions, although its stated intent is to preempt all state efforts.

Under OSHA's Medical Access Rule, an employer may not limit or deny an employee access to his or her own medical or exposure records. The current OSHA regulation, promulgated in 1980, grants employees a general right of access to medical and exposure records kept by their employer. Furthermore, it requires the employer to preserve and maintain these records for 30 years. There appears to be some overlap in the definitions of *medical* and *exposure* records, because both may include the results of biologic monitoring. Medical records, however, are in general defined as those pertaining to "the health status of an employee," whereas the exposure records are defined as those pertaining to "employee exposure to toxic substances or harmful physical agents."

The employer's duty to make these records available is a broad one. The regulations provide that on any employee request for access to a medical or exposure record, "the employer *shall* assure that access is provided in a reasonable time, place, and manner, but in no event later than 15 days after the request for access is made."

An employee's right of access to medical records is limited to records pertaining specifically to that employee. The regulations allow physicians some discretion as well in limiting employee access. The physician is permitted to "recommend" to the employee requesting access that the employee (a) review and discuss the records with the physician; (b) accept a summary rather than the records themselves; or (c) allow the records to be released instead to another physician. Furthermore, where information in a record pertains to a "specific diagnosis of a terminal illness or a psychiatric condition," the physician is authorized to direct that such information be provided only to the employee's designated representative. Although these provisions were apparently intended to respect the physician–patient relationship and

do not limit the employee's ultimate right of access, they could be abused. In situations in which the physician feels loyalty to the employer rather than the employee, the physician could use these provisions to discourage the employee from seeking access to his or her records.

Similar constraints do not apply to employee access to exposure records. Not only is the employee ensured access to records of his or her own exposure to toxic substances, but the employee is also ensured access to the exposure records of other employees "with past or present job duties or working conditions related to or similar to those of the employee." In addition, the employee has access to all general exposure information pertaining to the employee's workplace or working conditions and to any workplace or working condition to which he or she is to be transferred. All information in exposure records that cannot be correlated with a particular employee's exposure is accessible.

One criticism of the OSHA regulation is that it does not require the employer to compile medical or exposure information but merely requires employee access to such information if it is compiled. The scope of the regulation, however, should not be underestimated. The term *record* is meant to be "all-encompassing," and the access requirement appears to extend to all information gathered on employee health or exposure, no matter how it is measured or recorded. Thus, if an employer embarks on any program of human monitoring, no matter how conducted, he or she must provide the subjects access to the results. This access requirement may serve as a disincentive for employers to monitor employee exposure or health, if it is not clearly in the employer's interest to do so.

The regulations permit the employer to deny access to "trade secret data which discloses manufacturing processes or . . . the percentage of a chemical substance in a mixture," provided that the employer (a) notifies the party requesting access of the denial; (b) if relevant, provides alternative information sufficient to permit identification of when and where exposure occurred; and (c) provides access to all "chemical or physical agent identities including chemical names, levels of exposure, and employee health status data contained in the requested records."

The key feature of this provision is that it ensures employee access to the precise identities of chemicals and physical agents. This access is especially critical for chemical exposures. Within each "generic" class of chemicals, there are a variety of specific chemical compounds, each of which may have its own particular effect on human health. The health effects can vary widely within a particular family of chemicals. Accordingly, the medical and scientific literature on chemical properties and toxicity is indexed by specific chemical name, not by generic chemical class. To discern any meaningful correlation between a chemical exposure and a known or potential health effect, an employee must know the precise chemical identity of that exposure. Furthermore, in the case of biologic monitoring, the identity of the toxic substance or its metabolite is itself the information monitored.

Particularly in light of the public health emphasis inherent in the OSHAct, disclosure of such information does not constitute an unreasonable infringement on the trade secret interests of the employer. In general, chemical health and safety data are the least valuable to an employer of all the "proprietary" information relevant to a particular manufacturing process.

TSCA imposes substantial requirements on chemical manufacturers and processors to develop health effects data. TSCA requires testing, premarket manufacturing notification, and reporting and retention of information. TSCA imposes no specific medical surveillance or biologic monitoring requirements. However, to the extent that human monitoring is used to meet more general requirements of assessing occupational health or exposure to toxic substances, the data resulting from such monitoring are subject to an employer's recording and retention obligations.

The EPA has promulgated regulations requiring general reporting on several hundred chemicals, including information related to occupational exposure. The EPA administrator may require the reporting and maintenance of those data "insofar as known" or "insofar as reasonably ascertainable." Thus, if monitoring is undertaken, it must be reported. The EPA appears to be authorized to require monitoring as a way of securing information that is "reasonably ascertainable."

In addition to the general reports required for specific chemicals listed in the regulations, the EPA has promulgated rules for the submission of health and safety studies required for several hundred substances. A health and safety study includes "[a]ny data that bear on the effects of chemical substance on health." Examples are "[m]onitoring data, when they have been aggregated and analyzed to measure the exposure of humans . . . to a chemical substance or mixture." Only data that are "known" or "reasonably ascertainable" need be reported.

Records of "significant adverse reactions to [employee] health" must be retained for 30 years under Section 8(c). A rule implementing this section defines significant adverse reactions as those "that may indicate a substantial impairment of normal activities, or long-lasting or irreversible damage to health or the environment." Under the rule, human monitoring data, especially if derived from a succession of tests, would seem especially reportable. Genetic monitoring of employees, if some basis links the results with increased risk of cancer, also seems to fall within the rule.

Section 8(e) imposes a statutory duty to report "immediately . . . information which supports the conclusion that [a] substance or mixture presents a substantial risk of injury to health." In a policy statement issued in 1978, the EPA interpreted "immediately" in this context to require receipt by the agency within 15 working days after the reporter obtains the information. Substantial risk is defined exclusive of economic considerations. Evidence can be provided by either designed, controlled studies or undesigned, uncontrolled studies, including "medical and health surveys" or evidence of effects in workers. In the EPA's rule for Section 8(c), Section 8(e) is distinguished from Section 8(c) in that "[a] report of substantial risk of injury, unlike an allegation of a significant adverse reaction, is accompanied by information which reasonably supports the seriousness of the effect or the probability of its occurrence." Human monitoring results indicating a substantial risk of injury would thus seem reportable to the EPA. Either medical surveillance or biologic monitoring data would seem to qualify.

Section 14(b) of TSCA gives the EPA authority to disclose from health and safety studies the data pertaining to chemical identities, except for the proportion of chemicals in a mixture. In addition, the EPA may disclose information, otherwise classified as a trade secret, "if the Administration determines it necessary to protect . . . against an unreasonable risk of injury to health." Monitoring data thus seem subject to full disclosure.

In addition to the access provided by OSHA regulations, individual employees may have a limited right of access to medical and exposure records under federal labor law. Logically, the right to refuse hazardous work (see later), inherent in Section 7 of the NLRA and Section 502 of the Labor Management Relations Act, carries with it the right of access to the information necessary to determine whether or not a particular condition is hazardous. In the case of toxic substance exposure, this right of access may mean access to all information relevant to the health effects of the exposure and may include access to both medical and exposure records. These federal labor law provisions are clearly not adequate substitutes for OSHA access regulations, however, because there is no systematic mechanism for enforcing this right.

Collective employee access, however, is available to unionized employees through the collective bargaining process. In four cases, the NLRB has held that unions have a right of access to exposure and medical records so that they may bargain effectively with the employer regarding conditions of employment. Citing the general proposition that employers are required to bargain on health and safety conditions when requested to do so, the NLRB adopted a broad policy favoring union access: "Few matters can be of greater legitimate concern to individuals in the workplace, and thus to the bargaining agent representing them, than exposure to conditions potentially threatening their health, well-being, or their very lives."

The NLRB, however, did not grant an unlimited right of access. The union's right of access is constrained by the individual employee's right of personal privacy. Furthermore, the NLRB acknowledged an employer's interest in protecting trade secrets. Although ordering the employer in each of the four cases to disclose the chemical identities of substances to which the employer did not assert a trade secret defense, the NLRB indicated that employers are entitled to take reasonable steps to safeguard "legitimate" trade secret information. The NLRB did not delineate a specific mechanism for achieving the balance between union access and trade secret disclosure. Instead, it ordered the parties to attempt to resolve the issue through collective bargaining. Given the complexity of this issue and the potential for abuse in the name of "trade secret protection," the NLRB may find it necessary to provide further specificity before a workable industry-wide mechanism can be achieved.

The legal avenues for worker and agency access to information relevant to workplace exposures to toxic substances have been expanded substantially. Despite certain inadequacies in the current laws and despite current attempts by OSHA to narrow the scope of some of these even further, access to toxics data remains broader than it has ever been. By itself, however, this fact is of little significance. The mere existence of information transfer laws means little unless those laws are used aggressively to further the objective of the right to know: the protection of workers' health. The various rights and duties governing toxics information transfer in the workplace present workers, unions, and agencies with a magnificent opportunity. The extent to which they seize this opportunity over the next few years will be a true measure of their resolve to bring about meaningful improvement in the health of the American worker.

THE RIGHT TO REFUSE HAZARDOUS WORK

The NLRA and the OSHAct provide many employees a limited right to refuse to perform hazardous work. When properly exercised, this right protects an employee from retaliatory discharge or other discriminatory action for refusing hazardous work and incorporates a remedy providing both reinstatement and back pay. The nature of this right under the NLRA depends on the relevant collective bargaining agreement, if there is one. Nonunion employees and union employees whose collective bargaining agreements specifically exclude health and safety from a no-strike clause have the *collective* right to stage a safety walkout under Section 7 of the NLRA. If they choose to walk out based on a good-faith belief that working conditions are unsafe, they will be protected from any employer retaliation. Union employees who are subject to a comprehensive collective bargaining agreement may avail themselves of the provisions of Section 502 of the NLRA. Under this section, an employee who is faced with "abnormally dangerous conditions" has an *individual* right to leave the job site. The right may be exercised, however, only where the existence of abnormally dangerous conditions can be objectively verified. Both exposure and medical information are crucial here.

Under a 1973 OSHA regulation, the right to refuse hazardous work extends to all employees, *individually*, of private employers, regardless of the existence or nature of a collective bargaining agreement. Section 11(c) of the OSHAct protects an employee from discharge or other retaliatory action arising out of his or her "exercise" of "any right" afforded by the act. The Secretary of Labor has promulgated regulations under this section defining a right to refuse hazardous work in certain circumstances: where an employee reasonably believes there is a "real danger of death or serious injury," there is insufficient time to eliminate that danger through normal administrative channels, and the employer has failed to comply with an employee request to correct the situation.

Under the federal Mine Safety and Health Act, miners also have rights to transfer from unhealthy work areas if there is exposure to toxic substances or harmful physical agents, or if there is medical evidence of pneumoconiosis

ANALYSIS OF OSHA'S PERFORMANCE AND COMMENTARY ON NEW INITIATIVES

In the 1980s, OSHA turned to negotiated rule making allowed by the revisions to the Administrative Procedure Act. However, negotiation for the benzene standard failed, and, in 1983, OSHA issued a standard essentially the same as had been remanded by the Supreme Court, but with the required scientific/risk assessment justification. OSHA then promulgated negotiated standards for formaldehyde in 1992, methylenedianiline in 1992, and butadiene in 1996, but they were neither as protective as the law would have allowed, nor as technology forcing (see Caldart and Ashford, 1999).

Although OSHA standard-setting efforts continued in the latter part of the 1990s, its early commitment to worker protection has been further seriously compromised by both procedural requirements imposed by new legislation and by the chilling effect that this legislation has had on agency willingness to set stringent standards. This legislation—the Regulatory Flexibility Act, the Paperwork Reduction Act, the Unfunded Mandates Reform Act, the Small Business Regulatory Enforcement Fairness Act, and the National Technology Transfer and Advancement Act—has placed time-consuming burdens on the agency, contributing to a serious slowdown and resource intensiveness in the development of standards, compounding the effects of Executive (Presidential) Orders requiring the Office of Management and Budget to review OSHA's assessment of costs and benefits for major rules, defined as those having more than $100 million in costs per year.

Equally disturbing is the inadequacy of protection offered by some of the new health standards. The standard for the carcinogen methylene chloride was finally promulgated in 1997 after years of delay. The United Autoworkers Union (UAW) first petitioned OSHA in 1987 for a reduction of the permissible 8-hour exposure allowed by the prior PEL of 500 to 10 ppm. OSHA promulgated a standard of 25 ppm, without medical removal protection. That level was argued to present a lifetime cancer risk of 1 in 2,400, a risk considerably greater than that allowed in prior standards for individual carcinogens, such as vinyl chloride and benzene, and in sharp contrast to the level of 1 in 1 million required by the Clean Air Act of 1990 for environmental ambient air exposures to carcinogens. Originally challenging the standard in court as being too lax, the UAW negotiated a legal settlement with the opposing industry for a revision of the standard, retaining the 25 ppm level, but adding medical surveillance and removal requirements. Legislation introduced in Congress to veto the standard through a disapproval resolution, as allowed by the Paperwork Reduction Act, was unsuccessful.

As discussed earlier, OSHA has to make findings of fact with regard to both the significance of the risk and the feasibility of a proposed standard. Unfortunately, OSHA has pulled back from its historically protective determinations of these factors by (a) being content to regulate near the 1 per 1,000 lifetime risk, which was the *lower* bound of

significance suggested by the Supreme Court in its benzene decision; and (b) finding gratuitously that a proposed standard is feasible, rather than protecting workers to the extent feasible—that is, to the limits of feasibility, using its technology-forcing authority. A study undertaken by the now-defunct Congressional Office of Technology Assessment (OTA) examined the postpromulgation costs of past OSHA standards (including vinyl chloride, ethylene oxide, lead, cotton dust, and formaldehyde) and, in general, found them to be a fraction of the prepromulgation estimates. The OTA concluded that:

> OSHA's current economic and technological feasibility analyses devote little attention to the potential of advanced or emerging technologies to yield technically and economically superior methods for achieving reductions in workplace hazards . . . . Opportunities are missed to harness leading-edge or innovative production technologies (including input substitution, process redesign, or product reformulation) to society's collective advantage, and to achieve greater worker protection with technologically and economically superior means.
>
> [I]ntelligently directed effort can yield hazard control options—attributes that would, no doubt, enhance the "win-win" (for regulated industries and their workforces) character of OSHA's compliance requirements in many cases and support the achievement of greater hazard reduction.

Thus, OSHA in no way seems to be pushing regulation to its limits of technology (see U.S. Congress, Office of Technology Assessment, 1993).

OSHA ran into tremendous industry resistance to a proposed ergonomics standard, and capitulating to both industry and White House pressure, OSHA temporarily withdrew from the fray. It is now intent on formulating a new standard. In spite of the delay of OSHA in promulgating an ergonomics standard, the agency has made it clear the employer has obligations to protect workers from ergonomic hazards under the general duty clause and that enforcement activity will be applied in appropriate situations. The Review Commission upheld OSHA's authority to use the general duty clause in these circumstances. OSHA also experienced political difficulty in establishing standards for secondary tobacco smoke (environmental tobacco smoke) as part of its concern for indoor air quality. OSHA issued a proposed rule in 1994, but action is yet to be taken.

OSHA continues to plan the promulgation of new standards and the review/reconsideration of standards more than 10 years old as required by the Regulatory Flexibility Act. According to its published Regulatory Agenda, new standards expected in the near term include hexavalent chromium, metalworking fluids, silica, steel erection, prevention of work-related musculoskeletal disorders (ergonomics), respirator use, and occupational exposure to the bacteria that cause tuberculosis. OSHA is especially motivated to promulgate a rule for safety and health programs, which it sees as a cornerstone of worker protection. In addition, rules are planned for record keeping and reporting, PELs for air contaminants, and requirements to pay for personal protective equipment. Longer-term actions are planned for glycol ethers, indoor air quality, and beryllium (a rule was proposed in 1977, but was never issued), among others.

Four older standards are also being reviewed: lock-out/tag-out, ethylene oxide, cotton dust, and grain-handling facilities. So far, OSHA has determined that the first of these merits being continued. Also under consideration for revision is a rule for the Process Safety Management of Highly Hazardous Chemicals, to add other reactive chemicals to the rule and to bring it more in line with the EPA's Risk Management Plan. As provided by the Small Business Regulatory Enforcement Fairness Act, the effects of revisited standards on small business must be assessed, and that assessment is now reviewable by the circuit courts of appeal.

The standard-setting process for metalworking fluids is using a standards advisory committee authorized by the OSHAct. Negotiated rule making is being used for steel erection and shipyard fire protection rule

makings. Voluntary guidelines are proposed for workplace violence (the Workplace Violence Prevention Program). Paralleling efforts in other regulatory agencies such as the EPA, OSHA has gone out of its way to involve stakeholders early in the standard development process, holding open meetings on planned agency initiatives, often in several locations, to explain and discuss its approaches with the affected publics.

In the current antiregulatory climate, OSHA has, as have other regulatory agencies, shifted toward more voluntary initiatives, including the use of expert advisors, outreach, compliance assistance, consultation, and partnering with industry, trade unions, and workers. OSHA has designated Special Emphasis Programs and Initiatives on silicosis, mechanical power press injuries, lead in construction, nursing home accidents, and workplace violence. These programs and initiatives target a specific occupational hazard or industry and combine outreach and education with enforcement. OSHA has issued to its field staff a Directive on Strategic Partnerships for Worker Safety and Health. OSHA Strategic Partnerships are intended to establish cooperative efforts at improving health and safety. Motivated by both concerns for efficient use of agency resources and by new-wave volunteerism, OSHA is attempting to nationalize state-based, creative compliance programs (such as the Maine 200 Program) through its Cooperative Compliance Program (CPP) launched in 1997. Under this approach, OSHA uses an industry hazard classification system to target employers with high rates of injury and illness for an optional partnership with OSHA that allows firms to engage in self-improvement, rather be subject to priority inspection by OSHA inspectors, provided that the employers establish comprehensive safety and health programs. These programs provide for the participation of labor. Ironically, employers and industry groups are not uniformly supportive of CPP, because OSHA extracts commitments to these programs—such as an ergonomics plan—which are otherwise not currently required by law. The Chamber of Commerce obtained a stay of the CPP program from the District of Columbia Circuit Court of Appeals while the legality of the program was being decided. As discussed previously, in the meantime, a rulemaking process is moving forward to require firms to adopt comprehensive safety and health programs. These programs are commonplace in both Europe and Canada. It should be a source of embarrassment that the United States stands out in its slow, if not reluctant, approach to protect workers sufficiently with all the tools at its disposal.

OCCUPATIONAL HEALTH AND SAFETY IN BRITISH COLUMBIA[1]

The discussion in this chapter has focused on occupational health and safety in the United States. The system in British Columbia, Canada, is very different and provides another useful perspective.

Profile of British Columbia

British Columbia is Canada's third-largest province, with 1.4 million workers of a total population of 3 million people. Thirty-seven percent of the workers are unionized, compared with approximately 15% in the United States. Ninety-five percent of the firms have 50 or fewer workers, and 75% have five or fewer workers.

Administrative Structure

In British Columbia, the occupational safety and health regulation and enforcement activities and the workers' compensation system are part of the same administrative public corporation, the Workers' Compensation Board (WCB), and both are funded by assessed premiums on employers (see Box 11-2 in Chapter 11). The WCB is administered by a panel of administrators appointed by the Minister of Labour.

---

[1] As of June 1997.

The Prevention Division (formerly the Occupational Safety and Health Division) employs approximately 400 people, which would translate into 28,000 for the United States (compared with the actual number of approximately 2,000). The annual Division budget would be equivalent to a U.S. $1.5 billion budget for OSHA, five times larger than the amount actually allocated in the United States.

Legal/Structural Basis

Two provincial pieces of legislation—the Workers' Compensation Act (see Box 11-2 in Chapter 11) and the Workplace Act—provide the basis for the WCB's standard-setting authority. The federal Workplace Hazardous Materials Information System serves as the basis for provincial right-to-know activities. The Panel of Administrators adopts regulations, with the assistance of a tripartite Regulation Advisory Committee, including professionals from the Division, which was responsible for developing new regulations and revising older ones during the last extensive regulation review process. A Policy Bureau in the Division provides advice to the Panel of Administrators concerning the final regulations. Thereafter, there is no legal mechanism to challenge the regulations in the British Columbia system. Thus, the development of regulatory policy by the courts discussed for the U.S. system does not exist in British Columbia, for all practical purposes.

Enforcement

Historically, British Columbia standards have not been technology forcing. For example, until 1993 the lead standard permitted exposures up to 150 $\mu g/m^3$, compared with the U.S. standard of 50 $\mu g/m^3$. First-instance citations (mandatory citations on discovery of violations) exist only for a few, mostly safety, violations. There is pressure to include specific chemical exposures and failure of the employer to provide an adequate health and safety program/health and safety committee in the list of violations requiring first-instance citations. The Prevention Division can and does impose penalty assessments; criminal penalties are rarely issued. Labor participates in the WCB's enforcement and appellate process in a significant way.

Inspection activity is targeted by a combination of industry hazard classification, payroll, compensation claims, and inspector experience through a rational targeting system called WorkSafe. The construction and logging industries are targeted for special attention because of their high hazard nature and poor claims experience. The Prevention Division places serious emphasis on its data collection and analysis activities, which appear to be more useful than those of OSHA and the U.S. Bureau of Labor Statistics. Accident reporting, which is being computerized, increasingly provides the information needed to focus prevention activities, such as the cause of the accident, rather the cause of the injury. The Prevention Division is implementing the Diamond Project, which, like OSHA's Cooperative Compliance Program, is based on the Maine 200 Program, and seeks to shift responsibility to firms and workers when justified by a good record of occupational injuries (and disease).

Consultation

Most inspection activity results in warnings and corrective orders rather than monetary penalties on the employer. Some consultation and technical assistance is usually rendered by the inspector at the time of the inspection or closing conference. The Division provides engineering guidance and advice to employers in the form of technical bulletins and on-site consultation. The WCB also has an active first-aid certification program for workplace-based first-aid attendants, which is required by law. The WCB does not charge a fee for consulting advice or laboratory assistance/analysis.

Worker Participation

Workplace safety and health programs are required to be provided by all employers with a workforce of 50 or more employees (5% of the firms). For especially hazardous industries, the programs are required for employers with a workforce of 20 or more employees. Joint workplace safety and health committees are considered an essential part of these programs. There is pressure to expand the number of firms required to have such a program. Workers complain that they need more authority in the functions of the safety and health committees. They also complain of the inadequacy of the antidiscrimination provisions of the current law/structure, such as in relation to the right to refuse hazardous work.

Comment

Features of the British Columbia system suggest possible U.S. OSHA reforms, such as mandatory health and safety programs and committees, greater recognition of occupational disease, a streamlined standard-setting process, and a linkage of compensation and prevention activities. The period since 1970 has revealed both the strengths and weaknesses of the U.S. system, including the need to strengthen the connection between OSHA and the EPA through the OSHAct, TSCA, and the safety provisions of the Clean Air Amendments.

OCCUPATIONAL HEALTH AND SAFETY IN THE EUROPEAN COMMUNITY

Occupational health and safety legislation in individual European countries is in a great deal of flux after the formation of the European Community (EC), now the European Union (EU). The Single European Act establishing the EC was enacted in 1987. Article 118A of the Act addresses employment, working conditions, and occupational health and safety and provides a streamlined legislative process for the development of health and safety directives, and minimum health and safety standards, affecting approximately 150 million people. The EC directives have the force of law and set down general principles for the protection of workers. However, individual countries are obligated to adopt national legislation implementing these principles, with important technical details concerning enforcement and administration left to the EC Member States. Thus, programs may be expected to differ considerably among countries in the near future, although these differences may narrow as European integration becomes a reality. Therefore, it may be some time before innovations in health and safety regulatory approaches can be evaluated and serve as models for OSHA reform in the United States. Nevertheless, the EC experience may be important for the United States because (a) with the formation of a North American Free Trade Zone, the problems of harmonization of legislation may be similar; (b) the EC will be an important force in occupational safety and health; and (c) the EC will be a major trade competitor. The recent agreement between the EC and the European Free Trade Association countries to set up a free trade area means that the EC safety and health legislation is applicable in 19 countries in Europe.

Legal and Structural Basis

Regulatory activity within the EC can include regulations, decisions, directives, resolutions, and recommendations, varying from commitments in principle to legally enforceable mandates on the Member States. The European Commission, aided by expert groups, makes formal proposals to the EC Council of Ministers. The Council, in consultation with the Economic and Social Committee and the European Parliament, adopts, rejects, or modifies the proposals and issues directives by a qualified majority vote of 54 of a total of 76. Individual Member States can maintain or introduce more stringent measures for the protection of working conditions than those contained in the directives.

Until 1988, EC directives, such as those dealing with occupational exposure limits for vinyl chloride, lead, asbestos, and benzene were very detailed and prescriptive. STELs were also specified. After Article 118A was enacted, a more general Framework Directive 89/391/EEC "on the introduction of measures to encourage improvements in the safety and health of workers at work" was issued. This directive is the centerpiece of EC health and safety policy and establishes the guiding principles on which more specific directives are issued. There are now seven so-called daughter directives to the Framework Directive. Directive 90/394/EEC addresses carcinogens at work. Directive 88/642/EEC addresses risks related to exposure to chemicals and physical and biologic agents at work and has led to some 27 indicative limit values (TLVs), which are advisory only. The enforcement of those limits is left to the individual regulatory systems and styles of the various Member States. Nevertheless, there is a preferred hierarchy of control for "dangerous substances and products." In order of preference, these are substitution of dangerous substances by safe or less dangerous ones, the use of closed systems or processes, local extractive ventilation, general workplace ventilation, and personal protective equipment.

Other EC directives address biologic agents, asbestos, video display terminals, work equipment, personal protective equipment, and handling of loads. In 1988, the European Parliament adopted a Resolution on Indoor Air Quality, which is receiving attention for development into a directive.

All Commission proposals are submitted to the Advisory Committee on Safety, Hygiene and Health Protection at Work, composed of representatives of employers, workers, and governments. Initially, an expert scientific group evaluates all scientific data relevant to protecting workers from a particular substance. The Commission makes a proposal and solicits Advisory Committee opinion. The Technical Progress Committee votes on the proposal. The limit values may be adopted as indicative values by Commission directive. If the exposure limits are mandatory, they are adopted by the Council of Ministers as directives pursuant to Article 118A. Compared with the United States, relatively few health standards have been established, reflecting the slowness of the tripartite process of participatory standard setting envisioned by the EC.

The Framework Directive applies to all sectors of employment activity, both public and private. However, it excludes the self-employed and domestic workers. Employers have a general "duty to ensure the safety and health of workers in every aspect related to the work" (Article 5.1). Among the employer's specific duties are (a) to evaluate risks in the choice of work equipment, chemicals, and design of the workplace; (b) to integrate prevention into the company's operations at all levels; (c) to inform workers or their representatives of risks and preventive measures taken; (d) to consult workers or their representatives on all health and safety matters; (e) to train workers on workplace hazards; (f) to provide appropriate health surveillance; (g) to protect especially sensitive risk groups; and (h) to keep records of accidents and injuries.

Enforcement

Labor inspectorates in each Member State have the responsibility for ensuring employer compliance with health and safety requirements. However, beyond broad principles and duties, the EC Directives are often advisory, and not many specific requirements are enforceable through EC channels. Attempts to place binding obligations on national governments to establish the necessary institutional elements to support proper implementation of safety and health regulations, such as health and safety technical centers, have been unsuccessful. The Commission established a Committee of Senior Labor Inspectors in 1982 to facilitate infor-

mation exchange to encourage coordination of policy. The Commission also established the European Agency for Safety and Health at Work in Bilbao, Spain.

The Commission does have the authority to bring action against a Member State for failure to adhere to EC Directives, but the Commission does not yet have the institutional capacity to monitor compliance effectively. Action against a Member State has never been brought, however, even though some countries have not adopted national legislation to conform with specific mandatory exposure limits, such as for noise. No uniform policy on enforcement of standards, such as first-instance citations or penalty levels, exists and it is likely that intercountry variations will be allowed.

Worker Participation

The Framework Directive calls for "the informing, consultation, balanced participation . . . and training of workers and their representatives" to improve health and safety at the workplace (Article 1.2). The Directive gives workers the rights to consult in advance with their employers on health/safety matters, to be paid for safety activities, to communicate with labor inspectors, and to exercise the right to refuse dangerous work. Safety committees are not explicitly addressed by the Directive, although many European countries have required them in transposing the Directive into national law. Similarly, joint decision making is not mandated, but may occur in practice.

Comment

The health and safety policy of the EC is evolving. Although the general principles declared in EC legislation and specific directives are laudable, it remains to be seen what course implementation will take and how much variation will continue to exist among the different Member States. European regulatory systems tend to be more advisory. On the other hand, they are also more participatory, inviting decision making on a tripartite basis.

BIBLIOGRAPHY

Ashford NA. Promoting technological changes in the workplace: public and private initiatives. In: Brown M, Froines J, eds. Technological change in the workplace: health impacts for workers. Los Angeles: UCLA Institute of Industrial Relations, 1993:23–36.
A series of essays on the effects of technological change on workers, and suggestions for improvement of workers' health and safety.

Ashford NA. The economic and social context of special populations. In: Frumkin H, Pransky G, eds. Special populations in occupational health. Occup Med 1999;14(3):485–493.
Introductory chapter on changes in the nature of technology, work, and labor markets, and their impact on different worker populations at risk.

Ashford NA, Caldart CC. Technology, law and the working environment. Washington: Island Press, 1996.
A textbook of law and policy related to the workplace, with court cases, law review articles, and policy analysis.

Caldart CC, Ashford NA. Negotiation as a means of developing and implementing environmental and occupational health and safety policy. Harvard Environmental Law Review 1999;23(1):141–202.
A major review of negotiated rule making in both OSHA and EPA, assessing the strengths and weaknesses of regulatory negotiation.

Commission of the European Community. Social Europe: health and safety at work in the European Community. Brussels, Commission of the European Community, February 1990.
Text and explanation of EC directives and legislation pertaining to occupational safety and health.

Hecker S. Part 1: early initiatives through the Single European Act. New Solutions 1993;3:59–69; Part 2: The Framework Directive: whither harmonization? New Solutions 1993;4:57–67.
A critical look at health and safety policy in the European Union.

Rest KM, Ashford NA. Occupational safety and health in British Columbia: an administrative inventory of the WCB Prevention Division. Cambridge, MA: Ashford Associates, 1997.
A review of Occupational Health and Safety Administration in British Columbia.

U.S. Congress, Office of Technology Assessment. Preventing illness and injury in the workplace. Washington, DC: U.S. Government Printing Office, 1985.
A comprehensive assessment of earlier political problems faced in preventing occupational disease and injury.

U.S. Congress, Office of Technology Assessment. Gauging control technology and regulatory impacts in occupational safety and health: an appraisal of OSHA's analytic approach. Publication no. OTA-ENV-635. Washington, DC: U.S. Congress, Office of Technology Assessment, September 1995.
An analysis of the differences between preregulatory estimates of cost and postregulatory costs of OSHA regulation, focusing on the importance of taking technological change into account.

11 Workers' Compensation

Leslie I. Boden

Workers' compensation is a legal system that shifts some of the costs of occupational injury and illness from workers to employers. Workers' compensation laws usually require employers or their insurance companies to reimburse part of injured workers' lost wages and all of their medical and rehabilitation expenses. This chapter focuses on workers' compensation systems in the United States, discusses the role of the physician, and analyzes the very difficult problems associated with compensating victims of chronic occupational disease.

HISTORICAL BACKGROUND

Before passage of the first workers' compensation act in 1911, workers usually bore the costs of their work-related injuries. Injured workers and their families were forced to cope with lost wages and medical care and rehabilitation costs. Under the common law,[1] workers had to prove in a court of law that their injuries were caused by employer negligence to recover these costs.

For several reasons, it was extremely difficult for workers to win such negligence suits. The injured worker had the burden of

proof and had to show that the employer was negligent, that there was a work-related injury, and that the negligence caused the injury. To sustain this burden of proof, the worker had to hire a lawyer (which was costly) and often had to rely on the testimony of fellow workers (who, along with the suing worker, might be fired for their part in the suit). All of this was enough to deter most workers from bringing suit.

In addition, employers had three very strong common-law defenses that usually protected them from losing negligence suits when they were brought:

Doctrine of contributory negligence: If employees were found by judges to have contributed in any way to their injuries, they were barred from winning.
Fellow-servant rule: If fellow employees' actions were found by judges to have caused the injuries, employers were not considered responsible.
Assumption of risk: If injuries were found to be caused by common hazards or by unusual hazards of which workers were aware, they could not recover damages.

In the late 19th century, these defenses were widely used, and less than one-third of all employees who brought such negligence suits won any award. In one case, a New York woman lost her arm when it was caught between the unguarded gears of the machine she had been cleaning. Unguarded gears were in violation of New York State laws then, and before the accident she had com-

L. I. Boden: Department of Environmental Health, Boston University School of Public Health, Boston, Massachusetts 02118.

[1] The common law is a body of legal principles developed by judicial decisions rather than by legislation. Statutory law can override these judge-made laws.

plained to her employer about this hazard. Still, her employer refused to guard the machine. After the accident the worker sued her employer, but the court held that she could not be compensated; she had obviously known about the hazard and, of her own free will, had continued to work. This evidence showed that she had "assumed the risk" and that her employer was not responsible for the consequences.

The inability to hold employers responsible for their negligent actions persisted in the face of the high and increasing toll of occupational death and disability at the beginning of the 20th century. After a disabling injury, workers and their families were left largely to their own resources and to assistance from relatives, friends, and charities.

By 1910, some efforts had been made to provide better means of compensation to injured workers and their families. Some of the larger corporations had established private compensation schemes, and several states and the federal government had enacted employers' liability acts. These laws retained the basic common-law liability scheme, but reduced the role of the three common-law defenses.

Most injured workers, however, were not able to take advantage of these changes. There was growing support for a major change in the law from the social reformers of the Progressive Era and from major corporations. These pressures gave rise to the passage of the first workers' compensation law in New York State. Many other states rapidly followed suit, and by 1920 all but eight states had passed similar laws, although most did not cover occupational disease. Mississippi, in 1948, was the last state to establish a workers' compensation system.

DESCRIPTION OF WORKERS' COMPENSATION

Workers' compensation provides income benefits, medical payments, and rehabilitation payments to workers injured on the job as well as benefits to survivors of fatally in-

jured workers. There are 50 state and three federal workers' compensation jurisdictions, each with its own statute and regulations.

Although state and federal systems are different in numerous ways, they have several characteristics in common. Benefit formulas are prescribed by law. In general, medical care and rehabilitation expenses are fully covered, but lost wages are only partially reimbursed. Employers are legally responsible for paying benefits to injured workers. Some large employers pay these benefits themselves, but most pay yearly premiums to insurers, which process all claims and pay compensation to injured workers. Workers' compensation is a no-fault system. Injured workers do not need to prove that their injuries were caused by employer negligence. In fact, employers are usually required to pay benefits even if the injury is entirely the worker's fault.

The change to a no-fault system was established to minimize litigation. For a worker to qualify for workers' compensation benefits, only three conditions must be met: (a) there must be an injury or illness; (b) it must "arise out of and in the course of employment"; and (c) there must be medical costs, rehabilitation costs, lost wages, or disfigurement.

Clearly, these conditions are much easier for the injured worker to demonstrate than employer negligence. For example, if a worker falls at work and breaks a leg, all three conditions are easily met. Unusual cases sometimes arise in which the question of the relationship of an injury to employment is difficult to resolve, and there may be questions about when a worker is ready to return to work. Such issues may result in litigation, but they are the exception, not the rule. In most cases, a worker files a claim for compensation with the employer, and the claim is accepted and paid either directly by the employer or by the workers' compensation insurance carrier of the employer.

The following case is typical of the events that follow many minor claims for workers' compensation:

Mr. Fisher acquired a painful muscle strain while lifting a heavy object at work on Monday afternoon. He went to the plant nurse and described the injury. He was sent home and was unable to return to work until the following Friday morning. On Tuesday, the nurse sent an industrial accident report to the workers' compensation carrier and a copy to the state workers' compensation agency. Three weeks after he returned to work, Mr. Fisher received a check from the insurance company covering part of his lost wages as mandated by state statute and all of his out-of-pocket medical expenses related to the muscle strain.

Workers' compensation provides wider coverage than the common law system did. Under workers' compensation, workplace injuries and illnesses are compensable even if they are only partially work related. In general, injuries and illnesses are considered eligible for compensation if occupational exposure is the sole cause of the disease, is one of several causes of the disease, was aggravated by or aggravates a nonoccupational exposure, or hastens the onset of disability (Table 11-1). Suppose, for example, a worker with preexisting chronic low back pain becomes permanently disabled as a result of lifting a heavy object at work. In this case, the worker's preexisting condition might just as easily have been aggravated by carrying out the garbage at home, but the fact that the disabling occurred at work is sufficient for compensation to be awarded.

Cases in which an occupational injury or illness becomes disabling as a result of nonwork exposures are similar in principle. For example, a worker with nondisabling silicosis may leave a granite quarry job for warehouse work. Without further exposure, the silicosis will probably never become disabling. However, the worker may begin to smoke cigarettes and lose lung function until partial disability results. In most states, this worker should receive compensation from the owner of the granite quarry, if the work relationship can be demonstrated.

Several states, including California and Florida, allow disability to be apportioned between occupational and nonoccupational causes. Although at first this may seem like a sensible approach, apportionment creates some difficult decisions for workers' compensation administrators. Many disabilities are not the additive result of two separable exposures. With silica exposure or cigarette exposure alone, the worker in the previous example would not have become disabled. Often, as in the case of lung cancer caused by asbestos exposure and smoking, the contribution to disability or death of two factors is many times greater than that of one alone. Such issues make the apportionment of disability very difficult, if not impossible.

When workers' compensation was introduced, workers gained a swifter, more certain, and less litigious system than existed before. In return, however, covered workers waived their right to sue employers through common law. (See Box 11-1 for situations in which workers with occupational injuries and illnesses can sue.) They also accepted lower awards than those given by juries in negligence suits: workers' compensation provides no payments for "pain and suffering," as there might be in a common-law settlement. In addition, disability payments under workers' compensation are often much less than

TABLE 11-1. *Likelihood of compensation, by source of preexisting condition and source of ultimate disability*

	Source of preexisting condition	
Source of ultimate disability	Work related	Nonwork related
Work related	Compensable	Generally compensable
Nonwork related	Generally compensable	Not compensable

Adapted from Barth PS, Hunt HA. Workers' compensation and work-related illnesses. Cambridge, MA: MIT Press, 1980.

Box 11-1. Another Legal Remedy for Workers Injured at Work: Third-Party Lawsuits

Emily A. Spieler

Workers' compensation was designed as the exclusive remedy for injuries arising out of employment. In return for providing limited no-fault benefits for their injured employees, employers were shielded from common-law actions that might have been brought by these employees. Employers have largely retained this immunity, even when occupational injuries and diseases result from the employer's negligence. Therefore, employers still cannot be sued for damages, even if the worker's injuries could have been prevented through the exercise of reasonable care.

However, workers are not always limited to workers' compensation as a remedy. Not surprisingly, injured workers (and their lawyers) have continually tried to find legal strategies that provide damages beyond the limited benefits provided under workers' compensation programs. There are—and always have been—some exceptions to the broad grant of legal immunity to employers. In addition, the exclusivity of the workers' compensation remedy never applied to anyone other than the employer: third parties, including manufacturers of hazardous equipment and substances, can be sued. These lawsuits are governed by state law, and therefore may vary considerably from one state to another.

When Employers Are Subject to Lawsuits for Workplace Injuries and Illnesses

In most states, employers may be sued after a worker is injured:

- If the employer has not properly purchased workers' compensation insurance coverage.
- If the particular injury or illness is not compensable under the state's workers' compensation law. For example, if a state specifically excludes coverage for a specific occupational diagnosis such as stress-related mental illness or cumulative trauma disorders, the employer is not shielded

from lawsuit. To be successful, the worker must, however, prove that the employer was negligent; the no-fault principles of workers' compensation do not apply.
- If the claim is related to a specific employment law. Until relatively recently, employers of occupationally injured employees could pay workers' compensation benefits and discharge the worker without adverse legal consequences. This practice is often no longer legal. Actions brought by employees alleging violation of various state and federal employment laws are not affected by workers' compensation rules, usually including situations in which the workers allege mental injury as a result of illegal discrimination or retaliation.

In addition, when an injury is the result of intentional—not merely negligent—conduct by the employer, some states allow the worker to bring a common-law action against the employer. Most states still preclude these actions or require the worker to prove that the employer specifically intended to injure the worker: the specific intent standard. Under this standard, even employers who intentionally violate health and safety rules or are reckless in their approach to occupational safety are protected by workers' compensation immunity. In contrast, a growing minority of states now allow workers to bring common-law actions against an employer when the employer knowingly allows hazardous conditions to exist that are substantially certain to result in serious injury. In these states, an employer's knowledge that conditions were extremely hazardous is relevant to determining whether the employer will be held liable for damages above those provided by workers' compensation.

Lawsuits Against Third Parties

Although workers' compensation coverage bars workers from suing their employers at common law for negligently caused injuries, there is no such bar against suits against third parties. Not surprisingly, workers and their lawyers have sought substantial additional damages from a variety of third parties. Most commonly, people involved in automobile crashes at work have sued the drivers of the other vehicles. Also relatively common are suits against manufacturers of

Box 11-1 (*continued*)

equipment or substances that contributed or caused the injury. For example, if workers are injured by faulty machinery at work, workers' compensation does not bar them from suing the manufacturer of the equipment. Similarly, workers whose lung disease or cancer is related to workplace exposure to asbestos have successfully sued asbestos manufacturers.

These lawsuits always require that the worker show that the third party was negligent—the mere fact of the injury is never enough. In the case of third-party manufacturers of equipment, this negligence is tied to the failure to warn the users of the product (including the injured worker and in some cases his or her employer) regarding the safe handling or operation of the dangerous machinery or substances. Other third parties who have sometimes been found liable to workers for events surrounding workplace injuries range from workers' compensation insurance carriers (for bad faith dealings regarding the handling of claims or for negligent safety inspections of the workplace) to the Occupational Safety and Health Administration (for failing to cite a flagrant workplace safety hazard). Employers have also sometimes been sued, with varying success, by workers who claimed that the employer was not wearing "the employer hat" when the injury occurred. These lawsuits have involved a wide variety of situations, including when employers made or substantially altered the equipment causing the injury or when employers have negligently provided medical services to employees.

Differences between Lawsuits and Workers' Compensation Claims

It is not surprising that workers attempt to pursue common-law litigation instead of, or in addition to, workers' compensation claims. One study of asbestos compensation showed that average payments to workers from successful lawsuits were over twice as high as payments for workers' compensation claims (1); damages in a few cases have been very large. But the barriers to successful litigation of these claims is also high. Injured workers must prove in most cases that the defendant was negligent; the no-fault rules of workers' compensation do not apply. Lawsuits are tried in the regular civil courts rather than in specialized workers' compensation administrative courts. Trials in the civil courts are run under much more stringent rules of evidence, and proceedings tend to be much longer and require much more work by attorneys. Unlike workers' compensation claims, these cases cannot be pursued without a lawyer. The combination of the higher standard of proof and the greater legal expense mean that it is more difficult to find a lawyer to provide the necessary legal representation. The high legal expenses mean that the worker may collect a much lower proportion of the awarded damages in civil litigation than in workers' compensation proceedings. A 1984 study of lawsuits found that workers with asbestos-related disease received only 39% of the awarded damages, just slightly more than the amount paid to either defense or plaintiff attorneys (2).

lost income, especially for more severe injuries.

The United States does not have a unified workers' compensation law. Each state has its own system with its own standards and idiosyncrasies. In addition, federal systems cover federal employees, longshoremen and harbor workers, and workers employed in the District of Columbia. Except for Texas and New Jersey, all states require employers either to purchase insurance or to demon-

strate that they are able to pay any claims that might be made by their employees. In most states, private insurers underwrite workers' compensation insurance paid for by premiums from individual employers. In some states, a nonprofit state workers' compensation fund has been established; the state government therefore acts as an insurance carrier, collecting premiums and disbursing benefits. State funds seem to be very effective in delivering benefits: they disburse a higher percentage of premiums in the form of benefits than do private insurance carriers.

THE ROLE OF THE PHYSICIAN IN WORKERS' COMPENSATION

Workers' compensation is basically a legal system, not a medical system. The decision points for claims in this complex system are shown in Fig. 11-1. If a claim is rejected by the workers' compensation carrier or self-insured employer, it usually is necessary for the injured worker to hire a lawyer. The worker's lawyer may then bargain with the lawyers for the insurance carrier in an attempt to settle the dispute informally. If this bargaining does not result in agreement, the claim must either be dropped or taken before an administrative board—a quasijudicial body established by state statute—for a hearing. To the worker or to a physician who may be called to testify in such a hearing, the proceeding may be indistinguishable from a formal trial: witnesses are sworn, rules of evidence are followed, and testimony is recorded.

As a part of this legal proceeding, medical questions are often raised. There may be disagreement about the degree of disability of a worker, when an injured worker is ready to return to work, or whether a particular injury or illness is work related. To settle these disputes, physicians may be called on to give their medical opinions about employees' disabilities. Many physicians do not like to testify in such hearings, and most are not prepared by their training or experience to assume this role. Their expertise may be challenged; moreover, they may be confused by the different meanings of legal and medical terminology.

In workers' compensation, decisions are based on legal definitions, and the legal distinction between disability and impairment is often unclear to physicians (see Chapter 12). A physician called in to testify about whether a worker is permanently and totally disabled may understand total disability as a state of physical helplessness and may therefore testify that the injured worker is not totally helpless. However, this standard is not what a workers' compensation board would apply. The term *disability*, as used in workers' compensation proceedings, means that wages have been lost, whereas *total disability* means that the injured worker loses wages as a result of not being able to perform gainful employment. A relatively small impairment could result in a substantial disability. For example, airline pilots might be barred from working because of a level of visual impairment that might minimally interfere with other aspects of their lives. On the other hand, a worker who has been exposed to silica at work may have substantially reduced pulmonary function and therefore impairment. However, if the worker continues to work at the same job, no wages have been lost and therefore no disability payment is made. Many states, however, offer specified payments for disfigurement or losses of sight, hearing, or limbs, with compensation

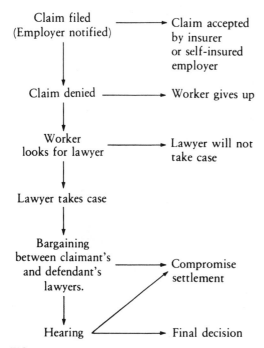

FIG. 11-1. Decision points for workers' compensation claims.

TABLE 11-2. *Income benefits for (in U.S. dollars) for scheduled injuries in selected jurisdictions (as of January 1, 1998)*[a]

Jurisdiction	Arm at shoulder	Hand	Thumb	First finger	Foot
Alabama	48,840	37,400	13,640	9,460	30,580
California	108,445	58,863	5,335	3,360	27,370
Connecticut	125,840	101,640	38,115	21,780	75,625
Delaware	98,115	86,341	29,435	19,623	62,794
Georgia	73,125	52,000	19,500	13,000	43,875
Illinois	244,524	154,865	57,056	32,603	126,337
Mississippi	55,956	41,967	16,986	9,793	34,972
New York	124,800	97,600	30,000	18,400	82,000
North Carolina	127,680	106,400	39,900	23,940	76,608
Washington	81,279	73,151	29,260	18,288	56,895
Wisconsin	89,500	71,600	28,640	10,740	44,750
Federal employees	424,292	311,818	101,993	62,556	278,782
U.S. longshore	250,722	203,921	62,680	38,444	171,326

[a] Amounts reflect maximum potential entitlement.
Reprinted with the permission of the Chamber of Commerce of the United States of America from *Analysis of Workers' Compensation 1998.* © 1998 Chamber of Commerce of the United States of America.

based on impairment and not on disability (Table 11-2).

Although physicians may feel that they have not been trained adequately for their role in workers' compensation, workers do need their support in this area. A lack of assistance may mean unnecessary financial hardship for the victim of an occupational injury or disease. The best way a physician can help identify a work-related disease or injury is by taking an occupational history (see Chapter 4). If the physician suspects a work-related disease or injury, the patient should be informed of the right to receive workers' compensation and the time limits on such claims. (The period for filling a claim usually begins when the patient is informed that the disease is work related.) Physicians should also suggest the possibility of seeking legal counsel, and they can provide direct help by completing any required reports, including descriptions of the illness or injury and why it is believed to be work related. The extent of probable disability should also be noted (see Chapter 12). None of these steps requires testimony before a workers' compensation board; most workers' compensation claims are either paid without contest or settled without a hearing.

A primary care physician may be the only person willing and able to provide documentation for an employee wishing to file a workers' compensation claim. Support for valid compensation claims not only assists injured workers, but helps to ensure that employers and their insurance carriers will approximately shoulder the costs that result from workplace hazards. If these costs are not paid under workers' compensation, they will be borne by workers and their families or by all of us through our share of the costs of third-party medical payments, welfare, Social Security, and other public support programs.

THE ADEQUACY OF WORKERS' COMPENSATION FOR OCCUPATIONAL INJURIES

The fundamental problems of the common-law scheme were that litigation was a necessary element of compensation and that it was very difficult for workers to win suits against their employers. Even when workers won negligence suits, payments were made long after they were injured, and a large amount of each settlement was diverted for legal fees. Today, workers with minor injuries covered by workers' compensation usually can expect to receive payments promptly and without contest. In fact, fewer than 10% of all claims for occupational injuries—as opposed to occupational diseases—are contested.

Under workers' compensation, insurance carriers or self-insured employers have the right to contest a claim. A claim may be contested because the injury is not considered work related, for example, or because the claim is for a larger settlement than the insurer is willing to pay. However, in most injury cases, the employer or insurance carrier has little incentive to contest because proof of eligibility is easy, and the potential gain to the insurer of postponing or eliminating small payments is not enough to offset the legal costs of pursuing a claim.

For expensive injury claims, such as permanent total disability and death claims, insurance companies are much more likely to contest. Even if they do not win, a contest enables them to keep the settlement money temporarily, invest it, and receive investment income until the case is closed and the injured worker paid. Because a contest delays the date of payment, this investment income is an incentive to contest even those cases the insurer is very likely to lose. The higher the potential settlement, the stronger the incentive to contest. Claims for permanent disability and death are contested 5 to 10 times more frequently than are claims for temporary disability.

Aside from the incentives to contest major claims, several other important problems can be cited in the more than 50 workers' compensation systems in the United States. In theory, workers' compensation should cover all employees; however, many states exempt agricultural employees, household workers, or state and municipal employees. Although compensation systems usually provide replacement for close to two thirds of lost wages for *temporary disability*, benefits for victims of *permanent disability* are characterized by low maximum weekly ceilings and low statutory limits on total benefits, with the worker usually receiving the smaller of two thirds of lost wages or the state maximum benefit.

The maximal weekly benefit provided varies widely among the jurisdictions: the highest, on January 1, 1998, was for federal employees ($1,360) and the lowest was for

TABLE 11-3. *Maximum weekly benefits for total disability provided by workers' compensation statues of selected states (as of January 1, 1998)*

Jurisdiction	Fraction of worker's wage	Maximum weekly benefit (to nearest dollar)
Alabama	2/3	493 (SAWW)
Alaska	4/5 of worker's spendable earnings	700
California	2/3	490
District of Columbia	2/3 up to 4/5 of worker's spendable earnings	775 (SAWW)
Florida	2/3	494 (SAWW)
Iowa	4/5 of worker's spendable earnings	873 (200% of SAWW)
Massachusetts	3/5	666 (SAWW)
Michigan	4/5 of worker's spendable earnings	553 (90% of SAWW)
Mississippi	2/3	280 (66% of SAWW)
New Hampshire	3/5	794 (150% of SAWW)
New York	2/3	400
North Carolina	2/3	532 (110% of SAWW)
Pennsylvania	2/3	561 (SAWW)
Rhode Island	3/4 of worker's spendable earnings	519 + 15 per dependent (SAWW)
Texas	7/10	508 (SAWW)
West Virginia	2/3	455 (SAWW)
Federal employees	2/3 or 3/4[a]	1,360 (66% or 75% of GS-15)[a]
U.S. longshore	2/3	836 (200% of NAWW)

[a] Maximum is 3/4 if one dependent or more.
SAWW, state's average weekly wage; NAWW, national average weekly wage.
Reprinted with the permission of the Chamber of Commerce of the United States of America from *Analysis of Workers' Compensation 1998*. © 1998 Chamber of Commerce of the United States of America.

Mississippi workers ($280) (Table 11-3). Many jurisdictions do not provide for cost-of-living adjustments; a person injured 30 years ago and still disabled may receive total disability payments of only $10 to $20 a week. Benefits also do not account for the increased wages that would have been earned if the employee had continued to work.

Studies have raised substantial concerns about the adequacy of workers' compensation benefits. Several suggest that a substantial number of all workers with occupational injuries never enter the workers' compensation system (3). Studies too recent to have yet been published show that in three states (California, Washington, and Wisconsin), many workers receive workers' compensation income benefits that are much lower than their lost earnings, especially workers with long-term impacts of injuries and those receiving permanent partial disability benefits.

WORKERS' COMPENSATION MEDICAL COSTS

Figure 11-2 shows that, during the 1980s, the annual rate of growth of workers' com-pensation medical costs was consistently above the growth of medical costs outside workers' compensation. From 1980 to 1990, there was a 265% increase in workers' compensation medical costs, compared with a 183% increase in medical costs outside workers' compensation. Between 1990 and 1992, there was no clear pattern. However, since 1992, workers' compensation medical costs have grown more slowly than other medical costs, actually falling by 8.3%, compared with the 18.5% increase in medical costs outside workers' compensation during the same period.

A 1990 study provides evidence that workers' compensation medical costs were higher than non–workers' compensation medical costs at that time. The Minnesota Department of Labor and Industry analyzed medical cost data from a matched sample of claims from Minnesota's largest workers' compensation insurer, the Liberty Mutual Insurance Company, and the major non–workers' compensation insurer, Blue Cross-Blue Shield (4) with the data on rapidly rising workers' compensation costs. The study found that workers' compensation medical costs were in general higher than Blue Cross charges

FIG. 11-2. Annual U.S. medical care cost growth: workers' compensation versus non–workers' compensation. (Sources: For 1980 to 1993, workers' compensation data are from Schmulowitz J. Workers' compensation: coverage, benefits, and costs, 1992–93. Social Security Bulletin 1995;58(2):51–57. For 1994 and 1995 workers' compensation data: National Academy of Social Insurance. Workers' compensation: benefits, coverage, and costs, 1994–95: new estimates. Washington, DC: National Academy of Social Insurance, 1997. For non-workers' compensation medical costs: Health Care Finance Administration web site at http://www.hcfa.gov/stats/nhe-oact/nhe.htm.)

for the same types of injuries. For all injuries in the Minnesota sample, workers' compensation charges were 75% higher on average than non–workers' compensation charges. Workers' compensation paid 2.5 times as much to treat back injuries, but only a little more to treat fractures, where treatment options were more limited.

Differences in the two systems may cause differential medical care utilization—that is, medical care providers may use more diagnostic tests on injured workers and may hospitalize them more frequently. On the other hand, the type and intensity of medical care services used may reflect differences in injury severity. We do not know if this is attributable to inherent differences between workers' compensation and non–workers' compensation medical care. Controlling for differential utilization may in part control for the systemic differences in the two systems. To measure these differences conservatively, data in Table 11-4 display workers' compensation medical costs relative to non–workers' compensation costs, controlling for both worker characteristics and medical care utilization.[2]

TABLE 11-4. *Ratio of workers' compensation medical charges to non–workers' compensation charges*[a]

Injury type	Ratio
All injuries	2.04
Back injuries	2.30
Sprains and strains	1.95
Lacerations	1.55
Fractures	1.00

[a] Charges controlled for patient characteristics and utilization.
From the Minnesota Department of Labor and Industry (1990), and Thornquist L. Health care costs and cost containment in Minnesota workers' compensation program. John Burton's Workers' Compensation Monitor 1990;3:3–26.

[2] Measures of utilization include the types of providers (e.g., physicians, chiropractors, physical therapists) and the quantity of medical services rendered (e.g., number of radiographs, blood tests, physical examinations, physical therapy visits, inpatient hospital days, operations).

The disparities between workers' compensation and non–workers' compensation medical costs appear to reflect a substantial difference between payments for the same injuries in the two systems. Alternatively, they might reflect unusually high workers' compensation medical costs in Minnesota relative to other states' systems. The Minnesota study suggests that this is not true (4). The authors compared Minnesota's workers' compensation medical costs with those of six other states (Iowa, Missouri, Illinois, Pennsylvania, Oregon, and Wisconsin) and found them similar. Also, Minnesota's average 1984 workers' compensation medical costs ranked 11th of 42 states; its 1980 to 1985 medical cost growth rate ranked as 21st of 43 states (5).

Another study, based on 1991 to 1993 workers' compensation claims in California, also found higher costs to treat workers' compensation injuries than to treat similar non–work-related injuries (6). The authors concluded that higher utilization caused the added costs of treating workplace injuries.

Minnesota's disparities between workers' compensation and non–workers' compensation medical costs thus provide additional grounds for concern that workers' compensation systems are facing medical cost problems even greater than those faced in the wider medical care system.

CAUSES OF HIGHER WORKERS' COMPENSATION MEDICAL COSTS

Factors specific to workers' compensation may have caused its medical costs to accelerate. Certain cost-control techniques are absent in workers' compensation, and others are more difficult to perform. For example, copayments and deductibles are used regularly outside workers' compensation to reduce the demand for medical care and, thus, its cost. Yet, workers' compensation systems traditionally have paid all medical costs resulting from covered injuries and illnesses.

Workers' compensation insurers or self-insured companies may find discounts for medical care difficult to negotiate for two reasons: (a) workers' compensation has only a small share of the medical care market (approximately 2%), and therefore has less bargaining power; and (b) workers have the legal right to choose a treating physician in approximately half the states, making it more difficult for employers to direct them to lower-cost providers. Evidence about the impact on costs of who chooses the treating physician is equivocal, with studies suggesting that employer choice does not reduce, and may even raise, costs (7).

Litigation also increases the medical costs of workers' compensation beyond those that might be incurred in other settings. Litigation can complicate care and delay recovery, prolonging the duration of medical treatment paid for by workers' compensation. In addition, most states allocate the expense of medical evaluations used to resolve legal disputes, which are often substantial, as medical costs.

WORKERS' COMPENSATION MEDICAL COST CONTROL

Rapidly increasing medical costs have driven many states to implement controls on medical costs. The most common of these are:

- Fee schedules that list maximum reimbursement levels for health care services or products
- Limited employee initial choice of medical care provider, or limitations on changing medical care providers
- Mandatory bill review for proper charges, usually tied to a fee schedule
- Mandatory utilization review of the necessity and appropriateness of admissions and procedures, length of hospitalization, and consultations by specialists before, during, or following an inpatient admission
- Managed care programs that seek to reduce the price and utilization of medical care (e.g., those of health maintenance or preferred provider organizations)

Over the past several years, many states have

adopted one or more of these methods of containing medical costs. From 1991 to 1997, 13 states added medical fee schedules, 14 states added hospital payment regulation, and six states mandated employers and insurers to provide managed care where none had done so in 1991.

The importance of these changes goes beyond their impact on medical costs. When they choose the medical provider or managed care organization, employers and insurers have a greater say about the medical treatment, but also about the provider's behavior in litigated cases. In many workers' compensation systems, treating providers furnish information about when a worker is ready to return to work (and temporary disability benefits may be terminated) or the worker's level of impairment (affecting permanent disability benefits). Because provider choice can affect income benefits, it would be in contention even if everybody agreed it did not affect medical costs.

MEDICOLEGAL ROADBLOCKS TO COMPENSATION FOR OCCUPATIONAL DISEASES

The burden of proving that occupational *injuries* arose "out of and in the course of employment" is usually straightforward. However, workers with occupational *illnesses* face a different situation (Table 11-5). The workers' compensation system expects a physician to say whether a worker's illness was caused by or aggravated by work. Physicians are asked, "Was this illness caused by workplace conditions?" This is a question for which medical science often does not have a simple answer.

Many aspects of occupational diseases make the disabled worker's burden of proof difficult to sustain. Physicians may not realize that their patients may have become ill as a result of workplace exposures. Many physicians are not able to identify occupational diseases because their medical training in this area has been inadequate—many have not even been trained in taking occu-

TABLE 11-5. *Roadblocks to compensation for occupational disease*

Limitations of medical science	Statutory limitations	Other limitations
Difficulty of differential diagnosis Lack of epidemiologic and toxicologic studies Multiple causal pathways Limitations of physician training	Time limits Burden of proof Restrictive definitions of disease	Lack of exposure records (duration and intensity)

pational histories. Furthermore, the signs and symptoms of most occupational diseases are not uniquely related to an occupational exposure. Medical and epidemiologic knowledge may be insufficient to distinguish clearly a disease of occupational origin from one of nonoccupational origin. For example, shortness of breath, an important symptom of occupational lung disease, is also associated with other chronic lung diseases (Fig. 11-3).

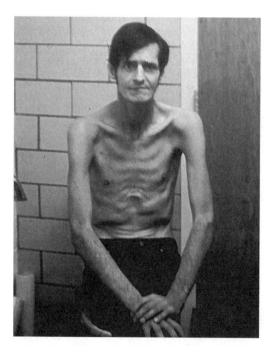

FIG. 11-3. Although workers with silicosis, like this rock driller, qualify for workers' compensation, most workers with chronic occupational diseases often find their claims denied. (Reprinted from Banks DE, Bauer MA, Castellan RM, Lapp NL. Silicosis in surface coal mine drillers. Thorax 1983;30:275.)

Another complicating factor is that a disease may have multiple causes, only one of which is occupational exposure. A worker who smokes and is exposed to ionizing radiation at work may contract lung cancer. Because both cigarette smoke and ionizing radiation are well-established risk factors for lung cancer, it may be impossible to say which of these two factors "caused" the disease. In many cases, occupational disease may develop many years after exposure began, and perhaps many years after exposure ceased. Consequently, memories of events and exposures may be unclear, and records of employment may not be available.

Some occupational injuries occur as a result of extended exposure to a hazard. These are "cumulative trauma" injuries, such as carpal tunnel syndrome, noise-induced hearing loss, and chronic low back pain. As with chronic occupational diseases, it may be difficult to prove the work-relatedness of these injuries. Moreover, records of exposure to occupational hazards are not often kept, so that even when a worker knows the type and duration of exposure, no written evidence of this can be presented.

These aspects of occupational disease mean that many victims do not even suspect that their disease is job related. For those who do and wish to make a claim, the causal relationship between disease and workplace exposures may be very difficult to establish. These are major reasons why so few claims for compensation for occupational disease are filed. A study of occupational disease in Washington and California revealed that, of the 51 probable cases of occupational respiratory conditions, only one was reported as a workers' compensation claim.

TABLE 11-6. *Percentage of alleged occupational disease cases controverted (contested) by category of disease*

Category	Percentage
Dust disease	88
Disorders due to repeated trauma	86
Respiratory conditions due to toxic agents	79
Cancers and tumors	46
Poisoning	37
Skin diseases	14
Disorders due to physical agents	10
Other	54
All diseases	63[a]

[a] In contrast, the percentage for all injuries is 10%. Adapted from Barth PS, Hunt HA. Workers' compensation and work-related illnesses. Cambridge, MA: MIT Press, 1980.

When claims for chronic occupational disease are filed, many are contested by the insurance carrier or self-insured employer (Table 11-6). Therefore, payments to disabled workers are delayed and uncertain (Table 11-7). Workers with chronic occupational diseases wait more than a year, on average, to receive compensation payments. In addition, administrative and legal costs absorb many of the resources devoted to compensating workers for their occupational diseases.

TABLE 11-7. *Delays in compensation for occupational disease by category of disease*

Category	Mean number of days from notice to insurer to first payment
Skin diseases	59
Dust diseases	390
Respiratory conditions due to toxic agents	389
Poisoning	111
Disorders due to physical agents	79
Disorders due to repeated trauma	362
Cancers and tumors	260
Other illnesses	180[a]

[a] In contrast, the mean delay for all injuries is 43 days. Adapted from Barth PS, Hunt HA. Workers' compensation and work-related illnesses. Cambridge, MA: MIT Press, 1980.

ESTABLISHING WORK-RELATEDNESS FOR COMPENSATION

The burden of proving that disease is occupational in origin lies with workers. They must find physicians who are convinced that their illnesses are occupational in origin or that their illnesses were aggravated or hastened by occupational exposures. Physicians must then be able to convince referees who hear the cases that the diseases are indeed work-related.

The burden of proof might at first seem to be impossible for those diseases that are not uniquely occupational in origin. For example, lung cancer may be caused by smoking, air pollution (although not definitely established), occupational or nonoccupational radiation exposure, or all of these factors.

Suppose that a worker with lung cancer has smoked cigarettes, has had diagnostic radiographs, and has also been occupationally exposed to ionizing radiation in a uranium mine. Because occupational lung cancer does not have distinctive clinical features, an expert medical witness, using clinical judgment, cannot say that the disease is, without question, occupational in origin. The expert witness cannot even say with certainty that occupational exposure to ionizing radiation was one of several causes or if it hastened the onset of the cancer. At best, all that can be determined is that the worker has an increased risk of lung cancer as a result of the job. In this case, the legal standard is that there must be a "preponderance of evidence" that the disease is occupational in origin, or the case is unlikely to be settled in favor of the disabled worker. A "preponderance of evidence" means that it is more likely than not (probability >50%) that the illness in question was caused by, aggravated by, or hastened by workplace exposure.

In some cases, workers' compensation laws have been written so that payment of a claim may be denied even though convincing evidence is presented that the illness was caused by or aggravated by the worker's em-

TABLE 11-8. *Time limits on filing occupational disease claims in selected jurisdictions (as of January 1, 1998)*

Jurisdiction	Time limit on claim filing
Alabama	Within 2 years after injury, death, or last payment; radiation—within 2 years after disability or death and claimant knows or should know relation to employment. Radiation or pneumoconiosis: exposure during at least 12 months over 5 years before last exposure.
California	Disability—within 1 year from injury or last payment; death—within 1 year after death and in no case more than 240 weeks after injury, except for asbestos-related disease claims. Date of injury is defined as when claimant is disabled and knows or should know of relation to employment.
Colorado	Within 2 years after commencement of disability or death; within 5 years in cases of ionizing radiation, asbestosis, silicosis, or anthracosis.
Massachusetts	Within 4 years of diagnosis or knowledge of the relationship to employment.
Michigan	Within 2 years after the claimant knows or should know of relation to employment.
New York	Within 2 years after disability or death, and within 2 years after the claimant knows or should know relation to employment.
North Carolina	Within 2 years after final disability determination or death. Within 6 years after death from an occupational disease. Asbestosis: disability or death within 10 years after last exposure. Lead poisoning: disability or death within 2 years after last exposure.
Oregon	Within 1 year of worker's discovery of the disease, after diablement, or physician informs claimant of disease.
Utah	Within 6 years after cause of action arose, but no later than 1 year after death. Notification must be given to employer or Industrial Commission within 180 days of this date.
Virginia	Within 2 years after diagnosis is first communicated to worker or within 5 years after exposure, whichever is first; within 3 years after death, occurring during a period of disability.
Federal employees	Within 3 years after injury, death, or disability and claimant knows or should know relation to employment; delay excusable.
U.S. longshore	Within 2 years of knowledge of relation to employment or 1 year after last payment.

ployment. Some states require that a disease not be "an ordinary disease of life." In other words, diseases such as emphysema and hearing loss may not be compensable because they often occur among people with no occupational exposure. More than 20 states have a related requirement that diseases are compensable only if they are "peculiar to" or "characteristic of" a worker's occupation.

All jurisdictions have a statute of limitations (often 1 or 2 years) on claims for workers' compensation. A 2-year statute of limitation means that the claim must be filed by the worker within 2 years of a given event. A time limit of 2 years after the worker has learned that a disease is work related imposes no particular hardship on those with occupational disease. In some states, however, the period begins when the disease becomes symptomatic, even if this takes place before the disease is diagnosed or determined to be work related. The latter policy for starting the statute of limitation may be a special problem if the worker's physician is not familiar with the occupational disease. The most burdensome statutes require that a claim be filed 1 or 2 years after exposure. Because chronic occupational diseases commonly do not manifest themselves until 5, 10, 20, or more years after exposure, such rules effectively eliminate the possibility of compensation for workers with these illnesses. In one study, occupational disease compensation among a group of asbestos insulators was only half as great in states with these restrictive statutes of limitation as in other states. Time limits for filing workers' compensation claims are described in Table 11-8.

THE PROBLEM OF COMPROMISE SETTLEMENTS

A workers' compensation claim that is denied by the employer or insurer does not automatically go to a hearing. The injured worker must first find a lawyer who will take the case. The lawyer's fee is often based on the portion of the award attributed to lost wages, which means that the lawyer's fee is small in a small award and that the lawyer receives nothing if the claim is denied. Thus, it is difficult for injured or ill workers to find lawyers to represent them when claims are small or success is unlikely.

Lawyers usually prefer to bargain informally with the defendant's lawyers rather than go to trial. If a compromise settlement can be reached before trial, no preparation for a hearing is necessary, and the lawyer will therefore have more time to work on other cases and earn additional income. A settlement reached outside the courtroom is called a compromise settlement because the amount paid to the injured worker usually is a compromise between the maximal and minimal amounts that the worker could receive in a court decision.

In the face of protracted litigation with uncertain results, a compromise settlement may seem very attractive to an injured worker who may have no wage income for a considerable period and may be facing large medical bills. The injured worker may therefore prefer a small settlement paid immediately to a much larger, but uncertain, settlement that would not be available for 1 or 2 years. Especially where the worker does not foresee a quick return to work, a settlement may be accepted that might seem quite small to an outside observer. Insurers may use their knowledge of the financial pressures on the claimant to obtain a small settlement; they will thus contest, delaying the time when the case is closed in the hope of obtaining a small compromise settlement.

The compromise settlement usually is paid in a lump sum to the injured worker and the attorney. This lump sum settlement takes the place of future payments for lost earnings and medical and rehabilitation costs. Many compromise settlements also release the insurer from future liability: if the worker's condition should change at a later date or if future medical needs or increased costs were inadequately estimated, the insurer would not incur the costs of any increased disability or medical or rehabilitation expenses. The injured worker who has accepted a compromise and release settlement may later need additional medical care, but not have the resources to pay for that care.

For example, a worker with a back injury was denied compensation by his employer, who claimed that the injury was not work related. He then took action that led to his being offered a lump sum settlement:

> I went to my union representative and filled out the forms for the industrial accident board, and about 3 weeks later they sent me an award which was about $600 . . . and I wouldn't take it. But then I applied for an attorney and talked to my attorney, and then filed suit. They turned around and told my attorney that they would consider [the injury] an industrial accident. So, I never did go to court. All they did was talk to my lawyer. They settled out of court. My lawyer told me while I was in the hospital that they wanted to settle it for $7,500. The fee for him [would be] $2,500.[3]

---

[3] Adapted from Subcommittee on Labor, Committee on Labor and Public Welfare, U.S. Senate. Hearings on the National Workers' Compensation Standards Act, 1974. Statement of Lawrence Barefield.

If the settlement of $7,500 was the result of a compromise and release agreement, the insurer or employer is not liable for any future disability or medical costs resulting from this injury.

RECOMMENDATIONS OF THE NATIONAL COMMISSION

As part of the Occupational Safety and Health Act of 1970, Congress established the National Commission on State Workmen's Compensation Laws to "undertake a comprehensive study and evaluation of state workmen's compensation laws in order to determine if such laws provide an adequate, prompt, and equitable system of compensation." In 1972, the Commission released its report, which described many problems of workers' compensation and made recommendations for improving state workers' compensation systems. This report, still relevant today, included these seven "essential" recommendations:

1. Compulsory coverage: Employees could not lose coverage by agreeing to waive their rights to benefits.
2. No occupational or numerical exemptions to coverage: All workers, including agricultural and domestic workers, should be covered, and all employers, even if they have only one employee, should be covered.
3. Full coverage of work-related diseases: Elimination of arbitrary barriers to coverage, such as highly restrictive time limits, occupational disease schedules, and exclusion of "ordinary diseases of life."
4. Full medical and physical rehabilitation services without arbitrary limits.
5. Employees' choice of jurisdiction for filing interstate claims.
6. Adequate weekly cash benefits for temporary total disability, permanent total disability, and fatal cases.
7. No arbitrary limits on duration or sum of benefits.

When this report was issued, some federal legislators threatened to establish federal minimum standards for state workers' compensation programs, or to supersede them with a national program. In the 1970s and 1980s, many states changed their statutes to follow some or all of the recommendations of the National Commission. By increasing coverage and raising benefits, they substantially improved the value of workers' compensation to injured employees. However, this trend ended in the 1990s. Moreover, general changes in coverage have done little to discourage the litigation of occupational disease claims and costly occupational injury claims.

THE FUTURE OF WORKERS' COMPENSATION

The common-law system in effect during the 19th century was time consuming, inefficient (because of litigation costs), and uncertain. Workers' compensation was designed to minimize litigation. Although it is a no-fault system, more than 80% of all compensation claims for chronic occupational diseases (see Table 11-6) and almost 50% of all injury claims for permanent total disability or death are contested, leading to delays of a year or more in settling workers' compensation claims (see Table 11-7). The settlements may be compromised and may thereby leave claimants seriously undercompensated. Also, legal fees, commonly 15% to 20% of the compensation award for lost wages, must be paid by the claimant. Among injured workers represented by attorneys in Wisconsin who were injured in 1989 or 1990, workers paid their attorneys an average of $3,200 of $20,900 in income benefits.

Two essential elements of workers' compensation that led to so many contested cases are (a) the necessity of proving that a disease is work related, and (b) the fact that large medical and wage replacement payments to compensated workers come directly from their employers or employers' insurance carriers. The first element leaves room for much legal argument; the second gives considerable incentive for employers or their insurance carriers to pursue such arguments. If contesting claims were less rewarding, more difficult, and less likely to be successful, employers might instead find it cheaper to pay them, thereby reducing the amount of litigation in the system.

In one proposed reform aimed at eliminating incentives for litigation, benefits would be paid from industry-wide funds rather than by individual employers or their insurance carriers. This plan would mean that individual employers or carriers would not gain financially by contesting an award and would thereby have little incentive to do so. A concern about such an arrangement, however, is that if workers submit claims for nonoccupational injuries and illnesses, insurers would have little incentive to screen out such claims.

A second proposed reform would establish expert medical boards to make decisions about the compensability of each contested case of alleged occupational disease as soon as a claim was denied by the employer's insurance carrier. If the medical board ruled that a disease was occupational in origin, the burden of proof would be lifted from the worker. The employer or insurer would be able to contest the decision of the medical board, but there would remain a strong presumption that the claim was valid.

Another way of reducing the burden of proof of causation would be to establish presumptive standards. A presumptive standard defines a level of evidence sufficient to demonstrate legally a causal relationship between occupation and disease. For example, a history of 10 years of work in a coal mine, combined with specific medical test results consistent with coal workers' pneumoconiosis ("black lung"), could be defined as legally sufficient evidence to presume that a specific disease is work related. An employer or insurance company would then have the burden of proving that the disease was not work related to avoid paying the claim.

The compensation of victims of occupational disease may also be improved through publicly funded programs such as Social Security Disability Insurance (SSDI). SSDI offers disability payments to people with total disability, regardless of whether that disability is work related. If SSDI were broadened to cover partial disability, those unable to receive workers' compensation disability payments would still receive income support. Similarly, a mandatory medical insurance program would ensure medical benefits for treatment of injuries and illnesses regardless of cause.

Other countries have workers' compensation systems with considerably less controversy surrounding compensation for occupational diseases (Box 11-2). These countries have social programs that provide medical benefits and disability benefits that are substantially greater than those provided to many American workers. Thus, a worker who does not receive workers' compensation still can pay for needed medical care and continue to help support the family through disability payments. Countries with national health systems, such as Great Britain and Sweden, provide universal medical care regardless of whether illnesses are occupational; Belgium and Denmark have excellent social insurance and disability programs that provide a significant amount of wage replacement for disabled workers, again regardless of whether disability is work related.

Because national medical care systems in these countries cover costs that might otherwise be paid by workers' compensation, they provide a medical safety net for victims of occupational disease. Still, in these countries, workers' compensation may not provide benefits for many occupational disease claims. As in the United States, physicians do not identify occupational diseases, workers are unaware of their exposures to workplace hazards, and, when they are, they find that exposures are difficult to document. Also, legislation and regulation may be restrictive in covering occupational diseases (9).

Box 11-2. Another North American System: Workers' Compensation in British Columbia[1]

Canadian workers' compensation systems are administered at the provincial level, just as most workers' compensation systems in the United States are state systems. The workers' compensation system in British Columbia is similar to U.S. systems in states with exclusive state funds. The Workers' Compensation Board (WCB) of British Columbia is the only organization allowed to offer workers' compensation insurance to employers in that province.

The British Columbia workers' compensation system offers the same types of benefits to injured employees as systems in the United States: temporary and permanent disability benefits, fatality benefits for survivors, medical benefits, and vocational rehabilitation benefits. Benefits to replace lost wages are high compared with most U.S. jurisdictions. This is, in part, because maximum benefits are high. Also, British Columbia has no waiting period for wage replacement benefits; in contrast, U.S. jurisdictions have waiting periods ranging from 3 to 7 days. Therefore, more workers are paid wage replacement benefits in British Columbia than in the United States.

British Columbia provides payments through its Medical Services Plan to cover health care for all its citizens, yet the workers' compensation system pays the medical expenses of injured workers. Providers are paid fees that are 10% higher than those in the private sector to cover the additional paperwork required by the workers' compensation system.

Four separate organizations within the Ministry of Labour and Consumer Services administer the workers' compensation act. The WCB provides insurance and administers the payment of claims. The Workers' Compensation Review Board (WCRB) adjudicates disputed claims, as do workers' compensation commissions and industrial accident boards in the United States. Two other agencies provide services that in

general have no parallel in U.S. systems. (Oddly, decisions of the WCRB can be reviewed by the Appeals Division of the WCB.) The Workers' Advisers Office (WAO) helps workers to bring claims and may represent them before the WCB or WCRB. The WAO also trains union personnel to represent their members in disputed claims. The Employers' Advisers Office (EAO) provides similar services to employers. An obmbudsman in the WCB responds to complaints. In addition, outside the Ministry of Labour and Consumer Services, the Ombudsman of British Columbia responds to complaints and provides oversight of other agencies.

The WAO, the EAO, and the Ombudsman of British Columbia have no counterpart in U.S. workers' compensation jurisdictions, although some workers' compensation agencies in the United States do provide information about the law to workers and employers. These agencies reflect a less litigious approach to workers' compensation, an approach that makes British Columbia's appeals rate similar to those in the least litigious U.S. states. Compared with U.S. jurisdictions, British Columbia has many fewer appealed claims. In 1994, workers or employers appealed only 4.4% of new claims filed, or 10.8% of claims with lost wages in British Columbia. In Wisconsin, a low-litigation state, 10% of claims with lost wages involve a request for hearing. This rate is 10% in North Carolina, 19% in Pennsylvania, 25% in Georgia, 42% in California, and 43% in Missouri.

In 1994, 74% of workers in appealed claims were represented, primarily by the WAO or union. It is likely that most of the unrepresented workers had consulted with their unions, the WAO, or somebody else before their hearings. Virtually all U.S. workers with appealed claims hire attorneys. This difference between the British Columbia system and the U.S. systems is probably due in part to the availability of assistance from the WAO and the EAO. A study of the British Columbia system suggests that another reason is "the strong posture in the Act and by the WCB that it should administer

---

[1] Unless otherwise noted, the source of the information presented here is Hunt HA, Barth PS, Leahy MJ. The workers' compensation system of British Columbia: still in transition. Kalamazoo, MI: W. E. Upjohn Institute for Employment Research, 1996.

Box 11-2 (*continued*)

the law in an inquiry, rather than an adversarial, manner."

Another interesting feature of the British Columbia workers' compensation system is the Medical Review Panel (MRP). When a medical issue is in dispute, a worker or employer can appeal that issue to an MRP. The provincial government (with advice from the British Columbia College of Physicians and Surgeons) appoints and keeps a list of private physicians who can chair panels. Chairs are chosen sequentially from this list. When a physician's name comes up, that physician is chosen to head a panel. The chair then sends the worker and employer a list of specialists in relevant disciplines from which they can each choose one. This panel of three physicians sees the worker, reads relevant records, and can order additional medical testing. The decision of the panel on medical issues is binding; it cannot be appealed. This feature probably would not survive a legal challenge in the United States. A Massachusetts law that established medical referees that could issue binding opinions was overturned in the Massachusetts state court because it limited the parties' access to appeals and therefore to due process [*Meunier's Case*, 319 Mass. 421, 66 N.E.2d 198 (1946)].

Medical payments are much lower in the British Columbia workers' compensation system than in U.S. systems. In 1994 medical payments were 39.4% of all workers' compensation benefits in the U.S. [8]; in that year they were 27.9% of all benefits paid in British Columbia. The reason for this is not clear. It may reflect generally lower Canadian medical costs. Perhaps higher litigation rates in the United States lead to more medical care or reduce the effectiveness of medical care, thus leading to greater utilization. Also, this ratio may be lower in British Columbia because wage-replacement benefits are high.

In British Columbia, unlike U.S. jurisdictions, the workers' compensation agency has authority to develop and enforce workplace safety and health regulations. WCB inspectors can require correction of hazards, recommend penalties, or issue 24-hour closure orders where imminent dangers exist. For firms above a certain size and hazard level, the law in British Columbia requires occupational health and safety programs, including labor–management health and safety committees. WCB inspectors review these programs, decide whether they are adequate, and provide advice on improving them.[2]

---

[2] Rest KM, Ashford NA. Occupational health and safety in British Columbia: an administrative inventory of the prevention activities of the Workers' Compensation Board. Cambridge, MA: Ashford Associates, 1997.

REFERENCES

1. Boden LI, Jones CA. Occupational disease remedies: the asbestos experience. In: Bailey E, ed. Regulation today: new perspectives on institutions and policy. Cambridge, MA: MIT Press, 1987:321–346.
2. Kakalik JS, Ebener PA, Felstiner WLF, Haggstrom GW, Shanley MG. Variation in asbestos litigation compensation and expenses. RAND publication R-3132-ICJ. Santa Monica, CA: The RAND Corporation, 1984.
3. Biddle J, Roberts K, Rosenman DD, Welch EM. What percentage of workers with work-related illnesses receive workers' compensation benefits? J Occup Environ Med 1998;40:325–331.
4. Thornquist L. Health care costs and cost containment in Minnesota's workers' compensation program. John Burton's Workers Compensation Monitor 1990;3:3–26.
5. Boden LI, Fleischman CA. Medical costs in workers' compensation: trends and interstate comparisons. Cambridge, MA: Workers' Compensation Research Institute, 1989.
6. Pozzebon S. Medical cost containment under workers' compensation. Industrial and Labor Relations Review 1994;48:153–167.
7. Johnson WG, Baldwin ML, Burton JF Jr. Why is the treatment of work-related injuries so costly? New evidence from California. Inquiry 1996;33:53–65.
8. National Academy of Social Insurance. Workers' compensation: benefits, coverage, and costs, 1994–95: new estimates. Washington, DC: National Academy of Social Insurance, 1997.
9. Mony AT. Compensation of occupational illnesses in France. New Solutions 1993;4:57–61.

BIBLIOGRAPHY

Barth PS, Hunt HA. Workers' compensation and work-related illnesses. Cambridge, MA: MIT Press, 1980. *The most complete description available of how work-*

ers' compensation programs handle occupational diseases. Describes how different states compensate occupational diseases and gives an overview of litigation
and settlement of workers' compensation disease
claims in the United States. Also reviews workers' compensation in other countries.

Boden LI. Workers' compensation in the United States:
high costs, low benefits. Ann Rev Public Health
1995;16:189–218.

A review of workers' compensation programs in the
United States, focusing on safety, medical costs, litigation, and benefit adequacy.

Chamber of Commerce of the United States. Analysis
of workers' compensation laws. Washington, DC:
Chamber of Commerce of the United States, 1998.

An annual review of workers' compensation laws in
the United States and Canada. Compensation statutes
are continually changing, and reading this review is
one way of keeping up to date on the coverage, payment
levels, and administrative arrangements of different
laws.

National Commission on State Workmen's Compensation Laws. Report and compendium on workmen's
compensation. Washington, DC: U.S. Government
Printing Office, 1972 and 1973.

The National Commission was created by the Occupational Safety and Health Act of 1970 to evaluate the
status of workers' compensation. It undertook a 2-year
study. The Compendium is a descriptive report on
workers' compensation in 1973, while the Report
makes recommendations to upgrade workers' compensation programs.

Selikoff IJ, ed. Disability compensation for asbestos-
associated disease in the United States. New York:
Mt. Sinai School of Medicine, 1983.

Reviews the asbestos occupational health problem
in detail and discusses the workers' compensation
and tort litigation experience of asbestos-exposed
workers.

Spieler EA. Perpetuating risk? Workers' compensation
and the persistence of occupational injuries. Houston
Law Review 1994;31:119–264.

A legal and political analysis of the most important
public policy debates concerning workers' compensation. Critical of efforts to contain costs that focus on
reducing benefits instead of reducing hazards.

12 Ability to Work and the Evaluation of Disability

Jay S. Himmelstein, Glenn S. Pransky, and Charles P. Sweet

Industrialized societies have taken two basic approaches to dealing with the problems of poverty and social isolation that often afflict people with disabilities who have been unable to achieve gainful employment. One approach, disability compensation, provides income support for those who are unable to work because of a disability. This approach is typified by the Social Security Disability Insurance (SSDI) and Supplemental Security Income (SSI) systems in the United States, described later in this chapter, and by various state and federal "workers' compensation" systems (see Chapter 11) designed to compensate those whose disability resulted from workplace injury or disease. A second approach promotes the independence of people with disabilities by supporting and enhancing their opportunities for employment. Vocational rehabilitation services facilitate or help maintain employment through training, accommodation, and advocacy. Federal and state laws are intended to remove barriers to employment by regulating employment practices and workplace

conditions that have tended to exclude people with disabilities. An example of such a federal law is the Americans with Disabilities Act (ADA; see chapter Appendix).

In both of these approaches, physicians and other health care providers play a key role in the generation and evaluation of medical information. This information often becomes the basis, in the first approach, for determining eligibility for disability-related programs (income replacement, medical benefits, and rehabilitation services) and, in the second approach, for entitlement to legal protection (employment, accommodations, and vocational services). Almost all health care providers become involved in medical issues related to their patients' employment.

This chapter offers clinicians a review of the theoretic basis and practical aspects of their role in evaluating work ability and disability. Clinicians' effectiveness in dealing with work ability and disability evaluations is enhanced by a clear understanding of (a) key definitions related to the evaluation process; (b) common features of insurance plans and antidiscrimination legislation affecting disabled workers; (c) the steps involved in the disability evaluation process and the clinician's role in the evaluation of work ability; and (d) unresolved controversies and potential role conflicts for the clinician.

J. S. Himmelstein: Center for Health Policy, Department of Family and Community Medicine, University of Massachusetts Medical Center, Worcester, Massachusetts 01655.

G. S. Pransky: Center for Disability Research, Liberty Mutual Research Center, Hompkinton, Massachusetts 01742.

C. Sweet: Department of Family and Community Medicine, University of Massachusetts Medical School, Worcester, Massachusetts 01655.

KEY DEFINITIONS

In reviewing the variety of compensation plans and the associated roles for the health care provider, it is important to recognize a few key concepts. Most important is the distinction between *impairment* and *disability*.

Impairment is commonly defined as the loss of function of an organ or part of the body compared with what previously existed. Ideally, impairment can be defined and described in purely medical terms and quantified in such a way that a reproducible measurement is developed—for example, severe restrictive lung disease with a total lung capacity of 1.6 L.

Disability, on the other hand, is usually defined in terms of the impact of impairment on societal or work functions. A disability evaluation, therefore, takes into account the loss of function (impairment) and the patient's work requirements and home situation. Certain agencies use a more restrictive definition of disability; for example, the Social Security Administration defines disability as "inability to perform any substantial gainful work." Often, private disability insurance policies define disability as an "inability to perform the essential tasks of the usual employment." The ADA defines a person with a disability as one who is "substantially limited in a major life activity." In theory, however, the determination of disability is always predicated on an objective assessment of impairment, followed by a determination of the loss in occupational or societal functioning that results from the impairment. In general, a health care professional (usually a physician) performs the determination of impairment; most often, nonphysician administrators use this information to determine the presence and extent of disability.

Disability compensation systems frequently request a determination of the *extent* and *permanence* of a disability condition. An injured worker who cannot do any work because of a medical condition is considered to be *totally disabled*. If this person can work but has some limitations and cannot do his or her customary work, a *partial disability* exists. Either type of disability is considered to be temporary as long as a resolution of the

disability is expected. When no significant functional improvement is expected, or a condition has not changed over a 1-year period, it is frequently inferred that a *medical end result* (sometimes called maximal medical improvement) has been achieved. At this point, a *temporary* (partial or total) *disability* would then be regarded by most systems as a *permanent disability*.

Workers' compensation insurance systems usually require determination of the work-relatedness of a disability. A *work-related* injury or disease refers to conditions that are thought to result from some exposure (physical, chemical, biologic, or psychological) in the workplace. In acute traumatic injuries, the causal relationship to the work-

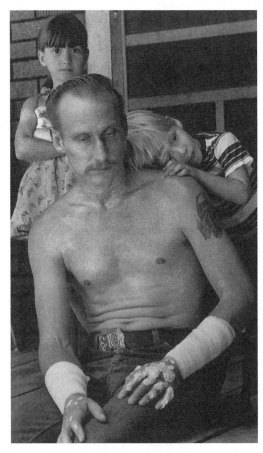

FIG. 12-1. Burn victim of a coal mine explosion in Virginia with his children. (Photograph by Earl Dotter.)

place is usually clear (Fig. 12-1). In chronic conditions, however, it may be difficult to be certain of the relationship between work and disease. The legal definition of cause may be less exacting than the medical definition, and most disability systems are based on the legal standard. One legal definition of a work-related condition is one "arising out of or in the course of employment" or "caused or exacerbated by employment." It is recommended that the physician's determination of work-relatedness be based on the evidence of disease, the exposure history, and the epidemiologic evidence linking exposure and disease (1).

DISABILITY COMPENSATION SYSTEMS

Some of the confusion regarding disability assessment stems from the multitude of disability compensation systems and plans because each may have its own definition of disability and criteria for assessing impairment. Different countries have designed varying approaches to providing income security to those who find their wage-earning capacity compromised by injury or disease (2). In the United States, most compensation systems fall into one of three major categories: workers' compensation insurance, Federal programs for the severely disabled, and private disability insurance (Table 12-1).

Occupational medicine specialists are most familiar with *workers' compensation insurance*, which provides coverage of most federal, state, and private employees. These plans compensate for medical expenses and a portion of lost wages due to work-related conditions (see Chapter 11).

The federal government sponsors two major *compensation programs for the severely disabled* through the Social Security Administration: SSDI and SSI. These programs pay a limited amount of compensation to those who are eligible and unable to achieve any gainful employment, regardless of the cause of disability.

Private disability insurance is often purchased by individuals or provided as an employer- or union-sponsored benefit. It is designed to (a) provide compensation for those who are unable to work at their regular jobs regardless of the cause of disability, or (b) supplement Social Security benefits.

Thus, a patient who can no longer work because of injury or illness might receive support from his or her employer's insurer, a federal or state agency, or an insurance policy that has been purchased privately.

Although each disability compensation system has different eligibility criteria and levels of payment, all share a few common features:

1. Every plan incorporates *shared risk*. Many individuals or employers at risk of financial losses contribute to a pool, from which a few individuals are reimbursed. The cost of entering the pool is partially determined by the actuarial risk of future events for that person or insured group. Thus, private disability insurance is much more expensive per year for a 55-year-old than for a 20-year-old because the older worker has a higher risk of disabling medical illness. Workers' compensation insurance is more expensive per employee for a construction company (higher risk of injury to employees) than for a stock brokerage firm.

2. Because payments into the pool are predictable, *finite resources* are available to all potential recipients of each plan. Therefore, eligibility criteria are structured so that the limited resources might go to those in greatest need. Workers' compensation plans often do not replace lost wages for fewer than 6 days of absence from work. The Social Security plans usually do not begin payments unless it is anticipated that the person has been or is anticipated to be unable to participate in gainful employment for at least 1 year.

3. Before medical evaluation of impairment, a potential recipient of benefits must first

TABLE 12-1. *Compensation systems*

Program	Eligibility	Source of benefits	Basis for claim	Clinician's role
Workers' compensation systems—"cause" of disability is determinant				
State	Private employees	Employer insurance	Work-related illness or injury	Evaluate work-relatedness, impairment, and disability
Federal	Fed. employees	Taxes		
Railroad workers	Railroad employees	Employer		
Black lung benefits	Coal miners	Tax on coal	Lung disease	Report chest radiograph, pulmonary function tests, and examination results only
Programs for severely disabled—inability to perform gainful activities is determinant				
Social Security Disability Insurance	Contributing workers	Workers' contributions	Severe disability	Evaluate and report impairment
Supplemental Security Income	The aged, blind, or severely disabled	Taxes		
Private disability plans—regardless of cause, protect income if unable to perform regular job				
Short-term disability	Enrolled workers	Payments by employee, employer, or union	Any illness preventing usual employment	Evaluate impairment and disability
Long-term disability				

demonstrate *administrative eligibility*. The basis for eligibility is different in each plan. For example, to be eligible for SSDI, a person must have worked and contributed to Social Security for 5 of the past 10 years. Workers' compensation covers only regular employees, not consultants or subcontractors. Private disability insurance often does not cover illness that occurs during the first 60 to 90 days of enrollment.

4. *Medical information* on impairment is utilized once a legal basis for a claim has been established. In every system, a medical diagnosis is necessary; in the workers' compensation system, physicians are often asked their opinions on the *work-relatedness* of employees' conditions, the prognosis for eventual return to work, and the restrictions or job accommodations that might be necessary to return the worker to employment.

5. The information from the physician, however, does not determine whether benefits are awarded or how much is paid; all of these systems are under *administrative control*. For example, in the Social Security system, an administrator–physician team reviews medical information from the evaluating physician and compares it with specific criteria for eligibility. In workers' compensation systems, benefits may be withheld by the insurance company if there is a significant discrepancy between the employer's report of an injury and the physician's report.

6. *Benefits are limited* and are intended to provide only a proportion of lost wages, medical expenses related to the specific impairment, and vocational rehabilitation. In rare circumstances, workers' compensation benefits for injured workers are enriched to punish gross negligence by an employer; in all other instances, *fault has no bearing on benefit levels*.

7. Applicants usually have a *right of appeal* of an administrative or medical decision, with review by a third party. In the Social Security system, applicants who are initially denied benefits can appeal to a sec-

ond administrator–physician team, then to a Social Security benefits coordinator, then to an administrative law judge, and finally to the federal courts, if desired. In most workers' compensation plans, the claimant can request an administrative hearing and be represented by an attorney. The agencies that provide benefits also conduct periodic reviews of cases to verify that continued eligibility (disability) exists.

8. There has been an increased emphasis on developing resources for *retraining and vocational rehabilitation*, closely allied with each system. Beneficiaries are often required to participate in programs to maximize their potential for return to gainful employment.

Workers' compensation insurance is reviewed in Chapter 11. The United States does not have a federal workers' compensation law, and each state has developed its own system. In addition, there are a number of workers' compensation programs that are occupation specific; these programs often have developed their own definitions of disability and related eligibility criteria. The Black Lung Program, for example, provides payments to coal miners with a documented work history and respiratory insufficiency who meet certain criteria. All disabling respiratory insufficiency is assumed to be related to mining if the miner meets a standard of number of years worked in the mines. Other examples of occupation-specific workers' compensation programs include those of railroad workers, longshoremen, military veterans, and municipal workers, such as police and firefighters.

The purpose of each plan is to reimburse workers for medical expenses, rehabilitation expenses, and a portion of lost wages that result from a work-related injury or illness. Systems in general are designed to be nonadversarial so that, in most cases, limited benefits are paid to injured workers without the necessity of a formal hearing. In most cases of acute traumatic injuries (e.g., fractures or lacerations occurring at work), the relationship to work is unquestionable and the system works reasonably well at compensating the injured worker. In many cases, however, the relationship to work is less clear, and the demand on the clinician more complicated, as the following case illustrates.

A 50-year-old truck driver followed by his physician for 6 years because of chronic low back pain came to the physician stating, "I cannot take it anymore." Although he could not recall a specific injury, he found that the requirements of driving a long-haul tractor-trailer caused him severe discomfort that was no longer relieved by rest or analgesics. His back discomfort improved while he was on vacation, but was clearly aggravated after more than 2 hours of driving or after any heavy lifting (at home or at work). He had been out of work for 1 week because of his discomfort and required a note from the physician before returning to work. Physical examination revealed a mild decrease in forward flexion. Radiographs were consistent with osteoarthritis of the spine. The patient wanted to know the physician's opinion on whether his back problems were due to his work as a truck driver, whether he should change his vocation because of his discomfort, and whether he should file a workers' compensation claim for work-related injuries.

This case illustrates some of the difficulties in evaluating and treating the patient with work incapacity. The patient went to the physician because his back discomfort was interfering with his ability to do his job. Like most patients with chronic low back pain, his symptoms and examination findings were nonspecific. The standard recommendations of rest and avoidance of exacerbating activities met with transient success, but his symptoms reappeared with his return to work. If the patient has not previously participated in a comprehensive back rehabilitation program aimed at maximizing conditioning, flexibility, and pain tolerance, this option should be considered because such programs have been shown to facilitate successful return to work.

It may also be appropriate for the provider to explore with the patient and employer

FIG. 12-2. Highly trained professionals take advantage of advanced technology in prosthetics and retaining equipment in worker rehabilitation to achieve the best possible function for an injured person, such as this man who is relearning his trade as a carpenter using a prosthetic arm. The immensity of the effort and its inherent limitations emphasize the importance of all possible efforts to prevent the initial injury. (Courtesy of Liberty Mutual Medical Service, Hopkinton, Massachusetts.)

possible job accommodations at work. Even modest accommodations can significantly improve the success of return to work after a rehabilitation program. Sometimes workers with severe injuries can still return to work with the support of rehabilitation professionals and, if necessary, advanced technology (Fig. 12-2). If no accommodations are possible, the patient might consider career changes consistent with his current physical abilities. For a variety of reasons, however, the patient may be reluctant to consider changing to another line of work, despite the discomfort associated with the current job.

With regard to causality, the high prevalence of nonspecific low back pain in the general population and the multifactorial etiology of this common condition make it impossible to say, with medical certainty, that this patient's back discomfort was caused entirely by his work. Several epidemiologic studies, however, have linked truck driving with a higher incidence of chronic disabling low back pain and have attributed this increase to excessive vibration, sitting, and heavy lifting. Despite medical uncertainty,

it is likely that most compensation systems would recognize this patient's low back pain as a condition that is aggravated by work, and that the patient's medical bills and lost wages related to his back pain would be covered by workers' compensation insurance.

The Social Security Administration, in the U.S. Department of Health and Human Services, administers two plans that provide benefits to those unable to work, regardless of the cause of disability. *SSDI* is a true insurance plan in that all nongovernmental employees in the United States contribute to the plan through mandatory deductions from wages. Administrative eligibility requires 5 years of contributions to the plan over the previous 10 years, and is determined by the federal Social Security Administration. *SSI* is funded through federal taxes—not by employee contributions—for blind, disabled, or elderly people who do not qualify for SSDI. Although medical criteria for disability are identical to those for SSDI, the SSI benefits are in general lower and do not begin until the claimant's assets and benefits from all other sources are exhausted.

State rehabilitation commissions are given the responsibility of determining whether the applicant's impairment qualifies for benefits. To qualify, an impairment must be the result of a documented medical condition and must be expected to result in at least 1 year of inability to work, or death; the impairment, however, does not have to be a consequence of work. The condition must be as severe as the standard description listed in the Social Security Administration's publication *Disability Evaluation under Social Security* (3). If a condition is not listed in the regulations, it must be medically equivalent to one that is listed. A sufficient impairment must result in inability to perform *any* gainful employment, not just the person's usual job. In cases in which the impairment does not meet the established criteria, the commissions also take into account age, education, and prior work history in determining the likelihood that an applicant would be able to find any future employment. Once a claim is accepted, there is a 6-month waiting period until benefits begin. Claims are reevaluated periodically to determine whether a severe impairment continues to exist, and updated medical information is requested.

The following case illustrates the basic medical considerations involved in the Social Security disability determination process:

A 60-year-old man asked his primary care physician to provide medical information in support of his application for SSDI. The patient had been a maintenance worker since 25 years of age and had moderate exposure to asbestos when he worked as a "fireman" on a Navy ship during World War II. He had smoked one to two packs of cigarettes a day for the past 42 years until quitting 6 months before this evaluation. He had been short of breath for 7 years. Merely dressing himself in the morning exhausted him. Physical examination revealed him to be thin, with a respiratory rate of 18 per minute at rest. His breath sounds were distant. There was no clubbing or cyanosis of the nail beds. A chest radiograph showed flattened diaphragms, emphysematous changes, and bilateral pleural plaques. His pulmonary function tests showed severe obstructive lung disease with a 1-second forced expiratory volume (FEV_1) of 1.1 L (35% of predicted), a forced vital capacity (FVC) of 3.1 L (68% of predicted), a total lung volume (TLV) of 5.4 L (80% of predicted), and an FEV_1/FVC ratio of 35%.

In this case, a patient with severe chronic lung disease was being evaluated for disability under SSDI. His exposure history was significant for occupational exposure to asbestos and nonoccupational exposure to cigarette smoke. His physical examination, chest radiograph, and pulmonary function tests were consistent with diagnoses of (a) severe obstructive lung disease, (b) possible restrictive lung disease, and (c) asbestos-related pleural plaques.

A number of systems have been developed for evaluation of severe pulmonary impairment. These systems are not identical in the level of impairment that is considered severe. For 40- and 60-year-old men who are each 70 inches tall, the FEV_1 values for "severe impairment" are different for each system

(Table 12-2). Comparison of this patient's pulmonary function tests with the guidelines for SSDI demonstrates that people of this height who have an FEV_1 of less than 1.55 L are considered disabled under this system.

The patient's occupational exposure to asbestos might have played an etiologic role in the development of pulmonary insufficiency, but this would have no effect on his application for SSDI.

Private disability insurance is available from many insurers in the United States and may be purchased individually or through an employer, union, or some other group purchaser. These programs provide benefits to supplement SSDI and have much less stringent criteria for acceptance of claims than SSDI. Usually, the claimant need only have his or her physician state that an impairment exists that prevents working at the usual job. Most short-term disability plans provide a percentage of regular income for the first 6 months; long-term disability plans pay for up to 2 years of disability if the claimant is unable to work at his or her usual job because of a medical condition. Afterward, full benefits are paid only if the claimant is totally disabled from *any* type of work; if the claimant can do work with lower wages than previously, the difference between current potential wages and prior wages may be used to determine the reimbursement level. Programs usually provide for maintenance of health insurance and certain other employee benefits, and benefits usually end at retirement age. Many variations and supplements exist that can be purchased by an employee.

The following case illustrates a typical situation that would be covered through private disability insurance:

A 60-year-old male maintenance worker, who formerly smoked cigarettes, was referred by his employer for a return-to-work evaluation. The patient had been out of work for 3 months after hospitalization for myocardial infarction. With an adequate medical regimen, the patient had fatigue after exertion but no chest pain. His home activities included 40 minutes of walking twice a day. On physical examination, he

TABLE 12-2. *Comparison of levels of FEV₁ required for severe impairment for 40- and 60-year-old white men, 70 in. (178 cm) tall, using different criteria*

Criteria	Actual FEV_1 and percentage of predicted value[a]	
	Age 40 yr	Age 60 yr
Social Security Administration (1996) (3)	1.55 (36%)	1.55 (40%)
American Medical Association (1993) (4)	1.68 (40%)	1.49 (40%)
Gaensler et al. (1966) (5)	1.59 (38%)	1.37 (37%)
Department of Labor Black Lung Benefits (1978) (6)	2.38 (56%)	2.06 (56%)

[a] Percentage predicted is based on Crapo RO, Morris AH, Gardner RM. Reference spirometric values using techniques and equipment that meet ATS recommendations. American Review of Respiratory Diseases 1981;123:659–666.
FEV_1, Forced expiratory volume in 1 second.

was in no acute distress, with a resting regular pulse of 80 beats per minute. His lungs were clear to auscultation, and his heart sounds were normal, without murmurs or gallops. A recent exercise thallium test had been discontinued after 8 minutes of a Bruce protocol because of fatigue, but there were no signs of ischemia or arrhythmia, and the ejection fraction was normal. His job as a maintenance worker involved walking long distances in the plant while pushing a 45-kg (100-lb) maintenance cart, and performing scheduled and unscheduled machinery repair.

In summary, this patient showed evidence of cardiopulmonary deconditioning after a myocardial infarction that might prevent him from performing his normal job tasks. It would therefore be appropriate to refer this patient to a cardiac rehabilitation program in an attempt to increase his exercise tolerance before making any judgment about permanent impairment. He would be supervised in a progressive exercise program, which might restore much of his exercise capacity.

Until fit enough to return to work, this patient would be eligible for continued short-term disability compensation from his company-sponsored plan. He would *not* be eligible for SSDI because the disability was not expected to last more than 1 year and did not meet the relevant criteria.

STEPS IN THE DISABILITY EVALUATION PROCESS

The following questions are central to the disability evaluation process:

1. What is the *medical basis* of the patient's disability claim?
   a. What is the patient's medical diagnosis?
   b. Does the patient have any objective evidence of impairment related to this diagnosis?
   c. If an impairment is present, is it temporary or permanent?
2. What is the *extent of any impairment* (Box 12-1)?
3. Is the patient's impairment or disease caused or aggravated by work?
4. What is the impact of this impairment on the patient's ability to obtain employment in specific occupations and to perform specific jobs? Alternatively, in the context of the ADA, what are this person's current capabilities? Might accommodations allow for employment?
5. What other sources of information on work capabilities or possible accommodations should be considered?
6. Given the information from the previous questions, what benefits (economic, medical, vocational assistance) are appropriate for the patient?

Physicians usually play a major role in answering the first three questions. The ADA encourages their participation in the fourth and fifth questions, along with that of other professionals. The answer to the fourth question often depends on the specialized skills of a vocational evaluation unit and input from the employer, and may be based on

Box 12-1. Standardized Measures of Impairment

Why bother to develop standardized measurements of impairment? First, because most reimbursement systems rely on objective medical data (physical examination findings and laboratory test results) to verify and quantify impairment, standardization might allow for a common method of reporting this information and might reduce variability between examiners. Second, consistent examination methods and reliance on objective findings rather than on subjective descriptions would eliminate much of the conflict over degree of functional loss and resulting compensation benefits. Third, quantifiable measurements of residual function in various organ systems should be linked with data on functional ability required by various jobs; this linkage might be helpful in designing individualized rehabilitation programs. Finally, objective measurements of function have shown promise as an aid in measuring progress in a rehabilitation program, or lack of progress when a medical end result has been reached.

The American Medical Association (AMA) has produced *Guidelines for the Evaluation of Permanent Impairment* (4) that are commonly used in state workers' compensation systems. The Social Security Administration has produced its own guidelines to disability evaluation under Social Security (3). Both organizations' guides organize evaluations by organ systems, specifying the medical information required for the evaluation. The Social Security Administration guide provides threshold criteria for impairments sufficient for eligibility; the AMA guidelines express impairment as a percentage of total body function that has been lost.

A number of problems frustrate attempts to develop measures of impairment that are reproducible, valid, and standardized. The absence of a "preinjury" baseline evaluation usually forces the examiner to use population-derived norms to predict degrees of functional loss. For example, a person's lung volume measurements are often compared with the distribution of test results in a population, standardized by age, sex, height, and race, to determine what percentage of a predicted value is present. A person who originally had large lung volumes and lost a significant percentage of function because of disease might have a substantial respiratory impairment but may still have "normal" lung volumes by testing. Conversely, another person may have had lung volumes below the population-derived norm, and a slight decrease on lung function tests may result in a misleading label of "abnormal lung function." Inaccuracies caused by poor cardiovascular conditioning, lack of understanding, fear of test procedures, or poor motivation can confound even the best test in providing estimates of actual functional ability. A determination of medical end result may be frustrated by an illness with a variable course of presentation in which signs and symptoms wax and wane.

Several devices have been developed to quantify musculoskeletal function. These devices measure ability to exert a rotational or linear force against a mobile or stationary object. Output of the device, which is usually reproducible, indicates the degree or distance of motion and the maximum force applied. Many programs have used these devices in assessment and rehabilitation of extremity injuries. These devices have not been shown accurately to predict ability to perform occupational tasks because the actions required to operate them are usually unrelated to job tasks. When considering purchase and use of these devices, the clinician should carefully review the scientific data supporting any advantages over existing forms of assessment and rehabilitation.

The following case is an example of an impairment and disability evaluation, where standardized measurement of impairment had been attempted:

A physician had been treating a 58-year-old clerk for apparently non–work-related carpal tunnel syndrome of both hands, mainly the right (dominant) hand. Her job requited filing, answering phones, and occasional typing. Six months after successful right carpal tunnel surgery, her pain and numbness were largely resolved, although she still complained of considerable weakness of the right hand. Physical examination showed a well healed scar and considerable loss of sensation in a median nerve distribution on the right, normal sensation on the left. Pinch strength by dynamometer was 5.4 kg (12 lb) with the left hand, and 2.7 kg (6 lb) with the right. She had been out of work for 6 months, and the physician believed that her condition had stabilized. She was concerned that "no one will hire me because I have hand problems" and wanted the physician to examine her for

Box 12-1 (*continued*)
Social Security Disability Insurance (SSDI) eligibility.

1. What was the patient's medical diagnosis?

Carpal tunnel syndrome: status post right carpal tunnel release, with residual median nerve damage.

2. What is the extent of the impairment?

There was loss of normal function of the right hand, and the information suggested that no further return of function was likely to occur. The clinician could first quantify the impairment using the AMA guides. History and examination revealed that the left hand was normal; however, there was evidence of considerable sensory loss of the right hand. She stated that she could do most of the activities of self-care but had difficulty with digital dexterity; physical examination revealed considerable loss of two-point discrimination, pain sensation, and sensation to fine touch over the affected area, as well as decreased oppositional strength between the thumb and other fingers. According to the charts in the AMA guides, she had a 2% loss of motor function in the upper extremity caused by median nerve motor injury (20% loss of median nerve motor function in the hand, multiplied by 10% maximal loss of upper extremity strength due to median neuropathy). Similar calculations revealed a 23% loss of function in the upper extremity from sensory deficit. These values are combined to yield a 25% impairment of the upper extremity, or a 15% impairment of the whole person. Using the AMA guides in this way is attractive as a means for reporting quantitative results for the purposes of the disability determination process. Readers should be aware, however, that there is limited scientific or medical evidence to support this rating system as accurately measuring impairment or predicting the impact of an impairment on functional capacity.

For the SSDI application, the physician must provide a description of the patient's functional status by history and examination; it would be helpful to describe her attempts to move objects in the office. The evaluation report should also include a copy of the nerve conduction studies and the statement that a medical end result had been reached.

3. Was the patient's impairment caused or aggravated by work?

No report of forceful or repetitive movements required at work was indicated. A careful review of the job tasks with the patient might have uncovered a history of unusual activities that might have exacerbated or even caused the carpal tunnel syndrome, such as folding large quantities of paper, typing on a manual typewriter, or using scissors extensively. Work relatedness, however, is not an issue in SSDI.

4. What was the impact of the impairment on the patient's ability to obtain employment or perform specific jobs?

The amount of impairment may have interfered with typing and filing but may not have interfered with other duties. If these activities were required on the job, then the physician might have had to ask the patient to demonstrate her ability to perform these skills; the exercise in the AMA guides would not have been helpful in determining disability. The actual abilities of the patient over an 8-hour workday are difficult to determine by a cursory office examination and are best determined by a vocational evaluation specialist with a thorough understanding of a patient's functional ability and job demands. The history of current household activities may also aid in assessment of functional ability.

5. To what, if any, economic benefits is this patient entitled?

It is unlikely that she would have initially received SSDI because the threshold in *Disability Evaluation under Social Security* (3) requires a process that "results in sustained disturbance of gross and dexterous movements in two extremities." If, on vocational case review, she demonstrated limited education and work experience, and impairment preventing performance of practically any job, her claim might have been accepted. The treating physician's opinion in this regard is not considered; however, the physicians' description of functional limitations is the basis for vocational assessment by a physician and vocational consultant for the Social Security disability determination services. However, if the patient were applying for private disability insurance, the physician might have to determine whether the essential job tasks could be performed by the patient.

legal–administrative criteria. Claims examiners and administrators using legal guidelines usually resolve the last question. Physical examination findings that support the degree of impairment and the stated diagnosis are important. For example, SSDI claims without positive objective physical findings (symptoms without physical or laboratory findings) are rejected. Evaluations often include measurement of strength and endurance, length of scar, degree of visual impairment, and other relevant items. Serial measurement of these findings over time provides an objective basis for deciding whether a medical end result has been achieved. Because the physician's office usually has insufficient resources fully to evaluate functional capacity, referral to an occupational therapist or vocational evaluation specialist for work capacity evaluation may be appropriate. A series of standardized tasks can be performed to document functional impairment.

The physician is often asked to determine whether the impairment is permanent or whether a medical end result has been achieved. For example, in the first case, it is likely that the patient would feel better once he had been away from work for a few weeks. However, his physician's experience showed that this patient consistently experienced back pain soon after returning to this type of work. Although some improvement of the symptoms is likely, the patient probably would not improve enough medically to allow him to continue working as a truck driver. Therefore, in terms of functional ability, a medical end result has been achieved; although his discomfort is not likely to be permanent, the medical limitations to his working as a truck driver probably will not change.

At times, insurance companies or lawyers ask for a determination of permanency that seems to require a crystal ball. In many cases, the prognosis for functional improvement is uncertain. Healthcare practitioners may fuel pressure from a lawyer or insurer to declare a medical end result so that a case can be settled. However, it is important to communicate uncertainty, both to the patient and to others involved with the case; patients usually do not benefit from a premature medical determination of permanent disability.

In workers' compensation and in private insurance disability cases, the physician is often asked whether the impairment is disabling and to describe how the impairment impedes the performance of usual job tasks. A clear job description is the basis for evaluating whether the employee can perform the essential functions of the job. Often, this cannot be determined without knowing what accommodations at work might be available. Thus, in the second case, it cannot be determined whether the patient is totally disabled until it is known whether any alternate work or accommodations are available. The same considerations apply to determining disability for private insurance. A visit to the workplace usually resolves the lack of clarity in standard job descriptions and may have an important role in encouraging an employer to provide accommodations for the disabled employee. In the uncomplicated case of a temporary injury or illness, in the absence of detailed knowledge of the job requirements or the ability of the employer to accommodate, the clinician can provide an opinion including the range of activities and job tasks the provider believes the worker can safely perform without interfering with recovery. A generic return-to-work form for use in this situation might include the following categories: lifting; carrying; sitting/standing; pushing/pulling; climbing/stooping; overhead work; repetitive activities of upper and lower extremities; operation of machinery; potentially hazardous exposures; duration of work; and duration of limitations.

Most insurance systems reimburse people for loss of earning capacity caused by objective impairment. It is often difficult to determine whether sufficient impairment exists for a person to qualify for benefits under a given plan. Physicians usually lack the experience, technical facilities, and ability to estimate vocational potential accurately; in these situations, early involvement with a qualified vocational rehabilitation specialist is worthwhile. Specialized skills and a broad database are required to predict residual earning capacity when employees are no longer able to return to their previous work. For example, factors related to worker autonomy, such as the availability of self-paced work, educational and experience levels, and self-employment, have been shown to be as important in determining disability status in patients with rheumatoid arthritis as the extent of medical findings (7).

In workers' compensation plans and in most private disability plans, the treating or reviewing physician is required only to determine that the impairment is sufficient to prevent work. However, in the Social Security, Veterans Administration, and Black Lung programs, there are often specific criteria for impairment that determine whether a person is eligible for benefits; these criteria vary from plan to plan. For example, the Black Lung and Social Security programs have threshold pulmonary function values; if an applicant's lung function is better than the threshold, then he or she does not qualify for disability. In the Veterans Administration system, the degree of lost function is expressed as a percentage of total lung function. Benefits are assigned based on the percentage of function lost; in contrast, the Social Security and Black Lung programs usually provide a fixed amount of benefits only if a worker is totally disabled according to the threshold criteria. Physicians are often frustrated with the arbitrary nature of the determination process. Under these criteria, some claimants with truly disabling impairments are refused compensation, whereas others capable of gainful employment receive benefits.

THE CLINICIAN'S ROLE

Within the disability evaluation process, three different and potentially conflicting roles for the clinician become clear: patient advocate, provider of information, and medical adjudicator. It is important to understand the requirements of each of these roles so that the patient can best be served. Clinicians must not neglect their role as patient advocate in treating patients with work incapacity and, when appropriate, assisting them in obtaining accommodations at work, or benefits to the extent entitled by a particular disability system. The clinician should not let personal feelings about a specific disability or impairment-rating system interfere with judgment in assisting patients.

As patient advocates, clinicians need to be aware of their role in preventing complications in patients with work disability. Patients frequently are limited financially and socially by their work incapacity, and social isolation frequently accompanies isolation from the workplace. Patients may be upset at how they have been handled by the "system," especially if benefits have been delayed or denied. They may be angry at the apparent insensitivity of their employer, insurer, or physician, and this anger often complicates their evaluation and treatment. Patients are often afraid of returning to the workplace where a serious injury occurred, or they may be afraid of being dismissed once they have successfully returned to work. Being aware of the complicated social, legal, and psychological state of disabled workers is an essential aspect of assisting their recovery. Appropriate referral for social and psychological support should be considered in every case of prolonged work incapacity. Mindful of the adverse psychological, physical, and economic consequences of disability, the clinician should be careful to avoid removing patients from gainful employment whenever possible.

As a patient gradually loses function because of a progressive disease process, the physician should anticipate the possibility that earning capacity may be lost and discuss this potential with the patient. Patients should be made aware of the potential loss of self-esteem and income, as well as the uncertainty of receiving benefits while out of work. Actively assisting patients in vocational rehabilitation and early selection of jobs that do not conflict with physical limitations may help the patient avoid unnecessary time out of work and economic dislocation. Physicians should learn of possible accommodations in the workplace by contacting the personnel manager or the patient's supervisor, and seek input from specialists in job accommodation. With this information, the physician can often help the employee return to work earlier, thus preserving earning capacity and often providing an additional stimulus to recovery.

In the routine care of patients, clinicians frequently are asked to provide information relating to their patient's medical condition for the purposes of determining ability to work, impairment, or disability. Such requests may originate from employers, insurance companies, state agencies, or patients. When such requests are accompanied by the patient's signed requests for release of information, it is appropriate to supply information that is relevant to the request. However, it would not be appropriate, for example, to comment on or release records about a patient's diabetes or epilepsy when the request was for information about impairment from an injury to the lower back. Because records relating to workplace injuries must be routinely supplied in workers' compensation cases, it may be worthwhile to make and provide office notes that are separate from the notes relating to other, non–work-related problems. Because a worker's reimbursement is frequently tied to the receipt of records from the attending clinician, it is important that clinicians be prompt in responding to such requests to minimize financial difficulties for their patients.

Several sources can aid in the provision of relevant information. For example, most private disability plans have a short guide on eligibility requirements for clinicians, state workers' compensation boards often publish free guidebooks, and the Social Security Administration office will provide a free copy of the publication, *Disability Determination under Social Security* (3).

A significant potential conflict arises between the primary clinician's traditional role as patient advocate and his or her gatekeeper or adjudicator role, brought about by a request from an employer, insurance company, or a public agency for professional evaluation of impairment. Patients often have significant concerns about the disability evaluation itself. Because they are aware that the outcome of the evaluation may determine their access to or continuation of benefits, they may feel the need to emphasize the extent of the disability to "prove" their case. They may have residual anger from previous examinations in which clinicians seemed unsympathetic or doubted their "true" disability. Increasingly, third parties are requesting and relying on so-called "independent medical evaluations" (IMEs) performed by physicians who are not involved in the medical care of the person being evaluated but have experience or specific training in the disability evaluation process. The "independence" and objectivity of IMEs requested by employers and insurance companies have been questioned by a number of authors (8,9), and there are increasing calls for establishing both ethical and technical standards for the performance of disability evaluations.

Primary care providers therefore need to be aware that an apparently simple request for an evaluation regarding the impairment or disability status of one of their patients can lead to hostility between patient and clinician because of unrealistic expectations and inexperience with disability systems and the determination process. Frequently, patients are not aware of the requirements of different systems and blame their clinicians if benefits are denied. Clinicians frequently share their patients' frustration with the arbitrary nature of a particular disability system. Clinicians occasionally resent their patients for "trying to take advantage" of an insurance system, and patients may rightly resent the need to "prove" their illness. All of these feelings may interfere with a satisfactory clinician–patient relationship.

Therefore, when a clinician is asked to provide information or act as an adjudicator involving one of their patients, it is important to clarify the purpose of the evaluation and the limitations of the clinician's situation. It may be appropriate for the clinician to refer the patient to a social worker, lawyer, or union representative for clarification of the social, legal, and financial issues surrounding application for disability. It may also be appropriate for a clinician to seek an independent opinion about impairment and disability from a colleague when there is a potential for conflict with a patient or significant uncertainty about the cause or extent of disability.

REFERENCES

1. Kusnetz S, Hutchinson MK, eds. A guide to the work-relatedness of disease, revised ed. NIOSH publication no. 79-116. Washington, DC: U.S. Government Printing Office, 1979.
2. Hadler NM. Disability backache in France, Switzerland, and the Netherlands: contrasting sociopolitical constraints on clinical judgment. J Occup Med 1989;31:823–830.
3. Social Security Administration. Disability evaluation under Social Security. Social Security Administration Office of Disability SSA publication no. 64-039, ICN 48 6600. Washington, DC: U.S. Government Printing Office, January 1998.
4. Committee on Mental and Physical Impairment, American Medical Association. Guidelines to the evaluation of permanent impairment, 4th ed. Chicago: American Medical Association, 1993.
5. Gaensler EA, Wright GW. Evaluation of respiratory impairment. Arch Environ Health 1966;12:146–189.
6. Richman SI, Smith CJ Jr. Legal aspects of impairment and disability in pneumoconiosis. Occup Med: State of the Art Reviews 1993;8(1):71–92.
7. Wolfe F, Hawley DJ. The longterm outcomes of rheumatoid arthritis: work disability: a prospective 18 year study of 832 patients. J Rheumatol 1998;25:2108–2117.
8. Dembe AE. Pain, function, impairment and disability: implications for workers' compensation and other disability insurance systems. In: Mayer TM, Gatchel RJ, Polatin PB, eds. Occupational musculoskeletal disorders: function, outcomes and evidence. Philadelphia: Lippincott Williams & Wilkins, 1999 (*in press*).
9. Pryor ES. Flawed promises: a critical evaluation of the American Medical Association's guides to the evaluation of permanent impairment. Harvard Law Review 1988;103:964–976.

BIBLIOGRAPHY

Carey TS, Hadler NM. The role of the primary physician in disability determination for Social Security and workers' compensation. Ann Intern Med 1986; 104:706–710.
A good overview of the physician's role in these systems.
Committee on Mental and Physical Impairment, American Medical Association. Guidelines to the evaluation of permanent impairment, 4th ed. Chicago: American Medical Association, 1993.
These guidelines offer a quantitative approach to the measurement of permanent impairment. They are easy to use and well illustrated. Although there is no documentation of the validity of the impairment ratings, the guidelines can be a useful starting point in impairment evaluations.
Engelberg AL. Disability and workers' compensation. Prim Care 1994;21:275–289.
Physicians are often requested to provide medical evaluations and opinions about disability. This article presents differing definitions of disability and impairment. It focuses on the specific disability systems known as workers' compensation systems, the roles physicians play in these systems, and the parties involved.
Harten JA. Functional capacity evaluation. Occup Med 1998;13:209–212.
The functional capacity evaluation (FCE) is a tool for assessing the extent of a patient's disability. Several factors help to determine whether an FCE should be conducted. Once underway, the evaluation includes an intake interview, musculoskeletal and functional assessments, validation of effort, and interpretation and recommendation.
Katz RT, Rondinelli RD. Impairment and disability rating in low back pain. Occup Med 1998;13:213–230.
This "how-to" guide for the examination of impairment and disability resulting from low back pain examines workers' compensation, Social Security, the Americans with Disabilities Act, and the American Medical Association's Guidelines to the Evaluation of Permanent Impairment. The medicolegal interface is addressed, and specific recommendations are made to assist the physician involved in an independent medical evaluation.
Social Security Administration. Disability evaluation under Social Security. Social Security Administration Office of Disability, SSA publication no. 64-039, ICN 48 6600. Washington, DC: U.S. Government Printing Office, January 1998.
A guide to SSA medical criteria for disability.

APPENDIX
A Human Rights Approach to People with Disabilities: The Americans with Disabilities Act

Jay S. Himmelstein, Glenn S. Pransky, and Christopher Kuczynski[1,2]

The employment provisions of the Americans with Disabilities Act (ADA) of 1990 seek to promote the independence of people with disabilities through equal opportunity for employment. As such, this act promotes

---

[1] Christopher Kuczynski is Assistant Legal Counsel and Director of the Americans with Disabilities Act Policy Division, U.S. Equal Employment Opportunity Commission. This appendix was cowritten by Kuczynski in his private capacity. No official support or endorsement by the Commission or any other agency of the U.S. Government is intended or should be inferred.

[2] David Fram, an attorney with the National Employment Law Institute, contributed to an earlier draft of this appendix.

a human rights approach, as opposed to a compensation approach, to the problems confronted by people with disabilities. This Appendix reviews the major employment provisions of the ADA and discusses opportunities for health professionals to support appropriate employment opportunities for people with disabilities and to thereby help *prevent* the tertiary consequences of disability such as poverty and social isolation.

Employment Provisions of the ADA

Title I of the ADA prohibits employment discrimination against a qualified person with a disability on account of his or her disability. A person with a disability is defined in the ADA as someone who (a) has a physical or mental impairment that substantially limits one or more major life activities, (b) has a record of such impairment, or (c) is regarded as having such an impairment. A person is qualified if she or he meets the basic job prerequisites and can perform the essential functions of the job with or without reasonable accommodation. This prohibition reaches to all aspects of the employment relationship, including application procedures, hiring, promotion, compensation, training, and discipline.

The ADA also requires employers to make reasonable accommodations for qualified people with disabilities. When a job applicant or an employee requests an accommodation and the disability or the need for accommodation is not obvious, an employer may require reasonable documentation of the existence of a disability or of the need for accommodation. Often this is provided by the applicant's or employee's own health care provider. However, if the applicant initially provides insufficient documentation from his or her health care provider, the employer may require that the person be examined by its own health care professional at the employer's expense (1).

Under the ADA, employers may not conduct medical examinations or make inquiries regarding the existence, nature, or severity of an applicant's disability before he or she is extended a conditional offer of employment. This prohibition applies to all applicants, regardless of whether they have disabilities (2,3). These are permitted after a person is offered a job, but before he or she starts work, if this applies equally to all applicants in the job category. Although an employer may obtain medical/disability–related information at this postoffer stage, the ADA does restrict employers subsequent actions based on this information. An employer may withdraw a conditional job offer because of a person's disability *only* if the employer can show one of the following:

1. The person cannot perform the essential functions of the job despite reasonable accommodations.
2. The person poses a direct threat to self or others in the position that cannot be eliminated or reduced to an acceptable level with a reasonable accommodation.
3. Other federal laws or regulations (e.g., those administered by the Federal Aviation Administration and the Department of Transportation) require the employer to withdraw the job offer because of the person's medical condition.

Therefore, to avoid liability under the ADA for withdrawing a conditional job offer for medical reasons, an employer must analyze the following key issues, frequently with input from medical personnel:

1. *Does the person have a disability protected by the ADA?* This requires the employer to examine whether the person has a physical or mental impairment, and whether the impairment substantially limits a major life activity. It also requires the employer to analyze whether the person has a record of or is regarded as having such an impairment.
2. *Is the person qualified to perform the job?* This requires the employer to examine whether the person can perform the essential functions of the job, with reasonable accommodation, if needed.

3. *Does the person pose a direct threat* and, *if so, can the risk of harm be reduced below the direct threat level through a reasonable accommodation?* This requires the employer to examine whether the person poses a significant risk of substantial harm to self or others, and whether that risk can be reduced through a reasonable accommodation.

Once a person becomes an employee, an employer may make disability-related inquiries or conduct medical examinations only if they are "job-related and consistent with business necessity." To meet this standard, the employer must demonstrate that it has a reasonable belief based on objective evidence that the employee will be unable to perform the essential functions of his or her job or would pose a direct threat because of a medical condition. Like the prohibition of disability-related inquiries and medical examinations of applicants, the limitations on inquiries and examinations of employees apply to employees with and without disabilities (4).

The Role of the Health Professional under the ADA

The ADA's requirements for employers are aimed at providing equal employment opportunities for people with disabilities. The role of medical professionals in this process is primarily to provide guidance on issues of whether a person has a covered disability, to determine whether the person poses a direct threat, and to make recommendations regarding workplace accommodations.

Does the Person Have a Covered Disability?

The health professional can assist the employer in determining whether a person has a physical or mental impairment and whether the impairment substantially limits a major life activity, without revealing information unrelated to the diagnosis of disability. In

the simplest case, this role can be fulfilled merely by providing information regarding an impairment that limits one or more major life activities. For example, a patient who has multiple sclerosis that substantially interferes with walking would have an impairment that substantially limits one or more major life activities and therefore would have a covered disability. Likewise, a potential employee who has fully recovered from back surgery but is limited by his or her inability to lift objects weighing more than five pounds from floor level may be considered to have a covered disability. A physician's determination that someone has no actual disability does not completely resolve the matter; a person may have a covered disability if an employer simply regards him or her as having a disability.

Does the Person Pose a Direct Threat?

An employer may require that a person not pose a direct threat to the health or safety of self or others in performance of the job. However, the requirements to meet the direct threat threshold as specified under the ADA are quite stringent. The risk and adverse outcome of concern must be specific, significant (severe), and highly probable; the risk factors must be based on objective medical data, not speculation or generalization from studies of large groups of similar people; the risk must be constant, not temporary; and, finally, the risk must be one that cannot be eliminated through reasonable accommodations. Each one of these areas should be addressed by an evaluating health professional in advising the employer whether the direct threat standard has been met. The consideration of whether a person constitutes a direct threat to self or others is probably the most complicated aspect of the ADA that is likely to be confronted by clinicians. We strongly recommend that clinicians familiarize themselves with the examples offered by the technical assistance manual published by the United States Equal Employment Opportunity Commission on this subject (5).

What Accommodations Are Appropriate?

An important and positive role for medical personnel is the recommendation of reasonable accommodations that would allow a person to perform the essential functions of the job. This, of course, requires that the evaluating clinicians has detailed information about the job requirements or has personal knowledge of the job through visiting the workplace. In the example of the patient with multiple sclerosis described previously, it may be that the only accommodation needed is wheelchair access to the desk so that the patient can perform his or her job as a data entry clerk. In the case of the patient after back surgery, the recommendations for accommodation may be more complex. For example, assuming that lifting is not an essential function of the job, reassigning the lifting duties to another employee would likely be an appropriate accommodation. If lifting is an essential function, then the clinician might suggest that certain equipment be obtained so that the employee can perform this function, despite his or her physical limitations. Regardless of the clinician's suggestion, it is ultimately up to the employer to provide the reasonable accommodation. It should be obvious from the previous description that active participation in enhancing employment of people with disabilities requires an expanded base of information for practicing clinicians. Several caveats are worth emphasizing:

1. *The evaluation of workers must be individualized to a specific person and a specific set of job tasks.* Physicians and employers are prohibited from relying on nonspecific physical standards unless it has been demonstrated that those who fail to meet the standards because of a disability constitute a direct threat to themselves or others or cannot perform essential job tasks with reasonable

accommodations. The only exceptions that are allowed occur where there are overriding health and safety requirements established under federal laws.

2. *A person's ability to do a job is often best determined by job simulation or job trial.* Many of the best methods for determining fitness are nonmedical in nature, and the role of the medical examination may be minimal in determining employability. Moreover, determining a person's future risk of injury or disease progression is especially complicated and filled with potential liability for the evaluating clinician and the employer. In relating to employers and employees, therefore, the physician should be careful to ensure that employers are aware of the very limited benefit of preplacement worker fitness and risk evaluations.

3. *Clinicians are health care providers, not lawyers or personnel managers.* Although health professionals need to become conversant with their role under the law, it is not their duty to act as legal counsels to employers or employees. Although clinicians should communicate with employers, they must remember that it is not their responsibility to make employment decisions. The physician's responsibility is to provide *medical opinions* and guidance regarding disability status, functional capability, and direct threat, and, where appropriate, to recommend modifications and accommodations to mitigate risk or enhance capabilities. The responsibility for making employment decisions or deciding whether it is possible to make a reasonable accommodation for a person with a disability lies with the employer. He or she is expected to incorporate insight from a variety of sources (including other health professionals and disability experts) in making employment decisions.

The ultimate impact of the ADA on the makeup of the American workplace will not be known for many years. Since its passage, a number of legal cases have explored the concepts of disability, direct threat, discrimination, and reasonable accommodations. In many instances, a clear consensus has not emerged. However, the presence of this landmark legislation has significantly affected the practices of employers and public and private agencies. Health care providers with their unique training and experience and their familiarity with medicine and demands of work, will play a distinct and vital role in enhancing the opportunities and safety at work of people with disabilities. To accomplish this, clinicians must shift from their traditional focus on diagnosing disability to recognizing and enhancing capability, consistent with the vision of occupational health as a true subspecialty of preventive medicine.

References

1. Equal Employment Opportunity Commission. EEOC guidance on the ADA and psychiatric disabilities. In: Americans with Disabilities Act manual. Washington, DC: Bureau of National Affairs, 1997, tab 70; 1281–1292.
2. *Roe v. Cheyenne Mountain Conference Resort*, 124 F.3d 1221 (10th Cir. 1997).
3. *Griffin v. Steeltek Inc.*, 8 AD Cases 1249 (10th Cir. 1998).
4. Equal Employment Opportunity Commission. A technical assistance manual on the employment provisions (Title I) of the Americans with Disabilities Act. Washington, DC: U.S. Government Printing Office, January 1992.
5. Equal Employment Opportunity Commission. Guidance on pre-employment disability-related inquiries and medical examinations under the ADA. In: Americans with Disabilities Act manual. Washington, DC: Bureau of National Affairs, 1995, tab 70: 1103–1109.

Occupational Health: Recognizing and Preventing Work-Related Disease and Injury, 4th ed.
Edited by Barry S. Levy and David H. Wegman, Lippincott Williams & Wilkins, Philadelphia © 2000.

13 Ethics in Occupational and Environmental Health

Kathleen M. Rest

Case 1: A team of university scientists (epidemiologists, physicians, industrial hygienists, toxicologists, and statisticians) is funded to conduct a large industry-sponsored epidemiologic study. The team wants to establish an independent advisory board to oversee the design and conduct of the study, as well as the data analyses and reporting and dissemination of study findings. The industry is reluctant to agree.

Case 2: A large industry is sponsoring a study of a community's exposure to two carcinogenic chemicals in its air emissions over the past 15 years, as well as a study of cancer incidence. Although the community is concerned about cancer, residents are much more concerned about birth defects and childhood respiratory illnesses, which they believe may be associated with many of the chemicals emitted from the plant.

Case 3: Hazardous wastes from industrial operations have contaminated the soil and water supply of a poor neighborhood with a large minority population. The government and responsible industrial parties have decided to incinerate the waste on site. Their contractor performs a risk assessment that indicates no health risk to the community. The residents want the waste removed from the neighborhood. They have used Environmental Protection Agency (EPA) funding to obtain their own technical advisor, who presents a risk assessment suggesting that incineration on site will indeed pose health risks to community residents.

Case 4: On petition from a labor union, the Occupational Safety and Health Administration (OSHA) is considering whether to lower the permissible workplace exposure limit of a chemical, because recently completed epidemiologic studies suggest neurotoxic effects at the current exposure level. The epidemiologists who conducted the studies are reluctant to get involved in the policy debate, but other experts from industry, labor, and academia are asked to comment about the need for such action. They present conflicting interpretations of the studies, different views about the findings from previous studies, and widely divergent opinions on the need for more stringent regulation.

Case 5: A U.S. company decides to move its pesticide manufacturing operations to another country because labor costs are cheaper there and because the other country's environmental and occupational regulations are less stringent and virtually unenforced. The other country's ministry of commerce and development welcomes the company.

Case 6: A hospital-based occupational health program has a contract to provide clinical and consultation services to a local furniture manufacturer. The company has many ergonomic problems, and many workers have experienced musculoskeletal and repetitive strain injuries. The company physician, nurse, and safety engineer have recommended ergonomic changes in the work environment many times, but the company has taken no action. Instead, the company has asked the nurse to institute a weight reduction program and a class on safe lifting techniques for the workers.

K. M. Rest: Occupational and Environmental Health Program, University of Massachusetts Medical School, Worcester, Massachusetts 01655.

Case 7: An occupational medicine resident conducts an "independent medical evaluation" of a person who has been out of work for 6 months with a work-related injury. The workers' compensation insurer requested the evaluation. The resident tells her preceptor that she cannot understand why the patient seems uncooperative and even hostile during the medical evaluation.

Case 8: A company contracted with a local family physician to perform preplacement physical examinations, periodic screening examinations, and fitness-for-duty and return-to-work evaluations. The company told the physician what it wanted to include in the preplacement and screening examinations. The physician has never visited the plant and has little information about workplace conditions and job demands. During the first 6 months of the contract, the physician is called many times by the company's human resource manager and asked for what she considers to be confidential medical information about specific employees' fitness for duty and return-to-work issues.

These cases typify the range of issues and types of ethical and moral questions that occupational and environmental health professionals, researchers, regulators, and others can expect to encounter. Literature from the United States and elsewhere frequently reflects on the ethical problems encountered by occupational and environmental health professionals in their work. Because the environments in which these health and safety professionals function can be characterized by competing goals and interests and differential power structures, thorny ethical issues are common.

As a field of study, *ethics* is a complex discipline that attempts to analyze, define, and defend the moral basis of human action. For our purposes, we can use the term *ethics* to refer to the rightness and wrongness of human behavior. Ethics entails a sense of "ought"; that is, ethics helps us decide how we ought to act or what ought to be done. Ethics is not law, social custom, personal preference, or consensus of opinion, although any of these may derive from or inform ethical considerations. Rather, ethics is

both an approach to moral reasoning and a collection of principles and rules to help guide judgment and action. Ethics can facilitate reflection that leads to consistent, informed, and justifiable decisions that can withstand close moral scrutiny.

In the field of occupational and environmental health, conflict and disagreement occur at many levels—from actions taken or not taken by regulatory and other governmental agencies in matters of public policy to decisions made by researchers and occupational and environmental health professionals in the context of their daily work. Difficult questions abound: Should a substance or condition in the workplace be regulated? At what level? How safe is safe enough? How clean is clean enough? Should medical screening or epidemiologic surveillance (or both) be instituted for a group of workers or residents of a contaminated community? When and how should information or concern about actual or potential health risks associated with present or past exposures be disclosed—to those exposed, to appropriate authorities, and to peers in the scientific community? How much information about their workers are employers entitled to have? How should tradeoffs among health, productivity, jobs, environmental protection, and economic development be made? Not all aspects of these conflicts and decisions are ethical in nature; some may reflect simple disagreements about facts, methods, processes, or desired outcomes. Many of the disagreements, however, have some underlying moral dimension, even when the arguments are framed in technical or economic terms.

The moral issues encountered in occupational and environmental health are socially constructed and therefore may vary temporally and geographically. These issues cannot be considered fully outside the social, cultural, institutional, political, and, increasingly, economic contexts from which they arise (1). In the workplace and in many environmentally-contaminated communities, an important contextual dimension is the power imbalance between employer and worker, polluter and resident. These power imbalances are reflected in significant differences in economic and technical resources, decision-making authority, access to powerful institutions, and the distribution of risks and benefits. A second important contextual dimension is the scientific uncertainty that attends many of the problems and issues encountered. Other contextual dimensions include the powerful economic forces generated by market competition and world trade and the very nature of the employer-employee relationship.

This chapter provides a brief and basic discussion of the ethical dimensions of common occupational and environmental health issues in public policy, scientific research, and practice. It also presents an array of principles, processes, and guidelines that have been suggested as aids to problem solving and decision making in situations that pose ethical dilemmas, conflicts, or problems. Probing the ethical dimensions of these situations can make a unique contribution to improved decision making and action.

REVIEW OF ETHICAL PRINCIPLES, PROCESSES, AND GUIDELINES

A lengthy and comprehensive discussion of moral philosophy and ethics is neither possible nor warranted in a textbook such as this. Yet it is important that the reader appreciate some of the concepts and constructs that are commonly applied when assessing or analyzing ethical problems. The literature of ethics and bioethics is replete with discussions of moral rights and their attendant moral obligations and duties; character traits and virtues of moral persons; moral principles against which actions can be judged; and necessary or established procedural and behavioral guidelines, such as ethical codes. This chapter considers a set of principles that has been prominent historically in bioethics, along with additional concepts that may further enrich ethical inquiry.

In 1994, Beauchamp and Childress provided a standard approach to bioethical issues with their focus on the principles of *autonomy* (which has come to mean individual self-determination but is also defined as respect for persons), *nonmaleficence* (the duty to do no harm), *beneficence* (the duty to do or promote good), and *justice* (the duty to be fair). Although these principles remain central, important, and helpful in moral reasoning, they have been criticized, as has the general approach of applying abstract moral principles to complex problems. It is suggested that they do not help in the most difficult of situations (i.e., when principles conflict); that they suggest application devoid of context; and that they are preoccupied with individual rights and removed from larger social and collective concerns (2). Additional guidance is offered by other scholars. Some emphasize the concept of *moral responsibility*, which arises from interpersonal relationships or from the special knowledge that one person has in relation to another person's welfare (3). Others suggest inquiry into the social construction of the moral issue—the vested interests and social, cultural, and institutional contexts that give rise to the problem (1).

ETHICAL ISSUES IN RESEARCH

Although science, in its pursuit of truth, is often held to be "objective," it is now generally acknowledged that individual, social, and cultural values influence scientists in many aspects of their work. These values may influence what researchers decide to study, how they frame research questions and design studies, what data they collect, how they analyze and interpret data, how they report study results, and how they participate in policy debates that involve the use of their findings. It is helpful when researchers are honest with themselves and with the public about the influence of their own values or those of relevant interest groups on their activities and pronouncements, because it adds an important dimension to public and private decision making. However, there have been cases in which science and scientists have been bought (directly or indirectly) to serve political or ideological interests under the guise of "objectivity" (4). In occupational health research, there are additional complexities. Because the workplace is the primary subject of the research, industry will always have an interest in the outcome, even if it does not fund the research. Powerful institutions may also have a strong interest in the conduct and results of environmental health research.

An often-discussed area of conflict in the research process is the reporting of results and dissemination of findings. There are both historic and current examples of situations in which occupational health professionals have failed to report relevant research findings or have experienced adverse personal and professional consequences after doing so (5,6). Questions relevant to this part of the research process include the following: How much control does the research sponsor have? Can the sponsor delay or even prevent presentation, publication, or dissemination of a study's findings? Are publications or presentations subject to approval of the sponsor? Can the sponsor edit, change, delete, or add to any manuscript prepared for publication?

These questions suggest consideration of the researcher's *autonomy* and *integrity*. Because research findings can have serious consequences for the interest groups involved, they may seek to influence the researcher and the work. In the past, industry sponsors have tried to influence the design, conduct, and interpretation of research and the dissemination and publication of results. For this reason, it would behoove independent researchers to institute structural safeguards before conducting a study sponsored by an industry or other interest group. Case 1 illustrates how one research team sought to protect the integrity and independence of their industry-sponsored research project. The sponsor's reluctance should signal the need for careful examination of the potential for interference or influence.

The questions listed also suggest consider-

ation of the researcher's professional responsibility. In many studies, researchers establish a direct or implied relationship with their subjects, from whom they obtain personal and medical information and, perhaps, biologic specimens. This special relationship confers a responsibility on researchers to at least respect and refrain from harming their subjects, whose welfare may be affected by inappropriate disclosure of personal information or by the researchers' failure to communicate the results of the study. Protection and promotion of worker and public health (beneficence) and fairness and justice for research subjects also merit deliberation.

A more subtle influence may be found in the way researchers themselves choose to interpret and report their findings. Will they refer to contradictory evidence and fairly discuss the significance of their own findings in this light? Will they fully discuss the limitations of their study—perhaps with reference to the study's statistical power? Although honesty and integrity demand attention to these matters, the researcher may be influenced by sources of previous, present, and potential future funding when findings are presented, interpreted, and discussed. Self-censorship can have a potent influence on the research process; it can affect any aspect, from the questions the researcher seeks to answer to the ways in which results are interpreted and reported.

The initial framing of the research question is also subject to influence. For example, a corporation concerned about the frequency of musculoskeletal injuries in its workforce may be willing to sponsor research on a device or method to identify susceptible workers, whereas the ergonomics researcher might prefer to study etiology or potential process, work practice, or product interventions. Industrial sponsors may directly influence toxicologists' or epidemiologists' decisions about which chemicals to study and which end points to investigate (case 2). Researchers concerned with the costs of workers' compensation may limit the focus of their studies to medical care cost-containment strategies, effectively precluding research on the prevention of workplace injury—perhaps because the sponsors view the problem solely in this light. Certainly, framing the research question is partly a function of the researcher's professional interests and abilities, but external factors and pressures can play a role during this phase of the research process. Important questions include the following: Who has defined the research question—the investigator, the sponsor, or both? If funding were not an issue, would this be the right question, the right issue, the right problem? Whose interests are being served by the way in which the question is framed?

Of course, the conduct of the study also poses ethical challenges. The concept of informed consent (justified by an appeal to *autonomy* and *respect for persons*) is especially important. Research subjects should be informed of the risks and benefits of their *voluntary* participation in any study. In occupational and environmental health research, there may be additional requirements. For example, subjects should be informed about the reasons for the study, the sponsors of the study, the timetable for completion of the study, and how the results will be used (7). Study subjects should also be informed about possible economic risks of participation, both for themselves and for their employers. Informed consent also has bearing on the conduct of medical examinations and screening and surveillance activities, performed by occupational and environmental health professionals, that are often carried out at the behest of third parties. This is discussed in a later section.

ETHICAL ISSUES IN SCIENCE AND POLICY

In the field of occupational and environmental health, most debates of science and public policy focus on the regulation and control of health and safety hazards. Beneath the technical and economic arguments regarding the basis, process, and content of regulation lie clear differences in values and in the deference given to widely held moral principles.

Several cases mentioned at the beginning of this chapter illustrate common ethical issues in science and policy. In these cases, scientists and experts have differing opinions about what a study means or which risk assessment reveals "the truth." These views may involve honest disagreements about methodology, but they may also reflect different understandings about the duties imposed by widely held moral principles (e.g., beneficence and autonomy), as well as differences in personal preferences, political ideology, and disciplinary biases.

In deciding whether to incinerate or remove the waste from a poor community (case 3), whether to lower the permissible exposure limit of a potential workplace neurotoxin (case 4), or whether to move a hazardous operation to a country with less stringent environmental controls (case 5), scientists, regulators, government officials, and employers are dealing with concepts of nonmaleficence, beneficence, autonomy, justice, and moral responsibility. Consideration of these moral concepts are particularly important in decisions involving health and safety issues and environmental regulations, which to a large extent transfer choice about "acceptable" risk from individuals to other entities, such as governments or corporations.

Although there is debate about the extent to which individuals are morally obligated to take positive action to contribute to the welfare of others, there is little disagreement that they should refrain from doing harm and, in some cases, take positive action to prevent harm from occurring. If the waste is incinerated, the health and well-being of community members may be harmed; if a chemical is not regulated more stringently, the health of workers may be endangered; if a hazardous facility is moved abroad, residents and workers in the new host country may be harmed. On the other hand, it could be argued that trucking the waste out of the community would create potential risks to others, that more stringent regulation of a chemical would adversely affect the financial resources of some manufacturers and might even cause some workers to lose their jobs, or that the economy of the other country would be improved by the migration of the potentially hazardous industry. How should these tradeoffs be made?

Many disagreements are likely to revolve around issues of scientific and technological uncertainty, in many ways an inherent element of occupational and environmental health research, policy, and practice. The effectiveness of technologies to clean up hazardous waste may be unknown. Risk assessments vary by orders of magnitude, depending on the models and assumptions made by the investigators. Most epidemiologic studies show association, not causation, and reported mortality risks are bounded by confidence intervals of varying widths. Furthermore, the models chosen, the assumptions made, and the designs employed may reflect the investigator's values or the values and interests of the study sponsors.

The comments of scientists and the decisions of policy makers often reflect their value-laden approaches to issues of uncertainty. What level of proof is needed to trigger action, and who bears the burden of proof? Should we wait for certain evidence of harm before we take action, or is a reasonable suspicion enough? Should consumers or workers be responsible for showing that a product is dangerous, or should the manufacturer or employer be required to show that the product is safe before it is allowed into the market or workplace?

Although science may contribute information and even define the parameters of the debate, the answers to these questions reflect policy judgments that invariably are influenced by values. Some scientists, regulators, and members of the public prefer to wait for additional evidence and accept the risk of future harm rather than expend potentially unnecessary resources to impose costly regulations now. These persons are likely to frame their arguments in economic terms

and to focus their critiques on the design flaws and technical limitations of individual studies bearing on the question.

Others prefer to err on the side of caution in protecting the health of workers and the public and seek to impose regulations or deny siting permits, even at the risk of being found wrong at some future date. Their arguments focus on human health or environmental impacts and are more likely to synthesize the available data, overlook the design flaws of individual studies, and give weight to the aggregate suggestive evidence. To what degree are these determinations "scientific" and to what degree are they driven by different views of one's duty to prevent harm, do good, or fulfill one's responsibility toward others in a personal or fiduciary relationship?

Cases 3 and 5 also illustrate an especially important dimension of *justice*. Many policy decisions reflect a utilitarian approach to policy making; that is, decisions are made to maximize benefit or to confer the greatest good on the greatest number. This approach, deeply rooted in ethical theory, often has salutary effects and can justify actions that harm a few while helping many. However, this approach often fails to consider issues of distribution. *Distributive justice* relates to fairness in the distribution of risks, costs, and benefits. Who benefits? Who bears the risks or costs? Who gets to make the decision? What is the relationship between those who bear the costs, those who reap the benefits, and those who make the decisions?

The environmental justice movement has provided evidence of environmental inequity, whereby economically disadvantaged and minority communities have borne more than their fair share of landfills, hazardous waste facilities, polluting industries, and environmentally related health problems, such as lead and pesticide poisoning (see, for example, ref. 8). In these cases, it seems that the most vulnerable members of society are asked to bear a disproportionate share of environmental pollution and associated health risks. Distributive justice requires us to examine and justify the allocation of risks, costs, and benefits.

Applications in Public Policy

The application of ethical principles to occupational and environmental health policy is more than just the fodder for interesting debate. We can see their reflection in many of the regulatory decisions taken by governmental agencies.

For example, the OSHA lead standard addresses issues of autonomy, justice, nonmaleficence, and beneficence. The regulatory debate on lead included scientific questions about safe airborne levels and the merits of using blood lead concentrations as the primary measure of compliance with the standard. An interesting part of the debate centered on the establishment of medical removal protection, with (hourly) rate retention for workers found to have elevated blood lead concentrations. There was little argument about the wisdom of removing such workers from exposure. Rather, the conflict arose over a proposal that obligated employers to maintain workers' hourly rates and seniority rights during the period of removal. The Lead Industries Association argued against this proposal on legal grounds. Workers' representatives focused on issues of autonomy and fairness, noting that workers would choose not to participate in a blood lead screening program that might threaten their livelihoods. Although medical removal was in the best interest of the workers' health, their representatives argued that it was unfair to penalize them by putting their wages and seniority benefits at risk in an exposure situation over which they have little, if any, control. The courts ruled in favor of the policy of medical removal with rate retention, but the ethical issues surrounding the regulation of lead did not abate. (More recently, OSHA's failure to include a medical removal protection requirement in its revised methylene chloride standard prompted the United Auto Workers Union to sue the agency. The Halogenated Solvents Industry Alliance also sued, but on economic grounds involving the costs of compliance. After the lawsuits were filed, the opposing parties worked together and proposed a settlement, which OSHA adopted in 1998. The final rule addressed some concerns for both parties. It included provisions for medical removal protection and extended compliance dates for certain employers to implement requirements for engineering controls and respiratory protection.)

Recognizing the reproductive effects of lead, some employers began to institute "fetal protection" policies. Such policies ostensibly sought to protect actual and potential fetuses of pregnant women or women of reproductive capacity. These policies involved the principles of beneficence and nonmaleficence, which in this case clashed with the principles of autonomy and justice. In weighing their options, employers who adopted such policies placed a higher value on protecting the fetus from harm (and themselves from future liability) than on providing autonomous choices to female workers, who could take steps to control their fertility. In addition, because lead is known to have toxic effects on both men and women, one could question the fairness of differentially protecting female and male employees.

The U.S. Supreme Court ruled against the use of fetal protection policies on the basis of sex discrimination (fairness), upholding the autonomy of women workers, in *United Auto Workers v Johnson Controls,* US 111 SupCt 1196 (1991). The Court stated that such policies force female workers "to choose between having a child and having a job" and that "it is no more appropriate for the courts than it is for individual employers to decide whether a woman's reproductive role is more important to herself and her family than her economic role." Rather, the Court found that "Congress has left this choice to the woman as hers to make." Thus, the Court gave overriding weight to the woman's autonomy.

Freedom and noninterference are tightly

guarded and highly cherished rights in the United States. In occupational and environmental health, these concepts are reflected in right-to-know laws and regulations (see Chapter 10). In this context, autonomy suggests that individuals have a right to know about their workplace or environmental exposures so that they can make decisions and take individual or organized action. Such decision making requires information—in this case, information about the exposures, their potential health effects, methods of protection, and, ideally, possible substitutes or alternatives.

In the United States, the concept of right-to-know has been embodied in both occupational and environmental legislation. The OSHA Medical Access Rule provides workers with access to their own medical records and to records pertaining to their exposures to toxic substances to the extent that the employer compiles such records. Under this rule, information is provided on request. The OSHA Hazard Communication Standard obliges employers to educate and train workers about the hazardous chemicals in the workplace. Employers must provide this information as a matter of course, even without a request. Congress extended the right to know about toxic hazards to communities in 1986 with the enactment of the Emergency Planning and Community Right-to-Know Act. Its many provisions provide local citizens with access to information about the location, use, and release of toxic chemicals in their communities (see Chapter 3).

In promulgating these right-to-know laws and regulations, the government recognized workers' and citizens' needs for information about toxic substances in their workplaces and communities, granted them a right to such information, and conferred on manufacturers and employers the duty to provide it. However, provisions were also enacted to balance these needs with the needs of employers and manufacturers to protect their trade secrets. The existing power imbalances in workplaces and communities have called into question the adequacy of these right-to-

know provisions. It is argued that, in addition to the right to know, workers and citizens need the right to act.

In enacting the Occupational Safety and Health Act of 1970, U.S. legislators recognized the potential harm that might befall workers who take action to protect their health and safety, for example by refusing hazardous work, calling unsafe conditions to the attention of their employers or fellow workers, or alerting authorities about health and safety problems at work. Section 11(c) of the Act seeks to protect workers in the exercise of these rights from retaliation and discrimination by their employers. A 1997 report issued by the Inspector General of the Department of Labor criticized the adequacy of OSHA's 11(c) program. The agency has taken steps to improve its performance in this area, enhancing protection of and justice for workers who take action to secure their health and safety on the job.

Applications in Corporate Policy

Public policy and regulation can go only so far in protecting workers and communities from occupational and environmental hazards. Corporate and business policies and decisions have the most immediate impact on health, safety, and the environment. They can promote good (beneficence) or result in significant harm. Although many employers have excellent health, safety, and environmental programs and function as responsible corporate citizens, there have been too many cases of business policies and practices that demonstrate wanton disregard for worker and community health (see, for example, refs. 9 and 10).

The public is well aware of and disillusioned by these corporate failures. Surveys indicate that the public considers industry the least trusted (but most knowledgeable) source of information about chemical risk (11). Recognizing the impact of this public perception, private industry has developed a variety of initiatives and policies to upgrade its practices and reassure a skeptical public.

For example, the Chemical Manufacturers Association instituted a voluntary program (Responsible Care) for its member companies, which provides codes for management practices and guidelines for community and employee outreach activities. Some companies participate in OSHA's Voluntary Protection Program, taking steps to enhance their own internal responsibility for worker health and safety. Individual employers also have refined their policies and practices to improve their health, safety, and environmental programs; facilitated right-to-know programs and worker training; and eliminated discriminatory practices against their workers. Business schools have instituted courses in Business Ethics.

Obviously, the employer community plays a significant part in creating ethical problems relating to occupational and environmental health. Employers need opportunities and incentives to enhance their abilities to recognize, understand, and respond to their ethical obligations and responsibilities and to respect the ethical duties and responsibilities of others.

Workers and their representatives, including organized labor, also have ethical duties and responsibilities to prevent harm, do good, be fair, and tell the truth. However, most workers in the United States and other countries are not in a position to enact and enforce policies and programs in the workplace. The conduct of individual workers, like that of everyone else, should be guided by moral considerations both on and off the job. Although individual workers (and labor organizations) can and sometimes do create ethical dilemmas for occupational health professionals, in most cases they are not a powerful and organized group, and they are not treated as such in this chapter.

OCCUPATIONAL HEALTH AND SAFETY PRACTICE

Most occupational and environmental health professionals routinely encounter conflicts and ethical problems that challenge their values as well as their professionalism. Those who provide occupational and environmental health services directly, such as occupational physicians, nurses, industrial hygienists, safety professionals, and occupational ergonomists, have clear professional and moral responsibilities that derive from their special knowledge of occupational and environmental health and safety and from the special relationships they develop with workers, employers, and communities. This section examines the ethical dimensions of activities commonly conducted or encountered by these professionals. Problem areas relate to dual agency, conflict of interest, confidentiality, professional competence, taking action in the face of scientific uncertainty or opposition, and responsibility for others.

In exploring these issues, it is important to appreciate the very real and personal consequences that these individuals may experience as a result of their work. Their actions can enhance their own reputation, status, and esteem or incur the wrath and distrust of employers, patients, and colleagues. Their decisions can affect their income, their employability, their standing in the professional community, and their respectability in the eyes of persons for and to whom they are responsible. Courageous and unpopular decisions may take a personal and emotional toll on these professionals and their families. The difficulty of "doing the right thing" in such situations should not be underestimated.

Working for Companies

Health and safety professionals who work for companies, whether full-time, part-time, contractually, or on a fee-for-service basis, frequently face a host of ethical issues that call their allegiance and values into question. Their goals, interests, and values may differ significantly from those of both employers and workers. The company's primary purpose and interest is to profitably manufacture a product or provide a service and to stay in business. The workers are primarily interested in earning a living, providing for their families, and finding some personal satisfaction in their work—without damage to their health. When the health and safety professional and the worker share the same employer and the interests of the employer and worker diverge, what is the role of the health care provider? Whose interests take precedence—those of the worker/patient or those of the employer/client?

Within this complicated structure, the practice of occupational medicine, nursing, safety, or industrial hygiene is inherently difficult and challenging. Employers may expect physicians and nurses to function as agents of social control, making determinations about when, where, and whether a person can work. A company's satisfaction with the services of these health care professionals may depend in large measure on their ability to get injured workers back to work quickly. Employers may limit the ability of their health and safety professionals to take preventive action regarding workplace hazards. At the same time, workers may expect occupational health professionals to protect their interests and function as their advocates when problems arise. The worker and the health professional may not always agree on issues related to return-to-work directives or job restrictions. When the employer/client's interests differ from those of the worker/patient, it is not surprising that skepticism, distrust, and hostility arise on all fronts. The occupational medicine resident in case 7 has probably failed to appreciate the significant economic and psychosocial factors, as well as the power dynamics, that often come into play when a worker is injured on the job and issues of compensation or return-to-work are raised.

The issue of confidentiality is perhaps the most frequently discussed ethical problem in the occupational health literature—and for good reason. Consider, for example, the almost unbelievable case of an employer installing hidden video equipment in a nurse's examining room—and then firing the nurses when they discovered and complained about it (12).

How much personal and medical information obtained by the health care provider is the company entitled to? This question frequently arises in the context of preplacement medical examinations, diagnosis and treatment of work-related injuries or illnesses, medical surveillance examinations, and fitness-for-work evaluations. Should employers be informed that a job applicant has diabetes or a history of back injury? Should the employer be told that the executive being considered for promotion has cardiac disease or is seeing a psychiatrist? What about liver abnormalities related to alcohol use discovered in a hazardous waste worker participating in a medical surveillance program?

As in case 8, physicians and nurses may encounter direct or indirect pressure to disclose or release this type of information, which the company believes will help protect its legitimate business interests. However, such disclosures invade workers' privacy and may threaten their job status. How should the health care provider reconcile these competing interests? Clarification of legal requirements and restrictions (especially in light of the Americans with Disabilities Act; see Chapter 12) may be a helpful first step. Reflection on the widely held professional ethic of maintaining patient confidentiality is also in order. In most cases, the employer does not need diagnostic or medical data but merely information about the employee's ability to work and the need for job modifications or work restrictions.

A larger and even more difficult problem relates to the extent to which occupational health and safety professionals are obligated to take action regarding suspected or known problems. The physician, nurse, and safety engineer in case 6 have expressed their concern about the company's ergonomic problems and have made numerous recommendations for correcting them. The company has taken no action. Having expressed their concern (perhaps even in writing), do these professionals have any further obligation to follow up on these problems? Do the principles of nonmaleficence and beneficence impose an ethical duty to do more than make recommendations? Workers have already been injured, and there is every reason to believe that such injuries will continue to occur. Should the physician, nurse, or safety officer notify OSHA? Try to organize the workers to take action? Quit? How far should the company-employed health and safety professional go to protect the workers? Similarly, if the company's environmental engineer is aware that the company has violated federal or state regulations on waste disposal, should he or she notify the appropriate authorities if the company continues to break the law?

The situation becomes even more complex when it is attended by uncertainty. Suppose, for example, that a hospital-based occupational health physician discovers several cases of serious disease (e.g., interstitial lung disease) among a group of workers (e.g., workers in the nylon flock industry), and that he suspects the disease may be related to unidentified exposures in the workplace. The physician commences an investigation, with the consent of the company. The physician wants to report his findings and suspicions about a new occupational disease to his colleagues at a professional meeting. If the flock manufacturer objects, citing a confidentiality agreement with the physician not to release trade secrets, and even the physician's hospital and affiliated university medical school object, what should the physician do? Continue to gather more scientific evidence? Provide the information to colleagues on an informal and ad hoc basis? Alert the workers and advise them to take action on their own? Take the personal and professional risk and report the findings anyway? (To find out what can happen in the real world of occupational medicine, see refs. 6, 13, 14, and 15.)

Worker Screening and Medical Examinations for Third Parties

Employers frequently ask physicians and nurses to perform medical examinations as part of various worker screening programs, for both predictive and preventive purposes. Preplacement examinations and medical surveillance programs are two examples—the latter possibly offered to comply with specific OSHA standards. Employers and insurers may also request examinations to assess work-relatedness, measure impairment or functional capacity, determine fitness for work, and make recommendations on return-to-work and job accommodation needs. Physicians have insisted and courts have agreed that no physician-patient relationship exists when a person is being examined on behalf of an employer or for another third party (16). This lack of a physician-patient relationship dilutes the legal duty of care a physician has for the patient and may relieve the physician of other traditional legal obligations in medical practice. In this situation, attention to the ethical aspects of these activities is of special importance.

Issues of privacy, confidentiality, fairness, informed consent or refusal, and professional competence and responsibility pervade almost every form of worker screening, as well as medical examinations required by third parties for other reasons. Ethical concerns may relate to the purpose and content of examination or screening programs or to the use of the results generated by these activities.

It is helpful to ascertain why the employer wants the workers screened or examined: To help ensure that the worker can do the job without injury to self or others? To comply with government regulations? To help evaluate the effectiveness of workplace controls? (For example, the employer may want to know whether the ventilation and respiratory protection programs are adequately protecting workers from chemical exposure.) Does the employer want to weed out applicants who may pose a future liability risk? To find medical reasons to justify the removal of a troublemaker or a frequently absent employee? Who is the intended beneficiary of the medical examination or screening program?

Decisions about what the screening program or examination should include can also be problematic. Sometimes these examinations and their content are mandated by OSHA regulations; sometimes the company has its own ideas about what the examination should include, such as back x-ray studies, strength testing, and drug screening. Although it is certainly within the employer's rights to require preplacement, random, or for-cause drug testing, ideally it is the health care provider who defines the clinical content of the medical examination, based on knowledge of the job requirements, workplace exposures, conditions, and risks and the diagnostic and preventive value of medical testing. Many worker screening programs are ill-conceived from both a scientific and an ethical point of view. Problems with test validity (sensitivity and specificity) and predictive value may weaken any appeal to beneficence. For example, with their low predictive value in forecasting future back injury, the use of lumbosacral x-ray studies to screen out persons with back problems provides no real benefit to the company and exposes the worker to unnecessary radiation, as well as the risk of job loss. The use of cardiovascular stress testing for healthy young adults applying for jobs in the hazardous waste industry is another questionable practice.

In all cases, the physician or nurse should be aware that a worker's participation in a screening program or consent to a medical examination does not necessarily reflect an autonomous and voluntary decision. Individuals may consent to these examinations simply because they need a job. During the examination, they may knowingly or unknowingly (through testing) divulge highly personal and sensitive information to the health care provider. They may not know how or even whether such information will be passed on to others and used to their benefit or detriment.

In case 8, the employer wants to dictate the contents of the preplacement and other periodic screening evaluations to a community physician, who has agreed to provide these services although she has never visited the plant. Can physicians exercise their professional and ethical responsibility to practice competently when, in making judgments about a worker's ability to perform a job, they have neither seen the job nor are well acquainted with its demands? Even if the physician understands the nature of the job, how will she balance fairness to the employer with protection of the prospective or current employee? Will the workers be informed that the results of the medical evaluation may adversely affect their employability, remuneration, or advancement in the company? Will they be informed of their test results, counseled about their meaning, or referred elsewhere for this type of follow-up? Will the results of medical evaluations be used to help improve workplace conditions or simply to weed out "unfit" employees? If the screening program is the employer's sole approach to controlling workplace illness and injury, health care providers should weigh their involvement very carefully.

These factors place stringent ethical obligations on physicians and nurses who participate in worker screening programs. They must decide whether the purpose and content of the proposed screening program is medically reasonable and ethically justifiable. They must decide how much information employers need and are entitled to. The concepts of nonmaleficence, beneficence, fairness, and professional competence and responsibility must also be considered in decisions about the use of screening information and the need for follow-up action. Any action or inaction may adversely affect the employer, the worker, and, possibly, the provider's own standing with the company.

In considering these and other issues related to worker screening, one group of authors suggested that such testing should be used only if (a) it is an appropriate preventive tool that addresses a specific workplace problem; (b) it is used in conjunction with environmental monitoring; (c) it is not used to divert attention and resources from reducing worker exposure to toxic substances and improving workplace conditions; (d) the tests are accurate, reliable, and have a high predictive value in the population screened; and (e) medical removal protection for earnings and job security is provided (17).

More recently, a bill of rights for persons who are subject to medical examinations at the direction of their employers has been proposed (16). This proposal suggests that each examinee should have the right (a) to be told the purpose and scope of the examination; (b) to be told for whom the physician works; (c) to provide informed consent for all procedures; (d) to be told how results will be conveyed to the employer; (e) to be told about confidentiality protection; (f) to be told how to obtain access to the medical information in the worker's file; and (g) to be referred for medical follow-up, if necessary.

Because workers involved in medical screening or medical examinations at the behest of third parties have few legal protections, occupational health professionals must take special care when engaging in these activities.

Workplace Health Promotion

In addition to the types of screening activities already mentioned, employers may ask health professionals to develop and deliver a variety of wellness or health promotion programs in the workplace. Moreover, health care professionals who work for a company may be approached by outside groups that sell these types of programs. Employers are attracted to wellness programs because of rising health care costs, concerns about declining productivity, and the recognition that behavioral risk factors have a significant impact on worker health and medical care costs.

In many ways, the workplace is an ideal site for such intervention programs. It is a place where large numbers of relatively healthy people spend a significant amount of time. The workplace is a potential locus for the exercise of a variety of social controls that may influence personal behavior, such as nonsmoking and vending-machine policies, contests and financial incentives, peer pressure, and accessible information and services. By helping workers change their unhealthful habits, these health promotion programs are buttressed by the principle of beneficence. Yet because they have the potential for misuse, they can present ethical challenges to the unwary health professional (18,19).

For example, there is concern that a focus on the lifestyle risks of the individual worker diverts attention away from the workplace risks that are under the direct control of the employer. This appears to be happening in case 6; the employer has ignored recommendations for ergonomic changes in the workplace and has decided to institute weight-reduction and safe-lifting classes for the workers instead. Is this the best way of preventing musculoskeletal injury in this workplace? Is it fair to place the sole burden on the workers? Should the nurse agree to conduct these health promotion programs in the absence of workplace ergonomic changes? Should the nurse encourage workers to participate in these wellness programs without also informing them of their substantial job-related risks and the actions the employer could take to reduce or eliminate them?

The debate has been framed as one between health promotion and health protection—or between the behavioralist approach and the environmentalist approach to public health. The priority and emphasis given these approaches by occupational health professionals may reveal something of their primary allegiance or, perhaps, their philosophy of health care. If the professional's primary role and expertise relates to *work-related* injury and illness, then perhaps a focus on behavioral risk factors should be a second-order concern.

This brief discussion is not meant to disparage workplace wellness programs. They can help address important public health problems and can be beneficial to and well received by both workers and employers. However, in the absence of effective health and safety programs targeted on workplace hazards, they should be subject to close ethical scrutiny.

The Business of Providing Occupational Health Services

The United States lags behind in requiring or providing occupational health services for its workers. Most European countries have legal, statutory requirements for the provision of these services. Of growing concern, however, is the increasingly competitive environment within which occupational health services must be delivered. Competition for contracts and business among external, community-based occupational health services and competition for resources within companies have raised challenges associated with business ethics for health care providers (20). These include issues relating to advertising and marketing; provider expertise, competence, and qualifications; commitment beyond contract periods; handing over records and responsibilities to new contractors; complying with regulations; and supervising trainees who help provide services. It has been suggested that there has been a shift from a professional ethic to a business orientation in the provision of medical care and that tension between the forces of business and professional practice gives rise to serious ethical conflicts. One study found widespread agreement on the need for a code of business ethics for occupational health services (20) (see Chapter 14).

A popular axiom for how to succeed in a competitive business environment is to provide client-focused, customer-friendly products and services that are responsive to customer needs. This approach may be problematic for the ethical provision of occupational health services in several ways. The customer (most often an employer) may have no appreciation for what is needed or what an occupational health service can provide. Activities and actions that are necessary to keep the company happy and retain its business may conflict with a provider's ethical duties or even the customer's (or provider's) legal obligations (e.g., recording or reporting cases of work-related injury or illness). The status of workers or patients and the provider's relationship and responsibility

to them is murky, at best. Competition on the basis of price or cost may limit the array or dilute the intensity of services that are needed to prevent work-related illness and injury. The movement of managed care organizations and techniques into the arena of workers' compensation medical care in the United States is another area that may present ethical challenges worthy of attention.

PROBLEM SOLVING AND DECISION MAKING: GUIDELINES, CODES, AND CONSULTATION

Codes of ethics are vehicles for articulating core values, establishing standards of ethical care and practice, and providing guidelines for conduct. Professional organizations of occupational and environmental health specialists have acknowledged the need for guidance in the face of ethical problems and have produced codes of ethics for their disciplines and members. In the United States, codes have been developed and adopted by the American College of Occupational and Environmental Medicine (ACOEM); the American Association of Occupational Health Nurses (AAOHN); four industrial hygiene associations (the American Conference of Governmental Industrial Hygienists, the American Board of Industrial Hygiene, the American Industrial Hygiene Association, and the American Academy of Industrial Hygiene); and the American Society of Safety Engineers (see appendix to this chapter). Professional groups in other countries have done the same. For example, in the United Kingdom, the Faculty of Occupational Medicine has issued *Guidance on Ethics for Occupational Physicians.* The International Commission on Occupational Health has adopted an international and interdisciplinary code of ethics for occupational health professionals. One U.S. group, the Association of Occupational and Environmental Clinics, has expressed its clear preference for the international code over the code promulgated by the largest professional organization of occupational and environmental medicine physicians in the United States. There is no paucity of written guidance or debate on the subject (see Bibliography).

Although these ethical codes address many of the same issues and problems, their articulation of principles varies in clarity, depth, emphasis, strength, and directness. The codes vary, for example, in what they say about the professional's primary purpose and what should be done when the needs, demands, or expectations of employers and workers conflict. The ACOEM code for occupational physicians directs them to "*accord the highest priority to the health and safety of individuals* in both the workplace and the environment." The hygiene code advises hygienists to "practice their profession following recognized scientific principles with the realization that the lives, health and well-being of people may depend on their professional judgment and that they are obligated *to protect the health and well-being of people.*" The nursing code enjoins nurses to "provide health care in the work environment with regard to human dignity and client rights . . . and promote collaboration with other health professionals and community health agencies in order *to meet the health needs of the workforce.*" The safety engineering code calls on members to "*Hold paramount the protection of people and property and the environment.*" The international code is perhaps the most straightforward; it states that "*the primary aim of occupational health practice is to safeguard the health of workers and to promote a safe and healthy working environment.*" (Italics were added in all instances.) A comparison of the provisions of various codes can be found in refs. 21 and 22.

Although they often are vague, these codes can be helpful. They could be appended to any contract or agreement that an occupational or environmental health professional enters into with a company or other organization. Ethical codes, however, cannot solve the moral dilemmas that arise in the day-to-day practice of occupational and environmental health. Studies suggest that occupational health professionals are often unaware of published guidelines and codes, or, if aware, seldom consult them (23,24). Fur-

ther, the codes and guidelines do not provide the protection that health professionals may need when they take action that is contrary to the wishes of their employer or other powerful interests. In the final analysis, these professionals must make their own decisions and live with the consequences.

Beyond codes, there are other concepts and methods that may help occupational health professionals in approaching ethical dilemmas. Light and McGee (1) suggested that insights can be gained from an examination of the social construction of the moral or ethical issue at hand, as well as consideration of the vested interests; the type of action involved (starting, stopping, continuing, or abstaining); the social, political, economic, and institutional contexts; the degree of volition; and the potential for various types of harms. For Philipp et al. (20), the tenets of professionalism are closely linked to ethics and can help guide conduct. These tenets include competence, a sense of dedication and purpose, responsibility, autonomy, accountability, a willingness to collaborate and work effectively with others, adherence to an ethical code, and conducting one's practice with personal integrity and for the public health. Weed (25) urged public health professionals to examine and proclaim their own philosophic commitments or values in the interest of making better choices. At the same time, he acknowledged that the examination and disclosure of economic interests, political ideologies, and other social forces may also be needed.

A clarification of legal responsibilities may be a helpful first step in trying to resolve ethical problems, but compliance with legal obligations alone may be insufficient. In many cases, consultation and deliberation with others can help clarify the underlying ethical dimensions of the conflict, help define a basis for decision making, suggest or help weigh ethical criteria and justifications for various courses of action, and share the moral responsibility for decision making. Questions that may stimulate reflection and discussion include the following: What makes this an ethical problem? Why does the health professional see it that way? Who will benefit or be harmed by each alternative? Who is the least advantaged or most affected person or group in this situation? Are the needs and preferences of this party known, and have they been given the appropriate weight? What are the long-term consequences of each action? What will happen if a particular action is not taken? What does the professional stand to lose or gain from each possible alternative? Does the professional feel pressured, or can independent judgment be exercised? Would the professional make a different decision or determination if the issue arose while practicing in another setting?

CONCLUSION

The fields of occupational and environmental health are charged with ethical problems and dilemmas that are not easy to resolve. Consideration of widely held moral principles—such as autonomy, respect for persons, nonmaleficence, beneficence, justice, responsibility, and the integrity of personal and fiduciary relationships—can help guide decision making. Honest and careful assessment of the social, economic, political, and institutional contexts of the problem may provide insight, as may meaningful deliberation and discussion with others. Although these steps can help point the way, they do not necessarily make hard choices any easier in the practical sense. Advances in medicine, science, and technology presage a growing number of complex ethical problems, as does the increasingly competitive market for health care, including occupational health services. The need for structural safeguards for occupational and environmental health professionals has never been more clear. Unless these professionals can somehow be insulated from personal and economic reprisals, it will remain difficult for them to make the bold decisions needed to ensure worker and community environmental health. The development of structural safeguards will require creativity and, most likely, legislative action. Constructive solutions also demand

honest dialogue and a clear understanding by all parties of how their actions and expectations can contribute to both the creation and the resolution of ethical issues in occupational and environmental health.

REFERENCES

1. Light D, McGee G. On the social embeddedness of bioethics. In: DeVries R, Subedi J, eds. Bioethics and society: constructing the ethical enterprise, 4th ed. Upper Saddle River, NJ: Prentice Hall, 1998:1–15.
2. Wolpe P. The triumph of autonomy in American bioethics: a sociological view. In: DeVries R, Subedi J, eds. Bioethics and society: constructing the ethical enterprise, 4th ed. Upper Saddle River, NJ: Prentice Hall, 1998:38–59.
3. Ladd J. The task of ethics. In: Reich WT, ed. Encyclopedia of bioethics. New York: The Free Press, 1978.
4. Soskolne CL. Epidemiology: questions of science, ethics, morality, and law. Am J Epidemiol 1989; 129:1–18.
5. Egilman DS, Hom C. Corruption of the medical literature: a second visit. Am J Ind Med 1998; 34:401–404.
6. Shuchman M. Secrecy in science: the flock worker's lung investigation. Ann Intern Med 1998;129: 341–344.
7. Ozonoff D, Boden L. Truth and consequences: health agency responses to environmental health problems. Sci Technol Hum Values 1987;12:70–77.
8. Bullard R. Dumping in Dixie: race, class and environmental quality. Boulder, CO: Westview Press, 1990.
9. Fagin D, Lavelle M. Center for Public Integrity. Toxic deception: how the chemical industry manipulates science, bends the law, and endangers your health, 2nd ed. Monroe, ME: Common Courage Press, 1999.
10. Robinson JC. Toil and toxics: workplace struggles and political strategies for occupational health. Berkeley: University of California Press, 1991.
11. McCallum DB, Covello VT. What the public thinks about environmental data. EPA J 1990;113:467–473.
12. Larsen S. Hidden camera triggers suit: IBP nurses fired after finding video equipment. The Dispatch and The Rock Island Argus. November 27, 1997:A1–A2.
13. Kern D, Crausman RS, Durand KTH, Nayer A, Kuhn C. Flock worker's lung: chronic interstitial lung disease in the nylon flocking industry. Ann Intern Med 1998;129:261–272.
14. Kern D. The unexpected result of an investigation of an outbreak of occupational lung disease. Int J Occup Med Environ Health 1998;4:19–32.
15. Davidoff F. New disease, old story. Ann Intern Med 1998;129:327–328
16. Rothstein MA. Legal and medical aspects of medical screening. Occup Med: State of the Art Reviews 1996;11:31–39.
17. Ashford NA, Spadafor CJ, Hattis DB, Caldart CC. Monitoring the worker for exposure and disease: sci-

entific, legal, and ethical considerations in the use of biomarkers. Baltimore: Johns Hopkins Press, 1990.

18. Walsh DC, Jennings SE, Mangione T, Merrigan DM. Health promotion versus health protection: employees perceptions and concerns. J Public Health Policy 1991;12:148–164.
19. Allegrante JR, Sloan RP. Ethical dilemmas in workplace health promotion. Prev Med 1986;15:313–320.
20. Philipp R, Goodman G, Harling K, Beattie B. Study of business ethics in occupational medicine. Occup Environ Med 1997;54:351–356.
21. Brodkin CA, Frumkin H, Kirkland KH, Orris P, Schenk M. AOEC position paper on the organizational code for ethical conduct. J Occup Environ Med 1996;38:869–881.
22. Rothstein MA. A proposed revision of the ACOEM code of ethics. J Occup Environ Med 1997; 39:616–622.
23. Aw TC. Ethical issues in occupational medicine practice: knowledge and attitudes of occupational physicians. Occup Med 1997;47:371–376.
24. Martimo KP, Antti-Poika M, Leino T, Rossi K. Ethical issues among Finnish occupational physicians and nurses. Occup Med 1998;48:375–380.
25. Weed D. Toward a philosophy of public health. J Epidemiol Community Health 1999;53:99–104.

BIBLIOGRAPHY

ACOEM Committee on Ethical Practice in Occupational Medicine. Commentaries on the code of ethical conduct: I. J Occup Environ Med 1995;37:201–206.

Brodkin CA, Frumkin H, Kirkland KH, Orris P, Schenk M. AOEC position paper on the organizational code for ethical conduct. J Occup Environ Med 1996; 38:869–881.

Brodkin CA, Frumkin H, Kirkland KH, et al. Choosing a professional code of ethical conduct in occupational and environmental medicine. J Occup Environ Med 1998;40:840–842.

Goodman KW. Codes of ethics in occupational and environmental health. J Occup Environ Med 1996; 38:882–883.

Rothstein MA. A proposed revision of the ACOEM code of ethics. J Occup Environ Med 1997; 39:616–622.

Teichman R, Wester MS. The new ACOEM code of ethical conduct. J Occup Environ Med 1994;36:27–30.

Teichman RF. ACOEM code of ethical conduct. J Occup Environ Med 1997;39:614–615.

These seven articles cover the debate about the ACOEM code of ethics for occupational health physicians.

Ashford NA, Spadafor, CJ, Hattis DB, Caldart CC. Monitoring the worker for exposure and disease: scientific, legal, and ethical considerations in the use of biomarkers. Baltimore: Johns Hopkins Press, 1990.

Koh D, Jeyaratnam J. Biomarkers, screening and ethics. Occup Med 1998;48:27–30.

Rothstein MA. Legal and medical aspects of medical screening. Occup Med 1996;11:31–39.

These three sources provide information on ethical issues in worker screening, genetic and otherwise.

Beauchamp TL, Childress JF. Principles of biomedical ethics. New York: Oxford University Press, 1994.

DeVries R, Subedi J. Bioethics and society: constructing the ethical enterprise. Upper Saddle River, NJ: Prentice Hall, 1998.

These two texts provide a general discussion of bioethics.

Bullard R. Dumping in Dixie: race, class and environmental quality. Boulder, CO: Westview Press, 1990.

Fagin D, Lavelle M, Center for Public Integrity. Toxic deception: how the chemical industry manipulates science, bends the law, and endangers your health, 2nd ed., Monroe, ME: Common Courage Press, 1999.

Robinson JC. Toil and toxics: workplace struggles and political strategies for occupational health. Berkeley: University of California Press, 1991.

These three books present accounts of historical failures described in this chapter.

Kern D, Crausman RS, Durand KTH, Nayer A, Kuhn C. Flock worker's lung: chronic interstitial lung disease in the nylon flocking industry. Ann Intern Med 1998;129:261–272.

Kern D. The unexpected result of an investigation of an outbreak of occupational lung disease. Int J Occup Med Environ Health 1998;4:19–32.

Shuchman M. Secrecy in science: the flock worker's lung investigation. Ann Intern Med 1998;129:341–344.

Davidoff F. New disease, old story. Ann Intern Med 1998;129:327–328.

These four articles describe a real-world example of ethical issues in occupational health.

McCunney RJ. Preserving confidentiality in occupational medical practice. Am Fam Physician 1996; 53:1751–1756.

Rischitelli DG. Licensing, practice, and malpractice in occupational medicine. Occup Med 1996;11:121–135.

Rothstein MA. Legal and medical aspects of medical screening. Occup Med 1996;11:31–39.

Tilton SH. Right to privacy and confidentiality of medical records. Occup Med 1996;11:17–29.

These three articles provide information about legal and ethical aspects of confidentiality of medical records in occupational health.

APPENDIX
Codes of Ethical Conduct for Major Occupational Health and Safety Professional Organizations in the United States

AMERICAN COLLEGE OF OCCUPATIONAL AND ENVIRONMENTAL MEDICINE CODE OF ETHICAL CONDUCT

Adopted October 25, 1993 by the Board of Directors of the American College of Occupational and Environmental Medicine (ACOEM).

This code establishes standards of professional ethical conduct with which each member of the ACOEM is expected to comply. These standards are intended to guide occupational and environmental medicine physicians in their relationships with the individuals they serve; employers and workers representatives; colleagues in the health professions; the public; and all levels of government, including the judiciary.

Physicians should:

1. Accord the highest priority to the health and safety of individuals in both the workplace and the environment.
2. Practice on a scientific basis with integrity, and strive to acquire and maintain adequate knowledge and expertise on which to render professional service.
3. Relate honestly and ethically in all professional relationships.
4. Strive to expand and disseminate medical knowledge and participate in ethical research efforts as appropriate.
5. Keep confidential all individual medical information, releasing such information only when required by law or overriding public health considerations, or to other physicians according to accepted medical practice, or to others at the request of the individual.
6. Recognize that employers may be entitled to counsel about an individual's medical work fitness but not to diagnoses or specific details, except in compliance with laws and regulations.
7. Communicate to individuals and/or safety groups any significant observations and recommendations concerning their health and safety.
8. Recognize those medical impairments in oneself and others, including chemical dependency and abusive personal practices, which interfere with one's ability to follow the above principles, and take appropriate measures.

(From Teichman R, Wester MS. The new ACOEM code of ethical conduct. J Occup Environ Med 1994;36:27–30.)

AMERICAN ASSOCIATION OF OCCUPATIONAL HEALTH NURSES CODE OF ETHICS AND INTERPRETIVE STATEMENTS (EXCERPTS)

Preamble

The AAOHN Code of Ethics has been developed in response to the nursing profession's acceptance of its goals and values, and the trust conferred on it by society to guide the conduct and practices of the profession. As a professional, the occupational health nurse accepts the responsibility and inherent obligation to uphold these values.

1. The occupational health nurse provides health care in the work environment with regard for human dignity and client rights, unrestricted by considerations of social or economic status, national origin, race, religion, age, sex, or the nature of the health status.
2. The occupational health nurse promotes collaboration with other health professionals and community health agencies in order to meet the health needs of the workforce.
3. The occupational health nurse strives to safeguard the employee's right to privacy by protecting confidential information and releasing information only on written consent of the employee or as required or permitted by law.
4. The occupational health nurse strives to provide quality care and to safeguard clients from unethical and illegal actions.
5. The occupational health nurse, licensed to provide health care services, accepts obligations to society as a professional and responsible member of the community.
6. The occupational health nurse maintains individual competence in occupational health nursing practice, recognizing and accepting responsibility for individual judgments and actions, while complying with appropriate laws and regulations (local, state, and federal) that impact the delivery of occupational health services.

7. The occupational health nurse participates, as appropriate, in activities such as research that contribute to the ongoing development of the profession's body of knowledge while protecting the rights of subjects.

(From Am Assoc Occup Health Nurses J, August 1991.)

CODE OF ETHICS FOR THE PRACTICE OF INDUSTRIAL HYGIENE (INTERPRETATIVE GUIDELINES NOT INCLUDED)

Objective

These canons provide standards of ethical conduct for industrial hygienists as they practice their profession and exercise their primary mission, to protect the health and well-being of working people and the public from chemical, microbiological and physical health hazards present at, or emanating from, the workplace.

Canons of Ethical Conduct and Interpretive Guidelines

Industrial Hygienists shall:

1. Practice their profession following recognized scientific principles with the realization that the lives, health and well-being of people may depend on their professional judgment and that they are obligated to protect the health and well-being of people.
2. Counsel affected parties factually regarding potential health risks and precautions necessary to avoid adverse health effects.
3. Keep confidential personal and business information obtained during the exercise of industrial hygiene activities, except when required by law or overriding health and safety considerations.

Professional Code of Ethics

4. Avoid circumstances where a compromise of professional judgment or conflict of interest may arise.

5. Perform services only in the areas of their competence.
6. Act responsibly to uphold the integrity of the profession.

(Developed jointly by the American Industrial Hygiene Association, the American Conference of Governmental Industrial Hygienists, the American Board of Industrial Hygiene, and the American Academy of Industrial Hygiene. From: http.//www. aiha.org/ethics.html.

AMERICAN SOCIETY OF SAFETY ENGINEERS CODE OF PROFESSIONAL CONDUCT

Fundamental Principles

As a member of the American Society of Safety Engineers, I recognize that my work has an impact on the protection of people, property and the environment.

In order to assume professional responsibility, I shall uphold and advance the integrity, honor and dignity of the safety, health and environmental profession by:

1. Enhancing protection of people, property and the environment through knowledge and skill.
2. Being honest, impartial and serving the public, employers and clients with fidelity.
3. Striving to increase my competence in and the prestige of the safety profession.
4. Avoiding circumstances where compromise of the professional conduct or conflict of interest may arise.

Fundamental Canons

In the fulfillment of my duties as a safety profession, I shall:

1. Hold paramount the protection of people and property and the environment.
2. Advise employers, clients, employees or appropriate authorities when my professional judgment indicates that the protection of people, property or the environment is unacceptably at risk.

3. Strive for continuous self-development while participating in the safety profession.
4. Perform professional services only in the area of my competence.
5. Issue public statements only in an objective and truthful manner and in accordance with the authority bestowed on me.
6. Act in professional matters as faithful agent or trustee and avoid conflict of interest.
7. Build my professional reputation on merit of service.
8. Ensure equal opportunities for individuals under my supervision.

Approved by the Assembly, June 1993.

(From the American Society of Safety Engineers web site. Available at: http://www. asse.org/hcode.htm.)

INTERNATIONAL CODE OF ETHICS FOR OCCUPATIONAL HEALTH PROFESSIONALS (EXCERPTS)

Basic Principles

The three following paragraphs summarize the principles of ethics on which is based the International Code of Ethics for Occupational Health Professionals, prepared by the International Commission on Occupational Health.

Occupational health practice must be performed according to the highest standards and ethical principles. Occupational health professionals must service the health and social well-being of the workers, individually and collectively. They also contribute to environmental and community health.

The obligations of occupational health professionals include protecting the life and the health of the worker, respecting human dignity and promoting the highest ethical principles in occupational health policies and programs. Integrity in professional conduct, impartiality, and the protection of the confidentiality of health data and of the privacy of workers are part of these obligations.

Occupational health professionals are ex-

perts who must enjoy full professional independence in the execution of their functions. They must acquire and maintain the competence necessary for their duties and require conditions which allow them to carry out their tasks according to good practice and professional ethics.

Duties and Obligations of Occupational Health Professionals

1. The primary aim of occupational health practice is to safeguard the health of workers and to promote a safe and healthy working environment. In pursuing this aim, occupational health professionals must use validated methods of risk evaluations, propose efficient preventive measures, and follow up their implementation . . . .
2. Occupational health professionals must continuously strive to be familiar with the work and the working environment as well as to improve their competence and to remain well informed in scientific and technical knowledge, occupational hazards and the most efficient means to eliminate or reduce the relevant risks . . . .
4. Special consideration should be given to rapid application of simple preventive measures which are cost-effective, technically sound and easily implemented. When doubt exists about the severity of an occupational hazard, prudent precautionary action should be taken immediately.
5. In the case of refusal or of unwillingness to take adequate steps to remove an undue risk or to remedy a situation which presents evidence of danger to health or safety, the occupational health professionals must make, as rapidly as possible, their concern clear, in writing, to the appropriate senior management executive, stressing the need for taking into account scientific knowledge and for applying relevant health protection standards, including exposure limits, and recalling the

obligation of the employer to apply laws and regulations and to protect the health of workers in their employment. Whenever necessary, the workers concerned and their representatives in the enterprise should be informed and the competent authority should be contacted.

6. Occupational health professionals must contribute to the information of workers on occupational hazards to which they may be exposed in an objective and prudent manner which does not conceal any fact and emphasizes the preventive measures . . . .

8. The objectives and the details of the health surveillance must be clearly defined and the workers must be informed about them. The validity of such surveillance must be assessed and it must be carried out with the informed consent of the workers . . . . The potentially positive and negative consequences of participation in screening and health surveillance programs should be discussed with the workers concerned.

9. The results of examinations . . . must be explained to the worker concerned. The determination of fitness for a given job should be based on the assessment of the health of the worker and on a good knowledge of the job demands and the worksite . . . .

10. The results of the examinations prescribed by national laws or regulations must only be conveyed to management in terms of fitness for the envisaged work or of limitations necessary from a medical point of view in the assignment of tasks or in the exposure to occupational hazards . . . .

14. Occupational professionals must be aware of their role in relation to the protection of the community and of the environment . . . .

Conditions of Execution of the Functions of Occupational Health Professionals

16. Occupational health professionals must always act, as a matter of priority, in the interest of the health and safety of the workers . . . .

17. Occupational health professionals must maintain full professional independence and observe the rules of confidentiality in the execution of their functions.

18. All workers should be treated in an equitable manner without any form of discrimination . . . . A clear channel of communication must be established and maintained between occupational health professionals and the senior management executive responsible for decisions at the highest level about the conditions and the organization of work and the working environment in the undertaking, or with the board of directors.

19. Whenever appropriate, occupational health professionals must request that a clause on ethics be incorporated into their contract of employment . . . .

21. Occupational health professionals must not seek personal information which is not relevant to the protection of workers' health in relation to work . . . .

(From International Commission on Occupational Health, 1992.)

Occupational Health: Recognizing and Preventing Work-Related Disease and Injury, 4th ed.
Edited by Barry S. Levy and David H. Wegman, Lippincott Williams & Wilkins, Philadelphia © 2000.

14 Occupational Health Services

Scott D. Deitchman

Occupational health services are those health care services provided to prevent, diagnose, or treat work-related injuries or illnesses or to promote rehabilitation from such an injury or illness. Other sections of this textbook provide information on specific work-related injuries and illnesses. This chapter describes the systems in which these ailments are cared for, including descriptions of the health care providers, the environment in which they work, and the various health care services they provide.

OCCUPATIONAL HEALTH CARE PROVIDERS

There is great diversity among providers of occupational health services. Although occupational health services are often provided by physicians, their training and backgrounds vary. Some physicians who designate their practice as occupational medicine have trained in occupational medicine residencies. Most occupational medicine physicians have trained in other specialties, although some also have passed the occupational medicine certifying examination offered by the American Board of Preventive Medicine. Of 7,018 current and retired members of the American College of Occupational and Environmental Medicine, only 1,486 (21%) are board certified in occupa-

tional medicine (Lanny Hardy, American College of Occupational and Environmental Medicine, personal communication, February 12, 1999).

Many physicians in other specialties also see patients with occupational illnesses. For example, patients with work-related musculoskeletal disease may be cared for by orthopedic surgeons or specialists in physical medicine and rehabilitation, and patients with occupational asthma may be referred to pulmonologists or allergists. Most cases of work-related injury or illness are seen in the offices of primary care physicians or in emergency departments; in these settings, however, the association between disease and work-related causes is often not recognized (1). This chapter focuses primarily on practice settings that are designated as occupational health clinics or are staffed by physicians and other providers who identify their primary practice as occupational medicine.

Occupational health care is also provided by occupational health nurses. Although not all nurses working in occupational health are specialty-trained, available training curricula include master's degrees in occupational health nursing. Certified occupational health nurses have completed requisite work experience and educational training and have passed a credentialing examination (2). Some occupational health nurses work in physician-staffed occupational health practices; others manage the routine functions of an occupational health clinic and refer cases needed medical management to consulting

S. D. Deitchman: National Institute for Occupational Safety and Health, Centers for Disease Control and Prevention, Atlanta, GA 30333.

physicians or serve as case managers in managed care organizations and insurance companies.

Working under the supervision of physicians, physician assistants provide occupational health care in many of the same settings as occupational health physicians and nurses. Approximately 3% of physician assistants in the United States work in occupational and environmental health services (Maryann Ramos, PA-C, MPH, American Academy of Physician Assistants in Occupational Medicine, personal communication, March 11, 1999). Other health professionals providing occupational health services include occupational therapists and occupational psychologists.

SETTINGS IN WHICH OCCUPATIONAL HEALTH SERVICES ARE PROVIDED

The diversity of occupational health practitioners is matched by the diversity of settings in which occupational health care is provided. For many years, it was common to find in large industrial plants a medical clinic staffed by both physicians and nurses. These were typically owned and operated by the corporation, and medical staff members, like the production workers they cared for, were employees of the corporation. Although many of these clinics still exist, other options have appeared. Some large plants maintain on-site clinics but have contracted to outside health care providers (e.g., hospitals, managed care organizations) to operate and staff them. In these plants, clinic staff members are not employees of the plant.

Other corporations have reduced the scope of practice of, or completely eliminated, the traditional plant medical clinic. In this regard, they have become more like smaller employers, many of whom never had an on-site clinic because they could not afford one or because the number of employees was too small to justify it. Some of these employers contract with off-site clinics to provide their occupational health services.

These clinics may be operated by and located at area hospitals; they may be satellite clinics operated by hospitals; or they may be free-standing clinics. In some cases, an employer makes a formal or informal arrangement with a physician in the area to see workers on an as-needed basis. Finally, workers initially may present with work-related illnesses or injuries to their family doctors because that is their preference, because there are no occupational health care providers available, because their health plan requires them to see their primary care provider first, or because the work-relatedness of their ailment is not recognized. Although practitioners in settings removed from the workplace may be equally as skilled as those in plant medical clinics, this distance may leave them less familiar with the particular exposures or hazards faced by their patients unless they make deliberate efforts to understand the workplace involved.

SCOPE OF OCCUPATIONAL HEALTH SERVICES

Occupational health services can be divided into four general categories: *preventive, curative, rehabilitative,* and *consultative.* Preventive health services are those that are intended to prevent the development of illness or injury, or to prevent or retard further progression of illness or injury that has already occurred. Preventive services are further divided into primary, secondary, and tertiary preventive services. As described in the following sections, many curative and rehabilitative services also have preventive functions and can be ranked within the hierarchy of preventive health services. In preventive, curative, and rehabilitative services, a physician-patient relationship exists in which the physician is working to ensure the patient's best interests and health, regardless of the source of the physician's compensation for these services. In consultative services, the physician is responsible for providing an accurate and informed judgment but may not be involved in the patient's care.

Primary Preventive Services

Primary preventive services are those intended to prevent illness or injury. Because occupational medicine is one of the preventive medical specialties, primary preventive services distinguish occupational health from other types of health care that address only curative services. Primary prevention is accomplished by eliminating hazardous exposures, protecting workers against remaining exposures, or protecting workers against the effects of those exposures.

Some primary preventive services are activities familiar to most providers of clinical care. They include immunizing workers against possible work-related infections, such as hepatitis B immunization for health care workers or rabies immunization for veterinarians and animal control workers. More often, primary preventive services involve modifying work environments to eliminate or contain hazards or supplying personal protective equipment to workers when hazardous exposures cannot otherwise be controlled. Although industrial hygienists, ergonomists, and safety specialists have the leading roles in these aspects of primary prevention (see Chapters 7 through 9), ideally occupational health care providers work closely with occupational hygienists and safety specialists to identify potential health hazards that require correction. The interaction between the various health and safety specialists is most clearly demonstrated in the linkage between primary, secondary, and tertiary prevention (see later discussion).

Secondary Preventive Services

Secondary prevention services are those intended to detect illness or injury at a relatively early stage, often before symptoms or clinical signs are noticed. When disease is detected at this early stage, it may be possible to take steps to arrest or reverse the disease process. Because the interventions are likely to be both clinical and workplace based, secondary prevention explicitly shows the need for occupational medicine clinicians to work with employers in a role that extends beyond their clinical role. For example, the physician may screen a worker at a battery manufacturing plant and discover that the worker has significantly elevated blood lead levels. The clinical response is to assess target organ function and determine whether chelation therapy is indicated. However, this case is a sentinel event that indicates excessive lead exposure in the workplace. In addition to providing clinical care, it is essential that the physician contact the employer to report the case. This information allows the employer to make workplace changes that can reduce or eliminate the hazardous exposures.

Secondary prevention usually addresses ailments that are not yet symptomatic, and these ailments typically are detected through screening examinations. Some screening examinations are required by the Occupational Safety and Health Administration (OSHA) for workers exposed to specific hazards, and OSHA standards regulating exposures to these substances often contain instructions regarding the medical examinations or tests and the frequency with which they must be performed. A list of substances for which OSHA has established standards for medical surveillance appears in Table 14-1. OSHA may also use the "general duty" clause of the Occupational Safety and Health Act to require medical surveillance for other occupational exposures (e.g., tuberculosis). Readers should contact OSHA for current requirements, recommendations, and interpretations (3). Other countries have different requirements for medical surveillance, and in some cases more specific guidance is provided regarding examinations, tests, recordkeeping, and other aspects of medical surveillance (4,5). Principles of medical surveillance and screening are presented in Chapter 4.

Tertiary Preventive Services—Curative and Rehabilitative

Tertiary preventive services are those provided after injury or illness has occurred or

TABLE 14-1. *OSHA standards mandating medical surveillance*

Substance	USA 29 CFR
Acrylonitrile	1910.1045(n)
2-Acetylaminofluorene	1910.1014[a]
4-Aminodiphenyl	1910.1011[a]
Arsenic	1910.1018(n)
Asbestos	1910.1001(l) and 1910.1001(m)(3)
Benzene	1910.1028(i)
Benzidine	1910.1010[a]
Bloodborne pathogens	1910.1030
1,3-Butadiene	1910.1051(k)
Cadmium	1910.1027(l)[b,c,d]
bis-Chloromethyl ether	1910.1008[a]
Coke oven emissions	1910.1029(j)[b]
Compressed air	1926.803(b)
Cotton dust	1910.1043(h)
1,2-Dibromochloropropane, 3-chloropropane	1910.1044(i)(5)
3,3'-Dichlorobenzidine (and its salts)	1910.1007[a]
4-Dimethylaminoazo-benzene	1910.1015[a]
Ethyleneimine	1910.1012[a]
Ethylene oxide	1910.1047(i)
Formaldehyde	1910.1048(l)
Hazardous waste and emergency response	1910.120(f)
Laboratories—hazardous chemical exposures	1910.1450(g)
Lead	1910.1025(j)[b]
Methyl chloromethyl ether	1910.1006[a]
4,4' Methylenedianiline (MDA)	1910.1050(m)[b]
Methylene chloride	1910.1052(j)
α-Naphthylamine	1910.1004[a]
β-Naphthylamine	1910.1009[a]
4-Nitrobiphenyl	1910.1003(g)
N-Nitrosodimethylamine	1910.1016[a]
Noise	1910.95(g)
β-Propiolactone	1910.1013[a]
Respiratory protection	1910.134(e)
Vinyl chloride	1910.1017(k)

[a] Medical surveillance regulations for these carcinogens have been grouped into 1910.1003(g).
[b] Also regulated in Contruction standards (29 CFR 1926).
[c] Also regulated in Maritime standards (29 CFR 1915).
[d] Also regulated in Agriculture standards (29 CFR 1928).
From: US Department of Labor, Occupational Safety and Health Administration website. Available at; http://www.osha-slc.gov/SLIC/medicalsurveillance/index.html. Accessed August 1999.

has become clinically apparent, and they are intended to prevent disability or further progression. Some tertiary services are curative, such as treatment for acute or chronic intoxication from a workplace exposure, orthopedic care for a fracture, or rehabilitation after surgery for a work-related injury. Tertiary preventive services include clinical care for occupational injuries and illnesses and are probably the most common occupational health services. Even clinical care, however, has aspects that extend beyond the clinic. Physicians who provide this care should be planning for the patient's ultimate return to work. This may require initial return to a modified or alternative job with reduced demands, with a graduated return to the original job (or as close to it as possible). It may also be necessary to modify the original job to correct ergonomic or other problems that would otherwise lead to reinjury or exacerbation of the ailment. The physician (or other clinician caring for the patient) should coordinate this process with an employer representative to ensure that needed workplace changes are made.

Other tertiary prevention strategies do not cure the illness but are intended to prevent its progression. Auditory screening in hearing conservation programs, for example, does not restore lost auditory acuity. Instead, the screening program identifies workers whose noise exposures must be reduced through control of noise emissions or use of hearing protection. These measures prevent further noise-related hearing loss but do not restore acuity that has already been lost.

Primary, Secondary, and Tertiary Prevention Interlinked

Many cases of occupational illness or injury indicate the presence of a particular hazard in the workplace. Lead intoxication in a working adult, for example, suggests occupational exposures to lead. Such diagnoses are termed *sentinel events* (6), and they illustrate that the three levels of prevention are interlinked in occupational health. If a worker is identified as having neurologic symptoms related to lead poisoning, chelation therapy can be given to reduce lead levels (tertiary prevention, since symptomatic disease is already present); other workers can be screened for possible subclinical lead intoxication (secondary prevention); and ventilation or respiratory protection can be used to reduce lead exposures for all workers (primary prevention).

Partners in Occupational Health Services—Liaison with the Workplace

Preventive health care can be practiced by the primary care clinician in the privacy of the office, with only the physician and the patient involved. Many of the issues discussed, such as diet, exercise, and tobacco use, are lifestyle risks that are under the patient's control. However, prevention in occupational medicine is a more complex matter. Occupational health and safety risks most often result from workplace hazards that are under the control of the employer, not the worker. There is a unique relation between occupational health services and the workplace because occupational health deals with the effects of workplace hazards. For this reason, occupational health services must address workplace risks. Strategies used by occupational medicine physicians may involve assessing workplace hazards; collaborating with primary prevention services, including occupational safety or occupational hygiene professionals; and reporting cases of illness to the employer or to regulatory agencies so that the hazards identified can be corrected.

Many clinical interventions are incomplete if they are not accompanied by needed workplace interventions. It would not be appropriate to use chelation therapy to reduce the body burden of lead in a lead-intoxicated worker and then send the worker back to an unchanged workplace for additional exposure. Similarly, a worker should not receive surgery and rehabilitation for carpal tunnel syndrome and then return to the same job with the same repetition rates and forces affecting the wrist. In both cases, the worker is at risk for reinjury, recurring illness, or exacerbation of existing disease.

Unlike purely clinical encounters, however, many of these preventive interventions are beyond the reach of the physician working alone. For this reason, clinicians treating occupational injuries or illnesses must work closely with other occupational health specialists. Cases of work-related disease should be reported to the employer, with appropriate caution to protect the worker's confidentiality and job security, so that causative workplace hazards can be controlled or eliminated. This may involve sharing information with occupational hygienists, safety specialists, ergonomists, or others qualified to assess and correct the identified hazard.

Conversely, the information provided by specialists in these related disciplines can provide valuable assistance to the occupational medicine physician. Industrial hygienists in a workplace can tell the physician which exposures are present, allowing the physician to recommend appropriate medical screening examinations. In addition, information on workplace exposures may be essential when considering a diagnosis of work-related disease or when selecting appropriate therapy after an acute exposure.

Occupational medicine specialists who are located at the workplace would seem to have the advantage in this situation, with more ready access to the workplace and its safety specialists. However, occupational medicine providers who are located elsewhere can include outreach to the workplace in their activities. This approach was used by a large managed care organization as it provided occupational health services under contract to employers (7).

Consultative Services

The services described previously are examples of care provided in a traditional physician-patient relationship, in which the physician is involved in the patient's welfare for an extended period. Some occupational health physicians also see patients for single or isolated visits in which they are not providing clinical care but instead are serving as medical consultants to the employer. One example of such a role is supervision of the testing of workers or job applicants for evidence of use of illegal drugs. The physician is paid by the employer to oversee the process of collecting the specimens (usually urine) and may be called on to use medical judgment in interpreting an uncertain result or ruling out other causes of a positive test, such as use of prescription medications. Although the physician has a duty to the patient to provide this service according to standards of confidentiality and scientific validity, no other relationship with the patient (e.g., medical treatment of addiction) is implied. Physicians performing this service are called *medical review officers.*

Another consulting role is that of the independent medical examiner. When a worker files for workers' compensation benefits, the state workers' compensation board or attorneys for the worker or the employer may request an independent examination of the patient to assess the degree of medical impairment (see Chapters 11 and 12). This examination is provided by a physician who is not otherwise connected with the case. The independent medical examiner takes a history from the patient and performs an examination to evaluate impairment. The physician reports this assessment to the requesting entity but has no further relationship with the patient. Other services that may be provided on a consultative basis include OSHA-mandated medical functions such as medical surveillance and clearance for workers who must wear respirators.

Health Risk Appraisal and Health Promotion

One other activity, health promotion (which can include health risk appraisal), often falls in the purview of the occupational health program, although it does not address work-related disease. Health risk appraisals are surveys, questionnaires, or interviews that help patients identify aspects of their lifestyle that are health risks. Survey instruments may ask about behaviors such as use of alcohol or tobacco, diet and exercise habits, or compliance with seatbelt laws. Health promotion activities seek to help patients adopt healthier behaviors, such as smoking cessation or

increased physical activity. They may use employer-funded programs, such as smoking cessation clinics, exercise programs, or nutritional counseling (8).

Health promotion activities are conducted in the workplace for several reasons. Because working people spend so much time in the workplace, it is a convenient place to conduct these activities. In addition, because the employer generally pays for employee health care, it is in the employer's interest to promote better health. Health promotion activities also may benefit the employer if improved workforce health results in increased productivity at work. Studies of whether health promotion activities are cost-effective have not shown a consistent benefit and suggest that target populations and interventions must be carefully selected (9,10). However, health promotion activities can complement occupational health activities; for example, workers with a history of asbestos exposure can be counseled about the interaction between that exposure and smoking (leading to a multiplied risk for lung cancer) and referred to a smoking cessation program if necessary.

ETHICS IN OCCUPATIONAL HEALTH SERVICES

Those who provide occupational health services have a unique obligation. Not only must they work to protect the patient's health, but they must do so in circumstances in which another party (the employer) pays for the care, has the capacity to address risk factors, and determines whether the patient is employed. The occupational health physician, nurse, or physician assistant therefore must seek the patient's consent, protect the patient's privacy, and be mindful of potential adverse effects on the patient's job status. A worker's medical information should remain confidential, and employers generally need to know only the employee's ability to work and whether job modifications (e.g., safety measures, exposure prevention) or work restrictions are needed. The reader is referred to Chapter 13 for more information about occupational health ethics.

FINANCING OCCUPATIONAL HEALTH SERVICES

Payment systems for occupational health services are diverse. Some care is directly underwritten by employers, including care given at on-site clinics by providers who are themselves plant employees and care given by on-site or off-site providers who either bill the employer for services or contract on a capitated basis. Sometimes this includes care for nonoccupational health conditions, if such care is delivered in an on-site clinic. Employers also pay for curative (although not consultative or preventive) occupational health care indirectly through contributions to state workers' compensation funds. In addition to payments by employers and workers' compensation systems, medical costs of occupational injuries and illnesses are also borne by other insurance systems—including health insurance for general health care, Medicaid and Medicare, and Social Security Disability Insurance—and payments workers make out of pocket (11). However, even employer-provided care involves a cost to workers, because workers' compensation fund contributions and health insurance payments are part of the employer's total resources for employee salaries and benefits; contributions to these insurance plans result in less available money for wages.

CHANGES IN THE DELIVERY OF OCCUPATIONAL HEALTH SERVICES: THE NEED FOR RESEARCH

Like other health services, occupational health services are undergoing rapid changes in their organization and delivery. In many places, the plant physician is being eliminated, replaced by contract providers or referral services. There are fewer plant clinics and increasing reliance on occupational health services provided by hospitals, clinics, managed care organizations, and other off-site providers (12). Most of these changes have been driven by efforts to reduce the cost of occupational health services. As a result, there is increasing competition among providers and provider organizations for employer contracts to provide employee health and workers' compensation care (13).

In general (nonoccupational) health care, the need to assess quality among different health care providers has prompted the development of performance measures. Measurement systems such as the Health Plan Employer Data and Information Set (HEDIS) allow patients and purchasers to compare managed health care plans on measures of the effectiveness of their care delivery (14). Efforts are now underway to develop similar performance measures of quality in the provision of workers' compensation care by managed care organizations (15). These measures may also be applicable to occupational health care provided in other settings, but occupational health services require additional, unique measures to assess their quality. A set of proposed quality indicators in occupational health care is presented in Table 14-2 (16,17).

These questions about the effectiveness of various providers and appropriate ways to measure performance illustrate the need for the relatively new field of occupational health services research. As defined by the Association for Health Services Research, health services research uses quantitative or qualitative methodology to examine the impact of the organization, financing, and management of health care services on access to, delivery, cost, outcomes, and quality of services (18).

There is great diversity in occupational health services, including the training and practice of the health care providers, the settings in which care is provided, the way that care is paid for, and the specific tests, therapies, and procedures used to screen for or to treat work-related injuries and illnesses. Different options may not be equally efficacious or cost-effective, and they therefore

TABLE 14-2. *Proposed quality indicators*
for occupational health care and possible specific measurements

---

1. Do patients have adequate and timely access to care?
   a. How long is the time needed to obtain an appointment for urgent and routine primary occupational health care?
   b. How long is the time needed to obtain a referral to specialty care?
   c. How long is the time from being authorized for a surgical procedure to receiving a scheduled time for surgery?
2. Primary prevention—how well is primary prevention carried out?
   a. What percentage of the contracted employers have had a review of their injury and illness experience and strategies for prevention?
   b. What percentage of contracted companies have decreasing rates of injury or illness?
   c. What percentage of sentinel health events have been reported to the employer?
   d. What percentage of health care workers have documented hepatitis B immunization?
3. How well are work-related illnesses recognized and diagnosed?
   a. What percentage of cases of work-related dermatitis or asthma were recognized in a report of first injury or illness?
   b. What percentage of lead-exposed workers received a blood lead level measurement?
4. How is the quality of clinical care?
   a. What percentage of cases of lumbar spine injury resulted in permanent partial disability?
   b. What percentage of occupational health hazardous exposure evaluations included a worksite visit?
   c. Appropriate utilization of surgical procedures: What percentage of patients receiving carpal tunnel release surgery had prior demonstration of delayed median-nerve conduction?
5. Are patients satisfied with their occupational health care?
   a. Survey question: "Overall, how satisfied are you with the service you received from your doctor?"
   b. Survey question: "Overall, how would you rate the care and services you received during this visit?"
   c. What is the complaint rate per 100 claims?
6. Are employers who purchase occupational health care satisfied with that care?
   a. What percentage of contracts were renewed?
   b. Mail survey: "Overall, how satisfied are you with the care and services your workers receive?"
7. What is the quality of care being provided, as assessed by the outcomes of that care?
   a. Using appropriate diagnostic category scores during treatment, how much time is lost from work for claims related to diagnoses of the lumbar spine and shoulder, or carpal tunnel syndrome?
   b. Comparing workers' compensation to group health claims, what is the average cost per claim in the lumbar spine diagnostic category?
   c. What percentage of patients with lost time greater than 3 days experience additional time lost due to reinjury after their initial return to work?
   d. What percentage of patients with lost time greater than 3 days are at their preinjury job or a modified job at 90 days after being released to return to work?
8. Are health care services being utilized appropriately?
   a. What is the rate of use of plain film and advanced imaging (computed tomography and magnetic resonance imaging) per low back diagnostic category claim?
   b. What is the rate of laminectomies per low back diagnostic category claim?
   c. What is the rate of treatment and rehabilitation visits (occupational therapy, work hardening, or physical therapy) per low back diagnostic category claim?
9. Are cases being managed appropriately?
   a. What percentage of filed claims are accepted as work-related?
   b. What percentage of cases require independent medical studies?
   c. What percentage of claims incur litigation?
10. Does the practice environment strive for quality in health care?
    a. Are clinical guidelines, including those for chemical exposure and surveillance, in use and updated annually?
    b. Is there a quality management program, including indicator tracking, and is it updated annually?
    c. What percentage of physicians are board-certified?

Adapted from occupational health quality measures proposed in Rudolph L. A Call for Quality. J Occup Environ Med 1996;38:343–344; and Feldstein A. Quality in Occupational Health Services. J Occup Environ Med 1997;39:501–503.

provide fruitful issues for study. The outcomes of occupational health care may be affected by provider characteristics (e.g., type of provider, provider organization, reimbursement system) or by employer characteristics (e.g., industry type, employer size, geographic location). Research addressing these issues will help to determine the most effective practices for providing occupational health services.

REFERENCES

1. Milton D, Solomon G, Rosiello R, Herrick R. Risk and incidence of asthma attributable to occupational exposure among HMO members. Am J Ind Med 1998;33:1–10.
2. American Board for Occupational Health Nurses, Inc., 201 East Ogden Road, Suite 114, Hinsdale, IL 60521-3652. Available at: http://www.abohn.org/certif.htm. Accessed August 1999.
3. U.S. Department of Labor, Occupational Safety and Health Administration Web site. Available at: http://www.osha-slc.gov/SLTC/medicalsurveillance/index.html. Accessed August 1999.
4. Straif K, Silverstein M. Comparison of U.S. Occupational Safety and Health Administration standards and German Berufsgenossenschaften guidelines for preventive occupational health examinations. Am J Ind Med 1997;31:373–380.
5. International Labor Office. Technical and ethical guidelines for workers health surveillance. Document MEHS/1997/D.2. Geneva, Switzerland: International Labor Organization, 1997.
6. Mullan R, Murthy L. Occupational sentinel health events: an up-dated list for physician recognition and public health surveillance. Am J Ind Med 1991;19:775–799.
7. Feldstein A, Breen V, Dana N. Prevention of work-related disability. Am J Prev Med 1998;14(3S): 33–39.
8. Tolsma D, Koplan J. Health behaviors and health promotion. In: Last JM, Wallace RB, eds. Public health and preventive medicine, 13th ed. Norwalk: Appleton & Lange, 1992:701–714.
9. Pelletier K. Clinical and cost outcomes of multifactoral, cardiovascular risk management interventions in worksites: a comprehensive review and analysis. J Occup Environ Med 1997;39:1154–1169.
10. Dishman R, Oldenberg B, O'Neal H, Shephard R. Worksite physician activity interventions. Am J Prev Med 1998;15:344–361.
11. Leigh JP, Markowitz SB, Fahs M, Shin C, Landrigan PJ. Occupational injury and illness in the United States. Arch Intern Med 1997;157:1557–1568.
12. Guidotti T, Cowell J. The changing role of the occupational physician in the private sector: the Canadian experience. Occup Med 1997;47:423–431.
13. Leone F, O'Hara K. The market for occupational medicine managed care. Occup Med: State of the Art Reviews 1998;13:869–879.
14. Committee on Performance Measurement. HEDIS 3.0. Washington, DC: National Committee on Quality Assurance, 1997.
15. Greenberg EL, Leopold R. Performance measurement in workers' compensation managed care organizations. Occup Med 1998;13:755–772.
16. Rudolph L. A call for quality. J Occup Environ Med 1996;38:343–344.
17. Feldstein A. Quality in occupational health services. J Occup Environ Med 1997;39:501–503.
18. Association for Health Services Research, 130 Connecticut Avenue, Suite 700, Washington, DC 20036. Available at: http://www.ahsr.org/hsrproj/define.htm. Accessed August 1999.

BIBLIOGRAPHY

Ashford NA, Rest KM. Shifting the focus of occupational health and safety services. In: Menckel E, Westerholm P, eds. Evaluation in occupational health practice. New York: Butterworth-Heinemann, 1999.
This is a thoughtful essay on the need for addressing technological change in improving occupational health and safety.

Committee on Performance Measurement. HEDIS 3.0. Washington DC: National Committee on Quality Assurance, 1997.
The HEDIS set of performance measures is used to assess performance of managed care organizations in general (nonoccupational) health care. The information and examples in the set provide useful perspectives on constructing and applying performance measures.

Harris J, ed. Managed care [special issue]. Occup Med: State of the Art Reviews 1998;13(4).
This issue focuses on managed health care and occupational medicine. The articles address many aspects of occupational health services, including practice guidelines, performance measurement, prevention of disability, and various insurance programs.

Harris J, ed. Occupational medicine practice guidelines: evaluation and management of common health problems and functional recovery in workers. Beverly Farms, MA: OEM Press, 1997.
A detailed compilation of practice guidelines for managing work-related injury and illness.

Occupational Health: Recognizing and Preventing Work-Related Disease and Injury, 4th ed.
Edited by Barry S. Levy and David H. Wegman, Lippincott Williams & Wilkins, Philadelphia © 2000.

PART III

Hazardous Exposures

15 Toxins

Howard Frumkin, with cases by
James Melius

A toxin is generally understood to be a substance that is harmful to biologic systems, but within this simple concept lies a great deal of variability. A substance that is harmful at a high dose may be innocuous or even essential at a lower dose. A toxin may damage a specific body system, or it may exert a general effect on an organism. A substance that is toxic to one species may be harmless to another because of different metabolic pathways or protective mechanisms. And the biologic damage may be temporary, permanent over the organism's lifetime, or expressed over subsequent generations.

Toxicology is the study of the harmful effects of chemicals on biologic systems. It is a hybrid science built on advances in biochemistry, molecular biology, genetics, physiology, pathology, physical chemistry, pharmacology, and public health. Toxicologists describe and quantify the biologic uptake, distribution, effects, metabolism, and excretion of toxic chemicals. A subfield, environmental toxicology, focuses on exposures to toxic substances in the atmosphere, in food

and water, and in occupational settings. These exposures have important effects both on humans and on the ecosystem in which we live.

The course of most toxicologic interactions takes the form of uptake → distribution → metabolism → excretion. Storage and biologic effects are other important events that may, but need not, occur. A knowledge of each of these steps is essential for a complete understanding of the effects of a chemical.

This chapter summarizes the principles of toxicology as applied to occupational health. It first presents five cases, with commentaries, that highlight aspects of some major categories of toxins. (These cases are presented in Boxes 15-1 through 15-5, on pages 310–318. The text continues on page 319.)

Box 15-1 presents cases of carbon monoxide poisoning among workers on an onion farm.

Box 15-2 presents two cases of birth defects in children of mothers exposed to chemicals at work.

Box 15-3 presents a case of a car painter with nonspecific central nervous system symptoms due to organic solvents.

Box 15-4 presents a case of recurrent stomach pains in a bridge repair worker exposed to lead.

Box 15-5 presents cases of acute poisoning among greenhouse workers exposed to pesticides.

H. Frumkin: Department of Environmental and Occupational Health, Rollins School of Public Health, Emory University, Atlanta, GA 30322.

J. Melius: New York State Laborers' Health and Safety Fund, Albany, New York 12211.

Box 15-1 Asphyxiants: Carbon Monoxide Poisoning among Workers at an Onion Farm

In December, a 50-year-old woman was brought to the emergency department of a small rural hospital after collapsing at work at an onion farm. She reported no previous episodes of syncope or chest pain and had no significant past medical history other than treatment for mild hypertension. She was doing her ordinary work at the farm's packing shed, preparing onions for shipment, when she suddenly became dizzy and passed out. Her electrocardiogram (ECG) showed mild ischemic changes, and she was admitted to the intensive care unit for observation.

The next afternoon, two other workers from the same farm were brought to the emergency department complaining of headaches, dizziness, and nausea. Blood samples were drawn for determination of carboxyhemoglobin concentration, and both workers had slightly increased levels (about 10%). Interpretation was complicated by the fact that more than 30 minutes had elapsed before the two patients reached the hospital from the farm, and it was unclear whether they had been treated with oxygen during that time. The emergency physician contacted the farm owner, who reported that he had called the gas company to check the propane heaters used in the barn. They had tested the barn with a "gas meter" and found no problem with carbon monoxide (CO) or other gases.

The two workers went back to work the next morning and again became ill. They returned to the emergency department. This time, their carboxyhemoglobin levels were between 14% and 16%. A nurse from a local occupational health program was notified and visited the farm that afternoon. In discussing the situation with the farmer and other workers, she found a number of potential problems. Temperatures in the barn were kept very cold, and there was little ventilation. Several small propane heaters provided some heat. More importantly, a propane-powered forklift was used intermittently in the barn. Because of weather conditions, the doors to the barn had been kept closed for the last several days.

The nurse requested that an industrial hygienist visit the facility to conduct further air sampling. He arrived the next day. Long-term personal samples taken that day showed acceptable CO levels—up to 24 ppm, compared with the Occupational Safety and Health Administration (OSHA) standard of 50 ppm. However, short-term samples showed levels up to 100 ppm at some locations, especially around the forklift. Doors in the facility were kept open during the day that sampling took place. Based on these findings, the farmer obtained a battery-powered forklift and took steps to improve ventilation in the facility.

CO, a byproduct of combustion, is the most common chemical asphyxiant. Most exposures can be related to a combustion source such as a heating device or a gasoline engine. CO has a very strong affinity for hemoglobin, forming carboxyhemoglobin, which interferes with oxygen transport and delivery to tissues. In acute exposure, the first symptom is usually headache, progressing to nausea, weakness, dizziness, and confusion. CO exposure should be considered in patients who collapse at work or report sudden headaches, lightheadedness, dizziness, or nausea. More severe poisoning can lead to unconsciousness and death.

The standard laboratory test for CO exposure is determination of the carboxyhemoglobin concentration in the blood; this reveals the proportion of hemoglobin that is bound to CO. Normal levels in nonsmokers range up to 4%, and smokers can have levels as high as 8%. Serious medical problems usually do not develop unless levels exceed 20%. However, patients with ischemic heart disease are especially susceptible to the effects of CO. Following CO exposures that raise their carboxyhemoglobin levels only slightly, exercise, can develop ischemic ECG changes. Interpretation of carboxyhemoglobin levels is challenging, because they return to normal within hours (even faster in patients who have been given oxygen). Therefore, if a patient collapses at work, is given oxygen, and is then brought to the emergency department, the carboxyhemoglobin level measured in the emergency department may be well below the peak level the patient reached during the exposure.

Hyperbaric oxygen therapy is used for severe CO poisoning, but even with such treatment permanent neurologic damage may occur.

Intermittent or episodic exposures to increased concentrations of CO can increase the risk of cardiovascular disease among groups such as tunnel workers and highway toll collectors. However, such exposures can be difficult to detect. In the case presented

Box 15-1 (*continued*)

here, the original testing by the "gas meter" might have occurred when the ventilation was especially good (e.g., with a breeze blowing through the barn), or the instruments might have been insensitive to slight CO elevations. Similarly, a worker's exposure from the forklift could vary with time and location in the facility. In this case, sampling with better instrumentation revealed the source of CO.

Asphyxiants usually are grouped into two major categories. *Simple* or *inert* asphyxiants, such as propane or hydrogen, act by displacing oxygen in the atmosphere. The most common scenario for this type of asphyxiation is work in a confined space, such as a manhole or a storage tank. OSHA requires special precautions for work in confined spaces, such as warning signs, air testing before entry, and the use of supplied-air respirators.

Chemical or *toxic* asphyxiants include a number of chemicals that interfere with the transport, delivery, or utilization of oxygen in the body. In addition to CO, common examples include hydrogen sulfide and hydrogen cyanide. Although these materials are sometimes used in a workplace, more commonly they are produced as a result of some other process, such as combustion or chemical mixing, and the asphyxiation occurs accidentally as a result of that process.

Hydrogen cyanide exposure may occur in several industries, including electroplating and production of certain specialty chemicals. Hydrogen cyanide is most commonly produced when acids come into contact with cyanide compounds. The burning of acrylonitrile plastics can also produce significant levels of hydrogen cyanide. This chemical acts by inhibiting the enzyme cytochrome oxidase, which is necessary for tissue respiration. Exposure to levels of about 100 ppm for 30 to 60 minutes can be fatal. Initial symptoms include headache and palpitations, progressing to dyspnea and then convulsions. Treatment with sodium nitrite, sodium thiosulfate, and amyl nitrite can be effective but must be started almost immediately. Blood cyanide levels can be used to monitor the effectiveness of treatment.

Acute hydrogen sulfide poisoning may occur in a number of workplace settings, including leather tanning, sewage treatment, and oil drilling. Hydrogen sulfide is a common cause of work-related fatalities in oil fields in the southwestern United States, where it occurs naturally as a contaminant of natural gas.

Hydrogen sulfide acts by interfering with oxidative enzymes, resulting in tissue hypoxia. Although at lower concentrations hydrogen sulfide has a characteristic "rotten egg" odor, at levels higher than 100 to 150 ppm olfactory sensation is diminished, which can provide a false sense of security. Initial symptoms of acute exposure include eye and respiratory irritation progressing to dyspnea and convulsions (from anoxia). As with hydrogen cyanide, rapid treatment with nitrites is effective. Delayed pulmonary edema has also been reported in some people after acute exposures.

The outbreak of CO poisoning described in this case, along with the consideration of other asphyxiants, highlights several important principles. First, not every toxic exposure is exotic. Such familiar items as a forklift can cause fatal exposures. Second, occupational medicine can be directly applicable to the general environment. For example, many cases of CO poisoning occur in the home and are caused by faulty heaters. Third, workers, when exposed to an asphyxiant or intoxicant, become less alert and less able to react briskly to hazards. This is a form of synergy, which increases the risk of injuries, further exposures, and other mishaps on and off the job. Fourth, in an environment with very high gas concentrations, every breath boosts the blood level of the gas, and toxicity can develop remarkably rapidly. Such acute toxicity is common in enclosed spaces, affecting not only the primary victims but coworkers who rush to provide assistance. Fifth, when a worker is found dead or unconscious after an unknown exposure, a blood sample should always be taken. Carboxyhemoglobin levels and evidence of other toxicities can be determined.

Perhaps the most important principle illustrated by the asphyxiants is the primacy of prevention. Asphyxiation can almost always be anticipated; the hazards of confined spaces, forklifts, and other sources are well recognized. Once anticipated, exposures can be prevented by some combination of usual measures: minimizing the formation of the asphyxiant, proper ventilation, personal protective equipment, proper work practices, and worker training.

Box 15-2. Teratogens: Birth Defects in Children of Mothers Exposed to Chemicals at Work

Case A: A 19-year-old woman, who worked in the reinforced plastics industry and whose husband was a 26-year-old carpenter in the same factory, gave birth to her first child, a 3,900-g, 54-cm boy, 18 days before her predicted delivery date. The child was found to have congenital hydrocephalus, anomaly of the right ear, and bilateral malformations of the thoracic vertebral column and ribs. Antibody tests for rubella, *Toxoplasma,* and *Listeria* in mother and child, and for mumps and herpes simplex in the mother, were negative.

In the third month of pregnancy the mother had had bronchitis, and she was given 3 days sick leave and treated with penicillin. Otherwise her pregnancy was normal, and she had taken no drugs except for iron and vitamin preparations. The mother worked regularly during pregnancy; she ground, polished, and mended reinforced plastic products and was exposed to styrene, polyester resin, organic peroxides, acetone, and polishes. In her second trimester, she was heavily exposed to styrene for about 3 days when she cleaned a mold without a face mask.

Case B: A 24-year-old woman, who worked in the reinforced plastics industry and whose husband was a 22-year-old welder-plater in the metal industry, gave birth to her first child, a 2,200-g, 47-cm girl, 6 weeks before her predicted delivery date. The baby died during delivery; anencephaly was diagnosed. Serologic tests of the mother for *Toxoplasma* and *Listeria,* and placental culture for *Listeria,* were all negative.

The pregnancy had been normal except for contractions during the second month. At that time, 10 mg of isoxsuprine was prescribed three times daily for 7 days. Slight edema occurred in the seventh month of pregnancy, and 50 mg of chlorthiazide per day was prescribed for 7 days. The mother worked during most of her pregnancy. In the third month of pregnancy she did manual laminating for about 3 weeks with no face mask and was then exposed to styrene, polyester resin, organic peroxides, and acetone. After this she did needlework in the same workshop for about 1 month and then did lamination again at varying intervals (1).

These two cases were identified during an investigation of congenital malformations in Finland. They were reported after the investigators found that workers in the reinforced plastics industry were overrepresented among parents of affected infants.

Teratogenesis caused by industrial chemicals may well have occurred in these cases. Styrene (vinyl benzene) is metabolized to styrene oxide, a known bacterial mutagen. Styrene is also a structural analogue of vinyl chloride, which is associated with lymphocyte chromosomal aberrations and hepatic angiosarcomas among exposed workers (see Chapter 16). These molecules are sufficiently fat-soluble to cross membranes and could have passed from the maternal to the fetal circulation. In both cases, the women had multiple chemical exposures, and the possibility of combined effects cannot be excluded.

In these cases, as is typical when a chemical exposure is clinically associated with teratogenic or carcinogenic effects, causation is difficult to establish. For any particular substance, it may be impossible ever to assemble a large enough group of exposed subjects to conduct an epidemiologic study that would yield statistically significant results (see Chapter 6). Health professionals therefore must use available toxicologic knowledge to evaluate case reports such as this one, identify potential hazards, and advise their patients regarding appropriate precautionary measures.

Box 15-3 Organic Solvents: A Car Painter with Nonspecific Central Nervous System Symptoms

During a routine medical examination, a 24-year-old man reported problems with concentration. He frequently lost his train of thought, forgot what he was saying in midsentence, and had been told by friends that he seemed to be forgetful. He also felt excessively tired after waking in the morning and at the end of his workday. He had occasional listlessness and frequent headaches. At work he often felt drunk or dizzy, and several times he misunderstood simple instructions from his supervisor. These problems had all developed insidiously during the previous 2 years. The patient thought that other employees in his area of the plant had complained of similar symptoms. He had noted some relief during a recent week-long fishing vacation. He denied appetite or bowel changes, sweating, weight loss, fever, chills, palpitations, syncope, seizures, trembling hands, peripheral tingling, and changes in strength or sensation. He was a social drinker and denied drug use and cigarette smoking.

The patient had worked for approximately 3 years as a car painter in a railroad car repair garage. On his physician's urging, he compiled a list of substances to which he had been exposed:

Paint Solvents	Paint Binders	Other Substances
Toluene	Acrylic resin	Organic dyes
Xylene	Urethane resin	Inorganic dyes
Ethanol	Bindex 284	Zinc chromates
Isopropanol	Solution Z-92	Titanium dioxide
Butanol		Catalysts
Ethyl acetate		
Ethyl glycol		
Acetone		
Methyl-ethylketone		

His plant had been inspected by OSHA 1 year previously, and only minor safety violations had been noted.

Physical examination, including a careful neurologic examination, was completely normal. The erythrocyte sedimentation rate was 3 mm/hour. Routine hematologic and biochemical tests, thyroid function studies, and heterophile antibody assay were all negative, except for slight elevations of serum γ-glutamyl transpeptidase (SGGT) and alkaline phosphatase (2).

This case illustrates some of the many problems that confront a health care provider in applying occupational toxicology. The patient reported vague, nonspecific symptoms, which a busy clinician might easily dismiss. However, many toxins have just such generalized effects. Furthermore, the patient had multiple chemical exposures, and no one toxin could readily be identified as the culprit. This patient was unusual in that he was able to provide a list of his exposures, but even this list had its limitations. Note the presence of two (fictional) trade names on the list; their identities are unknown and may be elusive even to an inquiring physician (see Appendix A). The absence of OSHA citations 1 year earlier may mean that all exposures were then at permissible levels, but one cannot be certain of this fact. The inspection might have been directed only at safety hazards, the plant may have been temporarily cleaned up for the inspector's benefit, conditions could have deteriorated in the subsequent year, and new production processes could have been initiated or new materials introduced. In any event, all the symptoms reported by this patient have been associated with "safe" levels of solvent exposure, so even a well-maintained plant might offer cause for concern.

Organic solvents are commonly used industrial chemicals. Exposures occur in a variety of workplace settings, including oil refining and petrochemical facilities, plastics manufacturing, painting, and building maintenance. Often several different solvents are used in a given product, and multiple products containing solvents may be used in a facility. Some products, such as paints, glues, and pesticides, are mixtures containing substantial portions of solvents. The formulation of products containing solvents has changed over time because of economic factors and concern about the toxicity of specific solvents. These factors may make identification of the specific exposure of an individual worker difficult to ascertain.

As illustrated in this case, many organic solvents target the nervous system, causing both acute effects (narcosis) and chronic neurobehavioral effects in some persons. In addition, several specific solvents, including carbon disulfide, n-hexane, and methyl n-butyl ketone, cause a peripheral neuropathy characterized by loss of distal sensation, progressing to include motor weakness and even paralysis. The disease may progress for several months after exposure has ceased, and permanent damage may occur (see also Chapter 29).

Box 15-3 (*continued*)

Long-term exposures to a number of solvents, including toluene, xylene, and trichloroethane, can cause chronic neurobehavioral changes, ranging from mild changes in affect and ability to concentrate to severe loss of intellectual function. Some persons appear to develop more severe disease than others, and the exact relation of disease to exposure history or to specific solvent exposures has been difficult to determine. Because most workers exposed to solvents are exposed to a mixture, the effects of individual solvents are difficult to separate.

Several organic solvents cause other toxic effects. Benzene was commonly used as an industrial and commercial solvent in the past. High exposures suppress bone marrow production, sometimes leading to anemia or pancytopenia. Benzene is also a potent carcinogen, leading to leukemia and other hematopoietic malignancies (Chapter 33). Because of this toxicity, benzene is used much less commonly today, although exposures continue to occur in the petrochemical industry and in some other industries. Some other hydrocarbons, including ethylene oxide, the chloromethyl ethers, and epichlorohydrin, are also known carcinogens. Many other solvents are suspected of being carcinogenic, including several halogenated compounds.

Several organic solvents are hepatotoxic. Carbon tetrachloride, chloroform, and tetrachloroethane can cause hepatic necrosis. Long-term exposure to carbon tetrachloride has been associated with the development of cirrhosis. Dimethyl formamide and 2-nitropropane have caused outbreaks of chemically induced liver disease in exposed workers (Chapter 34).

Several organic solvents, including the glycol ethers and ethylene oxide, have been shown to affect the reproductive system. Skin irritation also commonly results from solvent exposures. These chemicals dry the skin by removing natural skin oils. Many organic solvents are also acute respiratory irritants.

The diagnosis of health problems related to solvent exposure depends strongly on a thorough exposure history. Biologic monitoring may be helpful for ongoing exposures, but it is not useful for evaluating past exposures, because most solvents are metabolized and cleared from the body relatively quickly.

Control of solvent exposure is based on a careful evaluation of how the exposure occurs. Work procedures and practices are often important determinants of solvent exposure for many persons who work with organic solvents (e.g., painters). Personal protective equipment, a switch to a less toxic alternative, and/or changes in work practices may be needed to limit exposures. In some settings, traditional engineering methods such as ventilation may also be useful.

Box 15-4. Metals: Recurrent Stomach Pains in a Bridge Repair Worker

A 29-year-old laborer who worked intermittently for a construction firm that did bridge repair work complained to his family physician of intermittent stomach pains of several weeks' duration. The pain was not associated with meals. Onset had been gradual, and he had no associated systemic or gastrointestinal symptoms. He had not experienced any unusual stress at home or at work. He reported drinking one or two cans of beer per day. His physician treated him with antacids.

Approximately 9 months later, the patient saw his physician again with the same complaints. His earlier pains had resolved approximately 1 month after treatment, and he had been feeling fine until a few weeks earlier, when his stomach pains started to recur. This time, the pains were more severe and were associated with loss of appetite and generalized fatigue. There was no consistent association with mealtimes or with other activities. He had no other significant symptoms and reported no recent changes in his personal life or habits. His physical examination was normal. His doctor sent him for an upper gastrointestinal series that was scheduled approximately 1 week later. He was again treated with antacids and dietary restrictions.

The doctor saw the patient again approximately 1 week after the x-ray studies. The results had been normal, and the patient's symptoms had improved slightly over the past week. He was seen again 4 weeks later, with continued improvement. However, he still reported intermittent epigastric pains. At this time, his physician became concerned about possible exposures to lead from his occupation. Although the patient knew that lead had been an ingredient in gasoline, he was unaware that the paint on bridges could contain lead. He reported no other hobbies that might expose him to lead. The physician ordered a complete blood count and a blood lead level (BLL). The blood count showed slight anemia, and the BLL was 20 μg/dL. The physician continued antacid and dietary treatment.

Approximately 2 months later, the patient returned complaining of more severe epigastric pains, this time associated with abdominal cramping, headaches, and fatigue. He had been getting better but then started work at a new site, where he had used an oxyacetylene torch to remove paint from sections of an old bridge before welding. In reviewing the history of his episodes of pain, the patient reported that all three had occurred a few weeks after he started a similar type of job. After consultation with an occupational physician, the family physician obtained another BLL, which was 53 μg/dL. The patient stopped doing paint removal work, and his symptoms gradually improved. Within 2 weeks, his BLL was reduced to 43 μg/dL. The contractor arranged a ventilation system for use when paint was being removed from bridges, and quarterly monitoring of the patient's BLL showed a gradual decline.

Although the use of lead pigment was discontinued in most paints by the 1970s, older lead-containing paints still cover many interior and exterior surfaces in older buildings and continue to be used on steel structures such as bridges. Not only does this exposure account for many cases of childhood lead poisoning, but also painters and other workers conducting renovation work on buildings with lead paint can be significantly exposed to lead. Burning or torching of the surface to remove the paint produces a lead fume that is readily absorbed through the respiratory tract.

The time course of lead exposure is important. In this case, the worker's exposure was intermittent. At the time the first BLL was obtained, he had been unexposed for a few weeks, and the concentration had returned to a lower level. If the concentration had been taken at the time of the first visit, it would have been higher and the suspected occupational cause of the symptoms would have been confirmed. This illustrates the importance of a good occupational history, inquiring not only about usual activities but also about changes in workplace activities or possible exposures.

Most occupational lead exposures occur by inhalation, although ingestion may also contribute, especially through contamination of food or cigarettes at work. Lead is initially absorbed into the blood and then gradually stored in the bones. BLL determinations are a good indicator of recent exposure but may not reflect past exposures. Newer x-ray fluorescence techniques provide a better assessment of lead storage in bones, but they are not widely available.

Most metals, including lead, exert their biologic effects through enzyme ligand binding, and for many metals excretion can be hastened by chelation therapy with agents

Box 15-4 (*continued*)

such as dimercaptosuccinic acid (DMSA, succimer), dimercaprol (British antilewisite, BAL), or ethylenediaminetetraacetic acid (EDTA). Beyond these generalizations, however, metal toxicology is as varied as the metals themselves.

Lead affects a number of organ systems, including the hematopoietic, renal, and nervous systems. Typical early signs of exposure in adults include abdominal colic, headache, and fatigue. At higher levels of exposure, lead may cause a peripheral motor neuropathy with wrist or foot drop. Higher levels of exposure may also lead to an anemia related to the inhibition of several enzymes involved in hemoglobin production. Chronic exposure to lead can cause renal tubular damage and, eventually, renal failure. Lead exposure is also associated with adverse reproductive effects in both men and women.

The current OSHA lead standard is 50 $\mu g/m^3$ over an 8-hour day. Regular exposure at this level yields an average BLL of about 40 $\mu g/dL$. The standard requires routine BLL monitoring and removal of a worker from exposure if the BLL becomes elevated. This standard applies equally to general industry, where routine lead exposure occurs in such operations as battery manufacturing, and to the construction industry, where many lead poisoning cases are being reported, especially in workers who remove lead paint from highway bridges and similar structures.

Mercury is another important metal used in the manufacture of monitoring instruments and in certain industrial processes. It is important to distinguish the form of mercury (metallic, inorganic, or organic) when evaluating toxic effects. Metallic and inorganic mercury affect the nervous system and the kidneys. At high doses, exposed persons undergo personality changes such as irritability, shyness, and paranoia (a syndrome called *erethism*); tremor; and peripheral neuropathy. Lower doses cause more subtle forms of these problems, such as visual-motor changes on neurobehavioral testing and slowed nerve conduction velocity. The kidney toxicity can manifest as both tubular and glomerular dysfunction; patients show proteinuria and in severe cases impaired creatinine clearance. Gingivitis is another classic sign of severe mercury poisoning. Exposure to metallic mercury is usually monitored through determinations of urine mercury levels, although blood levels may also be useful.

Organic mercury compounds (usually methyl mercury) are sometimes encountered in workplace settings, but they are better known from outbreaks related to environmental contamination (usually human exposure to contaminated fish). These exposures have been associated with severe neurologic disease (both central and peripheral) and birth defects in children of pregnant women exposed to high levels of methyl mercury.

Arsenic is used in some industrial and chemical processes. Exposure also occurs in the smelting of some metal ores. Exposure to arsenic can cause a symmetrical distal polyneuropathy. High exposures cause liver damage and skin lesions. Arsenic is also carcinogenic, causing lung, liver, and skin cancer. Exposure to arsenic is usually monitored through urinary arsenic levels.

Cadmium exposure occurs in many different industrial processes. Its main effect is on the renal system, leading to renal tubular dysfunction as cadmium accumulates in the kidney. Cadmium also causes lung cancer. Cadmium exposure can be monitored with either urine or blood concentrations.

Beryllium is a metal used in electronics and some other industrial applications. Exposure leads to a fibrotic lung disease similar to—and often mistaken for—sarcoidosis. Lymphocyte transformation testing of blood or bronchoalveolar lavage can assist with the early diagnosis of this illness.

Other important toxic metals include nickel, which is carcinogenic and is a very common cause of contact dermatitis; chromium, which similarly causes contact dermatitis and is believed to be carcinogenic, but only in the hexavalent form; and manganese, which causes a neurologic condition similar to Parkinson's disease.

Prompt medical diagnosis is extremely important in the control of metal poisonings. Many current exposures occur in small businesses or involve exposures secondary to other work (e.g., lead exposure from removing lead paint). Biologic monitoring and a careful exposure history are critical for proper diagnosis and follow-up of people working in these industries. In larger industries, routine industrial hygiene control techniques are applied, including better ventilation and use of personal protective equipment.

Box 15-5. Pesticides: Acute Poisoning in Greenhouse Workers

A 38-year-old woman was seen in the emergency department of a rural hospital on Saturday evening, complaining of a severe rash. The rash had initially appeared on her forearms several weeks earlier, but during the last 2 weeks it had become more severe, spreading to her face and neck. She indicated that the itching from the rash had become so severe that she had hardly slept for the last three nights. She came to the hospital on a Saturday night because that was the only time she had off from work. She had no previous history of any skin problems. On questioning, the patient suspected that the rash might have resulted from chemical exposures at work. She worked at a greenhouse, where she had contact with pesticides, fungicides, fertilizers, and cleaning materials. Physical examination showed a severe maculopapular rash on her hands, forearms, face, and neck. The emergency room physician treated her with topical steroids and an antihistamine for the itching and referred her to a local community clinic for follow-up.

Two weeks later, the patient was seen at the community clinic. The rash was still quite severe. She had used up the medication provided at the hospital but was unable to fill her prescriptions because they were expensive. The physician asked her about the chemicals used at work, but she could not identify any of them by name. She did not apply pesticides or fungicides herself but was exposed to them when the greenhouses were sprayed before she arrived at work and when she handled the flowers. The physician provided her with medication for her dermatitis, advised her to return in 2 weeks, and asked her to try to get the names of the pesticides and fungicides that were used at work.

The patient did not return to the clinic until 4 weeks later. By this time, her dermatitis was much more severe. She reported that, although the medications had initially helped, many more pesticides had been used at work during the past week and several workers had become ill a few days ago. The greenhouse had been sprayed the night before, and they could smell a strong odor when they went to work. About 1 hour after entering the greenhouse, she and her two fellow workers became very ill with nausea and headaches. They left the greenhouse. After resting outside for an hour, they felt better and returned to work. By that time, the patient's dermatitis

was so bad that she decided to quit the job. She had requested information on the pesticides from the owner, but he told her that the information was too complicated for her to understand.

The physician from the community clinic treated the patient for the dermatitis, which slowly cleared up. After two more workers from the greenhouse came to the clinic reporting episodes of acute illness (headaches and nausea), the physician reported the problem to the state pesticide enforcement agency, which then inspected the facility. Although it found problems with labeling of the pesticides used at the facility and with disposal practices, no serious violations of current regulations were found. The owner did change some application practices, and the patient was later able to return to work in another area of the facility without problems.

Pesticides and fungicides include a wide range of chemicals used to control various undesirable species. Although most pesticide use occurs in agricultural settings, people may also be occupationally exposed from the use of these chemicals for structural pest control. Pesticides, as a class of chemicals, affect almost every organ system, but individual pesticides usually have a more limited and more specific toxicity.

Pesticides can be absorbed by all three routes: inhalation, ingestion, and skin absorption. Skin absorption is an important route for many pesticides, especially among workers who have extensive contact with sprayed plants or crops. Organophosphate pesticides are among the types most widely used in agricultural and structural applications. These compounds act by inhibiting the enzyme acetylcholinesterase at the nerve-to-nerve synapse or at the nerve-to-muscle motor end plate, leading to increased levels of the neurotransmitter acetylcholine at many different sites in the body.

The worker in this case probably was acutely exposed to one of the organophosphates. Any of these pesticides can be absorbed by the respiratory, percutaneous, and gastrointestinal routes. The exposure in this example probably included both inhalation (from the pesticide fogging) and dermal exposure from contact with pesticide-contaminated plants and surfaces. Once absorbed, organophosphates are metabolized by hepatic microsomal enzymes. For one of the most studied of these pesticides, parathion, the first major conversion it undergoes is

Box 15-5 (*continued*)

replacement of its sulfur by oxygen to form paraoxon, the actual anticholinesterase. Subsequent oxidation and hydrolysis result in detoxification.

Paraoxon binds with acetylcholinesterase molecules at cholinergic nerve endings, both centrally and peripherally. The organophosphates and the carbamates both act through this mechanism, but carbamate complexes dissociate spontaneously, whereas organophosphate complex formation is virtually irreversible. As a result, organophosphate poisoning causes a predictable constellation of muscarinic, nicotinic, and central nervous system symptoms. Severe cases can progress to coma and death.

Typical symptoms include miosis, salivation, sweating, and muscle fasciculation; at higher exposures, diarrhea, incontinence, wheezing, bradycardia, and even convulsions may occur. Cholinesterase inhibition can be measured with cholinesterase levels, but these tests are difficult to interpret for several reasons: people vary widely in their normal levels, laboratories vary widely in their measurements, and levels may quickly return to normal after exposure ceases. Cholinesterase levels are most useful for ongoing monitoring (if baseline levels are known) and in monitoring recovery from acute toxicity. Acute poisoning can be treated with atropine with or without pralidoxime.

A delayed neurotoxicity syndrome, with weakness, paresthesias, and paralysis of the distal lower extremities, has been found in persons exposed to organophosphate pesticides, and other chronic neurotoxicity syndromes also have been reported. These syndromes usually occur in people with chronic exposure or after a very severe acute exposure.

Another frequently used category of pesticides is the carbamates. Carbamates also inhibit the enzyme acetylcholinesterase, but this inhibition is more readily reversed than that caused by organophosphate pesticides. Hence, effects tend to be less severe. Because of the rapid reversal, serum and red blood cell cholinesterase levels tend to be less useful in the diagnosis of exposure to this type of pesticide.

Organochlorine pesticides, such as dichlorodiphenyltrichloroethane (DDT), were more widely used in the past, but their use has been limited owing to their persistence in the environment. However, some (e.g., lindane) are still commonly used. Organochlorine pesticides are metabolized very slowly and accumulate in fat and other tissues. Their major toxic effects involve the nervous system, leading to anorexia, malaise, tremor, hyperreflexia, and convulsions. In addition, evidence (mostly from animals) suggests that organochlorines and other persistent organic pollutants may have hormonal effects. If this is true, these agents could contribute to impaired reproductive function, developmental abnormalities in children, and increases in hormone-responsive cancers such as those of the breast and prostate. The endocrine disrupter hypothesis is the subject of intense research.

Many other individual pesticides have significant toxicity. Paraquat (an herbicide) can cause a severe pulmonary fibrosis. Dibromochloropropane (DBCP), used in the past to control nematodes, caused sterility among male workers exposed to high levels of this chemical (see Chapter 31). Many pesticides are carcinogenic. A series of studies found a high incidence of non-Hodgkin's lymphoma among midwestern U.S. farmers who used large amounts of herbicides. Dermatitis is also common among people working with pesticides, although some of this incidence is a result of exposure to other materials mixed with the pesticides.

Finally, although many fungicides and herbicides have low toxicity, others present a range of toxic effects. For example, skin problems may result not only from insecticides and the solvents used to dilute them for application, but also from fungicides and herbicides. In the case described, the patient had probably developed skin sensitization to fungicides used in the cultivation of the flowers.

The diagnosis of pesticide-related illnesses can be very difficult. A high index of suspicion and a very careful medical and exposure history are essential. Laboratory testing is helpful for some pesticides, as noted previously. Prevention strategies for control of pesticide-related health risks are discussed in Chapter 41.

CLASSES OF TOXIC SUBSTANCES

Toxic or harmful substances encountered in the workplace may be classified in various ways. A simple and useful classification is given here, along with definitions adopted by the American National Standards Institute (ANSI):

Dusts: Solid particles generated by handling, crushing, grinding, rapid impact, and detonation of organic or inorganic materials such as rocks, ore, metal, coal, wood, and grain. Dusts do not tend to flocculate except under electrostatic forces; they do not diffuse in air but settle under the influence of gravity.

Fumes: Solid particles generated by condensation from the gaseous state, usually after volatilization from molten metals, and often accompanied by a chemical reaction such as oxidation. Fumes flocculate and sometimes coalesce.

Mists: Suspended liquid droplets generated by condensation from the gaseous to the liquid state or by breaking up of a liquid into a dispersed state, as by splashing, foaming, or atomizing.

Vapors: The gaseous form of substances that are normally in the solid or liquid state and can be changed to these states by either increasing the pressure or decreasing the temperature. Vapors diffuse.

Gases: Normally formless fluids that occupy the space of enclosure and can be changed to the liquid or solid state only by the combined effect of increased pressure and decreased temperature. Gases diffuse.

This classification does not include the obvious categories of solids and liquids that may be harmful, nor does it encompass physical agents that cannot be considered "substances." Living agents such as bacteria and fungi constitute another group that would appear in a comprehensive classification of occupational health hazards.

ABSORPTION

In the workplace, people absorb chemicals by three main routes: the respiratory system, the skin, and the gastrointestinal tract. Other routes of absorption, usually of less importance, include mucous membranes and open lesions. The importance of each of these routes may be evaluated with specific questions during a careful occupational history (see Chapter 4).

Respiratory Tract

Inhalation is the major route of entry for gases, vapors, mists, and airborne particulate matter encountered in the workplace. To analyze this process we need to consider gases, vapors, and mists separately from particles.

Gases, Vapors, and Mists. Gases, vapors, and mists can damage the respiratory tract. They also can pass from the lungs to the bloodstream for distribution to other parts of the body. Both mechanisms may occur in the same patient.

Irritant gases are an important example of respiratory tract toxins (see Chapter 25). Their effects may be immediate or delayed. Gases that are very water soluble (e.g., hydrogen fluoride, ammonia, sulfuric acid) tend to dissolve in the moist lining of the upper respiratory tract, often producing immediate irritation, which forces an exposed worker to flee the area. Less soluble irritant gases (e.g., nitrogen dioxide, ozone, phosgene) reach the bronchioles and alveoli, where they dissolve slowly and may cause acute pneumonitis and pulmonary edema hours later. Long-term exposure at low concentrations may lead to chronic changes such as emphysema and fibrosis. Whether acute or chronic, the effects of irritant gases are seen mainly in the respiratory tract.

Asphyxiants are inhaled toxins that act by interrupting the supply or use of oxygen. *Simple asphyxiants* such as methane or nitrogen have relatively little direct physiologic effect, but by displacing oxygen in ambient air they can cause severe hypoxia. This is a common problem in enclosed spaces such as silos and storage tanks. *Chemical asphyxiants* block the delivery or use of oxygen at the cellular level through one of several mechanisms. One example is carbon monoxide, a product of incomplete combustion found in foundries, coke ovens, furnaces, and similar facilities. It binds tightly to hemoglobin, forming a carboxyhemoglobin complex (COHb) that is ineffective at oxygen transport. Another example is cyanide, which is used in plastics production, metallurgy, electroplating, and other processes. Cyanide inhibits the cytochrome oxidase enzymes, compromising oxidative metabolism and phosphorylation. A third example, hydrogen sulfide, a gas found in mines, petrochemical plants, and sewers, also inhibits cytochrome oxidase. Occupational fatalities from all these exposures remain common.

Gases, vapors, and mists that are fat soluble can cross from the alveoli into the bloodstream and migrate from there to binding or storage sites for which they have a special affinity. Substances that are readily absorbed after inhalation and exert their effects elsewhere include carbon disulfide, volatile aliphatic and aromatic hydrocarbons, volatile halogenated hydrocarbons, and aliphatic saturated ketones such as methyl ethyl ketone. Because of the impressive variety of substances involved and the wide spectrum of acute and chronic effects that may result, this is an important pathway of absorption in workplace settings.

The toxic action of some inhaled gases, vapors, and mists may be considerably enhanced when they are adsorbed to respirable particles. Presumably, the particles transport the toxins to deep parts of the respiratory tree, which would otherwise be inaccessible. An example is radon gas. This substance increases lung cancer incidence among uranium miners, but if it is experimentally inhaled in dust-free air, almost no radon is retained by the lung and much less carcinogenic effect is seen.

Several other variables influence the delivery of gases, vapors, and mists to the lower respiratory tract. With rapid, deep breathing, as occurs during strenuous exertion, the delivered dose increases. When a respirator is incorrectly chosen, poorly fitted, or inadequately maintained, significant amounts of airborne toxins can reach workers' lungs (see Chapters 7 and 25).

Particles. Inhaled particles, unlike most gases and vapors, are of interest primarily for their pathologic effects on the lungs. Their deposition, retention, and clearance are influenced by several well-defined factors.

One factor is inertia, which tend to maintain moving particles on a straight course. A second is gravity, which tends to move particles earthward, promoting the early settling out of larger and denser particles. A third is Brownian diffusion, the random motion that results from molecular kinetic energy. When these forces cause a particle to strike the airway wall or alveolar surface, deposition occurs.

Three factors influence the location and extent of deposition: anatomy of the respiratory tract, particle size, and breathing pattern. Branching, angling, and narrowing at each point in the airways define the local velocity and flow characteristics of the inspired air, which in turn influence particle deposition. For example, the nasopharynx features sharp bends, nasal hairs, and narrow cross sections with resulting high linear velocities, which together promote the impaction of inhaled particles.

This principle has some well-known applications. Nasal cancers are rare in the general population, but they occur with an increased incidence in certain occupational groups, probably because of nasal deposition of inhaled dusts that contain carcinogens. Examples include workers in furniture manufacturing, who are exposed to wood dust, and workers in leather shoe and boot manufacturing, who are exposed to leather, fiberboard, rubber, and cork dusts. A similar phenomenon is thought to occur farther along the airways. Most bronchogenic carcinomas arise at airway branch points, where particle deposition is promoted by locally turbulent air flow.

The effective anatomy of the respiratory

tract changes significantly with a simple shift from nose-breathing to mouth-breathing, as occurs normally during physical exertion. This bypasses the more efficient filtration of larger particles by the nasopharynx and results in greater deposition in the tracheobronchial tree. Through such a transition workers performing physical labor may lose the benefit of a major natural defense mechanism.

Because particle density and shape also influence deposition, these parameters are subsumed in a conceptual measure of size, the *effective aerodynamic diameter.* Comparisons among unlike particles, as if they were spherical and of unit density, are made possible with the use of this measurement. As Fig. 15-1 shows, deposition is very efficient for particles with effective aerodynamic diameters greater than several micrograms, reaches a minimal level at about 1.0 μm, and increases again for particles smaller than 0.5 μm. Particles with an effective aerodynamic diameter between 0.5 and 5.0 μm (the respirable fraction) can persist in the alveoli and respiratory bronchioles after deposition there; this is the first step in the development of pneumoconiosis. Smaller particles are cleared by macrophages, lymphatics, and the bloodstream, and larger particles are filtered out in the upper airways. The size of a particle is not always constant; hygroscopic particles (e.g., some salts) can expand significantly when hydrated in the upper airways. The result is a higher proportion of upper airway deposition.

As minute volume increases, the deposition of particles in the airways increases, especially for larger particles. In contrast, a change to rapid, shallow breathing diminishes the residence time of airborne particles in the lungs and hence the probability of deposition. Deep breathing, as during strenuous exertion, delivers a larger proportion of inhaled air to the distal airways and promotes alveolar deposition. Therefore, with some basic knowledge of the particles in question and the nature of the inhalation

exposure, one can evaluate, in general terms, the magnitude of a workplace exposure.

Filtration of particles in the upper airways and clearance of particles that do arrive distally are accomplished mainly by the mucociliary escalator. Particles that impact ciliated regions of the airways are carried toward the pharynx within hours and have little chance to dissolve or undergo leaching. However, particles deposited in nonciliated regions may be more persistent. If they are removed from surface sites by alveolar macrophages, they reach the mucociliary escalator and are cleared within 24 hours. (Materials that are carried to the pharynx by the mucociliary escalator are then swallowed, resulting in gastrointestinal exposure; this may explain the increased incidence of gastrointestinal cancer among asbestos workers.)

Alternatively, particles that reach nonciliated regions may penetrate into fixed tissue, such as connective tissue or lymph nodes, where they may reside for years, sometimes eliciting pathologic reactions. This penetration is especially likely when the performance of the mucociliary escalator is compromised, as it is in smokers and in some chronically exposed workers. In general, the mucociliary escalator is an effective defense against the retention of most inhaled particles. When it is overrun, however, particles are retained and the stage is set for subsequent pathology.

Another aspect of inhalation toxicology has also gained attention. Particulate air pollution, especially the fine particles with a diameter less than 2.5 μm, has been linked with increased cardiovascular mortality. This association played a major role in motivating the U.S. Environmental Protection Agency to formulate stricter standards for particulate air pollution in 1998 (A court decision blocked implementation of this standard). However, the mechanism of this toxicity remains poorly understood. One hypothesis holds that particulates cause an increase in blood viscosity, resulting in more ischemic events. Another theory proposes that partic-

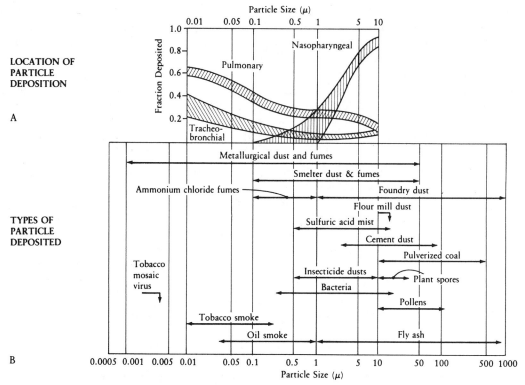

FIG. 15-1. A. Fractional deposition plotted against particle size (effective aerodynamic diameter) for three functional parts of the respiratory tree, based on the model of the International Committee on Radiation Protection. The broad bands reflect large standard deviations. **B.** Examples of inhaled particles, classified by size. By comparing (**A**) and (**B**), approximate predictions of the deposition pattern for each particle can be generated. (Adapted from Brain JD, Valberg PA. Aerosol deposition on the respiratory tract. Am Rev Respir Dis 1979;120:1325–1373.)

ulate exposure triggers changes in the autonomic nervous system, which in turn lead to cardiac arrhythmias. Understanding of this toxic effect is a major challenge for air pollution health scientists.

Skin

Human skin consists of three layers (the epidermis, dermis, and subcutaneous fat) and three kinds of glandular structures (sebaceous glands and apocrine sweat glands, which are both part of the pilosebaceous unit, and eccrine sweat glands) (see Fig. 27-1 in Chapter 27).

The outermost layer of the epidermis, the stratum corneum or horny layer, provides some of the skin's structural stability and

much of its chemical resistance. This layer consists mainly of thickened cell envelopes and a combination of sulfur-rich amorphous proteins and sulfur-poor fibrous proteins known as keratin.

There are two forms of percutaneous absorption: *transepidermal* (through the epidermal cells) and *appendageal* (through the hair follicles and sebaceous glands). The appendageal route offers greater permeability and plays an important role early in exposure and in the diffusion of ions and polar nonelectrolytes. However, the transepidermal route is generally more prominent because of its far greater absorbing surface.

Transepidermal transport occurs by passive diffusion, and several mechanisms have been hypothesized. According to one theory, the intracellular keratin matrix of the stratum corneum provides the main resistance to penetration by toxins, with polar and nonpolar molecules appearing to permeate through distinct channels. Alternatively, a two-phase series model postulates a protein-rich "cytoplasm" and a lipoidal "cell wall" to describe the passage of alcohols and steroids through the stratum corneum. A third possibility is that the stratum corneum limits the passage of polar molecules, while aqueous "boundary layers" prevent the diffusion of more lipophilic substances. Even very soluble substances encounter a relative barrier at the skin; in one experiment, mustard gas was found fixed in the epidermis and dermis of human skin 24 hours after application. (An interesting possibility is that the skin acts as a reservoir due to in situ fixation. After exposure ceases, toxins could continue to enter the bloodstream from skin stores. This process has been documented with some steroids.)

A tremendous variety of aliphatic and aromatic hydrocarbons, metals, and pesticides can undergo at least some percutaneous absorption. Generally, fat-soluble substances show a greater flux than water-soluble ones, and substances that are soluble in both fat and water show the greatest flux. For any substance, the relative importance of percutaneous absorption must be evaluated in light of other significant routes of entry. Although volatile solvents such as trichloroethylene and toluene are absorbed through the skin, inhalation would be far more significant in most occupational settings; on the other hand, for a substance such as benzidine, which is readily absorbed by the skin but has low volatility, percutaneous absorption may be the major route of entry.

A variety of factors can influence the extent of percutaneous absorption. Some are properties of the skin, such as wetness, location on the body, and vascularization. Previous damage, such as abrasion or dermatitis, can dramatically increase absorption. Certain organic solvents such as dimethyl sulfoxide (DMSO) and mixtures such as ethanol/ether or chloroform/methanol, are efficient delipidizing agents that compromise the barrier function of the stratum corneum. Occlusion often is used to promote the absorption of medications, but in the workplace occlusion can have untoward results. For example, if a worker wears rubber gloves that contain a solvent, the gloves may enhance absorption of the solvent. Other factors that influence percutaneous absorption are characteristics of the substance being absorbed; they include concentration, phase (aqueous or dry), pH, molecular weight, and vehicle. The best single determinant of the flux of a substance across the skin is its *partition coefficient* between stratum corneum and vehicle, which varies with the vehicle.

Skin absorption is necessary but not sufficient for the development of percutaneous toxicity. Cutaneous exposure may be followed by local effects (as discussed in Chapter 27), by systemic effects (as discussed later in this chapter), or by no effects at all.

Gastrointestinal Tract

Ingestion of hazardous substances is generally not a major route of workplace exposure, although there are important exceptions. Workers who mouth-breathe or who chew gum or tobacco can absorb appreciable amounts of gaseous materials during a workday. Inhaled particulates swept upward by ciliary action from the airways can be swallowed. And materials on the hands may be brought to the mouth during on-the-job eating, drinking, or smoking.

Several features of the gastrointestinal tract help to minimize toxicity by this route. Gastric and pancreatic juices can detoxify some substances by hydrolysis and reduction. Absorption into the bloodstream may be inefficient and selective. Food and liquid present in the gastrointestinal tract can dilute toxins and can form less soluble complexes with them. Finally, the portal circulation carries absorbed materials to the liver, where metabolism can begin promptly. As a result, the most serious workplace gastrointestinal exposures to consider are those with slowly cumulative action, such as mercury, lead, and cadmium. (Of course, nonoccupational gastrointestinal exposures, such as suicide attempts with acute toxins, can be catastrophic as well.)

DISTRIBUTION

Some toxins exert their effects at the site of initial contact. For example, the skin can be "burned" by strongly acidic or basic solutions or delipidized by some solvents. Inhalation of phosgene can lead to delayed pulmonary edema, and inhalation of cadmium fume can lead to pneumonitis. However, many toxins are taken up by the bloodstream and transported to other parts of the body, where they may exert biologic effects or be stored. The destination of a chemical is largely determined by its ability to cross various membrane barriers and by its affinity for particular body compartments.

Membranes are the main obstacle to the free movement of chemicals among the various compartments of the body. The mammalian cell membrane consists of a lipid layer sandwiched between two protein layers, with a combined thickness of about 100 Å. Large protein molecules, freely mobile in the plane of the membrane, can penetrate one or both faces, and small pores 2 to 4 Å in diameter traverse the membrane. There are therefore three mechanisms by which a chemical can cross a membrane: simple diffusion of lipid-soluble substances through the membrane, filtration through the pores by small molecules that accompany bulk flow of water, and binding by specialized carrier molecules, which transport polar molecules actively or passively through the lipid layer.

In a steady-state situation, the solute concentrations on the two sides of a membrane reach an equilibrium. This equilibrium is described by Fick's first law of diffusion, which states that the amount of solute (ds) that diffuses across an area (A) in time (t) is proportional to the local concentration gradient (dc/dx):

$$ds/dt = -DA(dc/dx)$$

The proportionality constant (D) is the *diffusion coefficient*. This constant is inversely proportional to the cube root of the molecular weight for a spherical molecule. Accordingly, many small molecules that ordinarily cross membranes without difficulty are unable to cross the same membranes once they are bound to large serum molecules. The diffusion coefficient also reflects the lipid/water partition coefficient. The result is an important generalization: Molecules that are fat-soluble cross membranes far more readily and far more rapidly than compounds that are water-soluble.

This generalization not only describes the distribution of toxins in the body; it has clinical applications as well. Manipulation of the pH on one side of a membrane can alter the extent of ionization of a solute, which, in turn, modifies its membrane solubility. This practice can be used to "trap" solute in a desired body compartment, in effect changing the flux by changing the diffusion coefficient. For example, patients who ingest large amounts of salicylates are treated with bicarbonate, in part because the resulting alkaline urine ionizes the filtered salicylate and prevents reabsorption by renal tubular cells.

Movement from one body compartment to another may not entail traversing of membrane bilayers. Each of three morphologic forms of capillaries—continuous, fenestrated, and discontinuous—provides an alternative way for solutes to leave the bloodstream. Therefore, various mechanisms exist by which membrane-insoluble substances and fluids can cross capillary walls, leave the vascular compartment, and reach target cells. The importance of these mechanisms is unclear, because most chemicals are thought to pass directly through capillary cells. However, in the face of local conditions such as acidosis, or in organs with discontinuous capillaries, such as the liver, these mechanisms may contribute significantly to the extravasation of toxins.

Two barriers to distribution deserve special attention. One is the *blood-brain barrier,* which impedes the passage of many toxins from central nervous system (CNS) capillaries into the brain. The blood-brain barrier has at least four anatomic and physiologic components. First, there are unusually tight junctions between the capillary endothelial cells. Second, the capillary endothelial cells have a transport mechanism, the multidrug-resistant (MDR) protein, that carries some toxins into the bloodstream. Third, glial connective tissue cells (astrocytes) surround the capillaries. Fourth, the CNS interstitial fluid has an unusually low protein concentration, which limits movement of water-insoluble chemicals that depend on protein binding. The blood-brain barrier is not uniform throughout the brain; for example, the cortex, lateral hypothalamic nuclei, area postrema, pineal body, and neurohypophysis are relatively less protected, perhaps because of a richer blood supply, a more permeable barrier, or both. Radiation, inorganic and alkyl mercury, and several other toxins and diseases can damage the blood-brain barrier and compromise its performance.

Chemical passage across the blood-brain barrier is governed by the same rules that characterize crossing of membranes elsewhere in the body. Only molecules that are not bound to plasma proteins can cross into the brain. Greater fat solubility helps a molecule permeate the barrier; therefore, nonionized molecules enter the brain at a rate proportional to their lipid/water partition coefficient, whereas ionized molecules are essentially unable to leave the CNS capillaries. The blood-brain barrier is distinctive mainly in its quantitatively greater protective effect compared with the body's other capillary beds.

The second barrier of particular interest is the *placenta.* In the placentas of humans and other primates, three layers of fetal tissue (trophoblast, connective tissue, and endothelium) separate the fetal and maternal blood. Some species such as the pig have six cell layers in the placental barrier, and others such as rat, rabbit, and guinea pig have only

one. These structural differences may correspond to differences in permeability, so animal data on transplacental transport must be interpreted with caution.

The placenta has active transport systems for vital materials such as vitamins, amino acids, and sugars. However, most toxins appear to cross the placenta by simple diffusion. In addition, a variety of nonchemical agents can cross the placenta, including viruses (e.g., rubella), cellular pathogens (e.g., the syphilis spirochete), antibody globulins (immunoglobulin G, but not immunoglobulin M), fibers (including asbestos), and even erythrocytes. There is evidence suggesting that the placenta actively blocks transport of some substances into the fetus, perhaps through chemical biotransformation mechanisms. Substances that do diffuse into the fetal circulation observe the laws discussed previously, and in a steady state their maternal and fetal plasma concentrations are equal. However, because maternal and fetal tissues have different affinities for some substances, the concentrations in specific kinds of tissues may differ. For example, the fetal blood-brain barrier is incompletely formed, and lead can accumulate in the fetal CNS far faster than in the maternal CNS; workplace exposure to lead therefore poses a special problem for pregnant women.

STORAGE

Most foreign substances that reach the plasma water are excreted fairly promptly, either intact or after chemical modifications. But some substances, although they are distributed as described previously, reach sites for which they have a high affinity, where they accumulate and persist. The result is storage depots.

These depots may occur at sites of toxic action. For example, carbon monoxide has a very high affinity for its target molecule, hemoglobin, and the herbicide paraquat, which causes pulmonary fibrosis, accumulates in the lungs. More commonly, however, the storage depot is different from the site of toxic action. For example, lead is stored in bone but exerts its toxic effects on various soft tissues. Because there is an equilibrium between plasma and depot concentrations of any stored substance, a depot can slowly release its content into the bloodstream long after exposure has ended.

One important storage depot is adipose tissue. Most chlorinated solvents (e.g., carbon tetrachloride) and chlorinated pesticides (e.g., DDT, dieldrin) migrate from the blood to the fat because of their high lipid solubility. Storage of such substances appears to involve simple physical dissolution in neutral fats, which constitute about 20% of a lean person's weight and up to 50% of an obese person's weight. The fat reservoir can therefore protect an obese person by sequestering a toxin from circulation; conversely, rapid loss of weight through illness or dieting might release toxic amounts of a stored substance back into the bloodstream.

A variety of proteins can bind exogenous molecules and thereby function as a storage depot. Probably the best known example is serum albumin, which binds and transports many pharmacologic agents. Evidence suggests that environmental agents such as DDE (the principal metabolite of DDT) and some dyes may compete for the albumin-binding sites. Molecules that are bound to albumin or to other plasma proteins (e.g., ceruloplasmin, transferrin) are unavailable for toxic action elsewhere as long as they remain bound. A more deleterious form of protein binding occurs with mercury and cadmium. Both metals complex with renal tubular proteins, and mercury complexes with CNS proteins as well, causing dysfunction at these sites. Finally, carcinogenic hydrocarbons appear to complex with proteins on contact with the skin and lung wall, and they initiate transformation by their continued presence (see Chapter 16).

Several important toxins are stored as insoluble salts. Lead, strontium, and radium form phosphates, which are deposited in bone. Lead has a toxic effect on marrow (see Chapter 33), and it can be mobilized during

acidosis to cause acute lead poisoning. Strontium, if deposited as a radioactive isotope (especially strontium 90), releases ionizing radiation, and radium is an alpha emitter; both are associated with osteosarcomas and hematologic neoplasms (see Chapter 16). Finally, fluoride is stored in the bones and teeth as an insoluble calcium salt; high levels can lead to bone and joint distortion, joint dysfunction, and mottling of dental enamel in children.

An unusual example of storage is the lymphatic accumulation of crystalline silica in the lung. Macrophages that engulf the silica migrate to lymph nodes, where they are destroyed. Continued irritation by the silica then induces inflammation and fibrosis (see Chapter 25).

BIOTRANSFORMATIONS

Between initial absorption and final excretion many substances are chemically converted by the body. Although many different metabolic conversions have been described, they can be characterized by a few simple generalizations and divided into a small number of categories.

Metabolic transformations are mediated by enzymes. The liver is rich in metabolic enzymes, and most biotransformation occurs there. However, all cells in the body have some capacity for metabolizing xenobiotics (chemicals foreign to the body). In general, metabolic transformations lead to products that are more polar and less fat-soluble, consistent with the eventual goal of excretion. This process usually entails a change from more toxic to less toxic forms. For example, benzene is oxidized to phenol, and glutathione combines with halogenated aromatics to form nontoxic mercapturic acid metabolites.

However, metabolic transformations sometimes yield products with increased toxicity. One example is the oxidation of methanol to formaldehyde. Another example is the solvent methyl-*n*-butyl ketone (MBK), which causes peripheral neuropathy (so-called cabinet-finisher's neuropathy) in exposed workers (see Chapter 29). Animal toxicology studies have revealed that a γ-diketone metabolite of MBK, 2,5-hexanedione, is probably the actual neurotoxin. *n*-Hexane is also oxidized to 2,5-hexanedione, which helps explain the neurotoxicity of that solvent (Fig. 15-2). In contrast, other hexacarbons such as methyl isobutyl ketone (MIBK) and methyl ethyl ketone (MEK) cannot give rise to 2,5-hexanedione and would therefore be preferable to MBK as solvents.

The concept of toxic metabolites is especially salient with regard to carcinogens (see Chapter 16). Vinyl chloride, a cause of liver, lung, lymphatic, and CNS tumors, is oxidized to a reactive epoxide intermediate, which is actually the proximate carcinogen. Similar transformations probably occur with trichloroethylene, vinylidene chloride, vinyl benzene, and chlorobutadiene. In fact, a major mechanism of carcinogenicity of aromatic compounds is conversion to reactive epoxides that, in turn, combine with cellular nucleophiles such as DNA and RNA.

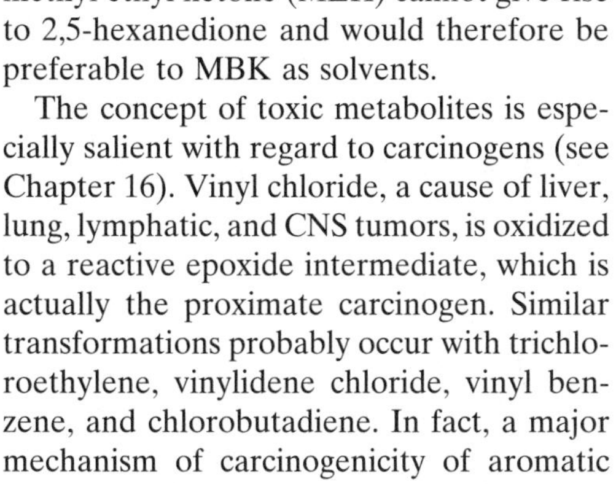

Classically, metabolic transformations are divided into four categories; oxidation, reduction, hydrolysis, and conjugation. The first three reaction types, which are known as phase I reactions, increase the polarity of substrates and can either increase or decrease toxicity. In conjugation, the only phase II reaction, polar groups are added to the products of phase I reactions. Most chemicals are handled sequentially by the two phases, but some are directly conjugated. The spectrum of reactions of each type can be found in any toxicology text, and only a few examples of occupational health interest are presented here.

Oxidation is the most common biotransformation reaction. There are two general kinds of oxidation reactions: direct addition of oxygen to the carbon, nitrogen, sulfur, or other bond; and dehydrogenation. Most of these reactions are mediated by microsomal enzymes, although there are mitochondrial and cytoplasmic oxidases as well. Figure 15-3 provides examples of oxidation. The thiophosphate insecticides (e.g., parathion) are relatively nontoxic until the sulfur is replaced by oxygen through this reaction.

Reduction is a much less common biotransformation than oxidation, but it does occur with substances whose redox potentials exceed that of the body. Figure 15-4 shows an example of azo reduction. The azo dyes,

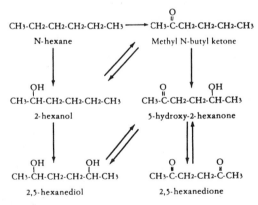

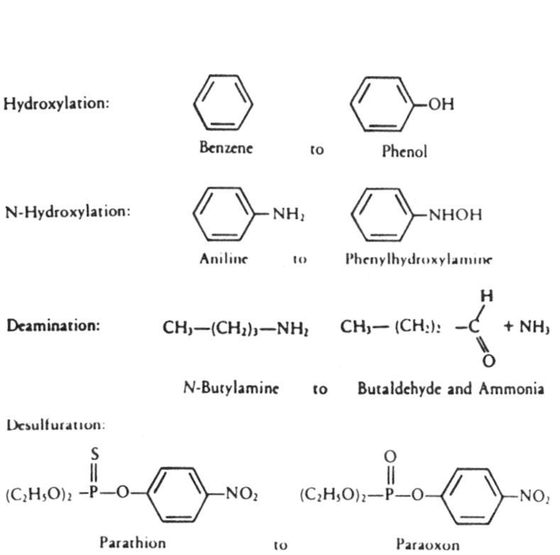

FIG. 15-2. The metabolic transformation of *n*-hexane. (From Spencer PS, Bischoff MC, Schaumburg HH. On the specific molecular configuration of neurotoxic aliphatic hexacarbon compounds causing central-peripheral distal axonopathy. Toxicol Appl Pharmacol 1978;44:17–28. Modified from Divincenzo GD, Kaplan CJ, Dedinas J. Characterization of the metabolites of methyl-*n*-butyl ketone, methyl iso-butyl ketone, and methyl ethyl ketone in guinea pig serum and their clearance. Toxicol Appl Pharmacol 1976;36:511–522.)

Hydroxylation:

Benzene   to   Phenol

N-Hydroxylation:

Aniline   to   Phenylhydroxylamine

Deamination:

$CH_3—(CH_2)_3—NH_2$ $CH_3—(CH_2)_2—\overset{H}{\underset{O}{C}}$ + NH_3

N-Butylamine   to   Butaldehyde and Ammonia

Desulfuration:

$(C_2H_5O)_2—P—O$—NO_2 $(C_2H_5O)_2—P—O$—NO_2

Parathion   to   Paraoxon

FIG. 15-3. Examples of oxidation.

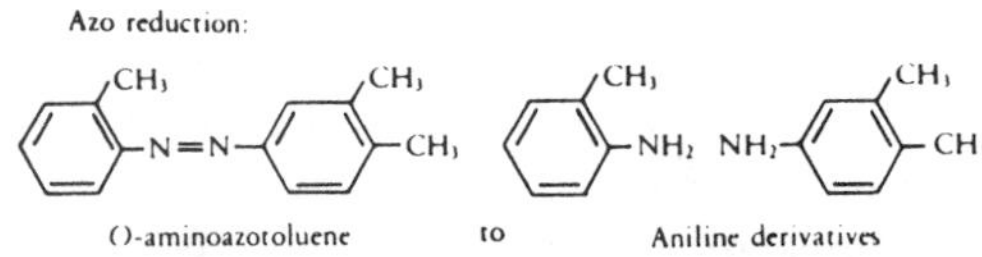

Azo reduction:

O-aminoazotoluene   to   Aniline derivatives

FIG. 15-4. Example of azo reduction.

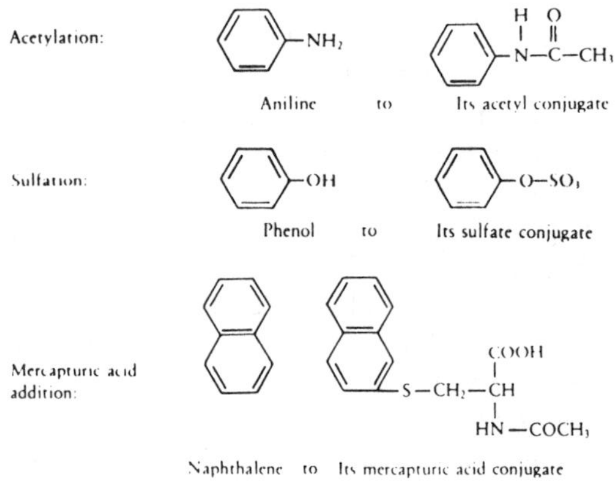

Acetylation:

Aniline   to   Its acetyl conjugate

Sulfation:

Phenol   to   Its sulfate conjugate

Mercapturic acid addition:

Naphthalene   to   Its mercapturic acid conjugate

FIG. 15-6. Examples of conjugation reactions.

based on the N=N bond, have been known since the 1930s to include many mutagens and carcinogens. Reduction of the azo bond yields aromatic amines, such as aniline and benzidine, which are probably the active carcinogens (Fig. 15-5). In mammalian test species, the principal target sites are the bladder and liver. Not all azo dyes are carcinogenic; extensive investigation of these compounds has yielded a rich body of information about structure-activity relations.

Conjugation involves combining a toxin with a normal body constituent. The result usually is a less toxic and more polar molecule that can be more readily excreted. However, conjugation can be harmful if it occurs in excess and depletes the body of an essential constituent. Figure 15-6 provides examples of conjugation reactions. Sulfation is the major means of preparing phenol for excretion. It is also used for alcohols, amines, and other groups. Sulfation and acetylation exemplify the sequential processing of substances by phase I and phase II reactions. Phenol and aniline can themselves be metabolites of other toxins, as previously illustrated; they are then conjugated and excreted. The addition of mercapturic acid (N-acetylcysteine) is a multistep process that proceeds through the addition of glutathione and subsequent cleavage to cysteine derivatives. This reaction is extremely important in handling reactive electrophilic compounds that result from exogenous exposure or endogenous metabolic processes. Polycyclic aromatic hydrocarbons and polyhalogenated

hydrocarbons are predominantly excreted in this way.

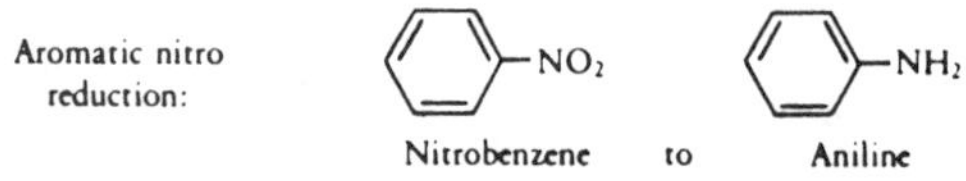

Hydrolysis is a common reaction in a variety of biochemical pathways. Esters are hydrolyzed to acids and alcohols, and amides are hydrolyzed to acids and amines.

As mentioned previously, various combinations of these reactions may be assembled in response to the same toxin. Metabolic strategies for a particular toxin vary widely among species, so an animal study, to be applicable to humans, must use a species with pathways similar to those of humans.

The most prominent enzyme system for performing phase I reactions is the cytochrome P-450 system, also known as the mixed-function oxygenase system. These enzymes are found in the endoplasmic reticulum of hepatocytes and other cells. Advances in molecular biology have greatly expanded our understanding of cytochrome P-450. Dozens of distinct P-450 genes have been identified, and many have been sequenced. They have been grouped into eight distinct families, and specific functions have been identified for many. For example, the enzyme CYP1A1 metabolically activates polycyclic aromatic hydrocarbons; the enzyme CYP2D6 is responsible for metabolizing such medications as β-blockers, tricyclic antidepressants, and debrisoquin; and the enzyme CYP2E1 bioactivates vinyl chloride, methylene chloride, and urethane.

Aromatic nitro reduction:

Nitrobenzene   to   Aniline

FIG. 15-5. Example of aromatic nitro reduction.

These insights, in turn, have helped explain why people can vary widely in their metabolic activity after similar exposures. Polymorphism in the genes that code for various P-450 proteins has been shown to result in different metabolic phenotypes. For example, people whose CYP2D6 phenotype makes them poor metabolizers of debrisoquin are at risk of various adverse drug reactions, whereas extensive metabolizers are at increased risk of lung cancer, probably because of carcinogenic metabolites they produce.

Any enzyme system has a finite capacity. When a preferred pathway is saturated, the remaining substrate may be handled by alternative pathways. (Most substrates can be metabolized by more than one enzyme system.) However, in some instances when a preferred metabolic pathway is saturated, the substrate may persist in the body and exert toxic effects. An example is dioxane metabolism. Dioxane (cyclic-$OCH_2CH_2OCH_2CH_2$) is a solvent with a variety of industrial applications, including painting, printing, and textile manufacturing. High exposure in rats causes hepatocellular and renal tubular cell damage as well as hepatic and nasal carcinoma. Based on rat studies, the principal human metabolite of dioxane has been identified as β-hydroxyethoxyacetic acid (HEAA). HEAA is found in the urine of exposed workers at more than 100 times the concentration of dioxane, suggesting that at low exposures dioxane is rapidly converted to HEAA. However, the metabolic pathway from dioxane to HEAA can be saturated by high-dose exposure, and in rats this event has been correlated with toxicity. Therefore, the effects of dioxane appear to be most pronounced when the ordinary metabolic pathway is saturated, allowing high levels of the solvent to accumulate.

A particular form of enzyme saturation is *competitive inhibition*. This may be a mechanism of toxicity, as when organophosphate pesticides compete with acetylcholine for the binding sites on cholinesterase molecules, or when metals such as beryllium compete with magnesium and manganese for enzyme ligand binding. However, competitive inhibition is also important in the metabolism of toxins. For example, methyl alcohol is oxidized by the enzyme alcohol dehydrogenase to the optic nerve toxin formaldehyde. This process can be blocked by large doses of ethanol, which competes for the binding sites of the enzyme and slows the formation of the toxic metabolite. The drug fomepizole acts in the same way, by selectively inhibiting alcohol dehydrogenase. This drug has been used to treat ethylene glycol poisoning, preventing the formation of the toxic metabolites glycolic acid and oxalic acid.

A less salutary example of competitive inhibition is the synergy demonstrated by two organophosphate pesticides, malathion and ethyl (*p*-nitrophenyl) phosphonothionate (EPN). Although they are structurally similar, EPN is far more toxic than malathion. When the two pesticides are present together, EPN competes for the enzyme that would ordinarily hydrolyze and thereby detoxify malathion. As a result malathion persists at unusually high concentrations, and the combined toxicity is far greater than would be expected. Other pairs of organophosphates that interact similarly are malathion with trichlorfon (Dipterex), and azinphos-methyl (Gusathion) with trichlorfon. Because most workplace exposures involve multiple substances, such synergistic effects are common but probably go unrecognized most of the time. This synergy is important to remember when evaluating reports on the toxicity of individual substances.

The enzyme systems that metabolize xenobiotics are not static. When demand is high, their synthesis can be enhanced through a process called *enzyme induction*. The resulting increase in enzyme activity helps the organism respond to subsequent exposures, not only to the original xenobiotic but to other similar substances. DDT and methyl cholanthrene are examples of substances known to induce metabolic enzymes.

People vary in their capacity for biotransformation in several ways. Two have already been mentioned: genetic differences and enzyme induction. Other factors that also account for interindividual differences in metabolism include general health, nutritional status, and concurrent medications.

PRINCIPLES OF BIOLOGIC EFFECTS

The aspect of toxic substances of greatest medical concern is their biologic effects. Several concepts have been developed to help classify and account for these effects.

Exposure to a toxic agent, and likewise the biologic response, may be either acute or chronic. An acute exposure often evokes an acute response, and a chronic exposure a chronic response, but neither sequence is invariable. For example, a short-term exposure to asbestos can result many years later in the development of mesothelioma, a fatal neoplasm.

A biologic effect may be either *reversible* or *irreversible,* independent of its time course. Chronic lead poisoning can cause hematologic, renal, neurologic, gastrointestinal, and reproductive effects, and all but the late renal and neurologic effects are reversible. Acute mercury poisoning can produce irritability, tremors, delirium, or outright psychosis, all of which may be irreversible. Reversibility is a function of the nature of the damage done and the regenerative capacity of the damaged tissue.

Generally, higher levels of toxic exposure lead to greater responses. They may occur within an individual, as when higher concentrations of inhaled carbon monoxide lead to higher concentrations of carboxyhemoglobin, or within a population, as when higher levels of benzidine dye exposure cause an elevation in the incidence of bladder cancer. Either relation may be depicted with the familiar *dose-response curve* (Fig. 15-7).

The dose-response curve quantifies the dependence of biologic effects on dose levels. Note that the curve in Fig. 15-7 has the common sigmoidal shape. At the lower end, the beginning of a linear increase reflects the existence of a *threshold dose,* below which

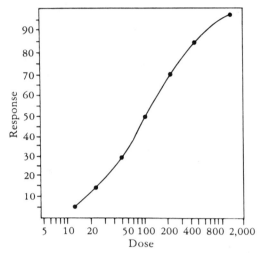

FIG. 15-7. The dose-response curve. (From Klaassen C, Amdur M, Doull J, eds. Casarett and Doull's toxicology, 3rd ed. New York: Macmillian, 1986, p. 18.)

variations in exposure presumably have no effect. At the upper end, the flattening of the curve reflects a *ceiling* level of maximal response that cannot be increased by greater doses. This level might correspond to death in an individual or to 100% cumulative incidence of disease in a population.

Several additional toxicologic concepts emerge from the dose-response curve. The LD_{50} is the dose that is lethal to 50% of a population; it is a measure of the *potency* of a compound, or the dose required to produce a certain effect. Potency should be distinguished from another pharmacologic measure, *efficacy,* which reflects the maximal effect a drug can produce (the ceiling on a dose-response curve). There are other standards that might be used. For example, cell killing by radiation is sometimes quantified by the "D_0" which corresponds to an LD_{63} on the exponential cell-kill curve.

Individual departures from the expected dose-response pattern can take several forms. *Hypersusceptibility* indicates an unusually high response to some dose of a substance. This term requires careful interpretation, however, because it is used in several different ways. It may refer to a genetic predisposition to a toxic effect, as discussed previously; it may indicate a statistically defined deviation from the mean; it may reflect an observer's subjective impression; or it may be used incorrectly as a synonym for hypersensitivity. *Hypersensitivity* is one form of hypersusceptibility; it is characterized by an acquired, immunologically mediated sensitization to a substance. In workplace settings this sensitization is most commonly manifested in pulmonary or dermatologic responses with features of all four immunologic reaction types described by Gell and Coombs (1). For example, toluene diisocyanate, a major ingredient in polyurethane manufacture, evokes asthmatic reactions in a small percentage of exposed workers even at permissible exposure levels (see Chapter 25), and workers exposed to nickel can develop a skin reaction that resembles chronic eczema (see Chapter 27). It is important to remember that such so-called hypersensitivities need not be aberrant or even unusual. Among epoxy workers, for instance, the incidence of skin sensitization to resins is approximately 50%, and at least 75% of long-term employees relate a history of dermatitis.

Hyposusceptibility, conversely, indicates an unusually low response to some dose of a substance, and it also is defined in a variety of ways. One form of hyposusceptibility is *tolerance,* a diminished response to a given dose of a chemical after repeated exposures. Tolerance classically involves an attenuation of the immune response to a specific antigen; it appears graphically as a shift to the right of the dose-response curve. In experimental animals tolerance can be induced by administration of antigen in certain forms or doses. This is the basis of desensitization treatments for allergies, and a similar phenomenon may well occur in workplace settings after chronic low-dose exposures. However, several equally conceivable events could yield an empirically similar result; these include impairment of absorption, induction of metabolizing enzymes, and enhancement of excretion. Typically, such acquired resistance to a chemically induced response can be overcome with sufficiently large doses. *Tachyphylaxis,* in contrast, is a rapidly acquired resistance that persists even with a large subsequent dose. In view of the variety of mechanisms of hypersusceptibility and hyposusceptibility, an individual worker may readily show a dose responsiveness that differs significantly from the population norm.

Many factors, some already mentioned, determine the localization of effect of a chemical exposure. These factors include translocation barriers (e.g., membranes) that limit migration of the chemical; chemical affinities that concentrate the chemical on certain molecules; and regional differences in metabolic activity, in susceptibility to a given toxic effect, or in capacity for repair.

A useful classification of chemical-induced toxic effects, based in part on the work of Loomis and Hayes (2), is as follows:

Normal (or expected) effects, depending on:
    Physical actions
    Nonspecific caustic or corrosive actions
    Specific toxicologic actions
    Production of pathologic sequelae
Abnormal (or unexpected) effects
    Immune mechanisms
    Genetic susceptibility

Normal effects can be repeatedly induced, do not require preconditioning, and occur with an incidence that is primarily dose-dependent. These range from simple physical phenomena, such as the displacement of oxygen by asphyxiants, to more elaborate pathophysiologic events. Immune mechanisms, in contrast, require preconditioning (sensitization) and may not be inducible in most members of a population simply by increasing the exposure. Finally, genetic susceptibility is in most cases not based on sensitization and may be nothing more than a quantitatively greater propensity for development of some "normal" response. An example is the elevated risk of hemolysis in people with glucose 6-phosphate dehydrogenase (G6PD) deficiency after exposure to oxidants such as naphthalene. Immune and genetic effects may coincide in heritable immunodeficiency diseases.

EXCRETION

Because biotransformation tend to make compounds more polar and less fat-soluble, the outcome of this process is that toxins can be more readily excreted from the body. The major route of excretion of toxins and their metabolites is through the kidneys (see Chapter 35). The kidneys handle toxins in the same way that they handle any serum solutes: passive glomerular filtration, passive tubular diffusion, and active tubular secretion. The daily volume of filtrate produced is about 200 L—five times the total body water—in a remarkably efficient and thorough filtration process.

Smaller molecules can reach the tubules through passive glomerular filtration, because the glomerular capillary pores are large enough (40 Å) to admit molecules as large as about 70,000 Da. However, substances bound to large serum proteins are excluded and must undergo active tubular secretion to be excreted. The tubular secretory apparatus apparently has separate processes for organic anions and organic cations, and, like any active transport system, these processes can be saturated and competitively blocked. Finally, passive tubular diffusion out of the serum probably occurs to some extent, especially for certain organic bases.

Passive diffusion also occurs in the opposite direction, from the tubules to the serum. As with any of the membrane crossings discussed previously, lipid-soluble molecules are reabsorbed from the tubular lumen much more readily than are polar molecules and ions; this explains the practice, already mentioned, of alkalinizing the urine to hasten the excretion of acids.

A second major organ of excretion is the liver (see Chapter 34). The liver occupies a strategic position because the portal circulation promptly delivers compounds to it after gastrointestinal absorption. Furthermore, the generous perfusion of the liver and the discontinuous capillary structure within it facilitate its filtration of the blood. Therefore, excretion into the bile is potentially a rapid and efficient process.

Biliary excretion is somewhat analogous to renal tubular secretion. There are specific transport systems for organic acids, organic bases, neutral compounds, and possibly metals. There are active transport systems with the ability to handle protein-bound molecules. Finally, re-uptake of lipid-soluble substances can occur after secretion, in this case through the intestinal walls.

Marked variations in biliary secretion can exist. Liver disease can compromise the process, and some chemicals such as phenobarbital and some steroids, in addition to inducing hepatic metabolic enzymes, can actually increase bile flow and hence biliary excretion. The effects of some chemicals have practical applications. For example, certain steroids have been demonstrated to decrease mercury toxicity in animals; this is attributed, at least in part, to their effect on biliary excretion.

Toxins that are secreted with the bile enter the gastrointestinal tract and, unless reabsorbed, are secreted with the feces. Materials ingested orally that are not absorbed and materials that are carried up the respiratory tree and swallowed are also passed with the feces. All of this may be supplemented by some passive diffusion through the walls of the gastrointestinal tract, but this is not a major mechanism of excretion.

Volatile gases and vapors are excreted primarily by the lungs. The process is one of passive diffusion, governed by the difference between plasma and alveolar vapor pressure. Volatiles that are highly fat-soluble tend to persist in body reservoirs and to take some time to migrate from adipose tissue to plasma to alveolar air. Less fat-soluble volatiles, on the other hand, are exhaled fairly promptly, until the plasma level has decreased to that of ambient air. The alveoli and bronchi can sustain damage when a vapor such as gasoline is exhaled, even if the initial exposure occurred percutaneously or through ingestion.

Other routes of excretion, although of minor significance quantitatively, are important for a variety of reasons. Excretion into mother's milk obviously introduces a risk to the infant, and because milk (pH 6.5) is more acidic than serum, basic compounds are concentrated in milk. Moreover, owing to the high fat content of breast milk (3% to 5%), fat-soluble substances such as DDT can also be passed to the infant (see Chapter 31). Some toxins, especially metals, are excreted in sweat or laid down in growing hair, which may be of use in diagnosis. Finally, some materials are secreted in the saliva and may then pose a gastrointestinal exposure hazard.

This chapter has presented basic principles of toxicology that are fundamental to understanding the effects of toxic substances and the treatment and prevention of these effects. In the chapters that follow (particularly Chapters 16 and 25 through 35), the types of biologic effects caused by workplace toxins are explored in detail. Readers interested in particular toxins should refer to Appendix A and the Index.

REFERENCES

1. Holmberg PC. Central nervous defects in two children of mothers exposed to chemicals in the reinforced plastics industry. *Scand J Work Environ Health* 1977;3:212.
2. Husman K. Symptoms of car-painters with long-term exposure to a mixture of organic solvents. *Scand J Work Environ Health* 1980;6:19.
3. Coombs RAA, Gell PGH. Classification of allergic reactions responsible for clinical hypersensitivity and disease. In: PGH Gell, RAA Coombs, PJ Lachmann, eds. Clinical aspects in immunology. Oxford: Blackwell, 1963:363.
4. Loomis TA, Hayes AW. Loomis's Essentials of Toxicology, 4th ed. New York: Academic Press, 1996.

BIBLIOGRAPHY

Brain JD, Valberg PA. Aerosol deposition in the respiratory tract. Am Rev Respir Dis 1979;120:1325–1373.
The classic article on airways deposition, with a discussion of major classes of inhaled particles and their effects on health.

Clayton GD, Clayton FE, eds. Patty's industrial hygiene and toxicology, Vol. II. New York: Wiley-Interscience, 1993–1994.
The mother of all industrial toxicology references. Volume II is dedicated to toxicology; it consists of six separate volumes, totaling over 5,000 pages, organized for the most part by classes of chemicals.

Klaassen CD, ed. Casarett and Doull's toxicology: the basic science of poisons, 5th ed. New York: McGraw Hill, 1996.
The standard toxicology text with chapters on general principles, individual body systems, and specific families of toxins.

Lauwerys RR, Hoet P. Industrial chemical exposure: guidelines for biological monitoring, 2nd ed. Boca Raton, FL: Lewis, 1993.
A review of the metabolism and excretion of various substances geared toward rational use of biological monitoring tests.

Lewis DFV. Cytochromes P450: structure, function, and mechanism. London: Taylor & Francis, 1996.
For more detail on the critically important cytochrome P-450 system.

Loomis TA, Hayes AW. Loomis' essentials of toxicology, 4th ed. Academic Press, 1996.
A basic, highly readable text.

Sullivan JB, Krieger GR. Hazardous materials toxicology: clinical principles of environmental health. Baltimore: Williams & Wilkins, 1992.
One of the best of the clinically oriented environmental toxicology texts.

Occupational Health: Recognizing and Preventing Work-Related Disease and Injury, 4th ed.
Edited by Barry S. Levy and David H. Wegman, Lippincott Williams & Wilkins, Philadelphia © 2000.

16 Carcinogens

Howard Frumkin and Michael Thun

A 67-year-old retired insulator develops a persistent cough and is diagnosed with lung cancer. He asks the chest surgeon whether his workplace exposure to asbestos over the years could have caused his lung cancer.

At a tire factory with 220 employees, five develop bladder cancer over a 6-year period. The plant manager calls the family physician, who also serves as plant physician, and poses a worried question: Is some exposure in our plant causing bladder cancer? What should we do about it?

A young couple brings their 18-month-old to the pediatrician's office for a routine checkup. They are concerned: "We have read that some pesticides in baby food, such as captan and carbaryl, might be carcinogenic. We have read that plasticizers in our baby's teething ring and other toys might be carcinogenic. We have read that when we fill up our car with gas, there are carcinogenic benzene fumes. We have even read that too much time playing outside in the sun can be carcinogenic. Should we be trying to limit our baby's exposure to these things? How should we do it?"

A computer company has a site with 1,200 employees, who engage in research and development, sales and service, and repairs. The human resources director knows that, as the workforce ages, the risk of cancer will rise. She also knows that the workplace provides an opportunity for early cancer

H. Frumkin: Department of Environmental and Occupational Health, Rollins School of Public Health, Emory University, Atlanta, GA 30322.

M. Thun: Department of Epidemiology and Surveillance Research, American Cancer Society, Atlanta, GA 30329.

detection through screening, such as mammography. She wonders which screening programs she should implement.

WHAT IS CANCER?

Cancer is the name given to a broad category of diseases that arise in various organs and tissues throughout the body. Cancer cells have in common damage to the genes that regulate cell growth. They grow and divide in a rapid, uncontrolled manner; they lose their differentiation; and they survive for abnormally long times. Because cancer originates with changes at the level of DNA, cancer is a genetic disease. And because something triggers these changes—an endogenous or exogenous chemical, a physical agent such as ionizing radiation, or a biologic agent such as viruses, *Helicobacter*, or aflatoxin—cancer is also an environmental disease. This chapter focuses on one class of environmental carcinogens: those occurring in the workplace.

Cancer is a major public health problem. Each year, approximately 1.2 million Americans are diagnosed with invasive cancer and slightly less than half this number die of various cancers. Cancer accounts for about one-fourth of all deaths. Among men, prostate cancer is the most common incident cancer, followed by lung cancer and colorectal cancer; among women, breast cancer is the most common diagnosis, followed by lung cancer and colorectal cancer. In both sexes, the three leading cancer sites account for more than half of new cases. Because survival is worse for lung cancer than for these other common types, lung cancer is the most common cause of cancer death among both men and women.

Cancer incidence and mortality patterns have shifted dramatically during the 20th century. In the United States and most developed countries, lung cancer has increased sharply, mostly because of smoking, and stomach cancer has declined to a low level, probably because of advances in food preservation and the increased availability of fresh fruits and vegetables. Cervical and uterine cancer mortality rates have declined because of early detection and treatment, and the mortality rate of colorectal cancer has declined substantially, especially in women. Incidence rates for some less common cancers, such as multiple myeloma, non-Hodgkin's lymphoma, melanoma, and brain cancer, have been increasing over time, but except for melanoma these appear to be stabilizing or even decreasing. Better diagnostic methods are part of the explanation, but because these cancers have been associated with workplace and environmental exposures, the increases are cause for concern. As cardiovascular disease mortality rates continue to decline, cancer will assume greater relative importance.

Cancer also has an enormous emotional impact. Perhaps because it often develops insidiously, it is often difficult to treat, or, as one author suggests (1), it offers an opportunity for attribution, blame, and compensation, cancer evokes deep concern from the American public. Nowhere is this more true than in the workplace setting, where occupational cancer has been a major part of the occupational health agenda.

The Process of Carcinogenesis

Our understanding of carcinogenesis is extensive but still incomplete. Based on experimental work with mouse skin in the 1940s, two broad stages in carcinogenesis were identified: initiation and promotion. *Initiation* was defined as the first event, when an irreversible change occurred in the cell's genetic material. *Promotion* consisted of one or more subsequent steps, when intracellular and extracellular factors allowed the transformed cell to develop into a focal proliferation such as a nodule, then to a locally invasive tumor, and finally to metastases. With advances in molecular biology, we now know that carcinogenesis is a more complex process, often involving multiple genetic alterations that compete with cellular repair and apoptosis (autodestruction of damaged cells). However, the concept of early and late stages remains helpful in understanding temporal relationships in carcinogenesis.

The development of a cancer involves damage to cellular DNA. This damage may lead to cell death, it may be repaired by ongoing processes in the cell nucleus, or, if the cell divides first, it may give rise to an irreversible, heritable genetic lesion termed a *mutation*. Mutations can be induced directly or indirectly, by carcinogens or by metabolic products of the cell itself. Mutations from endogenous metabolism account for many of the approximately 1 million lesions that occur in a person's DNA each day. Among the exogenous carcinogens, direct-acting agents require no metabolic or chemical changes to produce lesions in DNA. Bischloromethyl ether, for example, directly binds to DNA, creating DNA adducts that are known to be processed into mutations at a high frequency. Another direct-acting agent is ionizing radiation, which acts by causing DNA strand breakage. Other chemicals, such as the polycyclic aromatic hydrocarbons generated by coke ovens or foundry operations and β-naphthylamine, which is used to make dyes, must be converted to electrophilic metabolites before they can bind DNA and induce mutagenic changes.

As the genetic basis of cancer was increasingly recognized, the critical genes in cancer development were initially categorized into two classes: oncogenes and tumor suppressor genes. *Oncogenes*, or genes that accelerate cell growth, were originally recognized during studies of RNA viruses. Some of these viruses were found to code for genetic sequences (oncogenes) that, when inserted into host genomes, could cause malignant transformation. It later became clear that nascent forms of oncogenes, called *proto-oncogenes,* are common in many human and animal cells, encoding proteins that regulate normal cell growth and differentiation. If they are transformed into oncogenes, their products code for oncoproteins that act as growth factors, membrane receptors, protein kinases, or other factors that increase cell proliferation and, in some cases, dedifferentiation. One of the best-studied examples is the *ras* oncogene, which was first identified in rat sarcomas. This oncogene can be activated by polycyclic aromatic hydrocarbons, *N*-nitroso compounds, and ionizing radiation, and has been found in a wide variety of human cancers, including bladder, lung, and other cancers of occupational and environmental importance.

Tumor suppressor genes, or anti-oncogenes, are also important in carcinogenesis. Ordinarily these function to regulate cell growth and stimulate terminal differentiation. When inactivated, they fail to perform these functions, allowing neoplastic transformation to proceed. The most commonly identified example is the gene that produces the protein p53, which is located on chromosome 17. Mutations in the *TP53* gene have been identified in many cancers, including those of the colon, lung, liver, esophagus, breast, and reticuloendothelial and hematopoietic tissues, and in the Li-Fraumeni syndrome of familial multiple cancer susceptibility. Carcinogenic exposures such as aflatoxin and hepatitis B virus have been associated with specific mutations on the *TP53* gene, suggesting that some carcinogens may leave a unique genomic "signature." Although this may have clinical applications, as discussed later, present evidence suggests that these characteristic mutation patterns are probably more relevant in population research than to individuals.

In addition to oncogenes and tumor suppressor genes, genes that affect tumor progression are now also being recognized. The initiation, promotion, and progression of cancers comprises a sequence of genetic events that "activate" proto-oncogenes (to active oncogenes), "inactivate" tumor suppressor genes, and cause other cellular events that facilitate tumor progression. These events confer a growth advantage to the cell, leading to its proliferative clonal expansion. The population of clonally expanded mutant cells is then primed for subsequent mutations in other critical target genes. These mutations, in turn, lead to further growth of the cells and eventually to the transformed cell population that appears clinically as cancer. Any of the genetic lesions or metabolic processes along this pathway may be increased by exposure to carcinogens. It is therefore helpful to think of carcinogenesis as a series of probabilistic events rather than as a single occurrence.

All of these steps take time, sometimes many years. The term *latency* refers to the time that elapses between the initial damage caused by a carcinogen and the clinical detection of resulting cancers. This period presumably corresponds to the stages of initiation, promotion, and progression. Because of latency considerations, screening of workers who are at risk for cancer should focus on the time after the initiating lesion and before clinical presentation. If screening is conducted too soon after the onset of exposure, when no increase in risk is expected, the yield will be low, the benefit will not justify the cost, and negative test results may provide false reassurance. The correct timing of screening varies with the natural history of each cancer. The latency period for hematologic malignancies is in the range of 4 to 5 years, whereas that for solid tumors is measured in decades.

The existence of *threshold* levels of exposure to carcinogens has been controversial. A threshold is defined as a safe level of exposure to a carcinogen, below which carcinogenesis does not occur. The one-hit hypothesis, which grew out of early concepts of initiation and was consistent with the clonal nature of cancers, held that even a single mutation in a single cell could initiate a malignancy. As a result, the argument went, absolutely no exposure to a carcinogen could be considered safe. It remains difficult to generate definitive evidence on this point, because both epidemiologic and experimental data are inherently uninformative at very low exposure levels.

However, as our understanding of the molecular events of carcinogenesis has advanced, several arguments suggest that thresholds may exist, even if they cannot be empirically demonstrated. First, DNA damage occurs routinely, and numerous repair mechanisms have evolved to correct it. Second, certain carcinogens, such as trace elements, hormones, and oxygen free radicals from metabolism, are ubiquitous and even essential at low doses; it is argued that these substances are carcinogenic only at higher doses. Third, factors such as hormones may act epigenetically, stimulating cell division rather than directly affecting DNA. These typically have reversible effects, implying that a threshold exists.

In the absence of firm data on the effects of low-dose exposures to carcinogens, regulators have often adopted a conservative approach, assuming that no safe thresholds exist and banning such exposures altogether. More recently, the trend has been to define some acceptable level of risk, such as a one-in-a-million lifetime risk. Regulators then calculate an exposure level thought to confer less than that risk, and permit exposures at or below that level.

Interaction is another important concept in occupational carcinogenesis. This is defined as the joint effect of two or more carcinogens differing from what would have been predicted based on their individual effects. *Synergy,* in which joint effects exceed the combined individual effects, and *antagonism,* in which joint effects are less than combined individual effects, are two examples of interaction. A classic example is the relation between asbestos exposure and cigarette smoking, which was revealed in early studies by Selikoff and colleagues. The relative risk of dying from lung cancer was increased about 5-fold after asbestos exposure alone, about 10-fold after smoking alone, and about 80-fold after exposure to both substances (2).

Interaction is a complicated concept, in part because statisticians, biologists, and lay people approach it differently. For the statistician, interaction is defined by mathematical models; if a combined effect differs from that predicted by a model, then synergy (or antagonism) is said to exist. In modern epidemiologic analyses, logistic regression is commonly used, and synergy is signalled by the presence of *effect modification,* a term indicating statistically significant interaction. Because of the assumptions of logistic regression, effect modification implies departures from strictly multiplicative relations between different factors associated with cancer. On the other hand, simple additive relations among various causes of cancer may be more important in the public health context.

Endocrine Mechanisms

In recent years, environmental and occupational health researchers have increasingly focused on the health effects of hormonally active compounds. An environmental endocrine disrupter is defined as an exogenous agent that interferes with the synthesis, secretion, transport, binding, action, or elimination of natural hormones in the body (3). This concept is relevant to cancer because hormonal action, especially that of the sex hormones, influences several common cancers.

An early example of cancer from an exogenous estrogen was diethylstilbesterol (DES), which caused vaginal carcinomas in the daughters of women who took the drug during pregnancy. According to the endocrine disrupter hypothesis, other compounds may exert estrogenic effects, including some that are found in the workplace: polycyclic aromatic hydrocarbons; chlorinated organics such as polychlorinated biphenyls (PCBs), dioxins, and furans; phthalates; and some pesticides. These may exert direct hormonal effects on such outcomes as fertility, sperm production, and endometriosis. They also may influence the development of cancers, including those of the breast, testicles, endometrium, and perhaps prostate. The relation of industrial and agricultural chemicals to hormone-related cancers in humans remains controversial. One study showed an association between breast cancer and dichlorodiphenyldichloroethylene (DDE), but several others did not replicate that association. Some but not all studies of farmers show an increase in prostate cancer, which may be related to pesticide exposure, although studies of farmers' wives do not show increased risk of hormone-related cancers. Complicating the debate, a number of hormonally active compounds, especially naturally occurring plant phytoestrogens, may reduce cancer risk. Much remains to be learned about possible hormone-mediated environmental and occupational carcinogenesis.

CAUSES OF CANCER

Environmental and Occupational Causes of Cancer: Early Recognition

It is now so widely recognized that environmental (i.e., nongenetic) factors, including workplace exposures, can contribute to cancer that it is easy to forget how recently this awareness developed. In 1761, Dr. John Hill of London published the first modern clinical report of environmental carcinogenesis—a description of cancer of the nasal passages among tobacco snuff users. In 1775, another perceptive London physician, Dr. Percival Pott, became the first to recognize an occupational cancer—scrotal skin cancer among chimney sweeps, who were heavily exposed to soot in their work. In the 1800s, skin cancer was linked with occupational exposure to inorganic arsenic, tar, or paraffin oils (now known to contain polycyclic aromatic hydrocarbons), and bladder cancer was linked with workplace exposure to certain dyes. Experimental evidence of carcinogenesis first appeared in the early 20th century. In 1918, two Japanese scientists, Yamagiwa and Ichikawa, induced skin cancer in mice using coal tar.

In 1935, the first case report of lung cancer in a patient with asbestosis was published.

Over the years, other carcinogens have been identified, both in the workplace and in the general environment. Examples and the situations in which they were first recognized include vinyl chloride monomer (which caused a cluster of hepatic hemangiosarcomas in a chemical plant), benzene (which was linked with high leukemia rates in the Turkish shoe industry), and coke oven emissions (which were found to increase lung cancer risk among exposed steelworkers). In some industries, an increased incidence of cancer has been recognized from epidemiologic studies but no specific carcinogen has yet been identified. Definite occupational carcinogens are shown in Tables 16-1 and 16-2, and probable carcinogens in Tables 16-3 and 16-4. However, an Environmental Protection Agency (EPA) analysis found that of the approximately 3,000 chemicals ever produced in the United States at an annual volume of at least 1 million pounds, only 7% have been fully evaluated for toxicity, and a much smaller percentage with long-term carcinogenicity bioassays; therefore, the information in these tables must be regarded as incomplete.

TABLE 16-1. *Industrial processes involving exposure to established human occupational carcinogens (IARC[a] Group 1)*

Aluminum production
Auramine manufacturing
Boot and shoe manufacturing and repair
Coal gasification
Coke production
Furniture and cabinet making
Hematite mining (with radon exposure)
Inorganic acid mists, strong, containing sulfuric acid
Iron and steel founding
Isopropyl alcohol manufacturing (strong acid process)
Magenta manufacturing
Painting
Rubber industry

[a] International Agency for Research on Cancer.
Adapted and updated from Stellman JM, Stellman SD. Cancer and the workplace. Ca *Cancer J Clin* 1996;46:70–92. Up-to-date IARC evaluation data can be found at the IARC Web site, http://www.iarc.fr, or more specifically at the Monographs Database web page, http://www.193.51.164.11/.

TABLE 16-2. *Established human occupational carcinogens (IARC Group 1)[a]*

Exposures	Examples of occurrence	Target organ/comment
Aflatoxins	Grains, peanuts (farm workers)	Liver
4-Aminobiphenyl	Rubber industry	Bladder
Arsenic and its compounds	Insecticides	Lung, skin, hemangiosarcoma
Asbestos	Insulation, friction products	Lung, mesothelioma, respiratory tract, gastrointestinal system
Benzene	Chemical industry	Leukemia
Benzidine	Rubber and dye industries	Bladder
Beryllium and its compounds	Aerospace, nuclear, electric and electronics industries	Lung
bis-Chloromethyl ether and chloromethyl methyl ether	Chemical industry	Lung
Cadmium and its compounds	Metalworking industry, batteries, soldering, coatings	Prostate
Chromium (VI) compounds	Metal plating, pigments	Lung
Coal tar pitches	Coal distillation	Skin, scrotum, lung, bladder
Coal tars	Coal distillation	Skin, lung
Dioxin (2,3,7,8-tetrachlorodibenzo-p-dioxin)	Herbicide production and application.	All sites combined, lung
Erionite	Environmental (Turkey)	*See* asbestos
Ethylene oxide	Sterilant in health care settings; chemical component	Lymphoma, leukemia
Hepatitis B and C virus	Health care settings	Liver
Human immunodeficiency virus	Health care settings	Sarcoma
Mineral oils	Machining, jute processing	Skin
Mustard gas	Production, war gas	Lung
2-Naphthylamine	Rubber and dye industries	Bladder
Nickel compounds	Nickel refining and smelting	Nose, lung
Radon and its decay products	Indoor environments, mining	Lung
Schistosoma hematobium infection	Farming and other outdoor work in endemic areas	Bladder
Shale oils	Energy production	Skin
Silica, crystalline	Hard rock mining, sandblasting, glass and porcelain manufacturing	Lung
Solar radiation	Outdoor work	Skin
Soots	Chimneys, furnaces	Skin, lung
Sulfuric acid–containing strong inorganic acid mists	Metal, fertilizer, battery, and petrochemical industries	Larynx, lung , ? nasal sinus
Talc (with asbestiform fibers)	Talc mining, pottery manufacturing	*See* Asbestos
Vinyl chloride	Plastic industry	Hemangiosarcoma
Wood dust	Wood and furniture industries	Nose, sinuses

[a] Other carcinogens, including medications (especially cancer chemotherapeutic agents, a risk for health care workers), foods, tobacco, and viruses, are classified in Group 1 but are not listed here.

TABLE 16-3. *Industrial processes involving exposure to probable human occupational carcinogens (IARC Group 2A)*

Glass manufacturing
Hairdressing/barbering
Insecticide application (nonarsenicals)
Petroleum refining (certain exposures)

Testing and Evaluation of Carcinogens

Several methods have emerged to identify carcinogens (Table 16-5). These include epidemiologic studies, animal studies, *in vitro* test systems, and analysis of structure-activity relations.

Epidemiologic studies are potentially the most definitive source of information on hu-

TABLE 16-4. *Probable human occupational carcinogens (IARC Group 2A)*[a]

Exposures	Examples of occurrence
Acrylamide	Polyacrylamide manufacturing
Benz[a]anthracene	Coal distillation
Benzidine-based dyes	Dye industry
Benzo[a]pyrene	Coal and petroleum-derived products
1,3-Butadiene	Polymer and latex production
Captafol	Fungicide
Creosotes	Wood preservatives
Dibenz[a,h]anthracene	Coal distillation
Diesel exhaust	Motor vehicles
Diethyl sulfate	Petrochemical industry
Dimethyl carbamoyl chloride	Chemical manufacturing
1,2-Dimethylhydrazine	Rocket propellants and fuels, boiler water treatments, chemical reactants, medicines, cancer research
Dimethyl sulfate	Former war gas, now used in chemical industry
Epichlorhydrin	Resin manufacturing, solvent
Ethylene dibromide	Fumigant, gasoline additive
Formaldehyde	Chemical manufacturing; tissue preservative
4,4'-methylene bis(2-chloroaniline) (MOCA)	Resin manufacturing
N-nitrosodiethylamine	Solvent
N-nitrosodimethylamine	Solvent
Polychlorinated biphenyls	Electrical equipment
Propylene oxide	Chemical industry
Styrene oxide	Chemical industry
Tetrachlorethylene	Dry cleaning
Toluenes, α-chlorinated	Chemical manufacturing
Toluidine, para-chloro-ortho and its strong acid salts	Diazo dye manufacturing
Trichloroethylene	Metal degreasing
1,2,3-Trichloropropane	Pesticide, rubber manufacturing, solvent
tris-(2,3-Dibromopropyl) phosphate	Flame retardant, polystyrene foam manufacturing
Ultraviolet radiation A, B, and C	Outdoor work
Vinyl bromide	Plastic industry
Vinyl fluoride	Chemical industry

[a] Other probable carcinogens, including medications (especially cancer chemotherapeutic agents, a risk for health care workers), infectious agents, and foods, are classified in Group 2A but are not listed here.

man carcinogenicity, because they are based on human exposures in real-life situations. All major epidemiologic study designs are used, including ecologic studies, cohort studies, and case-control studies (see Chapter 6). However, several considerations complicate the interpretation of occupational cancer epidemiology findings. One is that prolonged exposure to suspect chemicals is uncommon, so that even in large workplaces the number

TABLE 16-5. *Relative advantages of the major methods of carcinogenicity testing*[a]

Advantage	In vitro bioassay	Animal bioassay	Epidemiologic study
Low cost	+++	++	+
Rapidity	+++	++	+
Ease of performance	+++	++	+
Ability to control for multiple exposures	+++	+++	+
Ability to demonstrate thresholds	0	0	0
Validity of results as predictor of human carcinogenesis	NA	+	+++
Validity of results in identifying human target organs	NA	+	+++

[a] The grading scale used in this table is qualitative only. Specific methodologic features and appropriate applications of each method are discussed in the text.

of cancer cases observed in highly exposed workers is small. A second challenge is cohort definition and follow-up; if a group of workers was exposed to a suspected carcinogen years ago, it can be difficult to enumerate its members and trace their health outcomes. Another limitation pertains to exposure assessment. Job designations and even direct measurements may not accurately reflect worker exposures, making it difficult to test whether differing levels of exposure are associated with differing levels of cancer risk. Still another limitation pertains to outcome information. Death certificates are the prime source of diagnostic information in occupational cancer epidemiology. They provide information on mortality but not incidence, have fairly high rates of misdiagnosis for some cancers, and do not ascertain the presence of many nonfatal diseases. Confounding is another problem. For many cancers of interest, both occupational and nonoccupational exposures play a role, but information on the relevant nonoccupational exposures is often unavailable. Many of these limitations bias results toward the null hypothesis, or a finding of no effect. Careful consideration is necessary in interpreting epidemiologic results.

Moreover, epidemiologic studies are expensive and time-consuming, and they provide evidence of carcinogenicity after the fact, when large numbers of workers have already been exposed to a substance for several decades. For these reasons, animal and *in vitro* studies have gained wide importance.

Animal studies for carcinogenicity use many species, from *Drosophila* to higher mammals. In accordance with National Cancer Institute guidelines, a substance is usually tested at two doses in both sexes of two strains of rodents, with at least 50 animals in each test group (2 doses × 2 sexes × 2 strains = 8 test groups). Therefore, at least 400 animals are usually studied, in addition to the 100 or more control animals that receive a placebo. The testing usually takes about 2 years, but 4 years can elapse from the initial planning of such a study until the

final report has been completed, at a cost of several hundred thousand to more than one million dollars.

The major question about animal bioassays is their accuracy in predicting human cancer. Skeptics argue that animal studies are of limited value because animals and humans often have different sensitivities to carcinogens, and because the ability to activate and detoxify carcinogens may vary markedly among species. Of the several hundred chemicals that have been formally bioassayed, about half have yielded positive results. Not all of these animal carcinogens have been shown to cause cancer in humans, but most have not been studied epidemiologically. On the other hand, every human carcinogen that has been adequately tested in animals is an animal carcinogen. Moreover, between 25% and 30% of established human carcinogens were first identified through animal bioassays. In general, therefore, properly conducted and replicated animal bioassays are reasonable predictors of human cancer risk.

Because cost limits the size of animal studies, small elevations in cancer incidence tend to be statistically insignificant. However, even a small elevation in incidence, when multiplied by a large exposed human population, could imply a substantial number of preventable cancers. Consequently, to ensure that positive results do not escape notice, every effort is made to detect the carcinogenic effect of a test substance. One way this is done is by using as one of the test doses the *maximum tolerated dose* (MTD)— the highest dose that does not kill or acutely poison the test animals. (Often the other test dose is one-fourth of the MTD.) Critics have noted that the MTD introduces an element of unreality to bioassays: such a large dose far exceeds human exposure levels, may create an unrepresentative biochemical milieu in test animals, and may overwhelm natural protective mechanisms such as detoxification and repair systems. Further, the MTD clearly causes cell death, which stimulates cell division in nearby surviving cells. This mitogenesis, rather than direct interaction of the carcinogen with DNA, could account for apparent carcinogenic effects. Cells with a limited amount of genetic damage may be intensely promoted by the repopulation that occurs after MTD-induced cell death, causing additional genetic changes critical to tumor formation. However, even with the use of MTDs, most chemicals tested show no carcinogenic activity. This finding is often cited to support the use of high doses and to refute the belief that "everything causes cancer."

Another controversial issue raised by animal testing is the existence of threshold levels of exposure. Because animal tests necessarily use high-dose exposures, their results must be extrapolated down to low doses to reach conclusions about the effects of more realistic exposures. Uncertainty about the mechanisms of carcinogenesis make this extrapolation speculative (Fig. 16-1). It is largely for this reason and for analogous considerations in epidemiology that no safe threshold for carcinogen exposure can be demonstrated with certainty.

Two related issues concern the organ and species specificities of animal bioassay results. The target organ in a test species may differ from that in humans, owing mostly to differing metabolic pathways. Benzidine, for example, causes bladder cancer in humans, hepatomas in mice, and intestinal tumors in rats. Increasingly, it is being recognized that some animal tumors, such as tumors of the rat forestomach after gavage, have no analog in humans. Similarly, different chemicals may vary across species in their ability to induce tumors. Positive animal bioassay results can help identify the most worrisome human exposures, but there is less basis to believe that they quantitatively predict human cancer incidence.

The results of animal bioassays are tabulated in an ongoing database, the Carcinogenic Potency Database, maintained by the Lawrence Berkeley Laboratory and the University of California, Berkeley. Results are published annually in the journal *Environ-*

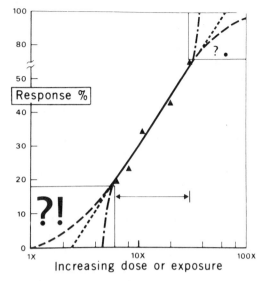

Figure 16-1. Hypothetical dose-response curve for animal cancer bioassay. Observations occur at high doses (*triangles*). The challenge is in extrapolating to low doses, the area at the lower left. As indicated, there are several ways to extrapolate, corresponding to different biologic models. A curve that intersects the x-axis at some positive value demonstrates a threshold—a dose or exposure that does not increase cancer in the test population. A curve that intersects the x-axis at 0, or does not intersect it at all, demonstrates no threshold. (From Maugh TH II. Chemical carcinogens: how dangerous are low doses? Science 1978;202:37–41.)

mental Health Perspectives, and they are available on the World Wide Web at http://potency.berkeley.edu/cpdb.html.

In vitro testing involves the use of bacterial or human tissue cultures. Suspected carcinogens are added to these systems, and end points that reflect DNA damage, such as DNA alteration, increased repair, or altered gene expression, are monitored. The prototype *in vitro* assay is the Ames test, first described in 1966. This test uses special strains of *Salmonella typhimurium* that cannot synthesize the amino acid tryptophan. Mutant strains synthesize tryptophan and show growth on a tryptophan-free medium, an end point that is easily detected by examining the medium for bacterial growth.

In vitro assays such as the Ames test determine *mutagenicity,* not carcinogenicity. However, both theoretic and empiric evidence suggests that mutagenicity and carcinogenicity are closely linked. The theoretic link lies in the molecular event that is common to both: alteration of DNA. The empiric link is that almost all known carcinogens are mutagenic. However, some reports have emphasized that many naturally-occurring substances not thought to be carcinogenic, including common plant products, food constituents, and hormones, are also mutagenic in the Ames test.

Numerous modifications of the Ames test have been made, including the addition of liver microsomes to metabolize substances that may not be mutagenic in their original form. Other short-term *in vitro* assays include the induction of chromosome aberrations, sister chromatid exchanges, and micronuclei and the malignant transformation of cultured cells that produce tumors when inoculated into an appropriate host. Short-term assays, in addition to suggesting potential human carcinogenicity tests to be performed in animals, are also used to help identify both mutagens in complex mixtures and active metabolites of mutagens in human body fluids and to study mechanisms of chemical carcinogenesis.

Finally, in analyzing *structure-activity relations,* investigators review chemical configurations for their similarities to known carcinogens. Such analyses can direct suspicion toward chemicals that may have carcinogenic potential; further testing usually is necessary.

Based on these four types of data, regulatory and research agencies have developed standardized methods for classification of chemical carcinogens. The most widely used scheme is that of the International Agency for Research on Cancer (IARC), which designates three categories. Group 1 includes chemicals and processes established by IARC as being human carcinogens, based on sufficient evidence, usually epidemiologic data. Group 2 includes chemicals and processes that are probably (Group 2A) or possibly (Group 2B) carcinogenic to humans.

Group 2A reflects limited evidence of carcinogenicity in humans and sufficient evidence of carcinogenicity in experimental animals; Group 2B reflects limited evidence in humans without sufficient evidence in animals, or sufficient evidence in animals without any human data. IARC policy has been to recommend treating Group 2 chemicals as if they presented a carcinogenic risk to humans. Group 3 includes agents that are not classified, and Group 4 includes agents that are probably not carcinogenic to humans. Using this classification, IARC has evaluated approximately 750 chemicals, industrial processes, and personal habits. More than 50 have been placed in Group 1 (Tables 16-1 and 16-2), and almost 250 have been placed in Group 2 (Table 16-3 and 16-4). The American Conference of Governmental Industrial Hygienists, the National Toxicology Program, and the National Institute for Occupational Safety and Health (NIOSH) have all adopted analogous systems.

How Much Cancer Is Occupational?

The workplace contribution to cancer has been a matter of considerable debate. Doll and Peto, in *The Causes of Cancer* (1981), estimated that between 2% and 8% of cancers at that time could be attributed to workplace exposures (4). In the United States, this would represent between 24,500 and 98,000 new cases of cancer each year. This proportion is higher in men (perhaps 5%) than in women (perhaps 1%), because men have historically sustained higher workplace exposures to carcinogens. The absolute workplace contribution to cancer is likely to decline in the United States as the manufacturing sector, where most exposures occur, decreases in size. However, occupational cancers may migrate with industrial plants to developing countries, increasing the role of the workplace in cancer in those countries.

The proportion of cancers attributed to a particular category of exposures (e.g., occupational) overlaps with other categories, so that the total of all these "attributable risks" exceeds 100%. For example, both asbestos and radon cause far more cancers through interaction with tobacco smoke than as isolated exposures.

Attributable risk estimates usually pertain to the entire population. Within high-risk groups such as heavily exposed workers, the proportion of cancers related to work may be substantially higher. The same is true of workers with few competing carcinogenic exposures, such as nonsmokers. Those cancers most closely associated with workplace exposures include both common types, such as lung cancer and skin cancer, and less common types, such as bladder cancer, leukemia, and brain cancer. Somewhat larger proportions of these latter cancers are likely to be work-related.

CANCER RISK ASSESSMENT

Risk assessment is a procedure for characterizing and quantifying the amount of harm expected to result from an exposure. This process was developed in the 1970s as regulatory agencies attempted to set permissible levels of exposure based on acceptable levels of risk, and to quantify the amount of benefit that would be expected from regulation at a particular level. Although risk assessment is a generic process that can be applied to any risk, including those posed by nonmalignant diseases, it is discussed in this chapter because it arose in the context of cancer risk (see also Chapter 3).

The basic components of risk assessment were codified by the National Research Council in 1983 (5). *Hazard identification* involves specifying what hazard will be studied. In *dose-response assessment,* epidemiologic and toxicologic data are used to define the dose-response relation. Because data are often available only for high-dose observations, this step frequently involves extrapolation, as discussed previously. In *exposure assessment,* the exposure to the target population is estimated. Finally, in *risk characterization,* the exposure level that leads to a particular magnitude of risk is estimated, us-

ing mathematical models. Different models, based on different biologic assumptions, yield different results. For example, the commonly used one-hit, linear, no-threshold model usually predicts higher risk for a given dose, whereas some nonlinear models predict lower risk. Because there is considerable uncertainty in risk assessment, uncertainty factors are often used to introduce margins of safety. More recently, physiologically-based pharmacokinetic (PBPK) models have been used to refine the predictions made when extrapolating animal data to humans and to assess the human relevance of certain animal tumors.

Risk assessment offers a quantitative approach to assessing the risk of exposures. If the public policy decision is to control rather than eliminate carcinogenic exposures, then risk assessment provides a framework for deciding how much exposure to permit as the basis of regulations. Risk assessment is transparent, in that the assumptions used are generally made explicit. However, critics point out that many of these assumptions, such as cross-species extrapolation and linear extrapolation to low doses, do not eliminate important scientific uncertainties. Moreover, risk assessment typically is performed on one substance at a time, whereas real exposures do not occur in isolation. Finally, risk assessment raises ethical concerns, because those who quantify and allocate levels of risk are usually not the ones who will bear the risk. Although risk assessment is an invaluable tool that will be with us for some time, it incorporates much of the uncertainty that characterizes occupational cancer causation generally.

CONTROLLING OCCUPATIONAL CANCER

Primary prevention is at the heart of the occupational health approach to workplace cancer. As described in Chapters 5 and 7, the preferred approach is to eliminate exposure altogether. For example, benzene, which causes leukemia, can be replaced by toluene in many uses. Similarly, exposure can be engineered out of a process. Exposure to vinyl chloride monomer in plastics factories was eliminated by enclosing the process. If necessary, personal protective equipment can be used to avoid exposure to carcinogenic materials such as asbestos (during cleanup jobs) and chromium. However, because personal protective equipment requires compliance and maintenance, can malfunction, and may be unpopular with workers, process changes are preferable.

Secondary prevention—early detection through screening—is discussed later and in Chapter 4. Other aspects of the control of occupational cancer are also important. These include worker training and product labelling to help build awareness of workplace hazards and how to minimize them, and regulatory controls to provide incentives to control exposures. In particular, regulations need to address the export of carcinogens from the United States to other nations, so as to control the transfer of risk to workers in those nations. Health care providers must be alert to the possibility of exposure to occupational carcinogens and take careful occupational histories, especially in cases of possibly work-related cancers (see Chapter 4). Research also is needed to recognize new occupational carcinogens, to understand the mechanisms of carcinogenesis, and to identify workers at risk. Finally, to control the overall cancer risk among working populations, it is important to control nonoccupational risks such as smoking.

OCCUPATIONAL CANCER AND THE CLINICIAN

As the examples at the beginning of this chapter illustrate, a clinician may face many questions about cancer in the workplace. Some of these relate to future risk: How do we prevent cancers, or how do we detect cancers early? Others arise after the fact: Was this case of cancer caused by workplace exposures? In each case, a firm grounding in science and a commitment to the doctor-patient relationship can help guide the clinician in providing answers.

Clinical Applications of Molecular Biology: Which Patients are at High Risk?

Insights gained from molecular biology may soon have a role in health care decisions for individual workers. One such application is in detecting markers of risk, and another is in detecting markers of exposure and early effect.

Markers of risk have been recognized for many years. For example, disease states such as xeroderma pigmentosum carry an unusually high risk of cancer because of defects in the ability to repair DNA damage. There is now a growing understanding of another dimension of increased risk, based on individual differences in metabolic enzymes. As noted previously, many workplace carcinogens are metabolized to active forms, usually electrophiles, that then bind covalently with DNA. The enzymes primarily responsible for this activation belong to the cytochrome P-450 system. Other enzymes, such as N-acetyltransferase, epoxide hydrolase, and glutathione S-transferase, may either increase excretion of toxic substances or, in other settings, form reactive electrophiles that can bond with DNA to create adducts and cause mutations.

Considerable variation in these metabolic functions has been noted, not only among species but among individuals. For example, people vary several thousand-fold in their levels of aryl hydrocarbon hydroxylase (AHH), an enzyme of the cytochrome P-450 system that helps metabolize polycyclic aromatic hydrocarbons. High levels of AHH have been associated in some studies with increased risk of lung cancer. Variations in AHH levels reflect genetic factors, but it is the interaction of these polymorphisms with environmental exposures such as cigarette smoking and diet that are thought to affect risk. Another highly variable factor is the cytochrome P-450 enzyme responsible for hydroxylating the antihypertensive drug debrisoquine. People who are "extensive hydroxylators" have several thousand times more enzyme activity than those who are "poor hydroxylators." The extensive hydroxylator phenotype has been associated with a markedly increased risk of lung cancer. A third example is *N*-acetyltransferase. Slow acetylators appear to have an increased risk of bladder cancer, presumably because they are less able to detoxify aromatic amines in the bladder. However, fast acetylators may have an increased risk of colorectal cancer, because the acetylated amines are carcinogenic to colonic mucosa. This polymorphism illustrates the truism that most gene activities are neither all good nor all bad.

What is the significance of differential susceptibility? Current knowledge is embryonic; researchers have only begun to identify genetic variations that may interact with environmental exposures to influence risk. Until much larger studies have replicated early findings, it cannot be known which associations are truly causal or biologically important. When such information does become available, health care practitioners and others will face ethical dilemmas of how to use the information.

Some have suggested that more susceptible workers should be kept from jobs that have carcinogenic exposures, and that less susceptible workers should preferentially work in such jobs. Although this approach has a certain logic, there are compelling counterarguments. First, the more definitive and therefore the preferred approach is to modify the workplace instead of the workforce (see Chapter 5). Second, the sensitivity and specificity of assays for most of the enzyme phenotypes now under study are too low to support confident predictions of cancer risk. Third, barring people of certain phenotypes from employment is discriminatory and probably illegal under the Americans with Disabilities Act (see Chapter 12). Despite these arguments, identification of high-risk and low-risk individuals may in the fu-

ture become part of the clinician's role, both in counselling prospective workers about job choices and in tailoring medical surveillance programs according to individual risk levels.

Clinical Applications of Molecular Biology: Which Patients Were Exposed?

Biomarkers of exposure are also likely to grow in importance in occupational health, both in clinical practice and in epidemiologic studies. For many years, carcinogens or their metabolites have been directly measured in biologic media such as blood and urine. For example, benzene exposure can be monitored directly in expired air and blood and through urinary phenol levels, and exposure to the aromatic amine 4,4'-methylenebis(2-chloroaniline) can be monitored through urinary levels of the chemical. More recently, assays have measured the level of mutagenicity in urine; the basis for this approach is that carcinogens, when metabolized to active forms and excreted, should be detectable in urine and should reflect the individual exposure. Perhaps the most promising approach to measurement of carcinogen exposure is to assess the dose at the ultimate target, DNA, by measuring DNA adducts.

DNA adducts may be long-lived or short-lived, and they may vary greatly in form. They can be measured by a variety of techniques. One is immunoassay, which uses specific antibodies against DNA adducts. Another is DNA digestion, in which individual bases of digested DNA are separated with two-dimensional chromatography after radiolabeling with phosphorus P 32; adducts are visualized with autoradiography. The study of DNA adducts has several limitations, including the instability and unknown clearance rates of DNA adducts and the lack of good dose-response data. Most importantly, the significance of the various kinds of adducts and their associated conformational changes in DNA are still not well understood. However, data from ethylene oxide workers, welders, hazardous waste workers, and others suggest that DNA adducts may reflect the biologically effective dose in exposed workers.

Another approach to biomarkers is to measure the level of chromosome aberrations or disturbances in the cells of exposed workers. Strictly speaking, this is a measurement of early effect rather than exposure. Cultured lymphocytes are the usual cells of choice, and the observations of interest may be sister chromatid exchanges, micronuclei, or mutations at so-called "reporter genes." Sister chromatid exchanges are four-stranded exchanges of genetic material that are efficiently induced by DNA adducts. Micronuclei are induced by gross chromosome damage that is not integrated into the nucleus at mitoses. Mutations induced *in vivo,* most commonly assessed at the hypoxanthine phosphoribosyltransferase (*HPRT*) gene locus, reflect DNA-based lesions and can also be quantified. Because the *HPRT* gene encodes a purine salvage pathway that is the primary source of purines for DNA synthesis, it can be used as a marker of induced mutations.

The precise relation of these phenomena to the exposure-induced genetic changes of interest—mutations in proto-oncogenes and tumor suppressor genes—remains unclear. As the genes directly involved in the carcinogenic process become better understood, the biomarkers available for use in occupational studies of cancer risk will become more varied and more directly associated with specific disease risk.

Biomarkers of exposure, whether simple blood levels of a chemical or sophisticated genetic measures, should never replace environmental monitoring as a check on workplace exposure levels. However, they have an important role as a backup means of monitoring exposure levels and identifying excessive exposures that require abatement.

Is This Chemical Carcinogenic?

This question is best answered according to defined criteria, such as those already discussed. The original reports of animal and epidemiologic studies on a chemical may be accessed through standard reference sources. Formal evaluations of carcinogenicity may be found in the IARC monographs on the Evaluation of the Carcinogenic Risk of Chemicals to Humans, in annual reports of the National Toxicology Program, and in other publications and databases.

I Am Exposed to a Carcinogen. What Should I Do?

This query has two components. One pertains to the carcinogenic exposure and one to the patient.

Ideally, there should be no workplace exposure to a carcinogen, because safe threshold exposure levels cannot currently be demonstrated. However, regulatory approaches permit exposures at levels thought to pose low levels of risk. A clinician who becomes aware of an ongoing carcinogenic workplace exposure should take appropriate steps to end that exposure if possible, or at least to lower it to below legal limits. This often involves contacting responsible persons at the workplace and at government agencies. Exposure may be ended through substitution of the carcinogen, enclosure of the process or other engineering techniques, or, when necessary, use of personal protective equipment.

The worker who reports being exposed to a carcinogen should be advised to terminate the exposure, to avoid concomitant carcinogenic exposures such as smoking, and to seek appropriate medical monitoring.

Did My Exposure Cause My Cancer?

Patients with cancer who have been exposed to carcinogens often ask whether their exposure caused the illness. The question may arise in the context of litigation, or it may simply reflect a patient's psychological need to explain a catastrophic life event. The issue of cancer causation in an individual patient is difficult to address because it entails the application to individuals of epidemiologic and statistical data that derive from groups.

Certain requirements must be met before it can be concluded that an exposure has causally contributed to a cancer. There must be evidence that the exposure has indeed occurred. The tumor type in question must be associated with the exposure, based on previous studies. Finally, the appropriate temporal relation must hold; in particular, a sufficiently long latency period must have elapsed between the onset of exposure and the diagnosis of cancer.

Suppose that the baseline incidence of lung cancer in unexposed adult men is 80 cases per 100,000 per year. Suppose further that a particular occupational exposure has been associated with a relative risk of lung cancer of 1.8. Therefore, the incidence among exposed men would be 144 cases per 100,000 per year. If an exposed man develops lung cancer and wonders whether his exposure caused his cancer, what should he be told?

The simplest analysis is qualitative. Any exposure that markedly increases risk may be considered to contribute to the development of cancer in an exposed person. The definition of a "marked" increase is not firm; relative risks as low as 1.3 have been considered in this category. By this analysis, the patient could be told that his exposure contributed to his cancer.

A second qualitative approach is to ask whether the patient's cancer would have occurred "but for" the exposure. In the example described, more than half of the cases of lung cancer in the exposed population would have occurred even without the exposure. It might then be concluded that any individual case is "more likely than not" to have occurred regardless of exposure. Similarly, a relative risk of 2.2 would lead to the conclusion that any individual case of cancer would not have occurred but for the exposure. Such a stochastic approach accounts for no cancer causation when the relative risk is below 2.0, and it accounts for all cancer causation when the relative risk exceeds 2.0. This violates commonsense notions of causation, and it places far too much weight on the precision of the relative risk estimate.

Finally, in the quantitative approach, causation is allocated to various causes, including occupational exposures. In the example, the patient might be told that the occupational exposure was "responsible" for 44% (0.8/1.8) of his lung cancer. On the other hand, smoking causes a tenfold (900%) increase in lung cancer risk; so if the patient were a smoker, he might be told that smoking accounted for 83% (9/10.8) of his cancer, the job exposure accounted for 7% (0.8/10.8), and baseline population risk factors accounted for 9% (1/10.8).

This approach has intuitive appeal because it confronts the multiplicity of exposures and attempts to quantify the relative importance of each. However, the data needed for this approach are rarely available. Interaction of multiple exposures (e.g., synergy) often occurs but is rarely quantitated. Consequently, even if a population relative risk can be estimated for an occupational exposure, the relative causal contribution of several factors in an individual worker is usually impossible to quantitate.

The latter two approaches outlined are often demanded in legal settings, but, as noted, they generally have inadequate scientific bases. Until further data or analytic methods become available, the first approach is recommended. It accords with commonsense, stays within the confines of available data, and is understandable to patients and their families.

How Should a Screening Program for Cancer Be Designed?

The theory and practice of screening have advanced in recent years. Cancer screening programs have three goals: (a) identifying susceptible individuals, presumably before exposure; (b) identifying markers of exposure (biologic monitoring); and (c) identifying early signs of disease (medical surveillance). Some screening tests combine aspects of the second and third categories by identifying physiologic changes that are related to exposure but of uncertain pathologic significance. The first two kinds of screening have already been discussed; this section focuses on the third type.

The issues raised by cancer surveillance in the workplace are perhaps best illustrated by the examples of bladder cancer and lung cancer. Bladder cancer surveillance has been extensively studied among workers exposed to β-naphthylamine, benzidine, and/or benzidine congeners such as o-tolidine. Two methods have been used: urinalysis for microscopic hematuria and urine cytology. The hematuria test is relatively sensitive in detecting both superficial and invasive bladder cancer, but its low specificity results in a high false-positive rate, leading to many invasive studies on healthy persons. Urine cytology has good sensitivity and specificity for invasive bladder cancer, but no firm evidence demonstrates a survival advantage for patients whose disease is detected through such screening. More advanced techniques, such as flow cytometry, quantitative fluorescence image analysis, and the use of a protein marker called urinary nuclear matrix protein 22 (NMP22), remain unvalidated but appear to have suboptimal sensitivity or specificity, or both. The International Conference on Bladder Cancer Screening in High-Risk Groups, sponsored by NIOSH in 1989, concluded that urinalysis and cytology might be appropriate methods, especially after high exposure to known or suspected bladder carcinogens, but that further research was necessary. Ongoing studies of high-risk groups should help to refine these recommendations.

Lung cancer surveillance consists of interval chest radiography or sputum cytology, or both. These approaches were evaluated in a series of trials at the Mayo Clinic, Johns Hopkins University, and the Memorial Sloan-Kettering Cancer Center in the 1970s. The combination of chest radiology and sputum cytology tests three times a year yielded a significant increase in lung cancer detection and resectability compared with controls (who were merely advised to be tested once a year). However, there was no significant decrease in lung cancer mortality. These results, in combination with other data, have supported the recommendation that no routine surveillance for lung cancer should be offered, even to high-risk populations.

Similarly, surveillance for other occupational cancers is generally not recommended. Most tests carry a high risk of false-positive findings, a low positive predictive value, high cost, worker unacceptability, morbidity, or some combination of these. The major exceptions are the tests recommended in general medical practice that might be provided conveniently in the workplace setting. These include Papanicolaou (Pap) smears for cervical cancer, sigmoidoscopy and testing for occult blood in stool for colorectal cancer, physical examination and mammography for breast cancer, and possibly digital examination and prostate-specific antigen (PSA) testing for prostate cancer.

We Have a Cluster of Cancer in Our Workplace. How Should We Respond?

Suspected excesses of cancer, clustered in time or space, may be noted by workers or by health care providers. Several of the most important occupational carcinogens were first recognized this way, including nickel, bis-chlormethyl ether, and vinyl chloride. Although there may still be some undiscovered occupational carcinogens, most clusters do not result in the recognition of a new carcinogen. They do, however, arouse a great deal of concern, and it is essential that they be handled systematically, openly, and professionally. A multidisciplinary approach is necessary, one that draws on the experience and knowledge of workers, physicians, epidemiologists, occupational hygienists, and managers. The following sequence is suggested.

First, the presence of a cluster should be confirmed or refuted. Each case of cancer should be confirmed, and tissue type, date of diagnosis, demographic data, and exposure data should be obtained. The worker population should be enumerated and subdivided into age, sex, and other pertinent categories. (If information on retirees is available, it should be included.) Age- and sex-specific cancer incidence rates for the state should be obtained from the state cancer registry, if one exists. With this information, an age-standardized cancer rate can be computed for the workplace and compared with the expected rate based on state data. Confidence intervals can be calculated for the workplace cancer rate; these usually will be broad because of the small numbers of cases involved. However, even a statistically insignificant elevation in the cancer rate in the workplace should prompt further evaluation.

Next, the tissue types of the cancer cases should be reviewed. An excess of unusual tumors, or tumors known to be environmentally induced, should prompt further concern.

Then the latency period of each cancer case should be reviewed. If many of the workers began their employment at the workplace only shortly before diagnosis, an exposure-related cluster is less plausible.

Next, confounders should be reviewed. Other factors that may have contributed to an elevation of cancer should be noted. However, care should be taken not to let the presence of confounders divert attention from an occupational carcinogen.

Next, the occupational histories of the affected workers should be reviewed. Detailed personnel histories may reveal that a particular job title is associated with cancer. It is important that jobs in the distant past, not just recent ones, be examined.

An occupational hygiene review should be made to determine whether any particular exposures are common among the affected workers. A variety of job titles may share a common chemical contact, which may help explain the cluster. Early exposures may need to be reconstructed, which may involve interviewing older workers or reviewing production records.

Next, the same analysis should be made with regard to worksites. If many of the cases arise from a single building or location, an environmental cause is suggested. Any worksites in question should be subjected to a thorough occupational hygiene evaluation, which should include both the production process and "incidental" exposures such as the heating, ventilation, and air-conditioning system and the drinking water.

Based on the results of these analyses, an initially suspected cancer cluster may be found to be (a) not actually present, (b) present but not consistent with occupational causation, (c) possibly related to occupational exposures, or (d) definitely related to occupational exposures. The results should be carefully and thoroughly communicated to all those concerned. If an occupational cause is implicated, aggressive corrective action is in order. Whatever the conclusion, careful ongoing surveillance of both the workplace and the workforce should continue.

A clinician who analyzes an apparent cancer cluster may require further assistance. The most appropriate sources are NIOSH, a state health department, or a qualified consultant group, such as a university-based occupational health program.

CONCLUSION

Occupational cancer is a dreaded consequence of certain workplace exposures. It differs from other occupational diseases in several ways: cancer elicits a unique fear among workers and the public, no absolutely safe level of exposure to carcinogens can be proven, many different forms of cancer exist, the long delay between exposure and onset of illness is confusing, most occupational cancer cannot be distinguished clinically from cancer of nonoccupational origin, and competing carcinogenic exposures are present in many cases. On the other hand, occupational cancer shares important features with other occupational diseases, in that there are large data gaps in relating exposure to disease and most cases are preventable.

The clinician's role in confronting occupational cancer is varied. A high index of suspicion of workplace causes should be maintained when treating cancer, especially lung, bladder, and brain cancer and leukemia. The clinician should work to identify past exposures, using the patient's knowledge, toxicologic resources, and consultants. If ongoing exposures are present, the clinician should assume a public health role, working to end such exposures. Finally, it is important to educate patients, employee and employer groups, and communities about the hazards of carcinogenic exposures and ways to prevent them.

REFERENCES

1. Veys CA. ABC of work related disorders: occupational careers. Brit Med J 1996;313:615–619.
2. Selikoff IJ, Hammond EC, Churg J. Asbestos exposure, smoking, and neoplasia. JAMA 1968;204:106–112.
3. US Environmental Protection Agency. Special report on environmental endocrine disruption: an effects assessment and analysis. EPA/630/R-96/012. Washington, DC: USEPA, February, 1997.
4. Doll R, Peto R. The causes of cancer. New York: Oxford University Press, 1981.
5. National Academy of Sciences. Managing the process: risk assessment in the federal government. Washington: National Academy Press, 1983.

BIBLIOGRAPHY

Bailar JS, Smith EM. Progress against cancer? N Engl J Med 1986;314:1226–1232.
A sobering argument that cancer treatment and prevention have not been very successful. Compare with Cole and Rocu's upbeat assessment 10 years later.
Bishop JM. Molecular themes in oncogenesis. Cell 1991;64:235–248.
An excellent review of the molecular biology of cancer.
Brandt-Rauf PW. New markers for monitoring occupational cancer: the example of oncogene proteins. J Occup Med 1988;30:399–404.
Reviews the role of oncogenes in occupational medicine.
Christiani DC, Monson RR. Cancer in relation to occupational and environmental exposures [special issue]. Cancer Causes Control 1997;8(3).
This special issue of Cancer Causes and Control *includes papers on most occupational and environmental carcinogens, and helps place each into the larger context of cancer occurrence.*
Cole P, Rocu B. Declining cancer mortality in the United States. Cancer 1996;78:2045–2048.
An upbeat assessment based on mortality data from the early 1990s. Compare with the Bailar and Smith paper.
Davis DL, Hoel D, eds. Trends in cancer mortality in industrial countries [special issue]. Ann NY Acad Sci 1990;609.
A collection of papers that discuss and analyze cancer rates in North America, Europe, and Japan.
Doll R, Peto R. The causes of cancer. New York: Oxford University Press, 1981.
A brief monograph reviewing and synthesizing epidemiologic and laboratory data.
Eddy DM. Screening for lung cancer. Ann Intern Med 1989;111:232–237.
A good summary of data on lung cancer screening.
Halperin W, Cartwright RA, Farrow GM, et al. Final discussion: where do we go from here? J Occup Med 1990;32:936.
A good summary of bladder cancer screening in the workplace.
Harris CC. Chemical and physical carcinogenesis: advances and perspectives for the 1990. Cancer Res 1991;51(Suppl):5023S–5044S.
An excellent review of mechanisms of carcinogenesis.
Harris CC. Interindividual variation among humans in carcinogen metabolism, DNA adduct formation and DNA repair. Carcinogenesis 1989;10:1563–1566.
A brief review of individual variability in cancer susceptibility.
Hart RW, Hoerger FD. Carcinogen risk assessment: new directions in the qualitative and quantitative aspects. Banbury Report 31. Cold Spring Harbor: Cold Spring Harbor Laboratory, 1988.
Proceedings of a conference on risk assessment, with papers that discuss methodologic difficulties and other problems.
Hemminki K. Occupational cancer and carcinogenesis. Scand J Work Environ Health 1992;18 (Suppl 1):1–117.
A monograph with papers on all aspects of occupational cancer.
Higginson J, Muir CS, Muñoz N. Human cancer: epidemiology and environmental causes. Cambridge: Cambridge University Press, 1992.
A brief but comprehensive general text on cancer and its causes.
Huff JE. Carcinogenesis results in animals predict cancer risks to humans. In: Wallace RB, ed. Maxcy-Rosenau-Last's public health and preventive medicine, 14 ed. Norwalk, CT: Appleton & Lange, 1998:543–550,567–69.
Reviews the correlation between animal bioassay results and human carcinogenicity.
International Agency for Research on Cancer. IARC monographs on the evaluation of carcinogenic risks to humans, suppl 7. Overall evaluations of carcinogenicity: an updating of IARC monographs volumes 1 to 42. Lyon, France: IARC, 1987.
The latest published in-depth summary of IARC monographs. The most recent IARC evaluations of carcinogenicity can be found at the IARC web site: http://www.iarc.fr.
National Academy of Sciences. Managing the process: risk assessment in the federal government. Washington: National Academy Press, 1983.

Explains the components of risk assessment as used by Federal agencies.

Office of Science and Technology Policy. Chemical carcinogens: a review of the science and its associated principles. Federal Register 1985;50:10372–10442. Reprinted in Environ Health Perspect 1986;67: 201–282.

Office of Technology Assessment. Identifying and regulating carcinogens. OTA-BP-H-42. Washington: US Government Printing Office, 1987.

These two sources explain the basis for Federal regulatory approaches.

Patterson JT. The dread disease: cancer and modern American culture. Cambridge: Harvard University Press, 1987.

A fascinating cultural history of cancer in the United States.

Rothman N, Hayes RB. Using biomarkers of genetic susceptibility to enhance the study of cancer etiology. Environ Health Perspect 1995;103(Suppl 8):291–295.

A brief overview of scientific (but not social) aspects of differential susceptibility to cancer.

Schottenfeld D, Fraumeni JF, eds. Cancer epidemiology and prevention, 2nd ed. Philadelphia: WB Saunders, 1993.

The best reference on cancer epidemiology, reviewing causal factors and clinical aspects.

Proceedings of the Second International Conference on Environmental Mutagens in Human Populations. Environ Health Perspect 1996;104(Suppl 3).

This collection of 51 papers includes work on biomarkers of susceptibility, exposure, and effect; on genetic mechanisms and gene-environment interactions in cancer susceptibility; and on the application of these concepts in clinical practice and epidemiology.

Siemiatycki J. Risk factors for cancer in the workplace. Boca Raton, FL: CRC Press, 1991.

Both a basic text and a report of a massive series of studies in Montreal.

Strauss GM. Measuring effectiveness of lung cancer screening: from consensus to controversy and back. Chest 1997;112(Suppl 4):216S–228S.

Suggests that there may be a role for lung cancer screening.

17 Ionizing Radiation

Arthur C. Upton

Chernobyl, USSR, April 27, 1986. An explosion and fire occurred in a large, uranium-fueled, graphite-powered reactor during a test in which the emergency cooling, regulating, and shutdown systems had been deliberately turned off. The resulting damage to the reactor caused the release of large quantities of radioactive fuel and fission products, which contaminated the plant site and areas downwind for hundreds of miles. Two plant workers died immediately after the accident from burns and traumatic injuries, and hundreds were later hospitalized for radiation sickness; 29 workers died within weeks after the accident.

No other radiation accident thus far has been as serious as the one at Chernobyl, but scores of other radiation accidents have occurred, some of which have caused fatalities (1). Even in the absence of accidents, mortality from cancer and other diseases has been increased among radiation workers in the past. Today, even the smallest doses are presumed to pose a potential risk of injury. In view of the large number of workers at risk (Tables 17-1 and 17-2), recognition, treatment, and prevention of radiation injury command an important place in occupational medicine.

NATURE AND MEASUREMENT OF IONIZING RADIATION

Ionizing radiations include x-rays, gamma rays, electrons, protons, neutrons, alpha par-

A. C. Upton: Department of Environmental and Community Medicine, Robert Wood Johnson Medical School, Piscataway, NJ 08854-5635.

ticles, and other corpuscular radiations of varying mass and charge (2). Such radiations are produced by the disintegration of naturally occurring unstable radioactive elements such as uranium, thorium, and radium and by the disintegration of elements that are disrupted by bombardment in an "atom smasher," nuclear reactor, or other such device.

X-rays and gamma rays travel with the speed of light and penetrate more deeply than do particulate radiations, which vary in initial velocity and penetrating power, depending on their energy and mass (2). As an ionizing radiation penetrates matter, it collides with atoms and molecules in its path, disrupting them and giving rise to ions and free radicals—hence the designation, *ionizing* radiation. Along the track of an impinging alpha particle, the collisions occur so close together that the radiation gives up all its energy in traversing only a few cells. The collisions along the track of an x-ray, on the other hand, tend to be distributed so sparsely that the radiation may traverse the entire body. The average amount of energy deposited per unit length of track (expressed in kiloelectron volts per micrometer, $keV/\mu m$) is called the linear energy transfer (LET) of the radiation.

Doses of ionizing radiation are expressed in terms of energy deposition (Table 17-3). Radiations of high LET, such as alpha particles, tend to cause greater injury for a given amount of energy deposited in the cell than do radiations of low LET, such as x-rays

TABLE 17-1. *Estimated numbers of U.S. workers exposed occupationally to ionizing radiation*

Type of work	Number of workers exposed annually	Average dose (mSv/yr)
Nuclear energy (fuel cycle)	62,000	8.4
Naval reactor	36,000	2.2
Healing arts	500,000	1.2
Research	100,000	1.2
Manufacturing and industrial	7,000,000	0.07

From Department of Health, Education, and Welfare. Interagency task force on the health effects of ionizing radiation: report of the work group on science. Washington, DC: Department of Health, Education, and Welfare, 1979. (Also see United Nations Scientific Committee on the Effects of Atomic Radiation: Sources and effects of ionizing radiation: report to the General Assembly, with annexes. New York: United Nations, 1994.)

TABLE 17-2. *Types of workers who may be exposed occupationally to ionizing radiation*

Airline crews	Oil assayers
Atomic energy plant workers	Petroleum refinery workers
Cathode ray tube makers	Physicians
	Pipeline oil flow testers
Dental assistants	Pipeline weld radiographers
Dentists	
Electron microscopists	Plasma torch operators
Fire alarm makers	Radar tube makers
High-voltage television repairmen	Radiologists
	Television tube makers
High-voltage vacuum tube makers	Thickness gauge operators
Industrial fluoroscope operators	Thorium-aluminum alloy workers
Industrial radiographers	Thorium-magnesium alloy workers
Inspectors and workers near γ-ray sources	Uranium mill workers
Liquid level gauge operators	Uranium miners
	X-ray aides
	X-ray diffraction apparatus operators
	X-ray technicians
	X-ray tube makers

Data from MM Key, et al., eds. Occupational diseases: a guide to their recognition. Washington, DC: NIOSH, 1977:471–472.

(2).The comparatively high potency, or high relative biologic effectiveness (RBE), of high-LET radiation results from the capacity of each densely ionizing particle to deposit enough energy in a critical site within the cell (such as a DNA molecule) to cause biologically significant molecular damage. Hence, although alpha particles generally travel too short a distance to cause injury if they are emitted outside of the body, they can be highly injurious to any cells that they penetrate (2).

The pattern of injury caused by an internally deposited radionuclide depends on the tissue distribution and retention of the radionuclide, which, in turn, varies with its physical and chemical properties. Radioactive iodine, for example, is normally concentrated in the thyroid gland, whereas strontium Sr 90 is deposited primarily in bone. After deposition of a given amount of radioactivity, the quantity remaining *in situ* decreases with time through both physical decay and biologic removal. The time taken for a radionuclide to lose one-half of its radioactivity by physical decay varies from a fraction of a second for some radionuclides to millions of years for others. For example, the physical half-life of iodine I 131 is 7 days, and that of plutonium Pu 239 exceeds 24,000 years. Biologic half-lives also vary, tending to be longer for bone-seeking radionuclides (e.g., radium, strontium, plutonium) than for radionuclides that are deposited in soft tissue (e.g., iodine, cesium, potassium, tritium) (2).

SOURCES AND LEVELS OF RADIATION IN THE ENVIRONMENT

Natural background radiation consists of (a) cosmic rays, impinging on the earth from outer space; (b) terrestrial radiation, emanating from radium, thorium, uranium, and other radioactive elements in the earth's crust; (c) internal radiation, emitted by potassium K 40, carbon C 14, and other radionuclides normally present within living cells; and (d) radiation from radon and its decay products in inhaled air. The average effective dose received annually from all four sources

TABLE 17-3. *Quantities and dose units of ionizing radiation*

Parameter measured	Definition	Dose unit[a]
Absorbed dose	Energy deposited in tissue (1 joule/kg)	Gray (Gy)
Equivalent dose	Absorbed dose weighted for relative biologic effectiveness of the radiation	Sievert (Sv)
Effective dose	Equivalent dose weighted for sensitivity of the exposed organ	Sievert (Sv)
Collective effective dose	Effective dose applied to a population	Person-Sv
Committed effective dose	Cumulative effective dose to be received from a given intake of radioactivity	Sievert (Sv)
Radioactivity	One disintegration per second	Becquerel (Bq)

[a] The units of measure listed are those of the International System, introduced in the 1970s to standardize usage throughout the world (see ref. 3). They have largely supplanted the earlier units, namely, the rad (1 rad = 100 ergs/g = 0.01 Gy), the rem (1 rem = 0.01 Sv), and the curie (1 Ci = 3.7×10^{10} disintegrations per second = 3.7×10^{10} Bq).

by a resident of the United States approximates 3 mSv, and the dose to the respiratory tract from inhaled radon greatly exceeds that from all other sources combined (2,3).

A nuclear plant worker who measured his radioactivity immediately on entering the plant found it to be unexpectedly high. On investigation, the source was found to be the radon in his home, which was present in concentrations thousands of times higher than average. The discovery that his house contained such high levels of radon prompted the worker to take steps to reduce the levels, by improving the ventilation of the house and blocking the entry of radon from the underlying soil. His discovery also prompted a survey of radon levels in other houses throughout the United States, results of which indicate radon levels to be well in excess of the recommended limits (4 to 8 pCi/L) in a significant percentage of houses (4).

People are exposed to radiation from artificial as well as natural sources. The average annual effective dose from medical and dental radiographic examinations in industrialized countries now amounts to a substantial fraction of that which is received from natural sources (2,3). Smaller doses are received from radioactive minerals in building materials, phosphate fertilizers, and crushed rock; radiation-emitting components of television sets, smoke detectors, and other consumer products; radioactive fallout from atomic weapons; and nuclear power plants (2,5).

In many occupations, workers are also exposed to radiation in the workplace (see Table 17-2), receiving doses that vary depending on their particular work assignments and operating conditions (see Table 17-1). The dose received occupationally by radiation workers in the United States averages less than 5 mSv (0.5 rem) per year, and it approaches or exceeds the maximum permissable limit (50 mSv) in fewer than 1% of workers in any given year (2).

TYPES OF RADIATION INJURY

Irradiation can cause many types of injury, depending on the dose and conditions of exposure (2). For purposes of radiation protection, mutagenic effects, carcinogenic effects, and certain teratogenic effects are viewed as *stochastic* effects, which vary in frequency but not severity with the dose, and lack thresholds (5). In contrast, erythema of the skin, cataract of the lens, impairment of fertility, depression of hematopoiesis, and various other tissue reactions are classified as *nonstochastic,* or *deterministic,* effects, because they have thresholds and vary both in frequency and severity with the dose (5).

Effects of Radiation on Cells

Any molecule in the cell can be altered by irradiation, but DNA is the most critical target because damage to a single gene can profoundly alter or kill the cell. A dose of radiation that is sufficient to kill the average dividing cell (e.g., 1 Sv) produces dozens of strand breaks and other changes in the cell's DNA molecules. Most such changes are repairable, but those caused by high-LET radiation are likely to be less reparable than those caused by low-LET radiation.

The susceptibility of cells to radiation-induced killing increases with their rate of proliferation; dividing cells are radiosensitive as a class. The percentage of cells surviving, as measured by their ability to proliferate, tend to decrease exponentially with increasing dose. A rapidly delivered dose of 1 to 2 Sv generally suffices to reduce the surviving fraction by more than 50%; however, if the same dose of low-LET radiation is divided into two or more exposures separated by several hours, it typically kills fewer cells because some of the sublethal damage is repaired between exposures (2).

Damage to Chromosomes

Radiation-induced changes in chromosome number and structure are among the most thoroughly studied effects of radiation. The frequency of chromosomal aberrations increases in proportion to the radiation dose in the low-to-intermediate dose range, approximating 0.1 per cell per sievert in human blood lymphocytes irradiated in culture. The frequency of such aberrations is also increased in radiation workers and other irradiated populations, in whom it can serve as a crude biologic dosimeter (2).

Damage to Genes

Mutagenic effects of radiation have been investigated extensively in many types of cells. The frequency of radiation-induced mutations averages about 10^{-6} per locus per sievert in human lymphocytes and 10^{-5} per locus per sievert in mouse spermatocytes and oocytes, depending on the conditions of irradiation (2,3). Heritable effects have not been detectable in the children of atomic bomb survivors, nor have heritable effects been detected as yet in other human populations (2,5). On the basis of the available data, however, the dose required to double the frequency of mutations in the human species is estimated to lie between 0.2 and 2.5 Sv (2,5).

Effects on Tissues

Effects of radiation include tissue reactions that vary markedly depending on the specific tissue or organ that is irradiated and the conditions of exposure (2). Tissues in which cells proliferate rapidly usually are the first to exhibit injury (2). Mitotic inhibition and cytologic abnormalities, for example, may be detectable immediately in irradiated tissues, whereas fibrosis and other degenerative changes may not appear until months or years after exposure. In tissues capable of rapid cell turnover, the killing of dividing cells by irradiation tends to elicit compensatory proliferation of surviving stem cells, so that a given dose causes less depletion of cells if it is spread out in time than if it is received in a single brief exposure (2).

Reactions that are particularly relevant to occupational radiation exposure include the following.

Skin. Because of the superficial location of the skin, its response to radiation has been investigated more thoroughly than that of any other organ. Erythema is the earliest outward reaction of the skin; it may occur within minutes or hours after exposure, depending on the dose. After rapid exposure to a dose of 6 Sv or more, the reaction typically lasts only a few hours and is followed 2 to 4 weeks later by one or more waves of deeper and more prolonged erythema. After a dose of 10 Sv or more, dry desquamation, moist desquamation, necrosis of the skin, and epilation may ensue, followed eventually by pigmentation. Sequelae, which may develop months or years later, include atrophy of the epidermis and its adnexae, telangiectasia, and dermal fibrosis (2).

Bone Marrow and Lymphoid Tissue. Hematopoietic cells show degenerative changes within minutes after a dose higher than 1 Sv. A dose of 2 to 3 Sv delivered rapidly to the whole body kills enough hemopoietic cells to interfere drastically with the normal replacement of aging leukocytes, platelets, and erythrocytes. As a result, the blood count declines gradually, leukopenia and thrombocytopenia becoming maximal in 3 to 5 weeks (Fig. 17-1). After rapid exposure to a dose greater than 5 Sv, fatal infection or hemorrhage is likely to ensue (2,4). Doses larger than 5 Sv can be tolerated only if they are accumulated gradually, over a period of months, or if they are delivered to only a small portion of the total marrow (2,4). However, repeated occupational exposure to doses higher than 0.5 Sv per year has been observed to cause a syndrome termed "chronic radiation sickness" (characterized by leukopenia, chronic fatigue, and depression) in some Russian nuclear plant workers (6).

Lymphocytes also are highly radiosensitive, degenerating rapidly after intensive irradiation. A dose of whole-body radiation in excess of 2 to 3 Sv causes severe lymphopenia, aplasia of lymphoid tissues, and depression of the immune response (2,4).

Gastrointestinal Tract. Dividing cells in the mucosal epithelium of the small intestine are highly radiosensitive; a dose of 10 Sv kills sufficient numbers of them to interfere with the normal renewal of the overlying epithelium. The resulting depletion of epithelial cells, if severe enough, may lead within a few days to ulceration and ultimately to denudation of the mucosa. The prompt delivery of a dose exceeding 10 Sv to a large part of the

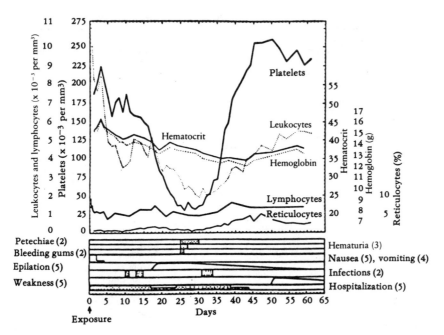

FIG. 17-1. Hematologic values, symptoms, and clinical signs in five men exposed to whole-body irradiation in a criticality accident. The blood counts are average values for the five men; the figures in parentheses denote the numbers showing the symptoms and signs indicated. (From Andrews GA, Sitterson EW, Kretchmar AL, Brucer M. Criticality accidents at the Y-12 plant. In: Diagnosis and treatment of acute radiation injury. Geneva: World Health Organization, 1961:27–48.)

small intestine, as may occur in a radiation accident, can cause a rapidly fatal dysentery-like syndrome (2,4).

Gonads. Because of the high radiosensitivity of spermatogonia, the seminiferous tubules are among the most radiosensitive organs of the body. Acute exposure of the testes to a dose as low as 0.15 Sv suffices to depress the sperm count for months, and a dose of more than 2 Sv can cause permanent sterility (2). Oocytes also are highly radiosensitive. Acute exposure of both ovaries to a dose of 1.5 to 2.0 Sv can cause temporary sterility, and a dose of 2.0 to 3.0 Sv can result in permanent sterility, depending on the age of the woman at the time of exposure (2).

Lens of the Eye. Irradiation of the lens can cause lens opacities that may not become evident until months or years later. The threshold for a vision-impairing opacity is estimated to vary from 2 to 3 Sv received in a single brief exposure, to 5.5 to 14.0 Sv received in repeated exposures over a period of months (2). In the 1940s, the occurrence of radiation-induced cataracts in a number of the pioneer cyclotron physicists provided the first indication of the high relative biologic effectiveness of neutrons for injury to the lens (2).

Radiation Sickness. Intensive irradiation of a major part of the hematopoietic system, the gastrointestinal tract, the lungs, or the brain can cause the *acute radiation syndrome* (Table 17-4). The associated prodromal symptoms characteristically include anorexia, nausea, and vomiting, which typically begin within hours after irradiation. Except with high doses, these symptoms usually subside within a day and are followed by a symp-

TABLE 17-4. *Major forms and features of the acute radiation syndrome*

Time after irradiation	Cerebral form (>50 Sv)	Gastrointestinal form (10–20 Sv)	Hemopoietic form (2–10 Sv)	Pulmonary form (>6 Sv to lungs)
First day	Nausea Vomiting Diarrhea Headache Disorientation Ataxia Coma Convulsions Death	Nausea Vomiting Diarrhea	Nausea Vomiting Diarrhea	Nausea Vomiting
Second week		Nausea Vomiting Diarrhea Fever Erythema Prostration Death		
Third to sixth weeks			Weakness Fatigue Anorexia Fever Hemorrhage Epilation Recovery (?) Death (?)	
Second to eighth months				Cough/Dyspnea Fever Chest pain Respiratory failure (?)

Modified from ref. 4.

tom-free interval before the onset of the main phase of the illness (see Table 17-4) (2,4).

> Within hours after performing maintenance work in a large industrial radiography facility that was subsequently found to have a defective safety interlock system, a pipefitter experienced transitory nausea and vomiting. One to 2 weeks later, generalized erythema developed, followed within several days by loss of hair, sore throat, bleeding from the gums, diarrhea, and weakness. On examination 3 weeks after the initial onset of his symptoms, the worker's lymphocyte count was 1,100/mm³, leukocyte count 2,200/mm³, platelet count 27,000/mm³, hematocrit 40%, and reticulocyte count 1%. On cytogenetic analysis, his blood lymphocytes revealed an increased frequency of chromosomal aberrations, consistent with whole-body ionizing irradiation. The symptoms subsided 2 to 3 weeks after treatment with platelet transfusions and antibiotics.

EFFECTS ON GROWTH AND DEVELOPMENT OF THE EMBRYO

Embryonal and fetal tissues are extremely radiosensitive. Rapid exposure to 0.25 Sv during a critical stage in organogenesis can cause malformations of many types in experimental animals (3), and comparable effects have been observed after larger doses in prenatally irradiated children (2). Mental retardation, for example, was greatly increased in frequency in children who were exposed to atomic bomb radiation at Hiroshima or Nagasaki between the 8th and 15th weeks of prenatal development (2,3).

EFFECTS ON CANCER INCIDENCE

Many types of cancer are induced by irradiation, depending on the conditions of exposure (2–6). The cancers resulting from irradiation do not appear until years or decades later, and they have no distinguishing features identifying them as having been induced by radiation. Hence, the occurrence of a given cancer cannot be attributed with certainty to previous irradiation.

The dose-incidence relation varies depending on the type of cancer; the dose, dose rate, and LET of the radiation; the age, sex, and genetic background of the exposed population; and other variables (2,3,6). With some types of cancer (e.g., chronic lymphocytic leukemia), no increase has been observed at any dose level, implying that such cancers are not inducible by radiation; with others (e.g., osteosarcoma), an increase in

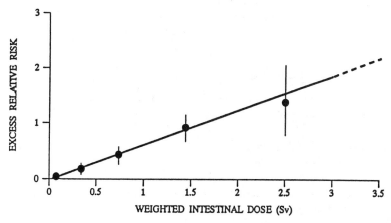

FIG. 17-2. Incidence of solid cancers, all types combined, in atomic bomb survivors, 1958–1987, in relation to the radiation dose. (From United Nations Scientific Committee on the Effects of Atomic Radiation. Sources and effects of ionizing radiation: report to the General Assembly, with annexes. New York: United Nations, 1994.)

frequency has been detectable only at high dose levels (more than 10 Sv). The existing data do not suffice to define the shape of the dose-response curve precisely for any neoplasm in the low-dose domain, but the following points are noteworthy: (a) The overall risk of solid cancer in Japanese atomic bomb survivors appeared to increase linearly with the dose up to 2.5 Sv (Fig. 17-2) and was elevated significantly at doses of only 5 to 50 mSv (7). (b) The risk of female breast cancer also appeared to increase as a linear-nonthreshold function of the dose, with essentially the same dose-dependent excess in women who received repeated fluoroscopic examinations of the chest during treatment for tuberculosis as in atomic bomb survivors or in women treated with x-rays for acute postpartum mastitis (2,3,6). (c) The pooled data from several large cohorts of radiation workers disclosed a dose-dependent excess of leukemia in this population (8) that was similar in magnitude to the excess in atomic bomb survivors, in whom the incidence appeared to increase as a linear-quadratic function of the dose (2,3,6). (d) The risk of thyroid cancer appeared to increase as a linear-nonthreshold function of the dose after acute irradiation in infancy or childhood and was significantly elevated by a dose of only about 100 mGy (2,3,6). Finally, (e) the risk of childhood cancer appeared to be increased significantly by prenatal exposure to only 10 mGy of x-radiation (9).

ADAPTIVE RESPONSES

There is growing evidence that a small "conditioning" dose of radiation may sometimes increase the resistance of an exposed cell or organism to a subsequent "challenge" dose; this has prompted many to question the plausibility of the linear-nonthreshold dose-response model. Some have even interpreted the data to suggest that the effects of low-level irradiation may be beneficial rather than harmful ("radiation hormesis") (6). Nevertheless, the weight of existing evidence suggests that for certain types of cancer and

for many of the biologic alterations that are precursors to cancer (e.g., mutations, chromosome aberrations), there is a linear-nonthreshold relation between risk and dose in the low-dose domain. Therefore, although alternative dose-response relations cannot be excluded, the linear-nonthreshold model appears to be the most plausible model, on the basis of present scientific knowledge (6).

COMPARATIVE MAGNITUDE OF RADIATION RISKS

On the assumption that the mutagenic and carcinogenic effects of radiation increase in frequency as linear-nonthreshold functions of the dose, the lifetime risk of a severe heritable disorder attributable to occupational irradiation is estimated to be approximately 0.8% per sievert (5), and the overall lifetime excess of cancer attributable to occupational irradiation is estimated at 5% per sievert (Table 17-5). From the foregoing, the average attributable risk of cancer in a radiation worker can be calculated to constitute less than 5% of the natural risk.

TABLE 17-5. *Estimated lifetime risk of fatal cancer attributable to low-level irradiation*[a]

Type or site of cancer	Excess cancer deaths per 10,000 person-Sv	
	No.	%[b]
Stomach	110	18
Lung	85	3
Colon	85	5
Leukemia (excluding chronic lymphocytic leukemia)	50	10
Urinary bladder	30	5
Esophagus	30	10
Breast	20	1
Liver	15	8
Gonads	10	2
Thyroid	8	8
Osteosarcoma	5	5
Skin	2	2
Other	50	1
Total	500	2

[a] Modified from ref. 5 (values rounded).
[b] Percentage by which the "spontaneous" baseline rate is increased.

RADIATION PROTECTION

From the beginning of this century, efforts have been made to prevent injury in radiation workers by limiting their occupational exposure. Initially, acute injuries were the chief concern, and attempts were made to set tolerance doses, or threshold limit values (TLVs) to prevent such injuries. Gradually, however, it came to be suspected that genetic, carcinogenic, and teratogenic effects might have no threshold. Hence the concept of a "tolerance" dose for such effects was eventually replaced by the concept of a "maximum permissible dose"—that is, a dose that is not expected to prevent such effects altogether but merely to limit their frequency to levels that are acceptably low. The present system of protection for radiation workers therefore involves two sets of dose limits, the first intended to restrict the risks of genetic, carcinogenic, and teratogenic effects, and the second to prevent other radiation effects altogether (by keeping the cumulative dose to any given organ from reaching one of the relevant thresholds) (5). Moreover, because no amount of radiation is assumed to be entirely without risk, the guiding rule in radiation protection is the *ALARA* principle, which holds that the dose should be kept *as low as reasonably achievable,* all social and economic costs considered (5).

To limit the risks of genetic, carcinogenic, and teratogenic effects, it is recommended that the effective dose from occupational irradiation not exceed 20 mSv (2 rem) per year averaged over any 5-year period, or 50 mSv (5 rem) in any single year; these dose limits are expected to keep the combined risks of such effects in radiation workers from exceeding a rate of 1 per 1,000 per year, a frequency of fatal work-related injuries encountered in many other occupations that are generally considered to be acceptably "safe" (5). In addition, to prevent the accumulated dose to any organ from reaching the threshold for a "deterministic" effect (such as radiation-induced depression of hemato-poiesis, impairment of fertility, or cataract of the lens), dose limits for each organ of the body also are recommended—namely, 0.15 Sv (15 rem) per year for the lens of the eye and 0.5 Sv (50 rem) per year for any other organ (5).

To minimize the radiation exposure of workers without unduly sacrificing their efficiency requires careful design of the workplace and work procedures, thorough training and supervision of workers, implementation of a well-conceived radiation protection program, and systematic health physics oversight and monitoring (Fig. 17-3).

Also needed are careful provisions for dealing with radiation accidents, emergencies, and other contingencies; systematic recording and updating of each worker's exposures; thorough labeling of all radiation sources and exposure fields; appropriate interlocks to guard against inadvertent irradiation; and various other precautionary measures (10–13).

General principles to be observed in every radiation protection program include the following:

1. Areas in which there is the potential for exposure to radiation should be monitored sufficiently to characterize spatial and temporal variations in the radiation level, to enable such areas to be properly posted and controlled.

2. Any work involving the potential for exposure to ionizing radiation should be authorized in advance, based on review of the tasks in question and the safety procedures to be observed in accomplishing the work.

3. Safety procedures should be designed to control internal as well as external exposure, should be spelled out clearly, and should be reviewed and updated periodically.

4. A personal dosimeter should be considered for any worker who is likely to receive more than 1 mSv in a given year; the device used should be capable of measuring the types and energies of the

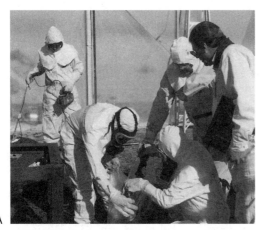

A   B

FIG. 17-3. A. Proper protection in the field is necessary during the collection and packaging of samples that may be contaminated with radioactivity. **B.** Monitoring radiation levels atop a nuclear reactor. (Photographs courtesy of RL Kathren, Batelle Pacific Northwest Laboratory.)

radiations to be encountered, which may vary depending on the nature of the radiation field (if the field is nonuniform, more than one device may be needed).

5. A permanent record of each monitored worker's exposure should be maintained, not only to document the exposure history but also to demonstrate compliance with guidelines and to assist in evaluating the effectiveness of radiation control measures.

6. A well-developed, well-rehearsed, and updated emergency preparedness plan should be in place to enable prompt and effective response in the event of a malfunction, spill, or other radiation accident.

7. Appropriate use should be made of shielding in facilities, equipment, and work clothing (e.g., apron and gloves).

8. Careful attention should be given to the appropriate selection, installation, maintenance, and operation of all equipment.

9. Because the intensity of exposure varies inversely with the square of the distance from the source, the time in the exposure field should be minimized and the distance between personnel and sources of radiation should be maximized.

10. Workers should receive appropriate training and supervision so that they can accomplish routine tasks with minimal exposure and cope safely with irregularities (13).

MANAGEMENT OF THE IRRADIATED WORKER

In any workplace where employees may be exposed accidentally to radiation or radioactive material, plans for coping with such contingencies should be made in advance. Such plans require delineation of lines of authority for managing accidents, knowledge in the workplace of local health care facilities that are capable of evaluating and treating radiation accident victims, plans for transporting radioactive victims, and an understanding by the workers of the hazards of radiation.

In caring for a radiation accident victim, good medical judgment and first aid should come first. Even if the worker has been heavily irradiated or contaminated, he or she must be evaluated also for other forms of injury, such as mechanical trauma, burns, and smoke inhalation. To guard against self-contamination, those handling or examining the victim should wear gloves, masks, and other protective clothing. The following general principles also should be observed (10–14):

A. General Emergency Medical Procedures
1. Emergency diagnostic and therapeutic maneuvers to ensure stability of airway, respiration, circulation, and so forth should be applied as necessary.
2. Appropriate transport should be summoned and the attendants alerted that the worker has been irradiated or contaminated.
3. The appropriate medical facility should be notified to expect the patient and informed that he or she has been irradiated or contaminated.
4. Detailed records should be kept concerning all examinations, measurements, findings, procedures, personnel, and the times involved.
B. Procedures for Radioactive Contamination
1. Clothing should be removed promptly if contaminated, isolated in a plastic bag, and labeled to denote radioactivity.
2. Any contaminated parts of the body should be isolated with plastic or paper from the rest of the body and from other surroundings and monitored for radioactivity.
3. Contaminated parts of the body should be thoroughly rinsed, the contaminated rinse water isolated as radioactive waste, and the part monitored again.
4. Care should be taken to avoid abrasion of contaminated skin during rinsing, to minimize hyperemia and further absorption of radioactivity.
5. To expedite elimination of any inhaled radioactivity, the victim should rinse the oral and nasal cavities by gargling and snorting water.
6. Secretions should be collected in plastic bags, labeled, and isolated for future examination.
7. The patient should be isolated from others who are not essential for emergency care.
8. Precautions should be taken to avoid contamination of other people, objects, and areas—for this purpose, the contaminated area should be sealed off as soon as possible.

REFERENCES

1. Lushbaugh CC, Fry SA, Ricks RC. Nuclear reactor accidents: preparedness and consequences. Br J Radiol 1987;60:1159–1183.
2. Mettler FA Jr, Upton AC. Medical effects of ionizing radiation. Philadelphia: WB Saunders, 1995.
3. National Academy of Sciences/National Research Council, Committee on the Biological Effects of Ionizing Radiation (BEIR V). Health effects of exposure to low levels of ionizing radiation. Washington, DC: National Academy Press, 1990.
4. United Nations Scientific Committee on the Effects of Atomic Radiation. Sources, effects and risks of ionizing radiation: report to the General Assembly, with Annexes. New York: United Nations, 1988.
5. International Commission on Radiological Protection. 1990 Recommendations of the International Commission on Radiological Protection. ICRP publication 60. Oxford: Pergamon, 1991. Annals of the ICRP 21:1–3.
6. United Nations Scientific Committee on the Effects of Atomic Radiation. Sources and effects of ionizing radiation: report to the General Assembly, with Annexes. New York: United Nations, 1994.
7. Cardis E, Gilbert E, Carpenter L, et al. Effects of low doses and dose rates of external ionizing radiation: cancer mortality among nuclear industry workers in three countries. Radiat Res 1995;142:117–132.
8. Pierce DA, Shimizu Y, Preston DL, et al. Studies of the mortality of atomic bomb survivors: report 12, part 1. Cancer: 1950–1990. Radiat Res 1996;146:1–27.
9. Doll R, Wakeford R. Risk of childhood cancer from fetal irradiation. Br J Cancer 1997;70:130–139.
10. National Council on Radiation Protection and Measurements (NCRP). Management of persons accidentally contaminated with radionuclides. NCRP report no. 65. Washington, DC: National Council on Radiation Protection and Measurements, 1980.
11. International Atomic Energy Agency (IAEA). What the general practitioner (MD) should know about the medical handling of overexposed individuals. IAEA-TECDOC-366. Vienna: International Atomic Energy Agency, 1986.
12. Mettler FA, Kelsey CA, Ricks RC. Medical management of radiation accidents. Boca Raton, FL: CRC Press, 1990.
13. National Council on Radiation Protection and Measurements (NCRP). Operational radiation safety program. NCRP report no. 127. Washington, DC: National Council on Radiation Protection and Measurements, 1998.
14. Kahn K, Ryan K, Sabo A, Boyce P. Ionizing radia-

tion. In: Levy BS, Wegman DH, eds. Occupational health. Boston: Little, Brown, 1983:189–206.

15. Andrews GA, Sitterson EW, Kretchmar AL, Brucer M. Criticality accidents at the Y-12 plant. In: Diagnosis and treatment of acute radiation injury. Geneva: World Health Organization, 1961:27–48.

16. AC Upton. Ionizing radiation. In: Levy BS, Wegman DH, eds. Occupational health: recognizing and preventing work-related disease, 2nd ed. Boston: Little, Brown, 1988.

BIBLIOGRAPHY

International Commission on Radiological Protection. 1990 Recommendations of the International Commission on Radiological Protection. ICRP publication 60. Oxford: Pergamon, 1991. Annals of the ICRP 21:1–3.
A detailed summary and explanation of the Commission's recommendations for limiting the exposure of workers and the public to ionizing radiation.

Mettler FA Jr, Upton AC. Medical effects of ionizing radiation. Philadelphia: WB Saunders, 1995.
A comprehensive review of the effects of ionizing radiation on human beings.

Lushbaugh CC, Fry SA, Ricks RC. Nuclear reactor accidents: preparedness and consequences. Br J Radiol 1987;60:1159–1183.
A review by recognized authorities on the causes, nature, and management of reactor accidents occurring in various countries between 1945 and 1987.

National Academy of Sciences/National Research Council, Committee on the Biological Effects of Ionizing Radiation (BEIR V). Health effects of exposure to low levels of ionizing radiation. Washington, DC: National Academy Press, 1990.
A comprehensive review of the biomedical effects of low-level irradiation, including estimates of the risks of genetic, carcinogenic, and teratogenic effects associated with occupational and environmental exposures.

United Nations Scientific Committee on the Effects of Atomic Radiation. Sources, effects and risks of ionizing radiation: report to the General Assembly, with annexes. New York: United Nations, 1988.
A comprehensive review of the sources and levels of ionizing radiation to which the population is exposed, and of the associated risks of health effects. Included is a review of the Chernobyl accident and its consequences.

United Nations Scientific Committee on the Effects of Atomic Radiation. Sources and effects of ionizing radiation: report to the General Assembly, with annexes. New York: United Nations, 1994.
A comprehensive review of recent epidemiologic data on the carcinogenic effects of ionizing radiation and a review of epidemiologic and experimental evidence of adaptive responses to radiation.

National Council on Radiation Protection and Measurements (NCRP). Operational radiation safety program. NCRP report no. 127. Washington, DC: National Council on Radiation Protection and Measurements, 1998.
A systematic presentation, explanation, and review of the purposes, principles, and requirements of an operational radiation safety program.

Occupational Health: Recognizing and Preventing Work-Related Disease and Injury, 4th ed.
Edited by Barry S. Levy and David H. Wegman, Lippincott Williams & Wilkins, Philadelphia © 2000.

18 Noise

Derek E. Dunn

The connection between noise exposure and a decline in hearing ability has been known for centuries. Almost 2,000 years ago, Pliney the Elder noted that people living near noisy waterfalls exhibited an accelerated and progressive hearing loss (1). Ramazzini directed attention to noise as a workplace hazard by documenting cases of occupational deafness in the 1700s (2). Early workplace noise tended to be mostly intermittent impact noise from the pounding of metalworkers and carpenters and affected mostly those performing the hammering. The dawn of the Industrial Revolution introduced continuous noise to the workplace, and the effect on hearing extended beyond the workers operating the machine to all those in the area. Today, noise is the most prevalent workplace hazard in the world.

The Occupational Safety and Health Administration (OSHA) has estimated that more than 7.9 million manufacturing workers in the United States are occupationally exposed to noise greater than 80 dBA (3).[1] The Environmental Protection Agency (EPA) estimated that more than 9 million U.S. workers in manufacturing are exposed to noise levels of 85 dBA or higher (4). Neither estimate included an additional 3 million workers employed in agriculture, mining, construction, transportation, or the federal government. More than 1 million workers in the U.S. manufacturing sector experience hearing loss from occupational noise, with half of them having moderate to severe hearing impairment (5). One worker in four exposed to 90 dBA noise over a working lifetime will develop a hearing loss that can be attributed to occupational noise exposure (6).

The largest number of workers exposed to levels of noise that are potentially damaging to their hearing are employed in manufacturing (7). Surveys indicate that more than half of industrial machines emit noise levels between 90 and 100 dB, and approximately 50% of industrial work environments have noise levels between 85 and 95 dB. Fewer than 6% of the machines surveyed produced noise levels lower than 85 dB (8).

Despite the prominence of noise in manufacturing, significant numbers of people in other industrial sectors (e.g., construction, agriculture, mining, transportation, the military) are exposed to hazardous noise (7–9). Studies indicate that, given the same noise exposures, women are less susceptible to noise-induced hearing loss than men, and African-Americans are less susceptible than Caucasians. The left ear shows a tendency to be more susceptible than the right to the effects of noise (10,11).

D. E. Dunn: Division of Biomedical and Behavioral Science, National Institute for Occupational Safety and Health, Robert Taft Laboratory, Cincinnati, OH 45226

[1] A common unit for measuring sound is the decibel (dB). Occupational noise is often assessed using a nonlinear filter (A-scale). The intensity of noise measured using the A scale is recorded as dBA.

The purposes of this chapter are to provide an overview of occupational noise and its effects on the ear and hearing and to review regulations and recommendations for effective programs to prevent job-related hearing loss. Although this chapter focuses on occupational noise-induced hearing loss and the consequences of excessive noise exposure, it should noted that workplace chemicals also can cause hearing loss (12,13); that increased mechanization has caused hearing loss from noise in the recreational, community, and home environments (14); that hearing loss occurs with aging (presbycusis) (15); and that noise can induce nonauditory effects (e.g., elevation of blood pressure, sleep disruption, stress, altered work performance) (16,17).

PROPERTIES OF SOUND AND NOISE

What we perceive as sound is the result of rapid fluctuations in the ambient air pressure caused by a vibrating object or a sudden expansion of gases, such as occurs in an explosion. These fluctuations in air pressure are called *sound waves* or *sound pressure.* Sound pressure waves are characterized by amplitude (loudness), frequency (pitch), and temporal pattern (18,19). The amplitude, measured in decibels (dB), represents the magnitude of change in the sound pressure wave relative to the ambient air pressure or a reference pressure. Human beings have an operational range up to approximately 130 dB before immediate damage results. The frequency, measured in hertz (Hz), is the speed with which the changes in the ambient air pressure occur per second. Humans perceive frequencies between 20 and 20,000 Hz. Amplitude and frequency may vary independently of one another. A pure tone is defined by a fixed frequency (e.g., 1,000 Hz) but may vary in amplitude. The temporal variations of these acoustic properties permit us to distinguish one sound from another and are the basis for our categorization of sounds as desirable or undesirable.

Noise generally comprises many pure-tone frequencies that interact with one another

to yield a sound with a complex mixture of loudness and pitch. A given noise with a relatively constant intensity is called *continuous* or *steady-state* noise. If a noise has occasional drops in intensity it may be referred to as *fluctuating* or *interrupted* noise. Noise characterized by a sudden rise followed by a decay in its intensity is referred to as *impact* or *impulse* noise.

The decibel (dB), the common unit for measuring the intensity of the noise or its sound pressure level (SPL), reflects the log of the ratio between the measured SPL, $p_{(1)}$, and a reference SPL, $p_{(0)}$. The reference SPL $(0.0002$ dynes/cm$^2)$ is the approximate lowest SPL that can be detected by the human ear. Consequently, SPL in decibels is a measure of decibels above 0.0002 dynes/cm^2. The formula for calculating a decibel SPL is:

$$SPL = 20 \log [p_{(1)}/p_{(0)}]$$

THE AUDITORY SYSTEM

The human peripheral auditory system can be divided into three parts: the external ear, the middle ear, and the internal ear (Fig. 18-1). The external ear consists of the auricle (which is often referred to as "the ear") and the external auditory meatus (ear canal). The middle ear is made up of the eardrum, the ossicles (three small bones), a number of suspension ligaments, two small muscles, the middle ear cavity, and the eustachian tube. The internal ear is the cochlea, which contains the sensory cells for detecting sound. This peripheral auditory system connects to the central auditory system, which consists of the eighth cranial nerve, the auditory pathways in the brain stem, and the auditory areas of the brain (19–21).

External Ear

Unlike animals with mobile pinnas, the human auricle does not play a major role in hearing or sound localization. The ear canal allows the more delicate structures of the auditory system to be recessed in the protec-

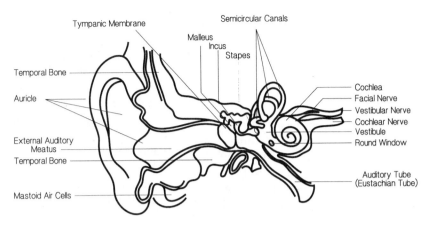

FIG. 18-1. Schematic drawing of the ear.

tive temporal bone of the skull. This part of the skull is the densest bone in the human body. The ear canal, somewhat like a cave, provides a stable temperature and humidity for the tissue paper–thin eardrum that seals its medial end. A very important characteristic of the ear canal is that, like any tube that is open on just one end, it has a resonant frequency. The resonance enhances the transmission of acoustic energy by approximately 15 to 20 dB for frequencies between 2,000 and 3,000 Hz and contributes to the fact that human auditory sensitivity is greatest for these frequencies.

Middle Ear

The principal role of the middle ear is to serve as an impedance matching and transmission device between the airborne acoustic vibrations in the ear canal and the fluid-filled cochlea.

The eardrum consists of a fibrous layer sandwiched between an outer dermal layer and an inner mucosal layer. A nonpathologic eardrum is transparent and permits visualization of middle ear structures. Sound waves traveling down the ear canal strike the eardrum and set it vibrating. These vibrations are transmitted to the ossicles (malleus, incus, and stapes), which are suspended in the middle ear cavity by ligaments. These are the smallest bones in the human body and

remain constant in size throughout life. The two smallest muscles in the human body (tensor tympani and stapedius) attach to the ossicles and, on contraction, reduce the amount of sound energy transmitted through the middle ear. The contraction of the tensor tympani and stapedius (acoustic reflex) can be triggered by sudden loud sounds, but this does not occur rapidly enough to prevent all the damaging energy produced by an impulsive noise.

The medial end of the stapes bone (the footplate) is attached to the fluid-filled cochlea. The eardrum is much larger than the footplate of the stapes; this results in improved energy transmission through the middle ear (area ratio). In addition, there is a difference between the long arm of the malleus and the long arm of the incus, which, like a crowbar, provides increased transmission of mechanical vibrations through the middle ear through a lever action. The area ratio and lever principle enhance the transmission of vibrations in the air of the ear canal to the fluid of the cochlea by as much as 30 dB. As a result of the external and middle ear, some sound transmissions are enhanced by as much as 50 dB. The eustachian tube connects the middle ear cavity and the upper throat cavity and serves to equalize pressure in the middle ear cavity with the external barometric pressure (e.g., when elevators or planes ascend or descend).

Muscles attached to the eustachian tube cause it to open during swallowing and yawning.

Internal Ear

The main function of the cochlea is to transduce mechanical vibrations in its fluid, generated by the stapes footplate, to neurally coded electrical impulses that are transmitted to the brain. The human cochlea has one row of inner hair cells and three to five rows of outer hair cells—a total of 20,000 to 25,000 hair cells. These hair cells sit on the basilar membrane, which vibrates in response to the pressure waves generated in the cochlear fluid. These mechanical movements produce an excitation of the sensory cells that causes them to stimulate neural fibers leading to the brain. The outer rows of hair cells have sometimes been considered the source of sensitivity for very-low-intensity sounds near the hearing threshold.

The rows of hair cells run the length of the cochlea. The hair cells responding to higher-frequency sounds are located nearer the basal end of the cochlea, where the footplate of the stapes is located, and the hair cells more sensitive to lower-frequency sounds are found more toward the opposite (apical) end of the cochlea. The spatial (tonotopic) differentiation of frequency coding in the cochlea is found also in the auditory pathways of the brain stem and the projections of the auditory system at cortical levels.

NOISE-INDUCED HEARING LOSS

Usually noise-induced hearing loss develops gradually as a result of damage to the sensory hair cells in the cochlea from prolonged loud exposures. The relative hazard to hearing from exposure to continuous, interrupted, impact or impulse noise depends on the intensity, duration, and frequency composition of the noise. Metabolic exhaustion and mechanical injury are the presumed mechanisms underlying cochlear damage from noise exposure. In metabolic exhaustion, the sensory cells are unable to keep pace with the energy demands placed on them. Most exposures to noise lack sufficient intensity or duration to result in immediate permanent damage to the auditory system. At these subtraumatic doses, metabolic exhaustion causes the auditory system to exhibit a temporary shift in hearing sensitivity that returns to normal with sufficient rest from noise exposure. However, repeated metabolic stress can result in damage or destruction of auditory hair cells and a permanent loss of hearing (22).

Mechanical stress on hair cells is thought to occur at higher noise levels and to result from a physical breakdown in cell structures caused by excessive vibratory forces transmitted into the cochlear fluids and basilar membrane by the stapes footplate (23–26). Noise exposures that tax cells' metabolic equilibrium can increase the their susceptibility to mechanical damage. The metabolic state of the hair cells is correlated with their ability to withstand or recover from physical stress.

Once the sensory hair cells that respond to a given frequency are destroyed, sounds at that frequency are no longer heard. Destroyed hair cells cannot repair themselves, nor can medical procedures restore normal function. Hair cell damage from noise usually affects the rows of outer hair cells before damaging the inner hair cells. Although the outer hair cells are more susceptible to damage from noise, the loss of an equal number of inner hair cells results in a much greater deficit in hearing.

Hair cells in the basal turn of the cochlea—those that respond to the higher-frequency tones—are more susceptible to noise-induced damage than are hair cells in the apical end of the cochlea. As a result, the early stages of noise-induced hearing loss usually are characterized by decreased ability to hear very soft, higher-frequency sounds. A common sign of noise-induced hearing loss is a greater decline in hearing acuity for tones between 4,000 and 6,000 Hz. The locus of damage to the hair cells along the basilar membrane is related to the physical properties of the noise. In most cases, the hearing loss is not perceptible in the early stages and may only be detected by audiometric tests. However, with continued overexposure to noise, the loss of sensitivity spreads from the higher frequencies into the speech frequencies (1,000 to 3,000 Hz). Once this occurs, considerable hearing loss already exists, and the ability to communicate or hear necessary sounds is significantly reduced.

The external or middle ear structures are damaged only by extremely intense noise, such as an explosive discharge. Severe damage to the external and middle ear can produce a reduction in hearing sensitivity of much as 30 to 50 dB as a result of loss of efficient transfer of sounds to the auditory sensory cells.

PREVENTING NOISE-INDUCED HEARING LOSS

Role of Government

Under the mandate of the 1970 Occupational Safety and Health Act, OSHA promulgated the 1971 noise standard for manufacturing companies engaged in interstate commerce (27). The 1971 noise standard covers most of manufacturing but does not cover workers in transportation, agriculture, construction, mining, or oil and gas servicing or drilling. However, mining is covered by noise standards enforced by the Mine Safety and Health Administration. Each of the enforced noise regulations incorporates an exposure limit based on a 90-dBA exposure for a duration of 8 hours with a 5-dB trading ratio. These noise standards also limit impulse noise to a peak SPL of 140 dB.

1971 Noise Standard

The 1971 noise standard requires that workers not be exposed to noise in excess of a 90-dBA time-weighted average (TWA) for 8 hours. The TWA is the average of the various exposure levels that occur during a period of measurement. Measurements are made

TABLE 18-1. *Permissible noise exposures*

Duration per day (hr)	Sound level (dBA slow response)
8	90
6	92
4	95
3	97
2	100
1.5	102
1	105
0.5	110
0.25 or less	115

using "the slow response" on the sound level meter. Higher dBA levels are permitted (not to exceed 115 dBA), but each increase of 5 dBA reduces the permitted exposure time by 50%; for example, 95 dBA for 4 hours or 100 dBA for 2 hours. This relation between noise level and exposure time is referred to as the 5-dB trading ratio or exchange rate; Table 18-1 gives an expanded list of allowable noise exposures. The OSHA noise regulation called for an "effective hearing conservation program" but does not specify the components of such a program. Exposure to impulse or impact noise is limited to peak pressures of 140 dB.

In 1983, based on the criteria for noise exposure recommended by the National Institute for Occupational Safety and Health (NIOSH) in 1971 (6), OSHA amended the 1971 noise standard. The 1983 amendment includes details of the components of a hearing conservation program and requires that a program be initiated when worker noise exposures exceed 85 dBA TWA. The OSHA-mandated hearing conservation program includes monitoring of noise exposure, periodic hearing testing of noise-exposed workers, noise abatement and/or administrative controls, provision of hearing protectors to employees, an employee education program, and record keeping (28).

Monitoring of Noise Exposure

The employer is required to identify workers whose noise exposure exceeds the limits defined in the 1971 noise standard. The identification of work areas with potentially hazardous noise levels is accomplished by assessing and monitoring noise levels. The noise data are reviewed as part of the process for determining compliance with noise regulations. Workers have the right to observe, or have a representative observe, the noise assessment procedures. Assessment of the noise exposure may be accomplished with the use of basic sound level meters, noise dosimeters, integrating sound level meters, or graphic level recorders (23). Measurement of noise with a sound level meter is appropriate when the levels are relatively steady and when the worker has a relatively stationary job. However, a noise dosimeter is required when noise levels are intermittent, vary greatly, or include impulse or impact noise. Many dosimeters offer options such as calculated dose for 3-dB or 5-dB exchange rates and selection of 8-hour criterion levels from 80 to 90 dBA.

As a general rule of thumb, potentially hazardous noise levels can be assumed to be present and hearing loss prevention strategies should be used if verbal communication is difficult at arm's length because of noise. However, decisions regarding which specific hearing loss prevention strategies to use require more objective assessments of the noise. Novices often are deceived into thinking that they can make accurate noise measurements because the devices appear simple to use: just point and note the position of the needle or digital readout. However, there are many variables and variations in the characteristics of both the physical environment (e.g., reverberation) and the measurement device (e.g., microphone selection, weighting, speed setting) that influence the accuracy of noise measurements. As the application of the noise data increases in importance (e.g., making decisions about effective engineering controls, developing a precise work area noise map, or determining whether permissible noise exposures are being exceeded), the services of professionals with training and experience in noise measurement should be obtained.

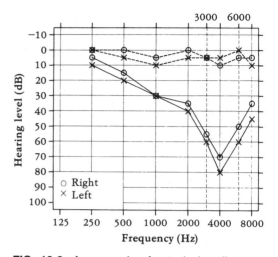

FIG. 18-2. An example of a typical audiogram from a person with normal hearing (*dashed lines*) and a person with bilateral sensorineural hearing loss resulting from excessive noise exposure. Note the maximum loss at 4,000 Hz and the spread of loss to the lower frequencies. (Audiogram provided by Robert I. Davis and Roger P. Hamernik.)

Periodic Hearing Tests

Hearing testing must be provided annually, at no cost, to workers exposed to 85 dBA TWA or more (Fig. 18-2). Specific procedures are explained for daily, yearly, and biannual audiometer calibration. Audiometry must be provided by a tester (audiologist, physician, or technician) or supervised by an audiologist, or an otolaryngologist, or other physician. Initial audiograms serve as a baseline reference against which subsequent audiograms are compared. An average shift of 10 dB at 2,000, 3,000, and 4,000 Hz is referred to as a standard threshold shift (STS). If an STS is determined to have occurred, the worker must be notified in writing of the STS, be retrained in the use of hearing protectors, and be refitted with hearing protectors.

Noise Abatement and Administrative Controls

Feasible engineering controls are required by the standard, but there is no clear definition of "feasible." Noise abatement encompasses efforts to reduce the workers' total noise exposure by means other than hearing protectors or to reduce the ambient noise. Administrative controls include actions such as locating lunch or break areas away from the noise or rotating workers out of noisy areas to reduce their total noise exposure. Although noise abatement and administrative controls require ingenuity and consideration for the specific conditions at a given work site, these approaches clearly reduce worker noise exposure and therefore are preferable to the less predictable performance of hearing protectors (24).

Hearing Protectors

Hearing protectors must be provided, at no cost to workers, before baseline audiometry if their noise exposure exceeds 85 dBA TWA or the allowable noise exposures, or if the worker experiences an STS (Fig. 18-3). The

FIG. 18-3. Many workers are exposed to loud noise at work. This fender stamping press operator at an auto parts plant in Michigan wears ear muffs to reduce his noise exposure. Noise production in a stamping operation may be difficult to control at the source. (Photograph by Earl Dotter.)

employee must be allowed to select from a variety of hearing protectors that reduce the occupational noise to exposures permitted by the noise standard. If an STS occurs, a hearing protector that reduces the noise exposure to 85 dBA TWA or less must be used.

Employee Education

All employees mandated to be in the hearing conservation program are required to receive education annually concerning the effects of noise on hearing, the purpose of audiometric testing, and the use of personal hearing protection.

Record Keeping

Records must be kept of noise assessments (for 2 years) and results of audiometric tests (for the period of employment). These records must be made available, on request, to employees or their representatives. The records are to be transferred to, and maintained by, new owners of a company.

1998 NIOSH Recommended Criteria for Noise Exposure

NIOSH reviewed its 1971 recommendations regarding noise exposure and hearing conservation in the light of data and research that was published after 1971. In 1998, NIOSH generated a new recommendation on noise exposure that focused on hearing loss prevention (avoiding a hearing loss) rather than hearing conservation (limiting the amount of hearing loss) (29). The new recommendation maintains the 85-dBA, 8-hour limit, but replaces the 5-dB trading ratio with a 3-dB trading ratio (exchange rate). There are substantial scientific data to support the 3-dB exchange rate; it is used by the military and by most industrialized countries.

There are numerous differences among the 1998 NIOSH noise recommendations, the 1971 NIOSH noise recommendations, and the OSHA noise standards of 1971 and 1983. The following section highlights aspects of the 1998 noise recommendations that differ from previous U.S. regulations or recommendations. However, the OSHA noise standard is still the enforced noise regulation.

Recommended Exposure Limit

The NIOSH recommended exposure limit (REL) for occupational noise exposure is 85 dBA, as an 8-hour TWA with a 3-dB exchange rate. Greater exposures are considered hazardous.

Audiometric Evaluation and Monitoring

NIOSH recommends pure-tone air-conduction threshold testing of each ear at 500, 1,000, 2,000, 3,000, 4,000, and 6,000 Hz. Testing at 8,000 Hz should be considered, because it may be beneficial in determining the cause of a hearing loss. A baseline audiogram (after 12 hours away from noise) should be obtained before employment or within 30 days of enrollment in the hearing loss prevention program (HLPP). However, the annual monitoring audiogram should be obtained during or at the end of the work shift. If efforts to prevent overexposure to noise are inadequate, the worker may show a temporary change in hearing and this change can serve as an alert to take action on behalf of the employee or to correct aspects of the HLPP.

The new proposal for a sentinel audiometric event is an STS, defined as a decrease in hearing ability of 15 dB or more at any required audiometric test frequency in either ear that is confirmed on retest. NIOSH does not recommend the use of age correction tables when determining whether an STS has occurred. The age correction data are group statistics that inappropriately evaluate the effect of aging on the hearing of anyone above or below the mean.

Hearing Protectors

NIOSH recommends mandatory provision of hearing protectors for employees exposed to an 8-hour, 85-dBA TWA or greater. If the exposure exceeds an 8-hour, 100-dBA TWA, a combination of ear plugs and ear muffs is suggested. The effective noise attenuation of the hearing protectors selected should reduce exposure below 85 dBA TWA for 8 hours. To determine the effective noise attenuation of a protector, one can use a subject fit method, or the current manufacturer's noise reduction rating for the protector can be derated (i.e., subtract 25% for ear muffs, subtract 50% for slow-recovery formable ear plugs, and subtract 70% for all other ear plugs).

Monitoring Hearing Loss Prevention Program Effectiveness

The effectiveness of an HLPP should be monitored in terms of both the hearing loss prevented for the individual employee and the overall rate of noise-induced hearing loss for the population of employees. For the individual worker, the comparison of the current monitoring audiogram with the baseline audiogram determines whether hearing loss has been prevented. Annual review of the worker's audiometric profile is probably the best quality assurance check for small programs.

It is advisable also to monitor trends in audiometric data for the overall population of employees in large HLPPs. High variability in sequential audiograms indicates a lack of integrity for the hearing data and suggests a need to improve the quality of audiometric test procedures, equipment calibration, or record keeping. NIOSH recommends comparing the incidence of STS among the monitored employees with the incidence found in workers not exposed to noise. NIOSH also recommends careful evaluation of the HLPP if the STS incidence considered significant for the noise-exposed population is 3% greater than the incidence found in a nonexposed population.

Role of Providers of Hearing Loss Prevention Services

Audiologists, industrial hygienists, and health care providers are called on to evaluate or make recommendations regarding hearing loss prevention programs. A common approach to this task is to review and cite the federal regulations concerning hearing conservation programs. Such a review results in identifying the major components of a hearing conservation program. However, meeting the minimal requirements for a hearing conservation program does not guarantee that the program is effective in protecting workers against hearing loss.

An effective HLPP is the result of careful planning and continuous monitoring to ensure that the goal of hearing loss prevention is reached. The employer cannot accomplish this goal alone. It is necessary to have the support of everyone concerned with worker safety, including workers, unions, and trade organizations. The full commitment of the employer to the HLPP is necessary to ensure that the program is effective. Commitment to effective hearing loss prevention is vital to obtaining the level of employee motivation, active participation, and communication that is necessary if the program is to be a success. The employer's commitment is evidenced by establishment and support of key policies that promote the effectiveness of the HLLP (30–33).

Advisors on safety and health have a unique opportunity to encourage company administrators to establish policies that will promote effective HLPPs. The employer should strive for excellence in the HLPP and not simply meet minimal requirements set forth by state or federal regulations.

It is important to integrate the HLPP fully into the total safety program. There are several benefits to this approach. First, the safety officer can combine several safety programs into the time allotted for educating and motivating workers. Separate safety programs often result in workers' weighing the relative importance of one program against another. Often, hearing safety is not regarded as being as important as those safety programs already in place to prevent immediate and observable injuries such as burns, lacerations,

or poisoning. Hearing loss occurs without pain, and it is not as immediate or dramatic as some other injuries.

In addition, the worker should be encouraged to use good hearing safety practices away from the job. There is growing evidence that when an effective HLPP reduces the hazard of workplace noise, much of a worker's hearing loss can be attributed to nonoccupational noise exposures (34–36). Making hearing protection available for off-the-job activities is one way to encourage good hearing safety outside the workplace.

Although an effective HLPP requires the full cooperation and participation of managers and workers, there should be one person (possibly assisted by a support team) who is responsible for ensuring the quality of the hearing conservation program. That person must serve as both a contact for all groups and a program coordinator. Absence of a contact person often leads to the perception that the employer is not committed to the HLPP and that the program really is not important. Physicians, nurses, safety officers, and union representatives frequently make good key personnel because their other activities generally involve contact and interaction with workers or committees.

Another policy should be to strive for simplicity and continuity in the HLPP. The more difficult it is for employees to understand and follow the rules, the more difficult it will be for employer to monitor the hearing conservation program.

The employer should have a policy of reviewing the program at regular intervals to determine whether the desired results are being obtained. The purpose of the periodic review is to identify problems and take corrective action. Consequently, there should also be a policy that requires careful modification of a program that does not adequately protect the employees.

There are many intangibles that are important to the success of an HLPP. Intangibles such as commitment, motivation, and continuity should not be overlooked when establishing or monitoring such a program. A sincere commitment by management to the key policies needed for a high-quality safety program, along with basic hearing loss prevention practices, usually results in successful incorporation of these intangibles into an effective HLPP.

REFERENCES

1. Bacon FL. Sylva sylvarum: or a natural history. London: W. Rawley, 1627, Seminars in Hearing, 1988;9:4.
2. Ramazzini B. De morbis atificum diatriba. (A Latin text of 1713 revised with translation and notes by WC Wright, 1940.) Chicago: University of Chicago Press, 1988;9:4.
3. 46 Federal Register 4078 (1981a). US Department of Labor: Occupational noise exposure. Hearing Conservation Amendment: final rule. (Codified at 29 CFR 1910.)
4. US Environmental Protection Agency. Noise in America: the extent of the noise problem. EPA Report No. 550/9-81-101. Washington, DC: EPA, 1981.
5. US Department of Labor, Occupational Safety and Health Administration. Final regulatory analysis of the Hearing Conservation Amendment. Report No. 723-860/752 1-3. Washington, DC: US Government Printing Office, 1981.
6. National Institute for Occupational Safety and Health. Criteria for a recommended standard: occupational exposure to noise. DHEW (NIOSH) Publication No. HSM 73-11001. Cincinnati: US Department of Health, Education, and Welfare, Health Services and Mental Health Administration, NIOSH, 1972.
7. National Institute for Occupational Safety and Health. National Occupational Hazard Survey. Publication No 74-127. Washington, DC, Department of Health, Education and Welfare, May 1974.
8. Karplus HB, Bonvallet, GL. A noise survey of manufacturing industries. Am Ind Hyg Assoc Q 1953; 14:235–263.
9. Suter AH, Von Gierke HE. Noise and public policy. Ear Hear 1987:8:188–191.
10. Berger EH, Royster LH, Thomas WG. Presumed noise-induced permanent threshold shift resulting from exposure to an A-weighted L_{eq} of 89 dB. J Acoust Soc Am 1978:64:192–197.
11. Ward WD: Endogenous factors related to susceptibility to damage from noise. In: Morata TC, Dunn DE, eds. Occupational hearing loss. Occup Med 1995;10:561–575.
12. Morata TC, Franks JR, Dunn DE. Unmet needs in occupational hearing conservation. Lancet 1994; 344:479.
13. Morata TC, Dunn DE, Kretschmer LW, Lemasters GK, Keith RW. Effects of occupational exposure to organic solvents and noise on hearing. Scand J Work Environ Health 1993:19:245–254.
14. Axelsson A, Clark W. Hearing conservation for nonserved occupations and populations. In: Morata TC, Dunn DE, eds. Occupational hearing loss. Occup Med 1995;10:657–662.
15. Rosenhall V, Redsen KE. Presbycusis and occupational hearing loss. In: Morata TC, Dunn DE, eds. Occupational hearing loss. Occup Med 1995; 10:593–607.
16. Cohen A. The influence of a company hearing conservation program on extra-auditory problems in workers. Journal of Safety Research 1976: 8:146–162.
17. Dunn DE, Marenberg ME. Noise. In: Rosenstock L, Cullen MR, eds. Textbook of clinical occupational and environmental medicine. Philadelphia: WB Saunders, 1994:673–680.
18. Kryter KD. The handbook of hearing and the effects of noise. San Diego: Academic Press, 1994:1–15.
19. Pickles JO. An introduction to the physiology of hearing. London: Academic Press, 1988:1–77.
20. Anson BJ, Donaldson JA. Surgical anatomy of the temporal bone and ear, 2nd ed. Philadelphia: WB Saunders, 1973:153–188.
21. Ward WD. Anatomy and physiology of the ear: normal and damaged hearing. In: Berger EH, Ward WD, Morrill JC, Royster LH, eds. Noise and hearing conservation manual. Akron, OH: American Industrial Hygiene Association, 1986.
22. Dunn DE: Noise damage to cells of the organ of Corti. Seminars in Hearing 1988:9:267–278.
23. Henderson D, Hamernik R. Impulse noise: critical review. J Acoust Soc Am 1986:80:569–584.
24. Lim DJ, Dunn DE. Anatomic correlates of noise-induced hearing loss. Otolaryngol Clin North Am 1979:12:493–513.
25. Lim DJ, Melnick W. Acoustic damage to the cochlea: a scanning and transmission electron microscopic observation. Arch Otolaryngol 1971:94:294–305.
26. Hunter-Duvar IM. Morphology of the normal and the acoustically damaged cochlea. Scanning Electron Microscopy 1977:2:421–428. 300.
27. Department of Labor. Occupational Noise Exposure Standard: Title 29, Chapter XVII, Part 1910, Subpart G, 1910.95. Federal Register 1971;36:10518.
28. Department of Labor, Occupational Health and Safety Administration. Occupational noise exposure: Hearing Conservation Amendment, final rule. Federal Register 1983;48(46):9738–9785.
29. National Institute for Occupational Safety and Health. Criteria for a recommended standard: occupational noise exposure. DHHS (NIOSH) Publication No. 98-126. US Department of Health, Education, and Welfare, Health Services and Mental Health Administration, NIOSH, 1998.
30. Royster LH, Royster JD. Education and motivation. In: Berger EH, Ward WD, Morrill JC, Royster LH, eds. Noise and hearing conservation manual. Akron, OH: American Industrial Hygiene Association, 1986:383–416.
31. Dunn DE. Making a commitment to effective hearing conservation. Applied Industrial Hygiene 1988; 3:F16–F18.
32. Stewart AP. The comprehensive hearing conservation program. In: Lipscomb DM, ed. Hearing con-

servation in industry, schools and the military. Boston: Little, Brown, 1988:203–230.

33. Royster JD, Royster LH. Hearing conservation programs: practical guidelines for success. Chelsea, MI: Lewis Publishers, 1990:7–22.

34. Clark WW, Bohl CD, Davidson LS, Melda KA. Evaluation of a hearing conservation program at a large industrial company. J Acoust Soc Am 1987;82(Suppl 1):S113(A).

35. National Hearing Conservation Association. Bang bang, you're deaf. Hearing Conservation News 1988;6(3):1–3.

36. Franks J, Davis R, Kreig E. Analysis of a hearing conservation program data base: factors other than workplace noise. Ear Hear 1989;10:273–280.

BIBLIOGRAPHY

Berger EH, Ward WD, Morrill JC, Royster LH, eds. Noise and hearing conservation manual. Akron, OH: American Industrial Hygiene Association, 1986.
A valuable reference for the person who wishes to have a thorough understanding of occupational noise effects and the approaches for protecting workers from noise-induced hearing loss.

Kryter KD. The handbook of hearing and the effects of noise. San Diego: Academic Press, 1994.
This book is the latest edition of the text that has been used by every serious acoustician and hearing conservationist. Some of the information may challenge the novice, but it remains a key reference because of its depth and breadth.

Morata TC, Dunn DE, eds. Occupational hearing loss. Occup Med: State of the Art Reviews 1995;10 (theme issue).
The acoustic, physiology, and epidemiologic information complied in this book covers a range of issues facing those trying to prevent occupational hearing loss from noise, chemicals, and other ototraumatic agents.

Pickles JO. An introduction to the physiology of hearing. London: Academic Press, 1988.
An excellent reference on the anatomy and physiology of the auditory system. Explanations that can be understood by novices are provided on complicated topics. These simple explanations are augmented by data from recent cutting-edge research.

Royster JD, Royster LH. Hearing conservation programs: practical guidelines for success. Chelsea, MI: Lewis Publishers, 1990.
This text goes well beyond listing the components of programs to prevent hearing loss by adding useful information on the "how to" of obtaining the desired objectives of the hearing conservation program. Very useful for the person responsible for work safety.

Occupational Health: Recognizing and Preventing Work-Related Disease and Injury, 4th ed.
Edited by Barry S. Levy and David H. Wegman, Lippincott Williams & Wilkins, Philadelphia © 2000.

19 Other Physical Hazards

Christopher T. Leffler and Howard Hu

A welder is diagnosed as having a cataract. He tells the physician that in the last 5 years, two other welders in his shop acquired cataracts, and asks whether welding may have caused his condition.

A patient who operates a pneumatic chipper complains of loss of grip strength and sensation in both hands.

A joint labor–management committee seeks to decrease the number of heat-related illnesses among the company's construction workers during the heat of the summer. The head of the committee calls an occupational health professional for advice.

An electrical worker is planning a pregnancy, and asks her physician if she should stop working during her pregnancy because of the possible adverse effects of electromagnetic fields.

In this chapter, several types of physical exposures and environments that may be hazardous to workers are discussed. These include *nonionizing radiation* [microwave/radiofrequency, ultraviolet (UV), visible light, infrared (IR), and laser radiation]; *electric and magnetic fields*; *vibration*, including *ultrasound*; and *atmospheric variations* [hot

C. T. Leffler: Department of Environmental Health and Medicine, Harvard School of Public Health, Boston, Massachusetts 02115.

H. Hu: Department of Environmental Health and Medicine, Harvard School of Public Health, Boston, Massachusetts 02115.

and cold environments and hyperbaric (compression/undersea) and hypobaric (high-altitude) environments]. Some of these have well recognized health effects, but others are controversial and are still being studied.

NONIONIZING RADIATION

Nonionizing radiation (see Chapter 17 for a discussion of ionizing radiation) refers to emissions from those parts of the electromagnetic spectrum where emitted photons usually have insufficient energy to produce ionization of atoms. These forms of radiation include microwaves, television and radiowaves, visible light, and IR and UV radiation, among others (Fig. 19-1). [Some portions of the UV spectrum are considered ionizing (1).] Laser radiation is an amplified form of nonionizing radiation.

All types of nonionizing radiation obey certain general laws of electromagnetic radiation. The equation that fundamentally characterizes electromagnetic radiation is:

$$\lambda = c/f$$

where λ = wavelength in meters, c = velocity (usually the velocity of light, 3×10^8 m/second), and f = frequency in cycles per second.

Nonionizing radiation has other important characteristics shared by all forms of electromagnetic radiation: (a) it travels in straight lines and can be bent or focused, (b) the energy delivered is directly proportional to the frequency (and therefore inversely pro-

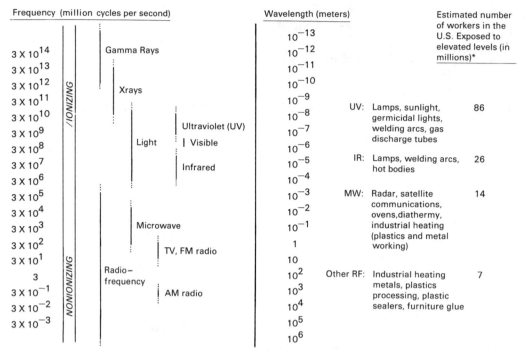

FIG. 19-1. The electromagnetic radiation spectrum. (From Key MM, et al. Occupational diseases: a guide to their recognition. Cincinnati, OH: National Institute for Occupational Safety and Health, 1977.)

portional to the wavelength), and (c) this energy occurs in small units called *quanta*.

When nonionizing radiation strikes matter, energy is absorbed. Nonionizing radiation is of lower frequency than ionizing radiation, and therefore contains less energy. Instead of causing ionization, this energy is usually transformed into heat, which accounts for many of its important physiologic effects. Absorption in the UV and visible portions of the spectrum can also produce photochemical reactions or fluorescence; this occurrence depends on the absorption spectrum of the molecule that has been struck and the efficiency of the specific radiation wavelength in producing this effect.

Ultraviolet Radiation

Ultraviolet radiation (UVR) is an invisible form of radiant energy produced naturally by the sun in a low-intensity form, and artificially by incandescent, fluorescent, and dis-

charge types of light sources. The UVR spectrum can be further subdivided, based on wavelength, into UV-A (4.0 to 3.2×10^{-7} m, black light) and UV-B (3.2 to 2.8×10^{-7} m) radiation, which are the two principal UV components in sunlight, and UV-C (2.8 to 2.0×10^{-7} m) radiation, which is germicidal. In industrial work settings, high-intensity UVR exposure occurs primarily from exposure to arc welding. Other industrial sources include plasma torches, which are used in heavy industrial cutting processes; electric arc furnaces; germicidal and black-light lamps; and certain types of lasers. UVR light is also being used in some coal processing and refining companies to detect coal tar residues on the skin of workers and equipment (because of coal tar's fluorescence under UVR).

The skin and the eyes absorb UVR and are particularly vulnerable to injury, primarily from UV-B radiation. The common sunburn is a well known result of acute overexposure to solar UVR. Long-term, low-intensity UVR exposure from the sun is also responsible for a number of other skin conditions, including solar elastosis (wrinkling) and solar keratoses (premalignant lesions). Sun exposure is the major etiologic agent for the most common skin cancers, basal cell and squamous cell carcinoma and malignant melanoma (see Chapter 27). All of these cancers are more common in areas with greater solar UVR exposure (few clouds, equatorial location, or high altitude), at sun-exposed anatomic sites, and in people with lighter skin, a greater sensitivity to sunlight, or a history of sunburn or sun-related skin damage (2). Whereas squamous cell carcinoma is associated with total and occupational sun exposure, basal cell carcinoma and malignant melanoma are more strongly associated with a history of multiple sunburns or other intermittent high exposures (2). Some epidemiologic studies have suggested that UVR in the form of fluorescent lighting can pose a slightly elevated risk for melanoma. The International Radiation Protection Association concluded that fluorescent light UVR did not pose a melanoma risk, but studies, including one that suggested moderately increased risks for men with increased cumulative exposure to fluorescent light UVR (3), indicate that further research is needed before this hypothesis can be completely dismissed.

Decreases in the earth's protective ozone layer due to chlorofluorocarbons have increased concern about the effects of UVR. Ozone losses will lead to increased UVR exposure and are expected to cause further increases in cataracts and basal cell and squamous cell skin cancer rates (4). The prospect of increased UVR also raises concern stemming from research suggesting the UVR can affect the immune system. UVR has been demonstrated to suppress certain aspects of cell-mediated immunity in mice. Langerhans cells (which activate T lymphocytes) are depleted in UV-irradiated human skin, and UVR is capable of reactivating both labial and genital herpes infections. UVR-related immunosuppression might provide a critical mechanism allowing the growth of skin cancers. The possibility that UVR-related immunosuppression may blunt cell-mediated immunity is particularly ominous for countries in which infectious diseases play a large role in morbidity and mortality.

Health effects of UVR on skin can be potentiated by several photosensitizing agents: locally applied coal tar, as well as many plants containing furmocoumarins and psoralens (e.g., figs, lemon and lime rinds, celery, and parsnips), are photosensitizers. When ingested, certain drugs (e.g., chlorpromazine, tolbutamide, and chlorpropamide) can increase susceptibility to UVR. Photosensitization can lead to an immediate sunburntype effect on exposure to a small dosage of radiant energy. People vary in their susceptibility to this reaction depending on their tendency to concentrate sufficient quantities of the photosensitizing agent in the skin.

Long-term, low-level UV-B exposure has also been clearly demonstrated to be a risk factor for the development of cortical cataracts (5), probably due to the generation of oxygen radicals that lead to photooxidation of proteins in the ocular lens. A doubling of UV-B would be projected to cause a 60% increase in cortical cataracts over current levels.

Covering skin with hats and long sleeves when possible, wearing UV-rated sunglasses, and avoiding prolonged exposures protects from the effects of low-intensity UVR. Benzophenone and anthranilate sunscreens absorb UVR and offer some protection to bare skin. Opaque sunscreens, such as titanium dioxide, offer the best protection and may be essential if photosensitization occurs. Some authors have postulated that sunscreen use might actually increase UVR exposure by failing to protect adequately against UV-A radiation and by increasing risk-taking behavior. Despite these theoretic concerns, randomized trials demonstrate that sunscreens reduce the frequency of premalignant solar keratoses (1). There is a consensus that for a given exposure, sunscreen use helps to protect the skin, and that broad-spectrum sunscreens that protect against UV-A radiation are ideal.

Ultraviolet radiation from arc-welding[1] operations has not been associated with skin cancer, probably because these high-intensity exposures are of an acute, short-term nature, and operators usually wear protective clothing. High-intensity acute exposures, however, can result in eye damage. It most commonly occurs with exposure to arc welding, but can also result from exposure to direct or reflected radiation from UVR lamps, such as those used in laboratories as bactericidal agents. The end result is conjunctivitis or keratitis (inflammation of the cornea), or both, a work-related condition commonly referred to as "ground-glass eyeball," "welder's flash," or "flash burn" (see Chapter 28). The welder or bystanders can be affected by just a brief exposure to unprotected or underprotected eyes. Prevention requires the isolation of high-intensity UVR sources and the use of goggles and other shields with proper filters.

Ultraviolet radiation also has an indirect impact on health through its ability to cause photochemical reactions. UVR, as encountered in welding operations, converts small amounts of oxygen and nitrogen into ozone and oxides of nitrogen, which are respiratory irritants (see Chapter 25). At certain wavelengths, UVR can also decompose halogenated hydrocarbon solvent vapors into toxic gases, such as perchloroethylene into hydrogen chloride, and trichloroethylene into phosgene. Control of these hazards requires proper local exhaust ventilation and isolation of high-intensity UVR sources from solvent processes.

Visible Radiation

Visible light plays an important role in determining working conditions. A consider-able amount of literature in the field of ergonomics deals with the proper use of different lighting systems for different tasks. Inadequate lighting may cause ocular fatigue and discomfort and, more important, contribute to work accidents. Types of illumination may also play a role in ocular health (see Chapter 28).

Infrared Radiation

In industry, significant levels of IR are produced directly by lamp sources and indirectly by sources of heat. The primary effect of IR on biologic tissue is thermal; it can cause skin burns, although IR exposure is usually detected by the sensation of heat before skin burns can occur. The lens of the eye, however, is particularly vulnerable to damage because the lens has no heat sensors and a poor heat-dissipating mechanism. Cataracts may be produced by chronic IR exposure at levels far below those that cause skin burns (see Chapter 28). Glass blowers and furnace workers have been shown to have an increased incidence of all types of cataracts (particularly posterior cataracts) after chronic IR exposure for over 10 years. Other workers potentially at risk for harmful IR exposure include those involved in handling molten metal (such as foundry workers, blacksmiths, and solderers), oven operators, workers in the vicinity of baking and drying heat lamps, and movie projectionists.

An epidemiologic investigation on the prevalence of cataract included 209 workers older than 50 years of age exposed to IR radiation in the Swedish manual glass industry for 20 years or more, and 298 non–IR-exposed control subjects. The risk that an IR-exposed worker will have to be operated for cataract is 12 times as high (95% confidence interval 2.6 to 53) (6). The left eye, which received higher IR exposures than the right eye, was more often affected.

Control of IR hazards requires shielding of the IR source and eye protection with

---

[1] Arcs are high-temperature sources generated between electrodes; during the welding process, they appear as blinding flashes of light and emit UVR.

IR filters. Maximizing the distance between workers and the IR source reduces the intensity of the exposure.

Laser Radiation

Laser is an acronym for *l*ight *a*mplification by the *s*timulated *e*mission of *r*adiation, a special category of human-made, nonionizing radiation. It is produced by forcing atoms of a particular gas or crystal to emit a stream of photons that are monochromatic and in phase with each other. These photons are emitted at specific wavelengths that depend on the source medium. The many and increasing applications of laser radiation derive from its ability to concentrate a large amount of energy in a small cross-sectional area in a highly coherent and minimally divergent manner. Lasers of various types are widely used as reference lines in surveying, instrumentation, and alignments; as a heating agent in welding; as a cutting instrument in microelectronics and microsurgery; in communications; in holography; and in the military, where their application is expected to increase dramatically.

The eye, including the retina, lens, iris, and cornea, is extremely vulnerable to injury from laser radiation in the near-UVR, near-IR, IR, and visible-light frequency ranges (see Chapter 28). Ocular damage can occur not only directly from intrabeam viewing, but indirectly by exposure to diffuse reflections from a high-power laser. Lasers of wavelengths that fall outside the visible portion of the electromagnetic spectrum can be particularly hazardous because exposure to the beam may not be readily apparent. Skin burns are caused by direct exposure to high-energy lasers.

Laser installations should be isolated wherever possible, and the laser beam should be terminated by a material that is nonreflective and fireproof. Special goggles can be helpful but must afford specific protection to the wavelength of the laser being used. Care must be paid to minimizing diffuse reflected radiation that may not be visually detectable.

Finally, proper worker education and a baseline eye examination must be included in the preventive program.

Microwave/Radiofrequency Radiation

Microwave/radiofrequency radiation (MW/RFR) encompasses a wide range of wavelengths used in radar, television, radio, and other telecommunications systems (see Fig. 19-1). Cellular telephones have emerged as the most important source of MW/RFR exposure in the general population (7). MW/RFR is also used in a variety of industrial operations, including the heating, welding, and melting of metals; the processing of wood and plastic (radiofrequency sealers); and the creation of high-temperature plasma. As many as 21 million workers in the United States may be exposed to MW/RFR, and its commercial and recreational use is expected to increase; thus, any health effect could have a very significant public health impact.

The heating effect of MW/RFR depends on the amount of energy absorbed, which, in turn, depends on the frequency of the radiation and the position, shape, and other properties of the exposed object. In general, detectable tissue heating requires relatively high-power MW/RFR, at power densities above 100 W/m² (10 mW/cm²). This thermal mechanism is responsible for the recognized adverse health effects of MW/RFR, some of which have been reported in humans, such as cataract formation, testicular degeneration, depressed sperm counts, focal areas of thermal necrosis, and, at extremely high intensities, death (8–10).

Recognition of the thermal effects of MW/RFR led to the current federal exposure guidelines, which limit MW/RFR power to 10 mW/cm² as averaged over any 6-minute period.

Whether adverse health effects can be caused by chronic exposure to MW/RFR at power densities *below* this standard is an area of ongoing research. Some laboratory studies of animals have demonstrated that MW/

RFR at intensities that are too low to cause thermal effects can cause biologic effects that are primarily of a neurologic and immunologic nature, such as alterations in electroencephalograms and behavior, and impairment of immune cell function (11). Studies suggest that MW/RFR can directly alter DNA sequences (1). It is possible that such effects could occur with low-level exposures by microheating—that is, heating localized to small regions of tissue (1). At present, there is no consensus on the adverse health effects related to typical human exposures (1,8, 9,12). Given the known potential for MW/RFR to cause thermal damage and the rapidly increasing low-level MW/RFR exposure to the general population from the proliferation of cellular telephones and their associated transmitters, a high priority must be given to establishing work guidelines, especially shielding and engineering controls that will bring compliance to the existing standard, and to performing research on the potential health effects of low levels of exposure. Arguments are being made in some communities to prevent, as a precaution, the current rush to install MW/RFR equipment for wireless communications. In the absence of conclusive evidence for or against the existence of significant low-level MW/RFR effects, these are understandably controversial policy debates.

Finally, microwaves and radiowaves, including those from cellular telephones, have the potential to interact with medical implants, particularly cardiac pacemakers. Current pacemaker leads are shielded to prevent electromagnetic interference; however, although unlikely, disruption is conceivable. Patients with pacemakers are advised to avoid carrying or placing an activated cellular phone or its antenna near the pacemaker (13). In cellular phone users and occupationally exposed people, estimations of the exposure and subsequent possibility of interference should be made in consultation with the (pacemaker) manufacturer and a cardiologist. Interference due to microwave ovens is not thought to be a problem given the shielding of both modern pacemaker leads and microwave ovens (14).

ELECTRIC AND MAGNETIC FIELDS

The ubiquitous use of electricity has led to exposure of the general population and occupational groups to high-voltage electricity. Although the hazards of electrocution and burns due to direct electrical exposure are well known, long-term effects, if any, due to exposure to low-frequency (50 or 60 Hz) electrical and magnetic fields associated with the use and transmission of electricity have not been determined.

An *electric field* describes the force between two electrically charged particles. A *magnetic field* describes the force between two objects carrying electric current. A variable electric field is always accompanied by a magnetic field; this interplay is often referred to as an *electromagnetic field* (EMF).

Occupational exposure to high-intensity EMFs may occur in people who work near high-voltage electrical lines (Fig. 19-2), such as streetcar and subway drivers, power station operators, power and telephone line workers, and other groups of technicians and electrical workers, including electricians, movie projectionists, and welders. Current occupational guidelines are designed to avoid acute electrical stimulation of neural and cardiac tissues (15).

A number of epidemiologic studies have suggested small increases in leukemia, brain cancer, and other cancers among occupational and residential groups exposed to higher levels of EMFs, but the findings have not been consistent (16–19). Criticism of the methodologies and interpretation of these studies has been intense. One of the most problematic aspects of these investigations is the tendency to find positive associations between cancer and surrogate exposure measurements, such as "electrical occupations" or wiring codes, but failure to find positive associations when actual field measurements of EMF are made. There is also some evidence of publication bias against studies that

FIG. 19-2. Telephone lineman repairing tornado damage in Texas. Utility line workers, who are exposed to high-voltage electrical lines, are at high risk for electrocution and are also being studied for disease potentially induced by electromagnetic fields. (Photograph by Earl Dotter.)

do not support an association between EMF exposure and cancer (16). Nevertheless, a meta-analysis that accounted for heterogeneity among studies, influence of individual studies, publication bias, and exposure type identified a consistent risk that could not be easily explained by random variation (20).

Electromagnetic fields may have an impact on health, given that endogenous electrical currents play crucial roles in physiologic processes such as neural activity, tissue growth and repair, glandular secretion, and cell membrane function. However, the fields induced within the body by environmental exposures are not large compared with those generated by human cells (17). Although it has been demonstrated that EMFs can exert biologic effects, it is unclear what, if any, adverse health outcomes might be produced.

In addition to the lack of unequivocal epidemiologic evidence, there is no conclusive evidence from animal or *in vitro* studies that EMFs are carcinogenic (17). More basic scientific and epidemiologic research is required to determine whether EMFs encountered in industry and the home pose a risk to health. Before future epidemiologic studies can generate meaningful results, more appropriate and precise exposure assessment metrics need to be established.

VIBRATION

A 40-year-old man came to see his physician with a chief complaint of hand problems. Approximately 12 years earlier, he started work as a copper stayer (metalsmith) using a pneumatic hammer weighing 12 lb with a frequency of 2,300 vibrations a minute (38 Hz[2]). He wore a glove on his left hand. After he had been working for 2 years, he noticed that his fingers would go white, starting with his left index finger, and

(Drawing by Nick Thorkelson.)

---

[2] 1 Hz = 1 hertz = 1 cycle per second; 1 kHz = 1,000 Hz.

along with this color change he noted a "pins-and-needles" sensation. Gradually, the right hand became involved and soon all fingers on both hands were involved. His hands became very awkward to use and he began to have difficulty in buttoning his shirt. It often took 2 hours for normal sensation to return to his hands. Initially, the syndrome occurred only in winter, but after 4 years it occurred in summer as well. After over 6 years' work with the hammer, he had to give up this occupation because of these symptoms. He is now working as a laborer in a boiler shop (21).

Vibration refers to the mechanical oscillation of a surface around its reference point. Health hazards of interest usually stem from vibration at frequencies of 2 to 1,000 Hz. Occupational health effects of vibration stem from prolonged periods of contact between a worker and the vibrating surface. It has been estimated that 8 million workers in the United States are exposed to occupational vibration, either whole-body vibration (WBV) or segmental vibration to a specific body part. Drivers of trucks, tractors, buses, and other vehicles, and certain heavy equipment operators, are subject to chronic WBV transmitted through the seat or floor of their work posts. This vibration is usually of lower frequency (2 to 100 Hz). Grinders and operators of chain saws, chipping hammers, and other pneumatic tools are exposed to segmental vibration, usually through the hands, that is of higher frequency (20 to 1,000 Hz).

Whole-Body Vibration

Less is known about the chronic effects of low-frequency WBV than about segmental vibration. Several epidemiologic studies have suggested an association with changes in bone structure, gastrointestinal disturbances involving secretion and motility, prostatitis (22), autonomic nerve dysfunction, and slowing of peripheral nerve conduction (23). Vibration at the resonating frequency of the eyeballs (60 to 90 Hz) has been associated with visual disorders. WBV exposure is associated with low back, neck, and shoulder pain among vehicle drivers

(24). Although it is difficult to isolate the effects of WBV from confounders, such as materials handling and posture, animal models lend biologic plausibility to this hypothesis. Thus, although definitive proof of causality is lacking, it would seem wise to limit exposure to WBV as much as possible.

Segmental Vibration

Chronic exposure to segmental vibration at low frequencies of 20 to 40 Hz, such as is encountered with the operation of heavy pneumatic drills, has been associated with degenerative osteoarticular lesions in the elbows and shoulders. These findings are believed to result from a "wear-and-tear" process with repetitive impulse loading; however, epidemiologic proof of causality has so far been lacking.

At frequencies between 40 and 300 Hz, the use of vibratory hand tools (e.g., chain saws, grinders, hammers, and drills) has been well known to elicit a neurovascular syndrome termed the *hand–arm vibration syndrome* (HAVS). HAVS has several features (25,26): (a) an abnormal peripheral circulation characterized by Raynaud's phenomenon (vasospastic attacks that cause finger blanching), the basis for alternate terms for HAVS, such as "vibration white finger" or "traumatic vasospastic disease" (27); (b) peripheral nerve sensory and motor changes characterized by decreased hand sensation and dexterity (23); and (c) muscular abnormalities, which manifest as decreased grip strength (28). A serious case of HAVS with frequent attacks can greatly interfere with a worker's performance and other activities. Moreover, many cases can progress to a chronic condition with permanent disability.

Other factors besides vibration may also contribute to the development of HAVS. Work under cold conditions, as is commonly experienced by loggers working with chain saws, increases the risk of this condition. Another important aggravating factor is tobacco use. Nicotine acutely increases peripheral vasoconstriction, which may precipitate an attack. Chronic use is associated with more advanced stages of the disease (29), but does not predict lack of improvement in established disease (28).

Although it is commonly agreed that arterial vasospasm is the cause of blanching, the mechanisms by which vibration damages the neurovascular and muscular tissues of the hands remain unknown. Microvascular abnormalities consistent with a local vasculitis may play a role in the pathogenesis (27). Early diagnosis is usually based on a history of typical symptoms, such as numbness or tingling of the fingers precipitated by exposure to vibration; these symptoms persist for progressively longer periods of time. Finger blanching is often the next symptom to occur with continued chronic exposure. Confirmation of the diagnosis and staging of disease severity have become increasingly reliant on objective vascular and neurologic testing, as well as the exclusion of other neuropathies (25,26). Thermal and vibratory sensation and two-point discrimination may be the most useful tests (26), although studies of the vascular response to cold provocation are also frequently used (25,26).

Hand–arm vibration in this range may contribute to carpal tunnel syndrome (25) and lead to scleroderma, especially when there is inhalation exposure to silica dust in addition to vibration (30).

Preventing these health effects requires minimizing exposure to segmental vibration and surveillance for early health effects. The National Institute for Occupational Safety and Health (NIOSH) has published a revised set of recommendations that includes (a) reduction in the intensity and duration of hand–arm vibration exposure; (b) exposure monitoring; (c) medical surveillance, including preplacement and periodic assessments of exposure history, possible vibration-related symptoms and signs, and discouraging tobacco use; (d) provisions for medical removal of symptomatic workers; (e) protective clothing against vibration and cold stress; and (f) worker training on the health hazards of vibration (31). Once vibration-

associated health problems arise, exposure must be terminated. The attacks of vasospasm are more likely to improve than the neurologic components (28). Severe disease and continued use of vibratory tools are associated with failure to improve (28).

No federal standard currently mandates limits on exposure to vibration or the use of vibrating tools. Thus, workers, managers, and clinicians have crucial roles in the early recognition of segmental vibration-induced problems and the prevention of their progression.

Ultrasound

The term *ultrasound* is used to describe mechanical vibrations at frequencies above the limit of human audibility (approximately 16 kHz[3]). These vibrations, like sound, are pressure waves and are unrelated to electromagnetic radiation. Ultrasound has many industrial and other uses, depending on the frequency of its generation. (With increasing frequency, ultrasound has a tendency to be absorbed by the transmitting medium and consequently has less penetrating ability.) Low-frequency (18 to 30 kHz), high-power ultrasound is used in industry for its penetrating and disruptive abilities to facilitate drilling, welding, and cleaning operations and to help in emulsifying liquids. It is also a component of noise generated by a variety of high-power engines. High-frequency (100 to 10,000 kHz) ultrasound has many analytic applications, and diagnostic ultrasound is increasingly used in many areas of medicine. No definitive evidence of any adverse effects of ultrasound on patients has been found (32). However, consensus documents recommend against excessive tissue heating, and note the theoretic possibility of cavitation-induced damage when ultrasound is used at gas–liquid interfaces (32).

Low-frequency ultrasound can cause a variety of health problems, either by its trans-

---

[3] 1 kHz = 1 kilohertz = 1,000 cycles per second.

mission through air or by bodily contact with the ultrasound generator or target, as when a worker immerses his or her hands in an ultrasound cleaning bath. A worker with such exposure may complain of headache, earache, vertigo, general discomfort, irritability, hypersensitivity to light, and hyperacusia (sensitivity to sound). Exposure through bodily contact with a source of ultrasound may eventually cause peripheral nervous system lesions, leading to autonomic polyneuritis or partial paralysis of the fingers or hands (see beginning of Vibration section).

High-frequency ultrasound may result in the same spectrum of problems; however, because high-frequency ultrasound is absorbed by air and poorly penetrating, excessive exposure is obtained only in cases of direct contact between the ultrasound generator and a part of the worker's body. Under these circumstances, the mechanism of harmful effects is similar to that of segmental vibration. Studies have suggested a reduction of vibration perception in the fingers of ultrasonic therapists (33).

Protection from the harmful effects of ultrasound is primarily an engineering problem requiring insulation and isolation of the ultrasound process. Direct contact between workers and the ultrasound source should be minimized. Ear protective devices are also helpful. Workers with ultrasound exposure should undergo yearly audiometric and neurologic examinations.

ATMOSPHERIC VARIATIONS

Heat

A 24-year-old laborer was brought to the emergency room after collapsing during work. Several hours before, he had become extremely irritable and, for no apparent reason, had provoked arguments with several of his coworkers. He had been working on an incentive basis next to a blast furnace in a metals factory on a day when the temperature had not been more than 90°F, but the humidity exceeded 60%. Physical exam-

ination showed a stocky man who was totally unresponsive with hot, flushed, moist skin. He had dilated, equally sized pupils that were unresponsive to light. Fundi were normal. Rectal temperature was greater than 41°C (105.8°F), pulse rate was 160, blood pressure was 80/0 mm Hg, and respirations were deep at a rate of 30 per minute. Despite rapid cooling in an ice bath with a decrease in core temperature and aggressive care in the intensive care unit, the patient had a series of grand mal seizures followed by marked metabolic acidosis, rhabdomyolysis, acute respiratory distress syndrome, and acute renal failure. After emergency hemodialysis, the patient had another grand mal seizure, followed by cardiac arrest, which was refractory to all attempts at resuscitation.

Heat is a potential physical hazard that can exist in almost any workplace, especially during summer. It is a particular hazard in tropical climates. Hot industrial jobs requiring heavy work afford the greatest potential for problems because they add to the worker's heat load by generating more metabolic heat (Fig. 19-3). When a worker's physiologic capacity to compensate for thermal stress is exceeded, heat can lead to impaired performance, an increased risk of accidents, and clinical signs of heat illness.

Internal temperature is regulated within narrow limits, predominantly by sweat production and evaporation. To a lesser extent, temperature regulation is accomplished by (a) physiologic control of blood flow from muscles and other deep sites of heat production to the cooler surfaces of the body, where heat is dissipated by convective exchange with air; and (b) by evaporative cooling from the lungs, to a minor degree.

The relationship between heat loss or gain variables and internal heat production is described by a simple heat balance equation:

$$S = (M - W) \pm C \pm R - E$$

in which S = net amount of energy gained or lost as heat by the body, M = heat produced by metabolism, W = external work performed, C = heat transfer by convection and conduction, R = heat transfer by radia-

FIG. 19-3. Foundry workers with exposure to excessive heat at work. (Photograph by Earl Dotter.)

tion, and E = body heat loss by evaporation (in kcal/hour). This formula can also be used to evaluate situations of extreme cold.

When homeostatic mechanisms are functioning properly, there is no net gain or loss of heat ($S = 0$). Should S be greater than zero, heat imbalance occurs, leading to manifestations of heat stress. *Convection* refers to heat transfer between the skin and the immediately surrounding air, assuming that air is being circulated. Its value is a function of the difference in temperature between skin and air, and the rate of air movement over the skin. *Conduction* refers to the transfer of heat from skin to a solid or liquid object resulting from direct contact. *Radiation* in this context refers to the radiative transfer of heat between surfaces of differing temperature, that is, the skin surface and the solid surroundings, such as an oven furnace or a cold floor.

The quantity ($M - W$) describes the total amount of body heat produced combining the gain from metabolic heat production and the loss due to external work effort. The C and R variables can be positive or negative

because convection/conduction and radiation can occur in both directions. Ordinarily, with ambient temperature lower than surface body temperature ($<95°F$ or $35°C$), convection and radiation promote transfer of body heat to the environment; however, if the environment is hotter than surface body temperature, radiation and convection can increase the body's heat load.

The heat loss/gain variables (C, R, and E) are influenced by environmental factors encountered in the workplace. In general, industrial heat exposures may be classified as either hot-dry or warm-moist. Hot-dry situations may prevail in a hot desert climate or near any furnace operation (e.g., in the metallurgical, glass, or ceramic industries) radiating high levels of heat. Heat absorption in these circumstances may overwhelm the cooling effect of sweat evaporation, leading to heat imbalance.

The most troublesome situation usually arises in warm-moist environments, as is found in tropical climates, and in such industries as canning, textiles, laundering, and deep metal mining. High humidity and still air impede evaporative and convective cooling. Heat imbalance may occur despite an only moderate increase in ambient temperature; when this happens, thermal stress begins and can result in a variety of illnesses.

The mildest form of heat stress is discomfort. Prolonged exposure to a moderately hot climate may also cause irritability, lassitude, decrease in morale, increased anxiety, and inability to concentrate.

Heat rash (prickly heat) is common in warm-moist conditions because of inflammation in sweat glands plugged by skin swelling. Affected skin bears tiny red vesicles and, if sensitive, can actually impair sweating and greatly diminish the worker's ability to tolerate heat. Heat rash can be treated with mild drying lotions, but it usually can be prevented by allowing the skin to dry in a cooler environment between heat exposures.

Prolonged exposure to heat may result in heat cramps, especially when there is profuse sweating and inadequate replacement of fluids and electrolytes. Heat cramps are alleviated by drinking electrolyte-containing liquids, such as juices.

As net retained body heat increases, heat exhaustion may occur, with pallor, lassitude, dizziness, syncope, profuse sweating, and clammy, moist skin. An oral temperature reading may or may not reveal mild hyperthermia, whereas a rectal temperature is usually elevated (37.5°C to 38.5°C, or 99.5°F to 101.3°F). Heat exhaustion usually occurs in the setting of sustained exertion in hot conditions with dehydration from deficient water intake.

Finally, heat stroke can occur. This is a medical emergency that often is found in a setting of excessive physical exertion. In some cases, the central nervous system inappropriately shuts down perspiration, leading to a loss of evaporative cooling; however, sweating may still be present in over half of exertional heat stroke cases (34). Furthermore, heat stroke may occur even in a healthy, acclimatized person when heat loss becomes less than endogenous heat produc-

tion. The uncontrolled accelerating rise in core temperature that follows is manifested by signs and symptoms that include dizziness, nausea, irritability, severe headache, hot, dry skin, and a rectal temperature of 40.5°C (104.9°F) or higher. This often leads quickly to confusion, collapse, delirium, and coma. Cooling of the body must be started immediately if vital organ damage and death are to be averted. Workers who are not acclimatized (see later) are especially prone to heat stroke; other predisposing factors include obesity, recent alcohol intake, and chronic cardiovascular disease.

Workers with heat exhaustion or heat stroke must be removed to a cooler environment immediately. Heat exhaustion can be treated with oral rehydration or, if more severe, by the administration of intravenous saline fluids and observation. Heat stroke, however, must also be treated with immediate cooling. Methods include immersion in ice water, ice packs to the groin and axillae, and spraying with cooler water while the victim is fanned to speed evaporation. Cooling efforts should be slowed down and the patient observed once the temperature reaches 39°C (102.2°F) to avoid overcooling.

Monitoring a workplace environment for heat exposure is a task that involves measuring heat-modifying factors and assessing workloads. Occupational hygienists commonly use a measurement called the wet bulb globe temperature (WBGT) index, which takes into account convective and radiant heat transfer, humidity, and wind velocity. It is calculated by integrating the readings of three separate instruments that indicate the "dry" bulb temperature, the natural "wet" bulb temperature (which takes into account convective and evaporative cooling), and the "globe" temperature (which, through the use of a black copper sphere, registers heat transfer by radiation). The WBGT index has been used by NIOSH and the American Conference of Governmental Industrial Hygienists as an indicator of heat stress in a recommended set of guidelines (Table 19-1).

TABLE 19-1. *Permissible heat exposure threshold limit values[a] (values are given in °C wet bulb globe temperature)*

Work–rest regimen	Workload		
	Light	Moderate	Heavy
Continuous work 75% Work	30.0	26.7	25.0
25% Rest, each hour 50% Work	30.6	28.0	25.9
50% Rest, each hour 25% Work	31.4	29.4	27.9
75% Rest, each hour	32.2	31.1	30.0

[a] Values are given in degrees Celsius wet bulb globe temperature.
From American Conference of Governmental Industrial Hygienists. Threshold limit values and biological exposure indices for 1985–86. Cincinnati: ACGIH, 1985:69.

These guidelines suggest maximally permissible standards of heat exposure, depending on the level of work performed in the hot environment (35,36). Even when these guidelines are followed, however, heat-related disorders may occur in exposed workers.

Ideally, prevention of heat stress should be accomplished by isolating workers from hot environments through engineering design. However, because this often is not possible, heat stress is usually averted by a combination of engineering controls, work practice changes, use of personal protective clothing, and the education of those who work under conditions of excessive heat.

In addition to guidelines regarding heat exposure, NIOSH has made other recommendations concerning heat exposure, including (a) medical surveillance supervised by a physician and paid for by the employer for all workers exposed to heat stress circumstances above the "action" level, to include comprehensive history taking, physical examinations, and overall assessment of a worker's ability to tolerate heat; (b) posting of warning signs in hazardous areas; (c) using protective clothing and equipment; and (d) providing information and training to workers. Workers should be trained to replace fluid losses systematically. During the course

of a day's work in a hot environment, a worker may lose each hour up to 1 L of fluid and electrolytes in sweat. This loss should be replaced by drinking fluids every 15 to 20 minutes, in greater amounts than are necessary to satisfy thirst, an inadequate stimulus for fluid replacement in stressful heat conditions. Other recommendations include (e) control of heat stress with engineering controls, and work and hygienic practices that specify time limits for working in hot environments, the use of a buddy system, provision of water, and other measures; and (f) maintaining and analyzing environmental and medical surveillance data.

One of the recommended work practices is an acclimatization program for new workers and for workers who have been absent from the job for longer than 9 days. Acclimatization is a process of adaptation that involves a stepwise adjustment to heat; it usually requires 1 week or more of progressively longer periods of exposure to the hot environment while working. Some workers may fail to acclimatize and consequently cannot work in such environments. Finding a short and easily administered screening test that can reliably predict heat intolerance is a research priority. NIOSH recommends (a) the reduction of peaks of physiologic strain during work; (b) the establishment of frequent, regular, short breaks for replacing water and electrolyte losses; and (c) preplacement and periodic medical examinations.

The finding that heat-acclimated workers lose less salt in sweat than previously thought, together with recognition of the high salt content in the average American diet and concern over exacerbating hypertension, has led experts to discourage the routine use of salt tablets to replace electrolyte losses. If salt replacement is required in a nonemergency situation, such as strenuous work on a hot, dry day, adding extra salt to food is recommended. Workers on a low-sodium diet because of cardiovascular or other reasons require close medical supervision if working in a hot environment.

Decreasing heat stress can involve engineering measures such as devising heat shielding, or insulating and ventilating with cool, dehumidified air. Workers should be provided with a cooled rest area close to the workstation. Although clothing impedes the evaporation of sweat, it can reduce heat exposure in situations in which the ambient air temperature is higher than the skin temperature and the air is relatively dry, or when a worker is near a strong heat-radiating source. Some types of industrial heat exposure require protective equipment, such as gloves, aluminized reflective clothing, insulated and cooling jackets, and even self-contained air-conditioned suits.

Cold

Cold stress is an environmental hazard that confronts cold-room workers, dry ice workers, liquefied gas workers, divers, and outdoor workers during cold weather. Other sources of cold exposure are the climates of the extreme northern and southern hemispheres, and even poorly heated indoor workplaces during the winter in temperate zones. The body maintains thermal homeostasis in the face of a cold environment by decreasing skin heat loss through peripheral vasoconstriction and increasing metabolic heat production through shivering. Harmful effects of cold include frostbite, trench foot, chilblains, and general hypothermia.

Frostbite is actual freezing of tissue due to exposure to extreme cold or contact with extremely cold objects. Symptoms and signs range from erythema and slight pain to painless blistering, deep-seated ischemia, thrombosis, cyanosis, and gangrene. Wind chill (loss of heat from exposure to wind) can play an important role in accelerating frostbite. Until a patient can receive definitive hospital treatment, it is imperative to protect the affected body part from trauma, including rubbing, to avoid rewarming if the possibility of refreezing exists. Emergency treatment includes warm-water immersion, elevation, débridement, analgesia, and tetanus prophy-

laxis. Long-term sequelae usually involve peripheral neurovascular changes.

Trench foot (immersion foot) is a condition that results from long, continuous exposure to damp and cold while remaining relatively immobile. It may involve not only the feet but the tip of the nose and ears; it typically occurs in a worker who has just experienced prolonged immersion in cold water or exposure to cold air. The clinical changes in affected tissue can progress from an initial vasospastic, ischemic, hyperesthetic (oversensitive), pale phase; to a later stage equivalent to a burn, with hyperemia, vasomotor paralysis, vesiculation, and edema; and finally to a gangrenous stage.

General hypothermia usually occurs in a person who is subjected to prolonged cold exposure and physical exertion. When a person becomes wet, either from exposure or sweating, body heat is lost even faster. Most cases occur in air temperatures between −1°C (30°F) and +10°C (50°F); however, they can also occur in air temperatures as high as 18°C (64°F) or in water at 22°C (72°F), especially in the setting of fatigue. As exhaustion sets in, the vasoconstrictive protective mechanism becomes overwhelmed, resulting in sudden vasodilation and acute heat loss. Mental status changes, arrhythmias, coma, and death can ensue rapidly. Those at the extremes of age and those with certain conditions, such as hypothyroidism, adrenal insufficiency, and malnutrition, are predisposed to hypothermia. The danger of hypothermia is also increased by the consumption of alcohol, phenothiazines, and other central nervous system depressants. Management of hypothermia requires removal of the victim from the cold environment, removing any wet clothing, and specific warming measures. The method of warming depends on the degree of hypothermia. Most patients with mild hypothermia can be treated with passive rewarming (warm, dry clothes and blankets in a warm environment). More severe cases require active external warming (heating blankets) and active core rewarming, including warmed oxygen, warmed intravenous fluids, and, in extreme cases, warmed dialysis fluids.

At temperatures below 15°C (59°F), the hands and fingers become insensitive long before these described cold injuries take place, thereby decreasing manual dexterity and increasing the risk of accidents. Many workers handle cold metal objects that can cause local freezing and metal–skin adhesions. The use of silk gloves makes it possible to handle cold metals up to −40°C (−40°F) without freezing to them while retaining good manual dexterity.

In general, cold hazards are well recognized and preventable. Insulated clothing and protective barriers should be provided and workers should be able to take adequate breaks inside warm shelters, with warm beverages and dry clothing changes available. Pain in the extremities and severe shivering should be recognized as warning signs of cold injury and hypothermia. The additional cooling that occurs with increased wind speed (wind chill) must also be taken into account when evaluating exposures.

Hyperbaric Environments

In a handful of environments, including certain caisson operations, underwater tunneling, diving, and tending patients in a hyperbaric chamber, the ambient air pressure exceeds one atmosphere absolute (1 ATA), the atmospheric pressure found at sea level. These environments usually range from 2 to 5 ATA, although commercial divers may dive to depths greater than 100 m at 11 ATA pressure (a 10-m increase in seawater depth adds 1 ATA pressure). In the United States, there are between 1.5 and 3.0 million certified divers, most of whom dive for recreation, with approximately 500 to 600 injuries and 75 to 100 deaths occurring each year.

The most common health problems stemming from working in a hyperbaric environment are caused by unequal distribution of pressure in tissue air spaces, incurred during compression (descent) or decompression (ascent). For example, when the eustachian

tube becomes blocked as a result of inflammation or failure of a diver to clear the ears, descent compression causes negative middle-ear pressure, with rupture of mucosal capillaries, bleeding into the middle ear, and even rupture of the tympanic membrane. The decongestant pseudoephedrine reduces the incidence of this condition, known as "ear squeeze." Sinus cavities also occasionally experience pressure imbalance during descent or ascent. Very rarely, the lungs can become compressed in a breath-holding dive. More commonly, during ascent (decompression), the lungs can expand beyond their capacity, leading to rupture and pneumothorax, mediastinal emphysema, or even arterial gas embolism. Pulmonary overinflation is associated with rapid ascent, or coughing or breath-holding during ascent, but may be unexplained. Cerebral air embolism may have a number of neurologic manifestations, including confusion, weakness resembling a stroke, or loss of consciousness. Arterial gas embolism usually occurs within the first 10 minutes after decompression, and must be treated with recompression (see later).

Another common problem arises when ascending too rapidly from depth. When the pressure of nitrogen dissolved in the tissues exceeds ambient pressure, bubbles form in the blood and other tissues. The bubbles impair circulation and lead to a variety of effects commonly known as decompression sickness (DCS). The same syndrome also may occur in air flight when there is a loss of cabin pressure (or in unpressurized aircraft). DCS usually has a rapid onset, but may be delayed. Type I DCS, "the bends," is characterized by dull, throbbing joint pain and deep muscle or bone pain, usually occurring within 4 to 6 hours. By definition, type II DCS involves neurologic or cardiopulmonary symptoms. Spinal cord lesions can lead to paralysis. Nitrogen bubbles in the pulmonary vasculature can cause cough and dyspnea and are known as the "chokes." People with air embolism or DCS, or both, require oxygen therapy and immediate recompression in a hyperbaric chamber, followed by prolonged decompression. Prolonged decompression is also the method of preventing the syndrome. Diving tables that provide sufficient decompression were developed by the military, and have been modified for recreational and commercial use.

Most of these problems have been minimized through appropriate decompression work practices. Procedures for medical surveillance and emergencies, involving divers and other workers in hyperbaric environments, are available (37).

Most decompression tables in current use were designed to avoid acute DCS, as opposed to chronic morbidity. Therefore, the evolving body of evidence suggesting chronic changes in divers or hyperbaric workers is not terribly surprising. Workers so exposed have been found to have aseptic bone necrosis (also termed *dysbaric osteonecrosis*), mostly of the articular heads of long bones. The pathophysiologic process is believed to involve the occlusion of small-bore arteries by nitrogen bubbles, with platelet aggregates acting as microemboli during decompression. Work from Norway has documented mild neuropsychological (38) and pulmonary function (39) abnormalities in commercial divers compared with control subjects with no history of diving. A history of DCS and extensive diving experience were associated with changes in cerebral perfusion on brain imaging (40). It is possible that frequent decompressions contribute independently to the risk of chronic disease, even when a person adheres to recommended decompression protocols. Decompression tables designed to avoid even asymptomatic intravascular bubbles (which can be detected with Doppler ultrasound) may be able to reduce the incidence of both acute and chronic decompression disorders.

Compressed-air medical examinations should evaluate the joints and musculoskeletal system, given the potential for aseptic bone necrosis. Routine bone radiography to identify aseptic bone necrosis may have merit, although it is not universally applied.

In addition to these mechanical hazards, workers in these environments encounter problems due to toxicity of the gas components of air at elevated partial pressures. When breathing air at 4 ATA or greater, gaseous nitrogen induces a narcosis ("rapture of the deep") marked by euphoria and disorientation. The experience of nitrogen narcosis is similar to that of alcohol intoxication, and limits air diving depth. Pulmonary toxicity associated with oxygen inhalation can occur at relatively low partial pressures, and can even affect oxygen-breathing patients at 1 ATA. Oxygen inhalation at a high partial pressure causes central nervous system toxicity involving paresthesias, tinnitus, visual changes, confusion, nausea, vertigo, and sometimes seizures. These reactions can be avoided by limiting the depth of descent, the time at depth, and the proportion of oxygen in the breathing gas. Obviously, the limits are more important for a free-swimming diver breathing oxygen because a seizure would likely be fatal.

It is important to recognize other hazards associated with hyperbaric environments. Compressed-air sources may be contaminated. Carbon monoxide is probably the most common contaminant of compressed air, particularly if a compressor is powered by a gasoline engine; other potential contaminants include carbon dioxide, oil vapor, and particulates. Finally, workers and patients must not take ignition sources or flammable products into a compressed air environment because of the risk of fire.

Hypobaric and Hypoxic (High-Altitude) Environments

A 41-year-old man was dead on arrival at the hospital. He had flown from 1,500 to 2,750 m, and during the next few days had climbed to 4,270 m, where he rapidly lost consciousness, dying 5 days after leaving 1,500 m. Necropsy approximately 5 hours after death revealed severe cerebral edema, multiple petechial hemorrhages throughout the brain, bilateral pulmonary edema, bilateral bronchial pneumonia, dilatation and hypertrophy of the heart, and moderate arteriosclerosis of coronary arteries (41).

Workers, skiers, and mountain climbers in

high-altitude regions and pilots in unpressurized cabins experience hypobaric (low-pressure) and hypoxic environments. At high altitudes, decreased pressure reduces the gradient for oxygen absorption, leading to a fall in arterial partial pressure of oxygen, despite the fact that the percentage of oxygen remains constant at 21%. Above 2,400 m, arterial oxyhemoglobin saturation falls below 90% in most people. Illness associated with high altitude is rarely experienced below 2,000 m; the incidence at 2,000 to 2,600 m ranges from 1.4% to 25%, and it becomes even more common above 3,000 m, especially in people who venture to this altitude without taking time for acclimatization.

The most common health problem at high altitudes is *acute mountain sickness*, an illness characterized by headache, nausea, vomiting, fatigue, and loss of appetite. Almost all people who undergo an abrupt altitude change by venturing into the mountains experience one of these symptoms to a degree within 24 hours; however acclimatization usually occurs in a few days, and symptoms disappear over 4 to 6 days of exposure. The pathophysiologic process has been suggested to involve the imbalance occurring between cerebral vasodilatation from hypoxia and cerebral vasoconstriction from hypocarbia (a decrease in the partial pressure of carbon dioxide in the blood). Acclimatization with few symptoms is best achieved by slow ascent (350 m or 1,000 feet per day) and gradually progressive activity for several days. Some climbers who are required to reduce acclimatization time have lessened acute mountain sickness symptoms by using acetazolamide, which results in a metabolic stimulus to increase the rate and depth of breathing. Dexamethasone is also effective as prophylaxis. Combination therapy is more effective prophylaxis than acetazolamide alone (42).

More serious is *high-altitude pulmonary edema*, which also strikes unacclimatized people, usually within 24 to 60 hours, and especially if they engage in vigorous activity soon after arrival at high altitude. Onset is usually insidious. Symptoms include shortness of breath, cough, weakness, tachycardia, and headache. This condition characteristically progresses to cough productive of bloody sputum, low-grade fever, and evidence of increasing pulmonary congestion, with rales and cyanosis. If the condition remains untreated, coma may ensue from hypoxia or cerebral edema (see later). The pathophysiologic process is unclear; clinical studies seem to suggest a process involving hypoxic vasoconstriction and pulmonary capillary leakage. It is treated ideally with descent to a lower altitude, rest, and administration of oxygen. Sometimes diuretics, positive-pressure breathing, and nifedipine have been used when immediate descent has not been possible. Lack of recognition of this syndrome is probably most responsible for serious morbidity and mortality.

Cerebral edema is a rare, but often fatal, consequence of high altitude. The presence of neurologic signs and symptoms of headache, confusion, ataxia, and hallucinations in a person who has had an altitude change should lead to suspicion of this diagnosis. Treatment, as in high-altitude pulmonary edema, centers on rapid descent and oxygen administration. Using dexamethasone in the same manner as for other forms of cerebral edema may be useful but is not a substitute for immediate descent.

Although most cases of acute high-altitude illness occur in people visiting the mountains for recreation or work, nonacute forms are occasionally found in those living at high altitude for longer periods. Chronic mountain sickness was first recognized among certain natives of the Peruvian Andes. An analogous disease characterized by polycythemia, pulmonary hypertension, and right heart enlargement has now been described in Tibet. It occurs almost exclusively among members of the Han population an average of 15 years after immigration to a high altitude, and is rare among the high-altitude natives. A subacute form with polycythemia, pleural effusion, right-sided congestive heart failure, and cardiomegaly has been recognized in a group of Indian soldiers stationed for approximately 11 weeks at 5,800 to 6,700 m (43). Although chronic mountain sickness improves with transfer to a lower altitude, subacute disease resolves within several weeks of the move to low altitude.

Most cases of acute altitude-related illnesses can be prevented by slow acclimatization. The physiologic mechanism of acclimatization involves an increase in red blood cell production and hematocrit, increases in intraerythrocytic 2,3-diphosphoglycerate (leading to a favorable shift in the oxyhemoglobin dissociation curve), and mild hyperventilation. Exercise can be a part of this program, especially if heavy physical work is anticipated. The lack of a previous history of altitude-related illness while at high altitudes does not preclude the possibility of its occurrence later.

REFERENCES

1. Verschaeve L, Maes A. Genetic, carcinogenic and teratogenic effects of radiofrequency fields. Mutat Res 1998;410:141–165.
2. English D, Armstrong BK, Kricker A, Fleming C. Sunlight and cancer. Cancer Causes Control 1997; 8:271–283.
3. Walter SD, Marrett LD, Shanon HS, From L, Hertzman C. The association of cutaneous malignant melanoma and fluorescent light exposure. Am J Epidemiol 1992;135:749–762.
4. Urbach F. Ultraviolet radiation and skin cancer of humans. J Photochem Photobiol B 1997;40:3–7.
5. Cruickshanks KJ, Klein BEK, Klein R. Ultraviolet light exposure and lens opacities: the Beaver Dam Eye Study. Am J Public Health 1992;82:1658–1662.
6. Lydahl E, Philipson B. Infrared radiation and cataract: II. epidemiologic investigation of glass workers. Acta Ophthalmol 1984;62:976–992.
7. Rothman KJ, Chou CK, Morgan R, et al. Assessment of cellular telephone and other radiofrequency exposure for epidemiologic research. Epidemiology 1996;7:291–298.
8. Michaelson SM. Biological effects of radiofrequency radiation: concepts and criteria. Health Physics 1991;61:3–14.
9. Yost MG. Occupational health effects of nonionizing radiation. Occup Med 1992;7:543–566.
10. Goldsmith JR. Epidemiologic evidence relevant to radar (microwave) effects. Environ Health Perspect 1997;105[Suppl 6]:1579–1587.
11. Izmerov NF. Current problems of nonionizing radiation. Scand J Work Environ Health 1985;11: 223–227.
12. Valberg PA. Radiofrequency radiation (RFR): the nature of exposure and carcinogenic potential. Cancer Causes Control 1997;8:323–332.
13. Hayes DL, Carrillo RG, Findlay GK, Embrey M. State of the science: pacemaker and defibrillator interference from wireless communication devices. Pacing Clin Electrophysiol 1996;19:1419–1429.
14. Anonymous. Microwaves and pacemakers. J Occup Med 1992;34:250.
15. Bailey WH, Su SH, Bracken TD, Kavet R. Summary and evaluation of guidelines for occupational exposure to power frequency electric and magnetic fields. Health Physics 1997;73:433–453.
16. Kheifets LI, Afifi AA, Buffler PA, Zhang ZW, Matkin CC. Occupational electric and magnetic field exposure and leukemia: a meta-analysis. J Occup Environ Med 1997;39:1074–1091.
17. Lacy-Hulbert A, Metcalfe JC, Hesketh R. Biological responses to electromagnetic fields. FASEB J 1998;12:395–420.
18. Linet MS, Hatch EE, Kleinerman RA, et al. Residential exposure to magnetic fields and acute lymphoblastic leukemia in children. N Engl J Med 1997;337:1–7.
19. Theriault G, Li C-Y. Risks of leukaemia among residents close to high voltage transmission electric lines. Occup Environ Med 1997;58:625–628.
20. Wartenberg D. Residential magnetic fields and childhood leukemia: a meta-analysis. Am J Public Health 1998;88:1787–1794.
21. Hunter D, McLaughlin AIG, Perry KM. Clinical effects of the use of pneumatic tools. British Journal of Industrial Medicine 1945;2:10.
22. Helmkamp JC, Talbott EO, Marsh GM. Whole body vibration: a critical review. American Industrial Hygiene Association Journal 1984;45:162–167.
23. Murata K, Araki S, Okajima F, Nakao M, Suwa K, Matsunaga C. Effects of occupational use of vibrating tools in the autonomic, central, and peripheral nervous system. International Arch Occup Environ Health 1997;70:94–100.
24. Pope MH, Magnusson M, Wilder DG. Kappa Delta Award. Low back pain and whole body vibration. Clin Orthop 1998;354:241–248.
25. Gemne G. Diagnostics of hand–arm system disorders in workers who use vibrating tools. Occup Environ Med 1997;54:90–95.
26. Lawson IJ, Nevell DA. Review of objective tests for the hand-arm vibration syndrome. Occup Med 1997;47:15–20.
27. Littleford RC, Khan F, Hindley MO, Ho M, Belch JJF. Microvascular abnormalities in patients with vibration white finger. QJM 1997;90:525–529.
28. Ogasawara C, Sakakibara H, Kondo T, Miyao M, Yamada S, Toyoshima H. Longitudinal study on factors related to the course of vibration-induced white finger. Int Arch Occup Environ Health 1997; 69:180–184.
29. Ekenvail L, Lindblad LE. Effect of tobacco use on vibration white finger disease. J Occup Med 1989;31:13–16.
30. Pelmear PL, Roos JO, Maehle WM. Occupational-induced scleroderma. J Occup Med 1992;34:20–5.
31. National Institute for Occupational Safety and Health. Criteria for a recommended standard: occupational exposure to hand–arm vibration. U.S. Department of Health and Human Service (NIOSH) publication no. 89-106. Washington, DC: U.S. Government Printing Office, 1989.
32. WFUMB symposium on safety of ultrasound in

medicine: conclusions and recommendations on thermal and non-thermal mechanisms for biological effects of ultrasound. Ultrasound Med Biol 1998; 24[Suppl 1]:S1–S58.

33. Lundstrom R. Effects of local vibration transmitted from ultrasonic devices on vibrotactile perception in the hands of therapists. Ergonomics 1985;28: 793–803.

34. Knochel JP. Heat stroke and related heat stress disorders. Disease-a-Month 1989;35:301–377.

35. National Institute for Occupational Safety and Health. Criteria for a recommended standard: occupational exposure to hot environments (revised criteria). Department of Health and Human Services (NIOSH) publication no. 86-113. Washington, DC: U.S. Government Printing Office, 1986.

36. American Conference of Governmental Industrial Hygienists. Threshold limit values and biological exposure indices for 1993–1994. Cincinnati, OH: American Conference of Governmental Industrial Hygienists, 1993.

37. U.S. Navy Diving Manual. Commander Naval Sea Systems Command publication 0994-LP-001-9010. Rev 3, vol 1. Washington, DC: U.S. Government Printing Office, 1993.

38. Todnem K, Nyland H, Kambestad BK, Aarli JA. Influence of occupational diving upon the nervous system: an epidemiological study. British Journal of Industrial Medicine 1990;47:708–714.

39. Thorsen E, Segadal K, Kambestad B, Gulsvik A. Diver's lung function: small airways disease? British Journal of Industrial Medicine 1990;47:519–523.

40. Shields TG, et al. Correlation between Tc-99(m)-HMPAO-SPECT brain image and a history of decompression illness or extent of diving experience in commercial divers. Occup Environ Med 1997; 54:247–253.

41. Houston CS, Dickinson J. Cerebral form of high-altitude illness. Lancet 1975;2:758–761.

42. Bernhard WN, Schalick LM, Delaney PA, Bernhard TM, Barnas GM. Acetazolamide plus low-dose dexamethasone is better than acetazolamide alone to ameliorate symptoms of acute mountain sickness. Aviat Space Environ Med 1998;69:883–886.

43. Anand IS, Malhotra RM, Chandrashekhar Y, et al. Adult subacute mountain sickness—a syndrome of congestive heart failure in man at very high altitude. Lancet 1990;335:561–565.

BIBLIOGRAPHY

It is difficult to offer a general bibliography given the wide assortment of hazards covered, their specific nature, and the rapid evolution of our understanding of many of these hazards. Reviews of specific topics that appear in the medical or environmental literature as well as technical monographs published by government agencies probably provide the best available overviews.

The U.S. Navy Diving Manual (see reference #37) provides detailed procedures for safe diving operations, and evaluation and treatment of diving-related medical conditions.

Other sources include the following:

Barrow MW, Clark KA. Heat-related illnesses. Am Fam Physician 1998;58:749–56.

Kanzenbach TL. Dexter WW. Cold injuries. Protecting your patients from the dangers of hypothermia and frostbite. Postgrad Med 1999;105:72–78.

These two reviews cover the symptoms, risk factors, treatment, and prevention of thermal injuries.

English DR, Armstrong BK, Kricker A, Fleming C. Sunlight and cancer. Cancer Causes Control 1997; 8:271–283.

Thoroughly reviews the epidemiologic association of sunlight exposure pattern and dose with dermal malignancies.

Gemne G. Diagnostics of hand–arm system disorders in workers who use vibrating tools. Occup Environ Med 1997;54:90–95.

Covers the pathophysiology, epidemiology, diagnosis, and exposure evaluation of vibration-induced disorders.

Kheifets LI, Afifi AA, Buffler PA, Zhang ZW, Matkin CC. Occupational electric and magnetic field exposure and leukemia: a meta-analysis. J Occup Environ Med 1997;39:1074–1091.

Wartenberg D. Residential magnetic fields and childhood leukemia: a meta-analysis. Am J Public Health 1998;88:1787–1794.

These two quantitative reviews demonstrate an association between electromagnetic fields and leukemia in workers and children. Sources of heterogeneity in the literature are identified.

Rothman KJ, Chou CK, Morgan R, et al. Assessment of cellular telephone and other radiofrequency exposure for epidemiologic research. Epidemiology 1996; 7:291–298.

This primer is essential for anyone interested in cellular telephone radiofrequency exposures.

Verschaeve L, Maes A. Genetic, carcinogenic, and teratogenic effects of radiofrequency fields. Mutat Res 1998;141–165.

Reviews the literature on radiofrequency radiation, from epidemiology to cellular studies, including a small body of recent work suggesting low-level effects on DNA.

WFUMB symposium on safety of ultrasound in medicine: conclusions and recommendations on thermal and non-thermal mechanisms for biological effects of ultrasound. Ultrasound Med Biol 1998;24 (Suppl 1):S1–S58.

This extensive consensus document is largely theoretical, but still references the clinical and epidemiologic literature where appropriate.

Yost MG. Occupational health effects of nonionizing radiation. Occup Med 1992;7:543–566.

This general review covers thermal effects.

20 Infectious Agents

Nelson M. Gantz

Work-related infectious diseases are caused by all categories of infectious agents: bacteria, viruses, fungi, protozoa, and helminths.

Workers in many different occupations are at risk (Table 20-1), but most cases occur in (a) health care workers, either with direct patient contact or laboratory exposure to infective material; and (b) workers having contact with animals or animal products, engaged in groundbreaking or earthmoving, or traveling to endemic areas. Table 20-2 presents the number of potentially work-related cases of selected infectious diseases in the United States reported to the Centers for Disease Control and Prevention (CDC) in 1997; it is not known how complete these data are or what percentage of these cases is actually work related. Additional types of work-related infectious agents are prevalent in developing countries (Box 20-1). Clinically, as with other types of disorders, the occupational history often provides critical information in making a diagnosis and in identifying the cause of a work-related infectious disease.

INFECTIOUS DISEASES IN HEALTH CARE WORKERS

The hazards of hospital-acquired (nosocomial) infectious diseases have been well rec-

N. M. Gantz: Department of Medicine and Division of Infectious Diseases, PinnacleHealth Hospitals, Harrisburg, Pennsylvania 19129.

ognized since the mid-19th century, when Semmelweis discovered the cause of puerperal fever. The risk of nosocomial infection exists both for hospitalized patients and for workers involved in their care. (Table 20-3 lists some infectious agents that have been transmitted from patients to health care workers.) Such problems exist not only for hospital workers, but for those in outpatient settings, such as dentists' offices (Fig. 20-1). The risk of infection is also present for personnel working in, among other places, outpatient renal dialysis centers, laboratories where workers have contact with blood, nursing homes, institutions for the retarded, and prisons.

Human Immunodeficiency Virus Infection

A 29-year-old nurse presents to the emergency department with 1 week of "flu-like" symptoms consisting of chills, fever, headache, myalgias, diarrhea, and a sore throat. She recalls that 1 month ago, she sustained a needlestick while drawing blood from a patient but failed to go to the employee health service because it was late in the day. On examination, her temperature was 102°F, pulse 96/minute, and blood pressure 110/70. She had an erythematous posterior pharynx with exudates. Small cervical and axillary nodes were palpable bilaterally. A diffuse maculopapular rash was also present. Laboratory studies revealed a white blood cell count of 2,200 cells/mm^3 with 50% polymorphonuclear leukocytes, 18% bands, and 23% lymphocytes with 9% atypical lymphocytes. A rapid test for group A *Strepto-*

TABLE 20-1. *Selected work-related infectious diseases, by occupation*

Occupation	Selected work-related infectious diseases
Bulldozer operator	Coccidioidomycosis, histoplasmosis
Butcher	Anthrax, erysipeloid, tularemia
Cat and dog handler	*Bartonella henselae, Pasteurella multocida* cellulitis, rabies
Cave explorer	Rabies, histoplasmosis
Construction worker	Rocky Mountain spotted fever, coccidioidomycosis, histo-plasmosis
Cook, food-processing worker	Tularemia, salmonellosis, trichinosis
Cotton mill worker	Coccidioidomycosis
Dairy farmer	Milker's nodules, Q fever, brucellosis, *tinea barbae*
Day care center worker	Hepatitis A, rubella, cytomegalovirus, other childhood infec-tious diseases
Delivery person	Rabies
Dentist	Hepatitis B, hepatitis C, AIDS
Ditch digger	Creeping eruption (cutaneous larva migrans), hookworm dis-ease, ascariasis
Diver	Swimming pool granuloma (*Mycobacterium marinum*)
Dock worker	Leptospirosis, swimmer's itch (*Schistosoma* species)
Farmer	Rabies, anthrax, brucellosis, Rocky Mountain spotted fever, tetanus, plague, leptospirosis, tularemia, coccidioidomyco-sis, ascariasis, histoplasmosis, sporotrichosis, hookworm disease
Fisherman, fish handler	Erysipeloid, swimming pool granuloma
Florist, nursery worker	Sporotrichosis
Forestry worker	California encephalitis, Lyme disease, Rocky Mountain spot-ted fever, tularemia, ehrlichiosis
Fur handler	Tularemia
Gardener	Sporotrichosis, creeping eruption (cutaneous larva migrans)
Geologist	Plague, California encephalitis
Granary and warehouse worker	Murine typhus (endemic)
Hide, goat hair and wool handler	Q fever, anthrax, dermatophytoses
Hunter	Lyme disease, Rocky Mountain spotted fever, plague, tulare-mia, trichinosis, ehrlichiosis
Laboratory worker	Hepatitis B, tuberculosis, salmonellosis
Livestock worker	Brucellosis, leptospirosis
Meat packer/slaughterhouse (abattoir) worker	Brucellosis, leptospirosis, Q fever, salmonellosis, *Staphylo-coccus aureus*
Mental retardation institute worker	Hepatitis A and B
Miner	Tuberculosis[a]
Nurse	Hepatitis B, rubella, tuberculosis, hepatitis C, AIDS, herpes simplex
Pet shop worker	*P. multocida* cellulitis, psittacosis, dermatophytoses
Physician	Hepatitis B, rubella, tuberculosis, hepatitis C, AIDS, Bacille Calmette-Guérin
Pigeon breeder	Psittacosis
Poultry handler	Newcastle disease, erysipeloid, psittacosis
Prison guard	Tuberculosis
Rancher	Lyme disease, rabies, Rocky Mountain spotted fever, Q fe-ver, tetanus, plague, tularemia, trichinosis, ehrlichiosis
Rendering plant worker	Brucellosis, Q fever
Sewer worker	Leptospirosis, hookworm disease, ascariasis
Shearer	Orf, tularemia
Shepherd	Anthrax, brucellosis, orf, plague
Soldier	Tularemia and other infectious diseases
Trapper, wild animal handler	Leptospirosis, Lyme disease, tularemia, rabies, Rocky Mountain spotted fever, ehrlichiosis
Veterinarian	Anthrax, brucellosis, erysipeloid, rabies, leptospirosis, *P. multocida* cellulitis, tularemia, salmonellosis, orf, psitta-cosis, *B. henselae*
Zoo worker	Psittacosis, tuberculosis

AIDS, acquired immunodeficiency syndrome.

[a] Silicotuberculosis (see Chapter 25) occurs among quarry workers, sandblasters, other silica processing workers, and workers in mining, metal foundries, and the ceramics industry.

TABLE 20-2. *Annual U.S. Incidence (1997) of selected infectious diseases that are sometimes work related*[a]

Disease	No. of reported cases
Anthrax	0
Brucellosis	75
Hepatitis, viral, type B	10,416
Leptospirosis	No data
Lyme disease	12,801
Plague	4
Psittacosis	37
Rabies	2
Rocky Mountain spotted fever	396
Tetanus	42
Tuberculosis	19,851
Tularemia	105
Typhus fever, murine	25–50 (estimate)

[a] Actual number of cases that are work related is unknown, data include both work- and non–work-related cases. Number of cases of typhus fever is an estimate because it is not a reportable disease.

From Centers for Disease Control and Prevention. MMWR Morb Mortal Wkly Rep 1997;52, 53:1259.

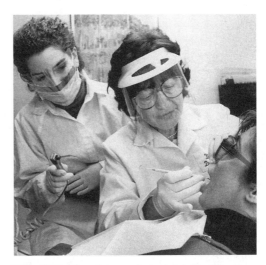

FIG. 20-1. Dentists and dental technicians require protection against pathogens in aerosols, blood, and saliva. The worker closest to patient has eye, but no respiratory, protection (no mask), whereas the other worker has a mask but lacks eye protection. Both workers should have masks and eye protection. (Photograph by Marvin Lewiton.)

Box 20-1. Infectious Diseases in Developing Countries

In developing countries, infectious diseases are highly prevalent. Some of these diseases are acquired in the workplace, from coworkers or other people or from other sources in the work environment. Others result indirectly from work, such as sexually transmitted diseases that may be prevalent among male workers who have left their families and traditional cultural supports in a rural area to seek employment in a large city.

Specific categories of infectious diseases that may be occupational in origin, organized according to their usual mode of transmission, are as follows:

1. Vectorborne diseases, such as malaria, which can be acquired by working near mosquito breeding sites.
2. Waterborne and foodborne diseases, such as gastrointestinal disorders caused by *Salmonella*, *Shigella*, *Giardia lamblia*, and other organisms, which can be acquired at work, especially if workers have direct contact with water or food that has been contaminated by these or other microorganisms.

3. Sexually transmitted diseases, including the acquired immunodeficiency syndrome, syphilis, gonorrhea, chlamydia, and other disorders, which can be transmitted at work. At high risk are commercial sex workers; at lower risk are clinical laboratory and health care workers.
4. Airborne diseases, which range from often benign viral respiratory infections to some more severe chronic respiratory infections, such as tuberculosis. Because HIV infection often leads to reactivation of latent infections with the tubercle bacillus, tuberculosis cases and deaths are increasing worldwide, including in developing countries.
5. Zoonoses, which include brucellosis, leptospirosis, anthrax, and rabies. These diseases are often prevalent in developing countries, especially in rural areas.
6. Other diseases, including viral hepatitis, type B (hepatitis B), which is highly prevalent in some developing countries.

From Levy BS, Choudhry AW. Endemic diseases. In: Jeyaratnam J, ed. Occupational health in developing countries. Oxford: Oxford University Press, 1992, pp. 314–325.

TABLE 20-3. *Some infectious agents that are occupational risks for health care workers*

Virus	Bacteria	Others
Adenovirus	*Bordetella* species	*Chlamydia psittaci*
B Virus	*Campylobacter* species	*Coxiella burnetti*
Creutzfeldt-Jakob agent	*Corynebacterium diphtheriae*	*Cryptosporidium* species
Cytomegalovirus	*Mycobacterium tuberculosis*	*Mycoplasma pneumoniae*
Ebola virus	*Neisseria meningitidis*	*Sarcoptes scabiei*
Hepatitis B	*Salmonella* species	
Hepatitis C	*Shigella* species	
Herpes simplex	*Yersinia* pestis	
Human immunodeficiency virus		
Influenza		
Lassa fever		
Measles		
Mumps		
Parainfluenza		
Parvovirus B19		
Poliovirus		
Respiratory syncytial virus		
Rotavirus		
Rubella		
Rubeola		
Varicella-zoster		

coccus and a mononucleosis spot test were negative. A human immunodeficiency virus (HIV) enzyme-linked immunosorbent assay was negative. Four days later, an HIV p24 antigen test was positive and the HIV RNA viral load determination using polymerase chain reaction (PCR) revealed 1,000,000 copies/mL. The patient was started on three antiretroviral drugs, zidovudine, lamivudine, and nelfinavir.

This patient had a primary HIV infection. Of the estimated 40,000 people in the United States who become infected with HIV yearly, approximately half manifest symptoms of primary HIV infection, which is also called the *acute retroviral syndrome* (1). The symptoms of the illness are similar to a mononucleosis-like or "flu-like" illness. The HIV antibody test is negative. The most sensitive test to establish the diagnosis during the so-called "window period" is to measure HIV RNA viral load. Patients with primary HIV infection typically have high levels of virus, usually in the range of 100,000 to 2 million copies/mL (2). Symptoms of primary HIV infection usually develop 3 to 6 weeks after exposure. In most people who seroconvert, HIV antibody develops within 2 months of exposure; in 95%, within 6 months of expo-

sure; and in all, within 1 year (3). In the United States, approximately 5,000 exposures to needlestick injuries from HIV-infected patients occur each year; with a transmission rate of 0.33%, approximately 15 health care workers become infected each year from an occupational exposure (4). An acute retroviral syndrome develops in approximately 75% of health care workers who seroconvert, as in this case.

Symptoms and signs of primary HIV infection include fever, lymphadenopathy, sore throat, rash, myalgias, arthralgias, headache, diarrhea, nausea, and vomiting. Common laboratory abnormalities are leukopenia, atypical lymphocytosis, anemia, thrombocytopenia, and abnormal liver function tests. The differential diagnosis includes Epstein-Barr virus infection, cytomegalovirus (CMV), toxoplasmosis, secondary syphilis, rubella, viral hepatitis, group A streptococcal infection, and primary herpes simplex virus infection. Symptoms of primary HIV infection usually last for 2 weeks, and then patients become asymptomatic for many months to years. Most experts recommend using three antiretroviral drugs to treat a patient with primary HIV infection. In addition, counseling is important to deal with the psychological consequences of such an event.

Guidelines exist to manage occupational exposures to HIV to prevent HIV transmission (5). The risk of acquiring HIV from a needlestick from an HIV-infected source is 0.33%; with a mucosal surface exposure, the risk is 0.09%. Although HIV transmission can occur with a skin exposure, the risk is even lower than reported for an exposure involving a mucous membrane. As of June 1997, there were 52 health care workers in the United States who were reported to have acquired occupational HIV infection (6). HIV infection occurred most often in nurses and laboratory technicians. Most transmissions involved blood or bloody body fluids. Laboratory workers had HIV seroconversion after an accident involving an HIV viral culture. To date, no HIV seroconversions have occurred with injuries from suture needles. Risk of HIV seroconversion is increased with deep punctures, visible blood on the device, hollow needles, needle used in a source patient's artery or vein, and a source with a high viral load. Risk is also greater with an exposure to a large volume of blood.

Postexposure antiretroviral prophylaxis is recommended for certain injuries based on the type of exposure and the viral load of the source. One study showed that zidovudine prophylaxis was associated with a 79% decrease in HIV transmission compared with placebo (7). Workers receive two or three antiretroviral drugs if there is a high risk of HIV transmission. Drug treatment should be started within 4 hours (preferably within 1 to 2 hours) of an exposure and continued for 4 weeks. Unfortunately, approximately 25% of patients discontinue the postexposure prophylaxis because of adverse effects of drugs. The source blood should also be evaluated for hepatitis B and C, the other major bloodborne pathogens. An HIV serologic test should be done initially and at 6 weeks, 3 months, and 6 months after exposure. Health care workers should receive counseling and information regarding primary HIV syn-

drome. They should be advised to practice safe sex for 6 months after the exposure. Confidentiality is critical in managing health care workers exposed to HIV.

Guidelines to prevent the transmission of HIV and hepatitis B virus (HBV) to health care workers have been developed (8,9). The CDC has also developed recommendations to prevent the transmission of HIV and HBV from an infected health care worker to a patient.

Universal precautions and work practices designed to prevent transmission of hepatitis B are also effective against the bloodborne pathogens, such as HIV. Health care workers should always take precautions to prevent needlestick injuries and exercise care when handling blood or tissues and touching surfaces of patients. To reduce the risk of acquiring any bloodborne pathogen, such as HIV or hepatitis B and C viruses, the CDC has recommended that blood and bloody body fluids of all patients be considered potentially infectious. These recommendations are known as *universal precautions*. To reduce further the risk of sharp injuries, safer products have been developed such as needleless systems. The CDC has published a statement on management of occupational exposure to HIV (5), and the Occupational Safety and Health Administration (OSHA) has published guidelines to prevent occupational exposure to bloodborne pathogens based on the CDC recommendations.

The risk of transmission of HIV from an infected health care provider to a patient during an invasive procedure is extremely low. In one dental practice in Florida, six patients acquired HIV infection presumably from an infected dentist. The mechanism of transmission is unknown. However, in 19,000 patients exposed to infected health care providers, no cases of HIV infection have been detected retrospectively.

Hepatitis B

Viral hepatitis, type B (hepatitis B), is another bloodborne, work-related infectious disease. It is a problem for physicians, nurses, dentists, and laboratory workers—especially those who have direct contact with blood. Evidence of past HBV infection in these health care workers in prevaccine surveys was over four times that reported for volunteer blood donors; rates were highest for pathologists and surgeons (10). With the increased use of hepatitis B vaccine for all health care providers and adherence to other preventive measures, cases of occupationally related hepatitis B have decreased by 90% since 1985 (11).

Blood is a major source of infective virus. Only minute amounts are required: 1 mL of blood from a chronic carrier diluted to 10^{-8} still retains infectivity. Hepatitis B surface antigen (HBsAg) is present not only in blood and blood products, but in saliva, semen, and feces; however, nonblood sources of infection are probably rare. The presence of HBsAg in serum correlates well, but not perfectly, with infectivity. (Infectivity is most highly correlated with HBeAg.) All patients who are HBsAg positive, however, should be considered as potentially infectious.

Transmission of HBV may occur from an accidental percutaneous stick from a contaminated needle or other instrument. Infection may also develop after contaminated blood enters a break in the skin, splatters onto a mucous membrane, or is ingested, such as in a pipetting accident. Airborne transmission has not been reported.

Health care workers who are positive for HBsAg or anti-HBs often have had substantial contact with blood or blood products. Patient contact seems to be less important than direct contact with patients' blood. The incidence of hepatitis B infection for health care workers is increased in certain work areas, including hemodialysis units, hematology and oncology wards, blood banks and clinical laboratories (especially where personnel have contact with blood), operating rooms, dental offices and oral surgery suites, and wash areas for glassware and other equipment.

In the hospital, the major sources of HBV are patients with acute hepatitis B infection, patients with chronic liver disease, patients receiving chronic hemodialysis, immunosuppressed patients, parenteral drug abusers, and recipients of multiple blood transfusions. In many hospitals, the chronic hemodialysis unit is the highest-risk area. Patients on chronic hemodialysis who become HBsAg positive have up to a 60% chance of becoming chronic carriers. Other people at high risk of becoming chronic carriers are male homosexuals and residents of institutions for the retarded. Parenteral drug abusers have an HBsAg carrier rate of 1% to 5%, compared with a rate of 0.1% for volunteer blood donors (10).

Although certain features help distinguish hepatitis B from other forms of acute viral or toxic hepatitis, often they are not distinguishable clinically, and serologic studies are required for a specific diagnosis. Some patients are asymptomatic; others have malaise, fatigue, anorexia, nausea, vomiting, distaste for cigarettes, fever, abdominal pain, dark urine, light-colored stools, and other symptoms. Laboratory studies show abnormally high serum aminotransferases, such as aspartate aminotransferase and alanine aminotransferase; often an increased serum bilirubin; and, in severe cases, a prolonged prothrombin time.

Hepatitis B surface antigen usually can be detected in the blood of a patient with hepatitis B 6 to 12 weeks after exposure and for approximately 1 to 6 weeks after onset of clinical illness. Of patients with hepatitis B, 5% to 10% become chronic carriers of HBV and serve as potential sources of infection for personnel. If a patient remains HBsAg positive for 5 months, the probability of becoming a chronic carrier increases to 88%. Most chronic HBsAg carriers are asymptomatic; a subgroup of these carriers have mild to moderate elevations in their liver function test results. Chronic active hepatitis occurs in 30% of chronic carriers and may result in cirrhosis.

Prevention of Hepatitis B Infection

Prevention of infections with bloodborne pathogens, including HBV, is detailed in the OSHA bloodborne pathogens standard.

Recommendations for preventing hepatitis B in health care workers include immunization with hepatitis B vaccine as well as surveillance of staff in high-risk areas, employee education, appropriate sterilization and disinfection procedures, designation of a specific person responsible for safety, use of protective clothing and gloves, and avoidance of eating or smoking in laboratory work areas. [The CDC has issued an update on hepatitis B prevention (12).] Health care personnel should minimize their contact with potentially infectious patient secretions. Patients and staff in high-risk areas, such as hemodialysis units, can be screened for HBsAg and anti-HBs. HBsAg-positive patients can be separated from HBsAg-negative patients and dialyzed by staff who have anti-HBs (or who are HBsAg positive). Personnel must carefully avoid needlesticks and contact of mucous membranes and skin with potentially contaminated blood or other secretions, such as by wearing gloves. Use of the various new needleless devices for injections is another means of reducing needlestick injuries. All blood specimens should be handled as if potentially infectious. HBV-contaminated reusable equipment should be autoclaved before reuse, and HBV-contaminated disposable material should be disposed of in appropriate containers (Fig. 20-2). Syringe and other sharps manufacturers are focusing attention on proper design of these currently unsafe products.

Work practices should be consistent with the CDC guidelines on invasive procedures (13).

Hepatitis B immune globulin (HBIG) is recommended for prophylaxis within 1 week after either parenteral or mucosal contact with HBsAg-positive blood. Clinical hepatitis developed in 2% of HBIG recipients after a needlestick exposure to HBsAg-positive blood, compared with a rate of 8% in immune serum globulin recipients. The HBsAg and anti-HBs status of exposed health care workers should be determined. If results of either test are positive, HBIG has no value and should not be given. If both tests are negative, HBIG should be given in two doses spaced 25 to 30 days apart.

Two types of hepatitis vaccines were developed for prophylaxis. A plasma-derived vaccine, which consists of a suspension of HBsAg particles that have been inactivated and purified, was approved by the Food and Drug Administration in 1981. Although this vaccine was safe and efficacious, fear of receiving a product prepared from plasma in the era of the acquired immunodeficiency syndrome somewhat limited its use by health care providers. In 1987, another vaccine became available that was prepared in baker's yeast, using recombinant DNA technology. Two recombinant vaccines are available, Recombivax HB (Merck & Co., West Point, PA) and Engerix-B (SmithKline Beecham, Philadelphia, PA); the plasma-derived vaccine, Heptavax B, is no longer made. The vaccine induces a protective antibody (anti-HBs). Three doses at 0, 1, and 6 months with hepatitis B vaccine are recommended. An alternate schedule using four doses of Engerix-B vaccine given at 0, 1, 2, and 12 months has been approved for more rapid induction of immunity. The vaccines are administered intramuscularly in the deltoid, not the buttock. The vaccine has been studied in high-risk, preexposure situations among male homosexuals and the staff of dialysis units. The results of the vaccine trials have been impressive, with efficacy rates from 71% to 92%. Lack of an antibody response may occur in those older than 50 years of age, smokers, and obese people. The vaccine has also been given in postexposure situations to sexual partners or spouses of those with hepatitis B and after needlestick exposures. In the United States, the three major target populations for the vaccine are people with occupational exposures, such as

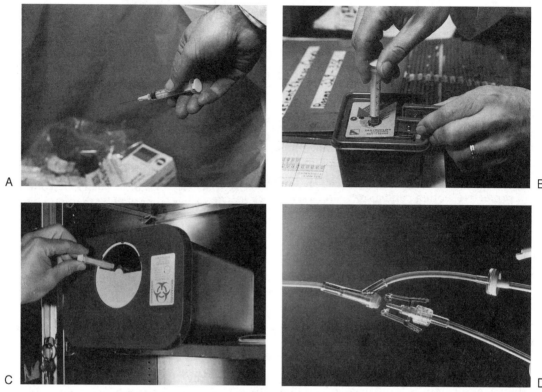

FIG. 20-2. Incorrect disposal of blood-contaminated needle in **(A)** a trash barrel and **(B)** a cutting device. **(C)** Correct method of disposal. Proper disposal of needles is an important means of reducing the risk for hospital and other health care workers of hepatitis B and hepatitis C infection. **(D)** New needleless device for intravenous injection systems. (Photographs by Marilee Caliendo.)

health care providers, all newborns, and those with sexual exposures. Target groups whose members, if susceptible, should receive hepatitis B vaccine are listed in Box 20-2.

In 1991, OSHA issued a regulation that hepatitis B vaccine be made available to all health care providers at risk. This regulation and the availability of the new hepatitis vaccines made with recombinant DNA technology have increased vaccine usage. The duration of immunity and the necessity for booster doses of vaccine need to be defined. Antibody levels decline after 7 years, but people may still be protected against clinical disease. Booster doses of hepatitis B vaccine usually are not necessary, and periodic serologic testing to monitor antibody levels is not recommended.

Recommendations for postexposure prophylaxis for percutaneous or a mucosal exposure to hepatitis B have been developed (14).

If the needle has been in contact with a known person ("donor"), then the HBsAg status of that person can be determined. If the donor is HBsAg positive or has a history of multiple blood transfusions and the worker is susceptible to hepatitis B (both HBsAg and anti-HBs negative), then the worker stuck by the needle is at risk for development of hepatitis B and should be given both HBIG and hepatitis B vaccine. If the status of the donor is unknown and the worker is susceptible, HBIG can still be given and a series of hepatitis B vaccine injections should be begun. Immunization is the most important preventive measure for hepatitis B (see Box 20-2).

Box 20-2. Groups Recommended for Hepatitis B Vaccine

1. People whose work involves contact with blood or blood-contaminated body fluids such as health care and public safety workers
2. Clients and staff of institutions for the developmentally disabled
3. Hemodialysis patients
4. Sexually active homosexual men
5. Users of illicit injectable drugs
6. Patients with clotting disorders who receive clotting factor concentrates
7. Household and sexual contacts of hepatitis B virus (HBV) carriers
8. People from countries of high HBV endemicity
9. Long-term inmates of correctional facilities
10. Sexually active heterosexual people, including sex workers
11. International travelers who plan to stay for more than 6 months in areas of high HBV endemicity, or short-term travelers who will have contact with blood, or sexual contact

From Centers for Disease Control and Prevention. Protection against viral hepatitis: recommendations of the Immunization Practices Advisory Committee. MMWR Morb Mortal Wkly Rep 1990;39(no. RR-2):14.

Hepatitis C

Hepatitis C virus (HCV) infection is a common cause of chronic hepatitis. HCV is bloodborne and the natural history of hepatitis C has been defined (15). Most cases (60%) of hepatitis C occur in intravenous drug users. In the past, blood transfusions accounted for a substantial number of infections, but since 1994, with better tests for hepatitis C in blood donors, transmission by this route rarely occurs. Approximately 10% to 20% of people with hepatitis C report exposure to an infected sexual partner or to having multiple partners in the absence of other risk factors. Some cases (10%) of hepatitis C are occupationally related and occur in health care providers, such as chronic hemodialysis personnel, and in emergency medical and public safety workers, such as law enforcement officers. In patients on chronic hemodialysis, the prevalence of hepatitis C antibody averages approximately 10% (16). With a needlestick or other sharp injury from an HCV-positive source, the risk of transmission is approximately 1.8% (range, 0% to 10%) (17).

Perinatal transmission can occur with a rate of infection of approximately 5% in infants born to HCV-positive mothers (18). In approximately 10% of patients, no source of exposure can be identified. (The risk of HCV transmission from an infected health care worker to a patient is very low.)

Diagnostic testing to detect HCV infection has improved. A number of tests are available. The enzyme immunoassay measures hepatitis C antibody, but a positive result occurs with an acute, chronic, or resolved infection. A more specific recombinant immunoblot assay is useful to identify false-positive enzyme immunoassay tests. Patients with a positive recombinant immunoblot assay should have HCV RNA measured using PCR, determining the number of viral copies per milliliter. This test can become positive 1 to 2 weeks after a viral exposure, even before the liver function tests are abnormal. Chronic HCV infection develops in most (75% to 85%) people infected with HCV. In approximately 30% of people with chronic HCV infection, liver function tests are normal. The course of chronic HCV infection is insidious, but often progresses to cirrhosis or hepatocellular carcinoma. Patients with hepatitis should be evaluated for chronic liver disease. Therapy of hepatitis C is evolving; the combination of interferon and ribavirin may be helpful. There are no data on prevention of HCV infection with postexposure prophylaxis. Immune serum globulin is not effective.

Hepatitis A

Infection with hepatitis A virus (HAV) is a risk for health care workers in institutions for the retarded where personal hygiene is poor and for certain animal caretakers, especially those with close exposure to chimpanzees. HAV is mainly transmitted by the fecal–oral route. Maximal viral shedding in the stool occurs during the 2-week period before onset of jaundice (and usually before diagnosis). HAV persists in the stool for approximately 7 to 15 days after the onset of jaundice (19). There is no chronic carrier state of HAV in the stool or blood.

Careful hand washing by health care personnel is probably the most important measure for preventing in-hospital spread of HAV. Immune serum globulin may be indicated for selected personnel working in institutions for the retarded within 1 to 2 weeks of contact with residents who have HAV infection. Nonhuman primates, which may be sources of HAV, should be quarantined for 2 months after importation. A vaccine to prevent hepatitis A is available.

Delta Hepatitis

Delta hepatitis infection is caused by a unique RNA virus that requires HBsAg for its replication. Hepatitis caused by the delta agent occurs only in patients with a simultaneous hepatitis B infection or as a superinfection in a chronic HBsAg carrier. Diagnosis can be established by demonstrating the presence of either immunoglobulin M or G, delta antibody (anti-HD), or delta antigen in the serum. Delta infection occurs most often in intravenous drug addicts and patients with hemophilia. Nosocomial transmission of the delta agent among patients undergoing hemodialysis has been reported and the potential for acquisition of delta hepatitis by health care personnel exists. Clinically, delta agent can cause fulminant hepatitis, and chronic carriage of the delta agent may be seen. Measures aimed at limiting the spread of hepatitis B should also limit the transmission of delta agent.

Tuberculosis

A resurgence of tuberculosis (TB) has occurred in the United States in the early 1990s secondary to the HIV epidemic, an increase in homelessness in the inner cities, and a change in immigration to the United States (20). Although the number of TB cases has decreased in the late 1990s, nosocomial transmission still occurs and health care workers are at risk. This problem has been further complicated by the identification of strains of *Mycobacterium tuberculosis* that are resistant to the standard antituberculous agents isoniazid (INH), rifampin, and ethambutol. These multidrug-resistant strains have been responsible for both community and nosocomial cases. These developments have prompted the CDC to issue guidelines to attempt to control this serious problem (21).

Transmission of TB from patients, often undiagnosed, to hospital workers and health care profession students remains a significant hazard. In one study, TB was not initially suspected in one half of the patients admitted with pulmonary TB and, in nearly one-third, the diagnosis was not established by time of discharge (22). The diagnosis was often missed because of a low index of suspicion of TB by today's physicians, an atypical clinical or radiographic presentation in people with HIV infection, and absence of tuberculin skin test positivity due to anergy. The incidence of tuberculin skin test positivity in U.S. physicians is at least twice the expected rate. Employees in nursing homes, mental health hospitals, and prisons are also at an increased risk for development of TB.

Prevention of TB in hospital employees requires that physicians have a high index of suspicion and institute respiratory isolation precautions pending confirmation of the diagnosis. Employees with nonreactive tuberculin skin tests should be screened with a tuberculin skin test once a year; tuberculin skin testing of exposed personnel is also indicated. Criteria for the interpretation of a positive tuberculin skin test depend on the host status. For an HIV-infected person, a positive test is an induration of 5 mm or more; for other immunosuppressed people, 10 mm or more; and for normal hosts, and induration of 15 mm or more. Indications for INH prophylaxis include household contacts of patients with active TB, recent skin test converters, tuberculin reactors with an abnormal chest radiographs, special clinical situations such as patients who are tuberculin reactors on high-dose corticosteroids, and tuberculin reactors younger than 35 years of age. INH is also indicated for HIV-infected people who (a) are tuberculin skin test positive, or (b) tuberculin skin test negative but at high risk for TB, such as intravenous drug abusers. Use of Bacille Calmette-Guérin (BCG) vaccination is not recommended for hospital personnel because it would make skin testing less helpful in identifying tuberculin reactors; in addition, some studies have shown a low level of effectiveness for the vaccine. Fundamentals of TB infection control (Fig. 20-3) have been described by the CDC. An effective TB control program requires early detection, isolation, and treatment of people with active TB. The primary emphasis of the TB infection control plan should be the achievement of these three goals. In all health care facilities, particularly those in which people who are at high risk for TB work or receive care, policies and procedures for TB control should be developed, periodically reviewed, and evaluated for effectiveness to determine the actions necessary to minimize the risk of TB transmission (21).

Rubella and Cytomegalovirus

Rubella infection may be transmitted from hospital employees to patients and from patients to susceptible personnel (23). In one outbreak, 47 cases of rubella occurred among hospital personnel (a dietary worker was the suspected index case); this outbreak resulted in one pregnancy being terminated and 475 lost workdays. The major hazard of rubella is infection in pregnant women, with the possibility of congenital rubella syndrome re-

FIG. 20-3. Health care workers can be protected from tuberculosis by proper isolation treatment of patients, use of enclosures, exhaust ventilation, and germicidal lamps. The last line of defense (illustrated) is the use of personal respiratory protection, one example of which (a powered air-purifying respirator) is illustrated. (Courtesy of the National Institute for Occupational Safety and Health, Washington, DC.)

sulting in the offspring. Schoolteachers are also at an increased risk of acquiring rubella because they are likely to have contact with people with the illness. Approximately 15% of women in the childbearing age group are susceptible to rubella. Rubella vaccine is an effective means of preventing disease, and susceptible personnel should be immunized. Also, the trivalent mumps-measles-rubella vaccine can be used. The vaccine is well tolerated and in the work setting results in minimal absenteeism. Pregnancy should be avoided for 3 months after vaccination.

Cytomegalovirus (CMV) is a potential threat for the developing fetus. Nurses who practice good personal hygiene are at no greater risk of acquiring this infection than their peers in the community (24). In contrast, day care workers may be at an increased risk of acquiring CMV (25). CMV is usually transmitted by sexual intercourse and through blood transfusions. Transmission of CMV appears to require prolonged, intimate contact in the hospital. The use of gloves when handling potentially contaminated wastes as well as hand washing after each patient contact are the appropriate infection control measures to prevent acquisition of CMV.

Measles

Since 1989, there has been a resurgence in cases of measles (26). Cases acquired in the medical setting account for 3.5% of all reported cases. The source of the measles virus is usually a patient in whom the illness was not recognized during the prodromal stage, when a rash is absent. Almost one third of cases in health care workers occur in those born before 1957, usually considered an immune group. In 1989, the Advisory Committee on Immunization Practices issued new guidelines for hospital employees: to provide proof of measles immunity or receive the measles vaccine (27).

Parvovirus B19

Parvovirus B19 is a cause of several clinical problems, including erythema infectiosum (also known as fifth disease), arthritis in adults, aplastic crisis, and fetal death in pregnancy. The virus has been detected in respiratory secretions and spread usually occurs by large particle aerosols or direct contact. Teachers and day care providers are at risk (28). In the hospital, pregnant health care providers should not care for patients with a known parvovirus B19 infection. Because the infection in the index patient may not be recognized, use of universal precautions with the wearing of gloves and frequent hand washing is mandatory.

Laboratory-Associated Infections

A 23-year-old hospital bacteriologist was admitted to the hospital with a 1-week history of chills, fever, headache, and diarrhea. She had a pulse of 80, temperature of 105°F, and a few macular areas on her abdomen. After blood cultures were obtained, she was given ampicillin and became afebrile. The blood cultures were positive for *Salmonella typhi*. The patient was diagnosed as having laboratory-acquired typhoid fever when further history revealed that she had been working with this organism 3 weeks earlier as part of a state laboratory proficiency testing program. She admitted to smoking in the laboratory and occasionally eating her lunch there.

The risk of infection from this organism can be reduced by immunizing laboratory personnel with typhoid vaccine as well as by enforcing basic laboratory safety procedures (29).

More than 6,000 cases of laboratory-associated infections with over 250 fatalities have been reported in the literature. The most frequently identified laboratory-associated infections are brucellosis, TB, tularemia, typhoid fever, hepatitis B, Venezuelan equine encephalitis, Q fever, coccidioidomycosis, dermatomycosis, lymphocytic choriomeningitis, and psittacosis. The CDC has classified infectious agents by risk hazard. Class 1 agents are the least hazardous; they include *Microsporum* and *Trichophyton* species (which cause tinea capitis, or ringworm of the scalp) and mumps virus. Class 4 agents pose the most risk and require the highest degree of containment; two examples are *herpesvirus simiae B* and Lassa fever virus. Smallpox, a major threat to health care workers in the past, remains a concern for the few laboratory workers who still work with the virus. Smallpox virus infection remains a threat for military troops as a possible agent for use in biologic warfare.

Laboratory-acquired infections usually result from accidents. Syringes and needles, either as a result of self-inoculation or causing a spray, have been involved in 25% of these infections; another 25% are related to spilling or spraying the infectious material. Other accidents result from injuries due to broken glass or other sharp objects, aspirating material by mouth-pipetting, and animal bites and scratches. The source of many laboratory-associated infections is obscure, although many of these are probably transmitted by aerosols. Air sampling techniques have shown that many laboratory procedures release organisms into the air.

Several approaches are available to prevent infection in laboratory personnel. Because inhalation of an infectious aerosol is assumed to be the major mode of transmission (although there are no data to support this contention), the use of biologic safety cabinets with filters and laminar air flow has been recommended to help protect against this hazard. Similarly, material being centrifuged should be put in tubes with sealable lids. Hand-pipetting devices should be used for pipetting infectious materials; mouth-pipetting should always be avoided.

Extreme care is required in the use and disposal of needles and syringes to decrease this frequent source of accidental infection. Experimentally infected animals are another important source of infection, and attention should be given to the animal quarters to minimize airborne spread of infection (see the next section). In addition, measures must be taken to prevent animal biting and scratching accidents; vaccination to prevent rabies, tetanus, or plague may be indicated for people at risk. Sera should be collected from employees and stored to be used diagnostically as acute-phase sera if an illness develops. Other measures to decrease the risk of infection include use of biohazard signs; limited access to laboratories; restrictions for pregnant employees working with CMV, herpesvirus, and rubella virus; and educational programs for personnel regarding possible hazards and preventive measures. Laboratory safety must be each employee's highest priority at all times.

Considerable interest has developed in the use of monoclonal antibodies for diagnosis as well as therapy of septic shock. Rat immunocytomas have been used in this work and the cell lines may be infected with various viruses. In one report, laboratory workers contracted hemorrhagic fever from contact with rat cell lines infected with hantavirus. Prevention of these infections requires screening of the rats before their use.

INFECTIOUS DISEASES IN NON–HEALTH CARE WORKERS

Most work-related infectious diseases among other workers are zoonoses (diseases primarily of animals that are transmitted to humans). Although these diseases occur infrequently, they are occupational hazards for workers who have contact with animals, such as farmers, veterinarians, butchers, and slaughterhouse workers. Some zoonoses also can result from recreational exposures. Examples of these diseases are listed in Table 20-4.

Bacterial Diseases

A sheep shearer became ill with fever and a headache. On examination, he was found to have a 2-cm ulcerative skin lesion on his right hand and enlarged right epitrochlear and axillary lymph nodes. A Gram stain of the skin ulcer drainage showed polymorphonuclear leukocytes but no organisms. Because of his occupation and the clinical presentation, tularemia was suspected. Isolation of the organism was not attempted because of the risk of creating an infectious aerosol in the bacteriology laboratory. The patient responded to oral tetracycline therapy and recovered in 10 days. The diagnosis of tularemia was later confirmed serologically by a fourfold rise in antibodies to tularemia agglutinins between the initial acute serum and convalescent serum.

Tularemia, like most other work-related bacterial diseases in non–health care workers, is an uncommon disease. However, the disease should be considered in the differential diagnosis of a patient with the appropriate work history who presents with a skin lesion and lymphadenopathy. Laboratory workers and others whose work requires re-

TABLE 20-4. *Examples of occupational zoonoses in the United States*

Disease	Common animal sources	Mode of acquisition	Workers at risk	Prevention
Bacterial diseases				
Anthrax	Cattle, sheep, horses, goats (usually in form of imported animal products)	Direct contact of inhalation of spores, souring of hides	Farmers, butchers, veterinarians	Vaccination, identification of infected animals
Brucellosis	Cattle, pigs, goats, sheep, dogs	Direct contact, ingestion, or inhalation	Meat packers, livestock workers, rendering plant workers, veterinarians	Animal vaccines, protective clothing and gloves
Cat-scratch disease	Cats, dogs	Direct contact	Veterinarians, cat and dog handlers	Avoidance of cat scratches
Erysipeloid (not to be confused with the acute cellulitis erysipelas)	Fish, other wild and domestic animals	Direct contact (often after skin abrasions)	Fishermen, butchers, fish handlers, poultry handlers, veterinarians, homemakers	Gloves, handwashing
Leptospirosis	Rodents, dogs, cats, cattle, pigs, wild animals	Contact with urine-contaminated soil or water, or direct contact with infected animal	Veterinarians, farmers (of sugar and rice), sewer workers, trappers, slaughterhouse workers	Animal vaccines, avoidance of contact with contaminated water, rat control workers, dock workers
Lyme	Mice, deer	Vectors of *Borrelia burgdorferi* are various *Ixodes* ticks	Forestry workers, hunters, ranchers	Protective clothing, insect repellents
Pasteurella multocida cellulitis	Cats and dogs (part of normal nasopharyngeal flora)	Animal bite	Veterinarians, pet shop workers	Prevention of animal bites
Plague	Ground squirrels, rabbits, hares, prairie dogs, rats, mice, coyotes	Direct contact with infected animal, rat flea bite, or respiratory droplets of infected patients	Farmers, ranchers, hunters, geologists in Southwest and West	Rat and vector control, vaccination
Salmonellosis	Poultry, cows, horses, dogs, cats, turtles	Direct contact with infected animal or its feces, or ingestion of infected food	Veterinarians, cooks, food processing and abattoir workers	Vaccination for laboratory workers, improved food processing and preparation
Swimming pool granuloma	Fish (marine and freshwater)	Direct contact	Fishermen, tropical fish store workers, drivers	Pool disinfection, gloves
Tularemia	Rabbits, hares, ticks	Direct contact with infected animal, ingestion, aerosolization, or tick bite	Trappers, fur handlers, ranchers, butchers, cooks	Vaccination, gloves, insect repellents
Viral diseases				
Encephalitis (e.g., California encephalitis)	Rodents, horses	Mosquito or tick bite	Agriculture and forestry workers, geologists, geographers, entomologists	Protective clothing, insect repellents

(continued)

TABLE 20-4. *Continued*

Disease	Common animal sources	Mode of acquisition	Workers at risk	Prevention
Rabies	Raccoons, dogs, cats, skunks, bats, faxes	Animal bite	Veterinarians, cave explorers, ranchers, trappers, farmers, wild animal handlers	Avoid animal bites, local wound care, preexposure and postexposure immunization
Rickettsial disease				
Murine typhus (endemic typhus)	Rats	Direct contact	Granary and warehouse workers	Rodent and vector control
Chlamydial disease				
Psittacosis	Parakeets, parrots, pigeons, turkeys, fowl	Inhalation of organism	Pet shop workers, pigeon breeders	Tetracycline, chemoprophylaxis of zoo workers, veterinarians, poultry handlers

peated exposure to tularemia are candidates for tularemia vaccine.

Other work-related bacterial diseases that affect non–health care workers include anthrax, brucellosis, erysipeloid, leptospirosis, Lyme disease, *Pasteurella multocida* cellulitis, plague, nontyphoid salmonellosis, and swimming pool granuloma. Prevention of bacterial zoonoses, which is summarized in Table 20-4, includes vaccination when available and use of protective clothing.

Viral Diseases

Viral infections acquired occupationally by non–health care workers fall into the following two categories:

1. Arthropod borne: yellow fever; Colorado tick fever; and Venezuelan, California, St. Louis, and western and eastern equine encephalitis are in this group. A human vaccine is available only for yellow fever; other preventive measures involve use of protective clothing, insect repellents, and vector control programs.
2. Nonarthropod borne: These diseases include rabies, orf, milker's nodules, and Newcastle disease. Orf is a skin disease caused by a poxvirus of sheep. Reddish-blue papules develop at sites of contact within 6 days of exposure to infected

sheep; spontaneous recovery occurs. Milker's nodules is another viral skin disease transmitted to the hands of milkers from the udders of infected cows; the nodules resolve within 4 to 6 weeks. Newcastle disease virus, which causes pneumoencephalitis in fowl, may produce a self-limited conjunctivitis in exposed workers.

Rabies is a good example of a possibly work-related viral disease for which prevention is available. In recent years, particularly in the northeastern United States, a marked increase has been seen in rabies cases in wild animals such as raccoons. High-risk groups include veterinarians, animal handlers, and laboratory personnel who work with the virus. Of course, not all cases are work related. Rabies prevention has been successful and has resulted in a decrease in the United States from an average of 22 cases per year in the 1946 to 1950 period to only one to five cases per year in the late 1990s. Prevention of rabies includes avoiding animal bites, local wound care, and immunization with vaccines or immune globulins. People at high risk of rabies should be immunized before exposure with multiple doses of human diploid cell rabies vaccine. Postexposure prophylaxis with human rabies immune globulin and the human diploid rabies vaccine is also recommended. To prevent rabies, the public must be educated in ways to reduce the risk of exposure to wild animals in affected areas, and the need to keep current the rabies vaccination status of pet dogs and cats.

Rickettsial Diseases

Rocky Mountain spotted fever, Q fever, and murine typhus can be work related. Rocky Mountain spotted fever is a potential risk for workers exposed to ticks in endemic areas such as the South Atlantic states; disease may occur, for example, in foresters, hunters, and construction workers. Q fever occurs commonly in cattle, sheep, and goats; dairy farmers, ranchers, stockyard and slaughterhouse workers, and hide and wool handlers are at risk for inhaling the organism and acquiring pneumonia or hepatitis. Murine typhus occurs in people working in rat-infested areas; granary and warehouse workers are at risk. (Most of the cases in the United States occur in Texas.)

Chlamydial Disease

Psittacosis, which is caused by a strain of *Chlamydia psittaci* and usually takes the form of a pneumonia, is an occupational hazard for pet shop employees, pigeon breeders, zoo workers, and veterinarians. In addition, several outbreaks have been reported in turkey-processing plants (30).

Parasitic Diseases

Some parasites are occupational hazards for non–health care workers in the United States; many more, not listed here, are hazards for workers in other countries. Schistosome dermatitis or "swimmer's itch" is an infrequent hazard for water workers such as skin divers, lifeguards, clam diggers, and rice field workers. Workers who have direct contact with the soil, such as barefoot farmers, ditch diggers, sewer workers, and tea plantation workers, may infrequently acquire hookworm disease, ascariasis, cutaneous larva migrans (creeping eruption), and visceral larva migrans. Mites, chiggers, and ticks that infest poultry and substances such as straw, dust, and grains may affect workers who handle them. Prevention of occupational parasitic diseases usually involves wearing protective clothing and shoes, health education, and measures to improve the standard of living.

Fungal Infections

Workers involved in earthmoving jobs, such as bulldozer operators, are at risk of acquiring deep fungal infections. Infection usually results from inhalation of spore-containing dust in endemic areas. Occupational fungal infections include histoplasmosis, coccidioidomycosis, and sporotrichosis. Superficial fungal infections can also result from occupational exposures; for example, ringworm in dogs, cats, cattle, and other animals is common and therefore animal handlers are at increased risk. Superficial candidal infection may be a hazard for workers such as bakers whose hands are often wet.

Histoplasmosis is a well recognized occupational pulmonary disease; the organism is found in certain soils, and its growth is stimulated by bat or bird guano. In the United States, the peak incidence of disease is in the Ohio and Mississippi River Valleys. At great risk are cave explorers, bridge scrapers, excavators, bulldozer operators, and grave diggers. Earthmoving activities at sites of chicken coops or bird roosts in endemic areas such as Tennessee or Kentucky should be preceded by appropriate soil cultures for fungi. A 5% formalin solution sprayed on contaminated soil before earthmoving operations may be effective in preventing disease where a disease risk has been documented.

Coccidioidomycosis is caused by *Coccidioides immitis*, a soil fungus found in the Southwest. Infection is associated with earthmoving activities. Rarely, infection may be transmitted through inanimate objects (fomites), such as fruits, cotton, and vegetables, containing spores that can become airborne. Archaeologists, geologists, bulldozer operators, and farm workers are at greatest risk. Prevention of naturally acquired disease is difficult, and workers involved in earthmoving activities should be aware of the risks. Before earthmoving in an endemic area, contractors should consider obtaining soil cultures (31,32).

Sporotrichosis infection usually results from traumatic inoculation of the organism into the skin. Sphagnum moss is often implicated as the source of the organism, although timber and other plant material are potentially a risk. Florists, nursery workers, farmers, berry pickers, and horticulturists are at greatest risk. Spraying sphagnum moss with a fungicidal solution may help prevent infection.

INFLUENZA CONTROL IN THE WORKPLACE

Influenza outbreaks continue to occur (Sometimes influenza-like symptoms may be due to other causes, as illustrated in Box 21-3). These outbreaks can be classified as (a) pandemics, which occur every 30 to 40 years and have an associated very high excess mortality; (b) epidemics, which occur more often with a lower excess mortality; and (c) sporadic outbreaks, which occur most often and have the lowest associated excess mortality. The explanation for these outbreaks involves the unusual capability of the influenza viruses to undergo antigenic changes. These changes in the virus can be classified as antigenic shifts and antigenic drifts.

Antigenic shifts result in major changes in one or both surface antigens, probably due to recombination of genetic material among different influenza viruses. Antigenic shifts are of great importance because the population usually lacks protective antibody against the new strain. Antigenic shifts can result in pandemics and epidemics.

Antigen drifts probably result from point mutations in the influenza virus genome leading to minor changes in the surface antigens. As a result of an antigenic drift, people have some protection against the new influenza virus but still require the new influenza vaccine to achieve optimal protection for that year.

Influenza viruses can be categorized as type A and type B viruses. Genetic recombination between the two types has not been demonstrated. Amantadine, a drug used for prophylaxis for people who have yet to receive the influenza vaccine, is effective only against influenza A virus.

Influenza and other respiratory viruses are responsible for considerable morbidity and lost workdays for workers in both health care and non–health care settings. Hospital personnel may transmit influenza to elderly patients; those with underlying heart, lung, or metabolic diseases; and immunosuppressed patients, resulting in morbidity and mortality.

Control measures include use of antiviral compounds, such as amantadine or ramantidine, and influenza vaccines. Use of influenza vaccine in the work setting for health care personnel should be considered, depending on the CDC recommendations for that year.

There are relatively few industry-based influenza immunization programs. Target groups who should receive yearly influenza vaccine are outlined in Box 20-4.

INTERNATIONAL TRAVELING WORKERS

A number of resources are available for health care providers and travelers to prevent diseases related to travel. The best source is the Centers for Disease Control and Prevention, Atlanta, GA 30333: (404) 639-3311 (workdays); (404) 639-2888 (after hours; emergency requests).

Malaria Branch: (770) 488-7788
Parasitic Diseases Drug Service: (404) 639-3356
Rabies Branch: (404) 639-1050
Travelers' Health Hotline: (877) 394-8747
    Fax: (888) 332-3299
Website: http://www.cdc.gov/travelershealth

Box 20-3. Fever and "Flu" May Not Be Infectious

Two work-related syndromes, metal fume fever and polymer fume fever, are characterized by fever and influenza-like symptoms but are noninfectious in origin.

Metal fume fever produces chills, increased sweating, nausea, weakness, headache, myalgias, and cough. It often begins with thirst and a metallic taste in the mouth. The white blood cell count is often elevated. Metal fume fever results from exposure to oxides of various metals, usually zinc, copper, or magnesium; aluminum, antimony, cadmium, copper, iron, manganese, nickel, selenium, silver, and tin have also been implicated. Welding, melting of copper and zinc in electric furnaces, and zinc smelting and galvanizing are work processes that have often been associated with this syndrome.

Polymer fume fever is characterized by dry cough, tightness in the chest, a choking sensation, and shaking chills. It is caused by exposure to unknown breakdown products of fluorocarbons, which are among the substances formed when polytetrafluoroethylene (PTFE, also known as Teflon or Fluon) is heated above 300°C. Since the first description of this syndrome in the 1950s, its control has been accomplished by preventing exposure to the heated fluorocarbon. In a workplace where there is PTFE exposure, however, cigarettes may become contaminated by PTFE on a worker's hands or in workplace air. In the cigarette, PTFE is heated to temperatures high enough to convert it to substances that are strong respiratory irritants. Therefore, even when the fluorocarbon is not directly heated to sufficient temperature, if smoking is allowed at the workplace the classic syndrome may still occur.

Both syndromes often occur after a delay of several hours from initial exposure. Both syndromes resolve within 24 to 48 hours with no known long-term sequelae.

For further information, see Gordon T, Fine JM. Metal fume fever. Occup Med 1993;8:505–518; and Shusterman DJ. Polymer fume fever and other fluorocarbon pyrolysis-related syndromes. Occup Med 1993;8:519–532.

Box 20-4. Target Groups for Influenza Vaccine

1. People older than 65 years of age
2. Residents of nursing homes and other chronic care facilities
3. Adults and children with chronic pulmonary or cardiovascular disease
4. Adults and children with chronic metabolic diseases (e.g., diabetes mellitus), renal disease, or hemoglobinopathies, or those who are immunosuppressed
5. Children and teenagers (6 months to 18 years of age) receiving long-term aspirin therapy
6. Health care providers
7. Members of households with high-risk people

REFERENCES

1. Schacker T, Collier AC, Hughes J, et al. Clinical and epidemiologic features of primary HIV infection. Ann Intern Med 1996;125:257–264.
2. Schacker TW, Hughes JP, Shea T, et al. Biological and virologic characteristics of primary HIV infection. Ann Intern Med 1998;128:613–620.
3. Cardo DM, Culver DH, Ciesielski C, et al. A case-control study of HIV seroconversion in healthcare workers after percutaneous exposure. N Engl J Med 1997;337:1485–1490.
4. Bell DM. Occupational risk of human immunodeficiency virus infection in healthcare workers: an overview. Am J Med 1997;102:9–15.
5. Centers for Disease Control and Prevention. Public health service guidelines for the management of healthcare worker exposures to HIV and recommendations for postexposure prophylaxis. MMWR Morb Mortal Wkly Rep 1998;47(No. RR-7):1–33.

6. Bartlett JG. IDCP notes on medical management of human immunodeficiency virus infection postexposure prophylaxis for healthcare workers. Infectious Diseases and Clinical Practice 1998;7:372–376.

7. Centers for Disease Control and Prevention. Case-control study of HIV seroconversion in healthcare workers after percutaneous exposure to HIV-infected blood: France, United Kingdom, and United States, 1988–August 1994. MMWR Morb Mortal Wkly Rep 1995;44:929–933.

8. Centers for Disease Control and Prevention. Recommendations for prevention of HIV transmission in health-care settings. MMWR Morb Mortal Wkly Rep 1987;36[Suppl 2S]:1S-18S.

9. Centers for Disease Control and Prevention. Recommendations for preventing transmission of human immunodeficiency virus and hepatitis B virus to patients during exposure-prone invasive procedures. MMWR Morb Mortal Wkly Rep 1991;40(no. RR-8):1–9.

10. Centers for Disease Control and Prevention. Inactivated hepatitis B virus vaccine. MMWR Morb Mortal Wkly Rep 1982;31:318.

11. Shapiro CN. Occupational risk of infection with hepatitis B and hepatitis C virus. Surg Clin North Am 1995;75:1047–1056.

12. Centers for Disease Control and Prevention. Protection against viral hepatitis: recommendations of the Immunization Practices Advisory Committee (ACIP). MMWR Morb Mortal Wkly Rep 1990; 39(no. RR-2):1–26

13. Centers for Disease Control and Prevention. Recommendations for preventing transmission of human immunodeficiency virus and hepatitis B virus to patients during exposure-prone invasive procedures. MMWR Morb Mortal Wkly Rep 1991; 40[Suppl no. RR-6].

14. Centers for Disease Control and Prevention. Immunization of healthcare workers: recommendations of the Advisory Committee on Immunization Practices (ACIP) and the Hospital Infection Control Practices Advisory Committee (HICPAC). MMWR Morb Mortal Wkly Rep 1997;46(no. RR-18):22–23.

15. Alter MJ, Margolis HS, Krawczynski K, et al. The natural history of community-acquired hepatitis C in the United States: The Sentinel Counties Chronic non-A, non-B Hepatitis Study Team. N Engl J Med 1992;327:1899–1905.

16. Centers for Disease Control and Prevention. Recommendations for prevention and control of hepatitis C virus (HCV) infection and HCV-related chronic disease. MMWR Morb Mortal Wkly Rep 1998;47(no. RR-19):1–39.

17. Puro V, Petrosillo N, Ippolito G. Italian Study Group on Occupational Risk of HIV and Other Blood-borne Infections: risk of hepatitis C seroconversion after occupational exposures in healthcare workers. Am J Infect Control 1995;23:273–277.

18. Granovsky MO, Minkoff HL, Tess BH, et al. Hepatitis C virus infection in the Mothers and Infants Cohort Study. Pediatrics 1998;102:355–359.

19. Krugman S, Ward R, Giles JP. The natural history of infectious hepatitis. Am J Med 1962;32:717–728.

20. Beck-Sagué C, Dooley SW, Hutton MD, et al. Hospital outbreak of multidrug-resistant *Mycobacte-rium tuberculosis* infections: factors in transmission to staff and HIV-infected patients. JAMA 1992; 268:1280–1286.

21. Centers for Disease Control and Prevention. Guidelines for preventing the transmission of Mycobacterium tuberculosis in health-care facilities. MMWR Morb Mortal Wkly Rep 1994;43(No. RR-13):1–132.

22. Counsell SR, Tan JS, Dittus RS. Unsuspected pulmonary tuberculosis in a community teaching hospital. Arch Intern Med 1989;149:1274–1278.

23. Polk BF, White JA, DeGirolami PC, Modlin JF. An outbreak of rubella among hospital personnel. N Engl J Med 1980;303:541–545.

24. Balcarek KB, Bagley R, Cloud GA, Paso RF. Cytomegalovirus infection among employees of a children's hospital: no evidence for increased risk associated with patient care. JAMA 1990; 263:840–844.

25. Adler SP. Cytomegalovirus and child day care: evidence for an increased infection rate among daycare workers. N Engl J Med 1989;321:1290–1296.

26. Atkinson WL, Markowitz LE, Adams NC, Seastrom GR. Transmission of measles in medical settings: United States, 1985–1989. Am J Med 1991; 91[Suppl 3B]:320S.

27. Centers for Disease Control and Prevention. Measles prevention: recommendations of the Immunization Practices Advisory Committee (ACIP). MMWR Morb Mortal Wkly Rep 1989;38(S9):10.

28. Cartter ML, Farley TA, Rosengren S, et al. Occupational risk factors for infection with parvovirus B19 among pregnant women. J Infect Dis 1991;163: 282–285.

29. Holmes MB, Johnson DL, Fiumara NJ, McCormack WM. Acquisition of typhoid fever from proficiency-testing specimens. N Engl J Med 1980; 303:519.

30. Hedberg K, White KE, Forfang JC, et al. An outbreak of psittacosis in Minnesota turkey industry workers: implications for modes of transmission and control. Am J Epidemiol 1989;130:569–577.

31. DiSalvo AF. Mycotic morbidity: an occupational risk for mycologists. Mycopathologia 1987;99: 147–153.

32. Kohn GJ, Linné SR, Smith DM, Hoeprich PD. Acquisition of coccidioidomycosis at necropsy by inhalation of coccidioidal endospores. Diagn Microbiol Infect Dis 1992;15:527–530.

BIBLIOGRAPHY

General References

Berenson AS, ed. Control of communicable diseases manual, 16th ed. Washington, DC: American Public Health Association, 1995.
A very useful handbook on the prevention and control of communicable diseases.

Guidelines

AIDS/TB Committee of the Society for Healthcare Epidemiology of America. Management of health-care workers infected with hepatitis B virus, hepatitis C virus, human immunodeficiency virus, or other blood-

borne pathogens. Infect Control Hosp Epidemiol 1997;18:349–362.
Guidelines for health care providers with hepatitis B virus, hepatitis C virus, or human immunodeficiency virus. Mandatory testing of providers is not indicated.

Centers for Disease Control and Prevention. Recommendations for prevention and control of hepatitis C virus (HCV) infection and HCV-related disease. MMWR Morb Mortal Wkly Rep 1998;47(RR-19):1–39.
Guidelines for diagnosis, management and prevention of hepatitis C.

Centers for Disease Control and Prevention. Public Health Service guidelines for the management of health-care worker exposures to HIV and recommendations for postexposure prophylaxis. MMWR Morb Mortal Wkly Rep 1998;47(no. RR-7):1–33.

Centers for Disease Control and Prevention and NIH. Biosafety in microbiological and biomedical laboratories, 3rd ed. Publication no. (CDC) 93-8395. Washington, DC: U.S. Department of Health and Human Services, 1993.
Safety guidelines for the laboratory.

Centers for Disease Control and Prevention. Immunization of healthcare workers: recommendations of the Advisory Committee on Immunization Practices (ACIP) and the Hospital Infection Control Practices Advisory Committee (HICPAC). MMWR Morb Mortal Wkly Rep 1997;46(no. RR-18):1–42.
A summary of vaccines for health care workers.

Centers for Disease Control and Prevention. Prevention and control of influenza: recommendations of the Advisory Committee on Immunization Practices (ACIP). MMWR Morb Mortal Wkly Rep 1998;47(no. RR-6):1–26.
Guidelines. Pneumococcal and influenza vaccines can be given simultaneously.

Garner JS, Hospital Infection Control Practices Advisory Committee. Guidelines for isolation precautions in hospitals. Infect Control Hosp Epidemiol 1996; 17:53–80.
Guidelines for health care workers for isolation of patients using standard precautions, a combination of universal precautions and body substance isolation, plus transmission-based precautions for pathogens transmitted by the airborne route or droplet nuclei.

Occupational Safety and Health Administration. Occupational exposure to blood-borne pathogens: final rule. 29 CFR 1010.1030. December 6, 1991.
Regulations designed to protect employees from bloodborne or other infectious material.

Infections in Health Care Workers

Bolyard EA, Tablan OC, Williams WW, et al. Guidelines for infection control in health-care personnel, 1998. Infect Control Hosp Epidemiol 1998;19: 407–463.
Comprehensive review to reduce the transmission of infections from patients to health care providers and vice versa.

Kelen GD, Green GB, Purcell RH, et al. Hepatitis B land hepatitis C in emergency department patients. N Engl J Med 1992;326:1399.
Twenty-four percent of patients were infected with hepatitis B virus, hepatitis C virus, or human immunodeficiency virus type 1, making use of universal precautions mandatory to prevent acquisition of these viruses.

Sepkowitz KA. Occupationally acquired infections in healthcare workers. Ann Intern Med 1996;125:826–834, 917–928.
A review of airborne and bloodborne occupationally acquired pathogens and prevention strategies.

Zoonoses

Palmer SR. Welsh Combined Centers for Public Health, Lord Soulsby, Simpson DIH. Zoonoses: biology, clinical practice, and public health control. Oxford: Oxford University Press, 1998.
This multidisciplinary book comprehensively covers the biology and epidemiology of zoonoses.

Sanford JP. Humans and animals: increasing contacts, increasing infections. Hosp Pract 1990;25:123.
A review of zoonoses.

Miscellaneous

Barry M, Russi MD, Armstrong L, et al. Brief report: treatment of a laboratory-acquired sabia virus infection. N Engl J Med 1995;333:294–296.
An unusual laboratory-acquired viral infection.

Khuder SA, Arthur T, Bisesi MS, Schaub EA. Prevalence of infectious diseases and associated symptoms in wastewater treatment workers. Am J Ind Med 1998;33:571–577.
Wastewater treatment workers have an increased frequency of gastrointestinal symptoms compared with control subjects.

BIBLIOGRAPHY FOR BOX 20-1

Berenson A, ed. Control of communicable diseases manual, 16th ed. Washington, DC: American Public Health Association, 1995.

21 Stress

Dean B. Baker and Robert A. Karasek

A 54-year-old taxi cab driver has unstable angina after a 10-year history of hypertension and two episodes of bleeding duodenal ulcers. He does not drink alcohol or smoke cigarettes.

A 21-year-old video display terminal operator has visual discomfort, headaches, backaches, irritability, and trouble sleeping.

A 46-year-old automobile assembly line worker calls in sick on a Monday for the fourth time in the past 2 months.

A 39-year-old investment banker with an exceptional record of achievement no longer seems able to complete projects on time. He reports feeling fatigue to the point where he has given up his regular recreational activities.

Considerable evidence exists that occupational stress contributes to a wide range of health effects such as those described in the preceding list (1,2). Job stress is a leading cause of worker disability in Europe and the United States (3). Stress contributes to the development of heart and cerebrovascular disease, hypertension, peptic ulcer and inflammatory bowel diseases, and musculoskeletal problems (1,4–6). Evidence suggests that stress alters immune function, possibly

facilitating the development of cancer (7–9). Anxiety, depression, neuroses, and alcohol and drug problems are associated with stress (1,10–12). Considered together, these disorders, which are affected by stress, are responsible for most of the mortality, morbidity, disability, and medical care utilization in the United States.

Awareness of occupational stress is increasing as managers, workers, and health care providers have to address issues such as increased job pressure, job insecurity, and feelings of powerlessness at work. There are indications that the modern technology used in production and international trade is contributing to an increase in stress risks throughout the world (4). Some of these stress issues occur at the level of the individual job, others have their source in complex organizational structures, and others operate at the level of the marketplace. The "free market" for labor, for example, means that any worker can be replaced at any time in the name of economic efficiency. The current work environment is associated with greater levels of stress even as traditional toxic exposures have come under better control.

Consideration of stress in the evaluation, prevention, and treatment of occupational diseases may be challenging to the health care provider. Unlike other occupational hazards, which tend to be specific for tasks, stress is associated to a varying extent with all work. It is a complex process involving social, psychological, and physiologic systems. Many causes of occupational stress

D. B. Baker: Center for Occupational and Environmental Health, University of California at Irvine, Irvine, California 92612.

R. A. Karasek: Department of Work Environment, University of Massachusetts Lowell, Lowell, Massachusetts 01854.

may combine to contribute to different effects. The health care provider should be able to recognize stressful working conditions, evaluate stress-related disorders, and manage stress in the individual patient.

DEFINITION OF STRESS

The National Institute for Occupational Safety and Health (NIOSH) defines job stress as "the harmful physical and emotional responses that occur when the requirements of the job do not match the capabilities, resources, or needs of the worker" (2). This definition emphasizes the development of health responses because of a mismatch between work conditions and the individual worker. This definition does not state exactly what conditions are responsible for causing stress, but these conditions include psychosocial, physical, and organizational factors.

Related terms have been used to define components of the stress process. A *stressor* is an environmental condition or *psychosocial* factor that results in stress. *Strain* is a short-term physiologic, psychological, or behavioral manifestation of stress, whereas a *modifier* is an individual characteristic such as coping style or environmental factor such as social support that may act on each stage of the stress process to produce individual variation in the stress response.

CONCEPTS OF OCCUPATIONAL STRESS ETIOLOGY AND MECHANISMS

Development of Occupational Stress Models

Because of the complexity of the stress phenomenon, stress research has led to the development of a number of conceptual models of how occupational stress arises. These models have evolved from earlier attempts at a stress theory from the field of cognitive psychology and from physiologic models of the stress response.

Contributions to a Stress Model from Cognitive Psychology

A central tenet of the cognitive stress models is that processes of perception and interpretation of the external world determine the development of psychological states and ensuing risk for chronic disease. Psychological stress effects are analyzed through the concept of mental workload. The cognitive psychological perspective defines workload in terms of total information load the worker is required to perceive and interpret while performing job tasks (13). "Overload" and stress occur when this human information processing load is too large for the person's information processing capabilities. These models emphasize the importance of information overloads, communication difficulties, and memory problems.

The psychological perspective tend to downplay the role of objective workplace stressors and emphasizes instead the importance of the worker's appraisal of the situation (14). Although this approach might be appropriate where there is little possibly of modifying the environment, it could harm workers in situations where the stressors are real and should be the target of change.

Physiologic Stress Theories

Physiologic stress response theories focus on two mechanisms: the adrenal medullary response involving epinephrine and norepinephrine and the adrenal cortical response involving cortisol. Cannon's (15) "fight–flight" response is most associated with stimulation of the adrenal medulla and epinephrine secretion. This pattern, occurring in conjunction with sympathetic arousal of the cardiovascular system is an active response mode where the organism is able to use metabolic energy to support both mental and physical exertion. Although this response mechanism is a basic element in all animal behavioral repertories, it is taxing in the short term, and long-term arousal of psychoendocrine mechanisms can lead to difficulties with relaxation and a state of chronic overarousal. Adrenal medullary mechanisms reflect the importance of sustained arousal conditions, threats to security, time pressures for increased performance, and a range of workplace social situations, including authority challenges.

The adrenal cortical response is a response of defeat and withdrawal—possibly occurring in a situation where the person faces stress over which he or she has little control. Henry and Stephens (16) described this behavior as defeat or loss of social attachments, leading to withdrawal and submissiveness in social interactions.

In the physiologic models, *stress* is a systemic concept referring to a disequilibrium of the system as a whole, in particular of the system's control capabilities. Biologic control systems include the nervous system, the cardiovascular system, and the psychoendocrine system. Stress-related diseases, including hypertension, are regarded as disorders of regulation in which the process of maintaining system equilibrium (homeostasis) is disturbed (16,17).

Integrated Occupational Stress Models

The integrated models take as their point of departure human behavior in complex environments, rather than psychological or physiologic brain functions. Behavioral models are almost always multidimensional to capture requisite complexity, and thus offer richer possibilities for understanding stressful job conditions. Two models of occupational stress have received the greatest amount of attention among stress researchers—the person–environment (PE) fit model and the job demand–control (D-C) model (1,4,18–21). A third model, the effort–reward imbalance (ERI) model, has received increasing attention in Europe (22–24).

Person–Environment Fit Model

The PE fit model states that strain develops when there is a discrepancy (a) between the demands of the job and the abilities of the person to meet those demands, or (b) between the motives of the person and the environmental supplies to satisfy the person's motives (20,25). Demands include workload and job complexity. Motives include factors such as income, participation, and self-utilization. Supplies refer to whether the job, for example, provides sufficient income to satisfy the person's motives. The model distinguishes the *objective* environment and person from the *subjective* environment and person, where *subjective* refers to the person's perceptions. It assumes that strain arises because of poor fit between the subjective person and subjective environment. The emphasis of the PE fit model on subjective perceptions is consistent with the earlier cognitive psychology model. As with that model, the PE fit model does not acknowledge the role of objective workplace stressors other than by their influence on a worker's perceptions.

A difficulty with the PE fit model is that it has demonstrated limited ability to predict what objective work conditions are likely to result in stress. Also, because interpretation of stressors is subjective, stress becomes a function primarily of individual perceptions. A strength of the model is its emphasis on the need for flexibility in job design and consideration of workers as individuals with varying abilities, motives, and perceptions.

Job Demand–Control Model

The demand–control model views strain as arising primarily because of the characteristics of work, rather than the subjective perceptions of the worker. It states that strain arises from an imbalance between demands and decision latitude (or control) in the workplace, where lack of control is seen as an environmental constraint on response capabilities (4,19,21). Decision latitude actually consists of two components that are highly correlated in job situations: personal control over decision making (autonomy), and skill level. The D-C model theorizes that the essential characteristic of a stressful work environment is that it simultaneously places demands and creates environmental constraints on a worker's response capabilities. Because of its hypothesis that the combination of high demands and low control leads to strain, the D-C model is also referred to as the *job strain* model.

The D-C model characterizes jobs by their combination of demands and control (Fig. 21-1). Jobs with high demands and low control, such as those of waiters, video display terminal operators, and machine-paced assemblers, result in strain (strain axis in Fig. 21-1). These characteristics typically are found in occupations with a high division of labor and deskilling of tasks. The model also hypothesizes that jobs in which psychological demands are accompanied by high control result in an active, high-motivation situation (activity axis in Fig. 21-1). This model is useful because it parsimoniously captures the two primary physiologic mechanisms described previously. The adrenal medullary response has been shown to correspond to increased job demands, or possibly increasingly active job situations, whereas increased adrenal cortical output is associated with deceased decision latitude, or possibly increased strain.

One omission of the D-C model is that social relations at the workplace are only indirectly included in the model. Therefore, much of the recent epidemiologic research on occupational stress and cardiovascular disease risk using the D-C model has been done with an expanded version of the model that includes social support as an additional dimension (19,21). The D-C model has also been criticized because of its focus on job task characteristics (22,24); however, the model can be conceptually expanded to include consideration of demands versus control over a range of physical and organizational factors at work. The amount of control over all aspects of the job is now recognized as a decisive factor in the development of occupational stress (2,3,26).

Effort–Reward Imbalance Model

Another, more recently developed model shares elements with the D-C model, but emphasizes an imbalance between the effort required for a job and the rewards provided by the job (22,23). In the effort–reward imbalance model, effort can be due to extrinsic factors, such as high workload, or intrinsic characteristics, such as the worker's "need for control." The model has more of a macrosociologic focus than the D-C model, and there is a greater emphasis on individual at-

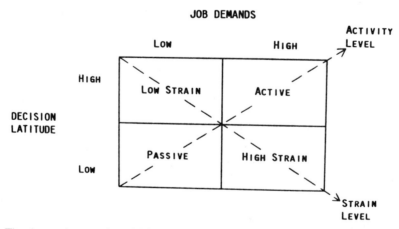

FIG. 21-1. The demand–control model focuses on the combined effects of job demands and decision latitude on strain. (From Baker DB. The study of stress at work. Am Rev Public Health 1985;6:367–381.)

tributes than in the D-C model (18,21). According to the ERI model, rewards can come from three main sources: money, esteem, and occupational status control. The concept of occupational status control relates to whether the person can control his or her job and career—including factors such as job stability, forced occupational change, downward mobility, or promotion prospects. In this model, the work role is considered a basic tool to link a person's important emotional and motivational needs, such as self-esteem and self-efficacy, with the opportunity structure. Thus, job stress arises both from the immediate conditions of work and from the broader context of career and the role of work in a person's life.

The ERI model has received increasing attention because it seems particularly relevant to the changes in the labor market described previously, in which an increasing number of jobs are becoming insecure as corporations continually restructure to compete in a global marketplace. A primary concern with the model is that it is "ambitiously broad" (18); it will be difficult empirically to test all parts of the model. On the other hand, initial studies based on the model have shown significant associations with cardiovascular disease and cardiovascular disease risk factors. A cohort study of British civil servants reported by Bosma and colleagues (24) found independent effects of components of the D-C model and the ERI model on coronary heart disease. More research is being done to explore the relationship between these models of job stress.

COMPONENTS OF THE STRESS PROCESS

Although these models are useful in understanding the development of stress-related disease, and epidemiologic studies have been able to demonstrate associations between stressors and stress responses, it is not possible to determine specific etiologic relationships between stressors and health effects for one individual or a small group of workers.

The researcher or clinician cannot analytically separate the effects of many coexisting stressors when examining one or a few individuals. Thus, from a clinical perspective, the relationship between stressors and health outcomes may appear nonspecific. The health care provider needs to assess each component of the stress process, including stressors, stress responses (strain) and long-term outcomes, and modifiers (Table 21-1).

Stressors

Stressors may be divided into those relating to the person's job, including the work schedule and pace, job content, and physical conditions; organizational factors such as role and organization structure; and extra-organizational factors.

Time Demands, Work Schedule, and Pace

Studies have reported associations between working excessive hours or holding down more than one full-time job and coronary heart disease morbidity and mortality (27). Lack of control over work hours exacerbates this stressor; required overtime has been associated with low job satisfaction and indices of poor mental health. Substantial evidence indicates that the temporal scheduling of work can have a significant impact on physical, psychological, behavioral, and social well-being. See Chapter 22 for more information on shift work.

An important factor is whether the work pace is controlled by the worker or is externally determined. *Machine pacing* means that the pace of the operation and the work output are controlled to some extent by a source other than the operator. In general, this work presents a deleterious combination of short-interval demands with lack of control. It requires vigilance, yet is monotonous and repetitive. As mentioned, research has indicated that machine-paced assembly line work is highly stressful (1,4). Control over work pace has been positively associated

TABLE 21-1. *Components of the stress process*

Stressors

Time demands, work schedule and pace—task demands, overtime, shift work, machine pacing, piecework
Task structure—lack of control, skill underutilization
Physical conditions—unpleasant, threat of physical or toxic hazard, ergonomic hazards
Organization—role ambiguity, role conflict, competition and rivalry
Extraorganizational—community, job insecurity, career concerns
Nonwork sources—personal, family, community

Outcomes

Physiologic
　Short term—catecholamines, cortisol, blood pressure increases
　Long term—hypertension, heart disease, ulcers, asthma
Psychological (cognitive and affective)
　Short term—anxiety, dissatisfaction, mass psychogenic illness
　Long term—depression, burnout, mental disorders
Behavioral
　Short term—job (absenteeism, productivity and participation), community (decreased friendships and participation), personal (excessive use of alcohol and drugs, smoking)
　Long term—"learned helplessness"

Modifiers

Individual—behavioral style and personal resources
Social support—emotional, value or self-esteem, and informational

---

with better pregnancy health among employed women (28).

In *piecework*, the worker's remuneration is based on the quantity of products produced. This system has been shown to induce stress responses similar to those of machine pacing (Fig. 21-2). People on piecework may increase their work pace even to the point of discomfort. For example, one study found that when invoicing clerks were put on piecework, as opposed to their usual hourly rate, they doubled their work rate, but also increased their urinary epinephrine and norepinephrine levels by approximately one-third.

FIG. 21-2. Garment workers, who often work on a piecework basis, are often under much stress at work. (Photograph by Earl Dotter.)

Task Structure

As discussed previously, research based on the D-C model has shown that the combination of high psychological demands and low decision latitude leads to psychological and physiologic strain reactions (Fig. 21-3) (4,19,21,29). Table 21-2 describes "good" and "bad" job characteristics based on an understanding of the D-C model of job content. Examples of high-demand task characteristics include having too much to do, time pressure or deadlines, or repetitious, fast-cycle workflow. Low decision latitude arises in tasks that are too narrow in content, lack stimulus variation, do not allow creativity or problem solving, and do not permit worker control over pace or work methods.

As predicted in the ERI model, changes in work organization may also be a source

FIG. 21-3. Secretaries are among the most highly stressed workers. (Photograph by Earl Dotter.)

The threat of a physical or toxic hazard can act as a stressor. Certain occupations are recognized to have increased risk of physical danger, such as police and firefighters. Chronic stress reaction among workers in these types of occupations is well recognized. Studies have also revealed that among workers in the trade and service sectors, the potential for abuse or violence from clients is a significant source of stress.

Physically demanding tasks, such as agricultural work (Fig. 21-4) and meat trimming (see Fig. 8-2 in Chapter 8), when repetitive and fast paced, can lead to increased risk of repetitive trauma injuries. Both psychological and physical risk factors are associated with this increasingly prevalent category of ergonomic injuries (see Chapters 9 and 26) (1,6,30,31).

Organization

Major organizational stressors include the worker's role in the organization, organizational structure, and interpersonal relationships. Role ambiguity results from lack of clarity concerning the requirements of the job; the worker does not know the objectives, scope, and responsibilities of the job. This stressor may lead to job dissatisfaction, feelings of tension, and lowered self-confidence. Role conflict occurs when conflicting demands are made on the worker by different groups in the organization or when the worker is required to do work that he or she dislikes or believes to be outside of the requirements of the job.

Competition and rivalry are sources of stress for many workers. Poor work relationships are "those which include low trust, low supportiveness, and low interest in listening to and trying to deal with problems that confront the organizational member" (32). Conflict tend to occur when inadequate information is provided to designate responsibilities, job roles, and the means of carrying them out. Sexual and gender harassment can also be an important workplace stressor (33,34). Poor interpersonal relationships can have a

of stress. Transfers, demotions, and even promotions can be stressful. The potential for change to act as a stressor is affected by the predictability of the event and the control the worker or work group has over the transition process.

Physical Conditions

Unpleasant working conditions include improper lighting, excessive noise, inadequate work space, depressing surroundings, and unsanitary conditions. Investigators have found a positive association between poor mental health and unpleasant working conditions. Investigations of complaints among office workers (Fig. 21-3) have found that uncomfortable workstations, crowding, and inadequate ventilation and temperature control are associated with stress responses.

TABLE 21-2. *Description of "Good" and "Bad" Job Characteristics based on Demand-Control Model*

"Bad" or high-strain jobs	"Good" jobs
Psychological demands There are long periods under intense time pressures, with the threat of unemployment at the end. Or there are long periods of boredom, but with the constant threat of crisis requiring huge efforts. There is great disorganization of work processes with no resources to facilitate order.	The job has routine demands mixed with a liberal element of new learning challenges, in a predictable manner. The magnitude of demands is mediated by interpersonal decision making between parties of relatively equal status.
Decision latitude: skill discretion Nothing is being learned, nothing is known of the product's destination. There is no hint of future development on the job. New technologies are difficult to understand, and knowledge is limited by secrecy requirements.	The job offers possibilities to make the maximum use of skill and provides further opportunities to increase skills on the job. New technologies are created to be effective tools in the workers' hands, extending their powers of production.
Decision latitude: autonomy The worker's minutest actions are prescribed and monitored by machine or by supervisors. There is no freedom independently to perform even the most basic tasks. New technologies restrict workers to rigid, unmodifiable information formats.	There is freedom from rigid "worker-as-child" factory discipline. Machine interfaces allow workers to assume control. Workers have influence over selection of work routines and work colleagues and can participate in long-term planning. It may be possible to work at home during flexible hours.
Social relations Workers are socially isolated from their colleagues. Random switching of positions prevents development of lasting relationships. Competition sets worker against worker.	Social contacts are encouraged as a basis for new learning and are augmented by new telecommunications technologies that allow contact when isolation was previously a necessity. New contacts multiply the possibilities for self-realization through collaboration.

Adapted from Karasek R, Theorell T. Healthy work: stress, productivity and the reconstruction of working life. New York: Basic Books, 1990.

FIG. 21-4. California farm workers stoop to harvest, wrap, and box iceberg lettuce for shipment direct from the farm field. The speed of the mechanical equipment sets the work pace. (Photograph by Earl Dotter.)

debilitating effect not only on individual workers but on the organization.

Extra-Organizational Stressors

Extra-organizational stressors are factors that are related to work but extend beyond the specific job or organization. These stressors include factors related to career security and advancement, unemployment, and issues of job security in relation to the global economy and free market. These factors are addressed in the ERI model of job stress.

Several conditions associated with career development or job future (e.g., lack of job security, underpromotion, overpromotion, and fear of job obsolescence) have been related to adverse behavioral problems, psychological effects, and poor physical health (1). Lack of job security is a primary stressor for many workers. Evaluation of career as a stressor must consider anticipated changes

A supervisor's job may be highly stressful due to its high degree of "boundariness." (Drawing by Nick Thorkelson.)

associated with the stage of the worker's career. Lack of change may be a stressor if it represents thwarted ambition. Retirement can be a stressor. It is an anticipated change that is modified by the extent that the worker requires work for self-image and has potential for constructive activity outside of work. Stress reactions may begin long before retirement.

Unemployment may have dramatic effects on stress-related illness, not only on those who become unemployed, but on those who remain working. Brenner and Mooney (35) observed that increasing unemployment is usually followed by increased cardiovascular mortality. Effects of unemployment on mental health (36), cortisol levels (37), and immune function (38) have been observed.

Nonwork Stressors

Finally, sources of stress other than work must be considered in an evaluation of the stress process. The boundary between work and nonwork stressors is not distinct, and certainly the two interact in causing stress responses.

Personal, family, and community factors can be stressors. These potential sources of stress should be more familiar to the reader and are not reviewed in this chapter.

Adverse Health Outcomes

Stress responses can be divided into three major categories: physiologic, psychological, and behavioral. Effects on the organization should also be considered.

Physiologic

Job stressors have been shown to induce a variety of short-term physiologic reactions (1,4,39,40). Substantial advances in understanding have accompanied the development of telemetry and biochemical measurement techniques. Bioelectric measures of stress reactions include the heart rate and rhythm, electromyogram, and galvanic skin response. Increases in galvanic skin response were found to be a sensitive index to behavioral changes related to boredom, fatigue, and monotony. Biochemical measures include catecholamines, corticosteroids (e.g., cortisol), cholesterol, free fatty acids, fibrinogen, thyroxine, and growth hormone. Stress produces changes in the level of antibodies in the blood and may alter cell-mediated immunity, although it is not known whether these changes are long-lasting and represent an adverse health effect (8,9).

The chronic pathophysiologic effects of stress are usually considered under the rubric of psychosomatic disorders. Depending on the perspective taken, this category may be as narrow as including only headache and gastritis or may encompass such diseases as cardiovascular disease, hypertension, ulcers, and musculoskeletal disorders.

Much research has focused on the potential effect of stress on cardiovascular disease. Studies since the mid-1970s based on the D-C model have shown significant associations between high-strain occupations and subsequent development of cardiovascular disease, after analytically controlling for other potential risk factors such as age, smoking, education, and obesity (5,19). Between 1981 and 1998, most of the more than 40 studies based on the D-C model have found significant, positive associations between job strain and cardiovascular disease. These studies used a variety of research designs and were performed in several countries (5,19,21). In general, the studies showed a stronger association between job strain and coronary heart disease than between job strain and biomedical risk factors such as serum cholesterol and smoking (41,42). However, studies using sophisticated ambulatory monitoring methods have found consistent associations between job strain and blood pressure (29,39). The associations appear to be stronger among blue-collar than white-collar occupations and stronger in studies of workers younger than 55 years of age (21). A smaller number of studies based on the ERI model have also found significant associations between job stress and cardiovascular disease (22–24) or risk factors (40).

A large number of studies have demonstrated an association between job stress and musculoskeletal problems, especially upper extremity disorders among office workers (1,6,30,31). The mechanisms for these associations may be that stress-related tension causes increased static loading of the muscles, or that workers under stress may alter their work behavior in a way that increases musculoskeletal strain (31). For example, workers may strike a keyboard with greater force or they may increase their work pace, leading to greater muscular strain.

Psychological

Research has demonstrated the association between job stress and adverse psychological effects, including anxiety, situational depression, and job dissatisfaction (11,12,43). Association with chronic mental illness is not as well documented. Nevertheless, data from workers' compensation claims indicate that psychological disorders due to chronic occupational stress are a major health problem (43). According to a study conducted by the National Council on Compensation Insurance, claims for "gradual mental stress" accounted for approximately 11% of all occupational disease claims (43,44).

Causes of depression and job dissatisfaction follow the predictions of the job stress models. Among blue-collar workers, major stressors are lack of job complexity, role ambiguity, and job insecurity. Particularly among clerical and service workers, low self-esteem arises from lack of respect from supervisors. For white-collar workers, major stressors are responsibility for people, high job complexity, and high and variable workload.

Behavioral

There is basic agreement among researchers that job stress affects behavioral outcomes such as absenteeism, substance abuse, sleep disturbances, smoking, and caffeine use (10,45). Quantitative workload has been associated with cigarette smoking and an inability to quit smoking. It also has been related to absenteeism, low work motivation, and lack of interest in contributing suggestions to management.

The work organization also is affected. Deleterious personal behaviors, such as absenteeism and accidents, have a negative impact on the organization. Stress decreases productivity through reduced output, production delays, and poor quality of work, which leads to lost time, equipment breakdown, and wasted material. Poor morale can lead to labor unrest, excessive number of grievances, sabotage, and increased turnover. Thus, both the individual and the organization suffer from the manifestations of job stress.

Specific Stress-Related Disorders

For the effects just described, job stress is one of many factors in the etiology of the conditions. For some disorders, stress is recognized as the primary etiologic agent. Examples include burnout and posttraumatic stress disorder.

Burnout is a phenomenon that has been recognized mostly in professional settings, although it can affect any worker (46,47). It may occur after years of high-quality performance when the person suddenly seems unable to perform his or her work. Behavioral manifestations include decreases in efficiency and initiative, diminished interest in work, and an inability to maintain work performance. Symptoms characterizing the syndrome are fatigue, intestinal disturbances, sleeplessness, depression, and shortness of breath. Other symptoms include irritability, decreased tolerance of frustration, blunting of affect, suspiciousness, feelings of helplessness, and increased risk taking. There is a tendency to self-medicate with tranquilizers, narcotics, and alcohol, all of which may lead to addiction. When away from the job, the person is unable to relax and often reports giving up recreation and social contacts.

Actually, burnout does not emerge spontaneously; it progresses in stages starting with the beginning of employment. Because burnout-prone people often start out as enthusiastic overachievers, the organization tend to heap more and more responsibilities on them. Failure to recognize the connection between constant overachievement and burnout and to take proper preventive measures can lead to an exhausted workforce whose achievement, desire, and creative talents are defunct.

Posttraumatic stress disorder is an anxiety disorder that occurs after a stressful or traumatic event (48). The disorder may occur among workers who have had occupational toxic exposure or injury. Any situation that evokes feelings of intense fear, helplessness, loss of control, or annihilation may precipitate the disorder. The trauma may be massive and discrete, like an accident, or may comprise episodes of exposure to a dangerous chemical or work process. The cardinal feature is repeated reexperiencing of the traumatic event. Typically, any stimulus that resembles the initial event evokes the reexperience of the event. Patients may believe they are allergic to everything. Other symptoms include emotional blunting, detachment from others, sleep disturbances, and trouble concentrating. The recurrent symptoms and anxiety experienced by people with posttraumatic stress disorder may be disabling. It is important to recognize the psychological effect of the initial trauma to avoid costly, unnecessary, and sometimes reinforcing medical diagnostic evaluations.

Modifiers

The association between stressors and stress responses is modified by characteristics of the person and environment. This notion is intrinsic to all models of stress.

Individual

Individual characteristics that affect vulnerability to stressors include emotional stability, conformity (versus inner-directedness), rigidity (versus flexibility), achievement orientation, and behavioral style. The following individual characteristics have received substantial attention as potential modifiers of the stress response (1). These factors are the major focus of stress management programs:

- Hardiness—a person's basic stance toward his or her place in the world that expresses commitment (related to a sense of purpose), control (feeling like one has control rather than being helpless), and challenge (understanding that change rather than stability is normal).
- Self-esteem—the favorability of a person's characteristic self-evaluations.
- Locus of control (LOC)—a person's tendency to believe that events in life are controlled by his or her own actions (internal LOC) or by outside influences (external LOC). Those with internal LOC believe they can exert control over life circumstances.
- Coping style—personality trait–like combinations of thoughts, beliefs, and behaviors that are used to reduce the negative impacts of stress. These styles can include problem-focused coping (e.g., information seeking and problem solving), emotion-focused coping (e.g., regulating emotions), and appraisal-focused coping (e.g., denial, acceptance, or redefinition of the problem) (49,50).

A behavioral style that has received much attention as a risk factor for coronary heart disease is type A coronary-prone behavior, characterized by a sense of competitiveness, time urgency, and overcommitment (51). Despite substantial research since Friedman and Rosenman (51) originated the concept in the mid-1970s, no one has been able to identify the precise aspects of the type A behavior pattern that engender the heart disease risk, and this concept is gradually losing favor among researchers (1).

Environment

Many investigators have concluded that the most important factor ameliorating the stress response is social support. Social support includes emotional, informational, and instrumental support. Informational support depends on the clarity and effectiveness of communication patterns in the organization. Instrumental support includes such factors as having adequate instructions and tools to complete the task. A large amount of research has demonstrated that social support can reduce the adverse health effects of stress (1,3,21). There has been disagreement, however, about whether the primary effect of social support is to modify or *buffer* the effect of stressors or whether social support has a direct effect in reducing stress (4). Regardless of whether social support directly reduces stress or acts as a modifier of the effect of stressors, a prime strategy of stress management is to encourage the development of supportive networks on the job.

MANAGEMENT OF STRESS

The phrase *stress management* typically denotes worksite programs that address employee stress by reducing individual vulnerability; however, management of work stress should occur in multiple settings, including the medical office or clinic. A comprehensive approach should encompass assessment, prevention, and treatment of stress in the individual worker and the organization.

Evaluating Stress in the Worker

Typically, evaluation of stress in a worker occurs when the person has already manifested initial strain reactions or a stress-related disorder. Therefore, the health care provider needs to assess sources of stressors, modifiers, and possible health effects for the worker to develop a plan for secondary prevention and treatment. Because many stress reactions appear to be nonspecific when evaluated at the individual level, it is often necessary to perform an assessment among a larger group of workers at the worksite to interpret findings for the individual worker.

Management of stress at the worksite is discussed in the following section.

Assessment

The initial step to manage stress in an individual worker is to assess potential stressors, stress reactions, and modifiers. Objective and subjective indicators should be measured simultaneously.

Knowledge of workplace exposures is an important aid in the interpretation of psychosocial stressors. The worker should be asked about the existence of toxic chemicals or physical hazards and the extent to which he or she is concerned about their effects.

Analysis of the work environment should include an assessment of tasks and the work organization. The worker should be asked about objective job structure (e.g., overtime, shift work, and machine pacing), task characteristics, and organizational structure. Standardized job stress questionnaires are available (52), but most of these are designed for epidemiologic studies rather than for assessment of an individual worker. Nevertheless, questions in these instruments can serve as a guide for the clinical evaluation.

Measuring Stress Reactions and Health Outcomes

Assessment of stress effects practically must be based on diagnostic skills already familiar to the health care provider. Many of the available stress diagnostic instruments are too expensive or time consuming to be of practical use outside of the research setting. Examples include measurement of urinary metabolites of catecholamines, galvanic skin response, or mental health status using structured interview instruments. The health care provider may have to make a diagnosis based on the results of several nonspecific but accessible indicators, rather than by relying on the results of one definitive test.

A minimum assessment should include the following observations: (a) complaints of "distress"—symptoms and motivation related to work conditions; (b) emotional reactions—anxiety, depression, and other reactions; (c) cognitive function and work performance—psychometric testing and performance evaluations at work; (d) behavioral changes such as sleep disturbances and drug and alcohol use; (e) physiologic function such as heart rate, blood pressure, and serum cholesterol; and (f) symptoms and diseases that may be due to stress. The clinician also should consider behavioral characteristics that may affect susceptibility or resistance and the worker's family, community, and cultural environment.

Diagnosis

The large number and complexity of factors involved in the stress process virtually preclude exhaustive documentation of each factor using definitive instruments. There is no simple, yet comprehensive instrument for measuring stressors and stress reactions. Consequently, the medical practitioner must inventory the range of factors discussed previously and make a judgment as to the likelihood that the identified stressors were responsible for the observed stress reactions. This judgment must take into consideration the following issues: (a) presence of potential workplace stressors; (b) presence of potential nonoccupational stressors; (c) physical, psychological, and behavioral health status (consistency with known patterns of stress reactions); (d) presence of other accepted risk factors for observed health conditions; (e) individual characteristics and social factors that may affect vulnerability to stress; and (f) presence of similar health conditions among coworkers.

Stress Treatment and Prevention for the Individual

Control of stress involves treatment for people already experiencing stress, secondary prevention for people who demonstrate early stress reactions, and secondary prevention through reducing individual vulnerability and increasing social support. Treatment of clinically apparent stress effects follows the traditional medical model, such as treatment of hypertension, depression, or excessive use of alcohol. At the same time, it is necessary to reduce exposure to stressors and increase individual resistance to prevent sequelae.

Reducing individual vulnerability to stress can be addressed by the health care provider or through group counseling, courses, and workshops. Common denominators of successful programs include training in self-awareness and problem analysis so that the person is better able to detect signs of increasing stress and identify the stressors that may be producing it. Another common denominator is training in assertiveness so that the person can become more active in controlling stressors. When the stress is clearly related to work organization, it is important for the person to understand the contribution of the environment. A professional can help a person realize that other workers in similar situations also have similar negative reactions. This realization can help the person overcome self-imposed blame for his or her psychosocial distress, recognizing that stress is not due to personal weakness. Also important are techniques, such as meditation, relaxation programs, exercise programs, and biofeedback, that reduce personally experienced stress to more tolerable levels (53).

Control of Stress at the Workplace and Through Work Reorganization

Research during has increasingly focused on worksite intervention strategies to reduce occupational stress (54–59). Hurrell and Murphy (55), Cooper and Cartwright (56), and others have discussed key strategies for preventing occupational stress. *Primary prevention* is designed to control stress by reducing job stressors, such as through job redesign and organizational change. *Secondary prevention* helps workers modify or control their appraisal of stressful situations. Stress management programs are concerned with the detection and management of stress by increasing awareness and improving the person's stress management skills through training and educational activities. These programs represent secondary prevention because they do not involve changes in job or organization stressors. *Tertiary prevention* includes programs, such as employee assistance programs and workplace counseling, that focus on treatment and rehabilitation of workers who have stress-related disorders.

Because intervention research on job stress is relatively recent, it is difficult to assess the effectiveness of these strategies. It appears that secondary prevention strategies, such as stress management, are effective in reducing the immediate impact of stress, but the beneficial effects are not long-lasting (56). In general, studies have shown that using a combination of techniques seemed to be more effective than those using single techniques (54,59). Primary prevention programs, such as work reorganization, can have a lasting effect on stress reduction. However, research on work reorganization interventions has had mixed findings (56,58). There are two primary explanations. First, it is difficult to develop good randomized study designs because the interventions require a willingness on behalf of the employer and employees to consider the possibility of substantial reorganization of work tasks and organization. Most research is based on observational or quasiexperimental study designs. Second, it is not clear that the intervention strategies used in some of the programs, such as participatory action research, are actually able to change workplace stressors.

NIOSH has developed a guideline for the prevention of work-related stress disorders that makes recommendations on job design (Table 12-3) (43). These recommendations are based on the findings of stress research. Scandinavian countries have implemented laws that mandate job characteristics similar to the NIOSH guidelines (e.g., the Swedish Work Environment Act of 1977, amended 1991, Chapter 2, Section 1).

TABLE 21-3. *National Institute for Occupational Safety and Health Recommendations for Job Design*

Work schedule—Work schedules should be compatible with demands and responsibilities outside of the job. . . . When schedules involve rotating shifts, the rate of rotation should be stable and predictable.
Workload—Demands should be commensurate with the capabilities and resources of individuals. Provisions should be made to allow recovery from demanding tasks or for increased job control under such circumstances.
Content—Jobs should be designed to provide meaning, stimulation, and an opportunity to use skills.
Participation and control—Individuals should be given the opportunity to have input on decisions or actions that affect their jobs and the performance of their tasks.
Work roles—Roles and responsibilities at work should be well defined. Job duties need to be clearly explained, and conflicts in terms of job expectations should be avoided.
Social environment—Jobs should provide opportunities for personal interaction both for purposes of emotional support and for actual help as needed in accomplishing assigned tasks.
Job future—Ambiguity should not exist in matters of job security and opportunities for career development

Adapted from Sauter S, Murphy LR, Hurrell JJ Jr. Prevention of work-related psychological disorders: a national strategy proposed by the National Institute for Occupational Safety and Health. Am Psychol 1990; 45:1146–1158.

NIOSH recommends a three-step process for implementing a stress prevention program (2) (Table 21-4). Before starting the process, it is essential to prepare the organization by raising awareness about job stress and securing commitment by top management to support the program. The first step is to identify the problem by using a stress diagnostic test, such as a worker attitude survey, that can provide information on stress levels and sources. Structured questionnaires are available to assess task and organizational characteristics. A Job Content Questionnaire based on the D-C model has been developed and validated (60). Other job stress questionnaires have been developed, some of which are widely used for stress evaluation (52).

Once the assessment has defined problem areas, interventions may include developing variable work schedules, job restructuring, supervisor training, changing management style, improving internal communications, and encouraging organizational development. Some stress managers state that the most important strategy is to increase social support through the organization of cohesive work groups, development of improved communication patterns, and provision of recreational facilities for employees. Workers' participation in analyzing problems and in planning programs can be an important contributor to success. Third, there should be an evaluation of the effort. Follow-up surveys permit comparison with baseline conditions.

International studies have shown that a key factor is that the work reorganization program must be based on a commitment to enhance the participation of workers and, thus, their decision latitude or control over job content and organizational issues. In 1991, the International Labor Organization commissioned 20 case studies of stress prevention programs in the worksite from nine industrialized and developing countries (61). At these workplaces, employers and workers attempted to address stress problems at their source by modifying the work situation rather than by attending to stress symptoms after the fact. Approximately 90% of the studies reported some evidence of success, indicating that elimination of work organizational sources of stress can succeed, and, in particular, that programs that engage workers in a participatory change process can succeed.

For a participatory process to occur, a feeling of trust must be created among lower-status employees that information they share openly about their feelings of job stress, and their ideas for work environment change, are protected from reprisals by management. A joint labor–management program may create the trust necessary for open communications, and is associated with program success.

Successful programs appeared to combine multiple intervention approaches covering both individually focused and environmen-

TABLE 21-4. *Recommended strategy to prevent job stress*

Prepare organization for a stress prevention program
- Build general awareness about job stress (causes, costs, and control).
- Secure top management commitment and support for the program.
- Incorporate employee input and involvement in all phases of the program.
- Establish the technical capacity to conduct the program, such as training of in-house staff or use of job stress consultants.

1. Identify the problem
- Hold group discussions with employees.
- Design an employee survey.
- Measure employee perceptions of job conditions, stress, health, and satisfaction.
- Collect objective data such as absenteeism, turnover rates, and performance.
- Analyze data to identify problem locations and stressful job conditions.

2. Design and implement interventions
- Target source of stress for change.
- Propose and prioritize intervention strategies.
- Communicate planned interventions to employees.
- Implement interventions.

3. Evaluate the interventions
- Conduct both short- and long-term evaluations.
- Measure employee perceptions of job conditions, stress, health, and satisfaction.
- Include objective measures such as absenteeism and health care costs.
- Refine the intervention strategy and return to Step 1.

Job stress prevention should be seen as a continuous process that uses evaluation data to refine or redirect the intervention strategy.

Adapted from National Institute of Occupational Safety and Health. DHHS, NIOSH publication no. 99-101. Washington, DC: US Government Printing Office, 1999.

tally focused activity, often in an ordered sequence. For example, stress education is often a first step, followed by group discussions of the problems and action planning sessions, and later by discussion of technical and economic resources issues. This process starts at the stress response in discussions generating awareness of the personal meaning of stress, moves to analysis of the task situation, and, finally, focuses on the level of work organization. The platform of raised awareness is necessary to gain support for job design and organizational issues.

A key feature of organizational intervention is to encourage an active process of joint labor and management participation that can lead to enhanced understanding of potential stressors and the possibilities for job redesign, changes in organizational function, and increases in buffering factors such as social support to reduce the level and effects of work-related stress.

REFERENCES

1. Sauter S, Hurrell J, Murphy L, Levi L, eds. Psychosocial and organizational factors. In: Encyclopaedia of occupational health, section 34. Geneva: International Labor Office, 1998:34.1–34.77.
2. National Institute for Occupational Safety and Health (NIOSH). Stress at work. U.S. Department of Health and Human Services, NIOSH publication no. 99-101. Washington, DC: US Government Printing Office, 1999.
3. Sauter S, Hurrell J, Murphy L, Levi L. Psychosocial and organizational factors (Introduction). In: Encyclopaedia of occupational health. Geneva: International Labor Office, 1998:34.2–34.3.
4. Karasek R, Theorell T. Healthy work: stress, productivity, and the reconstruction of working life. New York: Basic Books, 1990.
5. Schnall P, Landsbergis P, Baker D. Job strain and cardiovascular disease. Annu Rev Public Health 1994;15:381–411.
6. Punnett L, Bergqvist U. Visual display unit work and upper extremity musculoskeletal disorders: a review of the epidemiological findings. Arbete och Hälsa 1997;16:1–161.
7. Baker GHB. Psychological factors and immunity. J Psychosom Res 1987;31:1–10.
8. Ursin H. Immunological reactions. In: Encyclopaedia of occupational health. Geneva: International Labor Office, 1998:34.57.
9. Olff M, Brosschot JF, Benschop RJ, et al. Modulatory effects of defense and coping on stress-induced changes in endocrine and immune parameters. International Journal of Behavioral Medicine 1995; 2:85–103.
10. Hellerstedt WL, Jeffery RW. The association of job strain and health behaviours in men and women. Int J Epidemiol 1997;26:575–583.

11. Bourbonnais R, Brisson C, Moisan J, Vâezina M. Job strain and psychological distress in white-collar workers. Scand J Work Environ Health 1996;22:139–145.

12. Niedhammer I, Goldberg M, Leclerc A, Bugel I, David S. Psychosocial factors at work and subsequent depressive symptoms in the Gazel cohort. Scand J Work Environ Health 1998;24:197–205.

13. Saunders M, McCormic E. Human factors in engineering and design, 7th ed. New York: McGraw-Hill, 1992.

14. Lazarus RS. Cognition and motivation in emotion. Am Psychol 1991;46:352–367.

15. Cannon WB. Stresses and strains of homeostasis. Am J Med Sci 1935;189:1–14.

16. Henry JP and Stephens PM. Stress, health, and the social environment: a sociobiological approach to medicine. New York: Springer-Verlag, 1977.

17. Weiner H. Psychobiology of human disease. New York: American Elsevier, 1977.

18. Kasl S. The influence of the work environment on cardiovascular health: a historical, conceptual, and methodological perspective. Journal of Occupational Health Psychology 1996;1:42–56.

19. Karasek R. Demand/control model: a social, emotional, and physiological approach to stress risk and active behavior development. In: Encyclopaedia of occupational health. Geneva: International Labor Office, 1998:34.6–34.14.

20. Caplan R. Person-environment fit. In: Encyclopaedia of occupational health. Geneva: International Labor Office, 1998:34.15–34.17.

21. Theorell T, Karasek R. Current issues relating to psychosocial job strain and cardiovascular disease research. Journal of Occupational Health Psychology 1996;1:9–26.

22. Siegrist J. Adverse health effects of high-effort/low-reward conditions. Journal of Occupational Health Psychology 1996;1:27–41.

23. Siegrist J, Peter R, Jung A, Cremer P, Seidel D. Low status control, high effort at work and ischemic heart disease: prospective evidence from blue-collar men. Soc Sci Med 1990;31:1127–1134.

24. Bosma H, Peter R, Siegrist J, Marmot. Two alternative job stress models and the risk of coronary heart disease. Am J Public Health 1998;88:68–74.

25. French JR Jr, Caplan RD, Van Harrison R. The mechanisms of job stress and strain. Chichester, United Kingdom: John Wiley & Sons, 1982.

26. Johnson J. Conceptual and methodological developments in occupational stress research: an introduction to state-of-the-art reviews I. Journal of Occupational Health Psychology 1996;1:6–8.

27. Gardell B. Scandinavian research on stress in working life. Int J Health Serv 1982;12:31–41.

28. Wergeland E, Strand K. Work pace control and pregnancy health in a population-based sample of employed women in Norway. Scand J Work Environ Health 1998;24:206–212.

29. Schnall PL, Schwartz JE, Landsbergis PA, Warren K, Pickering TG. A longitudinal study of job strain and ambulatory blood pressure: results from a three-year follow-up. Psychosom Med 1998;60:697–706.

30. Hagen K, Magnus P, Vetlesen K. Neck/shoulder and low-back disorders in the forestry industry: rela-tionship to work tasks and perceived psychosocial job stress. Ergonomics 1998;41:1510–1518.

31. Lindstrom K, Leino T, Seitsamo J, Torstila I. A longitudinal study of work characteristics and health complaints among insurance employees in VDT work. International Journal of Human-Computer Interaction 1997;9:343–368.

32. French JR Jr, Caplan RD. Organizational stress and individual strain. In: Marrow AJ, ed. The failure of success. New York: AMACOM, 1972:30–66.

33. Piotrkowski C. Gender harassment, job satisfaction, and distress among employed white and minority women. Journal of Occupational Health Psychology 1998;3:33–43.

34. Goldenhar L, Swanson N, Hurrell J, Ruder A, Deddens J. Stressors and adverse outcomes for female construction workers. Journal of Occupational Health Psychology 1998;3:19–32.

35. Brenner MH, Mooney A. Relation of economic change to Swedish health and social well-being. Soc Sci Med 1987;25:183–195.

36. Starrin B, Larsson G, Brenner S-O, Levi L, Petterson I-l. Societal changes, ill health and mortality: Sweden during the years 1963–1983: a macro-epidemiological study (Swedish). Research report no. 13, Department of Community Medicine, County of Varmland, Karlstad, Sweden. Karlstad, Department of Comunity Medicine, 1988.

37. Brenner S-O, Levi L. Long-term unemployment among women in Sweden. Soc Sci Med 1987; 25:153–161.

38. Arnetz B, Wasserman J, Petrini B, et al. Immune function in unemployed women. Psychosom Med 1987;49:3–12.

39. Pickering TG, Devereux RB, James GD, et al. Environmental influences on blood pressure and the role of job strain. J Hypertens 1996;14[Suppl 5]:S179–S185.

40. Siegrist J, Peter R, Cremer P, Seidel D. Chronic work stress is associated with atherogenic lipids and elevated fibrinogen in middle-aged men. J Intern Med 1997;242:149–156.

41. Greenlund KJ, Liu K, Know S, McCreath H, Dyer AR, Gardin J. Psychosocial work characteristics and cardiovascular disease risk factors in young adults: the CARDIA study. Soc Sci Med 1995;41:717–723.

42. Netterström B, Kristensen TS, Damsgaard MT, Olsen O, Sjöl A. Job strain and cardiovascular risk factors: a cross sectional study of employed Danish men and women. British Journal of Industrial Medicine 1991;48:684–689.

43. Sauter S, Murphy LR, Hurrell JJ Jr. Prevention of work-related psychological disorders: a national strategy proposed by the National Institute for Occupational Safety and Health. Am Psychol 1990;45:1146–1158.

44. National Council of Compensation Insurance. Emotional stress in the workplace: new legal rights in the eighties. New York: NCI, 1985.

45. Shirom A. Behavioural outcomes. In: Encyclopaedia of Occupational Health. Geneva: International Labor Office, 1998:34.53–34.55.

46. Fielding JF. Corporate health management. Reading, MA: Addison-Wesley 1984:309–331.

47. Iverson RD, Olekalns M, Erwin PJ. Affectivity, or-

ganizational stressors, and absenteeism: a causal model of burnout and its consequences. Journal of Vocational Behavior 1998;52:1–23.

48. Schottenfeld RS, Cullen MR. Occupation-induced posttraumatic stress disorder. Am J Psychiatry 1985;142:198–202.

49. Lazarus R, Folkeman S. Stress: appraisal and coping. New York: Springer-Verlag, 1984.

50. Ingledew DK, Hardy L, Cooper CL. Do resources bolster coping and does coping buffer stress? An organizational study with longitudinal aspect and control for negative affectivity. Journal of Occupational Health Psychology 1997;2:118–133.

51. Rosenman RH, Brand RJ, Sholtz RI, Friedman M. Multivariate prediction of coronary heart disease during 8.5 year follow-up in the Western Collaborative Group Study. Am J Cardiol 1976;37:903–910.

52. Special section: The measurement of stress at work. Journal of Occupational Health Psychology 1998;3:291–401.

53. Warshaw JJ. Managing stress. In: Krinsky LW, Kieffer SN, Carone PA, Yolles SF, eds: Stress and productivity. New York: Human Sciences Press, 1984:15–30.

54. Murphy L. Stress management in work settings: a critical review of the health effects. American Journal of Health Promotion 1996;11:112–135.

55. Hurrell JJ, Murphy LR. Occupational stress intervention. American Journal of Industrial Medicine 1996;29:338–341.

56. Cooper CL, Cartwright S. An intervention strategy for workplace stress. J Psychosom Res 1997;1:7–16.

57. Cahill J. Psychosocial aspects of interventions in occupational safety and health. American Journal of Industrial Medicine 1996;29:308–313.

58. Reynolds S. Psychological well-being at work: is prevention better than cure? J Psychosom Res 1997;1:93–102.

59. Bellarosa C, Chen P. The effectiveness and practicality of occupational stress management interventions: a survey of subject matter expert opinions. Journal of Occupational Health Psychology 1997;2:247–262.

60. Karasek R, Brisson C, Kawakami N, Houtman I, Bongers P, Amick B. The Job Content Questionnaire (JCQ): an instrument for internally comparative assessments of psychosocial job characteristics. Journal of Occupational Health Psychology 1998;3:322–355.

61. International Labor Office. Conditions of work digest: preventing stress at work, vol 11, no. 2. Geneva: International Labor Office, 1992.

BIBLIOGRAPHY

International Labor Office. Conditions of work digest: preventing stress at work, vol 11, no. 2. Geneva: International Labor Office, 1992.
A unique compendium of resources for practitioners and researchers who want to reduce worker stress through work reorganization. Includes an introductory overview of the occupational stress problem, 19 case studies of preventive change from 9 countries, a summary analysis of the case studies, and a useful review of visual aides, articles, and books in the stress prevention field.

Karasek R, Theorell T. Healthy work: stress, productivity, and the reconstruction of working life. New York: Basic Books, 1990.
Combines explanations of stress development in the work process through the demand-control model with a proposal for work reorganization that could reduce psychosocial health risks. The model is presented at the task level, by occupation, at the psychophysiological level, and in epidemiologic studies. A psychosocial, skill-based productivity model, congruent with healthy work goals, is used to guide the book's second-half discussion through joint health and productivity topics: intervention processes, organizational issues, occupational strategies, and global economic challenges.

Sauter S, Hurrell J, Murphy L, Levi L, eds. Psychosocial and organizational factors. In: Encyclopaedia of occupational health, section 34. Geneva: International Labor Office, 1998:34.1–34.77.
Provides a comprehensive overview of job stressors, modifiers, and health outcomes, with extensive references to research through 1995.

The following journals have a strong emphasis on job stress, psychosocial factor at work, and stress management. Most issues have a relevant article, and some of the journals present special sections on job stress.
Journal of Occupational Health Psychology
Journal of Occupational and Organizational Psychology
Work and Stress
International Journal of Stress Management
British Journal of Psychology

22 Shift Work

Torbjörn Åkerstedt and Anders Knutsson

	Monday	Tuesday	Wednesday	Thursday	Friday	Saturday	Sunday
Week 1	Night	Night	Night	Night	Day off	Day off	Day off
Week 2	Morning	Morning	Day off	Afternoon	Afternoon	Afternooon	Morning
Week 3	Day off	Day off	Morning	Morning	Morning	Morning	Day off
Week 4	Afternoon	Afternoon	Afternoon	Day off	Night	Night	Night
Week 5	Day off	Day off	Day off	Day off	Day off	Day off	Day off

A 22-year-old man came to an occupational health clinic complaining of gastrointestinal symptoms and sleep disturbances. He worked at a paper mill, where for 3 years he had been working on a three-shift schedule that included night shifts, a rotating shift schedule in which the shifts changed at 6 a.m., 2 p.m., and 10 p.m., in the sequence shown above.

He had previously been well, with no history of sleep complaints or gastrointestinal problems. During the past 6 to 12 months, he had experienced increasing sleep disturbances, especially after night shifts, when he could sleep only 1 to 2 hours. He felt excessively tired at the end of his workday and on days off. He had also experienced epigastric pain, heartburn, and diarrhea. These problems had all developed insidiously over the past 6 months. He did not smoke or drink alcohol. He liked his work and his coworkers. He lived together with a female partner in a house in a small, quiet town, 25 miles from the paper mill. At home, he was not disturbed by noise during sleep. The physical examination was normal except for a tender epigastrium. There was no anemia. A month later, he was transferred to day work, and, 2 months later, his sleep problems and gastrointestinal disturbances disappeared, without any drug treatment.

This case illustrates the two most common disorders associated with shift work: disturbed sleep and wakefulness. Other major effects are accidents/injuries, gastrointestinal disorders, and cardiovascular disease. The central problem in shift-work disorders is that the work hours force the worker to adopt temporal patterns of biologic and social functioning that are at odds with those of the day-oriented environment. In particular, night work presents a problem. This chapter covers some of the major health problems in shift work, their mechanisms, and possible countermeasures.

T. Åkerstedt: Institutet För Psykosocial Medicin (IPM) and Department of Stress Research, Karolinska Institute, Stockholm.

A. Knutsson: Department of Occupational and Environmental Medicine, Umeå University Hospital, Umeå, Sweden.

SHIFT WORK

Shift work is an imprecise concept, although it usually refers to a work-hour system in which a relay of employees extends the period of production beyond the conventional daytime third of the 24-hour cycle. There are four major types of work hours: day work, permanently displaced work hours, rotating shift work, and roster work.

Day work involves work periods between approximately 7 a.m. and 7 p.m. *Permanently displaced* work hours require the person to work either a morning shift (approximately 6 a.m. to 2 p.m.), an afternoon shift (approximately 2 p.m. to 10 p.m.), or a night shift (approximately 10 p.m. to 6 a.m.). *Rotating shift work* involves alternation between two or three shifts. Two-shift work usually involves morning and afternoon shifts, whereas three-shift work also includes the night shift. Three-shift work is often subdivided according to the number of teams used to cover the 24 hours of the work cycle—usually three to six teams, depending on the speed of rotation (number of consecutive shifts of the same type).

Roster work is similar to rotating shift work but may be less regular, more flexible, and less geared to specific teams. It is used in service-oriented occupations, such as transport, health care, and law enforcement. In most industrialized countries, approximately one third of the population has some form of "non–day work" (shift work) (1). Approximately 5% to 10% have shift work that includes night work.

CIRCADIAN RHYTHMICITY AND SLEEP

A short introduction to circadian rhythms and sleep is useful because these concepts are necessary for an understanding of shift work problems. Almost all physiological and psychological functions are tied to a 24-hour *circadian* (from Latin, *circa dies*, "about one day") rhythm (2). Figure 22-1 illustrates the circadian rhythm of rectal temperature and

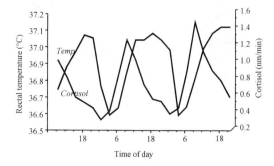

FIG. 22-1. Rectal temperature and urinary cortisol level during 64 hours of continuous activity (N = 12 subjects).

cortisol during 64 hours of continuous activity with intake of food and drink every 2 hours. Even though the basic circadian pattern is the same for most functions, the timing differs greatly. Thus, the hormone cortisol has its peak early in the morning, body temperature in the late afternoon, and the pineal hormone melatonin at night. On the whole, however, most rhythms peak during the daytime.

If a human being is deprived of normal time cues—so-called *synchronizers* or *Zeitgebers*—such as light, he or she will start to live on a 25.2-hour day/night schedule (3). The clock structures in the suprachiasmatic nuclei of the hypothalamus (4) receive neural input about environmental light levels from the retina, and this input is used to adjust the timing of the clock. Morning light (when the circadian rhythm is rising) advances the circadian system, whereas light exposure at night (when the rhythm is falling) delays the system (5).

An important aspect of circadian rhythms is their pattern of adjustment to shifted synchronizers (6). Because the spontaneous circadian period is 1 to 2 hours longer than that of the 24-hour astronomical day, it is very easy to go to bed 1 to 2 hours later than normal and rise 1 to 2 hours later than normal. This phase-delays the rhythm by 1 to 2 hours. In contrast, accomplishing a phase advance is much more difficult because this means going to bed 1 hour earlier—reducing the period length to 23 hours, 2 hours less than the spontaneous period.

For the shift worker, however, it appears that full adjustment never occurs to the night shift, not even in permanent night workers, mainly because morning or daytime light exposure advances the rhythm (instead of delaying it). At best, the adjustment is partial, but frequently only marginal (6), although exceptions may be found in groups isolated from the rest of society.

Sleep normally encompasses six stages (including wakefulness = stage 0). Stage 1 is unclear and accounts for only a small part of total sleep. Stage 2, "basic" sleep, usually accounts for half of all sleep. Stages 3 and 4 comprise approximately 20% and are called *slow-wave sleep* (SWS) because they are characterized by low-frequency, high-amplitude electroencephalographic (EEG) activity. Rapid eye movement (REM) sleep accounts for 25% of all sleep and involves dreaming. SWS seems most important for day-to-day functioning, whereas REM sleep seems important over the longer term. The

function of sleep is not completely understood. However, it seems that the normal amount of sleep (6 to 9 hours) is essential for sustaining waking activity in the short term, and life itself in the long term (7). The minimum amount of sleep necessary for reasonable daytime functioning in the long term seems to be approximately 6.5 hours. In the short term, day-to-day sleep reductions from 8 to 6 hours have only marginal effects. Reductions of 3 hours show effects on behavior, and total sleep loss for one night results in very clear reductions in performance capacity. Three nights of sleep loss result in almost complete inability to carry out normal tasks that involve perception, thinking, and decision making.

SHIFT WORK EFFECTS ON HEALTH AND WELL-BEING

Sleep

The dominant health problem reported by shift workers is disturbed sleep and wake-

(Drawing by Nick Thorkelson.)

fulness. At least three-fourths of shift workers are affected (8). EEG studies of rotating shift workers and similar groups have shown that day sleep is 1 to 4 hours shorter than night sleep. The sleep loss is primarily taken out of stage 2 sleep and REM sleep. Stages 3 and 4 ("deep" sleep) do not seem to be affected. Furthermore, the time taken to fall asleep (sleep latency) is usually shorter. Also, night sleep before a morning shift is reduced, but the termination is through artificial means and the awakening usually difficult and unpleasant. The level of sleep disturbances in shift workers is comparable to that seen in insomniacs. Interestingly, day sleep does not seem to improve much during a series of night shifts, although permanent night workers sleep slightly longer than rotating workers on the night shift. It is not clear if the sleep disturbances in shift work may develop into chronic sleep disturbances.

The main reason for short daytime sleep is the influence exerted by the circadian rhythm. When metabolism (rectal temperature) starts to increase from its trough at approximately 5 a.m., sleep becomes increasingly difficult, reaching a peak in the late afternoon (9). However, homeostatic influences also affect sleep, such that sleep length increases with preceding sleep loss. Thus, the time of sleep termination depends on the balance between circadian and homeostatic influences.

exhibit sleep incidents, most workers seem unaware of them. This suggests an inability to judge one's true level of sleepiness.

As may be expected, sleepiness on the night shift is reflected in performance (10). A classic study (11) showed that errors in meter readings over a period of 20 years in a gas works had a pronounced peak on the night shift (Fig. 22-2). There was also a secondary peak during the afternoon. Other studies demonstrated that telephone operators connected calls considerably more slowly at night, and that train engineers failed to operate their alerting safety device more often at night. Most other studies of performance have used laboratory-type tests and demonstrated, for example, reduced reaction time or poorer mental arithmetic on the night shift. Performance may be reduced to levels comparable with those seen in connection with considerable alcohol consumption (12).

If sleepiness is severe enough, interaction with the environment ceases. If this coincides with a critical need for action, an injury may occur. Such potential performance lapses due to night-work sleepiness were seen in several of the train drivers discussed earlier. The transport area is, in fact, where most of the available accident data on night-shift sleepiness have been obtained (10). The National Transportation Safety Board ranks fatigue as one of the major causes of heavy-vehicle accidents (13).

Alertness, Performance, and Safety

Night-oriented shift workers complain as much of fatigue and sleepiness as they do about disturbed sleep (8). The sleepiness is particularly severe on the night shift, hardly appears at all on the afternoon shift, and is intermediate on the morning shift. The maximum is reached toward the early morning (5 to 7 a.m.). Frequent incidents of falling asleep occur during the night shift, and this has also been documented through ambulatory EEG recordings in process operators, truck drivers, train drivers, pilots, and the like. Remarkably, even though one-fourth

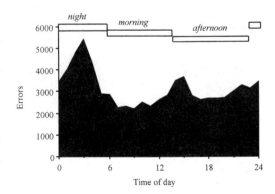

FIG. 22-2. Meter reading errors in three-shift workers.

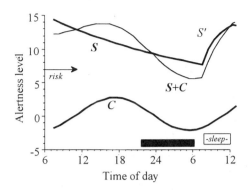

FIG. 22-3. Sleepiness model. Shaded area indicates the first night shift.

Very few relevant data are available from conventional industrial operations, but accidents in the automotive industry may exhibit night-shift effects (14). The Association of Professional Sleep Societies Committee on Catastrophes, Sleep and Public Policy (32) reports that the nuclear plant accidents at Chernobyl and Three Mile Island were due to fatigue-related errors during night work, as were the incidents at the David Beese reactor in Ohio and the Rancho Seco reactor in California, and the *Challenger* space shuttle disaster.

There are two obvious major sources of night-shift sleepiness, circadian rhythmicity and sleep loss. Their effects may be difficult to separate in field studies, but they are clearly discernible in laboratory sleep deprivation studies. Alertness falls rapidly after awakening, but gradually levels out as wakefulness is extended. The circadian influence appears as a sine-wave–shaped superimposition on this exponential decrease in alertness. These two components may be easily separated (Fig. 22-3). Space does not permit a discussion of the derivation of these functions, but the reader is referred to Folkard and Åkerstedt (16), in which the "three-process model" of alertness regulation is described. *Process C* represents sleepiness due to circadian influences and has a sinusoidal form. *Process S* is an exponential function of the time since awakening, which, at sleep onset, is reversed and called *S'* and recovers exponentially, leveling off toward an upper

asymptote. The estimated alertness (or sleepiness) is then expressed as the arithmetic sum (S+C) of the functions. Note that the figure illustrates the extension of wakefulness that occurs with the first night shift. Much of the night shift is worked below the critical level because it occurs at the circadian trough of alertness and after a long period of waking.

Other Effects on Health and Well-Being

Gastrointestinal complaints are more common among night-shift workers than among day workers. In a review of a number of reports covering 34,047 people with day work or shift work, ulcers occurred in 0.3% to 0.7% of day workers, in 5% of people with morning and afternoon shifts, in 2.5% to 15% of people with rotating shift systems with night shifts, and in 10% to 30% in former shift workers (17). Other gastrointestinal disorders, including gastritis, duodenitis, and digestive dysfunction, are more common in shift workers than in day workers (18).

The pathophysiologic mechanism underlying gastrointestinal disease in shift workers is unclear, but one possible explanation is that intestinal enzymes and intestinal mobility are not synchronized with the sleep/wake pattern. Intestinal enzymes are secreted with circadian rhythmicity, and shift workers' intake of food is irregular compared with intestinal function (19). A high nightly intake of food may be related to increased lipid levels (20), and eating at the circadian low point may be associated with altered metabolic responses (21). Another contributing factor to gastrointestinal diseases might be the association between shift work and smoking. A number of studies have reported that smoking is more common among shift workers (17,22).

Studies of alcohol consumption comparing day workers and shift workers have produced conflicting results (19,23,24), probably because of local cultural habits. One study, which used γ-glutamyltransferase as marker of alcohol intake, did not indicate that shift workers had a higher intake of alcohol than day workers (23).

A number of studies have reported a higher incidence of cardiovascular disease, especially coronary artery disease, in male shift workers than in men who work days [for reviews, see Kristensen (25) and Bøggild and Knutsson (26)]. In a study of 504 paper mill workers followed for 15 years, a dose–response relationship was found between years of shift work and incidence of coronary artery disease (27). A study of 79,000 female nurses in the United States gave similar results (28). As with gastrointestinal disease, a high prevalence of smoking among shift workers might contribute to the increased risk of coronary artery disease, but smoking alone cannot explain the observed elevated risk (29).

Most mortality studies concerned with occupational cohorts reveal standardized mortality ratios lower than 100, implying a healthy worker effect (30). Sickness absence is often used as a measure of occupational health risks. However, sickness leave is influenced by many unrelated factors and cannot be considered a reliable measure of true morbidity. Studies on sickness absence in day and shift workers have revealed conflicting results, and there is no evidence that shift workers have more sickness absence than day workers (30).

Only a few studies have addressed the issue of pregnancy outcome in shift workers. In one study of laboratory employees, shift work during pregnancy was related to a significantly increased risk of miscarriage (RR = 3.2) (31). Another study of hospital employees also demonstrated an increased risk of miscarriage (RR = 1.44, 95% confidence interval = 0.83 to 2.51) (32). Lower birth weight of infants of mothers who worked irregular hours has been reported (32,33). No teratogenic risk associated with shift work has been reported (33).

One of the major effects of shift work is the interference of work hours with various social activities (34). Thus, direct time conflict reduces the amount of time available to spend with family and friends or in recreation or voluntary activities. In addition, the alternating pattern of work hours and their anticipated interference may make participation in regular activities less worthwhile, resulting in passivity. Friends might also find the availability schedule of a shift worker too complicated and therefore refrain from contacts. The result is often social isolation and a reduced capacity to fulfill the various social roles expected by society. Freedom from work during daytime, however, carries certain advantages. The variable work hours also provide a method of child care for working couples.

FACTORS AFFECTING ADJUSTMENT

Which aspects of shift work are the most problematic? Which are most conducive to well-being? This section discusses the characteristics of shift systems and of individuals, and how possible guidelines for countermeasures may be developed.

Shift System Characteristics

The major problem of shift work is the night shift. As long as a night shift is involved in the work schedule, the problems of adjustment will remain. In comparison, other shift system characteristics are of minor importance. In contrast to the night shift, the problems of the morning shift are frequently overlooked. In fact, the amount of sleep loss before a shift that starts at 6 a.m. is often equal to the loss after a night shift (35). In addition, alertness is reduced.

Aside from the night shift *per se*, an important shift system characteristic is the number of night shifts in a row. Most studies indicate that the circadian system and sleep do not adjust (improve) much across a series of night shifts (36). Even in permanent night workers, no improvement is seen. Thus, a long series of night shifts (more than four) might be expected to be particularly taxing. One study found that rated alertness and general well-being in three-shift workers improved when a 2- to 3-day rotation was substituted for the usual 7-day rotation. On the whole, the advantage with rapid rotation is that the taxing night shift is not permitted to exert its influence for more than a limited time (37). On the other hand, if it is of major importance that performance capacity remain high during the night, it seems that a solution with permanent night shifts is preferable, in combination with other teams that work a two-shift system (with only morning and afternoon shifts) (38,39). The issue of speed of rotation is still debated, however, and there are arguments for the advantages of permanent night work.

As important as night work in causing difficulties is, in our opinion, the practice of "quick changes"—that is, reduced time between shifts, often only 8 hours. Obviously, if sleep is reduced between shifts, fatigue is increased in the second shift (40). The reason for the quick changes is usually that people like to compress the work week to obtain longer periods of time off.

With respect to the duration of shifts, researchers and practitioners believe that the prevalence of extended (to 10 to 12 hours) work shifts is increasing. Long shifts seem popular because they permit long sequences of free time and reduced commuting. Still, most would be tempted to assume that problems would increase with increasing duration, and there is scientific support for this notion (41). Similarly, a large number of consecutive work days gives rise to fatigue and overtime, no opportunity to take unscheduled rests, and paced jobs. Having a second job may exacerbate the effects of long shifts or lack of recovery days.

Another debated aspect of the shift schedule is its direction of rotation (37). Because the free-run period (the "day/night" length without the influence of time cues such as light, watches, and the like) of the human sleep/wake cycle averages 25 hours (instead of the normal "entrained" 24-hour day), and because it can be entrained by environmental time cues only to within ±2 hours, phase delays are easier to accomplish than phase advances. For the rotating shift worker, this implies that schedules that delay (i.e., rotating clockwise: morning-afternoon-night) are

preferable to schedules that rotate counterclockwise. There have been, however, very few practical tests of this theory, particularly in relation to sleepiness.

Individual Differences and Strategies

There have been many attempts to identify people who are more or less tolerant to shift work, but very few clear results have been obtained (42). However, health problems in shift workers usually increase with age and with increasing exposure to shift work. Being a "morning-type" person, as opposed to an "evening-type" person, is associated with poorer adjustment to shift work (43). Similarly, rigidity of sleep patterns is associated with difficulties in shift work (43). Gender is not in itself necessarily related to shift-work tolerance (42), although the extra burden of housework may put women at a disadvantage.

Good physical condition of the worker may facilitate shift work. One study had three-shift workers improve their physical fitness through a training program. This greatly reduced rated overall fatigue. A number of diseases have been considered incompatible with shift work, such as diabetes and peptic ulcer disease (44). There is, however, very little hard evidence that such diseases are exacerbated with shift work.

Preventive Measures

The preceding discussion suggests a number of preventive measures with respect to the *organization* of shift work. These are summarized in Table 22-1 at three levels of importance.

The most important *individual* preventive measure is good sleep hygiene: sleeping in a dark, cool, sound-insulated bedroom; using ear plugs; informing family and friends about one's sleep schedule; shutting off the phone during sleep hours; and hanging a "do not disturb" sign on the door.

Another important countermeasure is strategic sleeping. For night-shift work, the sleep period should be between 2 and 9 p.m.

TABLE 22-1. *Ways to Improve the Organization of Work-Shift Schedules*

Primary importance
1. Avoid night work (and morning work if possible).
2. Avoid quick changes.
3. Maintain daily rest at least at 11 h.
4. Avoid double shifts or other greatly extended work shifts.
5. Avoid very early morning shifts (starting before 6 a.m.).
6. Intersperse rest days during the shift cycle.

Clear improvements
7. Schedule naps during the night shift.
8. Provide long sequences of days off and few weekends with work.
9. Avoid having a morning shift immediately after a night or evening shift.

Probable improvements
10. Avoid long (more than three shifts) sequences of night or morning shifts (rotate rapidly).
11. Introduce permanent night work as an alternative under certain conditions.
12. Plan night shifts at the end of the shift cycle.
13. Give shift workers older than 45 years of age the right to transfer to day work.
14. Rotate shifts clockwise.

This would mean starting the night shift fairly refreshed from sleep, with only the circadian trough to combat. Sleep at the suggested hours may not be socially feasible, so the next-best alternative is to have a moderate morning sleep and then to add a 2-hour nap in the evening. This "split sleep" provides a solid ground of alertness for the subsequent night shift.

Very little is known about optimal food intake strategies in connection with shift work. Common sense suggests, however, that the worker should avoid intake of major meals during the night shift. Sleeping pills may improve daytime sleep but should be avoided in the long run. New methods for phase shifting the circadian system include bright-light exposure or intake of the pineal hormone melatonin. These methods, however, have not yet been proven effective.

REFERENCES

1. Maurice M. Shift work. Geneva: International Labor Office, 1975.
2. Moore-Ede MC, Sulzman FM, Fuller CA. The clocks that time us. Cambridge: MA: Harvard University Press, 1982.
3. Wever RA. The circadian system of man: results of experiments under temporal isolation. New York: Springer-Verlag, 1979.
4. Klein DC, Moorre RY, Reppert SM. Suprachiasmatic nucleus: the mind's clock. New York: Oxford University Press, 1991.
5. Czeisler CA, Johnson MP, Duffy JF, Brown EN, Ronda JM, Kronauer RE. Exposure to bright light and darkness to treat physiologic maladaptation to night work. N Engl J Med 1990;322:1253–1259.
6. Åkerstedt T. Adjustment of physiological circadian rhythms and the sleep-wake cycle to shift work. In: Monk TH, Folkard S, eds. Hours of work. Chichester, United Kingdom: John Wiley & Sons, 1985: 185–198.
7. Horne J. Why we sleep: the functions of sleep in humans and other mammals. Oxford: Oxford University Press, 1988.
8. Åkerstedt T. Shift work and disturbed sleep/wakefulness. Sleep Medicine Reviews 1998;2:117–128.
9. Czeisler CA, Weitzman ED, Moore-Ede MC, Zimmerman JC, Knauer RS. Human sleep: its duration and organization depend on its circadian phase. Science 1980;210:1264–1267.
10. Folkard S. Black times: temporal determinants of transport safety. Accid Anal Prev 1997;29:417–430.
11. Bjerner B, Holm Å, Swensson Å. Diurnal variation of mental performance: a study of three-shift workers. British Journal of Industrial Medicine 1955; 12:103–110.
12. Dawson D, Reid K. Fatigue, alcohol and performance impairment. Nature 1997;388:235–235.
13. National Transportation Safety Board. Factors that affect fatigue in heavy truck accidents. National Transportation Safety Board Safety Study. NTSB/SS-95/01. Washington, DC: National Transportation Safety Board, 1995.
14. Smith L, Folkard S, Poole CJM. Increased injuries on night shift. Lancet 1994;344:1137–1139.
15. Mitler MM, Carskadon MA, Czeisler CA, Dement WC, Dinges DF, Graeber RC. Catastrophes, sleep and public policy: consensus report. Sleep 1988; 11:100–109.
16. Folkard S, Åkerstedt T. A three process model of the regulation of alertness and sleepiness. In: Ogilvie R, Broughton R, eds. Sleep, arousal and performance: problems and promises. Boston: Birkhäuser, 1991:11–26.
17. Angersbach D, Knauth P, Loskant H, Karvonen MJ, Undeutsch K, Rutenfranz J. A retrospective cohort study comparing complaints and disease in day and shift workers. Int Arch Occup Environ Health 1980;45:127–140.
18. Koller M. Health risks related to shift work. Int Arch Occup Environ Health 1983;53:59–75.
19. Smith MJ, Colligan MJ, Tasto DL. Health and safety consequences of shift work in the food-processing industry. Ergonomics 1982;25:133–144.
20. Lennernäs M, Åkerstedt T, Hambraeus L. Nocturnal eating and serum cholesterol of three-shift workers. Scand J Work Environ Health 1994;20:401–406.
21. Hampton SM, Morgan LM, Lawrence N, et al. Postprandial hormone and metabolic responses in simulated shift work. J Endocrinol 1996;151:259–267.
22. Knutsson A, Åkerstedt T, Jonsson B. Prevalence of risk factors for coronary artery disease among

day and shift workers. Scand J Work Environ Health 1988;14:317–321.

23. Knutsson A. Relationships between serum triglycerides and γ-glutamyltransferase among shift and day workers. J Intern Med 1989;226:337–339.

24. Romon M, Nuttens MC, Fievet C. Increased triglyceride levels in shift workers. Am J Med 1992; 93:259–262.

25. Kristensen TS. Cardiovascular diseases and the work environment. Scand J Work Environ Health 1989;15:165–179.

26. Bøggild H, Knutsson A. Shift work, risk factors and cardiovascular disease. Scand J Work Environ Health (*in press*).

27. Knutsson A, Åkerstedt T, Jonsson BG, Orth-Gomér K. Increased risk of ischemic heart disease in shift workers. Lancet 1986;2:86–92.

28. Kawachi I, Colditz GA, Stampfer MJ, et al. Prospective study of shift work and risk of coronary heart disease in women. Circulation 1995;92:1–5.

29. Knutsson A. Shift work and coronary heart disease. Scand J Soc Med 1989;5(Suppl 44):1–36.

30. Harrington JM. Shift work and health: a critical review of the literature. London: Her Majesty's Stationery Office, 1978.

31. Axelsson G, Lutz C, Rylander R. Exposure to solvents and outcome of pregnancy in university laboratory employees. British Journal of Industrial Medicine 1984;41:305–312.

32. Axelsson G, Rylander R. Outcome of pregnancy in relation to irregular and inconvenient work schedules. British Journal of Industrial Medicine 1989; 46:306–312.

33. Nurminen T. Shift work, fetal development and course of pregnancy. Scand J Work Environ Health 1989;15:395–403.

34. Knauth P, Costa G. Psychosocial effects. In: Colquhoun WP, Costa G, Folkard S, Knauth P, eds. Shiftwork: problems and solutions. Frankfurt am Main: Peter Lang, 1996:89–112.

35. Kecklund G. Sleep and alertness: effects of shift work, early rising, and the sleep environment. Stress research reports no. 252. Doctoral dissertation. The National Swedish Institute for Psychosocial Factors and Health, 1996.

36. Åkerstedt T. Shifted sleep hours. Ann Clin Res 1985;17:273–279.

37. Knauth P. Speed and direction of shift rotation. J Sleep Res 1995;4[Suppl 2]:41–46.

38. Folkard S. Is there a "best compromise" shift system? Ergonomics 1992;35:1453–1463.

39. Wilkinson RT. How fast should the night shift rotate? Ergonomics 1992;35:1425–1446.

40. Kecklund G, Åkerstedt T. Effects of timing of shifts on sleepiness and sleep duration. J Sleep Res 1995;4[Suppl 2]:47–50.

41. Rosa R. Extended workshifts and excessive fatigue. J Sleep Res 1995;4[Suppl 2]:51–56.

42. Härmä M. Sleepiness and shiftwork: individual differences. J Sleep Res 1995;4[Suppl 2]:57–61.

43. Folkard S, Monk TH, Lobban MC. Towards a predictive test of adjustment to shift work. Ergonomics 1979;22:79–91.

44. Costa G. Special health measures for night and shift workers. In: Colquhoun WP, Costa G, Folkard S, Knauth P, eds. Shiftwork: problems and solutions. Frankfurt am Main: Peter Lang, 1996:143–154.

BIBLIOGRAPHY

Åkerstedt T. Shift work and disturbed sleep/wakefulness. Sleep Medicine Reviews 1998; 2: 117–128.
A good review article.

Colquhoun WP, Costa G, Folkard S, and Knauth P, eds. Shift work: Problems and solutions. Peter Lang, Frankfurt am Main, 1996.
A practical, useful publication.

Folkard S, Monk TH. The Hours of Work. Chichester and New York, Wiley, 1985.
This book is a carefully selected series of articles designed to address how to best design work hours to meet human needs as well as production goals.

Moore-Ede M. The Twenty-Four Hour Society. Reading, MA: Addison-Wesley, 1993.
This book summarizes the scientific findings on sleep/wake cycles, alertness, and fatigue, illustrating their effects and discussing ways in which one can protect against them.

23

Indoor Air Quality and Associated Disorders

Mark R. Cullen and Kathleen Kreiss

The focus of occupational medicine has been transformed in many ways by the increasing proportion of the workforce employed in offices and other kinds of public facilities. Once considered safe by crude comparison with industrial settings such as construction, mining, and agriculture, experience has proven that these indoor environments are not free of significant health hazards. Moreover, the workers engaged in these sectors are neither experienced with environmental risks, nor well prepared in general to think about hazards of work, as their industrial counterparts were even long before the modern regulatory era. Finally, because almost all previous attention has focused on the kinds of conditions and hazards that arise in more traditionally dangerous settings, the regulatory framework has not evolved forms of controls that ensure, at least in law, that work will be safe.

This chapter is divided into two sections. The first deals with the spectrum of problems that occur indoors in nonindustrial buildings, focusing on the common features that implicated facilities have. The reader is also referred to Chapter 43 on the health care setting, a specialized and uniquely hazardous nonindustrial setting.

M. R. Cullen: Occupational Health Program, Yale University School of Medicine, New Haven, Connecticut 06510.

K. Kreiss: Division of Respiratory Disease Studies, National Institute for Occupational Safety and Health, Morgantown, West Virginia 26505.

The second section of this chapter deals with the spectrum of clinical complaints related to low-dose (relative to doses that occur in industry) chemical exposures, which have received increasing attention. Although these problems of chemical sensitivity most often occur in association with indoor nonindustrial environments, they may also be seen in a range of other work settings as well as in the nonwork environment. Their distinguishing feature is the occurrence of symptoms or other clinical problems at levels that are far below those at which knowledge of toxicology would predict effects, and typically far below accepted standards in industry for human exposures (see Chapter 15). These somewhat vexing problems have challenged many of the cherished paradigms of occupational health about what is safe and what is not, and form a special challenge for the occupational specialist, as well as primary care providers whose patients may complain about chemicals at levels deemed "safe."

BUILDING-RELATED CONDITIONS

Sick Building Syndrome

Since the 1970s, office workers worldwide have frequently complained of mucous membrane irritation, fatigue, and headache when working in specific buildings, with improvement within minutes to an hour of leaving the building. This constellation of symptoms, with tight temporal association to building occupancy, is called *sick building syndrome.* It is the most frequent of the building-associated health complaints in industrialized countries, which also include diseases caused by infection, hypersensitivity, and specific toxins. Researchers have estimated that as many as 30% of office workers report symptoms attributed to poor air quality, and workers in buildings not known to have indoor air-quality problems have many complaints attributed to the indoor work environment.

Despite the impacts on productivity and employee morale when many of a building's workers have sick building syndrome, little progress has been made in understanding the causes of this syndrome. Early investigations of this phenomenon sometimes concluded that symptoms were caused by mass psychogenic illness because no specific contaminants were measured in concentrations that could account for symptoms. However, the endemic nature of complaints in specific buildings and the consistency of complaints from workers in sealed buildings across the world did not satisfy diagnostic criteria for mass psychogenic illness. Fortunately, such attribution to psychological cause is no longer common or acceptable, although work stress is associated with reporting of symptoms among occupants of specific buildings. Occupants of buildings with high levels of complaints are often angry and fearful because they may have encountered resistance of managers to investigation, inconclusive results, or ineffectual remediation for a syndrome whose scientific causes remain elusive.

The recognition of building-related complaints by public health authorities in the United States followed an energy crisis in which ventilation standards were lowered to 5 cubic feet of outdoor air per person per minute. This observation led to the hypothesis that the new building-related symptoms were attributable to lower rates of ventilation in relation to indoor contaminant sources. Some evidence exists, both in cross-sectional and experimental studies, that ventilation rates are related to sick building syndrome prevalence, particularly for ventilation rates below 30 cubic feet per person per minute. Indoor air-quality consultants commonly measure carbon dioxide levels in buildings with high complaint rates. However, human occupants, who are the source of increased concentrations of carbon dioxide in indoor air as opposed to outdoor air, are not the likely source of contaminants that would explain sick building syndrome. Carbon dioxide measurements simply reflect ventilation effectiveness in relation to human occupancy, and are not predictive of sick building syndrome.

The most interesting work on causes of sick building syndrome comes from epidemiologic studies of occupants of buildings selected without regard to known indoor air-quality complaints. These cross-sectional studies suggest that certain building features and occupant characteristics are related to sick building syndrome prevalence. The variation in prevalence of building-related complaints among buildings suggests remediable causes. Occupants of buildings with air conditioning have been shown to have higher rates of building-related symptoms than occupants of naturally ventilated buildings or buildings with mechanical ventilation that does not alter air temperature or humidity. This observation and other work suggest that the ventilation system itself may be the source of poor air quality in some buildings. However, measurable parameters of bioaerosols do not yet exist that consistently correlate with symptom rates, although this is a rapidly developing field of investigation. Other environmental correlates of sick building syndrome include carpeting, high occupancy load, and video display terminal use. Personal factors associated with building-related symptoms in many cross-sectional studies include female sex, job stress or dissatisfaction, and allergies.

The health care provider with the challenge of responding to indoor air-quality complaints must proceed without the benefit of complete scientific understanding of what may be a multifactorial syndrome. No single measurement establishes whether air quality is adequate or inadequate, and the acceptability of indoor air quality rests with the occupants, and not a laboratory. In the difficult situation of indoor air-quality complaints, a multidisciplinary approach allows attention to design and maintenance of air-conditioning systems, exclusion of obvious contaminant sources in the occupied space, and reassurance of occupants that sick building syndrome is a self-limited condition. Indoor air-quality investigations customarily assess the ventilation in relation to occupant load by measuring carbon dioxide, suggest remediation of ventilation system maintenance and cleanliness deficiencies, and examine smoking policies. On the multidisciplinary team alongside industrial hygienists and ventilation engineers, health care providers have an important role to exclude the possibility of less common, but more medically serious, building-related diseases that nearly always occur with a background of sick building syndrome complaints among other workers, such as asthma and hypersensitivity pneumonitis.

Building-Related Allergic Disease

A 48-year-old social services eligibility technician began working in the implicated office building in October. She had a history of sinus symptoms and 15 pack-years of cigarette smoking, having been an ex-smoker for 10 years. In January, she began to have insidious onset of dry cough, which in March was diagnosed as asthma. Skin prick tests were negative to aeroallergens, and she was referred to an occupational medicine clinic in August because she noted deterioration during the workday, when she needed to use inhaled bronchodilators, and recovery in the evenings and on weekends, when she did not. Her asthma became much worse when she manipulated dusty records while her desk was being moved. She performed peak-flow measurements with a peak-flow meter, which showed reproducible, striking air-flow limitation shortly after entering the building, with partial recovery during lunch breaks outside the building and full recovery on weekends (Fig. 23-1). Methacholine challenge testing in September and November, before and after a 16-day vacation, found provocative concentra-

tions (PC20) for a 20% decrement in forced expiratory volume in 1 second (FEV_1) of 0.29 and 0.47 mg/mL, respectively (normal PC20 > 15 mg/mL). These results confirmed a diagnosis of asthma and suggested slight improvement in airway hyperreactivity with a short work absence. Although she had notified her employer, her relocation to another building was delayed until late February, after her third course of prednisone treatment. After this relocation, her work-related airflow limitation (documented by peak-flow measurements), her symptoms, and her need for asthma medications all resolved. Her PC20 normalized to above 25 mg/mL 3 months after her relocation.

Nine months later, she was moved back to the original building into a set of offices that shared no ventilation system with the offices that she had previously occupied. Over the next 6 weeks, she experienced increasing symptoms and airflow limitation, once again requiring daily medication, and her PC20 fell to 0.22 mg/mL. She was medically restricted from the implicated building, with resolution of her work-related decrements in peak flows, decrease of her medication requirements, and increase in her PC20 to 5.19 mg/mL over the following 6 weeks. She has had no further difficulty with clinical asthma.

This building was built into an earthen bank, and workers reported musty odors and visible mold growth on the interior wall that abutted the bank. *Aspergillus* species of fungi were detected in the interior air but not in simultaneous measurements of outdoor air, suggesting amplification and dissemination of this bioaerosol indoors. The presumed source of the woman's asthma was fungal bioaerosols associated with moisture coming in from the earthen bank.

Building-related asthma is infrequently recognized by physicians, although it can lead to chronic irreversible illness, unlike sick building syndrome. Early recognition and removal from the building, as in this woman's case, can result in cure of asthma. Permanent asthma can result when recognition of occupational etiology is delayed and asthma becomes severe before the patient leaves the implicated exposure. Such sentinel cases of asthma imply risk for other workers. In this case, public health investigation after

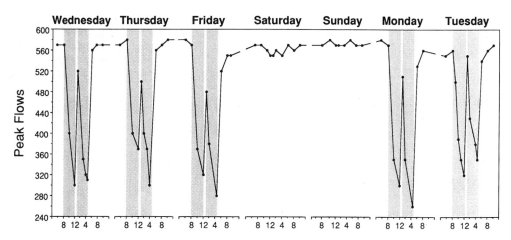

FIG. 23-1. Peak-expiratory flow measurements by hour and day in a case of office building-related asthma. Stippled columns indicate time at work.

two sentinel cases showed that coworkers had nearly five times the prevalence of physician-diagnosed asthma with onset or exacerbation since building occupancy compared with workers in another social service agency (1).

Building-related asthma occurs in water-damaged buildings and in relation to microbially contaminated humidifiers or biocides used in them. The biologic aerosols containing mold spores and perhaps bacteria are thought to be the sensitizing agents. Characterization of bioaerosols is difficult because few laboratories have expertise in identifying saprophytic fungi, in contrast to fungi that cause human infection. Despite the difficulty in characterizing the exposure, the history and peak-flow measurements can be valuable in documenting the occupational nature of building-related asthma. Cases of building-related asthma may occur along with cases of hypersensitivity pneumonitis in water-damaged buildings.

> A 46-year-old pediatrician had been followed by an allergist for 10 years for upper respiratory and chest complaints after moving into an office suite. At first, he complained of sinus drainage and a sore feeling in his nose and throat. Over the years, he had acquired achiness in his chest associated with fever, productive cough, chest tightness, wheezing, fatigue to exhaustion,

and shortness of breath on exertion. His forced vital capacity (FVC) fell within 3 years of building occupancy, consistent with a restrictive pattern. He had been treated with nasal cromolyn, inhaled steroids, bronchodilators, theophylline, antibiotics, and intermittent oral corticosteroids, without receiving a diagnosis. A year before his referral to an occupational medicine specialist, he had noted exacerbation in his chest symptoms and wheezing when he returned to his office suite after a week away from work. He then began to suspect an office-related cause to his symptoms, with increased cough, chest tightness, and achiness when he entered his suite, and resolution over hours after leaving and improvement on weekends. He noted a musty smell and fungal discoloration of wall board in the suite bathroom, which resulted from leaking pipes.

> On referral, he was found to have basilar rales, bronchial hyperreactivity on histamine challenge test, and reduced exercise tolerance with excessive respiratory rate at rest and excessive minute ventilation for oxygen consumption. The chest radiograph was normal, but a high-resolution computed axial tomography scan showed fine centrilobular nodules. Bronchoalveolar lavage showed a lymphocytic alveolitis compatible with hypersensitivity pneumonitis, and a transbronchial lung biopsy showed a mild, patchy lymphocytic interstitial pneumonitis. His symptoms resolved with prednisone and removal from the office suite.

> However, 2 months later, chest aching,

exertional shortness of breath, profound fatigue, and chilly feelings recurred within 45 minutes of using a musty restaurant bathroom that had been recurrently water damaged from roof leaks. He had a prolonged recovery time, requiring systemic steroids for 7 months. A year after this acute exacerbation, he again had a recurrence of chest symptoms, within hours of handling medical records from his previous office suite that had become wet when a hot water heater broke in his basement. He again required months of prednisone use and did not fully recover his health until a year later.

This case of building-related hypersensitivity pneumonitis illustrates the typical medical delay in suspecting and diagnosing a building-related etiology for symptoms. Few physicians are aware that office settings can be associated with diseases related to organic antigens. In contrast to building-related asthma, however, there are many published case reports and epidemic investigations of hypersensitivity pneumonitis and humidifier fever (2). Typically, people with hypersensitivity lung diseases may not be able to reoccupy a building in which they were sensitized to biologic aerosols from contaminated humidifiers, ventilation systems, or water-damaged materials on which fungal growth occurred. Even after remediation of the conditions that led to sensitization and disease, low levels of exposure can trigger recurrent symptoms. Because hypersensitivity pneumonitis can lead to irreversible lung fibrosis after recurrent acute episodes or prolonged exposure, early recognition and restricting affected people from the implicated building is the best means of preventing progression. Remediation can prevent cases in coworkers who are not yet sensitized. Occupational health practitioners can encourage specialists to proceed with diagnostic tests before development of classic late-stage abnormalities, such as those shown on chest radiographs. The history of this pediatrician suggests that he was sensitized to an antigen that was not unique to his water-damaged office setting.

Cases of hypersensitivity pneumonitis are often accompanied by systemic symptoms of myalgias, fever, and profound fatigue. These symptoms are not usually present in asthma, although both diseases commonly share chest symptoms, such as cough, chest tightness, and wheezing. In contrast to asthma and hypersensitivity pneumonitis, sick building syndrome alone is not accompanied by chest symptoms. When indoor air-quality complaints exist, health care providers should ensure that building-related asthma and hypersensitivity pneumonitis are not occurring, in addition to the more common complaints of mucous membrane irritation, headache, and fatigue. The occurrence of building-related chest disease dictates evaluation for sources of fungal and bacterial growth and means of dissemination from areas of water damage or from the ventilation system. Chest disease also requires more aggressive medical restriction from the building to prevent chronic disease.

Many patients report that they have building-related nose and sinus symptoms. It is likely that allergic rhinosinusitis can occur, in a way analogous to the response of airways and lung tissue to building-related antigen exposure. Little research has been done on this common clinical complaint to document its occurrence epidemiologically, to distinguish it from mucous membrane complaints in sick building syndrome, or to link it to exposures in implicated buildings.

Building-Related Infection

In 1976, 182 cases of a mysterious pneumonia occurred among members of the American Legion attending a convention in Philadelphia. After months of laboratory investigation, a newly discovered bacterial organism, *Legionella pneumophila*, was found responsible. We now know that this common environmental organism frequently grows in warm waters of building cooling towers in the absence of vigorous attempts to eradicate it. When contaminated cooling tower mists are entrained in air intakes of large buildings, cases of infection with the organism (legionellosis) can occur. Outbreaks have also been recognized as a result of contaminated industrial water sprays, hospital shower heads, and hot tubs.

In addition to pneumonia, *Legionella* organisms have been associated with another building-related disease called Pontiac fever, which is characterized by fever, chills, headache, and myalgia. This disease was first described in 1968, in a building-related epidemic of 144 cases in a county health department in Michigan. The attack rate was nearly 100%, with an average incubation period of 36 hours.

In addition to infections that cannot be spread to other people, such as *Legionella* pneumonia, building ventilation characteristics are important to the spread of infections that can be passed on to other people, such as viral respiratory infection. Military studies have shown that types of housing with different ventilation characteristics, such as air-conditioned buildings compared with tents or naturally ventilated barracks, are associated with incidence of respiratory symptoms and signs of communicable disease in troops. Other airborne infections, such as tuberculosis, pneumococcal disease, varicella, and measles, may be affected by ventilation rates. A major concern in hospitals, prisons, and shelters is control of tuberculosis, for which ventilation and air disinfection techniques are critical. (See also Chapter 20).

Building-Related Complaints Due to Specific Toxins

Health professionals responding to building-related complaints must also consider specific exposures or toxins as a possible explanation. This is particularly important when complaints differ from those of sick building syndrome or occur in epidemic, rather than endemic, fashion. For example, complaints of headache and nausea dictate consideration of carbon monoxide poisoning, which can occur when air intakes entrain fumes from loading docks, parking garages, or boiler stack emissions. Building-related itching without rash can occur with fibrous glass

exposure, which can result when air-duct lining is entrained in the airstream entering the occupied space. Epidemic coughing, dry throat, and eye irritation can result from detergent residues after the misapplication of carpet cleaning products. In instances of building-related complaints associated with specific exposures, a careful evaluation of types of symptoms, their prevalence, and their temporal onset may point investigators to the cause and to remediation resources.

Environmental tobacco smoke may contribute to the irritant symptoms of sick building syndrome. In many buildings, environmental tobacco smoke is circulated throughout the building as air is recirculated, with modest dilution from outdoor air ventilation. In buildings with indoor air-quality complaints, restriction of smoking to areas with separate exhaust ventilation may result in improved air quality for the remainder of the building.

Asymptomatic Problems

Sometimes building-related exposures do not lead to occupant symptoms, but nonetheless pose a health risk. For example, radon gas emitted from building materials, water, and soil surrounding foundations poses increased risk of cancer. Similarly, asbestos in insulation and some building materials in older buildings poses risks of cancer of the lung and other sites as well as nonmalignant lung disease if it is disturbed during occupant activities or renovation. Occupational health specialists and other health professionals are often called to help communicate risks of such exposures to building occupants or the public during removal of asbestos from older buildings.

MULTIPLE CHEMICAL SENSITIVITIES

Since the 1980s, a new clinical syndrome has been recognized in occupational and environmental health practice characterized by occurrence of multisystem symptoms after exposure to low levels of synthetic chemicals. Unlike any other building-related illness, this disorder recurs in affected people in a diverse array of environmental situations and cannot be readily reversed by attention to any single exposure situation. The following is a representative example of what is now most widely referred to as multiple chemical sensitivities (MCS).

A 46-year-old library worker enjoyed good health until the onset of eye, nose, and throat irritation and recurrent headache associated with a renovation of the library where she worked. She and many coworkers complained primarily of dust and paint fume exposures, which were initially poorly controlled. After several weeks of effort, the employer succeeded in establishing temporary ventilation for the work area and conducting most of the construction activities at night. Almost all of the patient's coworkers improved dramatically after these changes were instituted. She, however, felt no better and began experiencing similar symptoms in her car, at various stores, and whenever she was around anything "scented," especially experiencing these symptoms in the office. She believed she was "reacting" to the small residual levels of construction-related exposures, but temporary transfer to another part of the library brought no relief. New symptoms, including difficulty breathing, muscle and joint aches, and confusion occurred both at work and at home, triggered by an increasing list of offensive odors, irritants and products. Efforts to clean her house of such materials, as well as a trial leave of absence from work (without the benefit of workers' compensation), resulted in only minimal improvement.

On clinical evaluation, the patient appeared well and had no physical findings. Laboratory tests, including workup for respiratory and central nervous system abnormalities, were unrevealing. Consultants in pulmonary medicine, rheumatology, and neurology were unhelpful, as were the various inhalers, nonsteroidal antiinflammatory agents, and migraine therapies. Because of the disparity between complaints and findings, the patient was referred to a psychiatrist who confirmed some depressive features, but could not explain the patient's complaints. A trial of selective serotonin reuptake inhibitor antidepressants was not tolerated by the patient, who discontinued the drug after 3 days.

Finally, frustrated by unsympathetic physicians and her employer, the patient took advice she obtained from the internet and sought an "environmental" physician, who advised total avoidance of all chemical exposures (including her job) and a variety of nontraditional remedies based on results of blood and hair tests, which purported to demonstrate organic chemicals and heavy metal "poisons," as well as immunologic "reactions" to a range of widely found chemicals such as formaldehyde. She remains highly symptomatic.

Although this case occurred in the setting of building-related illness, MCS may develop in any occupational setting and in people who have experienced a single episode or recurring episodes of a chemical injury, such as solvent or pesticide poisoning. Once the problem begins, however, many types of environmental contaminants in air, food, or water may elicit the symptoms at doses well below those that clinically affect others. Although there may not be measurable impairment of specific organs, the complaints are associated with dysfunction and disability. Although MCS is not common, it is prevalent enough to have generated substantial controversy. However, research has not yet elucidated its cause and pathogenesis, nor ways to treat or prevent it.

Multiple Chemical Sensitivities: Definition and Diagnosis

There is no general consensus on a definition for MCS, but certain features are sufficiently characteristic to raise suspicion and differentiate it from other occupational and nonoccupational health problems. Its major features are as follows:

• Symptoms usually occur after an occupational or environmental inhalation or toxic exposure. This precipitating event may be a single episode, such as an exposure to a pesticide spray, or recurrent, as in the case presented previously. Often the initial event or reaction is mild and may merge without clear demarcation into the syndrome that follows.

- Symptoms resembling those associated with the preceding exposure begin to occur after exposures to surprisingly lower levels of various materials, including chemicals, perfumes, and other common work and household products, especially materials that have a pungent odor or are irritating.
- Symptoms appear referable to many organ systems. Central nervous system problems, such as fatigue, confusion, and headache, occur in almost every case.
- Complaints of chronic symptoms, such as fatigue, cognitive difficulties, and gastrointestinal and musculoskeletal disturbances, frequently complicate the temporal relationship between specific exposures and effects. These more persistent symptoms may even predominate over acute reactions to chemicals in some cases.
- Objective impairment of the organs that would explain the pattern or intensity of complaints is typically absent.
- No other diagnosis easily explains the range of responses or symptoms. Although the patient may, in fact, have other physical or emotional ailments, such as allergy or anxiety, MCS should be considered if it better explains the overall clinical picture.

Of course, not every patient meets these criteria precisely. But because the diagnosis of MCS is, in the end, based on subjective information, each point should be carefully considered. Each serves to rule out other clinical disorders that MCS may resemble, such as generalized anxiety disorder, classic sensitization to environmental antigens (e.g., occupational asthma), late sequelae of organ system damage (e.g., reactive airways dysfunction syndrome after a toxic inhalation), or systemic disease (e.g., systemic lupus erythematosus). On the other hand, MCS is not typically diagnosed by exclusion of all other possibilities, and exhaustive testing is not required in most cases.

In practice, diagnostic problems are seen in two clinical situations. Early in the course of the disorder, it is often difficult to distinguish MCS from occupational or environmental health problems that may have preceded it. For example, patients who have experienced symptomatic reactions to pesticide spraying indoors may find that their reactions are persisting even when they avoid direct contact with these chemicals. In this situation, a clinician might assume that significant exposures could still be occurring and may focus entirely on altering the environment further, which usually does not relieve the recurrent symptoms. This is especially troublesome in an office setting, where MCS may develop as a complication of sick building syndrome (see earlier). Although most coworkers improve after steps are taken to improve air quality, the patient who has acquired MCS continues to experience symptoms despite the lower exposures involved.

Later in the course of MCS, diagnostic dilemmas arise because of the chronic aspects that may obscure the patient's intolerance to common odors and chemicals. After many months, the patient with MCS is often depressed, anxious, and frustrated about his or her health. Physical inactivity, often with weight gain, sleep disturbances, and significant social dysfunction, are common. These phenomena demand considerable attention therapeutically and may also make it hard to focus on the patient's perceived strong intolerance to chemicals and odors.

Pathogenesis

The sequence of pathologic events that leads from apparently self-limited episodes of an environmental exposure to the development of MCS in certain people is not known. There are several current theories.

A group of "environmental" physicians, initially called "clinical ecologists," have promulgated the view that MCS is a form of immune dysfunction caused by insidious accumulation of exogenous chemicals over a lifetime. Susceptibility factors in this view may include nutritional deficiencies (e.g., vitamins and antioxidants), the presence of subclinical infections (e.g., candidiasis), or other host factors. In this theory, the precipitating exposure or exposures are important because of their contribution to lifelong "chemical overload."

Another biologically oriented theory is that MCS represents an atypical biologic sequela of chemical injury, such as a new form of neurotoxicity due to solvents or pesticides, or injury to the respiratory tract after an acute inhalational episode. In this theory, MCS is seen as a final common pathway of different primary disease mechanisms.

A more recent concept has focused on the relationship between the mucosa of the upper respiratory tract and the limbic system, especially the close anatomic proximity of the two in the nose. Under this view, relatively small stimulants to the nasal epithelium could result in amplified limbic responses (as occurs, for example, in addicted people to the substances to which they are addicted), explaining the dramatic and sometimes stereotypic responses to low-dose exposures. This theory also may explain the prominent role of stimuli with strong odors, such as perfumes, in triggering responses in many patients.

Many investigators and clinicians with experience have invoked primarily psychological mechanisms to explain MCS, linking it to other anxiety or affective disorders. Some believe that MCS is a variant of posttraumatic stress disorder or a conditioned response to a toxic experience. One group has suggested MCS is a late-life response to early childhood traumas, such as sexual abuse. In these theories, the precipitating illness plays a more symbolic than biologic role in the pathogenesis of MCS. Host susceptibility is obviously very important in these theories, particularly the predisposition to somaticize psychological distress.

Although there is much published literature, few clinical or experimental studies have been presented or published to support strongly any of these views as the single best explanation for MCS. Research has been hampered by variously defined study populations, inappropriately matched control groups, and lack of "blinding" of subjects and investigators. As a result, most available data are descriptive. Perhaps most difficult of all, debate over the etiology of MCS has been heavily dominated by strong dogmas. Because major financial decisions, such as patient benefit entitlements and physician reimbursement, may depend on how MCS is viewed, many physicians have very strong opinions that make the scientific validity of their observations questionable. Treating MCS patients also requires awareness of the possibility that these theories may be well known to patients as well in the internet era, and they may also have very strong views.

Epidemiology

Detailed information about the epidemiology of MCS is not available. Estimates of prevalence in the U.S. population range as high as several percent, but the scientific basis for these estimates is questionable. Best evidence suggests that although many people find chemicals and other odors objectionable, and may even report discomfort when around them, MCS in clinically overt form is uncommon. Although most available data come from case series by various practitioners who treat patients with MCS, some general observations appear recurrently in the reports:

Multiple chemical sensitivities occurs most commonly in midlife, although patients of virtually all ages have been described.

Workers in higher socioeconomic status jobs seem more often affected, whereas economically disadvantaged workers seem underrepresented; this may be an artifact of differential access to occupational and environmental health services, or a diagnostic bias.

Women are more frequently affected than men.

Some host factor or susceptibility is important because mass outbreaks have been uncommon, and only a small fraction of victims of chemical overexposures acquire MCS or anything like it. Although most host factors have not been adequately studied, common atopic allergic disorders do not appear to be an important risk factor for MCS.

Several classes of chemicals have been commonly implicated in the initial presentation of MCS, specifically organic solvents, pesticides, and respiratory irritants. This may be a function of the widespread exposure to these materials. The other commonplace setting in which many cases occur is in the "sick building" situation, with some patients evolving from sick building syndrome into MCS, as in the patient described in the previous case. Although the two illnesses have much in common, their epidemiologic features serve to distinguish them: sick building syndrome usually affects a high proportion of people sharing a common environment, whereas MCS occurs sporadically and without tight temporal association with one environment.

Finally, there is great interest in whether MCS is a new disorder or a new presentation of an old one. Views on this are divided, much as is opinion on the pathogenesis of MCS. Those favoring a biologic role for chemicals argue that MCS is a 20th-century disease with rising incidence related to widespread chemical usage. Those who support psychologic mechanisms see MCS as an old somatoform disorder with a new societal metaphor—the social perception of chemicals as agents of harm.

Natural History

Multiple chemical sensitivities has not yet been studied enough to delineate its clinical course completely, although reports of large series of patients have provided some clues. The general pattern is early progression as the process evolves, followed by less predictable periods of small improvements and exacerbations. These modest changes are often perceived by the patient in relation to environmental factors or treatments, but no scientific basis for such relationships has been established.

Two important observations have been made. First, there is little evidence that MCS is a progressive disorder. Patients do not get worse from year to year in any demonstrable physical way, or have resultant complications, such as organ system failure, unless there is intercurrent illness. MCS is not lethal—perceptions of patients notwithstanding, a basis for a hopeful prognosis and reassurance. Unfortunately, it has been equally clear from clinical series that complete remissions are unlikely, given current treatment (or lack thereof). Although significant improvement may occur, this is usually related to better patient function and sense of wellbeing. The underlying tendency to react to chemical exposures persists, although symptoms may become tolerable enough to allow a normal lifestyle.

Clinical Management

There remains no specific treatment for MCS. Many traditional and nontraditional strategies have been tried, although few have been subjected to the usual scientific standards to document success or failure, such as a blinded clinical trial. Approaches to treatment of the disorder have followed theories of pathogenesis. Those who believe that MCS is caused by biologic consequences of large burdens of exogenous chemicals have focused attention on avoidance of further exposures through the use of "natural" products and the radical alteration of lifestyle. Diagnostic tests of unproved significance, including body fluid assays for trace organic chemicals and antibodies to common chemicals, have been developed as a basis to attempt to "desensitize" patients. Dietary supplements, such as vitamins and antioxidants, have been recommended to improve host resistance to chemical effects, again without evidence of efficacy. A more radical treatment involves elimination of toxins from the body by chelation or accelerated turnover of fat, where lipid-soluble pesticides, solvents, and other organic chemicals may be concentrated.

Those who take to a psychological view of MCS have tried approaches consistent with these theories. Supportive individual or group therapies and behavioral modification techniques have been described, although the efficacy of these therapies remains unproved. These patients tend to be intolerant to pharmacologic agents used to treat affective and anxiety disorders, making treatment plans much more difficult.

Despite limitations of current knowledge, certain treatment principles can be synthesized:

To the extent possible, the search to "get to the bottom" of MCS in an individual patient should be minimized—it is counterproductive to starting support and treatment. Many patients have already had considerable medical evaluation by the time MCS is first recognized, and further evaluation, unless necessary to exclude treatable diseases, is often a distraction.

Whatever the particular beliefs of the clinician, the existing knowledge and uncertainty about MCS should be explained to the patient, including the fact that its cause is unknown.

The patient must be reassured that consideration of psychological complications that commonly arise does not mean that the illness is not real, serious, and worthy of treatment.

The patient may also be reassured that MCS is neither progressive nor fatal, but that complete cures are not likely with current modalities.

Uncertainty about pathogenesis aside, it is most often necessary to modify the patients' work environments to remove them from triggers of symptoms. Although radical avoidance is counterproductive to the goal of enhancing function, regular and severe symptomatic reactions must be limited to allow the patient to begin the supportive care he or she needs in a trusting doctor–patient relationship. Often this requires a job change. Workers' compensation may be appropriate in the perspective of MCS as a complication of a work exposure, which often appears to be the case.

The goal of all therapy must be improvement of function because the underlying problem cannot be changed given current knowledge. Psychological problems, such as adjustment difficulties, anxiety, and depression, must be treated, as should coexistent clinical disorders, such as atopic allergies. Because patients with MCS do not tolerate chemicals in general, nonpharmacologic approaches may be necessary. Most patients need direction, counseling, and reassurance to adjust to life with an illness such as MCS. Whenever possible, patients should be encouraged to increase activities to their premorbid level. Passivity and dependence, common responses to the disorder, should not be reinforced.

Prevention and Control

Primary prevention strategies cannot be developed without knowledge of the pathogenesis of the disorder or the host risk factors that predispose some people to become affected. However, reduction of opportunities in the workplace for the overexposures that seem to lead to MCS in some people, including especially respiratory irritants, solvents, and pesticides, may reduce the occurrence of MCS. Certainly better ventilation in offices and other nonindustrial workplaces would also help, and there are more than enough reasons to recommend such an approach.

Secondary prevention would appear to offer some greater control opportunities, although no specific interventions have been studied. Because psychological factors may play a role in victims of occupational overexposures, careful and early management of people who get exposed is advisable even when the prognosis from the exposure itself is good. Patients seen in clinics or emergency departments immediately after acute exposures should be assessed for their reactions to the events and should probably receive very close follow-up when undue concerns of long-term effects or persistent symptoms are noted. Obviously, efforts should be made for such patients to ensure that preventable recurrences do not occur because this may be an important risk factor for MCS by whatever mechanism is causal.

REFERENCES

1. Hoffman RAE, Wood RC, Kreiss K. Building-related asthma in Denver office workers. Am J Public Health 1993;83:89–93.
2. Kreiss K, Hodgson MJ. Building-associated epidemics. In: Walsh PJ, Dudney CS, Copenhaver ED, eds. Indoor air quality. Boca Raton, FL: CRC Press, 1984:87–106.

BIBLIOGRAPHY

Cone JE, Hodgson MJ, eds. Problem buildings: building-associated illness and the sick building syndrome. Occup Med 1989;4.
A compilation of papers offering diverse opinions regarding aspects of indoor air quality issues. The chapter by Kreiss has a more comprehensive review of the literature pertinent to building-related complaints, with an exhaustive list of references.

Mendell MJ. Nonspecific symptoms in office workers: a review and summary of the epidemiologic literature. Indoor Air 1993;3:227–236.
Describes the conclusions about risk factors for sick building syndrome that can be drawn from the literature through the lenses of methodologic strength of design and consistency of findings among investigations.

Mendell MJ, Smith AH. Consistent pattern of elevated symptoms in air-conditioned office buildings: a reanalysis of epidemiologic studies. Am J Public Health 1990;80:1193–1199.
A useful paper that allows some sense to be made of the seemingly disparate epidemiologic findings of studies looking for building-related and personal risk factors for sick building syndrome.

Seiber WK, Stayner LT, Malkin R, et al. The National Institute for Occupational Safety and Health indoor environmental evaluation experience: part three. Associations between environmental factors and self-reported health conditions. Applied Occupational and Environmental Hygiene 1996;11:1387–1192.
A report of systematic investigations of 2,435 respondents in 80 buildings with indoor air quality complaints, summarizing the environmental risk factors that were associated with symptom prevalence.

Sparks PJ, Daniell W, Black DW, et al. Multiple chemical sensitivity: a clinical perspective. I. Case definition, theories of pathogenesis, and research needs. J Occup Med 1994;36:718–730.

Sparks PJ, Daniell W, Black DW, et al. Multiple chemical sensitivity: a clinical perspective. II. Evaluation, diagnostic testing, treatment, and social considerations. J Occup Med 1994;36:731–737.
These two papers comprise a referred review of MCS.

PART IV

Injuries and Disorders by Organ System

24 Injuries

Dawn N. Castillo, Timothy J. Pizatella, and Nancy Stout

Occupational injuries are caused by acute exposure in the workplace to physical agents such as mechanical energy, electricity, chemicals, and ionizing radiation, or from the sudden lack of essential agents, such as oxygen or heat. Examples of events that can lead to worker injury include motor vehicle crashes, assaults, falls, being caught in parts of machinery, being struck by tools or objects, and submersion. Resultant injuries include fractures, lacerations, abrasions, burns, amputations, poisonings, and damage to internal organs.

Occupational injuries are a serious public health problem. In 1996, more than 6,100 workers died from occupational injuries (1) and more than 6 million workers sustained nonfatal injuries, based on a survey of employers (2). This latter estimate is conservative because it relies on employer reporting and excludes important groups of workers, such as the self-employed, workers on small farms, and government employees. The annual societal cost of occupational injuries in the United States has been conservatively estimated at more than $145 billion (3).

CAUSES OF INJURIES

Although the immediate cause of injury is exposure to energy or deprivation from es-

D. N. Castillo, T. J. Pizatella, and N. Stout: Division of Safety Research, National Institute for Occupational Safety and Health, Morgantown, West Virginia 26505.

sential agents, injury events arise from a complex interaction of factors associated with materials and equipment used in work processes, the work environment, and the worker. These factors include physical hazards in the workplace or setting, hazards and safety features of machinery and tools, the development and implementation of safe work practices, the organization of work, the design of workplaces, the safety culture of the employer, availability and use of personal protective equipment (PPE), demographic characteristics of workers, experience and knowledge of workers, and economic and other social factors.

The case described in Box 24-1 illustrates how the occurrence of occupational injury events can be influenced by a variety of factors and circumstances. Some of the contributory factors are clear, others are surmised. The victim did not have experience doing this type of work, and may not have fully recognized all the fall hazards associated with the job, or the importance that fall protection be used. Social and economic factors may have contributed to the crew being asked and agreeing to do work they had never done before—for example, the need for the income from the job and their desire to be seen as employees who would do what was asked of them. A number of factors may have accounted for the absence of a comprehensive safety and training program, including the employer's perceptions that workers know how to conduct work safely and do

Box 24-1. Laborer Dies After 41-Foot Fall from Roof under Construction

A 22-year-old male laborer worked as part of a four-man construction crew that normally performed floor, truss, and deck work. On the day of the incident, the crew was asked by their supervisor to roof a building because the regular roofing crew had been sent to another job. No one from the crew had ever done roofing work before, nor were they provided with fall protection equipment or any training in the recognition and avoidance of fall hazards. At approximately 7:00 a.m., the crew began installing sheets of plywood on roof trusses of a steep, pitched roof of an apartment building. Each member of the crew was working on a separate area of the roof carrying, laying, and nailing sheets of plywood to the trusses. The employer did not have an on-site safety program specifying hazards routinely faced by workers, such as hazards faced during roofing, or safety practices mandated by the employer to ensure work was done safely. Late in the afternoon, the victim fell 41 feet to the ground below from an unprotected roof edge. Emergency medical services arrived shortly after the incident, and pronounced the victim dead at the scene. The coroner listed the cause of death as blunt trauma injury to the head and chest.

National Institute for Occupational Safety and Health, Division of Safety Research. Fatality assessment and control evaluation (FACE) report 98-17. Morgantown, WV: National Institute for Occupational Safety and Health, Division of Safety Research, 1998.

not need guidance or training, employer and worker perceptions that fall protection decreases productivity, costs to hire safety and health expertise if not available with current staff, and costs associated with providing safety equipment and PPE. The fact that the victim did not speak English would have been a barrier to his taking part in weekly safety training meetings, held in English, by the general contractor for subcontractor workers.

This case illustrates that injury events can arise from a complex array of factors. Not all factors carry the same weight in contributing to injury events. In addition, the responsibilities for a safe work environment and safe work practices are not borne equally by all involved parties. The greatest responsibilities are borne by employers, who are responsible for providing a safe work environment, including the identification of potential safety hazards as well as the implementation of hazard controls and safe work practices and procedures to keep workers safe. Workers are responsible for following established procedures and for reporting safety hazards to employers.

EPIDEMIOLOGY OF INJURIES

Occupational injuries are not random events. They cluster or are associated with specific types of workplaces and jobs, workplace exposures, and worker characteristics. Violence in the workplace is one type of exposure contributing to occupational injuries (see Box 24-2). Because occupational injuries are not random, they can be anticipated and steps can be taken to prevent them.

Epidemiologic data allow those involved in injury prevention efforts to target groups and settings with high numbers or rates of occupational injuries, and to anticipate and take steps to prevent injury in specific workplaces or settings. Epidemiologic data on fatal and nonfatal occupational injuries differ and thus are addressed separately. Both categories of injuries require attention—fatal injuries, because they represent the most severe consequence of occupational injury and are devastating to families, communities, and workplaces; and nonfatal injuries, because of the sheer volume and aggregate costs to workers, families, employers, and society.

Fatal Injuries

The distribution and risks for fatal occupational injury differ by demographic characteristics of workers. Men account for more than 90% of occupational fatalities, and have fatality rates approximately 10 times higher

Box 24-2. Violence in the Workplace

The role of violence as a cause of work-related injury and death had been largely overlooked until the early 1990s, when the National Institute for Occupational Safety and Health (NIOSH) reported findings from its National Traumatic Occupational Fatality Surveillance System (NTOF).

Reviewing death certificate data for the 1980s, NIOSH found that approximately one in eight work-related fatalities in the United States is a homicide. The high prevalence of work-related homicides has continued into the 1990s. Among women, homicide is the largest single cause of death in the workplace, accounting for almost half of these deaths. Victims of occupational homicide are young (primarily 25 to 44 years of age). Although 73% are white, the rate of occupational homicide among African-Americans and other minority workers is more than twice that for whites.

Because of the availability of death certificates, the discovery of the importance of violence was primarily based on the surveillance of occupational fatalities. However, these data probably *underestimate* occupational violence. The problem is clearly more widespread when nonfatal effects of violence, ranging from injuries due to violent acts to sexual harassment on the job, are considered. Approximately 1 million workers are physically attacked at work in the United States each year. Data from the United Kingdom indicate that one in eight health care workers suffers a physical attack each year. Studies in the United States suggest that those responsible for violent events are most commonly patients receiving health care, customers, or strangers, with coworkers, relatives, and intimates accounting for fewer cases. The typical victim of occupational homicide is a regular worker employed in a work setting that allows continued exposure to hazards of crime and violence. Retail trade and service workers are those most frequently killed by violent acts at work; these workers accounted for 55% of occupational homicides in the 1980s and early 1990s. In this context, some of the most hazardous workplaces were taxicabs, liquor stores, gas stations, detective/protective services, and justice/public order establishments; some workers with the most hazardous occupations were taxicab drivers (with a risk of occupational homicide greater than 30 times the average), law enforcement officers, gas station workers, security guards, stock handlers/baggers, store owners/managers, sales clerks, and bartenders.

The circumstances of these fatalities indicate robbery as a primary motive, with some being caused by disgruntled workers and clients. NIOSH has summarized the major risk factors known to increase probability for occupational homicide: (a) exchange of money with the public, (b) working alone or in small numbers (Fig. 24-4), (c) working late-night or early-morning hours, (d) working in high-crime areas, (e) guarding valuable property or possessions, and (f) working in community settings, as taxicab drivers and police do.[1]

NIOSH has recommended preventive measures that can be quickly introduced to reduce the risk of occupational homicides, especially in high-risk establishments and occupations:

- Make high-risk areas visible to more people and install good external lighting.
- To minimize cash on hand, use drop safes, carry small amounts of cash, and post signs stating that limited cash is on hand.
- Install silent alarms and surveillance cameras.
- Increase the number of staff on duty and have police check on workers routinely.
- Provide training in conflict resolution and nonviolent response as well as the importance of avoiding resistance during a robbery.
- Provide bullet-proof barriers or enclosures.
- Close establishments during high-risk hours (late at night and early in the morning).

[1]National Institute for Occupational Safety and Health. Current Intelligence Bulletin 57: violence in the workplace. DHHS (NIOSH) publication no. 96-100. Cincinnati, OH: U.S. Department of Health and Human Services (DHHS), Centers for Disease Control and Prevention (CDC), National Institute for Occupational Safety and Health, 1996.

than those for women. Approximately 80% of occupational fatal injuries are among white workers (including Hispanic workers), 10% among black workers, and 3% among Asian or Pacific Islander workers (1,4). Black workers have slightly higher fatality rates than white workers (4). One analysis suggested that most, but not all, of the difference in fatality rates between black and white men could be attributed to the types of jobs held by each (5). Differences in employment patterns have also been suggested as the reason for the sex disparity. Sixty-eight percent of fatal occupational injuries occur to workers between 25 and 54 years of age, with approximately 10% of the fatalities among workers younger than 25 years of age and 22% of the fatalities among workers 55 years of age and older. Rates of fatal occupational injury begin to increase at approximately 45 years of age, with the highest rates among workers 65 years of age and older (1,4). Decreased ability to survive injuries may account for some of the increased fatality rates among older workers.

Eighty percent of occupational injury deaths are among wage and salary workers; the remainder are among the self-employed, who have fatality rates approximately 2.5 times greater than wage and salary employees. The types of jobs held by the self-employed explain some of this difference (1). For example, higher proportions of the self-employed work in the agriculture and construction industries (two industries with the highest risk of fatal injury) than wage and salary workers.

Transportation-related events account for many occupational injury deaths each year: 42% of the 6,112 occupational injury deaths in the United States in 1996. These events involve motor vehicles and mobile equipment (e.g., tractors and forklifts), occur on and off the highway, and include pedestrians or bystanders as well as operators or drivers. Assaults and violent acts accounted for 19% of fatalities in 1996, with most of these events involving homicides, and some involving suicides and animal attacks. Contact with ob-

jects or equipment accounted for 16% of the fatalities, including being struck by falling objects (e.g., materials falling from cranes and trees falling down while being cut), being caught in running equipment or machinery, and being caught in or crushed by collapsing materials (e.g., in trench cave-ins or collapsing buildings). Falls, mostly to a lower level, accounted for 11% of fatalities. Exposure to harmful substances or environments (e.g., electric current, temperature extremes, hazardous substances, and oxygen deficiency) accounted for 9% of fatalities, with more than half of these being electrocutions. Fires and explosions accounted for 3% of the fatalities (1). There are some variations by demographic characteristics; for example, homicide typically accounts for higher proportions of deaths, and is frequently the leading cause of death for women, minorities, and the self-employed (1,4).

There is tremendous variability in the incidence of occupational injury deaths by industry division (Table 24-1), and among specific industries within industry divisions. The occupational injury fatality rate averaged across all industries in the United States in 1996 was 4.8 per 100,000 workers (1). There are dozens of specific industries with injury rates far in excess of the average for all indus-

TABLE 24-1. *Number and rate of fatal occupational injuries by industry division, 1996*

Industry division	No. of fatalities	Fatality rate[a]
Mining	152	26.8
Agriculture/forestry/fishing	798	22.2
Construction	1,039	13.9
Transportation/public utilities	947	13.1
Wholesale trade	267	5.4
Manufacturing	715	3.5
Retail trade	672	3.1
Services	767	2.2
Finance/insurance/real estate	114	1.5
Total	6,112	4.8

[a] Rate per 100,000 workers.

From Bureau of Labor Statistics. Fatal workplace injuries in 1996: a collection of data and analysis. Report 922. Washington, DC: U.S. Department of Labor, Bureau of Labor Statistics, 1998.

tries—information on these industries can be accessed through routine and summary publications of occupational injury fatality data (1,6). Some of these high-risk industries are in industry divisions with relatively low rates of fatal injury—specifically, manufacturing, wholesale trade, services, and retail trade.

There is also considerable variability in the incidence and patterns of injury death by occupation. Table 24-2 provides information on the incidence and patterns of fatal injury for select detailed occupations. In some occupations, there is a predominant type of injury event, such as aircraft crashes among military occupations; in other occupations, such as those in farming, a variety of events contribute to injury death. Not included in Table 24-2 is information on several specific occupations with fewer numbers of deaths each year, but very high annual fatality rates, such as fishers (178.4 deaths per 100,000 workers) and structural metal workers (85.2 deaths per 100,000 workers) (1,6).

Nonfatal Injuries

Although not as dramatic as for fatal injuries, differences are also seen across demographic categories for nonfatal injuries. Men account for approximately 70% of nonfatal work-related injuries, and, based on data from emergency department visits, have rates from 1.6 to 1.8 times higher than those for women

TABLE 24-2. *Fatality rate and most frequent events leading to occupational injury death for select occupations, 1996*

Occupation	No. of deaths	Rate[a]	Most frequent events
Timber cutting and logging	118	157.3	Contact with objects and equipment (mostly struck by)—78% Transportation—14%
Airplane pilots and navigators	100	87.7	Transportation (all aircraft)—100%
Construction laborers	291	35.7	Transportation—31% Contact with objects/equipment—30% Falls—22% Exposure to harmful substances/environment—13%
Truck drivers	785	26.0	Transportation—79%
Farming occupations	589	24.8	Transportation—43% Contact with objects/equipment—26% Assaults and violent acts—10% Falls—10% Exposure to harmful substances/environment—10%
Laborers, except construction	213	15.9	Transportation—30% Contact with objects/equipment—30% Falls—13% Assaults and violent acts—12% Exposure to harmful substances/environment—11%
Electricians	98	12.8	Exposure to harmful substances/environment (mostly electrocutions)—51% Falls—18% Transportation—12% Contact with objects and equipment—10%
Police and detectives	114	11.9	Assaults and violent acts (mostly homicides)—52% Transportation—42%
Military	123	9.5	Transportation (mostly aircraft)—79%
Sales, supervisors and proprietors	225	5.0	Assaults and violent acts (mostly homicides)—70% Transportation—23%
Cashiers	94	3.3	Assaults and violent acts (mostly homicide)—92%

[a] Rate per 100,000 workers.

From Bureau of Labor Statistics. Fatal workplace injuries in 1996: a collection of data and analysis. Report 922. Washington, DC: U.S. Department of Labor, Bureau of Labor Statistics, 1998.

TABLE 24-3. *Number and rate of nonfatal occupational injuries by industry division, 1996*

Industry division	No. of injuries[a]	Injury rate[b]
Manufacturing	1,952.9	10.6
Construction	483.8	9.9
Transportation/public utilities	514.4	8.7
Agriculture/forestry/fishing	113.0	8.7
Retail trade	1,117.5	6.9
Wholesale trade	412.9	6.6
Services	1,466.8	6.0
Mining	33.3	5.4
Finance/insurance/real estate	144.3	2.4
Total	6,238.9	7.4

[a] Number × 1,000.
[b] Rate per 100 full-time workers.
From Bureau of Labor Statistics. Workplace injuries and illnesses in 1996. Bulletin 97-453. Washington, DC: U.S. Department of Labor, Bureau of Labor Statistics, 1997.

(7,8). Data from the annual survey of employers conducted by the Bureau of Labor Statistics (BLS) suggest that 53% of injuries and illnesses with lost workdays are among white, non-Hispanic workers, with non-Hispanic blacks and Hispanic workers each accounting for approximately 9% of lost workday cases (9). An analysis of emergency department data that did not separate out Hispanic ethnicity found that black workers had injury rates, approximately 1.3 times higher than white workers (8). Workers 25 to 54 years of age account for more than 73% of nonfatal injuries, those younger than 25 years of age account for approximately 16%, and those older than 54 years of age approximately 8% (9). Based on emergency department data, workers 18 to

TABLE 24-4. *Incidence and most frequent events of nonfatal occupational injury and illnesses requiring days away from work for select occupations, 1995*

Occupation	Estimated no. of injuries	Most frequent events
Truck drivers	151,338	Overexertion (mostly lifting)—29% Contact with objects/equipment (mostly struck by)—20% Transportation—12% Fall to same level—10%
Laborers, except construction	115,545	Contact with objects/equipment (mostly struck by)—34% Overexertion (mostly lifting)—31%
Nursing aides, orderlies, and attendants	100,596	Overexertion (mostly lifting)—59% Fall to same level—10%
Assemblers	55,537	Contact with objects/equipment (mostly struck by)—34% Overexertion (mostly lifting)—27% Repetitive motion—15%
Janitors and cleaners	52,582	Overexertion (mostly lifting)—30% Contact with objects/equipment (mostly struck by)—25% Fall to same level—14%
Construction laborers	43,496	Contact with objects/equipment (mostly struck by)—37% Overexertion (mostly lifting)—22% Fall to lower level—11%
Cooks	35,440	Contact with objects/equipment (mostly struck by)—30% Exposure to harmful substances—21% Overexertion (mostly lifting)—19% Fall to same level—18%
Carpenters	35,044	Contact with objects/equipment (most struck by)—38% Overexertion (mostly lifting)—23% Fall to lower level—14%
Stock handlers and baggers	34,711	Overexertion (mostly lifting)—36% Contact with objects/equipment (mostly struck by)—29%
Cashiers	30,177	Overexertion (mostly lifting)—28% Contact with objects/equipment (mostly struck by)—29% Fall to same level—20%

From Bureau of Labor Statistics. Occupational injuries and illnesses: counts, rates, and characteristics, 1995. Bulletin 2493. Washington, DC: U.S. Department of Labor, Bureau of Labor Statistics, 1998.

19 years of age have the highest rates of injury, with injury rates decreasing with increasing age (7).

Information from the BLS annual survey of employers on employee tenure are available only for lost workday cases, and injuries are not separated from illnesses in published data (illnesses represent only approximately 7% of the cases). Thirteen percent of all cases occurred among employees with less than 3 months of service with the employer, 18% among employees with 3 to 11 months of service, 31% with 1 to 5 years of service, and 27% with more than 5 years of service with their employer (9).

The magnitude and risk of nonfatal injuries by industry division vary substantially from those for injury deaths (Table 24-3). The occupational injury rate averaged across all industries in 1996 was 7.4 per 100 workers. A substantial number of specific industries have injury rates far in excess of the average rate. Most are within the manufacturing industry division, with many in the lumber and wood products industries, primary metal industries, and the manufacture of transportation equipment.

Table 24-4 provides information on the estimated incidence (numbers, not rates) and patterns of nonfatal injury for select specific occupations (10). Many nonfatal injury events are common across a variety of occupations, with fewer examples of differences in occupation-specific injury event patterns. Injury patterns among occupations are important for focusing prevention efforts.

CLINICAL PRESENTATION AND COURSE OF INJURIES

Data from the 1998 National Health Interview Survey indicate that 34% of workers with occupational injuries are treated in emergency departments. The remainder are treated on-site, at private physicians' offices or clinics, or in other medical treatment facilities. Table 24-5 provides information on diagnoses and anatomic sites of occupational injuries treated in emergency departments in the United States in 1996. Sprains and strains accounted for 27% of the injuries, followed by lacerations (22%), and contusions/abrasions/hematomas (20%). Thirty percent of injuries were to the hand or finger (7). Among the 1996 work-related emergency department visits, wound care was provided to 34% of patients, extremity radiographs were ordered or provided to 30% of patients, and orthopedic care was provided to 21% of patients. Approximately 1.5% of injuries resulted in hospital admission (8).

Additional information on the course of

TABLE 24-5. *Occupational injuries treated in emergency department by diagnosis and anatomic site, 1996*

Diagnosis	Estimated no.	Part of body affected					
		Trunk, back, groin	Leg, knee, ankle	Arm, wrist, shoulder	Head, face, neck	Hand/ finger	Other
Sprain or strain	885,000	44%	22%	20%	6%	4%	3%
Laceration	731,000	<1%	6%	10%	15%	68%	1%
Contusion, abrasion, or hematoma	660,000	14%	17%	15%	21%	21%	12%
Dislocation or fracture	220,000	11%	15%	22%	4%	34%	15%
Burn	132,000	5%	8%	18%	38%	26%	6%
Other	674,000	10%	5%	10%	29%	29%	18%
Total	3,302,000	18%	13%	15%	17%	30%	8%

From National Institute for Occupational Safety and Health. Surveillance for nonfatal occupational injuries treated in hospital emergency department. *Morb. Mortal Wkly Rep* 1998;47:302–306.

injuries is available from the BLS annual survey of employers (9). Of the estimated 1.9 million injuries and illnesses with lost workdays in 1996, the median days away from work was five. Median days away from work were highest for amputations (20 days) and fractures (17 days). With respect to fatally injured workers, in 1996, 84% died the day they were injured; 97% died within 30 days (1).

PREVENTION OF INJURIES

Hierarchical Approach to Occupational Injury Control

Over the years, a number of models for occupational injury control have evolved. Many of these models categorize worker protection strategies based on a hierarchical approach, such as the five-tier model presented in Table 24-6. Other variations of this hierarchical approach have been published (11). Haddon proposed 10 basic strategies for injury prevention that have a number of similarities to the hierarchical approach, such as hazard elimination, hazard reduction, and using barriers for protection (12). Haddon also introduced the concept that injury causation was a chain of multifactorial events, each of which provided opportunities for intervention. Linn and Amendola (13) suggested an approach that combines the public health model with safety engineering analysis for injury prevention. The disciplines of epidemiology, safety engineering, biomechanics, ergonomics, psychology, safety management, and others form a multidisciplinary approach that is useful for identifying injury

TABLE 24-6. *Safety hierarchy*

Priority rank	Safety action
1	Eliminate hazard or risk
2	Apply safeguarding technology
3	Use warning signs
4	Train and instruct
5	Use personal protective equipment

Adapted from Barnett RL, Brickman DB. Safety hierarchy. Journal of Safety Research 1986;17:49–55.

risk factors and developing control strategies (see Chapters 5, 6, 8, 9, 21, and 26).

Primarily, the hierarchical approach focuses on eliminating a hazard through design; using safeguards that eliminate or minimize worker exposure to a hazard; providing worker training in safe work practices and procedures; and using PPE to prevent or minimize worker exposure to hazards, or to reduce the severity of an injury if one occurs. In general, there are three main categories of control strategies: engineering control, administrative control, and the use of PPE.

Control Strategies

The optimal injury control strategy should be to eliminate a hazard completely. Many times, hazard elimination can be accomplished through equipment design. Consider the example illustrated in Box 24-3. Although a number of factors contributed to this fatal event, one of the National Institute for Occupational Safety and Health recommendations suggested that equipment and tool manufacturers design a unique coupling system to prevent the use of unsuitable hydraulic hoses on booms, aerial buckets, or aerial bucket attachments. In this situation, the unique design of the hose coupling mechanism could eliminate the hazard of metal-reinforced hoses on bucket truck booms, allowing only the appropriate type of hydraulic hose to be installed.

Because hazard elimination is not always possible, other control strategies in the hierarchy must be implemented to achieve worker protection. If a hazard cannot be eliminated completely, then the next control level should be to eliminate or minimize worker exposure through protective safeguarding technologies (commonly referred to as *engineering controls*). Effective engineering controls are designed into equipment, workstations, and work systems to provide protection without direct worker involvement—that is, passive control. Typically, engineering controls do not eliminate the hazard *per se*; rather, they prevent or

Box 24-3. Electrical Lineman Dies after Falling 35 Feet to the Ground from a Burning Aerial Bucket

A 37-year-old electrical lineman was sagging (adjusting slack in) the center phase of a three-phase, 12,400-volt, energized power line. A metal-reinforced rubber hydraulic hose was attached to an impact wrench the lineman was using. When the hose simultaneously contacted two phases of the power line, the heat generated by the electric current in the metal reinforcement caused the hose to melt and rupture. When the hydraulic fluid from the ruptured hose contacted the power line, the fluid ignited and the aerial bucket became engulfed in flames. The lineman attempted to jump to an earthen bank approximately 15 feet from the side of the bucket, but caught his foot on the bucket lip and fell 35 feet to the ground. A field mechanic had installed the metal-reinforced hose 5 months before the fatal incident. During later interviews, the mechanic stated that he knew he was installing the wrong type of hose but did not understand the hazards involved.

National Institute for Occupational Safety and Health. Request for assistance in preventing injuries and deaths from metal-reinforced hydraulic hoses. DHHS (NIOSH) publication no. 93-105. Cincinnati, OH: U.S. Department of Health and Human Services (DHHS), Centers for Disease Control and Prevention (CDC), National Institute for Occupational Safety and Health, 1993.

minimize worker exposure to a hazard, as long as the control is in place and functions properly. Experience has shown that effective engineering controls must be well designed so as not to interfere adversely with the work process and introduce additional hazards.

Administrative controls also seek to eliminate or minimize worker exposure to hazards, but require behavioral actions by workers to be most effective. Administrative controls are usually defined as management-directed work practices or procedures that, when implemented consistently, reduce the risk of injury. Administrative controls are sometimes referred to as *active controls* because they require worker involvement to be effective.

Personal protective equipment is usually viewed as the last control option in the hierarchy. PPE consists of devices used to protect workers by reducing the risk that exposure to a hazard will injure the worker, or reducing the severity of an injury if one does occur. Although the hazard still exists, the potential for worker injury is mitigated through the PPE.

A comprehensive approach to worker injury prevention efforts inevitably includes all tiers of a control hierarchy to achieve maximum worker protection. The following sections provide some examples of each type of control.

Engineering Controls

Many types of industrial equipment require power transmission units that include belts, pulleys, gears, shafts, and other mechanisms necessary for the equipment to function. Workers can be exposed to serious, or even fatal, injury hazards if they come into contact with these rotating or moving components. A fixed barrier guard that completely encloses the power transmission unit is an engineering control that protects workers from these types of hazards. As long as the barrier guard remains in place, the worker is protected from injury. Another engineering control is an optical sensor (also called a *light curtain*) used to protect the point of operation (area where the machine operation takes place) on a mechanical power press (Fig. 24-1). In this example, the sensor is integrated into the press control mechanism so that if any part of the worker's body breaks the plane of light in front of the hazardous point of operation, either the down-

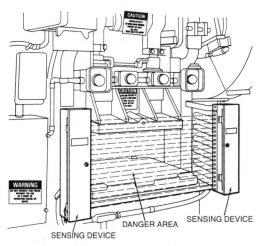

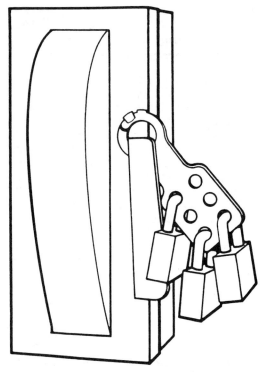

FIG. 24-1. Photoelectric (optical) sensor installed on a mechanical power press to protect the point of operation. (From Occupational Safety and Health Administration. Concepts and techniques of machine safeguarding. OSHA 3067. Washington, DC: U.S. Department of Labor, Occupational Safety and Health Administration, 1980.)

FIG. 24-3. Lock-out hasp on an electrical control panel provides a method for applying a lock (lock-out) to the panel during maintenance or repair to ensure that the equipment is not energized until the work has been completed. The control panel should also be tagged (tag-out) with a label indicating that work is being performed. Workers should be provided with individually keyed locks, and only the worker who applied the lock should remove it. (From Occupational Safety and Health Administration. Concepts and techniques of machine safeguarding. OSHA 3067. Washington, DC: U.S. Department of Labor, Occupational Safety and Health Administration, 1980.)

ward motion of the press ram cannot be initiated or it is automatically disengaged if motion has begun.

Many engineering controls are *interlocked* to ensure that they cannot be removed without disabling the machine or equipment. An

FIG. 24-2. Example of poor housekeeping on a construction site. Loose bricks, lumber, and other debris create a potential tripping hazard for workers.

interlock is a device that is integrated into the control mechanism of a machine or work process to prevent the work cycle from being initiated until the interlock (usually, an electrical or mechanical control) is closed, signaling the equipment that the work cycle can be initiated. Interlocks need to be designed so that they are not easily bypassed or defeated.

Although engineering controls should be viewed as a primary tier of prevention, it is not always possible to develop such controls

FIG. 24-4. Violent injuries are particularly a problem for workers employed in isolated work settings, such as this gas station. (Photograph by Marvin Lewiton.)

for all potentially hazardous work situations. Thus, administrative controls are the next tier for reducing or minimizing worker exposure to hazards.

Administrative Controls

An administrative control in the form of good housekeeping procedures requires that spills or debris be cleaned up to reduce the potential for a slip, trip, or fall injury. When spills or debris are cleaned up quickly, the potential hazard of a slippery work surface or tripping hazard is removed or minimized (Fig. 24-2).

The implementation of a hazardous energy control policy during maintenance activities is an administrative control in the form of safe work practices and procedures to prevent worker injuries due to inadvertent energization of equipment. Lock-out/tag-out procedures are important components of a hazardous energy control policy (Fig. 24-3). However, to be effective, the procedures must be written and consistently implemented, and workers trained in their use (14). Other examples of administrative controls are policies in retail establishments that increase visibility of cash exchange areas (through limited use of displays on storefronts) and minimize the amount of cash that

employees have access to as a deterrent to robberies that can be associated with assaults on workers (Fig. 24-4). Worker training, which is essential to the implementation of safe work practices and procedures, is also considered by many safety and health professionals to be an administrative control. The use of training as an injury prevention strategy is discussed in further detail in the section on Training, later.

Personal Protective Equipment

The use of PPE is common in many work environments, and in many situations, essential for worker protection. However, PPE is usually viewed as the last tier of the protection hierarchy. If hazardous exposures cannot be eliminated or controlled through equipment design or the application of engineering or administrative controls, then PPE provides another opportunity for worker protection. Examples of PPE designed for reducing worker injuries include protective hard hats, eye wear and face shields, steel-toed safety shoes, fall restraint devices, and personal flotation devices. When worn properly and consistently, these devices can prevent or at least reduce the severity of traumatic injury. Fall restraint devices (e.g., lanyards and body harnesses) do not prevent a worker from falling, but protect workers from a more serious injury or fatality from a fall event when working at elevations, such as on a roof (Fig. 24-5).

In most work environments, a combination of engineering controls, administrative controls, and PPE is required to have a complete and effective injury prevention program. The following examples illustrate how the combined application of controls can be used to achieve an enhanced level of worker protection.

Tractors equipped with a rollover protective structure (ROPS) significantly reduce the risk that the operator will be injured in a rollover event (Fig. 24-6). However, additional protection can be achieved if a seat belt is worn to keep the operator within the protective envelope of the ROPS. A similar

FIG. 24-6. Tractor with a two-post protective rollover protective structure (ROPS) frame installed. ROPS are designed to reduce the risk of injury or death by preventing the tractor from rolling onto and crushing the operator. A properly fastened seat belt greatly improves the chances that the operator will stay within the protective envelope provided by the ROPS (i.e., the seat). (From National Institute for Occupational Safety and Health. Safe grain and silage handling. DHHS (NIOSH) publication no. 95-109. Cincinnati, OH: National Institute for Occupational Safety and Health, 1995.)

example is the increased protection afforded by the use of seat belts, mandated in company safety policies and programs, in motor vehicles that are also equipped with air bags.

Training

Training refers to methods to assist workers in acquiring knowledge (safety information), changing attitudes (perceptions and beliefs regarding safety), or practicing safe work behaviors (organizational, management, or worker performance). Despite a paucity of data on the direct relationship between training and injury, there is evidence showing a positive impact of training on establishing safe working conditions (15). Research indicates that training is one of the key factors accounting for differences between companies with low and high injury rates, and in developing and implementing effective hazard control measures (15,16). Training increases hazard awareness, knowledge and adoption of safe work practices, and other workplace safety improvements.

Characteristics of effective training pro-

FIG. 24-5. Ironworker using a full-body harness and lanyard attached to a rope grab. This worker has the flexibility to move up the structure while maintaining fall protection. (Photograph courtesy of the Construction Safety Council.)

grams include assessing training needs specific to the work task; developing the training program to address these needs specifically; setting clear training goals; and evaluating the posttraining knowledge and skills and providing feedback of these results to the workers (16). There are also indications that training is more effective when coupled with organizational support and management commitment to safety, and when it emphasizes early indoctrination and follow-up instruction and reinforcement (15,16).

Unique characteristics of the specific workforce must be considered when developing or implementing safety training programs. Language, literacy, cognition, and cultural issues may diminish the effectiveness of training when programs are not tailored to account for unique or diverse characteristics of the workforce. Workplace safety training appears to be most effective when it includes active learning experiences stressing job site applications, and when it is placed in the context of a broader, workplace-based prevention approach (15).

Standards

There are many standards aimed at protecting workers from traumatic injury. These standards cover a multitude of hazards and address the work environment, work practices, equipment, PPE, and worker training. There are primarily two types of worker protection standards: mandatory standards, such as those promulgated by a regulatory agency such as the Occupational Safety and Health Administration (OSHA); and voluntary standards, such as those developed through independent organizations like the American National Standards Institute through a consensus process involving various stakeholders in an industry, such as representatives from labor, management, and government. There are also numerous specifications, codes, and guidelines for machinery, equipment, tools, and other materials that can assist engineers and designers in developing safer products and systems, many of which have application in the occupational setting. Examples include the National Electric Code, published by the National Fire Protection Association, and numerous consensus standards from the American Society of Mechanical Engineers and the American Society for Testing and Materials.

The OSHA standards are the primary federal regulations governing workplace safety and health for all industries except mining. The OSHA standards are codified under Title 29 (U.S. Department of Labor). Title 29 CFR (Code of Federal Regulations), Part 1910 *Occupational Safety and Health Standards*, covers General Industry, Maritime and Agriculture, whereas 29 CFR 1926 includes *Safety and Health Regulations for Construction*. Title 29 also provides standards under Chapter V, Wage and Hour Division, that address child labor issues. Safety and health standards for the mining industry are promulgated by the Mine Safety and Health Administration under Title 30 CFR 1 to 199, Chapter I. (See Chapter 10.)

Injury Control: Roles and Responsibilities

Occupational injury prevention is not the sole responsibility of a single person or group. Employers, workers, regulators, and policy makers each share in the responsibility for prevention. A multidisciplinary approach involving interaction among diverse groups in an organization and active participation by both management and workers are crucial to an effective safety program.

Employers are responsible for establishing written safety policy, developing a comprehensive safety program, and effectively implementing that program at the worksite. A competent person or committee should be designated with responsibility for company safety policy. This person or committee should have knowledge of safety policy, standards, regulations, and hazard abatement, and should actively participate with management and employees in overseeing the safety program.

An effective safety program strives to identify hazards through job safety analysis or other methods of systems safety analysis, and eliminates or controls identified hazards through the various approaches previously discussed. Workers, managers, and safety specialists should work together to analyze the job and potential hazards, and to recommend changes or controls to abate them before an injury event occurs. In industries or jobs where the worksite is not constant, site hazard assessments should be performed before beginning work in any new environment. Occupations such as farming, logging, construction, and mining are characterized by frequently changing worksites, and require a site hazard assessment before commencing work in any new or changed environment. This is particularly important in construction, where worksites change not only from job to job, but from day to day, even hour to hour, with constant potential for new hazards.

Employers are also responsible for ensuring proper maintenance of vehicles, equipment, and machinery, and their safety features, such as machine guarding, interlocks, and barriers. Where job hazards cannot be eliminated or controlled, employers are responsible for providing appropriate PPE, such as fall restraint systems, respirators, hearing protection, hard hats, or eye protection.

Employers must also ensure that workers receive appropriate training in minimizing their risk, including safety policy and practice, hazard recognition and control technologies, and the appropriate use of PPE. Enforcement of safety policy is also a critical employer responsibility. Management's demonstrated commitment to safety has been recognized as a major factor in successful workplace safety experience (17,18). Employers who demonstrate concern and support for safety activities have top managers personally involved in safety activities, and routinely involve workers in safety matters and decision making. These employers are more likely than others to have successful safety programs. As part of a comprehensive safety program, employers should require systematic reporting or surveillance of occupational injuries, and assessment of these data for use in corrective action to prevent similar occurrences.

Workers also play a vital role in workplace safety. Workers share in the responsibility for complying with safe work practices and policies, maintaining a safe workstation, and using appropriate PPE as required by their employers. Workers should also participate in company-sponsored safety training and reporting unsafe conditions for corrective action. Participation of workers in the workplace safety program is essential; as the experts in their jobs, workers should be involved in safety analysis and development of safe solutions. Worker input into recommended design or modification of safety controls, processes, or technology, and into the development of safe work practices, increases the acceptance of positive changes and, thus, the success of the safety program.

An effective workplace safety program that minimizes injuries results from a multidisciplinary effort that actively involves every level of the workforce, from the employer and upper-level managers to employee representatives and hourly workers. Each must assume some responsibility for safety and must work together interactively to achieve the common goal of preventing injuries.

Occupational injuries continue to exert too large a toll on the workforce. Reductions in fatal and nonfatal injuries from earlier periods demonstrate that progress can be made (1,2,4,9,19). However, progress requires concerted and consistent efforts from multiple parties using multiple strategies. In addition to the primary stakeholders in the workplace, additional groups can help reduce occupational injuries. These groups include manufacturers and distributors of industrial equipment and tools who design and promote safety features of equipment, insurers who provide monetary incentives for good safety records, and health care providers who give their patients information on preventing workplace injuries.

REFERENCES

1. Bureau of Labor Statistics. Fatal workplace injuries in 1996: a collection of data and analysis. Report 922. Washington, DC: U.S. Department of Labor, Bureau of Labor Statistics, 1998.
2. Bureau of Labor Statistics. Workplace injuries and illnesses in 1996. Bulletin 97-453. Washington, DC: U.S. Department of Labor, Bureau of Labor Statistics, 1997.
3. Leigh PJ, Markowitz SB, Fahs M, Shin C, Landrigan PJ. Occupational injury and illness in the United States: estimates of cost, morbidity, and mortality. Arch Intern Med 1997;157:1557–1568.
4. Jenkins EL, Kisner SM, Fosbroke DE, et al. Fatal injuries to workers in the United States, 1980–1989: a decade of surveillance; national profile. DHHS (NIOSH) publication no. 93-108. Cincinnati, OH: U.S. Department of Health and Human Services (DHHS), Centers for Disease Control and Prevention (CDC), National Institute for Occupational Safety and Health, 1993.
5. Loomis D, Richardson D. Race and risk of fatal injury at work. Am J Public Health 1998;88:40–44.
6. Fosbroke DE, Kisner SM, Myers JR. Working lifetime risk of occupational fatal injury. Am J Ind Med 1997;31:459–467.
7. Centers for Disease Control and Prevention. Surveillance for nonfatal occupational injuries treated in hospital emergency departments. MMWR Morb Mortal Wkly Rep 1998;47:302–306.
8. McCaig L, Burt CW, Stussman BJ. A comparison of work-related injury visits and other injury visits to emergency departments in the United States, 1995–1996. J Occup Environ Med 1998;40:870–875.
9. Bureau of Labor Statistics. Lost-worktime injuries and illnesses: characteristics and resulting time away from work, 1996. Bulletin 98-157. Washington, DC: U.S. Department of Labor, Bureau of Labor Statistics, 1998.
10. Bureau of Labor Statistics. Occupational injuries and illnesses: counts, rates, and characteristics, 1995. Bulletin 2493. Washington, DC: U.S. Department of Labor, Bureau of Labor Statistics, 1998.
11. Hammer W. Occupational safety and management and engineering, 4th ed. Englewood Cliffs, NJ: Prentice-Hall, 1989.
12. Baker SP, O'Neill BO, Ginsburg MJ, Li G. The injury fact book, 2nd ed. New York: Oxford University Press, 1992.
13. Linn HI, Amendola AA. Occupational safety research: an overview. In: Stellman, JM, ed. Encyclopaedia of occupational health and safety. Geneva: International Labor Office, 1998:60.2–60.5.
14. National Institute for Occupational Safety and Health. Request for preventing worker injuries and fatalities due to the release of hazardous energy. DHHS (NIOSH) publication no. 99-110. Cincinnati, OH: U.S. Department of Health and Human Services (DHHS), Centers for Disease Control and Prevention (CDC), National Institute for Occupational Safety and Health, 1999.
15. National Institute for Occupational Safety and Health. Assessing occupational safety and health training. DHHS (NIOSH) publication no. 98-145. Cincinnati, OH: U.S. Department of Health and Human Services (DHHS), Centers for Disease Control and Prevention (CDC), National Institute for Occupational Safety and Health, 1999.
16. Johnston JJ, Cattledge GH, Collins JW. The efficacy of training for occupational injury control. Occup Med 1994;9:147–158.
17. Cohen A. Factors in successful occupational safety programs. Journal of Safety Research 1977; 9:168–178.
18. Shannon HS, Walters V, Lewchuck W, et al. Workplace organizational correlates of lost-time accident rates in manufacturing. Am J Ind Med 1996; 29:258–268.
19. Stout NA, Jenkins EL, Pizatella TJ. Occupational injury mortality rates in the United States: changes from 1980–1989. Am J Public Health 1996;86:73–77.

BIBLIOGRAPHY

Baker SP, O'Neill BO, Ginsburg MJ, Li G. The injury fact book, 2nd ed. New York: Oxford University Press, 1992.
A reference document that includes information on injury epidemiology and prevention strategies. Includes a chapter on occupational injury as well as chapters addressing injury causes relevant to occupational settings and overview chapters.

Hammer W. Occupational safety and management and engineering, 4th ed. Englewood Cliffs, NJ: Prentice-Hall, 1989.
A good overall reference on occupational safety and health issues. Provides an overview of standards and codes for workplace safety; identifying and controlling hazards; analyzing safety hazards and conducting incident investigations; and developing and implementing workplace safety programs. Addresses both the engineering and management aspects of occupational injury and disease prevention.

Jenkins EL, Kisner SM, Fosbroke DE, et al. Fatal injuries to workers in the United States, 1980–1989: a decade of surveillance; national and state profiles. DHHS (NIOSH) publication no. 93-108S. Cincinnati, OH: U.S. Department of Health and Human Services (DHHS), Centers for Disease Control and Prevention (CDC), National Institute for Occupational Safety and Health, 1993.
Provides a comprehensive compilation of surveillance data for fatal occupational injuries for the period 1980 to 1989. The document contains detailed data by sex, race, age, cause of death, industry, and occupation, both nationally and by state.

Occupational Safety and Health Administration. Concepts and techniques of machine safeguarding. OSHA 3067. Washington, DC: U.S. Department of Labor, Occupational Safety and Health Administration, 1980.
An excellent reference for identifying potential hazards when working with industrial machinery. The publica-

tion also provides general principles of machine safeguarding to protect workers from injury.

Stout N, Borwegan W, Conway G, et al. Traumatic occupational injury research needs and priorities: a report by the NORA Traumatic Injury Team. DHHS (NIOSH) publication no. 98-134. Cincinnati, OH: U.S. Department of Health and Human Services (DHHS), Centers for Disease Control and Prevention (CDC), National Institute for Occupational Safety and Health, 1998.

Provides a broad framework of research needed to begin filling gaps in knowledge and furthering progress toward the prevention of traumatic occupational injury in the United States. The recommendations target government agencies, academic institutions, public and private research organizations, labor groups, profes- *sional societies, and individual researchers, who could use the document as a basis for planning and prioritizing research efforts. The report is a product of deliberations of a team of experts representing industry, labor, academia, and government, representing a variety of scientific disciplines and organizational perspectives.*

Wallerstein N, Rubenstein H. Teaching about job hazards: a guide for workers and their health providers. Washington, DC: American Public Health Association, 1993.

This comprehensive manual provides guidance for health and safety education to workers, including guidance specific to health care providers, as well as information for occupational safety and health training resources.

Occupational Health: Recognizing and Preventing Work-Related Disease and Injury, 4th ed.
Edited by Barry S. Levy and David H. Wegman, Lippincott Williams & Wilkins, Philadelphia © 2000.

25 Respiratory Disorders

David C. Christiani and
David H. Wegman

A 60-year-old man, who had been a sand-blaster for 23 years, was hospitalized for the third time in the past 4 months for shortness of breath. Three years ago, he began having respiratory problems, initially mild shortness of breath and increased heart rate when walking in snow and climbing steps, and with heavy exertion at work. These symptoms increased moderately over the next several months. He was seen by the company physician, who told him that he had "bad lungs," but gave him no treatment. Two years ago, he sought therapy at a community hospital due to increasing shortness of breath while walking at normal speed on the level ground for one to two blocks. He was hospitalized. Resting room-air arterial blood gases were Pa_{O2} = 87 mm Hg and Pa_{CO2} = 31 mmHg. A chest radiograph showed multiple interstitial nodules without evidence of hilar disease. Pulmonary function tests revealed a reduced forced vital capacity (FVC; 73% of predicted) with a normal diffusing capacity. Tuberculosis smear, culture, and cytology of bronchial washings were all negative. He was sent home without therapy, and was told not to return to work. He has not worked since.

Seven months ago, he acquired a cough occasionally productive of thin, clear-to-grayish sputum. Three more hospital admissions for increasing shortness of breath occurred with no new findings reported. Since the last hospitalization, 1 month ago, he has been on oxygen continuously and stays in bed most of the day. He has also had dysuria and some trouble initiating his urinary stream, which seems to make his shortness of breath worse.

The patient had smoked one pack of cigarettes per day for 5 years, until he quit 20 years ago. He has no history of asthma, pneumonia, surgery, or allergies.

His occupational history revealed a 23-year period of operating a sandblasting machine located in a basement room (20 × 40 ft). Dust escaped continuously through crevices of the sandblasting unit; every time he opened the door to remove and install a piece to be blasted, much fine dust escaped. The windows were closed; there was an exhaust fan in the wall that did not seem to remove any dust. A room fan, installed to circulate the air in the room, was often out of order. The patient wore a helmet with a cloth apron on the bottom, covering his shoulders, and, when the room was especially dusty, a compressed air supply.

Physical examination revealed a thin man in moderate respiratory distress, sitting hunched over, gasping for breath, with grunting expirations. Pulse was 110, respiratory rate 40, blood pressure 110/80, and temperature 98°F. Pulmonary and cardiac examinations were normal, except for a systolic ejection murmur and an increased second heart sound over the pulmonic area. His extremities revealed clubbed fingernails and cyanosis. The rest of the examination was normal. Resting room-air arterial blood gases revealed Pa_{O2} = 39 mmHg and

D. C. Christiani: Occupational Health Program, Harvard School of Public Health, Boston, Massachusetts 02115, and Pulmonary and Critical Care Unit, Massachusetts General Hospital/Harvard Medical School, Boston, Massachusetts 02214.

D. H. Wegman: Department of Work Environment, University of Massachusetts Lowell, Lowell, Massachusetts 01854.

$PaCO_2$ = 38 mmHg. Chest radiography showed diffuse, interstitial, small, rounded densities throughout both lung fields with hilar fullness. These were judged to be "q"-sized with a 2/2 profusion in all lung fields, using the International Labor Organization (ILO) nomenclature for chest radiographs. The diagnosis of silicosis was made. He remained completely disabled and died 3 months later.[1]

This case is an example of a severe occupational respiratory disease—in this instance, pneumoconiosis. However, workplace exposure responsible for such chronic disabling lung disease occurs gradually over long periods of time; initially, exposures do not result in any obvious acute symptoms, but once symptoms do appear, often little can be done beyond making the worker comfortable. Unless discovered very early in their course, most work-related respiratory diseases are not curable. Disease prevention is therefore critically important.

Occupational lung disease is recorded in accounts of ancient history. Case reports exist in the writings of Hippocrates, and evidence of silicosis is present in pictographs from Egypt. Yet, some of those chronic diseases remain important problems for workers today. Estimates of the prevalence and incidence of occupational respiratory disease suggest that less than 5% of chronic occupational respiratory disease is correctly identified as associated with work.

Pneumoconiosis and occupational asthma are two work-related respiratory diseases that often are not correctly diagnosed. For example, approximately 5% of Americans have what physicians diagnose as asthma, but a much larger proportion of people report either having asthma or episodes of wheezing; physicians who see workers who report wheezing should determine if a work-related bronchoconstrictive response is occurring.

---

[1] Case courtesy of Stephen Hessl, M.D., Daniel Hryhorczuk, M.D., and Peter Orris, M.D., Section on Occupational Medicine, Cook County Hospital, Chicago, Illinois.

EVALUATION OF INDIVIDUALS

Evaluations of pulmonary response to toxic exposures are important because work-related respiratory disease is frequently a contributory cause—and commonly a primary cause—of pulmonary disability. Usually performed in a physician's office, they include a minimum of four elements: (a) a complete history, including occupational (direct and indirect) and environmental (including home and recreational) exposures, a cigarette-smoking history, and a careful review of respiratory symptoms (see Chapter 4); (b) a physical examination with special attention to breath sounds; (c) a chest radiograph with appropriate attention to parenchymal and pleural opacities; and (d) pulmonary function tests.

History

Review of symptoms should include questions on chronic cough, chronic sputum production, shortness of breath (dyspnea) on exertion compared with peers or usual level of activity, wheezing unrelated to respiratory infections, chest tightness, pain, and reports of allergic or asthmatic responses to work or nonwork environments. For example, one peculiar characteristic of several types of occupational asthma and of pulmonary edema is that symptoms may peak in intensity approximately 8 to 16 hours after exposure has ended. The symptoms often occur at night as shortness of breath or cough. In assessing acute airway disease, the clinician should question the patient about the principal symptoms: chest tightness, wheezing, dyspnea, and cough. Symptom periodicity or timing is critical. For example, respiratory symptoms (cough, wheeze, chest tightness) occurring on work days (or nights) with improvement on weekends or holidays strongly suggest a workplace-induced condition. A formal survey questionnaire for systematic respiratory effect studies, the American Thoracic Society (ATS) Respiratory Symptom Questionnaire, is available (1).

I. **Size and Shape of Small Opacities**

ROUNDED OPACITIES		IRREGULAR OPACITIES	
p	≤ 1.5 mm diameter	s	fine linear opacities > 1.5 mm width
q	1.6 - 3.0 mm diameter	t	medium opacities 1.6 - 3.0 mm width
r	3.1 - 10.0 mm diameter	u	coarse, blotchy opacities 3.1 - 10.0 mm width

· Size recorded by two letters to distinguish single type from mixed type. For example, q/q
 if only q opacities are present, but q/t if q opacities predominate but t are also present.

II. **Concentration (Profusion) and Distribution**

SMALL OPACITIES

	Major Categories	Minor Divisons			Distribution·	
0	Small Opacities Absent or Less Than Category 1 Normal Lung Markings Visible	0/-	0/0	0/1	RU	LU
1	Small Opacities Present but Few Normal Lung Markings Usually Visible	1/0	1/1	1/2	RM	LM
2	Small Opacities Numerous Normal Lung Markings Partially Obscured	2/1	2/2	2/3	RL	LL
3	Small Opacities Very Numerous Normal Lung Markings Totally Obscured	3/2	3/3	3/+		

LARGE OPACITIES

A	One or More Opacities with Greatest Summed Diameter 1 - 5 cm
B	One or More Opacities Larger or More than Category A. Total Area < Equivalent of Right Upper Zone
C	One or More Opacities. Total Area Exceeds Equivalent of Right Upper Zone

· Recorded by dividing lungs into 3 regions per side and checking all regions containing the designated small opacities

III. **Pleural Thickening**

	WIDTH·		EXTENT·		CALCIFICATION·
a	Maximum Up to 5 mm	1	Up to 1/4 Lateral Wall	1	One or Several Regions Summed Diameter ≤ 2 cm
b	Maximum 5 - 10 mm	2	1/4 - 1/2 Lateral Wall	2	One or Several Regions Summed Diameter 2 - 10 cm
c	Maximum > 10 mm	3	Exceeds 1/2 Lateral Wall	3	One or Several Regions Summed Diameter > 10 cm

· **Width** estimated only if seen in profile. **Extent** estimated as maximum length of thickening (profile or face on).
 Calcification site (diaphragm, wall, other) and extent are noted separately for two sides.

FIG. 25-1. Schematic of International Labor Organization classification system for chest radiographs. In addition to these scores, the reader is guided in scoring technical quality of the radiograph (good, acceptable, poor, unacceptable) and in identifying other relevant features (e.g., bullae, cancer, abnormal cardiac size, emphysema, fractured rib, pneumothorax, tuberculosis).

Physical Examination

The physical examination is helpful when the results are abnormal. The most remarkable finding in most patients with occupational lung diseases is the relative absence of physical signs; however, certain conditions are associated with physical signs, and the presence or absence of such abnormalities should be noted.

Auscultation can reveal important diagnostic clues. Fine rales at the bases, often at end-inspiration, are more common in asbestosis than in other interstitial lung diseases. Wheezes and their relationship to exposure are helpful in evaluating a suspected case of work-related asthma. A pleural rub can occur with pleural reaction due to acute, chronic, or long-ago asbestos exposure.

Clubbing of the digits is a nonspecific sign that occurs rarely and usually in relatively advanced lung diseases, including asbestosis, and therefore usually appears after other evidence of the disease has become apparent. This finding is nonspecific and cannot be used as a reliable clinical indication for the diagnosis of asbestosis. It does not usually occur in other mineral pneumoconioses or in hypersensitivity pneumonitis. The most common nonoccupational causes of clubbing are bronchial carcinoma and idiopathic pulmonary fibrosis.

Examination of the heart is important because left ventricular failure can present as dyspnea alone, and right ventricular failure may indicate severe lung disease.

Chest Radiography

A chest radiograph should be taken and, in addition to a standard interpretation, it should, if possible, be interpreted according to the ILO system for pneumoconiosis by a trained reader (Fig. 25-1) (2). Although this classification was developed for epidemiologic studies, and not for clinical evaluation *per se*, it may be an important function in the common posteroanterior chest radiograph interpretation. This scheme is not useful for evaluation of occupational asthma, but is relevant for suspected pneumoconiosis. The standard technique permits semiquantitative interpretation of radiographs to identify early evidence and progression of parenchymal and pleural disease; it focuses on size, shape, concentration, and distribution of small parenchymal opacities as well as distribution and extent of pleural thickening or calcification. For example, rounded opacities in the upper lung fields are usually associated with silicosis, whereas linear (irregular) opacities in the lower lung fields are usually associated with asbestosis (Fig. 25-2). Deviations from these patterns are common; for example, silicosis and coal workers' pneumoconiosis (CWP) can be associated with irregular opacities. Moreover, workers exposed to mixed dusts, such as silica and asbestos, can present with mixed, rounded, irregular opacities in any or all lung fields. The ILO system has the advantage of using a standardized set of comparison radiographic films, which can be used to classify radiographs at one point in time or to follow an individual or a population for

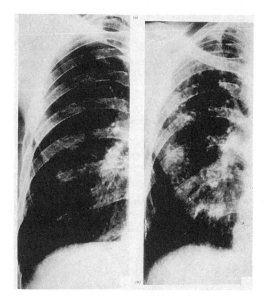

FIG. 25-2. Progression of discrete nodules of silicosis over 10 years in a slate quarry worker. (From Parkes WR. Occupational lung disorders, 2nd ed. London: Butterworths, 1982.)

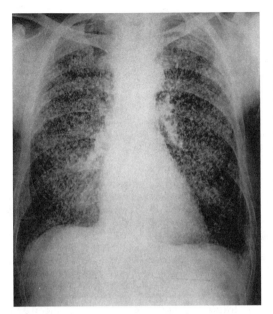

FIG. 25-3. Chest radiograph demonstrating stannosis, the benign pneumoconiosis due to the inhalation of tin oxide, in a man who worked as a furnace charger in a smelting works for 42 years. (From Parkes WR. Occupational lung disorders, 2nd ed. London: Butterworths, 1982.)

change over time. Even though chest radiographs present evidence of abnormality, they do not provide information on disability or impairment and do not necessarily correlate well with pulmonary function test findings. A person with severe obstructive disease may show little evidence of it on chest radiography. In contrast, a person exposed chronically to iron oxide or tin oxide may show a dramatically abnormal chest radiograph, although little, if any, pulmonary inflammatory reaction or lung function abnormality (Fig. 25-3).

Pulmonary Function Tests

A critical element in determining respiratory status is an evaluation of pulmonary function. In a well equipped pulmonary function laboratory, spirometry, lung volume determinations, gas exchange analyses, and exercise testing can be performed with relative ease. In a physician's office, only spirometry is readily and inexpensively performed; it does, however, provide a surprising amount of information. Pulmonary function tests, required for medical surveillance by some Occupational Safety and Health Administration standards, are commonly used and are easy to perform, reliable, and reproducible. Most lung disease yields abnormal results or accelerated declines within the "normal" ranges before onset of clinical symptoms, especially if patients are followed at regular 1- to 3-year intervals. Although these tests may demonstrate several patterns of abnormalities, alone they are not capable of determining etiology. Hospital-based tests (i.e., lung volume determinations, gas exchange analyses, exercise tests, and bronchial challenge tests) can contribute to a refined diagnostic evaluation once an abnormality is suspected.

The basic tests of ventilatory function can be obtained with a simple portable spirometer. Test results are derived from the forced expiratory curve (Fig. 25-4). Many types of equipment are marketed to provide these tests, yet several have been inadequately standardized; the ATS has evaluated spirometers and can provide information on which ones are most reliable and accurate (3). Although many measures can be derived from the forced expiratory curve, the simplest and generally the most useful ones for evaluating work-related respiratory disease are FVC, forced expiratory volume in the first second of a forced vital capacity maneuver (FEV_1), and the ratio of these two measurements (FEV_1/FVC). A simple scheme for the interpretation of these tests is shown in Table 25-1 and Fig. 25-4. Results are compared with expected values, derived from a normal population of nonsmoking adults, and are expressed as a percentage of the expected value.

Race-specific expected values are now available for nonwhite Americans. Criteria

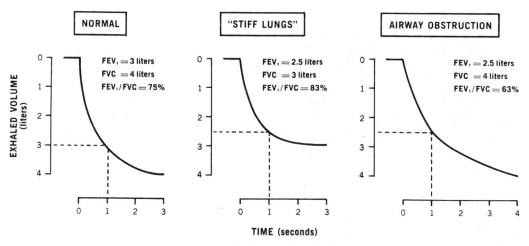

FIG. 25-4. Spirographic results in normal and disease states. [Adapted from Nadel JA. Pulmonary function testing. Basis of Respiratory Disease (American Thoracic Society) 1973;1:2.]

for the proper performance and evaluation of spirometry are based on ATS recommendations (4–6).

Pneumoconioses, such as silicosis and asbestosis, are considered restrictive diseases because there is reduction in lung volume. In the absence of significant airways disease, flow rates are maintained and may be above normal because of decreased lung compliance with increased elastic recoil. CWP, on the other hand, is more often associated with an obstructive pattern, with decreased airflow and normal or increased lung volumes. Occupational asthma is also considered an obstructive disease because there is obstruction of airflow without reduction in lung volume. With multiple environmental exposures (including tobacco smoking), a mixed

condition is frequently present. Moreover, some mineral dusts, such as asbestos and coal, have been shown to cause abnormalities in both the airways and the interstitium. Nevertheless, the basic distribution of ventilatory function abnormalities is useful in considering the general characteristics of work-related respiratory disease (see Table 25-1).

EVALUATION OF GROUPS

If the physician is able to examine several workers from the same work environment, careful attention should be directed toward evaluation of the grouped results in addition to those of each worker. For an individual, emphasis is on the work history and collection of information to explain specific symp-

TABLE 25-1. *Spirometry interpretation*

| Type of response | Percentage predicted[a] | | | Response to inhaled bronchodilators |
	FEV$_1$	FVC	FEV$_1$/FVC %	
Normal	≥80%	≥80%	≥75%	—
Obstructive	<80%	≥80%[b]	<75%	±
Restrictive	≥80%	<80%	≥75%	—
Mixed	<80%	<80%	<75%	±

FEV$_1$, forced expiratory volume in 1 second; FVC, forced vital capacity.

[a] Predicted FEV$_1$ and FVC based on RJ Knudson, Lebowitz MD, Holberg CJ, et al. Changes in the normal maximal expiratory flow volume curve with growth and aging. American Review of Respiratory Diseases 1983;127:725.

[b] Severe obstruction can result in reduction of FVC also.

TABLE 25-2. *Major types of occupational pulmonary disease*

Pathologic function process	Occupational disease example	Clinical history	Physical examination	Chest radiograph	Pulmonary function pattern
Fibrosis	Silicosis	Dyspnea on exertion, cyanosis, shortness of breath	Clubbing	Nodules	Restrictive or mixed obstructive and restrictive
	Asbestosis	Dyspnea on exertion, cyanosis, shortness of breath	Clubbing densities, rales	Linear pleural plaques, calcifications	D_{LCO} normal or decreased
Reversible airway obstruction (mucous plugging, asthma)	Byssinosis, isocyanate, asthma	Cough, chest tightness, shortness of breath, asthma attacks	↑ Respiratory rate, wheeze	Usually normal	Normal or obstructive with bronchodilator improvement Normal or high D_{LCO}
Emphysema	Cadmium poisoning (chronic)	Cough, sputum, dyspnea	↑ Respiratory rate, ↑ expiratory phase	Hyperaeration, bullae	Obstructive Low D_{LCO}
Granulomata	Beryllium disease	Cough, weight loss, shortness of breath	↑ Respiratory rate	Small nodules	Usually restrictive with low D_{LCO}
Pulmonary edema	Smoke inhalation	Frothy, bloody sputum production	Coarse, bubbly rales	Hazy, diffuse Air space disease	Usually restrictive with decreased D_{LCO} Hypoxemia at rest

D_{LCO}, diffusing capacity of lung for carbon monoxide.

toms and signs. The absence of basilar rales, however, does not exclude asbestosis; wheezes do not necessarily diagnose occupational asthma; opacities on chest radiography do not specify the underlying pathologic process; and pulmonary function tests may be falsely considered normal because of the wide variation in standard populations. It may not be until a group of coworkers is evaluated that pulmonary disease can be recognized as associated with work.

Group evaluations enable subdivision of results according to duration of work or types of exposure. Chest radiographic findings, pulmonary function tests, and symptom histories can be examined by subgroups to evaluate previously unrecognized work effects (see examples in Chapter 6). Furthermore, the average value of a group of tests has less variability than an individual test result. For example, individual measurements of FEV_1 and FVC that vary between 80% and 120% of

the population standards are still considered normal; a group of 10 or 20 actively working people, however, should have a mean result much closer to the standard values (100%). If the average population difference is as little as 5% lower—that is, 95% of the predicted value—then an adverse health effect in that population should be seriously considered (6).

Comparisons with baselines should be performed whenever possible to permit evaluation of change over time in individuals or a group compared to a known—not a predicted—value. Accelerated decrements in lung function, accelerated development of respiratory tract symptoms, or recognition of subtle chest radiographic abnormalities are far more significant when the comparison is based on earlier examinations rather than on expected population experience. Any worker potentially exposed to respiratory hazards at work should have a baseline ventilatory function test before being exposed.

The major types of respiratory response to external agents discussed in this chapter are summarized in Table 25-2. Occupational lung cancer and work-related infectious diseases of the respiratory tract are discussed in Chapters 16 and 20, respectively.

ACUTE IRRITANT RESPONSES

Irritation in the upper respiratory tract, in contrast to the middle or lower tract, is frequently associated with work-related symptoms. Acute symptoms are often due to regional inflammation, which a patient perceives as irritation. With nasal and paranasal sinus irritation, there is congestion that can result in violent frontal headache, nasal obstruction, runny nose, sneezing, and occasionally nosebleed. Throat inflammation is commonly reported as a dry cough. Laryn-geal inflammation can cause hoarseness and, if severe, may result in laryngeal spasms associated with glottal edema, dramatic anxiety, shortness of breath, and cyanosis.

In the mid-respiratory tract, the acute reaction is characteristically bronchospasm. The extreme case is asthma, which is histologically distinguished by a thickened basement membrane, increased number of goblet cells with secretions, mucus plugging, and increased smooth muscle at preterminal bronchioles. Asthma associated with work is being recognized with increasing frequency. Precipitating agents number over 250 and include isocyanates, detergent enzymes, and Western red cedar dust (Table 25-3). In addition to asthma caused by exposure to agents listed in this table, many irritant substances not usually associated with asthma can produce bronchial hyperreactivity when high

TABLE 25-3. *Selected causes of occupational asthma*[a]

Agents	Occupations
High–molecular-weight compounds	
Animal products: dander, excreta, serum, secretions	Animal handlers, laboratory workers, veterinarians
Plants: grain, dust, flour, tobacco, tea, hops, latex	Grain handlers, tea workers, bakers and workers in natural oil manufacturing and in tobacco, food processing, and health care workers
Enzymes: *B. subtilis,* pancreatic extracts, papain	Bakers and workers in the detergent, pharmaceutical, trypsin, fungal amylase, and plastic industries
Other: crab, prawn	Crab and prawn processors
Low–molecular-weight compounds	
Diisocyanates: toluene diisocyanate, methylene-diphenyldiisocyanate	Polyurethane industry workers, plastics workers, workers using varnish, and foundry workers
Anhydrides: phthallic and trimellitic anhydrates	Epoxy resin and plastics workers
Wood dust: oak, mahogany, California redwood, western red cedar	Carpenters, sawmill workers, and furniture makers
Metals: platinum, nickel, chromium, cobalt, vanadium, tungsten carbide	Platinum- and nickel-refining workers and hard-metal workers
Soldering fluxes	Solderers
Drugs: penicillin, methyldopa, tetracyclines, cephalosporins, psyllium	Pharmaceutical and health care industry workers
Other organic chemicals: urea formaldehyde, dyes, formalin, azodicarbonamide, hexachlorophene, ethylene diamine, dimethyl ethanolamine, polyvinyl, chloride pyrolysates	Workers in chemical, plastic, and rubber industries; hospitals; laboratories; foam insulation, manufacture; food wrapping; and spray painting

[a] Mechanism believed to be IgE mediated for high–molecular-weight compounds and for some low–molecular-weight compounds. The immunologic mechanism for asthma from many low–molecular-weight substances remains undefined.

Adapted from Chan-Yeung M, Lam S. Occupational asthma. American Review of Respiratory Diseases 1986;133:686–703.

levels of exposure have occurred. Single high-dose exposure and episodic low-dose exposure to irritants such as ammonia or chlorine can result in nonspecific bronchial reactivity, referred to by some authors as reactive airways dysfunction syndrome or irritant asthma, which may persist for months to years or may never fully resolve (7).

The conditions deriving from acute irritation of the deep respiratory tract are pulmonary edema and pneumonitis. With pulmonary edema, there is extravasation of fluid and cells from the pulmonary capillary bed into the alveoli. Primary pulmonary edema is due to direct toxic action on the capillary walls. For example, exposure to ozone or oxides of nitrogen, common in industrial settings, can cause pulmonary edema—either acutely when a trapped worker cannot escape exposure, or in a more delayed fashion when overexposures are not too high. Pneumonitis, on the other hand, is an inflammation of the lung parenchyma in which cellular infiltration rather than fluid extravasation predominates. Beryllium and cadmium are metals that can cause acute pneumonitis.

Factors Involved in Toxicity

The most widespread causes of acute responses are irritant gases. Water is a major constituent of the respiratory tract lining, and solubility of these gases in water is the most significant factor influencing their site of action. Gases with high solubility act on the upper respiratory tract within seconds. For example, fatal epiglottic edema has been associated with irritants of high solubility, such as ammonia, hydrogen chloride, and hydrogen fluoride. The moderately soluble gases act on both the upper and lower respiratory tract within minutes. Chlorine gas, fluorine gas, and sulfur dioxide are irritants of this type, producing upper respiratory irritation as well as symptoms of bronchoconstriction. The low-solubility irritants are most insidious. There are fewer warning signs, and they penetrate to the deep portions of the respiratory tract and act predominantly on

the alveoli 6 to 24 hours after exposure. Because of the considerable delay in onset of symptoms, large doses can be delivered without any irritant symptoms to serve as warnings. Pulmonary edema is the major effect of overexposures to materials such as ozone, oxides of nitrogen, and phosgene.

Other factors influencing the site of action of an irritant gas are intensity and duration of exposure. The amount of exposure depends not only on air concentrations but on work effort: a worker with a sedentary job exposed to a given concentration of a respiratory irritant receives a much lower dose than one with an active job requiring rapid breathing and a high minute ventilation (tidal volume × respiratory rate).

A final element that influences the site of action is interaction—both synergism and antagonism. Sulfur dioxide and water droplets are synergistic; they combine to deliver a sulfuric acid–like vapor to the respiratory tract. Ammonia and sulfur dioxide, however, are antagonistic and together produce less response than either can individually. The presence of a carrier, such as an aerosol, may increase the effect of an irritant gas: sulfur dioxide may cause a moderate effect and a sodium chloride aerosol no effect on the respiratory tract, but the two combined may result in a marked effect because the aerosol delivers the sulfur dioxide more deeply into the lung.

Highly Soluble Irritants

Primary examples of highly soluble irritants are (a) ammonia, used as a soil fertilizer, in the manufacture of dyes, chemicals, plastics, and explosives, in tanning leather, and as a household cleaner; (b) hydrogen chloride, or hydrochloric acid, used in chemical manufacturing, electroplating, and metal pickling; and (c) hydrogen fluoride, or hydrofluoric acid, used predominantly for etching and polishing of glass, as a chemical catalyst in the manufacture of plastics, as an insecticide, and for removal of sand from metal castings in foundry operations.

The primary physical effects of highly water-soluble irritants are first the odor and then eye and nose irritation; throat irritation is slightly less frequent. In high doses, the respiratory rate can increase and bronchospasm can occur. Lower respiratory tract effects, however, do not occur unless the person is severely overexposed or trapped in the environment. The irritant effects are powerful and usually provide adequate warning to prevent overexposure of people free to escape from exposure. The history and physical examination are the most important parts of irritant exposure evaluation. Reflex bronchoconstriction may be evident on pulmonary function tests shortly after exposure. Chest radiographs are not helpful unless there is pulmonary edema.

Management of reactions to these irritants is immediate removal of the worker and, if breathing is labored or hypoxemia is present, administration of oxygen. If severe exposure or loss of consciousness occurs, observation in a hospital for development of pulmonary edema is advisable.

Prevention of exposures relies on proper industrial hygiene practices with local exhaust ventilation as an essential component. Respirators should be used only as a temporary control measure in an emergency. If respirators are required to prevent overexposure, workers must be trained in their proper use and maintenance.

A 25-year-old man came to the emergency room with acid burns. Before taking a job as an electroplater 5 weeks before admission, he was in perfect health. On the first day at this job, he developed itching. Subsequently, he developed sores, which healed with scars, at sites of splashes of workplace chemicals. After 4 days on this job, he had a runny nose, throat irritation, and a productive cough. He also noted some shortness of breath at work.

His work involved dipping metal parts into tanks containing chrome solutions and acid. He wore a paper mask disposable respirator, rubber gloves, and an apron, but no eye protection. Although heavy fumes were present in the $60 \times 20 \times 14$-foot room, no ventilation was provided. Apparently, none of the other eight workers in the room had similar medical problems.

Past history revealed three prior hospitalizations for pneumonia, but not asthma or allergies. He smoked about four cigarettes per day.

From age 16 to 18 years, he worked as a sheet metal punch-press operator for a tool and die company. At age 18 years, he worked as a drip-pan cleaner for a soup company. He was a student in an auto mechanics' school from age 19 to 21 years. From age 21 to 24 years, he occasionally worked as a gas station attendant.

Physical examination was normal, except for multiple areas of round, irregularly shaped, depigmented, 1-mm atrophic scars on both forearms and exposed areas of the anterior thorax and face; a 4-mm, rounded, punched-out ulcer, with a thickened, indurated, undermined border and an erythematous base on his left cheek; an erythematous pharynx; and bilateral conjunctivitis. There was no perforation of the nasal septum. Patch tests with dichromate, nickel, and cobalt were all negative. A chest radiograph was normal.

The diagnoses were irritation of the upper respiratory tract and an irritant contact dermatitis, both due to chromic acid mist. His symptoms resolved with removal from exposure. Periodic medical surveillance was advised to provide early diagnosis of a possible malignancy of the nasal passages for which he may be at risk as a result of the chromium exposure. Finally, a follow-up industrial hygiene survey of the workplace was initiated to control exposures for the other exposed workers.[2]

Many small electroplating firms have no local ventilation over open vats of chromic and other acids. Frequently, a high level of chrome or other metals in the fumes is liberated when metal parts that are being plated are immersed. Chrome and chromic acid mist are local irritants. Primarily in hexavalent forms, chromium is considered to be a carcinogen; epidemiologic studies have shown an elevated lung cancer risk among exposed workers.

Moderately Soluble Irritants

The moderately soluble irritants commonly encountered in industrial settings are chlo-

---

[2] Case courtesy of Stephen Hessl, M.D., Daniel Hryhorczuk, M.D., and Peter Orris, M.D., Section of Occupational Medicine, Cook County Hospital, Chicago, Illinois.

rine, fluorine, and sulfur dioxide. Chlorine is widely used in the chemical industry to synthesize various chlorinated hydrocarbons, whereas outside the chemical industry its major use is in water purification and as a bleach in the paper industry. Fluorine is used in the conversion of uranium tetrafluoride to uranium hexafluoride, in the development of fluorocarbons, and as an oxidizing agent. Fluoride is used in the electrolytic manufacture of aluminum, as a flux in smelting operations, in coatings of welding rods, and as an additive to drinking water. Sulfur dioxide is commonly used as a disinfectant, a fumigant, and a bleach for wood pulp, and is formed as a by-product of coal burning, smelter processes, and the paper industry.

These irritants, like the highly soluble ones, initially cause mucous membrane irritation, often manifested by a persistent cough. Acute symptoms are usually of short duration. Low levels of continuous exposures, which are better tolerated than exposures to highly soluble irritants, may cause obstructive respiratory disease.

In addition to causing respiratory symptoms, these irritants lead to other health problems. Chlorine gas contributes to corrosion of the teeth, whereas fluorine is a sig-

nificant cause of chemical skin burns. Chronic exposure to fluoride is associated with increased bone density, cartilage calcification, discoloration of teeth in the young, and possibly rheumatologic syndromes. Sulfur dioxide, in particular, is associated with bronchospasm, especially in people with asthma; it may also cause chronic obstructive pulmonary disease.

Management and prevention are similar to those for highly soluble irritants. Pulmonary function tests, especially the FEV_1, are recommended in surveillance programs for workers with chronic exposure.

Irritants of Low Solubility

Usually, the effects of irritants with low solubility are mild throat irritation and occasionally headache. Much more significant is pulmonary edema, which manifests itself 6 to 24 hours after exposure, preceded by symptoms of bronchospasm—chest tightness and wheezing (Fig. 25-5). Ozone and oxides of nitrogen are the two low-soluble irritants most commonly encountered. Both are present in welding fumes, and therefore are found in many work environments. Ozone is used as a disinfectant; as a bleach in the food, textile,

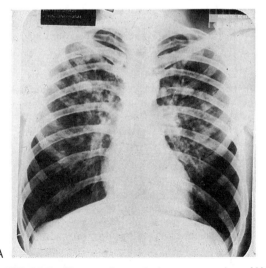 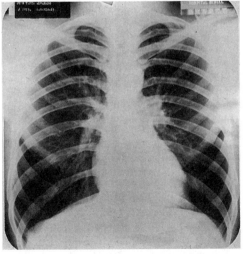

FIG. 25-5. Chest radiographs in a copper miner. **(A)** Twenty-four hours after overexposure to oxides of nitrogen. Pulmonary edema is evident. **(B)** One week after exposure, there is resolution of pulmonary edema. (Courtesy of the late Benjamin G. Ferris, M.D., Harvard School of Public Health, Boston, Massachusetts.)

and pulp and paper industries; and as an oxidizing agent. Oxides of nitrogen are used in chemical and fertilizer manufacture and in metal processing and cleaning operations.

Chronic exposure to oxides of nitrogen may result in bronchiolitis obliterans. An acute obstructive defect is revealed by tests. The chest radiograph may show early pulmonary edema and the appearance of bronchiolitis obliterans.

A specific syndrome associated with oxides of nitrogen is silo filler's disease, which results from exposures to this gas in the upper chambers of grain silos, where it forms in the anaerobic fermentation of green silage. The brownish color of nitrogen dioxide is an important warning sign for farmers. Numerous instances of acute overexposure and death have resulted from inadequately ventilated silos.

Although management and prevention are similar to those for highly soluble irritants, overnight observation of patients is frequently necessary when excess exposure has occurred because of the insidious onset of pulmonary edema.

OCCUPATIONAL ASTHMA

An 18-year-old woman arrived at an emergency department complaining of shortness of breath. Eight weeks previously, she had consulted her physician about daytime wheezing and cough productive of white phlegm. She was treated with antibiotics and an expectorant and remained at home for 3 days with significant improvement. A week later, a cough and shortness of breath again developed. Again, she was treated with antibiotics, an expectorant, and bed rest with significant improvement. She had an exacerbation of coughing, shortness of breath, and cyanosis of her fingertips the day before her visit.

Her occupational history revealed that she had begun working at a tool supply and manufacturing company 9 weeks previously, a week before her symptoms began. Her usual job there was grinding carbide-steel drill bits. In her work, she used one of four machines that sharpened drill bits. Her machine generated much metal dust,

often covering the machines and her face, hands, and clothes. There was no exhaust system to draw dust away from her breathing zone, and no respiratory protection was provided.

After being treated for the first time 8 weeks previously, she was temporarily assigned to cleaning drill bits in a solvent bath. On this job she felt lightheaded but had no difficulty breathing. After a long holiday weekend, she was again assigned to drill bit grinding and after several hours acquired a cough. The next day, the cough increased and she experienced shortness of breath, prompting a second visit to her physician. When she improved from that episode, she returned to work again and experienced exacerbation of coughing and shortness of breath. This prompted her emergency department visit.

Past medical history revealed occasional seasonal rhinitis as a child but no asthma, eczema, or other allergies. There was no family history of allergies or asthma.

Physical examination revealed a pulse rate of 128 and a respiratory rate of 40. She had cyanosis of the lips and fingertips. Chest examination revealed diffuse bilateral wheezes and use of accessory muscles for breathing.

Arterial blood gases in room air at rest revealed a Pao_2 of 39 mmHg. Spirometry showed a normal FVC, but a markedly abnormal FEV_1 (53% of predicted). Chest radiography was normal. White blood cell count was 11,200 cells/mm^3, with 10% eosinophils.

She was treated with oxygen, bronchodilators, and steroids. She improved clinically; by the second day, her FEV_1 had improved to 82% of predicted.

A later call by her physician to the state occupational safety and health agency revealed that carbide-steel bit alloys contain nickel, cobalt, chromium, vanadium, molybdenum, and tungsten. Grinding such bits can produce cobalt and tungsten carbide dusts, which are recognized pulmonary sensitizers.

The diagnosis in this case was occupational asthma. No specific agent was proved responsible, but the presence of tungsten carbide and cobalt dusts suggest probable agents. Since changing jobs, she has felt well and has not had further bronchospasm.[3]

---

[3] Case courtesy of the late James Keogh, M.D., University of Maryland School of Medicine, Baltimore, Maryland (unpublished curriculum materials).

Many occupational asthma cases are not seen by physicians or other health care providers, probably because workers recognize the association between exposure and asthma and thus avoid further contact. There are serious economic and employment costs as a consequence of occupational asthma.

Individual responses may be so clear and occur so early in a new job that those workers who respond adversely may leave quite soon after being hired. Thus, in population surveys, very few workers may be identified with immediate sensitivity because most of those who had experienced adverse effects had already left the job to avoid the asthma-producing exposure. A wide variety of materials and circumstances have been shown to cause occupational asthma (see Table 25-3).

Diagnosis of Occupational Asthma

Diagnosis of occupational asthma depends greatly on the occupational history. Major or minor constituents of substances as well as accidental byproducts can incite attacks. Many people with occupational asthma have a history of atopy, especially when the exposure is to high–molecular-weight compounds. However, those without such a history may become sensitized after exposure to specific environmental agents such as diisocyanates. Suspicion of this diagnosis should be aroused even when a worker has had no previous history of asthma. Often the worker reports wheezing, chest tightness, shortness of breath, or severe cough developing in the evening or at night with recovery overnight or over a weekend away from work. However, if exposure and its effects have been prolonged, the symptoms may persist at home or over the weekend. Specific questioning about nocturnal symptoms may elicit responses otherwise not volunteered. The physical examination of an acutely ill worker reveals wheezing and rhonchi.

A particularly useful test for bronchoconstriction of occupational origin is the FEV_1 before and after a work shift. A drop of at least 300 mL, or 10%, of the FEV_1 (measured as the mean of the two best of three acceptable results each time) between the beginning and end of the first shift of the work week suggests a work-related effect. An acute drop in FEV_1 as large as 1.8 L has been measured without the worker reporting symptoms. Serial measurements of peak flow, such as four times daily, both on days at and days away from work, with a simple, inexpensive peak-flow meter can be extremely valuable in detecting work-associated declines in airflow (8). Peak-flow monitoring has also become a mainstay of asthma management. Excessive eosinophils in the sputum or blood may distinguish asthma from bronchitis. Reliance should not be placed on skin tests for diagnosing allergic reactions because skin and bronchial responses do not always correlate well. Specific bronchoprovocation with suspected offending agents is usually not needed for diagnosis and can be dangerous. Nonspecific bronchoprovocation testing may be necessary to confirm reversible airways disease. Because virtually any chemical substance can precipitate an asthma attack, physicians should rely heavily on the patient's medical and work histories even in the absence of a documented association between a given exposure and asthma. Clinical guidelines for diagnosis and management of occupational asthma have been formulated recently and published (9).

Acute care of those with attacks of occupational asthma is the same as for any case of asthma. Long-term management, however, almost always requires removal from exposure, because after sensitization even very low levels of exposure can trigger an asthmatic response. Close monitoring of symptoms and lung function should be maintained for a person who must continue exposure to a suspected offending agent (10).

An important, and increasingly prevalent challenge, is the recognition, management, and prevention of occupationally exacerbated asthma (i.e., workplace triggering of symptoms and airflow obstruction in a person with otherwise controlled asthma). All people with asthma are at risk, and the inciting conditions may be chemical, biologic, or physical (11).

HYPERSENSITIVITY PNEUMONITIS

Hypersensitivity pneumonitis refers to reactions associated with the most picturesque of all occupational disease names (Table 25-4). This response results from organic materials, commonly fungi or thermophilic bacteria, that are present in a surprising variety of settings. In contrast to asthma, this response is more focused in the lung parenchyma (respiratory bronchioles and alveoli). Characteristics of this kind of reaction include antibodies (precipitins) in serum and the collection of lymphocytes in pulmonary infiltrates. Activation of pulmonary macrophages with an increased number of T lymphocytes and probably a change in their function appear to be the underlying cellular mechanisms. The end result can be fibrosis, yet the responses are much less dose dependent than those for primary fibrosis due to inorganic dusts. Once hypersensitivity is established, small doses may trigger episodes of alveolitis. This disease is a complex inflammatory response, often due to bacterial or fungal material—neither an infection nor a true allergic response. Therefore, the commonly used clinical term *hypersensitivity pneumonitis* is inaccurate. Research has focused on the etiologies, pathophysiology, treatment, and prevention of this condition.

Outbreaks of hypersensitivity pneumonitis have been described in workers using metal fluids (coolants) (12,13).

The worker with hypersensitivity pneumonitis experiences shortness of breath and nonproductive cough. In contrast to asthma, wheezing is not a prominent component. In acute episodes, the sudden onset of the respiratory symptoms along with fever and chills is dramatic. Physical examination may show rapid breathing and fine basilar rales. Pulmonary function tests can show marked reduction in lung volumes consistent with restrictive disease. The FEV_1 is reduced, but in proportion to the decreases in FVC and total lung capacity; in general, there is a normal or increased FEV_1/FVC ratio. Arterial blood gas measurements show an increased alveolar–arterial oxygen difference $[P(A - a)O_2]$ and a reduced lung diffusing capacity for carbon monoxide (D_{LCO}). A chest radiograph

TABLE 25-4. *Examples of hypersensitivity pneumonitis*

Disease	Antigenic material	Antigen
Farmer's lung	Moldy hay or grain	Thermophilic actinomycetes
Bagassosis	Moldy sugar cane	
Mushroom worker's lung	Mushroom compost	
Humidifier fever	Dust from contaminated air conditioners or furnaces	
Maple bark disease	Moldy maple bark	*Cryptostroma* species
Sequoiosis	Redwood dust	*Graphium* species, Pallurlaria
Bird breeder's lung	Avian droppings or feathers	Avian proteins
Pituitary snuff user's lung	Pituitary powder	Bovine or porcine proteins
Suberosis	Moldy cork dust	*Penicillium* species
Paprika splitter's lung	Paprika dust	*Mucor stolonifer*
Malt worker's lung	Malt dust	*Aspergillus clavatus* or *Aspergillus fumigatus*
Fishmeal worker's lung	Fishmeal	Fishmeal dust
Miller's lung	Infested wheat flour	*Sitophilus granarius* (wheat weevil)
Furrier's lung	Animal pelts	Animal fur dust
Coffee worker's lung	Coffee beans	Coffee bean dust
Chemical worker's lung	Urethane foam and finish	Isocyanates (e.g., toluene diisocyanate), anhydrides

can be helpful in acute episodes by revealing patchy infiltrates or a diffuse, fine micronodular shadowing.

If the person is removed from exposure, symptoms and signs usually disappear in 1 to 2 weeks. If repeated exposures are experienced, especially at levels low enough to result in only mild symptoms, a more chronic disease may ensue. The worker may be unaware of the work association because the low-level effects may appear symptomatically like a persistent or intermittent respiratory "flu." Over a period of months, however, there is a gradual onset of dyspnea, which can be accompanied by weight loss and lethargy. The physical examination is similar to that in the acute episode, although the patient may appear less acutely ill and may demonstrate finger clubbing. The chest radiograph, however, is more suggestive of chronic interstitial fibrosis, and the pulmonary function tests show a restrictive defect. The disease may progress to severe dyspnea and the end result resembles, even histologically, chronic interstitial fibrosis of unknown etiology. Sometimes an asymptomatic patient without an episode of acute pneumonitis in the past acquires interstitial fibrosis.

Prevention rests on removal from exposure. This can be more readily accomplished than with asthma because environmental controls can focus on the elimination of conditions that foster bacterial or fungal growth. Process changes may also be necessary to prevent antigen production, and local exhaust ventilation, rather than personal protective equipment (masks), should be used.

BYSSINOSIS AND OTHER DISEASES CAUSED BY VEGETABLE DUSTS

Some types of airway constriction are believed to be due not to sensitization, but to direct toxic effects on the airways. This has been referred to as *pharmacologic bronchoconstriction*. For byssinosis, however, the pathogenesis is still poorly understood.

Byssinosis (meaning *white thread* in Greek) is associated with exposure to cotton, hemp, and flax processing. It has been popularly called *brown lung* (a misnomer because the lungs are not brown), by analogy to the popular term *black lung* used to describe CWP.

Byssinosis has been shown to develop in response to dust exposure in cotton processing. It is especially prevalent among cotton workers in the initial, very dusty operations where bales are broken open, blown (to separate impurities from fibers), and carded (to arrange the fibers into parallel threads). A lower prevalence of disease occurs in workers in the spinning, winding, and twisting areas, where dust levels are lower. The lowest prevalence of byssinosis has been found among weavers, who experience the lowest dust exposure. Processing of cloth is practically free of cotton dust, as in the manufacture of denim, which is washed during dyeing, before thread is spun. Byssinosis has also been described in other than textile sectors where cotton is processed, such as cottonseed oil mills, the cotton waste utilization industry, and the garnetting, or bedding and batting, industry. The same syndrome has been shown to occur in workers exposed in processing soft hemp, flax, and sisal.

Byssinosis is characterized by shortness of breath and chest tightness. These symptoms are most prominent on the first day of the work week or after being away from the factory over an extended period of time ("Monday morning tightness"). No previous exposure is necessary for symptoms to develop.

Symptoms are often associated with changes in pulmonary function. One characteristic of the acute pulmonary response to cotton dust exposure is a drop in the FEV_1 during the Monday work shift or the first day back at work after at least a 2-day layoff. Because workers do not normally lose lung function during a workday, an acute loss of at least 10% or 300 mL (whichever is greater) in an individual, or 3% or 75 mL (whichever is greater) in a group of 20 or more workers, can be considered significant enough to require further investigation. Over time, cotton dust workers have an accelerated decrement in FEV$_1$ consistent with fixed airflow obstruction and chronic obstructive lung disease. Diagnosis is based mainly on symptoms; no characteristic examination or chest radiographic findings are associated with byssinosis. Therefore, the patient should be questioned systematically about symptoms.

It is assumed that the disease progresses if duration of exposure to sufficiently high dust levels is prolonged. Mild byssinosis probably is reversible if exposure ceases, but long-standing disease is irreversible. People with severe byssinosis are rarely seen in an industrial survey because they are too disabled to be working. Byssinosis seems more severe when it is associated with chronic bronchitis. The end stage of the disease is fixed airway obstruction with hyperinflation and air trapping. Cigarette smokers are at increased risk of irreversible byssinosis.

Much research has been done on possible etiologic mechanisms and effects. Extracts of cotton bract have been shown to release pharmacologic mediators, such as histamine, as well as prostaglandins. It seems likely that the mechanism of byssinosis involves stimulation of the same inflammatory receptors by endotoxin and by cotton dust. Gram-negative bacterial endotoxin contaminates cotton fiber, and aqueous extracts of endotoxin have produced acute symptoms and lung function declines.

Two other respiratory conditions are associated with work in the cotton industry:

1. *Mill fever*—This self-limited condition usually happens on first exposure to a cotton dust environment. It lasts for 2 or 3 days and has no known sequelae. It is characterized by headache, malaise, and fever. A flu-like illness, it has symptoms similar to metal fume fever and polymer fume fever (see Box 20-3 in Chapter 20). Mill fever is probably related to gram-negative bacterial material in mill dust; it usually affects workers only once, but after prolonged absence from a mill, reexposure may trigger another attack.
2. *Weaver's cough*—Weavers have experienced outbreaks of acute respiratory illness characterized by a dry cough, although their dust exposure is comparatively low. It may result from sizing material or from mildewed yarn that is sometimes found in high-humidity weaving rooms.

Other organic/vegetable materials are associated with respiratory diseases, including flax (baker's asthma), swine confinement buildings (acute airflow obstruction), and wood dust (asthma, chronic airflow obstruction). Evidence is accumulating that chronic exposure to organic dusts can result in both acute and chronic lung disease (14).

CHRONIC RESPIRATORY TRACT RESPONSES

Pneumoconiosis

Pulmonary fibrosis is a well documented work-related chronic pulmonary reaction. This condition, which varies according to inciting agent, intensity, and duration of exposure, is generally referred to as a *pneumoconiosis*. It is usually due to an inorganic dust or coal that must be of respirable size (<5 μm) to reach terminal bronchioles and alveoli; dust of this size is not visible and so its presence may not be recognized by a worker. There are two basic types of fibrosis: (a) localized and nodular, usually peribronchial fibrosis; and (b) diffuse interstitial fibrosis. The clinical features of all pneumoconioses are similar: initial nonproductive cough, shortness of breath of increasing severity, and, in the later stages, productive cough, distant breath sounds, and signs of right heart failure. The pneumoconioses are often associated with obstructive airways disease caused by the same agents.

Silica-Related Disease

Crystalline silica (SiO_2) is a major component of the earth's crust. Therefore, exposure occurs in a wide variety of settings, such as mining, quarrying, and stone cutting (see Fig.

8-4 in Chapter 4); foundry operations; ceramics and vitreous enameling; and in use of fillers for paints and rubber.

Estimates of the prevalence of silicosis in the United States vary, ranging from 30,000 to 100,000 current cases. No distinct clinical features can be cited beyond the ones already listed, but there is distinct pathologic process. Silicosis occurs more frequently in the upper rather than the lower lobes, with nodules varying in size from microscopic to 6 mm in diameter. In severe cases, nodules aggregate and become fibrotic masses several centimeters in diameter. Nodules are firm and intact with a whorled pattern, and rarely cavitate (Fig. 25-6). Microscopically, the nodules are hyalinized, with a well organized circular pattern of fibers in a cellular capsule. The amount of fibrosis appears proportional to the free silica content and to the duration of exposure. One notable characteristic of this disease is that fibrosis progresses even after removal from exposure. Except in acute silicosis, symptoms usually do not occur until 10 to 20 years after initiation of exposure. Evidence of pathologic response to silica exposure exists well before symptoms occur.

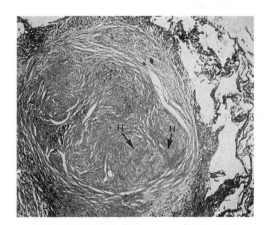

FIG. 25-6. Microscopic section of a typical silicotic nodule showing the concentric (onion skin) arrangement of collagen fibers, some of which are hyalinized (H); lack of dust pigmentation; and peripheral cellularity. The lesion is clearly demarcated from adjacent lung tissue, which is substantially normal. (From Parkes WR. Occupational lung disorders, 2nd ed. London: Butterworths, 1982.)

Evaluation of workers exposed to silica includes lung function tests (which may show reduced FVC or total lung capacity, or mixed obstructive and restrictive patterns), a chest radiograph (which may appear more abnormal than the lung function tests), and determination of (a reduced) hemoglobin oxygen saturation on exercise. As the disease progresses, there can be decreased oxygen saturation at rest and reduced total lung capacity. The radiograph usually shows rounded opacities, localized initially to the upper lung fields (see Fig. 25-2). The size and distribution of these opacities increase over time, and "eggshell" calcification of hilar lymph nodes occurs in a few cases.

Chronic silicosis is classified either as a simple or complicated, although there is a continuum between these two forms of the disease. The simple form is noted on the chest film by the presence of multiple, small, round opacities, usually in the upper zones. The concentrations of these opacities are used in classifying simple silicosis (categories 1 to 3) (2). Although simple silicosis alone is not a common cause of disability, it can contribute to disability as well as progress to complicated silicosis. In progressive massive fibrosis (PMF), several of the simple nodules appear to aggregate and produce larger conglomerate lesions, which enlarge and encroach on the vascular bed and airways (ILO categories A, B, and C). The extent of lung function impairment appears directly related to the radiographic size of the lesions and is most severe in categories B and C.

An important complication of silicosis is tuberculosis (TB), which persists today as an added hazard peculiar to this pneumoconiosis. The association between silicosis and pulmonary TB has been known for decades. More recent publications also show an increased incidence of TB among workers in the mining, quarrying, and tunneling industries, and steel and iron foundries. Workers exposed to silica may be at increased risk of TB even in the absence of radiographic evidence for silicosis. Infections with atypical mycobacteria such as *Mycobacterium kan-*

sasii and *Mycobacterium avium-intracellulare* can also occur and are related to the geographic distribution of these organisms. Treatment of such cases may require more vigorous drug treatment than TB without silicosis. No relationship has yet been shown between silicosis and cigarette smoking.

Another potential complication of silicosis exposure is lung cancer. Epidemiologic studies have shown a link between silicosis and lung cancer, and the International Agency for Research on Cancer (IARC) has classified silicosis as a class 1 human carcinogen.

Prevention of silicosis focuses on reduction of exposure through wet processes, isolation of dusty work, and local exhaust ventilation. Annual TB screening by purified protein derivative (PPD) skin testing or, if the PPD is positive, chest radiography is essential in silica-exposed workers. There is an ongoing national effort to eliminate silicosis in all sandblasting operations. Elimination of silicosis from each work practice would reduce the at-risk population substantially.

Acute silicosis, a distinct entity, is a devastating disease. It is due to extraordinarily high exposures to small silica particles (1 to 2 μm). These exposures occur in abrasive sandblasting and in the production and use of ground silica. Symptoms include dyspnea progressing rapidly over a few weeks, weight loss, productive cough, and sometimes pleuritic pain. Diminished resonance on percussion of the chest and rales on auscultation can be found. Lung function tests show a marked restrictive defect, with an impressive decrement in total lung capacity. The radiograph has a diffuse ground-glass, or miliary TB–like appearance, rather than the classic nodular silicosis. The pathologic process in this disease is characterized by a widespread fibrosis, with a diffuse interstitial, rather than nodular, macroscopic appearance, and a microscopic appearance and chemical constituency resembling pulmonary alveolar proteinosis, but with doubly-refractile particles of silica lying free within the alveolar exudate. Disease onset usually occurs 6 months to 2 years after initial exposure. Acute silicosis is often fatal, usually within 1 year of diagnosis.

Diatomaceous earth is an amorphous silica material mined predominantly in the western United States. It is used as a filler in paints and plastics, as a heat and acoustic insulator, as a filter for water and wine, and as an abrasive. In contrast to the various forms of crystalline silica, amorphous silica has relatively low pathogenicity. However, some processes using diatomaceous earth include heating (calcinating) it to remove organic material. This heating process can produce up to 60% crystalline silica as cristobalite, which is highly fibrogenic. Exposure to this form of diatomaceous earth, therefore, must be treated the same as exposure to crystalline silica.

Silica appears in a wide variety of minerals in different combined forms known as silicates. Many of these silicates, such as asbestos, kaolin, and talc, also cause pneumoconiosis, but the forms they produce have features distinct from those of silicosis. Asbestos is the most widespread and best known of the silicates and is responsible for asbestosis as well as several types of cancers (see Chapter 16).

Asbestos appears in nature in four major types (chrysotile, crocidolite, amosite, and anthophyllite) that produce similar chronic respiratory reactions. All four types are characterized by being fibrous and are indestructible at temperatures as high as 800°C. Use and production of these materials has greatly increased in the past century; more than 3,000,000 tons of asbestos are produced in the world annually. Over 30 million tons have been used in construction and manufacture in the United States alone. Asbestos is used in a variety of applications: asbestos cement products (tiles, roofing, and drain pipes), floor tile, insulation and fireproofing (in construction and ship building), textiles (for heat resistance), asbestos paper (in insulating and gaskets), and friction materials (brake linings and clutch pads). Probably the most hazardous current exposures occur in repair and demolition of buildings and ships

FIG. 25-7. Brake mechanic exposed to asbestos fibers while using compressed air to clean brake drum. (Photograph by Nick Kaufman.)

and in a variety of maintenance jobs where exposures may be unsuspected by the workers (Fig. 25-7). In the United States, the construction industry is the major source of asbestos exposure to workers, mainly from asbestos products in place.

As with silicosis, the predominant symptoms of asbestosis are shortness of breath, which may be more severe than the appearance of the chest radiograph might indicate, and cough. Although not common, pleuritic pain or chest tightness may occur, and these are more frequent than in other pneumoconioses. In 20% of those affected, basilar rales are present, heard best at the end of inspiration or early expiration, and pleural rubs and pleural effusions can occur. Pleural effusion in a person with a history of asbestos exposure even many years earlier should be evaluated for mesothelioma, although benign asbestos effusions also occur.

Pathologically, the lung appears macroscopically as a small, pale, firm, and rubbery organ with a fibrotic adherent pleura. The cut surface shows patchy to widespread fibrosis, and the lower lobes are more frequently affected than the upper. The microscopic appearance is characterized by interstitial fibrosis. Chest radiography shows widespread irregular (linear) opacities more common in the lower lung fields, in contrast to the round opacities seen in silicosis, which occur first in the upper lung fields.

A great deal of attention has focused on asbestos (or ferruginous) bodies in sputum and lung tissue. These are dumbbell-shaped bodies 20 to 150 μm in width that appear to be fibers covered by a mucopolysaccharide layer. Iron pigment (from hemoglobin breakdown) makes them golden-brown. They are not diagnostic of asbestos-related disease, but when present even in small numbers in sputum or tissue sections, they indicate substantial occupational exposure to airborne fibers. Most urban dwellers in industrialized countries have a measurable asbestos burden, but the concentrations of asbestos bodies in the nonoccupationally (or paraoccupationally) exposed populations are orders of magnitude lower than in those with known occupational exposures. Pathology studies have shown that in the "background" population of urban dwellers, 50 to 100 microscopic sections of lung would have to be searched to find a single asbestos body, whereas people with very early asbestosis have asbestos bodies in nearly every section and those with more severe asbestosis have scores of asbestos bodies per section. Asbestos bodies may also be found in other parts of the body besides the lungs; they form round fibers that are transported by lung lymphatics into the circulation.

A particular feature of asbestos exposure, unlike other pneumoconioses, is the frequent presence of asbestos-induced circumscribed pleural fibrosis, known as pleural plaques, which are sometimes the only evidence of exposure. These plaques, which can calcify, may be bilateral, and are located more commonly in the parietal pleura. In fact, the evidence for prior asbestos exposure or the explanation of abnormal pulmonary function tests may sometimes be found because of the calcified pleural plaques seen on chest radiography (Fig. 25-8).

Pleural plaques are one manifestation of the rather marked pleural reaction to asbestos fibers. Other such evidence seen on the chest radiograph is a "shaggy"-appearing

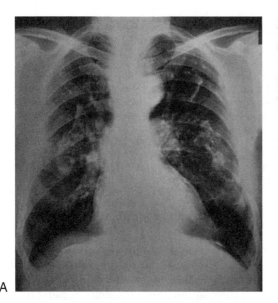
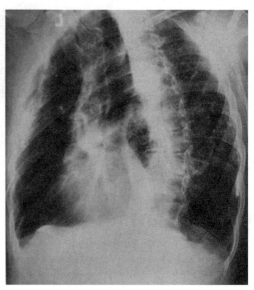

A   B

FIG. 25-8. Bilateral calcified pleural plaques on chest walls and diaphragm. **(A)** Note irregular outline and variable density of the large legion seen *en face* and the rim of calcification along the left cardiac border. The small, rounded lesions also represent calcification in plaques and are not intrapulmonary. **(B)** The large plaque in the right lung field on the posteroanterior film is seen end-on against the chest wall (left field). There is no evidence of diffuse interstitial pulmonary fibrosis in either film. The patient was an ex-insulation worker (1925 to 1932), 65 years of age. There were no crackles in the lungs, and lung function testing showed severe airflow obstruction and hyperinflation only. (From Parkes WR. Occupational lung disorders, 2nd ed. London: Butterworths, 1982.)

cardiac or diaphragmatic border. An early, nonspecific sign is a blunted costophrenic angle. Diffuse pleural thickening also occurs, probably less commonly than the more specific pleural plaques. Asbestos-induced diffuse visceral pleural fibrosis may also occur and may impair lung function. Advanced pleural fibrosis may act like a cuirass, severely constricting breathing and leading to respiratory failure.

The evaluation of a worker suspected of having asbestosis includes determining if there has been a history of exposure; a physical examination to ascertain if rales are present; a chest radiograph, which may show irregular linear opacities and a variety of pleural reactions; and pulmonary function tests, which may show evidence of an interstitial type of abnormality—that is, restrictive disease and a diminished D_{LCO}. In addition, the peribronchiolar fibrosis may have an obstructive component. Hence, in both nonsmokers and smokers with asbestosis (as with all pneumoconioses), a mixed restrictive–obstructive pattern may be seen.

Asbestosis, like silicosis, may progress after removal from exposure. Asbestos exposure even without asbestosis carries with it the added risk of cancers of the lung, pleura and peritoneum (mesotheliomas), gastrointestinal tract, and other organs (see Chapter 16). Prevention focuses on substitution with materials such as fibrous glass, use of wet processes to reduce dust generation, local exhaust ventilation to capture the dust that is generated, and respiratory protection.

Exposed patients who smoke should be advised to stop smoking for the rest of their lives.

Talc is a hydrated magnesium silicate that occurs in a variety of natural forms. The two major types are nonfibrous and fibrous. The nonfibrous forms, such as those found in Vermont, are free of both crystalline silica and fibrous asbestos tremolite; the fibrous forms, such as those found in New York State, can contain up to 70% fibrous material, including amphibole forms of asbestos. Talc exposures occur mainly during its use as an additive to paints and as a lubricant in the rubber industry, especially in innertubes. Evidence suggests that high doses of nonfibrous talc or moderate doses of fibrous talc accumulated over a long time result in chronic respiratory disease known as *talcosis*, with the same symptoms as other pneumoconioses.

Pathologically, the macroscopic appearance of the lung is characterized by poorly structured nodules, unlike the firm nodules of silicosis and the diffuse fibrosis of asbestosis. The microscopic appearance consists of ill-defined nodules with some diffuse interstitial fibrosis. Evaluation of people exposed to talc includes pulmonary function tests and a chest radiograph. The chest radiograph may show both nodular and linear opacities and also pleural plaques. Studies addressing the possibility of a cancer risk associated with fibrous talc exposure found a fourfold increased risk of lung cancer in New York State talc miners.

Kaolin (China clay) is a hydrated aluminum silicate found in the United States (in a band from Georgia to Missouri), India, and China. It is used in ceramics; as a filler in paper, rubber, paint, and plastic products; and as a mild soap abrasive. Kaolin is not particularly hazardous in the mining processes because it is usually a wet ore and mined by jet-water mining techniques.

The pneumoconiosis resulting from chronic exposures to kaolin dust produces no unique clinical features. Pathologically, the macroscopic appearance is one of immature silicotic nodules, although conglomerate nodules may appear. Pleural involvement occurs only if the lung is massively involved. The microscopic appearance consists of nodules with randomly distributed collagen.

Coal Workers' Pneumoconiosis

In the United States until the 1960s, coal workers' respiratory disease was considered a variant of silicosis and was often known an *anthracosilicosis*. It is now clear that CWP is an etiologically distinct entity that can be induced by both coal dust and pure carbon. CWP exists both in uncomplicated and complicated forms; the latter, known as progressive massive fibrosis (PMF), is the most severe form of the disease. Although exposure to coal dust occurs most commonly in underground mines, there is also some exposure in handling and transportation of coal. Significant exposure also occurs in the trimming or leveling of coal in ships when preparing material for transport.

Uncomplicated CWP increases the likelihood for future development of the complicated form, which is generally agreed to be a disabling condition. The diagnosis of CWP has relied primarily on the chest radiograph, which shows nodular opacities of less than 1 cm (mostly <3 mm) in diameter. PMF, in contrast, is seen on chest radiography as the development of conglomerations of these small opacities to sizes greater than 1 cm in diameter.

In the early stages, CWP is asymptomatic. The initial symptoms are dyspnea (breathlessness) on exertion with progressive reduction in exercise tolerance. As nodular conglomeration begins and PMF is diagnosed, symptoms become more severe, with marked exertional dyspnea, severe disability, or total incapacity. There is general agreement that PMF leads to premature disability and death. No such agreement, however, exists for the impact of simple CWP of grade 2 or less.

Coal dust also contributes independently to the disability observed in coal workers through the production of chronic bronchitis, airways obstruction, and emphysema. The bronchitis and pulmonary function loss is dose related to coal dust in both smokers and nonsmokers. The greater the intensity and duration of exposure (cumulative exposures), the more likely that a miner will get any of these diseases, as well as silicosis if the quartz content of the coal is high. Moreover, the diseases may present in any combination.

Pathologically, CWP appears as soft, black, indurated nodules. Microscopic observation shows dust in and around macrophages near respiratory bronchioles. Nodules show random collagen distribution and the lung shows centrilobular emphysema. Chest radiography shows widely distributed, small, round opacities.

In PMF, the large conglomerate masses have variable shapes and do not respect the architecture of the lung. The surfaces are hard, rubbery, and black, and cavitation often occurs (Fig. 25-9). Copious, black sputum is often produced. Microscopically, the appearance is not distinct from the simple nod-

ules. Chest radiography shows large conglomerate opacities (Fig. 25-10). A separate condition, called Caplan's syndrome, or rheumatoid CWP, is PMF accompanied by rheumatoid arthritis. It has a different pathologic appearance, with alternate black and gray-white bands of material in the conglomerate masses. The conglomerate masses frequently cavitate or calcify. Whether there is a different clinical course for people with PMF accompanied by rheumatoid arthritis is not known.

Although evaluation for CWP is the same as for the other pneumoconioses, a particular feature affecting evaluation is the federal Mine Safety and Health Act of 1977, which prescribes what types of abnormalities make a person eligible for disability benefits (see Box 10-1 in Chapter 10). Because these are subject to continuous revision, consultation with the Mine Safety and Health Administration in the U.S. Department of Labor is advisable. Miners enjoy special rights to a low-dust environment with increased medical

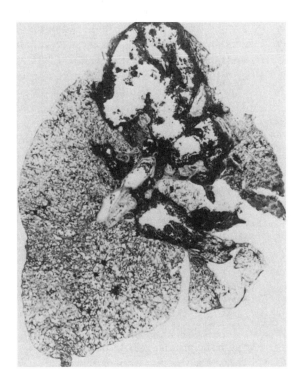

FIG. 25-9. Gough section of lung of coal worker with 18 years of mining experience completed 20 years before death. It shows cavitation as well as centrilobular emphysema, which was present in both lungs. (Courtesy of J. C. Wagner, MRC Pneumoconiosis Unit, Llandough Hospital, Penarth, Wales, United Kingdom.)

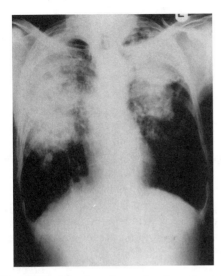

FIG. 25-10. Chest radiograph of coal worker whose lung section appears in Fig. 25-9, taken 2 weeks before death. The appearance is classic for progressive massive fibrosis with larger conglomerate masses in both lung fields. (Courtesy of J. C. Wagner, MRC Pneumoconiosis Unit, Llandough Hospital, Penarth, Wales, United Kingdom.)

monitoring if they are found to have CWP, and they have the right to permanent removal from the high-dust environment with wage retention. These rights are unique among American workers, although such an approach should be applied in the prevention of all pneumoconioses.

Miscellaneous Inorganic Dusts

Fibrous glass and related products, referred to as synthetic vitreous fibers (SVF), man-made vitreous fibers, or very fine vitreous fibers, have been used for insulation purposes for over 60 years. More recently, they have played an important role as an asbestos substitute. SVF are amorphous silicates with a length-to-diameter ratio of greater than 3:1. They are made mainly from rock, slag, glass, or kaolin clay, and can be divided into three main groups: mineral wool, fibrous glass, and ceramic fiber.

Synthetic vitreous fibers can induce skin and upper respiratory tract irritant responses. There have been few case reports of pulmonary disease due to SVF exposure. Prevalence studies of chest radiographic findings, respiratory symptoms, and lung function in exposed workers have in general been negative. However, limited studies of workers exposed to fine-diameter fibers have revealed evidence of irregular opacities consistent with pneumoconiosis. No abnormalities in lung function were reported in these studies. There is growing concern about the possible carcinogenicity of these very fine fibers. The IARC has classified glass wool, rock wool, slag wool, and ceramic fiber as possibly carcinogenic to humans; continuous glass fiber as not classifiable as to human carcinogenicity. Epidemiologic studies have suggested some chronic pulmonary effects, including lung cancer, are associated with SVF exposures, but a number of studies are still in progress, and longer-term follow-up is necessary. Long-term studies of employment in industries using the respirable sizes of SVF are now being performed. Although few data are available with regard to neoplastic diseases in workers exposed to SVF, in chronic inhalation studies, ceramic fibers produce an increase in the incidence of mesothelioma in rats and hamsters and an increased incidence of lung tumors in rats. Because of persistent uncertainties, occupational exposures to SVF should be lowered as much as possible with engineering controls, proper worker training, and safe work practices.

Individual exposures to iron dusts, particularly those resulting from steel-grinding operations, welding, or foundry work, are common. The only clinical effect of pure iron oxide exposure is a reddish-brown coloring of the sputum. Lung function tests show no clinical abnormality, whereas the chest radiograph shows many small (0.5 to 2.0 mm) opacities without confluence (as with stannosis; see Fig. 25-3). Lung sections show macrophages laden with iron dust but without fibrosis or cellular reaction. With removal from further iron oxide dust exposure, the radiographic abnormalities slowly resolve.

Similar results can be seen in exposures to tin, barium, and antimony.

Chronic Bronchitis

Probably the most common of the chronic responses of the respiratory tract is chronic bronchitis, which results from excessive mucus production in the bronchi. Diagnosis is made strictly on clinical grounds. *Chronic bronchitis* is a formally defined diagnosis that must meet ATS criteria: recurrent productive cough occurring four to six times a day at least 4 days of the week, for at least 3 months during the year, for at least 2 years. The definition of simple bronchitis—the production of phlegm on most days for as much as 3 months of the year—can be used to distinguish those with probably important symptoms from those without. The excess mucus production associated with bronchitis is often associated with airflow obstruction. Chronic bronchitis is not a unique occupational pulmonary response; it is frequently superimposed on other respiratory diseases due to occupational toxins and most often cigarette smoke. Occupational toxins that can cause chronic bronchitis include mineral dusts and fumes (e.g., from coal, fibrous glass, asbestos, metals, and oils), organic dusts, irritants (e.g., ozone and oxides of nitrogen), plastic compounds (e.g., phenolics and isocyanates), acids, and smoke (e.g., as experienced in firefighting).

Emphysema

Emphysema is a chronic response that depends more specifically on a pathologic description: it is the enlargement of air spaces distal to terminal (nonrespiratory) bronchioles that includes destruction of the alveolar walls and results in air trapping. Occupational examples of this response are not well studied, but evidence suggests that fixed airway obstruction is the end stage of disease due to chronic coal dust or chronic cadmium exposure.

Granulomatous Disease

Another type of chronic response not commonly described as work related is granuloma formation. In a granuloma, many cells responding to an inciting agent become surrounded by bundles of collagen. The foreign body granuloma in the skin is an analogous kind of tissue reaction. The best occupational example of pulmonary granulomas is chronic beryllium disease; workers who make metal alloys containing beryllium are exposed when dust control is poor. The disease appears as a restrictive pneumoconiosis, although the pulmonary reaction is out of proportion to the amount of metal dust in the lungs. It is very similar to sarcoid and can be impossible to distinguish without measuring tissue levels (in lung and lymph nodes) of beryllium. A more specific test, available at a few academic centers, is the lymphocyte blast transformation test (LTT) on peripheral or lavaged lymphocytes. The LTT has been useful in the early diagnosis of beryllium disease. In this test, the proliferation of lymphocytes cultured in the presence of a beryllium salt is assessed by measuring the incorporation of radiolabeled thymidine. This *ex vivo* test is both highly specific and sensitive.

REFERENCES

1. Ferris BG. Epidemiology standardization project. Part 2 of 2. American Review of Respiratory Disease 1978;118:1–120.
2. International Labor Office. Guidelines for the use of the ILO international classification of pneumoconioses. Occupational Safety and Health Series 22, Rev. 80. Geneva: International Labor Office, 1980.
3. Enright PL, Hyatt RE. Office spirometry: a practical guide to the selection and use of spirometers. Philadelphia: Lea & Febiger, 1987.
4. American Thoracic Society. Lung function testing: selection of reference values and interpretation strategies. Official statement of the American Thoracic Society. American Review of Respiratory Disease 1991;188:1208–1218.
5. European Respiratory Society. Official statement, standardized lung function testing: lung volumes and forced ventilatory flows, 1993 update. Report of the Working Party, Standardization of Lung Function Tests. Eur Respir J 1993;6[Suppl 16].

6. Hankinson JL, Odencratz JR, Fedan KB. Spirometic reference values from a sample of the general U.S. population. Am J Respir Crit Care Med 1995;159:179–187.
7. Picrila PL, Nordman H, Korhonen OS, Winblad I. A thirteen year follow-up of respiratory effects of active exposure to sulfur dioxide. Scand J Work Environ Health 1996;22:191–196.
8. National Asthma Education Program. Guidelines for diagnosis and management of asthma, August 1991. DHHS (PHS) Publication no. 91-3042. Washington, DC: U.S. Department of Health and Human Services, Public Health Service, National Institutes of Health, 1991.
9. Wagner GR, Wegman DH. Occupational asthma: prevention by definition. Am J Ind Med 1998;33:427–429.
10. Gassert TH, Hu H, Kelsey KT, Christiani DC. Long-term health and employment outcomes of occupational asthma and their determinants. J Occup Environ Med 1998;40:481–491.
11. Milton DK, Solomon GF, Rosiello RA, Herrick RF. Risk and incidence in asthma attributable and occupational exposure among HMO members. Am J Med 1998;33:1–10.
12. Fox H, Anderson H, Moen T, et al. Metal working fluid-associated hypersensitivity pneumonitis: an outbreak investigation and case–control study. Am J Ind Med 1999;35:58–67.
13. Kreiss K, Cox-Ganser J. Metalworking fluid-associated hypersensitivity pneumonitis: a workshop summary. Am J Ind Med 1997;32:423–432
14. Christiani DC. Organic dust exposure and chronic airflow obstruction. Am J Respir Crit Care Med 1996;156:833–835.

BIBLIOGRAPHY

American Thoracic Society (ATS) Committee of the Scientific Assembly on Environmental and Occupational Health. Adverse effects of crystalline silica exposure. Am J Respir Crit Care Med 1997;155: 761–768.
Expert committee review and statement. Up to date and concise.
Banks DE, Parker JE. Occupational lung disease: an international perspective. London: Chapman and Hall, 1998.
An excellent reference text. Up to date and global in perspective.
Harber P, Schenker M, Balmes J, eds. Occupational and environmental respiratory disease. St. Louis: Mosby-Year Book, 1995.
A well written review of occupational and environmental respiratory diseases.
Parkes WR. Occupational lung disorders, 3rd ed. London: Butterworths, 1992.
Excellent, detailed summary of occupational respiratory disease. Includes clinical and pathologic details. Some terminology is British, but this does not cause a significant problem. Best used as a reference.
Rom WN, ed. Environmental and occupational medicine, 3rd ed. Boston: Little, Brown, 1998.
Excellent general reference text with strong chapters on occupational lung diseases.
Schenker M, Christiani DC, Husman K, et al: ATS Committee on the Scientific Assembly on Environmental and Occupational Health. Respiratory health hazards in agriculture. Parts 1 and 2. Am J Respir Crit Care Med 1998;158.
Excellent, in-depth review of the topic.

26 Musculoskeletal Disorders

Gunnar B. J. Andersson, Lawrence J. Fine, and Barbara A. Silverstein

Part I. Low Back Pain

Gunnar B. J. Andersson

Work-related musculoskeletal disorders commonly involve the back, cervical spine, and upper extremities. Understanding of these problems has developed rapidly during the past decade. The two sections of this chapter provide an overview of these problems and a framework for recognizing and preventing them.

Low back pain is one of the oldest occupational health problems in history. In 1713, Bernardino Ramazzini, the "founder" of occupational medicine, referred to "certain violent and irregular motions and unnatural postures of the body by which the internal structure" is impaired. Ramazzini examined the harmful effects of unusual physical activity on the spine, such as the sciatica caused by constantly turning the potter's wheel, lumbago from sitting, and hernias among porters and bearers of heavy loads.

In addition to being one of the oldest occupational health problems, low back pain is one of the most common. Approximately 80% of workers experience low back pain sometime during their active working life. At any given moment, 10% to 15% of the adult U.S. population experiences low back pain (1,2). Back pain is the most frequent cause of activity limitation in people younger than 45 years of age, the second most frequent reason for physician visits, and the third ranking reason for surgical procedures. Approximately 11% of Americans report low back impairment or a reduced ability to function. Every year, about 2% of the employed population lose time from work because of low back pain, and approximately half of these people receive compensation for lost wages. Lost time from low back pain averages 4 hours per worker per year, and among medical reasons for work absence it is second only to upper respiratory tract infections.

Low back pain is clearly the most costly occupational health problem. Although 16% to 20% of all workers' compensation cases involve low back pain, these cases are responsible for 34% to 40% of the total costs. More than $16 billion (in direct costs) is spent each year on the treatment and compensation of low back pain in the United States (1). Including indirect costs, the total expenses for these back problems may be as

G. B. J. Andersson: Department of Orthopedic Surgery, Rush-Presbyterian-St. Luke's Medical Center, Chicago, Illinois 60612.

L. J. Fine: Division of Surveillance, Hazard Evaluations, and Field Studies, National Institute for Occupational Safety and Health, Cincinnati, Ohio 45226.

B. A. Silverstein: Safety and Health Assessment and Research for Prevention (SHARP), Olympia, Washington 98504-4330.

Many different types of workers are prone to back problems. (Drawing by Nick Thorkelson.)

high as $50 to $80 billion (3). However, back pain expenses are not equally distributed; they are highly biased toward the more expensive cases—25% of low back pain cases account for more than 90% of the expenses. Most cases are relatively inexpensive. Psychological impairments accompany many of the more expensive cases, either preceding or in response to the physical disability. Therefore, in cases of chronic low back pain, both the medical and the psychological aspects must be addressed.

PATHOPHYSIOLOGY

Pain in the lumbosacral spine can result from inflammatory, degenerative, neoplastic, gynecologic, traumatic, metabolic, or other types of disorders. However, the great majority of low back pain is nonspecific and of unknown cause. Many theories regarding the origin of nonspecific low back pain have been proposed, but so far no one has been able to prove how and where the pain arises.

The intervertebral disc as a source of low back pain has attracted much attention, mainly for two reasons. Discs herniate and

the herniated nucleus pulposus (HNP) is such an obvious, dramatic, and reputed cause of back pain and sciatica that most workers know, or know of, someone with this diagnosis. Furthermore, most workers know that a herniated disc may require surgical treatment and that the results vary. The second reason for attention to the disc is that it degenerates, and the degenerative process is readily visible on magnetic resonance images (MRI) and often on radiographs.

The significant focus on the disc as a source of back pain is unwarranted. In reality, herniated discs are responsible for only a minor share of back problems (1% to 5%). A herniated disc is a specific clinical entity characterized by pain radiating into one leg (sometimes both legs) and usually accompanied or preceded by low back pain. Physical examination often reveals the presence of one or more objective neurologic changes, such as reflex asymmetry, sensory change in the distribution of a nerve root, or muscle weakness. Additionally, the clinical diagnosis requires the presence of positive nerve root tension signs, for which the straight-leg raising test is most commonly used. Fewer than

1% of patients who have lumbar disc herniations have a massive extrusion of nuclear material sufficient to interfere with nerve control of bladder and bowel function (cauda equina syndrome). Although it is not common, the cauda equina syndrome is a true surgical emergency, because failure to decompress the lesion may result in permanent loss of bladder and bowel control. Most patients who meet the clinical criteria for HNP recover spontaneously from acute symptoms and have minimal residual functional or work capacity impairment.

Disc degeneration is not synonymous with back pain and, indeed, it is as common in those without as in those with low back pain. Disc degeneration, however, may predispose to herniations and to other clinical syndromes, such as "spinal instability" and spinal stenosis. All human spines degenerate with time. Most autopsy specimens show the onset of gross and microscopic evidence of intervertebral disc degeneration by the third decade of life. These pathologic changes are accompanied by alterations in the chemical composition of the disc, such as decrease in water content, increase in collagen, and decrease in proteoglycans. The onset of changes occurs earlier in life in men and occurs most commonly in the L4–5 and L5–S1 discs.

Radiographic changes indicative of disc degenerations, such as disc space narrowing and spinal osteophytes, lag behind the histologic and chemical events. The prevalence of these degenerative changes is the same in patients with and without low back pain. The presence of a narrowed intervertebral disc is not correlated with the risk for, or the presence of, a disc herniation. Nor does it provide an explanation of the patient's pain in most cases. This issue is of even greater importance in the interpretation of imaging studies such as computed tomographic (CT) scans and MRI studies. Disc degeneration is often present on MRI scans of workers in their 20s and 30s. As the disc degenerates, it usually bulges. The presence of disc bulging is equally common in those with and without a history of back pain. Furthermore, 20% to 30% of asymptomatic patients have evidence of disc herniation on these structural examinations, emphasizing the need to carefully correlate the patient's symptoms and signs with imaging studies. A positive image without appropriate clinical symptoms and signs is insufficient to make a clinical diagnosis.

Back sprains and strains are probably the most common causes of occupational low back pain. A strain is defined as a muscle disruption caused by indirect trauma, such as excessive stretch or tension. Sprains are actually specific to ligaments, but the terms *sprain* and *strain* are loosely interchanged. There is currently no available method to specifically diagnose a sprain or strain injury; these injuries usually are diagnosed by exclusion of other possible causes of pain. Animal studies have shown that strains heal rapidly (within weeks) and that the healing process, after the first few days, is positively influenced by controlled activation, defined as smooth movements within the normal range of the affected joint. The previously common recommendation of prolonged rest has been abandoned. Controlled activity leads to a more rapid recovery.

Other spinal diagnoses of relevance to occupational low back pain include spinal stenosis, spinal instability, facet syndrome, internal disc disruption, and spondylolysis/spondylolisthesis. A brief description of those entities is provided here; for detailed descriptions see references 3 and 4.

Spinal stenosis is defined as a narrowing of the spinal canal or nerve root foramina, or both. Diffuse narrowing of the spinal canal (central spinal stenosis) has many causes, but most commonly the stenosis results from degeneration with posterior osteophytes projecting into the spinal canal, hypertrophy of the articular facets, and buckling of the ligamentum flavum. Neurogenic claudication is a common symptom, as is pain on extension relieved by flexion. Reflex, motor, and sensory changes are often completely absent. Unlike lumbar disc herniations, nerve root tension signs (e.g., straight-leg raising test) are often negative.

A subgroup of patients with spinal stenosis

have predominantly lateral recess or nerve root canal stenosis. This often is caused by a combination of facet hypertrophy and disc bulge, which reduces the exit space available at the affected disc level or levels and compromises the nerve root. These patients may present with disc hernia–like symptoms, particularly if only one level is affected. It is important to recognize the relation between spinal stenosis and disc herniations. When the spinal canal and lateral nerve root canals are narrowed, a relatively small disc herniation can produce clinically significant symptoms, such as radiculopathy, which would not have occurred if the canal had been of adequate dimensions. Therefore, preexistence of spinal stenosis can contribute to the development of symptoms caused by a disc herniation.

Spinal instability is an ill-defined entity characterized by recurrent episodes of low back pain and/or sciatica triggered by minor mechanical overloads. For definite diagnosis, a shift in the alignment of vertebrae observed on flexion-extension radiographs or other provocation radiographs is required. A special class of spinal instability comprises the gross instabilities caused by fractures, tumors, and infections. In these patients, flexion-extension radiographs usually are not necessary for diagnosis and can, in fact, be harmful, causing damage to the nerve structures.

Facet osteoarthritis can contribute to back pain. The role of the so-called facet syndrome has been deemphasized in recent years, because it appears not possible to classify it clinically with any certainty. Furthermore, specific treatment aimed at the facet joints has been largely unsuccessful.

Internal disc disruption is another syndrome for which the classification remains uncertain. Injections into the disc with contrast media (discography) should reproduce the patient's pain and also demonstrate disruption of the disc architecture. Although internal disc disruption probably exists, it should be considered a rare cause of occupational low back pain.

Spondylolysis (a defect in the neural arch) and *spondylolisthesis* (a forward displacement of one vertebra in relation to the underlying vertebra or sacrum), occurs most commonly at the L5–S1 level; it develops during adolescence, with little further risk of slippage in adulthood. It is generally believed to represent a fatigue fracture of the neural arch, occurring at a specific region (the pars interarticularis), that has failed to heal. Workers with spondylolysis are no more at risk for back pain than those without, whereas spondylolisthesis may render the worker somewhat more susceptible to low back pain. This is particularly the case when

TABLE 26-1. *Classification of Low Back Disorders*[a]

Classification	Symptoms
1	Pain without radiation
2	Pain + radiation to extremity, proximally
3	Pain + radiation to extremity, distally
4	Pain + radiation to upper/lower limb neurologic signs
5	Presumptive compression of a spinal nerve root on a simple roentgenogram (i.e., spinal instability or fracture)
6	Compression of a spinal nerve root confirmed by specific imaging techniques (i.e., computerized axial tomography, myelography, or magnetic resonance imaging) or other diagnostic techniques (e.g., electromyography, venography)
7	Spinal stenosis
8	Postsurgical status, 1–6 months after intervention
9	Postsurgical status, >6 months after intervention 9.1 Asymptomatic 9.2 Symptomatic
10	Chronic pain syndrome
11	Other diagnoses

[a] In addition to symptom classification, low back disorders are classified by duration of symptoms (<7 days, 7 days to 7 weeks, >7 weeks) and by working status at the time of the evaluation (W, working; I, idle).

From Spitzer WO, LeBlanc FE, Dupuis M, et al: Scientific approach to the assessment and management of activity-related spinal disorders: a monograph for clinicians. Report of the Quebec Task Force on Spinal Disorders Spine 1987;12(Suppl 7):S1–S59.

the olisthesis is grade 2 or greater—that is, the slip is more than 25% of the vertebra below.

A variety of classification systems have been developed for low back pain. The most comprehensive system is based on symptoms and was developed by the Quebec Study Group (5). Table 26-1 outlines the system, which is applicable to all anatomic regions of the human spine. Table 26-1 also notes two other important means for classifying low back disorders: duration of symptoms and working status. The Quebec Study Group states that acute symptoms are those that last 7 days or less; subacute symptoms last from 7 days to 7 weeks, and chronic symptoms last longer than 7 weeks. This is a convenient division that reflects the normal recovery period of patients with back pain. Work status is also important because of its influence on prognosis.

DIAGNOSIS

The cornerstones of diagnosis are the history and the physical examination (3,4). In most patients, this is the only evaluation necessary. Patients with recurrent or chronic symptoms, those with more severe back pain of an unrelenting nature, and those with neurologic deficits may require imaging, electrodiagnostics, or laboratory studies. The use of these tests should be based on specific indications derived from the history and physical examination, and the results must be correlated to the history and physical examination. Patients with chronic symptoms sometimes also need a psychological evaluation.

The history should contain information about present and previous symptoms, significant other medical diseases, and use of medications. The onset of the symptoms should be explored in detail, including recent trauma. Pain is the most important symptom. Its pattern, intensity, site, and distribution should be determined, as well as the factors that accentuate and relieve it. The time of day when pain is most severe should be ascertained. Worsening through the day suggests mechanical low back pain and is the most common pattern. Pain on arising in the morning, with improvement during the day, suggests an inflammatory condition. Pain that is most severe at night and awakens the patient is a sign of possible malignancy or infection. Pain distribution is critical for correct classification of patients, particularly when pain radiates down the leg. Dermatomal pain must be separated from referred or nonspecific leg pain.

Neurologic symptoms should be identified, including sensory changes such as numbness and tingling, subjective sense of lower extremity weakness, and changes in bladder and bowel control or sexual function. Loss of ability to initiate voiding and urinary or fecal incontinence are symptoms of a cauda equina syndrome that must be further evaluated with urgency. Progressive lower extremity weakness and "foot drop" are other symptoms requiring further evaluation.

Over time a series of symptoms and signs have been identified as "red flags" (Table 26-2). They are helpful in deciding which patients need emergency diagnostic tests or immediate referral.

The physical examination consists of inspection, palpation, range of motion measurements, and neurologic tests. Body movements, gait observations, and inspection of the patient's standing posture provide information on the severity of symptoms and allow an estimate of functional limitations. The range of motion, including flexion-extension, lateral bending, and axial rotation, are observed and measured. Examination of the lower extremities serves to determine the presence or absence of any significant joint deformities and to assess neurologic function. Hip motion should also be evaluated, because hip and spine problems are sometimes difficult to differentiate. Knee (quadriceps) and ankle (Achilles) reflexes should be obtained in the sitting or supine position. The knee reflex can be affected by an L4 root compression; the ankle reflex is medi-

TABLE 26-2. *The "Red Flags" of low back pain*

Possible condition	History	Physical examination
Fracture	Major trauma	Intense tenderness
	Minor trauma (older patient)	
Tumor	Age <15 or >50 yr	
	Known cancer	
	Unexplained weight loss	
	Night pain	
Infection	Recent fever or chills	Intense tenderness
	Recent bacterial infection (urinary tract infection)	Fever
	Intravenous drug use	
	Immune suppression	
	Unrelenting pain	
Cauda equina syndrome	Saddle numbness	Weak anal sphincter
	Urinary retention, incontinence	Perianal sensory loss
	Severe (progressive) neurologic deficit in legs	Major motor weakness

ated primarily by the S1 nerve root. Nonspecific loss of sensation must be differentiated from well-defined dermatomal loss. Motor function is also grossly evaluated. The strength of the extensor hallucis longus is typically affected by an L5 nerve root compression. Because more than 90% of all neurologic low back pathology involves the L4, L5, and S1 nerve roots, the two reflexes (knee and ankle), one strength test (extensor hallucis longus), and a sensory examination are sufficient for screening purposes.

The evaluation of nerve root tension signs is important. The straight-leg raising test is the test most commonly used. It is positive when sciatica (posterior leg pain) is reproduced. The degree of elevation at which the symptoms occur is recorded.

At the completion of the history and physical examination, decisions regarding the need for further diagnostic tests can be made and an initial therapeutic plan formulated. For most patients, further diagnostic studies are not required. Additional evaluation of

the patient depends on symptom response to treatment and the severity of any remaining symptoms. The diagnostic tests under consideration fall into two categories: those performed to detect physiologic abnormalities and those performed to detect anatomic abnormality and provide anatomic definition. Tests of physiologic dysfunction include laboratory tests, bone scans, and electrodiagnostic studies (electromyography). Anatomic tests include radiology and other imaging studies (CT, MRI, and myelography). In selected cases a psychological evaluation may also be required.

Radiographs are commonly used, but as a screening tool they have limited value. Degenerative changes are nonspecific and often are unrelated to the patient's pain. A variety of other imaging techniques, as indicated previously, are now available to assess patients who fulfill criteria for their use. These tests allow assessment of the anatomy of the spinal canal and its contents. These tests should not be used routinely in patients with low back pain, but only when indicated, based on the patient's history and physical examination. In other words, these tests are used to confirm a clinical suspicion. The basic principle is that anatomic images are only as valid as their correlation with clinical signs and symptoms. This is important because imaging abnormalities are common in people who have never had low back symptoms or sciatica. Further, in addition to clinically significant abnormalities, a number of other findings of questionable or unknown importance are commonly observed. These include disc bulging and disc degeneration, which are both nonspecific. Other diagnostic tests are more rarely indicated (5). Laboratory tests are rarely helpful. A bone scan is an excellent screening test when there is clinical suspicion of tumor, fracture, or infection. Electrodiagnostic tests (electromyography, somatosensory evoked potential, and nerve conduction velocity) are indicated only when there is strong suspicion of neurologic involvement, but the neurologic examination and imaging studies are unclear.

RISK FACTORS—HIGH-RISK JOBS

It is difficult to determine the relation between occupational factors and low back pain because (a) low back pain is not easily defined and classified; (b) sickness absence and other disability data are influenced not only by pain but also by physical and psychological work factors, social factors, and insurance systems; (c) there is a poor relation between tissue injury and disability; (d) the "healthy worker effect" influences data; and (e) exposure is difficult to determine. The retrospective assessment of exposure in most studies limits casual interpretation.

The seven most frequently discussed factors are listed in Table 26-3 (2,6). The six physical work factors have been experimentally associated with the development of injuries in spinal tissues. The seventh, psychological and psychosocial work factors, is probably more related to back disability than to an actual back injury.

There is presently an enormous body of data implicating heavy work in increasing the risk of back pain, sciatica, and disc herniations. Most investigators are using sickness absence and injury reports as their sources; therefore, their studies reflect not only back pain but also disability caused by back pain. However, some studies are based on questionnaires, interviews, and even disc hernia operations. Several countries report similar data. One study of occupational safety and health data from the United States found significantly higher rates of back sprain or strain among workers in heavy industries and in physically demanding occupations (7). These data were confirmed in other popula-

TABLE 26-3. *Occupational factors associated with an increased risk of low back pain*

Heavy physical work
Static work postures
Frequent bending and twisting
Lifting, pushing, and pulling
Repetitive work
Vibration
Psychological and psychosocial stress

tions (2,8). In another study, injury rates were related to predicted spine compression forces in 55 industrial jobs. Back pain was twice as common if the predicted disc compression was greater than 6,800 N (1,500 lb) (9).

Static work postures include primarily long-term sitting, which appears to increase the risk of low back pain and which, in combination with driving, increases the risk of disc herniation (2,10-12) Frequent bending and twisting are usually associated with lifting when reported as causes of back injuries. However, one study found low back pain to be associated with asymmetric postures in a car assembly plant even when lifting was not performed (13).

Lifting is a well-known triggering event for back pain (2,14-17). One study compared workers who performed heavy manual lifting with sedentary workers; the odds ratio was 8 (18). Insurance company data indicate that a worker is three times more susceptible to compensable low back pain if he or she performs excessive manual handling tasks (19). Associations between prevalence of low back pain and lifting have also been established by others (15-17,20). Another study found that disc herniations were more frequent among subjects who had to lift, particularly when lifting was performed in a bent and twisted posture (odds ratio, 6) (21).

Low back pain is more frequent in vehicle drivers than in control subjects, implicating vibrations (perhaps in combination with the sitting posture) as a cause. One study found truck drivers to have a fourfold increased risk of disc herniations, whereas simple car commuting increased the risk by a factor of 2 (22). Other studies also indicate an increased risk of low back pain with vibration (2). Another study found that the risk of being hospitalized for HNP in Finland was particularly high for professional motor vehicle drivers (23). Studies of vehicle drivers have disclosed that radiographic changes occur over time (2).

Psychological and psychosocial work factors have received increasing attention because of their effect on low back disability. Monotony has been identified as a risk factor for back pain, (15) as has poor work satisfaction (14,15). Investigators concluded, based on the Boeing study, that psychological work factors were more important than physical factors as risk indicators of filing worker's compensation claims for low back pain (14,24).

MANAGEMENT

The management of patients with low back problems has undergone major changes over the last decade (3,4,25). In general, patients recover from an acute episode in days to a few weeks and require little treatment (Fig. 26-1) (5). If symptoms continue, physical activity is now recommended, whereas rest was often prescribed in the past.

Early treatment should include (a) information about back pain and its excellent prognosis; (b) help with control of symptoms, including medication and advice on activity; and (c) a reassessment plan. Many patients are concerned about a back problem when it first arises and have misconceptions about its natural history and effect on their future lives. This fear can be eliminated by appropriate information. Symptom control is typically achieved by medication, but manip-

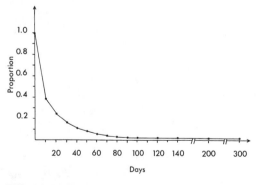

FIG. 26-1. The recovery rate after a back injury is rapid. The figure shows the proportion of people who were absent from work at different times after first reporting sick. (From: Andersson GB, Svensson HO, Oden A. The intensity of work recovery in low back pain. Spine 1983;8:880–884.)

ulation can also have beneficial effects in the acute stages of the problem. Analgesics and nonsteroidal antiinflammatory drugs (NSAIDs) are the medications that are most frequently used. Muscle relaxants can cause drowsiness and appear to be no more effective than NSAIDs.

Activity advice is critical. Some degree of activity restriction is often necessary to reduce symptom severity. This can often be accomplished by modifying work activities and work postures, while allowing the patient to continue working. Bed rest should be advised only in more severe cases, and then for no more than 2 to 3 days (26). Early return to work is critical to prevent prolonged disability and should be emphasized as being part of the recovery process. Activity recommendations (restrictions) may include limitations on lifting, twisting, and bending, and on the duration of sitting, standing, and walking (3). Restrictions should remain in place only for a short period. The beneficial natural history should allow unrestricted return to work after several weeks.

Many treatment alternatives have been and are being used in the treatment of acute low back pain, but few have been rigorously studied for treatment effect (5,25,27). There is no evidence that physical therapy modalities influence the natural history of recovery, although the patient many sometimes feel temporary relief. The medical literature also does not support benefits from the use of transcutaneous electrical nerve stimulation (TENS), corsets, traction, acupuncture, or biofeedback in the treatment of acute low back pain. Corset use may allow the patient to return to physical activities (e.g., lifting) earlier, but it does not affect long-term results. Back schools provide an efficient method of education but not an effective treatment method. Manipulation appears to hasten recovery in patients with acute low back pain without radiculopathy, when it is used in the first 4 weeks of symptoms. Manipulations should not be used in patients with neurologic deficits and should not be continued if a few attempts are unsuccessful.

Surgery is almost never indicated in the first 6 to 12 weeks, but it should then be considered for patients with severe persistent sciatica that has not responded to conservative treatment and when the findings on clinical examination and confirmatory diagnostic tests indicate nerve root compromise (28). The cauda equina syndrome remains the exception: it should be addressed surgically without unnecessary delay. Simple disc surgery does not prevent return to physically demanding jobs.

PROGNOSIS

Nonspecific low back pain is a self-limited disorder that resolves rapidly in more than 90% of cases (see Fig. 26-1). Approximately 40% of the patients recover within 1 week, 80% within 3 weeks, and 90% within 6 weeks, regardless of treatment. The rate of recurrence, however, is very high—estimates range up to 50% in the year following the first episode. A British study indicates that the probability of low back pain is almost four times greater in people who have had previous episodes of low back pain (29). A Swedish study found that industrial workers who have a first-time episode of low back pain have a 28% chance of a new episode during the next year (30).

At the onset of low back pain, it is difficult to predict which patients will require longer than 6 weeks to recover. Among factors that negatively influence recovery are the presence of leg pain, compensation, older age, a specific diagnosis, psychological factors, physical work factors, and socioeconomic factors. Because the interactions among these factors are complex, none of them is a particularly useful predictor of chronicity. If workers have been off the job for longer than 6 months, it is estimated that there is only a 50% possibility of their ever returning to productive employment; if disabled for longer than 1 year, only a 25% chance; and 2 years, almost none (31). The last decade has witnessed the development of comprehensive treatment programs that have improved on these data using various types of

pain treatment and activation programs, but the prognosis after long work absence still remains poor (32).

Early return to work is now considered an important part of the therapy. However, there are many reasons why patients do not go back to work. The obstacles to early return to work include (a) lack of patient motivation, (b) illness behavior, (c) the problem of identifying and providing modified work, (d) unwillingness by employers to accept workers back unless they are fully recovered (despite evidence that accepting a partially recovered worker may be less costly and more effective in promoting recovery than continued disability), (e) difficulties imposed by rigid work rules, (f) inappropriate treatment by practitioners or prolonged use of ineffective treatment, and (g) legal advice to accept a "lump sum" settlement instead of a rehabilitation program designed to return the patient to work. The physician should be aware of these situations and recognize their relative importance in each case of low back pain.

PREVENTION AND CONTROL

Prevention of work-related low back pain is a complex challenge (33,34). Low back pain prevention in work settings is best accomplished by a combination of measures that are listed in Table 26-4. Low back pain can be controlled by reducing the probability of the initial episode, reducing the severity of the symptoms, reducing the length of disability, and reducing the chance of recurrence.

It is estimated that good ergonomic design can eliminate up to one-third of cases of compensable low back pain in industry (see Chapter 9) (20). Not only can good job design reduce the probability of initial and recurring episodes; it also allows the worker with moderate symptoms to stay on the job longer and permits the disabled worker to return to the job sooner. Good ergonomic design reduces the worker's exposure to the risk factors of low back pain through the following:

TABLE 26-4. *Prevention of low back pain*

Job design (ergonomics)
  Mechanical aids
  Optimum work level
  Good workplace layout
  Sit/stand work stations
  Appropriate packaging
Job placement (selection)
  Careful history
  Thorough physical examination
  No routine radiographs
  Strength testing
  Job-rating programs
Training and education
  Training workers
    Biomechanics of body movement (safe lifting)
    Strength and fitness
    Back schools
  Training managers
    Response to low back pain
    Early return to work
    Ergonomic principles of job design
  Training labor union representatives
    Early return to work
    Flexible work rules
    Reasonable referrals
  Training health care providers
    Appropriate medication
    Prudent use of radiographs
    Limited bed rest
    Early return to work (with restrictions, if necessary)

FIG. 26-2. Automotive worker at risk of low back injury. (Photograph by Earl Dotter.)

1. Mechanical aids (powered or manual) to assist with heavy weights and forces (Figs. 26-2 and 26-3)
2. Optimum work level to reduce unnecessary bending and stretching (Fig. 26-4)
3. Good workplace layout to reduce unnecessary twisting and reaching (Fig. 26-5)
4. Sit/stand work stations to reduce prolonged sitting and standing
5. Appropriate packaging to match object weights with human capabilities
6. Good work organization to reduce repetitive high loading and fatigue
7. Good chairs to support the back properly.

The National Institute for Occupational Safety and Health (NIOSH) has provided guidelines for evaluating and designing manual lifting tasks (35,36). Guidelines for other manual-handling tasks, such as pushing, pulling, or carrying, have also been developed (37).

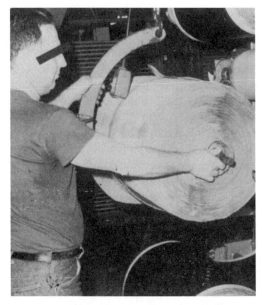

FIG. 26-3. Overhead hoists and lifting attachments. Overhead hoists with specially designed lifting attachments to fit the objects being handled eliminate the awkward manual handling tasks. This C-shaped attachment allows quick and easy placement of heavy rolls on a tire-building machine. (From Liberty Mutual Research Center, Hopkinton, Massachusetts.)

FIG. 26-4. Scissors lift. Hydraulic scissors lifts have a variety of uses in industry to position the workpiece so that a minimum of effort must be used to perform the operation. They can place the workpiece at the proper height to convert lifting and lowering tasks to carries or pushes. (From Liberty Mutual Research Center.)

Considerable attention and hope has been placed on the use of "lifting belts" as a means to reduce back injuries from lifting. Currently, there are insufficient data on the use of these belts to make recommendations on a scientific basis. The mechanical effect of the belts is unclear, but with respect to spinal loading it appears to be minor. Studies suggest that an initial beneficial effect on back injury reporting may occur, but the data remain preliminary and short-term. As we understand today, there are no direct negative effects of belt use, although long-term effects on trunk muscles have been discussed as a possible disadvantage (38–40).

Although job design may be applicable to many manufacturing operations, there are other jobs that are difficult to design and control, such as firefighting, police work, and certain construction and delivery operations. These jobs require greater dependence on preplacement testing and selection of workers (3,32,41). In the past the preemployment medical examination was used in many industries, especially since the enactment of workers' compensation laws. The introduction of the Americans with Disabilities Act changed this situation. Preemployment examinations are no longer acceptable, but preplacement examinations continue to be used to determine whether the worker, once hired, is capable of performing a specific job.

Many authorities believe that the medical history is the most important part of the medical examination for identifying workers who are susceptible to future low back pain (41). Use of this information to prevent back pain occurrence is difficult, however, because back pain is so common and its effect on an individual so variable.

Routine radiographs of the lumbar spine have often been part of the preplacement medical examination, but the preponderance of evidence indicates that the small yield does not justify the radiation exposure or increased cost. According to guidelines issued by the American College of Occupational and Environmental Medicine, lumbar spine radiologic examinations should not be used as a risk assessment procedure for back problems.

FIG. 26-5. Work dispensers are self-leveling devices designed to keep parts at a particular height, eliminating the need to bend over to pick up or release an item. Models are available that use trays or pans or provide for bulk-dispensing of small parts. The dispensers are portable and may be wheeled from station to station, eliminating the need to lift, carry, and lower trays and individual parts. (From Liberty Mutual Research Center.)

Several studies have demonstrated the relation between strength or fitness and the incidence of low back pain (9,32,42). Studies of isometric strength testing have shown that the probability of a musculoskeletal disorder is up to three times greater when job-lifting requirements approach or exceed the worker's isometric strength capability (9). A Danish study revealed that men who experienced low back pain for the first time had lower isometric endurance of the back muscles measured up to 1 year before the episode (42). In women, however, such lower endurance had no predictive value. Although isometric strength testing may be an effective selection technique, it should be used only for jobs that are difficult to design or control and for which a careful ergonomic evaluation has been made (3). Strength testing should never be used as a substitute for good job design. Furthermore, it should be used only when the job has been thoroughly analyzed so that the test truly reflects the demands of the work. Strength testing under a general protocol does not appear to be effective in preventing back injury reports (14).

Training and education are the oldest and most commonly used approaches for reducing low back pain in industry (3,32). Safety and personnel departments have typically used education and training to instruct employees in proper methods and work procedures. For example, training of workers in the biomechanics of safe lifting has been a part of safety programs in industry for more than 50 years. However, according to NIOSH, the value of training programs in safe lifting is open to question because there have been no controlled studies showing consequent decreases in rates of manual-handling accidents or back injuries (35). A major problem is compliance; even after training, most workers do not lift correctly because it is a more difficult way to lift. Greater compliance with training occurs with workers who have (or have had) low back pain. Exercises to increase strength and fitness have been a part of low back pain treatment programs for many years, but only recently have strength and fitness programs been advocated in industry to reduce or prevent the onset of low back pain.

The back school is an attempt to educate the worker in all aspects of back care; it represents a much more comprehensive approach to back care that includes the previous topics of safe lifting, strength, and physical fitness. The original concept of the back school was to educate patients who already had (or had recently had) low back pain—that is, it was a form of treatment. A more recent use of the back school has been to educate workers on how to prevent low back pain. However, no controlled study has shown the effectiveness of such a school as a preventive technique for workers (33,34).

Training of managers is as important as training of workers, both for primary and for tertiary prevention. Long-term disability is associated with adversary situations, litigation, hospitalization, and lack of follow-up and concern. Many of these situations can be alleviated by training foremen, supervisors, and upper-level managers in appropriate responses to worker complaints of low back pain. Also important is providing modified, alternative, or part-time work as a means of returning the worker to the job as quickly as possible.

A program in which management was trained in positive acceptance of reports of back pain has been described (43,44). An atmosphere was created in which workers were encouraged to report all episodes of low back pain—even minor episodes—to the company clinic. Immediate and conservative in-house treatment, including worker education, was provided by the company nurse. Attempts were made to keep the worker on the job—often with modified duties or a redesigned job. If necessary, referrals were made to the company physician, who closely monitored treatment and progress. Over a 3-year period, annual workers' compensation costs for low back pain claims were reduced from more than $200,000 to less than $20,000. Although this was not a controlled study, the results were impressive.

In addition to training for managers, efforts to incorporate union leaders into prevention efforts is important. Early return to work should be encouraged as an important part of the treatment of low back pain. Unions can often assist in the recovery of their members by allowing early return to work through flexible work rules and referrals to clinicians and lawyers who will not unnecessarily prolong the disability.

Company medical personnel should be trained in the benefits of early intervention, conservative treatment, patient follow-up, and job placement techniques. Both physicians and nurses should be familiar with recent literature that objectively evaluates various types of treatment for low back pain. Medical personnel should also become familiar with the physical demands of jobs performed in the company so that they can adequately place both injured workers and new employees.

The effectiveness of a standardized approach to the diagnosis and treatment of low back pain in industry has been demonstrated (32,45). This approach was used to monitor the course and treatment of low back pain. If there was any disagreement between the investigators (who were orthopedic surgeons) and the treating physician, they discussed the case together in detail. Usually, they reached an agreement; if they did not, another physician was consulted for an independent opinion. This program dramatically decreased the number of low back patients, the number of days lost, and the number of patients sent to surgery.

Although knowledge of low back pain is limited, enough is already known to adequately control the problem in industry. Instead of waiting for a major medical breakthrough to occur, emphasis should be placed on applying the knowledge that is already available. Low back pain control requires the combined efforts of workers, managers, unions, nurses, physicians, and others.

Part II. Work-Related Disorders of the Neck and Upper Extremity

Lawrence J. Fine and
Barbara A. Silverstein

A 31-year-old, right-handed man had been employed in a variety of automobile manufacturing jobs for 13 years. Two years ago, he switched to a new plant and was assigned to a job that required him to manipulate a spot-welding machine beneath cars moving overhead. He had 1 minute to complete four welds on each car. The spot welder, which had metal handles, required substantial force for appropriate positioning, and it had to be repositioned four times for each car. The worker's wrists were in complete extension for a substantial portion of the job cycle.

When the worker started on this job, the weekday work shift was 9 hours long and Saturday work was required in most weeks. After 3 weeks on the job, he noted that he had pain in both wrists. He also noted numbness and tingling in the first four fingers of his left hand, at first only at night, a few nights each week, after he had fallen asleep. When he awoke at night with the numbness, he would get up and walk around shaking his hands; in about 10 minutes he would be able to go back to sleep. Gradually, over the next several months, the numbness and pain worsened in both frequency and intensity. His left hand would feel numb by the end of the work-shift, and any time he was driving his hands would become numb. Because he liked his job and did not want to be placed on restriction, which would mean he could not work overtime, he decided to visit his private physician rather than the company physician. He also was not sure that the company physician would be very sympathetic to his complaints.

The physician found on physical examination that the worker had decreased sensitivity to light touch in the left index and middle fingers and a positive Phalen's test of the left hand. She suspected carpal tunnel syndrome (CTS) and believed that the disorder might be work-related because the patient was young, male, and had no other risk factors, such as diabetes, past history of wrist fracture, or recent trauma to the wrist. The physician discussed job changes with the patient. She also prescribed wrist splints to be used at night.

The splints relieved some of the night-time numbness for a period. However, over the next 6 months, the patient's symptoms began to be present all of the time, and he thought that his left hand was becoming weaker. Similar symptoms also developed in his right hand.

The patient felt he could no longer do his job and returned to his physician. She referred him to a hand surgeon and ordered nerve conduction tests. The tests showed slowing of sensory nerve impulse conduction in the median nerve in the region of the carpal tunnel, more so on the right than the left.

One year after the problem was first noted, the worker had surgery, first on the left hand and then on the right. After surgery, the company placed him in a transitional work center for a 3-month period, where he worked at his own pace and had no symptoms. He then returned to the assembly line with the restriction that he not use welding guns or air-powered hand tools. When he worked on the line, he occasionally had symptoms, but they were substantially less intense and less frequent than before.

He later transferred to a warehouse, because he felt that he would have a better chance of avoiding long layoffs there. His job required use of a stapling gun to seal packages. Three weeks after beginning this job, his symptoms began to return with their former intensity. Through ordinary channels, he immediately sought and was given a transfer to a position driving a fork lift truck. This change reduced, but did not eliminate, his symptoms. Currently, he has numbness, tingling, and pain in the fingers of both hands about twice a month. Playing volleyball usually triggers a severe attack. With the use of nighttime splints, he can sleep through most nights without awakening. Although he believes that his hands are weaker than before the symptoms developed, he still is able to perform his job. He has decided that he will continue working as long as the symptoms remain at no more than the present level.

This case illustrates the intermittent and progressive nature of most work-related disorders of the upper extremity, and particularly of carpal tunnel syndrome (CTS) (46), the best known of the common work-related disorders of the upper extremity. Other examples of these disorders that may be related to work include de Quervain's disease (47), epicondylitis, shoulder supraspinatus tendinitis (48,49), and tension neck syndrome (50). This family of disorders may involve muscles (tension neck syndrome), tendons (supraspinatus), joints (degenerative joint disease), skin (calluses), nerves (CTS), or blood vessels (hand-arm vibration syndrome, or Raynaud's phenomenon of occupational origin). (See appendix to this chapter.) (51,52)

MAGNITUDE AND COST OF THE PROBLEM

The only national source of data about the magnitude of work-related musculoskeletal disorders is the Bureau of Labor Statistics (BLS) Annual Survey of Occupational Injuries and Illnesses (53). This is a federal/state program in which employer reports are collected from about 165,000 private industry establishments. The survey excludes the self-employed, farms with fewer than 11 employees, private households, and employees in all governmental agencies. Each episode is classified as an illness or injury.

Illnesses or injuries that result in days away from work are classified on several bases, including (a) the associated events or exposures, such as a fall or overexertion; (b) the nature of the injury or illness, such as sprain, strain, or CTS; and (c) the source of injury or illness, such as machinery, worker motion, or worker position. Some of these categories probably represent illnesses and injuries caused by physical stressors at work, such as repetitive movements of the hands or lifting of objects. Decisions about the event or exposure that results in injury or illness are based on the validity of the employer's description of the associated events and decisions made by the coder and therefore should not be considered scientific causal inferences. Because the survey is confidential, which promotes accurate recording by employers, it is useful for identifying surveillance associations. A worker's or employer's decision to report may be influenced by personal, cultural, administrative, peer pressure, or economic factors (54). The accuracy of the survey depends on the interplay of complex factors, making it difficult to interpret time trends. The rate of illnesses or injuries that result in days away from work has declined for 7 consecutive years. This trend has occurred among most types of injuries and illnesses, including musculoskeletal disorders.

The BLS survey also tracks the number of cases of certain illnesses and injuries. Between 1992 and 1996, injuries and illnesses resulting in days away from work declined about 20%, and the number of CTS cases dropped slightly more than 10% (to 29,900 cases) over the same period. The decrease was less for amputations and fractures. Some of the decline in the overall trend for cases with days away from work may be explained by a modest increase in the rate of cases involving work restrictions (e.g., shortened hours, no heavy lifting). There is little evidence that underreporting has increased. No single factor explains the reduction. However, increases in health and safety activities by private-sector firms, brought about after increases in workers' compensation costs and changes in Occupational Safety and Health Administration (OSHA) activities, may have contributed to these declines (54).

All disorders and injuries involving lost time identified in the annual BLS survey are classified based on the following question: What was the employee doing just before the incident occurred? Repetitive motion is one of the many narrowly-defined categories that trained survey coders record based on the employer's response to this question. Cases are classified as resulting from repetitive motion if a bodily motion such as typing or key entry, repetitive use of tools, or repeated grasping of objects other than tools is considered by the coder to be related to the illness or injury. Another event category that is relevant to estimating the number of work-related musculoskeletal disorders is overexertion, as in maneuvering especially heavy or bulky objects, such as cartons of soft drinks. In 1996, the BLS survey reported that about 74,000 (3.9%) of the almost 1,900,000 cases were coded as being associated with repetitive motion, most commonly in the upper extremities (shoulder, hand, and wrist) (53,55). In comparison, 311,900 cases (16.6%) were coded as being associated with overexertion in lifting, pulling, pushing, or other activities; approximately two-thirds of these were back injury cases. The survey also collects information on the duration of absences from work. For all cases, the median is 5 days; for CTS cases, it is 25 days. Although these numbers do suggest that back injuries caused by lifting and other manual-handling activities are a larger problem, upper-extremity disorders and injuries are clearly major problems. In 1994, there were about 48,000 cases of shoulder injury associated with overexertion from lifting, carrying, pulling, or pushing objects, and about 51,000 wrist cases were associated with repetitive motion (55).

The BLS data not only provide information on the magnitude of the work-related musculoskeletal problem, but they also can be useful in identifying industries that contain many jobs with higher levels of exposure to repetitive tasks and manual-handling activities (55). In 1994, several industries had rates of overexertion disorders that were significantly higher—more than three times higher than the average rate for private industry (7.6 cases per 1,000 workers). These include nursing and personal care facilities; scheduled air transportation; manufacturers of travel trailers and campers, food products machinery, soft drinks, and mattresses; wholesaling of alcoholic beverages; and coal mining. Similarly, several industries had significantly increased rates of injuries and disorders associated with repetitive motion— more than seven times higher than the average rate for all private industry (1.2 cases per 1,000 workers). These industries include a number of clothing manufacturing sectors (e.g., knit underwear mills, male workclothes) and other manufacturing industries, including potato chips, motor vehicles, and meat-packing plants. More than 3.5 million workers were employed in these high-risk industries, but not all of them were in high-exposure jobs. However, other industries, such as trucking and courier services, with more than 1.6 million workers, had rates of overexertion disorders that were almost three times higher than the average rate for all private industries.

The BLS survey, like most surveillance data systems, has some degree of misclassification or error in the information on exposure and health outcomes. Nevertheless, industries with consistently increased rates are likely to contain a substantial number of high-risk jobs. Overexertion in lifting, other types of overexertion, and repetitive motion were associated in 1996 with about 600,000 illnesses and injuries, most involving the back or upper limbs.

More information about the prevalence of some specific disorders of the upper extremity and the costs of these disorders is provided by workers' compensation data from Washington State (56). Workers' compensation data for 1989 to 1996 were used to iden-

tify all the accepted musculoskeletal claims; these claims were categorized specifically (rotator cuff syndrome, epicondylitis, CTS, and sciatica), based on physician diagnoses. Other, less specific musculoskeletal disorders of the upper extremity and back were also identified. An effort was made to determine whether the onset was acute (e.g., after a fall) or gradual. Gradual-onset disorders represented 67% of the back claims, and 36% of the upper-extremity claims. The claims incidence rate for gradual-onset back disorders was 16 per 1,000 workers, and for gradual-onset upper-extremity problems it was 9 per 1,000. The claims incidence rates for the specific disorders of the upper limb were approximately 2 per 1,000 workers for rotator cuff syndrome, 1 per 1,000 for epicondylitis, and 3 per 1,000 for CTS. The percentage of all claims that were for musculoskeletal disorders remained constant over the 8-year period. The rates from these data are higher than those from the BLS data, suggesting some degree of underreporting in the latter program. Workers in industries with widespread manual-handling tasks, such as nursing homes, several construction activities, and garbage collection, had the highest incidence rates for back and shoulder disorders. Workers in industries with highly repetitive work, such as wood product manufacturing or fruit and vegetable packing, were at highest risk for CTS and epicondylitis. One group of workers who appeared to be at increased risk were those in the rapidly growing temporary-help industry. The data suggest that these workers are being placed in high-risk jobs.

Overall, the three specific diagnoses for upper limb disorders accounted for about one-third of upper-extremity injuries and illnesses. The Washington State data suggest that, in addition to CTS, both epicondylitis and particularly rotator cuff injuries are also common disorders that deserve more study and greater efforts at prevention. This finding underlines the conclusion that, although back disorders are more common than upper limb disorders, work-related upper limb disorders are often serious enough to result in workers' compensation claims—in about 1 of every 100 workers per year.

The Washington State data also allow the development of a national estimate of the cost of gradual-onset disorders of the back and upper extremity. For the 96 million workers in the United States covered by OSHA, the estimated direct cost for gradual-onset back disorders ranges from $1 to $17 billion, depending on whether one uses the average or the median cost for each Washington State claim. For upper-extremity disorders, the range is from $291 million to $6.7 billion (56). These figures do not consider the indirect costs to employers for lost productivity or training of replacement workers—nor do they consider the costs to workers and their families that are not covered by workers' compensation.

Although the precise cost and prevalence of work-related musculoskeletal disorders of the upper extremity is unknown, they are among the most common occupational disorders. Typical annual incidence rates for CTS, epicondylitis, and rotator cuff syndrome in workplaces with an average level of risk should be less than 1 per 100 workers. Both the Washington State and BLS data show an overall decline for all illnesses and injuries and a trend toward a higher proportion of workers with injuries and disorders remaining on the job. The overall decline results from the interplay of several factors, such as changes in the nature of work and more effective prevention efforts. The trend for more affected workers to be able to remain on the job may have resulted from either changes in treatment or greater use of work restrictions and alternative work placements (54).

Pathophysiology

Clinical, laboratory, and epidemiologic studies all have contributed to the current understanding of the pathophysiology of work-related musculoskeletal disorders of the upper extremity and neck.

Although the current level of knowledge is incomplete, it nevertheless guides both treatment and preventive strategies. Five workplace physical factors are important in the etiology of these disorders: repetitive motions, forceful motions, mechanical stresses, static or awkward postures, and local vibration (Fig. 26-6). The effects of these physical load factors probably can be exacerbated by workplace psychosocial factors, such as the perception of intense workloads, monotonous work, and low levels of social support at work (55). In assessing the role of workplace factors, it is important to consider the duration, frequency, and intensity of the individual and combined factors.

Repetition and Force

Repetitive motions of the hands, wrists, shoulders, and neck commonly occur in the workplace. A data-entry operator may perform 20,000 keystrokes per hour; a worker in a meat-processing plant may perform 12,000 knife-cuts per day; and a worker on an assembly line may elevate the right shoulder above the level of the acromion 7,500 times per day. Such repetitive motions may eventually exceed the ability of the individual to recover from the stress, especially if forceful contractions of muscles are involved in the repetitive motions.

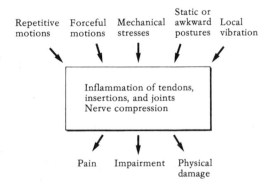

FIG. 26-6. Schematic representation of pathophysiology of work-related disorders of the upper extremity and neck.

Failure to recover usually implies some type of tissue damage or dysfunction ("tissue reaction to injury"), which may represent acute inflammation and may be totally reversible. Tissue damage may even lead, over time, to improved function—a "training effect." Acute damage to muscle from overexertion often leads to muscle hypertrophy. In work-related musculoskeletal disorders, the sites of likely tissue damage are most commonly tendons, tendon sheaths, and tendon attachments to bones, bursae, and joints. It is probable that, over time, these tissue changes can in some cases lead to nerve compression (CTS), chronic fibrous reaction in the tendon, tendon rupture (rotator cuff injury), calcium deposits, or fibrous nodule formations in a tendon, leading to a trigger finger (57).

Abrupt increases in the number of repetitive motions performed by a worker each day are well recognized clinically as a cause of tendinitis (58). Too many forceful contractions of muscles can cause corresponding tendons to stretch, compressing the microstructures of the tendons and leading to ischemia, microscopic tears in tendons, progressive lengthening, and sliding of tendon fibers through the ground substance matrix. All of these events can cause acute inflammation of tendons. The process by which these acute changes evolve over time to the more common chronic picture of damage, such as that observed during carpal tunnel surgery (59), is not well understood. Both laboratory and epidemiologic studies have provided substantial evidence that high levels of exposure to the combination of repetitive and forceful movements is strongly associated with several musculoskeletal disorders of the upper extremity (55,60).

An important but inadequately investigated question is whether the repetitive stresses to the joints of upper limbs that occur in some occupations lead to accelerated development of localized osteoarthrosis. Because localized osteoarthrosis is not a specific disease but the final common pathway of biomechanical and pathologic changes in cartilage, subchondral bone, and bone surrounding joints, it is reasonable to assume that repetitive stresses accelerate its development (51).

Repetitiveness has a number of components to be considered, including the velocity and acceleration of movement and the amount of recovery time within any repetitive cycle or task.

Posture, Stress, and Vibration

In addition to repetitive and forceful motions, three other exposure variables that influence the development of work-related musculoskeletal disorders are external mechanical stress, work performed in awkward or static postures, and segmental (localized) vibration.

Mechanical stress in tendons results from muscle contractions or from compression of a tendon or other tissues by contact between the body and another object. One of the major determinants of the level of the mechanical stress is the force of the muscle contractions. For example, a pinch that is very forceful is more stressful than one that is not very forceful. Another source of mechanical stress results from work surfaces or handheld tools with hard, sharp edges or the ends of short handles that press on soft tissues. The tool exerts just as much force on the hand as the hand does on the tool. These stresses can lead, for example, to neuritis associated with the forceful contact between the edge of scissors handles or bowling ball holes with the sides of the fingers or thumb, and to cubital tunnel syndrome in microscopists who must position their elbows on a hard surface for long periods. Short-handled tools (e.g., needle-nosed pliers) that dig into the base of the palm exert as much force on the hand, particularly on the superficial branches of the median nerve, as the hand does on the tool.

Work performed in awkward or static postures is another important influence on the development of work-related musculoskeletal disorders. The level of mechanical stress produced by a muscle contraction varies with the posture of a joint. For example, the amount of force that can be exerted in a power grip is greatly reduced as the wrist moves from a neutral to a flexed position. Work with the arm elevated more than 60 degrees from the trunk is more stressful for the rotator cuff tendons than work performed with the arm at the trunk. Work performed in static postures that requires prolonged, low-level muscle contractions of the upper limb or trapezius muscle may also trigger chronic localized pain by an unknown mechanism (perhaps decreased blood flow to the muscle).

Segmental vibration is transmitted to the upper extremity from impact tools, power tools, and bench-mounted buffers and grinders. The mechanism by which localized vibration from power tools contributes to the development of work-related Raynaud's phenomenon is not clear. Nevertheless, this syndrome has been associated with several types of power tools, including chain saws, rock drillers, chipping hammers, and grinding tools.

Most of the effects thus far described center on the tendon structures. The chronic effects of repetition and other risk factors on muscles are not as well understood as the effects on tendons. Chronic or intermittent pain originating in muscles may be important in understanding several disorders, including tension neck syndrome (costoscapular syndrome) and overuse injuries in musicians (61). Two types of muscle activity may be important in work-related disorders: low force with prolonged muscle contractions (e.g., moderate neck flexion while working on a video display terminal for several hours without rest breaks) and infrequent or frequent high-force muscle contractions (e.g., intermittent use of heavy tools in overhead work). Sustained static contractions can lead to increases in intramuscular pressure, which in turn may impair blood flow to cells within the muscle.

Motor nerve control of the working muscle may be important in sustained static contrac-

tions, because even if the relative load on the muscle as a whole is low, the active part of the muscle can be working close to its maximal capacity. Therefore, small areas of large muscles, such as the trapezius, may have disturbances in microcirculation that contribute to or cause the development of muscle damage (red ragged fibers), reduction of strength, higher levels of fatigue, and sensitization of pain receptors in the muscle leading to pain at rest (62). High levels of tension (strong contractions) can lead to muscle fiber Z-line rupture, muscle pain, and large delayed increases in serum creatine kinase. If not prolonged, these changes are reversible and completely repaired, often making the muscle stronger.

It is hypothesized that if damage occurs daily from work activity, the muscle may not be able to repair the damage as fast as it occurs, leading to chronic muscle damage or dysfunction. The mechanism of this damage at the cellular level is not fully understood. Work activities that lead to sustained relatively low-level muscle activity or higher-level muscular contractions may be a causal factor in some work-related musculoskeletal disorders.

Nonoccupational Factors

In addition to occupational risk factors or exposures such as repetitive work, personal risk factors may influence the risk of developing these work-related disorders. For example, forceful repetitive activities, such as wrist extension, can occur in some recreational activities and contribute to the development of work-related disorders; however, factors related to some specific disorders, such as rotator cuff tendinitis, have not been adequately studied.

The nonoccupational factors for CTS that have been most thoroughly studied include coexisting medical conditions such as rheumatoid arthritis, diabetes mellitus, pregnancy, and acute trauma, especially after a Colles' fracture of the wrist. Additional possible nonoccupational risk factors are age, gender, obesity, and carpal tunnel size or shape. Carpal canal size, either small or large, has been proposed as a risk factor based on limited evidence (63). For nonoccupational CTS, men have increasing risk with age; in women, the risk peaks at approximately 50 years of age (64). Both men and women may experience occupationally-related CTS at an earlier age (65). Although most clinic- or community-based studies report three times as many CTS cases among women as among men, studies from workplaces have not tended to observe so large a gender difference. In a study of occupational CTS in Washington State, based on workers' compensation data, the female-male ratio was 1.2:1 (65). Several studies have identified obesity as an independent risk factor (66,67). Both pregnancy and cigarette smoking have been also identified as risk factors (68,69). Few, if any, personal factors are useful and strong predictors of susceptibility to work-related disorders of the upper extremity.

Psychosocial Factors

In addition to the physical factors described, psychosocial factors may be important in both the initial development of these disorders and the subsequent long-term disability that sometimes occurs (see Chapter 21). Few studies have rigorously investigated both psychosocial and physical factors or their combined effects (56,60,70). In the future, this type of study is likely to become more common. At least one such study has been done on low back pain; it found that specific aspects of the physical and psychosocial demands of work were strong and independent predictors of low back pain, after adjusting for individual factors (71). In a study of aluminum smelter workers, both forearm rotation and low degree of latitude in decision making were associated with shoulder and elbow disorders in carbonsetters (72).

Psychosocial factors relate to the way in which work is organized. The effects of psychosocial factors may operate indirectly by altering muscle tension or other physiologic processes and, through the latter, may also influence the perception of pain (73). Psychological factors may be particularly important in determining whether specific musculoskeletal disorders evolve into chronic pain syndromes. Overall, psychosocial factors appear to be more important in disorders of the neck and shoulder muscles than in tendon-related disorders of the forearm and the hand (55). Epidemiologic studies of upper-extremity disorders suggest that the perception of an intense or stressful workload, monotonous work, and low levels of social support at work increase the risk of upper limb disorders (55). Different studies use a variety of measures to define intense or stressful workloads, such as lack of control over how work is done, perceived time pressure, deadlines, work pressure, or workload variability (55). The causal role of psychosocial factors probably is not limited to particular jobs, such as the use of computers in the office setting.

Studies that have addressed psychosocial factors have often used the demand-control-support model originally introduced by Karasek and Theorell (73). In this model, high levels of psychological job demands may contribute to the development of work-related musculoskeletal disorders when they occur in an occupational setting in which the worker has little ability to decide what to do or how to do a particular job task and little opportunity to use or develop job skills. Further, these adverse effects are hypothesized to occur more frequently in a work environment in which there is little social support from coworkers or supervisors. Nonoccupational psychosocial factors could also be important. Further studies are needed to better understand the complex interactions among physical occupational factors and occupational psychosocial factors.

A model has been developed to incorporate both physical and psychosocial factors in the development of work-related musculoskeletal disorders (Fig. 26-7) (74). Exposures may be either to physical work factors or to psychosocial factors. Dose also may be either

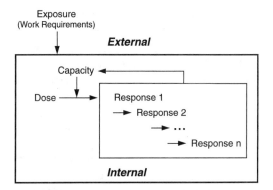

FIG. 26-7. The proposed model contain sets of cascading exposure, dose, capacity, and response variables, such that the response at one level can act as a dose at the next level. In addition, the response to one or more doses can diminish (impairment) or increase (adaptation) the capacity for responding to successive doses.

physiologically or psychologically defined. Capacity may physiologically defined (e.g., strength of a specific tendon) or psychologically defined (e.g., level of self-esteem). The internal processes resulting from a specific dose could occur in the spinal neural circuits or in the muscles of the forearm.

In summary, excessive repetitive or forceful motions, high levels of mechanical stress, work in static or awkward postures, and the use of vibrating power tools can—either singularly or, more often, in combination—increase the risk of upper-extremity disorders and cause inflammation and chronic localized pain in the upper limb and neck. Precise dose-response relations have not been developed for these risk factors, and the precise role of the psychosocial factors needs to be further elucidated. At present, it appears that when the exposure to several physical factors is high, then the risk of these disorders is substantially increased. When the level of exposure to physical factors is more moderate, then the overall level of risk may appear to depend more on the combination of personal attributes, physical factors, and psychosocial factors. As with many occupational exposures, the risk is influenced also by nonoccupational factors (55,60).

DIAGNOSIS

This broad group of work-related disorders of the neck and upper extremity has a diverse set of symptoms and physical findings. The evaluation of a patient for a suspected work-related disorder should have three major components: obtaining a history of present illness from the worker, performing a physical examination of the upper extremity and the neck, and assessing the work setting and tasks (75,76).

The *history* of the present illness should fully characterize the symptoms by determining the location, radiation, duration, evolution, time patterns, and exacerbating factors. The worker's description of work activities is useful. The worker should describe the nature of specific work tasks by risk factors (forceful exertions, repetitive activities, and other adverse exposures). For example, a worker who for 8 hours a day uses a vibrating hand tool to perform a task that is repeated every 30 seconds may be at high risk for wrist tendonitis or CTS. Similarly, a repetitive job that requires the arms to be held overhead during most of the work shift may increase the risk of a rotator cuff shoulder tendinitis. Because specific job tasks can vary within even a high-risk occupation, a careful history of specific job tasks is important.

When a worker who has been performing the same job for a considerable period develops a disorder, the history should be directed not only at the chronic stable exposures, but also at acute factors. For example, if the worker uses a power screwdriver, perhaps the symptoms started when screws from a "bad" batch were used, requiring more force for proper insertion. Other common acute risk factors are changes in work pace or length (longer or more frequent overtime, either by lengthening of the normal workday or by decreasing the number of days off); such changes may reduce the opportunity for recovery from fatigue and occult injury.

Despite the conscientious efforts of the employee and careful interviewing by the physician, description of work tasks may not be sufficient. In general, direct observation of the work process provides the most accurate view of the risk factors associated with specific job tasks. In addition to visiting the workplace, review of representative videotapes of job tasks, written descriptions of job tasks based on industrial engineering data, and results of ergonomic job analyses can all provide helpful information. Although direct observation of the work is often required to determine more precisely the level of risk factor exposure in specific job tasks, descriptions by workers may identify many high-risk exposures with sufficient accuracy for a correct diagnosis.

Determining whether the patient has a predisposing medical condition (e.g., previous injury to the symptomatic area) is also important. In a study of all cases of CTS in Rochester, Minnesota, from 1961 through 1980, the following conditions were associated with CTS in more than 4% of the cases: Colles' fracture, other acute trauma, collagen vascular disease, arthritis of the wrist including rheumatoid arthritis, hormonal agents or oophorectomy, diabetes mellitus, and pregnancy (68). Nonoccupational exposure to risk factors can be a potential confounding influence and should be elicited during the worker interview. For these nonoccupational exposures to be significant as causal factors, they must be similar in intensity and frequency to the known occupational exposures.

Surveillance and epidemiologic studies have identified several industries and occupations associated with risks of CTS or other upper-extremity disorders. Awareness of these findings can alert physicians to the industries and occupations in which adverse exposures are more common. A few illustrative examples are listed in Table 26-5.

Finally, clinical experience, surveillance, and epidemiologic research suggest that many jobs with substantial exposure may adversely affect closely related muscles, tendons, or joints at the same time. As a result, workers in these high-risk jobs may present with any of several disorders, such as CTS and de Quervain's disease or epicondylitis.

The *physical examination* is an important part of evaluation of the patient with work-related musculoskeletal disorders. An examination of the upper extremity typically involves inspection, palpation, assessment of the range of motion, and evaluation of peripheral nerve function.

One of the main objectives of the physical examination is to determine the precise structure or structures in the upper extremity that are the anatomic source of the symptoms. Numbness and paresthesias often result from peripheral nerve compression, but there are many other reasons why there might be numbness and tingling in the fingers. Increased pain on resisted movements (e.g., resisted wrist extension) often results from lesions in a tendon or at the insertion of a tendon. In some cases, it is not possible to determine the precise source of the pain in the upper extremity; in others it is possible to determine the specific disorder that is present (e.g., CTS, de Quervain's stenosing tenosynovitis). The severity of these disorders ranges from very mild, with no significant impairment of the ability to work, to very severe. The symptoms and physical signs of several of these disorders are described in the appendix to this chapter (77–80).

In addition to the disorders with specific findings on physical examination, workers in certain occupations (e.g., keyboard operators, musicians, newspaper reporters) often have an increased rate of complaints of pain in the upper extremity or neck. These symptoms are similar to those of low back pain because a specific anatomic source of the pain often cannot readily be identified on clinical evaluation. As with low back pain, these pains are common, often intermittent in nature, and sometimes lead to substantial disability and impairment (61).

The diagnosis of a work-related musculoskeletal disorder is based on a three-step process. First is the determination of whether the patient has a specific disorder, such as flexor tendinitis of the forearm. This is usually based on the history and physical examination.

TABLE 26-5. *Illustrative examples of industries and occupations with high-risk job tasks*

Occupations and industries	Disorders (Ref)
Occupations	
Seafood packers	Carpal tunnel syndrome (21)
Carpenters	'' '' ''
Invasive cardiologists (removal of intraaortic balloons)	Carpal tunnel syndrome (32, 77)
Metal platers	Carpal tunnel syndrome (19)
Sausage makers	Epicondylitis (3)
Rock blasters	Shoulder tendinitis (4)
Dentists	Cervical spondylosis (5)
Data-entry operators	Tension neck syndrome (5)
Instrumental musicians	Focal dystonias (16)
Industries	
Manufacture of knit underwear, hats, bras, girdles, house slippers, men's and boys' workclothes	Disorders associated with repetitive motion (11)
Manufacture of potato chips	'' '' ''
Automobile manufacturing	'' '' ''
Meatpacking plants	'' '' ''
Temporary help, assembly	Hand/wrist disorders, gradual-onset (12)
Seafood canneries	'' '' ''
Meat and poultry dealer, wholesale	'' '' ''
Wallboard installation	Elbow disorders, gradual-onset (12)
Roofing	'' '' ''
Temp help assembly	'' '' ''
Wallboard installation	Shoulder disorders, gradual-onset (12)
Fence erection	'' '' ''
Temporary help, assembly	'' '' ''
Glass installation	'' '' ''
Garbage installation	'' '' ''

Second, there should be evidence from a detailed occupational history, or, better yet, from direct observation of the workplace, of substantial exposure to specific occupational risk factors. Analysis of health surveillance data, such as OSHA logs or workers' compensation records from the specific workplace, may be particularly helpful in confirming that a particular job is associated with an increased risk of a work-related musculoskeletal disorder. Some employers, to facilitate return-to-work evaluations, now provide physicians with a videotape of the job that the worker normally performs. This may be useful in determining the approximate level of exposure.

Third, nonoccupational causes should be considered as possible primary causal factors or as extenuating factors based on the history and physical examination. Review and analycsis of surveillance and epidemiologic data of similar work may provide information on the relative contributions of occupational and nonoccupational factors in the causation of a specific work-related musculoskeletal disorder in the patient's selected occupation and industry (see Table 26-5). With the exception of tests for abnormalities in nerve conduction, elaborate diagnostic or laboratory studies often are not necessary unless the patient has a history of trauma or symptoms suggestive of underlying systemic disease or fails to improve with conservative treatment.

The most difficult part of the diagnosis of work-related musculoskeletal disorder is determination of the relative contribution of occupational factors in the etiology of the disorder. As with other diagnostic evaluations of work-relatedness, the critical question is: Was the exposure of sufficient intensity, frequency, and duration to have caused the injury or illness? Because intense periods of high exposure as short as weeks in duration can cause lateral epicondylitis or other work-related musculoskeletal disorders, attention should be directed to estimating the intensity and frequency of exposure. It is not uncommon for there to be exposure to multiple risk factors at the same time—for example, repetitive and forceful exertions of the hands, shoulder abduction, and exposure to vibration from hand tools. There are no simple rules for assessing whether exposures has been of sufficient intensity and frequency to cause a disorder.

One study of CTS was based on patients referred for an independent medical examination for both diagnostic evaluation and determination of work-relatedness. The investigators used typical workers' compensation criteria and a standardized approach to assess exposure (frequently based on plant visits or reviews of videotapes). Approximately 60% of the confirmed cases of CTS were found to be work-related (80). Fifteen percent of the patients were found to have not CTS but rather a different work-related disorder of the upper limb, such as localized muscle "fatigue/myalgias." An additional 21% had non–work-related disorders. This study emphasizes the need for thorough diagnostic evaluation to identify the specific disorder and to assess exposure as carefully as possible.

TREATMENT AND PROGNOSIS

The goals of treatment are elimination or reduction in symptoms and impairment and return of the employee to work under conditions that will protect his or her health. These goals can be most easily achieved by early and conservative treatment. Treatment of work-related musculoskeletal disorders early in the course has several advantages: such treatment is less difficult and less costly, surgical procedures can be avoided, periods of absence from work or stressful exposures

are shorter, and the effectiveness of treatment is greater (50,81).

The initial goals of treatment are to limit further tissue damage, dysfunction, and inflammation (if present) and to assist the repair of any tissue damage. Symptomatic relief is provided by the use of antiinflammatory medications, rest (sometimes facilitated by splints), and application of heat or cold. Physical therapy techniques are used to assist in symptom relief, to ensure normal joint motion (stretching), and to recondition muscles after periods of rest or reduced use. If these more conservative measures fail to reduce symptoms and impairment for some conditions such as CTS, steroid injections or surgical treatments can be helpful (75,82,83). Surgery, even in CTS, may be ineffective if the worker is returned to the old job without an effort to reduce the occupational exposures that were present. Because few scientifically valid studies have evaluated the long-term effectiveness of the treatment of work-related musculoskeletal disorders of the limb and neck, an empiric approach is indicated (83).

Resting of the symptomatic part of the upper extremity is the most important part of the treatment program. Reducing or eliminating worker exposure to the known risk factors can achieve this. In addition to engineering changes, restricted duty, job rotation, or temporary transfer may be effective. In order for job transfer or rotation to be effective, the new job duties must result in a net reduction in the level of exposure. It is often necessary to conduct an evaluation of the new duties to determine whether a reduction in exposure will occur. The magnitude of reduction required to facilitate recovery often is not known. In general, the more severe the disorder, the greater the reduction in magnitude and duration that will be required. Because of the adverse consequences of complete removal from the work environment, this step should be taken only in severe cases or after less drastic measures have failed.

Splints and other immobilization devices

may provide rest to the symptomatic region. However, they may increase the level of exposure if the worker must resist the device in order to carry out regular job tasks (e.g., frequent wrist flexion while wearing neutral-position wrist splints for CTS). Workers may also adapt to wearing a splint by altering their work activities in a way that leads to substantial stress on another region of the upper extremity, such as the elbow or shoulder. Immobilization or prolonged rest may have direct adverse effects if either leads to muscle atrophy. As a result, careful monitoring of the worker who is on restricted duty or job transfer or is wearing an immobilization device is indicated. In addition, because it is difficult to predict the clinical course of these conditions and because the empiric basis of many of the treatments is poorly understood, frequent follow-up is desirable. Failure of the treatment plan to produce improvement over several weeks should lead to thorough re-evaluation of the plan and its underlying assumptions. Many of these conditions resolve within a few weeks with early treatment. The prognosis is generally good with early treatment and reduction in exposure.

Sometimes CTS and other conditions of the upper extremity follow a course similar to that of chronic severe low back pain. With conservative treatment and appropriate adjustments in the work setting, most cases should improve enough so that the patient can successfully return to work, but a small minority become chronic and very difficult to treat successfully with conventional approaches. In these cases of work-related disorders of the back and upper limb, the physical capabilities of the worker, the work demands, and the psychosocial factors related both to the worker and the employer are all-important in determining whether the worker successfully returns to work (84). The ways in which these factors interact are complex. The recognition that psychosocial factors—such as job satisfaction or negative self-fulfilling beliefs on the part of the patient, the employer, or the health care provider—are important should not lead to

ignoring the role of occupational physical exposures or to "blaming the victim" (85). When the latter occurs, delayed recovery is often attributed to personal weakness, low job satisfaction, or desire for secondary gain (86). Critical to prevention of these persistent cases is early intervention—an important reason to eliminate barriers to early reporting of symptoms.

It is increasingly recognized that comprehensive approaches that directly address all facets of the patient's situation may prove to be most effective in returning the patient to work and reducing future impairment. As a consequence, there has been a rapid development of comprehensive programs that ideally address the physical reconditioning of the worker, psychosocial factors, and workplace factors such as ongoing exposure (84,86). A contract between the patient and the health care provider should be established early in the treatment process, with the explicit aim of returning the worker safely to work. The diagnosis and treatment of severe or chronic work-related musculoskeletal disorders is sometimes challenging. Identification of the level of exposure by patient history is difficult, and usually direct observation of work is the preferred approach. There is substantial uncertainty about how to best measure exposure in some occupational settings, especially in the office. There is a danger of both overdiagnosis and underdiagnosis—for example, assuming either that every case of CTS is work-related or that no case of CTS is work-related. Not only is the assessment of exposure difficult, but scientific understanding of the relation between exposure and disease is still limited and imprecise. The danger exists of not recognizing when a case is becoming chronic and severe and when a multidisciplinary approach should be considered. Several observations are helpful when one is faced with challenges of diagnosing and treating these work-related conditions. A careful history and physical examination are important. An extensive objective assessment of the work environment may be required. In most cases, conservative treatment that preserves normal physical conditioning, that relies on a reduction of the level of occupational exposure while the patient remains at work, and that incorporates careful monitoring of the patient is a reasonable initial approach and is effective.

Prevention of these disorders requires the successful identification and remediation of adverse exposures.

PREVENTION

Preventive strategies are largely experience-based and have not been comprehensively evaluated by scientific studies. The principles outlined here must be adapted to fit the specific characteristics of each working environment. They should be viewed as a guide rather than a blueprint, and they require ongoing scientific evaluation (87,88). Three standard preventive strategies can be considered: (a) a reduction of the exposure to suspected occupational risk factors, such as vibrating hand tools; (b) a conditioning process that increases the tolerance of workers to the suspected occupational risk factors; and (c) development of a replacement process that is highly predictive and reliable to identify those persons who are at unusually high risk for development of an upper-extremity disorder. The remainder of this chapter discusses the first strategy in detail. Before this discussion, however, brief comment is in order for the other two strategies.

Development of a replacement process is the least desirable of these strategies because there are no scientifically valid screening procedures to identify which persons are at high risk, and because this shifts the cost of reducing the incidence of symptoms onto the workers (who are denied employment or placement) and increases the costs of the hiring and replacement processes. A prospective study comparing workers who were without symptoms of CTS but had abnormally slow sensory median nerve conduction with workers from the same industries with normal conduction concluded that workers with asymptomatic slowed conduction did

not have an increased risk of developing CTS over an 18-month period (89).

The second approach, a conditioning process that provides a period of time during which workers can gradually adapt their muscles and tendons to the new demands, could be a useful approach for workers in forceful- or repetitive-action jobs. Training of new workers in the most efficient and least stressful ways of performing their jobs may also be useful, provided that the work tasks can be done in alternative ways that are both less stressful and at least as fast. Similarly, workers with symptoms may, with training, be able to adapt an equally efficient, but less stressful, work method. Training activities have not been evaluated specifically. Several employers, perceiving long-term benefits from a "phasing-in period," have established transitional or training areas where employees may work at a reduced pace for a limited time. In a survey of 5,000 employers in Washington State, among those who took prevention steps, a larger percentage reported decreased number and severity of musculoskeletal disorders with engineering and administrative measures (e.g., task variety, reduced overtime) than with strictly personal controls (e.g., exercise programs, personal protective equipment) (90).

Reduction in exposures, the standard preventive approach that directs attention to control of occupational factors, has the most promise. This approach often requires changes in the workstation, work process, or use of tools. Sometimes administrative changes, such as work restrictions, use of personal protective equipment (e.g., palm pads), or job rotation are useful alternatives, either as preventive or as therapeutic interventions.

In order for work restrictions to be effective in the treatment of injured workers, the health care provider must be specific about the type of work activity that should be avoided or reduced. For example, it is better to limit repetitive hand activities to "fewer than 10 movements per minute for more than 2 hours per day" than to prescribe "no repetitive hand movements during the work shift." Developing specific recommendations for work restrictions is facilitated by viewing videotapes of the usual job of the worker or by obtaining detailed job descriptions from the employer. As a preventive intervention, job rotation of workers among jobs that require different types of motions of the upper extremity may simply expose an even greater number of workers to a considerable degree of risk.

To reduce exposure, the first step required for instituting changes in workstations or work processes is to analyze the specific characteristics of suspected high-risk jobs. Although the job review can be conducted by an industrial engineer or occupational health professional with training in ergonomics, the involvement of those persons who are most knowledgeable about the job is important. Experience has shown that operators and supervisors with limited technical training can successfully identify many of the hazardous aspects of a specific job, and that specific solutions may not be effective or accepted without the involvement of such persons in the job review and development of solutions.

REDUCING EXPOSURE TO RISK FACTORS

A job analysis performed for the patient with the spot-welding job in the case at the start of this section would have identified several exposure factors. The job was repetitive and required forceful gripping of the handles of the welding gun. The wrists were extended through most of the job cycle.

After a job analysis has identified the potentially hazardous exposures associated with a particular job, specific solutions should be solicited from those who are knowledgeable about the job. With limited training in the control principles (discussed in the next section), engineers, production employees, and front-line supervisors often propose the most useful methods for eliminating hazardous risk factors. If several factors are present, it can be difficult to determine which is the most detrimental. Where

possible, integrated solutions should be developed that reduce multiple risk factors at the same time.

Control of repetitiveness, forcefulness, awkward posture, mechanical stress, vibration, and cold are often possible, as illustrated in the following examples.

Control of Repetitiveness

1. Use mechanical assists and other types of automation. For example, in packing operations, use a device, rather than the hands, to transfer parts.
2. Rotate workers among jobs that require different types of motions. Rotation must be viewed as a temporary administrative control, one used only until a more permanent solution can be found.
3. Implement horizontal work enlargement by adding different elements or steps to a job, particularly steps that do not require the same motions as the current work cycle.
4. Increase work allowances or decrease production standards. This control strategy is rarely looked on favorably by management.
5. Design a tool for use in either hand and also so that fingers are not used for triggering motions.

Control of Forcefulness

1. Decrease the weight held in the hand by providing adjustable fixtures to hold parts being worked on. Many conventional balancers are available to neutralize tool weight. Articulating arms are used in many plants to hold and manipulate heavy tools into awkward positions.
2. Control torque reaction force in power handtools by using torque reaction bars, torque-absorbing overhead balancers, and mounted nut-holding devices. Control the time that a worker is exposed to torque reaction by using shut-off rather than stall power tools. Avoid jerky motions by hand-held tools.

3. Design jobs so that a power grip rather than a pinch can be used whenever possible. (Maximum voluntary contraction in a power grip is approximately three times greater than in a pinch).
4. Increase the coefficient of friction on hand tools to reduce slipperiness, for example, by use of plastic sleeves that can be slipped over metal handles of tools.
5. Design jobs so that slides or hoists are used to move parts, to reduce the amount of handling or carrying of parts by the worker.

Control of Awkward Posture

The primary method for reducing awkward postures is to design adjustability of position into the job. Wrist, elbow, and shoulder postures required on a job often are determined by the height of the work surface with respect to the location of the worker. A tall worker may use less wrist flexion or ulnar deviation than a shorter worker. Additionally, awkward postures can be reduced by the following procedures.

1. Alter the location or method of the work. For example, in automotive assembly operations, changing the line location at which a particular part is installed may result in easier access.
2. Redesign tools or change the type of tool used. For example, when wrist flexion occurs with a piston-shaped tool that is used on a horizontal surface, correction may involve use of an in-line type tool or lowering of the workstation.
3. Alter the orientation of the work.
4. Avoid job tasks that require shoulder abduction or forward flexion greater than 30 to 45 degrees, elbow flexion greater than 110 degrees, wrist flexion or extension greater than 20 degrees, or frequent neck rotation.
5. Provide support for the forearm when precise finger motions are required, to reduce static muscle loading in the arm and shoulder girdle.

Control of Vibration

1. Do not use impact wrenches or piercing hammers.
2. Use balancers, isolators, and damping materials.
3. Use handle coatings that attenuate vibrations and increase the coefficient of friction to reduce strength requirements.

Control of Mechanical Stress

1. Round or flare the edges of sharp objects, such as guards and container edges.
2. Use different types of palm button guards, which allow room for the operator to use the button without contact with the guard.
3. Use palm pads, which may provide some protection until tools can be developed to eliminate hand hammering.
4. Use compliant cushioning material on handles or increase the length of the handles to cause the force to dissipate over a greater surface of the hand.
5. Use different-sized tools for different-sized hands.
6. Avoid narrow tool handles that concentrate large forces onto small areas of the hand.

Control of Cold and Use of Gloves

1. Properly maintain power tool air hoses to eliminate cold exhaust air leaks onto the workers' hands or arms.
2. Provide a variety of styles and sizes of gloves to ensure proper fit of gloves. Although gloves may protect the hands from cold exposures, they often decrease grip strength (requiring more forceful exertion), decrease tactile sensitivity, decrease manipulative ability, increase space requirements, and increase the risk of becoming caught in moving parts.
3. Cover only that part of the hand that is necessary for protection. Examples include use of safety tape for the fingertips with fingerless gloves and use of palm pads for the palm.

Conclusion

In summary, work-related low back pain and disorders of the upper extremity are together among the most common occupational health problems. Although scientific knowledge often limits our ability to determine precisely the role of occupational and nonoccupational factors in the diagnosis of these conditions, substantial progress can be made in reducing their severity by applying existing knowledge about the role of physical factors in these disorders, including forceful repetitive hand work and frequent lifting of heavy objects. Work should be designed to reduce exposure to the known physical risk factors. Encouragement of prompt and appropriately conservative medical evaluation of workers with such disorders can contribute to secondary prevention. Finally, for the minority of workers with disorders that do not respond to conservative treatment, including reduction in the level of exposure, treatment programs that address all aspects of the problem, both the psychosocial and the physical, probably have the greatest chance of preventing permanent disability from these disorders.

REFERENCES

1. American Academy of Orthopaedic Surgeons. Musculoskeletal conditions in the United States. Park Ridge, IL: AAOS, 1992.
2. Andersson GBJ. The epidemiology of spinal disorders. In: Frymoyer JW, ed. The adult spine: principles and practice, 2nd ed. Philadelphia: Lippincott-Raven Press, 1997:93–141.
3. Pope MH, Andersson GBJ, Frymoyer JW, Chaffin DB. Occupational low back pain: assessment, treatment and prevention. St. Louis: Mosby –Year Book, 1991:1–325.
4. Frymoyer J. The adult spine: principles and practice. Philadelphia: Lippincott-Raven, 1997.
5. Spitzer WO, LeBlanc FE, Dupuis M, et al. Scientific approach to the assessment and management of activity-related spinal disorders: a monograph for clinicians. Report of the Quebec Task Force on Spinal Disorders. Spine 1987;12:S1–S59.
6. Andersson GBJ. Epidemiologic aspects on low back pain in industry. Spine 1981;6:53–60.
7. Klein BP, Jensen RC, Sanderson LM. Assessment of workers' compensation claims for back strains/sprains. J Occup Med 1984;26:443–448.
8. Frymoyer JW, Pope MH. Epidemiologic insight into the relationship between usage and back disorder. In: Hadler NH, ed. Current concepts in regional musculoskeletal illness. Orlando, FL: Grune & Stratton, 1987:263–279.
9. Chaffin DB, Herrin GD, Keyserling WM. Preemployment strength testing: an updated position. J Occup Med 1978;20:403–408.
10. Kelsey JL. An epidemiological study of acute herniated lumbar intervertebral discs. Rheumatol Rehab 1975;14:144–159.
11. Kelsey JL, Githens PB, O'Connor T, et al. Acute prolonged lumbar intervertebral disc: an epidemiologic study with special reference to driving automobiles and cigarette smoking. Spine 1984;9:608–613.
12. Magora A. Investigation of the relation between low back pain and occupation. Ind Med Surg 1972;41:5–9.
13. Keyserling WM. Postural analysis of the trunk and shoulders in simulated real time. Ergonomics. 1986;29(4):569–583.
14. Bigos SJ, Battie MC. Risk factors for industrial back problems. Seminars in Spine Surgery 1992;4:2–11.
15. Svensson H-O, Andersson GBJ. Low back pain in forty to forty-seven year old men: work history and work environment factors. Spine 1983;8:272.
16. Svensson H-O, Andersson GBJ. The relationship of low-back pain, work history, work environment, and stress: a retrospective cross-sectional study of 38 to 64 year old women. Spine 1989;14:517–522.
17. Frymoyer JW, Pope MH, Costanza MC, Rosen JC, Goggin JE, Wilder DG. Epidemiologic studies of low back pain. Spine 1980;5:419–423.
18. Chaffin DB, Park KS. A longitudinal study of low-back pain as associated with occupational weight lifting factors. Ind Hyg Assoc J 1973;34:513–525.
19. Snook SH, Campanelli RA, Hart JW. A study of three preventive approaches to low back injury. J Occup Med 1978;20:478–481.
20. Snook SH. Low back pain in industry. In: White AA, Gordon SL, eds. Symposium on idiopathic low back pain. St. Louis: Mosby, 1982:23–28.
21. Kelsey JL, Githens PB, White AA, et al. An epidemiologic study of lifting and twisting on the job and risk for acute prolapsed lumbar intervertebral disc. J Orthop Res 1984;2:61–66.
22. Kelsey JL, Hardy RJ. Driving of motor vehicles as a risk factor for acute herniated lumbar intervertebral disc. Am J Epidemiol 1975;102:63.
23. Heliovaara M. Occupation and risk of herniated lumbar intervertebral disc or sciatica leading to hospitalization. J Chronic Dis 1987;3:259–264.
24. Bigos SJ, Spengler DM, Martin NA, Zeh J, Fisher L, Nachemson A. Back injuries in industry: a retrospective study. III: Employee-related factors. Spine 1986;11:252–256.
25. Bigos S. Acute low back problems in adults. Clinical Practice Guidelines No. 14. Agency for Health Care Policy and Research Publ. No 95-0642, Washington, DC: U.S. Department of Health and Human Services, Public Health Service, Agency for Health Care Policy and Research, 1994:1–160.
26. Deyo RA, Diehl AK, Rosenthal M. How many days of bed rest for acute low back pain? A randomized clinical trial. N Engl J Med 1986;315:1064–1070.
27. Deyo RA. Conservative therapy for low back pain:

distinguishing useful from useless therapy. JAMA 1983;25:1057.

28. Weber H. Lumbar disk herniation: a controlled, perspective study with ten years of observation. Spine 1983;8:131–140.

29. Dillane JB, Fry J, Kalton G. Acute back syndrome: a study from general practice. Br Med J 1966;2:82–84.

30. Bergquist-Ullman M, Larsson U. Acute low back pain in industry. Acta Orthop Scand 1977;48 (Suppl. 170): 1–117.

31. McGill CM. Industrial back problems: a control program. J Occup Med 1968;10:174–178.

32. Battie MC. Minimizing the impact of back pain: workplace strategies. Seminars in Spine Surgery 1992;4:20–28.

33. King PM. Back injury prevention programs: a critical review of the literature. J Occup Rehabil 1993;3:145–157.

34. Lahad A, Maller AD, Berg AO, Deyo RA. The effectiveness of four interventions for the prevention of low back pain. JAMA 1994;272:1286–1291.

35. US Department of Health and Human Services. Work practices guide for manual lifting. NIOSH publication no. 81-122. Washington, DC: DHHS, 1981.

36. Waters TR, Putz-Anderson V, Garg A, Fine LJ. Revised NIOSH equation for the design and evaluation of manual lifting tasks. Ergonomics 1993; 26:749–776.

37. Snook SH. The design of manual handling tasks. Ergonomics 1978;21:963–985.

38. Lavender SA, Thomas JS, Chang D, Andersson GBJ. Effect of lifting belts, foot movement, and lift asymmetry on trunk motions. Hum Factors 1995;37:844–853.

39. Reddell C, Congleton J, Huchingson D, Montgomery J. An evaluation of weight lifting belt and back injury prevention training class for airline baggage handlers. Appl Ergon 1992;25:319–329.

40. Walsh NE, Schwartz RK. The influence of prophylactic orthoses on abdominal strength and low back injury in the workplace. Am J Phys Med Rehabil 1990;69:245–250.

41. Himmelstein JS, Andersson GBJ. Low back pain: risk evaluation and preplacement screening. Occup Med 1988;3:255–269.

42. Biering-Sorensen F. Physical measurements as risk indicators for low-back trouble over a one-year period. Spine 1984;9:106–119.

43. Fitzler SL, Berger RA. Attitudinal change: the Chelsea back program. Occup Health Saf 1982;51:24–26.

44. Fitzler SL, Berger RA. Chelsea back program: one year later. Occup Health Saf 1983;52:52–54.

45. Wiesel SW, Feffer HL, Rothman RH. Industrial low back pain: a prospective evaluation of a standardized diagnostic and treatment protocol. Spine 1984;9:199.

46. Phalen G. The carpal tunnel syndrome: clinical evaluation of 598 hands. Clin Orthop 1972;83:29.

47. Moore JS. De Quervain's tenosynovitis. Stenosing tenosynovitis of the first dorsal compartment. J Occup Environ Med 1997;39:990–1002.

48. Kurppa K, Viikari-Junta E, Kuosma E, Huuskonen M, Kivi P. Incidence of tenosynovitis or peritendin-

itis and epicondylitis in a meat-processing factory. Scand J Work Environ Health 1991;17:32–37.

49. Stenlund B, Goldie IN, Hagberg M, Hogstedt C. Shoulder tendinitis and its relation to heavy manual work and exposure to vibration. Scand J Work Environ Health 1993;19:43–49.

50. Hagberg M, Wegman DH. Prevalence rates and odds ratios of shoulder-neck disease in different occupational groups. Br J Ind Med 1987;44:602–610.

51. Hadler NM, Gillings DB, Imbus HR, et al. Hand structure and function in an industrial setting. Arthritis Rheum 1978;21:210–220.

52. Taylor W, Pelmar PL. The hand-arm vibration syndrome: an update. Br J Ind Med 1990;47:577–579.

53. Bureau of Labor Statistics. Lost-worktime injuries and illnesses: characteristics and resulting time away from work, 1996. Washington, DC: Bureau of Labor Statistics, US Department of Labor, April 23, 1998.

54. Conway H, Svenson J, Occupational injury and illness rates, 1992–1996: why they fell. Monthly Labor Review, November 1988:36–58.

55. National Institute for Occupational Safety and Health. Musculoskeletal disorders and workplace factors: a critical review of epidemiological evidence for work-related musculoskeletal disorders of the neck, upper extremity, and low back. DHHS (NIOSH) Publication No. 97-141. Cincinnati: US Department of Health and Human Services, Public Health Service, Centers for Disease Control and Prevention, NIOSH, 1997.

56. Silverstein B, Kalat J. Work-related disorders of the back and upper extremity in Washington State, 1989–1996. Technical Report 40-1-1997. Olympia: Safety and Health Assessment and Research for Prevention (SHARP) Program, Washington State Department of Labor and Industries, January 1998.

57. Gorsche R, Wiley JP, Renger R, Brant R, Gemer TY, Sasyniuk TM. Prevalence and incidence of stenosing tenosynovitis (trigger finger) in a meat-packing plant. J Occup Environ Med 1998;40:556–560.

58. Thompson AR, Plewes LW, Shaw EG. Peritendinitis crepitans and simple tenosynovitis: a clinical study of 544 cases in industry. Br J Ind Med 1951;8:150–160, 1951.

59. Gross AS, Louis DS, Carr KA, Weiss SA. Carpal tunnel syndrome: a clinicopathologic study. J Occup Med 1995;37:437–441.

60. Steering Committee for the Workshop on Work-Related Musculoskeletal Injuries. Work-related musculoskeletal disorders: the research base. Workshop summary and papers. Washington, DC: National Academy Press, 1999.

61. Lacewood AH. Medical problems of musicians. N Engl J Med 1989;320:221–227.

62. Armstrong TJ, Buckle P, Fine LJ. A conceptual model for work-related neck and upper-limb musculoskeletal disorders. Scand J Work Environ Health 1993;19:73–84.

63. Hagg GM. Static work load and occupational myalgia: a new explanation model. In: Anderson P, Hobart D, Danoff J, eds. Electromyographical kinesiology. Amsterdam: Elsevier Science Publishers, 1991:141–144.

64. Hagberg M, Morgenstern H, Kelsh M. Impact of occupations and job tasks on the prevalence of car-

pal tunnel syndrome. Scand J Work Environ Health 1992;18:337–345.

65. Stevens JC, Sun S, Beard CM, O'Fallon WM, Kurland LT. Carpal tunnel syndrome in Rochester, Minnesota, 1961 to 1980. Neurology 1988; 38:134–138.

66. Franklin GM, Haug J, Heyer N, Checkoway H. Occupational carpal tunnel syndrome in Washington State, 1984–1988. Am J Public Health 1991;81: 741–746.

67. Werner S, Albers JW, Franzblau A, Armstrong TJ. The relationship between body mass index and the diagnosis of carpal tunnel syndrome. Muscle Nerve 1994;17:1492.

68. Nathan PA, Keniston RC, Myers LD, Meadows KD. Obesity as a risk factor for slowing of sensory conduction of the median nerve in industry: a cross-sectional and longitudinal study involving 429 workers. J Occup Med 1992;34:379–383.

69. Stevens JC, Beard CM, O'Fallon WM, Kurland LT. Conditions associated with carpal tunnel syndrome. Mayo Clin Proc 1992;67:541–548.

70. Tanaka S, Wild DK, Cameron LL, Freund E. Association of occupational and nonoccupational risk factors with the prevalence of self-reported carpal tunnel syndrome in a national survey of the working population. Am J Ind Med 1997;32:550–556.

71. Theorell T, Harms-Ringdahl K, Ahlberg-Hulten G, Westin B. Psychosocial job factors and symptoms from the locomotor system: a multi-causal analysis. Scand J Rehabil Med 1991;23:165–173.

72. Kerr MS, Frank JW, Shannon HS, et al. Independent biomechanical and psychosocial risk factors for low back pain at work. Personal communication.

73. Hughes RE, Silverstein BA, Evanoff BA. Risk factor for work-related musculoskeletal disorders in an aluminum smelter. Am J Ind Med 1997;32:66–75.

74. Theorell T, Nordemar R, Michelsen H, Stockholm Music Study Group. Pain thresholds during standardized psychological stress: relation to perceived psychosocial work situation. J Psychosom Res 1993; 37:299–305.

75. Rempel DM, Harrison RJ, Barnhardt S. Work-related cumulative trauma disorders of the upper extremity. JAMA 1992;267:838–842.

76. Rempel D, Eranoff B, Amadio PC, et al. Consensus for the classification of carpal tunnel syndrome in epidemiologic studies. Am J Public Health 1998;88:1447–1451.

77. Stevens K. The carpal tunnel syndrome in cardiologists [Letter]. Ann Intern Med 1990;112:796.

78. Dawson DM, Hallet M, Millender LH. Entrapment neuropathies, 2nd ed. Boston: Little Brown, 1990.

79. Herington TN, Morse LH, eds. Occupational injuries: evaluation, management, prevention. St. Louis: Mosby, 1995.

80. Moore JS. Clinical determination of work-relatedness in carpal tunnel syndrome. J Occup Rehab 1991;1:145–158.

81. Hales TR, Bertsche PK. Management of upper extremity cumulative trauma disorders. AAOHN J 1992;40:118–128.

82. American Academy of Orthopaedic Surgeons. Clinical policies: carpal tunnel syndrome. Park Ridge, IL: American Academy of Orthopaedic Surgeons, 1991.

83. Katz JN, Keller RB, Simmons BP, et al. Maine carpal tunnel study: outcome of operative and nonoperative therapy for carpal tunnel syndrome in a community-based cohort. J Hand Surg [Am] 1998; 23:697–710.

84. Feurstein M. A multi-disciplinary approach to the prevention, evaluation and management of work disability. J Occup Rehab 1991;1:5–12.

85. Niemeyer LO. Social labeling, stereotyping, and observer-patient interaction on outcome. J Occup Rehab 1991;1:251–267.

86. Parenmark G, Malmkvist AK. The effect of an outpatient rehabilitation program on occupational cervicobrachial disorders. J Occup Rehab 1992; 2:67–72.

87. Silverstein B, Fine LJ, Armstrong SJ. Carpal tunnel syndrome: causes and a preventive strategy. Semin Occup Med 1986;1:213–221.

88. National Institute for Occupational Safety and Health. Elements of ergonomics programs. DHHS (NIOSH) Publication No. 97-117. Cincinnati, OH: US Department of Health and Human Services, Public Health Services, Centers for Disease Control and Prevention, NIOSH, 1997.

89. Werner RA, Fransblau A, Albers JW, Buchele H, Armstrong TJ. Use of screening nerve conduction studies for predicting future carpal tunnel syndrome. Occup Environ Health 1997;54:96–100.

90. Foley M, Silverstein B. Musculoskeletal disorders, risk factors and prevention steps: a survey of employers in Washington State. Technical Report 53-1-1999. Olympia: Safety and Health Assessment and Research for Prevention (SHARP) Program, Washington State Department of Labor and Industries, January 1999.

BIBLIOGRAPHY

Cailliet R. Hand pain and impairment, 2nd ed. Philadelphia: Davis, 1981.
A good but old introductory text with excellent illustrations. (Look at the Seminars in Occupational Medicine for a better example.)

Deyo RA. Conservative therapy for low back pain: distinguishing useful from useless therapy. JAMA 1983;250:1057.
Reviews the evidence supporting commonly used conservative therapies for low back pain.

Nachemson A. Advances in low back pain. Clin Orthop 1985;200:266.
An objective and thorough review of current knowledge about low back pain. Topics include epidemiology, etiology, biomechanics, and treatment. Emphasis is placed on the benefits of patient education and motion and the disadvantages of prolonged bed rest (inactivity) and repeat surgery.

National Institute for Occupational Safety and Health. Musculoskeletal disorders and workplace factors: a critical review of epidemiological evidence for work-related musculoskeletal disorders of the neck, upper extremity, and low back. DHHS (NIOSH) Publication No. 97-141. Cincinnati: US Department of Health and Human Services, Public Health Service, Centers for Disease Control and Prevention, NIOSH, 1997.

Reviews epidemiologic studies and discusses the clearest delineation of the association of upper-extremity disorders with work exposures.

Nordin M, Andersson GBJ, Pope MH, eds. Musculoskeletal disorders in the workplace: principles and practice. St. Louis: Mosby, 1997.
Covers basic concepts of pain, biomechanics, psychosocial factors, clinical evaluation, and patient care as well as workplace modifications.

Quinet RJ, Hadler NM. Diagnosis and treatment of backache. Semin Arthritis Rheum 1979;8:261.
A comprehensive, objective, and well-written review of the etiology, diagnosis, and treatment of low back pain. The authors discuss traction, drugs, diathermy, heat, cold, manipulation, corsets, braces, injection therapy, exercises, chemonucleolysis, and surgery.

Rempel DM, Harrison RJ, Barnhardt S. Work-related cumulative trauma disorders of the upper extremity. JAMA 1992;267:838–842.

A concise summary of the causes, diagnosis, and principles of treatment for common work-related disorders of the upper extremity.

Rowe ML. Backache at work. Fairport, NY: Perinton, 1983.
Describes the results of a 20-year clinical study of low back pain at a large company. Included in the study are 1,500 cases. The effectiveness of current selection techniques in preventing low back pain is discussed. Emphasis also is placed on preventing the disability.

Wiesel SW, Feffer HL, Rothman RH. Industrial low-back pain: a prospective evaluation of a standardized diagnostic and treatment protocol. Spine 1984;9:199.
Describes a standardized approach to the diagnosis and treatment of low back pain in employees at a large company. Two orthopedic surgeons monitored the course and treatment of employee low back pain and intervened when they believed it necessary. The program dramatically decreased the number of patients sent to surgery.

APPENDIX

WORK-RELATED DISORDERS OF THE UPPER EXTREMITIES

Disorder	History	Physical examination
Wrist		
Carpal tunnel syndrome	Pain, tingling, or numbness in medial sensory distribution of the hand; nocturnal exacerbation; problems with dropping things	Positive Phalen's test; positive Tinel's test; thenar atrophy in severe cases; rule out pronator teres syndrome, cervical root syndrome
De Quervain's disease	Pain in anatomic snuffbox; may radiate up forearm; no history of radial or wrist fracture	Positive Finkelstein's test with sharp pain rather than just pulling sensation; rule out radial nerve entrapment
Trigger finger	Finger locks in extension or flexion; requires assistance in unlocking; nodule or tendon	Nodule at base of digit palpable; locking on flexion or extension of digits
Ulnar nerve compression (Guyon canal syndrome)	Burning, tingling, or numbness in fourth and fifth digits; clumsiness in fine movements	Positive Tinel's sign at Guyon canal; positive Phalen's test in ulnar distribution; decreased pinch strength; weakness on resisted abduction and adduction of digits; rule out cervical root disorder, thoracic outlet syndrome, cubital tunnel syndrome
Tendinitis, tenosynovitis	Localized pain and swelling over muscle-tendon structure	Pain exacerbated by resisted motions; fine crepitus on passive range of motion (ROM) possible; no pain on passive ROM; pronounced asymmetric grip strength
Elbow/Forearm		
Lateral epicondylitis (tennis elbow)	Pain at lateral epicondyle during rest or active motion of wrists and fingers	Pain on resisted extension of wrist with fingers flexed; no pain or limitation on full passive ROM; pain at epicondyle on palpation; pain on resisted radial deviation; rule out radial nerve entrapment
Medial epicondylitis (golfer's elbow)	Pain at medial epicondyle during rest or active motion of wrist and fingers	No pain on passive ROM; pain on resisted wrist flexion and resisted forearm pronation; pain at medial epicondyle on palpation
Olecranon bursitis	Pain and swelling at olecranon	No pain on passive or resisted ROM; swelling around olecranon on palpation; rule out rheumatoid arthritis
Pronator teres syndrome	Burning pain in first three digits of hand and forearm	Increased pain in forearm by resisted pronation with clenched fist and flexed wrist (Mill's test); sensory impairment of thenar eminence; rule out carpal tunnel syndrome

WORK-RELATED DISORDERS OF THE UPPER EXTREMITIES (Continued)

Disorder	History	Physical examination
Shoulder		
Rotator cuff tendinitis (mainly supraspinatus)	Dull ache generally localized to deltoid area without neck or arm radiation; no symptoms of distal paresthesia; nocturnal exacerbation; subject may not "catch" on movement.	Diffuse tenderness over shoulder, especially over humeral head and lateral to acromion; if tenderness is localized, it is most often over supraspinatus insertion; weakness uncommon *Supraspinatus:* shrugs shoulder on abduction, painful arc at 70–90 degrees; passive ROM normal; pain on resisted abduction *Infraspinatus:* pain on resisted external rotation; painful arc; rule out rheumatoid arthritis
Bicipital tenosynovitis	Pain localized to bicipital groove area that may radiate to anterior aspect of arm; no distal paresthesia; nocturnal exacerbation; subject able to use forearm when upper arm held against chest; subject notes pain on abduction and rotation	Positive Yergason's test (resisted supination), or positive Speed's test (resisted wrist flexion); normal passive and active ROM
Degenerative joint disease—acromioclavicular joint	Generalized aching shoulder pain exacerbated by motion; least difficulty in morning but worse as day progresses	Limitation is similar on active and passive ROM; most discomfort is with mild abduction; crepitus common; tenderness on palpation directly over acromioclavicular articulation; pain reproduced as arm is abducted more than 90 degrees; pain on shoulder shrug
Degenerative joint disease—glenohumeral joint	Pain is very diffuse and nocturnal	Tenderness to palpation along joint line; no deltoid or suprespinatus pain; passive ROM full but painful; active ROM retarded on flexion and extension (normal is 240 degrees in youth, 190 degrees at age 70; normal abduction in youth is 166 degrees, 116 degrees at age 70)
Neck/Scapula		
Tension neck syndrome (costalscapular syndrome)	Neck pain or stiffness; no history of herniated cervical disc, injury, or ankylosing spondylitis	Muscle tightness, palpable hardening and tender spots; pain on resisted neck lateral flexion and rotation
Cervical root syndrome	Pain radiating from neck to one or both arms with numbness in one or both hands; exacerbated by cough	Limited passive or active ROM; radiating pain on passive motions; positive foreminal test; decreased pinprick in dermatome; absence of joint findings

Occupational Health: Recognizing and Preventing Work-Related Disease and Injury, 4th ed.
Edited by Barry S. Levy and David H. Wegman, Lippincott Williams & Wilkins, Philadelphia © 2000.

27 Skin Disorders

Michael E. Bigby, Kenneth A. Arndt, and
Serge A. Coopman

A cutaneous abnormality caused directly or indirectly by the work environment is an occupational skin disorder. Work-related cutaneous reactions and clinical syndromes are as varied as the environments in which people work. Skin disorders are the second most frequently reported type of occupational disease (after disorders associated with repeated trauma). A basic understanding of occupational skin disorders is therefore essential for everyone involved in occupational health.

An occupational skin injury is defined as an immediate adverse effect on the skin that results from instantaneous trauma or brief exposure to toxic agents involving a single incident in the work environment (1). The National Institute for Occupational Safety and Health (NIOSH) used the National Electronic Injury Surveillance System for surveillance of nonfatal occupational injuries treated in hospital emergency departments to estimate occupational injury rates in the United States in 1996. Sprains and strains accounted for 27% of injuries, followed by lacerations (22%); contusions, abrasions and hematomas (20%); dislocations and fractures

(7%); and burns (4%). The hands and fingers sustained the most injuries (2).

Occupational skin diseases or illnesses also result from exposure to toxic agents or environmental factors at work. In contrast to occupational skin injuries, occupational skin diseases require prolonged exposures and involve longer intervals between exposure and occurrence of disease (1). In comparison with other occupational health problems, skin disorders are often more easily diagnosed and recognized as being work-related.

The average annual reported incidence of occupational disease in the United States in 1997 was 5 cases per 1,000 full-time workers. Sixty-four percent of the workplace illnesses were disorders associated with repeated trauma, and 13% were skin diseases (3). It is estimated that the actual incidence of occupational skin diseases is 10 to 50 times higher than the reported incidence (5). Occupational skin disease also causes significant time lost from work. In a study of occupational skin disease in Washington State, based on filed claims from 1989 to 1993, of 7,445 total claims, 675 (9%) involved more than three lost workdays (6).

The annual cost of occupational skin disease in the United States is great: at least 200,000 lost workdays, an estimated direct economic cost (lost wages or productivity) of $9.6 million, and a total cost (adding the costs of replacement workers, indemnity, medical costs, and insurance) of $20 to $30 million. If this estimate is multiplied 10 to

M. E. Bigby and K. A. Arndt: Department of Dermatology, Harvard Medical School, and Beth Isreal Deaconess Medical Center, Boston Massachusetts 02215.

S. A. Coopman: Eevwfeestclinic, Antwerp, Belgium.

50 times to compensate for underreporting, the actual annual cost may be $200 million to $1.5 billion (5).

Occupational skin disorders are unevenly distributed among industries. Workers in manufacturing and services account for the majority of cases (7). In other countries, specific occupational groups account for the majority of cases: in England, miners; in Germany, steelworkers; and in Italy, bricklayers (8).

SKIN STRUCTURE AND FUNCTION

To understand skin disorders, a basic review of skin structure and function is helpful. Because skin represents the boundary between a person and the environment, it is often the first site exposed to environmental insults. Skin weighs 3 to 4 kg, constitutes 6% of body weight, and covers about 20 square feet (about 2 m^2) in the average adult (9). It consists of three principal layers.

The *epidermis* is the most superficial layer. Its outermost compartment, the anucleate stratum corneum or horny layer, acts as the primary barrier that retains water and interferes with the entrance of microorganisms and toxic substances. This barrier is impermeable to hydrophilic substances. Percutaneous absorption is greater with lipophilic compounds, through inflamed or abraded skin, or after occlusion by water-permeable materials (e.g., waterproof clothes or rubber gloves) (see Chapter 15). Melanocytes of neural crest origin lie within the epidermis and synthesize the pigment melanin, which protects against ultraviolet (UV) radiation. Langerhans cells, dendritic antigen-presenting cells[1] that are important in the development of allergic contact dermatitis, also reside within the lower layers of the epidermis.

The *dermis* consists primarily of the fibrous protein collagen in a glycosaminoglycan[2] ground substance, both of which protect against trauma and envelop the body in a strong and flexible wrap (Fig. 27-1). Also within the dermis are blood vessels, lymphatics, nerves, and the epidermal appendages: eccrine and apocrine sweat glands, sebaceous glands, and hair follicles. The epidermal appendages, especially the pilosebaceous unit (hair follicle and sebaceous gland), are important portals of entry for chemical irritants and allergens.

The third layer is the *subcutaneous tissue.* The thick, fatty subcutaneous tissue helps conserve the body heat and serves as an additional shock-absorbing buffer.

Certain aspects of normal or altered skin are particularly important for workers in an industrial environment. Induced or inherent alterations of barrier-layer function increase susceptibility to the effects of workplace exposures and open the skin to further damage. This phenomenon occurs in contact dermatitis, psoriasis, and atopic dermatitis. Workers with atopic dermatitis are estimated to have a 13-fold higher risk of developing occupational irritant dermatitis (10). Patients with psoriasis may develop psoriasis in areas that are exposed to trauma or irritation (Koebner's phenomenon). These two common skin diseases, if noted on preplacement examination, may be sufficient reason to place a worker where exposure to irritating or sensitizing chemicals or to physical trauma does not occur.

Poorly pigmented skin is far more susceptible to UV light damage. Workers who tan poorly or not at all and who sunburn easily are more likely to develop basal cell and squamous cell skin cancers as a result of chronic exposure to sunlight. Such fair-skinned workers should protect their skin from UV damage by using sunscreens and wearing hats and long sleeves, especially if

---

[1] These cells are dendritic cells that have the capacity to process antigens (i.e., engulf, partially metabolize, and display antigens on their surface) and to present antigens to lymphocytes.

[2] This substance has a polysaccharide (glycan) structure that contains hexosamines (glycosamine), hyaluronic acid, dermatin sulfate, and chondroitin sulfates.

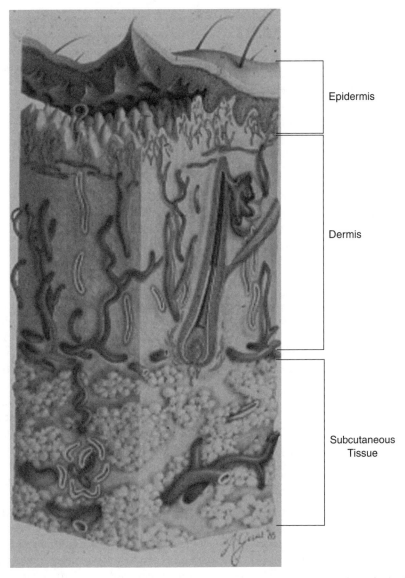

Epidermis

Dermis

Subcutaneous
Tissue

FIG. 27-1. Cross-section of human skin. The outermost stratum corneum is the principal barrier to chemical absorption.

they work in outdoor occupations such as agriculture, fishing, forestry, or construction.

Excessive eccrine (sweat) gland function results in hyperhidrosis (excessive perspiration), which may make it difficult to grasp objects. In the metal industry, malfunction of metal components occurs because of the problem of "rusters"—workers whose palmar sweat has a tendency to cause corrosion of metal objects. Another problem exacerbated by sweating results when otherwise harmless dusts, such as soda ash, become hazardous when they are dissolved after being deposited on wet (but otherwise normal) skin. Disorders affecting the pilosebaceous units, such as acne, are worsened after exposure to heavy oils, grease, and hot and humid working conditions.

CAUSES OF OCCUPATIONAL DERMATOSES

Workplace agents that may induce skin disorders can be arbitrarily divided into categories (Table 27-1). Chemical agents produce the great majority of occupational dermatoses by inducing either irritant or allergic contact sensitivity reactions. Approximately two-thirds of occupational contact dermatitis cases are of the irritant type (11). Common

TABLE 27-1. *Examples of workplace agents that induce skin disorders[a]*

Chemical agents	Bacteria
Rhus oleoresin (poison ivy and oak)	Secondary superinfection
Acids	Impetigo
Alkalis	Furuncles
Solvents	Anthrax (infected hides)
Oils	Brucellosis (animals)
Soaps and detergents	Erysipeloid (fish and poultry)
Plastics	
Resins	Mycobacteria infection (animals and fish tanks)
Paraphenylenediamine	
Chromates	Tularemia (rodents)
Acrylates	Cat-scratch disease (cats)
Nickel compounds	
Rubber chemicals (including latex)	Lyme disease (tick bites)
Petroleum products not used as solvents	**Fungi**
	Candida infection (moist conditions)
Glass dust	Dermatophytosis (animals, soil, and humans)
Plant and wood substances	
Pink rot celery	Sporotrichosis (soil)
Citrus fruit	Blastomycosis (inhalation)
Physical agents	
Ionizing and nonionizing radiation	Coccidiodomycosis (inhalation)
Wind	**Ectoparasites**
Sunlight	Cutaneous larva migrans
Temperature extremes	Scabies
Humidity	Grain itch (mites)
Viruses	Bites (ticks, fleas, mites, and spiders)
Orf (sheep)	
Milker's nodule (cows)	Swimmer's itch (schistosomes)
Warts	
Herpes simplex (patients)	**Biting animals**
Rickettsiae	**Mechanical factors**
Rocky Mountain spotted fever (ticks)	Pressure
Murine typhus (fleas)	Friction
Tick typhus (ticks)	Vibration
Rickettsialpox (mites)	
Scrub typhus (mites)	

[a] Agents of transmission or most common sources appear in parentheses.

causes of irritant contact dermatitis in the workplace include acids, soaps, detergents, and solvents. Common agents that cause allergic contact dermatitis in the workplace include formaldehyde, nickel, epoxy resins, chromates, and *p*-phenylenediamine (12).

Plant and wood substances may induce contact dermatitis or a light-activated photocontact dermatitis, which occurs when contact with the photosensitizing substance is followed by exposure to sunlight. Dermatoses caused by vegetation are most commonly found among outdoor workers, especially farmers, construction laborers, lumber workers, and firefighters.

Among the physical agents, UV radiation can cause both acute and chronic skin disorders. Acute exposures to UV radiation can result in sunburn that may vary from erythema to severe blistering accompanied by fever, chills, and malaise. Short-wave UV radiation (UVB) damages epidermal cells. The subsequent release of inflammatory mediators causes vascular dilatation and influx of inflammatory cells in the dermis. Longerwave UV light (UVA) can penetrate into the dermis and directly damage cells in the dermis. Acute UV exposure also causes increased production and dispersion of melanin, which results in darkening of the skin (tanning).

Chronic exposure to UV radiation causes solar elastosis, actinic keratoses, basal cell carcinomas, and squamous cell carcinomas. Solar elastosis is characterized by dryness and cracking of the stratum corneum, hyperpigmentation, decreased elasticity of the skin, and telangiectasia. Actinic keratoses are premalignant, erythematous, rough, scaly plaques that may develop into squamous cell carcinomas if left untreated. Solar UV radiation is responsible for the great majority of new skin cancers each year. All of the chronic effects of UV radiation occur most commonly and are most severe in lightly pigmented persons who tan poorly and sunburn easily.

Ionizing radiation can also induce basal cell and squamous cell skin cancers. Chronic exposure to x-irradiation can cause a peculiar abnormality of the skin known as poikiloderma (areas of atrophy, telangiectasia, hyperpigmentation, and hypopigmentation) (see Chapter 17).

Exposure to cold temperatures may precipitate pernio (chilblains) or Raynaud's phenomenon, or it may cause frostbite. Working in hot and humid environments predisposes to the development of intertrigo and miliaria (heat rash) (see Chapter 19). Low ambient humidity leads to dry skin and pruritus.

Biologic agents are said to be the second most common cause of occupational skin diseases. In contrast to most occupational skin disorders, it is often difficult to recognize that a skin disease caused by a biologic agent is work related. Making the correct diagnosis requires insight and thoroughness. Examples of occupational skin diseases caused by biologic agents include herpetic infections of the hands, as seen in dentists and respiratory therapists, and orf in sheep handlers.

Mechanical factors can cause calluses and fissures. Fibers that are too small to see with the naked eye, such as those of fibrous glass, may become embedded in the skin and cause pruritus and excoriation. The use of pneumatic devices, such as jackhammers and chain saws, is an often overlooked cause of occupational Raynaud's phenomenon.

REACTION PATTERNS IN OCCUPATIONAL DERMATOLOGY

Occupational skin diseases can be classified into several clinical patterns, including contact dermatitis, infections, pilosebaceous unit abnormalities, pigment disorders, and neoplasms. This system of classification can serve as a tool for remembering a wide variety of work-related clinical syndromes and can also serve as a reminder of the causes of occupational skin diseases.

Contact Dermatitis

The most common skin reaction seen in the industrial setting, contact dermatitis ac-

counts for more than 90% of all occupational dermatoses. In a study based on the Register of Occupational Diseases in Denmark, 98% of occupational skin diseases were eczematous and two-thirds of them were irritant contact dermatitis (11). Contact dermatitis can be produced either by irritants or by allergic sensitizers. It occurs at the site of contact with the irritant or sensitizer, usually on exposed surfaces, especially the hands. Acute contact dermatitis is characterized by redness, swelling, vesicle and blister formation, and exudation leading to crusting and scaling. Thickening (lichenification), excoriations, and often hypopigmentation or hyperpigmentation characterize chronic contact dermatitis.

Primary irritant contact dermatitis is a non-allergic reaction of the skin caused by exposure to irritating substances (Fig. 27-2). Any substance can act as an irritant, provided the concentration and duration of contact are sufficient. Most irritants are chemical substances, although physical and biologic agents can produce a similar clinical picture. Approximately two-thirds of occupational contact reactions are of the irritant type.

Irritants can be classified as mild (relative or marginal) irritants, which require repeated or prolonged contact to produce inflammation including soaps, detergents, and most solvents, or as strong (absolute) irritants, which injure skin immediately on contact including strong acids and alkalis. Exposure to strong acids or alkalis constitutes a medical emergency that should be treated by washing the area with copious amounts of cool water to limit tissue damage. Hydrofluoric acid (HF) is particularly damaging to the skin and can cause rapid, deep, and extensive tissue necrosis (Fig. 27-3). Therapy for HF burns may require injection of 5% calcium gluconate solution or application of a calcium gluconate gel or ointment after washing (13). Follow-up care to prevent superinfection is essential for all chemical burns.

If exposure to mild irritants is constant, normal skin in some workers may become hardened or tolerant of this trauma, and con-

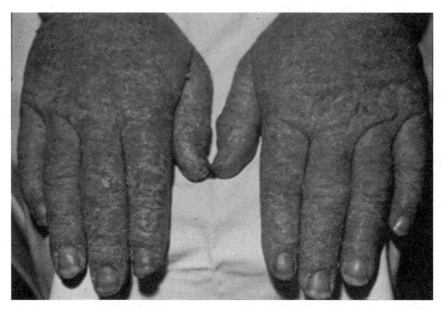

FIG. 27-2. Chronic irritant hand dermatitis. Skin is thickened, scaling, and inelastic to touch and shows deep fissures. (From American Academy of Dermatology set of teaching slides on occupational dermatitis.)

tact can be continued without further evidence of inflammation.

Allergic contact dermatitis is a manifestation of delayed-type hypersensitivity and results from the exposure of sensitized persons to contact allergens (Fig. 27-4). Most contact allergens produce sensitization in only a small percentage of those exposed. *Rhus* antigens, which induce sensitization in more than 70% of those exposed to poison ivy, oak, or sumac, are marked exceptions to this rule. The incubation period after initial sensitization to an antigen is 5 to 21 days, and the reaction time after subsequent reexposure is 6 to 72 hours. The normal reaction of a sensitized person after exposure to a moderate amount of poison oak or ivy is the appearance of a rash in 2 to 3 days and clearing within 1 to 2 weeks; with larger exposure, lesions appear more quickly (6 to 12 hours) and resolve more slowly (2 to 3 weeks). Contact sensitizers less commonly lead to other cutaneous reaction patterns, such as urticaria (hives). Since the emergence of the human immunodeficiency virus (HIV), health care workers have increased their use of rubber gloves to protect themselves from body fluids of patients. This practice has led to an increase in the incidence of allergic contact dermatitis caused by rubber or latex (14).

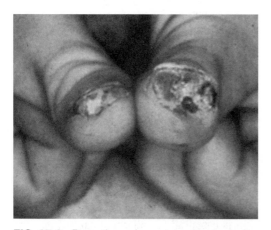

FIG. 27-3. Deep tissue loss caused by hydrofluoric acid burn. (From American Academy of Dermatology set of teaching slides on occupational dermatitis.)

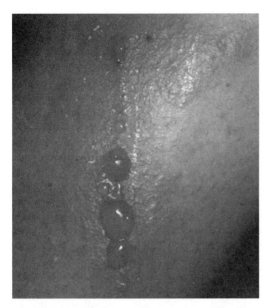

FIG. 27-4. Allergic contact dermatitis due to poison ivy. Note linear array of vesicles and bullae. (Photograph by Michael E. Bigby.)

Latex allergy has reached epidemic proportions in the United States and has its highest occupational incidence among health care workers and latex industry workers (15). Prevalence of sensitization to latex in most studies of health care workers has ranged from 5% to 12% (16).

Irritants and allergens cause itching as the primary symptom of contact dermatitis. Irritants also can cause an inelastic skin, discomfort due to dryness, and pain related to fissures, blisters, and ulcers. Strong irritants can cause blistering and erosions; mild irritants or allergens may cause dermatitis with erythema, microvesicles, and oozing. Allergens most typically cause grouped or linear tense vesicles or blisters. Edema can occur and may be severe, particularly on the face and the genital areas. Most often contact dermatitis can be distinguished from other types of eczema or dermatitis not by the specific morphology of lesions but by their distribution and configuration—that is, the rash occurs in the exposed or contact areas, often with a bizarre or artificial pattern of sharp margins, acute angles, and straight lines. Air-

borne dermatitis is caused by chemicals dispersed in the air in the form of vapors (e.g., formaldehyde), droplets (e.g., hair sprays, pesticides), or solid particles (e.g., industrial powders). The upper eyelid is frequently involved.

Infections

Pyodermas induced by streptococci or staphylococci are the most common bacterial skin infections. These infections may occur as a result of trauma or as a complication of other occupational dermatoses. Barbers and cosmeticians who have contact in their work with customers who have contagious skin diseases are particularly at risk for bacterial and fungal infections. Lesions range from the superficial, such as impetigo, to the deep, such as folliculitis, carbuncles, cellulitis, and secondary lymphangitis. Diagnosis is made by clinical appearance, Gram staining, and culture. A less common infection is erysipeloid among meat and fish handlers and veterinarians; even rarer is anthrax among sheep handlers and animal hide and wool workers (see Chapter 20). A large outbreak of sporotrichosis in workers exposed to sphagnum moss was reported in 1988 (17).

Fungal infections with dermatophytes (ringworm) or the yeast-like fungus *Candida albicans* are often found in a local environment of moisture, warmth, and maceration; they therefore occur frequently in body folds and during warm seasons. The lesions of ringworm are annular, red, and scaling at the periphery with clearing in the center, whereas those of *Candida* are beefy red, weeping areas with satellite papules or pustules. Both organisms may involve the nail, but the nailfold is most often affected by *Candida,* particularly in workers whose hands are often exposed to water (e.g., dishwashers, hairdressers, canning industry workers who handle fruit). In these persons, the interdigital spaces of the hands are also predisposed. Demonstration of septate hyphae or budding yeasts and pseudohyphae on potassium hydroxide (KOH) examination confirms the presence of fungal or yeast infection, and culture delineates the type of organism.

Warts and *Herpes simplex* are frequently identified work-related viral skin infections. Occupational handlers of meat, poultry, and fish have a higher prevalence of warts than other workers (18). The prevalence is as high as 42% in some slaughterhouses (19). Health care workers, especially those who are exposed to oral secretions (e.g., dental technicians, nurses, anesthetists), may develop painful infections in their fingertips and around their nails (herpetic whitlow). These are often confused with felons or bacterial paronychial infections. Primary cutaneous infection by *Herpes simplex* virus is also frequently observed in wrestlers and rugby players (herpes gladiatorum). Diagnosis is confirmed by the finding of viral giant cells on microscopic examination of scrapings from the vesicle floor (Tzanck's preparation), direct fluorescence antibody test, or culture.

Pilosebaceous Unit Abnormalities

Acne vulgaris, a multifactorial disorder involving the pilosebaceous unit, is usually first noted in teenage years and subsides by early adulthood. Heavy oils, such as insoluble cutting oils and greases used by machine-tool operators or machinists, may aggravate idiopathic acne or cause comedone-like follicular plugging and folliculitis. Skin under saturated clothing is most at risk, but lesions are also found on the dorsal surfaces of the hands and on extensor surfaces of the forearms (Fig. 27-5). Oil boils may develop as the result of entrapment of surface bacteria.

Some halogenated aromatic hydrocarbons induce chloracne, an acneiform eruption consisting of comedones, hyperpigmentation, and pathognomonic oil cysts on exposed skin sites. These disfiguring lesions can last for as long as 15 years. Though chloracne is not itself a disabling illness, it is important evidence of percutaneous absorption of chloracnegens, chlorinated hydrocarbons that have been associated with skin, lung, and bladder cancer (20). Workers at risk include herbicide manufacturers as well as cable splicers and others exposed to polychlorinated biphenyls (PCBs) in dielectric applications (electrical insulation). Chloracne has been noted as a side effect in instances of massive environmental contamination by herbicides, such as occurred in

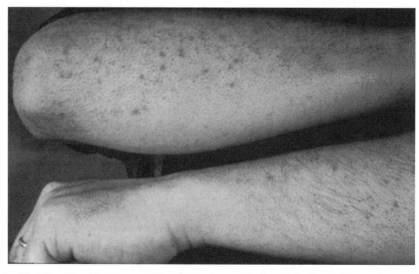

FIG. 27-5. Oil folliculitis. These acne-like lesions were induced by insoluble cutting oils that had saturated work pants. (From American Academy of Dermatology set of teaching slides on occupational dermatitis.)

Seveso, Italy, in 1976, when an explosion in a nearby chemical plant sent a toxic cloud containing dioxin into the air. Dioxin, a trace contaminant of the chlorinated hydrocarbon 2,4,5-trichlorophenoxyacetic acid (2,4,5-T) is one of the most toxic substances known, second only to the neurotoxin botulin and certain nerve gas and chemical warfare components. The dioxin-containing cloud settled on Seveso, a rural community of 5,000 people. There were 187 confirmed cases of chloracne, many in children; the long-term effects of this explosion remain to be seen. Another exposure to dioxin occurred among inhabitants of Vietnam and military personnel sent there between 1962 and 1971, when Agent Orange, which contains 2,4,5-T, was used as a defoliant.

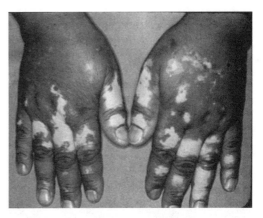

FIG. 27-6. Depigmentation caused by monobenzyl ether of hydroquinone. This chemical was used as an antioxidant in workers' gloves. (From American Academy of Dermatology set of teaching slides in occupational dermatitis.)

Pigment Disorders

Occupational exposures can produce many pigments on the skin. Some are stains adhering to the stratum corneum, and some are a result of systemic absorption or inoculation (tattoo) of heavy metals, but most are caused by altered melanin pigmentation.

Postinflammatory hyperpigmentation and hypopigmentation are the most common pigment disorders. Any dermatitis or other trauma to the skin, such as thermal or chemical burns, may lead to temporary increase or decrease in melanin pigmentation in that area. Persons who are more heavily pigmented show these findings most notably and with slower reversibility. If damage has been severe, melanocytes may have been destroyed and permanent depigmentation may occur, sometimes with scarring. Exposure to the sun or other sources of UV light in the presence of photosensitizes such as tar, pitch, or psoralen leads to enhanced erythema and, later, tanning.

An antioxidant used in rubber manufacturing, monobenzyl ether of hydroquinone, is the chemical most notorious for inducing permanent work-related loss of pigment (leukoderma) (Fig. 27-6). Metastatic lesions of hypopigmentation not in sites of direct contact may also be found. Other phenolic compounds used as antioxidants or germicidal disinfectants have also been found to produce pigment loss. This phenomenon results from the structural resemblance of these substances to tyrosine, an amino acid precursor of melanin synthesis.

Neoplasms

Occupational neoplasms may be benign, premalignant, or cancerous. Foreign-body granulomas or granulomatous inflammatory nodules caused by beryllium or silica are important examples of benign lesions. Exposure to silica dust has also been suggested as a factor in the induction of scleroderma (21). Foreign-body granulomas and fistulas occur in the interdigital spaces of barbers when the skin is penetrated by small fragments of hair. Keratoses produced by exposure to shale oils or to UV light are examples of premalignant lesions. Examples of cancerous lesions include squamous cell carcinoma and basal cell carcinoma caused by chronic exposure to UV light (Fig. 27-7), and squamous cell carcinoma, often on the scrotum, occurring historically among chimney sweeps and cotton spinners and currently among metal machinists exposed to carcinogens in lubricating oils

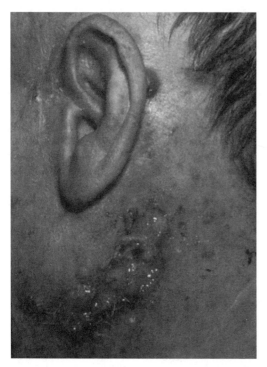

FIG. 27-7. Basal cell carcinoma. This worker had years of exposure to sunlight. (From American Academy of Dermatology set of teaching slides on occupational dermatitis.)

(22). The incidence of melanoma is higher among professional, technical, and white-collar workers, presumably because of their higher recreational sun exposure. However, a higher incidence of melanoma has also been described among workers in basic chemical production, in the printing industry, in breweries and malt-processing industries, and in shoe fabrication from leather and skins (23). Oncogenic agents include ionizing radiation; UV light; some insoluble oils and greases, especially shale oils; and degradation products of the incomplete combustion of wood, oils, and tar. A diagnostic skin biopsy is always necessary when a skin cancer is suspected.

Disorders of Hair and Nails

Alopecia (hair loss) may be caused by exposure to thallium-containing rodenticides, chloroprene, boric acid, or high doses of ionizing radiation. Nail abnormalities may follow mechanical or chemical injury to the nail matrix, or they may result from infection with *Candida* or dermatophytes.

Other Occupational Skin Disorders

The development of scleroderma-like lesions has been observed in those who work with polyvinyl chloride, particularly reactor cleaners who have been exposed to vinyl chloride, and in other workers after exposure to perchloroethylene, a solvent used in dry cleaning. An Eastern European study of 61 patients with systemic sclerosis found that 28% had undergone significant exposure to organic solvents (24). An earlier Japanese study drew attention to an association between scleroderma-like skin changes and occupational exposure to epoxy resin (25). The occurrence of scleroderma in South African coal miners with silicosis has been known for a considerable time.

DIAGNOSIS OF OCCUPATIONAL SKIN DISORDERS

Occupational skin disorders are diagnosed by an accurate, detailed, and discerning history; a careful physical examination; laboratory tests; and tests for allergic contact dermatitis. The history is of utmost importance in diagnosing occupational skin disorders. The history should include a detailed description of the patient's job and a complete list of chemical, physical, and biologic agents and mechanical factors to which the patient is exposed. The onset of the skin disorder in relation to starting or changing job duties is important. Occupational skin disorders often improve on weekends and during vacations. Up to one-third of skin disorders seen in workers are unrelated to their jobs but instead are caused by exposure to irritants and allergens in the patient's home or other sites. Finally, all patients with contact dermatitis should be questioned about a past history of atopic dermatitis (see Chapter 4).

In the physical examination, emphasis should be placed on recognizing the patterns of occupational skin disorders. These disorders predominantly affect the hands, wrists, and forearms. The face, eyelids, and ears may also be involved. Examination of the entire cutaneous surface is recommended. Attention to spared areas as well as involved areas may provide clues to the cause of the skin disorder. For example, phototoxic eruptions tend to spare areas covered by clothing, the upper eyelids, under the chin, and behind the ears. The appearance of acne lesions outside the typical acne distribution may be helpful in establishing the diagnosis of occupationally induced acne.

Laboratory tests may include smears and cultures for viruses, bacteria, and fungi. A skin biopsy may be useful. Descriptions of the laboratory tests used to help establish a diagnosis of skin disorders can be found in several references in the Bibliography.

Patch testing is used to document and validate a diagnosis of allergic contact sensitization and to identify the causative agent (Fig. 27-8). It may also be of value as a screening procedure in patients with chronic or unexplained dermatitis to assess whether contact allergy is playing a causative role. The patch test is a unique means of *in vivo* reproduction of disease on a small scale. Because sensitization affects the whole body, allergic reactions may be elicited at any cutaneous site. Patch testing is easier and safer than the "use" test because potential allergens can be applied in low concentrations on small areas of skin for short periods. Patch testing is of no value in diagnosing irritant dermatitis except to exclude allergic contact dermatitis as a primary or contributing cause. Toxic chemicals and strong irritants should not be tested, although open patch tests are sometimes used for relative irritants, such as shampoos or other detergents. Proper performance and interpretation of patch tests require considerable experience. Possible side effects include a severe local reaction, secondary autosensitization reaction, and actual sensitization of the patient to the test compound. Patch-test allergens are often available commercially; sometimes they must be prepared from the suspected occupational agent. These substances must be nonirritating and nontoxic. The selection of optimal patch-test preparations is the subject of much debate and varies from country to country according to their agricultural and industrial economies (27). Patch-testing procedures and techniques for proper dilution of reagents can be found in several references in the Bibliography.

MANAGEMENT OF SKIN DISEASES

The treatment of occupational skin diseases generally depends on accurate diagnosis of the clinical reaction pattern. Although it may be possible to treat some disorders success-

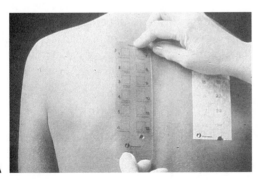 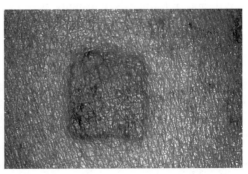

A   B

FIG. 27-8. A. Patch testing can be conveniently and reliably performed with a test that is taped to the patient's skin for 48 hours. **B.** A positive reaction appears as an acute, eczematous dermatitis.

fully without having an exact etiologic diagnosis, an exact diagnosis should always be made for purposes of workers' compensation and prevention of the disorder in other workers. There are no differences in the management of skin diseases caused by occupational, as opposed to nonoccupational, factors. The following is a brief summary of the more commonly used therapies. For detailed management strategies, texts on dermatologic therapy should be consulted.

Contact Dermatitis

Acute exudative and vesicular contact dermatitis should be treated with compresses or dressings such as Burow's solution and topical antiinflammatory agents. Topical corticosteroids are the antiinflammatory agents of choice; they usually hasten resolution of the dermatitis and reduce itching. If the eruption is severe and is accompanied by marked edema, a short course of oral corticosteroids may be necessary. Antihistamine drugs may reduce itching, and oral antibiotics are effective in eliminating secondary infection. Chronic contact dermatitis requires adequate lubrication and the skillful use of topical corticosteroids. Attempts to induce clinically relevant hyposensitization with purified allergens have not been successful.

Infections

Systemic antibiotics are usually warranted for cutaneous pyodermas. For superficial infections, topical antibiotics may be effective (e.g., mupirocin against staphylococcal and streptococcal infections, silver sulfadiazine against many gram-negative organisms). The topical imidazole drugs miconazole and clotrimazole are almost always effective in eliminating ringworm (dermatophyte) or yeast (*C. albicans*) infection. Fungal infections of the scalp or the nail plate, however, should be treated with oral antifungal agents. Acyclovir, famciclovir, and valacylovir are available for the treatment of selected patients with *Herpes simplex* infection.

Pilosebaceous Follicle Abnormalities

Acne and chloracne respond to topical and systemic treatment, but chloracne is far more recalcitrant and responds less completely. In most instances, topical creams and gels containing tretinoin or benzoyl peroxide are the agents of choice. Some patients may need treatment with oral antibiotics or isotretinoin. Lesions of chloracne often last months to years despite therapy.

Pigment Disorders

Postinflammatory pigment changes slowly resolve with time. That part of postinflammatory hyperpigmentation reflective of increased epidermal melanocyte activity can be treated with hydroquinone-containing creams. However, bleaching creams usually are of limited value. The leukoderma caused by monobenzyl ether of hydroquinone or phenolic disinfectants is not amenable to treatment.

Neoplasms

Premalignant keratoses can be treated with topical 5-fluorouracil (5-FU) or with liquid nitrogen cryosurgery, electrosurgery, or curettage. Cryosurgery, scalpel excision, and radiation therapy are appropriate for basal or squamous cell carcinomas. The best therapy in any instance depends on factors such as the size and site of the lesion and the patient's age and history of cutaneous carcinomas.

PROGNOSIS

The prognosis of occupational skin disorders depends on several variables, including the type of cutaneous reaction pattern, its exact cause (if determined), duration of the eruption before diagnosis, type and effectiveness of treatment, patient compliance, and adequacy of preventive measures. Those workers with a specific allergic contact sensitivity may do well if they can avoid the allergen in work and home environments; in industrial settings, it is almost always necessary to transfer such workers to another area of the plant. Workers with irritant contact reactions may be able to continue at work if the duration and intensity of exposure to contactants is decreased by environmental or protective measures. Those with atopy or a long history of dermatitis have a particularly dismal prognosis.

Persistence of what is presumed to be a work-related eruption even after the worker has been removed from the putative cause is not uncommon and may occur in the majority of cases (28,29). For example, in a 2-year follow-up study of 230 workers with occupational skin disease, Holness and Nethercott found that 35% had lost at least 1 month of work and 43% had applied for workers' compensation benefits (29). However, persistence of a work-related skin disorder should raise several potential questions: Was the correct diagnosis made and the best treatment prescribed? Has the patient conscientiously carried out the treatment plan? Has work exposure been eliminated, and are there other possible sources of contact in second jobs or at home? Are psychological factors playing a role, and is there any evidence of malingering? These questions are often difficult to answer definitely, just as it is often difficult to delineate the specific cause of an occupational dermatosis.

The American Medical Association has published guidelines on evaluating the degree of permanent impairment induced by occupational skin disorders; these guidelines are shown in Table 27-2. As opposed to permanent skin impairment, functional loss is best evaluated by assessing the degree of itching, scarring, and disfigurement.

TABLE 27-2. *Guidelines for evaluating permanent skin impairment[a]*

Category	Impairment (%)	Comments
Class 1	0–9	Signs and symptoms of skin disorder are present or only intermittently present; **and** There is no limitation or limitation in the performance of *few* ADL, although exposure to certain chemical or physical agents might increase limitation temporarily; **and** *No* treatment or intermittent treatment is required.
Class 2	10–24	Signs and symptoms of skin disorder are present or intermittently present; **and** There is limitation in the performance of *some* of the ADL; **and** Intermittent to constant treatment may be required.
Class 3	25–54	Signs and symptoms of skin disorder are present or intermittently present; **and** There is limitation in the performance of *many* of the ADL: **and** Intermittent to constant treatment may be required.
Class 4	55–84	Signs and symptoms of skin disorder are *constantly* present; **and** There is limitation in the performance of *many* of the ADL that may include intermittent confinement at home or other domicile; **and** Intermittent to constant treatment may be required.
Class 5	85–95	Signs and symptoms of skin disorder are *constantly* present; **and** There is limitation in the performance of *most* of the ADL, including occasional to constant confinement at home or other domicile; **and** Intermittent to constant treatment may be required.

[a]The signs and symptoms of disorders in classes 1 and 2 may be intermittent and not present at time of examination. The impact of the skin disorder on daily activities should be the primary consideration in determining the class of impairment. The frequency and intensity of signs and symptoms and the frequency and complexity of medical treatment should guide the selection of an appropriate impairment percentage and estimate within any class.

ADL, activities of daily living.

From American Medical Association. Guides to the evaluation of permanent impairment, 4th ed. Chicago: AMA, 1993:280. Copyright © 1993, American Medical Association.

PREVENTION

Occupational skin diseases are often preventable by a combination of environmental, personal, and medical measures.

Environmental cleanliness is paramount in preventing occupational dermatoses. Maintenance of a clean workplace involves frequently cleaning floors, walls, windows, and machinery; recognizing hazardous materials and either providing substitutes or altering or eliminating them from the workplace; ensuring proper ventilation, temperature, and humidity; and using exhaust hoods, splash guards, and other protective devices and systems.

Personal cleanliness is a key element in preventing occupational skin disease. Washing facilities with hot and cold running water, towels, and proper cleansing agents must be easily accessible and strategically placed. The mildest soap that will clean the skin should always be used. Appropriate cleansing agents might include waterless hand cleaners for oils, greases, or adherent soils. Safety showers must be available if highly corrosive chemicals are being handled. If strong irritants are in the work environment, it is necessary for workers to shower at the end of each shift, or possibly more often. In some industries, it is appropriate to supply clothing and laundering to ensure both the use of proper types of material and daily clothing changes. If water is not easily accessible for washing, waterless hand cleansers can be used. Solvents such as kerosene, gasoline, and turpentine should *never* be used for skin cleansing; they are damaging to the skin because they "dissolve" the cutaneous barrier and can either induce an irritant contact dermatitis or predispose to a cumulative-insult contact dermatitis. A skin moisturizer should be used after hand washing, especially if frequent washing is necessary.

Protective clothing is often all that is needed to prevent a dermatitis by blocking contact of chemicals with the skin. The clothing should be chosen based on the skin site needing protection and the type of chemical involved (some solvents can dissolve certain fabrics or materials). Natural rubber gloves are impervious to most aqueous compounds but deteriorate after exposure to strong acids and bases. Very common allergens, such as nickel salts, penetrate rubber but not polyvinyl chloride glove material. Synthetic rubbers are more resistant to alkalis and solvents; however, some are altered by chlorinated hydrocarbon solvents. [In Sweden, a database has been developed that contains test data on protective effects of gloves against chemicals (30).] It is always useful to wear absorbent, replaceable soft cotton liners inside protective gloves to make them more comfortable. Commercially available gear includes gloves of different lengths, sleeves, safety shoes and boots, aprons, and coveralls composed of materials such as plastic, rubber, glass fiber, metal, and combinations of these materials. Clothing that might become caught in machines must be avoided for safety reasons.

Protective creams, referred to as "barrier" creams, afford much less protection than does clothing. However, these creams can be valuable when gloves would interfere with the sense of touch required to perform a job or when use of a face shield might be awkward. Barrier creams should be applied to clean skin, removed when the skin becomes excessively soiled or at the end of each work period, and then reapplied. Proper use of a barrier cream not only provides some degree of protection but induces the worker to wash at least twice during the work shift.

There are four types of protective creams. *Vanishing creams* contain detergents that remain on the skin and facilitate removal of soil when washing. *Water-repellent creams* leave a film of water-repellent substances, such as lanolin, petrolatum, or silicone, on the skin to help prevent direct contact with water-soluble irritants (e.g., acids and alkalis). *Solvent-repellent creams* repel oils and solvents and may leave either an ointment film or a dry, oil-repellent film on the skin surface. *Special creams* include sunshades; sunscreens that absorb UVA, UVB, or both spectra of UV light; and insect repellents. Some common constituents of protective creams (e.g., lanolin, propylene glycol, sunscreens) can also induce contact dermatitis.

Careful screening of new workers in preplacement examinations decreases the incidence of job-related dermatoses. Persons with a history of atopic dermatitis should not work in occupations involving frequent exposure to harsh chemicals or water, such as certain machining, cooking, bottle-washing, and operating-room jobs. Those with psoriasis of the hands would do poorly in the same situations and furthermore may respond with a Koebner's reaction, in which psoriatic lesions develop in sites of heavy trauma. Such trauma includes scratches, abrasions, and cuts that disrupt the epidermis as well as rough handwork or continual kneeling.

Workers with dermatographism, at high risk for annoying pruritic responses to trauma or to foreign bodies such as fibrous glass, should avoid these occupational exposures; workers with acne should not be employed in hot and humid workplaces or where they would be exposed to oil mists, heavy oils, or greases; and fair-skinned or sunlight-sensitive persons should not work in intense UV light (see Chapter 19) or around potentially photosensitizing chemicals such as tar, pitch, or psoralens. Job applicants should be questioned concerning previous skin diseases, including childhood eczema and atopic diseases, contact dermatitis, psoriasis, fungal infections, and allergic reactions to drugs or other agents.

Preplacement patch testing is generally not advised; although it may occasionally identify a previously allergic person, the yield is very low and the test carries the greater risk of inducing contact sensitization. Chemical agents used in the workplace should undergo toxicologic testing to detect irritancy, allergenicity, acnegenicity, carcinogenicity, and other properties. When potentially hazardous substances are detected, they should be properly labeled, and workers should be educated about these hazards and how to avoid them.

Workers with severe, chronic, or unremitting allergic dermatoses should be transferred to other plant areas, if possible. By definition, allergy implies sensitivity to very low levels of an antigen, and it usually is not possible to continue working in the same site even though careful precautions are taken. Those with irritant dermatoses are often, but not always, able to continue working by decreasing the duration and intensity of exposure to irritants. Relocation of workers is more easily accomplished within large workplaces; however, only one third of the workforce is employed in such workplaces. Most cases of occupational dermatitis occur in small workplaces with poorly developed preventive services and less sophisticated or no supervisory or medical personnel.

REFERENCES

1. Centers for Disease Control and Prevention. Leading work-related diseases and injuries. MMWR Morb Mortal Wkly Rep 1986;35:561–563.
2. Surveillance for nonfatal occupational injuries treated in hospital emergency departments, United States. MMWR Morb Mortal Wkly Rep 1998; 47:302–306.
3. Workplace injuries and illnesses in 1997 [news release]. Available at: http://stats.bls.gov/os/osnr0007.pdf. Accessed December 17, 1998.
4. Bureau of Labor Statistics. Nonfatal occupational illnesses by category of illness, private industry, 1993–1997. Available at: http://stats.bls.gov/os/ostb0628.pdf. Accessed December 17, 1998.
5. Mathias CGT. The cost of occupational skin disease. Arch Dermatol 1985;121:332–334.
6. Kaufman JD, Cohen MA, Sama SR, Shields JW, Kalat J. Occupational skin diseases in Washington State, 1989 through 1993: using workers' compensation data to identify cutaneous hazards. Am J Public Health 1998;88:1047–1051.
7. Bureau of Labor Statistics. Nonfatal occupational illnesses by category of illness, private industry, 1997. Available at: http://stats.bls.gov/os/ostb0629.pdf. Accessed December 17, 1998.
8. Rycroft RJG. Occupational dermatosis. In: RH Champion, JL Burton, FJG Eblingo. Textbook of dermatology, 5th ed. Oxford: Blackwell, 1991.
9. Goldsmith LA. My organ is bigger than your organ. Arch Dermatol 1990;126:301–302.
10. Shmunes E, Keil JE. Occupational dermatoses in South Carolina: a descriptive analysis of cost variables. J Am Acad Dermatol 1983;9:861–866.
11. Halkier-Sorensen L. Occupational skin diseases: reliability and utility of the data in the various registers; the course from notification to compensation

and the costs. A case study from Denmark. Contact Dermatitis 1998;39:71–78.

12. Bernstein DI. Allergic reactions to workplace allergens [review]. JAMA 1997;278:1907–1913.

13. Trevino MA, Herrman GH, Sproue WL. Treatment of severe hydrofluoric acid exposures. J Occup Med 1983;25:861–863.

14. Bubak ME, Reed CE, Fransway AF, et al. Allergic reactions to latex among health-care workers. Mayo Clin Proc 1992;67:1075–1079.

15. Woods JA, Lambert S, Platts-Mills TA, Drake DB, Edlich RF. Natural rubber latex allergy: spectrum, diagnostic approach, and therapy [review]. J Emerg Med 1997;15:71–85.

16. Liss GM, Sussman GL. Latex sensitization: occupational versus general population prevalence rates. Am J Ind Med 1999;35:196–200.

17. Coles FB, et al. A multistate outbreak of sporotrichosis associated with sphagnum moss. Am J Epidemiol 1992;22:444–448.

18. Kilkenny M, Marks R. The descriptive epidemiology of warts in the community. Australas J Dermatol 1996;37:80–86.

19. Aziz MA, Bahamdan K, Moneim MA. Prevalence and risk factors for warts among slaughterhouse workers. East Afr Med J 1996;73:194–197.

20. Rosestock L, Cullen MR. Textbook of clinical occupational and environmental medicine. Philadelphia: WB Saunders, 1994.

21. Haustein UF, Ziegler V, Herrmann K, Mehlhorn J, Schmidt C. Silica-induced scleroderma. J Am Acad Dermatol 1990;22:444–448.

22. Calvert GM, Ward E, Schnorr TM, Fine LJ. Cancer risks among workers exposed to metalworking fluids: a systematic review. Am J Ind Med 1998; 33:282–292.

23. Linet MS, Malker HS, Chow WH, et al. Occupational risks for cutaneous melanoma among men in Sweden. J Occup Environ Med 1995;37:1127–1135.

24. Czirjak L, Bokk A, Csontos G, Lorincz C, Szegedi G. Chemical findings in 61 patients with progressive systemic sclerosis. Acta Dermatol Venereol 1989; 69:533–536.

25. Yamakage A, Ishikawa H, Saito Y, Hattori A. Occupational scleroderma-like disorders occurring in men engaged in polymerization of epoxy resins. Dermatologica 1980;161:33–44.

26. Móroni P, Peirini F. Patch testing and detective work in the workplace. Clin Dermatol 1992; 10:195–200.

27. Keczkes K, Bhate SM, Wyatt EH. The outcome of primary irritant hand dermatitis. Br J Dermatol 1981;105(Suppl 21):65–70.

28. Burrows D. Prognosis and factors influencing prognosis in industrial dermatitis. Br J Dermatol 1981; 105(Suppl 21):65–70.

29. Holness DL, Nethercott JR. Work outcome in workers with occupational skin disease. Am J Ind Med 1995;27:807–815.

30. Mellstrom GA, Lindahl G, Nahlberg JE. DAISY: reference database on protective gloves. Semin Dermatol 1989;8:75–79.

BIBLIOGRAPHY

Adams RM, ed. Occupational skin disease. Occup Med: State of the Art Review 1986;2:199–360.
A concise, well-written treatise on this subject, covering the basics of occupational skin diseases. The role of atopy in occupational skin disease, vibration syndromes, and AIDS in the workplace are among the interesting topics covered.

Adams RM, ed. Occupational skin disease, 3rd ed. Philadelphia: WB Saunders, 1999.
A thorough and comprehensive general reference book. An essential book for the serious student of occupational skin disorders.

American Medical Association. Guides to the evaluation of permanent impairment, 4th ed. Chicago: ANA, 1993:277–289.
A valuable resource for evaluating permanent impairment of the skin.

Arndt KA. Bowers KE, Chuttani JR. Manual of dermatologic therapeutics: with essentials of diagnosis, 5th ed. Boston: Little, Brown, 1994.
A practical manual. Discusses the pathophysiology, diagnosis, and treatment of common skin disorders seen in ambulatory patients. Patch testing is described.

deGroot AC. Patch testing. Amsterdam: Elsevier, 1986.
This book provides recommendations for test concentrations and vehicles for 2,800 allergens.

Rietschel RL, Fowler JF, eds. Fisher's contact dermatitis, 4th ed. Philadelphia: Williams & Wilkins, 1995.
Essential, detailed reference work on contact dermatitis. Contains a glossary that describes the proper patch-test concentrations of many common antigens.

28 Eye Disorders

Paul F. Vinger and David H. Sliney

Case 1: A 52-year-old metal worker was polishing a brass fixture on a buffing wheel. The fixture was torn from his grasp by the wheel and struck his left eye. In addition to a full-thickness laceration of the left upper eyelid, a severe laceration of the globe of the eye was apparent when he was evaluated at a hospital soon afterward. It was discovered then that the patient's best-corrected vision in the uninjured right eye enabled him only to count fingers because of dense amblyopia ("lazy eye" secondary to unilateral uncorrected high myopia, or nearsightedness). Despite extensive surgical procedures, vision was lost in the injured left eye. Attempts to improve the vision in the uninjured right eye by optical means were not successful. He is now legally blind with no hope of recovery of useful vision. He can no longer drive or perform his usual work.

Case 2: A 46-year-old man was using a band saw to cut stainless steel rings off the end of a furnace roll. A ring piece broke off and struck the left lens of safety glasses that conformed to the requirements of ANSI Z87.1-1989 for spectacles with removable lenses. The 2.0-mm plano polycarbonate lens remained intact but was driven through the frame from the impact at the lower left corner of the lens. The result was a corneal laceration, commotio retinae, hyphema, iritis, and cataract with dislocated lens. Final

best-corrected vision is 20/200 because of a macula scar.

Case 3: An 18-year-old arc-welding student stared at an electric welding arc while another welder was welding a piece of aluminum. He was outside a protective curtain and about 200 cm (about 6.5 ft) from the arc, yet was able to stare at the arc with his right eye for approximately 10 minutes. He sustained a retinal injury that initially resulted in marked visual loss. This loss slowly resolved over 16 months, leaving him with normal visual acuity and a residual, partially pigmented foveal lesion. (The term *foveal* refers to the central portion of the retina, which is required for normal reading and driving visual acuity and accurate color vision).

Case 4: A scientist was working with a relatively weak neodymium-yttrium-aluminum-garnet (neodymium:YAG) laser without safety goggles. He was not looking directly at the beam but heard a popping sound inside his eye accompanied by almost immediate obscuring of his vision. The laser burn, between the fovea and the optic nerve of his left eye, resulted in a large, permanent blind area in his visual field.

Every working day, more than 2,000 preventable job-related eye injuries occur among workers in the United States. During the 6 months from March 15 to September 15, 1985, there were 3,185 eye trauma visits to the Massachusetts Eye and Ear Infirmary emergency department. Almost half of these injuries were incurred during automobile repair and probably would have been prevented by use of appropriate safety eyewear. Although only 5% of the total injuries were

P. F. Vinger: Department of Ophthalmology, Tufts University School of Medicine, Concord, Massachusetts 01742, and Lexington Eye Associates, Emerson Hospital, Concord, Massachusetts 01742.

D. H. Sliney: Laser Program, Center for Health Promotion and Preventive Medicine, Aberdeen Proving Ground, Maryland 21010-5422.

deemed serious, they were responsible for $2 million in hospital bills, 30 person-years of work lost, and a 9% incidence of litigation (1). Occupational vision programs, including preplacement examinations and mandatory use of *appropriate* eye protectors (the worker in case 2 should have been wearing goggles), can prevent the majority of work-related eye injuries. Such programs could have prevented loss of vision in the cases described.

CAUSES OF OCCUPATIONAL EYE INJURIES

Workers have enormously varied exposures to eye injuries. The employee who drives to work or drives as part of the job can have an automobile accident in which eye injury results, either from the accident itself or from deployment of an airbag. At work there may be exposure to impact, dust, chemicals, heat, radiation, or bright light. The employee may play basketball or racquetball at lunchtime or after work. There are softball games at company picnics, where there is also the probability of sunburn. All of these risks are especially serious for the worker who has only one useful eye or a visual field defect.

Direct Trauma

Direct trauma is the most common cause of occupational eye injuries. Jobs in which there is both a risk of high-speed flying particulate material and a tradition of working without eye protectors (e.g., automobile mechanics) present the highest risk. Eye injuries from direct trauma are almost totally preventable with protective eyewear. Prevention of eye injury from direct trauma requires a combination of making equipment safer with proper shields; positioning workers so that flying particles from one worker's area do not enter the space of a nearby worker; prescribing, dispensing, and maintaining protective eyewear; and ensuring that safety eyewear is worn at all times.

Chemicals

Many different types of chemicals are commonly involved in industrial eye injuries (2).

Alkalis are especially dangerous because of their ability to rapidly penetrate into the interior of the eye, with severe consequences. (Ammonia penetrates into the interior of the eye within seconds and sodium hydroxide within minutes.) The severity of ocular injury is proportional to the alkalinity and not to the specific cation. Many workers do not realize that wet plaster containing lime is sufficiently alkaline to result in blindness if not immediately and thoroughly removed from the eye.

Acids can burn the eye. Acid burns usually are less severe than alkali burns because acids do not penetrate into the eye as readily. However, because sulfuric acid is one of the most frequently used compounds in industry, permanent corneal scarring from acids is not rare.

Carbon dioxide poisoning may result in retinal degeneration, photophobia, abnormalities of eye movements, constriction of peripheral visual fields, enlargement of the blind spots, and deficient dark adaptation.

Alkyl esters of sulfuric acid are intensely irritating to the conjunctiva, cornea, and eyelid. Many esters are irritating to the conjunctiva.

Hydrogen sulfide causes a keratoconjunctivitis that is responsible for colored rings around lights and increased photophobia—a possible warning sign of early poisoning.

Quinone vapor and hydroxyquinone dust cause corneal and conjunctival pigmentation that can result in significant visual loss.

Ketones can be irritating to the eyes and mucous membranes.

Methanol can cause total blindness from optic atrophy.

Concentrated nitric acid, when splashed into the eye, produces immediate opacification of the cornea. Severe nitric acid burns cause blindness, symblepharon (adhesion of the eyelids to the globe of the eye), and phthisis (blind, soft eye).

Although silver results in argyrosis—a dramatic staining of the conjunctiva, cornea, and, rarely, the lens—significant associated visual loss has not been reported.

Organic tin compounds can cause intense chemical conjunctivitis associated with severe itching.

Light, Laser, Heat, and Ionizing Radiation

Optical radiation is considered nonionizing radiation because the photon energies for ultraviolet (UV) wavelengths longer than approximately 180 nm are insufficient to individually ionize atoms found in important biologic molecules. In contrast to the non-threshold biologic effects of ionizing radiation (e.g., x-rays), a threshold appears to exist for each biologic effect of optical radiation. In the UV and visible regions of the spectrum, photochemical damage mechanisms are demonstrable. Thermal injury mechanisms dominate for most pulsed-laser exposures and infrared (IR) radiation exposures.

The optical spectrum includes the UV, visible (light), and IR regions of the electromagnetic spectrum (see Fig. 19-1 in Chapter 19). Lasers, which produce wavelengths in all parts of the spectrum, are unique sources of optical radiation with extremely high brightness.

Three factors enter into any analysis of light and laser hazards: (a) the type of laser or light source and the potential hazards of associated equipment, (b) the environment, and (c) potentially exposed individuals. Because many combinations are possible, numerous, rigid laser safety regulations should be avoided (3).

Bright, continuous visible-light sources elicit a normal aversion or pain response that can protect the eye from injury. Therefore, visual comfort can often be used as an approximate hazard index for the design of goggles and other hazard controls. Almost all conceivable accident situations require a hazardous exposure to be delivered within the period of the blink reflex. Few arc sources are sufficiently large and bright enough to pose a retinal burn hazard under normal viewing conditions, but if the arc or tungsten filament is greatly magnified by an optical projection system, it is possible for hazardous irradiances to be imaged on a sufficiently large area of the retina to cause a burn. If an arc were initiated at close viewing range (a few meters for all but the most powerful xenon searchlights, or a few inches from a welding arc, most movie projection equipment, or movie lamps) a retinal burn could result. Several hazard-reduction options are available to prevent individuals from viewing the source at close range.

The probability of hazardous laser retinal exposure is almost always remote. The pencil beam from most lasers is so small that direct entry into the 2- to 7-mm pupil of the eye is unlikely unless deliberate exposure occurs or unless an extremely careless atmosphere exists in the laser work area. However, the perception of a low likelihood of injury is probably the greatest single problem that exists with laser safety programs. When workers who do not follow precautions or who do not wear eye protectors are not injured, overconfidence, a lack of trust in health and safety professionals, and a continued disregard for safety programs result. The only solution is a sound program of education, coupled with strict enforcement of company safety regulations. If workers understand that the laser hazard is somewhat similar to Russian roulette, they are more likely to take precautions. Better-educated workers may also be less likely to attribute all eye irritation or vision changes they experience to work with the lasers.

Eye injuries can occur from essentially all portions of the electromagnetic spectrum; however, some sources of electromagnetic energy are far more hazardous to the eye because (a) the eye is more sensitive to the particular wavelengths, (b) the energy dose is very high, or (c) the energy is delivered in a very brief interval.

Ionizing radiation is rarely a cause of industrial eye injury because standard precautions for exposure to ionizing radiation give adequate eye protection.

Exposure to excess heat can cause thermal burns to the lids and eye. Cataracts occur in unprotected glass blowers and furnace workers. Heat-absorbing or heat-reflecting protective eyewear is available. Other protective clothing or head or face shields may be indicated, depending on the severity of heat exposure.

Although most workers are aware of the potential for lasers to cause eye injury, more conventional light sources, especially those with high output in the UV portion of the spectrum (e.g., welding arcs, sun lamps), can also be extremely hazardous.

There are at least five separate types of potential hazards to the eye and skin from lasers and other, more conventional optical sources:

1. UV photochemical injury to the skin (erythema and carcinogenic effects) and to the cornea (photokeratitis) and lens (cataract) of the eye (180 to 400 nm)
2. Thermal injury to the retina of the eye (400 to 1,400 nm)
3. Blue-light photochemical injury to the retina of the eye (principally 400 to 550 nm)
4. Near-infrared thermal hazards to the lens (approximately 800 to 3,000 nm)
5. Thermal injury (burns) of the skin (approximately 340 nm to 1 mm) and of the cornea of the eye (approximately 1,400 nm to 1 mm).

The hazards posed by lasers can be categorized as follows:

Class 4: High power can produce hazardous diffuse reflections and may also present a fire hazard or significant skin hazard.
Class 3: Medium power cannot produce hazardous diffuse reflections but does present a hazard when the beam is viewed directly.
Class 2: Low power (less than 1 mW), visible lasers are safe for momentary viewing (less than 0.25 seconds); the eye's aversion response to bright light normally precludes a person from staring into the light.

Class 1: Nonhazardous lasers emit less than the recommended limits for intrabeam viewing for the maximal reasonable viewing duration.

Laser Pointers

Laser pointers fall into class 2 or 3a. These laser systems, which have a "CAUTION" (class 2) or "DANGER" (class 3a) label, would not injure the eye if viewed for only momentary periods (within the aversion response time) with the unaided eye, but may present a greater hazard if they are viewed with the use of collecting optics. Although class 3a lasers are capable of exceeding permissible levels for the eye in 0.25 second, laser pointers in this class still pose a low risk of injury. Most laser light pointers require a fixed gaze at the laser for 8 to 10 seconds to produce an injury. Laser pointers are safe for their intended purpose and pose no eye hazard to a member of the audience if the laser is accidentally pointed into the visual axis. A risk of the laser pointer is the intentional prolonged viewing of the beam by children who consider the laser pointer a toy. Another risk is the distraction of a driver or operator of dangerous equipment by an involuntary aversion response to a laser pointer intentionally directed by someone in an aggressive manner.

Computer Terminals

There is no evidence that prolonged use of computer terminals is harmful to the eyes.

TREATMENT

Recognize Serious Eye Injury

Symptoms of serious eye injury indicating the need for immediate referral are the following (4,5):

1. Blurred vision that does not clear with blinking
2. Loss of all or part of the visual field of an eye

3. Sharp stabbing or deep throbbing pain
4. Double vision.

Signs of eye injury that require ophthalmologic evaluation are the following:

1. Black eye
2. Red eye
3. An object on the cornea
4. One eye that does not move as completely as the other
5. One eye protruding forward more than the other
6. One eye with an abnormal pupil size, shape, or reaction to light, compared with the other eye
7. A layer of blood between the cornea and the iris (hyphema)
8. Laceration of the eyelid, especially if it involves the lid margin
9. Laceration or perforation of the eye.

Keep First Aid to a Minimum

There are two rules governing first aid for significant eye injuries: (a) for chemical injuries, copiously irrigate the eye first, and then call for medical help; (b) for all other injuries with signs or symptoms of serious eye injury, patch the injured eye lightly with a dry, sterile eye pad. If laceration of the eye is suspected, add a protective shield over the sterile eye pad. Instruct the patient not to tightly squeeze the eye shut because doing so greatly elevates the intraocular pressure. Calmly transport the patient to the ophthalmologist or hospital emergency department. Never put eye ointment in an eye about to be seen by the ophthalmologist. The ointment makes clear visualization of the retina very difficult.

Immediate emergency action followed by immediate referral to an ophthalmologist is indicated for all chemical injuries. Any chemical splashed into the eye must be considered a vision-threatening emergency. Forcibly keep the patient's eyelids open while irrigating with water from any source for at least 5 minutes; then refer the patient to an ophthalmologist. Inform the ophthalmologist of the nature of the chemical contaminant.

Injuries that are treatable at the site include conjunctival foreign body, dislodged contact lens, and spontaneous nontraumatic asymptomatic subconjunctival hemorrhage. Conjunctivitis, with normal vision and a clear cornea, can be treated with an antibiotic eye ointment for several days; if there is no improvement, referral to the ophthalmologist is indicated. Never give a patient a topical anesthetic to relieve pain, such as from a flash burn. The prolonged use of topical anesthetics can result in blindness from corneal breakdown. Never treat a patient with a topical steroid unless directed by the ophthalmologist. Topical steroids can make several conditions worse, including *herpes simplex* keratitis, fungal infections, and some bacterial infections. If in doubt as to the severity of an ocular symptom or sign, always err on the side of caution and refer the worker to an ophthalmologist for diagnosis and treatment.

PREVENTION

Provide Preplacement Eye Examinations

The purpose of the preplacement examination is to check for preexisting eye disorders, identify workers who are functionally one-eyed, prescribe appropriate protective eyewear, and ascertain that workers such as truck drivers meet standards for visual performance. The examination must be accompanied by education to teach the employee the potential for ocular injuries on the job, the importance of protection, and workplace rules regarding eye safety.

In any work setting where there is risk of eye injury, it is important that the employer document basic visual skills and rule out significant preexisting eye disease. In addition to guiding decisions about job placement and eye protection, this documentation protects both the employee and the employer by providing baseline data that can be compared with findings after an eye injury has occurred or after the employee has terminated employment. For most occupations, satisfactory visual screening includes central visual acuity (distant and near), muscle balance, stereopsis, color discrimination, and horizontal peripheral visual field. Any trained person can do these tests in a few minutes using one of a variety of available binocular test instruments. These instruments are simple to operate, easy to transport, relatively inexpensive, and rarely require maintenance.

A person is functionally one-eyed if loss of the better eye would result in a significant change in lifestyle owing to poor vision in the remaining eye. Because loss of the ability to drive legally in most states would be a significant handicap, we suggest that an employee be considered functionally one-eyed if the best-corrected vision in the poorer eye is less than 20/40. Every employee who tests less than 20/40 (with glasses, if worn) on the preplacement examination must be evaluated by an optometrist or an ophthalmologist to determine whether the subnormal vision is simply caused by a change in refraction.

If the best-corrected vision in either eye is less than 20/40 after refraction, ophthalmologic evaluation to obtain a definitive diagnosis is indicated. Effective eye protection is possible only when the employee understands the risks and is willing to cooperate with protective measures. Usually it is fairly easy to reach an agreement among the employer, the employee, and the ophthalmologist as to whether the employee is functionally one-eyed and the level of extra protection needed to decrease the risk of eye injury on and off the job. Most jobs and other activities that pose a high risk to unprotected eyes can be made quite safe with the use of appropriate protective devices. The employee deserves a careful explanation of the eye injury risk in the proposed job category, both with and without various types of eye protectors.

Teach Employees Workplace Rules for Eye Safety

Every workplace that has jobs carrying a risk of eye injury must have written, enforced rules concerning eye safety. The employee must know both the rules and the consequences for not obeying them. A *laissez-faire* attitude on the part of the safety officer will result in noncompliance with the eye safety program, leading to eye injuries, disability, and litigation.

Chemicals must be copiously and thoroughly irrigated from the eye. All employees who might receive ocular chemical burns through their work must be taught the principle and method of prompt irrigation for any chemical that comes in contact with the eye. Emergency eyewash fountains must be located conveniently in chemical laboratories and in industrial chemical facilities. Eyewash fountains must be tested regularly to make sure that they function properly.

Occupational health nursing personnel must be taught the symptoms and signs of serious eye injury and workplace first-aid policies. The nurses must have constant access to immediate ophthalmologic consultation by prearranged agreement with a hospital emergency department that has ophthalmologic coverage or an independent ophthalmologist (or group of ophthalmologists) who agrees to give full-time coverage for serious eye injuries.

Analyze the Work Environment

A person who is in a hazardous area without eye protection is at risk for eye injury. Check the work area. Are all power tool safety devices operational? Is there appropriate space between workers? Is there a safe area for passersby and observers? Is the lighting adequate, or are there areas of poor illumination, excess brightness, or radiation hazard? Are dust and chemical controls maintained? Are all workers and visitors wearing appropriate eye protection? Are warnings posted for hazardous areas? Is it easy to find eyewash and shower stations? Failure to provide a safe work area is the prime sin of omission for an employer or safety officer.

Full-site evaluation and determination of visual and safety needs may require input from an occupational health team that includes occupational medicine and primary care physicians, ophthalmologists, nurses, optometrists, opticians, industrial hygienists, laser safety officers, and/or other safety personnel.

The occupational health team evaluates and preserves vision and ocular health by assessing:

The visual requirements needed to perform the job

The worker's visual skills

The safety of various job tasks

The illumination and visual/ergonomic conditions of the worksite

The intrinsic safety of the worksite

The availability and suitability of protective eyewear

Medical access and a system for first aid and definitive care in the event of injury

Obviously stated safety rules easily seen by all workers at the site

A definite statement concerning enforced company penalties for violations of safety rules.

Evaluate the Performance of Employees with Compromised Vision

Various job categories require different visual skills. A receptionist could be totally blind, whereas a precision machine worker might require better than average skills for near vision. Some highly motivated but visually impaired employees may, with the help of low-vision aids, function quite well in jobs that usually require good visual skills; they may do so as long as their placement on the job does not cause undue hazard to themselves or to other employees. As these examples show, visual guidelines are important, but there may be exceptions.

Basic guidelines for visual skills associated with successful performance of various job tasks are available (6), but these standards must be combined with information gathered at the workstation. This information includes observation of the employee's performance on the job so as not to discriminate against a visually handicapped but highly motivated employee who may be performing the job

in a more satisfactory manner than a less-motivated coworker with perfect vision.

A person with only one useful eye can perform most work tasks, including working with power tools and operating machinery (e.g., forklifts). This statement is contrary to some guidelines, but it has proved true for many of our patients who have lost an eye or had a disease that compromised vision in one eye. Individuals with one good eye can train themselves, by motor memory and monocular clues, to do most tasks, with the possible exception of fine close work that requires true stereopsis. Employees who have compromised or no vision in one eye and normal vision in the other should be allowed to take a performance test to determine whether they can safely do a particular task. Any employee who has compromised vision should undergo performance testing annually, or sooner if warranted. Eye protection must be worn at all times to protect the remaining eye. The safety supervisor should be aware of the uncorrected and best-corrected vision levels in each eye of every employee. Employees with decreased vision should be evaluated by an eye care professional for spectacles. If the vision is not corrected with spectacles, further evaluation by an ophthalmologist is indicated. Some workers cannot perform well with compromised vision. Performance testing is essential.

Prescribe Appropriate Protective Eyewear

Prescribing appropriate protective eyewear requires knowledge of the potential on-the-job risks and the best available means of protection. If there is any doubt, the advice of an ophthalmologist, optometrist, or safety consultant who is experienced in eye hazards and protective eyewear should be sought.

All industrial safety eyewear must meet the American National Standards Institute (ANSI) Z87.1 standard requirements. Polycarbonate plastic is the best material for a protective device unless special filtration or optical considerations preclude its use. Industrial safety lenses made of glass or allyl-

A

B

FIG. 28-1. Safety wear with side shields, for use (**A**) by workers who require a prescription lens or (**B**) by workers who require eye protection but do not need a spectacle lens for better vision.

resin plastic have only a fraction of the impact resistance of polycarbonate.

Side shields are always indicated on safety eyewear (Fig. 28-1) unless the eyewear is to be used only in an office environment. Because the total protection is decreased by up to 25% when side shields are removed (Fig. 28-2), dispensing safety eyewear without side shields cannot be justified for any work situation with a potential risk of flying particles or a significant risk of class 3 or class 4 laser exposure.

Goggles are indicated (a) for work in which there is a high potential for many fine flying particles, (b) for use with certain lasers, (c) for use over streetwear spectacles, (d) for

use with chemicals (splash goggles), and (e) for welding that does not require a full-face shield (Fig. 28-3 and Fig. 28-4).

Face shields are required for arc welding (Fig. 28-5) and for use of tools that can project particles that pose an injury potential to the face, including the eyes. Face shields are secondary protectors and must always be worn in conjunction with safety spectacles or goggles.

Laser eye protection is designed to provide the greatest visual transmission along with an adequate optical density at the laser wavelengths. Most laser hazard controls are com-

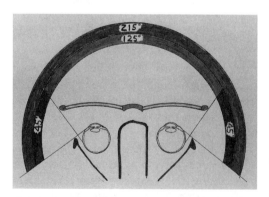

FIG. 28-2. View of the area of coverage of safety glasses without side shields. Note that the eyes are unprotected over a range of 90 degrees of and are exposed to possible penetration by a fragment.

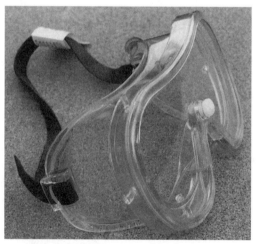

FIG. 28-3. Goggles, designed to protect eyes from chemical splash, suitable for use alone or over other spectacles.

FIG. 28-4. Dust and impact goggles that could support protective lenses for impact, welding, or laser use.

monsense procedures designed to keep personnel away from the beam path or to keep the primary and reflected beams away from occupied areas. The degree of control must be correlated with the injury potential of the laser in use.

Selection of appropriate protective eyewear for laser use requires knowing (a) at least three output parameters of the laser— maximum exposure duration, wavelength, and output power (or energy); (b) applicable safe corneal radiant exposure; and (c) envi-

ronmental factors, such as ambient lighting and the nature of the laser operation.

The laser wavelength (or wavelengths) for which an eye shield was designed should be specified. Commercial protective eyewear is designed to greatly reduce or essentially prevent particular wavelengths from reaching the eye. Although most lasers emit only one wavelength, many emit more, in which case each wavelength must be considered. It is seldom adequate merely to mark on the goggle that it protects against radiation from a particular laser or against the wavelength corresponding to the greatest output power. For example, a helium-neon laser may emit 100 mW at 632 nm and only 10 mW at 1,150 nm; however, safety goggles that absorb at the 632-nm wavelength may absorb little or nothing at the 1,150-nm wavelength. Hence, the wavelength range of use must be specified.

Eye protection filters for glass workers, steel and foundry workers, and welders are specified by shades (logarithmic representations of visual transmission). Typical shade values are the following: acetylene flames,

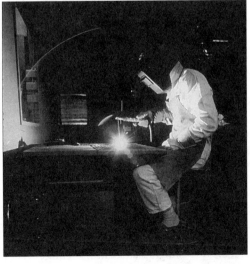

A   B

FIG. 28-5. Use of proper eye protection in welding operations is necessary to protect against permanent eye damage. **A.** Inadequate protection. (Photograph by Barry Levy.) **B.** Adequate protection. (Photograph by Denny Lorentzen, from Newsletter, National Swedish Board of Occupational Safety and Health, January 1981.)

shade 3 or 4; electric welding arcs, shades 10 to 13; viewing the sun or plasma cutting/spraying, shade 13 or higher. These densities greatly exceed those necessary to prevent retinal burns but are required to reduce the luminance (brightness) to comfortable viewing levels. The user of the eye protection therefore should be permitted to choose the shade most personally desirable for the particular operation. Actinic UV radiation from welding arcs is effectively eliminated by all standard welding filters. Few people use welding goggles for arc welding because a welding helmet or shield is necessary to protect the face as well as the eyes from radiant energy.

Contact lenses give no protection from UV keratitis (flash burn). Despite cases in the media reporting the "welding" of contact lenses to the cornea, a contact lens cannot be welded to the eye. All cases investigated have proved to be false reports. In all cases the pain was caused by UV keratitis from failure to wear proper occupational eye injury protection gear over the contact lenses.

Contact lenses give no eye protection and are contraindicated when the employee is exposed to chemicals, especially alkalis. The contact lens can be very difficult to remove when the employee has a chemical corneal burn that involves extreme uncontrolled eyelid closure. Its presence makes irrigation far less effective because chemicals or caustic particles may become trapped beneath the lens, away from the flushing stream. Contact lenses are relatively contraindicated in areas where there is a large amount of dust, because particles trapped beneath the lens may increase the chance of corneal injury.

Consider the Use of Sunglasses

Sunglasses are beneficial for outdoor workers with long exposure to bright light or glare. However, sunglasses can be dangerous if they are worn in lower light levels (i.e., indoors), and they should never be worn to drive at night. Sunglasses should fulfill three basic requirements: impact protection, protection from potentially harmful UV and blue radiation, and comfort. Sunglasses must meet the impact requirements for the work or sport activity and should be certified to meet the appropriate impact standard. Comfort includes adequate light attenuation to avoid discomfort from excessive brightness and possible polarization to decrease glare in certain conditions, such as working on the water.

Provide Protection For Workplace-Organized Sports and Exercise Programs

Many workplaces have physical fitness and sports programs that are important to the well-being, happiness, and ultimate on-the-job productivity of employees. However, many sports carry a potential for eye injury that can exceed the on-the-job risk (7). Hockey and the racket sports have very high eye-injury risks. Eye protection that meets American Society of Testing Materials (ASTM) standards should be mandated for these sports. Baseball, softball, basketball, and soccer have relatively high eye-injury risks. Safety glasses with polycarbonate lenses are indicated for these sports.

Recognize Safety Enforcement as a Management Policy

It is not possible to overstate the importance of eye injury prevention in terms of worker pain and suffering, medical costs to society, and direct and indirect costs to industry. Safety is as much a concern of management as it is of the occupational health team. Management must ensure that safety rules are followed, that overall planning includes safety considerations, and that safety activities are promoted through newsletters or membership in the Wise Owl Club, which is sponsored by Prevent Blindness.

REFERENCES

1. Schein OD, Hibberd PL, Shingleton BJ, et al. The spectrum and burden of ocular injury. Ophthalmology 1988;95:300–305.

2. Grant WM, Schuman JS. Toxicology of the eye. Springfield: Charles C Thomas, 1993.
3. Sliney D, Wolbarsht M. Safety with lasers and other optical sources. New York: Plenum, 1980.
4. American Academy of Ophthalmology, Interprofessional Education Committee. The worker's eye. San Francisco: American Academy of Ophthalmology, 1998.
5. Duke-Eider S, MacFaul PA. System of ophthalmology injuries. St. Louis: Mosby, 1972.
6. Good GW, Maisel SC, Kriska SD. Setting an uncorrected visual acuity standard for police officer applicants. J Appl Psychol 1998;83:817–824.
7. Vinger PF. The eye and sportsmedicine. In: Duane TD, Jaeger EA, eds. Clinical ophthalmology, vol 5. Philadelphia: JB Lippincott, 1994:1103.

BIBLIOGRAPHY

American National Standards Institute (ANSI) Safety Standards

ANSI Z80.5-1997 Requirements for ophthalmic frames.
ANSI Z80.1-1995 Prescription ophthalmic lenses recommendations.
ANSI Z80.3-1996 Requirements for nonprescription sunglasses and fashion eyewear.
These are safety standards for dress eyewear, also called streetwear spectacles. The test requirements are minimal and are geared to the desire for a diversity of styles in fashion eyewear. Streetwear spectacles are not appropriate for work or for sports with impact potential. Use of lenses made of polycarbonate, a highly shatter-resistant material, can reduce injuries from shattered spectacle lenses.
ANSI Z87.1-1989 Practice for occupational eye and face protection.
The industrial safety standard is being revised. In its present form, the standard allows for removable lenses that shatter with relatively little energy. Polycarbonate lenses should be used unless there is a specific reason for using another lens material.
ANSI Z358.1-1998 American national standard for emergency eyewash and shower equipment.
Engineering controls are the primary defense against eye injuries by chemicals. Splash goggles are essential if the safety and engineering controls fail and exposure occurs. Eyewash equipment should meet this standard specification.
ANSI Z136.1 Safe use of lasers.
ANSI Z123.3 Safe use of lasers in health care facilities.
ANSI Z49.1 Welding and cutting.

American Society for Testing and Materials (ASTM) Safety Standards

ASTM F803-97 Eye protectors for selected sports (racket sports, women's lacrosse, baseball).
ASTM F513-95 Eye and face protective equipment for hockey players.
ASTM F659-92 High impact resistant eye protective devices for alpine skiing.
ASTM F910-92 Face guards for youth baseball.
ASTM F776-97 Eye protectors for use by players of paintball sports.
ASTM F587-96 Head and face protective equipment for ice hockey goaltenders.
ASTM sports standards are extremely effective. There have been no reported blinding eye injuries to any player wearing an appropriate protector certified to one of these standards, despite extensive exposure.

Laser Institute for America (LIA) Guidelines

LIA laser safety guide.
LIA guide to medical laser safety.
LIA guide for the selection of laser eye protection.
These publications give a good summary of the laser safety guidelines.

29 Disorders of the Nervous System

Edward L. Baker, Jr.

A 29-year-old man was seen after 8 years of employment in a chloralkali plant, where he was primarily employed in maintenance and operation of the electrolytic cells. Four years after beginning work in the plant, he began to notice increased nervousness and irritability. His nervousness continued for 2 years; he then began to experience episodes of severe depression. At that time, he also experienced a tremor of the hands, bleeding gums, easy fatigability, increased salivation, and loss of appetite. He sustained an injury to his left Achilles tendon and was away from work for 7 months, during which time most of his symptoms improved, but tremulousness, nervousness, and depression remained.

This man and his wife reported that before his employment at the plant he was outgoing, calm, and patient. He had been a military policeman in the U.S. Marines and did not experience emotional upsets during this tour of duty despite significant stress.

Urine mercury monitoring, which had been performed by his employer during the entire period of employment, had demonstrated numerous values over 500 μg/L, the highest of which was 736 μg/L in his 5th year of employment (normal range in the general population, 5 to 30 μg/L).

Physical examination performed at the end of the 7-month removal from work showed no evidence of tremor, a mild loss of pinprick sensation on the dorsal aspect of his arms, and an otherwise normal neurologic examination. Lines of increased pig-

mentation were observed at the gingival margins of several teeth.

Neuropsychological testing showed normal levels of intellectual functioning. The worker showed mild defects in his ability to perform mental calculations and in his immediate verbal and visual memory. Written spelling was particularly impaired, with an inability to copy simple sentences. He could not concentrate on various tasks and, as a result, his performance was erratic, with incorrect answers to simple questions and correct answers to more difficult ones. He was emotionally labile in the test situation, appearing anxious and depressed. He displayed average performance on tests of manual dexterity.

This patient's illness was manifested primarily by emotional disturbances that had secondary effects on standardized tasks of psychological performance. He showed no particular deficits in memory, psychomotor performance, learning ability, or recall of current events. His most striking deficit was one of impaired concentration, which resulted in erratic performance on various tests. These effects were still detected months after he was removed from mercury exposure.

Increasing concern has been focused on the occurrence of neurobehavioral disorders among workers in various occupations. In many instances, as in this case study, specific chemical substances have been identified that are responsible for characteristic pathologic processes within the nervous system. In other cases, groups of substances, such as solvents, have been associated epidemiologically with manifestations of nervous system disease. Although exposure to industrial toxins has been known for more than 100 years

E. L. Baker, Jr.: Public Health Practice Program Office, Centers for Disease Control and Prevention, Atlanta, GA 30333.

to affect behavior, studies performed during the past three decades have applied quantitative methods to the study of behavioral aberrations after toxin exposure and have demonstrated a wide range of clinical and subclinical effects for numerous substances (1). Neuroimaging techniques such as positron emission tomography (PET) and single photon emission computed tomography (SPECT) hold promise as additional objective tools for evaluating the impact of neurotoxic agents on the central nervous system (CNS) (2). Nerve conduction studies can be useful in quantifying peripheral nervous system (PNS) dysfunction (2). Many neurotoxic agents produce a dose-related spectrum of impairment, ranging from mild slowing of nerve conduction velocity or prolongation in reaction time to neuropathy and frank encephalopathy.

Disorders with predominantly psychiatric manifestations have been described in some workers; these range from acute psychosis and mass psychogenic illness to chronic neurasthenia. Although specific chemical substances have been identified that may be associated with certain of these psychiatric syndromes, etiologic mechanisms are unclear. (Psychiatric disorders are discussed in Chapter 30.)

With the introduction of new substances into the workplace, some neurologic disorders are being newly recognized. Such a discovery occurred in the 1970s, when an industrial catalyst, dimethylaminopropionitrile (DMAPN), was found to be associated with bladder neuropathy in workers producing polyurethane foam (3). Another such discovery occurred when peripheral neuropathy was diagnosed in employees in a coated-fabrics plant and traced to the introduction of a neurotoxic solvent, methyl *n*-butyl ketone (MBK) (4).

NEUROLOGIC DISORDERS

Pathophysiology

Peripheral Nervous System Effects. Two basic forms of damage to peripheral nerves have been identified as responsible for the peripheral neuropathies associated with occupational exposure to neurotoxins (5). *Segmental demyelination* results from primary destruction of the neuronal myelin sheath, with relative sparing of the axons. This process begins at the nodes of Ranvier and results in slowing of nerve conduction. Characteristically, there is no evidence of muscle denervation, although disuse atrophy may occur if paralysis is prolonged. As remyelination begins during the recovery phase, recovery is rapid and usually complete in mild to moderate neuropathies.

Axonal degeneration is associated with metabolic derangement of the entire neuron and is manifested by degeneration of the distal portion of the nerve fiber. Myelin sheath degeneration may occur secondarily. Nerve conduction rates are usually normal until the condition is relatively far advanced. Distal muscles show changes of denervation. Recovery may occur by axonal regeneration, but it is very slow and often incomplete.

In some instances, axonal degeneration and segmental demyelination coexist, presumably as a result of secondary effects derived from damage to each system. Therefore, although the classic descriptions of these syndromes hold in experimental models, the clinical manifestations of neuropathy in exposed workers may represent a combination of both pathologic processes.

Central Nervous Systems Effects. Investigations of lead, chlordecone (Kepone), carbon monoxide, and other chemicals have shown significant disruption of neurotransmitter metabolism, affecting dopamine, norepinephrine, gamma-aminobutyric acid (GABA), and serotonin, which correlates with behavioral aberrations in experimental animals. Furthermore, many industrial solvents cause acute depression of CNS synaptic transmission, resulting in drowsiness and weakness (6). Such mechanisms may be responsible for the manifestations of CNS toxicity induced by workplace substances.

Combined Peripheral and Central Nervous System Effects. Certain industrial neurotoxins cause distal degeneration of axons in both the CNS and PNS (5). This form of axonal degeneration was originally described as "dying back" neuropathy. In view of the association of CNS and PNS degeneration, it has been suggested that this process be referred to as "central-peripheral distal axonopathy." Substances associated with this effect include acrylamide, *n*-hexane, MBK, carbon disulfide, and organophosphorus compounds, most notably triorthocresyl phosphate (TOCP).

Characteristically, distal degeneration occurs within the long nerve fiber tracts of both the CNS and PNS. Once degeneration begins peripherally, it becomes more severe in the initially affected nerve segments while progressing centrally to involve more proximal segments of nerve fibers. Within the spinal cord, the long ascending and descending tracts (the spinocerebellar and corticospinal tracts) appear to be the most severely affected. Involved fiber tracts demonstrate axonal swellings, which are often focal and are associated with neurofilament accumulation within the axon. Although the length of the axon is a key determinant of fiber susceptibility, fiber diameter may also be important: large-diameter, myelinated fibers are more frequently affected.

The precise locus of the metabolic derangement that is responsible for these manifestations of axonal damage is unknown. Chemical substances may bind to the inactive intraaxonal enzyme systems required for maintenance of normal axonal transport mechanisms.

Manifestations

Peripheral Nervous System. Virtually all of the industrial toxins that affect the PNS cause a mixed sensorimotor peripheral neuropathy. The initial manifestations of this disorder consist of intermittent numbness and tingling in the hands and feet; motor weakness in the feet or hands may develop somewhat later and progress to the development of an ataxic gait or an inability to grasp

heavy objects. Although the distal portion of the extremities is involved initially and to a greater degree, severely affected patients may also have proximal muscle weakness and muscle atrophy. Nerve biopsies in affected persons have shown axonal swellings and paranodal myelin retraction. Extensor muscle groups usually manifest weakness before flexors do.

Although the manifestations are somewhat similar from one toxin to another, certain specific characteristics are unique to individual agents (Table 29-1). Painful limbs and increased sensitivity of the feet to touch are particularly characteristic of arsenical neuropathy. Sensory involvement predominates in the relatively rare neuropathy seen with alkyl mercury poisoning. Both motor and sensory disorders are observed in the neuropathies associated with exposure to *n*-hexane, MBK, and acrylamide.

The peripheral neuropathy associated with lead exposure is unusual because only the motor system is involved. The most characteristic early manifestation of lead neuropathy is wrist extensor weakness. Reports of involvement of the lower extremities resulting in ankle drop were made during the 1930s, when cabaret dancers consumed lead-contaminated illicit whiskey and developed lead neuropathy in the muscles that they used most actively. Overt wrist drop, which was a characteristic manifestation of lead neuropathy in reports of many years ago, is rare today.

The development of these syndromes is usually insidious. Very slow development of numbness and tingling of the fingers and toes occurs over several weeks and may then be followed by motor weakness. With several toxins, including acrylamide, *n*-hexane, and MBK, the neuropathy may progress even after the worker is removed from exposure. This deterioration may continue for 3 to 4 weeks; at that point, recovery may begin. The duration of the recovery process is proportional to the degree of severity of neuropathy: less severely affected patients may experience total resolution in 3 to 6 months,

TABLE 29-1. *Peripheral nervous system effects of occupational toxins[a]*

Effect	Toxin	Comments
Motor neuropathy	Lead	Primarily wrist extensors Wrist drop and ankle drop rare
Mixed sensorimotor neuropathy	Acrylamide	Ataxia common Desquamation of hands and soles Sweating of palms
	Arsenic	Distal paresthesias earliest symptom Painful limbs, especially in calves Hyperpathia of feet Weakness prominent in legs
	Carbon disulfide	Peripheral neuropathy rather mild CNS effects more important
	Carbon monoxide	Only seen after severe intoxication
	DDT	Only seen with ingestion
	n-hexane and methyl n-butyl ketone	Distal parethesias and motor weakness Weight loss, fatigue, and muscle cramps common
	Mercury	Predominantly distal sensory involvement More common with alkyl mercury exposure
	Organophosphate in-secticides (se-lected agents)	Delayed onset following single exposure (usually nonoccupational)

[a] Includes most but not all of the neurotoxic substances associated with listed conditions.

whereas those with advanced disease may continue to have signs and symptoms 1 to 2 years later.

Physical examination of affected workers shows a characteristic distribution of sensory loss, particularly to pain and temperature discrimination (Fig. 29-1). Frequently vibra-

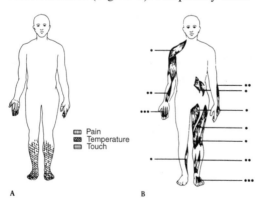

FIG. 29-1. A. Pattern of sensory loss in a severe case of MBK neuropathy. **B.** Distribution of muscle weakness in MBK neuropathy. The degree of weakness is proportional to the number of asterisks shown. (From Allen N. Solvents and other industrial organic compounds. In: Vinken PK, Bruyn GW, eds. Handbook of clinical neurology: intoxications of the nervous system, part I. Vol 36. Amsterdam: Elsevier North-Holland Biomedical Press, 1979.)

tion sensation is impaired and touch perception, particularly with acrylamide poisoning, is lost. Tremor of the hands is particularly common in several types of chemical intoxication; in most instances, it is a resting tremor that is not increased with movement. The tremor seen with chlordecone poisoning is a common manifestation of the disease and has characteristic features: it is irregular, it is nonpurposive, and it is most severe when the limb is static but unsupported against gravity. In contrast, the tremor seen with mercury poisoning is fine and affects the eyelids, tongue, and outstretched hands. Motor weakness in toxic neuropathies is characteristically found in distal muscles of the arms and legs (see Fig. 29-1). Intrinsic muscles of the hands and feet are particularly affected in neuropathies caused by n-hexane, MBK, and acrylamide. Extensor weakness of the forearms is characteristic of lead neuropathy. Impaired coordination is often seen in persons with motor weakness in the extremities; cerebellar pathology need not be present for these manifestations to occur. In summary, distal sensory and motor impairment characterized by numbness and weakness of the hands and feet is followed by more proximal involvement as the toxic neuropathy develops.

Other Neurologic Manifestations. A wide variety of additional manifestations may be seen that are specific to individual toxins (Table 29-2). Movement disorders that resemble Parkinson's disease have been reported in persons exposed to carbon disulfide, carbon monoxide, or manganese; hypotonia, dystonia, and other disorders of locomotion occur in persons with excessive exposure to these

TABLE 29-2. *Other neurologic manifestations of occupational toxins[a]*

Manifestation	Agent
Ataxic gait	Acrylamide
	Chlordane
	Chlordecone (Kepone)
	DDT
	n-Hexane
	Manganese
	Mercury (especially methyl mercury)
	Methyl *n*-butyl ketone
	Toluene
Bladder neuropathy	Dimethylaminopropionitrile
Constricted visual fields	Mercury
Cranial neuropathy	Carbon disulfide
	Trichloroethylene
Headache	Lead
	Nickel
Impaired visual acuity	*n*-Hexane
	Mercury
	Methanol
Increased intracranial pressure	Lead
	Organotin compounds
Myoclonus	Benzene hexachloride
	Mercury
Nystagmus	Mercury
Opsoclonus	Chlordecone (Kepone)
Paraplegia	Organotin compounds
Parkinsonism	Carbon disulfide
	Carbon monoxide
	Manganese
Seizures	Lead
	Organic mercurials
	Organochlorine insecticides
	Organotin compounds
Tremor	Carbon disulfide
	Chlordecone (Kepone)
	DDT
	Manganese
	Mercury

[a] Includes most but not all of the neurotoxic substances associated with listed conditions.

substances. In the case of manganese toxicity, significant improvement after drug therapy can be seen in the characteristic mask-like facies of a patient with occupational exposure to this toxin (Fig. 29-2).

A characteristic abnormality of eye movements called *opsoclonus* can be caused by exposure to chlordecone. It consists of irregular bursts of involuntary, abrupt, rapid jerks of both eyes simultaneously; these movements usually are horizontal, but in severely affected persons they are multidirectional.

Seizures are often seen in workers with acute excessive exposure to industrial toxins. Organochlorine insecticides, such as dichlorodiphenyltrichloroethane (DDT) and chlordane, have been associated with seizures after acute ingestion of a large dose. Seizures are a rare manifestation of lead encephalopathy in adults.

Cranial nerve involvement is uncommon with peripheral neurotoxins. However, trichloroethylene has a predilection for the trigeminal nerves and has been associated with facial numbness and weakness. Carbon disulfide exposure is also associated with cranial neuropathies.

An unusual manifestation of neurotoxicity was seen in a group of workers exposed to DMAPN. This substance caused a neuropathy in the bladder, resulting in urinary retention, urinary hesitancy, and sexual dysfunction. Although the symptoms improved after removal of the affected workers from exposure to this substance, symptoms and signs persisted in some for at least 2 years.

Diagnosis

Electrophysiologic tests that assess peripheral nerve function, including electromyograms (EMGs) and nerve conduction measurements, are important tools for assessing the extent and severity of neurologic disorders in workers exposed to industrial toxins. These techniques are often useful in the evaluation of individual patients. Noninvasive techniques that measure sensory thresholds for vibration and temperature have been de-

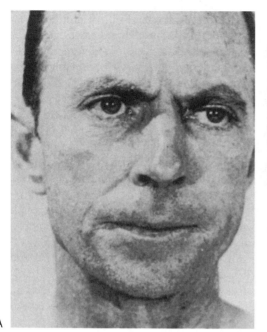
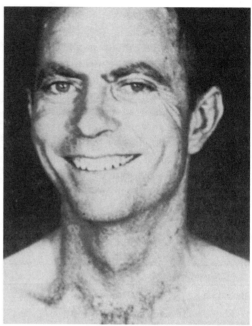

A

B

FIG. 29-2. A. Mask-like faces of patient with manganese toxicity before therapeutic trial with L-dopa. **B.** Full facial expression of same patient being maintained on L-dopa. (From Rosenstock HA, Simons DG, Meyer JS. Chronic manganism: neurologic and laboratory studies during treatment with levodopa. JAMA 1971;217:1355. Copyright © 1971, American Medical Association.)

veloped to monitor diabetic patients for the occurrence of sensory neuropathy; these are also efficient tools for reliable screening of workers with significant exposure to neurotoxic agents or with early sensory symptoms. In addition to detection of toxic neuropathy, these instruments may be useful in detection of compression neuropathies, such as carpal tunnel syndrome.

EEGs have also been used in the evaluation of workers exposed to neurotoxins, but these tests usually are not as useful as nerve conduction tests. The EEG may be of value as an adjunct in the assessment of altered states of consciousness of unknown cause. A more promising extension of EEG use is the measurement of cortical-evoked potentials after auditory or visual stimuli; for example, prolonged latency of visually evoked responses has been reported in workers chronically exposed to *n*-hexane.

BEHAVIORAL DISORDERS

Manifestations

Excessive exposure to industrial toxins may result in behavioral effects ranging from mild symptoms of fatigue to persistent impairment of nervous system function. In view of the nonspecific nature of many behavioral manifestations of neurotoxin exposure, standardized psychometric testing has greatly facilitated the evaluation of these disorders. In general, neurotoxins particularly affect psychomotor performance by causing slowness in response time, impaired eye-hand coordination, and diminished concentration ability. Emotional effects are also seen; they consist of irritability and, at times, emotional lability. Recent memory may be disrupted; this is manifested in testing situations as an inability to learn new material. Aspects of cognitive functioning that usually are not af-

TABLE 29-3. *Behavioral effects of occupational toxins*[a]

Manifestation	Agent
Acute psychosis or marked emotional instability	Carbon disulfide
	Manganese
	Toluene (rare)
Acute intoxication	Organic solvents
	Carbon monoxide
Chronic behavioral symptoms	Acrylamide
	Arsenic
	Lead
	Manganese
	Mercury
	Methyl *n*-butyl ketone
	Organotin compounds
Chronic toxic encephalopathy	Organic solvents
	Styrene
	Lead
	Carbon disulfide

[a] Includes most but not all of the neurotoxic substances associated with listed conditions.

fected by toxins include remote memory and fund of general information.

Although few toxins have unique behavioral effects, several substances deserve particular attention (Table 29-3). Carbon disulfide affects all levels of the CNS and may result in bizarre clinical syndromes including acute psychosis. Excessive exposure has been associated with suicide in at least one worker. Neurotoxins may cause both behavioral effects and peripheral neuropathy in the same person.

Most chlorinated hydrocarbon solvents in current use in industry cause a relatively brief "high" after exposure to significantly elevated concentrations in air. Intentional abuse of industrial solvents by workers desiring these intoxicating effects has been reported to cause permanent damage to the PNS and CNS (7). Such misuse obviously should be prevented, in view of the potentially severe consequences.

Diagnosis

Standardized psychometric testing, using measures of memory, intelligence, attention, dexterity, reaction time, personality, and general psychomotor function, is very useful in evaluating exposed individuals as well as groups of workers (8). These neurobehavioral tests have been adapted for computer administration to facilitate reproducibility in testing of groups and to improve data-handling efficiency. Computerized testing has been used in epidemiologic and clinical research to evaluate health effects of nervous system toxins. For correct evaluation of the etiologic role of toxin exposure, interpretation of test results must take into account confounding factors, such as age, education, alcohol consumption, and preexisting neurologic disease. The most important feature of the diagnostic process is a carefully obtained occupational history that identifies specific neurotoxins and assesses the magnitude and duration of exposure to each. The work history is particularly important in evaluating behavioral disorders, because these conditions are often attributed to factors unrelated to work.

MANAGEMENT AND CONTROL

Management of occupationally-induced neurologic problems consists primarily of identification of the offending agent and removal of the worker from continued exposure. In some instances (e.g., DMAPN exposure), removal of the offending agent from the workplace may prevent the development of new cases. Some workers with known exposure may develop mild, early symptoms of neurotoxicity; objective demonstration of functional impairment on standardized tests is essential in the management of individual cases. Workers with evidence of toxin-related symptoms or functional impairment should be removed from exposure until these deficits resolve and exposure in the workplace is terminated.

Prevention of occupationally-induced neurologic disorders can be accomplished through workplace medical and environmental control programs. The goal of environmental control is to reduce concentrations of neurotoxic substances in the worker's environment by various manipulations. Medical strategies designed to reduce neurologic morbidity include preplacement evaluation and periodic medical monitoring. The goal of preplacement evaluation as it relates to neurologic disorders is to avoid placement of workers with preexisting disease (e.g., peripheral neuropathy) in jobs with exposures that might exacerbate these conditions. Conditions that might impair a worker's ability to perform a job, such as uncontrolled epilepsy in a person operating hazardous machinery, would be grounds for medical exclusion from such jobs.

Periodic medical monitoring programs are becoming more common in industries where neurotoxins are used. An important element of such programs is measurement of the neurotoxic agent in biologic fluids. The most common such application occurs in industries where lead or mercury is used.

Periodic monitoring of lead-exposed workers should include occupational and medical histories; a physical examination with special attention to the nervous system; blood and urine studies to evaluate hematologic and renal effects of lead exposure; and, most importantly, determination of the blood concentrations of lead and zinc protoporphyrin (ZPP). The content of such examinations is mandated by the Occupational Safety and Health Administration (OSHA) standard on occupational exposure to lead, in which specific guidelines are provided for job transfer of workers with excessive blood lead levels (BLLs). The OSHA standard for inorganic lead exposure requires that employers make routine BLL monitoring available to all employees who are exposed to lead at concentrations greater than the action level of 30 μg per cubic meter of air, regardless of whether respirators are worn. Specific actions must be taken depending on the results of BLL testing. Any worker removed from a job because of an elevated BLL is protected by the medical removal protection provision of the OSHA lead standard. This provision requires an employer to "maintain

the worker's earnings, seniority, and other employment rights and benefits (as though the worker had not been removed) for a period of up to 18 months."

The evaluation of mercury-exposed workers is similar, with three exceptions. First, urine mercury determinations are used rather than blood measurements. Second, there is no enzymatic test, such as the ZPP test, that measures the metabolic toxicity of mercury exposure. Finally, there is no comprehensive OSHA standard for occupational mercury exposure that prescribes the content of periodic medical evaluations and medical action levels for job transfer.

Workers exposed to cadmium and arsenic should be monitored periodically with urinary determinations of these metals in addition to standard medical evaluations.

Workers chronically exposed to solvents should have periodic medical histories and physical examinations with attention to the nervous system. Measurement of urinary metabolites of solvents is sometimes helpful as an adjunct to other medical monitoring techniques.

Pesticide-exposed workers, particularly those using organophosphate insecticides, should be periodically evaluated, including red blood cell cholinesterase levels, to assess their degree of pesticide exposure. Although some recommendations have been made that periodic nerve conduction testing should be performed in addition to a standard medical history and physical examination, this test is not suitable for routine monitoring of asymptomatic workers.

Treatment of occupational neurologic disease beyond removal of the worker from continued toxic exposure may consist of the administration of drugs designed to remove the offending agent or counteract its effects. Chelating drugs, such as ethylene diamine tetraacetic acid (EDTA), dimethylsuccinic acid (DMSA), and penicillamine, are given as treatment for symptomatic poisoning by lead and other heavy metals. These drugs should not be given prophylactically to lower blood levels of the metal; they have known toxicities, which may add to the toxic effects of the metal and also may increase gastrointestinal absorption of the metal. Workers should be removed from exposure to the offending agent before initiation of drug therapy.

Treatment of organophosphate insecticide poisoning is accomplished primarily by giving atropine, a pharmacologic antagonist of the pesticide. If patients are seen very soon after exposure, other drugs (i.e., oximes) can be given to regenerate inhibited cholinesterase enzyme.

Ultimately, prevention of occupational diseases of the nervous system rests on adequate testing of chemicals before their introduction into the workplace and on environmental measures designed to reduce exposure. The Toxic Substances Control Act (TSCA) addresses the issue of premarket testing, and the Environmental Protection Agency (EPA), which administers this Act, has specified criteria for neurologic evaluation of chemical substances. Biologic assays of organophosphate compounds have successfully predicted those substances that are neurotoxic to humans. Substances such as *n*-hexane and MBK, which produce an axonal neuropathy in exposed humans, have been shown to produce similar effects in animals, and the neurologic disorder associated with chlordecone toxicity was seen in experimental animals several years before it was reported in exposed humans. Therefore, testing of industrial substances by administration of toxins to experimental animals is essential in the identification of substances with neurotoxic potential.

In rare instances, structural similarity alone has proved useful in predicting toxicity. *n*-Hexane and MBK are metabolized to 2,5-hexanedione, which is thought to be responsible for the neurotoxic manifestations of these two industrial chemicals. Investigation of structure-activity relations therefore may be of value in identifying substances with potential neurotoxicity. In those instances in which neurotoxicity is suspected because of the chemical structure

of the compound, animal tests are still required.

As standardized tests of neurologic function become increasingly available, field studies of industrial toxins using these techniques will be used to determine the appropriate levels of industrial exposure. These levels, referred to as threshold limit values (TLVs) or permissible exposure limits (PELs), have in the past been based on informed opinions of experts. As epidemiologic studies become more precise in assessing the neurotoxic hazard of these substances, control measures and specific exposure standards will be based on much more objective information.

EFFECTS OF SELECTED NEUROTOXINS

Lead

The most commonly encountered workplace substance with clearly recognized neurotoxic effects is lead. (See also Chapters 3, 15, 31, 33 and 35.) The National Institute for Occupational Safety and Health (NIOSH) has estimated that more than 1 million U.S. workers are daily exposed to lead (Fig. 29-3). The manifestations of lead neurotoxicity as currently encountered differ significantly from those seen in reports from the earlier part of the 20th century, when more overt disorders were observed. The most common neurologic finding is impaired CNS function, manifested by symptoms of fatigue, irritability, difficulty in concentrating, and inability to perform tasks requiring sustained concentration. These symptoms are associated with abnormalities on standardized neuropsychological testing that indicate impairment of verbal intelligence, memory, and perceptual speed. Symptoms of arm weakness, characteristically affecting extensor muscle groups, are also seen in the early phases of lead toxicity. Often, weakness occurs before abnormalities are seen on nerve conduction testing. Such abnormalities tend to develop in person with BLLs in the range of 60 μg/dL,

A   B

FIG. 29.3. Current work practice rules require significant personal and environmental protection in situations of lead exposure. **A.** Lead battery worker is protected mainly by local exhaust ventilation. **B.** Automobile lead grinder is protected by air-supplied hood and floor exhaust ventilation. (Photographs by Earl Dotter.)

and they become more apparent as the BLL rises. After removal from exposure, these symptoms and abnormalities resolve slowly over weeks to months, the duration depending on their initial intensity and other factors.

Neurologic abnormalities caused by lead exposure usually occur after hematologic toxicity, as manifested by an elevated ZPP and a reduced blood hemoglobin concentration. Permanent renal damage occurs much later than neurologic dysfunction and characteristically develops only after at least 5 years of lead exposure. In contrast, neurologic abnormalities may develop within 2 to 3 months after the onset of work in a lead-contaminated environment, particularly where exposure is relatively poorly controlled. Abnormalities of nerve conduction tend not to occur before at least 6 to 8 months of chronic exposure to lead.

Mercury

Although disease as striking as that experienced by Lewis Carroll's Mad Hatter no longer occurs in workplaces in the United States, behavioral effects of exposure to elemental mercury are still seen. Erethism, a set of behavioral symptoms classically associated with mercury toxicity, is characterized by unusual shyness, irritability, and other symptoms. Standardized memory tests are affected in persons with urine mercury concentrations of 200 μg/L or higher. A fine tremor of the hands is associated with mercury poisoning, and computer-assisted analysis of EMGs has shown a shift in the frequency of normal forearm tremor as an early manifestation of mercury toxicity. Peripheral neuropathy is not a recognized feature of elemental mercury poisoning. Measurement of mercury in urine and blood is a useful tool in the assessment of workplace exposure. Clinical signs of mercury poisoning usually do not occur when the urinary concentration of mercury is kept below 300 μg/L.

Organic mercurials, particularly alkyl mercury compounds such a methyl mercury, have a strong affinity for the CNS, and severe neurologic effects have been associated with excessive exposure. The best-described episode of organic mercury poisoning occurred in Minamata, Japan, where early symptoms of poisoning consisted of distal paresthesias, cerebellar disorders, visual impairment, deafness, and mental disturbances. Sensory deficits were seen, with loss of position sense, impaired two-point discrimination, astereognosis, and mild hypalgesia. Visual impairment characteristically consisted of constriction of visual fields. Mental disturbances were characterized by agitation alternating with periods of stupor and mutism. The more severely affected patients exhibited dystonic flexion postures. Peripheral neuropathies were not seen in this group of patients.

The devastating and usually irreversible effects on the nervous system of organic mercury poisoning should be prevented through restriction of the use of mercury. To that end, the practice of treating seed grain with organic mercurial fungicides has been curtailed by the EPA, following outbreaks of neurologic disease in the United States and Iraq among persons ingesting food inadvertently contaminated with these substances.

Organophosphate Insecticides

Acute organophosphate insecticide poisoning is characterized by the inhibition of acetylcholinesterase, with resultant overactivity of cholinergic components of the autonomic nervous system, inhibition of conduction across myoneural junctions in skeletal muscle, and interference with CNS synaptic transmission. Manifestations of acute toxicity include meiosis, blurring of vision, chest tightness, increased bronchial secretion, and wheezing. Gastrointestinal effects are also seen, including abdominal cramps, nausea, and vomiting. Increased sweating, salivation, and lacrimation are additional characteristic features.

Atropine is the drug of choice for treatment of the acute manifestations of organophosphate insecticide poisoning. Repeated doses are given to the point of atropinization, and subsequent doses of the drug may be required since the duration of action of atropine is less than that of organophosphate

insecticides. Because organophosphate compounds bind irreversibly to cholinesterase, reactivation of the enzyme system occurs only through synthesis of additional cholinesterase molecules. Therefore, recovery of normal cholinesterase concentrations in red blood cells is slow, and repeated exposure may result in cumulative depression of cholinesterase stores. Recovery after an acute episode of poisoning is usually complete within 7 days unless anoxia has occurred during the acute phase of the episode. Measurement of red blood cell cholinesterase concentrations is valuable during the acute intoxication episode and is also used for surveillance of occupationally exposed workers. Plasma cholinesterase concentrations are of less value in occupational settings because they can be altered by many factors.

A syndrome of delayed neurotoxicity has been reported with certain organophosphate compounds (although not with organophosphate pesticides currently used in the United States). This syndrome develops 8 to 35 days after exposure. Progressive weakness begins in the distal lower extremities, and toe and foot drop often develop; finger weakness and wrist drop follow the lower-extremity manifestations. Sensory loss is minimal. Deep tendon reflexes are frequently depressed. The disease may progress for 1 to 3 months after onset, and recovery is very slow.

Studies of workers occupationally exposed to organophosphates have revealed some evidence of psychomotor impairment and abnormal EEG findings. Further studies are required to assess the extent and nature of these disorders. Evaluation of patients exposed to organophosphate insecticides should include, in addition to manifestations of autonomic nervous system dysfunction, measurement of the red blood cell cholinesterase concentration. Cholinesterase levels correlate reasonably well with manifestations of clinical toxicity. Migrant workers are at risk for the acute and chronic effects of exposure to organophosphate insecticides used in agriculture. This population has not been adequately studied, and often migrant workers do not receive adequate protection from pesticide exposure (see Chapter 41).

Organic Solvents

Exposure to organic solvents occurs daily for more than 1 million U.S. workers. The most frequently used solvents are toluene, xylene, trichloroethylene, ethanol, methylene chloride, and methyl chloroform. Although chemically heterogeneous, these compounds are often discussed as a group because of toxicologically similar effects and the high frequency of exposure to various combinations of these substances (Fig. 29-4).

Acute intoxication, with symptoms of dizziness, lightheadedness, or feeling "high," occurs after exposure to excessive concentrations of solvent vapors. Exposure to very high concentrations of solvent vapors or fumes may lead to narcosis with total loss of consciousness.

To facilitate the characterization of persistent health effects of solvent exposure, a nomenclature has been developed by a World Health Organization (WHO) working group

FIG. 29.4. Printer cleaning type with organic solvent is exposed via skin through permeable cloth gloves and air via evaporation from work surface and open bottle. (Photograph by Barry S. Levy.)

(10) and by a workshop of invited experts held in the United States (11). The mildest form of effect, *organic affective syndrome* (or type 1 solvent health effect), is characterized by symptoms of irritability, fatigability, difficulty in concentrating, and loss of interest in daily events. This type of effect is typically reversible. In the second type of effect, mild chronic toxic encephalopathy, abnormalities on neurobehavioral testing are observed, and symptoms that are similar to those of the type 1 effect are reported. Sustained personality or mood change (type 2a effect) or impairment in intellectual function (type 2b) may be seen at this level, either singly or in combination. Some individuals with type 2 effect may experience reversal of signs and symptoms; some do not, leading to permanent cognitive impairment. Severe *chronic toxic encephalopathy* (type 3 effect) is characterized as a type of dementia with global deterioration of memory and other cognitive functions that are unlikely to be reversible. Workers exposed to solvents may exhibit any of these three syndromes, depending on the intensity and duration of their exposure. Epidemiologic studies have frequently shown decreases in reaction time, dexterity, speed, and memory among workers with prolonged exposure to solvents. Relatively few abnormalities have been demonstrated in PNS function; nerve conduction abnormalities were reported in one study of mixed solvent exposure. Measurement of urinary metabolites, such as hippuric acid in toluene exposure, may be of value in monitoring of exposed populations.

Clinical diagnosis of chronic toxic encephalopathy caused by exposure to organic solvents is made by obtaining a careful occupational and clinical history and by performance of standardized neurobehavioral testing (12). Government agencies have concluded that workers exposed to solvent fumes at excessive levels (those sufficient to cause acute intoxication frequently) for more than 10 years are clearly at risk for development of toxic encephalopathy (13,14). Although some studies (15) have shown that workers excessively exposed to solvents for 5 to 10 years are at risk for toxic encephalopathy, there is little epidemiologic evidence that exposure for less than 5 years is sufficient to place a worker at increased risk. Therefore, any worker with more than 5 years of excessive solvent exposure should be considered potentially at risk of solvent encephalopathy and provided a thorough medical and neurobehavioral assessment to determine whether symptoms or signs of neurobehavioral dysfunction are present.

REFERENCES

1. Johnson BL, Baker EL, El Batawi M, et al. Prevention of neurotoxic illness in working populations. New York: John Wiley & Sons, 1987.
2. Feldman RG. Occupational and environmental neurotoxicology. Philadelphia: Lippincott-Raven, 1999.
3. Keogh JP, Pestronk A, Wertheimer D, Moreland R. An epidemic of urinary retention caused by dimethylaminopropionitrile. JAMA 1980;243:746–749.
4. Allen N, Mendell JR, Billmaier DJ, Fontaine RE, O'Neill J. Toxic polyneuropathy due to methyl *n*-butyl ketone. Arch Neurol 1975;32:209–212.
5. Spencer PS, Schaumberg HH, eds. Experimental and clinical neurotoxicology. Baltimore: Williams & Wilkins, 1980.
6. Gerr F, Letz R. Solvents. In: Rom WN, ed. Environmental and occupational medicine, 2nd ed. Boston: Little, Brown, 1994.
7. Morton HG. Occurrence and treatment of solvent abuse in children and adolescents. Pharmacol Ther 1987;33:449–469.
8. White RF, Feldman RG, Proctor SP. Neurobehavioral effects of toxic exposure: clinical symptoms in adult neuropsychology. In: White RF, ed. The practitioner's handbook. Amsterdam: Elsevier, 1992:1–51.
9. Baker EL, White RF, Pothier LJ, et al. Occupational lead neurotoxicity: improvement in behavioral effects after exposure reduction. Br J Ind Med 1985;42:507–516.
10. World Health Organization and Nordic Council of Ministers. Chronic effects of organic solvents on the central nervous system and diagnostic criteria. Copenhagen: WHO Regional Office for Europe, June 1985.
11. Cranmer JM, Goldberg L. Workshop on neurobehavioral effects of solvents. Neurotoxicology 1987;7:1–95.
12. White RF, Proctor SP. Solvents and neurotoxicity. Lancet 1997;349:1239–1243.
13. Lundberg I, Hogstedt C, Liden C, Nise G. Organic solvents and related compounds. In: Rosenstock L, Cullen MR, eds. Textbook of clinical occupational and environmental medicine. Philadelphia: WB Saunders, 1994.
14. New Zealand Department of Labor, Occupational Safety and Health Service. Chronic organic solvent neurotoxicity: diagnostic criteria. Wellington, New Zealand: Occupational Safety and Health Service, Department of Labor, 1992.
15. Mikkelsen S, Jorgensen M, Browne E, Gyldenstead C. Mixed solvent exposure and organic brain damage. Acta Neurol Scand 1988;78(Suppl 118):1–96.

BIBLIOGRAPHY

Baker EL, Feldman RG, French JG. Environmentally related disorders of the nervous system. Med Clin North Am 1990;74:325–345.
A review of neurologic disorders caused by clinical and physical factors encountered in the workplace or the general environment.

Chang YC. An electrophysiological follow-up of patients with *n*-hexane polyneuropathy. Br J Ind Med 1991;48:12–17.
A report of an electroneurographic and evoked-potential study of 11 Taiwanese printers that showed persistent abnormalities of PNS and CNS electrophysiology even though most workers regained motor and sensory functions.

Cherry N, Gautrin D. Neurotoxic effects of styrene: further evidence. Br J Ind Med 1990;47:29–37.
A report of a study of 70 Canadian factory workers that showed neurobehavioral and neurologic effects of chronic styrene exposure.

Cranmer JM, Goldberg L, eds. Proceedings of the Workshop on Neurobehavioral Effects of Solvents. Neurotoxicology 1987;7:1–95.
A comprehensive review of solvent-related health effects developed by a workshop of international experts.

Feldman RG. Occupational and environmental neurotoxicology. Philadelphia: Lippincott-Raven, 1999.
A new, authoritative textbook with exclusive clinical descriptions.

He FS, Zhang SL, Wang HL, et al. Neurological and electroneuromyographic assessment of the adverse effects of acrylamide on occupationally exposed workers. Scand J Work Environ Health 1989;15:125–129.
A study of 71 Chinese acrylamide workers that showed PNS and cerebellar dysfunction.

Heyman A, Pfeiffer JB, Willett RW, Taylor HM. Peripheral neuropathy caused by arsenical intoxication. N Engl J Med 1956;254:401.
Largest series of cases of arsenical neuropathy. Careful discussion of prognosis and treatment.

Johnson, BL, Baker EL, El Batawi M, et al, eds. Prevention of neurotoxic illness in working populations. New York: John Wiley & Sons, 1987.
A comprehensive overview of neurotoxic illness.

Kurland LT, Faro SN, Siedler H. Minamata disease. World Neurol 1960;1:370.
Extensive discussion of historic outbreak of methyl mercury poisoning.

Landrigan PJ. Current issues in the epidemiology and toxicology of occupational exposure to lead. Toxicol Ind Health 1991;7:9–14.
An authoritative update on health effects of occupational lead exposure.

Linz DH, Barrett ET, Pflaumer JE, Keith RE. Neuropsychological and postural sway improvement after Ca EDTA chelation in mild lead intoxication. J Occup Med 1992;34:638–648.

An interesting case report illustrating the use of neurodiagnostic techniques in evaluating occupational lead intoxication.

Namba T, Nolte CT, Jackrel J, Grob D. Poisoning due to organophosphate insecticides: acute and chronic manifestations. Am J Med 1971;50:475.

A clinical review with excellent discussion of treatment.

Reels HA, et al. Assessment of the permissible exposure level to manganese in workers exposed to manganese dioxide dust. Br J Ind Med 1992;49:25–34.

A report of a comprehensive epidemiologic study of 92 exposed workers combined with an assessment of the PEL for manganese.

Sharp DS, Eskenazi B, Harrison R, Callas I, Smith AH. Delayed health hazards of pesticide exposure. Annu Rev Public Health 1986;7:441–471.

A review of the delayed health effects of organophosphate insecticides and related compounds.

Snoeij NJ, Penninks AH, Seinen W. Biological activity of organotin compounds: an overview. Environ Res 1987;44:335–353.

A review of organotin neurotoxicity.

Spencer PS, Schaumberg HH, eds. Experimental and clinical neurotoxicology. Baltimore: Williams & Wilkins, 1980.

In-depth discussion of the pathophysiology of neurotoxic-induced disease.

Taylor JR, Selhorst JB, Houff SA, Martinez AJ. Chlordecone intoxication in man. Neurology 1978;28:626.

A complete description of a severe outbreak of occupational neurologic disease.

Vinken PJ, Bruyn GW, eds. Handbook of clinical neurology: intoxications of the nervous system, parts I and II. Vols 36 and 37. Amsterdam: Elsevier North-Holland, 1979.

A collection of comprehensive monographs on various neurotoxins. An excellent reference work.

White RF, Proctor SP. Research and clinical criteria for development of neurobehavioral test batteries. J Occup Med 1992;34:140–148.

A comprehensive discussion of the issues related to the use of neurobehavioral testing in the evaluation of workers at risk for toxic encephalopathy.

World Health Organization, Nordic Council of Ministers. Chronic effects of organic solvents on the central nervous system and diagnostic criteria. Copenhagen: WHO Regional Office for Europe, 1985.

An important summary of a consensus WHO workshop on the nature of organic solvent neurotoxicity.

30 Psychiatric Disorders

Nancy F. Fiedler and Elise Caccappolo

The impact of psychiatric disorders in the workplace can be approached from several perspectives, thus raising a number of questions. From the public health perspective, are there occupations that produce psychiatric illness regardless of individual characteristics, or do vulnerable individuals develop psychiatric disorders in response to work conditions that most can tolerate? Although a large literature explores stress in the workplace (see Chapter 21), much less is known about the relation between specific work factors and the incidence of psychiatric disorders (1). From the industrial/organizational perspective, what are the direct and indirect costs to the organization and how can the organization deal with affected workers? For example, psychiatric disorders may impair productivity through several avenues, such as reduced labor supply, absenteeism, poor morale, and reduced quality and quantity of work. For workers with psychiatric disorders, what is the responsibility of the workplace to prevent, treat, or accommodate the disabilities that may accompany psychiatric illness? Although work is acknowledged as being essential for mental health, defining those characteristics of the work environment that promote mental health has received relatively less attention than work stress.

N. F. Fiedler, E. Caccappolo: Environmental and Occupational Health Sciences Institute, Robert Wood Johnson Medical School, University of Medicine and Dentistry of New Jersey, Piscataway, NJ 08855

In this chapter, an overview of the prevalence of mental disorders in various occupations is presented along with discussion of the management of these disorders in the workplace. Work-related risk factors for mental disorders are reviewed, as are the data documenting the impact on productivity of various psychiatric disorders. Finally, worksite intervention programs, including employee assistance programs (EAPs) and accommodations arising out of the Americans with Disabilities Act (ADA) are discussed (see also Chapter 12).

PSYCHIATRIC DISORDERS IN THE WORKPLACE

Few epidemiologic studies of the prevalence of psychiatric disorders in the workplace exist. One strategy has been to identify occupations with higher rates of admission to psychiatric treatment facilities (2). For example, Sauter et al. (3) cited several studies documenting higher rates of suicide and hospital or mental health center admissions among health care workers. However, increased risk in these professions could be confounded by demographic differences (e.g., gender) or by variability in the awareness and acceptability of mental health issues rather than work-related differences in the incidence of psychiatric disorders.

More recent studies used structured interview data from the Epidemiologic Catchment Area community survey to diagnose psychiatric disorders according to the criteria

(Drawing by Nick Thorkelson.)

of the *Diagnostic and Statistical Manual—III* (DSM-III) of the American Psychiatric Association and controlled for demographic differences. Of the 104 occupational groups included, lawyers, teachers and other counselors (excluding college), and secretaries had elevated rates of major depression compared with the overall rate for all employed persons (4). Movers (freight, stock, and material); workers in transport and material-moving occupations; handlers, equipment cleaners, and laborers; janitors and cleaners; and waiters and waitresses had elevated rates of alcoholism (5). The relative odds of an alcoholism diagnosis increased for managers and construction laborers when they were unemployed rather than employed. Another study of white collar workers found no association between prevalence of major depression or alcohol abuse or dependence and work-related variables such as managerial or professional status, length of employment, hours worked per day, and supervisory responsibility. Rather, non–work-related factors (e.g., marital status) were of significance for both diagnoses. However, a nationwide

study of Swedish men found that a combination of low work control, low work demands, and low work social support was related to later alcoholism after controlling for other risk factors (6).

The mechanisms for higher rates of psychiatric disorders in some occupations are unclear. Status at work appears to be an important risk factor: employees in lower positions demonstrate higher rates of psychiatric disorders than those in higher positions (7). Workers in lower-status positions often experience job demands with little control over decisions; these work characteristics are associated with job strain (8,9). Although workload is perceived to be a risk factor for psychiatric illness, it is often difficult to separate workers' perceptions of job characteristics from individual negative affective traits; for example, depressed employees probably perceive their workload as heavier than healthier workers do. Moreover, high workload is not consistently associated with increased risk for all psychiatric disorders. For example, both Mandell et al. (5) and Hemmingsson and Lundberg (6) reported greater risk of alcoholism with lower or no work load. Other psychosocial factors that may contribute to stress and psychiatric disorders among workers include conflicting demands and poor social support, both of which may have more of a detrimental effect than high work pace (see Chapter 21) (9,10). In fact, Stansfeld et al. (9) found that both high work social support and skill discretion were protective against absence due to psychiatric illness. Social support consisted of high levels of support from colleagues and supervisors coupled with clear and consistent information from supervisors, and skill discretion referred to job variety and the opportunity to use skills at work.

The association of occupations with psychiatric disorders seems to vary depending on the method used to assess the disorder, the demographic characteristics of the sample surveyed, and employment status at the time of the survey. Optimally, studies involving a large cross-section of occupations and assessment not only of occupational categories but also of work and nonwork factors, such as hours worked and family duties, provides more precise targeting of those factors and occupations in which increased risk occurs and prevention programs are needed.

WORKPLACE HAZARDS AND PSYCHIATRIC DISORDERS

Chemical Hazards

Studies have documented psychiatric symptoms in response to occupational exposures to neurotoxicants such as lead, organic solvents, carbon monoxide (CO), and mercury (see Chapters 15 and 29). Lead exposure, whether acute or chronic, may result in nonspecific symptoms often found in psychiatric disorders, such as fatigue, decreased libido, restlessness, and depression (11,12). Exposure to organic solvents has resulted in a variety of psychiatric symptoms, ranging from mild mood disturbances to severe psychoses. Acute solvent exposure is most often

followed by mood changes (13), transient euphoric reactions, or complaints of mental confusion, whereas long-term exposure can lead to symptoms of posttraumatic stress disorder (14), somatoform disorder (15), schizophreniform disorder (16), and panic disorder (17). The organic affective syndrome, identified by the World Health Organization and by the National Institute for Occupational Safety and Health (NIOSH) (18,19), is associated with chronic solvent exposure and is characterized by symptoms such as irritability, poor concentration, and loss of interest— symptoms also seen in several psychiatric disorders.

Sources of CO that may cause poisoning include exhaust fumes from motor vehicles, inhaled smoke, malfunctioning heating systems (20), and cleaner fuels such as propane and methane (21). Psychological symptoms of CO poisoning include fatigue, apathy, emotional lability accompanied by lowered frustration tolerance, impulsivity, irritability (22), and, at times, psychosis (23). Psychological symptoms are often delayed, occurring anywhere from 3 to 240 days after recovery from acute intoxication (24). Approximately 50% to 75% of exposed patients recover from this delayed syndrome within 1 year (25).

Mercury is well known for the clinical syndrome seen in the late 19th century in European hatting industry which was characterized by the phrase, "mad as a hatter." Significant exposure to mercury, which occurred when hatmakers used mercury to process felt, was found to result in chronic depressed mood with apathy and social withdrawal (26). Chronic mild exposures may lead to irritability, nervousness, fatigue, and depression (27).

Despite the reported associations between psychiatric symptoms and neurotoxicant exposure, the Epidemiologic Catchment Area study did not report higher prevalence rates of depression among workers in occupations expected to have greater chemical exposures, such as construction laborers and metal workers (4). The causal relation between neurotoxic exposure and psychiatric symptoms is unclear. Because cognitive deficits and various physical symptoms are also associated with exposure, it is possible that recognition of these impairments may lead to reactive depression. Alternatively, it has been suggested that psychiatric responses may reflect central nervous system dysfunction within the frontal, temporal, and limbic regions resulting directly from exposure. Additional work must be done to investigate the pathogenesis of psychiatric symptoms, such as depression, that result from work-related neurotoxicant exposures.

Physical and Psychosocial Hazards

Exposure to physical agents within the workplace may also lead to psychiatric symptoms. For example, factory workers exposed to high levels of noise demonstrated depressive symptoms such as insomnia, anxiety, and weight loss (see Chapter 18) (28). Shift work represents another source of potential psychological symptoms for employees (see Chapter 22). Little evidence exists regarding a causal relation between shift work and psychiatric disorders. Nonetheless, shift workers report lower subjective levels of physical health and well-being (29–31). In addition, they have higher rates of alcohol and substance abuse compared with daytime workers (32) and high rates of neuroticism (33,34). At present, studies investigating the prevalence of depression in shift workers are contradictory and inconsistent, reflecting the necessity for further research to determine more thoroughly the potential psychological consequences involved with shift work.

The proliferation of conditions characterized by nonspecific symptoms has expanded the range of potential mental health concerns within the workplace. For example, a number of workers who rely on video display units have reported hypersensitivity to electromagnetic fields and radio-frequency waves, with symptoms such as eye irritation, headache, fatigue, dermatologic problems (including itching, redness, and swelling), memory impairment, nausea, and mucosal irritation (35). It is difficult to determine the cause of nonspecific symptoms and whether they are from an occupational source. Factors that may influence nonspecific symptom reporting, in addition to hazardous workplace exposures, include personality style, individual attitudes and belief systems, premorbid psychiatric status, and social pressures such as employees' perceptions of managers' competence (36). Likewise, the tendency to experience negative, distressing emotions (neuroticism) has been associated with symptomatic responses to poor indoor air quality and with the clinical syndrome known as multiple chemical sensitivity (37–39) (Chapter 23). These clinical examples serve to highlight the complex interaction between occupational exposure and individual psychological and physical susceptibility.

Trauma and the Workplace

A range of traumatic events can occur in the workplace and may have significant economic and psychological consequences (see Chapters 8 and 24). Traumatic events include death or serious injury to self or others and workplace violence, ranging from threats and harassment to murder. For example, NIOSH data reveal that homicide is the second leading cause of occupational death, after work-related motor vehicle accidents, and surpasses machine-related deaths (40). Workplace factors—such as exchanging money, interacting with the public, working at night or in the early morning, delivering goods, and working alone—place employees at high risk for violent attacks (40).

Controlling the risk of psychological trauma in the workplace presents a challenge. Research has presented guidelines for identifying potentially violent employees within the workplace. There is no specific profile of a violence-prone employee, but there are possible early warning signals of a potentially violent employee, including paranoid behavior, desperation over recent personal problems, an inability to accept criticism of job performance, blaming others for problems, and direct or subtle threats of harm (41). The work environment may increase the risk of workplace violence; for example, an environment in which the dignity of the employees is not respected (e.g., one with frequent invasions of privacy or high levels of secrecy) or an authoritarian management style may play a role in leading a potentially violent employee to commit a violent act.

A potential consequence of trauma in the workplace is posttraumatic stress disorder (PTSD). PTSD occurs after exposure to a traumatic event involving threatened death or serious injury to self or others (e.g., witnessing the death of a coworker). The individual's response to such an event involves feelings of intense fear, helplessness, and horror. Symptoms of PTSD include intrusive thoughts or dreams and recollections of the trauma, reexperiencing of the trauma, and avoidance of stimuli that arouse recollection of the trauma. PTSD was first recognized among veterans of the Vietnam War, a unique occupational setting. The extreme traumas of war are not often seen on such a scale within nonmilitary workplaces, although workers in certain occupations, such as police personnel, firefighters, and emergency medical technicians, are at high risk for psychological trauma (42).

MANAGEMENT OF PSYCHIATRIC DISORDERS IN THE WORKPLACE

Employee Assistance Programs

In recognition of the impact that untreated psychiatric disorders have on productivity and morale in the workplace, more than 20,000 U.S. companies offer EAPs to detect and treat employees with psychiatric disorders (43). Initially, EAPs were developed in response to alcohol-related problems, but it became clear that personal problems beyond substance abuse could also interfere with work performance. Moreover, early detection, before recognition of impaired work performance by a supervisor, is preferable. Therefore, EAPs broadened their scope to include evaluation and referral services for personal problems such as marital, individual psychiatric, and financial issues. Where broad-brush programs are offered, employees are encouraged to seek services on their own or as self-referrals rather than waiting until their job performance suffers. Table 30-1 provides a description of the primary elements and options available as part of an EAP.

Despite the proliferation of EAPs and widespread claims that EAPs are cost-effective, few data are available to address this issue. Evaluation of broad-brush EAPs by several investigators has revealed significant improvements in indicators such as absenteeism, lost time, warnings, and supervisor ratings of performance (44–48) after institution of counseling services. As managed care has increased, EAPs are being asked not only to evaluate and refer troubled employees but also to act as case managers and gatekeepers for utilization of mental health benefits.

Psychiatric Treatment and Productivity

From an organizational perspective, psychiatric disorders impair productivity through several avenues, including reduced labor supply, absenteeism, poor morale, and reduced quality of work. Statistics validating the overall economic impact of mental or psychiatric disorders within the United States estimate that depression alone annually costs $43.7 billion (49). This amount includes the costs of lost productivity and health care. Although the economic burden of alcoholism is frequently cited, in a random undiagnosed sample of employees, drinking behavior (e.g., coming to work hung over) was also associated with problems at work (50,51). In short, the costs of behavioral problems and psychiatric disorders affect not only workers and their families but also managers, employers, and insurance companies.

An extensive literature documents the efficacy of psychiatric treatment. Yet, there is a paucity of literature that specifically addresses the effects of treatment on work-

TABLE 30-1. *Elements of employee assistance programs (EAP)*

Program Type	Eligible Participants	Referral to EAP
Internal: EAP staff are company employees External: outside consultant or organization provides EAP	Employees only Employees and eligible dependents	Self-refer; voluntary Supervisor referral: voluntary or involuntary based on documented poor job performance
Problem Type	**Service Type**	**Supervisor Training**
Substance abuse or dependence "Broad brush": substance abuse, psychological, marital, financial, elder care	800-number telephone evaluation One to three evaluation sessions; referral recommendation; crisis intervention Short-term treatment (e.g., up to 10 sessions)	Documentation of work performance and referral procedures Prevention (e.g., stress management skills)

related variables. Mintz et al. (52) reviewed the literature addressing the effect of psychiatric treatment on the capacity to work for those diagnosed with drug addiction, alcoholism, anxiety or affective disorders, gambling, or schizophrenia and concluded that most attention has been given to work outcomes for substance abusers. These authors found that long-term treatment was not more beneficial than standard alcohol treatment regimens and that successful treatment aimed at reducing abusive drinking increased productivity (52). In a separate review, data on occupational outcomes from 10 treatment studies for depression were analyzed. Symptom improvements occurred more rapidly than did improvements in work-related variables such as missed time, lower productivity, and interpersonal problems. These improvements were not affected by treatment duration, although work outcomes improved as treatment duration increased, with maximum benefits achieved at 4 to 6 months. For schizophrenia, neuroleptic drugs reduce symptoms, but some studies suggest that they may also adversely affect work capacity by interfering with the learning process (53). Overall, the most striking finding has been the lack of attention in psychiatric outcome research to the effects of treatment on functional work capacity, despite the stated importance of occupational impairment inherent in the DSM-IV criteria for most psychiatric disorders (54).

In sum, psychiatric disorders and associated behavioral problems, such as alcohol consumption, significantly affect productiv-

ity—regardless of their cause or their relation to worksite factors and stressors. From the data available through program evaluation of EAPs or in the general treatment outcome literature, it appears that when employed persons are treated, their work improves. This finding is encouraging and further supports the importance of health insurance benefits that include psychiatric treatment.

Fitness for Duty

When an employee has been out of work because of treatment of psychiatric disorders such as depression or anxiety, or a question arises about the employee's ability to function on the job, a fitness-for-duty evaluation may be requested (see Chapter 12). Fitness for duty is defined as the ability of the individual to perform a job based on the specific job requirements. A detailed understanding of the job duties is required, and this can often be problematic, because job descriptions are not necessarily informative or sufficiently behaviorally oriented. Ancillary information, such as interviews with workers in similar positions or with supervisors, may be needed to understand the essential behaviors expected on the job. Fitness for duty can never be based solely on a psychiatric diagnosis; rather, it must be based on a behavioral analysis of the employee's abilities. Past job performance is the best predictor of future job performance. Further, a global assessment of functioning can be useful as a behavioral guide for the individual's current level of functioning and ability to perform daily tasks related to work (55). Overall, matching of an assessment of the employee's current behavioral functioning with the essential functions required to perform a job, along with consideration of the employee's premorbid level of function on the job, yields the best prediction of the employee's fitness for return to the job.

Accommodation in the Workplace

Since the ADA was passed in 1990, employers have been under increasing pressure to hire and accommodate workers with disabilities, including psychiatric illness. The number of discrimination claims against employers based on emotional or psychiatric impairment has also increased since the passage of this legislation. For example, in 1997 the Equal Employment Opportunity Commission reported that 15% of discrimination claims were related to emotional or psychiatric impairment—the largest category of claims in that year. The need to properly evaluate an individual's ability to perform a job and the ability to make reasonable accommodations is a growing concern among employers in the United States.

The ADA prohibits discrimination based on disability and provides that employers must make "reasonable accommodations" to the disabilities of "qualified" applicants so long as this does not impose "undue hardship." "Qualified" means that the individual can perform the essential functions of the job, except for the disability. "Reasonable accommodation" refers to any modification or adjustment to a job or work environment that allows the qualified employee with the disability to perform the job functions. "Undue hardship" refers to "an action requiring significant difficulty or expense" (56). Employers are not allowed to inquire about a disability before hiring, and the applicant does not have to reveal a psychiatric history at the time of hire. Moreover, if a long-term employee who was previously performing the job develops a psychiatric disorder, the employer is obligated to make accommodations (57).

For people who are hospitalized for psychiatric diagnoses such as schizophrenia, employment rates have traditionally been low (less than 20%) (57). For those who are chronically mentally ill, the best predictors of future work performance seem to be ratings of work adjustment in a sheltered job site, ability to function socially with others, and previous employment history (58,59). Therefore, type of diagnosis alone (e.g., psychotic versus nonpsychotic) is not as predictive of work capacity as is assessment of objective behavioral performance. Although these findings apply specifically to the psychoses, the same guideline seems to be applicable for any physical or psychiatric illness.

The Mental Health Law Project's guidebook on the ADA (60) provides a helpful document outlining reasonable accommodations for persons with psychiatric disabilities. Accommodations include analysis of the individual employee's behavioral problems (e.g., anxiety, sensitivity to criticism) and development of accommodations based on individual needs. For example, it has been suggested that supervisors be trained to offer positive feedback, along with critiques of performance, to a sensitive employee returning from hospitalization.

In summary, the literature associating occupations or specific work factors with the prevalence of psychiatric disorders is inconclusive. What is clear is that a significant number of individuals within workplaces have diagnosable psychiatric conditions, particularly depression, at any given time, and that some occupations show higher rates of such disorders than others. Moreover, some occupations appear to place workers at greater risk for traumas that result in psychiatric disorders, such as PTSD. Whatever the cause, psychiatric illness will continue to affect the workplace, and therefore must be recognized, treated, and accommodated—rather than dismissed or ignored.

REFERENCES

1. Kasl SV. Surveillance of psychological disorders in the workplace. In: Keita GP, Sauter SL, eds. Work and well-being: an agenda for the 1990s. Washington, DC: American Psychological Association, 1992: 69–95.
2. Colligan MJ, Smith MJ, Hurrell JJ Jr. Occupational incidence rates of mental health disorders. Journal of Human Stress 1977;3:34–39.
3. Sauter SL, Murphy LR, Hurrell JJ Jr. Prevention of work-related psychological disorders: a national strategy proposed by the National Institute for Occupational Safety and Health (NIOSH). In: Keita GP, Sauter SL, eds. Work and well-being. Washington, DC: American Psychological Association, 1992: 17–40.
4. Eaton WW, Anthony JC, Mandel W, Garrison R. Occupations and the prevalence of major depressive disorder. J Occup Med 1990;23:1079–1087.
5. Mandell W, Eaton WW, Anthony JC, Garrison R. Alcoholism and occupations: a review and analysis of 104 occupations. Alcohol Clin Exp Res 1992; 16:734–746.
6. Hemmingsson T, Lundberg I. Work control, work demands, and work social support in relation to alcoholism among young men. Alcohol Clin Exp Res 1998;22:921–927.
7. Hotopf M, Wessely S. Stress in the workplace: unfinished business. J Psychosom Res 1997;43:1–6.
8. Karasek RA. Job demands, job decision latitude, and mental strain: implications for job redesign. Adm Sci Q 1979;24:285–308.
9. Stansfeld SA, Fuhrer R, Head J, Ferrie J, Shipley M. Work and psychiatric disorder in the Whitehall II Study. J Psychosom Res 1997;43:73–81.
10. Hammar N, Alfredsson L, Johnson JV. Job strain, social support at work, and incidence of myocardial infarction. Occup Environ Med 1998;55:548–553.
11. Schottenfeld RS, Cullen MR. Organic affective illness associated with lead intoxication. Am J Psychiatry 1984;141:1423–1426.
12. Eskanazi B, Maizlish N. Effects of occupational exposure to chemicals on neurobehavioural functioning. In: Tarter RE, VanThiel DH, Edward KL, eds. Medical neuropsychology: the impact of disease on behavior. New York: Plenum, 1988:223–263.
13. Johnson BL, Baker EL, El Batawi M, Gilioli R, Hanninen H, Seppalainen AM. Prevention of neurotoxic illness in working populations. New York: John Wiley & Sons, 1987.
14. Morrow LA, Ryan CM, Goldstein G, Hodgson MJ. A distinct pattern of personality disturbance following exposure to mixtures of organic solvents. J Occup Med 1989;31:743–746.
15. Schottenfeld RS, Cullen MR. Recognition of occupation-induced post traumatic stress disorders. J Occup Med 1986;28:365–369.
16. Goldblum D, Chouinard G. Schizophrenia psychosis associated with chronic industrial toluene exposure: case report. J Clin Psychiatry 1985;46:350–351.
17. Dager SR, Holland JP, Cowley DS, Dunner DL. Panic disorder precipitated by exposure to organic solvents in the work place. Am J Psychiatry 1987; 144:1056–1058.
18. National Institute for Occupational Safety and Health. Organic solvent neurotoxicity. Bulletin 48. Washington, DC: US Department of Health and Human Services, NIOSH, 1987.
19. World Health Organization, Nordic Council of Ministers. Chronic effects of organic solvents on the central nervous system and diagnostic criteria. Copenhagen, WHO Regional Office for Europe, 1985.
20. Meredith T, Vale A. Carbon monoxide poisoning. Br Med J 1988;296:77–79.
21. Ely EW, Moorehead B, Haponik EF. Warehouse workers' headache: emergency evaluation and management of 30 patients with carbon monoxide poisoning. Am J Med 1995;98:145–155.
22. Lezak MD. Neuropsychological assessment, 3rd ed. New York: Oxford University Press, 1995.
23. Min SK. A brain syndrome associated with delayed neuropsychiatric sequelae following acute carbon monoxide intoxication. Acta Psychiatr Scand 1986; 73:80–86.
24. Ernst A, Zibrak JD. Carbon monoxide poisoning. N Engl J Med 1998;339:1603–1608.
25. Choi IS. Delayed neurologic sequelae in carbon monoxide intoxication. Arch Neurol 1983;40: 433–435.
26. Maghazaji HI. Psychiatric aspects of methylmercury poisoning. J Neurol Neurosurg Psychiatry 1974; 37:954–958.
27. Gross LS, Nagy RM. Neuropsychiatric aspects of poisonous and toxic disorders. In: Yudofsky SC, Hales RE, eds. American psychiatric press textbook of psychiatry, 2nd ed. Washington, DC: American Psychiatric Press, 1992:541–561.
28. Bing-shuang H, Yue-lin Y, Ren-yi W, Zhu-bao C. Evaluation of depressive symptoms in workers exposed to industrial noise. Homeostasis in Health Disease 1997;38:123–125.
29. Akerstedt T. Psychological and psychophysiological effects of shift work. Scand J Work Environ Health 1990;16:67–73.
30. Frese M, Semmer N. Shiftwork, stress, and psychosomatic complaints: a comparison between workers in different shiftwork schedules, non-shiftworkers, and former shiftworkers. Ergonomics 1986;29: 99–114.
31. Verhaegen P, Dirkx J, Maasen A, Meers A. Subjective health after twelve years of shift work. In: Haider M, ed. Night and shift work: longterm effects and their prevention. Frankfurt Am Main: Verlag Peter Lang, 1986:67–74.
32. Costa G, Apostali P, D'Andrea F, Gaffuri E. Gastrointestinal and neurotic disorders in textile shift workers. In: Reinberg A, Vieux N, Andlauer P, eds. Night and shift work: biological and social aspects. Oxford: Pergamon Press, 1981:187–196.
33. Tasto DL, Colligan MJ, Skjel EW, Polly SJ. Health consequences of shift work. SRI Project URU-4426. Washington, DC: US Department of Health, Education and Welfare, NIOSH, 1978.
34. Harma M, Illmarinen J, Knauth P. Physical fitness and other individual factors relating to the shiftwork tolerance of women. Chronobiol Int 1988;5:417–424.
35. Arnetz BB, Wilhom C. Technological stress: psychophysiological symptoms in modern offices. J Psychosom Res 1997;43:35–42.

36. Spurgeon A, Gompertz D, Harrington JM. Non-specific symptoms in response to hazard exposure in the workplace. J Psychosom Res 1997;43:43–49.
37. Stenberg B, Wall S. Why do women report "sick building symptoms" more often than men. Soc Sci Med 1995;40:491–502.
38. Ryan CM, Morrow KA. Dysfunctional buildings or dysfunctional people: an examination of the sick building syndrome and allied disorders. J Consult Clin Psychol 1992;60:220–224.
39. Fiedler N, Kipen H, DeLuca J, Kelly-McNeil K, Natelson B. A controlled comparison of multiple chemical sensitivity and chronic fatigue syndrome. J Psychosom Med 1996; 58:38–49.
40. National Institute of Occupational Safety and Health. Violence in the workplace: risk factors and prevention strategies. Bulletin 57. Washington, DC: US Department of Health, Education and Welfare, NIOSH, 1997:1–22.
41. Trafford C, Gallichio E, Jones P. Managing violence in the workplace. In: Cotton P, ed. Psychological health In the workplace: understanding and managing occupational stress. Carlton, Australia: The Australian Psychological Society, 1996:147–158.
42. Williams T. Trauma in the workplace. In: Wilson JP, Raphael B, et al, eds. International handbook of traumatic stress syndromes. New York: Plenum Press, 1993:925–933.
43. Adamson DW, Gardner MD. Employee assistance programs and managed care: merge and converge. In: Sauber SR, ed. Managed mental health care: major diagnostic and treatment approaches. Bristol, PA: Brunner/Mazel, 1997:67–82.
44. Cooper CL, Sadri G. The impact of stress counselling at work. Journal of Social Behavior and Personality 1991;6:411–423.
45. Mitchie S. Reducing absenteeism by stress management: valuation of a stress counselling service. Work and Stress 1996;10:367–372.
46. Ramanathan CS. EAP's response to personal stress and productivity: implications for occupational social work. Soc Work 1992;37:234–239.
47. Walsh DC, Hingson RW, Merrigan DM, et al. A randomized trial of treatment options for alcohol-abusing workers. N Engl J Med 1991;325:775–782.
48. Guppy A, Marsden J. Assisting employees with drinking problems: changes in mental health, job perceptions and work performance. Work and Stress 1997;11:341–350.
49. Greenberg PE, Stiglin LE, Finkelstein SN, Berndt ER. The economic burden of depression in 1990. J Clin Psychiatry 1993;54:405–418.
50. Harwood JJ, Kristiansen P, Rachal JV. Social and economic costs of alcohol abuse and alcoholism. Issue Report No. 2. Research Triangle Park, NC: Research Triangle Institute, 1985.
51. Ames GM, Grube JW, Moore RS. The relationship of drinking and hangovers to workplace problems: an empirical study. J Stud Alcohol 1997;58:37–47.
52. Mintz J, Mintz LI, Arruda MJ, Hwang SS. Treat-ments of depression and the functional capacity to work. Arch Gen Psychiatry 1992;49:761–768.
53. Hogarty GE, McEvoy JP, Munetz M, et al. Dose of fluphenazine, familial expressed emotion, and outcome in schizophrenia: results of a two-year controlled study. Arch Gen Psychiatry 1988;45:797–805.
54. American Psychiatric Association. Diagnostic and statistical manual of mental disorders, 4th ed. Washington, DC: American Psychiatric Association, 1994.
55. Sperry L. Psychiatric consultation in the workplace. Washington, DC: American Psychiatric Press, 1993.
56. United States Department of Justice. The Americans with Disabilities Act: questions and answers. Washington, DC: US Department of Justice, Civil Rights Division, 1991.
57. Carling PJ. Reasonable accommodations in the workplace for individuals with psychiatric disabilities. Consulting Psychology Journal 1993;45:46–62.
58. Anthony WA, Jansen MA. Predicting the vocational capacity of the chronically mentally ill. Am Psychol 1984;39:537–544.
59. Massel HK, Liberman RP, Mintz J, et al. Evaluating the capacity to work of the mentally ill. Psychiatry 1990;53:31–43.
60. Mental Health Law Project. Mental health consumers in the work place: how the Americans with Disabilities Act protects you against employment discrimination. Washington, DC: Mental Health Law Projects, 1992.

BIBLIOGRAPHY

Carling P. Reasonable accommodations in the workplace for individuals with psychiatric disabilities. Consulting Psychology Journal 1993;45:46–62.
This article reviews information related to accommodation for persons with psychiatric disabilities and the methods to assist employers and employees in complying with the Americans with Disabilities Act.

Mintz J, Mintz L, Arruda M, Hwang S. Treatments of depression and the functional capacity to work. Arch Gen Psychiatry 1992;49:761–768.
This review summarizes the literature on the effects of antidepressants and psychotherapy treatment on the functional capacity to work.

Sauter SL, Murphy LR, Hurrell JJ. Prevention of work-related psychological disorders. Am Psychol 1990; 45:1146–1158.
This article reviews the scientific literature regarding work stress and its impact on psychological and emotional well-being.

Schottenfeld RS. Psychological sequelae of chemical and hazardous materials exposures. In: JB Sullivan, GR Kreiger, eds. Hazardous materials toxicology: clinical principles of environmental health. Baltimore: Williams & Wilkins, 1992.
This chapter provides an overview of the psychiatric issues that arise as a consequence of exposure to toxic substances.

Occupational Health: Recognizing and Preventing Work-Related Disease and Injury, 4th ed.
Edited by Barry S. Levy and David H. Wegman, Lippincott Williams & Wilkins, Philadelphia © 2000.

31 Reproductive Disorders

Maureen Paul and Linda Frazier

Prevention of reproductive system disorders is an important public health priority. These problems affect both men and women. In the United States, approximately one in seven married couples is involuntarily infertile. Between 10% and 20% of pregnancies end in clinically recognized spontaneous abortion, and rates of very early periimplantation loss are even higher. Among newborns in the United States, approximately 7% are of low birth weight (less than 2,500 g), and 3% have major congenital malformations.

Historically, several dramatic events have highlighted the issue of reproductive hazards. The drug thalidomide, prescribed as an antiemetic and sedative to pregnant women, was linked to limb malformations and other defects in newborns. *In utero* exposure to diethylstilbestrol (DES) resulted in anomalies of the reproductive tract and later development of vaginal cancer in daughters of the treated women. In Japan, contamination of fish with methyl mercury and cooking oil with polyhalogenated biphenyls led to serious development toxicity in children exposed prenatally. These tragedies served to break the prevailing belief that the placenta acted as a protective barrier for the fetus. Moreover, in 1977 occupational exposure to dibromochloropropane (DBCP) was linked to

male infertility (Box 31-1), demonstrating that reproductive toxicity could affect both women and men.

REPRODUCTIVE AND DEVELOPMENTAL TOXICOLOGY

Reproductive processes in humans are complex and incompletely understood. In a precisely regulated hormonal milieu, normal human reproduction proceeds from formation and transport of the germ cells through fertilization, implantation, and prenatal and postnatal development. Toxic agents can act at one or many sites to disrupt this chain of events, resulting in reproductive dysfunction or adverse pregnancy outcomes.

Numerous agents have been shown to have adverse reproductive or developmental effects, although for many substances the available evidence is limited to studies in experimental animals. Most workplace exposure limits were formulated to protect against adverse health outcomes other than reproduction, such as acute toxicity or cancer. Because some research suggests that the reproductive system is more sensitive than other organ systems, it is important that exposure limits formulated in the future take reproductive toxicity into consideration.

Genotoxicity

Genotoxicity can occur in either men or women. Most numeric chromosomal abnormalities (too many or too few chromosomes) are incompatible with survival. Infants born

M. Paul: Planned Parenthood, Boston, Massachusetts 02315.

L. Frazier: Department of Preventive Medicine, University of Kansas School of Medicine, Wichita, Kansas 67214.

Box 31-1. DBCP: A Potent Male Reproductive Toxicant

In 1977, a small group of men in a northern California pesticide formulation plant noticed that few of them had recently fathered children. Investigation of the full cohort of production workers found a strong association between decreased sperm count and exposure to DBCP, a brominated organochlorine that had been used as a nematocide since the mid-1950s. The spermatotoxic effects in some of the exposed men were sufficient to render them sterile. Testicular biopsies showed the seminiferous tubules to be the site of action and spermatogonia to be the target cell. The relation between reduced sperm count and exposure to DBCP, both in its manufacture and in its use, has been confirmed in studies of other plants in the United States and abroad. Follow-up of

workers after cessation of exposure shows that spermatogenic function is eventually recovered in those less severely affected. However, many of the azoospermic men have remained so for many years after cessation of exposure.

Much DBCP has been exported by U.S.-based multinational corporations to many developing countries. A substantial amount of this pesticide was exported even after DBCP was banned in the United States. In most instances, workers exposed to DBCP in developing countries were not informed of its hazards, trained in its use, or provided personal protective equipment to safeguard themselves adequately (Fig. 31-1). In one study of approximately 26,400 DBCP-exposed workers in developing countries who sued US companies, 24% were azoospermic and 40% were oligospermic (1).

A   B

FIG. 31-1. Many workers in developing countries became sterile from exposure to dibromochloropropane (DBCP), even after it was banned in the United States. **A.** Simulation of worker pouring DBCP solution, which he has mixed with a stick in a 55-gallon drum, into an applicator. **B.** Simulation of worker injecting DBCP solution around the roots of a banana tree. (Photographs by Barry S. Levy.)

with chromosomal abnormalities often have physical, behavioral, and intellectual impairments. Structural chromosomal changes may have no adverse effects, or they may be associated with mental retardation, anomalies, reduced fertility, or malignancy.

With the use of bacterial assays, increased

mutagenic activity has been detected in the urine of workers exposed to such substances as anesthetic gases, chemotherapeutic agents, and epichlorohydrin. Increased frequencies of chromosomal aberrations have been reported in radiation workers and in workers exposed to chemicals such as ben-

TABLE 31-1. *Selected occupational agents with suspected effects on male reproductive function*

Adverse effects	Examples[a]
Decreased libido, hormonal alterations	Lead, mercury, manganese, carbon disulfide, estrogen agonists (e.g., polychlorinated biphenyls and organohalide pesticides); workers manufacturing oral contraceptives
Spermatotoxicity[b]	Lead, dibromochloropropane (DBCP), carbaryl, toluenediamine and dinitrotoluene, ethylene dibromide, plastic production (styrene and acetone), ethylene glycol monoethyl ether, welding, perchloroethylene, mercury, heat, military radar, Kepone, bromine, radiation (Chernobyl), carbon disulfide, 1,4-dichlorophenoxy acetic acid (2,4-D).
Spontaneous abortion in partner	Solvents, lead, mercury; workers in rubber and petroleum industries
Altered sex ratio in offspring	Dibromochloropropane (DBCP)
Congenital malformations in offspring	Pesticides, chlorphenates, solvents; firefighters, painters, welders, auto mechanics, motor vehicle drivers, sawmill workers and workers in aircraft, electronics and forestry and logging industries
Neurobehavioral disorders in offspring	Alcohols, cyclophosphamide, ethylene dibromide, lead, opiates
Childhood cancer in offspring	Solvents, paints, pesticides, petroleum products; welders, auto mechanics, motor vehicle drivers, machinists and workers in aircraft and electronics industries

[a] Some human evidence, albeit limited, is available for all examples listed except those associated with neurobehavioral disorders in offspring; animal evidence is available for these paternal exposures.
[b] NIOSH has included these agents on its list of male reproductive hazards (www.cdc.gov/niosh/malerepro.html).

zene, styrene, ethylene oxide, epichlorohydrin, arsenic, chromium, and cadmium. Although these assays are useful as biologic markers of exposure to genotoxicants, they do not predict specific reproductive health effects in individual workers.

The risk of adverse pregnancy outcome after preconception exposure of men to toxic agents is an area of active research (2–7). A number of mechanisms have been postulated for these male-mediated effects, including germ cell mutagenesis. Although findings have yet to be replicated or precise mechanisms clarified, increased rates of pregnancy loss have been reported among the wives of men exposed to lead, inorganic mercury, organic solvents, and other agents (Table 31-1). Some studies suggest that certain paternal occupations pose an increased risk for congenital malformations and childhood cancers, but more research is needed to explore this issue.

Spermatogenesis

Alterations in sperm count or semen quality have been documented for a number of occupational exposures. Occupational agents can disrupt sperm production either directly, by injuring testicular cells, or indirectly, by interfering with the hormonal regulation of spermatogenesis. Toxic agents may also impair sexual function by reducing libido or by inhibiting erection and ejaculation. As long as the stem cell precursors are spared, spermatogenic damage may be reversible over time. This appears to be the case with most substances studied so far, although few substances have been studied thoroughly. Table 31-1 lists some occupational agents that are known or suspected to affect male reproductive function adversely (2–7).

Pesticides. Perhaps the best known spermatoxin is DBCP, the first substance discovered to cause infertility in American workers (see Box 31-1). Although the manufacture of DBCP has been banned in the United States, it remains a low-level groundwater contaminant in some states. Another pesticide, ethylene dibromide, has been associated with post-testicular effects including decreased sperm velocity, motility, and viability.

Heavy Metals. Among the heavy metals, lead is the best-studied spermatotoxin (7). In investigations of workers exposed to lead in battery manufacture, blood lead levels (BLLs) higher than 40 μg/dL have been associated with decreased sperm counts and aberrant sperm motility and morphology. Evidence suggests that lead has a direct toxic effect on the gonads and may also act at the level of the hypothalamus and pituitary to impair endocrine function. Agents that affect the central nervous system or that cause severe debilitation may affect sexual function. For example, decreased libido has been associated with severe lead or manganese poisoning.

Glycol Ethers. The ethylene glycol ethers, 2-methoxyethanol and 2-ethoxyethanol, and their acetates are organic solvents used in multiple industrial applications. These agents cause testicular atrophy and disruption of the seminiferous tubules in several laboratory animal species, and they have been associated with decreased sperm counts in exposed workers (8). The ethylene glycol ethers target meiotic spermatocytes; at high doses, effects on spermatogonia and late spermatids have also been reported. The ethylene glycol ethers are metabolized in the body to alkoxyacetic acids, which are responsible for their reproductive toxicity.

Hormonally Active Compounds. Gynecomastia and decreased libido have been reported in men involved in the manufacture of oral contraceptives. Some chemicals, such as the polyhalogenated biphenyls and organohalide pesticides, are structurally similar to the reproductive sex steroid hormones, raising the possibility that they could disrupt male reproduction by binding to endogenous hormone receptors. Several studies have suggested that average sperm counts in men have fallen during the past several decades, but other studies have not confirmed this finding. More research is needed to clarify the potential role of endocrine-disrupting chemicals on male fertility and reproductive outcomes (9).

Oogenesis

Toxicity to oocytes may occur from occupational exposure received by a woman worker. In addition, exposures received by a female fetus while her mother is working

could theoretically affect her fertility during adulthood. In adults, disturbances in ovulation manifest clinically as infertility or menstrual dysfunction. Ovarian toxicity can result in premature menopause, as tobacco research has demonstrated.

Menstrual disorders have been reported among women in various occupations, including athletes and dancers, agricultural workers, and those formulating oral contraceptives. Reduced fertility has been reported in semiconductor workers and in dental assistants exposed to high levels of metallic mercury vapor or nitrous oxide. The probability of conception in each menstrual cycle was almost 60% lower among women exposed to unscavenged nitrous oxide for 5 or more hours per week than among unexposed women (10).

Pregnancy

With rare exceptions, contemporary studies from industrialized nations reveal better pregnancy outcomes among women in the workforce compared with unemployed women. This finding may be related in part to the healthy worker effect (see Chapter 6) and to the economic and health care benefits derived from the work experience. On the other hand, a number of work exposures may be associated with adverse pregnancy outcomes (Table 31-2) (11).

The preimplantation phase of development is often referred to as the "all-or-none" period. This terminology derives from studies showing that sufficiently high doses of ionizing radiation may cause death of the conceptus but sublethal doses are unlikely to result in teratogenic effects because of effective cellular repair processes.

The embryo (from the 17th to the 56th day after conception) is acutely sensitive to teratogenic insult. The second and third trimesters are marked by significant growth of the conceptus and by the continued differentiation and maturation of some organ systems. Therefore, exposure to toxic agents after the first trimester can still cause prob-

TABLE 31-2. *Selected occupational agents with suspected effects on pregnancy*

Agent (illustrative exposures)	Reported effects	Recommendations
Physical agents		
Strenuous work (standing >6 hr/wk per shift, working >40 hr/wk)	Preterm delivery. Women with medical or obstetric conditions predisposing to preterm delivery may be at particular risk	Lifting aids; limited duty options; rest breaks; preterm delivery prevention programs
Ionizing radiation (x-rays, radionucleotides such as P32)	At high doses, growth deficits, CNS malformations, mental retardation; possible low risk of genetic defects, childhood cancer at doses <5 rem.	Exposure monitoring; shielding; minimize time of exposure; increase distance from source; occupational limit <0.5 rem total dose during pregnancy
Noise	Possible fetal hearing loss (beyond the fifth month of pregnancy), possible increased risk for preterm birth	ACGIH has proposed that risk to fetal hearing may occur at >115 dBC 8-hr TWA or at peak exposure of 155 dBC
Electromagnetic fields (fields derived from flowing electric currents)	No reproductive effects from fields generated by video display terminals; further research is needed on exposures of very high field strength.	Most exposures do not exceed recommended limits; higher exposures found in certain electrical utility jobs or worksites with very large amounts of flowing electrical current
Chemical agents		
Heavy metals	Neurobehavioral deficits in infants (with prenatal lead exposure, deficits reported at cord blood lead levels as low as 10–20 μg/dL)	Standard hierarchy of controls; air and blood lead monitoring; medical removal protection
Organic solvents (glycol ethers, toluene, xylene)	Spontaneous abortion; fetal loss rates increased in semiconductor workers exposed to EGEE/EGME; modestly increased risk of birth defects for mixed solvent exposure; toluene abuse (fetal solvent syndrome)	Standard hierarchy of controls; exposure monitoring; OSHA proposed limits for EGEE/EGEEA, 0.5 ppm, and for EGME/EGMEA, 0.1 ppm of special concern is exposure causing eye, respiratory, or neurologic symptoms
Antineoplastic agents (cis-platin, doxorubicin, fluorouracil, methotrexate)	Spontaneous abortion	Follow OSHA *Work Practice Guidelines for Personnel Dealing with Cytotoxic Drugs*
Other pharmaceuticals (antivirals such as ribavirin; estrogenic or antiestrogenic compounds such as tamoxifen; immunosuppressive agents such as cyclosporine)	Spontaneous abortion, sperm effects in male animals and teratogenesis in female animals have been noted at high doses	Follow procedures used for antineoplastic agents
Carcinogens and mutagens (ethidium bromide, aflatoxin B_1)	Human data limited; sperm effects noted in male animals, and teratogenesis and cancer in offspring of exposed female animals	Reduce exposure levels through standard hierarchy of controls
Waste anesthetic gases (nitrous oxide), [N_2O]	Spontaneous abortion	Standard hierarchy of controls; air monitoring; NIOSH recommends that exposures to N_2O not exceed 25 ppm.
Sterilants and disinfectants (ethylene oxide, formaldehyde)	Spontaneous abortion	Standard hierarchy of controls; exposure monitoring; NIOSH recommended limit for ethylene oxide, <0.1 ppm.
Polychlorinated biphenyls (PCBs) (chlorodiphenyls, chlorobiphenyls)	Congenital PCB syndrome at high doses; excreted efficiently into breast milk; low-level dietary exposure related to mild neonatal growth and neurobehavioral deficits in some studies	FDA tolerance limit: 1.5 ppm in cow's milk and dairy products; exposure risk for workers who repair PCB machinery; measure serum levels when excessive exposure is suspected

(*continued*)

TABLE 31-2. *Continued*

Agent (illustrative exposures)	Reported effects	Recommendations
Pesticides (organochlorines such as lindane; organo-phosphates such as chlorpyrifos; *n*-methyl carbamates such as carbaryl; fungicides such as benomyl; herbicides such as 2,4-D)	Both male and female reproductive effects have been noted in animal studies for a number of pesticides; there are also some positive studies in workers; review data for each compound.	Follow EPA minimum field re-entry intervals after pesticide application; standard hierarchy of controls; biologic monitoring when available (e.g., blood cholinesterase assays)
Biologic agents		
Hepatitis B virus (HBV)	Neonatal carrier state; chronic liver disease and mortality	Universal screening of pregnant women; vaccine (not contraindicated during pregnancy); postexposure HBIG for susceptible workers
Human immunodeficiency virus (HIV)	Morbidity and mortality for infected pregnant women and neonates	Universal precautions; certain anti-HIV medications should be given during pregnancy if mother is infected
Cytomegalovirus (CMV)	Neonatal death, malformations, developmental deficits; seroconversion rates in health care workers using adequate precautions not increased compared with community controls	Universal precautions; serology
Rubella virus	Spontaneous abortion, stillbirth, congenital defects with infection during first 16 wk of pregnancy	Isolation precautions; document immunity by serology preplacement; vaccine (>3 mo before pregnancy); susceptible workers should not care for patients with rubella infection
Varicella-zoster virus	Serious maternal pneumonia; malformation risk approximately 5% for infection in first half of pregnancy; neonatal morbidity/mortality if maternal infection <5 days before or <2 days after delivery	Isolation precautions; vaccine (>3 mo before pregnancy); limit patient contact based on history of immunity—not pregnancy *per se*. Post-exposure VZIG prophylaxis for susceptible workers
Human parvovirus B19	Nonimmune fetal hydrops; fetal death	Serology; maternal serum α-fetoprotein and ultrasounds in B19 immunoglobulin M–positive women; community viral reservoir, so removal of susceptibles may lessen, but not eliminate, risk of infection; workers may return to job 21 days after last reported case

ACGIH, American Conference of Governmental Industrial Hygienists; FDA, United States Food and Drug Administration; TWA, time-weighted average; EGEE/EGEEA, ethylene glycol monoethyl ether and its acetate; EGME/EGMEA, ethylene glycol monomethyl ether and its acetate; HBIG, hepatitis B immune globulin; VZIG, varicella-zoster immune globulin.

lems. Certain exposures may reduce fetal growth, result in functional or neurobehavioral abnormalities in offspring, or increase the risk of pregnancy complications such as preeclampsia or preterm birth.

Transfer of chemicals across the placenta occurs primarily by passive diffusion. Chemicals that are lipophilic and of low molecular weight cross the placenta readily. Clinical manifestations of developmental toxicity depend on the properties of the agent, the timing and dose of exposure, genetic susceptibility, and other factors. Some agents, such as thalidomide and DES, affect the embryo at doses far below those that induce maternal toxicity. Other agents, such as methyl mercury and cyclophosphamide, are harmful to the conceptus only at doses that are toxic to the mother.

Lead. Low-level exposure to lead *in utero* can cause subtle neurobehavioral deficits during the early years of life (Fig. 31-2). The Occupational Safety and Health Administration (OSHA) lead standard recommends that BLLs of prospective parents and the fetus or newborn not exceed 30 μg/dL. Cognitive deficits have been noted, however, with prenatal exposures as low as 10 to 20 μg/dL (as measured by cord BLLs at birth). Follow-up studies suggest that these developmental delays do not persist in school-aged children unless the postnatal BLL remains elevated. Associations with preterm delivery, fetal growth deficits, and minor

FIG. 31-2. Pregnant workers in seemingly clean work environments may be exposed to reproductive hazards. This quality control worker in a Hungarian leaded-glass factory licks her fingers between inspections of glasses that may have lead dust on them. (Photograph by Barry S. Levy.)

malformations are less consistent. Although lead can induce spontaneous abortion at high doses, a study of women residing near a lead smelter in the former Yugoslavia found no increased risk of spontaneous abortion with BLLs ranging from 5 to 40 μg/dL (12).

Solvents. In a study of semiconductor workers, an elevated risk of spontaneous abortion was found among female fabrication workers exposed to photoresist/developer solutions containing glycol ethers, xylene, or *n*-butyl acetate (13). Other studies have noted increased rates of miscarriage and congenital malformations among solvent-exposed women (14–16). Symptomatic exposure appears to predict higher risk for malformations (16).

Pesticides. Although many of the highly toxic chlorinated hydrocarbon pesticides have been banned from production or use in the United States, these chemicals persist as environmental contaminants and are still in widespread use in developing nations. Information on the effects on human developmental of the approximately 21,000 pesticides registered for use in the United States is sparse. Many pesticides are mutagenic or teratogenic in laboratory animals. Limited epidemiologic investigations have reported increased risks for spontaneous abortion, stillbirth, or birth defects among farm workers and residents living in communities with high pesticide use.

Strenuous Work. Evidence suggests that women who perform strenuous work may be at increased risk of preterm delivery (17). Although definitions of strenuous work differ in these studies, findings are most consistent for women whose jobs involve prolonged standing (more than 5 to 6 hours per shift), long work shifts (more than 8 hours), or long work weeks (more than 40 hours). Physically stressful postures and heavy lifting or carrying are risk factors in some studies. If two or more ergonomic risk factors are present, risk appears to increase.

Many women can tolerate physical job demands during pregnancy, especially if they are physically fit before pregnancy. Women who have predisposing risk factors for early delivery, such as an incompetent cervix or a history of preterm birth, may benefit from job modifications such as avoidance of heavy lifting, reduced hours, or reduced time standing during the third trimester. In addition, the patient should be counseled about the signs and symptoms of preterm labor and should have periodic cervical examinations to monitor for cervical dilatation.

Advanced pregnancy may modify physical work capacity. Therefore, work changes may be indicated during the second half of pregnancy for women whose jobs involve prolonged sedentary postures or tasks that require delicate balance, such as heavy lifting or climbing. Pregnant workers may also be intolerant of prolonged standing or strenuous work in hot, humid environments. Reduction of cardiac output created by decreased venous return, coupled with peripheral vasodilation to dissipate the heat generated by both the fetus and the employee's own increased metabolic rate, may lead to dizziness or syncope in these settings.

Biologic Agents. Biologic agents present hazards to a wide range of workers, including health care workers, veterinarians, public safety personnel, hazardous waste workers, day care workers, and school teachers. Rubella virus, cytomegalovirus, and *Toxoplasa* pose a teratogenic risk, whereas human immunodeficiency virus, hepatitis B virus, and perinatal varicella infection may result in significant infant morbidity and mortality. In addition, infection during pregnancy with human parvovirus B19, the etiologic agent of fifth disease, has been associated with nonimmune fetal hydrops and pregnancy loss. (See also Chapter 20.)

Ionizing Radiation. Depending on the gestational timing of exposure, high-dose ionizing radiation can cause death of the embryo, growth retardation, or birth defects. The most sensitive window of vulnerability for induction of mental retardation is between about 8 and 15 weeks of gestation. Prenatal x-ray exposures in the range of 2 rem have also been associated with a modest

increased risk for childhood leukemia. Although patients may be very concerned about working near radiation sources during pregnancy, usual occupational exposures are far below those expected to produce adverse effects. The Nuclear Regulatory Commission stipulates that the total dose to the embryo or fetus from occupational exposure of a declared pregnant woman not exceed 0.5 rem. (See Chapter 17.)

EVALUATION AND CONTROL OF PERSONAL RISK

Steps in the Clinical Work-up

Common clinical situations that require knowledge of occupational reproductive hazards include preconception counseling, evaluation of the infertile couple, and assessment of workers who are pregnant or who have experienced an untoward pregnancy outcome. In all of these situations, it is essential to answer the following four questions:

To What Agents Is the Patient Potentially Exposed?

Because reproductive disorders have multiple causes, the medical history, work history, and history of environmental exposures all help the clinician to assess the reproductive risk profile of a given worker (Chapters 3 and 4). This information should be gathered for both the male and the female partner. Physical as well as chemical and biologic exposures should be noted. No physical findings are pathognomonic for work-related reproductive disorders; however, the examination may identify signs of exposure (e.g., dermatitis in a solvent-exposed worker) or pathology contributing to the problem (e.g., uterine fibroids in a woman with menstrual irregularity).

Is the Patient Actually Exposed to the Agents and, If So, What Are the Timing and Dose of Exposure?

Working with an agent is not necessarily the same as being exposed to the agent. Through the occupational history, the clinician can estimate whether the likelihood of internal body exposure to the agent is high, low, or negligible. The history can help to determine whether exposures are episodic or chronic in nature. Use of protective measures at work, such as engineering controls or personal protective equipment, may have reduced the exposure (Chapters 5 and 7). Because teratogens exert their effects during specific critical periods of organogenesis, every effort should be made to establish gestational age at the time of exposure precisely. Abnormal semen parameters should prompt a search for gametotoxic exposures occurring several months before the onset of the problem, since spermatogenesis takes about 2 months to complete. Occasionally, exposures in the more distant past are important. For example, 90% of absorbed lead is stored in bone; conditions that increase bone turnover, such as menopause and perhaps pregnancy, may increase the BLL.

Worksite walk-throughs, the assistance of industrial hygienists, and the use of available biologic markers can greatly enhance understanding of actual exposures. However, in only a few instances in the United States have occupational standards been established with reproductive risks in mind (lead, ethylene oxide, DBCP, and ionizing radiation). Therefore, exposure data should not be considered simply in terms of whether regulatory limits are exceeded (18). The recommended exposure limits (RELs) promulgated by the National Institute for Occupational Safety and Health (NIOSH) do consider available data on reproductive and developmental effects, but they are not updated often.

Is There Evidence to Suggest That the Agents Cause Adverse Reproductive or Developmental Effects?

Even after a comprehensive history has been taken, questions frequently remain about the precise identity of chemicals handled on the job. Material safety data sheets (MSDSs)

contain essential information about hazardous product ingredients and should be carefully reviewed. However, reproductive and developmental toxicity data on the MSDSs may be sparse or missing entirely. Of almost 700 MSDSs reviewed for lead and the ethylene glycol ethers, 62% failed to mention reproductive system effects; those that did were 18 times more likely to mention pregnancy-related toxicity than adverse male reproductive effects (19).

Additional data on reproductive and developmental toxicities of occupational agents are available from computerized databases, toxicology hotlines, reference books, and government agencies. The bibliography at the end of this chapter lists several textbooks that provide further information on literature searching and on the toxicities of specific agents (see also Appendix A).

Given the Information Collected, Does the Patient's Exposure to the Agents Pose a Reproductive or Developmental Risk?

The final step in the work-up involves assimilating exposure and health effects data to estimate the degree of risk to the patient. In addition to the properties of the agent and the characteristics of the exposure, the health professional must consider biologic factors that modify risk, such as age, nutritional status, and preexisting medical or reproductive problems. For example, the risk of fetal chromosomal abnormalities increases in women 35 years of age and older. Cigarette smoking is associated with subfertility, earlier age at menopause, spontaneous abortion, and fetal growth deficits. Not every reproductive problem is work-related. On the other hand, the presence of a personal risk factor does not rule out contribution by a work exposure.

Risk Prevention and Management

Intervention is clearly warranted when exposure to any chemical or physical agent exceeds regulatory exposure limits. Because few legally mandated exposure limits are designed to protect against reproductive system effects, exposures at or below mandated limits deserve attention when the agent is a known or suspected reproductive hazard. Health professionals can prevent or reduce work-related health risks through patient education and counseling and by advocating for workplace change to decrease or eliminate deleterious exposures.

Counseling

Reproductive problems are intensely personal; patients who believe that their workplace exposures may be contributing to their reproductive problem typically feel anxious and vulnerable. Counseling of these patients requires time, attentive listening, and a compassionate attitude. Information should be conveyed in an understandable and nonjudgmental fashion. Because patients may not retain all information conveyed in the counseling session, it is helpful to summarize key points in a follow-up letter.

Uncertainties or limitations in the data must be clearly conveyed. Frequently, patients' perceptions of risks are distorted, and counseling can help to place the risk in proper perspective. For example, a pregnant woman with a BLL of 20 μg/dL may fear that her child will be mentally retarded; the clinician should explain that, although close follow-up of the infant is advisable, the developmental delays associated with low-level prenatal lead exposure are subtle and may not persist. In addition, the concept of relative risk should be put in perspective. For example, a two-fold increased risk for fetal loss may elevate the miscarriage rate from 15% to 30%, whereas a doubling of the risk of a malformation that occurs with a baseline frequency of 1 per 1,000 live births would result in only 1 additional affected infant in 1,000.

In all cases, the clinician should provide information on background rates of adverse reproductive outcomes and their multifactorial causes. Frequently, the potential risks from a workplace exposure are small compared with spontaneous risks or with risks attributable to nonoccupational causes. Through this type of comprehensive counseling, patients are better able to make informed choices and to accept clinicians as educators and advocates, while understanding that a perfect outcome can never be guaranteed.

Controlling Exposure

When exposures are of concern, clinicians should make every effort to eliminate potential hazards through consultation with employers and regulatory agencies. The traditional industrial hygiene hierarchy that emphasizes engineering controls is especially important for chemical exposures (see Chapters 7 and 15). Many types of personal protective equipment can be used, but bulky protective clothing and respirators may be uncomfortable during the later stages of pregnancy. Short-term use of personal protective equipment while more effective controls are being developed may provide a satisfactory alternative to removal of workers from their jobs.

Safety-oriented employers are likely to welcome these measures. On the other hand, some patients may fear employer retaliation or job loss if they raise health concerns, and the clinician's recommendations may not be supported by legislative protections for workers. Given the complexities of these problems, the uncertainty of risk in many cases, and the limited options available to some workers, decisions ultimately belong to the well-informed patient.

In addition to control of chemical exposures, attention must be directed to control of ergonomic risk factors that particularly affect the pregnant woman (Box 31-2).

Protective Legislation

In nations such as France and Finland, liberal maternity and paternity leaves allow broad latitude in addressing reproductive hazards in the workplace. In the United States, how-

Box 31-2. The Impact of Ergonomic Risk Factors on Reproduction

Laura Punnett

There is reason to expect poor tolerance of the pregnant worker's uterus and the fetus to ergonomic stressors such as heavy physical work, prolonged standing, and static non-neutral trunk postures. There is a physiologically plausible common mechanism for the effect of these exposures on pregnancy. A wide variety of stressors, including fatigue and psychosocial strain, affect the sympathetic nervous system and cause the release of hormones (e.g., catecholamines, prostaglandins) into the maternal circulation. Catecholamines, such as norepinephrine, tend to stimulate uterine contractility; uterine irritability causes cervical changes; the two together initiate the onset of labor. Furthermore, epinephrine levels are elevated in the presence of anxiety or stress. Norepinephrine and epinephrine increase blood pressure and decrease uteroplacental blood flow (UPBF) and placental function. The resulting decreased progesterone production leads to an increase in prostaglandin and to cervical changes.

During exercise, catecholamine and prostaglandin levels also increase, with similar effects on uterine contractility. In addition, the circulatory response to exercise involves an increase in sympathetic nervous system activity, which decreases circulation to the uterus in order to increase it to the skeletal muscle. Therefore, exercise is associated with visceral vasoconstriction. Animal studies suggest that the uterus is not protected during this reaction and that UPBF is decreased as a result. Because fetal nutrition depends on UPBF, it is likely that prolonged strenuous exercise leads to fetal deprivation. It appears that this effect is largely independent of maternal nutritional status; it is not known how it can be modified by general physical fitness. It is biologically plausible that ergonomic strain is relevant even very early in pregnancy, because hypoxia in the first trimester could affect fetal enzymes. A placenta compromised by hypoxia from the beginning of pregnancy might support fetal growth only to a limited stage of development.

It is also important to distinguish static from dynamic exercise in terms of cardiovascular requirements. Local exercise does not have the same effect as whole-body work on redistribution of the circulation, but local static exertion has been shown to produce marked strain on the cardiovascular system. For this reason, there may be more reason to expect an effect on UPBF from whole-body exercise, but the importance of prolonged static exertion of local muscle groups needs further investigation. The relative importance of static and dynamic exercise remains confusing. For example, regular physical exercise also increases plasma volume. Moderate aerobic exercise has been positively associated with fetal growth among well-nourished, low-risk, physically fit women, but rest during the last 6 weeks of pregnancy also increases fetal weight. There is no increase in preterm delivery in women who participate regularly in sports.

Prolonged standing, which affects venous return, decreases plasma volume, and activates the sympathetic nervous system, would be expected to have a deleterious effect on placental perfusion. In addition, growth of the uterus is associated with loss of the normal lumbar spine curve. As the spine is projected forward, the major blood vessels are compressed, resulting in decreased UPBF. Prolonged standing, especially in forward flexed postures, could further aggravate the decrease in UPBF. The effects of such non-neutral trunk postures might be more likely to occur later in pregnancy, when there is decreased room for the pregnant uterus.

The epidemiologic findings for studies of ergonomic risk factors and pregnancy outcome are mixed. They are complicated, in part, by the need to include the many other known risk factors for low birthweight and prematurity and by the selection of higher-risk women out of the workforce. In some studies, heavy physical work has been shown to be associated with preterm delivery and low birthweight for gestational age. In particular, frequent heavy lifting (e.g., loads greater than 25 lb lifted more than 50 times per week) is associated with uterine contractions, spontaneous abortions, prematurity, and low birthweight. Long hours at work (more than 45 hours per week), especially

Box 31-2 (*continued*)

in physically strenuous jobs, has been linked to decreased gestational age. Finally, although it is not a uniform finding, some studies have shown increased risk of low birthweight in women who stand for most of the workday.

Variable shift work, in addition to operating as a psychosocial stressor, may affect circadian rhythms (e.g., body temperature, hormone secretion) and thus reproductive hormones. Anovulation, menstrual disor-

ders, and spontaneous abortion are postulated to result. Shift work has been associated with an increase in first-trimester miscarriages, preterm deliveries, and decreased birthweight for gestational age.

An additional impact of ergonomic issues might be through interaction with chemical exposures. Physical exertion results in increased ventilation rates (as does pregnancy itself). Therefore, a higher tissue dose might result from the same environmental concentration of a chemical.

---

ever, leave options are more limited. The OSHA lead and cadmium standards are the only occupational regulations that specifically allow for compensated removal from the job of workers with medical conditions that may be exacerbated by exposure.

Pregnancy is not considered a disabling condition under the Americans with Disabilities Act (see Chapter 12). Some states and many private employers have short-term disability insurance programs for workers with physical or medical conditions that interfere with their ability to perform their jobs. However, disability benefits are frequently denied to workers with reproductive concerns, on the basis that speculative fetal risk or conditions such as infertility are not strictly work-disabling. Although such interpretations are arguable from a legal standpoint, many workers lack the resources to challenge denial of benefits. Workers' compensation is also an inadequate remedy, because it also applies to job-disabling conditions and requires proof of the work-relatedness of illnesses and injuries—a criterion difficult to fulfill given the scientific uncertainties and multifactorial causes of most reproductive system disorders. As a last resort, employees who leave their jobs because of hazardous conditions may be eligible for unemployment benefits (20).

SPECIFIC CLINICAL ENCOUNTERS

A 27-year-old man complains of inability to conceive with his spouse for 13 months.

He is employed at an automobile radiator repair shop. His job involves cleaning radiators with a caustic soda solution, disassembling them with an oxyacetylene torch, and repairing and then resoldering the units with lead-tin solder. His workroom has two windows that are left open on warm days. He wears overalls that are laundered at home, gloves when handling the cleaning solutions, and appropriate shields when using the torch. A one-pack-a-day smoker, he often smokes and eats in his work area. His wife, a 25-year-old woman who has never been pregnant, is employed as a waitress. An infertility work-up by her gynecologist reveals no abnormalities. The histories are otherwise unremarkable, and his physical examination is normal.

A semen analysis reveals a sperm count of 18 million sperm per milliliter, with mildly abnormal motility and morphology. His other laboratory test results include a BLL of 63 µg/dL, a normal complete blood count, and normal renal function tests. His wife has a BLL of 22 µg/dL, a zinc protoporphyrin concentration of 65 µg/dL, and blood indices consistent with a mild iron deficiency anemia.

Management of infertility requires assessment of all potential contributing factors, including reduction of workplace exposures that may impair reproductive function. Often, the only way to determine whether an occupational agent is responsible for the problem is to see whether abnormalities, such as low sperm count, improve after cessation of exposure. Exposure abatement at the workplace is the best solution; obtaining disability leave in these circumstances is problematic, because causation is often uncertain and infertility does not interfere with job performance *per se*.

This case involves lead poisoning caused by inhalation of lead fumes during soldering and contamination of the workspace with lead dust. The worker's habit of eating and smoking in contaminated areas may increase lead exposure through ingestion and volatilization of lead. Lead poisoning may manifest as reproductive system impairment with few or no other systemic symptoms. Home laundering of workclothes laden with lead dust can result in exposure to family members and probably accounts for the modestly increased BLL in this worker's spouse. Elevation of zinc protoporphyrin reflects heme enzyme inhibition, which, in her case, probably results from both lead absorption and iron deficiency.

This patient is eligible for medical removal protection under the OSHA lead standard, which requires temporary removal of workers with BLLs of 50 μg/dL or higher without loss of wages or benefits. This rule also applies to workers with medical conditions that might be exacerbated by exposure to lead, and it therefore may be instrumental in protecting pregnant or reproductively impaired workers who have lower BLLs.

The OSHA lead standard also calls for institution of control measures to decrease exposure. Before returning to work, this radiator mechanic should be counseled about the hazards of lead and ways to minimize exposures through safe work practices and personal hygiene measures, including smoking cessation. Reporting this case of lead poisoning to OSHA would trigger a workplace inspection and better ensure that effective control measures are put in place to protect all of the workers.

The BLL should be normalized in both the man and the woman before conception; therefore, contraceptive counseling is important. A finding of BLLs greater than 10 μg/dL before or during pregnancy should prompt action to identify and remove sources of exposure. Correction of iron deficiency helps to lessen lead absorption. Treatment for the mechanic could include chelation therapy, and follow-up would include serial semen analyses to monitor recovery of sperm parameters.

A 26-year-old woman, at 7 weeks' gestation with her first pregnancy, has been employed at a small nonunionized furniture manufacturing plant for 4 years. Her job involves applying an adhesive to the backs of furniture cushions with a medium-sized brush and setting them into couch and chair frames. Other workers in her immediate area finish the wooden furniture frames with solvent-based cleaners and lacquers. The only ventilation in the room consists of a ceiling fan, and no personal protective equipment except gloves is available. Review of MSDSs reveals that the adhesive is toluene-based and that the other chemicals used by nearby workers include acetone, xylene, and methylene chloride. During a recent OSHA inspection, the woman's 8-hour time-weighted average exposure to toluene was 80 ppm; OSHA's 8-hour permissible exposure limit (PEL) is 100 ppm, but the American Conference of Governmental Industrial Hygienists recommends an 8-hour exposure limit of 50 ppm. Airborne concentrations of the other solvents ranged from one third to one half of the OSHA PELs for those substances. The worker is concerned about the effects of these chemicals on her pregnancy. However, she is reluctant to voice her concerns because of fear of employer harassment or job loss. Although her husband works, the couple depend on her job for adequate income and health benefits. No disability insurance plan is available through the state or through the employer.

This case is among the most difficult that the clinician is likely to encounter, because it involves both scientific uncertainty and limited employment and benefit options for the worker. Some studies suggest that mixed organic solvent exposure during pregnancy may increase the risk for spontaneous abortion. In addition, both toluene and xylene at high doses are fetotoxic in laboratory animals, and methylene chloride is metabolized to carbon monoxide *in vivo*.

From a public health perspective, even the

uncertain and limited data on these organic solvents would warrant minimization of exposures during pregnancy. However, the degree to which this goal is achievable may vary according to the resources and goodwill of the employer, as well as the financial status of the worker. In stable, large industries, employers are often willing to temporarily transfer exposed workers while implementing more effective controls to reduce exposures. In small firms, like the one in this case, transfer options are limited and employers may be unable to bear the cost of new engineering controls. This employer is not violating OSHA standards, so it is unlikely that OSHA would compel the employer to decrease exposures further. At the same time, OSHA limits for these organic solvents do not necessarily protect against adverse developmental outcomes, so a potential, albeit uncertain, risk to the pregnancy remains.

Workers who are employed in low-wage, nonunionized jobs with few protections or benefits face difficult choices. This worker may ultimately decide that attempts at intervention in the workplace are worth the potential risks of employer retaliation; she may choose to leave work and pursue unemployment benefits; or she may decide to stay on the job, preferring uncertain occupational risks to the adverse health and financial consequences of unemployment. Whatever she chooses, the health care provider should ensure that her decisions are well informed. If she remains at work, counseling regarding safe work practices and simple control measures (e.g., respirator use) is crucial. In addition, referral to legal professionals may be warranted to ensure that her rights are fully protected.

CONCLUSION

Workers are turning increasingly to health care providers with concerns about potential reproductive hazards. Evaluation of patients includes identifying possible harmful exposures, defining the characteristics of expo-

sure, and making estimates of risk based on the best available data. Clinicians can play vital roles in encouraging employers to decrease hazards and in providing supportive advocacy and counseling to workers.

In the United States, there is a tremendous need for policy reform in the area of reproductive health hazards. Workers have won important legal cases involving gender discrimination in the workplace and denial of disability or unemployment benefits to pregnant women who are exposed to potential hazards. However, these cases are extremely costly and beyond the resources of most workers. As with many public health issues, addressing the problem of occupational reproductive hazards will require not only scientific advances but also more stringent regulation of reproductive hazards and fundamental changes in the policies that govern the protections and accommodations available to affected workers (20). Health care providers can play a critical role in the critique and reformulation of policies that affect patients with occupational reproductive concerns.

REFERENCES

1. Levy BS, Levin JL, Teitelbaum DT, eds. Symposium: DBCP-induced sterility and reduced fertility among men in developing countries: a case study of the export of a known hazard. In J Occup Environ Health 1999;5:115–153.
2. Colie CF. Male mediated teratogenesis. Reprod Toxicol 1993;7:3–9.
3. Olshan AF. Male-mediated developmental toxicity. Annu Rev Public Health 1993;14:159–181.
4. Colt JS, Blair A. Parental occupational exposure and risk of childhood cancer. Environ Health Perspect 1998;106(Suppl 3):909–925.
5. Nelson BK, Moorman WJ, Schrader SM. Review of experimental male-mediated behavioral and neurochemical disorders. Neurotoxicol Teratol 1996; 18:611–616.
6. Lahdetie J. Occupation- and exposure-related studies on human sperm. J Occup Environ Med 1995; 37:922–930.
7. Winder C. Reproductive and chromosomal effects of occupational exposure to lead in the male. Reprod Toxicol 1989;3:221–233.
8. Occupational Safety and Health Administration. 29 CFR Part 1910. Occupational exposure to 2-methoxyethanol, 2-ethoxyethanol and their ace-

tates (glycol ethers): proposed rule. Federal Register 1993;58:15526–15632.

9. Daston GP, Gooch JW, Breslin WJ, et al. Environmental estrogens and reproductive health: a discussion of the human and environmental data. Reprod Toxicol 1997;11:465–481.

10. Rowland AS, Baird DD, Weinberg CR, Shore DL, Shy CM, Wilcox AJ. Reduced fertility among women employed as dental assistants exposed to high levels of nitrous oxide. N Engl J Med 1992;327:993–997.

11. Scialli AR, ed. Pregnancy and the workplace. Semin Perinatol 1993;17:1–57.

12. Murphy M, Graziano J, Popovac D, et al. Past pregnancy outcomes among women living in the vicinity of a lead smelter in Kosovo, Yugoslavia. Am J Public Health 1990;80:33–35.

13. Schenker MB, Gold EB, Beaumont JJ, et al. Association of spontaneous abortion and other reproductive effects with work in the semiconductor industry. Am J Ind Med 1995;28:639–659.

14. Taskinen H, Kyyronen P, Hemminki K, Hoikkala M, Lajunen K, Lindbohm ML. Laboratory work and pregnancy outcome. J Occup Med 1994;36:311–319.

15. McMartin KI, Chu M, Kopecky E, Einarson TR, Koren G. Pregnancy outcome following maternal organic solvent exposure: a meta-analysis of epidemiologic studies. Am J Ind Med 1998;34:288–292.

16. Khattak S, K-Moghtader G, McMartin K, Barrera M, Kennedy D, Koren G. Pregnancy outcome following gestational exposure to organic solvents: A prospective controlled study. JAMA 1999;281:1106–1109.

17. Marbury MC. Relationship of ergonomic stressors to birthweight and gestational age. Scand J Work Environ Health 1992;18:73–83.

18. United States Congress, General Accounting Office. Reproductive and developmental toxicants: regulatory actions provide uncertain protection. Washington, DC: General Accounting Office, 1991.

19. Paul M, Kurtz S. Analysis of reproductive health hazard information on material safety data sheets for lead and the ethylene glycol ethers. Am J Ind Med 1994;25:403–415.

20. Clauss CA, Berzon M, Bertin J. Litigating reproductive and developmental health in the aftermath of *UAW v. Johnson Controls*. Environ Health Perspect 1993;101(Suppl 2):205–220.

BIBLIOGRAPHY

American College of Occupational and Environmental Medicine. ACOEM reproductive hazard management guidelines: committee report. J Occup Environ Med 1996;38:83–90.

This ACOEM committee report reviews general principles for assessment and management of workers who may be exposed to reproductive hazards. After providing an overview of strengths and limitations of the reproductive toxicity literature, procedures are recommended for the assessment of reproductive and developmental health risks. A sample reproductive health questionnaire is provided. Then practical reproductive health management options are described.

Frazier LM, Hage ML, eds. Reproductive hazards of the workplace. New York: John Wiley & Sons, 1998.

The first six chapters of this book provide a framework for managing reproductive hazards in the workplace or in clinical settings. Most of the book provides detailed reproductive health data for a wide variety of exposure agents, including male effects and the effects of female exposures on preconception and conception, early pregnancy, middle and late pregnancy, and breast feeding, as well as postnatal effects. More than 100 chemical compounds are reviewed, as are physical, biologic, and other exposures. Appendix 1 provides occupational exposure limits for each chemical reviewed in the book; appendix 2 provides examples of forms used for patient assessment.

Paul M, ed. Occupational and environmental reproductive hazards: a guide for clinicians. Baltimore: Williams & Wilkins, 1993.

This book begins with sections on basic principles of reproductive toxicology, research methods, and an overview of clinical assessment and management of the worker. Then detailed sections are provided on the major reproductive toxicants. Clinical case studies provide insight into management of exposed persons, and the appendix shows differences in recommended exposure limits for known and suspected reproductive and developmental hazards.

Schardein JE. Chemically induced birth defects, 2nd ed. New York: Marcel Dekker, 1993.

This book provides data on teratogenic effects of exposures among female animals and women; about 75% of the text is devoted to pharmaceuticals and 25% to chemicals. Of particular use are the extensive tables, which provide overviews of effects in many species for a large number of compounds.

32 Cardiovascular Disorders

Gilles Thériault and Devendra Amre

A 23-year-old refrigeration repairman developed palpitations and lightheadedness while servicing an air conditioning unit in a large commercial building. He was sent to a local emergency department where he was evaluated for atrial fibrillation. Exposure measurements at the workplace revealed high levels of fluorocarbons (1).

A 50-year-old welder developed bradycardia with ectopic ventricular beats while performing gas-tungsten arc welding. These effects resulted from the malfunctioning of his pacemaker. He was exposed to elevated electromagnetic fields during the welding.

Two young men employed in the mineral assay industry developed noninflammatory cardiomyopathy. By review of clinical findings, elicitation of occupational and environmental histories, worksite evaluations, and ascertainment of tissue cobalt levels, public health authorities confirmed these cases to be caused by occupational cobalt exposure. Hair and heart cobalt levels were elevated for the two patients, but control samples had no detectable cobalt (2).

One of the great achievements of medicine in developed countries over the last three decades has been the decline of cardiovascular disorders (CVDs) and, in particular, coronary heart disease (CHD) among people aged 45 to 64 years. Several reasons are invoked to explain this decline, some preventive (such as changes in dietary habits, smok-

ing, and exercise) and some curative (such as improvement in the management of heart diseases). Nevertheless, CVDs remain the leading cause of death worldwide. In the industrialized countries, 15% to 20% of all working people experience a CVD at some time during their working lives. Among people 45 to 64 years of age, these diseases cause more than one-third of the deaths in men and more than one-fourth in women.

Because of their complex etiology, only a small proportion of CVD causes are attributable to occupation. Working conditions and job demands play an important role in the multifactorial process that leads to these diseases, but ascertaining individual causal factors is difficult. Nevertheless, because CVDs cause so much mortality, preventing even a small increase in risk due to occupation can protect many people and be a significant public health measure.

In order to quantify the impact of work on CVDs, some authors have estimated the *etiologic fraction*—the proportion of cases of the disease that would not have occurred had the risk factor not been present in the population. Table 32-1 presents such a risk assessment for Danish workers (3). According to these estimates, 51% of premature cardiovascular deaths among men and 55% among women would not have occurred had none of the occupational risk factors existed. (These findings are 16% in men and 22% in women when sedentary work is not included among risk factors.) These estimates are remarkably high and emphasize the impor-

G. Thériault and D. Amre: Joint Departments of Epidemiology and Biostatistics and Occupational Health, McGill University, School of Medicine, Montreal, Quebec, Canada H3A 1A2.

TABLE 32-1. *Etiologic fractions of occupational factors in premature cardiovascular diseases in Denmark*

Risk factor	Prevalence of exposures (%)		Relative risk	Etiologic fractions (%)	
	Men	Women		Men	Women
Monotonous, fast-paced work	6	16	2.0	6	14
Shift work	20	20	1.4	7	7
Noise	7	4	1.2	1	1
Chemical exposures	Low	Low	>1.0	0–1	0
Passive smoking	12	13	1.3	2	2
"Sedentary" work	90[a]	90[a]	2.0	42	42
All occupational risk factors				51	55
All occupational risk factors except "sedentary" work				16	22

[a] Only those 72% who are also physically inactive during leisure time will gain any benefit from physical activity at work.

From Olsen O, Kristenen TS. Impact of work environment on cardiovascular diseases in Denmark. J Epidemiol Commun Health 1991;45:4–10.

tance of preventing deleterious occupational exposures.

RISK FACTORS FOR CARDIOVASCULAR DISORDERS

Risk factors for CVDs can be broadly classified as follows: (a) somatic factors, such as high blood pressure, lipid metabolism disorders, obesity, and diabetes mellitus; (b) behavioral factors, such as smoking, poor nutrition, lack of physical activity, type A personality, high alcohol consumption, and drug abuse; and (c) environmental factors, such as exposures in the occupational, social, and personal environment.

These factors act synergistically, so that a smoker with high blood pressure and high serum cholesterol is at eight times greater risk of developing CHD than a nonsmoker who has normal serum cholesterol and normal blood pressure (4). Genetic factors also play a role in the causation of CHD. In certain families, the risk of CHD is high and is correlated with the number of blood relatives who have developed the disease at an early age (5).

Although the association between personal risk factors and CVD is well documented, knowledge of the role of occupational risk factors in the causation of CVD is still limited. Table 32-2 lists some known occupational hazards associated with CVD.

TABLE 32-2. *Known occupational hazards associated with cardiovascular disorders and degree of scientific evidence*

Hazard	Cardiovascular disorders	Strength of the scientific evidence
Carbon monoxide	Arteriosclerosis	Weak
Carbon disulfite	Arteriosclerosis	Strong
Certain aliphatic nitrates (such as nitroglycerin, ethylene glycol dinitrate)	Coronary spasm Arteriosclerosis	Strong Satisfactory
Lead	Hypertension (renal)	Weak
Cadmium	Hypertension (renal)	Insufficient
Arsenic	Coronary heart disease	Insufficient
Cobalt	Cardiomyopathies	Satisfactory
Physical inactivity	Coronary heart disease	Strong
Noise	Transient high blood pressure	Strong
	Long-term high blood pressure	Insufficient
Shift work	Coronary heart disease	Weak
Halogenated solvents	Arrhythmia	Satisfactory
Chronic hand-arm vibration	Vibration white finger	Strong
Electromagnetic fields, radiofrequency radiation	Malfunction of electrical implants	Satisfactory
Work in aluminum production industry	Telangiectasis	Satisfactory

PHYSICAL HAZARDS

Noise

High levels of noise—those exceeding 85 dBA (decibels on the "A" scale)—are common in workplaces (see Chapter 18). There are few factories, smelters, or mines where hazardous noise is not prevalent. The association between chronic noise exposure and hearing loss is well documented. Evidence that long-term exposure to noise can contribute to CVD is generally lacking. However, in a study of female textile workers in China, it was observed that an increase in the noise level from 70 to 100 dBA raised the risk of hypertension by a factor of 2.5. This study population comprised only nonsmoking women; the risks for men could be even higher (6). In Western industrialized countries, hearing protection programs that limit noise levels to 85 to 90 dBA are widely implemented. As a consequence, few extraauricular effects of noise are being reported.

Acute exposures to high levels of noise initiate cardiovascular responses that mimic the effects of acute stress, including increased blood pressure, heart rate, blood levels of catecholamines and lipids, and vascular tone of the peripheral vessels. These changes are transitory, however, and disappear shortly after exposure ceases (7).

Vibration

Chronic exposure of the upper limbs to vibration from tools such as pneumatic drills, hammers, chisels, riveters, metal grinders, and chainsaws has been associated with a vascular syndrome affecting the fingers (see Chapter 19). This syndrome, known as Raynaud's phenomenon, vibration white finger, or hand-arm vibration syndrome, manifests as an episodic whitening of the fingers accompanied by numbness and loss of sensation. On recovery, there is reddening and tingling of the affected areas accompanied by pain. In forestry, the prevalence of this phenomenon has been estimated to be more than 30% (8). After several years of expo-

sure, the syndrome becomes so disabling that the affected worker is forced to leave the job. Recovery takes place slowly once exposure has ceased.

Physical Inactivity

Of all the occupational factors that have been associated with work, inactivity is the one that seems to carry the strongest association with CHD. Physical inactivity is casually related to an excess risk of both CVD and CHD (9). When inactive people are compared with active ones, the relative risk for CVD is estimated to be 2.0. Modern technological developments give rise to more sedentary work. Only health promotion programs can reduce the effects of sedentary work on CVD.

Electromagnetic Fields and Radiofrequency Radiation

Concerns about the potential health effects of exposure to electromagnetic fields, microwaves, and radio frequencies (cellular telephones) has raised interest in the phenomenon of "electromagnetic interference," whereby workers implanted with ferromagnetic materials such as pacemakers, are susceptible to malfunctioning of the implants because of interference from electromagnetic fields in the workplace (see Chapter 19). One study that evaluated several work environments having the potential for electromagnetic interference with pacemakers found that electric-arc welding operations (up to 225 amp, without high-frequency voltage, with cables uncoiled) did not interfere with the device; however, welding operations, such as gas tungsten arc welding, interfered with the pacemaker if the worker was within 1 m of the weld. Also, industrial work involving large degaussing coils interfered with pacemaker function when the worker was within 2 m of the coils. There are reports of possible effects of radiofrequency radiation generated by cellular telephones on the functioning of pacemakers. These findings imply that workers with such implants should be advised regarding specific precautions to be taken when working in environments with high exposure to electromagnetic fields or radiofrequency radiation.

Heat and Cold

Although the effects of acute exposures to extreme heat and cold (see Chapter 19) on the circulatory system have been well documented, there is little, if any, epidemiologic evidence to support the hypothesis that the risk of CVD is higher in occupational groups with long-term exposures to these hazards. Some studies carried out among metalworkers, glassworkers, underground miners, and metallurgical workers exposed to elevated temperatures for long periods have indicated that the risk for hypertension and ischemic heart disease is higher among such workers. On the other hand, some studies carried out among steelworkers and other heat-exposed workers reported lower risks for CVD. Although slaughterhouse workers and fishery workers are likely to be exposed to very low temperatures for extended periods, there is no epidemiologic evidence to suggest that they are at higher risk for CVD.

CHEMICAL HAZARDS

Several studies have investigated the role of chemicals in the etiology of CVD. Although the findings are inconclusive, the contribution of chemicals to CVD is probably low, as suggested by the estimated etiologic fraction in the Danish population (less than 1%) (see Table 32-1). (See also Chapter 15.)

Carbon Monoxide

The potential for exposure to carbon monoxide (CO) in industry is high. This odorless and colorless gas is produced in most processes that involve fire, combustion, or oxidation. High exposures may occur in many workplaces, such as steel and iron foundries, petroleum refineries, pulp and paper mills, and plants where formaldehyde and coke are produced. One of the most common and in-

sidious sources is the internal combustion engine; workers in garages and enclosed parking spaces may be chronically exposed to fairly high levels of CO. Firefighters, apart from the usual hazards of their work, may be exposed to excessively high levels of CO in smoke. CO causes a variety of signs and symptoms depending on the level of exposure. Exposure to high concentrations of CO (more than 1,500 ppm) can result in death by anoxia. Exposure to low concentrations decreases myocardial oxygen consumption, concomitantly increases coronary flow and heart rate, and lowers exercise tolerance in healthy persons. People who already suffer from a certain degree of coronary insufficiency may develop into increased stress test segment depression on electrocardiograms (ECGs) studies, the onset of angina pectoris (11), and occasionally the development of an acute myocardial infarction (MI).

The influence of chronic exposure to low levels of CO on the development of coronary atherosclerosis leading to CHD has yet to be shown. The literature indicates that intermittent exposure to low levels of CO accelerates the development of arteriosclerosis in laboratory animals when combined with a diet rich in saturated fats (12). The few epidemiologic studies that have been conducted among working groups have been unable to show a distinct relation between chronic exposure to CO and development of CHD. The known association of CHD with cigarette smoking, combined with observations that workers intermittently exposed to peaks of CO have a higher risk of heart disease, keeps this question open to further research.

It should be noted that exposure to methylene chloride leads to formation of carboxyhemoglobin in the body, with effects similar to those caused by direct exposure to CO.

Carbon Disulfide

Of all the chemicals for which an association with heart disease has been studied, the evidence is most convincing for carbon disulfide (CS₂). Although CS₂ is used mostly as a solvent in the production of organic chemicals, paints, fuels, and explosives, it was its use in the viscose rayon industry that revealed this association (Fig. 32-1). Mortality studies of viscose rayon workers showed that they were at a two to five times greater risk of dying from heart disease than were unexposed workers. Reduction of exposures lowered the risk to workers. In one study, the excess mortality declined from a relative risk of 4.7 to a risk of 1.0 over a 15-year period after implementation of exposure-reduction measures (13).

Although the exact mechanism by which CS_2 causes CVD is unclear, it is known to interfere with lipid and carbohydrate metabolism, with thyroid functioning (triggering hypothyroidism), and with the coagulation metabolism (promoting thrombocyte aggregation and inhibiting plasminogen and plasmin activity). There is evidence for a direct cardiotoxic effect. There is also evidence that CS_2 can cause atherosclerosis, which is prob-

ably the cause of the increase in CHD associated with long-term exposure to CS_2.

Nitroglycerin and Other Aliphatic Nitrates

Some aliphatic nitrates are potent vasodilators of coronary vessels; this property has long been exploited for the treatment of angina pectoris. However, some workers exposed continuously to nitroglycerin, and in particular to ethylene glycol dinitrate, during the manufacturing of explosives have developed angina pectoris on withdrawal from exposure (Fig. 32-2). This phenomenon occurs on weekends or vacations and disappears on return to work. The mechanism involved is thought to be a coronary spasm that develops after the chemically induced vasodilatation has ceased. Reversal of this vascular spasm by the administration of nitroglycerin has been directly observed under angiography during the withdrawal period. Studies have reported elevated risk for CHD after some 20 years of exposure, which seems to indicate that nitro compounds are not only responsi-

FIG. 32-1. A worker tends machines that spool rayon thread from carbon disulfide. Worker exposure to carbon disulfide was high until this process was enclosed, which reduced worker exposure and, by recycling of the carbon disulfide, saved the company a substantial amount of money (Liang YX, Qu DZ. Cost benefit analysis of the recovery of carbon disulfide in the manufacturing of viscose rayon. Scand J Work Environ Health 1985;11(Suppl. 4):60–63. (Photograph by Barry S. Levy.)

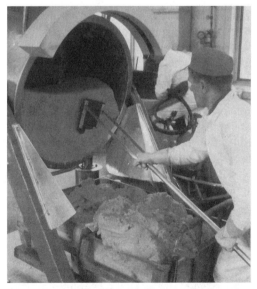

FIG. 32-2. Dynamite kneading involves exposure to nitrates; this can result in rebound vasospasm of coronary arteries on days off from work. Note local exhaust and protective clothing. (Photograph courtesy of C. Hogstedt.)

ble for acute vasospastic reactions but also increase the risk of CHD after long exposure. The mechanism by which nitro compounds generate atherosclerosis remains unknown.

Metals

There has been much speculation regarding the potential for lead, cadmium, and arsenic to cause CVD among workers with long-term chronic exposure (14).

Much research has been conducted on the relation between lead exposure and high blood pressure. A pattern has emerged to the effect that workers exposed to lead have high mortality from cerebrovascular diseases, very likely mediated by hypertensive diseases. However, no clear association with CHD has been reported. Most studies show an increased mortality from chronic renal diseases among lead-exposed workers. (See Chapters 15 and 35.)

For several years in the 1960s and early 1970s, cadmium was believed, on the basis of animal studies, to cause high blood pressure among exposed workers. Most of the more recent studies have not confirmed this association, and today it is recognized that cadmium exposure does not cause an increased risk of CHD among exposed workers.

Arsenic has been associated with "blackfoot disease," a unique peripheral vascular disease that is endemic in some regions of Taiwan, where residents consume arsenic-contaminated well water (15). Similar peripheral lesions have been reported in other countries, such as Chile and Mexico, and also have been attributed to exposure to arsenic in drinking water. One study demonstrated a dose-dependent increase in mortality from ischemic heart disease with long-term exposure to environmental arsenic (16). Some occupational epidemiologic studies have suggested an increased risk for CHD mortality among workers exposed to arsenic, such as copper smelter workers, chimneysweeps, and glassworkers. However, in most of these studies, confounding from other exposures cannot be ruled out, and therefore compel-

ling evidence on the cardiovascular effects of workplace arsenic exposure is still lacking.

In the mid-1960s, an epidemic of fatal cardiomyopathy was reported among heavy beer drinkers after the addition, by some breweries, of a foam-stabilizing, cobalt-containing substance to the beer. It was suggested that the synergistic effects of alcohol, cobalt, and a protein-poor diet were at the root of the cobalt-induced cardiomyopathy, with symptoms that resembled those of thiamine deficiency (17). The problem disappeared after withdrawal of cobalt from the beer. Since then, four cases of cobalt cardiomyopathy have been reported among workers in the hard metal and cobalt-using industries, the last two of which were identified in the mineral assay industry (2). This may indicate that such a disease has long existed among cobalt-exposed workers but has been misdiagnosed or overlooked.

Workers in the primary aluminum production industry, most specifically those working with the Soderberg smelting process, have shown a high prevalence for superficial skin angiomas. These manifest as numerous maculae (workers have referred to them as "red spots") on the chest, upper back, neck, and arms. These are clusters of telangiectases. Apart from their unaesthetic appearance, they do not seem to have any health significance. The affected workers were heavily exposed to coal tar pitch volatiles and fluorides. Neither the mechanism involved nor the causal chemical is known, although it is proposed that a fluoride element bound to a hydrocarbon molecule excreted by the sweat may account for the phenomenon (18).

Solvents

Acute exposures to some halogenated hydrocarbons and nonhalogenated industrial solvents (e.g., toluene, xylene, chloroform, trichloroethylene) and to fluorocarbon aerosol propellants have been associated with sudden death. Halogenated hydrocarbons are said to act as myocardial irritants, which can lead to fatal arrhythmia or ventricular fibrillation. These problems result from sensitization of the heart myocardium to epinephrine. This potential of the chemical to cause sensitivity depends on the degree of halogenation and the type of halogen. Chlorinated hydrocarbons are more sensitizing than fluorinated hydrocarbons. Short-chain, nonsubstituted hydrocarbons have higher toxicity than those with longer chains. The triggering factor is usually a sudden, stressful situation among people heavily exposed. Epidemiologic studies, however, have not revealed increases of CVD mortality among painters or other groups of workers exposed to organic solvents.

BIOLOGIC HAZARDS

Bacteria, fungi, viruses, and protozoa are among the biologic hazardous materials that can harm the cardiovascular system. A number of occupations entail exposures to such biologic agents, especially in the fields of health care, farming, forestry, and sewage and waste handling and treatment. Although there are some reports of increased prevalence of CVD (endocarditis, vasculitis, myocarditis, pericarditis, and pancarditis) among these workers, epidemiologic studies have not yet explored the problem extensively.

STRESS AND OTHER OCCUPATIONAL PSYCHOSOCIAL FACTORS

Stress

Among less well-defined risks of CVD is a wide array of psychological factors. The most widely studied is the behavior of type A individuals, who chronically struggle to achieve an unlimited number of goals in the shortest possible time, often in competition with other people or opposing forces in the environment (see Chapter 21) (19).

An extensive review of the evidence regarding the association between CVD (especially CHD) and type A behavior (20) concluded that such a personality entails an increased risk of CHD among healthy working men. However, this does not hold for recurrent events or for mortality among men who have had a first heart attack or who have angina. Subsequent studies have pointed out that anger and hostility seem to play an important role in this association.

Contrary to previous beliefs, white collar workers who are exposed to the more stressful psychological environment of the decision-making process have lower mortality and incidence rates of CHD than blue collar workers do. This may reflect the high socioeconomic status of white collar workers; it could also indicate a risk associated with the physical demands and the deleterious exposures of the blue collar working environment.

In the early 1980s, the job strain model was proposed (21). It classified work situations bidirectionally according to the degree of monotony of a job and the level of control over the job that could be exercised by the worker. Jobs characterized by high monotony, understimulation, high predictability, and lack of control entail the highest risk of CHD. The risk of CHD for workers exposed to strainful work situations can range between 1.3 and 4.0. It is proposed that the mechanism behind this phenomenon is an increase in circulating hormones (e.g., plasma testosterone, noradrenaline) and an increase in blood pressure. Under the same intervention, cortisol in plasma increased more among passive than among active workers.

Shift work

Several epidemiologic studies have suggested that shift workers have a higher risk for CVD, especially CHD (see Chapter 22) (22). In Denmark, it was estimated that 7% of CVD in men and women could be traced to shift work (3). Studies have shown that shift work induces increased blood pressure and increased levels of triglyceride or serum cholesterol, or both, which along with other risk factors such as smoking and obesity (commonly found in shift workers) can cause increased mortality and morbidity from atherosclerotic disease. Though the exact mechanisms of the association between increased CVD and shift work have not been identified, it is recommended that CVD prevention programs be directed to these workers.

CARDIOVASCULAR DISORDERS AND CIGARETTE SMOKING

Although antitobacco campaigns have succeeded in decreasing the number of smokers, many workers still smoke cigarettes. It is therefore appropriate to stress the relation between smoking and CHD.

Most studies have estimated the risk of CHD among smokers compared with nonsmokers to be on the order of 2.5. This risk is associated more closely with the number of cigarettes smoked per day than with the number of years of smoking. Studies demonstrate that this risk is reversible after smoking ceases; 10 years after cessation, the risk decreases to the same level as in nonsmokers (23).

SCREENING FOR CARDIOVASCULAR DISORDERS IN ASYMPTOMATIC WORKERS

Exercise stress testing has been proposed as a means of screening out from strenuous jobs those people who are at higher risk for development of ischemic heart disease. This concept stemmed from the results of studies who during an exercise stress test have a lowering of the ST segment on ECG are three to five times more likely to develop CHD after 5 years than those without this ECG change (24).

This type of screening may seem attractive, particularly if people are working under conditions that may represent a higher risk of CHD (e.g., regular exposure to low levels of CO, strenuous jobs). In some Western countries such as Canada, this type of screening is performed for workers whose occupations entail risk to the general public (e.g., pilots, flight engineers, air traffic controllers) (25). (See screening section in Chapter 4.)

Many sound arguments militate against screening for CVD in asymptomatic workers, including (a) the low reliability of exercise stress testing in predicting the development of CHD; (b) the discrimination in hiring of workers on unproven grounds; (c) the unavailability of preventive measures for those identified as being at higher risk; (d) the association of risk with numerous other factors; and (e) the existence of risk associated with the testing itself.

RETURN TO WORK AFTER MYOCARDIAL INFARCTION

There is 5% to 10% mortality rate during the first year after an MI; after 1 year, the rate is approximately 3% annually. Factors associated with poor prognosis are the extent of the damage caused to the myocardium (expressed by a poor left ventricular ejection fraction), ventricular irritability with persistent ectopic activity, persistent angina, and the presence of associated diseases such as diabetes mellitus or high blood pressure.

One of the challenges of modern medicine is to adequately match the worker who has experienced a heart attack and is left with a handicap to a work activity commensurate with residual capacities. There are no magic formulas here, and much is left to the discretion of the attending physician. Some authors have provided guidelines to be followed when evaluating the working capacity of patients with MI (26). In general, for patients who are free of clinically significant cardiac dysfunction at rest, it is recommended that impairment be evaluated by symptom-limited exercise testing on a motor-driven treadmill or cycle ergometer.

In the commonly used metabolic equivalence system (METS), the amount of oxygen that the patient can consume before showing signs of cardiac stress on the treadmill is measured. Oxygen consumption is expressed in METS: 1 MET = oxygen consumption of 3.5 mL/kg body weight per minute. Based on this value, the physician refers to an occupational equivalence table that classifies working activities according to their energy requirements and matches the residual capacity of the patient with the energy requirements of the particular job. A rule of thumb is that patients should demonstrate an exercise performance free of significant cardiac dysfunction that is at least twice the average energy requirements and 20% more than the expected peak energy requirements encountered on the particular job. For workers who have an acceptable functional evaluation with standard laboratory techniques but experience symptoms on the job, on-site evaluation of the cardiovascular responses can be done by ambulatory monitoring of ECG (looking for arrhythmias and ST segment responses) and blood pressure. Such monitoring is not recommended for general use.

MYOCARDIAL INFARCTION AND WORKERS' COMPENSATION

Although CHD is essentially a personal disease, MI cases sometimes lead to a lawsuit, with allegations that the MI resulted from the stress or strain of a specific work activity. The assumption is that work has contributed in an appreciable degree to the development of the MI or has aggravated a preexisting condition. In certain circumstances, such cases can be recognized as legally work-related.

In most North American workers' compensation programs, MI is considered to be an injury caused by an accident rather than an occupational disease (see Chapter 11). The presence of an accident originally meant that some "unusual exertion" was a necessary precondition, but this definition has varied considerably with time and place (27).

It is reasonable to assume that most courts of law will accept cases when four conditions are met: (a) the asserted heart pathology is well demonstrated; (b) the MI occurred after an exertive activity not encountered normally in the execution of the work (often this is related to an emergency situation); (c) the MI occurred immediately, or within a reasonable period of time, after the effort; and (d) a physician has stated that the exertion, more probably than not, triggered the attack.

There seems to be general agreement that firefighters and police officers are at high risk for MI and, accordingly, several states have favorably considered compensation cases for these workers. The reason for such a consensus has been concern about the physical and emotional stresses of both occupations, and, in addition, the chemical hazards encountered in fighting fires. However, such assumptions have not been confirmed by epidemiologic findings.

REFERENCES

1. Ferry GF. Occupational medicine forum. J Occup Environ Med 1995;37:122–123.
2. Jarvis JQ, Hammond E, Meier R, Robinson C. Cobalt cardiomyopathy: a report of two cases from mineral assay laboratories and a review of the literature. J Occup Med 1992;34:620–626.
3. Oslen O, Kristensen TS. Impact of work environment on cardiovascular diseases in Denmark. J Epidemiol Community Health 1991;45:4–10.
4. Feinleib M. Risk assessment, environmental factors and coronary heart disease. J Am Coll Toxicol 1983;2:91–104.
5. Barrett-Connor E, Khaw KT. Family history of heart attack as an independent predictor of death due to cardiovascular disease. Circulation 1986; 69:1065–1069.
6. Zhao Y, Zhang S, Selvin S, Spear RC. A dose-response relationship for occupational noise induced hypertension. Schriftenr Ver Wasser Boden Lufthyg 1993;88:189–207.
7. Delin CO. Noisy work and hypertension. Lancet 1984;2:931.
8. Thériault GP, De Guire L, Gingras S. Raynaud's phenomenon in forestry workers of the Province of Quebec. Can Med Assoc J 1982;126:1404–1408.
9. Kristensen TS. Cardiovascular diseases and the work environment: a critical review of the epidemiologic literature on nonchemical factors. Scand J Work Environ Health 1989;15:165–179.
10. Marco D, Eisinger G, Haynes DL. Testing of work environment for electromagnetic interference. Pacing Clin Electrophysiol 1992;15:2016–2022.
11. Anderson EW, Andelman RJ, Strauch JM, Fortuin NJ, Knelson JH. Effect of low-level carbon monoxide exposure on onset and duration of angina pectoris. Ann Intern Med 1973;79:46–50.
12. Weir FW, Fabiano VL. Re-evaluation of the role of carbon monoxide in production or aggravation of cardiovascular disease processes. J Occup Med 1982;24:519–525.
13. Nurminen M, Hernberg S. Effects of intervention on the cardiovascular mortality of workers exposed

to carbon disulfide: a 15 year follow-up. Br J Ind Med 1985;42:32–35.

14. Kristensen TS. Cardiovascular diseases and the work environment: a critical review of the epidemiologic literature on chemical factors. Scand J Work Environ Health 1989;15:245–264.

15. Tseng WP. Effects and dose-response relationships of skin cancer and blackfoot disease with arsenic. Environ Health Perspect 1992;97:110–119.

16. Chen CJ, Chiou HY, Chiang MH, Lin LJ, Tai TY. Dose-response relationship between ischemic heart disease mortality and long term arsenic exposure. Arterioscler Thromb Vasc Biol 1996;16:504–510.

17. Morin Y, Daniel P. Quebec beer drinkers cardiomyopathy: etiological considerations. Can Med Assoc J 1967;97:925–928.

18. Thériault GP, Harvey R, Cordier S. Skin telangiectases in workers at an aluminium plant. N Engl J Med 1980;303:1278–1281.

19. Dorian B, Taylor CB. Stress factors in the development of coronary artery disease. J Occup Med 1984;26:747–756.

20. Matthews KA, Haynes SG. Type A behavior pattern and coronary disease risk: update and critical evaluation. Am J Epidemiol 1986;123:923–960.

21. Karasek R, Baker D, Marxer F, Ahlbom A, Thorell T. Job decision latitude, job demands and cardiovascular disease: a prospective study of Swedish men. Am J Public Health 1981;71:694–705.

22. Knutsson A, Hallquist J, Reuterwall C, Theorell T, Akerstedt T. Shiftwork and myocardial infarction: a case-control study. Occup Environ Med 1999; 56:46–50.

23. Rosenberg L, Kaufman DW, Helmrich SP, Shapiro S. The risk of myocardial infarction after quitting smoking in men under 55 years of age. N Engl J Med 1985;313:1511–1514.

24. Hltaky MA. Exercise stress testing to screen for coronary artery disease in asymptomatic persons. J Occup Med 1986;28:1020–1025.

25. Canadian Guidelines for the Assessment of Cardiovascular fitness in Pilots, Flight Engineers and Air Traffic Controllers. Ottawa: Civil Aviation Medicine (AARG), Medical Services Branch, Health Canada, 1994.

26. Haskell WL, Brachfeld N, Bruce RA, et al. Task Force II: determination of occupational working capacity in patients with ischemic heart disease. J Am Coll Cardiol 1989;14:1025–1034.

27. Barth PS, Hunt MA. Occupational disease in the law. In: Workers' compensation and work-related illnesses and diseases. Cambridge, MA: MIT Press; 1980:92–116.

BIBLIOGRAPHY

Epstein FP. The epidemiology of coronary heart disease. J Chronic Dis 1965;18:735–774.
A thorough review of the epidemiology of CHD, including mortality, incidence and prevalence, and the importance of reduction of known risk factors. This is an old reference, but its findings are still applicable today.

Kristensen TS. Cardiovascular diseases and the work environment: A critical review of the epidemiologic literature on (1) non-chemical and (2) chemical factors. Scand J Work Environ Health 1989;15:165–179 and 245–264.
A comprehensive review (more than 2,000 references) of all identifiable epidemiologic literature on cardiovascular diseases and work; two remarkable articles.

Scott A. Employment of workers with cardiac disease. J Soc Occup Med 1985;35:99–102.
Despite advances in the management of cardiac disorders, many patients do not return to gainful employment. The author discusses this isssue in the socioeconomic context of modern workplaces.

Stamler J, Wentworth D, Neaton JD. Is relationship between serum cholesterol and risk of premature death from coronary heart disease continuous and graded? JAMA 1986;256:2823–2828.
A very large study that illustrates remarkably well the risks associated with personal risk factors, age, serum cholesterol, high blood pressure, and smoking.

Trends and determinants of coronary heart disease mortality: international comparison. Int J Epidemiol 1989;18(Suppl 1).
The papers in this supplement provide information presented at a a workshop of the same title. It is a broad, comprehensive look at CHD mortality from the view of international comparisons of mortality, morbidity, risk factors, and medical care.

Turino GM. Effect of carbon monoxide toxicity: physiology and biochemistry. Circulation 1981;63:253A–259A.
A critical review of the evidence regarding the effects of carbon monoxide toxicity on the cardiorespiratory system.

33

Hematologic Disorders

Bernard D. Goldstein and
Howard M. Kipen

The hematologic system is a primary end point of effect for a variety of occupational health problems. It is also a conduit of unwanted material to other organ systems and often an early indicator of important effects in other tissues. In addition, a number of the most important problems in the workplace are due to chemical agents that may produce more than one hematologic effect (1,2).

AGENTS THAT INTERFERE WITH BONE MARROW FUNCTION

Benzene: Aplastic Anemia and Acute Myelogenous Leukemia

The potent hematotoxicity of benzene was first described in workers in the 19th century. Life-threatening aplastic anemia appears to be an inevitable consequence of exposure to high levels of benzene. Acute myelogenous (or myeloid) leukemia (AML) in association with benzene exposure was originally reported in 1927, and the causal relationship

B. D. Goldstein: Department of Environmental and Community Medicine, Robert Wood Johnson Medical School, and Environmental and Occupational Health Sciences Institute, Piscataway, New Jersey 08854.

H. M. Kipen: Division of Occupational Health, Robert Wood Johnson Medical School, and Division of Occupational Health Environmental and Occupational Health Sciences Institute, Piscataway, New Jersey 08854.

has since been established through epidemiologic research (3).

The control of benzene exposure in the workplace (Fig. 33-1) and in the general environment has been a subject of much interest and controversy since the mid-1980s. A major reason for the controversy surrounding the regulatory control of benzene is that, among the organic chemicals that are known to be human carcinogens, it is produced in the highest volume (8 million tons in the United States in 1998) and has the greatest potential of exposure for workers and the general public. The National Institute for Occupational Safety and Health (NIOSH) estimates that up to 2 million American workers may be exposed. Benzene is an integral component of petrochemical feedstocks and is present in gasoline in the United States in the range of 1.0% to 1.5%. It is a useful intermediate in organic synthesis and is frequently used in research and commercial laboratories. Benzene is also formed during coke oven operations. Although an excellent solvent, it can and should be largely replaced in this role by much less toxic solvents, such as toluene. Benzene, however, is likely to remain a ubiquitous component of our society for the near future.

In 1991, three physicians from Brazil described an occupational health disaster that had developed among workers in the steel industry outside of Sao Paulo. They reported that in 1984 a total of 680 workers were removed from work because of hema-

FIG. 33-1. Benzene worker with respiratory protection and impermeable gloves. (Photograph by Earl Dotter.)

tologic abnormalities, and that this had progressed beyond 1,000 workers as of 1991. Data were presented on 95 workers with repeated neutrophil counts below 2,000/mm³ who also had bone marrow biopsies or aspirates (4).

The average white blood cell (WBC) count was 3,770, the average hemoglobin was 14.5 g/dL, hematocrit, 44.6%, and mean corpuscular volume (MCV), 93 fL. Bone marrow examinations showed total hypocellularity in 78%, granulocytopenia in 83%, megakaryocytopenia in 65%, and erythropenia in 51% of the subjects. Thus, the dominant effects in this group appeared to be on leukocytes, although the sample was selected based on WBC counts. The authors were aware of three cases of acute leukemia, two of aplastic anemia, and three of myelodysplastic syndrome.

Given that no formal cohort was constructed and followed, it is not clear whether this represented complete case ascertainment.

Benzene exposures in this working group

occurred in a coke byproducts production plant. Limited sampling information described the benzene exposures as being at least 10 parts per million (ppm) on a chronic basis. This may be an underestimate in light of the widespread bone marrow depression in those affected. Surveillance of populations in the United States since the 1950s has not yielded similar examples of such widespread hematologic effects of benzene (5).

The current U.S. occupational standard for benzene is 1 ppm TWA (time-weighted average). In 1977, after discovery by NIOSH of a group of rubber workers with a fivefold increased risk of AML, the Occupational Safety and Health Administration (OSHA) attempted to lower the workplace standard for benzene to 1 ppm, first by an emergency temporary standard and then through a formal rule-making process. Both approaches were overturned by the courts, the latter in a five-to-four U.S. Supreme Court decision, which, in part, called on OSHA to calculate the benefits of its action. In 1987, using updated information on the rubber worker cohort, OSHA established the 1 ppm standard (6). (See also Chapter 10.)

The mechanism by which benzene produces hematotoxicity has not been determined. Benzene does not produce bone marrow damage; one or more metabolites do. Approximately 50% of inhaled benzene is excreted by exhalation; the remainder is metabolized to a variety of products, particularly phenol. Measurement of urinary phenol has been used as a marker of benzene exposure. Although this is a useful technique to confirm relatively high-level exposure, such as may occur in spills or accidents, the variation in background levels of urinary phenol, which are presumably from dietary sources, appears to preclude use of this assay as a technique to determine benzene exposure at levels present in a modern, well regulated workplace. Benzene can also be measured in blood or in exhaled breath, both assays being in equilibrium with total body benzene. Although useful for determining instantaneous benzene body burden, the relatively steep slope of the disappearance curve of benzene after after exposure complicates interpretation as an indicator of quantitative exposure. Studies in humans have determined that the half-life of benzene is relatively short, benzene being no longer detectable in the body approximately 1 week after exposure. Other potentially useful urinary markers of benzene exposure, such as muconic acid and S-phenyl mercapturic acid, are undergoing evaluation.

In laboratory animals, the metabolism of benzene has been reported to be inhibited by simultaneous exposure to toluene, resulting in protection against benzene-induced hematotoxicity. However, this protection does not occur at relatively low exposure levels in animals; in addition, toluene does not appear to inhibit benzene metabolism in humans exposed to usual workplace concentrations. In animal studies, benzene hematotoxicity is potentiated by ethanol and lead; however, the relevance of these findings to humans has yet to be determined. A potential interaction of note is the report that Japanese atomic bomb survivors who contracted leukemia were more likely also to have had an occupational exposure to benzene (7). Studies in Chinese workers heavily exposed to benzene indicated that variations in the activity of enzymes involved in benzene metabolism are of importance in determining susceptibility to its hematologic effects (8).

Studies of the toxicology of benzene in laboratory animals have clearly demonstrated benzene-induced aplastic anemia in a variety of species. However, only recently has it been possible to demonstrate tumorigenicity in laboratory animals, predominantly carcinomas and lymphatic tumors, in addition to a weak leukemic effect. Although various lymphatic tumors have been associated with benzene exposure in humans, causality does not yet appear to be fully proven (Table 33-1) (9). The evidence that benzene is causally related to human nonhematologic neoplasms is minimal.

Benzene not only destroys the pluripotential stem cell responsible for red blood cells,

TABLE 33-1. *Relationship of benzene exposure to hematologic disorders*

Causality proven
  Pancytopenia: aplastic anemia
  Acute myelogenous leukemia and variants
    (including acute myelomonocytic leukemia,
    acute promyelocytic leukemia, and erythro-
    leukemia)
  Myelodysplasia
Causality reasonably likely
  Paroxysmal nocturnal hemoglobinuria
  Acute lymphoblastic leukemia
  Multiple myeloma
  Lymphoma: lymphocytic and histiocytic
  Myelofibrosis and myeloid metaplasia
Causality suggested but unproven
  Chronic lymphocytic leukemia
  Hodgkin's disease
  Chronic myelogenous leukemia
  Thrombocythemia

platelets, and granulocytic WBCs, it causes a rapid loss in circulating lymphocytes in laboratory animals and humans. Based on studies in animals and the lymphocytopenic effects in humans, the potential for a benzene effect on the immune system in workers has been suggested but not demonstrated. There is evidence that benzene results not only in a decrease in number, but in structural and functional abnormalities of circulating blood cells. An increase in red cell MCV and a decrease in lymphocyte count may be useful parameters for surveillance of potentially exposed workers. Cytogenetic abnormalities are common in significant benzene hematotoxicity and, with advances in techniques, may become useful in surveillance.

Aplastic anemia is frequently a fatal disorder, with death usually occurring because of infection related to leukopenia or hemorrhage due to thrombocytopenia. Studies of groups of workers with overt evidence of benzene hematotoxicity have often been initiated after the observation of a single person with severe aplastic anemia. For the most part, those people who did not succumb relatively quickly to aplastic anemia demonstrated recovery after removal from benzene exposure. A follow-up study of a group of workers who previously had significant benzene-induced pancytopenia revealed minor residual hematologic abnormalities 10 years later (10). An occasional patient in these groups with initially relatively severe pancytopenia was observed to proceed from aplastic anemia through a myelodysplastic preleukemic phase to the eventual development of AML. This development is not unexpected because people with aplastic anemia from any cause appear to have a higher-than-expected likelihood of development of AML. Accordingly, it has been suggested that AML is not a direct consequence of a benzene effect on the genome, but rather occurs indirectly through the production of aplastic anemia or at least significant pancytopenia. If true, this would imply that workplace control of benzene to a level that prevents pancytopenia would be sufficient also to preclude a risk of myelogenous leukemia. However, based on available evidence, there is no reason to alter the current prudent public health approach of assuming that there is no threshold for the carcinogenic effect of benzene. In other words, we assume that every molecule of benzene to which a person is exposed has some finite risk, albeit small, of producing a somatic mutation that may result in AML. Findings in large cohorts of heavily exposed workers in China are leading to reconsideration of the risk assessment for benzene hematotoxicity and leukemia (11).

Review of leukemia cases that have been reported to be associated with exposure to benzene reveals not only AML, but a number of variants of this disorder, including acute myelomonocytic leukemia, promyelocytic leukemia, and erythroleukemia, the last of which is a relatively rare variant that has been disproportionately reported. Even in cases not specifically designated as erythroleukemia, evidence of erythroid dysplasia is relatively common in the bone marrow. A possibly related observation is the increased MCV of erythrocytes as a fairly common and early manifestation of benzene hematotoxicity.

Myelodysplasia refers to a clonal expansion of dysplastic myeloid stem cells. This relatively rare disorder occurs *de novo* in older people, where it can slowly progress to frank AML. It is a not infrequent occurrence in people who contract AML after exposure to bone marrow toxins, including benzene as well as chemotherapeutic agents.

A typical case report of AML in a worker heavily exposed to benzene follows.

A 29-year-old white man went to his physician with complaints of nonspecific malaise and bleeding gums. On physical examination, he was noted to be febrile and to have exquisite sternal tenderness. Laboratory findings included hematocrit, 32%; WBC, 28,000/mm³; undifferentiated blast forms, 40%; and platelet count, 28,000/mm³. The bone marrow findings were diagnostic of AML. Cytogenetic studies were not performed until after the patient had undergone chemotherapy. Although abnormalities were observed, the findings could not clearly be related to the etiology of the acute leukemia.

The patient had been working in the chemical industry since 20 years of age, when he received an associate degree in laboratory technology. Since that time, he had taken courses at two universities to earn a bachelor's degree in chemistry. Control of chemical exposure in the student laboratories of all three academic institutions was negligible to modest. Benzene was present in each. In only one laboratory course was there any specific instruction concerning the control of chemical hazards in the laboratory. His initial job was in the control room of a petrochemical plant, where he primarily checked vials. He rarely was outdoors or in the chemical area, but did participate in the cleanup of occasional spills.

From 23 to 27 years of age, he worked as a technician in the quality control laboratory of a chemical company that produced chlorinated hydrocarbon pesticides. On approximately two-thirds of workdays, his job was to test the extent of chlorination of the product. The procedure included a solvent extraction in which 50 mL of benzene was added to each of 30 samples on a laboratory bench, and the samples were then placed into a hood containing a heating device.

After a specified time, the 30 beakers were removed from the hood and, while still hot and bubbling, placed on a laboratory bench. After titration and analysis, the residual material was dumped into a sink and the viscous residue was washed vigorously with hot water. He reported that the smell of benzene was particularly notable during the washing procedure, and on occasion he would become lightheaded. This feeling would clear in a few minutes if he left the room or stood by an open window.

During this period, his supervisor, who did most of the same procedures and had also been exposed to lindane (γ-benzene hexachloride), contracted fatal aplastic anemia. Benzene was used in the laboratory as a general solvent and kept in bottles out of the hood.

Two years before the development of AML, he began work in another chemical plant as a laboratory technician. In this laboratory, he also performed chloride analysis, but acetone was used as the extractant and benzene was kept in the hood. In addition, the latter laboratory regularly received visits from industrial hygienists with monitoring equipment, and they presented lectures about laboratory safety.

Other Hematologic Disorders Due to Benzene

Benzene has also been associated with other myeloproliferative disorders, including chronic myelogenous leukemia, myelofibrosis, and myeloid metaplasia; however, evidence is less convincing for a causal relationship with benzene exposure than it is for AML. It is not surprising that cases of the relatively rare disorder paroxysmal nocturnal hemoglobinuria have been reported in benzene-exposed workers. This paraneoplastic disorder is related both to aplastic anemia and to AML. Lymphoproliferative disorders, including Hodgkin's and non-Hodgkin's lymphoma, acute and chronic lymphatic leukemia, and multiple myeloma, have also been reported in association with benzene exposure (12). Again, the evidence of a causal relationship with benzene ranges from suggestive to highly likely but not conclusive.

Other Causes of Aplastic Anemia and Hematologic Neoplasia

Other agents in the workplace have been associated with aplastic anemia, the most notable being radiation. The effects of radiation on bone marrow stem cells have been used in the therapy of leukemia. Similarly, a variety of chemotherapeutic alkylating agents produce aplasia and pose a risk to workers responsible for producing or administering these compounds. Radiation, benzene, and alkylating agents all appear to have a threshold concentration below which aplastic anemia does not occur. Protection of the production worker becomes more problematic when the agent has an idiosyncratic mode of action in which minuscule amounts may produce aplasia; chloramphenicol is an example of such an agent. A variety of other chemicals have been reported to be associated with aplastic anemia, but it is often difficult to determine causality. An example is the pesticide lindane. Case reports have appeared, usually after relatively high levels of exposure, in which lindane is associated with aplasia. This finding is far from being universal in humans, and there are no reports of lindane-induced bone marrow toxicity in laboratory animals treated with large doses of this agent. Bone marrow hypoplasia has also been associated with exposure to ethylene glycol ethers, trinitrotoluene, pentachlorophenol, and arsenic.

Hematologic neoplasms have been reported in other occupational situations. There is some evidence that (a) farmers have a higher risk of lymphoma; (b) production workers exposed to ethylene oxide, used primarily as a sterilant, have an increased incidence of AML (although this increase was not found in one large cohort of sterilization workers); and (c) much more controversially, workers exposed to electromagnetic fields may possibly be at greater risk of leukemia (13–15). Multiple myeloma has been associated with radiation and solvent exposure. The reported association of toluene with aplastic anemia and leukemia almost certainly is incorrect because this represents contamination of toluene with benzene.

AGENTS THAT AFFECT RED BLOOD CELLS

Agents That Interfere with Hemoglobin Oxygen Delivery

Certain toxins produce adverse effects by interfering with the orderly delivery of oxygen from red cell hemoglobin to tissues (see Chapters 15 and 32).

Carbon Monoxide

Carbon monoxide (CO) binds relatively firmly with the oxygen-combining site of hemoglobin, thereby preventing the uptake of oxygen from the lungs. This effect is magnified because once one or more of the four oxygen-combining sites on a hemoglobin molecule is occupied by CO, it becomes more difficult for the oxygen on the molecule to be released at the tissue level. This "shift to the left" of the oxygen dissociation curve accounts for the potential lethality of carboxyhemoglobin (COHb) concentrations as low as 25% to 35% of total hemoglobin. The affinity of hemoglobin for CO is more than 200 times greater than it is for oxygen; that is, a gas mixture of 1,000 ppm CO and 20% oxygen (200,000 ppm) would result in approximately one-half oxyhemoglobin and one-half COHb at equilibrium. At the normal rate of respiration, equilibration occurs slowly, requiring approximately 8 to 12 hours. The rate of change in the COHb level depends on the rate of respiration, thereby putting active workers at greater risk.

There is a natural background level of approximately 0.5% COHb because of the formation of CO during metabolism. A pack-a-day cigarette smoker achieves COHb levels of approximately 5%. The U.S. workplace standard for CO is 50 ppm TWA, which results in approximately 8% to 9% COHb at the end of an 8-hour workday. There is also a 400-ppm short-term exposure limit—a 15-minute time-weighted average not to be exceeded at any time during a workday. Cigarette smoke and ambient exposures have a roughly additive effect on COHb levels.

In a workplace, CO poisoning is usually caused by incomplete combustion coupled with improper ventilation. CO is also formed through the metabolism of certain exogenous agents, most notably methylene chloride.

There is some evidence that COHb levels in the range of 5% are associated with a minimal decrement in psychomotor function, suggesting that at these or higher levels there may be an increased risk of performing tasks wrong, leading to industrial accidents (16). People with preexisting cardiovascular disease have been reported to have a decrease in exercise tolerance and earlier development of acute angina or intermittent claudication, with effects occurring at levels ranging down to perhaps 2.5% COHb. Fetal hemoglobin has a greater affinity for CO than does adult hemoglobin, putting the fetus at greater risk (17).

Detection of CO poisoning depends primarily on clinical suspicion. Complaints include headache, weakness, lassitude, and mental obtundation. In severe cases, the blood is a characteristic cherry red. Treatment includes removal from the contaminated air and ventilation with oxygen.

Methemoglobin-Producing Compounds

Methemoglobin is another form of hemoglobin that is incapable of delivering oxygen to the tissues. Hemoglobin iron must be in the reduced ferrous state to participate in the transport of oxygen. Under normal conditions, oxidation to methemoglobin, with iron in the ferric state, goes on continuously, resulting in approximately 0.5% of total hemoglobin in the form of methemoglobin in the steady state. Reduction to ferrous hemoglobin occurs through the activity of a nicotinamide-adenine dinucleotide hydrogenase

(NADH)-dependent methemoglobin reductase.

Clinically significant methemoglobinemia has been a not uncommon event in industries using aniline dyes. Other chemicals that frequently cause methemoglobinemia in the workplace have been nitrobenzenes, other organic and inorganic nitrites and nitrates, hydrazines, and a variety of quinones (Table 33-2) (18).

Cyanosis, confusion, and other signs of hypoxia are the usual symptoms. In chronically exposed people, blueness of the lips may be observed at levels of methemoglobinemia (approximately 10% or greater) that are without other overt consequences. The blood is a characteristic chocolate brown. Treatment consists of avoiding further exposure. With significant symptoms, usually at methemoglobin levels greater than 40%, therapy with methylene blue or ascorbic acid can accelerate reduction of the methemoglobin level.

There are inherited disorders that lead to persistent methemoglobinemia, either due to heterozygosity for an abnormal hemoglobin or homozygosity for deficiency of red cell NADH-dependent methemoglobin reductase. People who are heterozygous for this enzyme deficiency are not able to decrease elevated methemoglobin levels caused by chemical exposures as rapidly as people with normal enzyme levels.

In addition to oxidizing the iron component of hemoglobin, many of the chemicals that cause methemoglobinemia, or their metabolites, are also relatively nonspecific oxidizing agents that at high levels can cause a Heinz body hemolytic anemia. This process is characterized by oxidative denaturation of hemoglobin, leading to the formation of punctate, membrane-bound red cell inclusions known as Heinz bodies, which can be identified with special stains. Oxidative damage to the red cell membrane also occurs. Although this may lead to significant hemolysis, the compounds listed in Table 33-2 primarily produce their adverse effects through the formation of methemoglobin, which may be life threatening, rather than through hemolysis, which is usually a limited process. In essence, two different red cell defense pathways are involved: (a) the NADH-dependent methemoglobin reductase required to reduce methemoglobin to normal hemoglobin, and (b) the nicotinamide-adenine dinucleotide phosphate hydrogenase (NADPH)-dependent process through the hexose monophosphate shunt, leading to the maintenance of reduced glutathione as a means to defend against oxidizing species capable of producing Heinz body hemolytic anemia (Fig. 33-2). Heinz body hemolysis can be exacerbated by the treatment of patients with methemoglobinemia with methylene blue because it requires NADPH for its methemoglobin-reducing effects. Hemolysis is also a more prominent part of the clinical picture in patients with deficiencies in one of the enzymes of the NADPH oxidant defense pathway or an inherited unstable hemoglobin. Except for glucose-6-phosphate dehydrogenase (G6PD) deficiency, described later in this chapter, these are relatively rare disorders.

Another form of hemoglobin alteration produced by oxidizing agents is a denatured species known as sulfhemoglobin. This irreversible product often can be detected in the blood of people with significant methemoglobinemia produced by oxidant chemicals. Sulfhemoglobin is also the name given, and more appropriately, to a specific product formed during hydrogen sulfide poisoning (see Chapter 15).

TABLE 33-2. *Selected agents implicated in environmentally and occupationally acquired methemoglobinemia*

Nitrate-contaminated well water
Nitrous gases (in welding and silos)
Aniline dyes
Food high in nitrates or nitrites
Moth balls (containing naphthalene)
Potassium chlorate
Nitrobenzenes
Phenylenediamine
Toluenediamine

$$GSH + GSH + (0) \xrightarrow[\text{Peroxidase}]{\text{Glutathione}} GSSG + H_2O$$

$$GSSG + 2NADPH \xrightarrow[\text{Reductase}]{\text{Glutathione}} 2GSH + 2NADP$$

$$\text{Glucose-6-Phosphate} + NADP \xrightarrow{\text{G6PD}} \text{6-Phosphogluconate} + NADPH$$

$$Fe^{+++} \text{Hemoglobin (Methemoglobin)} + NADH \xrightarrow[\text{Reductase}]{\text{Methemoglobin}} Fe^{++} \text{Hemoglobin}$$

FIG. 33-2. Red blood cell enzymes of oxidant defense and related reactions.

Hemolytic Agents: Arsine

The normal red blood cell survives in the circulation for 120 days. Shortening of this survival time can lead to anemia if not compensated by an increase in red cell production in the bone marrow. There are essentially two types of hemolysis: (a) intravascular hemolysis, in which there is an immediate release of hemoglobin in the circulation; and (b) extravascular hemolysis, in which red cells are destroyed in the spleen or the liver.

One of the most potent intravascular hemolysins is arsine gas (AsH₃). Inhalation of a relatively small amount of this agent leads to swelling and eventual bursting of red blood cells in the circulation. It may be difficult to establish the causal relation of workplace arsine exposure to an acute hemolytic episode, partly because of the delay between exposure and onset of symptoms but primarily because the source of exposure is often not evident. Arsine gas is made and used commercially, often in the electronics industry. However, most of the published reports of acute hemolytic episodes have documented the unexpected liberation of arsine gas as an unwanted byproduct of an industrial process, such as if acid is added to a container made of arsenic-contaminated metal. (A characteristic untoward exposure is described later.) Any process that reduces arsenic, such as acidification, can lead to the liberation of arsine gas. Because arsenic can be a contaminant of many metals and organic materials, such as coal, arsine exposure can often be unexpected. Stibene, the hydride of antimony, appears to produce a hemolytic effect similar to that of arsine.

NIOSH has recommended an immediate danger to life and health level of 6 to 30 ppm for 30 minutes, and the National Academy of Sciences Committee on Toxicology has recommended an emergency exposure limit of 1 ppm for 1 hour. Death can occur directly because of complete loss of red blood cells. (A hematocrit of zero has been reported.) However, a major concern at arsine levels less than those that produce complete hemolysis is acute renal failure due to the massive release of hemoglobin in the circulation (see Chapter 35). At much higher levels, arsine may produce acute pulmonary edema and possibly direct renal effects. Hypotension may accompany the acute episode. There is usually a delay of at least a few hours between inhalation of arsine and the onset of symptoms. In addition to red urine due to hemoglobinuria, the patient frequently complains of abdominal pain and nausea, symptoms that occur concomitantly with acute intravascular hemolysis from a number of causes (18).

Treatment is aimed at maintenance of renal perfusion and transfusion of normal blood. Because the circulating red cells affected by arsine appear to some extent to be doomed to intravascular hemolysis, an exchange transfusion in which arsine-exposed red cells are replaced by unexposed cells appears to be optimal therapy. As in severe, life-threatening hemorrhage, it is important that replacement red cells have adequate 2,3-diphosphoglyceric acid levels to be able to deliver oxygen to the tissue.

> Two maintenance workers were assigned to clean up a clogged drain with a mixture of sodium hydroxide, sodium nitrate, and aluminum chips, which together act to form hydrogen gas. Arsine was formed from the combination of hydrogen with arsenic, which was residual in the drain from a use 5 years before. Both men noted the development of a sewer-like odor during the 2 to 3 hours during which they worked on the drain. One man noted a headache, followed by numbness of tongue and cheeks, weakness, and nausea while at work. After being home for a few hours he sought medical attention. The second patient first noted abdominal discomfort at the end of the workday and while at home experienced nausea, vomiting, and hematuria. He was treated at an emergency department that evening with "stomach medicine" and after being anuric all night was admitted the next morning to a local hospital. Both patients were treated for acute renal failure with partial recovery in one and a need for maintenance dialysis in the other patient (19).

A variety of other hemolytic agents are found in the workplace. As discussed before, for many the toxicity of concern is methemoglobinemia. Other hemolytic agents include naphthalene and its derivatives. In addition, certain metals, such as copper, and organometals, such as tributyl tin, shorten red cell survival, at least in animal models.

Mild hemolysis can also occur during traumatic physical exertion (march hemoglobinuria); a more modern observation is elevated WBC counts with prolonged exertion (jogger's leukocytosis). The most important metal that affects red cell formation and survival in workers is lead.

Hematologic Aspects of Lead Poisoning

Occupational exposure to dusts and fumes that contain lead compounds is a frequent concern of medical personnel who evaluate workers in a variety of occupational and en-

FIG. 33-3. Overexposure to lead occurs in iron workers. The man at left is exposed to lead as he cuts apart a ship. The man at right is obtaining an air sample for lead determination. Protection against lead exposures is essential for all its uses. (Photograph courtesy of National Institute for Occupational Safety and Health.)

vironmental settings. In addition to occupations with well recognized risks, such as battery makers, secondary lead smelter workers, and foundry workers, the list of potentially lead-exposed occupations includes more commonly encountered groups, such as firing range personnel, jewelry and pottery workers, bridge workers and other iron workers (Fig. 33-3), welders, and painters. The last three groups are at risk when sanding or burning through lead pigment–containing paints, particularly in a marine setting. Exposure usually is from inhalation of dust or lead oxide fumes. Skin absorption occurs readily only with organic lead compounds, such as tetraethyl lead (a gasoline antiknock additive no longer used in the United States); poor work habits and hygiene can predispose to ingestion and inhalation of lead on cigarettes. The finer the lead particle generated at work (as in lead fume from burning), the greater the proportion of lead absorbed from the respiratory tract. An understanding of the hematologic aspects of lead poisoning is important, in part, because the erythrocytic cells of the bone marrow and peripheral blood represent a major target for lead toxicity as well as an easily sampled window into its systemic effects (see also Chapters 3, 15, 29, 31, and 35).

Pathophysiology of Lead Hematotoxicity

Although lead is known to inhibit most of the enzymes in the heme biosynthetic pathway, its most pronounced effect is inhibition of the final enzymatic reaction in heme formation, in which heme synthetase (ferrochelatase) catalyzes incorporation of ferrous iron into the heme ring (Fig. 33-4). The resultant decrease in intraerythroid levels of iron-containing heme releases earlier reactions in the biosynthetic pathway from feedback inhibition by the end product, particularly for the irreversible first reaction, the synthesis of Δ-amino-levulinic acid (Δ-ALA) from succinyl-coenzyme A and glycine, catalyzed by Δ-ALA synthetase. Increased production of Δ-ALA and decreased incorporation of iron

into protoporphyrin IX causes accumulation of abnormally high levels of all constituents of the pathway, particularly of the penultimate protoporphyrin IX. These high intracellular levels of erythrocyte protoporphyrin provide the biochemical basis for one group of diagnostic tests in the evaluation of lead toxicity. This increased porphyrin production is an example of a well documented type of toxic porphyria.

Interference with heme formation leads to a series of predictable abnormalities in red cell maturation. Bone marrow examination of the lead-intoxicated patient, although rarely clinically indicated, reveals erythroid hyperplasia and sideroblastic changes in which the more mature cells display iron-staining granules in a perinuclear arc. Of greater clinical relevance are the peripheral red cell abnormalities. Because of a combination of decreased red blood cell production and reduced survival time (from membrane abnormalities), normochromic microcytic or normochromic normocytic anemia is the classic finding, although anemia is expected only with the combination of high blood lead levels (BLLs)—over 50 μg/dL—and chronic exposures. Examination of the peripheral smear may demonstrate increased punctate basophilic stippling of red blood cells (stippled cells) and reticulocytosis. Decreased osmotic fragility may also be demonstrated in the laboratory.

It is of utmost importance that neither absence of the aforementioned hematologic changes nor absence of other signs and symptoms of clinical lead poisoning (e.g., neurologic dysfunction) be used to rule out the presence of significantly high lead burdens in either individuals or populations. Development of overt hematologic disease from lead may be expected to result only after serious breaches of safety procedures and of recommended exposure limits, such as those of the 1978 OSHA lead standard and its 1993 update to include originally excluded construction industries. Air concentrations are limited to a TWA of 50 μg/m^3, personal hygiene standards are set, and mandatory bio-

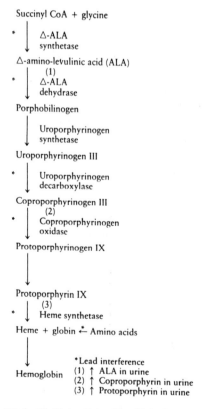

FIG. 33-4. Multiple sites of lead interference with hemoglobin synthesis. (From Rom WN. Environmental and occupational medicine. Boston: Little, Brown, 1982.)

logic monitoring of BLLs is now the law for those with significant exposures to lead (20,21). Ideally, elevation of biologic indices, such as BLL and zinc protoporphyrin (ZPP), the latter of which was not specified in the 1978 OSHA standard but added in the 1993 construction update, will be recognized before overt hematologic disease supervenes.

In chronic lead intoxication, the marrow produces erythrocytes with a modestly decreased survival due to membrane abnormalities, and thus there is some hemolytic component to the resulting anemia. The unusual occurrence of acute lead intoxication due to massive inhalation of dust, as has occurred in some painters when sanding paint from older homes, may result in acute hemolysis similar to that encountered after exposure. The diagnosis is usually evident in the occupational history.

Hematologic Tests for Diagnosis of Lead Poisoning

The most commonly used test is direct measurement of the BLL, routinely performed in clinical laboratories by atomic absorption. Only a few milliliters of whole venous blood is required; however, significant unreliability of laboratory values has been demonstrated. Because lead is a ubiquitous contaminant of industrialized societies, 1990 background levels in U.S. adults approximate 1.5 to 4.5 μg/dL, although neurologic dysfunction has been suggested, especially in children, at levels as low as 10 μg/dL. Values greater than 20 μg/dL commonly indicate significant exposures. The OSHA lead standard is designed to control airborne exposures so that most workers' BLL will be less than 40 μg/dL, although it does not specify worker removal from exposure until the BLL reaches 50 μg/dL.

Blood lead levels best reflect recent exposure, that is, exposure within 2 to 3 weeks. Large amounts of lead are readily incorporated into bone and maintained in equilibrium with blood and soft tissue levels. Thus, chronic or intermittent exposure may lead to relatively moderate elevations of the BLL (30 to 60 μg/dL) in the face of a substantial total body burden. In situations in which people have the opportunity for chronic or intermittent exposures, BLL may be an incomplete index of both body burden and toxicity, and the following two alternatives are indicated to supplement this evaluation for BLLs that exceed 30 μg/dL.

As implied earlier, lead inhibition of heme synthetase results in large accumulations of various heme precursors in developing erythrocytes, the most prevalent of which is protoporphyrin IX. Increases in erythrocyte protoporphyrins can be measured by extraction of the porphyrins from whole blood, followed by spectrophotometry or fluorometry. They also can be measured in urine, a common approach in the workup of other abnormalities of porphyrin metabolism. A more recent and highly useful test is the fluorometric assay for intraerythrocytic ZPP, which has been shown to correlate almost perfectly with free erythrocyte protoporphyrin determinations (22). Zinc is the most abundant intracellular heavy metal, apart from iron, and readily complexes with excess erythrocyte protoporphyrins, forming a specific fluorescent product. During extraction, zinc is usually removed—hence the term *free erythrocyte protoporphyrin*. The direct assay of ZPP with a hematofluorometer can be performed in the field using only microcapillary amounts of whole blood. Rather than assaying a level of lead, it represents a sensitive indicator of biochemical dysfunction. Elevated protoporphyrin measurements are not specific for lead inhibition of heme synthesis; this parameter is commonly elevated in iron-deficiency anemia. Thus, all abnormal values must be interpreted in the context of BLL and serum iron determinations.

It is important to recognize differences between the interpretation of the BLL and ZPP (23). ZPP measures intraerythrocytic abnormalities developed during bone marrow maturation; because the red cells of even a lead-intoxicated person circulate for over 100 days, ZPP reflects the cumulative average inhibition of heme synthesis over the preceding 3 to 4 months. It remains elevated for the life of an individual red cell. The BLL, however, reflects more recent lead exposure. Thus, in an intermittently exposed worker, ZPP may be elevated from exposures 1 to 2 months previously, whereas the BLL could be normal after initially rising and subsequently falling as lead is redistributed to bone and excreted slowly by the kidney. Although no clear guideline has been established, in many laboratories a ZPP greater than 100 μg/dL is said to be beyond the usual range. Even lower levels may be of concern in protecting against lead-induced neurologic dysfunction, especially in children. The important caveat is that ZPP is relatively insensitive when the BLL has not exceeded 20 to 30 μg/dL.

Occult lead intoxication, especially if exposures have taken place more than 3 months before, may be demonstrated by a diagnostic challenge with calcium ethylenediaminetetraacetic acid. This metal chelator binds lead cations in bone and soft tissue. Excretion of greater than 600 μg of lead in a 24-hour urine specimen suggests previous significant exposure. In the presence of renal insufficiency, 72-hour urine collections may be required (24).

X-ray fluorescence (XRF) testing is being shown to be a valuable tool in assessing bone lead levels, as an indicator of cumulative lead exposure (25).

OTHER HEMATOLOGIC DISORDERS

Porphyrias

Various clinicians have raised the hypothesis that mild cases of porphyria, uncommon inborn or inherited disorders of heme synthesis (porphyrin metabolism), may account for medically unexplained illnesses, including multiple chemical sensitivities. Although a number of occupational chemical exposures are known to cause porphyrinopathies, most notably lead and the fungicide hexachlorobenzene, there is no convincing evidence that the symptoms of multiple chemical sensitivities or other medically unexplained syndromes are mediated by an abnormality of heme synthesis such as a porphyria (see Chapter 23) (26).

White Blood Cells

There are a variety of drugs, such as propylthiourea, that affect the production or survival of circulating polymorphonuclear leukocytes relatively selectively. In contrast, nonspecific bone marrow toxins affect the precursors of red cells and platelets as well. Workers engaged in the preparation or administration of such drugs should be considered at risk. There is one report of complete granulocytopenia in a worker poisoned with dinitrophenol. Alteration in lymphocyte number and function, and particularly of subtype distribution, is receiving more attention as a possible subtle mechanism of effects due to a variety of chemicals in the workplace or general environment, particularly chlorinated hydrocarbons, dioxins, and related compounds. Validation of the health implications of such changes is not complete.

Coagulation

Similar to leukopenia, there are many drugs that selectively decrease the production or survival of circulating platelets, which could be a problem in workers involved in the preparation or administration of such agents. Otherwise, there are only scattered reports of thrombocytopenia in workers. One study implicates toluene diisocyanate as a cause of thrombocytopenic purpura. Abnormalities in the various blood factors involved in coagulation usually are not noted as a consequence of work. People with preexisting coagulation abnormalities, such as hemophilia, often have difficulty entering the workforce. However, although a carefully considered exclusion from a few selected jobs is reasonable, such people are usually capable of normal functioning at work.

HEMATOLOGIC SCREENING AND SURVEILLANCE IN THE WORKPLACE

Markers of Susceptibility

In part because of the ease in obtaining samples, more is known about inherited variations in human blood components than about those of any other organ. Extensive studies sparked by recognition of familial anemias have led to fundamental knowledge concerning the structural and functional implications of genetic alterations. Of pertinence to occupational health are those inherited variations that might lead to an increased susceptibility to workplace hazards. A number of such testable variations have been considered or actually used for the screening of workers. The rapid increase in knowledge concerning human genetics makes it a certainty that in the future we will have a better understanding of the inherited basis of variation in human response, and we will be more capable of predicting the extent of individual susceptibility through laboratory tests (see Chapters 4 and 37).

Before discussing the potential value of available susceptibility markers, the major ethical considerations in the use of such tests in workers should be emphasized. It has been questioned whether such tests favor exclusion of workers from a site rather than maintaining a focus on improving the worksite for the benefit of the workers. Some distinction has been made between the use of susceptibility markers as a screening device at a preplacement examination and such use as part of an ongoing evaluation of employed workers. At the very least, before embarking on the use of a susceptibility marker at a workplace, the goals of the testing and consequences of the findings must be clear to all parties (see Chapter 13).

The two markers of hematologic susceptibility for which screening has taken place most frequently are sickle cell trait and G6PD deficiency. The former is at most of marginal value in rare situations, and the latter is of no value whatsoever in most of the situations for which it has been advocated.

Sickle cell disease is a fairly common disorder in which the person is homozygous for hemoglobin S (HbS). It is a relatively severe disease that often, but not always, precludes entering the workforce. The HbS gene may be inherited with other genes, such as HbC, which may reduce the severity of its effects. The basic defect in patients with sickle cell disease is the polymerization of HbS, leading to microinfarction. Microinfarction can occur in episodes, known as sickle cell crises, and can be precipitated by external factors, particularly those leading to hypoxia and, to a lesser extent, dehydration. With a reasonably wide variation in the clinical course and well-being of those with sickle cell disease, employment evaluation should focus on the individual case history. Jobs that have the possibility of hypoxic exposures, such as those requiring frequent air travel, or those with a likelihood of significant dehydration, are not appropriate.

Much more common than sickle cell disease is sickle cell trait, the heterozygous condition in which there is inheritance of a gene for HbS and one gene for HbA. People with this genetic pattern have been reported to undergo sickle cell crises under extreme conditions of hypoxia. Some consideration has been given to excluding people with sickle cell trait from jobs where hypoxia is a common risk, probably limited to the jobs on military aircraft or submarines, and perhaps on commercial aircraft. People with sickle cell trait do very well in almost every other situation, however. For example, athletes with sickle cell trait had no adverse effects from competing at the altitude of Mexico City (2,200 m, or 7,200 ft) during the 1968 Summer Olympics. Accordingly, with the possible few exceptions described previously, there is no reason to consider exclusion or modification of work schedules for those with sickle cell trait.

Another common genetic variant of a red blood cell component is the A form of G6PD deficiency. It is inherited on the X chromosome as a sex-linked recessive gene and is present in approximately 1 in 7 black men and

1 in 50 black women in the United States. Like sickle cell trait, G6PD deficiency provides a protective advantage against malaria. Under usual circumstances, people with this form of G6PD deficiency have red blood counts and indices within the normal range. However, because of the inability to regenerate reduced glutathione, their red blood cells are susceptible to hemolysis after ingestion of oxidant drugs and in certain disease states. This susceptibility to oxidizing agents has led to workplace screening on the erroneous assumption that people with the common A variant of G6PD deficiency will be at risk from the inhalation of oxidant gases. In fact, it would require exposure to levels many times higher than those at which such gases would cause fatal pulmonary edema before the red cells of G6PD-deficient people would receive oxidant stress sufficient to be of concern (27). G6PD deficiency increases the likelihood of overt Heinz body hemolysis in people exposed to aniline dyes and other methemoglobin-provoking agents (see Table 33-2), but in these cases the primary clinical problem remains the life-threatening methemoglobinemia. Although knowledge of G6PD status might be useful in such cases primarily to guide therapy, this knowledge should not be used to exclude workers from the workplace.

There are many other forms of familial G6PD deficiency, all far less common than the A variant. People with certain of these variants, particularly those from the Mediterranean basin and Central Asia, have much lower levels of G6PD activity in their red blood cells. Consequently, the affected person can be severely compromised by ongoing hemolytic anemia. Deficiencies in other enzymes active in defense against oxidants have also been reported, as have unstable hemoglobins that render the red cell more susceptible to oxidant stress in the same manner as G6PD deficiency.

Surveillance

One of the most difficult tasks in occupational medicine is to distinguish between statistical abnormality and clinical abnormality in the interpretation of laboratory tests obtained as part of surveillance. Surveillance differs substantially from clinical testing in both the evaluation of ill patients and the regular screening of presumably healthy people. In ill patients, a laboratory value just outside normal limits is unlikely to explain the cause of the illness. Similarly, physicians evaluating a patently healthy person tend to discount an unsupported finding of one laboratory test just outside normal limits. In contrast, the laboratory tests chosen for a surveillance program are in general those for which any abnormality is a matter of concern because of the nature of the job. These tests are not intended for diagnostic purposes, although they may be useful to supplement normal diagnostic testing. In an appropriately designed surveillance program, the aim is to prevent overt disease by picking up subtle early changes through the use of laboratory testing. Therefore, a slightly abnormal finding should automatically trigger a response—or at least a thorough review—by physicians.

To respond appropriately, the statistical basis of the "normal" laboratory value must be understood. For blood counts, it has been traditional to describe the "normal" range as including 95% of the distribution of blood counts in healthy people. This range implies that 1 of 20 clinically normal people will be statistically abnormal in any one blood count, with 1 in 40 being higher than normal and 1 in 40 being lower than normal. This situation presents a problem when reviewing blood counts in that clinical information can be obtained by findings beyond either end of the normal range. For most hematotoxins, such as benzene, the primary concern is with the lower end of the range. The problem of making a distinction between statistical abnormality and clinical abnormality is compounded by the number of tests in use and the relatively large number of workers who may be under surveillance. Consider that in a normal person, blood counts for platelets, white cells, and red cells might each have a 1 in 40 chance of being below the normal range. In a study of 100 healthy people with no known exposure to hematotoxins, it would not be surprising if perhaps 6 were found to have a statistically abnormal low count for one of these parameters. This finding would be the baseline of false-positive results if a physician were evaluating 100 workers potentially exposed to benzene.

In the initial review of hematologic surveillance data in a workforce potentially exposed to a hematotoxin such as benzene, there are two major approaches that are particularly helpful in distinguishing false-positive results. The first is the degree of the difference from normal. As the count gets further removed from the normal range, there is a rapid drop-off in the likelihood that it represents just a statistical anomaly. Second, the clinician should take advantage of the totality of data for that person, including normal values, keeping in mind the wide range of effects produced by benzene. For example, there is a much greater probability of a benzene effect if a slightly low platelet count is accompanied by a low-normal WBC count, a low-normal red cell count, and a high-normal MCV. Conversely, the relevance of this same platelet count to benzene hematotoxicity can be relatively discounted if the other blood counts are at the opposite end of the normal spectrum. These same two considerations can be used in judging whether the person should be removed from the workforce while awaiting further testing and whether the additional testing should consist only of a repeat complete blood count.

If there is any doubt as to the cause of the low count, the entire complete blood count should be repeated. If the low count is due to laboratory variability or some short-term individual biologic variability, it is less likely that the blood count will again be low. Comparison with preplacement or other available blood counts should help distinguish those people who have an inherent tendency to be on the lower end of the distribution. Detection of an individual worker with an effect due to a hematologic toxin should be considered a sentinel health event, prompting careful investigation of working conditions and of coworkers (see section on Surveillance in Chapter 4).

The wide range in normal laboratory values for blood counts can present an even greater challenge because there can be a substantial effect while counts fall within the normal range. For example, it is possible that a worker exposed to benzene or ionizing radiation may have a fall in hematocrit from 50% to 40%, a fall in the WBC count from 10,000 to 5,000/mm^3, and a fall in the platelet count from 350,000 to 150,000/mm^3—that is, more than a 50% decrease in platelets; yet all these values are within the "normal" range of blood counts. Accordingly, a surveillance program that looks solely at "abnormal" blood counts may miss significant effects. Therefore, blood counts that decrease over time while staying in the normal range need particular attention.

Another challenging problem in workplace surveillance is the detection of a slight decrease in the mean blood count of an entire exposed population—for example, a decrease in mean WBC count from 7,500 to 7,000/mm^3 because of a widespread exposure to benzene or ionizing radiation. Detection and appropriate evaluation of any such observation require meticulous attention to standardization of laboratory test procedures, the availability of an appropriate control group, and careful statistical analysis. For instance, cigarette smoking increases WBC counts by approximately 25%, and thus smoking habits should be accounted for in making intergroup comparisons.

REFERENCES

1. Wintrobe MM, et al. Clinical hematology. Philadelphia: Lea & Febiger, 1993.
2. Williams WJ, et al. Hematology. New York: McGraw-Hill, 1990.
3. Laskin S, Goldstein BD, eds. Benzene toxicity: a critical evaluation. J Toxicol Environ Health 1977;[Suppl 2]. 69–105.
4. Ruiz MA, Vassallo J, De Souza C. A morphological study of the bone marrow of neutropenic patients exposed to benzene of the metallurgical industry of Cubatao, Sao Paulo, Brazil [Letter]. J Occup Med 1991;33:83.
5. Rinsky RA, Hornung RW, Landrigan PJ, et al. Benzene and leukemia. An epidemiologic risk assessment. N Engl J Med 1987;316:1044–1050.
6. Ishimaru T, Okada H, Tomiyaso T, Tsuchimoto T, Hoshino T, Ichimaru M. Occupational factors in the

epidemiology of leukemia in Hiroshima and Nagasaki. Am J Epidemiol 1971;93:157–165.

7. Rothman N, Smith MT, Hayes RB, et al. Benzene poisoning, a risk factor for hematological malignancy, is associated with the NQO1 609C-T mutation and rapid fractional excretion of chlorzoxazone. Cancer Res 1997;57:2839–2842.

8. Yin SN, Hayes RB, Linet MS, et al. A cohort study of cancer among benzene-exposed workers in China: overall results. Am J Ind Med 1996; 29:277–235.

9. Goldstein BD. Clinical hematotoxicity of benzene. In: Carcinogenicity and toxicity of benzene, vol 4. Princeton, NJ: Princeton Scientific, 1983:51–61.

10. Hernberg S, Savilahti M, Ahlman K, Asp S. Prognostic aspects of benzene poisoning. British Journal of Industrial Medicine 1966;23:204–209.

11. Goldstein BD. Is exposure to benzene a cause of human multiple myeloma? Trends in cancer mortality in industrial countries. Ann NY Acad Sci 1990; 609:225–234.

12. Steenland K, et al. Mortality among workers exposed to ethylene oxide. N Engl J Med 1991; 324:1402–1407.

13. Hogstedt C, Aringer L, Gustavsson A. Epidemiologic support for ethylene oxide as a cancer-causing agent. JAMA 1986;255:1575–1578.

14. Daniell WE, Stockbridge HL Labbe RF, et al. Environmental chemical exposures and disturbances of heme synthesis. Environ Health Perspect 1997; 105[Suppl 1:37–53].

15. Savitz DA, Pearce NE. Occupational leukemias and lymphomas. Seminars in Occupational Medicine 1987;2:283–289.

16. National Academy of Sciences, Committee on Medical and Biologic Effects of Environmental Pollutants. Carbon monoxide effects on man and animals. Washington, DC: National Academy of Sciences, 1977:68–167.

17. Smith RP. Toxic responses of the blood. In: CD Klaassen Casarett and Doull's toxicology: the basic science of poisons, 5th ed. New York: McGraw-Hill, 1996.

18. Fowler BA, Wiessberg JB. Arsine poisoning. N Engl J Med 1974;291:1171–1174.

19. Parish GG, Glass R, Kimbrough R. Acute arsenic poisoning in two workers cleaning a clogged drain. Arch Environ Health 1979;34:224–227.

20. U.S. Department of Labor, Occupational Safety and Health Administration: Occupational exposure to lead: final standard. Federal Register 1978;43: 52952–53014 and 43:54353–54616.

21. U.S. Department of Labor, Occupational Safety and Health Administration. Lead exposure in construction: interim final rule. Federal Register 1993; 58:34218.

22. Lamola AA, Joselow M, Yamane T. Zinc protoporphyrin (ZPP): a simple, sensitive, fluorometric screening test for lead poisoning. Clin Chem 1975;21:93–97.

23. Fischbein A, Thornton JC, Lilis R, et al. Zinc protoporphyrin, blood lead and clinical symptoms in two occupational groups with low-level exposure to lead. Am J Ind Med 1980;1:391–399.

24. Lifis R, Fischbein A. Chelation therapy in workers exposed to lead. JAMA 1976;235:2823–2824.

25. HVH, Rabinowitz M, Smith D. Bone lead as a biological marker in epidemiologic studies of chronic toxicity: conceptual paradigms. Environ Health Perspect 1998;106:1–8.

26. Daniell, WE, Stockbridge HL, Labbe, RF, et al. Environmental chemical exposures and disturbances of heme synthesis. Environ Health Perspect 1997;105(Suppl 1):37–53.

27. Amoruso MA, Ryer J, Easton D, Witz G, Goldstein BD. Estimation of risk of glucose 6-phosphate dehydrogenase-deficient red cells to ozone and nitrogen dioxide. J Occup Med 1986;28:473–479.

BIBLIOGRAPHY

Cullen MR, Robins JM, Eskenazi B. Adult inorganic lead intoxication: presentation of 31 new cases and a review of recent advances in the literature. Medicine (Baltimore) 1983;62:221–247.
A clinically oriented review with a good perspective on the relevant issues.

Ernst A, Zibrak J. Current concepts: carbon monoxide poisoning. N Engl J Med 1998;339(22):1603–8.
Contemporary article.

Lilis R, Fischbein A, Eisinger J, et al. Prevalence of lead disease among secondary lead smelter workers and biological indicators of lead exposure. Environ Res 1977;14:255–285.
A complete clinical and laboratory evaluation of a cohort of highly exposed workers that clearly describes the relationship between zinc protoporphyrin, blood lead, and clinical findings.

Linet MS. The leukemias: epidemiological aspects. New York: Oxford University Press, 1985.
An excellent in-depth review of the epidemiology of leukemia. Contains valuable summary tables. The approach to the existing literature is both informative and critical.

Nielsen B. Arsine poisoning in a metal refining plant: fourteen simultaneous cases. Acta Medica Scandinavica Supplementum 1969;1–496.
Contains five papers describing various aspects of an episode of arsine poisoning at a metal refining plant, including industrial hygiene studies, the clinical picture, and examination of the renal circulation in affected workers.

Savitz DA, Andrews KW. Review of epidemiologic evidence on benzene and lymphatic and hematopoietic cancers. Am J Ind Med 1997;31:287–295.
An excellent review on the subject.

34 Hepatic Disorders

Jia-Sheng Wang and John D. Groopman

The liver normally functions to maintain homeostasis by processing nutrients, such as dietary amino acids, carbohydrates, lipids, and vitamins. This organ is involved in the phagocytosis of particulate material in the splanchnic circulation, synthesis of serum proteins, formation of bile, and biliary excretion of endogenous products and xenobiotics. Further, the liver is a major site of biotransformation and detoxification of drugs, chemicals, and circulating metabolites. Thus, the central role of the liver in total health status makes it susceptible to a wide variety of occupational hepatotoxins.

Occupational hepatic disorders can be divided into three etiologic categories: viral hepatitis, chemically induced injuries, and physically induced disorders (1,2). Viral hepatitis usually affects health care workers who are at risk by virtue of blood contact or needlesticks (see Chapters 20 and 43). Chemically and physically induced hepatic disorders are also a hazard of a number of occupations. Many of these disorders are difficult to diagnose clinically and treat because of the similarity of the presenting symptoms. People with acute hepatic disorders frequently have nonspecific clinical manifestations, and those with chronic hepatic disorders are often asymptomatic until disease progresses to its end stage. Sometimes hepatic disorders are inferred in certain occupational settings by epidemiologic or exposure studies or when liver function screening tests have been performed. Simultaneous exposure to agents such as hepatitis B virus (HBV) and alcohol may present a complicated picture compared with hepatic disorders caused by specific occupational hepatotoxins. Further, host susceptibility factors, including genetic polymorphisms of metabolic and detoxifying enzymes, complicate the direct interpretations of descriptive epidemiology studies. In the future, the application of molecular biomarkers specific to occupational hepatotoxins and specific gene-based biomarkers for the hepatic disorders may help resolve these difficulties of etiology and diagnosis.

This chapter focuses only on chemically and physically induced hepatic disorders. Chapters 20 and 43 include discussions of biologically induced hepatic disorders. Table 34-1 summarizes agents that cause occupational hepatic disorders.

CHEMICALLY INDUCED HEPATIC DISORDERS

Hepatic disorders induced by chemicals have been clinically recognized for more than a century. Early documentation of hepatic deposition of lipids after exposure to yellow phosphorus was well described in the literature. Clinical cases and experimental studies of hepatic lesions produced by diverse agents, such as arsphenamine, carbon tetra-

J-S. Wang and J. D. Groopman: Department of Environmental Health Sciences, Johns Hopkins University, School of Public Health and Hygiene, Baltimore, Maryland 21205.

TABLE 34-1. *Agents that caused occupational hepatic disorders*

Type of agents	Example	Hepatic disorders
Chemical agents		
Naturally occurring chemicals		
Mycotoxins	Aflatoxins	HCC, acute and chronic hepatitis
Plant toxins	Cycasin, safrole	Acute hepatic injury
Bacterial toxins	Exotoxins and endotoxins	Acute cholestatic injury
Mushroom toxins	Phalloidins	Acute cholestatic injury
Algae toxins	Microcystin	Acute and chronic hepatic injury
Synthetic compounds		
Alcohol	Ethyl alcohol	Fatty liver, cirrhosis
Pesticides and insecticides	Dichlorodiphenlytrichloroethane (DDT), Kepone, chlorobenzene	Fatty liver, chronic hepatic injury
Aromatic amines	Methylene dianiline (MDA)	Acute cholestatic injury
	2-Acetylaminofluorene	Hepatic injury
Halogenated hydrocarbons	Carbon tetrachloride (CCl_4,), trichloroethane (TCA)	Fatty liver, cirrhosis
	Vinyl chloride	HAS, hepatic sclerosis
	Chloroform	Acute hepatic injury
Chlorinated aromatics	Polychlorinated biphenyls (PCBs), tetrachlorodibenzo-*p*-dioxin (TCDD)	Acute and chronic hepatic injury
Nitroalkane and nitroaromatics	Trinitroluene (TNT), dimethylnitrosamine (DMN), trinitrophenol (TNP)	Fatty liver, acute and subacute hepatic injury
Metals and metalloids	Arsenic	Acute hepatic injury, cirrhosis, HAS
	Beryllium, copper	Hepatic granulomata
	Chromium	Cholestatic injury
	Phosphorus	Acute hepatic injury, cholestatic injury
Physical agents	Heat stroke	Cholestatic injury
	Ionizing radiation	Subacute hepatic injury, fibrosis
Biologic agents		
Hepatitis viruses	A, B, C, D, E	Acute and chronic hepatitis, HCC (associated with hepatitis B and C)
Parasites	Schistosomiasis	Acute hepatic injury, cholestatic injury, fibrosis, cirrhosis
Bacteria	Listeriosis	Hepatic injury, hepatic granuloma

HAS, hepatic angiosarcoma; HCC, hepatocellular carcinoma.

chloride (CCl_4), trinitrotoluene (TNT), dimethylnitrosamine, chloroform, and the polychlorinated biphenyls (PCBs), were reported in the first-half of the 20th century (2,3).

Hepatotoxic chemicals are encountered in a variety of industries, including painting, textiles, munitions, rubber, cosmetics, perfume, food processing, refrigeration, and insecticide, herbicide, pharmaceutical, plastic, and dye manufacturing. More than 100 chemicals have been found to be toxic to the liver by epidemiologic studies of occupationally exposed workers. Hepatotoxic exposures can also occur domestically when chemicals such as cleaning agents, paints, or paint removers are inappropriately used. Rare instances of hepatic disorders associated with massive environmental contamination have been reported, such as cooking oil heavily contaminated with PCBs in Japan, wheat with hexachlorobenzene in Turkey, and flour with 4,4'-diaminodiphenylmethane in England (4). Some industrial chemicals, despite premarket safety testing, have been found to be hepatotoxic in workplace settings long after their initial release because of insufficient appreciation of their toxicity or inappropriate use. Although exposures to many common hepatotoxic chemicals have been reduced through regulation, education, and testing, many new chemicals are being found to damage the liver. For example, a clinical trial for the promising drug fialuridine, being evaluated for use as a therapeutic agent for chronic hepatitis, was terminated in 1993 when a number of patients suddenly died from hepatic failure (5).

Classification of Chemical Hepatotoxins

Hepatotoxic chemical agents encompass both naturally occurring and synthetic compounds. The naturally occurring toxins include bacterial toxins, mycotoxins, mushroom toxins, and algae toxins. Among these species, aflatoxins, phalloidin, and microcystin are potent hepatotoxins for both humans and animals. The synthetic hepatotoxic chemicals include therapeutic drugs, ethanol, pesticides, metals, and other industrial agents. Among the industrial chemicals, aromatic hydrocarbons, halogenated hydrocarbons, chlorinated aromatic compounds, and various nitro compounds have been found to be toxic to the liver.

Hepatotoxins have also been categorized as either intrinsic or idiosyncratic, based on their presumptive mechanisms of action. Most hepatotoxins are intrinsic toxins—that is, their hepatotoxicity is a predictable property of the substance itself, and most people are sensitive if the dose is high enough. An intrinsic, or direct, toxin is defined as a substance, or its metabolite, that directly injures the liver rather than indirectly interfering with metabolic pathways. Acute and subacute injury by intrinsic toxins induces varying degrees of hepatocellular injury, with necrosis and steatosis. A few hepatotoxins, such as beryllium, are idiosyncratic—they cause liver injury that is sporadic and usually not dose related, possibly because of a hypersensitivity or immunologic reaction.

Clinical Disorders Induced by Chemical Hepatotoxins

Chemically induced hepatic disorders can be acute, subacute, or chronic hepatic injuries as classified by their clinical presentations. Specific hepatotoxins may induce both acute and chronic lesions. Pathologically, chemical hepatic injury manifests in different ways. Acute injury often results in an accumulation of lipids (steatosis) and the appearance of degenerative processes, leading to cell death (necrosis). The necrotic process may affect small groups of isolated parenchymal cells (focal necrosis); groups of cells located in zones, as in centrilobular, midzonal, or periportal necrosis; or virtually all the cells in a hepatic lobule (massive necrosis). Although acute injury may cause both fatty accumulation and necrosis, they are not the consistent features for making a diagnosis. Chronic injuries usually present with piecemeal and bridging necrosis, influx of chronic inflammatory cells accompanied by hepatocyte regeneration, and fibrous septum formation.

Cholestatic Injury

Cholestatic lesions result in diminution or cessation of bile flow and retention of bile salts and bilirubin (6). The appearance of jaundice is a significant characteristic. Acute cholestatic hepatic injury is rare, but has been reported after exposure to the chemical methylene dianiline (MDA), an aromatic amine used as an epoxy resin hardener. In 1965, an epidemic of cholestatic jaundice known as *Epping jaundice* occurred in Epping, England after bread was made from flour contaminated with MDA (7). Similar cases were reported in workers after occupational exposure during the manufacture or handling of MDA. In addition to jaundice, clinical symptoms included abdominal pain, pruritus, and fever. Laboratory studies showed elevated bilirubin, alkaline phosphatase (AP), and aminotransferase levels. Liver biopsies revealed bile stasis, portal inflammation, and variable hepatic necrosis.

Chronic cholestatic hepatic disorder with systemic illness was reported with the so-called "Spanish toxic oil syndrome" associated with accidental large-scale ingestion of denatured rapeseed oil (8). Laboratory studies included transient increases of serum aminotransferase activity to chronically increased aminotransferase, AP, and bilirubin levels for up to 30 months after exposure. Liver biopsies revealed chronic biliary disease, fibrosteatosis, and chronic hepatitis.

Fatty Liver (Steatosis)

Steatosis, fatty change in the liver, was first characterized in patients with alcohol-related liver disease. Steatosis is defined morphologically as greater than 5% of hepatocytes containing fat, or, quantitatively, as greater than 5 g lipid per 100 g hepatic tissue. This syndrome also occurs in other disorders, including diabetes mellitus, hypertriglyceridemia, and obesity. Some degree of steatosis usually accompanies acute hepatocellular necrosis; however, marked steatosis is more commonly seen after chronic exposure to a hepatotoxic agent.

Fatty liver associated with chemicals was first described as a result of yellow phosphorus poisoning, with pronounced steatosis and necrosis found at autopsy. Steatosis has also been associated with occupational exposure to styrene, toluene, trichloroethane (TCA), and other aromatic compounds. Similar cases of acute massive necrosis and steatosis have been reported with TNT exposure in munitions industries, arsenical pesticide use in vintners, and the use of certain chlorinated aliphatic solvents (e.g., CCl_4, methyl chloroform, and TCA). More subtle microsteatosis was described after short-term, low-level exposure to dimethylformamide (DMF) in a fabric-coating factory (9). Chronic exposure to chlorinated solvents, such as CCl_4, causes varying degrees of steatosis and hepatocellular injury. Several studies have found steatosis in workers chronically exposed to nonchlorinated solvents, including DMF, toluene, and mixed aliphatic and aromatic solvents.

Hepatoportal Sclerosis

Hepatoportal sclerosis is a rare form of noncirrhotic periportal fibrosis that can lead to portal hypertension. This syndrome has been associated with occupational exposure to the vinyl chloride monomer in polyvinyl chloride polymerization manufacturing plants, inorganic arsenicals, and thorium compounds. Liver biopsies have revealed hyperplasia of hepatocytes and sinusoid cells, with dilatation of sinusoids and progressive subcapsular, portal, perisinusoid, and, occasionally, intralobular fibrosis, accompanied by portal hypertension and splenomegaly.

Fulminant Hepatic Failure and Necrosis

Fulminant hepatic failure is a severe liver disorder in which hepatic insufficiency progresses from the onset of symptoms to hepatic encephalopathy within 2 to 3 weeks, resulting in liver necrosis and liver failure. This disorder was reported after exposure to TNT, used extensively in munitions manufacture during World Wars I and II. The symptoms included jaundice, hepatomegaly, and severe liver necrosis. Even in people who survived the acute phases of the disease, there was often a later development of postnecrotic cirrhosis or aplastic anemia. This severe hepatic disorder can also be induced by CCl_4 and chloroform after inhalation exposure in an enclosed space. Onset of symptoms develops 2 to 4 days after exposure, often accompanied by renal failure in severe cases. Those who survive the acute stages recover in 2 to 4 weeks, but repeated subclinical exposure can induce cirrhosis.

As many as 10% of workers exposed for more than 30 work-years to TCA reportedly become jaundiced. On autopsy, subacute or massive hepatic necrosis and postnecrotic scarring have been noted. Exposure to trichloroethylene also can result in hepatic necrosis similar to that seen with CCl_4. Similar findings were reported after occupational exposure in an enclosed space to an epoxy resin coating containing 2-nitropropane. In several cases this resulted in the death, and some of the workers who recovered had persistent serum aminotransferase elevations (10).

Cirrhosis

Cirrhosis is a chronic, irreversible condition in which the normal lobular architecture is replaced by fibrous tissue and regenerating nodules derived from the remaining hepatocytes. Although cirrhosis is most commonly due to chronic viral infection and alcohol abuse, increased morbidity associated with cirrhosis has been noted in several studies of shipyard workers and printers who were exposed daily to a variety of organic solvents. In addition, other workers exposed to dimethylnitrosamine, TNT, TCA, various pesticides, and hydrazines also have increased rates of cirrhosis. Cirrhosis and other hepatic disorders have been reported to be more prevalent among anesthesiologists than other hospital personnel. In one study, morticians exposed over the long term to formaldehyde had a greater prevalence of cirrhosis than an unexposed control population, although ethanol was a possible confounding factor. Increased prevalence of cirrhosis has also been associated with arsenic exposure, such as to vineyard workers using arsenic compounds as a pesticide. An increased death rate for cirrhosis was also reported in a retrospective cohort study of over 2,500 workers in two plants where PCBs were used in the manufacture of electrical capacitors.

Granulomatous Hepatic Disorder

Granulomatous hepatic disorder has been reported in occupational exposure settings when beryllium and copper were used. In beryllium injury, the histopathologic appearance of the liver biopsy specimen can be indistinguishable from sarcoidosis. Berylliosis may include associated granulomas in the spleen, bone marrow, and lungs, as well as the usual granulomatous interstitial lung disease. Thirty workers with vineyard sprayer's lung were also found to have liver damage with inclusion of copper in biopsy tissue. The hepatic disorder included proliferation and swelling of the Kupffer cells, sarcoid-like granulomas, fibrosis, micronodular cirrhosis, hepatic angiosarcomas, and idiopathic portal hypertension (11).

Porphyria Cutanea Tarda and Related Abnormalities

A chronic hepatic disorder of porphyrin metabolism was described in 36 workers with vinyl chloride–induced hepatic injury from long-term industrial exposure. This abnormality was apparently induced by the inhibition of a number of hepatic enzymes in the porphyrin biosynthesis pathway. Porphyrinuria due to vinyl chloride exposure is rarely diagnosed (12). The disorder has also been described in association with methyl chloride poisoning, dioxin exposure, hexachlorobenzene, and other kinds of polyhalogenated aromatic hydrocarbon–induced liver injury (13). Occupational exposure to 2,4,5-trichlorophenoxyacetic acid (2,4,5-T) and PCBs can also cause porphyria. After exposure to tetrachlordibenzo-*p*-dioxin (TCDD), porphyria cutanea tarda seems to be a specific disorder, producing increased urinary concentrations of uroporphyrin. Eleven of 55 workers who had been exposed during the manufacture of TCDD were diagnosed and then were followed for 10 years because of liver abnormalities.

Other Hepatic Disorders Caused by Chemical Exposures

Transiently increased liver function test values were recorded in several case reports after occupational exposure to methylene chloride (14). Increased aminotransferase values were found in DMF-exposed workers who had microvesicular fat and hepatocellular changes in liver biopsy specimens (15). Liver biopsies of workers manufacturing the pesticide Kepone (chlordecone) showed increased fat, numerous dense bodies, and proliferative smooth endoplasmic reticulum. These workers had severe neurologic symptoms. Jaundice and mild transient liver necrosis have been reported in workers exposed to chromium during chrome plating operations.

Transient liver function abnormalities were also found in association with TCDD exposure. Ten percent of the Seveso (Italy) population exposed to TCDD after an industrial explosion had modest elevations of γ-glutamyltranspeptidase (GGT). Workers exposed to tetrachlorophenol or TCDD had prolonged prothrombin times, elevated plasma lipids, and elevated liver aminotransferase values. Mild steatosis, periportal fibrosis, activation of Kupffer cells, and porphyria cutanea tarda were reported in workers who manufactured TCDD. In Japan and Taiwan, more than 2,000 people who ingested cooking oil contaminated with PCBs and related compounds had abnormal liver function tests, hepatomegaly (in severe cases), and alterations in the endoplasmic reticulum and mitochondria in biopsy samples seen in electron microscope studies. A variety of hepatic abnormalities have also been reported after occupational PCB exposure (16). In one study of electrical workers who had PCB levels in blood of 41 to 1,319 μg/kg, 16 of 80 workers had hepatomegaly and variably increased GGT, aspartate aminotransferase (AST), and alanine aminotransferase (ALT) values.

Residents near a toxic waste dump with leachate-contaminated drinking water sources exhibited statistically significant increases in liver function test values (AP and AST) and a greater prevalence of hepatomegaly compared with area residents drinking uncontaminated water (17). Increased hepatic enzyme values (especially GGT and ALT) were seen in human populations who consumed water from a reservoir contaminated with a heavy bloom of the toxic blue-green alga, *Microcystis aeruginosa*, compared with an adjacent population who drank water from other sources (18). An outbreak of toxic hepatitis that occurred in India in 1974 and was associated with a high mortality rate among the exposed villagers and animals, was traced to food contaminated by aflatoxin and other mycotoxins (19).

HEPATIC DISORDERS INDUCED BY PHYSICAL AGENTS

Hyperthermia (heat stroke) can cause acute hepatic injury characterized by centrilobular necrosis and cholestasis (20). Exposure to a cumulative dose of ionizing radiation in excess of 3,000 to 6,000 cGy gives rise to radiation-induced hepatitis within 2 to 6 weeks.

Those who survive frequently acquire cirrhosis with progressive fibrosis and obliteration of the central veins, with centrilobular congestion. Radiation-induced hepatitis has been reported after accidental intense exposure. For example, a group of Japanese fishermen exposed to radioactivity from hydrogen bomb experiments in the Bikini Islands in 1955 had liver fibrosis and proliferation of bile canaliculi in biopsy specimens (4,21).

MALIGNANT HEPATIC DISORDERS

There are two types of human malignant hepatic disorders associated with occupational hepatotoxicants, hepatic angiosarcoma (HAS) and hepatocellular carcinoma (HCC). HAS, also called endothelial cell sarcoma, is a rare malignant tumor in the general population, but has been found to be associated with chronic exposure to vinyl chloride monomer, arsenic, anabolic steroids, and Thorotrast (an obsolete scintigraphy contrast agent that contained colloidal thorium dioxide, an emitter of alpha particle ionizing radiation) (22). Many of the reported HAS cases have appeared in workers who had been exposed to vinyl chloride for many years in a few specific plants (23). HAS has also been diagnosed in vineyard workers and others who have used arsenicals, including Fowler's solution (1% potassium arsenite), or copper as a pesticide. In addition, several case reports have indicated that long-term ingestion of arsenic-contaminated well water can cause HAS.

Hepatocellular carcinoma is one of the leading causes of cancer mortality in Asia and Africa. In the People's Republic of China, this disease is the third leading cause of cancer mortality and accounts for at least 150,000 deaths per year, with an annual incidence rate in some areas of the country approaching 100 cases per 100,000. In contrast, the annual HCC incidence in the United States is approximately 1.5 cases per 100,000. Thus, the incidence of this malignancy varies worldwide by at least 100- to 1,000-fold (21,24). In the United States and most Western countries, excessive ethanol consumption, HBV and hepatitis C virus (HCV) infections, and possible exposure to chlorinated hydrocarbons have been estimated to account for as many as 75% of HCC cases. In addition, a slightly increased risk is thought to be associated with use of oral contraceptives or androgenic anabolic steroid hormones. A number of epidemiologic studies have also examined the association between occupation and risk of HCC. In Texas, a statistically significant excess risk of HCC mortality has been described in certain occupations such as oil refinery workers, plumbers, pipefitters, butchers and meat cutters, textile workers, and longshoremen—all with odds ratios of at least 2.0 (25). Another study reported an increased risk of HCC in the chemical or petrochemical industry. A case–control study of HCC in New Jersey found an association with road builders, manufacturers of automobiles and plastics, workers exposed to anesthetics, gas station workers, and (even after adjustment for level of ethanol consumption) workers in eating and drinking places (26). Studies have also found increases in HCC in the highway construction industry (especially among workers who have been exposed to asphalt), synthetic abrasive manufacture, and the automobile industry (27). Many of these studies, however, did not adjust for other known risk factors, such as ethanol consumption and HBV and HCV infection.

In contrast to North America, the major risk factors for HCC in Africa and Asia, including China, are chronic hepatitis virus infection and exposure to aflatoxins (28,29). Aflatoxins are mycotoxins produced by *Aspergillus flavus* and *Aspergillus parasiticus*. Aflatoxin B_1 (AFB$_1$) is a potent dietary hepatocarcinogen and has been classified as a group I human carcinogen by the International Agency for Research on Cancer (IARC) (30). Since the late 1960s, there have been extensive efforts to investigate the association between aflatoxin exposure and human HCC. Several epidemiologic studies have found that increased aflatoxin ingestion

corresponded to increased HCC incidence (28). The relationship between aflatoxin exposure and development of human HCC is further highlighted by studies of the *p53* tumor suppressor gene, the most commonly mutated gene detected in cases of human cancer (24). The initial results, from three independent studies of *p53* mutations in HCCs occurring in populations exposed to high levels of dietary aflatoxin, revealed high frequencies of G → T transversions, with clustering at codon 249. On the other hand, studies of *p53* mutations in HCCs from Japan and other areas, where there is little exposure to aflatoxin, revealed no mutations at codon 249. These studies provided the circumstantial linkage between this signature mutation of *p53* and aflatoxin exposure in HCC from China and Southern Africa.

Hepatocellular carcinoma tissues from two different areas in China—Qidong, where exposure to HBV and AFB$_1$ is high, and Beijing, where exposure to HBV is high but that of AFB$_1$ is low—showed distinct differences in the pattern of *p53* mutations at codon 249, suggesting that AFB$_1$ or other environmental carcinogens may contribute to this difference.

The observation of the codon 249 mutation in *p53* with aflatoxin exposure is not limited only to China and Southern Africa. Senegal is a country where the incidence of HCC is one of the highest in the world and where people are exposed to high levels of aflatoxins. One study found mutations at codon 249 of the *p53* gene in 10 of 15 tumor tissues tested, the highest rate thus far described. Another study examined the role of AFB$_1$ and *p53* mutations in HCCs and in normal liver samples from the United States (negligible AFB$_1$ exposures), Thailand (low), and Qidong, China (high). The frequency of the AGG to AGT mutation at codon 249 paralleled the level of AFB$_1$ exposure, further supporting the hypothesis that aflatoxin has a causative, and probably early, role in human hepatocarcinogenesis (24).

Results from experimental studies have also implicated aflatoxin as a causative agent in the described *p53* mutations. Previous work had shown that AFB$_1$ exposure causes almost exclusively G $\rightarrow$ T transversions in bacteria, and that aflatoxin-epoxide can bind to codon 249 of *p53* in a plasmid *in vitro*. Further study examined the mutagenesis of codons 247 to 250 of *p53* by rat liver microsome-activated AFB$_1$ in human HCC HepG2 cells, and found that AFB$_1$ preferentially induced the transversion of G $\rightarrow$ T in the third position of codon 249; however, AFB$_1$ also induced G $\rightarrow$ T and C $\rightarrow$ A transversions in adjacent codons, albeit at lower frequencies. A study of the mutability of codons 247 to 250 of *p53* with AFB$_1$ in human hepatocytes, using the same strategy, found that AFB$_1$ preferentially induced the transversion of G $\rightarrow$ T in the third position of codon 249, generating the same mutation that is found in a large fraction of HCCs from regions of the world with AFB$_1$-contaminated food. These experimental results support AFB$_1$ as an etiologic factor for HCCs in AFB$_1$-contaminated areas (21,24).

PREVENTION OF HEPATIC DISORDERS

Prevention of occupationally related hepatic disorders follows the general principles of primary and secondary prevention. Strategies for primary prevention include identifying and removing (or reducing) hepatotoxic agents in the work environment, and minimizing worker expose to known hepatotoxic agents. Hepatotoxic agents can be identified by review of the chemicals and physical agents used or produced in specific industrial process, followed by further industrial hygiene assessment. Hepatotoxic exposure can be minimized by changing the manufacturing process or improving the working conditions, such as by improving ventilation and using personal protective equipment. Preplacement screening is useful to rule out factors such as alcohol or barbiturate abuse or other existing hepatic disorders that could make the worker particularly susceptible to the effects of hepatotoxic exposure.

Secondary prevention involves the screening of workers actively exposed to known or suspected hepatotoxic agents. This approach is particularly important to identify hepatic disorders at an early, reversible stage when exposure is unpredictable or unavoidable—not amenable to primary prevention. Chemoprevention has proven effective in several settings of exposure to hepatotoxic agents and is being developed for many hepatic disorders.

Despite the knowledge base for prevention, the incidence of hepatic disease continues to rise in the United States and elsewhere. The evolving endemic presence of HCV infection portends an even steeper increase in cases. It is therefore critical for occupational health workers to maintain vigilance in the workplace to help reverse these trends.

REFERENCES

1. Redlich C, Brodkin CA. Gastrointestinal disorders: liver diseases. In: Rosenstock L, Cullen MR, eds. Textbook of clinical occupational and environmental medicine. Philadelphia: WB Saunders, 1994: 423–436.
2. Moslen MT. Toxic responses of the liver. In: Klaassen CD, ed. Cosarett and Doull's toxicology: the basic science of poisons, 5th ed. New York: McGraw-Hill, 1995;403–416.
3. Reynolds ES, Moslen MT. Environmental liver injury: halogenated hydrocarbons. In: Farber E, Fisher MM, eds. Toxic injury of the liver. New York: Marcel Dekker, 1980:541–596.
4. Fleming LE, Beckett WS. Occupational and environmental disease of the gastrointestinal system. In: Rom WN, ed. Environmental and occupational medicine, 2nd ed. Boston: Little, Brown, 1992: 633–647.
5. Macilwain C. NIH, FDA seeks lessons from hepatitis B drug trial deaths. Nature 1993;364:275.
6. Oelberg DG, Lester R. Cellular mechanisms of cholestasis. Annu Rev Med 1986;37:297–317.
7. Kopelman H, Scheuer PJ, Williams R. The liver lesion of the Epping jaundice. QJM 1966;35: 553–564.
8. Velicia R, Sanz C, Martinez-Barredo F, Sanchez-Tapias JM, Bruguera M, Rodes J. Hepatic disease in the Spanish toxic oil syndrome. J Hepatol 1986;3:59–65.
9. Fleming LE, Shalat SL, Redlich CA. Liver injury in workers exposed to ethylformamide. Scand J Work Environ Health 1990;16:289–292.
10. Harrison R, Letz G, Pasternak G, Blanc P. Fulminant hepatic failure after occupational exposure to 2-nitropropane. Ann Intern Med 1987;107:466–468.

11. Pimentel JC, Menezes AP. Liver diseases in vineyard sprayers. Gastroenterology 1977;72:275–283.
12. Tamburro CH. Relationship of vinyl monomers and liver cancers: angiosarcoma and hepatocellular carcinoma. Semin Liver Dis 1984;4:158–169.
13. Doss M, Lange C-E, Veltman A. Vinyl chloride induced hepatic coproporphyrinuria with transition to chronic hepatic porphyria. Klin Wochenschr 1984;64:175–178.
14. Cordes DH, Brown WD, Quinn KM. Chemically induced hepatitis after inhaling organic solvents. West J Med 1988;148:458–460.
15. Redlich CA, Beckett WS, Sparer JS, et al. Liver disease associated with occupational exposure to the solvent dimethylformamide. Ann Intern Med 1988;108:680–686.
16. Higuchi K, ed. PCB poisoning and pollution. Tokyo, Kodansha, Ltd. and Academic Press, 1976.
17. Meyer CR. Liver dysfunction in residents exposed to leachate from a toxic waste dump. Environ Health Perspect 1983;48:9–13.
18. Falconer IR, Beresford AM, Runnegar MTC. Evidence of liver damage by toxin from a bloom of the blue-green alga, *Microcystis aeruginosa*. Med J Aust 1983;1:511–514.
19. Krishnamachari KAVR, Bhat RV, Nagarajan V, Tilak TBG. Hepatitis due to aflatoxicosis. Lancet 1975;1:1061–1063.
20. Bianchi L, Ohnacker H, Beck K, Zimmerli-Ning M. Liver damage in heatstroke and its regression. Hum Pathol 1972;3:237–248.
21. Wang JS, Groopman JD. Toxic liver diseases. In: Rom WN, ed. Environmental and occupational medicine, 3rd ed. Philadelpia: Lippincott–Raven, 1998:831–841.
22. Creech JL, Johnson MN. Angiosarcoma of the liver in the manufacture of polyvinyl chloride. J Occup Med 1981;16:150–151.
23. Forman D, Bennett B, Stafford J, Doll R. Exposure to vinyl chloride and angiosarcoma of the liver: a report of the register of cases. British Journal of Industrial Medicine 1985;42:750–753.
24. Groopman JD, Wang J-S, Scholl P. Molecular biomarkers for aflatoxins: from adducts to gene mutations to human liver cancer. Can J Physiol Pharmacol 1996;74:203–209.
25. Suarez L, Weiss NS, Martin J. Primary liver cancer death and occupation in Texas. Am J Ind Med 1989;15:167–175.
26. Stemhagen A, Slade J, Altman R, Bill J. Occupational risk factors and liver cancer: a retrospective case control study of primary liver cancer in New Jersey. Am J Epidemiol 1983;117:443–454.
27. Wegman DH, Eisen E. Causes of death among employees of a synthetic abrasive product manufacturing company. J Occup Med 1981;11:748–753.
28. Wogan GN. Aflatoxins as risk factors for hepatocellular carcinoma in humans. Cancer Res 1992;52[Suppl]:2114s–2118s.
29. Groopman JD, Cain LG, Kensler TW. Aflatoxin exposure in human populations and relationship to cancer. Crit Rev Toxicol 1988;19:113–145.
30. IARC Working Group on the Evaluation of Carcinogenic Risk to Humans. Some naturally occurring substances: food items and constituents, heterocyclic aromatic amines and mycotoxins. Lyon, France: IARC Press, 1993:245–395.

BIBLIOGRAPHY

Arias IM, Boyer JL, Fausto N, Jakoby WB, Schachter D, Shafritz DA, eds. The liver: biology and pathobiology, 3rd ed. New York: Raven Press, 1994.

Schiff L, Schiff ER, eds. Diseases of the liver, vols I and II, 7th ed. Philadelphia: JB Lippincott, 1993.

Zakim D, Boyer TD, eds. Hepatology: a textbook of liver disease, vols 1 and 2, 2nd ed. Philadelphia: WB Saunders, 1990.

Zimmerman HJ, Hepatotoxicity: the adverse effects of drugs and other chemicals on the liver. New York: Appleton-Century-Crofts, 1978.

Excellent general references. The book edited by Arias and colleagues gives a biologic and pathologic view of the liver and its diseases. The books edited by Schiff and Schiff and Zakim and Boyer describe almost every known liver diseases, including their etiologies, pathophysiologic mechanisms, diagnosis, and management. Zimmerman's book, although published more than 20 years ago, remains a comprehensive source on hepatotoxicity induced by drugs and chemicals.

Klaassen CD, ed. Casarett and Doull's toxicology: the basic science of poisons, 5th ed. New York: McGraw-Hill, 1996.

Meeks RG, Harrison SD, Bull RJ, eds. Hepatotoxicology. Boca Raton, FL: CRC Press, 1991.

Plaa GL, Hewitt WR, eds. Toxicology of the liver, 2nd ed. Washington, DC: Taylor & Francis, 1998.

Klaassen's book deals with all aspects of modern toxicology and is an authoritative reference for toxicologists. Several chapters in the book (e.g., 13, 24, and 33) are specifically related to occupational hepatic disorders. The books edited by Meeks and colleagues and Plaa and Hewitt are excellent references for chemically induced liver injury.

Davidson CS, Leevy CM, Chamberlayne EC. Guidelines for detection of hepatotoxicity due to drugs and chemicals. NIH publication No. 79–313. Washington, DC: U.S. Department of Health, Education and Welfare, National Institutes of Health, 1979.

An important regulatory document detailing the methods and standards for detection of drug- and chemically induced hepatotoxicity.

Dossing M, Skinhoj P. Occupational liver injury. Int Arch Environ Health 1985;56:1–21.

An excellent review of the characterization and identification of occupational hepatotoxins.

35 Renal and Urinary Tract Disorders

Richard P. Wedeen

Occupational renal diseases due to lead, cadmium, mercury, arsenic, chromium, uranium, organic solvents, and silica are well recognized (Table 35-1), as are cancers of the urinary tract due to specific carcinogens.

Cause and effect are relatively easy to demonstrate when the toxic agent is known and the renal damage is acute and dramatic, as in acute renal failure. Establishing the contribution of an occupational toxin to chronic kidney disease is considerably more difficult when there is a long latent period and etiology is multifactorial. When renal disease is a consequence of long-term, low-dose, asymptomatic exposure modulated by complex interactions with nutritional factors, systemic disease, lifestyle, and genetic susceptibility, the etiology may remain obscure. The impact of occupational nephrotoxins on the incidence of renal disease is therefore unknown.

In 1996, more than 280,000 Americans were treated for end-stage renal disease (ESRD) by hemodialyisis or peritoneal dialysis at a cost of $14.6 billion (1). The incidence of ESRD in that year in the United States was 268 per million and the prevalence rate was 1,041 per million. Despite extensive data collected for the U.S. Renal Data System, information on the causal role of occupational toxins in the induction of renal disease is sparse. In the United States, causes

TABLE 35-1. *Occupational and environmental renal diseases*

Exposure	Tubulointerstitial nephritis	Glomerulonephritis
Lead	+++	
Cadmium	+++	
Mercury	(+++)	+++
Arsenic, chromium, uranium	(+++)	
Solvents	(+++)	+++
Silica	+	+++

(+++), Tubular proteinuria usually without clinically important interstitial nephritis.

of ESRD are reported to be: diabetes, 39%; hypertension, 28%; glomerulonephritis, 11%; and interstitial nephritis/pyelonephritis, 4%—almost 85% of ESRD is accounted for by four diagnoses. Solvent-induced glomerulonephritis is hidden in the category "glomerulonephritis." Interstitial nephritis and hypertension include most of the clinically important renal diseases due to occupational nephrotoxins, including lead, cadmium, and mercury. The contribution of occupational nephrotoxins to ESRD in the United States, consequently, is also unknown.

TUBULAR VERSUS GLOMERULAR DISEASE

Regardless of etiology, renal diseases are usually divided into acute (onset over days or weeks) and chronic (onset over months

R. P. Wedeen: Research Service, Department of Veterans Affairs New Jersey Health Care System, East Orange, New Jersey 07018.

or years). Further characterization is based on anatomic site. If the major initial injury is to the glomerulus, the renal disease is termed *glomerulonephritis*. If the major initial injury is to the renal tubules, the disease is termed *tubulointerstitial nephritis*. Most occupational nephrotoxins, including lead, cadmium, mercury, and organic solvents, are selectively accumulated in the proximal tubule, where they produce intracellular damage, sometimes leading to acute tubular necrosis. If the person survives the initial insult, residual tubulointerstitial nephritis may remain after incomplete recovery. Prolonged exposure to doses insufficient to produce acute tubular necrosis can also cause tubulointerstitial nephritis. These histopathologic designations are particularly useful in the early stages of disease. As the end stage is approached, much of the unique histopathologic appearance that might signify primary glomerular or tubular injury is lost. Tissue damage, which is distinctly tubulointerstitial or glomerular at the outset, may be indistinguishable at the end stage, when all structures in the kidney show distortion and fibrosis.

Occupational nephrotoxins that induce primary glomerular disease (mercury, silica, and organic solvents) do so through immunologic mechanisms. Clinically, glomerular disease is readily recognized by the presence of albuminuria. Simple laboratory tests to detect glomerular proteinuria due to leaky glomerular capillaries have been available since the early 19th century. Nonglomerular disease (tubulointerstitial nephritis) is characterized by the absence of heavy albuminuria in its early phase and is therefore more difficult to detect until a substantial fraction of kidney function is lost. Not until the glomerular filtration rate (GFR) is reduced by more than 60% are the blood urea nitrogen (BUN) and serum creatinine concentrations elevated.

TUBULAR PROTEINURIA

Normally only a small fraction of circulating albumin and other high–molecular-weight proteins (HMWPs) pass through the glomerular capillary filter. Much of the protein that passes through the glomerulus is reabsorbed and catabolized in the proximal tubule, so that less than 300 mg of albumin per day appears in the urine. Nevertheless, slightly increased albuminuria (30 to 300 mg/day), although within "normal" limits, often heralds the future development of kidney disease in patients with diabetes mellitus.

The appearance of minute quantities of low–molecular-weight proteins (LMWPs) in the urine is an early effect of nephrotoxins. LMWPs, which are found in urine in micrograms-per-liter concentrations, must be distinguished from the heavy albuminuria in grams-per-liter concentrations associated with glomerular disease. In contrast to HMWPs, most circulating LMWPs pass through the glomerular filter, but only microgram quantities reach the final urine because of reabsorption and intracellular catabolism in the proximal tubule. Larger quantities are excreted in the presence of tubular injury, but still only in milligrams-per-liter amounts. Lysosomal enzymes, which are also HMWPs, may appear in the urine after proximal tubule damage, but these usually originate in damaged tubular cells rather than from the circulation.

Characteristic urinary excretion patterns of LMWPs, enzymes, certain growth factors, and prostaglandins have been found when workers with specific known exposures have been studied. There is good evidence that increased urinary excretion of β_2-microglobulin (B2M) or retinol-binding protein (RBP) signals the development of renal failure after cadmium absorption. In the absence of acute tubular necrosis, tubular proteinuria induced by other heavy metals or solvents is almost always reversible.

LEAD NEPHROPATHY

Acute Lead Nephropathy

Acute lead nephropathy is associated with severe acute lead poisoning, characterized by colic, encephalopathy, peripheral neuropathy, and anemia (Chapters 3, 15, 29, 31, and 33) (2). Renal findings include Fanconi's syndrome, a proximal tubular reabsorption defect characterized by aminoaciduria, glycosuria, hyperphosphaturia, and hypercalciuria. It has become evident that tubular proteinuria accompanies these proximal tubular reabsorptive defects. Histologically, the proximal tubules show characteristic acid-fast intranuclear inclusion bodies (Fig. 35-1), which have also been identified in other tissues. The renal findings are minor compared with the often overwhelming neurologic symptoms of lead encephalopathy. Acute lead poisoning occurs most frequently in children younger than 6 years of age as a result of ingesting nonfood materials (pica), usually leaded paint chips. Blood lead levels (BLLs) usually exceed 100 μg/dL.

Removal of lead by chelation therapy reverses the proximal tubule reabsorptive defect and removes the intranuclear inclusion bodies. If the child survives without chelation therapy, chronic interstitial nephritis may develop decades later. An epidemic of interstitial nephritis in young adults who had sustained lead poisoning in childhood at the turn of the century was identified in Queensland

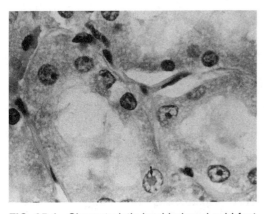

FIG. 35-1. Characteristic lead-induced acid-fast intranuclear inclusion body in proximal tubule of a lead worker who died of lead encephalopathy. (From Wedeen RP. Injury to the kidneys from heavy metals and radiation. In: Jamison RL, Wilkinson R, eds. Nephrology. New York: Chapman and Hall, 1997:732–741, with permission.)

Australia and is termed *Queensland nephritis* (2).

Chronic Lead Nephropathy

Lead poisoning has been recognized as a cause of kidney disease since the 19th century. Symptomatic lead poisoning from occupational exposure or contaminated alcoholic beverages permitted the identification of lead as an etiologic agent of interstitial nephritis.

Typically, renal failure is evident only after years of intense exposure and is frequently associated with hypertension and gout. Fifty percent of patients with lead nephropathy have *saturnine gout*. Although hyperuricemia is universal in renal failure, gout is rare in renal disease unrelated to lead. However, in the absence of renal failure, gout cannot usually be attributed to lead exposure, despite coexisting hypertension. As with other forms of interstitial nephritis, albuminuria is initially meager but increases as renal failure progresses.

In chronic lead nephropathy, the kidney shows the characteristic morphology of relatively acellular tubulointerstitial nephritis. Intranuclear inclusion bodies are usually absent when there is no current exposure. The appearance of arteriolar nephrosclerosis before hypertension and the relatively short duration of hypertension preceding renal failure suggest that the primary injury from lead is to the renal microvasculature (3,4). This view is consistent with the observation that creatinine clearance decreases with increasing BLL in the general population, an association that is independent of blood pressure (5).

Advanced lead nephropathy is not reversed by chelation. No improvement in renal function can be anticipated when the serum creatinine exceeds approximately 3 mg/dL. However, lead-induced interstitial nephritis may be accompanied by an acute, reversible functional component that is, at least in part, prerenal in origin. Chronic volume depletion and hyporeninemic hypoaldosteronism may contribute to the reversible renal dysfunction.

Even low-level lead absorption may contribute to renal failure. In Germany, people without unusual lead exposure but with modest renal failure (mean serum creatinine 2.5 mg/dL) had significantly higher chelatable lead levels than did control subjects without renal failure (6), even though renal failure *per se* does not appear to increase chelatable lead (3). Both groups had mean chelatable leads under 200 μg over 4 days, well below the upper limit of normal of 600 μg over 3 days. Patients in China with comparable modest renal failure and low body lead burdens showed a significantly decreased rate of progression of renal failure after chelation therapy with 8 weekly infusions of 1 g CaNa$_2$-ethylenediaminetetraacetic acid (EDTA) compared with untreated patients (7). The effect of low-level lead absorption on GFR is further supported by the Department of Veterans Affairs Normative Aging Study in Boston, where a rise in BLL of 10 $\mu g/dL$ was associated with a decrease in creatinine clearance of 10.4 mL/minute in an essentially normal population of men older than 43 years of age (8).

In the absence of symptoms of acute lead poisoning or known excessive lead exposure, the diagnosis of lead nephropathy is difficult. After exposure has ceased, the BLL falls with a biologic half-life approximating 2 months. BLL thus reflects recent, rather than past, cumulative lead absorption. However, 95% of the body lead stores are retained in bone with a biologic half-life approximating 20 years. Lead in bone is in slow, continuous equilibrium with blood so that as BLL falls, the contribution of bone stores to the BLL increases. Cumulative lead absorption is assessed by the EDTA lead mobilization test. Because renal failure reduces the rate of lead–chelate excretion, urine collection during the lead mobilization test is increased to 3 to 4 days after the administration of CaNa$_2$EDTA (4). The degree of renal failure that precludes performance of the lead mobilization test has not been determined; the

test is probably inappropriate when the creatinine exceeds approximately 6 mg/dL.

In vivo tibial x-ray–induced fluorescence (XRF) is a safe, noninvasive method for the measurement of the bone lead concentration (9). Because lead is stored in bone for decades, XRF provides more direct information on cumulative past absorption than either the BLL or the lead mobilization test.

Hypertension

Hypertension has been associated with lead poisoning since blood pressure measurements were first made in the late 19th century (2). The early view that lead poisoning causes hypertension has gained increasing support as careful epidemiologic studies have shown that BLL predicts blood pressure, even when both are within the accepted normal range (9). Bone lead levels determined by *in vivo* tibial XRF have been found to be higher among hypertensive than among normotensive subjects in the Normative Aging Study of 2,280 people followed for more than 30 years (10). A role for lead in the induction of hypertension or gout with renal failure (serum creatinine >1.5 mg/dL) has also been suggested by use of the EDTA lead mobilization test (11,12). Hypertensive patients with elevated levels of chelatable lead are designated *essential hypertensives* if the chelation test is not performed because of the absence of the acute symptoms of lead poisoning. Mortality data indicate that hypertensive cardiovascular disease is a more frequent cause of death among lead-exposed workers than among the general population (13). Some small epidemiologic studies have not found an association between lead and blood pressure, probably reflecting the reduced statistical power of small cohorts and the presence of numerous confounding variables in the determination of blood pressure.

The role of lead in the induction of the excess of essential hypertension and renal failure in African-American men remains to be elucidated. The coexisting highest BLLs and highest incidence of ESRD due to hypertension in African-American men has not been systematically investigated. African-American boys younger than 6 years of age have the highest BLLs in the United States, whereas the point prevalence of ESRD attributed to hypertension in black men is over eightfold higher than in white men (1).

Prevention

Acute childhood lead poisoning with BLLs over 100 μg/dL has been largely eliminated in the United States through educational and lead paint abatement efforts. Since the late 1970s, there has been a dramatic drop in blood levels in the United States, paralleling the removal of lead from gasoline. However, chronic low-level lead poisoning remains a problem in the United States and other countries.

Occupational exposure has also been reduced, but high exposures still occur in some occupations. The Occupational Safety and Health Administration (OSHA) requires medical monitoring for workers exposed to airborne lead concentrations greater than 30 μg/m^3. If the BLL is greater than 40 μg/dL, repeat blood measurements must be made every 2 months and a complete history and physical examination are required. If the BLL exceeds 50 μg/dL, medical removal protection for up to 18 months and follow-up are mandatory.

CADMIUM NEPHROPATHY

Cadmium is widely used in the manufacture of plastics, pigments, glass, alloys, and electrical equipment. An increased incidence of prostatic and lung cancer has been reported in cadmium workers, but the interpretation of these findings is debatable (14). After absorption, cadmium is transported albumin-bound to the liver, where it induces the synthesis of a carrier protein, metallothionein, within 24 hours. The cadmium–thionein complex is released from the liver into the blood, passes through glomerulus, is accumu-

lated in proximal tubules by pinocytosis, and transferred to lysosomes. Catabolism of the cadmium–thionein complex with release of unbound cadmium into the cytoplasm is believed to contribute to the proximal tubule injury as well as to the continuous resynthesis of metallothionein and prolonged retention in the renal cortex. The biologic half-life of cadmium–metallothionein is several days. The biologic half-life of cadmium in the kidney is approximately 30 days. The cadmium–metallothionein complex is approximately 15 times more nephrotoxic than either cadmium alone or the zinc–thionein complex. Total body stores of cadmium in nonoccupationally exposed adults range from 10 to 30 mg, with roughly one-third in the kidneys and one-third in the liver.

Proximal Tubule Dysfunction

The kidney is considered the "critical organ" for cadmium because the metal is accumulated and produces its most prominent toxic effect in the proximal tubule. When the intrarenal concentration reaches approximately 200 μg/g tissue, tubular injury is manifest by the appearance of low–molecular-weight proteinuria, enzymuria, aminoaciduria, renal glucosuria, hypercalciuria, and increased urinary excretion of cadmium. These early renal findings predict the later development of tubulointerstitial nephritis.

Although urinary cadmium excretion is normally less than 2 μg/day, after the critical concentration has been exceeded in the kidney, urinary cadmium in excess of 10 μg/day is usual. Clinically important abnormalities of proximal tubular function are associated with urinary cadmium excretion in excess of 30 μg/day. Blood levels greater than 1 μg/dL, as well as urine creatinine concentrations exceeding 10 μg/g, are considered evidence of excessive exposure.

Increased urinary excretion of LMWPs, such as B2M, *N*-acetyl glucosaminidase (NAG), or RBP, and human intestinal alkaline phosphatase are early indicators of cadmium nephrotoxicity. B2M (15) has been the most extensively examined LMWP in cadmium nephropathy, but because of its instability in acid urine, measurement of urinary RBP or NAG is more reliable. Proteinuria in cadmium workers rarely exceeds a few hundred milligrams per day and does not approach nephrotic levels (>3.5 g/day). Immunologic techniques are required for specific protein identification and measurement.

Renal calcium wasting is responsible for the high incidence of urinary tract stones among cadmium workers, which may reach 40%. Osteomalacia with pseudofractures and severe bone pain was the major clinical manifestation of environmental cadmium poisoning identified in Japan after World War II. This syndrome, known as *itai-itai byo* ("ouch ouch disease"), arose from consumption of rice contaminated by cadmium from rivers polluted by metal manufacturing operations. The disease affected primarily postmenopausal, multiparous women who had subsisted on calcium-deficient and vitamin D–deficient diets for decades. The victims manifested a waddling gait, shortened stature, anemia, glycosuria, and elevated serum alkaline phosphatase. Hypertension was absent. Excretion of B2M often exceeded the normal maximum (1 mg/g creatinine) by 100-fold, and GFRs were substantially reduced in severely affected patients.

In vivo measurement of liver and kidney cadmium content by neutron capture gamma-ray analysis has proved to be a safe, accurate, noninvasive, and portable method for assessing cumulative cadmium retention in the liver and kidney. Once renal failure is clinically apparent, the renal cortex begins to lose cadmium; therefore, in renal failure, neutron activation shows stable or falling, rather than increasing, cadmium content, even when exposure is continued. As renal cadmium content decreases, liver cadmium continues to increase.

Cadmium nephropathy cannot be reversed once urine cadmium exceeds approximately 30 μg/g creatinine or when tubulointerstitial disease has progressed sufficiently to produce renal failure. When urine cad-

mium is less than approximately 20 μg/g creatinine, cessation of exposure can prevent progression and reverse tubular dysfunction (16). Chelation therapy with EDTA is ineffective after cadmium has accumulated in the kidney.

Prevention

In 1992, OSHA set the maximum permissible exposure level for airborne cadmium in the workplace at 5 μg/m^3 8-hour time-weighted average. At airborne exposure levels of half this value, air monitoring must be performed regularly and respirators provided for workers. If the airborne cadmium level exceeds 2.5 μg/m^3 for more than 30 days per year, medical surveillance is mandated, including medical examinations and determination of blood cadmium, urine cadmium, and urine B2M. The goal is to maintain urine cadmium less than 3 μg/g creatinine, urine B2M less than 300 μg/g creatinine, and blood cadmium less than 0.5 μg/dL.

MERCURY NEPHROTOXICITY

The principal symptoms after environmental or occupational exposure to mercury are neurologic, although acrodynia is still occasionally encountered in infants after the application of mercurial ointments for skin rashes.

Occupational exposure to elemental mercury with urine concentrations exceeding 50 μg/L is associated with increased human intestinal alkaline phosphatase and NAG excretion, but little increase in other LMWPs, other renal enzymes, or prostaglandins. Tubular proteinuria after exposure to elemental mercury is reversible (17). There is no evidence that tubular proteinuria predicts the development of the nephrotic syndrome or renal failure after exposure to mercury.

Metabolism

Inorganic mercury is selectively accumulated in the kidney, where it is retained with a half-life of approximately 2 months (Fig. 35-2). The effect of mercury on the kidney is deter-

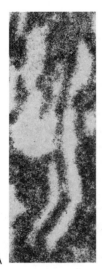

A   B

FIG. 35-2. Autoradiograph **(A)** and periodic acid-Schiff stain of underlying tissue **(B)** from a dog that received a diuretic dose of Hg^{203}-labeled chlormerodrin. Mercury is selectively accumulated in convoluted and straight segments of proximal tubules. (From Littman E, Goldstein MH, Kasen L, Levitt MF, Wedeen RP. Intrarenal distribution of Hg^{203}-chlormerodrin. Fed Proc 1964; 23:362, with permission.)

mined by its chemical form and the genetic background that modulates susceptibility to autoimmune disease. Neither elemental mercury nor the mercurous salt (Hg_2Cl_2, calomel) produce sustained renal tubular injury despite induction of tubular proteinuria. Tubular proteinuria occurs only after more than 100 μg/L of mercury appears in the urine. Biotransformation of inorganic mercury to mercuric ions is believed to account for renal accumulation. However, mercuric chloride ($HgCl_2$), at parenteral doses of more than 1 mg/kg body weight, regularly produces acute tubular necrosis. Divalent mercury binds avidly to metallothionein and sulfhydryl groups in the kidney. Recovery from tubular necrosis may be incomplete, leaving calcified tubular remnants, persistent tubulointerstitial nephritis, and chronic renal insufficiency.

The inorganic and organic mercurials bind avidly to sulfhydryl groups in circulating proteins and amino acids as well as to intracellular glutathione, cysteine, and metallothionein. Selective accumulation of the mercuric

ion in the proximal tubules is accomplished by absorptive endocytosis from the luminal side of mercury bound to amino acids or proteins. Mercury is released into the cytosol by intralysosomal enzymatic degradation (18). It is less clear if mercury moves across the proximal tubular epithelium in the secretory direction, entering proximal tubular cells from the peritubular surface. Excretion is primarily through bile in the feces. Enzymuria from elemental, methyl, or phenyl mercury is useful as an indicator of exposure for the prevention of the neurologic effects, but whether such low–molecular-weight proteinuria signifies clinically important renal damage remains to be determined. Potassium-wasting nephropathy has been described from accidental poisoning with methyl mercury fungicide (19).

Immunologically Mediated Glomerulonephritis

Case reports of the nephrotic syndrome developing after occupational or therapeutic exposure to mercury suggest idiosyncratic immunologic reactions in humans. Kidney biopsies frequently show immune complex deposits in glomerular basement membranes indicative of membranous glomerulonephritis. A specific antigen has not been identified in the immune complexes, and both normal glomeruli and anti-glomerular basement membrane antibody disease have been reported. Low–molecular-weight proteinuria, even when accompanied by minimal albuminuria, should not be confused with the massive albuminuria of the nephrotic syndrome; very different pathophysiologic mechanisms appear to be involved.

Subcutaneous doses of $HgCl_2$ too low to induce acute tubular necrosis have been shown to induce membranous glomerulonephritis in rodents with defined genetic backgrounds. This mercury-induced, immunologically mediated glomerulonephritis may serve as a model for understanding mercury-induced glomerulonephritis in humans. The disease has been produced with various forms of mercury, different routes of administration, and doses as low as 0.005 mg per 100 g body weight. The autoimmune response is under precise genetic control and is actually biphasic (20). Anti-glomerular basement membrane antibodies are found in glomeruli after 1 week. These are replaced by granular IgG deposits in arteriolar walls and mesangium, as well as in the glomerular basement membrane (GBM), 2 and 3 weeks after disease onset: The changes in the GBM are those of immune complex glomerulonephritis. Immunoglobulin localization in the glomeruli is associated with heavy proteinuria, circulating immune complexes, and polyclonal B-cell activation owing to anti-self Ia autoreactive T cells. The glomerular disease can be transferred to T-cell–depleted rats by T cells and T helper cells taken from $HgCl_2$-treated rats of the same genetic background. The mercury-induced nephrotic syndrome disappears spontaneously after termination of exposure. British antilewisite (BAL) is an effective chelating agent for mercury when administered at an initial dose of 5 mg/kg intramuscularly.

URANIUM

The injection of uranyl nitrate into experimental animals is a standard technique for the induction of experimental acute tubular necrosis. Extensive experience in humans connected with development of the atomic bomb in the Manhattan Project during World War II made it clear that the kidney is the primary site of acute toxicity (21). U^{6+} readily enters the bloodstream after inhalation and is filtered at the glomerulus as a bicarbonate complex. The bicarbonate complex breaks down to UO_2, which binds to intracellular proteins, producing necrosis in the second and third parts of the proximal tubule. Catalase, alkaline phosphatase, and B2M excretion are increased in uranium workers, but chronic renal failure induced by uranium has not been described in humans.

SILICA (SILICON DIOXIDE)

Crystalline silicon has a metallic luster and grayish sheen and is therefore sometimes considered metallic—although with a specific gravity of only 2.3, it is not a heavy metal. Silicosis (see Chapter 25), the major effect of inhalation of silicon dusts that occurs in miners, sandblasters, and glass manufacturers, stimulates an autoimmune response characterized by the presence of circulating anti-nuclear antibodies, rheumatoid factor, and immune complexes. In its severe form, nodular pulmonary fibrosis occurs, often complicated by tuberculosis and systemic manifestations of connective tissue disease simulating scleroderma or lupus erythematosus. Focal glomerulosclerosis and glomerular immune complex deposition have been described. Rapidly progressive crescentic glomerulonephritis in association with elevated anti-nuclear antibodies (ANA) may complicate a fulminant form of silicosis known as silicoproteinosis. Anti-nuclear cytoplasmic antibodies (ANCA)-positive glomerulonephritis and Wegener's granulomatosis have been reported after exposure to silica with little evidence of pulmonary silicosis (22). The odds ratio for patients with ESRD due to Wegener's granulomatosis who worked as sandblasters is 3.8 compared with matched control subjects. Tubular proteinuria is found in workers exposed to silica dust (23), suggesting a toxic effect on the proximal tubule that appears to be distinct from the immunologically mediated glomerular disease.

ARSENIC

Chronic low-level exposure from contaminated well water has been responsible for an epidemic of peripheral neuropathy, hyperpigmentation, and hyperkeratosis of the palmar surfaces of the hands in India that was first recognized in 1984 (24). An excessive incidence of cancers of the skin, bladder, kidney, and lung has been reported in populations heavily exposed to arsenic in drinking water (25). In Taiwan, gangrene of the extremities similar to Buerger's disease, known as *blackfoot disease*, has been attributed to arsenic in water obtained from artesian wells (26). Chronic renal disease has, however, rarely, been reported after occupational or environmental exposure to arsenic (27).

Arsenic is a common ingredient of insecticides, which may be ingested accidentally or deliberately. Severe acute poisoning results in cardiovascular collapse and acute renal failure. Acute tubular necrosis has also resulted from exposure to arsine gas (AsH_3) in industrial accidents. Arsine, used as a poisonous gas in World War I, is a colorless, odorless gas evolved when arsenicals are mixed with acid. Inhalation produces massive hemolysis, hematuria, jaundice, and abdominal pain within a few hours, and acute tubular necrosis within a few days. Hemodialysis is required, and exchange transfusion may be lifesaving, removing the arsenic–hemoglobin complex from the circulation. Incomplete recovery from acute tubular necrosis has resulted in persisting chronic tubulointerstitial nephritis in surviving victims. Incomplete recovery from patchy cortical necrosis after consumption of illicitly distilled whiskey ("moonshine") contaminated with arsenic has also been reported to result in persisting renal disease (see also Chapters 15, 16, 27, 29 and 33) (28).

CHROMIUM

Acute oliguric renal failure and tubular necrosis occur after massive absorption of hexavalent chromium as the chromate or dichromate. Renal failure is not produced by trivalent chromium. Chromium is selectively accumulated in the proximal tubule, but there is little evidence of chronic renal disease resulting from usual occupational exposure. Minimal tubular proteinuria in the absence of reduced glomerular filtration has been reported in chromeplaters when urine chromium exceeds 15 $\mu g/g$ creatinine, but negative findings have also been reported (29). The finding of an odds ratio of 2.7 for occupational exposure to chromium in a case–control study of chronic renal failure nevertheless warrants further evaluation of the association of environmental exposure to chromium with chronic renal disease (30).

ORGANIC SOLVENTS

Halogenated hydrocarbons have often been implicated in the induction of acute tubular necrosis or Fanconi's syndrome in both humans and experimental animals. Low-level occupational absorption by inhalation of volatile hydrocarbons or absorption through the skin may also induce tubular proteinuria, which does not necessarily signify the presence of clinically important renal disease, glomerular damage, or immune system activation (31).

Light Hydrocarbon Nephropathy

Toxicologic studies of the effects of gasoline distillates have identified an effect of these complex mixtures on the renal tubule of male rats. Referred to as *light hydrocarbon nephropathy*, the tubular injury is induced by exposing Fischer 344 male rats to petroleum hydrocarbon vapors from a few hours up to a few years. Hyaline droplets form within epithelial cells of proximal tubules. Sustained renal failure with permanently reduced GFR has not been reported in light hydrocarbon nephropathy in humans or experimental animals.

Light hydrocarbon nephropathy, which occurs regularly in heavily exposed humans and experimental animals, should not be confused with the relatively rare glomerular disease in humans referred to as *solvent nephropathy*. Although such tubular proteinuria is common, the massive albuminuria of solvent nephropathy is distinctly unusual.

Severe neurologic toxicity has been reported after glue sniffing among teenagers seeking intoxication. Permanent as well as transient neurologic and hepatic damage has resulted. Fanconi's syndrome, caused by proximal tubular reabsorptive defects, has been observed after recreational glue sniffing, apparently the result of toluene, mixed with acetone, isopropyl alcohol, ethyl acetate, and trichloroethylene. Immunologically mediated glomerulonephritis and the hemolytic–uremic syndrome have also been reported after glue sniffing. In addition, distal renal tubular acidosis and myoglobinuria have been reported with acute renal failure and residual chronic tubulointerstitial nephritis.

Toluene is metabolized to benzoic acid and then to hippuric acid, which are selectively accumulated in, and secreted by, the proximal tubules. These organic acids may contribute to high–anion-gap metabolic acidosis because they are "unmeasured anions" in the blood. However, significant renal disease has not been found in cross-sectional studies of industrial workers regularly exposed to styrene, toluene, and xylene, or in oil refinery workers. The finding of increased low–molecular-weight proteinuria in cohorts exposed to solvents, oils, or hydrocarbons suggests an effect on the renal tubule that is distinct from solvent-induced glomerulonephritis, an immunologically mediated disease.

Solvent Nephropathy

At least 40 epidemiologic studies have examined the relationship between glomerulonephritis and exposure to organic solvents. A number of these concluded that patients with chronic glomerulonephritis have been exposed to both aliphatic and aromatic organic solvents more frequently than patients with other diseases. Initially, solvent nephropathy was associated with Goodpasture's syndrome (anti-glomerular basement membrane antibody–mediated glomerulonephritis and pulmonary hemorrhage). Subsequently, a wide variety of glomerular diseases have been associated with excessive exposure to solvents in the workplace.

The etiologic role of solvents remains controversial because the dose and composition of industrial solvents are usually unknown. Moreover, of the thousands of workers exposed, very few contract glomerular disease. The genetic and environmental factors that make specific individuals susceptible to solvent nephropathy have not been delineated.

On the other hand, it has been proposed that solvent exposure may contribute to the progression of preexisting kidney disease in susceptible people (32).

BLADDER CANCER

In 1885, Rehn reported three cases of bladder cancer among 45 German workers exposed to aniline dyes. Historically, this was one of the first occupational cancers recognized, after Percival Pott's 1775 description of scrotal cancer among English chimney sweeps. By 1912, it was established that the incidence of bladder cancer was 23 times higher in dye workers than among the general male population (33). As was the case with the chimney sweeps, knowledge of etiology did not result in prevention. At least seven clusters of bladder cancer involving over 750 workers have been identified in the United States since 1930. In 1992, a 25-fold, dose-dependent increase in bladder cancer was discovered among 1,972 benzidine-exposed workers in China (34). An increased incidence of bladder cancers has also been reported to occur in populations exposed to excessive arsenic in drinking water (see Chapter 16) (25).

Benzidine, 4-aminobiphenyl, and β-naphthalene are the major carcinogenic chemicals responsible for bladder cancer. These carcinogens are found in cigarette smoke as well as in the paper, textile, leather, and dye manufacturing industries. Benzidine-based dyes are metabolized to benzidine, which is classified as a confirmed human carcinogen by the American Conference of Governmental Industrial Hygienists, and is particularly associated with bladder cancers. These dyes are no longer manufactured in the United States. 4-Aminobiphenyl, β-naphthalene, and other heterocyclic arylamines are activated to potent carcinogens by O-acetylation and oxidation to N-hydroxylarylamines by cytochromes, such as P-450. The highly mutagenic products are deactivated by N-acetylation by cytosolic transacetylases located in the bladder mucosa. Glucuronide formation in the liver may also reduce carcinogenicity of activated intermediates before they reach the bladder. The metabolites are particularly carcinogenic in the bladder because they are concentrated in urine, have a relatively long dwell time, and are further activated by acid urine. The mutagenic intermediates also appear to increase the incidence of laryngeal, colorectal, hepatic, and breast cancer.

The *N*-acetylation reaction that reduces the carcinogenicity of the aniline dye group of chemicals is an inherited trait involved in the *in vivo* detoxification of a number of pharmacologic agents, including hydralazine, procainamide, isoniazid, and sulfamethazine. Genetic control of *N*-acetylation can be detected by measuring the rate of appearance of acetylated sulfamethazine in urine. Slow acetylators have an increased risk of bladder cancer compared with fast acetylators exposed to the same chemicals. The acetylation reaction thus provides the potential basis for ecogenetic investigations to make the workplace safe. It also provides genetic information on individual susceptibility, which can be abused by excluding slow acetylators from the workforce.

KIDNEY CANCER

Renal cell carcinoma most frequently presents with flank pain, hematuria, and a flank mass. However, a variety of manifestations, including fever, weight loss, hepatic dysfunction, amyloidosis, and anemia may first bring the patient to clinical attention. Dramatic paraneoplastic syndromes may occur, including erythrocytosis, hypertension, hypercalcemia, and feminization—presumably due to erythropoietin-, renin-, parathyroid hormone-, and gonadotropin-like activity of tumor proteins. A variety of chemicals have been reported to produce renal adenocarcinomas in experimental animals, including aromatic hydrocarbons, aromatic amines and amides, aliphatic compounds, *N*-nitroso compounds, and arsenic. Most rats fed a 1% lead acetate diet acquire renal cancers after 1 year, but lead does not appear to be carcin-

ogenic in humans. An increased risk for renal cell cancer has sometimes been found among workers exposed to dry cleaning solvents (e.g., perchloroethylene), petroleum products, and asbestos (see Chapter 16) (34).

REFERENCES

1. U.S. Renal Data System. USRDS 1998 annual data report. Bethesda, MD: National Institutes of Health, National Institute of Diabetes and Digestive and Kidney Diseases, 1998.
2. Wedeen RP. Poison in the pot: the legacy of lead. Carbondale, IL: Southern Illinois University Press, 1984.
3. Batuman V, Landy E, Maesaka J, et al. Contribution of lead to hypertension with renal impairment. N Engl J Med 1983;309:17–21.
4. Wedeen RP, Maesaka JK, Weiner B, et al. Occupational lead nephropathy. Am J Med 1975;59:630–641.
5. Staessen J, Lauwerys RR, Bernard A, et al. Renal function is inversely correlated with lead exposure in the general population. N Engl J Med 1994;327:151–156.
6. Koster J, Ehrhardt A, Stoeppler M, et al. Mobilizable lead in patients with chronic renal failure. Eur J Clin Invest 1989;19:228–233.
7. Lin J-L, Ho H-H, Yu C-C. Chelation therapy for patients with elevated body lead burden and progressive renal insufficiency. Ann Intern Med 1999;130:7–13.
8. Payton M, Hu H, Sparrow D, et al. Low-level lead exposure and renal function in the Normative Aging Study. Am J Epidemiol 1994;140:821–829.
9. Wedeen RP. Lead, the kidney, and hypertension. In: Needleman H, ed. Human lead exposure. Boca Raton, FL: CRC Press, 1992:170–189.
10. Hu H, Aro A, Payton M, et al. The relationship of bone and blood lead level to hypertension. JAMA 1996;275:1171–1176.
11. Batuman V, Maesaka JK, Haddad B, et al. The role of lead in gout nephropathy. N Engl J Med 1981;304:520–523.
12. Sanchez-Fructuoso AI, Torralbo A, Arroyo M, et al. Occult lead intoxication as a cause of hypertension and renal failure. Nephrol Dial Transplant 1996;12:1775–1780.
13. Nuyts GD, Dalemans RA, Jorens G, et al. Does lead play a role in the development of chronic renal disease? Nephrol Dial Transplant 1991;6:307–315.
14. Friberg L, Elinder CG. Kjellstrom T, et al, eds. Cadmium and health: a toxicological and epidemiological appraisal. Boca Raton, FL: CRC Press, 1986.
15. Roels HA, Bernard AM, Cardenas A, et al. Markers of early renal changes induced by industrial pollutants: III. application to workers exposed to cadmium. British Journal of Industrial Medicine 1993;50:37–48.
16. Roels HA, Van Assche FJ, Oversteyns M, et al. Reversibility of microproteinuria in cadmium workers with incipient tubular dysfunction after reduction of exposure. Am J Ind Med 1997;31:645–652.

17. Ellingsen DG, Barregard L, Gaarder PI, et al. Assessment of renal dysfunction in workers previously exposed to mercury vapour at a chloralkali plant. British Journal of Industrial Medicine 1993; 50:881–887.

18. Zalups RK, Lash LH, Advances in understanding the renal transport and toxicity of mercury. Journal of Environmental Toxicology and Health 1991; 42:1–44.

19. Szylman P, Benzakin A, Szjnader Y, et al. Potassium wasting nephropathy in an outbreak of chronic organic mercurial intoxication. Am J Nephrol 1995; 15:514–520.

20. Goldman M, Baran D, Druet P, et al. Polyclonal activation and experimental nephropathies. Kidney Int 1988;34:1411–1450.

21. Dounce AL. The mechanism of action of uranium compounds in the animal body. In: Voegtlin C, Hodge HC, eds. Pharmacology and toxicology of uranium compounds, division VI, vol 1. New York: McGraw-Hill, 1949:951–991.

22. Kallenberg CGM. Renal disease: another effect of silica exposure. Nephrol Dial Transplant 1995;10:1117–1119.

23. Hotz P, Lorenzo J, Fuentes E, et al. Subclinical signs of kidney dysfunction following short exposure to silica in the absence of silicosis. Nephron 1995; 70:130–138.

24. Subramanian KS, Kosnett MJ. Human exposures to arsenic from consumption of well water in West Bengal, India. Int J Occup Med Environ Health 1998;4;217–230.

25. Smith AH, Goycolea M, Biggs ML. Marked increase in bladder and lung cancer mortality in a region of Northern Chile due to arsenic in drinking water. Am J Epidemiol 1998;147:660–669.

26. Lin T-H, Huang Y-L, Wang M-Y. Arsenic species in drinking water, hair, fingernails, and urine of patients with blackfoot disease. Journal of Environmental Toxicology and Health 1998;53:85–93.

27. Prasad GCR, Rossi NF. Arsenic intoxication associated with tubulointerstitial nephritis. Am J Kidney Dis 1995;26:373–376.

28. Gerhardt RE, Crecelis A, Hudson JB. Moonshine related arsenic poisoning. Arch Intern Med 1977; 140:211–213.

29. Wedeen RP, Haque S, Udasin I, et al. Absence of tubular proteinuria following environmental exposure to chromium. Arch Environ Health 1996; 51:329–332.

30. Nuyts GD, Van Viem E, Thys J, et al. New occupational risk factors for chronic renal failure: defined and undefined occupational risk factors of chronic renal failure. Lancet 1995;346:711.

31. Roy AT, Brautbar N, Lee DBN. Hydrocarbons and renal failure. Nephron 1991;58:385–392.

32. Hotz P, Pillod J, Bernard A, et al. Hydrocarbon exposure, hypertension and kidney function tests. Int Arch Occup Environ Health 1990;62:501–508.

33. Bi W, Hayes RB, Feng P, et al. Mortality and incidence of bladder cancer in benzidine-exposed workers in China. Am J Ind Med 1992;21:481–489.

34. Mandel JS, McLaughlin JK, Schlehofer B, et al. International renal-cell cancer study: IV. occupation. Int J Cancer 1995;61:601–605.

BIBLIOGRAPHY

Nakagawa H, Nishijo M. Environmental cadmium exposure, hypertension and cardiovascular risk: review in depth. J Cardiovasc Risk 1996;3:11–17.
A review of the controversies surrounding the renal-cardiovascular effects of cadmium.

Sheehan HE, Wedeen RP, eds. Toxic circles: environmental hazards from the workplace into the community. New Brunswick, NJ: Rutgers University Press, 1993.
Historical perspectives on occupational urinary tract cancers from aniline dyes and scrotal cancers from petroleum products. Other occupational exposures in the past that have become environmental exposures in the present in New Jersey are also examined.

Wedeen RP. Nephrotoxicity secondary to environmental agents and heavy metals. In: Schrier R, Gottschalk C, eds. Diseases of the kidney, 6th ed. Boston: Little, Brown, 1997:1231–1247.
An in-depth review of occupational renal diseases.

Yaqoob M, Bell G. Review: occupational factors and renal disease. Ren Fail 1994;16:425–434.
A review of the contribution of hydrocarbon exposure (organic solvents) to end-stage renal disease.

PART V

Selected Groups
of Workers

36 Women and Work

Margaret M. Quinn, Susan R. Woskie, and Beth J. Rosenberg

Most women in the United States today perform wage-earning work outside of the home. There are more than 60 million working women, comprising nearly half (46%) of the U.S. workforce (1). Work offers many economic and social advantages to a woman and her family, and it has been shown that women who work outside of the home live longer than women who do not (2). Although wage-earning work has many advantages for women, as for men, there are associated hazards. The reporting of occupational health hazards and symptoms, as well as illness and injury rates, vary by sex. These differences are primarily the result of a constellation of social and economic factors that affect working men and women differently. In certain situations, there is evidence that the biologic differences between women and men contribute to their having different occupational health experiences. In most work situations, however, it is impossible to separate the effects of biologic differences from the complex social and economic forces that lead

M. M. Quinn: Department of Work Environment, University of Massachusetts Lowell, Lowell, Massachusetts 01854.

S. R. Wolskie: Department of Work Environment, University of Massachusetts Lowell, Lowell, Massachusetts 01854.

B. J. Rosenberg: Department of Family Medicine and Community Health, Tufts University School of Medicine, Boston, Massachusetts 02111.

men and women into different work experiences (2,3).

Social and economic factors determine many important aspects of a person's work experience, including the type of job likely to be held, the likelihood of being employed and remaining employed, income, opportunities for job advancement, and the degree of authority in workplace decisions. Despite women's recent progress in entering some jobs that have traditionally been held by men, most of the work that women and men do in industrial societies is still highly segregated. Most of the leading 20 professions for women in the United States in 1997, which are shown in Table 36-1, have 70% or more women employed in them. This U.S. list is similar to those of other industrially developed nations. In Sweden, for example, it is estimated that only 10% of all employees are working in nonsegregated occupations (defined as occupations employing between 40% and 60% of either sex) (3). Thus, most women do different jobs than men.

In cases where a woman does the same job or a comparable job as a man, she is still likely to have a different experience. Even within the same job title, women may be assigned to different tasks and may be treated quite differently by supervisors, coworkers, and clients. For example, among commercial cleaners, light cleaning tasks are assigned to women and heavy cleaning tasks to men. Light work is not considered as important as heavy work, and this view is re-

TABLE 36-1. *Twenty leading occupations of employed women, 1997 annual averages*

Occupations	Employed (no. in thousands)	Percent women
Total, 16 years and over (all employed women)	59,873	100.0
Secretaries	2,989	98.5
Cashiers	2,356	78.3
Managers and administrators	2,237	30.2
Registered nurses	1,930	93.5
Sales supervisors and proprietors	1,780	38.4
Nursing aides, orderlies, and attendants	1,676	89.4
Bookkeepers, accounting and auditing clerks	1,602	92.3
Elementary school teachers	1,571	83.9
Waiters and waitresses	1,070	77.8
Sales workers, other commodities	1,014	69.2
Receptionists	970	96.5
Accountants and auditors	921	56.7
Machine operators, assorted materials	913	31.5
Cooks	888	41.8
Textile, apparel, and furnishing machine operators	781	72.1
Janitors and cleaners	756	34.0
Investigators and adjusters, excluding insurance	735	74.8
Administrative support occupations	702	76.8
Secondary school teachers	685	58.4
Hairdressers and cosmetologists	676	90.4

From U.S. Department of Labor, Bureau of Labor Statistics, Women's Bureau, Washington, DC, 1999.

flected in the wage scale and in the social attitudes of supervisors and coworkers toward the different tasks and the workers doing them (2).

The best documented socioeconomic man-ifestation of different gender experiences for comparable work is in wages. Despite a rhetoric of equality, as of 1997, average earnings for women in the United States were 74 cents to every dollar earned by a man. This wage

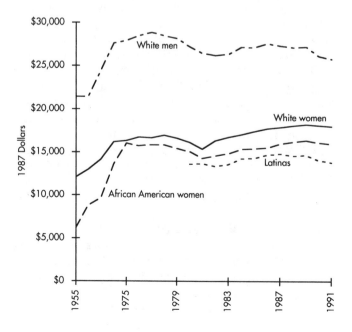

FIG. 36-1. Median annual income of women and white men, full-time, year-round workers, 1955 to 1991 (in 1987 dollars). (From Amott T. Caught in the crisis: women and the U.S. economy today. New York: Monthly Review Press, 1993.)

discrepancy widens as the worker ascends the career ladder. For Latinas and African-American women, the overall gap is wider (Fig. 36-1). Women comprise the majority of low-wage workers. More than 75% of working women earned less than $25,000 a year in 1995, compared with 54% of men (4).

Six of every 10 mothers of children younger than 6 years of age are in the labor force (5). In Boston in 1995, housing, food, clothing, and transportation costs for a family with one preschool and one school-age child were $13,000. The additional costs of working nearly double the minimum amount needed to maintain the family, without even considering taxes. Such costs of working include child care ($646 per month), transportation to work ($106 per month), and health care coverage, even assuming the employer pays 30% of the premiums ($214 per month). The total needed for these basic payments is $25,500 per year (6). However, a woman whose wage equaled the nationwide single-mother average of $9.71 per hour earned only $20,200 a year (6).

The number of families maintained by women grew by almost 90% between 1970 and 1985. In 1993, women headed 12 million families in the United States. The percentage of families headed by black women has grown dramatically, more than doubling between 1970 and 1993. In 1993, 48% of all black families were headed by women, compared with 25% of Latino families and 14% of white families. Over half of children in families with a female head of household lived below the poverty level. Over two-thirds of African-American and Latino children whose mothers supported them lived in poverty, primarily because African-American and Latino women maintaining families had lower median earnings, lower median ages, and higher unemployment rates than white women maintaining families.

Women who work outside of the home still do most of their household work as well. Added to the pressures of long hours of work inside and outside the home are the time conflicts that emerge when the person is a

homemaker, the primary family caretaker, and a wage earner. The responsibilities of caring for sick children, children on school holidays, and ill or dependent elderly relatives often fall on women. Inflexible work schedules and the lack of decent, accessible, and affordable child care and elder care exacerbate the burden. Although the restructuring of work to include more part-time and "temporary" work has had some advantages for women, there are serious drawbacks as well. Not only is there a loss of income from the shorter work week, but part-time work also pays less and has fewer benefits (6). In addition, part-time workers are more likely to be at a disadvantage for professional development because they are not considered to be as serious or dedicated to their work as those working full time.

On the other end of the spectrum, women are moving into heavy industrial jobs (Fig. 36-2), the building trades, and professional fields in small but increasing, numbers. These women experience a new set of occupational

FIG. 36-2. Many women now work in jobs that were traditionally held only by men. (Photograph by Earl Dotter.)

health concerns as they adapt to a physical environment and work organizational structure that was designed for the average man.

SOURCES OF OCCUPATIONAL STRESS

Stress can be defined as a physical or psychological stimulus that produces strain or disruption of the person's normal physiologic equilibrium. Men and women may experience a wide range of stress reactions, including adverse health effects (see Chapter 21).

Problems of Multiple Roles

Although men are assuming an increasing domestic role, particularly with regard to child care (3), working mothers still spend more time on household labor than their male counterparts. Working mothers sleep less, get sick more, and have less leisure time than their husbands (7). A study on working women found that the subjects focused on sleep, how little they could survive on, which of their friends needed how much sleep— and then they apologized for the amount that they needed. The author concluded, "These women talked about sleep the way a hungry person talks about food" (7). A 1993 study found that single mothers with one child younger than 5 years of age spend an average of 25 hours per week on unpaid domestic labor, whereas a mother with a spouse and a young child spends twice as much time as the man—34 versus 17 hours a week—doing housework, child care, and household shopping (8). These extra hours spent in working at home have two profound implications. One is that the stress and fatigue from balancing work life and home life present a serious problem for women and their partners and their children. The other is that when a woman becomes incapacitated because of a workplace injury, such as repetitive strain injury, which is associated with many women's jobs (Box 36-1), the effects on her home life are tremendous because of the household's dependence on her free labor. A large Italian women's group, formed by a coalition of women from political parties ranging from conservative to progressive, has emphasized that the structure and organization of work in industrialized societies is built around the lives of men. The Italian women's group developed an alternative, detailed proposal for organizing work and related commercial, government, and social support services around the life cycle and needs of the working mother. The group argues that work organized in this way would accommodate all members of society because working mothers have the most diverse social needs resulting from the combined demands of workplace, household, and child care (9).

The Structure of Women's Work: Part-time, Temporary, and Home Work

Employers are increasingly using part-time and temporary workers ("temps" or contract workers). Often, this arrangement provides the desired flexibility in working hours that women need. There are costs, however. On average, part-time workers earn only 60% as much as full-time workers on an hourly basis. Not only are they paid less, but their benefits, such as health insurance, pensions, paid sick leave, and vacation, are substantially less than those of full-time workers. Fewer than 25% of part-time workers have employer-paid health insurance, compared with nearly 80% of full-time workers. Sixty percent of full-time workers have pensions, whereas only 25% of part-time workers have this coverage. In 1990 in the United States, there were nearly 5 million part-time workers who preferred to be employed full time. Women make up more than two-thirds of all part-time workers in the United States, and over half of them work less than full time, contrary to their wishes (10). Two of every three temporary workers are female (11). Temporary workers often live with the stress of not knowing when they will be working. They also tend to work more overtime because they are often hired for "crunch periods" when intense work needs to be done to meet

Box 36-1. Commonly Asked Questions about Musculoskeletal Disorders and Gender[1]

Laura Punnett

1. Do women get more musculoskeletal disorders (MSDs) of the upper extremity or low back than men?

Being female is often described as a "risk factor" for many MSDs in that prevalences in the general population appear to be approximately twice as high among women as among men. However, in a review of the differences in MSD frequency between men and women, it was found that after adjusting for sex differences in job demands, there was no obvious increase in the risk of upper extremity disorders among women compared with men (12). Being female was associated with double or more the risk of upper extremity MSDs in only approximately one third of the studies analyzed. For low back pain, the risk was more often lower for women than for men. Information is still very limited, however, as to whether women have a higher MSD risk than men in the few situations where they have similar exposures.

2. What causes MSDs in women? Is it biology, the types of jobs, or the psychosocial conditions under which most women work?

The causes of MSDs in female workers are largely similar to the causes for male workers (see Chapters 9 and 26), although the specific types of occupational ergonomic stressors experienced by women are often different. Men are less likely than women to work in sedentary jobs with repetitive hand motion, such as clerical work or stitching. Other "women's jobs" involve a lot of whole-body motion and lifting, such as nursing aides, industrial cleaning, and laundry work. Both patterns of demands are associated with musculoskeletal strain and injury; the stereotype of women's work as "light" is not justified on the basis of the MSD risks incurred by female workers. In

one large study, workers with repetitive jobs had a very high risk of carpal tunnel syndrome (CTS), whereas the risk attributed to sex, although elevated, was 20 times smaller (13). Preventive measures should not and need not be discriminatory to be effective.

The effects of psychosocial job conditions on MSDs are less well studied but are highly plausible: monotony, low control over decision making, and poor social relations at work are associated with physiologic indicators of strain such as increased cortisone levels, reduced circulation to the musculoskeletal tissues, and excess muscle tension. In a Finnish study, monotonous work content and low social support at work were associated with MSDs among blue-collar workers of both sexes (14). Low work control at baseline predicted 10-year musculoskeletal morbidity of the neck/shoulder in white-collar women and of the low back in blue-collar women, although it had no predictive value among men.

If there is a sex difference in MSD occurrence—which is not certain, its interpretation is challenging. Different occupational exposures might be responsible, either because women are more often found in repetitive, monotonous jobs, because many worksites fail to accommodate female body size and shape, or because women sustain or perceive higher levels of psychosocial job strain than men. Housework may be a factor; for example, women more often care for children, and having young children at home has been associated with increased risk of musculoskeletal pain, perhaps because of lifting and carrying the children. There has been little study of the ergonomic features of housework or their effects on MSD risk.

In traditionally male-dominated jobs, women workers often feel that they need to work harder or take more risks than their male colleagues take to prove themselves to their supervisors and coworkers (15). On this issue, the psychosocial aspects of the work environment, especially a lack of social support, have potential consequences for the physical exposures, if overcompensation leads to workers injuring themselves, such as in lifting too heavy loads.

Another possible factor may be a greater willingness by women to report symptoms to workplace supervisors or to medical care providers. Although some have accused female workers of overreporting symptoms and malingering, there is little evidence that MSD symptoms reported by women are less

[1] A fuller set of citations for material in this Box may be found in Punnett L, Herbert R. Work-related musculoskeletal disorders: is there a gender differential, and if so, what does it mean? In: Goldman MB, Hatch MC, eds. Women and health. San Diego, CA: Academic Press, 1999.

Box 36-1 (*continued*)

valid than those reported by men. A few studies have suggested that men may be more likely to deny MSD symptoms than women, even when they have similar findings on physical examination.

Last, there are some physiologic factors that differ by sex and may predispose to the development of MSDs. These include body size, muscle strength and muscle fiber type distribution, endocrine hormones, and body changes associated with pregnancy (see below).

3. What, if any, role do hormonal factors have in the development of MSDs in women?

The possible role of female hormones in development of MSDs is poorly understood. Tendons and ligaments have receptors for female sex hormones, but the implications of these findings are unknown. It is possible that hormones fluctuating during the menstrual cycle or pregnancy may contribute to differences in regulation of ligament and tendon function or in susceptibility to tissue inflammation. Sex hormones may also influence biochemical changes related to joint cartilage degeneration through regulation of the inflammatory response.

Use of oral contraceptives and surgical removal of the ovaries, both of which affect endocrine hormone levels, are sometimes cited as risk factors for CTS among women, although the epidemiologic literature is inconsistent as to their importance. In a few studies, low back pain has been associated with menstruation, oral contraceptive use, induced abortion, number of live births, menopausal symptoms, and lower age at menopause, but the evidence is sparse and the mechanisms of these associations are largely unknown. Data are also lacking as to whether such indicators of hormonal status interact with the effect of ergonomic exposures at work on MSDs.

4. Does pregnancy increase the risk of MSDs? Are there any MSD preventive actions recommended during pregnancy in particular?

A number of physical changes during pregnancy may place women at least temporarily at increased risk for MSDs (see section on The Pregnant Worker). Increased body weight and changes in body weight distribution change the fit between the worker's body and the workplace layout. For example, pregnant women sit further from work tables than do nonpregnant women, with increased compensatory trunk flexion and arm flexion and extension. Postural strain is markedly decreased when the work surface height can be adjusted by the worker to accommodate her changing body dimensions.

Muscle strength and physical fitness are not substantially affected by pregnancy, although whole-body lifting capacity is altered in the later months as the center of gravity moves and as increased body size prevents objects from being lifted close to the body. In addition, the ligaments and muscles of the back and abdomen are stretched, making lifting potentially more hazardous. Pregnant women also have altered connective tissue function, including increased peripheral joint laxity, possibly because of release of relaxin or other hormones. There is some evidence of increased risk of low back injury for pregnant women performing specific tasks such as heavy lifting, standing, and frequent climbing of stairs, although there has been little effort to distinguish low back pain secondary to pregnancy *per se* from an increase in susceptibility to occupational ergonomic factors during pregnancy.

Carpal tunnel syndrome during pregnancy is a well described phenomenon. The cause has not been completely elucidated, although it has been hypothesized that fluid retention in pregnancy may be sufficient to lead to nerve compression at the wrist. It is not known if pregnant women working in high-risk jobs are at greater risk for development of CTS than pregnant women who do not sustain ergonomic exposures. In general, patients with CTS recover spontaneously after delivery, but the possible role of postpartum occupational exposures in preventing the usual recovery has not been investigated. Although pregnancy accounts only for a small portion of women's work lives, the relationship between the potential increased risk of MSD development in pregnancy and subsequent inability to heal due to sustained biomechanical insult in the workplace, even after pregnancy has ended, is troublesome, particularly in nations such as the United States that do not provide for prolonged maternity leave from work.

Box 36-1   (*continued*)

5. Do women with MSDs have different outcomes than men with MSDs (such as absenteeism, disability, interference with social and family life, financial losses?)

Despite the widespread belief that MSDs disproportionately affect women, the outcomes of these conditions have been examined primarily among men. One study found that return to work after CTS treatment was markedly lower for women than men, whereas another showed that a rehabilitation program for chronic back pain benefited only the female participants. However, other studies have found no sex differences in the clinical or occupational outcomes of MSDs.

Women workers differ from their male counterparts in several important ways that might influence their outcomes. First, as noted, women's jobs are more repetitive and offer less decision-making autonomy than men's jobs; thus, women may have more difficulty than men in remaining at work after a musculoskeletal injury because it interferes with job performance or the job reaggravates the injury. Second, women usually have more responsibility for household work and care of family members, so the family burden may have more influence on return-to-work decisions by women than by men; further, the cost of lost household services greatly outweighs wage losses for women. Physicians' perceptions of the work demands in men's and women's jobs may also differentially affect medical treatment or the success of a compensation claim and thus the financial option to stay out of work.

6. How do women compare with men in the strength and flexibility needed for heavy lifting or other physical demands? Is strength testing an accurate and fair method to determine whether a woman can handle a certain level of physical work?

Women's total body strength is, on average, approximately two thirds that of men's; however, the ratio in static strength (ability to move a stationary weight) ranges from 35% to 85%, depending on the tasks and muscles involved. Women's average strength is relatively lower in the upper extremity and closer to men's for leg exertions, dynamic lifting, pushing and pulling activities, and manual handling of smaller containers. The difference is also smaller when men and women have similar industrial experience or athletic training. There is also substantial overlap in the strength distribution between men and women, as much as 50% or more for certain muscle groups. In all, the factors of sex, age, weight, and height explain only approximately one third of the variability in human strength data.

The musculoskeletal health implications of this strength differential are not clear. In a group of Swedish students followed from 16 to 34 years of age, several measures of physical capacity in adolescence, such as flexibility and strength, were found to influence the development of neck, shoulder, and low back symptoms in adulthood in both sexes (16). However, several other studies have been inconclusive as to whether muscle strength is protective against low back pain, and at least one has shown high muscle force capacity to be a risk factor. Although stronger muscles are capable of generating higher internal forces, they do not imply greater strength in other soft tissues, such as nerves and spinal discs. The low predictive value of muscle force capacity may also be related to the fact that women are more often employed in repetitive, low-effort work, where the limiting factor may be physiologic tolerance for prolonged static work rather than the ability to lift a heavy object.

Strength testing has been proposed as a gender-neutral hiring criterion that might protect both men and women from MSDs. Because there is a large overlap between population distributions of the static strength of men and women, selection of individuals able to lift a given weight will not exclude all prospective female employees from heavy jobs. However, because it is not clear that strength is protective against MSDs, using such criteria might erroneously exclude people who would not incur injury and include those who would. Some have argued that there is a sex bias inherent in standard strength testing procedures because motion patterns observed in the workplace are typically used to determine the relevant testing maneuvers, but women of-

Box 36-1   (*continued*)

ten use different strategies to perform physically demanding tasks (17). Furthermore, strength tests do not measure other impor-tant physiologic capacities such as dynamic (aerobic) endurance or range of motion, which differ less between the sexes or even favor women, and which may be as or more important to the risk of injury.

---

a deadline. Neither part-time workers nor temps receive equal protection under government laws, including occupational safety and health regulations, unemployment insurance, and pension regulations. Few are represented by labor unions (10). A case study commissioned by the Occupational Safety and Health Administration of contract labor in the petrochemical industry showed that contract workers get less health and safety training and have higher injury rates than noncontract workers (18).

Another cost-cutting measure affecting women is the rise in work conducted in workers' homes—"homework." In 1949, Congress passed a law making industrial homework illegal because it was so difficult to enforce labor standards, such as the minimum wage. The Reagan Administration made it legal again, which caused a growth in this type of work in the 1980s (10). Typical female homeworkers are garment workers; costume jewelry, microelectronics, and other manufacturing assembly workers; clerical workers; independent contract workers; and entrepreneurs. Home garment and assembly workers and some clerical workers are usually paid by piece rates rather than by hourly rates. Piece-rate payments lead to increased speed of performance of job tasks, long hours, and, combined with a repetitive or sustained posture, a high rate of injuries, many of which are due to poor workplace or workstation design (see Chapters 9 and 26).

Homework can be related to both occupational and environmental problems because the handling and disposal of chemicals and other materials often is not adequately controlled. For example, in semiconductor manufacturing homework, workers and their family members are exposed to hazardous chemicals, which can also contaminate residential sewage systems. In addition, children can be exposed to hazardous agents used in the home, and it is very difficult to enforce child labor laws with homework. Trade unions, which have played a major role in improving working conditions, are weakened with homework because workers are isolated from each other.

Job Control

A method to identify stressful components of jobs and work organizational structures, the "demand and control" model, has been validated in the United States and other countries in Europe and Asia (see Chapter 21) (19). Using this model, jobs are analyzed according to the degree of control a worker has in decision making and in the required skill level. Women more frequently report low authority over decisions and low skill discretion in their work—that is, uncontrollable, hectic, and monotonous work (19). A Swedish population-based study of five counties found that the excess risk of coronary heart disease associated with hectic and monotonous work was higher for women than for men. The authors concluded that, despite the much lower incidence of coronary heart disease in working women, this type of job strain is associated with risks for women at least as strong as those for men (19).

One study examined the widely held opinion that certain people are more prone to stress and that they are responsible for causing their own stress problems. Workers in typically female jobs were found to have much less control over decision making than those in typically male jobs. Female-dominated occupations, such as clerical work, electronics manufacturing, garment work, and poultry processing, are characterized by tedium, ergonomic hazards, and low job control. It may be the concentration of women in these jobs that accounts for the higher prevalence of stress-related disorders in women, rather than their lack of ability to cope (20).

Recognizing the social, economic, and physical determinants of health effects related to occupational stressors, instead of focusing solely on personal pathology, is a first step in the complete and long-term management of stress-related problems. Although many women may benefit from programs that provide individual coping and relaxation exercises, workplace stress management programs should also address the broader social and economic constraints that provide the context for the daily lives of working women (see Chapter 21).

Violence in the Workplace

Violence is about power and control. It is a process that involves using force or deliberately trying to intimidate someone; the abuser may feel frustration or powerlessness or be fighting loss of control (21). Violence includes verbal intimidation as well as assault. It has been institutionalized in our society in the forms of sexual harassment, racial discrimination, and age discrimination. Although workplace violence is a serious public health hazard, it has been minimized, like "family violence," or tolerated as "part of the job" (21). The social denial of violence occurs, at least in part, because most violent actions, whether in the home or in the workplace, are against women.

In 1997, women comprised 46% of the workforce, yet only 8% of occupational fatalities. However, homicides accounted for 31% of the occupational deaths of women, compared with 12% for male workers (22). Industries with the highest proportion of deadly violent assaults on women included retail trade (24% of all deaths) and service occupations (24%). However, when the number of women working in various industries is accounted for, the death rate per 100,000 workers was highest among female construction workers (1.8), miners (1.7), and agriculture/forestry/fishing workers (1.6) (23).

Fifty-seven percent of nonfatal violent assaults were experienced by women workers (24). The largest proportion of these victims worked as caregivers in nursing homes and hospitals, where they experienced hitting, kicking, or beating by their patients to a degree that resulted in a median of 4 days away from work.

A framework to analyze and address workplace violence has been developed by the Health and Safety Executive (HSE) in Britain (21). The HSE, as well as other workplace violence researchers, have identified that the main risk factor is working with "the public." Nearly all the leading 20 jobs held by women (see Table 36-1) involve interfacing with the public. As a first step in developing effective control methods, the HSE recommends the identification of those aspects of workplace violence that are common across job types and organizations. The starting point is a five-element framework including a profile of (a) *the assailant*: personality, temporary conditions (e.g., drug or alcohol abuse, illness, or personal stress), presence of negative/uncertain expectations, age and maturity; (b) *the employee*: appearance (e.g., physical build and wearing of a uniform), health, age, experience, sex, personality and temperament, attitudes, and expectations; (c) *the interaction*: type of contact (may be direct, as in nursing or child care, or indirect, as in service work over the telephone), and potential for someone to believe that the worker, or system the worker represents, is being unfair or unreasonable; (d) *the work environment*: the total context of the job (e.g., details of work and features of the organizational structure, culture, and physical environment), relations between coworkers and managers, working alone, job location, handling cash, waiting, time of day, and privacy; and (e) *the outcome*: physical injury, threats with a weapon, verbal abuse, and angry behavior (21).

Workplace prevention strategies include safer cash handling procedures, physical separation of workers from the public, better workplace design to limit access and improve lighting and visibility, and use of security systems and personal protective devices. Administrative controls include improvements in staffing patterns, clear systems for communication among staff regarding client/patient aggressive behaviors, development of workplace policies, and procedures around workplace violence and workplace training (25).

Discrimination

Racial, gender, and age discrimination all contribute to the stress experienced by working women (see Chapters 37 and 39). These types of discrimination take many forms and affect women in economic terms (lower pay than men who do comparable work), social terms (isolation from supervisors and coworkers), and personal terms (low self-esteem and reduced creative growth). Women who work in occupations in which most workers are male are often seen as unwelcome intruders, and many experience gender-based harassment. These working women may also feel excessive pressure to perform faultlessly to prove that they are good enough. These experiences, which are likely to be even more extreme for women who are part of racial or ethnic minorities, can have serious consequences for health and safety.

Sexual Harassment

Women are the predominant targets of sexual harassment at work. Any unwanted verbal or physical sexual advance constitutes harassment, which can range from sexual comments and suggestions, to pressure for sexual favors accompanied by threats concerning the person's job, to physical assault, including rape. Studies indicate that 40% to 80% of women have experienced some form of sexual harassment at work (20). In a study of more than 500 cases of sexual harassment, 46% of the women said that it interfered with their work performance and 36% reported physical ailments that they associated with the harassment, including nausea, vomiting, depression, headaches, and drastic weight change (20).

Sexual harassment is not only harmful to women's health, it is costly to business in terms of job turnover, sick time, impaired productivity, and the cost of legal claims. It was estimated that in 1988 the U.S. Government spent $189 million on the cost of sexual harassment. Almost one-third of the 500 largest U.S. companies spend a total of $6.7 million a year because of this problem (26).

Sexual harassment is illegal. It is a violation of rights under Title VII of the Civil Rights Act and of many state fair employment practice laws. A woman who is fired or resigns as a result of a harassment situation may be entitled to unemployment compensation. Organized protest against managers for sexual harassment is protected activity under the National Labor Relations Act.

A woman can seek help from her union, coworkers, and local women's organizations. Talking to other people and getting assistance can help the woman feel less isolated and frustrated and is probably necessary to deal effectively with the situation. It may be necessary to seek legal help and to go to a local or state agency that deals with fair employment practices, or the closest office of the federal Equal Employment Opportunity Commission. Unfortunately, legal remedies cannot fully address the devastating effects of sexual harassment. The incidence of sexual harassment may decrease when the balance of power is more equal between men and women in the workplace.

ERGONOMIC HAZARDS

Many workplaces are a haphazard layout of tools, machines, and workstations. Little thought is given to the fit of tools, the nature of the lifting tasks, or the fit of personal protective equipment (PPE) to the worker. This lack of thought results in injuries that could be prevented if basic guidelines for lifting and for job and tool design were used. Although these issues are not particular to women, this group often bears the brunt of poor workplace design because most tools, workstations, and PPE were developed for use by the "average" man.

Repetitive, forceful, and awkward motions have been associated with a number of occupational musculoskeletal disorders (see Chapters 9 and 26). For example, among nurses and nurse's aides, those aged 20 to 29 years who spent more time lifting had a higher prevalence of low back pain than those whose jobs did not require heavy lifting (27). Neck, shoulder, and upper limb disorders have been associated with a number of jobs held largely by women, including keyboard operators of typewriters, telex and calculating machines, computers, and telephone exchangers; cash register operators; film roller and capper workers; cigarette rolling and packing workers; and scissors manufacture and assembly workers. Tenosynovitis has been reported among female assembly line packers in a food production factory, female poultry-processing plant workers, and female scissors manufacture and assembly workers. Carpal tunnel syndrome (CTS) has been found among female garment workers, hotel cooks, maids, workers in the boning department of a poultry-processing plant, and cash register operators, including checkout counter workers who use bar-code readers. Packaging operations are particularly apt to produce musculoskeletal disorders because of the repetitive nature of the work. Because women are concentrated in jobs that require repetitive motions, such as bench assembly and small parts manufacturing, it is difficult to determine whether there are aspects of this disease that are solely related to sex (see Box 36-1). However, reduction of major musculoskeletal injuries can be accomplished through careful consideration of the ergonomics of the work process. Improvements of tool and workstation design can contribute significantly to preventing musculoskeletal disorders (see Chapters 9 and 26).

One company made improvements in the working conditions of cash register operators, including shortened operating time, development of a work rotation system, and change from mechanical to electronic registers with lighter key touches. The result was a decrease in the workload on the arms and hands and a significant drop in symptoms in the hands, fingers, and arms (28).

When video display terminal (VDT) workers in an office were given an adjustable workstation that they could adapt to their preferred settings for a number of factors, including keyboard and screen height, viewing angle, and screen distance, they reported significantly fewer muscle complaints and significantly less impairment in the neck, shoulder, back, and wrist (29).

PROBLEMS WITH PERSONAL PROTECTIVE EQUIPMENT

Personal protective equipment should not be the primary method of controlling a worker's exposure to a workplace hazard, but it can be important in temporary situations, emergencies, or situations that cannot be controlled in other ways. However, for PPE to work, it must fit. In addition, PPE, such as gloves, that does not fit can increase the risk of injuries and increase the strength requirements for a task. For women, it is often difficult to find manufacturers who make equipment in women's sizes. Often small, medium, and large sizes of equipment are available, but even the small size is designed for small men, not small women. A survey of over 350 companies found that, in women's sizes, only 14% provided ear protection, 58% hand protection, 18% respirators, 14% head and face protection, 50% body protection, and 59% foot protection (30). Many women end up purchasing men's PPE and trying to modify it to fit, or they purchase women's equipment that is not up to safety standards. Until PPE is available in a variety of sizes for women, it may actually be contributing to the risk of workplace injuries.

REPRODUCTIVE HEALTH HAZARDS

Occupational reproductive hazards are often viewed as a woman's personal problem. As a result, the individual woman is left to bear the burden of the social, economic, and health consequences of reproductive hazards, while the hazards faced by men are ignored. In fact, almost all occupational crises with documented adverse reproductive effects, such as exposure to dibromochloropropane, chlordecone (Kepone), exogenous estrogens, and dimethylaminoproplonitrile, have involved men (see Chapter 31). In situations in which men have been at risk, control of the hazard was achieved by elimination of the exposure. However, when even the potential for a reproductive health problem has existed for female workers, control of the "hazard" has often been achieved by eliminating women from the job, particularly if they are seeking a job in an industry that has not traditionally hired women. Both men and women will benefit if it is recognized that reproductive hazards may seriously compromise the health of all workers as well as their children.

There has been increasing concern over the role of chemicals that may be recognized by the endocrine systems of both men and women as a hormone, particularly estrogen. *Endocrine disrupters* or *environmental estrogens* have been proposed as risk factors for breast cancer in women (31) and reproductive disorders, such as low sperm counts, in men (32). Chemicals suspected of estrogenic activity are used or produced in the occupational setting and then dispersed in the environment as products or effluents in air and water. These chemicals include (a) phthalates, which are used as plasticizers in plastics used for food wraps and that may constitute as much as 30% to 40% of the total weight of some plastic products such as polyvinyl chloride blood bags and intravenous bags

and tubing; (b) dioxins, which may be formed under certain conditions during the incineration of polyvinyl chloride and other plastics; and (c) possibly certain pesticides, polychlorinated biphenyls, and organic solvents (33). It is difficult to study the health effects of these substances in human populations because they are diffused widely throughout industrialized populations in air, water, foods, and occupational settings. The Environmental Protection Agency is now funding the development of assays to identify environmental estrogens, and is recommending testing of suspected agents.

In addition to chemical agents, ergonomic stressors can be related to reproductive health problems. There is some evidence that heavy lifting, heavy industrial cleaning, long periods of standing at work, and other strenuous exertions are associated with spontaneous abortions, premature birth, and low birth-weight (see Box 31-1 in Chapter 31). The effect of ergonomic stressors on the pregnancies of women working in the home has not been evaluated.

THE PREGNANT WORKER

Many organ system and musculoskeletal changes occur during pregnancy to accommodate the needs of the developing fetus. The most evident changes are modifications in cardiovascular, respiratory, and metabolic functions as well as shifts in the center of gravity associated with weight gain. These changes are normal and healthy, and may or may not affect a woman's ability to work. Pregnant women can usually continue to perform the physical activities to which they have been accustomed; however, pregnancy may not be the time to change to a new or unfamiliar level of work activity unless the woman undertakes a carefully supervised program of physical conditioning. Some medical conditions can be compromised by pregnancy; others predispose the pregnant woman to an increased likelihood of complications during pregnancy. The American Medical Association has developed guidelines for the continuation of various levels of work during pregnancy (34).

As with pregnancy, data about postpartum readiness to resume work are lacking. The American College of Obstetricians and Gynecologists stated the following in its *Guidelines on Pregnancy and Work*: "The normal woman with an uncomplicated pregnancy and a normal fetus in a job that presents no greater potential hazards than those encountered in normal daily life in the community may continue to work without interruption until the onset of labor and may resume working several weeks after an uncomplicated pregnancy" (35). A woman's ability to work during pregnancy and return to work after pregnancy should be determined by the woman and her health care provider, considering the requirements of her job, her health status at the time she becomes pregnant, and her pregnancy experience.

The First Trimester

During the first trimester, many women experience nausea and vomiting as well as fatigue and breast swelling and tenderness. There is a large increase in total blood volume, which results in an increased heart and metabolic rates. The ventilation rate begins to increase (from a nonpregnant average of 7 L/minute to an average of 10 L/minute by term). Glomerular filtration increases by approximately 50% above the nonpregnant level. These physiologic changes may affect a working woman in several ways. The vomiting and fatigue may decrease her capacity for work. Some women are particularly sensitive to bad odors that may increase their nausea. The increased metabolic rate raises body temperature, increasing sensitivity to hot and humid environments. The increased blood volume decreases the total percentage of red blood cells in the circulating blood and leads to an anemia, which is normal for a pregnant woman. However, it is especially important that a pregnant woman does not already have anemia, which can be caused by workplace exposures such as lead or benzene. Because the percentage of hemoglobin-carrying red blood cells is decreased, she is also more vulnerable to environmental agents that interfere with the blood's ability to carry oxygen, such as carbon monoxide. The increased ventilation rate can result in increased absorption of any toxic materials in the air. Increased kidney function results in increased urination, making access to suitable bathroom facilities particularly important (35).

The Second Trimester

Although the body changes mentioned persist throughout pregnancy, many of the symptoms disappear and women often find they feel better during the second trimester. By the end of this period, there is a weight gain of approximately 7 kg (15 lb.) and the uterus grows approximately 28 cm (11 in.) above the pelvis. As the uterus enlarges, its bulk tilts the body forward and the lower spine curves inward. The pelvic joints also become increasingly mobile. In addition to aggravation of any preexisting back problems, women often experience low back discomfort and stiffness (see Box 36-1). The change in the center of gravity may result in decreased balance. Tolerance to physical exertion varies widely, and it may be difficult to sustain high exertion levels if nausea or other health problems develop. In general, women who are in good physical condition before pregnancy have fewer problems and are capable of greater exertion. Along with these changes in body structure, there may be a greater tendency for the blood to pool in the legs. This can lead to dizziness and fainting with prolonged standing or working in hot environments, which increase the pooling. Varicose veins may also develop under these work conditions (35).

The Third Trimester

During the third trimester, the uterus continues to enlarge and total body weight gain increases to an average of 11 kg (24 lb.). Peripheral edema is common because there may be a decrease in venous return from the legs owing to the pressure of the uterus on the pelvic veins. As the third trimester progresses, many women also experience increasing fatigue, insomnia, and shortness of breath. These symptoms may be caused by the uterus pushing on the diaphragm, increased respiratory demands with a tendency toward oxygen debt, and discomfort associated with weight gain. Near delivery, the uterus pressing on the bladder may cause women to be incontinent or to urinate frequently (35). Prolonged standing or jobs requiring balance, endurance, exertion, and work in hot environments or in locations remote from bathroom facilities may become increasingly difficult.

Breastfeeding

Chemical exposures in the workplace may present a hazard to women and the infants they breastfeed. Toxic substances are passively transferred from plasma to breast milk if they are lipid soluble, polarized at body pH, and have a low molecular weight. These include many drugs, alcohol, some components of cigarette smoke, and many occupational and environmental toxins such as lead, mercury, halogenated hydrocarbons, and organic solvents. The dose and duration of exposure to the infant also depend on how quickly the substance is metabolized or excreted by the mother. For chemicals such as dichlorodiphenyltrichloroethane (DDT), polychlorinated biphenyls (PCBs), and related halogenated hydrocarbons, which are stored in body fat, breast milk can be a major route of excretion. The infant's dose may thus be as high as the mother's even after she has been removed from the exposure. However, solvents, although lipid soluble, are also metabolized or excreted through the liver, lung, and kidneys as well as breast milk, so maternal body burden rapidly decreases after removal from exposure (36).

What the Health Care Provider Can Do

Information on a pregnant woman's work activity and that of her partner, including wage-earning work and work done at home, and chemical and physical exposures should be an essential part of the comprehensive perinatal health history. This information should be obtained from her at the first prenatal visit and reconfirmed by inquiry at each subsequent visit. This inquiry can often best be done by means of a questionnaire (34). In some instances, it may be important to augment the questionnaire information, with the woman's permission, with information from the physician, nurse, or industrial hygienist in her employee health unit; from the plant safety director; or from the union representative. (See information on the occupational history in Chapter 4.)

Legal Rights of Pregnant Workers

When I told my new employer that I was pregnant, they lowered the amount of money they had offered to pay me, actually telling me I was worth less to them now. When my husband told his boss he was going to be a parent, he got a raise.

The 1978 Pregnancy Discrimination Act, an amendment to Title VII of the 1964 Civil Rights Act, requires that women affected by pregnancy and related conditions must be treated the same as other employees and applicants for employment when an employer determines their probable ability or inability to perform a job. This law protects a woman from being fired or refused a job or promotion merely because she is pregnant. A woman unable to work for pregnancy-related reasons is entitled to disability benefits, sick leave, and health insurance (except for abortions), just like employees disabled for other medical reasons. Under the law, pregnant workers temporarily unable to perform their jobs must be treated in the same manner as other disabled employees, such as by modifying the task, changing the work assignment, or granting disability leave or leave without pay. An employer who assigns a pregnant woman to another job because she cannot perform her regular work or because the job represents a hazard, however, can reduce her pay to that of the new job (Fig. 36-3).

The employer cannot enforce a rule prohibiting a return after childbirth for a set period of time. The woman's ability to return to work is the only test. Unless the employee on leave has informed the employer that she does not intend to return to work, her job must be held open on the same basis as jobs are held open for employees on sick or disability leave for other reasons. During her pregnancy-related leave, the employee must receive equal credit for seniority, vacations, or pay raises.

Pregnancy cannot have a role in the decision to hire an employee. If the applicant can perform the major functions required by the position, pregnancy-related conditions cannot be considered. The employer cannot refuse to hire her because of any real or imagined preference of coworkers, customers, or suppliers. In addition, the woman's request for maternity leave must be honored if the employer grants employees the right to leaves of absence for such purposes as travel or education.

If a woman believes she has been discriminated against because of childbirth or pregnancy-related conditions, she may sue the employer. The law requires filing a complaint first with the state agency dealing with discrimination. If this state agency does not proceed with informal mediation or legal action under state law after 60 days, the woman can file a complaint with the federal Equal Employment Opportunity Commission. The complaint must be filed within 6 months of the discriminatory event (37). The Family and Medical Leave Act of 1993 provides for a total of 12 weeks' leave from work without pay and return to work at the same job or a job with equal pay, status, and benefits. Those eligible are men and women caring for a newborn or adoptive or foster child, or for a sick child, spouse, or parent. The worker requesting the leave must have been employed with the same company for at least 12 months and have worked at least 1,250 hours in the past year. In addition, the company must have at least 50 employees and have 50 employees who work within 75 miles of the work location where the leave is requested. This law provides only limited protection because many women are employed in workplaces with fewer than 50 employees.

FIG. 36-3. Adequate control of lead exposure for all workers requires a high level of engineering control and may also include the need for personal protective equipment. Because U.S. law prohibits exclusion of pregnant women from lead work, controls must be sufficient to protect the fetus as well (see also Chapters 3, 13, 29, 31, and 35). (From Zenz C. Occupational medicine: principles and practical applications. Chicago: Year Book, 1975. Reproduced with permission.)

SPECIFIC OCCUPATIONS

Most U.S. women are employed in a only few occupational categories, such as clerical, service, health care, and manufacturing assembly. At the same time, a major shift in the employment of women has been into jobs that have traditionally been held by men, or "nontraditional jobs." Nontraditional jobs for women are defined by the Women's Bureau of the U.S. Department of Labor as those in which women make up 25% or less of the total number of workers. The number of women entering traditionally male occupations increased, at least in part, because of federal legislation and affirmative action programs.

Office Work

"For a year-and-a-half, I worked 3 feet from a copier . . . . The fumes got so bad, I felt like I was being poisoned . . . . It's important that I have won a workers' compensation claim . . . but I cannot place an order for another pair of lungs and I cannot wipe the worried look from the faces of the members of my family when I can't breathe, can't stop coughing, when I have to sit up to sleep . . . ."—Former secretary before the House Subcommittee on Health

There are over 18 million clerical workers in the United States, and nearly 80% of them are women. As illustrated by the preceding statement, despite its image of safe, clean, white-collar work, there is a growing awareness of the hazards associated with office work. Among the more obvious safety hazards are slippery floors, open file cabinet drawers, electrical cords strung across the floor, swinging doors, and movement of bulky and heavy objects, such as cases of paper and office furniture.

The design of the office and its workstations is important in determining the extent to which noise, lighting, and ventilation are problems. The use of fluorescent light can produce overillumination and flicker that can result in headaches and light sensitivity. By using desk blotters and task lighting controlled by the worker, the shadows and glare that are often the source of eye strain can be reduced. Once the ventilation system, copiers, fax machines, printers, and telephones are operating in a modern, open office space occupied by many people, it is not uncommon for the noise level to exceed the 45 to 55 dBA recommended for easy office and phone conversation (38).

Many office buildings have serious indoor air pollution (see Chapter 23). The combination of poor ventilation design, sealed buildings, the buildup of chemicals from building materials and office products, office machines, cleaning product residue, air fresheners, and cigarette smoke has resulted in an office smog in many buildings. Photocopiers can produce ozone, nitropyrenes, and ultraviolet light. Duplicating machines emit ethanol, methanol, and ammonia. Formaldehyde is found in carbonless paper, building materials, carpets, and draperies. Several solvents, including trichloroethane, tetrachloroethylene, and trichloroethylene, are used in liquid eraser products. Pesticides used in the building can remain in the air for extended periods (38).

In buildings constructed primarily between 1930 and 1976, asbestos was used as insulation for ducts and pipes, as fire retardant on the structural steel of the building, and as a spray-on ceiling material. Office workers can be exposed to asbestos when it degrades from age, water damage, or disruption during renovations and when computer, phone, or electrical lines are installed between floors in a building. Motor vehicle exhaust is a frequent air contaminant in buildings that have air ducts near busy streets, parking garages, or loading docks. Buildings made of granite, bricks, or cement may accumulate surprisingly high levels of radon emitted from these materials. Offices in the basements of buildings may also have radon gas exposures from the soil. Microorganisms can flourish in the air-conditioning and humidifying systems, evaporative condensers, and cooling towers in many office buildings. The result may be allergies and respiratory infections, such as Legionnaires' disease, that sometimes can reach epidemic proportions. Cigarette smoke can increase the level of respirable particles in the air to five times that of a nonsmoking office. Because research has linked the cigarette smoking of a spouse with the increased lung cancer risk of a nonsmoking spouse, nonsmoking office workers may also be at risk.

Controlling the indoor air pollution of an office involves curtailing, alternating, or substituting some processes; isolating those sources of toxins that must be used; and improving the fresh-air ventilation of most buildings. For example, photocopiers should be kept in a separate room and vented, and substitutes should be used for carbonless copy paper containing sensitizing agents. Most office buildings have reduced fresh-air circulation to cut energy costs. The American Society of Heating, Refrigeration, and Air Conditioning Engineers recommends 20 cubic feet per minute (cfm) of fresh outdoor air per person where smoking is permitted (5 cfm where it is not) (39). In addition, the relative humidity should be between 40% and 60%. By improving ventilation, the buildup of office smog can be avoided. Workplace smoking policies that prohibit smoking in office buildings significantly reduce office air pollution.

With the introduction of computers and VDTs into the office, a series of health problems have occurred. Among these problems are eye strain, headaches, neck and shoulder pain, and symptoms of CTS and tenosynovitis.

Most VDTs in use today do not emit detectable levels of ionizing radiation (x-rays or gamma radiation). Nonionizing radiation, in the form of very–low-frequency and extremely–low-frequency radiation, can be emitted from VDTs that are not shielded; however, there is much scientific debate about the health significance of exposure to this form of radiation (see Chapter 17). In the early 1980s, a series of miscarriage and birth defect clusters were reported among VDT users. There was some concern that they might be related to radiation leaks from VDTs. Since then, several studies have found inconsistent or inconclusive results regarding the association between VDT use and birth defects or miscarriages. However, these stud-

ies have been small and thus limited in their sensitivity.

Of the many recommendations made regarding VDTs, perhaps the most important is the need to provide workers with some control over their work patterns and environment. Many VDT data entry positions have been set up with required hourly keystroke rates that are constantly monitored. The combination of this pressure, poor workplace design, and few, if any, breaks contributes to the health problems experienced by these workers. NIOSH has recommended a 15-minute break for every 2 hours of VDT work, or 15 minutes every hour if the workload or visual demands are high (40).

A review of the literature on VDT design suggests that there is no consensus on the best ergonomic parameters for a VDT workstation. Adjustable chairs and VDT tables with separate and adjustable platforms for keyboard and screen should be used so that each worker can find the best fit of viewing angle, distance, and keyboard height. Supports for palm, hand, wrist, and arm may be desirable for prolonged typing. Chairs should provide a high seat back with proper support for both the lower and upper back. The VDT screen should have a glare reduction device, where necessary, and background lighting should be low with a separate light for the hard copy. The VDT unit should be positioned in the office to minimize glare on the screen. Each worker should be able to adjust the lighting and layout of the workstation to his or her own comfort (Fig. 36-4).

Hospital Work

Over 2.3 million women work in hospitals and another 4.5 million in other health services. Women make up over 80% of the total workforce of these industries but less than 24% in "health diagnosing" occupations, such as physician. In all health care occupations, women earn approximately 80% of what men earn, with the percentage decreasing as the skill level of the occupation increases (see Chapter 43).

FIG. 36-4. A model video display terminal (VDT) workstation. In addition to specific features (indicated by letters), it has adequate ventilation, no excess noise or crowding, adequate privacy, relaxing colors and nonglare surfaces, and windows with blinds or curtains, and it allows adequate social contact with coworkers. The terminal should be regularly serviced and cleaned, and records should be kept easily accessible. The printer should be in a separate area; if located near the work area, it should be equipped with a noise shield. A, Indirect general lighting; moderate brightness (can be turned off if desired); B, screen approximately 1 to 2 feet away with midpoint slightly below eye level; characters are large and sharp enough to read easily; brightness and contrast controls present; adjustable height and tilt; glareproof screen surface; no visible flicker of characters; C, if necessary, special glasses for VDT viewing distance; D, adjustable back rest to support small of back; E, easily adjustable seat height and depth; F, swivel chair; safer with five-point base and casters; G, feet firmly resting on the floor; foot rest available; H, thighs approximately parallel to the floor; I, movable keyboard on surface with adjustable height; arms approximately parallel to the floor; J, copy holder at approximately same distance as screen; adjustable space for copy holder and other materials; K, direct, adjustable task lighting. [From Office Technology Education Project (OTEP), Somerville, Massachusetts. Drawn by Beth Maynard.]

Stress is a major problem for health care workers, who are often overworked and yet have responsibility for human life (see Chapter 21). Six of the leading 27 occupations with the highest rate of mental health disorders

are in the health care field (20). Managed care and other cost-containment programs have resulted in even more understaffing, as well as equipment maintenance problems and supply shortages. Rotating shifts, which are common among hospital workers, are another source of stress that can affect reproductive and cardiovascular health (see Chapter 22). Nursing and personal care workers have an injury rate equal to that of agricultural workers. Hazards contributing to this high injury rate include needlesticks, the transport of equipment and patients, slippery floors, violent attacks from patients, and electrical hazards from the extensive use of equipment to monitor and treat patients (Fig. 36-5). Hospitals use a wide variety of chemicals for cleaning, sterilizing, laboratory analysis, and chemotherapy. A hospital worker can be exposed to chemicals or infectious agents through direct handling, cross-contamination of hospital areas due to poor ventilation design, the transport and disposal of waste, and the maintenance and cleaning of equipment or rooms. Control methods have been developed for many of the hazardous substances to which hospital workers are exposed. Ventilation, substitution with less harmful products, and changes in work practices can minimize many hospital hazards. For example, inexpensive scavenger systems are available to reduce exposures to anesthe-

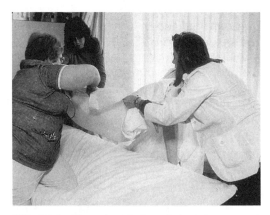

FIG. 36-5. Many female hospital workers are at risk of back strains from lifting patients. (Photograph by Marilee Caliendo.)

tic gases. Substitution and changes in work practices can eliminate formaldehyde (formalin) exposures in autopsy, surgical pathology, and histology laboratories, and in renal dialysis units. The use of control procedures, such as complete evacuation of sterilizer units before opening, use of aeration cabinets and catalytic converters, and careful equipment maintenance, can reduce ethylene oxide exposures from sterilizers. Guidelines developed to minimize exposure to chemotherapeutic drugs include provisions for careful labeling of materials, use of biologic safety cabinets in drug preparation, and use of gowns and gloves. Radiation exposures can result from portable x-ray machines, other diagnostic tests or therapies using radiation, and patients emitting radiation after therapeutic implants or diagnostic tests. Shielding and distance from the radiation source are the best protection along with regular maintenance, personal monitoring, reductions in unnecessary procedures, and careful labeling of materials, patients, and waste.

Hospital workers are exposed to a wide variety of infectious agents, including those that cause hepatitis B and C, rubella, influenza, tuberculosis, meningitis, toxoplasmosis, and cytomegalovirus disease, as well as human immunodeficiency virus. To prevent infections, all blood, excretions, and secretions should be treated as infectious, and personal hygiene should be strictly maintained (see Chapters 20 and 43) (41).

Latex allergy has become a serious hazard for nurses, other health care providers, and for some patients. The allergic reactions are to proteins found in products that contain natural rubber latex, such as examination gloves and other types of gloves, catheters, condoms, medication vials, tourniquets, band-aids, adhesive tape, clothing, and dental dams. The degree of allergic reactions may range from skin irritation and itching, to severe asthma-like symptoms, to anaphylactic shock and death. It is thought that frequent exposure to latex may lead to the development of latex allergies and the

worsening of reactions in those already sensitized. Several studies found that from 5% to 12% of health care workers have latex sensitivity, and the prevalence is as high as 40% to 65% in children with spina bifida and patients who undergo frequent medical procedures. Latex exposures in health care can be reduced by selecting latex-free gloves and other medical products. In situations where latex gloves cannot be replaced, the gloves should be powder free. Powder increases the potential for latex exposure because latex proteins adhere to the powder particles, which become airborne when the gloves are donned. Because the powder particles are in the respirable size range, they remain airborne for considerable periods of time and are easily inhaled. Some hospitals have created latex-free examination and operating rooms (42).

Microelectronics

The microelectronics industry includes a wide array of processes and products centered around computers, consumer electronics, telecommunications, medical and industrial electronics, and electronic accessories and components. This section focuses on the manufacture of semiconductor integrated circuits and the production and assembly of printed circuit boards that are used in computers, communication, and consumer products. A high percentage of women work in the final assembly processes of microelectronics and run process equipment in semiconductor manufacturing. Men tend to work in maintenance-related tasks, engineering, and management (43). The manufacture of semiconductors involves a sequence of processes that must be repeated many times before the integrated circuit, which contains millions of transistors of doped silicon, is completed. Printed circuit board production includes preparation of the board for assembly, attachment of electronic components by soldering, and board cleaning. Increasingly, these processes are being concentrated in the Pacific Rim countries of Asia (Fig. 36-6).

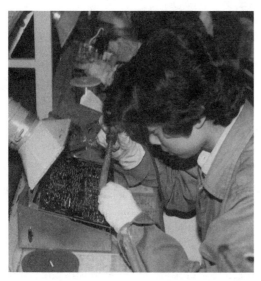

FIG. 36-6. Local exhaust ventilation successfully captures soldering fumes in a semiconductor assembly operation in China. (Photograph by Barry S. Levy.)

Many of the processes in printed circuit board assembly and semiconductor manufacturing involve repetitive motions of the wrists, hands, and arms, as well as sustained awkward postures from poorly designed workstations. In a study of semiconductor workers, 7 of 12 measures of musculoskeletal symptoms were significantly higher in fabrication room workers, with these symptoms occurring with greater prevalence among women (44). However, the incidence of lost workday cases of CTS reported among U.S. semiconductor workers was less than half the rate for all manufacturing (43). Whether this reflects lower risk or lack of reporting by a nonunionized, often immigrant, workforce of women is unknown.

Circuit board assembly operations use a variety of solvents in their cleaning processes. Metals and inorganic compounds are present in soldering operations and acids are used in etching and electroplating processes. Each of these processes and chemicals may present health hazards (43). One study found that dysmenorrhea occurred more frequently in women with high exposure to tolu-

ene in an electronics assembly plant, although other worksite conditions besides the solvent exposures may have contributed to this finding (45). In addition to chemical exposures, electronics workers are often exposed to low-frequency (60 Hz) magnetic fields from power sources. Exposures to these 60-Hz magnetic fields have been associated with a higher risk of breast cancer (46).

The semiconductor industry manufactures the chips used in electronic equipment. The chip manufacturing process involves the use of many toxic chemicals, including metals, such as arsenic, antimony, boron, phosphorus; many inorganic acids; and solvents, such as xylene, glycol ethers, and n-butyl acetate. Exposures to 60-Hz magnetic fields are also present in semiconductor facilities, as are radiofrequency exposures. The fabrication process takes place in highly controlled clean rooms where ventilation rates are very high and some areas have special yellow lighting that lacks the normal ultraviolet wavelengths. In clean rooms, workers must wear special clothing. In some cases, workers are required to wear totally enclosed "bunny suits" with their own air supply.

A series of epidemiologic studies of the reproductive effects of semiconductor work found increased risk of spontaneous abortions among fabrication facility workers, in particular those in the photolithography areas of the facilities. Exposures to fluoride and photoresist solvents were associated with increased spontaneous abortions. There were also small but statistically insignificant differences in measures of menstrual cycle patterns among fabrication facility workers. (47). Semiconductor fabrication workers reported more respiratory symptoms, dermatitis, and alopecia than nonfabrication workers, and these symptoms were more prevalent among women (48).

Construction Trades Work

Women who work in nontraditional jobs face different health and safety hazards than women who work in jobs where the workforce is predominately female. This is in part because the work is different from traditionally female occupations, and in part because of the stress of being an unwelcome minority. In 1997, approximately 5,048,000 women worked in the construction trades, compared with 800,000 in 1981. Despite this increase in the total number, women make up only 2.7% of all employees working in the construction trades (49). The most common trades for women are painters and laborers (50). It is still common for a tradeswoman to be the only woman on a job site employing hundreds of construction workers. As such, isolation, animosity, and sexual harassment are often part of a tradeswoman's daily working conditions. Job security, personal safety, and the need to fit in all present sources of significant stress for female construction workers (51,52).

When interviewed, tradeswomen express concerns about safety hazards, such as slips, falls, and lacerations; exposure to physical agents, such as noise, temperature extremes, and vibration; chemical agents such as paints and solvents; high dust levels; and injuries to the back and eyes. Many of these concerns are similar to those of male construction workers, and depend in part on the trade of the worker (see Chapter 42). In addition, tradeswomen often highlight the lack of training, the unavailability of adequately fitting PPE and tools, and the lack of onsite toilets and washing facilities (51,52).

Mortality rates among white tradeswomen are elevated for lung cancer, traumatic injuries, suicide, bladder cancer, nonmalignant respiratory disease, and connective tissue cancer. For black women in construction, mortality rates from traumatic injuries and lung cancer are elevated (53). When specific trades are examined, women employed as dry wall installers, paperhangers, plumbers, roofers, carpenters, and construction laborers all have elevated mortality rates (54). Chemical and physical hazards vary by the type of construction and the trades involved; however, asbestos, crystalline silica, and diesel exhaust are all known carcinogens and are frequently present on construction sites (see Chapter 42).

Cumulative trauma disorders and sprains and strains are the most common injuries among female construction workers. In addition, back injuries and neck and shoulder pain are frequently reported. These injuries are more prevalent in construction than in any other industry except transportation (50). Among tradeswomen, a common concern is the strength requirements for construction work. Tradeswomen are concerned about injuring themselves, but at the same time they do not want to seek "special" treatment. In some cases, heavy lifting can be avoided by adapting tools or work practices. These adaptive methods are often learned from older male construction workers, many of whom have been injured. In other cases, union agreements may define maximum lifting guides. Many tradeswomen lift weights or do other strength-conditioning exercises to prepare for work.

For pregnant construction workers, the heavy physical demands of nontraditional work may necessitate transfer to a job where the needs for good balance and stamina are not as great. The worker, in conjunction with her health care provider, can determine if and when this is desirable. No research has been done to evaluate the reproductive outcomes of construction workers, although some of the agents, such as noise, vibration, heat, cold, paints, and solvents, are thought to have potential adverse reproductive effects.

Household Work

The single largest category of work done by women in our society is that of unpaid household work. Despite the increasing number of women who work for wages, approximately one third of married women in the United States are still full-time housewives, and most wage-earning women are responsible for work in their own households. In addition, over 1 million paid household workers are employed as servants, nannies, housekeepers, janitors, or cleaners. The exact number of paid household workers is very difficult to determine, however, because much of this work is undocumented. Most of these workers are female and more than one third are of racial or ethnic minorities. Few household workers, paid or unpaid, get sick days and many have no health insurance, Social Security, or workers' compensation.

Household work is thought of as "women's work," and is often not regarded as work at all. This perception devalues the labor involved and implies that there are no health and safety concerns of importance. In fact, a wide range of industrial substances and equipment is used and stored in the home under relatively uncontrolled conditions. There is growing recognition that certain home environmental conditions may pose important health problems, and that homes are significant small-source generators of hazardous waste. Among the products used routinely in the home are pesticides, drain cleaners, chlorine bleaches, scouring powders, ammonia, oven cleaners containing lye, furniture polish, furniture or paint strippers containing organic solvents, glues, paints, epoxies, and air fresheners. There is an aerosol product containing ammonia, chlorine, organic solvents, acids, or detergents for every surface of the bathroom. All of these products are potential hazards and may become a problem, particularly when used in an enclosed space, such as an unventilated bathroom or attic, closet, crawl space, or the space beneath cabinets.

In well insulated homes, potential health hazards may arise from building, insulating, and decorating materials, such as formaldehyde emitted from particle board, plywood, carpeting, fabrics, and foam insulation; and asbestos in pipe lagging or furnace and boiler insulation. Gas and wood stove emissions can cause respiratory problems. Long-term radon exposure in homes has been associated with lung cancer. Biologic agents found in the home are related to asthma and other respiratory problems. Chief among these are molds, fungi, and bacteria, all of which can grow in air-conditioning systems, damp wallboard, and other building materials; dust mites, particularly in carpeting and bedding; and cockroaches (see Chapter 23).

In addition to hazardous chemical, physical, and biologic agents in the home, some of the social factors that define the nature of housework can contribute to poor mental health. For example, sexual harassment and abuse can be serious problems for women in the home. Although many household workers enjoy their work and acquire a wide range of skills, their skills are largely unrecognized in the broader labor market, and they may have little long-term opportunity for job advancement. An important step toward addressing the health and safety issues of women who work in the home is their recognition as legitimate workers with corresponding rights.

REFERENCES

1. National Center for Health Statistics. Women: work and health. Series 3, no. 31. Atlanta: U.S. Department of Health and Human Services, Centers for Disease Control and Prevention, December, 1997.
2. Messing K. One-eyed science: occupational health and women workers. Philadelphia: Temple University Press, 1998.
3. Kilbom A, Messing K, Thorbjörnsson CB, eds. Women's health at work. Solna, Sweden: National Institute for Working Life, 1998.
4. U.S. Department of Commerce, Bureau of the Census. Money income in the United States. Washington, DC: U.S. Department of Commerce, 1995:60–193.
5. U.S. Department of Labor, Women's Bureau. Work related child care statistics. Washington, DC: U.S. Department of Labor, 1998.
6. Albelda R, Tilly C. Glass ceilings and bottomless pits: women's work, women's poverty. Boston: South End Press, 1997.
7. Hochschild A. The second shift: working parents and the revolution at home. New York: Viking Press, 1989.
8. Zukewich GN. Les femmes sur le marche du travail. Ottawa, Ontario, Statistique Canada, 1993.
9. Quinn M, Buiatti E. Women changing the times: an Italian proposal to address the goals and organization of work. New Solutions: A Journal of Environmental and Occupational Health Policy 1991; 1:48–56.
10. Amott T. Caught in the crisis: women and the U.S. economy today. New York: Monthly Review Press, 1993.
11. National Committee on Pay Equity. Face the facts about wage discrimination and equal pay. National Committee on Pay Equity Newsnotes Winter 1996.
12. Punnett L, Herbert R. Work-related musculoskele-

tal disorders: is there a gender differential, and if so, what does it mean? In: Goldman MB, Hatch MC, eds. Women and health. San Diego, CA: Academic Press, 1999.

13. Silverstein BA, Fine LJ, Armstrong TJ. Occupational factors and carpal tunnel syndrome. Am J Ind Med 1987;11:343–358.

14. Leino PI, Hänninen V. Psychosocial factors at work in relation to back and limb disorders. Scand J Work Environ Health 1995;21:134–142.

15. Goldenhar LM, Sweeney MH. Tradeswomen's perspectives on occupational health and safety: a qualitative investigation. Am J Ind Med 1996;29:516–520.

16. Barnekow-Bergkvist M, Hedberg GE, Janlert U, Jansson E. Determinants of self-reported neck-shoulder and low back symptoms in a general population. Spine 1998;23:235–243.

17. Stevenson JM. Gender-fair employment practices: developing employee selection tests. In: Messing K, Neis B, Dumais L, eds. Invisible: issues in women's occupational health/La santé des travailleuses. Charlottetown, PEI: Gynergy Books, 1995:306–320.

18. John Gray Institute. Managing workplace safety and health: the case of contract labor in the U.S. petrochemical industry. Beaumont, TX: John Gray Institute, Lamar University, 1991.

19. Karasek R, Theorell T. Healthy work: stress, productivity, and the reconstruction of working life. New York: HarperCollins, 1990.

20. Chavkin W, ed. Double exposure: women's health hazards on the job and at home. New York: Monthly Review Press, 1984.

21. Wigmore D. Taking back the workplace, workplace violence: a hidden risk in women's work. In: Messing K, Neis B, Dumais L, eds. Invisible: issues in women's occupational health. Charlottetown, PEI: Gynergy Books, 1995;321–352.

22. U.S. Department of Labor, Bureau of Labor Statistics. National census of fatal occupational injuries, 1997, USDL publication no. 98-336. Washington, DC: U.S. Department of Labor, 1998.

23. Jenkins EL. Occupational injury deaths among females: the US experience for the decade of 1980 to 1989. Ann Epidemiol 1994;4:146–151.

24. Toscano G. Workplace violence: an analysis of Bureau of Labor Statistics data. Occup Med 1996;11:227–235.

25. National Institute for Occupational Safety and Health. Violence in the workplace: risk factors and prevention strategies. Current Intelligence Bulletin 57. U.S. Department of Health and Human Services, Public Health Service, Centers for Disease Control and Prevention publication no. 96-100. Atlanta: Centers for Disease Control and Prevention, 1996.

26. Spangler E. Sexual harassment: Labor relations by other means. New Solutions 1992;3:24–29.

27. Videman T, Nurminen T, Tolas S, Kuorinka I, Vanharanta H, Troup JD. Low back pain in nurses and some loading factors of work. Spine 1984; 9:400–404.

28. Ohara H, Aoyama H, Irani T. Health hazards among cash register operators and the effects of improved working conditions. Journal of Human Ergonomics 1976;5:31–40.

29. Grandjean E, Hunting W, Nishiyama K. Preferred VDT workstation settings, body posture and physi-

cal impairments. Journal of Human Ergonomics 1982;11:45–53.

30. Murphy DC, Henifin MS, Stellman JM. Personal protective equipment for women: results of a manufacturers' and suppliers' survey. In: Transactions of the 43rd annual meeting of the American Conference of Governmental Industrial Hygienists. Cincinnati, OH: American Conference of Governmental Industrial Hygienists, 1981:62–72.

31. Aschengrau A, Coogan P, Quinn M, Cashins L. Occupational exposure to estrogenic chemicals and the occurrence of breast cancer: an exploratory analysis. Am J Ind Med 1998;34:6–14.

32. Colburn T, vom Saal F, Soto A. Developmental effects of endocrine disrupting chemicals in wildlife and in humans. Environ Health Perspect 1993; 101:378–384.

33. Soto A, Sonnenschein C, Chung K, Fernandez M, Olea N, Serrano F. The E-SCREEN assay as a tool to identify estrogens: an update on estrogenic environmental pollutants. Environ Health Perspect 1995;103:113–122.

34. Kipen H, Stellman J. Core curriculum: reproductive hazards in the workplace. White Plains, NY: March of Dimes, 1985.

35. National Institute for Occupational Safety and Health. Research report: guidelines on pregnancy and work. The American College of Obstetricians and Gynecologists, U.S. Department of Health, Education and Welfare. Rockville, MD: National Institute for Occupational Safety and Health, 1977.

36. Welch LS. Decisionmaking about reproductive hazards. Seminars in Occupational Medicine 1986; 1:97–106.

37. Goerth CG. Pregnant workers have discrimination protection. Occup Health Saf 1983;6:22–23.

38. Stellman J, Henifin M. Office work can be dangerous to your health. New York: Pantheon, 1983.

39. American Society of Heating, Refrigeration, and Air Conditioning Engineers (ASHRAE). Standard: ventilation for acceptable indoor air quality. Publication no. 62-1981. Atlanta: ASHRAE, 1981.

40. Murray WE, Moss CE, Parr WH. Potential health hazards of video display terminals. Publication no. 81-129. Rockville, MD: National Institute for Occupational Safety and Health, 1981.

41. Patterson WB, Craven DE, Schwartz DA, Nardell EA, Kasmer J, Nobel J. Occupational hazards to hospital personnel. Ann Intern Med 1985;102: 658–690.

42. U.S. Department of Health and Human Services. NIOSH alert: preventing allergic reactions to natural rubber latex in the workplace, NIOSH publication no. 97-135. Washington, DC: U.S. Department of Health and Human Services, 1997.

43. Stellman, JM, ed. Encyclopaedia of occupational health and safety. Chapter 83: Microelectronics and semiconductors. Williams ME, chapter ed. Geneva: International Labor Organization, 1998.

44. Pocekay D, McCurdy SA, Samuels SJ, Hammond SK, Schenker MB. A Cross-sectional study of musculoskeletal symptoms and risk factors in semiconductor workers. Am J Ind Med 1995;28:861–872.

45. Ng TP, Foo SC, Yoong T. Menstrual function in workers exposed to toluene. British Journal of Industrial Medicine 1992;49:799–803.

46. Coogan PF, Clapp RW, Newcomb PA, et al. Occupational exposure to 60 hertz magnetic fields and risk of breast cancer in women. Epidemiology 1996;7:459–464.
47. Schenker MB, Gold EB, Beaumont JJ, et al. Association of spontaneous abortion and other reproductive effects with work in the semiconductor industry. Am J Ind Med 1995;28:639–660.
48. McCurdy SA, Pocekay D, Hammond SK, Woskie SR, Samuels SJ, Schenker MB. A cross-sectional survey of respiratory and general health outcomes among semiconductor industry workers. Am J Ind Med 1995;28:847–860.
49. U.S. Department of Labor. Nontraditional occupations for employed women in 1997. Washington, DC: U.S. Department of Labor, 1998.
50. Center to Protect Workers' Rights. The construction chart book. Washington, DC: Center to Protect Workers' Rights, 1998.
51. Goldenhar LM, Sweeney MH. Tradeswomen's perspectives on occupational health and safety: a qualitative investigation. Am J Ind Med 1996;29:516–520.
52. National Institute for Occupational Safety and Health. Women in the construction workplace: providing equitable safety and health protection. NIOSH publication no. 98-135. Washington, DC: U.S. Department of Health and Human Services, 1998.
53. Robinson CF, Burnett CA. Mortality patterns of US female construction workers by race, 1979–1990. J Occup Med 1994;36:1228–1233.
54. Robinson C, Stern F, Halperin W, et al. Assessment of mortality in the construction industry in the United States: 1984–1986. Am J Ind Med 1995; 28:49–70.

BIBLIOGRAPHY

Headapohl DM. Women workers. Occup Med 1993.
Presents a variety of medical and social issues related to women and the work environment. In addition to discussing reproductive risks, attention is given to cardiovascular, ergonomic, psychological, and psychosocial risks. Also considered are sexual harassment, gender discrimination, disability, problems associated with economic circumstances, and issues related to the dual roles of worker and homemaker.

Kilbom A, Messing K, Thorbjörnsson CB, eds. Women's health at work. Solna, Sweden: National Institute for Working Life, 1998.
Describes the broad social, economic, and health experience of working women in Sweden and other countries. In addition to comprehensive chapters on musculoskeletal disorders, stress, and cardiovascular illness, it has several chapters on new topics, including one on work-related skin disease from a gender perspective

and one on developing indicators for measuring the health of working women.

MassCOSH Women's Committee. Confronting reproductive health hazards on the job. Boston: Massachusetts Coalition for Occupational Safety and Health, 1993.
A practical handbook for working women that focuses on reproductive hazard identification and problem solving.

Messing K. One-eyed science: occupational health and women workers. Philadelphia: Temple University Press, 1998.
A precise and comprehensive analysis of the scientific and social factors that affect working women in North America, beginning with the question, "Is there a women's occupational health problem?" The author is a member of an interdisciplinary group of researchers that has produced some of the leading work on women and occupational health. Also contains a compelling foreword by Jeanne Mager Stellman that gives a concise, historical overview of women's occupational health issues and direction for the future.

Messing K, Neis B, Dumais L, eds. Invisible: issues in women's occupational health. Charlottetown, PEI: Gynergy Books, 1995.
A thoughtful collection of technical, social, and policy analyses related to women and work. Contains several detailed case studies to illustrate the theoretical analyses. The chapter on workplace violence presents a very far-reaching analytic framework followed by a thorough discussion of possible solutions.

Paul ME. Occupational and environmental reproductive hazards: a guide for clinicians. Baltimore: Williams & Wilkins, 1993.
The major guide for health care providers on workplace reproductive hazards. Covers medical issues related to specific hazardous agents such as solvents and has a useful chapter on legal and policy issues such as fetal protection and a regulatory framework.

Three texts are somewhat dated, but remain classics in the field:

Chavkin W, ed. Double exposure: women's health hazards on the job and at home. New York: Monthly Review Press, 1984.
A well documented selection of the social, economic, and scientific issues of women at work. Includes topics that are often overlooked, such as household work, farm work, sexual harassment, and the role of trade unions.

Hunt VR. Work and the health of women. Boca Raton, FL: CRC Press, 1979.
Physical, chemical, and biologic hazards are discussed as well as the hazards of specific industries. Presents a concise summary of male reproductive hazards.

Stellman JM. Women's work, women health. New York: Pantheon, 1977.
An overview of the history of working women, health hazards on the job, and policy issues concerning protective legislation.

Occupational Health: Recognizing and Preventing Work-Related Disease and Injury, 4th ed.
Edited by Barry S. Levy and David H. Wegman, Lippincott Williams & Wilkins, Philadelphia © 2000.

37 Minority Workers

Andrea Kidd Taylor[1] and Linda Rae Murray

Despite a strong U.S. economy with an increase in the availability of jobs and a decrease in unemployment, the job market and income status for most minority workers[2] has not improved substantially. In 1998, median earnings for black men working at full-time jobs were 76% of the median income for white men, and black women's earnings were 88% of those for their white counterparts. Median earnings overall for Hispanics who worked full time were lower than those for blacks and whites. Among the major occupational groups, people employed full time in managerial and professional occupations had the highest weekly earnings; the largest proportion of white men and white women are employed in this occupational group. Men and women in service occupations and farm jobs, mainly minority workers, earned the least (1).

A. K. Taylor: U.S. Chemical Safety and Hazard Investigation Board, Washington, DC 20037.

L. R. Murray: Winfield Moody Health Center, Chicago, Illinois 60610.

[1] The views and opinions expressed in this document are solely those of the authors and do not represent official views and policies of the U.S. Chemical Safety and Hazard Investigation Board, unless they are expressly identified as such.
[2] Minority workers, or workers of color, as presented in this chapter, include African Americans (blacks), Asian Americans (and Pacific Islanders), Latino Americans (Hispanics), and Native Americans (American Indians). The word *minority* is synonymous with *nonwhite*.

Minority workers in general have less education, lower income levels, inferior housing, worse health status, and less access to services such as health care, compared with white workers. U.S. Census data on the civilian labor force from 1990 indicated that the proportion of workers who had not completed high school was 14% for whites, 26% for blacks, and 45% for Hispanics. The proportion of workers completing college in the 1990 census data was 24% for whites, 13% for blacks, and 9% for Hispanics (2). Such disparities in income, education, and employment status indicate the need continuously to address the health and safety problems faced by minorities in the workplace.

Most of the examples of minority workers in this chapter refer to African-American workers because, as poor as the data are for African Americans, very few occupational health data exist for the other groups of minority workers. In addition, African Americans today constitute most of the minority workers in the United States, and the occupational health problems they face are likely to represent the general experience of other minority groups.

Each immigrant group that arrived in the United States historically worked in the most dangerous jobs in industry, succeeded within the next one or two generations by the next immigrant group. Racially motivated discriminatory hiring and employment patterns in many industries have been used to prevent African Americans, Latino Americans, and

Native Americans from moving out of these entry-level jobs, in contrast to many of the white immigrant groups. Consequently, as far back as the 1920s, minority workers have been concentrated in the worst jobs in many industries. In addition, discriminatory job placement practices have resulted in elevated cancer incidence and death rates in African Americans exposed to occupational carcinogens (3).

PLACEMENT PATTERNS OF MINORITY WORKERS

Minorities continue to work disproportionately in higher-risk occupations. These jobs are also the jobs with lower pay and lower status. Nonwhite workers are concentrated in many of the most hazardous industries in the manufacturing sector; for example, they are concentrated in the logging industry, "other primary iron and steel industries," and the meat products industry, all of which have high injury and illness rates. Minorities also tend to be overrepresented in the lead industry, such as lead smelting and storage battery manufacture. Over 50% of black and Hispanic workers are employed in the lead smelting industry as laborers, operatives, and service workers—the jobs with the greatest potential for lead exposure (4).

Minority workers have been documented to face greater on-the-job hazards. Forty-seven percent of African-American workers report themselves as exposed to at least one hazard, compared with 37% of white workers; the average number of significant hazards reported by blacks is 1.6, compared with 1.0 for whites (5). Of 13 hazards studied in one study, African Americans were observed to be exposed significantly more often than whites working in situations with extreme temperature, dirty conditions, loud noise, and risk of disease (5).

Ever since African Americans were brought to the United States as slaves, they have been employed in the lowest-paid and least-desirable jobs. They have in general been denied entrance to many industries or skilled trades and are often assigned the most dangerous jobs in the industries in which they have worked.

African-American women and other women of color, on average, have the lowest-paid jobs and are mainly employed in domestic and service positions and blue-collar jobs. There are industries in the United States and elsewhere where large numbers of women of color are employed:

- Of the 240 poultry, catfish, and meat processing plants in the United States employing over 150,000 workers, nearly 75% are located in the southern part of the United States, where communities are poor and many African-American women are employed. Thus, conditions similar to those found in the Mexican *maquiladoras*[3] exist. Working conditions are unsafe and unhealthy, minimum wages are paid, there are no health care benefits, and most of the plants are nonunion (6).
- Latino Americans constitute 53% of the farmworkers in California, Nevada, and Arizona; 80% of those in the East; approximately 90% of those in the Midwest; and almost 100% of those in the Southwest. Seasonal farmworkers are 71% Latino, and the migrant worker population has been estimated to be 80% to 95% Latino (7).
- Asian immigrant women, working in poorly lit and crowded sweatshops, make up 85% of the San Francisco Bay Area's 20,000 garment workers (Fig. 37-1). The story is similar in Los Angeles and New York, where subcontractor sweatshops pay women at or below the legal minimum wage and, when profit margins dwindle, they close their doors and pay nothing for work already done (8).

Often one particular minority group is concentrated in performing the unskilled and

---

[3] *Maquiladoras* are factories along the Mexican side of the U.S.–Mexican border that usually subcontract from United States-based corporations, hire mostly women, and offer inadequate, unpleasant, and hazardous working conditions.

FIG. 37-1. Asian workers in the United States sometimes work in low-paying, monotonous jobs, such as in the garment industry, as shown here. (Photograph by Earl Dotter.)

semiskilled jobs in a particular industry in the United States. For example, African Americans have been concentrated in these jobs in the poultry and meat processing industries in the South, and Latino Americans have been concentrated in these jobs in copper smelting and agricultural production in the West and Southwest. Native Americans have been concentrated in semiskilled jobs in uranium mining in the Southwest, and Asian Americans in such jobs in the garment, hotel, and restaurant industries on the West and East Coasts.

Not only are minority workers concentrated in the most dangerous sectors of the U.S. economy, they are overrepresented in the more dangerous occupations in these industries, even after controlling for education and experience. The most significant occupational health problems faced by minority workers are related to patterns of their placement in the least desirable, semiskilled and unskilled jobs within industry. It has been estimated that African-American workers have a 37% greater chance than white workers of having an occupational injury or illness, and a 20% greater chance of dying from one (3).

Evidence indicating that minority workers

are specifically concentrated in jobs where elevated rates of occupational disease have been documented is incomplete. Three research practices common to many epidemiologic studies of occupational disease and injury are responsible for the paucity of solid data on occupational disease among minority workers: (a) focusing on white workers only because they have a better national comparative database; (b) controlling for race so that no race-specific data are presented; and (c) including race as a variable in study design, but not in the analysis of study results. Each of these practices can be defended on methodologic grounds, but they have not been complemented by other practices that would yield race-specific risk information.

The collection of U.S. government statistics also contributes to the problem—and similar government practices may contribute to the problem in other countries. Although census and employment data are collected according to race on broad groups of workers by industry, such as steelworkers, data by detailed job classification, such as "coke oven workers," are not collected by any organization or agency. This practice means that when epidemiologic studies associate a particular occupational disease with a particular job classification, such as lung cancer with coke oven workers, there are no national data to estimate the number of workers in that job classification.

EXPOSURE TO OCCUPATIONAL HAZARDS: EVIDENCE OF ELEVATED RISK IN MINORITY-INTENSIVE JOBS

The following summary of selected race-specific epidemiologic studies, lawsuits, and industrial disasters that have attracted widespread public attention provides compelling evidence of direct links between job placement patterns and occupational disease patterns among minority workers. The examples that follow describe situations in which mainly African Americans and some Latino Americans were concentrated in particular jobs that are now recognized to be linked with occupational disease. In some of the situations described, however, direct evidence is lacking that work exposure actually caused disease. Exposure data were not collected and appropriate comparisons were not performed. It is important, however, to examine this kind of evidence because most people are not aware that such exposures occurred in such a manner.

The 1930s and 1940s

Five thousand workers, mostly African American, were recruited to drill a tunnel through a mountain in West Virginia for transporting river water to a power plant. These workers were exposed to high concentrations of silica dust. A congressional committee later discovered that 1,500 of these workers had become disabled and another 476 died from silicosis (9). In Alabama, during the same period, more than half of the bituminous coal miners were African American, as were approximately 25% of all coal miners in southern Appalachia (10). Many of the workers were displaced during the Depression and never received compensation or medical treatment for the coal workers' pneumoconiosis they had developed (see Chapter 25).

The U.S. Public Health Service, in an investigation to determine the factors that contribute to high rates of pneumonia among steelworkers, found that African Americans hired in the blast furnace, coke oven, and open hearth departments had disproportionately high rates of pneumonia. The researchers believed that this problem was due to heat exposure, but steel industry representatives suggested that African Americans have been "predisposed to the disease" (11).

A total of 108 African-American steelworkers in Indiana sued subsidiaries of a large steel company in 1935 for failing to provide healthful working conditions. The worker plaintiffs were mostly furnace cleaners and coke oven workers who had been given the dirtiest jobs in the mill. The workers charged that their jobs had caused tuberculosis, silicosis, and other lung disease. In 1938, the suit was settled out of court for an undisclosed amount of money (12).

So many African-American workers were hired in the foundry department of automobile companies in the 1940s that it became known as the "black department." Subsequently, a study has documented an increased cancer rate among foundry workers during this period (13).

During World War II, African-American workers were hired in record numbers in the shipbuilding and munitions industries. After the war, most were laid off (14). Most African-American workers exposed to asbestos and other hazardous substances during the war have not received compensation for cancer and other occupational diseases they incurred.

After World War II

A U.S. Public Health Service study of the chromate industry uncovered an occupational lung cancer epidemic. In this industry, African-American workers were found to have 80 times the expected number of respiratory cancers—significantly higher than the 29-fold excess found for all chromate workers. The cause was identified as exposure to chromium-bearing dusts in the "dry end" of processing, where 41% of the African-American workers, compared with 16% of the white workers, were employed (15).

A series of long-term mortality studies among American steelworkers began to be published in the 1960s, showing that African Americans had been concentrated in the most dangerous areas of steel plants and had had disproportionately high rates of lung cancer. Of the African Americans assigned full time to coke oven departments, 19% were employed in the most dangerous job of top-side coke oven worker, compared with 3% of the white coke oven workers; 80% of the full-time top-side coke oven workers were African American. Full-time top-side coke oven workers were shown to have 10 times the expected risk for development of lung cancer (16).

Long excluded because of discriminatory hiring practices, over 90,000 African-American workers entered the textile mill industry between 1960 and 1979 (Fig. 37-2). An epidemiologic study of one of the largest textile corporations revealed that (a) the employment of African Americans was being concentrated in the high-dust work areas, and (b) although African-American workers had been employed in the textile mills for fewer years, they were at higher risk than white workers for development of byssinosis (17).

A study of the rubber industry showed that employment of African Americans was concentrated in the compounding and mixing areas, where elevations in cancers of the stomach, respiratory system, prostate, bladder, and blood, and lymph nodes were found. In addition, African-American workers in these areas were found to have particularly high rates of respiratory and prostate cancer compared with white workers in the same work area (18).

In another situation, African-American workers in lead smelters and battery plants

FIG. 37-2. Minority workers often are employed in the least desirable and most hazardous jobs, such as this worker opening bales of cotton—a job that has traditionally caused a high risk of byssinosis. (Photograph by Earl Dotter.)

in three major cities instituted legal actions against their employers, charging them with responsibility for excessive exposure of workers to lead, falsification of medical monitoring records, and unethical and dangerous use of chelating drugs in attempts to prevent lead poisoning (19).

On behalf of African-American workers employed in a shipyard, the National Association for the Advancement of Colored People instituted a class action suit. The suit alleged that (a) virtually all workers assigned to sandblasting (with hazardous exposure to silica dust) were African American; and (b) workers were being systematically denied workers' compensation benefits for resultant cases of silicosis.

A study in California showed that Latino-American and African-American workers are at greater risk of occupational illness and injury than white workers (20). Latino-American men had more than double the risk of work-related illness and injury than white men, and African-American men had 1.4 times the risk of white men; for Latino-American women, the risk was 1.5 that of white women, and for African-American women it was 1.3 that of white women. At least part of the difference in risk cannot be explained by differences in years of education and work experience among racial and ethnic groups.

For both sexes, African Americans have greater rates of disabling injuries than whites. Fifteen percent of African-American workers are unable to work because of permanent or partial disabilities (21). There has been a dramatic narrowing of racial differences in exposure to occupational hazards for male workers since the 1960s, but no such narrowing for female workers. Although they work in inherently safer jobs, minority women workers now have approximately the same rate of occupational injury as white men.

The risk for occupational disease among Latino-American workers is aggravated by several factors, including federal worker protection laws not being enforced in certain job

FIG. 37-3. Migrant farmworkers throughout the world, who are often members of minority groups, face many work-associated hazards, including physical stresses and exposure to pesticides. In the United States, migrant farmworkers are often Latinos. (Photograph by Ken Light.)

classifications in which a large proportion of Latino-American workers are employed. For example, the Occupational Safety and Health Administration field sanitation standard for farmworkers excludes farms that employ 11 or fewer workers, leaving an estimated 89% of agricultural establishments and approximately 64% of all farmworkers—largely Latino-American workers—uncovered by the standard (7). In the United States, pesticides annually cause over 300,000 cases of toxicity and 1,000 deaths among farmworkers, also disproportionately affecting Latino-American workers (Fig. 37-3) (2).

UNEMPLOYMENT

Nonwhite workers experience higher unemployment rates than white workers. During the 1970s, they had approximately double the national unemployment rate, and tended to remain unemployed longer than white workers. Several studies on unemployment in the late 1970s and early 1980s linked job loss with elevations in blood pressure, adverse changes in mental health, and excess morbidity and mortality (22–24).

According to more recent statistics, the black unemployment rate has for some time been approximately twice the white unemployment rate. The causal chain of unemployment as a risk factor for morbidity and mortality operates in both directions. Although ill people are more likely to become unemployed, unemployment and its consequences play an important role in contributing to poor health status (2).

GENERAL HEALTH STATUS AND HEALTH CARE ISSUES

In the United States, minority workers have dramatically higher morbidity and mortality rates than white workers for most major diseases. Almost 15% of African-American workers are unable to work because of permanent or partial disabilities. The average life expectancy for African Americans is approximately 7 years less than that for whites, and between the ages of 25 and 44 years, hypertension-associated mortality rates are approximately 16 times higher among African Americans than among whites. Age-adjusted death rates for African Americans are twice those of whites for pneumonia, influenza, diabetes mellitus, cirrhosis, and cerebrovascular disease, and five times those of whites for tuberculosis. In the working age population (between 17 and 64 years of age), the prevalence of heart disease is higher among nonwhites. Age-adjusted death rates for all malignancies for men are approximately 25% higher for nonwhites than whites, and for women approximately 10 times higher for nonwhites than whites.

Although the causes of these large differences between white and nonwhite mortality and morbidity rates are not entirely understood, the causes of many common diseases are multifactorial (2). Although workplace exposures may contribute to disease, so may a range of genetic, social, and lifestyle factors, as well as environmental factors beyond the workplace. Thus, if African Americans and other minority groups have an increased "baseline" risk for some of these illnesses, then workplace and other environmental exposures could pose additional risks of disease for members of these groups.

Inadequate health care contributes to the reduced health status of minority workers. Inadequate health care in minority communities is well documented. Minorities are less likely to have access to health care and health insurance (25,26). They are also less likely to have a regular source of medical care, and less likely to receive preventive services such as cancer screening. Nonwhite workers are also less likely than white workers to participate in voluntary worksite health promotion programs. Members of minority groups are more likely to die—and die earlier—of diseases affecting them than whites (2). Specific data documenting that this pattern extends to occupational illnesses and injuries, however, do not exist.

WORKERS' COMPENSATION

Although race-specific data are not kept by any state workers' compensation program, some data suggest that minority workers are less likely than white workers to receive workers' compensation benefits for work-related injuries, even though minority workers as a group experience higher rates of occupational injury and disease (see Chapter 11). Workers with severe job-related disabilities are more likely to receive welfare benefits than disability payments from workers' compensation programs, which suggests that workers as a group may be having difficulty receiving compensation for work-related injury and illness.

It is reasonable to assume that minority workers, concentrated in low-paying jobs

with high job turnover, are more likely to experience special difficulty receiving workers' compensation benefits, particularly for occupational disease. A study of chronic disease and work disability supported this assumption, finding that nonwhites were less likely to report disability than whites, but 1.5 times more likely than whites to be severely disabled—although they were, on average, younger (21). A long-term study of steelworkers indicated that employers were more likely to have information about the health status of white retirees than of African-American retirees, a finding that may further reflect the problems nonwhites may encounter in obtaining workers' compensation benefits (27).

WHAT CAN BE DONE

Gaps in knowledge hinder the reduction of risks of occupational illnesses and injuries for minority workers. Most epidemiologic studies have excluded minority workers. With an increasing body of evidence indicating that minority workers are disproportionately exposed to hazards on the job, better research methodologies, including the collection of race-specific data, need to be implemented so that the patterns of occupational disease and injury among minority workers can be better understood. The knowledge gained from such studies could help develop policies and programs that may reduce, and ultimately eliminate, the disparities in occupational health between minority and nonminority workers.

Preventive interventions should be designed and implemented with the needs of minority workers in mind. Interventions should initially focus on the most hazardous occupations, particularly those with the largest representation of minority workers. In addition, training and risk communication must be directed toward minority workers, paying special attention to cultural and language needs (2).

Because minority workers are at an increased risk of occupational disease, there is also the need to recruit more people of color into the occupational and environmental health professions. People respond better to professionals with the same ethnic and cultural backgrounds as themselves. Occupational health professionals should also be more alert to the possible work-related problems among minority workers.

Labor unions have represented their members on health and safety matters by bargaining with employers for agreements aimed at improving working conditions. Such efforts have clearly had an impact in creating stronger occupational safety and health legislation. A traditional tactic of employers has been to use marginalized and minority workers as strikebreakers, thereby creating resentment of better-paid union workers, most of whom are usually white workers.

In addition to health and safety issues, organized labor must appropriately and adequately address social and civil rights issues as part of its efforts to organize minority workers (see Chapter 40). Labor's survival greatly depends on its ability to organize all workers, including workers of color. Acknowledging, understanding, and recognizing the problems of racism, sexism, and other forms of discrimination can be useful for identifying and resolving issues that may be the root causes of hostility among different racial and ethnic populations. The realization that an injury to one is an injury to all should be the rallying theme for all workers.

To achieve further gains in the occupational health of minority workers, public policies must be developed and implemented to ensure access to health care and to occupational health and safety services for all workers. Minority workers must be active participants in efforts that seek to improve their health and safety. In addition, public policies must be developed and implemented that help overcome continuing discrimination in the labor market and thereby achieve social and economic equity and improved health status for all workers.

REFERENCES

1. U.S. Department of Labor. Usual weekly earnings of wage and salary workers: third quarter 1998, USDL Publication No. 98–425. Washington, DC: U.S. Department of Labor, Bureau of Labor Statistics, 1998.
2. Frumkin H, Walker ED, Friedman-Jiménez G. Minority workers and commnities. Occup Med: State of the Art Reviews 1999;14(3):495–517.
3. Davis ME. The impact of workplace health and safety on black workers: assessment and prognosis. Labor Law Journal 1980;31:723–732.
4. Alexander D. Chronic lead exposure: a problem for minority workers. American Association of Occupational Health Nursing Journal 1989;37:105–108.
5. Robinson JC. Racial inequality and the probability of occupation-related injury or illness. Milbank Memorial Fund Quarterly 1984;62:567–90.
6. Lillie-Blanton M, Martinez R, Taylor AK, Robinson BG. Latina and African-American women: continuing disparities in health. Int J Health Serv 1993;23:555–584.
7. Friedman-Jiménez G, Ortiz JS. Occupational health. In: Molina CW, Aguirre-Molina M, eds. Latino health in the US: a growing challenge. Washington, DC: American Public Health Association, 1994:341–389.
8. Needleman R. Organizing low-wage workers. Working USA 1997;1(1):45–59.
9. Cherniack M. The Hawk's Nest incident: America's worst industrial accident. New Haven, CT: Yale University Press, 1986.
10. Northrop H. Organized labor and the Negro. New York: Harper, 1944:156–157.
11. U.S. Public Health Service. Frequency of pneumonia among iron and steel workers. Public Health Bulletin no. 202. Washington, DC: UPHS, November 1932.
12. Berman D. Death on the job. New York: Monthly Review Press, 1978:29–30.
13. Kotelchuck D. Occupational injuries and illnesses among black workers. Health PAC Bulletin 1978;81–82:33–34
14. Weaver RC. Negro labor. Port Washington, NY: Kennikat Press, 1949 [Reprinted in 1969].
15. Gafafer W. Health of workers in the chromate producing industry: a study. Washington, DC: U.S. Public Health Service, 1953.
16. Lloyd W. Long-term mortality study of steelworkers: V. respiratory cancer in coke plant workers. J Occup Med 1971;13:59–68.
17. Martin C, Higgins J. Byssinosis and other respiratory ailments: a survey of 6,631 cotton textile employees. J Occup Med 1976;18:455–62.
18. McMichael AJ, Spirtas R, Gamble JF, Tousey PM. Mortality among rubber workers: relation to specific jobs. J Occup Med 1976;18:178–185.
19. Chicago Area Committee for Occupational Safety and Health News 1976;4:1 and 1977;4:1.
20. Robinson JC. Exposure to occupational hazards among Hispanics, blacks, and non-Hispanic whites in California. Am J Public Health 1989;79:629–630.
21. Robinson JC. Trends in racial inequality and exposure to work-related hazards, 1968–1986. American

Association of Occupational Health Nurses Journal 1989;37(2):56–63.
22. Kasl S, Cobb S. Blood pressure changes in men undergoing job loss: a preliminary report. Psychosom Med 1970;76:184–188.
23. Navarro V. Race or class or race and class? Growing mortality differentials in the United States. Lancet 1990;336:1238–1240.
24. Lerner M, Henderson LA. Income and race differentials in heart disease mortality in Baltimore City, 1979–81 to 1984–86. In: Health status of minorities and low-income groups, 3rd ed. U.S. Department of Health and Human Services publication no. 40085. Washington, DC: Government Printing Office, 1991:271–848.
25. Blendon RJ. Access to medical care for black and white Americans: a matter of continuing concern. JAMA 1989;261:278–281.
26. Short PF, Cornelius LJ, Goldstone DE. Health insurance of minorities in the United States. J Health Care Poor Underserved 19901(1):9–24.
27. Redmond CK, Ciocco A, Lloyd JW, Rush HW. Long term mortality study of steelworkers: VI. mortality from malignant neoplasms among coke oven workers. J Occup Med 1972;14:621–629.

BIBLIOGRAPHY

Coles R. Migrants, sharecroppers, mountaineers: children of crisis, vol 2. Boston: Little, Brown, 1971.
Florida Rural Legal Services. Danger in the field. 1980.
United States Commission on Civil Rights. The working and living conditions of mushroom workers. Washington, DC: U.S. Government Printing Office, 1977.
University of Wisconsin. Health care needs of Hispanic population in Dane, Doge, and Jefferson Counties. Madison, WI: Department of Rural Sociology, University of Wisconsin Extension, Dane County Mental Health Center, 1977.
Important references on migrant and seasonal agricultural workers. Robert Coles studied the children of migrant workers who travel the East Coast of the United States. The survey from Florida Rural Legal Services, one of the best in the past 20 years, studies health status as self-reported and its relationship to pesticide overspraying. The mushroom workers studied by the U.S. Commission on Civil Rights are almost all Spanish-speaking; they are among the lowest-paid, most poorly housed, and most medically impoverished groups in the United States. The survey from the University of Wisconsin focuses on the health needs of permanent, year-round residents of the Hispanic community in south-central Wisconsin.
Davis ME. The impact of workplace health and safety on black workers: Assessment and prognosis. Labor Law Journal 1980;31:723.
Overview of general health status, stress-related disease, and occupational cancer among black workers. Includes a discussion of the policy issues and suggested strategies raised by the occupational health problems encountered by black workers.
Davis ME, Rowland AS. Occupational disease among black workers: an annotated bibliography. Berkeley: Labor Occupational Health Program, University of California, 1980.
A useful source book.
Rowland AS. Black workers and cancer. LOHP (Labor Occupational Health Program) Monitor 1980;8:14.
Discussion of the possible role of occupational exposures in the rise of cancer death rates among blacks. Links industrial placement patterns of black workers in the 1930s and 1940s with the rise in cancer incidence today. Available from the Labor Occupational Health Program, Institute for Industrial Relations, University of California, Berkeley, CA 94720.
Friedman-Jimenez G. Occupational disease among minority workers: a common and preventable public health problem. American Association of Occupational Health Nursing Journal 1989;37:64–70.
Friedman-Jimenez G, Ortiz JS. Occupational health. In: Molina CW, Aguirre-Molina M, eds. Latino health in the US: a growing challenge. Washington, DC: American Public Health Association, 1994:341–389.
Frumkin H, Walker ED, Friedman-Jiménez G. Minority workers and communities. Occup Med: State of the Art Reviews 1999;14(3):495–517.
Lillie-Blanton M, Martinez R, Taylor AK, Robinson BG. Latina and African-American women: continuing disparities in health. Int J Health Serv 1993;23:555–584.
Morris LD. Minorities, jobs, and health: an unmet promise. American Association of Occupational Health Nursing Journal 1989;37:53–55.
Needleman R. Organizing low-wage workers. Working USA 1997;1(1):45–59.
Robinson JC. Racial inequality and the probability of occupation-related injury or illness. Health and Society 1984;62:567–590.
Robinson JC. Trends in racial inequality and exposure to work-related hazards, 1968–1986. American Association of Occupational Health Nursing Journal 1989;37:56–63.
Taylor AK. Organizing marginalized workers. Occup Med: State of the Art Reviews 1999;14(3):687–695.
These nine references further document the greater occupational health risks of minority workers.

38 Child and Adolescent Workers[1]

Letitia Davis, Dawn N. Castillo, and David H. Wegman

- A 15-year-old working in a restaurant was severely burned while changing the oil in a fryer. The youth slipped while carrying the hot oil and sustained second- and third-degree burns over 14% of his body.
- A 15-year-old was killed in a tractor roll-over incident on a tobacco farm. The youth, along with his brother and another boy, was using a tractor to plow a field. As he tried to turn the tractor around near the edge of a small ravine, one of the wheels went over the ravine and the tractor over-turned. It landed on the boy, and he died immediately.
- A 17-year-old dietary aide in a hospital was blinded for 2 weeks after the chemical she was using to wash pans splashed into her eyes.
- A 17-year-old underwent amputation of both legs after he fell inside a paper baler at a resource recovery center. Investigators found that he was employed in violation of child labor laws and that the energy source to the baler was not turned off as it should have been.
- A 17-year-old working in a sandwich shop was sexually assaulted and robbed at gun-point. The crime occurred while the youth was working alone after 11:00 p.m.

Millions of adolescents and children in the United States work. Although work can provide important benefits for youth—enhanced self-esteem, job skills, and income—it also poses substantial health risks. Young workers, here defined as workers younger than 18 years of age, routinely confront safety and health hazards on the job, and each year in the United States tens of thousands are injured, hundreds are hospitalized, and at least 70 are killed. In addition, working more than 20 hours per week while going to school has been linked with a number of adverse psychosocial outcomes such as increased daytime fatigue and substance abuse, issues important to consider within a broad definition of child and adolescent health (1).

This chapter provides an overview of youth employment and what is known about occupational injuries and illnesses. It discusses factors that raise special health and safety concerns about youth in the workplace and describes child labor laws, which establish extra protections for working youth. Finally, it discusses innovative opportunities for prevention.

L. Davis: Massachusetts Department of Public Health, Boston, Massachusetts 02108.

D. N. Castillo: Division of Safety Research, National Institute for Occupational Safety and Health, Morgantown, West Virginia 26505.

D. H. Wegman: Department of Work Environment, University of Massachusetts Lowell, Lowell, Massachusetts 01854.

[1] Adapted from Castillo D, Davis LK, Wegman DH. Young workers. Occup Med: State of the Art Reviews 1999;14(3):519–536.

Box 38-1. International Child Labor: The Impact of Economic Exploitation on the Health and Welfare of Children

David L. Parker

Social policy questions concerning child labor are neither unique to the current era nor the Western world. Perhaps the earliest regulation of child labor dates to 1284, when a statute of Venetian glass makers forbade the employment of children in certain dangerous aspects of the glass trade (2). Efforts at reform have been and continue to be impaired by a lack of substantive data on the effects of child labor on health and development.

In 1919, the first meeting of the International Labor Organization (ILO) fixed a minimum age for the employment of children at 14 years. Subsequently, many international conventions and regulations have been adopted by the ILO and many nations have set a minimum age for employment. However, as judged by educational data, a substantial number of nations report large numbers of children leaving school below the nation's specified minimum age.

The right of children to an education and freedom from exploitation are clearly stated in the Convention on the Rights of the Child adopted by the United Nations General Assembly in 1989. The passage of this declaration represents a commitment on the part of member nations to work toward a future in which the rights of all children are respected. International law also provides a clear and generally accepted consensus on the nature and definition of child labor.

The Magnitude of Child Labor

In the 1970s, the ILO estimated that there were approximately 150 million working children in the world; more recently, the ILO has estimated that the number may be as high as 250 million (3). Even the latter estimate may be low. There are an estimated 1.1 billion children between 5 and 16 years of age in developing and the least developed nations in the world. In many nations, fewer than 50% of children complete primary school. If most of these children work, the number of working children may be closer to 500 million (Figs. 38-1 and 38-2).

In spite of the lack of comprehensive data,

it is clear that working children in developing nations spend long hours at work and often have little or no time away from the

FIG. 38-1. Child carpet weaver in India. (Photograph by David Parker.)

FIG. 38-2. Young brick workers in Colombia. Thousands of children work as forced labor in brick kilns, rock quarries, or mines. (From the International Labor Office, Geneva, Switzerland.)

Box 38-1 (*continued*)

workplace. In one study of 210 Malaysian children, the children worked an average of 10 hours per day. Thirty-eight children reported that they worked 7 days per week, and 132 had only a half-day off per week (4). Although sometimes combined with school, similar numbers of hours worked have been reported from Sudan, Turkey, and Nigeria.

The Health Effects of Child Labor

The impact of child labor should be seen in four parts: the effect of work on growth and development; job-specific hazards as they relate to injury and illness morbidity; the effect of latency on the future health of children; and sexual and emotional abuse.

Perhaps the most obvious impact of child labor is on intellectual development, and child labor has been frequently associated with adult illiteracy. Studies have also demonstrated the impact of child labor on lost educational opportunity. For example, in one study in Bangladesh, children of women with no education had a four- to fivefold risk of severe malnutrition compared with children of mothers with a university education (5). Studies have also demonstrated the general poorer health of child workers (4).

It is not surprising that there are few studies on the impact of specific work-related exposures on the health of children. First, children are often working illegally. Second, children are rarely the beneficiaries of any type of labor contract. Third, such studies are expensive and difficult to conduct. What is known is that working children can be found around the world cutting rock in stone quarries, working in heavy construction, tanning leather, electroplating metals, scavenging garbage for food, tending goats and sheep, and any of hundreds of menial tasks. Too often, work in developing nations is performed without adequate protections and, when available, personal protective equipment has been designed for use by adults and is virtually useless for a child.

Young workers in many developing nations are at substantial risk of developing both work-related and non–work-related illness. Most data indicate that the health status of young workers is poor. These health problems are compounded by the all-too-often intolerable work conditions. Although data on the toxic effects of occupational exposures to children are limited, existing disease models amply support the hypothesis that children are likely to acquire disease at an early age as a result of hazardous work. For example, young children are known to be more susceptible than adults to the adverse health effects of lead.

The focus of the chapter is limited to the health and safety risks faced by young workers in the United States. Many of the issues are relevant to youth employment in other industrialized countries, but less so to child labor in newly industrializing and developing nations, where there are many basic human rights as well as health and safety concerns. (Box 38-1 provides a discussion of child labor in these countries.) Exploitative child labor continues to be a reality for some young people in the United States. In considering the health and safety concerns of young workers in the United States, however, it is important to distinguish between children employed in frankly exploitative situations, such as sweatshops in the apparel industry, and what is here called *youth employment*. The former typically involves minority children working out of economic necessity and hidden from public view. It involves a comparatively small number of young people in the United States. Youth employment, on the other hand, is the norm in American society. Failure to differentiate between frankly exploitative child labor and youth employment enables society at large to distance itself from the problems—to focus on the extreme conditions and overlook risks faced by youth in common, everyday jobs. Both issues need to be addressed (6).

YOUTH EMPLOYMENT IN THE UNITED STATES

Youth employment is more extensive than often realized. According to the U.S. Department of Labor, an average of 34% of 16- to 17-year-olds—over 2.6 million teens—were employed at any given point during 1997. An additional 10% were in the workforce looking for work (7). These official estimates undercount the true number of employed youths. For example, they exclude 14- and 15-year-olds, who are allowed to work under child labor laws, as well as younger children, who can legally work under federal law as news carriers and on family farms and in other family businesses. Surveys of youth themselves indicate that 80% have worked for pay by the time they leave high school. Although working youth historically contributed to the support of their families, most youth today report that they primarily work for discretionary income. Although comparisons with other countries are difficult, available data suggest that more children in the United States work while going to school than in other industrialized nations (1).

Young workers are typically employed in part-time, low-paying jobs, and move in and out of the workforce. When employed, they spend substantial numbers of hours at work. In a nationwide survey, 18% of high school students reported working more than 20 hours per week during school (8). As shown in Fig. 38-3, close to half of working youth are employed in the retail sector, predominantly at grocery stores and restaurants, where they constitute approximately 10% of the workforce. More than one-fourth of working youth are employed in service industries, such as at nursing homes, and approximately 8% are employed in agriculture. Youth from low-income families and minority youth are less likely to be employed and to reap the potential benefits of work. When they do work, they are more likely to be engaged in higher-risk occupations such as agriculture, manufacturing, and construction (9). Although official statistics on the numbers of children employed in frankly exploitative situations are not available, it has been estimated that as many as 13,000 children in the United States may be employed in sweatshops (10).

OCCUPATIONAL HEALTH OF YOUNG WORKERS

Concern about young workers in the United States cannot simply be relegated to the past. There is considerable evidence that occupational injuries to working youth are a significant public health problem today (Figs. 38-4 and 38-5). Each year, at least 70 workers younger than 18 years of age are killed on the job. Close to half of those killed are younger than 16 years of age. The leading causes of these deaths include traffic-related incidents, homicides, and machine-related deaths.

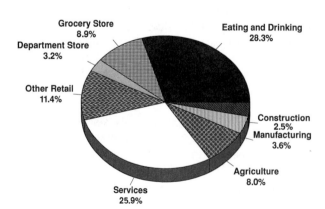

FIG. 38-3. Distribution of working youths, 15 to 17 years of age, by industry in the United States, 1996. (From Bureau of Labor Statistics, 1997).

FIG. 38-4. An underage supermarket meat packer unjams store machinery. Jobs in meat packing place adolescent workers at high risk for serious injuries. (Photograph by Earl Dotter.)

FIG. 38-5. Adolescent worker in a dietary department of a hospital. Burns, sprains and strains, and lacerations are common injuries in food service jobs. (Photograph by Elise Morse.)

Agriculture, which has a number of special exemptions under the child labor laws, accounts for more fatalities than any other industry. Half of these fatalities involve youth working on family farms. The overall fatality rate for young workers has been found to be similar to that for adults, despite the fact that child labor laws prohibit youth from working in particularly dangerous jobs (11).

Nonfatal injuries far outnumber fatalities. In New York State, from 1980 to 1987, more than 1,200 youths annually received compensation for occupational injuries resulting in 8 or more lost workdays; 44% of the injured youth sustained permanent disability (12). In Washington State, which tracks all injuries regardless of lost work time, over 4,400 adolescents received workers' compensation benefits each year from 1988 to 1991 (13).

Workers' compensation data, however, reveal only part of the problem because many young workers who are injured may never enter workers' compensation systems. In a Massachusetts study, 7% to 13% of all medically treated injuries among 14- to 17-year-olds occurred at work. In 1996, an estimated 105,000 youth were treated in emergency departments nationwide for work-related injuries. Surveys of young workers themselves underscore the extent of the problem. Between 7% and 16% of teens who have worked report being injured at work seriously enough to seek medical care (1).

The numbers of teen work injuries alone raise concern. Because youths usually work in part-time, temporary jobs, these numbers lead to estimates of high injury rates *per hour worked*. Several studies indicate that the overall injury rate per hour worked for teens is actually higher than that for adults. In 1996, for example, the rate of work-related injuries treated in emergency departments for 16- to 17-year-old workers was 4.9 per 100 full-time equivalent workers, compared with the rate of 2.8 for all workers aged 16 years and older.

These rates are crude and do not take into account the different types of jobs held by teens and adults; further research is necessary to determine whether teens have higher injury rates than adults doing comparable work.

As would be expected, most nonfatal injuries occur in those industries in which the greatest numbers of youth are employed—in retail trades, predominantly restaurants and grocery stores. Other leading locations include general merchandise stores, nursing homes, and farms. Agricultural injuries are especially common among youth younger than 16 years of age, and it appears that injuries sustained in agriculture are more severe than injuries in other industries. Lacerations, sprains, and strains are the most common nonfatal work injuries to youth. Similar to adults, half of the sprains and strains involve the back. Burns and fractures are also common. As shown in Fig. 38-6, the types of injury vary by industry.

Although little is known about the long-term disability associated with these injuries, the impact should not be underestimated. Such injuries, identified through emergency department records and workers' compensation claims, are serious enough to result in medical care. The longer-term human and economic costs associated with these injuries need to be documented.

Although surveillance studies have greatly improved our understanding of occupational *injuries* to youth, little is known about the extent to which youth experience acute or chronic *illnesses* as a result of work-related exposures. There are incidence reports of young workers with acute occupational disease. For example, youth were among the ill in an outbreak of green tobacco sickness in Kentucky. There is also evidence that youth are potentially exposed to a number of hazardous conditions at work that can contribute to latent illness. In a North Carolina survey, for example, 27% of youth who had worked reported exposure to very loud noises, 24% reported working with gasoline, and 19% reported working with pesticides or other chemicals. Because youth typically work part time, their exposures may not exceed existing standards that assume a lifetime of 8-hour workdays and 40-hour workweeks. Their short periods of exposure may reduce the likelihood of long-term health impacts. On the other hand, the fact that they are exposed at all raises important research questions about the potential increased susceptibility of young workers, the effects of age at first exposure, and the combined effects of multiple exposures over a working lifetime. Youth employment also raises important policy considerations about what are acceptable health risks for working youth (1). Washington State, for example, has revised its child labor laws to prohibit youth

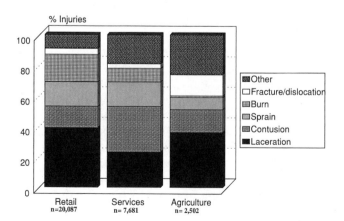

FIG. 38-6. Type of injury by industry in youth 14 to 17 years of age treated in emergency departments for work-related injuries in the United States, July to December, 1992. (From Layne LA, Castillo DN, Stucte N, Cutlip P. Adolescent occupational injuries requiring hospital emergency department treatment: National Representative Sample. Am J Public 1994:84:657–660.

from working in occupations involving potential exposure to carcinogens or bloodborne pathogens.

RISK FACTORS SPECIFIC TO YOUNG WORKERS

In considering the risk factors specific to young workers, it is critical that the focus on unique attributes of youth should not detract attention from the work environment itself. Young workers, like adult workers, are injured on the job because there are hazards in the jobs where they work. Notably, many of the industries in which youth are commonly employed—grocery stores, nursing homes, farms—have high injury rates for workers of all ages. The primary prevention goal for workers of all ages is the elimination and control of hazards in the workplace.

A number of factors do, however, raise special concerns about the health and safety of children and adolescents at work. Like all *new* workers, young workers are at increased risk of injury simply from lack of experience. Inexperienced workers are unfamiliar with the requirements of work, are less likely to recognize hazards, and are commonly unaware of their legal rights on the job. Lack of health and safety training is also a problem for young workers, even more so in light of their inexperience. In several surveys of working teens, approximately half report that they have not received health and safety training on the job. In addition, teens typically do not receive training about workplace health and safety or labor laws in school, except for some training in vocational education. Inadequate supervision is likewise a concern.

Developmental characteristics—physical, cognitive, and psychological—may also place young workers at increased risk. There is tremendous variability in the size of adolescents, for example, and smaller youth may not be able to reach parts of machines or may lack strength required to do the tasks demanded of them. Some organ systems, such as the musculoskeletal, reproductive, and endocrine systems, undergo periods of rapid growth or activity during adolescence. It is not known if this introduces unique youth susceptibilities to chemical or ergonomic insults at work.

Adolescence is well recognized as a period of profound psychological change. The psychological transition typically lags behind physical maturation, and psychological immaturity may be obscured by the physical appearance of the adolescent who may be assigned tasks for which he or she is neither emotionally nor cognitively prepared. Because adolescence is a time of exploration and risk taking, occupational injuries in this age group are commonly attributed to irresponsible acts associated with risk-taking behavior. Interviews with adolescents injured at work in Massachusetts, however, suggest the opposite is often the case. It is frequently young workers trying to act responsibly and demonstrate new independence and skills who are injured at work. Many of adolescents' positive traits—energy, enthusiasm, a need for increased challenge—combined with a reluctance to ask questions or make demands on employers can result in adolescents taking on tasks they are not capable of doing safely.

Young workers face the added challenge of balancing work and school. Federal child labor laws allow 14- and 15-year-olds to work 18 hours during school weeks; there are no federal restrictions on hours for older teens, although such time restrictions do exist in some states. Heavy part-time work schedules of adolescents have been linked with inadequate sleep and increased daytime fatigue, which may increase risk of injury in other arenas as well as work. Working more than 20 hours per week has also been associated with increased substance abuse—increased smoking and use of alcohol and illicit drugs (1).

THE CHILD LABOR LAWS

Early concerns about child labor led to passage of state child labor laws and, eventually,

TABLE 38-1. Hazardous jobs under the Fair Labor Standards Act

---

Nonfarm work
Seventeen hazardous nonfarm jobs are prohibited under the Fair Labor Standards Act. In general, youth younger than 18 years of age cannot do work involving the following:
Manufacturing or storing explosives
Driving a motor vehicle and being an outside helper on a motor vehicle
Coal mining
Logging and sawmilling
Power-driven wood-working machines[a]
Exposure to radioactive substances and to ionizing radiations
Power-driven hoisting equipment
Power-driven metal forming, punching, and shearing machines[a]
Mining, other than coal mining
Meat packing or processing (including power-driven meat slicing machines)[a]
Power-driven bakery machines
Power-driven paper products machines[a]
Manufacturing brick, tile, and related products
Power-driven circular saws, band saws, and guillotine shears[a]
Wrecking, demolition, and ship-breaking operations
Roofing operations[a]
Excavation operations[a]

Farm work
Children working on their parents' farms are exempt from the prohibitions of the Fair Labor Standards Act. For other children younger than 16 years of age working in agriculture, the following occupations/tasks are prohibited:
Operating tractors with horsepower greater than 20 power take-off (PTO).
Operating corn pickers, cotton pickers, grain combines, hay mowers, forage harvesters, hay balers, potato diggers, mobile pea viners, feed grinders, crop dryers, forage blowers, auger conveyors, nongravity-type self-unloading wagons or trailers, power post-hole diggers, power post drivers, or nonwalking-type rotary tillers
Operating trenchers or earthmoving equipment, forklifts, potato combines, or power-driven saws
Handling breeding animals, sows with suckling pigs, cows with newborn calves
Felling, bucking, skidding, loading, or unloading timbers with a butt diameter of more than 6 inches
Using ladders or scaffolds more than 20 feet high
Driving a bus, truck, or car while transporting passengers or riding as a passenger or helper on a tractor
Working inside fruit, forage, or grain storage units, silos, or manure pits
Exposure to agricultural chemicals classified as category I or II of toxicity
Working with explosives
Being exposed to anhydrous ammonia

---

[a] Limited exemptions are provided for apprentices and student-learners under specified standards.

in 1938, of federal law to protect the educational opportunities, health, and well-being of young workers. These laws reflect the long-standing societal viewpoint that there are different levels of acceptable risk for youth in the workplace. They provide an additional layer of protection, above and beyond occupational safety and health standards that apply to all workers.

The child labor laws establish minimum ages for employment, limit the hours and times of day youth can work, and prohibit employment of youth in certain jobs deemed to be too hazardous. Most state laws require young workers to obtain work permits, typically issued by schools. The hours of work and prohibited jobs vary significantly for children working in agricultural and nonagricultural occupations (Table 38-1). The basic minimum age of employment is 14 years in nonagricultural occupations, except that under the federal law, children of any age may work in family businesses as long as they are not engaged in the prohibited jobs. Children working on family farms are completely exempt from the federal child labor laws, as are news carriers. State laws vary widely and may be less or more protective than the federal law. Federal law applies only to businesses that are engaged in interstate commerce or have an annual gross income of $500,000 or more. Therefore, many small businesses are exempt from federal regulations. When both federal and state laws are applicable, the most stringent law applies.

Substantial revisions in federal and many state laws have not been made in decades. Consequently, many laws do not reflect changes in patterns of youth employment and education, changes in the nature of work, and new knowledge about occupational health and safety risks. For example, at the federal level, virtually all of the prohibited occupations for youth in nonagricultural jobs are prohibited on the basis of safety hazards; none of the restrictions addresses health hazards in the workplace; nor do they address issues of violence in the workplace. In addition, enforcement of the laws is limited. In 1998, the National Research Council issued a report calling for federal limits on hours of work for 16- and 17-year-olds, elimination of the differences between the restrictions for children working in agricultural and nonagricultural jobs, and updating of the list of prohibited occupations in an effort to address new and emerging technologies and working conditions (1). Additional research and input from public health professionals are needed to provide the scientific basis for updating these regulations.

OPPORTUNITIES FOR INTERVENTION

Although the presence of youth in the workplace raises special health and safety concerns, there are also special opportunities for prevention of occupational injuries and illnesses in this population. These include not only use of the federal and state child labor laws to protect young workers, but innovative efforts that involve the broader set of stakeholders for youth. Multiple adults, in addition to employers, have important roles to play in safeguarding the health and safety of young workers. Parents and guardians retain legal and social responsibility for their children's well-being. Educators also play a

role in the work lives of youth: in approving work permits, providing or facilitating work experiences, and preparing students for the world of work. In some states, health care professionals are required to sign off on work permits.

This expanded set of stakeholders provides an important opportunity for using the community at large to promote the health and safety of working youth. Examples of ongoing efforts include incorporation of health and safety education into school-to-work initiatives or middle and high school curricula; development of peer leadership programs that include occupational health and safety among the list of topics addressed by peers; and information dissemination about health, safety, and workers' rights to both young workers and their parents through community organizations and the local media. Health care professionals who provide services to children and adolescents also have a potentially important role to play in providing guidance to young patients about work (Table 38-2). It remains the ultimate responsibility of the adult employer, who profits economically from the labor of youths, to provide a safe, appropriate work environment, including adequate supervision and training. Community resources can be marshaled to shift community norms and help ensure that this responsibility is met.

Preventing occupational injuries and illnesses among young workers ultimately requires a comprehensive approach that includes regulation and enforcement, engineering advances to control hazards in the workplace, and education of adults and youth. It calls for new alliances between occupational health experts and other stakeholders, including pediatricians and maternal and child health professionals, child labor regulators, educators, and community leaders. Although safeguarding youth at work poses unique challenges, it also has tremendous potential for influencing the safety and health of the next generation. Health and safety education that goes beyond task-specific safety training can provide young work-

TABLE 38-2. *Advice to health care professionals about protecting working teens*

As health care professionals, you have an excellent opportunity to counsel teenage patients during the high-risk transition period from childhood to adulthood. You can play an important role by providing them with information, promoting safe work practices, and encouraging them to know their rights and to speak up when there are problems.

- Ask your teenage patients whether they work, and if so, where.
- Ask if they or their friends have ever been injured at work.
- Ask how many hours they work in a week, especially during the school year, and discuss whether the number of hours interferes with other activities and contributes to fatigue.
- Ask about work tasks, both regular and occasional. Are the tasks appropriate to your patients' developmental and physical abilities? Are they prohibited by the child labor laws?
- Encourage your patients to follow safety rules at work, including using protective clothing and equipment as required.
- Encourage your patients to tell someone (parent, boss, older coworkers) if they encounter problems at work.
- Provide material to teens and their parents or guardians about child labor laws and resources for more information. These materials are available in many states.
- Contact the Occupational Safety and Health Administration or the relevant state agency about workplaces where you believe young workers may be at risk of serious injury.

Adapted from educational materials prepared by the Massachusetts Department of Public Health with funding provided by the National Institute for Occupational Safety and Health.

ers with transferable knowledge and skills— hazard recognition and an understanding of the principles of hazard control as well as legal rights and responsibilities—that they will carry with them throughout their working lives. This education is especially relevant in looking toward intervention models that focus on joint labor–management health and safety efforts and greater involvement of workers who are empowered to effect change in the workplace. Working with youth now to provide them with knowledge and skills in occupational safety and health will better enable them to be active participants in creating and ensuring safe and healthful workplaces of the future.

ACKNOWLEDGMENTS

The authors thank Robin Dewey, Elise Morse, and Ellen Frank for their helpful comments.

REFERENCES

1. National Research Council. Protecting youth at work: health, safety, and development of working children and adolescents in the United States. Washington, DC: National Academy Press, 1998.
2. Wiener M. The child and the state in India. Princeton NJ: Princeton University Press, 1991.
3. Bellamy C. The state of the world's children. Oxford: Oxford University Press. 1997.
4. World Health Organization. Children at work: special health risks, WHO Techincal Report Series 756. Geneva: WHO, 1997.
5. Islam MA, Rahem MM, Mahalanabis D. Maternal and socioeconomic factors and the risk of severe malnutrition in a child: a case control study. Eur J Clin Nutn 1994;48:416–424.
6. Davis L. Youth employment versus exploitative child labor. Public Health Rep 1998;113(1):3–4.
7. Bureau of Labor Statistics. Employment and earnings, issue 1. Washington, DC: U.S. Department of Labor, Bureau of Labor Statistics, January, 1998.
8. Resnick MD, Bearman PS, Blum RW, et al. Protecting adolescents from harm: findings from the national longitudinal study on adolescent health. JAMA 1997;278:823–832.
9. U.S. General Accounting Office. Child labor: characteristics of working children. Publication no. GAO/HRD 91-83BR. Washington, DC: U.S. General Accounting Office, 1991.
10. Kruse D. Illegal child labor in the United States. Paper prepared for the Associated Press, November 1997 [available from Douglas Kruse, Rutgers University].
11. National Institute for Occupational Safety and Health. NIOSH alert: preventing deaths and injuries of adolescent workers. U.S. Department of Health and Human Services (NIOSH) publication no. 95-125. Cincinnati, OH: National Institute for Occupational Safety and Health, 1995.
12. Belville R, Pollack SH, Godbold JH, Landrigan PJ. Occupational injuries among working adolescents in New York State. JAMA 1993;269:2754–2759.
13. Miller M, Kaufman JD. Occupational injuries among adolescents in Washington State, 1988–1991. Am J Ind Med 1998;34:121–132.

BIBLIOGRAPHY

Castillo D, Davis LK, Wegman DH. Young workers. Occup Med: State of the Art Reviews 1999; 14(3):519–536.
Contains a comprehensive review of the literature on occupational illnesses and injuries among young workers and sets out a research agenda for the future.

Children's Safety Network at Education Development Center, Inc. and the Massachusetts Occupational Health Surveillance Program. Protecting working teens: a public health resource guide. Newton, MA: Education Development Center, Inc., 1995.
Targets state public health practitioners who want to address the issue of health and safety for teen workers. It includes a description of data sources available for surveillance of teen work injuries and examples of prevention efforts initiated in various states.

National Committee for Childhood Agricultural Injury Prevention. Children and agriculture: opportunities for safety and health. Marshfield, WI: Marshfield Clinic, 1996.
Includes recommendations developed by a broad-based coalition for public and private efforts to address prevention of childhood agricultural injuries.

National Institute for Occupational Safety and Health. NIOSH alert: preventing deaths and injuries of adolescent workers. U.S. Department of Health and Human Services (NIOSH) publication no. 95-125. Cincinnati, OH: National Institute for Occupational Safety and Health, 1995.
A summary of risks faced by adolescent workers that includes recommendations for parents, employers, educators, and adolescents.

National Institute for Occupational Safety and Health. Promoting safety work for young workers: a community based approach, NIOSH Publication No. 99–141. Cincinnati, OH: NIOSH, 1999.
A how-to resource guide for anyone interested in protecting the health and safety of young workers. Based on the actual experiences of three NIOSH-funded community-based projects.

National Research Council. Protecting youth at work: health, safety, and development of working children and adolescents in the United States. Washington, DC: National Academy Press, 1998.
Provides a comprehensive review of the literature on both health and psychosocial outcomes of youth employment in the United States and the available data systems that provide information about teen employment and work injuries. Includes social policy and research recommendations to protect working children and promote healthful work experiences for youth.

BIBLIOGRAPHY FOR BOX 38-1

Parker DL. Stolen dreams: portraits of working children. Minneapolis: Lerner Publications, 1998.
A photographic expose of child labor around the world.

Wiener M. The child and the state in India. Princeton NJ: Princeton University Press, 1991.
Perhaps the most comprehensive contemporary work on the history of child labor and its regulation.

Bureau of International Affairs, U.S. Department of Labor. By the sweat and toil of children: the use of child labor in American imports. Washington, DC: U.S. Department of Labor, 1994–1999.
Four volumes of interviews with hundreds of governmental and nongovernmental workers around the world.

Occupational Health: Recognizing and Preventing Work-Related Disease and Injury, 4th ed.
Edited by Barry S. Levy and David H. Wegman, Lippincott Williams & Wilkins, Philadelphia © 2000.

39

Older Workers

David H. Wegman

Increasing attention is being given to the occupational health of older workers and the relationship between work and aging. Important issues include how age affects workers' abilities to meet job demands, how job-related factors affect the aging process, how to help older workers maintain and update working skills and knowledge, how age determines retirement decisions, how better to use the expertise and wisdom of older workers, and how to design work to enable people to retire in optimal health. Despite the importance of these issues, the knowledge base on them is limited.

No standard definition of an "older" worker exists. Many definitions focus on the workers 55 years of age or older, and some focus on "younger" older workers—those who are 45 years of age or older (1,2). Relevant issues vary by age: for those older than 55 years, attention is directed more to issues related to retirement, whereas for those younger than 55 years, focus can be on workplace interventions that prevent disability and promote the person's continued development in an occupation.

Factors that convert advancing age into a handicap are mostly related to working conditions that impose constraints that outstrip actual human capabilities, and work organization that prevents employees from

making significant contributions to their own growth in their jobs (3).

TRENDS IN THE 20TH CENTURY

An unusual bulge in the population pyramid—the "baby boomers" born between 1946 and 1964 in the United States, Western Europe, and Japan—represents a large number of workers, and a substantial number of them are nearing retirement. The size of the aging workforce (aged 45 to 64 years) and older adults (65 years and older, most of whom will be retired) will have an important impact at least through the first quarter of the 21st century (Fig. 39-1).

Interest in life after work has increased as life expectancy has increased worldwide (4), from less than 40 years at mid-century to 65 years in 1995. Life expectancy in industrialized countries is now over 75 years; in developing countries, 64 years; and even in the least developed countries, 52 years.

Before World War II, work did not arbitrarily terminate at a fixed age. Just after the war, however, age-barrier retirement was introduced in many industrialized nations. In the United States, there was a sharp decrease in male workers older than 55 years of age between 1950 and 1993; smaller changes occurred among female workers (5).

Age-barrier retirement policies enabled governments to reduce unemployment by transferring older workers to pensions, thus opening new positions for younger workers. These developments also facilitated employ-

D. H. Wegman: Department of Work Environment, University of Massachusetts Lowell, Lowell, Massachusetts 01854.

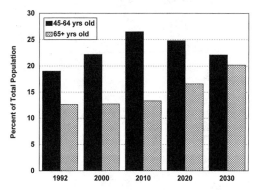

FIG. 39-1. Population of the United States: estimated for 1992 and projected for 2000 to 2030. (From U.S. Bureau of the Census. Current population reports, series P25-1092: Population projections of the United States by age, race, and Hispanic origin: 1992 to 2050. Washington, DC: U.S. Government Printing Office, 1992:Table 2.)

ers' efforts to "downsize" (with the short-term goal of increasing productivity), and offered workers the opportunity to retire early rather than work until the end of life.

For a time, growth in Social Security benefits delivered on the promise and also provided great opportunity for workers to retire early or to benefit from programs providing new access to disability retirement. This social commitment and capacity resulted from economic growth in the industrialized world starting in the 1950s, matched by the belief that the society could, and should, enable older citizens to enjoy activities of their own choosing for a number of years after working life had ended (the "Third Age").

The decrease in retirement age has now begun to be questioned, especially in Europe. The simultaneous aging of the population and the increased costs of providing retirement benefits for this population have now led most European nations to conclude that the costs are too great. Increasingly, Western European nations are restructuring the eligibility requirements for retirement and promoting continuation of work until at least 65 years of age. Somewhat similar actions have occurred in the United States, driven by concerns about future solvency of the Social Security system. Moreover, in con-

trast to Europe, the 1986 amendment to the Age Discrimination in Employment Act of 1967 now prohibits forced retirement of workers at any age in the United States.

By the early 1990s, trends in the United States indicated that the average age of retirement for men was no longer declining and for women was continuing to rise, probably not only because of the abolition of mandatory retirement, but also (a) Social Security benefit changes making work at later ages more attractive, (b) transition to defined-contribution retirement plans in the private sector (eliminating age-specific work disincentives), and (c) the increasing use of bridge jobs and part-time work to move gradually into retirement (6). In the United States, more older than younger workers work part time: approximately 25% of women 25 to 54 and 40% of those 55 years of age or older, and approximately 11% of men 25 to 54 and 23% of those 55 years of age or older. Over 50% of all those 65 years of age or older work part time (7,8).

AGING AND ACCELERATED AGING

There is no easily available functional definition of age or aging. Old age is not necessarily a disabling condition; it is an increase in the likelihood of small changes in performance parameters (9). We therefore need to study how age-related changes in particular subsystems interact to alter efficiency in carrying out complex everyday tasks.

Most of us mean *chronologic* age when we speak of age, but people of the same age can have strikingly different functional characteristics. Ways have been explored to measure *biologic* or *functional* age and summarize the balance of characteristics that yield interindividual differences, but, so far, an appropriate integrated measure has not been developed. Other features that are not a matter of change in physical or cognitive capacities have been described (10). For example, *psychological aging* represents the evolution of self-image and social relations; as workers age, they generally appear to experience a loss of self-image despite gains they have made in expertise or wisdom. *Social aging* places age in its social context; many older workers "accept" age discrimination because, in some sense, they expect it and consider it justified.

How Aging Might Affect Working Capacities

Many stereotypes deal with capabilities of older workers, but very few are documented. Stereotypes have been summarized into five myths: older workers are (a) in poorer health and have decreased capacity and loss of stamina; (b) at greater risk of injury, lost workdays, and higher insurance and medical costs; (c) less productive; (d) rigid and will not learn new skills; and (e) a poor investment to try to retrain (11). It is useful, therefore, to review what aspects of aging are appropriately seen as problematic for work.

Workers meet the demands of work through the use of a combination of resources. The most obvious of these can be grouped according to physical and mental resources. These primary components are supplemented by social capacities and features of motivation and experience (12).

Physical Resources

The best studied of the capacities relevant to work are those related to physical ability. Maximal physical strength for both male and female workers is reached between the ages of 20 and 30 years, followed by a very gradual decrement over the remainder of life (Fig. 39-2) (12–15). The decrements are larger for the lower limbs than the upper (14), and the rate of change increases after approximately 40 years of age. Similarly, maximum oxygen uptake peaks at 20 years and decreases to approximately 70% of that maximum by 65 years of age (13).

The variance among individuals at any given age is quite substantial. Some of this variance is genetically determined, but much can be attributed to differences in active ver-

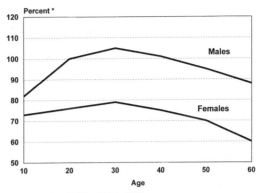

FIG. 39-2. Mean muscle strength by age and sex: sum of 25 muscle groups. (From de Zwart BC, Frings-Dresen MH, van Dijk FJ. Physical workload and the aging worker: a review of the literature. Int Arch Occup Environ Health 1995;68:1–12.)

sus sedentary lifestyles, degree of conditioning exercise apart from work, and weight gain or body mass, which are amenable to change throughout life (13).

Some data in the 11-year follow-up of Finnish municipal workers provide evidence for the value of physical exercise to promote maintenance of physical capacities (1,16). Laboratory-based investigations from Japan support this finding as well (17). However, the benefits achieved by regular physical exercise do not appear to be achieved by most physically demanding jobs; for example, meat cutters who perform work with high demands for hand strength have been noted to lose hand strength *more* rapidly than those without such demands (18).

Maximum physical capacities describe potential for use in work, a potential that may be limited by musculoskeletal changes that also occur with aging. Joint mobility and body posture gradually change between the ages of 20 and 60 years, and arthritis tend to present more frequently after 45 years of age. These changes have to be considered when determining how much of maximum capacity can actually be used in a given job (14).

Few jobs, however, require maximum capacity or maximum effort. Most jobs have continuous demands at a submaximal level such that older workers can meet the demands, although they will likely work closer to their maximal capacity (Fig. 39-3). Health care professionals can guide people toward health behaviors that promote their best possible physical state at any age with advice concerning body weight, exercise, and active lifestyles.

There are, however, adverse effects on physical capacity that result from the long-term effects of work, such as cumulative trauma disorders or repetitive strain injuries (see Chapter 26). These disorders are believed to result from continued exposure to poor job factors such as awkward postures,

repetitive motions, and excessive forces (19). The larger the dose, the greater the risk of injury. Because older workers naturally have accumulated larger doses of exposure, cumulative trauma disorders or repetitive strain injuries appear more prevalent among older workers. Older workers also have a higher prevalence of low back pain, although the incidence of low back pain is the same among older and younger workers (20).

For characteristics of work that might be considered to cause accelerated aging, intervention at the workplace is necessary. Such intervention can take advantage of the extensive ergonomic knowledge about prevention of musculoskeletal disorders. Many ergonomic interventions fail, however, not so much for absence of a correct understanding of risk, but more significantly because ergonomists emphasize *health* and *comfort,* whereas managers emphasize *productivity, quality,* and *efficiency* (21). Therefore, to make a difference in the control of workplace risks, occupational health professionals probably need to work directly with managers to find the best ways to expand the latitude within which the professionals will be able to act.

FIG. 39-3. A heat-stressed farmworker in Mississippi takes a water break with the temperature approaching 105°F. The changed physiology of older workers places them closer to maximum tolerance when working in conditions such as extremes of temperature. (Photograph by Earl Dotter.)

Mental Resources

In industrialized economies, mental capacity may be of greater importance to workers than physical capacity. There is, however, no measure of maximum mental capacity. Much of what we know comes from studies of component capacities, especially cognitive functioning in laboratory settings. Consequently, we know much about changes in test scores, but little about how workers balance their strengths and weaknesses in performing tasks.

Cognitive processes, such as information processing, and fluid abilities, such as drawing inferences and understanding implications, decline with age; cognitive products, such as verbal abilities and crystallized knowledge (accumulated information), however, remain stable with aging (14,22,23).

Compared with younger workers, older workers show more difficulty in the processing of complex and confusing stimuli and in paying attention to task-relevant information (24). Older workers seem to take longer to retrieve unfamiliar information and for tasks that place demands on short-term memory (24).

Older workers seem to be less competent than younger ones in processing information, especially with "fluid" intelligence, such as in problem-solving tasks, paying selective attention to subtasks within a series of tasks, using active or working memory, and carrying out several tasks simultaneously. These are illustrated by the age trends in performance on components of the Woodcock-Johnson Test of Cognitive Abilities (Fig. 39-4). Decline in the measures illustrated are greater after 50 years of age, but the declines begin earlier. Variance in most of these measures increases with age, suggesting that there is a wide range of cognitive functioning among older workers (Fig. 39-5).

Although there is ample evidence of the age-related decline in specific tests of cognitive functioning, declines do not seem to affect work performance. There appear to be at least two reasons for this finding: problems with slower mental processing are important only when tasks are complex; and compensation, such as using environmental cues, usually balances functional declines (25).

Attitude and Motivation

A person's attitude toward work and motivation at work are significant determinants of effective and productive work. Unfortunately, there is little information on these characteristics, let alone how they might relate to age. A review of 30 studies supports

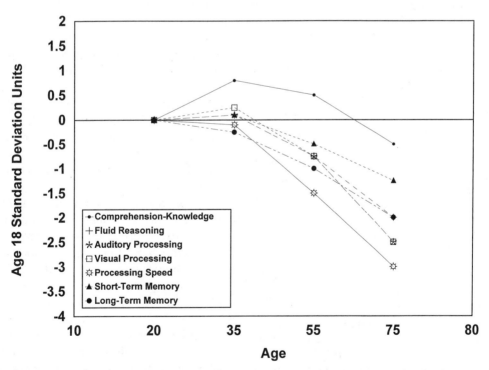

FIG. 39-4. Age trends in composite measures of several cognitive abilities taken from the Woodcock-Johnson Test of Cognitive Abilities. Values are plotted in standard deviation units based on the average 18-year-old test values. [From Salthouse TA. Implications of adult age differences in cognition for work performance. Arbete och Hälsa 1997;29(1):15–28.]

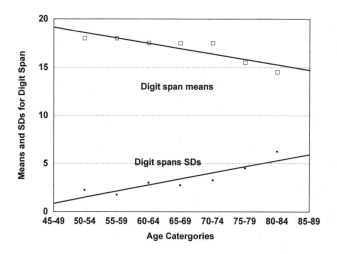

FIG. 39-5. Digit-span memory test scores by age. Values are plotted as means and standard deviations. (From McDaniel MA, Schmidt FL, Hunter JE. Job experience correlates of job performance. J Appl Psychol 1988;73:327–330.)

a relationship between older age and overall job satisfaction (22/28 studies), job involvement (18/21), organizational commitment (17/21), and low job turnover (10/15) (26). The determinants of job satisfaction, however, are poorly understood. For example, it could be that older workers are more satisfied because they have more desirable jobs, have more realistic expectations of work, or have not given work as high a priority in their lives as younger workers (27,28).

Motivation is a complex combination of interests and needs, including whether greater importance is assigned to affiliation and recognition or to material gain or power. Motivation probably plays an important part in determining occupational effectiveness, but there is little research on motivation, age, and work (25).

Experience and Expertise

Experience and expertise are frequently cited as reasons why older workers, despite any loss in fundamental capacities, remain effective and desirable employees. Again, little has been done to study these factors other than to examine job tenure and performance. Several studies have documented a strong positive association between experience and job performance (29–32). Because age and experience are correlated, age might be ex-

pected to be positively associated with job performance as well. For example, a study of speed and skill demonstrated that a significant correlation between age and performance became nonsignificant after controlling for years of experience (33). Whether there is added value to experience (years in the job) after the first 5 or 10 years is less well understood (34). It is important to distinguish expertise from experience.

Expertise or wisdom represents "procedural" knowledge, as distinct from experience that might be expected to result simply from repeated practice of a task. Expertise, then, is distinct from experience alone. Some efforts have been made to examine job-related expertise in occupational studies. There is evidence that relevant job knowledge is associated with higher supervisor ratings (35). Expertise is domain specific and is not likely to be transferable to jobs that require different types of knowledge. Expertise among experienced workers has probably been gained through developing somewhat individualized automatic behaviors (short cuts) to carry out tasks more efficiently than novices. A simple model of the interaction of decline in capacities with age and the offsetting effects of experience is provided in Table 39-1 (36).

Both experience and expertise probably

TABLE 39-1. *Four categories of job activity and expected relationships to performance with age during the years of labor market participation*

Task category	Basic capacities exceeded with increasing age	Performance enhanced by experience	Expected age relationship	Illustrative job content
Age enhanced	No	Yes	Positive	Knowledge-based judgments without time pressure
Age neutral	No	No	Zero	Relatively unde-manding activities
Age counteracted	Yes	Yes	Zero	Skilled manual work
Age impaired	Yes	No	Negative	Continuous paced data processing

From Warr P. Age and employment. In: Triandis HC, Dunnette MD, Hough LM, eds. Handbook of industrial and organizational psychology, vol 4, 2nd ed. Palo Alto, CA: Consulting Psychologists Press, 1994:485–530.

serve older workers in maintaining and advancing their effectiveness in many occupational settings. More studies targeted to measuring these characteristics and examining them with respect to age and occupation should provide valuable knowledge to guide occupational training programs and effective job placement for older workers.

HEALTH EXPERIENCE OF OLDER WORKERS

The best data available on age and health effects of work concern injury experience. In general, injury frequency decreases whereas injury severity, including likelihood of death, increases with age (14,37–39). Reduced injury frequency with age probably is due to job experience and familiarity with tasks. Increased severity with age is less easily explained. "Severity" is usually measured by time off the job. Greater injury severity among older workers could result from diminishing resilience as the neurologic and musculoskeletal systems age. Older workers may also have a reduced propensity to report minor injuries. In addition, managers and physicians may give them more time off work when they are injured (38).

More detailed studies of injury experience within occupation reveal a less clear pattern. For example, age-related injury risk has been shown to vary by specific job, as does age-related employment (38). Thus, it is not sur-

prising to find that both injury type and source vary by age, with the rate of back injuries higher in older workers, but the rate of eye and hand injuries higher in younger workers. Similarly, the agents of injury differ by age, with, for example, hand tools a more common agent in younger workers and working surfaces a more important agent in older workers. Little is known about actual magnitude of occupational disease in the population, especially in older workers. In general, disease is probably quite prevalent in older age groups, and many of these diseases are likely to be made worse by work exposures or occupational settings. Although many older workers have to retire early because of illness, many more are working with diagnosed diseases. The reasons for continuing to work may be that the disease is not disabling or that the job plays too important a role economically or socially for the person to seek early retirement. An estimate of the magnitude of the working ill comes from the Finnish study of municipal workers (40). Approximately 65% of workers aged 44 to 55 years had some diagnosed disease at the beginning of the study, and 11 years later over 80% did.

This study provides some further details on prevalence of diagnosed diseases among older workers in a cross-section of different types of work. The prevalent conditions are the same as would be seen in the population at large: musculoskeletal, cardiovascular, respiratory, and mental disorders. Disease prevalence for all reported conditions was higher for workers receiving old age or disability retirement benefits, but even those still at work had surprising levels of disease (40). For example, among workers still active at the 11-year follow-up, the prevalence of musculoskeletal disease was 44%—essentially double the baseline rate. Older people experience more arthritis and difficulty with joint mobility, but many work settings may add unfavorably to the "background" risk. Every effort should be made to control or eliminate exposures for any age group (see Chapters 7 to 9), but older workers are obviously at added risk because of underlying disease or disorders.

Finally, shift work is one specific feature of many of today's jobs, and tolerance of shift work does appear to differ according to age. Studies of shift work, defined as work other than that performed during the normal day work hours 7:00 a.m. to 6:00 p.m., have shown an adverse impact on sleep, alertness, performance, and safety; induction of strain from social and domestic sources; and some relationship to a variety of subjective health complaints such as gastrointestinal symptoms (41). When age is examined as a factor in these studies, the most important finding is the increased tendency to earlier waking, better performance earlier in the workday, and poorer tolerance for evening or night shifts (42). Given the opportunity or choice, older workers should probably be guided toward early rather than late shifts (see Chapter 22).

PERFORMANCE, PRODUCTIVITY, AND OLDER ADULTS

Two basic questions face older working adults and the health care professionals advising them: will advancing age affect work performance and productivity, and will changing capacities make older people less desirable as workers? Age stereotypes have been the source of the impressions among many managers that older workers perform

less well (34,43), and that they do not have adequate skills and cannot learn new ones (43,44). These attitudes are more often expressed about older workers from lower socioeconomic classes (43). It is particularly unfortunate that this discrimination appears to be accepted by older workers (Fig. 39-6) (10).

Efforts to study performance and productivity of workers have had mixed success. Problems center around difficulty in selecting appropriate measures of either *performance* or *productivity*.

As for performance, the most common criterion, supervisor evaluation, suggests only a very weak association with age (34). Emphasis frequently is placed on functional capacities, whereas little attention is directed at the specific characteristics of the work itself (45). Development and application of an effective measure of work capacity—what the Finnish researchers call *work ability* (1,16)—is essential to determining which older workers should stay in the workforce

FIG. 39-6. Stereotypes of older workers may result in their assignment to dead-end jobs in isolated settings, like this warehouse worker in Michigan. (Photograph by Earl Dotter.)

and which should be encouraged to leave work.

As for the importance of experience, studies in several work settings suggest that early during employment (approximately the first 5 years), there is a good correlation between experience and performance, but that after that the association is less well established (34). The importance of experience *in the occupations studied* has largely been shown during the early years of employment, regardless of age when the job is started.

The examination of age differences in learning and the transfer of learning to job settings has been an active topic of research. Older employees are less likely to participate in learning and, when they do, they are less effective learners than younger people. In addition, they seem to learn less, especially when cognitive demands are increased by rapid pacing. Not only are older workers less likely to take part in educational programs, they are often excluded from such programs (44).

These limitations are not insurmountable if efforts are made to motivate older workers to participate in learning programs, create a positive learning climate with encouragement from senior management, promote learning confidence with feedback and pre-training, provide additional learning time, and use newer teaching methodologies (e.g., active problem solving) to help address the learning strategies used by older people. Efforts to involve older workers in training programs to maintain and upgrade skills need to overcome some recognized resistance. Older people are relatively inactive as learners, they usually have lower educational qualifications and lack recent learning experiences, and they see learning as difficult, unrewarding, and not part of their typical lifestyle. Organizational culture may also interfere if managers believe that there is better financial return from training younger workers and do not promote learning opportunities among the older workforce.

Although the aforementioned evidence provides little support for older workers being less productive, employers still seem to believe in myths. A national survey in the United Kingdom asked employers what might discourage their recruitment of older people (46). The most common reasons they gave were lack of appropriate skills or qualifications and inadequate payback period on training. The fact that level of training and skills play such a large role in managers' beliefs is notable because these are primarily remediable factors that can be addressed by training of older workers. Moreover, the reliability and commitment associated with older workers could well offset the shorter payback period on training (47).

A study sponsored by the American Association of Retired Persons surveyed high-level managers from a sample of 400 companies ranging in size from 50 to over 1,000 employees (48). Considering the impact of downsizing with layoff of older workers, approximately half of the respondents indicated that this resulted in a workforce with less of a work ethic and with fewer seasoned, experienced workers. More than 70% of these managers considered older workers to have solid and reliable performance records and experience, whereas more than 80% considered older workers to have good attendance and punctuality, commitment to quality, and dependability in a crisis. The negative aspects of older workers reported by some respondents included lack of flexibility and lack of comfort with new technology. Despite these formal responses, focus groups held in association with the survey seemed to suggest that human resource decision makers believe that younger managers do not want to employ older employees no matter how good their skills.

DISABILITY AND RETIREMENT

Many workers must retire because of disability or, by the time of retirement, have experienced some form of disability that affects the quality of their lives during retirement. Disability retirement is more common among blue-collar than white-collar workers

FIG. 39-7. Older workers, as illustrated by this retired coal miner who had multiple finger amputations, often experience the physical, mental, and social effects of acute and chronic injuries accumulated over many years of work. (Photograph by Earl Dotter.)

(Fig. 39-7). Class, educational level, and marital status are also associated with disability retirement (49). Few studies have examined the differential impact of various diseases or injuries on disability in different occupations in older workers. Reports on two different Finnish populations provide some useful details that could help target prevention and intervention actions.

The first of these reports is based on an 11-year longitudinal study of Finnish municipal workers aged 44 to 58 years, whose work ranged widely—from construction workers, bus drivers, and nurse's aides to teachers, administrators, and physicians. This study was performed to determine the basis, if any, for the wide differences in age-related retire-

ment rules that existed among different occupations. For example, bus drivers and firefighters were eligible to retire at 55 years, whereas nurses and nurse's aides were eligible between 57 and 60 years of age. Other occupations required work until 65 years of age. Although the workers included were public employees, the nature of the work studied was comparable with that in the private sector. The cohort has been examined initially and 4 and 11 years later with respect to work ability, current illnesses, old age retirement, disability retirement, and death (1,16)

Disability pensions had been awarded to approximately one-third of the workers by the end of the 11-year follow-up, when their average age was 61 years (50). This high proportion of disability pensions reflects Finnish social policy on work disability. Overall, the types of disability were similar to those seen among comparable workers in the United States, but the threshold for awarding a work disability pension was lower for Finnish workers. Occupations were categorized as physically demanding, mentally demanding, or both. For physically demanding work, the disability retirements were approximately equal to regular retirements. However, disability retirements were less common for those in mentally demanding work, and much less common for those in work that required both physical and mental demands.

After 11 years of follow-up, more than two-thirds of these older workers had left work. Only 20% of workers in occupations characterized either as physically demanding (with or without mental demands) were still employed; only 30% of those in occupations characterized predominantly by mental demands were still employed (Fig. 39-8).

In this same study, "work ability" was estimated by using a questionnaire-based index (Work Ability Index or WAI) (51) covering (a) present work ability compared with lifetime best, (b) work ability in relation to physical and mental demands of work, (c) disease diagnosed by a physician, (d) subjective estimation of work impairment from this dis-

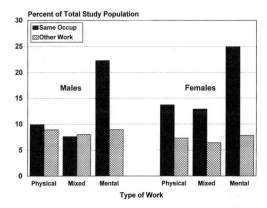

FIG. 39-8. Occupational changes after 11 years by sex and type of work among Finnish municipal workers. (From Tuomi K, Ilmarinen J, Klockars M, et al. Finnish research project on aging workers in 1981–1992. Scand J Work Environ Health 1997;23[Suppl 1]:7–11.)

ease, (e) sickness absences in the past year, (f) prognosis of work ability 2 years hence, and (g) psychological resources and support. The questionnaire was completed by all subjects at the start and end of the study; disability retirements 11 years later were then examined according to the WAI score assigned at the start of the study (Fig. 39-9) (50). The most important predictor of a disability retirement was self-estimated work ability at

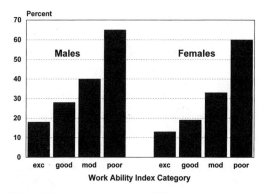

FIG. 39-9. Percentage disability pension awards among retired Finnish municipal workers after 11 years by baseline Work Ability Index category. (From Tuomi K, Ilmarinen J, Jahkola A, Katajarinne L, Tulkki A. Work ability index, 2nd revised ed. Helsinki: Finnish Institute of Occupational Health, 1997.)

the start (Ilmarinen J, personal communication, 1998).

Another population-based longitudinal study from Finland examined work determinants of disability for subjects who were enrolled in the Kuopio Ischemic Heart Disease Risk Factor Study. The population was a 30% random sample of adults from eastern Finland who worked in either the private or the public sector. The subjects, 42 years of age or older, were studied 4.2 years after enrollment to examine risk factors for disability retirement (52). When only age at the start of the study was considered, those 54 years of age had the highest risk of disability retirement compared with those 42 years of age (odds ratio = 3.6). After adjusting for age and socioeconomic variables, the important significant factors predicting disability retirement were heavy physical work, noise exposure, physical strain, mental strain, job dissatisfaction, and supervisor support. In the absence of longitudinal studies in the United States examining specific work factors on aging or retirement, the findings from Finland provide important insights for U.S. workplaces.

CONCLUSION

The general population and the workforce are aging. It makes sense to keep older workers in the workforce for as long as they are willing and able. However, older workers are different from younger ones, and popular misconceptions about competence, knowledge, and work capacity play a large role in determining whether older workers are likely to remain employed. One particular concern is the increasing tendency to see older workers as good contingent or part-time workers (48), which permits employers to take advantage of their experience, reliability, and work ethic while eliminating the need to pay costly benefits. In Western Europe, policy makers have realized that the loss of older workers through downsizing and age-discriminating policies is threatening the overall strength of the workforce (44). There, the "baby boom" has been followed by a "baby bust," and in a number of specific geographic regions or particular occupations there is, and will be, a shortage of younger workers to replace the skilled older workers. In the United States, the baby bust appears less evident—probably because of higher immigration rates than in Western Europe and higher fertility rates among immigrants. Although these immigrants may relieve U.S. industry from experiencing a shortage of future workers, many older workers want to continue working and others may need to continue working to maintain adequate income.

Health care providers have a special responsibility to use their skills and knowledge to make certain that inappropriate health or capacity considerations are not the basis for closing work opportunities for the older members of our population. We need to especially consider in job placement, matching of work tasks with a worker's personal characteristics and skills, while improving workplace conditions for all age groups (53).

REFERENCES

1. Ilmarinen J, ed. The aging worker. Scand J Work Environ Health 1991;17[Suppl 1]:1–141.
2. Kilbom A, Westerholm P, Hallsten L, Furåker B. Work after 45? Proceedings from a scientific conference held in Stockholm 22–25 September 1996. Arbete och Hälsa 1997;29(1,2).
3. Devezies P. The development of the functional condition of people in relation to their age, and the consequences for work activity. In: Ageing at work: proceedings of a European colloquium, Paris, 12 June 1991. Dublin: European Foundation for the Improvement of Living and Working Conditions, 1991:25–29.
4. Kinsella K, Gist YJ. Older workers, retirement and pensions: a comparative international chartbook. U.S. Department of Commerce, Bureau of the Census. Washington, DC: U.S. Government Printing Office, December, 1995, publication no. IPC/95-2RP.
5. U.S. Bureau of the Census. Current population reports, special studies P23-190: 65+ in the United States. Washington, DC: U.S. Government Printing Office, 1996:4-1.
6. Quinn JF. Retirement trends and patterns in the 1990s: the end of an era? Public Policy and Aging Report 1997;8(summer):10–14, 19.
7. Rones PL, Ilg RE, Gardner JM. Trends in hours of work since the mid-1970s. In: Monthly labor review.
Washington, DC: U.S. Department of Labor, Bureau of Labor Statistics, April, 1997.
8. Tilly C. Reasons for the continuing growth of part-time employment. In: Monthly Labor Review. Washington, DC: U.S. Department of Labor, Bureau of Labor Statistics, March, 1991.
9. Rabbitt P. Management of the working population. Ergonomics 1991;34:775–790.
10. Hjort PF. Age and work: good or bad for whom? Arbete och Hälsa 1997;29(1):3–13.
11. Peterson DA, Coberly S. The older worker: myths and realities. In: Morris R, Bass SA, eds. Retirement reconsidered: economic and social roles for older people. New York: Springer, 1988:16–28.
12. Ilmarinen J. A new concept for productive aging at work. In: E Heikkinen, J Kuusinen, I Ruoppila, eds. Preparation for aging. New York: Plenum Press, 1995:215–222.
13. de Zwart BC, Frings-Dresen MH, van Dijk FJ. Physical workload and the aging worker: a review of the literature. Int Arch Occup Environ Health 1995; 68:1–12.
14. Garg A. Ergonomics and the older worker: an overview. Exp Aging Res 1991;17:143–155.
15. World Health Organization (WHO) Study Group. Report: aging and work capacity. WHO Technical Report Series no. 835. Geneva: WHO, 1993.
16. Tuomi K, ed. Eleven-year follow-up of aging workers. Scand J Work Environ Health 1997;23[Suppl 1]:1–71.
17. Nakamura E, Moritani T, Kanetaka A. Biological age versus physical fitness age. Eur J Appl Physiol 1989;58:778–785.
18. Olofsson G. Firms and the older workforce: the case of Sweden in the 1990s. Arbete och Hälsa 1997; 29(2):243–249.
19. Bernard BP, ed. Musculoskeletal disorders and workplace factors: a critical review of epidemiologic evidence for work-related musculoskeletal disorders of the neck, upper extremity, and low back. U.S. Department of Health and Human Services (NIOSH) publication no. 97-141. Cincinnati, OH: U.S. Department of Health and Human Services, Public Health Service, Centers for Disease Control and Prevention, National Institute for Occupational Safety and Health, 1997.
20. de Zwart BC, Broersen JP, Frings-Dresen MH, van Dijk FJ. Repeated survey on changes in musculoskeletal complaints relative to age and work demands. Occup Environ Med 1997;54:793–799.
21. Winkel J, Westgaard RH. Editorial: a model for solving work related musculoskeletal problems in a profitable way. Applied Ergonomics 1996;27:71–77.
22. Baltes PB, Baltes MM. Psychological perspectives on successful aging: the model of selective optimization with compensation. In: PB Baltes, MM Baltes, eds. Successful aging: perspectives from the behavioral sciences. Cambridge: Cambridge University Press, 1990:1–34.
23. Salthouse TA. Implications of adult age differences in cognition for work performance. Arbete och Hälsa 1997;29(1):15–28.
24. Salthouse T. Speed of behavior and its implications for recognition. In: Birren JE, Schaie KW, eds.

Handbook of the psychology of aging, 2nd ed. New York: van Nostrand Reinhold, 1985:400–426.

25. Warr P. Age, competence and learning at work. In: Å Kilbom, ed. Aging of the workforce. Key-note presentations from a workshop held in Brussels March 23–24. Arbetslivsrapporter 1998:24. ISSN 1401-2928. Report series published by National Institute for Working Life, Sulna, Sweden, pp. 23–62.

26. Rhodes SR. Age-related differences in work attitudes and behavior: a review and conceptual analysis. Psychol Bull 1983;93:328–367.

27. Schooler C, Caplan L, Oates G. Aging and work: an overview. In: Schaie KW, Schooler C, eds. Impact of work on older adults. New York: Springer, 1998:1–20.

28. Sterns HL. Commentary: decision to retire or work. In: Schaie KW, Schooler C, eds. Impact of work on older adults. New York: Springer, 1998:134–135.

29. Avolio B, Waldman DA, McDaniel MA. Age and work performance in nonmanagerial jobs: the effects of experience and occupation type. Academy of Management Journal 1990;33:407–422.

30. Jacobs R, Hofman DA, Kriska SD. Performance and seniority. Human Performance 1990;3:107–121.

31. McDaniel MA, Schmidt FL, Hunter JE. Job experience correlates of job performance. J Appl Psychol 1988;73:327–330.

32. Schmidt Fl, Hunter JE, Outerbridge AN, Goff S. Joint relation of experience and ability with job performance: test of three hypotheses. J Appl Psychol 1988;73:46–57.

33. Giniger S, Dispenzieri A, Eisenberg J. Age, experience and performance in speed and skill jobs in an applied setting. J Appl Psychol 1983;68:469–475.

34. Sterns HL, McDaniel MA. Job performance and the older worker. In: Rix SE, ed. Older workers: how do they measure up? An overview of age differences in employee costs and performances. Public Policy Institute report no. 9412. Washington, DC: American Association of Retired Persons, November, 1994.

35. Borman WC, While LA, Pulakos ED, Oppler SH. Models of supervisory performance ratings. J Appl Psychol 1991;76:863–872.

36. Warr P. Age and employment. In: Triandis HC, Dunnette MD, Hough LM, eds. Handbook of industrial and organizational psychology, vol 4, 2nd ed. Palo Alto, CA: Consulting Psychologists Press 1994:485-530.

37. Kisner SM, Pratt SG. Occupational fatalities among older workers in the United States: 1980–1991. J Occup Environ Med 1997;8:715–721.

38. Laflamme L, Menckel E. Aging and occupational accidents: a review of the literature of the last three decades. Safety Science 1995;21:145–161.

39. Mitchell OS. The relation of age to workplace injuries. In: Monthly Labor Review. Washington, DC: U.S. Department of Labor, Bureau of Labor Statistics, July, 1988.

40. Seitsamo J, Klockars M. Aging and changes in health. Scand J Work Environ Health 1997;23 [Suppl 1]:27–35.

41. Akerstedt T, Knutsson A. Shiftwork. In: Levy BS, Wegman DH, eds. Occupational health: recognizing and preventing work-related disease, 3rd ed. Boston: Little, Brown, 1995:407–417.

42. Reilly T, Waterhouse J, Atkinson G. Aging, rhythms of physical performance, and adjustments to changes in the sleep–activity cycle. Occup Environ Med 1997;54:812–816.

43. Walker A, Taylor P. Ageism versus productive aging: the challenge of age discrimination in the labor market. In: Caro PG, Chen Y-P, eds. Achieving a productive aging society. London: Auburn House, 1993.

44. Walker A. Combating age barriers in employment: European research report. Dublin: European Foundation for the Improvement of Living and Working Conditions, 1997.

45. Robertson A, Tracey CS. Health and productivity of older workers. Scand J Work Environ Health 1998;24:85–97.

46. Taylor PE, Walker A. The aging workforce: employers' attitudes towards older people. Work, Employment and Society 1994;8:569–591.

47. Shephard RJ. Human rights and the older worker: changes in work capacity with age. Med Sci Sports Exerc 1987;19:168–173.

48. American Association of Retired Persons. American business and older workers: a road map to the 21st century. Washington, DC: American Association of Retired Persons, March, 1995.

49. Höög J, Stattin M. Who becomes a disability pensioner? The Swedish case 1988 and 1993. Arbete och Hälsa 1997;29(1):166–176.

50. Tuomi K, Ilmarinen J, Klockars M, et al. Finnish research project on aging workers in 1981–1992. Scand J Work Environ Health 1997;23[Suppl 1]:7–11.

51. Tuomi K, Ilmarinen J, Jahkola A, Katajarinne L, Tulkki A. Work ability index, 2nd revised ed. Helsinki: Finnish Institute of Occupational Health, 1997.

52. Krause N, Lynch J, Kaplan GA, et al. Predictors of disability retirement. Scand J Work Environ Health 1997;23:403–413.

53. Westerholm P, Kilbom A. Aging and work: the occupational health services perspective. Occup Environ Med 1997;54:777–780.

BIBLIOGRAPHY

de Zwart BC, Frings-Dresen MH, van Dijk FJ. Physical workload and the aging worker: a review of the literature. Int Arch Occup Environ Health 1995;68:1–12. *A fine review and summary of what is known about the physical effects of aging that relate to working ability.*

Kilbom A, Westerholm P, Hallsten L, Furker B. Work after 45? Proceedings from a scientific conference held in Stockholm 22–25 September 1996. Arbete och Hälsa 1997;29(1,2). *A meeting summary with excellent keynote addresses and papers presenting brief reports from a variety of relevant and interesting studies. Arbete och Hälsa is a scientific report series published by the National Institute for Working Life in Sweden. Copies can be purchased or downloaded from the Institute's web site: http://www.niwl.se.*

Robertson A, Tracey CS. Health and productivity of older workers. Scand J Work Environ Health 1998;24:85–97.

A report based on one of a series of comprehensive working papers on Issues of an Aging Workforce undertaken by the Centre for Studies of Aging at the University of Toronto. In addition to the review on which this article is based, the series includes separate literature reviews on the aging workforce with respect to displaced workers, women, and work organization.

Schaie KW, Schooler C, eds. Impact of work on older adults. New York: Springer, 1998.

An important monograph that presents a series of articles from the fields of sociology and psychology with critical commentary from members of both fields. The articles address a variety of aspects of work and aging in an interesting and challenging way.

Ilmarinen J, ed. The aging worker. Scand J Work Environ Health 1991;17[Suppl 1]:1–141.

Tuomi K, ed. Eleven-year follow-up of aging workers. Scand J Work Environ Health 1997;23[Suppl 1]: 1–71.

Two extensive reports on the Finnish studies of aging workers, the single most extensive longitudinal study of the impact of aging on work and work on aging. The first supplement presents the results at 4 years.

World Health Organization (WHO) Study Group. Report: aging and work capacity. WHO Technical Report Series no. 835. Geneva: WHO, 1993.

Probably the best available compact summary of the issues and knowledge base regarding work and aging.

40 Labor Unions and Occupational Health

Michael Silverstein and Franklin E. Mirer

Tragedies such as the Triangle Shirtwaist fire of 1911, the Gauley Bridge silicosis disaster of the 1930s, the Farmington Mine explosion of 1968, and the malignant legacy of asbestos exposures taught generations of workers that organization for health and safety protection is critical (1).

The Occupational Safety and Health Act (OSHAct) of 1970 and the mine safety acts that preceded it first established the right to a safe and healthful workplace at the national level, replacing less effective state legislation (see Chapter 10). But even this protection has been insufficient. Working people, therefore, have turned to their unions and other organizations for protection from the chemical, physical, and other hazards that accompany earning a living.

Personal and union involvement in health and safety programs is a moral imperative for workers. It is also a practical necessity for occupational health professionals who seek to prevent work-related illness and injury. Just as clinical medicine cannot be practiced effectively without the participation of informed patients, occupational health and safety cannot be pursued successfully without the active involvement of the workers whose lives and health are at stake. Workers possess unique information about working conditions that is vital to the diagnostic process. Moreover, the health professional lacks the independent ability to intervene to correct problems on the job once they have been identified. Without this preventive component, occupational medicine is reduced to minor treatment, fitness-for-duty examinations, disability evaluations, and drug testing. Many occupational health professionals fail to recognize the potential for a mutually rewarding alliance with workers and their unions (see Box 40-1). This chapter addresses the role of unions and their commitment to health and safety to facilitate such alliances.

VEHICLES FOR WORKER INVOLVEMENT IN HEALTH AND SAFETY

Labor unions are the major organizations pursuing the collective interests of workers in health and safety. In addition to assisting their members with day-to-day needs, unions actively work for legislative and regulatory remedies for health and safety problems. Therefore, although only approximately 14% of workers in the United States are unionized, their influence extends far beyond the workplaces where their members are employed.

Labor unions take two basic forms. Industrial model unions, such as the United Steelworkers of America or the Service Employees International Union (SEIU), represent employees in a specific workplace or indus-

M. Silverstein: Department of Labor and Industries, State of Washington, Olympia, Washington 98504.

F. E. Mirer: International Union, United Automobile Workers, Detroit, Michigan 48214.

Box 40-1. Checklist for Health and Safety Work with Unions

Health professionals who provide health and safety services for a union-organized workplace cannot be fully effective until they establish a working relationship with unions based on trust and mutual respect. The following steps will help prepare health professionals for this:

1. Determine which union(s) represents workers. Identify the local union leaders (presidents or chairs) and key representatives, such as health and safety or benefits representatives. Identify the representatives of the international union (servicing representatives or business agents) who are assigned to work with the local leaders.
2. Find out what kind of help the international union provides the local union on health and safety problems. Is there a health and safety department? Who is its director? Does the international union have staff professionals, such as industrial hygienists, safety engineers, or health educators?
3. Determine whether the local union or the international union has arrangements with outside experts for help on health and safety, such as those at academic institutions, COSH groups, and the Workers Institute for Safety and Health (WISH).
4. Establish communications with local union leadership before specific problems arise. The first contact should be with the elected leaders of the local, who can then make introductions to other key representatives. Ask about any outstanding issues of concern to the union.
5. Become knowledgeable about the nature of labor-management relations at the workplace. Read the collective-bargaining agreement with particular attention to any language on health and safety. Are there other written guidelines or procedures covering health and safety matters? For example, there should be a written respirator program under the terms of OSHA Standard 1910.134. How does the grievance procedure work for health and safety complaints? Does the union have the right to strike over health and safety? Is there a joint health and safety comimittee, a quality of work life program, or an employee assistance program?
6. Learn what procedures guard the confidentiality of workers with medical problems and move immediately to strengthen them if they are inadequate.
7. Find out what types of health and safety training programs are provided for employees and seek to become directly involved with them.
8. Obtain and read copies of any health and safety studies that have been done at the workplace (industrial hygiene surveys, ventilation or other engineering studies, or medical surveillance or other epidemiologic reports). Make sure these have been made available to the union in accordance with legal requirements.
9. Visit the shop floor early and often. Establish a presence in the plant, independent of management or labor, but also be sure to tour the plant frequently while accompanied by union representatives. While observing jobs, be sure to talk with workers and listen carefully to their concerns.

---

try. Members of craft unions, such as the International Brotherhood of Carpenters or the International Longshoremen's Association, typically move among different employers by virtue of their union membership status, sometimes dispatched from a hiring hall. Employees at a specific workplace may be organized into several different bargaining units and unions based on their type of work. At a construction site, as many as a dozen unions may represent groups of workers in the various building trades, such as the International Brotherhood of Painters and Allied Trades. Craft unions typically manage a health and pension fund covering many employers. For most industrial union members, typically the employer manages insurance and pension with union input. The American union movement is increasingly based in the public sector, and oriented to white-collar and service jobs.

At the individual workplace, unions are called *local unions* or *lodges*. These local unions are usually part of a national or international union (representing workers in Canada and the United States). Most international unions in the United States are, in turn, affiliated with the American Federation of Labor–Congress of Industrial Organization (AFL-CIO), headquartered in Washington, DC, and composed of unions who represent over 13 million members. On a regional or city-wide basis, local unions may work together in a group called a *central labor council.*

Local unions and their internationals enter directly into collective bargaining with employers. The AFL-CIO and local labor councils do not usually engage in collective bargaining, but rather focus on political action and other public policy initiatives on behalf of workers.

Labor laws govern relationships between employers and unions, including the National Labor Relations Act (Wagner Act, 1935), the Labor–Management Relations Act (Taft-Hartley Act, 1947), and the Labor–Management Reporting and Disclosure Act (Landrum-Griffin Act, 1959). Public employment law is governed by state legislation. These laws are designed to make employers bargain fairly with unions, to protect the rights of unions and union members in their relationships with employers, and to ensure democratic procedures and sound fiscal practices in unions. Labor law requires that employers bargain in good faith about concerns related to wages, hours, and conditions of employment. Mandatory subjects of bargaining include "conditions of employment" related to health and safety, such as the provision of ventilation, the use of personal protective equipment, or the operation of a plant medical clinic (Fig. 40-1). Labor law also forbids "employer domination" of labor organi-

FIG. 40-1. After the breakdown in local negotiations over health and safety conditions, a United Automobile Workers local union went on strike at a California automobile assembly plant in an effort to secure improvements in the working environment. (Photograph by Robert Gumpert. Courtesy of United Automobile Workers.)

zations. It would be illegal, for example, for an employer to establish a health and safety committee, set the agenda, and appoint the employee representatives. However, enforcement of labor law is limited and must be triggered by an employee complaint.

Unions have represented their members on health and safety matters in several ways:

1. Unions bargain with employers for specific improvements in working conditions, such as ventilation or safety equipment.
2. Unions bargain for representation and systems for improving conditions, such as health and safety committees, union health and safety representatives, rights to refuse unsafe work, grievance procedures for members to use in pressing specific complaints, and the right to strike during the life of a contractual agreement.
3. Unions provide technical assistance to members facing health or safety dangers. Many international unions have health

and safety departments with professionals, such as industrial hygienists or safety engineers. Other unions secure technical aid from Committees on Occupational Safety and Health (COSH groups; see later), the Center to Protect Workers Rights (a technical support group established by the Building Trades Department of the AFL-CIO), academic programs, and government agencies. The AFL-CIO Executive Council includes a health and safety committee, and also maintains a staff health and safety subcommittee comprising the health and safety directors of the affiliated unions, which meets bimonthly.

4. Unions conduct educational and training programs so that members can better understand their legal and contractual rights and can more effectively recognize hazards and work for their elimination.

5. Unions work politically for laws, standards, and regulations designed to improve working conditions and worker health. Unions have been the critical force behind most major Occupational Safety and Health Administration (OSHA) standards, providing evidence in the rule-making record and initiating litigation to force rule making and defend rules against industry opposition.

Workers, unions, and sympathetic health professionals have worked with a variety of organizations independent of unions to educate and enlist the energies of workers on behalf of health and safety reforms, and to apply pressure to employers, government agencies, and the scientific community. One of the earliest was the Workers' Health Bureau of America, which sought to assist labor unions with occupational health investigations, clinical services, education, and public policy agitation during the 1920s (2).

In the 1960s, the Black Lung Association, a coalition of mine workers and their families, union and community activists, and health professionals, raised awareness of the urgent need to eliminate the extreme hazards

in the coal mines by education, demonstrations, lobbying, and use of the media (Fig. 40-2). Special attention was focused on the risk of pneumoconiosis to underground coal miners. In conjunction with the United Mine Workers of America (UMWA), the Black Lung Association was instrumental in the passage of the federal Coal Mine Safety and Health Act of 1969 and later the Black Lung Benefits Reform Act. Dr. Lorin Kerr, director of the UMWA Occupational Health Department and for many years organized labor's only physician, was a major force in these proceedings. In a similar fashion, the Brown Lung Association and the Textile Workers Union [now the Union of Needletrades, Industrial and Textile Employees (UNITE)] drew public attention to the hazards of byssinosis in the cotton textile industry and successfully pushed OSHA to promulgate its cotton dust standard in 1978.

Since the early 1970s, a loose network of COSH groups has provided most nonunion support for health and safety. The Chicago area COSH group (CACOSH) was an early

FIG. 40-2. Protest by union members that was part of the movement that led to the passage of the black lung legislation (Coal Mine Safety and Health Act) in 1969. (Photograph by Earl Dotter.)

prototype for these coalitions of local union activists with supportive health professionals, lawyers, and students. CACOSH generated excitement and action through worker education programs, provisions of technical support to local unions, and political pressure for workers' compensation reform and stronger OSHA enforcement. A network of approximately 25 COSH groups across the country had developed by the late 1970s, and continues in the 1990s. COSH groups have expanded staffing and union base by successfully competing for health and safety training grant funds for projects emphasizing communities and unorganized workers. Training-oriented government requirements such as hazardous waste, asbestos, and lead abatement have indirectly supported COSH group structure. Many of the health and safety professionals in the early organizations progressed to union, government, or academic positions.

The right-to-know movement initiated in the late 1970s by the Philadelphia area COSH group (PHILAPOSH) was a significant outgrowth of the COSH groups. It brought national attention to the denial of full and accurate information on the composition and the hazards of chemicals to workers. After a tumultuous Philadelphia City Council hearing orchestrated by PHILAPOSH and the Delaware Valley Toxics Coalition, the nation's first local right-to-know ordinance was enacted in 1981. This city ordinance provided broad worker and community access to material safety data sheets (MSDSs) and other information on workplace chemicals.

As unions and COSH groups continued their successful work on state and local right-to-know bills, industry groups reluctantly decided that it would be preferable to have a single, uniform regulation rather than the evolving patchwork of local provisions. OSHA, under the Carter Administration, responded with a proposed federal regulation. The OSHA Hazard Communication Standard, eventually issued in late 1983 during the Reagan Administration, was very much

a compromise regulation. Organized labor has concentrated on the provisions that require employers to provide training on hazard recognition and control for all employees. Many unions, through collective bargaining, have secured the right to participate directly in the development and delivery of this training, which would otherwise be a unilateral employer prerogative.

HEALTH AND SAFETY PRINCIPLES UNDERLYING THE UNION AGENDA

Union health and safety activities rest on a set of principles about rights and responsibilities in the workplace. Health professionals who intend to work with unions will be more effective if they understand and respect these premises.

Premise 1: Workers and their union representatives are entitled to participate fully in the development and implementation of health and safety policies and programs because their lives and well-being are directly affected by the decisions made.

Convictions about workplace democracy and the right to "a seat at the table" underlie worker demands for participation. Worker participation in health and safety takes many forms: health and safety committees, review of new technology and equipment, and access to technical information and reports. National collective bargaining agreements with major automobile manufacturers provide for union representation on hazardous materials control committees, which approve or disapprove proposed new chemicals before use.

Premise 2: Working conditions shall be safe for all employees; employers have an obligation, both legally and morally, to provide such conditions without resorting to discriminatory hiring and placement practices based on sex, ethnicity, genetic predispositions, or physical handicaps.

This premise means that employers must make a good-faith effort to design or alter the job to fit the worker rather than to limit jobs to those who are judged to be "fit."

The most prominent example has been the United Automobile Workers (UAW) campaign to prevent exclusion of women from jobs with exposure to lead, which led to the U.S. Supreme Court prohibiting this practice. Unions argued that engineering controls and personal protective equipment should be used to reduce exposures sufficiently so that fertile or pregnant employees could work without discrimination. A more mundane example is the demand that the size and shape of tool handles should be adjustable so that workers of all sizes and shapes can work without fear of cumulative trauma disorders. Return-to-work programs should be aimed at placing a worker back on the original job, even if the job must be adjusted to allow this (see Chapters 9 and 12).

This principle of equal opportunity has several corollaries. First, medical examination programs should be used to guide environmental interventions rather than to determine hiring and personnel decisions. Second, procedures should be established to permit an employee to refuse to work without penalty on a job that is honestly believed to be unsafe until appropriate investigations and corrections can be made. Third, the employer should have a "medical removal protection" program that entitles a worker to an alternative placement without loss of seniority, earnings, or other employment rights and benefits in the event that all feasible protections have been built into a job and a medical examination determines that it is still unsafe for a specific worker.

Premise 3: Workers should not be forced to try to "work safe" in an unsafe environment.

This principle refutes the myth, unsupported by data, that 90% of injuries are caused by "unsafe acts." Unions are critical of behavior-based safety approaches, injury incentive programs, or more benign notions of safety awareness or training as dominant strategies for protecting workers. Personal respirators are not an acceptable alternative to local exhaust ventilation to reduce harmful chemical exposure. Workers should not be lectured to "be careful" around auto-

mated powered machinery by an employer who wants to substitute warning signs for mechanical enclosures that completely prevent worker entry into danger zones.

Premise 4: Workers have a fundamental need for and right to all information that is known to the employer, vendors, or suppliers about chemical, physical, and other hazards on the job.

The OSHA Hazard Communication Standard is only a partial solution to this need. For example, containers should be labeled with full chemical identities of all ingredients. Full disclosure of chemical ingredients should not be compromised by trade secret claims of chemical manufacturers.

Employers have argued that workers need only to be told how to protect themselves and can be provided this information without disclosure of detailed chemical data that are technically beyond worker comprehension. Unions respond that such policies are not only demeaning but that workers cannot afford to defer vital judgments about their safety to company representatives whose interests may conflict with their own.

Premise 5: Workers who have been subjects in scientific studies are entitled to full notification about the results along with advice and support services to help them cope with their problems if they were found to be at high risk.

Worker notification is a particularly serious problem for thousands of workers who are the surviving members of cohort studies that were conducted by employers, universities, and government agencies and were found at high risk of cancer mortality. Unless these survivors are notified, they will be unlikely to seek whatever early diagnosis and treatment services may be available, they will not have the opportunity to make informed choices about their remaining years, they will not be able to take prompt advantage of any forthcoming advances in cancer prevention, and they will be kept ignorant of the need for protections on the job if the high-risk conditions persist.

Premise 6: Workers who are temporarily or

permanently disabled as a result of workplace injuries or illnesses deserve full, prompt compensation for any lost earnings and associated pain or suffering.

State workers' compensation laws are antiquated. State-to-state differences in payment levels, diagnostic criteria, and filing and appeal procedures require substantial legislative reform, including federal provisions that ensure equal treatment for all workers (see Chapter 11).

Premise 7: Workers should not work with chemicals that have not been adequately tested for toxicity. There is a need for a substantial increase in occupational health and safety research and a national commitment to apply the results of research to the workplace.

It is not acceptable to unions that thousands of potentially toxic chemicals are introduced into industrial use without adequate premarket toxicologic testing.

Premise 8: Workers need a combination of health and safety laws and collective bargaining agreements to achieve maximum protection on the job. Neither legislation nor contracts alone is sufficient.

Many serious health and safety problems are not covered by OSHA standards and require direct agreements with employers. For example, there are no federal standards governing control of ergonomic risk factors that are associated with musculoskeletal disorders. Most OSHA health standards consist only of a permissible exposure limit (PEL) and contain no provisions for worker training, medical surveillance, process design, and work practices.

On the other hand, even the strongest union contracts can reach their full potential only in the context of well designed labor and public health laws. The UAW, for example, worked with a major employer to develop two training programs for hourly employees. One was hazard communication training and the other was safety training for skilled trades workers. After the first year of implementation, 95% of eligible employees had received the hazard communication training, which was required by law, but less than 50%

had received the skilled trades training, which was not required by law.

Premise 9: Obsolete OSHA and voluntary standards are obstacles to protection. Vigorous efforts are needed to bring many existing standards in line with current scientific knowledge.

Almost all existing OSHA permissible exposure limits (PELs) for chemical exposure were adopted directly from the threshold limit value (TLV) list of the American Conference of Governmental Industrial Hygienists (ACGIH) in effect in 1968. Although many of the TLVs, which are essentially consensus standards, have been revised, the TLVs are also unprotective. However, most TLVs are now stricter than their corresponding OSHA-mandated PELs. The National Institute for Occupational Safety and Health (NIOSH) has published numerous documents recommending more stringent PELs for dozens of chemicals.

The current situation is harmful to workers because most employers strongly resist demands that exposure levels be further reduced as long as the OSHA standards have been met, even though most scientists agree that many legal exposures, which are below the OSHA PEL, but above the ACGIH and NIOSH limits, are harmful.

Premise 10: Workers are entitled to strict protections of confidentiality and privacy when they participate in occupational medical programs. This premise is particularly important because a violation of confidentiality can result in discrimination or job loss.

All medical records should be maintained in locked files and restricted to the use of medical personnel unless authorized in writing by the employee or otherwise required by law. Unions strongly supported the passage of the OSHA Regulation on Access to Employee Exposure and Medical Records, which prohibits release of medical records to anyone without written authorization from the employee. However, this regulation makes it clear that personal air-sampling and biologic monitoring tests, such as blood lead levels, are measures of environmental conditions and are to be more widely available.

Unions support Principle 6 of the American College of Occupational and Environmental Medicine Code of Ethical Conduct, which permits physicians to counsel employers about the medical fitness of people to work but asks them not to provide the employer diagnoses or other clinical details (see Chapter 13) (3). There is a widespread perception among workers, however, that these guidelines are frequently violated by company physicians and nurses and that confidential medical information often ends up in the hands of supervisors and other management personnel. Substantial testimony was made to this effect during the public hearings on the OSHA access regulation.

UNION HEALTH AND SAFETY ACTIVITIES: EXAMPLES

Collective Bargaining

Before the passage of the OSHAct in 1970, collective bargaining agreements typically contained general statements about health and safety. Perceived problems were handled by the grievance procedure. Specific issues regarding conditions were negotiated locally. The renewal of worker consciousness about health and safety in the late 1960s brought recognition that these agreements were insufficient and that stronger, more comprehensive contract language would be necessary to force employers to address serious hazards in a serious way. With a few unions leading the way, most notably the Oil, Chemical and Atomic Workers (OCAW), a new generation of health and safety agreements began to emerge—often only in response to strike actions or other determined struggles between labor and management. The most fully developed agreements today address five areas, detailed in the following sections.

Responsibility

The OSHAct states employers' legal obligation to maintain a safe and healthful working environment. Union contracts often restate this employer obligation and amplify it by enumerating specific responsibilities, such as providing appropriate medical examinations. Unions typically commit to cooperate with and participate in programs aimed at fulfilling the employer's responsibility. The legal duty of the union is to represent its members fairly with respect to the employer obligations and workers' rights spelled out in the contract. Courts have generally found that this duty does not make the union liable for injuries arising from unsafe conditions or the failure to correct them.

Representation

Many agreements, especially those that cover relatively small workplaces, assign existing union representatives to handle health and safety matters along with their other duties. In industrial settings, these representatives are usually paid "lost time" to carry out such duties. Increasingly, unions have been able to negotiate additional representation time for health and safety matters, and part-time or full-time positions for local union health and safety representatives. These representatives receive technical training, conduct investigations, and handle health and safety complaints before their introduction in the grievance procedure (Fig. 40-3). UAW national and local agreements have created additional positions for industrial hygiene technicians, ergonomics analysts, and health and safety trainers.

Many unions have bargained for joint health and safety committees to which management and the union each appoint members. These committees are usually advisory in nature and have little authority to make environmental changes, enforce agreements, or shut down hazardous operations. (See Box 5-5 in Chapter 5.)

Information and Participation

Some access to information is guaranteed by OSHA regulation, including the right of workers and union representatives under

FIG. 40-3. A union health and safety representative (*left*) shows a supervisor (*center*) a dangling line that could get caught and cause a wheel to explode in a worker's face. This inspection led to the company correcting the problem and prohibiting use of the machine until this was done. (Photograph by Russ Marshall. Courtesy of United Automobile Workers.)

various circumstances to obtain MSDSs, industrial hygiene and biologic monitoring results, the OSHA 200 logs (of work-related injuries and illnesses that are maintained by employers), company medical records, and copies of any scientific analyses and reports prepared by or for the employer (see Chapter 10). Legal rights to safety information (injury prevention rather than health information) are limited. Unions have won the right to review plans for the introduction of new processes or equipment so that potential health and safety problems can be identified and prevented before reaching the plant. Others have secured agreements to use air-monitoring equipment, to accompany company industrial hygienists during sampling, or to use the employer's computerized toxic materials database.

Many unions have secured company agreement to provide training for employees in hazard recognition and control beyond that required by law. The most comprehensive agreements to date were initiated in 1984 and 1985 between the UAW and General Motors (GM), Ford, and Chrysler. The centerpiece of these agreements is a fund for training activities that is jointly administered by the company and the union. At GM, for example, 4 cents is put in this fund for every hour worked by a UAW member. The funds originally developed and delivered training programs for high-risk workers, such as skilled trades workers performing mainte-

nance tasks and materials-handling vehicle drivers, and for local joint health and safety committees. At present, most job-related safety training for employees in the automotive industry is jointly developed and delivered.

Grievance Procedures

Collective bargaining agreements contain grievance procedures for the union to follow when it believes that management has violated the agreement, including provisions regarding health and safety. Grievance procedures typically prescribe steps by which progressively senior management and union representatives attempt to settle a complaint. Some agreements establish special expedited procedures for health and safety complaints. In the event the parties find it impossible to agree, the contract directs that disputes may be resolved by binding arbitration or by strike within the life of the agreement. Health and safety conditions are rarely subject to arbitration, although job refusals are typically subject to arbitration.

Environmental Controls and Other Specific Protections

Local agreements frequently go beyond general rights by specifying the way that particular problems will be handled. The company and union may agree that a new ventilation system will be installed, that a chemical will be eliminated, that medical examinations will be offered, or that a research project will be undertaken.

Research

Unions have pressed employers, academic institutions, and government agencies to conduct high-quality occupational health research and to eliminate the inadequate and often self-serving research that has plagued the field for years. Two early leaders in this regard were the OCAW and the United Rubber Workers (URW).

During the 1970s, OCAW was aggressively engaged in nearly all elements of health and safety activity, including campaigns for better standards and enforcement, access to information for workers, and improved research. With an OSHA New Directions grant, OCAW was able to hire several health professionals who provided technical assistance to local unions and who worked to stimulate needed research on potential hazards faced by OCAW members. For example, OCAW was able to challenge research of dubious value being conducted by petrochemical companies and to press NIOSH for the improvement and expansion of its health hazard evaluation and industry-wide studies activities.

In 1970, the URW negotiated a research program of great vision and importance with the major tire manufacturers. Research funds were established with each employer contributing a sum of money, ranging from 0.5 to 2 cents for each hour worked by a URW member. These funds, under joint labor–management control, were used to support university research into the hazards of the rubber industry. A series of epidemiologic and industrial hygiene studies was conducted by two schools of public health. These studies demonstrated the relationship between solvent exposures in the rubber industry and both stomach cancer and leukemia. Most important, manuals were prepared

with information about how to reduce exposures to the chemicals of likely danger.

In the 1980s, the UAW identified occupational health research as an area of major importance, expanding some of the earlier union initiatives. The UAW developed its own in-house epidemiology program to respond to concerns of patternmakers and diecast workers. It established a mortality surveillance system based on union job history and public death record, helped local unions do their own investigations, and completed a series of substantial mortality and morbidity studies (4,5). The work-related health risks identified included associations between stomach cancer and work with metalworking fluids, between both lung cancer and nonmalignant respiratory disease and foundry work, and cumulative trauma disorders and various assembly jobs.

Because of the pressure generated by independent activity, the UAW reached agreements on research with employers. In 1982, the UAW and GM agreed to establish an Occupational Health Advisory Board of mutually acceptable, university-based scientists to assist in the development of research activities. The Board's first major task was to sponsor a competitive peer-review process to consider proposals for a comprehensive investigation of the health effects of exposure to metalworking fluids. This process resulted in the first of almost two dozen studies of mortality, respiratory effects, measurement, and control technology of metalworking fluids. Starting in 1984, the UAW negotiated occupational health research funds with GM, Ford, and Chrysler totaling over $5 million for each 3-year contract period. The two leading achievements, identification of respiratory and carcinogenic hazards of water-based metalworking fluids, have spurred work on new occupational health standards. The validation of ergonomic risk factor identification and abatement methods under this research umbrella provides much of the basis for industrial ergonomics programs and standards. Additional areas of work were mortality studies identifying cancer risks among foundry workers, electronics, and stamping and assembly processes. A promising current area is injury prevention research.

Occupational Safety and Health Administration Standards

Unions have consistently been the chief advocates for protective OSHA standards, pressuring successive generations of OSHA administrators who have moved slowly on their own—and in some cases, not at all. The first major OSHA health standard, asbestos, was started with a union petition. The United Steelworkers of America pioneered union methods for getting a standard: petition for emergency action, participate in public hearings, go to court to stop administrative delays, and file complaints when employers fail to implement the provisions of new standards. The Steelworkers pressure enabled OSHA to incorporate medical removal protection and multiple physician review in the lead standard. These regulations protect workers from economic loss as a result of cooperating with medical programs, and have been included in most OSHA chemical standards.

The Textile Workers Union (now UNITE) demonstrated the importance of enlisting public support in the campaign for a cotton dust standard.

The UAW championed a safety standard for energy lock-out to workers during service and repair of machinery and equipment. The UAW had determined that the largest single circumstance of fatal injuries among its members was skilled workers doing service and repair. This march to this broad safety standard started with an American National Standards Institute committee chaired by a UAW staff member, and was then negotiated with automobile manufacturers; these programs were the foundation for a comprehensive OSHA rule.

By contrast, proposals for standards with little union interest or support, such as OSHA's ill-fated seat belt rule, have had lit-

tle success. OSHA's 1984 attempted PEL Update demonstrated the need for union support for a standard. OSHA proposed to adopt very modest improvements in several hundred chemical exposure limits adopted in 1970. Unions responded that these modest improvements would be an excuse not to set more protective standards. The few industries affected by these changes filed a lawsuit against the rule, claiming it was too strict; labor challenged it as not protective enough, and eventually the courts wiped out the entire PEL update. OSHA rule making for chemicals has stagnated since this episode.

The union regulatory agenda shifted as labor's attention was increasingly paid to the service and public sectors. For example, SEIU and the American Federation of State, County and Municipal Employees (AFSCME) were leading forces behind the bloodborne pathogens standard and have continued to press for regulatory protection against tuberculosis. Many unions support an indoor air quality rule that goes beyond "no smoking" to good ventilation. Substance-specific OSHA rules remain a union priority, including rules for silica, metalworking fluids, and hexavalent chromium. The most important issue that unites all sectors is the need for an OSHA ergonomics standard (see Chapters 9 and 26). Because musculoskeletal disorders comprise from 40% to 65% of all injuries in all sectors of industry, all unions have pressed OSHA to adopt such a standard. Unions have also supported innovative approaches to standard setting that would regulate large groups of similar materials, such as chlorinated solvents or pesticides, at one time rather than the current inefficient and time-consuming substance-by-substance approach (see Chapter 10).

Training and Technical Support

Since passage of the OSHAct in 1970, union activities have dramatically expanded.

In the late 1970s, OSHA established a New Directions grant program to support expansion of union training and technical support, including industrial hygiene expertise, in dozens of unions. For a time, this program was a balance to the free consultation provided to industry as part of OSHA's compliance assistance effort. OSHA phased this program down and replaced it with short-duration, targeted training grants that go to union and management groups for training on specific issues.

Although OSHA funding declined after 1980, a significant expansion in union health and safety activity was supported by Hazardous Waste Worker training grants funded by the Superfund law, and later supplemented by Department of Energy waste worker training grants. These grants primarily targeted cleanup and emergency response workers, particularly among the construction and firefighters unions. These programs include technical support for chemical hazard control and innovative adult education techniques, often with university affiliations.

The construction trades have established the Center to Protect Workers' Rights, originally focusing on asbestos and lead abatement, now expanding into ergonomics (Appendix B). The Laborers International Union and International Union of Operating Engineers have incorporated health and safety training in their joint labor–management training programs, including apprenticeship.

Jointly administered, management-funded, labor–management training in the industrial sector, including the automotive industry–UAW and Boeing–Machinists Union centers, has greatly expanded.

Occupational Safety and Health Administration Enforcement Support

Unions participate actively in the OSHA enforcement process, particularly in designing and implementing abatement procedures. For example, in 1987, the United Food and Commercial Workers (UFCW) discovered that a meat packing employer was not recording musculoskeletal disorders on the OSHA 200 log as required. The local union filed a complaint, leading to a large penalty citation for management's willful failure to keep a record of injuries. The correct records revealed the gravity of the ergonomic problems, which led to a major citation under the OSHA general duty clause. Management contested the citations. The UFCW elected to become a formal participant in the appeal process, and aggressively participated in the settlement discussions. Eventually, company-wide ergonomics-based abatement programs were negotiated. These were, in turn, the basis of the OSHA red-meat guidelines. These programs may eventually be the template for an OSHA ergonomics standard. (See Box 9-1 in Chapter 9.) Similar campaigns regarding ergonomics issues were conducted by the SEIU in the nursing home sector and the UAW in the automotive industry.

Legislation

Unions were deeply involved in activities necessary to secure passage of the federal Coal Mine Health and Safety Act in 1969 and the OSHAct in 1970. Since then, much union energy has gone toward implementing these acts; protecting them from erosion; defending the budgets of OSHA, the Mine Safety and Health Administration, and NIOSH; and supplementing the statutory protections with collective bargaining protections. There is still a need, however, for additional legal protection for workers. Unions have identified priorities for improvement of the OSHAct to address better the needs of protecting the health and safety of working people. Targets for improvement include coverage for public employees; the requirement for written, comprehensive health and safety programs at every workplace; mandatory employee health and safety representatives and joint labor–management health and safety committees; employee appeal rights in OSHA enforcement equal to those of management; increased criminal sanctions for rules violations; mechanisms for accelerated rule making; and reduction in the PELs. These changes are not immediately feasible at the federal level because the primary emphasis of Congress has been regulatory relief for employers.

REFERENCES

1. Cherniack M. The Hawk's Nest incident: America's worst industrial accident. New Haven, CT: Yale University Press, 1986.
2. Rosner D, Markowitz G. Safety and health on the job as a class issue: the Workers' Health Bureau of America in the 1920s. Science and Society 1984;48:466–482.
3. American College of Occupational and Environmental Medicine. Code of ethics. Chicago: American College of Occupational and Environmental Medicine, 1993.
4. Silverstein M, Maizlish N, Park R, et al. Mortality among workers exposed to coal tar pitch volatiles and welding emissions: an exercise in epidemiologic triage. Am J Public Health 1985;75:1283–1287.
5. United Automobile Workers. The case of the workplace killers: a manual for cancer detectives on the job. Detroit: United Automobile Workers, 1981.

BIBLIOGRAPHY

American Federation of Labor and Congress of Industrial Organizations (AFL-CIO). Death on the job: the toll of neglect. Washington, DC: AFL-CIO, 1998.
The AFL-CIO's annual report on safety and health.

American Federation of Labor and Congress of Industrial Organizations (AFL-CIO). Manual for shop stewards. Washington, DC: AFL-CIO, 1984.
A basic guide for union representatives, including information about rights and responsibilities.

Ashford N, Caldart C. Technology, law and the working environment. Washington, DC: Island Press, 1996.
Includes numerous documents on union involvement in OSHA rulemaking and related policy issues.

Babson S. Working Detroit: the making of a union town. New York: Adama, 1984.
An excellent chronicle of workers and unions in America's industrial heartland.

Cherniack M. The Hawk's Nest incident: America's worst industrial disaster. New Haven, CT: Yale University Press, 1986.
A historical and epidemiologic reconstruction of the scandalous epidemic of acute silicosis among workers building the Hawk's Nest Tunnel 50 years ago to provide hydroelectric power for a Union Carbide Corporation plant in Gauley Bridge, West Virginia.

Dembe A. Occupation and diseases. New Haven, CT: Yale University Press, 1996.
An excellent and provocative examination of how social factors affect the conception of work-related disorders. Major sections examine the social and political history of noise-induced hearing loss, back pain, and cumulative trauma disorders of the hand and wrist.

Deutsch S, ed. Theme issue: occupational safety and health. Labor Studies 1981;6.
A collection of articles providing more details about several of the subjects covered in this chapter, including health and safety committees, collective bargaining, COSH groups, and workplace democracy.

Dotter E. The quiet sickness: a photographic chronicle of hazardous work in America. Washington, DC: American Industrial Hygiene Association, 1998.
A remarkably compelling vision of workplace safety and health in America since the mid-1970s.

McAteer JD. Miner's manual: a complete guide to health and safety protection on the job. Washington, DC: Crossroads, 1981.
A good example of a health and safety training publication designed with rank-and-file workers in mind.

Melkin D, Brown M. Workers at risk: voices from the workplace. Chicago: The University of Chicago Press, 1984.
A powerful series of first-person accounts by workers faced with chemical hazards on the job.

Page J, O'Brien M. Bitter wages. New York: Grossman, 1972.
The best discussion available about the origins of the OSHAct, including substantial material about the political activity of labor unions in health and safety during the pre-OSHA era.

Rashke R. The killing of Karen Silkwood: the story behind the Kerr-McGee plutonium case. Boston: Houghton Mifflin, 1981.
An investigative reporter's probe into the tangled story of union activism, radiation hazards, the nuclear fuel industry, liability law, and, in the view of some, murder.

Stein L. The triangle fire. New York: Carroll & Graf, 1962.
A historical recreation of a workplace disaster that was linked to industry negligence and that led to important health and safety reforms in the early 20th century.

Occupational Health: Recognizing and Preventing Work-Related Disease and Injury, 4th ed.
Edited by Barry S. Levy and David H. Wegman, Lippincott Williams & Wilkins, Philadelphia © 2000.

41 Agricultural Workers

Richard A. Fenske and Nancy J. Simcox

Agriculture is the world's largest economic activity, involving an estimated 63% of the population in developing countries (1). It is also among the most dangerous and physically demanding occupations. Its practice ranges from highly mechanized operations employing state-of-the-art technology to maintenance of subsistence plots. The agricultural workplace may be a well-defined commercial enterprise resembling an industrial setting, comprising a workforce conducting specific tasks with assembly-line regularity; but it also may be virtually indistinguishable from the habits of everyday life, occurring in residential and community space, with family members helping in all phases of production. The diversity of global agricultural activities represents a challenge to occupational health practitioners. The identification of occupational health hazards and the development of methods to evaluate, mitigate, and ultimately prevent occupational exposures and illnesses is labor intensive and requires knowledge of almost all aspects of occupational medicine. This chapter discusses the primary hazards, the health problems associated with these hazards, and the controls that are employed to reduce health risks for agricultural workers.

WORKFORCE CHARACTERISTICS

In most countries, a substantial proportion of agricultural activity is controlled by farmers—that is, individuals who both own and work the land. Although they may hire workers to assist in various aspects of production, the core of their workforce is the family. Although this chapter focuses primarily on agricultural workers who are not owners of the means of production, many of the health risks discussed here are also applicable to family farmers.

The Hired Farmworker

Hired farmworkers are not adequately protected by federal laws, regulations, and programs; therefore, their health and well-being are at risk. . . . Furthermore, . . . their children—who may work in the fields because the families need the money or lack access to child care facilities—are subject to educational disadvantages and health risks from injuries and pesticides.
—U.S. Government Accounting Office, 1992 (2)

The number of agricultural workers in the world is difficult to determine. Estimates of the U.S. workforce range from 2.5 to 4.1 million, composed primarily of immigrants from Mexico, Puerto Rico, Haiti, Jamaica, Central America, and indigenous Native Americans and African-Americans (3). During the past

R. A. Fenske: Department of Environmental Health, University of Washington School of Public Health and Community Medicine, Seattle, WA 98195.

N. J. Simcox: Field Research and Consultation Group, University of Washington School of Public Health and Community Medicine, Seattle, WA 98195.

10 years, the demographics of the farm labor population has become increasingly male (80%), increasingly foreign-born (70%), and very young (two-thirds younger than 35 years of age) (4). From 1990–1991 to 1991–1992, the proportion of workers 17 years of age and younger doubled, from 4% to 8%, primarily because of an influx of young U.S.-born whites into the workforce. Women's participation in farm work has declined over the last several years, from 25% to 19%. Approximately one-half (51%) of all women farmworkers were born in the United States, compared with one-fourth of male farmworkers.

Agricultural labor has traditionally been treated as a "special case," both in the United States and elsewhere. In the United States, agricultural workers are excluded from key labor laws, from many state and federal occupational health and safety laws, and, in one half of the states, from workers' compensation laws (5). This treatment of agriculture has even extended to basic rights, such as workplace sanitation. Since 1971, all other U.S. workers have been guaranteed rights to the provision of basic sanitation (hand washing, potable water, and portable toilets). Not until 1987, however, did the Occupational Safety and Health Administration (OSHA) issue a field sanitation standard for agricultural fieldworkers. In many parts of the world, this same dichotomy between agriculture and other industries remains.

The occupational health issues for agricultural workers are not easily separated from other political, social, and ethical concerns related to the status and treatment of this labor group. In industrial societies, the living conditions of many agricultural workers more closely resemble those of families in developing countries. Deficiencies in nutrition, housing, sanitation, education, and access to health care all contribute to the general health status of these families and individuals and frequently exacerbate occupational health risks (Fig. 41-1).

Housing in or near fields can expose workers and their families to pesticide spray drift. Inadequate sanitation is associated with an

FIG. 41-1. Migrant farmworkers typically suffer from housing and other socioeconomic problems. These workers are sleeping under orange trees because no housing was provided. (Photograph by Ken Light.)

increased prevalence of parasitic and infectious disease among farmworkers. The inaccessibility of potable water in the fields and in some housing forces workers to drink irrigation water. Such water can be contaminated with pesticide runoff and in some cases is used as a direct method of pesticide application. Lack of education can make it difficult or impossible for workers to read pesticide labels and posted signs. Finally, the same economic and geographic factors that limit access to medical care for agricultural workers may also delay or prevent appropriate treatment for job-related injuries and illnesses.

Child Labor

Child labor became an important occupational and public health concern in the 1990s, and particularly so in agriculture (see Chapter 38) (6-8). Agriculture employs about 8% of all working adolescents (8). The use of children as agricultural laborers can jeopardize their education and development and place them at risk for injury, illness, and chemical exposures. Approximately 300 children and adolescents die each year from farm injuries in the United States, and 23,500 suffer nonfatal trauma (9). In 1993, there were 68 injury deaths among farmworkers younger than 18 years of age. Agriculture accounted for 25% of all occupational injury deaths of those age 16 and 17 years, 41% of those age 14 and 15 years, and 78% of those younger than 14 years of age (10). Injuries included amputations, lacerations, and crush accidents from machinery, trauma from animals, and vehicular crashes. In addition to the physical hazards posed by machinery, noise produced by such equipment has been linked to early hearing loss in young agricultural workers (11).

In the United States, the minimum age for employment is 16 years, and a child may not be employed in hazardous occupations or activities (e.g., mining, logging, roofing) until 18 years of age. Under the Fair Labor Standards Act, however, the minimum age for agricultural labor is 12 years, and 16-year-old children can participate in hazardous activities. For children on family farms, a full exemption has been granted from these age requirements. Allowable activities that are considered hazardous include operating tractors, using heavy machinery, climbing ladders, and mixing or applying acutely toxic pesticides. In the mid-1990s, several states passed stricter state regulations to limit children's exposure to hazardous activities in agriculture. A report by the Committee on the Health and Safety Implications of Child Labor strongly recommended that "the current distinctions between hazardous orders [regulations] in agriculture and nonagricultural industries should be eliminated from child labor laws. Furthermore, the minimum age of 18 should apply for all hazardous occupations, regardless of whether the adolescent is working in an agricultural or nonagricultural job. . . . No health and safety justification for the distinction between agricultural

and nonagricultural settings appears to exist" (8).

Children younger than 16 years are estimated to constitute about 16% of the agricultural workforce in the United States. Children of hired farmworkers assist their parents with picking crops, carrying produce in bags and buckets, climbing ladders, caring for animals, and operating farm machinery. The risks posed by farm hazards may be increased for young workers because of their small size, inexperience, and inadequate safety training.

Few injury and illness data are available regarding pesticide exposure. Children contact pesticides in much the same way as adults, and they are exposed to the same concentrations of pesticide residues. A U.S. National Academy of Sciences report, *Pesticides in the Diets of Infants and Children,* discussed differences in pesticide-related health risks for adults and children (12). Differences in metabolic activity and a higher ratio of surface area to weight may place children at greater risk than adults after equivalent exposures. Also, research in animal models suggests that developing hormonal systems may be more susceptible to the toxic effects of some compounds. In general, however, the physiologic and clinical effects of pesticide exposure in children and adolescents remain undocumented.

Cross-Cultural Medicine

Diversity in race, culture, language, and religion among the hired farmworker population brings forth special cross-cultural issues that should be considered during medical treatment. Medical schools and residency curricula are beginning to include cross-cultural training in their programs (13,14). In 1996, the American Academy of Family Physicians adopted a set of core curricula on cultural sensitivity and competent health care. Physicians working in diverse communities are beginning to extend beyond the biomedical approach to the treatment of disease (15). In treating the hired farmworker population for occupational diseases and illnesses, the same principles should apply and should be useful to avoid miscommunication and misunderstanding.

OCCUPATIONAL HYGIENE IN AGRICULTURE

The fundamental concepts of occupational hygiene and modern preventive medicine have only recently been applied systematically to agriculture. The principles of recognition, evaluation, and control, for example, are as applicable to agricultural hazards as they are to more traditional industrial hazards but have not been used effectively. The relative neglect of agricultural health and safety is a multifaceted problem. In the United States, those federal agencies with direct responsibility for workplace evaluation (the National Institute for Occupational Safety and Health [NIOSH]) and regulation (OSHA) have only recently begun to focus their efforts on agricultural production. This neglect also may be a result of a prevailing agrarian myth which has focused almost exclusively on the "family farm," portraying work in agriculture as a healthy lifestyle (16). Family farm health and safety was given long overdue attention by a 1991 national conference convened by the U.S. Surgeon General, by infusion of both foundation funds (W.W. Kellogg) and federal funds (NIOSH) into research and surveillance programs, and by a variety of outreach initiatives. These efforts are now focusing on the unique nature of U.S. farming, including not only the family farmer but also the large hired farmworker population. Across the United States, nine regional centers for agricultural health and safety research and education have been established by NIOSH.

Morbidity and Mortality in Agriculture

Any discussion of illness and injury in agriculture must be prefaced with the caveat that most of available statistics probably underestimate true incidence and prevalence. This

occurs because (a) 95% of farms in the United States have 10 or fewer employees, and most are therefore exempt from OSHA regulations, including the reporting of illnesses and injuries (16); and (b) rural communities lack trained professionals who specialize in occupational health (e.g., occupational hygienists, occupational health nurses, occupational health physicians) who can respond to the unique problems and issues in agriculture.

Encouragingly, states are taking action to change their agricultural health and safety regulations to increase the numbers of workers protected. In 1996, the State of Washington became the first to remove the exemption of 10 or fewer employees, so that the state's agricultural regulations now apply to all farm operations, regardless of size (17). This action is consistent with the view expressed in the recent National Academy of Sciences report on child labor:

> Congress should undertake an examination of the effects and feasibility of extending all relevant Occupational Safety and Health Administration regulations to agricultural workers, including subjecting small farms to the same level of OSHA enforcement as that applied to other small businesses (8).

Statistics available from the United States indicate that injury and death rates rank agriculture consistently among the three most hazardous industries. The death rate for agricultural workers in 1997 was 20 per 100,000 workers, compared with 13 and 24 per 100,000 workers in construction and mining, respectively (18). In 1994, the Traumatic Injury Surveillance of Farmers (TISF) survey (19) found that nonfatal lost-time work injury rates were the highest among nursery operations (9.4 injuries per 100,000 hours worked); dairy operations (6.8); vegetable, fruit, and nut operations (5.0); and beef, hog, or sheep operations (4.6). The leading causes of lost-time work injuries on the 11,630 farms participating in this survey were livestock (20%); machinery, excluding farm tractors (19%); and working surfaces (8%). Farm tractors accounted for 5% of these nonfatal injuries. Active surveillance programs, such as TISF, are needed to assist agricultural workplaces with appropriate injury prevention programs (Fig. 41-2).

As with injury rates, illness rates place agriculture among the most hazardous occupations. Occupational illnesses per 10,000 full-time U.S. workers in 1996 were 52 for all industries combined, 144 for manufacturing, 36 for agriculture, 16 for construction, and 15 for mining (18). Although injury rates have been declining in the United States, occupational disease rates have not, and they represent a rising proportion of all morbidity among agricultural workers. Agriculture had the highest rate of all industry divisions for skin diseases and disorders (18). In public health terms, these statistics provide evidence of an environmental risk transition in developed industrial societies toward predominant patterns of chronic disease. Specific health risks highlighted in this chapter include skin disorders, musculoskeletal and soft tissue disorders, and pesticide-related illnesses.

FIG. 41-2. Farmworkers, as illustrated by this grape picker, face physically demanding jobs that often predispose them to serious injury. (Photograph by Ken Light.)

Skin Disorders

Skin disorders are the most frequently reported category of occupational disease in agriculture, as is true in many occupations. In 1996, the incidence rate per 10,000 full-time U.S. workers for skin diseases and disorders was 16.8 for agriculture, 14.1 for manufacturing, 3.3 for construction, and 1.6 for mining (18). Several large studies have been conducted regarding the prevalence of dermatitis in general populations from Western Europe, the United States, and the Netherlands (20).

Most agricultural dermatoses are caused by plant exposures, although pesticide-related skin disorders often require extended periods of disability leave. It is usually difficult to distinguish the clinical manifestations of plant dermatoses from those of pesticide dermatoses, be they irritant contact dermatitis, allergic contact dermatitis, or a photosensitivity reaction. Agricultural workers exposed to pesticides are four times more likely to develop a skin rash than the average worker (21). Environmental conditions, such as heat, sweating skin, clothing, and skin damage (sunburn), can exacerbate skin irritation. Pesticide-related dermatitis outbreaks have been reported in several U.S. regions (20,22).

Differentiation of plant dermatosis from pesticide dermatosis is accomplished primarily through the history of temporal association with work in a certain field or with a certain crop, the agricultural cycle and chemical applications, and symptoms. For example, a recurrent rash in the early part of June may be related to a weed that flourishes then; a rash occurring in July may result from exposure to an herbicide used on the crop at that stage of the cycle each year. Distribution of a rash on the extremities may also provide a clue; contact with plants may produce irritation only on the forearms, hands, and ankles, whereas pesticide residues dislodged from foliage often irritate the face and neck as well. Patch testing with plant extracts or pesticide samples frequently permits a definitive diagnosis in cases of allergic contact dermatitis (see Chapter 27).

Musculoskeletal Disorders

Farmworkers are exposed to many of the risk factors that are associated with musculoskeletal diseases. Occupational factors include heavy lifting and carrying, forward bending, kneeling, bent-over positioning, and excessively fast-paced work. The physical strain and repetitive motions involved in farm work can lead to traumatic injuries, irritation of joint tissues, and accelerated degeneration of the joints. In the 1980s, health surveys in the United States documented that farmworkers have a higher prevalence of arthritis than do white collar workers, blue collar workers, service workers, or all workers combined. Musculoskeletal conditions are the most commonly reported ailments among farmers and farm managers, and the rate of musculoskeletal disease among farmers is more than 50% higher than among farm managers (23). By 1996, statistics regarding these disorders were more readily available and showed that overexertions accounted for 19% of all agricultural nonfatal occupational injuries and illnesses involving lost workdays (18). Musculoskeletal disorders, such as back injuries, sprains, strains, and overexertions, are a significant problem in agriculture from both an incidence and a cost perspective, but there has been little application of ergonomic research and intervention in agricultural workplaces (24,25).

The significance of this evidence of widespread musculoskeletal disease is often ignored by clinicians, either because they look predominantly for signs of pesticide exposure or because they assume that musculoskeletal disease is an unavoidable result of farm labor. In fact, ergonomic strain associated with farm work can be minimized or entirely prevented with the appropriate redesign of equipment and labor practices. In forestry and construction occupations related or similar to some in agriculture, such changes have significantly reduced the ergonomic problems of many tasks (see Chapters 9, 26, and 42).

Few studies exist documenting the risk factors of musculoskeletal conditions in agricultural populations. Agricultural working postures can be analyzed with the use of ergonomic field methods (26). In one study, for example, women bagging pears had a higher prevalence rate of physical symptoms, such as fatigue and pain in neck and shoulders, than women bagging apples. Risk factors associated with overhead work in bagging pears were found to be different than those for apples (27). Recognition of the risk factors for musculoskeletal disorders is required for many more agricultural tasks to determine cost-effective ergonomic interventions. Much more emphasis needs to be placed on ergonomic risk factors and associated musculoskeletal disorders among agricultural workers. Several intervention studies among greenhouse workers are currently underway in California and the State of Washington (28).

Traditional agricultural tools and labor practices exact a toll on the musculoskeletal system of lifelong farmers and workers. Modernization, in turn, has frequently brought more extensive and physiologically less adaptive use of traditional tools such as "el cortito," the short-handled hoe (29). Replacing or modifying a tool can have a significant effect on the health of workers; for example, a 34% decrease in sprain and strain injuries occurred among California farmworkers as use of the short-handled hoe declined. Although California banned use of the short-handled hoe in 1975, other states did not do so until the 1980s.

The industrialization of agriculture has also introduced new equipment for harvesting and on-field packaging of many fruits and vegetables. This equipment has frequently been designed without the benefit of ergonomic analyses and produces new forms of musculoskeletal disease and risks of injury for farmworkers and equipment operators. Federal and state governments in the United

States are developing ergonomic standards to address the increasing numbers of musculoskeletal disorders. It is not clear whether agriculture will be exempt or included in these new regulations. However, international guidelines on ergonomics in agriculture, addressing developing countries involved in large- and small-scale farming, are being prepared in collaboration with the Silsoe Research Institute in the United Kingdom (30).

Other Occupationally Related Disorders

In addition to the occupational diseases and disorders already discussed, agricultural work entails risks associated with the use of equipment, exposure to animal-borne infectious diseases (see Chapter 20), stress due to heat (see Chapter 19) and other extreme conditions of weather, and hearing loss (see Chapter 18) from the operation of heavy equipment such as tractors and combines. Both acute and chronic lung disease have been associated with workplace exposures in agriculture (see Chapter 25). Airborne dusts, particularly those with high organic content, can produce acute hypersensitivity pneumonitis and organic dust toxic syndrome, a febrile illness associated with myalgias, malaise, dry cough, chest tightness, and headache (31). Of special concern are exposures to mycotoxins and bacterial endotoxins associated with many agricultural processes and products.

Reproductive hazards exist in agriculture, but few studies have found persuasive associations between adverse birth outcomes and parental occupation in agriculture. Nonetheless, the increased use of synthetic chemicals in modern agriculture has led many investigators to explore causal hyphotheses related to congenital malformations and fecundability (32). Certain pesticide groups, such as fungicides and chlorophenoxy herbicides, have received the most attention. Results from these studies are not conclusive because of limitations in study design, misclassification of maternal occupation, and lack of knowledge of the degree of gestational exposure to specific agents. Pesticides and other environmental contaminants, such as DDT, DDE, and dioxins (including TCDD), are currently under review as potential endocrine disruptors in the United States and elsewhere (33). Little information is presently available to evaluate the impact of these chemicals on agricultural workers. This issue is likely to play a central role in future research agendas aimed at elucidating the linkage between pesticides and health.

AGRICULTURAL WORKER EXPOSURE TO PESTICIDES

The use of hazardous chemicals to control pests has become a pervasive aspect of agricultural production both in the United States and internationally. The term *pesticide* refers to a functional rather than a chemical category and encompasses a heterogeneous group of compounds developed to control a variety of pests. The term can be applied to microorganisms with biocidal properties as well as to common household products. Therefore, in the investigation of health hazards associated with pesticide use, it is essential that the physical, chemical, and toxicologic properties of the specific compound be identified. Pesticides are generally categorized according to the type of pest for which they have shown efficacious action. The primary categories, based on volume of use, are insecticides, herbicides, and fungicides. Many other categories (e.g., termiticides, rodenticides, miticides) are also in use. An excellent resource describing the various pesticide classes is called *The Pesticide Book* (34).

Pesticides are manufactured and sold in various formulations. Formulations may be designed to improve effectiveness or to enhance storage or safety. Persistence of the pesticide in the environment can also be affected by the formulation. Formulations may be liquid or powder sprays (e.g., emulsifiable concentrates, wettable powders), dusts, aerosols, or granular materials. The

TABLE 41-1. *EPA Labeling toxicity categories by hazard indicator*

Hazard indicators	Toxicity categories			
	I (Danger-Poison)	II (Warning)	III (Caution)	IV (Caution)
Oral LD_{50}	$\leq$50 mg/kg	50–500 mg/kg	500–5,000 mg/kg	>5,000 mg/kg
Inhalation LD_{50}	$\leq$0.2 mg/L	0.2–2 mg/L	2–20 mg/L	>20 mg/L
Dermal LD_{50}	$\leq$200 mg/kg	200–2,000 mg/kg	2,000–20,000 mg/kg	>20,000 mg/kg
Eye effects	Corrosive; corneal opacity not reversible within 7 days	Corneal opacity reversible within 7 days; irritation persisting for 7 days	No corneal opacity, irritation reversible within 7 days	No irritation
Skin effects	Corrosive	Severe irritation at 72 hr	Moderate irritation at 72 hr	Mild or slight irritation at 72 hr

From: EPA Pesticide Programs, Registration and Classification Procedures, Part II. Federal Register 40:28279.

formulation may be designed to volatilize quickly or to act in a slow-release or controlled-release manner. Knowledge of formulation can often prove useful in understanding the potential for worker exposures.

The U.S. Environmental Protection Agency has primary responsibility for regulation of pesticides in the United States. It has devised a toxicity category classification system to assist in setting standards and labeling requirements (Table 41-1.)

As can be seen from the table, this system is designed to warn users of the acute hazards associated with these compounds. The toxicity categories are based primarily on the acute systemic toxicity of the compound, as determined by animal testing. However, acute dermal and ocular toxicity are also considered in the chemical classification. Chronic health effects of pesticides are regulated through a process called "special review." Special review compounds believed to be carcinogenic or teratogenic, for example, are subject to quantitative risk assessments. Therefore, although they can be very helpful in determining the likelihood of an acute intoxication, these toxicity categories provide no information regarding chronic health effects.

Global Pesticide Issues

Pesticide use patterns have changed significantly during the past 20 years. Developed countries predominantly use herbicides and the more selective insecticides and fungicides, whereas developing countries employ primarily insecticides, many of which are acutely toxic (1). Highly toxic fumigants also are used in agriculture and are a serious hazard in some parts of the world. In terms of amounts applied per hectare, pesticide consumption has been greatest in Japan, Europe, the United States, and China. Other major users of pesticides are Brazil, Mexico, Malaysia, Colombia, and Argentina.

The size of the population at risk for pesticide exposure is largest in developing countries, where most economically active members of the population work in agriculture. A report by the Pesticides Trust (35) considers that pesticide use is most hazardous in countries with the following characteristics: high rates of illiteracy, hot and humid climates, lack of protective equipment, absence of medical facilities, lack of washing water, and lack of trained workforce.

The World Health Organization (WHO) has identified Africa as the fastest growing pesticide market. This fact has serious implications for countries such as Benin, Senegal, and Togo, because they lack pesticide legislation, have insufficient information on pesticide health effects and controls, and have no monitoring system to evaluate the health impact. Although some trade control measures were implemented during the late

1980s, developing countries continue to rely on pesticides that have been restricted or banned for use in developed countries, such as carbofuran, methamidophos, parathion, and paraquat. Several explanations can be offered for the continued use of these compounds: they are less expensive than the more selective pesticides; they are broad spectrum and can be used for a variety of pest problems; many chemical companies promote their safety through strong marketing campaigns; information regarding health effects is not always provided to the importing nation; and knowledge pertaining to integrated pest management (IPM) practices and least toxic alternatives is inadequate in importing nations.

Developing nations are now experiencing many of the pesticide health and exposure problems experienced by industrialized countries during the 1970s. According to the few case reports and surveys from developing countries, pesticide poisoning is a major public health concern today. (See Box 31-1 in Chapter 31.) This situation is well illustrated by a report from Costa Rica, in which workers' compensation data were used to estimate rates of agrichemical-related injuries in three banana-producing provinces in 1990 (36). In this study, the overall annual injury rate was 5.3 per 100 workers, with chemical burns accounting for more than 60% of the injuries. Statistics such as these are generally viewed as underestimates of true morbidity, because they rarely include mild poisonings or chronic effects. Cooperation among governments (especially those exporting pesticides), the international agrochemical industry, workers and their organizations, other nongovernmental organizations, and institutions such as the World Bank is crucial to progress in the reduction of pesticide-related illness.

International Code of Conduct

United Nations Food and Agriculture Organization (FAO) adopted the International Code of Conduct on the Distribution and Use of Pesticides in 1985 as a means of addressing the growing number of illnesses, deaths, and environmental hazards caused by pesticides (37). The Code, which received support from major pesticide manufacturers, governments, and nongovernmental organizations, delineated a comprehensive set of international guidelines to assist governments with the registration and safe use of pesticides. Pesticide-manufacturing companies participate on a voluntary basis. An additional provision passed in 1989, called Prior Informed Consent, granted governments the right to refuse importation of pesticides that are already banned or severely restricted by other countries for health and environmental reasons. Countries also can apply this provision to those pesticides that have been identified as causing health or environmental problems in their own country. As of 1999, a total of 19 pesticides, including DDT, chlordane, ethyl parathion, toxaphene and 2,4,5-T, had been granted Prior Informed Consent status, with others under consideration (35).

A new industry-based program, known as the Global Safe Use Initiative, is aimed at training pesticide users in the developing nations (35). The initiative is promoted by the industry-sponsored Global Crop Protection Federation (formerly Groupement International des Associations Nationales de Fabricants de Produits Agrochemiques [GIFAP]) and is initially focused on three countries: Kenya, Thailand, and Guatemala (38). Its ultimate success appears to depend in large part on the industry's ability to integrate workers and public health scientists into the design, implementation, and evaluation stages of the project.

Acute Pesticide Poisonings

WHO estimates that acute pesticide poisonings have doubled in developing countries since the 1970s and are likely to increase further as the use of more toxic, less persistent pesticides become widespread (1). The use of organophosphates (a class of insecticides with neurotoxic properties) has re-

---

Box 41-1. Worker Reentry Poisoning in California, 1998

In July 1998, a total of 34 workers entered a cotton field that had been sprayed 2 hours earlier with carbofuran, a cholinesterase-inhibiting carbamate pesticide. The EPA restricted entry interval (REI, the minimum time between application and worker reentry) for field work in cotton for this compound is 48 hours. After approximately 4 hours of weeding, the workers began to feel ill and stopped working. More than 80% reported nausea and headache, eye irritation, dizziness, and muscle weakness. Thirty of the workers were transported directly to a medical clinic, and the other four eventually sought medical treatment. Important steps taken to manage this epidemic included the following:

1. *Worker decontamination:* All workers reporting to the clinic were decontaminated by clothing removal and showering.
2. *Clinical observation:* Workers were sent from the clinic to nearby hospitals for further evaluation; most were sent home the same day, but one was kept overnight because of new-onset atrial fibrillation.
3. *Treatment:* Atropine was administered to patients with severe symptoms.
4. *Removal from work:* Twenty-eight of the 34 workers did not return to work for at least 1 day.

---

[a] Centers for Disease Control and Prevention. Worker reentry poisoning in California, 1998. MMWR Morb Mortal Wkly Rep, February 19, 1999.

5. *Diagnosis and documentation of pesticide-related illness:*
   A. *Cholinesterase monitoring:* Blood samples were drawn on the day of the incident and assayed for plasma and red blood cell cholinesterase activity. No baseline activity values were available, so results were compared with laboratory normal range values. All results were within the normal range, but investigators found that blood samples had not been placed on ice and that analytical procedures were not uniform. Subsequent retesting of 10 workers under a stricter protocol demonstrated that all had low cholinesterase activity compared with the normal range.
   B. *Urinary metabolite analysis:* Urine samples were collected within 11 days of exposure and analyzed for carbofuran metabolites; 18 (58%) of 31 had detectable levels.
   C. *Foliar residue sampling:* Leaf samples were collected from the field on the same day that workers reported to the clinic and were found to contain carbofuran residues consistent with a recent application.

This poisoning outbreak demonstrates the hazard inherent in applying acutely toxic chemicals to fields that require hand labor activities. The editors of MMWR note that "failure to adhere to an REI can result in substantial morbidity among exposed workers," and that because "sole reliance on these control measures may be inadequate, the substitution of safer, less toxic alternative pesticides should be adopted when feasible."

---

sulted in more occupational acute pesticide poisonings and deaths than any other class. Few state and national registries that document and report pesticide health problems exist. In most cases, estimates for acute pesticide poisoning are based on hospital visits and community-based surveys from various regions of the world.

Mortality and morbidity statistics related to pesticide exposure on a global level are difficult to obtain. In 1990, WHO estimated that 1 million unintentional severe acute poisonings occur annually worldwide, resulting in 20,000 fatalities (1). WHO also estimates that for every 500 symptomatic cases, there are 11 hospital admissions, and 1 death. Although the United States has arguably the most advanced regulatory requirements related to pesticide use, the full dimensions of pesticide-related illness are not well documented. The EPA estimates that 10,000 to 20,000 cases of acute pesticide poisoning occur annually in the United States (39). Only a few states require mandatory reporting of pesticide illnesses. In 1990, almost 2,000 pesticide-related illnesses related to occupational pesticide exposures were reported in California; more than 50% involved exposure during use (e.g., mixers, loaders, applicators) and 25% involved exposure to residues (40).

Large crew poisonings in the United States are less common today than two decades ago, but they do still occur despite well-defined regulations. A case in point is a 1998 worker reentry poisoning in California caused by carbofuran (Box 41-1) (41). In countries where health and safety laws are virtually nonexistent and health care is often inaccessible, pesticide use can result in severe epidemics.

Chronic Health Effects of Pesticides

Public health investigators in industrialized countries are increasingly focusing their attention on long-term or chronic health effects of occupational pesticide exposure. Exposure levels that were considered acceptable a decade ago are now being carefully evaluated for health risks. Current knowledge on the effects of such exposures was thoroughly reviewed in an issue of *Occupational Medicine: State of the Art Reviews* entitled "Human Health Effects of Pesticides" (42).

Most epidemiologic studies have focused on manufacturing workers, pesticide applicators, and farmers rather than hired farmworkers. Studies have indicated associations between pesticides and cancers of the lymphatic and hematopoietic system, connective tissue, brain, prostate, skin, stomach, colon, and lip. Noncancer end points are also of concern, including reproductive, nervous, and other system effects (43). Several studies employing neurobehavioral and neuropsychological test batteries have indicated that persons with acute poisoning histories who appear normal on clinical evaluation exhibit

deficits in one or more aspect of neurologic function (44-45). The U.S. National Cancer Institute has initiated a large prospective cohort study to investigate health effects among pesticide handlers in Iowa and North Carolina (46), and the Canadian government has undertaken a similar study. These studies provide the first real opportunity to evaluate the long-term effects of pesticide exposures in agricultural populations.

PESTICIDE EXPOSURE ASSESSMENT

The evaluation of pesticide use and exposure in agriculture is complex for several reasons. First, the physical nature of the chemical can vary greatly: Workers may be exposed to pesticides in their concentrated form, as a dilute aqueous spray, as granules or pellets, or as a residue on crops and foliage. Second, workers are exposed to a wide variety of chemicals over a single season: An acutely toxic organophosphorus compound may be applied one day, a skin-irritating sulfur material the next day, and a mutagenic fungicide on the following day; or several compounds may be applied simultaneously. Third, chemical exposures occur under uncontrolled environmental conditions: Factors such as wind, rain, and sunlight can dramatically alter exposure potential. Finally, unlike exposures in many industrial settings, the major route of exposure to pesticides in agriculture is dermal rather than respiratory. For this reason, traditional occupational hygiene approaches, such as air sampling and adherence to guidelines (e.g., threshold limit values), have limited relevance.

In considering the evaluation and control of agricultural pesticide exposures, it is useful to distinguish between workers who directly handle pesticides during mixing and application and those who are exposed primarily to residues after application.

Mixers and Applicators

Workers who mix, load, and apply formulated pesticides can be referred to generally as pesticide handlers. Pesticide handlers are considered to be at highest risk for acute intoxications. They also represent a high-exposure group for chronic health effects, but the extent of exposure depends on such factors as frequency of application (times per season) and exposure duration (years of application). Several databases have been developed for this worker population to better understand the extent and variability of such exposures (47). A 1995 international workshop publication also provided an excellent review of these issues (48).

Exposure during specific pesticide handling events can be modified by several important factors:

1. *Type of equipment:* Mixing and loading of pesticides can result in substantial exposures over brief periods. Closed systems have been developed to mitigate such exposures, and, when properly used and maintained, they have proved effective. Applications with such equipment as airblast (speed) sprayers produce much higher exposures than applications with ground boom equipment.
2. *Formulation:* Different formulation types can produce very different exposure patterns. For example, a liquid concentrate can be spilled during mixing, resulting in permeation of clothing and skin contact; emergency washing facilities are often required in proximity to mixing stations to prevent overexposures. Wettable powders can become airborne during mixing, producing both a respiratory hazard and exposures to the face and eyes.
3. *Chemical protective clothing:* Protective clothing, such as chemical-resistant gloves and coveralls, is often required during pesticide handling and application. Label requirements may also call for respiratory protection. Use of such personal protective equipment can dramatically reduce skin contact and inhalation exposures.
4. *Hygienic behavior:* Worker care in regard to pesticide handling can also have substantial impact on exposure. Workers who avoid mixing and spraying during windy

conditions can reduce their exposure. Proper use and maintenance of protective clothing are also important behaviors associated with reduced chemical exposures.

Studies in the 1950s in the United States demonstrated that the primary exposure pathway for pesticide handlers is skin exposure and absorption. Investigators in Washington State developed what has come to be known as the "patch technique," using absorbent pads as collection devices on various parts of the body to determine levels of deposition. Hand exposure was measured by rinsing the hands with water or alcohol and concentrating the residue collected. These methods were subsequently used throughout the world to assess worker exposure to pesticides, and they form the basis of the current WHO guidelines (49).

These methods have provided sufficient information to draw several important conclusions. First, the highest exposures usually occur during mixing and handling of the concentrated material. Second, wind is the single most important environmental factor determining dermal exposure during application. Third, in most cases, exposure to the hands constitutes a major fraction of the total exposure. Fourth, the proper use of protective clothing can substantially reduce total dermal exposure. Pesticide exposure evaluation can be greatly aided by the use of fluorescent tracers to visualize and quantify patterns of pesticide deposition on the skin (Fig. 41-3), allowing evaluation of protective clothing performance and effective worker education (50–52).

Efforts to reduce mixer and applicator exposures have focused on placing a barrier between the worker and the source of exposure. The primary method to reduce exposure during pesticide mixing is an engineering control: the closed system that transfers the pesticide from its container to the mixing tank without direct handling by the worker. When functioning properly, these systems can reduce exposure consider-

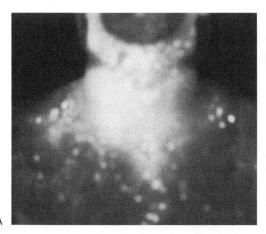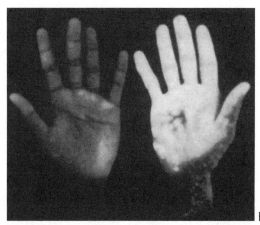

A   B

FIG. 41-3. A. Fluorescent tracer evaluation of a pesticide applicator reveals deposition on the neck and on the chest beneath coveralls. The area around the mouth was protected by a respirator. **B.** The use of gloves while handling pesticides can reduce exposure dramatically. In this photograph done with fluorescent tracer, the right hand was not protected by a glove but the left hand was. (Photographs by Richard A. Fenske.)

ably, but if a system failure occurs potential exposure is very high.

The most common engineering control for exposure reduction during application of pesticides is the closed-cab tractor. The closed cab normally provides an effective barrier to dermal contact, but the efficiency of the air filtration system is critical. If the cab is left partially open or if improper filters are employed, the utility of this approach is reduced drastically. Also, exposure may actually be higher inside the cab than outside if the worker enters the cab with boots and workclothes contaminated during mixing procedures. An engineering control aimed at hand spraying, developed in Brazil, involves mounting the spray apparatus on bicycle wheels, which are pushed ahead of the applicator. As the distance between the operator and the spray is increased, exposure is decreased substantially (53).

A supplemental, but very important, preventive measure for mixers and applicators is training; knowledge of the operation of closed systems, the relative toxicity of compounds, the proper handling and disposal of concentrated material, and when to spray are all required. (Because windy conditions increase exposures and off-target drift substan-

tially during application, many agricultural regions prohibit pesticide spraying at these times.)

The final control strategy for mixers and applicators is protective clothing. Protective gear, such as gloves, face shields, aprons, and boots, can effectively reduce exposure. This strategy may prove practical if mixing is only an intermittent and short-term activity, but skin contamination can still occur during removal of the gear, if clothing is not properly cleaned before reuse, or if the barrier properties of the material fail. Regulatory agencies have increasingly turned to protective clothing as a primary strategy for exposure reduction in agriculture, and personal protection plays a prominent role in the EPA's Worker Protection Standard related to agricultural pesticides (39). Personal protection has always been considered a control of last resort by occupational hygienists because (a) such a control strategy requires continual training of personnel in the proper use and maintenance of clothing; (b) the clothing must be inspected periodically and replaced when necessary; and (c) use of chemical protective clothing often reduces the comfort, agility, and dexterity of the worker and may contribute to heat stress under agricultural conditions. For these reasons, an *effective* control program based on protective clothing may prove to be more costly and more difficult to monitor than equally effective engineering and administrative approaches.

Field Workers

Pesticide exposure among agricultural workers who enter treated fields has been a public health concern since the early 1950s and remains a central issue in agricultural health and safety. Residues on crops can become airborne or be transferred to the skin during field labor activities. On any given day the pesticide exposure of a field worker may be lower than that of a pesticide handler, but the frequency of exposure (days per season) for a field worker may be substantially greater, resulting in a relatively high cumulative exposure. For example, it is estimated by the EPA that strawberry harvesters in California work between 80 and 120 days per season, whereas applications occur over only a few days.

The pesticide hazard that agricultural field workers confront takes the form of residues on fruit, foliage, or soil. This hazard is complicated by the general unawareness among workers of their potential for exposure and the consequent health risks they face. Because many of these workers are migratory and are not involved with other farm operations, they may not know what pesticides have been used or when they were sprayed. Furthermore, a substantial number of studies have demonstrated that residue levels on foliage are difficult to predict. In arid regions, such as California and the southwestern United States, high residue levels can remain for many weeks. Under certain environmental conditions, a number of organophosphorus compounds can be transformed into their more toxic "oxon" derivatives; for example, the oxon derivative of parathion, paraoxon, is 10 times more acutely toxic than parathion.

The thinning and harvesting operations that field workers perform require direct contact with foliage, and significant dermal exposure to any pesticide residues on the foliage is largely unavoidable. The hard physical labor and high temperatures typically encountered make protective clothing an unrealistic method for prevention of exposure. Several decades of research point to the conclusion that the only practical means of minimizing exposure for field workers is to make certain that toxic levels of residues have degraded or dissipated before workers are allowed into the fields. To achieve this goal, *restricted entry intervals* are derived from repeated studies of pesticide residue decay on specific crops. Under the Worker Protection Standard, the EPA has established a minimum restricted entry interval of 12 hours for all pesticide applications, superseding the previous minimum interval of waiting "until sprays have dried and dusts have settled." A 48-hour restricted interval is now required for Toxicity I compounds and a 24-hour interval for Toxicity II compounds. In California, where extended reentry is an issue because of the arid climate and persistent residues, restricted entry intervals may extend up to 60 days for particular pesticide and crop combinations.

Traditional engineering controls used in occupational health usually are not feasible as a means of reducing pesticide residue exposures in agriculture. However, several prevention strategies with engineering aspects are useful. These include product substitution, in which a less toxic pesticide replaces a more hazardous one; alternative pest control technologies (e.g., biologic control) that reduce pesticide use; alternative cultivation practices (e.g., crop rotation, cultivation of mixed varieties) that reduce the need for heavy pesticide use; and provision of handwashing facilities for the removal of pesticide residues before eating or using the bathroom and at the end of the workday.

Agricultural Worker Families

Increased attention has been directed at spouses of agricultural workers and children living in residences where pesticides are used. In 1995, NIOSH reviewed children's exposures to environmental health hazards associated with parental occupation, including pesticides (54).

Reported associations between childhood cancer and parental pesticide exposures or household pesticide use have been reviewed, with a general finding that such associations exist across a wide range of studies (55). Elevated rates of birth defects have also been associated with occupational pesticide use (56). In almost all of these studies, however, exposure metrics were based on pesticide use records or parental recall rather than actual measurements of pesticide exposure in children.

The exposure potential for children of agricultural families may be higher than for other child populations, because concentrated formulations of pesticides are used in high volume near the home. Pesticides used during work may also be introduced into the home inadvertently by various pathways. Poor hygienic practices among pesticide formulators have been associated with measurable blood levels of pesticides (e.g., chlordecone, or Kepone) in family members (57). Classic organophosphorus pesticide exposure symptoms in spouses and children of greenhouse workers have been reported (58). Several studies have shown that agricultural workers bring contaminated clothing into the home (59,60).

A study in Washington State found evidence that children living with agricultural workers and in proximity to tree fruit orchards have more opportunity for exposure than children living in homes without such risk factors (61). These findings were supported by an additional study, which measured pesticide metabolites in children's urine (62).

Current attempts to control such exposures are aimed at reducing track-in of pesticides from the outdoors, proper handling and cleaning of workclothes, and possible restrictions on children's activity during or after pesticide applications. Also, agricultural workers should be cautioned regarding the dangers inherent in the use of acutely toxic pesticides in residential environments. An incentive program aimed at reducing this practice should be considered by appropriate public health agencies.

PREVENTION STRATEGIES FOR PESTICIDE-RELATED ILLNESS

Public health practice recognizes three levels of illness prevention. Many of the strategies discussed in the previous section are considered primary prevention—that is, reduction or elimination of exposure through engineering or administrative controls. Secondary and tertiary prevention strategies, such as medical surveillance and clinical management of pesticide poisonings, are discussed in detail here.

Clinical Management of Pesticide Poisonings

Acute pesticide-related illnesses are frequently recognized during the initial presentation of patients to clinicians, although the specific chemical may be unknown. Most common among acute pesticide poisonings today are those caused by organophosphates, which have fairly typical symptoms. As emergency life-support procedures are instituted, samples of urine and blood should be taken and the patient should be rapidly decontaminated. Also, care must be taken to protect health workers from exposure caused by handling of body fluids or contaminated clothing. Such exposures may be substantial, and emergency department personnel have become ill while assisting pesticide-poisoned patients. Specific clinical treatments for acute and emergency pesticide exposure are comprehensively detailed in the book, *Recognition and Management of Pesticide Poisonings* (63) and in a 1997 review article (64).

The clinical diagnosis of moderate or severe organophosphate poisoning is confirmed when a test dose of atropine does not result in symptoms of atropinization, including flushing, rapid heartbeat, large pupils, and dryness of the mouth. Chemical tests for the presence of pesticide residues or their metabolites permit the identification of organophosphates and other compounds for medicolegal purposes, although usually not early enough to affect the course of clinical treatment. Because farmworkers who become ill at work often must obtain medical assistance on their own, it is important to inquire whether other workers were potentially exposed, so that public health workers can attempt to locate them both to investigate the incident and to offer medical care.

Mild acute organophosphate effects—that is, those not severe enough to require treatment with atropine—have been associated with low-level, occupational pesticide exposures. These effects manifest as one or more central nervous system symptoms that are nonspecific, including headache, fatigue, drowsiness, insomnia and other sleep disturbances, mental confusion, disturbances of concentration and memory, anxiety, and emotional lability. It is difficult to demonstrate a clinical association between symptoms and exposure under these circumstances, because the symptoms often occur at relatively low levels of cholinesterase depression. Individual cases of such mild effects typically go undiagnosed and therefore unreported.

Medical Surveillance

Measurements of cholinesterase inhibition have been used since the 1950s to evaluate acute poisonings and chronic exposure among pesticide applicators, and more recently to evaluate low-level exposure among farmworkers. Although plasma cholinesterase measurements are more widely used, the red blood cell cholinesterase values are a more valid indicator of a pesticide's physiologic effect on the nervous system. There is a significant degree of intraindividual and interindividual variation for both cholinesterases. Consequently, the reduction in cholinesterase activity required to diagnose

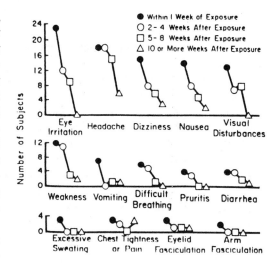

FIG. 41-4. The time course of symptoms reported by crew members exposed to organophosphates over the follow-up period (22-29 subjects examined at different times). (From Coye MJ, Barnett PG, Midtling JE, et al. Clinical confirmation of organophosphate poisoning of agricultural workers. Am J Ind Med 1986;10:399–409.)

pesticide-induced inhibition is relatively large, even when preexposure baseline values are available for comparison.

Without baseline values, plasma activity levels usually must be 30% or more below the laboratory normal range to achieve clinical significance. When a baseline value is available, a 20% decline in plasma and a 15% decline in red cell activity is significant. Mild but persistent and disabling symptoms may occur at levels of inhibition far less impressive than these. In a large study of California lettuce harvesters exposed to mevinphos (phosdrin), for example, moderately severe symptoms were reported despite a average plasma cholinesterase inhibition of 16% and an an average red blood cell cholinesterase inhibition of 6% (Fig. 41-4.). A 1997 review article provided a thorough discussion of the role of cholinesterase monitoring for worker protection (65).

Other Prevention Strategies

The treatment of agriculture historically as a special case in regard to occupational safety and health has resulted in a failure to adequately protect the health of many agricultural workers. Health professionals have a unique opportunity and responsibility to redress this inequity by offering these workers adequate information about occupational hazards and medical care appropriate to their circumstances. The experience of many agricultural workers has led them to believe that they must accept pesticide exposures as a price of employment. Without a thorough review, clinicians are never justified in assuming that workers are adequately informed about the hazards of their work or that they are adequately protected.

Several techniques and approaches that may contribute to a pesticide exposure prevention strategy for agricultural workers can be highlighted here in brief:

1. *Reduced use or elimination policies for highly toxic pesticides:* The World Bank has promoted a shift to relatively low-toxicity pesticide formulations in Bank-financed projects. This recommendation is based on the concept that complete protection of workers cannot be expected in hot conditions (1).

2. *Integrated pest management programs:* IPM programs are gaining widespread acceptance in agriculture and represent a shift in management practices from traditional chemical control to a mixture of control approaches. This approach recognizes pest control as a process requiring substantial knowledge of insect behavior within a broad agri-ecosystem context.

3. *Pesticide use and incident reporting systems:* Until pesticide-related illnesses can be enumerated and linked to specific pesticide use patterns, regulatory interventions will continue to be instituted on an ad hoc basis. Systems such as the California Pesticide Incident Reporting System can serve as models for such programs.

4. *Routine medical surveillance:* Increased medical surveillance will be possible in many parts of the world as field-based systems for measuring cholinesterase inhibition are validated and made widely available. A field portable cholinesterase test kit has performed well in field trials and reduces the cost of such assays considerably (66).

5. *Improved hazard communication:* Workers should be provided with meaningful health and safety information. Fluorescent tracer evaluation of dermal exposures is one promising and inexpensive tool for improved worker education. Such an approach is particularly effective because it draws on the worker's own knowledge in analyzing the sources of exposure and the means for reducing exposures (51).

6. *Improved analytical procedures:* Residue monitoring and occupational exposure assessments can be simplified and costs reduced by use of new analytical techniques. Immunoassays for pesticide residue analysis show great promise for measuring low amounts of pesticides and their metabolites with great specificity.

7. *Safe use promotion programs:* Programs such as the Global Safe Use Initiative, discussed earlier, can contribute significantly to the prevention of pesticide-related illness in developing countries. However, such programs are likely to be successful only if workers and the public health community are included in the design, implementation, and evaluation phases of the effort.

REFERENCES

1. World Health Organization. The public health impact of pesticides used in agriculture. Geneva: WHO, 1990.
2. Hired farmworkers: health and well-being at risk (GAO/HRD-92-45). Washington, DC: U.S. General Accounting Office, 1992.
3. Commission on Agricultural Workers. Report of the Commission on Agricultural Workers. Washington, DC: U.S. Government Printing Office, 1992.
4. Mines R, Gabbard S, Seirman A. A profile of U.S. farm workers: Office of Program Economics research report no. 6, prepared for the Commission on Immigration Reform. Washington, DC: U.S. Department of Labor, April 1997.

5. Moses M. Pesticide-related health problems and farmworkers. Am Assoc Occup Health Nursing J 1989;37:115–130.

6. Pollack SH, Landrigan PJ, Mallino DL. Child labor in 1990: prevalence and health hazards. Annu Rev Public Health 1990;11:359–375.

7. Wilk V. Health hazards to children in agriculture. Am J Ind Med 1993;24:283–290.

8. National Research Council Committee on the Health and Safety Implications of Child Labor. Protecting youth at work. Washington, DC: National Academy Press, 1998. Available at: http://www.nap.edu.

9. Rivara F. Fatal and nonfatal farm injuries to children and adolescents in the United States. Pediatrics 1985;76:567–573.

10. National Institute for Occupational Safety and Health. Preventing deaths and injuries of adolescent workers, Cincinatti, OH: NIOSH Publication No. 95-125. 1995.

11. Broste SK, Hansen DA, Strand RL, Steuland DT. Hearing loss among high school farm students. Am J Public Health 1989;79:619–622.

12. National Research Council. Pesticides in the diets of infants and children. Washington, DC: National Academy Press, 1993.

13. Rothschild SK. Cross-cultural issues in primary care medicine. Disease Monthly 1998;44:293–319.

14. The EthnoMed Home Page. Harborview Medical Center, Seattle WA. Available at: http://www.hslib.-washington.edu/clinical/ethnomed/index.html.

15. Richards HN. Some good sources on cross-cultural medicine [Letter]. Minn Med 1996;79:2.

16. Schenker MB. Preventive medicine and health promotion are overdue in the agricultural workplace. J Public Health Policy 1996;17:275–305.

17. Langley R, McLymore RL, Meggs WJ, Roberson GT. Safety and health in agriculture, forestry, and fisheries. Rockville, MD: Government Institutes, Inc, 1997.

18. National Safety Council. Accident facts, 1998 ed. Chicago: National Safety Council, 1998.

19. Meyers J. Injuries among farm workers in the United States 1994. DHHS Publication No. 98-153. Cincinnati: U.S. Department of Health and Human Services, Centers for Disease Control and Prevention, National Institute for Occupational Safety and Health, July 1998.

20. Mobed K, Gold EB, Schenker MB. Occupational health problems among migrant and seasonal farm workers. West J Med 1992;157:367–373.

21. Wilk V. The occupational health of migrant and seasonal farmworkers in the United States. Washington, DC: The Farmworker Justice Fund, 1986.

22. Outbreak of severe dermatitis among orange pickers—California. MMWR Morb Mortal Wkly Rep 1986;35:465–467.

23. National Institute for Occupational Safety and Health. Musculoskeletal disease in agricultural workers. Cincinnati: NIOSH, 1983. [Internal Document. Information extracted for this chapter courtesy of Shiro Tanaka, M.D., Industrywide Studies Branch, Division of Surveillance, Hazard Evaluations and Field Studies.]

24. Meyers J, Miles J, Faucett J, Janowitz I, Tejeda D, Kabashima J. Using ergonomics in the prevention of musculoskeletal cumulative trauma injuries in agriculture. J Agromed 1995;2:11–24.

25. Meyers J, Miles J, Faucett J, Janowitz I, Tejeda D, Kabashima J. Ergonomics in agriculture: workplace priority setting in nursery industry. Am Ind Hyg Assoc J 1997;58:121.

26. Pinzke S. Observational methods for analyzing working postures in agriculture. J Ag Safety Health 1997;3:169–194.

27. Sakakibara H, Miyao M, Kondo T, Yamada S, Nakagawa T, Kobayashi F. Relation between overhead work and complaints of pear and apple orchard workers. Ergonomics 1987;30:805–815.

28. Janowitz I, Meyers J, Tejeda D, et al. Reducing risk factors for the development of work-related musculoskeletal problems in nursery work. Appl Occup Environ Hyg 1998;13:9–14.

29. Murray DL. The abolition of "el cortito," the short handled hoe: a case study in social conflict and state policy in California agriculture. Social Problems 1982;30:26–39.

30. International Labor Office Web site. Available at: http://www-ilo-mirror.who.or.jp/public/english/90travai/sechyg/agrivfer.htm.

31. Von Essen S. Airborne dusts. Papers and Proceedings of the Surgeon General's Conference on Agricultural Safety and Health, USDHHS/CDC/ NIOSH, April 30–May 3, 1991. DHHS (NIOSH) Publication No. 92-105. Washington, DC: U.S. Government Printing Office, 204–215.

32. Sever LE, Arbuckle TE, Sweeney A. Reproductive and developmental effects of occupational pesticide exposure: the epidemiologic evidence. Occup Med 1997;12:305–326.

33. Kavlock RJ, Daston GP, DeRosa C, et al. Research needs for the risk assessment of health and environmental effects of endocrine disruptors: a report of the U.S. EPA sponsored workshop. Environ Health Perspect 1996;104(Suppl 4):715–740.

34. Ware GW. The Pesticide Book, 4th ed. Fresno, CA: Thomson Publications, 1994.

35. Dinham B. The pesticide hazard: a global health and environmental audit. New Jersey: The Pesticides Trust, 1993.

36. Vergara AE, Fuortes L. Surveillance and epidemiology of occupational pesticide poisonings on banana plantations in Costa Rica. Int J Occup Environ Health 1998;4:199–201.

37. United Nations Food and Agriculture Organization. International code of conduct on the distribution and use of pesticides (amended version). Rome: FAO, 1990.

38. Global Crop Protection Federation Web page. Available at: http://www.gcpf.org/.

39. Environmental Protection Agency. Worker protection standard, hazard information, hand labor tasks on cut flowers and ferns exception; final rule, and proposed rules (40 CFR Parts 156 and 170). Federal Register 1992;57:38102–38176.

40. California Environmental Protection Agency, Department of Pesticide Regulation. Summary of illnesses and injuries reported by California physicians as potentially related to pesticides in 1990. HS-1666. Sacramento, CA: California Environmental Protection Agency. March 1, 1993.

41. Farm worker illness following exposure to carbofuran and other pesticides Fresno County, California, 1998. MMWR Morb Mortal Wkly Rep 1999;48:113–116.

42. Keifer MC, ed. Human health effects of pesticides [theme issue]. Occup Med 1997;12(2):203–411.

43. Maroni M, Fait A. Health effects in man from long-term exposure to pesticides. Toxicology 1993; 78:1–180.

44. Savage EP, Keefe TJ, Mounce LM, Heaton RK, Lewis JA, Burcar PJ. Chronic neurological sequelae of acute organophosphate pesticide poisoning. Arch Environ Health 1988;43:38–44.

45. Rosenstock L, Keifer M, Daniell WE, McConnell R, Claypoole K. Chronic central nervous system effects of acute organophosphate pesticide intoxication. Lancet 1991;338:223–226.

46. Alavanja MCR, Sandler DP, McMaster SB, et al. The agricultural health study. Environ Health Perspect 1996;104:362–369.

47. van Hemmen JJ. Agricultural pesticide exposure data bases for risk assessment. Rev Environ Contam Toxicol 1992;126:1–85.

48. Curry PB, Iyengar S, Maloney PA, Maroni M, eds. Methods of pesticide exposure assessment. NATO Committee on the Challenges of Modern Society. New York: Plenum Press, 1995.

49. World Health Organization. World Health Organization field surveys of exposure to pesticides standard protocol. Toxicol Lett 1986;33:223–235.

50. Fenske RA. Use of fluorescent tracers and video imaging to evaluate chemical protective clothing during pesticide applications. In: Mansdorf SZ, Sager R, Nielsen AP, eds. Performance of protective clothing: 2nd symposium. STP 989:630-639. Philadelphia: American Society for the Testing of Materials, 1988.

51. Fenske RA. Visual scoring system for fluorescent tracer evaluation of dermal exposure to pesticides. Bull Environ Contam Toxicol 1988;41:727–736.

52. Fenske RA, Birnbaum SG. Second generation video imaging technique for assessing dermal exposure (VITAE System). Am Ind Hyg Assoc J 1997; 58:636–645.

53. Machado N, Matuo JG, Matuo YK. Dermal exposure of pesticide applicators in staked tomato crops: efficiency of a safety measure in the application equipment. Bull Environ Contam Toxicol 1992; 48:529–534.

54. National Institute for Occupational Safety and Health. Report to Congress on workers' home contamination study conducted under the Workers' Family Protection Act. Cincinnati: U.S. Department of Health and Human Services, NIOSH, 1995.

55. Daniels JL, Olshan AF, Savitz DA. Pesticides and childhood cancers. Environ Health Perspect 1997;105:1068–1077.

56. Garry VF, Schreinemachers D, Harkins ME, Griffith J. Pesticide appliers, biocides, and birth defects in rural Minnesota. Environ Health Perspect 1996;104:394–399.

57. Cannon SB, Veazey JM, Jackson RS, et al. Epidemic kepone poisoning in chemical workers. Am J Epidemiol 1978;107:529–537.

58. Richter ED, Chuwers P, Levy Y, et al. Health effects from exposure to organophosphate pesticides in workers and residents in Israel. Isr J Med Sci 1992;28:584–598.

59. Chiao-Cheng JH, Reagan BM, Bresee RR, Meloan CE, Kadoum AM. Carbamate insecticide removal in laundering from cotton and polyester fabrics. Arch Environ Contam Toxicol 1989;17:87-94.

60. Clifford NJ, Nies AS. Organophosphate poisoning from wearing a laundered uniform previously contaminated with parathion. JAMA 1989;262:3035–3036.

61. Simcox NJ, Fenske RA, Wolz S, Lee I, Kalman D. Organophosphate pesticide residues in residential housedust and soil in the tree fruit region of Washington state. Environ Health Perspect 1995; 103:1126–1134.

62. Loewenherz C, Fenske RA, Simcox NJ, Bellamy G, Kalman D. Biological monitoring of organophosphorous pesticide exposure among children of agricultural workers. Environ Health Perspect 1997; 105:1344–1353.

63. Morgan D. Recognition and management of pesticide poisonings, 4th ed. EPA-540/9-88-001. Washington, DC: Environmental Protection Agency, 1989.

64. Wagner SL. Diagnosis and treatment of organophosphate and carbamate intoxication. Occup Med 1997;12:239–250.

65. Wilson BW, Sanborn JR, O'Malley MA, Henderson JD, Billitti JR. Monitoring the pesticide-exposed worker. Occup Med 1997;12:347–364.

66. McConnell R, Cedillo L, Keifer M, Palamo M. Monitoring organophosphate insecticide exposed workers for cholinesterase depression: new technology for office or field use. J Occup Med 1992; 1:34–37.

BIBLIOGRAPHY

Environmental Protection Agency. Worker protection standard, final rule, and proposed rules. Federal Register 1992;57:38102–38176.
This document provides a comprehensive rationale for current U.S. regulations designed to protect agricultural workers from pesticide-related health risks.

World Health Organization. The public health impact of pesticides used in agriculture. Geneva: WHO, 1990.

Dinham B. The pesticide hazard: a global health and environmental audit. New Jersey: The Pesticides Trust, 1993.
These two monographs provide a global review of pesticide use. Work-related aspects of pesticide exposures are a small but important section of these monographs.

Keifer MC, ed. Human health effects of pesticides [theme issue]. Occup Med 1997;12(2):203–411.
This volume is the best recent review of health effects associated with pesticide use.

Maroni M, Fait A. Health effects in man from long-term exposure to pesticides. Toxicology 1993;78: 1–180.
This article reviews the available scientific literature

through about 1990 and serves as an excellent source of references.

Morgan D. Recognition and management of pesticide poisonings, 4th ed. EPA-540/9-88-001. Washington, DC: Environmental Protection Agency, 1989.
The most authoritative source for clinical aspects of pesticide poisonings, this book is used throughout the world as a handbook.

National Research Council. Protecting youth at work. Washington DC: National Academy Press, 1998.
This report provides a thorough review of child labor in the United States and lists public health recommendations. Agriculture is one of the key industries discussed in the report. Available on line at http://www.nap.edu.

Occupational Health: Recognizing and Preventing Work-Related Disease and Injury, 4th ed.
Edited by Barry S. Levy and David H. Wegman, Lippincott Williams & Wilkins, Philadelphia © 2000.

42 Construction Workers

Knut Ringen, Anders Englund,
Jane L. Seegal, Michael McCann, and
Richard A. Lemen

NATURE AND CHARACTERISTICS OF CONSTRUCTION WORK

Construction workers build highways, stadiums, industrial plants, office buildings, and homes. They also repair or renovate roads, bridges, and other structures and demolish or clean up former building sites. In the United States, construction work also includes cleanup of hazardous waste sites. The work is hard physical labor, often under difficult conditions, including hot, cold, and wet weather (Fig. 42-1).

ORGANIZATION OF THE WORK

Several factors exacerbate safety and health problems in construction, related to how the industry operates or how the work is performed.

Construction rarely provides steady em-

K. Ringen: The Center to Protect Workers' Rights (former director), Seattle, WA 98166.

A. Englund: Swedish National Board of Occupational Safety and Health, Solva, Sweden.

J. L. Seegal: The Center to Protect Workers' Rights, Washington, DC 20001.

M. McCann: The Center to Protect Workers' Rights, Washington, DC 20001.

R. A. Lemen: National Institute for Occupational Safety and Health (retired), Duluth, Georgia 30155.

ployment; construction workers are always working themselves out of their jobs. Although some projects may last several years, many last only a few months. And many assignments on a project, such as roofing or painting, last only a few days each, with several trades working on a site simultaneously. Therefore, a construction worker may have five or more employers in a year.

Just as the work assignments change throughout a construction project, so do the topography of the worksite and the cast of employers. Each trade on a site may work for a different contractor.

The universe of contractors is marked by high turnover. The U.S. Census counted 634,000 construction establishments in 1995, excluding the self-employed. (Some companies may have been counted more than once, depending on the number of business offices they have.) Most of the establishments were "mom-and-pop" operations; 82% had one to nine employees. An estimated 35% of all firms belonged to a construction organization, such as the National Erectors Association or the Associated General Contractors.

These features create public health problems. With so many job changes and small and short-lived firms, it is difficult to monitor an individual's work history. It is even more difficult to monitor injuries or exposures to hazards (see the following section). In the United States, the recording of injuries (ex-

FIG. 42-1. Construction workers building a large sewage treatment plant. Women remain a small minority of the construction workforce in most countries. (Photograph by Marvin Lewiton.)

cept those that can be handled with first aid alone) in a log—as required by the Occupational Safety and Health Administration (OSHA)—has proved virtually unenforceable. (And the requirement does not apply to the 40% of construction workers who are self-employed or work for firms having fewer than 11 employees.) A 1987 study commissioned by the National Academy of Sciences noted the difficulties in obtaining injury and illness data for construction. Among other things, the study cited poorly defined responsibility for reporting injuries and illnesses (1).

The constantly changing worksite has another marked effect on safety and health. In contrast to an industrial setting, where the tasks are often repetitive and controlled by the location of machinery, the construction site allows and requires extensive movement by the worker from moment to moment. This means that the worker is much more responsible for self-protection.

In addition to risks of injuries and exposures to hazardous substances, construction workers face long-term risk from the stress of episodic employment. Stress may be caused by the fear of not having a paycheck. And because construction jobs can be few and far between, construction workers may have to travel long distances daily to work or may need to move their families often.

A lack of comprehensive employer organizations to work with hampers some public health efforts. Labor-management organizations have served as the vehicles for successful safety and health programs in the Netherlands, Sweden, and Ontario, Canada. (Some 24% of blue-collar construction workers in the United States—including 40% of those in public-sector construction—belong to unions or are covered by union contracts.) For all these reasons, it is difficult to implement preventive safety and health programs, including training, in the construction industry.

Although the causes of work-related injuries are well defined, the risks of chronic work-related musculoskeletal disorders are poorly defined, as are the relations between exposures and chronic diseases.

EPIDEMIOLOGIC OVERVIEW

The construction industry in the United States employs about 6% of the labor force—about 8.3 million workers (or 7.9 million full-time equivalents). In 1997, however, this industry had 18% of the fatal injuries and more than 10% of the lost-workday cases resulting from injuries. The death rate is 13.9 per 100,000 full-time construction workers, or more than 1,000 deaths per year; on average, four or more workers are killed by injuries sustained on the job each workday.

The rate of lost-time injuries, which increased slightly from 1975 to 1990, declined by 40% from 1990 to 1996, the most recent year of injury reporting by the Bureau of Labor Statistics. During that time, injury rates for other industries declined as well, although to a lesser degree in some cases.

No apparent changes in construction industry practices can account for such a dramatic decline.

Other factors in injury rates continue to hold. Generally, the largest firms have the lowest injury rates. For workers, experience is a factor, with the rate of injuries decreasing substantially as length of service increases (2). Familiarity with a job site also is a consideration. For laborers, who have some of the highest injury rates, 12% of lost-time injuries occur during the first day on a job site; this pattern appears to hold for most of the trades (occupations).

The workforce, drawn largely from immigrants and other low-income groups, faces predictable occupational ailments. These include the white finger of the jackhammer operator, silicosis of the tunnel builder, low back pain of the bricklayer, dermatitis of the cement mason, carpal tunnel syndrome of the ironworker or electrician, solvent-induced kidney ailments of the painter and roofer, lead poisoning of the bridge rehabilitation worker, asbestosis of the building demolition worker, and heat stress of the hazardous waste cleanup worker (from wearing "moon suits").

For a mix of reasons—work-related and not, many still poorly understood—the average age at death for construction workers in the United States is substantially lower than for other groups such as teachers or physicians (Fig. 42-2). Standardized mortality ratios (SMRs) for selected occupations in California show that many construction-related occupations have SMRs that are much greater than the average for all professions (100), and many times higher than for low-risk occupations.

Protection of the construction workforce in some industrial countries appears to be more effective than in the United States. International comparisons are difficult to make, but the death rates for construction workers in the United States, Australia, and Japan are high for industrial countries, at about 14 or 15 deaths per 100,000 full-time-equivalent workers. By contrast, the death

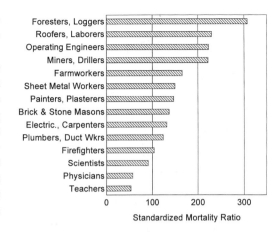

FIG. 42-2. Standardized mortality ratio for selected occupations. (Data from California Department of Health Services. California Occupational Mortality, 1979–1981. Table 7A. Sacramento, March 1987.)

rates for Sweden and the Netherlands, countries that have long-standing safety and health programs, are at about half that level (3). And, since 1970, the rate of deaths from worksite injuries in Sweden has been reduced by 75%.

The magnitude and severity of occupational health and safety hazards in construction work are greater in developing countries (Box 42-1).

Nonparticipants

In Sweden, the construction unions and employers established a comprehensive program focusing on safety and health, Bygghälsan, which monitored more than 225,000 construction workers for about 25 years. The rate of cardiovascular disease—the leading cause of death for workers and the general population—was lower among the construction workers. Cancer incidence was at 90% to 95% of the rate for the overall population, with some exceptions (Table 42-1). Bygghälsan attributes the differences partly to selection mechanisms but also to preventive programs it implemented on hypertension, smoking, diet, and exercise.

One striking finding involved a cohort of

Box 42-1. Construction Work in Developing Countries: The Experience in East Africa

Occupational health and safety hazards facing construction workers in developing countries are even greater than those facing their counterparts in industrialized countries. The experience in the East African countries of Kenya, Tanzania, and Uganda is illustrative.

In these developing countries, the hazards that are inherent in construction work are compounded by hot and humid conditions, inadequate training of workers, and shortages or inadequacies of construction tools, scaffolding, and helmets and other basic personal protective equipment. There is a great variety of inconsistency in equipment and processes. Among major construction hazards cited in Kenya, for example, are excavation work, inadequate scaffolds (Fig. 42-3) and electric equipment, and lack or inadequacy of lifting equipment. In these and other countries, hand tools and diesel engines at construction sites generate much noise; dusts and fumes are also major problems, especially in confined spaces.

In East African countries, well over half of construction workers are casual workers who are employed on a daily basis and do not belong to unions, as many other construction workers do. In these countries, construction workers are often paid on a piece-rate basis, which encourages rapid work and often unsafe work practices. Clean water and nutritious food are often not available to construction workers in the countries of East Africa. The same is true for basic health and safety services.

Data for work-related injuries and illnesses of construction workers in East Africa are even more incomplete than comparable data for these workers in industrialized countries. Nevertheless, available data provide some useful descriptions of the situation. Reporting is highly variable; for example, in Zanzibar, Tanzania, 44 construction incidents were reported between 1983 and 1989, with the number varying from 0 to 19 annually. In Kenya in 1989, according to workers' compensation data, there were 1,117 reported injuries among the approximately 60,000 building and engineering construction workers. Of these injuries, 43 were fatal and 52 caused permanent total incapacity.

The rate of reported fatal injuries for construction workers (72 per 100,000 workers per year) was three times the estimated rate in the United States and 12 times the rate in Denmark. Of the 618 injuries that could be classified, 40% were caused by stepping on or being struck by objects and 31% were caused by transportation vehicles.

Ministries of labor in the East African countries, with assistance from the International Labor Organization, the Finnish Institute of Occupational Health, and other organizations and groups, are attempting to improve safety and health in the construction industry by improving training of workers, establishing better means of surveillance and evaluation of hazards, and instituting better measures for prevention, often using low-cost, readily available materials and methods.

FIG. 42-3. Construction laborers work, without personal protective equipment, on dangerous bamboo scaffolding in Kenya. (Photograph by Barry S. Levy.)

(Adapted from the East African Newsletter on Occupational Health and Safety, December 31, 1990.)

TABLE 42-1. *Selected excess mortality and incidence for cancer and other disorders, Swedish construction workers*[a,b]

Trade	Cause of death (n)	SMR	Incident cancers (n)	SIR
Bricklayer	—	—	Peritoneum (2)	12.5
Carpenter	—	—	Nose (11)	2.2
Driver (truck/heavy equipment)[c]	—	—	Lip (6)	3.6
			Multiple myeloma (7)	2.7
Electrician	Bladder cancer (13)	2.3	[Bladder (39)]	1.3
Insulator	Pneumoconiosis (2)	40	Peritoneum (2)	200
Laborer	Accidental falls (67)	1.5	Lip (41)	1.8
			Stomach (152)	1.2
Maintenance worker[c]	Other accidents (7)[d]	8.4	Colon (11)	2.5
	Drowning (5)	4.7		
Plumber	Pneumoconiosis (4)	4.4	Pleura (15)	6.3
			Lung (105)	1.3
Sheet metal worker	Accidental falls (11)	2.4	Lung (26)	1.6
Tunnel worker	Prostate cancer (20)	1.9	[Prostate (33)]	1.1
	Other accidents (10)[d]	6.1		
	Violent death (69)	1.5		

SMR, standardized mortality ratio; SIR, standardized incidence ratio.

[a] Ratios are based on comparison with the white male Swedish population. The cohort of more than 225,000 workers was established in 1971 to 1979. Mortality was followed through 1988. Cancer incidence was followed through 1987.

[b] All values are statistically significant unless in brackets.

[c] Compared with the other construction trades, drivers and maintenance workers have a risk of death from heart disease of 1.37 and 1.40, respectively.

[d] Other accidents were predominantly work-related injuries.

Data from Göran Engholm, Bygghälsan, 1992.

48,754 male union painters and certified plumbers and insulators. The group was divided almost equally into workers who participated in voluntary medical examinations and those who did not, either because they chose not to or because they were no longer employed in construction. By 1988, nonparticipants showed mortality rates 72% higher than those of participants. By 1987, similar results were obtained for cancer incidence.

The differences between nonparticipants and participants were especially noteworthy for diseases associated with poor health behaviors. A disparity in mortality rates from cancer of the larynx probably could be assigned to differences between the two groups in alcohol and tobacco consumption; SMRs for nonparticipants and participants were 3.12 and 1.11, respectively. Excess mortality from alcoholism, cirrhosis of the liver, and cardiovascular disease could also be tied to behavioral differences.

A second distinguishing factor is analogous to the healthy worker effect, whereby the employed population tends to be healthier overall than the total population. (This is particularly true in occupations with heavy manual labor, including most construction work.) For example, a higher mortality rate from diabetes among nonparticipants may reflect the inability of workers who initially were part of the cohort to remain employed, because of illness.

These findings point up the risk of basing epidemiologic conclusions on the results of a voluntary screening program. Although voluntary screenings can suggest whether a problem exists, results based on those who participate in a voluntary program would show a large selection bias. In this case, the findings would paint much too rosy a picture of the health status of construction workers generally.

Female Workers

Female workers are only beginning to join the skilled trades in substantial numbers in most countries; in the United States, they accounted for 2.5% of blue collar workers in 1996 (4). Injury rates for women in construction have been lower than those for men, at least partly because women have been restricted from the most dangerous and highest-paying jobs (5,6). Women's safety and health risks in construction are tied to gender discrimination (7,8). Discrimination can take the form of a lack of site orientation or cooperation by coworkers, sexual harassment, or outright threats to safety (e.g., a brick dropped from overhead). One obvious result is increased stress. In addition, female construction workers face problems tied to anatomy. Some job equipment (e.g., hand tools) or personal protective equipment (e.g., respirators) may not properly fit women, who tend to be smaller than many men. A lack of private sanitary facilities may lead women to avoid urinating during a workshift, with consequent discomfort and health problems (see Chapter 36).

TABLE 42-2. *Distribution of lost-time injuries (fatal and nonfatal) among roofers and laborers*

Cause of injury	Roofers	Laborers
Falls from elevations	23%	11%
Overexertion	23%	22%
Struck by an object	14%	25%
Contact with temperature extremes	9%	2%
Struck against	7%	10%
Falls from same level	6%	7%
Bodily reaction[a]	5%	3%
Caught in/under/between (including cave-ins)	3%	8%
Rubbed/abraded	3%	7%
Contact with radiation, caustics, etc.	2%	3%
Other[b]	5%	2%

[a] Includes, for instance, slipping and twisting body to catch oneself or twisting an ankle while climbing a ladder.
[b] Includes transport and nonclassifiable injuries.
Labor injury data from Bureau of Labor Statistics. Injuries to construction laborers. Bulletin 2252. Washington, DC, March 1986:8. Roofer injury data from Martin E. Personick, Bureau of Labor Statistics, based on workers' compensation data, selected states.

INJURIES AND ILLNESSES ASSOCIATED WITH CONSTRUCTION WORK

Traumatic Injuries

Construction workers are at great risk of injury partly because of where they work, from scaffolding and roofs hundreds of feet up to trenches. Specific hazards and overall risk vary by trade (Table 42-2). Based on what is known in the United States, ironworkers have the highest risk of work-related deaths, more than twice the rate of the next-highest group, laborers; in other countries, roofing may be the most dangerous trade because of the danger of falls and exposures to hot tar (9,10).

The rankings of causes of fatal and nonfatal injuries appear to differ, however. For instance, falls from heights tend to be so serious that they are responsible for most of the traumatic work-related deaths. For nonfatal injuries, most studies list overexertion, followed by "struck by" injuries (see Chapters

8 and 24) (11). (In regions where a large proportion of the construction labor force consists of immigrants, workers' inability to understand the national language may increase the risk of injury.)

Musculoskeletal Disorders

Some musculoskeletal disorders result from traumatic injuries, but many others develop incrementally. These stem from repetitive tasks, awkward body positions, and materials handling on the job. The bricklayer lifts an estimated 3 to 4 tons daily, with 1,000 trunk-twist flexions. The ironworker tying intersections of the perpendicular rods used to reinforce concrete may bend over for more than half the workday, repeatedly twisting the wrist under pressure. In building construction, much of the finishing work involves areas above shoulder height or below knee level. Wallboard sizes typically used in the United States (4 ft × 12 ft × 5/8 inch) weigh more than 100 pounds apiece and are often packaged in pairs. (Arbouw, the Dutch institute on safety and health in construction, recommends that wallboard weigh less than 20 kg [45 lb]).

Although musculoskeletal disorders are not fatal, they are significant. Disorders of the back and upper extremities account for roughly one-third of all workers' compensation claims (see Chapters 11 and 26) (12). Very little detailed analysis has been performed on the problems of each trade. A Bygghälsan questionnaire from 1989 to 1992 was completed by more than 83,000 construction workers, including more than 19,000 carpenters. Responses to questions about the proportion of time spent in certain work postures and about the locations of the most prevalent musculoskeletal complaints suggest a correlation between the two.

The most detailed scientific report on musculoskeletal disorders was done by Holmström, who produced a series of studies, one of them on a randomized, cross-sectional sample of 1,773 construction workers in Malmö, Sweden (13). Holmström found low back pain to be correlated with increasing age, construction trade, personal habits, and psychosocial factors. Only 8% of the workers studied reported no musculoskeletal problems in the preceding year; low back pain was reported by 72%, knee problems by 52%, and neck-shoulder pain by 37%.

The prevalence of most musculoskeletal symptoms increased with age. Low back pain correlated with frequent handling of hand-held equipment, handling of bricks and roofing materials, and awkward postures such as stooping or kneeling for more than 1 hour per day. The prevalence and type of disorder varied by trade; roofers, carpet and tile layers, and scaffolding erectors had the highest prevalence of low back pain.

Reported low back pain was 2.7 times more prevalent for smokers than for non-smokers. Psychosocial factors also contributed significantly to "explain" low back pain, when other factors were kept constant. Workers who reported no low back problems were generally in better physical condition, were more involved in recreational activities, smoked less, and had a more positive out-look. They reported fewer psychosomatic symptoms and were more active participants in worksite decision making. The average length of employment in construction for this group was 15 years. Workers reporting severe low back problems had significantly reduced back muscle endurance. Both groups of workers were found to have the same maximum abdominal and back muscle strength.

Illnesses

Some work-related illnesses appear to be correlated with specific construction trades (Table 42-3). In addition, some illnesses may develop as a result of the location of work. For instance, in the United States, construction workers who work outdoors where Lyme disease is found have an increased risk of developing the disease, compared with the general population (14). Listeriosis, histoplasmosis, malaria, yellow fever, and rashes from poisonous plants are other ailments tied to location. For several reasons, including long latencies, it has generally been difficult to relate chronic diseases to an individual worker's employment history.

Pulmonary Diseases, including Lung Cancers, and Bystander Exposures. Construction sites are generally dusty; powdered bags of cement are emptied for mixing, wood is sawed, heavy machinery lumbers across uneven terrain, and pneumatic tools are used on drywall, concrete, and rock that contains quartz. Fumes are also produced by such activities as welding, roofing, and paving. So, in several ways, construction workers' lungs are exposed to toxic hazards.

Asbestos and silica are the two best-documented hazards. The National Institute for Occupational Safety and Health (NIOSH) (15) found that the highest proportionate mortality ratios (PMRs) for white male construction workers younger than 65 years of age were for asbestosis (393) and silicosis (327). These findings, derived from underlying cause-of-death codes on death certificates for 1984–1986, compared with a PMR for falls of 177 (see Chapter 25).

TABLE 42-3. *Common toxic hazards on the construction site*

Substance	Key source of exposure	Substance	Key source of exposure
Dusts		**Solvents**	
Asbestos[a]	Demolition, maintenance, insulation	Benzene[a]	Hazardous waste cleanup, petrochemical plant sites
Cement	Foundations, sidewalks, floors	Methylene chloride	Paint strippers
Synthetic vitreous fibers, other insulation	Insulation on pipes, air conditioning	Toluene	Varnishes, paints, adhesives, cleaners
Silica	Sandblasting, tunneling	Trichloroethylene	Varnishes, paints, adhesives, cleaners
Wood dust[a]	Remodeling, demolition, sawing	**Other chemicals** Epoxy resins	Impermeable paints, wood floor primers
Metals (dusts and fumes)			
Cadmium[a]	Welding, cutting pipe	Polyurethanes (isocyanates)	Seam sealers, insulation, electrical wire coats
Hexavalent chromium[a]	Welding, cutting pipe		
Copper	Welding, cutting pipe	Coal tar pitch[a]	Roofing, road work
Lead	Demolition, work on lead-paint sur-faces		
Magnesium	Welding, cutting pipe		
Zinc	Welding, cutting pipe		

[a] Human carcinogen. Asbestos is a disease risk only when asbestos is disintegrating (friable).
Adapted from Workplace Hazard and Tobacco Education Project. Construction workers' guide to toxics on the job, Berkeley: California Public Health Foundation, 1993.

Studies have documented a pattern of lung cancer among insulation (asbestos) workers, as well as the risk for other neoplasms and the likelihood that others besides workers producing or handling asbestos are also at risk (16). The research showed an almost seven times greater risk of death from bronchogenic carcinoma and mesothelioma, compared with the general U.S. population. The higher the exposure, the greater the risk for developing asbestos-related disease; nonetheless, any worker near asbestos work is at risk of developing disease through "bystander exposure." Although the spray application of asbestos insulation has been banned in the United States since 1973 and most uses of asbestos are controlled, construction workers involved in demolition

continue to be exposed to asbestos-containing materials that were installed years ago. There is no scientific basis for application of differing regulations for different asbestos fiber types (17–19).

Similarly, although OSHA has set permissible exposure limits for respirable silica, new cases of silicosis are still reported in the United States; however, the disease is underdiagnosed and therefore underreported. Those at risk include tunnel workers, sandblasters, workers in trades working with concrete and mortar (laborers, masons, concrete finishers, tile setters, and plasterers), and bystanders (20). Depending partly on the percentage of silica in the materials used, a wide range of tasks can prove hazardous, including drilling holes, grinding concrete surfaces, power-cleaning concrete forms, and cutting through concrete block, walls, or pipe. The risks have been documented for decades and are not limited to new construction. For instance, powered grinders may be used to remove mortar for restoration.

A health examination survey of construction workers formerly employed at U.S. Department of Energy facilities found evidence of positive results on lymphocyte proliferation tests, a clinical indication of beryllium exposure. Although work that may lead to beryllium exposure is not widespread in construction, such exposure should be considered particularly during maintenance, demolition, and cleanup of nuclear facilities, other types of power generating plants, and aerospace facilities, where beryllium may have been used.

Other Cancers. Exposures to carcinogens are possible in all types of construction. Some sources are well known, such as polycyclic aromatic hydrocarbons in roofing tar. Welding can produce carcinogenic fumes, such as fumes of cadmium, nickel, or hexavalent chromium.

More than 60 million tons of asphalt (bitumens) are produced worldwide annually. Although studies have not yet established that asphalt exposure leads to human cancers, general agreement exists, on the basis of experimental studies, that asphalt may pose such a risk (21). When working with asphalt, skin exposures should be avoided; work in poorly ventilated areas should be done only with engineering controls and use of personal protective equipment.

Some carcinogenic exposures are of relatively recent origin. The use of plastics has been multiplying, and the health effects of their use remain unknown. Specialty paints may include metals and dangerous solvents. Among resins, epichlorhydrin in epoxies and isocyanates in polyurethanes pose potential but sometimes poorly documented risks. Benzene and vinyl chloride are among the substances commonly found at Superfund cleanup sites.

Central Nervous System Disorders. Lead poisoning continues to be a particular concern for construction workers. Although lead has been restricted to trace amounts in residential paints since 1978 in the United States, it is still allowed for industrial uses, including signs, road paints, and steel structures. The California Occupational Lead Registry found that construction workers accounted for 18% of the workers who had peak blood lead levels of 80 μg/dL (22). (This is well above the 50 μg/dL level requiring medical removal.) Exposures to lead occur during rehabilitation or demolition of lead-painted structures, including housing built before 1950. Lead exposures can occur during scraping, cutting with torches, sandblasting, and welding. (See also Chapters 3, 15, 29, 31, 33, and 35.) Welders may be exposed to fumes containing lead, but also manganese, nickel, and decomposition products of epoxy, polyurethane, and polyvinyl chloride coatings. In California, a welder's Parkinson's disease has been tied to work-related manganese exposures for workers' compensation (see Chapter 29.)

Threats to the nervous system commonly found by hazardous waste cleanup workers include toluene, trichloroethylene, tetrachloroethylene, arsenic, benzene, lead, and mercury.

Skin Disorders. Bricklayers, cement masons, and others who handle wet cement are prone to allergenic and irritant dermatitis on the hands and other exposed areas. The symptoms can be severe enough to necessitate leaving the trade. The irritant dermatitis is believed to be caused by the abrasive nature and high pH of wet cement (see Chapter 27).

Hearing Loss. Noise levels on construction sites represent a long recognized and inadequately addressed problem. Among other things, limited hearing endangers workers who cannot hear an approaching vehicle or warnings of immediate danger. Reports from Sweden, Germany, the United States, and British Columbia, Canada, show extensive hearing loss, apparently work related, among construction workers. In the

United States, a 1997–1998 study of construction sites in Washington State found that 40% of workers were commonly exposed to noise levels greater than 85 dB (23). A jackhammer can exceed 105 dB; a bulldozer, 95 dB. Bygghälsan found in 1974 that bilateral normal hearing among construction workers decreased gradually with age, so that only about 8% of those examined at 50 years of age had hearing in the normal range. At the same time, about 12% of construction workers at that age had bilateral severe high-tone hearing loss. (See also Chapter 18.)

Temperature Extremes. Some of the most difficult hazards faced by construction workers are those presented by extremes of temperature, particularly in summer and winter. Consequences of overexposure to heat are similar to those documented for other working groups (see Chapter 19). Although the same is true for the consequences of cold temperatures, there is the additional hazard of increased risk of traumatic injury that can result from contact with frozen surfaces or reduced ability to handle tools and equipment properly.

Family-Contact Disease. Asbestos fibers reportedly have been transported in asbestos products carried home to show family members and inadvertently on clothing, in hair, and on shoes. A similar risk exists for lead poisoning among construction workers' families, particularly children, who are more vulnerable to lead.

APPROACHES TO PREVENTION

Adaptation of Measurement Technology

Industrial hygiene exposure measurements in construction have some special aspects. To measure exposures in manufacturing, integrated samples are collected to determine 8-hour time-weighted averages. Such an approach may not give an accurate picture about health effects in construction, however. Brief, high-level exposures are common and may have different and significant

health effects compared with longer-term, low-level exposures. Air sampling based on tasks may be more relevant. There is a need for researchers to take on the special issues in construction and develop new methods, such as real-time monitoring, that may help address this problem. One goal would be to develop exposure profiles of individual tasks and possibly of trades.

Use of exposure profiles is limited, however, in that individual construction workers or workers in a given trade do not perform the same tasks each day. The lack of consistent exposures and varying ambient conditions (e.g., wind, rain, temperature) limit the effectiveness of air-sampling measurements for routine monitoring.

Construction workers can also be subject to bystander exposures. Because construction workers all move about a site, each worker's position in relation to exposure sources changes often. At some moments, a worker may be directly exposed while using a hazardous substance, but at other times the worker may be exposed to another substance as a bystander 10 feet downwind. It is therefore difficult to anticipate all the substances and degrees of exposure an individual may encounter on a given day or in a career.

In addition, air sampling does not measure skin absorption or ingestion of toxic substances, which contribute an unknown, but not necessarily insignificant, amount of exposure, particularly on worksites where sanitary facilities for hand washing are nonexistent.

New Technologies

Technological improvements have reduced the risks of musculoskeletal and other health problems. Many of the changes are straightforward. For instance, a two-handed screwdriver with a longer handle, now used in Sweden, increases torque and reduces stress on the wrists.

To make lifting easier for the bricklayer, bricks in Germany now are designed with

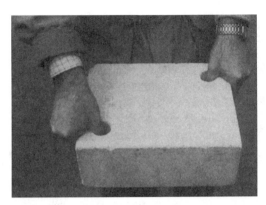

FIG. 42-4. A variety of handholds can be designed to assist in lifting bricks of different sizes and shapes. (Photograph by Bau-Berufsgenossenschaft, Frankfurt.)

holes or handles (Fig. 42-4). Regulations require that bricks weighing more than 25 kg (55 lb) be lifted only by machine. In the Netherlands, brick manufacturers, unions, and management developed a system in which bricks are packaged in sets that are easily moved about the worksite on dollies and lifted by levers to a height that is convenient for the worker.

Tower crane cabins are being redesigned in Germany and Sweden. One change extends the window to the cabin floor. This enables operators to see below without having to lean forward constantly. As a result, they report less chronic neck pain.

To reduce the risk of silicosis by about 90%, researchers in Germany have turned to wet blasting using water to dampen sand as it is being sprayed and thereby minimize dust. Nevertheless, throughout Europe, silica is being banned, except where its use is essential.

In Denmark, Finland, Sweden, and Germany, researchers found that adding small amounts of ferrous sulfate to cement changes water-soluble hexavalent chromium to trivalent chromium. This modification was applied beginning in the mid-1980s and appears to explain the substantial decline in allergic dermatitis in the Nordic countries since that time. The change costs about $1 per ton.

Some efforts to improve worker health combine technologies and training. By 1986–1990, such efforts by Bygghälsan had reduced noise-related hearing loss by half among workers 30 to 50 years old, the ages for which hearing loss could have been affected by Bygghälsan's interventions.

Personal Protective Equipment

Engineering or work-practice changes are preferable to relying on personal protective equipment. Workers may not know when they need to use protective equipment. If they do know, they may lack proper equipment or the needed training. And use of some controls may create problems. For instance, although construction workers often perform as teams, respirators may prevent coworkers from communicating with each other. Full-body protective clothing can also contribute to heat stress.

Having protective gear without knowing its limitations can do more harm than good, by giving the worker or employer the illusion that the worker is protected. For instance, no gloves can protect for longer than 2 hours against methylene chloride, which is present in many paint strippers. And some solvent mixtures can seep through the best gloves in minutes, even though the components may each have breakthrough times of 4 hours or longer.

A lack of eating and sanitary facilities may also present problems. Often, workers cannot wash up before meals and must eat in the work zone. A lack of showering and changing facilities may result in transport of contaminants from the workplace to a worker's home.

Regulation

Since 1989 in the United States, OSHA has implemented standards that could improve safety and health in construction. Among these are: Process Safety Management of Highly Hazardous Chemicals (29 CFR 1926.64), 1992; Hazardous Waste Operations and Emergency Response (29 CFR 1926.65), 1993; Lead Exposure in Construction (29 CFR 1926.62), 1993; and a proposed revised Subpart R Steel Erection standard.

Recent studies have confirmed, however, what critics have long believed to be a major deficit in OSHA's enforcement of its construction regulations. OSHA tends to repeatedly inspect the largest worksites maintained by the largest companies, where unsafe conditions are least likely to exist (24). To direct its limited resources to more hazardous sites, OSHA has developed a "focused inspection" program. If a worksite has an effective safety and health program that includes worker training and participation, OSHA inspectors are authorized to focus on the hazards known to cause 80% or more of deaths from injuries: falls, and "struck by," "caught in-between," and electrical hazards. If a focused inspection finds no violations of OSHA standards, inspectors normally move on to other worksites. Some OSHA standards do not apply to construction, however. Such as a mandatory safety and health program (29 CFR 1900.1) and a proposed ergonomics standard.

Ergonomics is beginning to be addressed by other agencies. Since 1996, the U.S. Army Corps of Engineers has had an ergonomics standard for its construction contractors. A Swedish ergonomics standard in place since 1998 applies to construction and requires training for workers; the standard, for instance, notes that 90-cm-wide gypsum wallboard is 25% lighter than traditional, larger board and should be used whenever feasible (25).

Sometimes regulation has unintended negative effects. Regulation and the threat of lawsuits regarding use of asbestos in industrial countries have led to its exportation to developing countries, where it is not controlled. In Africa, Asia, and Latin America, asbestos is used mainly not for insulation but in asbestos cement, to build homes and produce drinking-water pipe. Workers installing asbestos in developing countries generally are not warned of potential health hazards to themselves or their families. Instead, the industry promotes asbestos as a safe material. Even if asbestos use is immediately banned in such countries as Zimbabwe, China, the Philippines, Thailand, and Brazil, the hazards from asbestos already in place will continue for decades. Furthermore, potential substitutes for asbestos must be monitored for possible similar health risks, and, if necessary, regulated.

Site Safety and Health Planning and Management

There is widespread agreement that planning is key to improved safety and health at the worksite. It is most effective to require initially a written plan by a project owner who monitors its use; such a plan should include accountability, coordination among subcontractors, and active worker participation. In the United States, written site safety plans have long been required and enforced by the Army Corps of Engineers, which reports an injury rate one fourth of that for the entire industry. Such plans are increasingly being used, at least by the largest multinational firms; as of 1998, contractor fees from the U.S. Department of Energy are based partly on execution of site safety plan requirements.

Education and Training

Safety training and worker and manager education have long been provided by some companies and some unions. The training programs cover such topics as rigging, trenching, stretching exercises, and substance-abuse recovery.

In the United States, many training programs include instruction about dangerous substances, which is mandated under OSHA's Hazard Communication Standard. The Building and Construction Trades Department, representing the 15 construction unions in the AFL-CIO, has produced training on safety and health in more than 10 one-hour modules, on topics ranging from confined spaces to ergonomics.

In Germany, the Gefahrstoff-Informationssystem der Berufsgenossenschaften der

Bauwirtschaft (GISBAU) program works with manufacturers to learn the content of all substances used on construction sites. The information is provided in a form to suit the differing needs of health staff, managers, and workers through training programs, in print, and on CD-ROM. The software, known as WINGIS, can be used on construction sites. It tells how to substitute for some risky substances and how to safely handle others.

ROLE OF THE CLINICIAN

The major safety and health need in construction remains the broad implementation of what may be considered best practices, particularly by smaller employers. Few of the risks and successful interventions that have been documented through increased research on construction safety and health are new discoveries. An obscure presentation in 1942 by M.F. Trice, an industrial hygienist for the State of North Carolina, listed most of the major risks known today and proposed solutions that are still accepted (26). For whatever reasons, occupational safety and health professionals shied away from the construction industry and its occupations.

The occupational health professions can have a major impact in the construction industry. To be effective, however, health professionals must first understand the organizational and sociologic aspects of construction work, so that effective health care delivery systems can be developed. Second, health professionals must understand the risks that workers face on the job. And third, health professionals must develop protocols and programs appropriate to the needs of construction workers and their families.

Delivery Systems

There are numerous opportunities to provide improved health programs for construction workers through health and welfare plans, workers' compensation carriers, employers, and unions. Structured care systems

are needed that are based on close cooperation among clinicians and experts in physical therapy, occupational hygiene, and safety engineering.

Few employers provide their own medical staff. Therefore, existing community or academic occupational health clinics generally have a large volume of building trades activities, seeing individual workers who have been referred or self-referred and investigating special problems in the worker population. This pattern, however, does not reflect industrywide health care use. Because of the episodic nature of construction employment and the high cost of health care, most construction workers do not have continuous medical care or a long-term relationship with a medical provider who knows them—except, perhaps, where there is universal health care or national health insurance.

A partial exception in the United States has been in the unionized sector, which since the late 1940s has provided health insurance through health and welfare funds that are jointly trusteed (with employers). There are about 750 such funds of varying size in the United States, most of them local. To address episodic employment, the health and welfare funds have established hour banks in which workers can accumulate hours worked to qualify for coverage. Workers can thereby maintain group coverage through as much as 3 to 6 months of unemployment by drawing on their bank reserves. But this system has its limitations. With erratic employment, some workers still are not able to build up reserve hours. And, even with good coverage, medical care utilization has been poor.

Despite this spotty record, construction may be unique in the strong incentives it offers employers and workers to support preventive health care. Although workers' compensation costs in the industry in the United States have leveled off and are being reduced, in some cases, through state-approved collective bargaining, construction premiums remain excessive. For instance, the workers' compensation premium for ironworkers in 1998 averaged $37.64 per $100 of payroll, and the rate for general carpenters was $18.08 (27); the rates vary enormously, depending on trade, jurisdiction, and type of work done. The high costs mean that most safety and health programs can have positive economic impacts.

Preventive Services. Targeted preventive services are needed in health insurance plans that cover construction workers and their families. These should include identification of each patient's occupation, and then discussion with the patient regarding precautions to take around the hazardous substances most likely to be encountered in that trade.

Development of Protocols. Few occupational medicine protocols have been developed and validated for use for construction groups, even for preventive medicine generally. Issues include whether there is a role for chest radiographs in periodic preventive medical examinations and whether liver function tests have a useful role in predicting fitness for work in hazardous waste cleanup.

Targeted Medical Monitoring. Because of the difficulties of reliably establishing work histories and exposures, medical monitoring is especially important, particularly in hazardous waste cleanup and lead abatement. Only a few types of exposures require medical monitoring in the United States, whereas German law specifies medical monitoring for many potentially toxic substances in the work setting. In Germany, checkups are required before work begins and at specified intervals. For some known carcinogens, checkups are required at regular intervals even after exposure ends. The preemployment checkup provides a baseline for future examinations. Information from the checkups that can be used to improve workplace conditions is given to employers. Employers are responsible for continuing the medical monitoring according to a schedule begun with the previous employer. However, if a worker is unemployed for longer than the scheduled interval for checkups, the examination schedule usually begins over again

with a baseline examination at the time of next employment.

Screening for Disease. Although there is a long history of screening programs for asbestos-related diseases, relatively little is known about the prevalence or incidence of most musculoskeletal disorders, hearing loss, dermatitis, silicosis, or other chronic diseases. As the first and second waves of asbestos disease end, it will also be useful to study whether the overall incidence is in decline. (In industrial countries, the first wave of asbestos disease early in the 20th century was related to asbestos mining, milling, and manufacturing; the second wave accompanied end-product use.) A third wave of asbestos disease is beginning to appear as buildings and materials containing asbestos introduced during the second wave deteriorate or are demolished.

Determination of Disability and Support for Rehabilitation

Medical Panels. Better medical support to determine disability under workers' compensation and help workers return to work early are needed. Closed medical panels for disability determination can help make the system more responsive to the needs of disabled workers. The occupational physician can also help identify light-duty tasks that match a worker's level of disability. (See Chapter 12.)

Standards Setting. Occupational health physicians can help with the setting and implementation of health-related standards. In the United States, most OSHA standards are inadequate in terms of medical monitoring requirements for construction workers. Medical monitoring requirements in the OSHA Hazardous Waste Operations Standard, for instance, are vague. The OSHA Lead Exposure in Construction Standard has clear medical requirements for monitoring and actions, but it is impractical. For instance, a worker with a high blood lead level is supposed to be provided another job removed from the lead exposure and at the same pay. But if a project has been completed, the contractor no longer is required to provide a job. The worker then may have difficulty finding another job until the blood lead level decreases.

Future Research Needs

Issues such as the measurement of and consequences of toxic exposures are poorly understood, partly because of inadequate past research on work-related safety and health in construction. For health professionals interested in research, construction is largely unexplored terrain. Areas that have not been addressed are exposure characterization, epidemiology, health services, interventions, and policy.

Exposure Characterization. Differences among construction sites hamper efforts to predict exposures. In response to this challenge, researchers have been trying to develop an exposure assessment model for construction work (Fig. 42-5). The intent is to efficiently collect descriptive and quantitative data that can predict exposures before jobs begin and present the information in a form that workers can easily use on site.

Given the limits of existing technology, the best approach may be to develop estimates of the range of exposures for given tasks (e.g., rod tying, drywall sanding, welding), taking into account such factors as the substances used, duration of use, and ventilation. For welding, for instance, an estimate would consider the welding method, the welding rod used, and the materials welded, including any coatings. These estimates should include exposures via ingestion and skin absorption. Exposures worth special attention include noise, dusts, solvents, vibration, manual lifting, and work postures.

Epidemiology. Epidemiologic surveillance systems cover few workers in the United States, and estimates based on different data sets show major inconsistencies. Work-related illnesses are grossly underreported, for a mixture of reasons. There is a shortage of reliable data on the mortality and morbidity patterns of the various trades. Virtually no research has been done on patterns of disability for the trades. Descriptive studies are needed. So are longitudinal cohort studies of construction workers to determine the magnitude of many of these problems.

Health Services. No research has been done on the delivery of occupational or general medical services for construction workers in the United States, and little is known about construction workers' patterns of health care use. If one accepts that continuity of care is a key determinant of the effectiveness of medical care, the episodic nature of construction appears to be a major barrier.

Interventions. Studies are needed regarding four types of intervention. First, various approaches to the delivery of preventive services should be tested. Second, information is needed on a host of preventive measures, including training, certification of workers and contractors, and use of technologies, work practices, and personal protective

FIG. 42-5. Worker, spraying fibrous glass insulation on a building under construction, illustrates the difficulties of assessing and controlling construction-site hazards. (Photograph by Ken Light.)

equipment that may reduce exposures. Third, no studies have been performed on the best systems for monitoring the health of construction workers, such as exposure monitoring, medical monitoring, or the tracking of workers in this transient industry. Fourth, studies should be performed to reduce the sequelae of disability and to return injured workers to gainful employment.

Policy. Little consideration has been given to the economics of improved safety and health in construction. There have been no valid studies to examine the effects of disability on workers and their families. In addition, there have been no systematic attempts to characterize the construction industry's composition in a defined geographic area. Much more needs to be done to understand better what types of interventions work in an industry in which employers are mostly very small contracting companies, employment is transient, and worksites are temporary and constantly changing—and with different employers working together without adequate coordination.

Acknowledgment

The National Institute for Occupational Safety and Health (NIOSH) supports work by the Center to Protect Workers' Rights through grants CCU310982 and CCU312014. The contents of this chapter are the responsibility of the authors and do not necessarily represent the official views of NIOSH.

REFERENCES

1. Pollack ES, Keimig DG, eds. Counting injuries and illnesses in the workplace: proposals for a better system. Washington, DC: National Academy Press, 1987.
2. Culver C, Marshall M, Connolly C. Construction accidents: the workers' compensation data base 1985–1988. Office of Construction and Engineering, Occupational Safety and Health Administration, 1992:13.
3. The construction chart book: the U.S. construction industry and its workers, 2nd ed. Washington, DC: The Center to Protect Workers' Rights, 1998:31a. Available at: http://www.cpwr.com.
4. The construction chart book: the U.S. construction industry and its workers, 2nd ed. Washington, DC: The Center to Protect Workers' Rights, 1998:18a. Available at: http://www.cpwr.com.
5. Toscano G, Windau JA, Knestaut A. Work injuries and illnesses occurring to women. Compensation and Working Conditions 1998;3(2):16–22.
6. Robinson CF, Burnett CA. Mortality patterns of US female construction workers by race, 1979–1990. J Occup Med 1994;36:1228–1233.
7. Women in the construction workplace: providing equitable safety and health protection: a working paper of the Health and Safety of Women in Construction Workgroup [Mimeo, 27 pgs]. Washington, DC: Occupational Safety and Health Administration, U.S. Department of Labor, March 4, 1997.
8. Eisenberg S. We'll call you if we need you: experiences of women working in construction. Ithaca, NY: ILR Press, 1998.
9. Helander MG. Safety hazards and motivation for safe work in the construction industry. Int J Ind Economics 1991;8:205–223.
10. The construction chart book: the U.S. construction industry and its workers, 2nd ed. Washington, DC: The Center to Protect Workers' Rights, 1998:31c.
11. U.S. Bureau of Labor Statistics, ftp web site, ostb0377 (table R4).
12. Silverstein B, Kalat J. Work-related disorders of the back and upper extremity in Washington state, 1989–1996. Technical Report 40-1-1997. Olympia, WA: Washington State Department of Labor and Industries, 1998.
13. Holmström E. Musculoskeletal disorders in construction workers. Lund, Sweden: Lund University, Department of Physical Therapy, 1992.
14. Parkinson DK, DeVito A, Dattwyler RJ, Luft B, Kennedy JM. Lyme disease prevalence among construction workers on Long Island, New York. Washington, DC: The Center to Protect Workers' Rights, 1996.
15. Robinson C, Stern F, Halperin W et al. Assessment of mortality in the construction industry in the United States, 1984–1986. Am J Ind Med 1995;28:49–70.
16. Selikoff IJ, Churg J, Hammond EC. Asbestos exposure and neoplasia. JAMA 1964;188:22–26.
17. Stayner L, Dankovic D, Lemen R. Occupational exposure to chrysotile asbestos and cancer risk: a review of the amphibole hypothesis. Am J Public Health 1996;86:179–186.
18. Stayner L, Smith R, Bailer J, et al. Exposure-response analysis of risk of respiratory disease associated with occupational exposure to chrysotile asbestos. Occup Environ Med 1997;54:646–652.
19. Occupational exposure to asbestos: final rule 29 CFR 1910 et al. Occupational Safety and Health Administration, U.S. Department of Labor, Part II, Federal Register August 10, 1994;59(153, Part II):40964–41062.
20. Lofgren DJ. Silica exposure for concrete workers and masons. Appl Occup Environ Hyg 1993; 8:832–836.
21. Finklea J. Asphalt. In: JM Stellman, ed. Encyclopaedia of occupational health and safety, vol III, 4th

ed. Geneva: International Labor Office, 1998; Chapter 93:51.

22. Waller K, Osorio AM, Maizlish N, Royce S. Lead exposure in the construction industry: results from the California Occupational Lead Registry, 1987 through 1989. Am J Public Health 1992;82:1669–1671.

23. Neitzel R. An assessment of occupational noise exposures in four construction trades [MS thesis]. Seattle, WA: Department of Environmental Health, University of Washington, 1998.

24. David Weil, assistant professor of economics, School of Management, Boston University. Personal communication. February 26, 1999.

25. Ergonomics for the prevention of musculoskeletal disorders. Stockholm: Swedish National Board of Occupational Safety and Health (Arbetarskyddsstyrelsen), 1998:35.

26. Safeguarding health in construction work. Durham, NC: J Appl Occup Environ Hyg 1942;12:648–649. Available at: http://www.cpwr.com.

27. Programs cut into workers comp cost [based on figures from Marsh & McLennan Inc, Insurance Brokers]. Engineering News-Record 1998(Sep 28):32–33.

BIBLIOGRAPHY

Construction Safety Association of Ontario. Construction health and safety manual. 343 pages. Toronto: Construction Safety Association of Ontario, 1992.
This is the most complete, concise compendium of recommended safety and health practices published in English for construction. Guidelines provided include general information about responsibilities onsite, personal protective equipment, and ways to avoid back injuries, as well as trade-specific safety and health information. Versions are available for general building
trades (carpenters, drywallers, and others) and the mechanical trades.

Ringen K, Englund A, Welch L, Weeks JL, Seegal JL, eds. Construction safety and health. Occup Med: State of the Art Reviews 1995;10(2).
This publication presents a comprehensive discussion for health professionals of safety and health issues; it includes Swedish longitudinal data on construction worker mortality and cancer incidence.

Construction. In: JM Stellman, ed. Encyclopaedia of occupational health and safety, vol III, 4th ed. Geneva: International Labor Office, 1998; Chapter 93:51.
For the first time, the encyclopedia contains a chapter focusing on construction. The 52-page chapter was produced with input from experts in 10 countries and covers topics ranging from cement dermatitis to asphalt and gravel.

The construction chart book: the U.S. construction industry and its workers, 2nd ed. Washington, DC: The Center to Protect Workers' Rights, 1998.
This publication characterizes the industry through the use of 128 charts in five categories—industry summary, labor force characteristics, employment and income, education and training, and safety and health—based on a wide range of government and private-sector sources. The format is easy to use and the text explains how data were derived for each chart. (It is available for downloading at http://www.cpwr.com.)

Waitzman NJ, Smith KR. Unsound conditions: work-related hearing loss in construction, 1960–75. Washington, DC: The Center to Protect Workers' Rights, 1998.
This is the first comparative multivariate analysis of hearing test data from national surveys done in 1960–1975 (the most current data on adults). Blue collar workers were found generally to have more noise-induced hearing loss than white collar workers, and the rate of such hearing loss was markedly higher among construction workers compared with other blue collar workers.

Occupational Health: Recognizing and Preventing Work-Related Disease and Injury, 4th ed.
Edited by Barry S. Levy and David H. Wegman, Lippincott Williams & Wilkins, Philadelphia © 2000.

43 Health Care Workers

Jane Lipscomb and Bill Borwegen

Almost one of every 10 workers in the United States is a health care worker. Characterized as people committed to promoting health through treatment and care for the sick and injured, health care workers, ironically, confront perhaps a greater range of significant workplace hazards than workers in any other sector. In addition to exposure to airborne and bloodborne infectious agents, typical exposures include workplace assault, ergonomic hazards, toxic drugs and other chemicals, radiation, and work stress, often caused or exacerbated by inadequate staffing. For these reasons, health care workers often struggle to provide quality and compassionate care in an inherently dangerous work environment.

According to data from the Bureau of Labor Statistics (BLS), the number of injury and illness cases recorded among health services workers increased 130% between 1983 and 1993, while total employment in the category grew only 46% (1). In 1997, employers reported that 652,800 health care workers had an Occupational Safety and Health Administration (OSHA)–recordable injury or illness (1). Furthermore, in the health care work environment, unlike the situation in many other industries, workers are not the only ones who are affected when occupa-

tional safety and health threats are not adequately identified and addressed: patient care also deteriorates.

In 1989, the health care industry for the first time exceeded the injury and illness rate for all private industries, and it has maintained this dubious status ever since (Fig. 43-1). Possible explanations for the dramatic increase in reported injuries and illnesses in the health care industry beginning in 1986 are discussed later in the chapter and include health care restructuring, the related reduction in staffing levels, and a consequent increase in musculoskeletal injuries, assaults, and needlestick injuries. In 1996, when private sector injury and illness rates decreased 5% to a rate of 7.4 per 100 employees, hospital employers reported an injury and illness rate increase of almost 10%. For the first time, the rate (11.0 per 100 workers) eclipsed rates for workers employed in mining (5.4), manufacturing (10.6), and construction (9.9). The nursing home segment of the health care industry has consistently reported injury and illness rates significantly higher than those for the most hazardous industries—as high as 18.2 per 100 as recently as 1995 (1). Keep in mind that actual numbers and rates of injuries and illnesses experienced by health care workers are considerably higher than those reported, in part because of gross underreporting of needlestick injuries. It is estimated that 64% to 96% of the 600,000 to 1,000,000 needlesticks occurring each year in the United States go unreported (2,3).

Yet despite these reported high injury and

J. Lipscomb: School of Nursing, University of Maryland School of Nursing, Baltimore, Maryland 20201.

B. Borwegen: Service Employees International Union, AFL-CIO, Washington, DC 20005.

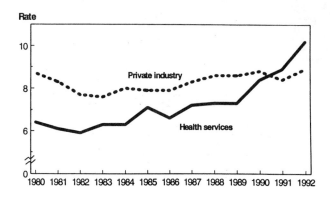

FIG. 43-1. Work Injuries and illnesses per 100 full-time workers in the health services industry and all private industry, 1980–1992. (U.S. Department of Labor, Bureau of Labor Statistics, 1994.)

illness rates, health care workers have received relatively little attention from occupational health and safety professionals compared with workers in industries traditionally viewed as hazardous. This chapter focuses on the leading hazards faced by health care workers and discusses these hazards within the historical, social, and cultural contexts in which health care is provided.

HISTORICAL PERSPECTIVE

In the United States, the practice of occupational health and safety dates from the late 1800s. However, as health care institutions focused on improved patient care, concerns about occupational health and safety in the health care industry were largely ignored or denied. As recently as the 1950s, there was still no consensus regarding the occupational risk of exposure to tuberculosis (TB), in part because of fears that young women would avoid nursing if they knew the risks involved and that liability issues might surface. It was not until TB declined significantly in the general population while remaining elevated among medical workers that it was fully recognized as an occupational hazard (4).

The health care industry has been slow to recognize and respond to the epidemic of injuries and illnesses facing its own workforce. Explanations given for this lack of attention or concern include the following:

1. *A false perception that the industry is self-regulated:* The Joint Commission on the

Accreditation of Health Care Organizations (JCAHO), developed in response to the Medicare Act of 1965 as a means of compliance with Medicare and Medicaid program reimbursement requirements, conducts preannounced inspections of most hospitals every 3 years. Accreditation is primarily directed at assessing the quality of patient services. As a consequence, JCAHO inspectors are poorly trained in occupational health and safety and direct little attention toward workplace exposures and hazards during inspections.

2. *The idea that "an industry that employs mainly females must be a safe industry":* Seventy-six percent of hospital workers, 83% of nursing home workers, and 93% of home care workers are female.

3. *Focus on curative rather than preventive medicine:* Health care institutions are more concerned and better prepared to respond to the more dramatic aspects of curative medicine rather than to preventive medicine and public health, including occupational health and safety.

4. *A low unionization rate within the health care sector:* Health care workers, compared with workers in more heavily unionized industries, have little voice and power to effectively negotiate for and improve workplace health and safety conditions. It was not until 1974, almost 30 years after the passage of the National Labor Relations Act, that private-sector health

care workers finally joined other private-sector workers with coverage under this Act. Today, only 13.5% of hospital workers, the most unionized segment of the health care industry, are covered by a collective bargaining agreement.

5. *Lack of attention by governmental agencies responsible for health and safety:* Little research has been conducted and few governmental standards have been issued for the hazards causing most injuries to health care workers. OSHA continues to operate largely in an industrial sector mindset, failing to recognize and respond to a major shift in the economy from the manufacturing to the service sector. OSHA continues to conduct 85% of inspections within the manufacturing and construction sectors, whereas 58% of all injuries and illnesses now occur in the service sector, including health care. When OSHA does inspect health care facilities, such as under its 1997 nursing home initiative, inspectors receive little, if any, training on how to do inspections in the health care sector. In addition, OSHA gives inadequate attention to cumulative trauma, including back injuries, because it has no ergonomics standard. The National Institute for Occupational Safety and Health (NIOSH) has earmarked funds for sector-specific research programs for miners, construction workers, and agricultural workers, but not for health care workers.

CHARACTERISTICS OF THE WORKFORCE AND WORK SETTING

The health care sector employs more than 9 million health care workers, having grown by 3 million workers between 1980 and 1997. BLS projections for the 10 fastest growing occupations in 1996–2006 include six health-related occupations—personal and home care aides, physical and corrective therapy aides, home health aides, medical assistants, physical therapists, and occupational therapists—all projected to increase by 70% to 85% during this period (5). As is reflected in

these projections, health care workers are increasingly being employed outside the acute care or hospital setting. The term *health care workers* includes both health care professionals and support staff members working in hospitals, outpatient clinics, nursing homes, home health care settings, medical laboratories, dental offices, and veterinary settings.

Health care facility operations that have a health impact on workers include all patient care and treatment areas, sterilization areas, pharmacies, support laboratories, housekeeping, maintenance, and waste disposal areas. It should be noted that the generation and disposal of biologic, chemical, and radiologic wastes pose risks to the communities surrounding health care facilities and beyond, in particular if these facilities incinerate their waste on site. The widespread use and resulting incineration of plastics containing chlorine compounds (e.g., polyvinyl chloride) have the potential to create and release into the atmosphere dioxins, which are among the most toxic substances known (see Chapter 3).

The degree of health care worker exposure and the risk of various hazards vary by health care setting. For example, nursing home workers are at increased risk for back injuries compared with hospital-based nursing staff—not only because of the frequency of patient handling activities and the load associated with them, but generally because nursing homes have fewer nursing staff or lifting teams to assist with hazardous lifts.

HAZARDS OF HEALTH CARE WORK

Hazards facing health care workers include biologic hazards associated with airborne and bloodborne exposures to infectious agents (see Chapter 20); chemical hazards—especially those found in hospitals, including waste anesthetic and sterilant gases, antineoplastic drugs and other therapeutic agents, mercury, and industrial-strength disinfectants and cleaning compounds (see Chapters 15 and 16); physical hazards, including ioniz-

TABLE 43-1. *Selected hazards, health effects, and control strategies in health care*

Hazards	Health effects	Control strategies
Biologic		
Viral (hepatitis B virus, hepatitis C virus)	Acute febrile illness, liver disease, death	Safer needle devices, hepatitis B vaccine
Bacteria (*Mycobacterium tuberculosis*)	TB infection, TB illness, multiple drug resistance, death	Isolation of suspect patients, respirators, ultraviolet light, negative pressure rooms
Natural rubber latex proteins (and rubber chemical additives)	Range from type IV delayed hypersensitivity to rubber additives to type I immunologic response, anaphylactic shock, death	Substitution with low latex protein powderless gloves or nonlatex gloves and supplies
Chemical		
Ethylene oxide	Peripheral neuropathy, cancer, reproductive effects	Substitution, enclosed systems, aeration rooms
Formaldehyde	Allergy, nasal cancer	Substitution, local ventilation
Glutaradehyde	Mucous membrane irritation, sensitization, reproductive effects	Substitution, local ventilation
Antineoplastic drugs	Cancer, mutagenicity, reproductive effects	Class 1 ventilation hoods, isolation of patient excreta
Waste anesthetic gases	Hepatic toxicity, neurologic effects, reproductive effects	Scavenging systems, isolation of off-gassing patients
Mercury	Neurologic effects, birth defects	Substitution with electronic thermometers
Physical		
Patient handling	Back pain, injury	Patient handling devices, lifting teams, training
Static postures	Musculoskeletal pain and injury	Rest breaks, exercise, support hose and shoes
Ionizing radiation	Cancer, reproductive effects	Isolation of patients, shielding and maintenance of equipment
Lasers	Eye and skin burns, inhalation of toxic chemical and pathogens, fires	Local exhaust ventilation, equipment maintenance, respirators and face shield
Physical assault	Traumatic injuries, death	Alarm systems, security personnel, training
Psychosocial/Organizational		
Violence threat and physical assault	Traumatic injury, death, post-traumatic stress disorder	Training, post-assault debriefing
Restructuring	Mental health disorders, exacerbation of musculoskeletal injuries, traumatic injuries, burn out	Acuity-based staffing, employee involvement in restructuring activities
Work stress (other than above)	Mental health disorders, burn out	Stress prevention and management programs
Shift work	Gastrointestinal disorders, Sleep disorders	Forward, stable and predictable shift rotation

ing and non-ionizing radiation (see Chapters 17 and 19); safety and ergonomic hazards that can lead to a variety of acute and chronic musculoskeletal problems (see Chapters 8, 9, and 26); violence (see Chapter 24); psychosocial and organizational factors including psychologic stress (see Chapter 21) and shift work (see Chapter 22); and the many health consequences associated with changes in the organization and financing of health care (Table 43-1).

Musculoskeletal Injuries

Musculoskeletal injuries rank second among all work-related injuries, with the greatest number occurring among health care workers. Exposures include the requirements to lift, pull, slide, turn, and transfer patients, move equipment, and stand for long hours (see Chapters 8, 29, and 26). Among all occupations, hospital and nursing home workers experience the highest number of occupational injuries and illnesses involving lost workdays due to back injuries. Nurse's aides report a greater percentage of their injuries as back injuries than workers in any other occupation. A 3-year review of BLS annual survey data indicates that nursing personal care facilities have an occupational musculoskeletal injury and illness rate of 4.62 per 100 workers per year, the highest among all three-digit Standardized Industrial Classification codes (1).

Back injuries continue to take a huge economic and personal toll within the health care sector. The nursing home industry alone spends more than $1 billion each year in workers' compensation premiums, even though the implementation of engineering and administrative controls (e.g., safe staffing levels, lifting teams, use of newer mechanical patient handling devices) has been shown to reduce both injury rates and workers' compensation premiums dramatically. For example, a Maine nursing home, with 245 residents and 270 workers, had 573 lost workdays due to back injuries in 1991 and paid a workers' compensation premium of $1.5 million. After the employer purchased 12 mechanical patient lifts for a total cost of $60,000 and implemented a policy banning the lifting of residents unless more than one worker was present to assist, its lost workdays decreased to 12 in 1996, and its workers' compensation premium dropped by 50%.

By contrast, one study of a large nursing home found that nursing assistants experienced four episodes of low back pain on average in a 3-year period, but that assistive devices (a mechanical lift and transfer belt) were used for fewer than 2% of all transfers. Patient safety and comfort, lack of accessibility, physical stresses associated with the devices, lack of skill, increased transfer time, and lack of staffing were some of the reasons cited for not using these devices. Ironically, when patients and residents were surveyed it was found that they actually preferred mechanical lifts, which made them feel more secure (6).

Similarly, the actual use of lifting teams in hospitals or other health care settings is limited despite published studies demonstrating the effectiveness of this control strategy. In a large, acute-care public hospital in Northern California, annual lost-time injuries decreased from 16 to 1 one year after deployment of a lifting team, with a savings of $144,000 (7).

Musculoskeletal injuries in the health care industry, other than those affecting the back, are less well understood. Laboratory workers are at increased risk for cumulative trauma disorders of the hand and wrist related to repetitive work, such as pipetting. Operating room workers who must maintain static postures for long periods and those involved in overhead work (e.g., holding instruments overhead during lengthy surgeries) experience neck and shoulder pain and injury.

Workplace Violence

The health care sector also leads all other industry sectors in incidence of nonfatal workplace assaults. In 1993, 38% of all nonfatal assaults against workers resulting in lost workdays in the United States occurred in the health care sector. Among all nonfatal workplace assaults resulting in lost workdays, nursing aides and orderlies had the highest proportion—27%, compared with 7% for police and guards. In 1994, BLS reported the rate of nonfatal assaults among workers in "nursing and personal care facilities" as 38 per 10,000, compared with a rate of 3 per 10,000 in the private sector as a whole—an almost 13-fold difference (1). Among these assault victims, 30% were government employees, even though they make up only 18% of the workforce. Across all industries, in 45% of reported cases of nonfatal workplace assault, the perpetrator of the assault was a health care patient. In contrast, only 8% of these injuries were perpetrated by a coworker; yet, much more media and employer attention is disproportionately directed to coworker violence (see Chapter 24) (1).

In a Washington State psychiatric facility, 73% of staff surveyed had reported at least a minor injury related to an assault by a patient during the past year. Only 43% of those reporting moderate, severe, or disabling injuries related to such assaults had filed for workers' compensation. The survey found an assault incidence rate of 437 per 100 employees per year, whereas the hospital incident reports indicated a rate of only 35 per 100 (8).

Emergency department personnel face a significant risk of fatal injuries from assaults by patients or their families. Weapon-carrying in emergency departments, reported to be as high as 25% in major urban hospitals, creates the opportunity for severe or fatal injuries (9). Although mental health and emergency departments have been the focus of attention and research on the subject, no department within a health care setting is immune from workplace violence. Consequently, all departments should have violence prevention programs.

A number of environmental risk factors have been associated with assaults by patients, including poor security, inadequate staffing patterns, time of high activity, and containment activities. Inadequate training and a lack of clear policies and procedures have also been identified as contributing to the incidence of assaults.

In 1993, after the murder of a state employee, Cal OSHA (the California OSHA program) published the first set of OSHA guidelines describing the components of a comprehensive workplace violence prevention program. After several years of pressure from a multiunion task force on workplace violence, federal OSHA followed California's lead in 1996, issuing a similar set of guidelines, entitled *Guidelines for Preventing Workplace Violence for Health Care and Social Service Workers*. As described by OSHA, the essential elements in developing a violence prevention program mirror those for any comprehensive safety and health program and include the following:

1. *Management commitment* must be evident in the form of high-level management involvement and support for a written workplace violence prevention policy and its implementation.
2. Meaningful *employee involvement* in policy development, joint management-worker violence prevention committees, postassault counseling and debriefing, and follow-up are all critical program components.
3. *Worksite analysis* includes regular walk-through surveys of all patient care areas and the collection and review of all reports of worker assault. A successful job hazard analysis must include strategies and policies for encouraging the reporting of all incidents of workplace violence, including verbal threats that do not result in physical injury.
4. *Hazard prevention and control* includes the installation and maintenance of alarm systems in high-risk areas. It may also include the training and posting of security personnel in emergency departments. Adequate staffing is an essential hazard prevention measure, as is adequate lighting and control of access to staff offices and secluded work areas.
5. *Training and education* must include preplacement and periodic, educationally-appropriate training regarding the risk factors for violence in the health care environment and control measures available to prevent violent incidents. Training should also include skills in aggressive behavior identification and management, especially for staff working in the mental health and emergency departments.

Needlestick Injuries

The most prevalent, least reported, and largely preventable serious risk health care workers face comes from the continuing use of inherently dangerous conventional needles. Such unsafe needles transmit blood-borne infections to health care workers employed in a wide variety of occupations.

This is an example of an "active" safety mechanism, requiring the healthcare worker to pull the sheath over the needle after use.

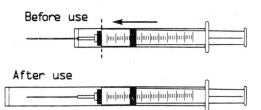

This is an example of a "passive" safety mechanism, where the needle retracts automatically into the barrel when the plunger is depressed after use.

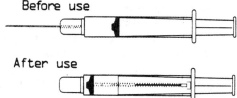

FIG. 43-2. Active and passive safety features on safer needle-bearing devices.

Elimination of unnecessary sharps and the use of safer needles can dramatically reduce needlestick injuries (see Fig. 20-2 in Chapter 20) (10,11).

Safer needle devices have integrated safety features built into the product that prevent needlestick injuries. The term *safer needle device* is broad and includes many different devices, from those that have a protective shield over the needle to those that do not use needles at all. Needles with integrated safety features are categorized as passive or active. Passive devices offer the greatest protection because the safety feature is automatically triggered after use, without the need for health care workers to take any additional steps. An example of a passive device is a spring-loaded retractable syringe or self-blunting blood collection device (Fig. 43-2). An example of an active safety mechanism is an employee-activated self-sheathing needle. Use of conventional needles in the health care environment today has been compared with the use of unguarded machinery decades ago in the industrial workplace.

Despite this existing and widely available technology, it is estimated that 600,000 to 1,000,000 health care workers continue to get stuck by these older, obsolete, conventional needles each year. Once stuck, not only do health care workers run the risk of acquiring a number of serious infectious diseases, but they and their families must deal with the emotional stress of waiting a minimum of 6 months to find out, through antibody testing, whether a particular needlestick injury will

cause a potentially life-threatening disease. Health care workers must also avoid exposing others to their body fluids, including practicing "safe sex" during this period. In addition, those who take prophylactic drugs in hope of preventing human immunodeficiency virus (HIV) infection, may have serious side effects.

As of 1998, the U.S. Food and Drug Administration had approved more than 250 safer needle-bearing products. The Centers for Disease Control and Prevention (CDC) reported in 1997 that an eight-hospital study demonstrated that, during such high-risk procedures as drawing blood, needlesticks could be reduced up to 76% with the use of safer needles (10,11); nevertheless, fewer than 10% of the needles being purchased in 1998 by health care employers had these integrated safety features.

After a needlestick injury, the risk of developing occupationally acquired hepatitis B virus (HBV) infection for the nonimmune health care worker ranges from 2% to 40%, depending on the hepatitis B e antigen status of the source patient (12). The risk of transmission from a positive source for hepatitis C is between 3% and 10% (12), and the average risk of transmission of HIV is 0.3% (13). However, the risk of transmission increases if the injury is caused by a device visibly contaminated with blood, if the device is used to puncture the vascular system, or if the stick causes a deep injury. All of these diseases are associated with significant morbidity and mortality and only hepatitis B can be prevented by vaccine. Health care workers, laundry workers, and housekeeping workers are all too often engaged in duties that create an environment for these high-risk needlestick injuries.

The above data translate into the estimate that each year more than 1,000 health care workers contract a serious infection, such as HBV, hepatitis C virus (HCV), or HIV infection, from an occupational needlestick injury. HCV infects 560 to 1,120 health care workers in the United States each year (14), with 85% becoming chronic carriers. It is thought that one health care worker per week will eventually die from occupational exposure to HIV (2). Some groups of workers, such as phlebotomists, experience a particularly high risk of needlesticks and subsequent infections. According to a 1994 survey of this group of workers, 24% of health care workers who drew blood were stuck by a needle in the previous year.

Hepatitis B can be prevented by the administration of a vaccine that is now widely used. However, before the promulgation of the 1991 OSHA Bloodborne Pathogen Standard, which mandated free availability of this vaccine to potentially exposed workers, hepatitis B was the leading needlestick-transmitted bloodborne infectious disease affecting health care workers (15).

Latex Allergy

Despite the success of the Bloodborne Pathogen Standard and related guidance from CDC and professional associations, a very significant health problem has emerged that can be attributed, in part, to the increased use of examination and surgical gloves required by this standard. An epidemic of latex allergy is now affecting health care workers and others exposed. The prevalence of latex allergy among health care workers is estimated to be between 5% and 12%, with atopic workers at even greater risk (16–18). Manifestations of this exposure range from type IV delayed hypersensitivity to rubber additives, which manifests as contact dermatitis, to type I immunologic responses to residual proteins in gloves and other medical devices. In 1997, NIOSH made recommendations for prevention of allergic reactions to natural rubber latex in the workplace and for controlling exposure (available at: http://www.cdc.gov/niosh). Specifically, NIOSH recommended the use of latex gloves only when protection from infectious agents is needed for food-handling and other occupational tasks. Most importantly, NIOSH recommended the use of powderless, low-protein latex gloves for protection from bloodborne pathogens in health care and other settings. NIOSH is currently evaluating the effectiveness of these recommendations in a series of clinical intervention trials. Some institutions have reduced or eliminated the use of even powderless latex gloves, instead using gloves made from vinyl, nitrate, and other nonlatex materials.

Chemical Hazards

Health care workers are exposed to a wide range of chemical disinfectants, anesthetic waste gases, and chemotherapetic drugs that are known to cause human health effects, as well as others for which no or inadequate testing has been conducted. NIOSH estimates that the average hospital contains 300 chemicals, twice the number of the average manufacturing facility. Among disinfectants, formaldehyde is a probable human carcinogen and has been linked to occupational asthma in the hospital setting. A more recent cold sterilant substitute, glutaraldehyde, has been associated with mucous membrane irritation, allergic contact dermatitis, and adverse reproductive effects. Ethylene oxide, a gas sterilant, is a neurotoxin, a carcinogen, and a reproductive health hazard. Thousands of health care workers were exposed to harmful levels of this gas before the 1984 OSHA standard for ethylene oxide was issued. This chemical continues to be of concern to central supply hospital workers because of leaks from distribution lines, especially when gas cylinders are being changed.

Anesthetic agents, used in large amounts in hospitals, pose a threat to health care workers when operating room scavenging systems are poorly maintained. Health care workers are also exposed when patients are transferred to the recovery room and exhale anesthesia gases. Specially designed nonrecirculating general ventilation systems with adequate room air exchanges are necessary in these areas.

Therapeutic agents associated with human health effects among workers who handle and administer them include antineoplastic agents, which are known to cause reproductive effects and cancer. Also associated with the handling of antineoplastic agents are acute and chronic effects associated with the therapeutic effects of these drugs. Safe-handling guidelines were published in the mid-1980s by the National Institutes of Health, and later by OSHA, to control dermal and inhalation exposures associated with the mixing and administration of these drugs. Use of proper glove material is critical, because most of these substances easily penetrate regular latex gloves. A survey conducted in the late 1980s found variations in staff compliance with the OSHA guidelines; workers who mixed the drugs were most protected, and those who handled patient excretions were least protected. Aerosolized medications pose unique threats because of how these drugs are administered; ribavirin is a potential human teratogen.

Organization of Work

Organization of work refers to management and supervisory practices as well as production processes and their influence on the way work is performed. Perhaps no other single factor influences worker injury and illness rates more than the manner in which work is organized and staffing decisions are made. Few industries in the United States have undergone more sweeping changes over the past decade than the health care industry.

Macro-level changes in the organization of the work of health care delivery have included organizational mergers, downsizing, changes in employment arrangements (e.g., contract work), job restructuring and redesign, and changes in worker-management relations. Many of these changes have accompanied the emergence of managed care, the priority given to cost containment, and conversions to for-profit health care institutions.

The widespread concern regarding adequacy of nursing staffing levels in health care facilities led to a 1993 Congressional mandate that the Institute of Medicine of the National Academy of Sciences study the adequacy of nurse staffing in hospitals and nursing homes. The Institute found a paucity of data linking staffing patterns to nurses' health and well-being, although there were some indications that reductions in staffing lead to increased back injury rates among nurses (19).

Since that time, additional information on the subject has become available. Between 1981 and 1993, total hospital employment grew steadily, but the number of nursing personnel declined by 7%, while adjusting for the severity of the illness of the patients under their care. The decline in nursing personnel was 20% to 27% in Massachusetts, New York, and California, all of which had high managed care penetration rates (19).

Shifts in nursing staffing patterns and greater severity of inpatient illnesses are increasing the risks to health care workers and the nosocomial infection risks to patients in hospitals in the United States. The CDC estimates that 90,000 patients each year die from nosocomial infections—more than twice the number of people who die from motor vehicle accidents. In 1996, two thirds of 5,000 registered nurses polled reported an increase in the number of patients assigned to them; three fourths also reported increased severity of illnesses in patients under their care. Staffing reductions, combined with the replacement of registered nurses by unlicensed personnel, has led to work speed-up for many nurses. Two fifths of responding nurses said they would not want a family member to receive care at the institution at which they worked (21). The Minnesota Nurses Association examined OSHA-200 worker injury and illness logs at 86 Minnesota hospitals over a 4-year period; it found a 65% increase in injuries and illness reported by nurses, while nursing staff at the same time was reduced by 9%. Needlestick and back injuries contributed most to the increase in reportable incidents (22).

HOME HEALTH CARE: A SETTING DESERVING SPECIAL ATTENTION

One outcome of the restructuring of the health care industry described in the previous section has been shorter hospital stays and a resulting increase in the type and volume of health care provided in the home. Home health care workers today provide services ranging from care of acutely ill medical and surgical patients to care of the chronically-ill elderly and disabled population. As projected in the BLS figures referenced earlier, the industry that supplies nursing and personal care in patients' homes is adding jobs faster than any other segment of the U.S. economy, having doubled its workforce from 250,000 employees in 1989 to 500,000 in 1994. It probably will employ 1.25 million workers by the year 2005. A consequence of this growth and of providing care in an environment within which the worker has little or no control is injury rates 50% higher than in hospitals and 70% higher than the national private-sector rate. In 1994, most of the industry's 18,800 injury cases involved home health care nurses and aides; these injuries resulted on average in 1 to 2 weeks of absence from work. The comparatively high injury rate in the home health industry reflects, in part, the relatively large number of highway-related injuries sustained by home care personnel while making house calls. In fact, home care workers have higher rates of motor vehicle injuries than truck drivers do. When highway-related injuries are excluded from industry totals, the

home health care lost-worktime injury rate (398 cases per 10,000 workers) still exceeds the corresponding rate in hospitals by almost 25% and the national rate by almost 50% (23).

Despite these trends, the home health care setting has rarely been the focus of research in the prevention of exposures to occupational hazards. Instead, the hazards and attendant risks of the traditional acute health care environment are usually simply extrapolated to health care work in the home environment. Although these health care workers face many of the same hazards as their counterparts in the hospital environment, there are additional factors, including unpredictability and less control of the physical work environment; lack of policies, equipment, and training specific to this environment; and the differing social roles of caregiver and client. These factors may intensify the occupational risk for illness or injury in the home care setting. For example, hand-washing and waste disposal facilities, especially for needles and other sharps, may be inadequate or unavailable, and coworkers may not be available to assist with patient handling or in emergency situations.

Threatening to compound these differences are legal precedents that provide the home health care environment with less protection than that afforded in an institutional setting. For example, in 1993, the U.S. Court of Appeals for the 7th Circuit upheld the home health care industry's position that the employer should not be held responsible for ensuring the implementation of the specific provisions of the OSHA Bloodborne Pathogen Standard regarding the use of protective clothing and equipment.

LEGISLATIVE AND REGULATORY ACTIONS TO PROTECT HEALTH CARE WORKERS

Legislation, regulations, and even voluntary guidelines to protect health care workers have been slow in coming and inadequate in their coverage. In 1958, the American Medical Association and American Hospital Association issued a joint statement in support of worker health programs in hospitals. In 1977, NIOSH published criteria for effective hospital occupational health programs. In 1982, the CDC published the *Guideline for Infection Control in Hospital Personnel,* which focused on infections transmitted between patient care personnel and patients, not exclusively on health care workers' risks of contracting infectious diseases. CDC guidelines for blood and body fluid precautions (1982) and universal precautions (1987) were published to provide guidance to health care workers. In 1984, OSHA promulgated its first health care worker–specific standard, covering the use of ethylene oxide. In 1987, the U.S. Department of Labor and the U.S. Department of Health and Human Services issued a joint advisory notice entitled *Protection Against Occupational Exposure to HBV and HIV.* In 1988, NIOSH published comprehensive guidelines for protecting the safety and health of health care workers. In 1991, OSHA promulgated the Bloodborne Pathogen Standard, which required the observance of universal precautions, offering of the hepatitis B vaccine by employers, and evaluation of engineering controls to protect workers against the health hazards related to blood-borne pathogens. OSHA was scheduled to publish a final tuberculosis standard in 1999. Health care workers continue to await OSHA actions on an ergonomics standard, a workplace violence standard, and a standard requiring the use of safer needle devices.

REFERENCES

1. Bureau of Labor Statistics. U.S. Department of Labor, Surveys of occupational injuries and illnesses, Washington, DC: U.S. Department of Labor, 1995–1999.
2. Ippolito G, Pura V, Petrosillo N, et al. Prevention, management and chemoprophylaxis of occupational exposure to HIV. Charlottesville, VA: International Health Care Workers Safety Center, University of Virginia, 1997.
3. Cardo D, Culver D, Srivastava P, Campbell S. Results from the first phase of the national surveillance system for hospital health care workers (NaSH) [Abstract]. Am J Infect Control 1997;25:131.
4. Sepkowitz KA. Tuberculosis and health care workers: an historical perspective. Ann Intern Med 1994;120:71–79.
5. Bureau of Labor Statistics. U.S. Department of Labor, Survey of Occupational Injuries and Illnesses. 1996–2006 Employment Projections 1997, USDL publication no. 97-429. Washington, DC: U.S. Department of Labor
6. Garg A, Owen BD. Reducing back stress to nursing personnel: an ergonomic intervention in a nursing home. Ergonomics 1992;35:1353–1375.
7. Charney W. The lift team method for reducing back injuries: a 10-hospital study. Am Assn Occup Health Nurses J 45:300–304.
8. Bensley L, Nelson N, Kaufman J, Silverstein B, Kalat J, Walker J. Injuries due to assaults on psychiatric hospital employees in Washington State. Am J Ind Med 1997;31:92–99.
9. Wassenberger J, Ordog GH, Kolodny M, Allen K. Violence in a community emergency room. Arch Emerg Med 1989;6:266–269.
10. Centers for Disease Control and Prevention. Evaluation of safety devices for preventing percutaneous injuries among health-care workers during phlebotomy procedures—Minneapolis-St. Paul, New York City, and San Francisco, 1993–1995. MMWR Morb Mortal Wkly Rep 1997;46:21–25.
11. Centers for Disease Control and Prevention. Evaluation of blunt suture needles in preventing percutaneous injuries among health-care workers during gynecologic surgical procedures—New York City, March 1993–June 1994. MMWR Morb Mortal Wkly Rep 1997;46:25–29.
12. Gerberding JL. Prophylaxis for occupational exposures to bloodborne viruses. N Engl J Med 1995: 332:444–455.
13. Centers for Disease Control and Prevention. Recommendations for preventing transmission of human immunodeficiency virus and hepatitis B virus to patients during exposure-prone invasive procedures. MMWR Morb Mortal Wkly Rep 1991;40 (RR-8):1–9.
14. Alter MJ. The detection, transmission and outcome of hepatitis C virus infection. Infectious Agents and Disease 1993;2:155–166.
15. Gibas A, Blewett DR, Schoenfeld DA, Dienstag JL. Prevalence and incidence of viral hepatitis in health workers in the prehepatitis B vaccination era. Am J Epidemiol 1992;136:603–610.
16. Turjanmaa K. Incidence of immediate allergy to latex gloves in hospital personnel. Contact Dermatitis 1987;17:270–275.
17. Arellano R, Bradley J, Sussman G. Prevalence of latex sensitization among hospital physicians occupationally exposed to latex gloves. Anesthesiology 1992;77:905–908.
18. Liss GM, Sussman GL. Latex sensitization: occupational versus general population prevalence ratio. Am J Ind Med 1999;35:196–200.
19. Institute of Medicine. Nursing staff in hospitals and nursing homes: is it adequate? Washington, DC: National Academy Press, 1996.
20. Aiken LH, Sochalski J, Anderson GF. Downsizing

the hospital nursing workforce. Health Aff (Millwood) 1996;15:82–92.

21. Shindu-Rothchild J, Berry J, Long-Middleton E. Where have all the nurses gone? Final results of our patient care survey. Am J Nurs 1996;96:25–39.

22. Shogren E. Restructuring may be hazardous to your health. Am J Nurs 1996;96:64–66.

23. Bureau of Labor Statistics. U.S. Department of Labor, Survey of Occupational Injuries and Illnesses. Injuries to caregivers working in patients homes 1997. Summary 97-4. Washington, DC: U.S. Department of Labor, 1997.

BIBLIOGRAPHY

American College of Occupational and Environmental Medicine (ACOEM) Section on Medical Center Occupational Health. Guidelines for Employee Health Services in Health Care Facilities. 1998. Available at: http://www.occenvmed.net/ehsg/.
This set of guidelines focuses on the overall management of an employee health program in the health care setting. The Web site provides links to government and professional sites addressing the field.

American Nurses Association (ANA) Web site. Available at: www.nursingworld.org.dlwa/osh/.
This Web site offers information on cutting-edge issues of primary concern to American nurses. It contains ANA informational brochures on such topics as latex allergy, workplace violence, and pollution prevention in health care. It provides links to relevant Web sites.

Charney W. Ed. Essentials of modern hospital safety, vol 4. Boca Raton, FL: Lewis Publishers, CRC Press, Inc, 1999.
This text provides an in-depth discussion of current hazards facing hospital workers and was authored by experts in the field. The book was written by public health professionals with an activist approach to health and safety. The text offers survey forms and other useful tools for eliminating or reducing workplace hazards.

NIOSH. Guidelines for protecting the safety and health of health care workers. Washington, DC: U.S. Department of Health and Human Services, 1988.
These comprehensive guidelines provided a very useful resource at their time of publication. NIOSH continues to receive numerous requests for copies of the guidelines despite the fact that they are outdated owing to the dramatic changes in the field over the past 10 to 15 years. NIOSH has recently issued a request for information on how to proceed to update the 1988 document. See their Web site for this notice—available at: http://www.cdc.gov/niosh/hcw-fr.html.

U.S. Department of Labor, Occupational Safety and Health Administration. Guidelines for Preventing Workplace Violence for Health Care and Social Service Workers. 1996. Available at: http://www.osha.gov.
This document provides a succinct discussion of the background of the problem and a detailed description of the critical elements of a violence prevention program. The documents provides excellent examples of how to respond to these performance-based guidelines, including a staff assault survey, checklists, and forms.

Occupational Health: Recognizing and Preventing Work-Related Disease and Injury, 4th ed.
Edited by Barry S. Levy and David H. Wegman, Lippincott Williams & Wilkins, Philadelphia © 2000.

APPENDICES

Appendix A: How to Research the Effects of Chemical Substances or Other Workplace Hazards

Will Forest, Charleen Kubota, Stephen Zoloth, David Michaels, and Jim Cone

STEP 1: SET PRIORITIES

It may be impractical to research every substance, physical hazard, or process to which your patients are exposed. A few common-sense guidelines can help focus your efforts on the exposures most likely to cause disease. The two main determinants of risk are the magnitude of exposure and the potency and nature of toxicity.

Consider the Magnitude of Exposure. The substances to which your patient has the most intensive, frequent, or prolonged exposures should be high priorities for investigation.

Consider Toxicologic Information. Your background reading will alert you to the toxicity of certain classes of chemicals; it should help you be selectively suspicious and direct

W. Forest: Occupational Health Branch, California Department of Health Services, Oakland California, 94612.

C. Kubota: University of California, Berkeley Public Health Library No. 7360 Berkeley, CA 94720-7360.

S. Zoloth: Hunter College, New York, NY 10021.

D. Michaels: Department of Community Health and Social Medicine, The City University of New York Medical School, New York, NY 10031.

J. Cone: Department of Medicine, University of California, San Francisco, Berkeley, California 94709, and Occupational Health Branch, California Department of Health Services, Oakland, California 94612.

your research accordingly. For example, pesticides, heavy metals, chlorinated organic compounds, polycyclic compounds, aromatic amines, anhydrides, glycol ethers, and gases should generally be given priority in your research, because they are often highly toxic. Even low-dose exposure to a carcinogen or a reproductive hazard may be cause for concern (see Chapters 16 and 31).

Consider Clinical-Toxicologic Correlations. If the onset of symptoms in your patient dates from the introduction of a new chemical at the workplace, that product certainly deserves further investigation.

Consider Epidemiologic Correlations. If your patient and coworkers share the same symptoms or diseases, discover what exposures they have in common.

Ask the Patient for Guidance. Workers are often well aware of the toxic substances or other hazards to which they are exposed. In many cases, workers have already obtained useful information about chemical hazards, from container labels, material safety data sheets (MSDSs), their unions, libraries, Internet searching, or other sources, and they will often share it with you. Furthermore, by observing health and disease patterns among themselves, workers are often able to identify clusters of work-related disease, most often for acute disease occurring shortly after exposure, but sometimes for chronic disease as well. Ask if your patients suspect that their health problems, or those of coworkers, are work related.

Alert Your Patient to Right-to-Know Laws and Regulations. Most workers in the private sector in the United States have the right, under the Occupational Safety and Health Administration (OSHA) Hazard Communication Standard, to obtain information from their employer about the chemicals with which they work (see Chapters 10 and 15). Many public-sector workers have similar rights under right-to-know laws in effect in 25 states. Under these regulations, employers must provide workers with information on all hazardous chemicals used in the workplace, through proper labeling of containers, distribution of MSDSs, and training programs. The MSDS for a product, generally prepared by its manufacturer, is useful because it lists the product's hazardous ingredients; it also discusses the product's toxicity and gives recommendations for safe use and other important information.

Your patient has the right to see and copy the MSDS for any product he or she is exposed to at work. As your patient's agent, you may also request an MSDS, but you should never contact the employer without your patient's permission—contacting the employer could jeopardize your patient's job.

Do Not Rely on Material Safety Data Sheets Alone. MSDSs are very often incomplete, inaccurate, or out of date. MSDSs can help you identify the chemical constituents of a product, but the toxicity information they contain should be checked, when possible, by other means. If your patient is no longer exposed to the substance, it may be difficult to obtain the appropriate MSDS. More up-to-date information can be obtained by calling the telephone number listed on the product label and speaking with the company's toxicologist (see below). Many on-line databases of MSDSs provide up-to-date information about specific chemicals. These can be located by searching (for "MSDS") with an Internet search engine or by using an MSDS Web site such as MSDS-Search (http://www.msdssearch.com/), which provides free access to more than 750,000 MSDSs. The Canadian Centre for Occupa-

tional Health and Safety (CCOHS) produces an MSDS database available on CD-ROM or on-line formats. Finally, on many states and localities, community residents, firefighters, and others have the same rights to information about chemicals used by local employers, under community right-to-know regulations.

STEP 2: RESEARCH THE CHEMICAL

Many products are known by their trade names. It is extremely difficult to investigate a product without knowing the generic or chemical names of its ingredients.

Check Reference Books. Several sources of generic ingredient information are available. The National Institute for Occupational Safety and Health (NIOSH) Registry of Toxic Effects of Chemical Substances, commonly known as RTECS (available on CD-ROM from various database vendors; last hard copy edition: 1985–1986) provides a comprehensive listing of more than 130,000 chemicals, extensively cross-referenced to synonyms and trade names. RTECS is best used to identify references to the toxicologic literature. *The Clinical Toxicology of Commercial Products* (1), available in most medical libraries, lists the ingredients of 15,000 trade-name products. If neither of these are sufficient, try one of the several chemical dictionaries or collections of synonyms that list common trade names. Pesticides' active ingredients are identified in the Environmental Protection Agency *Catalog of Pesticide Chemical Names and Their Synonyms* (2), the *Farm Chemicals Handbook* (3), updated annually, or Hayes' *Handbook of Pesticide Toxicology* (4).

Contact the Manufacturer. You should not call your patient's employer without your patient's permission, but you may choose to contact the manufacturer of the product. Product labels usually contain the manufacturer's name and telephone number and often an address as well. If the address is not listed, use Thomas' *Register of American*

Manufacturers (5), available at many public libraries. Request MSDSs on the substances of interest. Calls from physicians and other health care providers generally receive quick responses from manufacturers, who may be concerned about potential liability.

Contact the Regional Poison Control Center. Poison control centers are a vital source of data on both the generic ingredients and the acute toxic properties of chemical substances, and one is located in every region of the United States.

Contact NIOSH. NIOSH has a computerized databank of the ingredients of trade-name substances from both the National Occupational Hazard Survey (1972–1974) and the National Occupational Exposure Survey (1981–1983). Although these databanks are somewhat out of date, NIOSH can provide the generic chemical ingredients of approximately one-third of the trade-name products requested. Contact the Hazard Surveillance Section, NIOSH, Mail Stop R19, 4676 Columbia Parkway, Cincinnati OH 45226; telephone: 800-35-NIOSH.

Check Chemical Fact Sheets. Basic information on common chemicals can be found in chemical fact sheets. Many state health and labor departments publish chemical fact sheets or serve as sources for information on hazardous chemicals. For example, the health departments of New York State (http://www.health.state.ny.us/index.htm), California (http://www.ohb.org/hesfact.htm), and New Jersey (http://www.state.nj.us/health/eoh/rtkweb/rtksfs.htm) have all developed well-researched fact sheets on industrial chemicals, many of which are available on their respective Web sites. Some state health departments also provide responses by telephone or in writing to inquiries from workers, employers, and clinicians within their individual states via telephone hotlines. Contact your state health department to determine whether such a service exists in your state.

Check NIOSH's Pocket Guide to Chemical Hazards (6). This guide lists 677 major industrial chemicals and provides very brief statements on their toxicity, exposure limits, and protective equipment. NIOSH has also published *Occupational Health Guidelines for Chemical Hazards* (7). This three-volume set of 5-page fact sheets covers more than 300 substances for which there are federal occupational safety and health regulations. Each fact sheet provides a more detailed summary of toxicology data, a description of symptoms and signs of exposure, and recommended surveillance protocols.

Check Criteria Documents. NIOSH has published a CD-ROM with a complete set of "Criteria Documents," critical evaluations of the toxicologic and epidemiologic literature on which NIOSH has based its recommended exposure limits (RELs) for individual substances. Most of the criteria documents were published during the 1970s or early 1980s, but they remain valuable for their identification of minimal-risk exposure levels.

Check Toxicological Profiles. The Agency for Toxic Substances and Disease Registry (ATSDR) publishes the "Toxicological Profile" series, which provides information concerning the health effects of more than 275 hazardous substances that are commonly found at Superfund sites—and in many workplaces—and that pose the most significant potential threats to human health. These profiles, also available in a complete collection on CD-ROM format, undergo periodic review and are updated frequently. The ATSDR toll-free telephone number for general information is 888-42-ATSDR or 888-422-8737. Its Web site is: http://atsdr1.atsdr.cdc.gov:8080/atsdrhome.html.

Consult Textbooks. Several chapters of this book review the effects of chemicals on specific organ systems (Chapters 25–35), and the bibliographies in these chapters are good starting points for obtaining information. Major textbooks of industrial hygiene, toxicology, and pharmacology usually have chapters devoted to specific chemicals or families of chemicals (see bibliographies in Chapters 1, 7, and 15).

Consult Electronic Resources. An ever-widening array of computerized databases is available for research on the toxic properties of chemicals. There exist more than 39 on-line toxicology databases provided by commercial and governmental agencies worldwide, categorized as factual, full-text, or bibliographic databases (8,9). See Table A-1 for a list of database vendors, and Table A-2 for a list of databases with their format and content.

Factual databases contain concise factual or numeric data on a chemical. Perhaps the best example of a factual database is the Hazardous Substances Databank (HSDB), maintained by the U.S. National Library of Medicine. Full-text databases contain the full text of articles, books, or reports. Examples include the NIOSH CD of Criteria Documents and Reports and the ATSDR Toxicological Profile CD. Bibliographic databases provide bibliographic references about a specific subject, usually with specific information sources to which the user may refer to obtain an entire article, book, or report.

Bibliographic databases, such MEDLINE and TOXLINE, sponsored by the National Library of Medicine (NLM), contain references to the published medical and toxicologic literature. These databases are available free to anyone with access to the Internet at: http://www.igm.nlm.nih.gov, http://www.ncbi.nlm.nih.gov/PubMed, or http://medlineplus.nlm.nih.gov/medlineplus/. MEDLINE contains approximately 11 million citations from about 4,000 biomedical journals and provides health professionals rapid access to the most recently published information on any biomedical subject. Biochemical, pharmacologic, physiologic, and toxicologic effects of drugs and other chemicals are accessible through TOXLINE. TOXLINE and its backfile, TOXLINE65, contain more than 2.5 million bibliographic citations from secondary sources that are searchable from 18 subfiles.

TOXNET is a comprehensive database containing both factual information and references on several thousand chemicals. One advantage of TOXNET is that statements on toxicity and biomedical effects undergo scientific peer review. TOXNET provides a free Web-based search and retrieval interface (http://toxnet.nlm.nih.gov) to an integrated system of databases: Chemical Carcinogenesis Research Information System (CCRIS), Hazardous Substances Databank (HSDB), Integrated Risk Information System (IRIS), and Genetic Toxicology (GENE-TOX).

NIOSH maintains its own occupational safety and health database, NIOSHTIC, which covers more than 160 journals and includes toxicology, epidemiology, ergonomics, and other information about health and safety in the workplace. Abstracts of all NIOSH health hazard evaluations and testimony before committees of the U.S. Congress are also included. Free searches of NIOSHTIC can be arranged through NIOSH by telephoning 800-35-NIOSH. Online access or a CD-ROM of the entire data-

TABLE A-1. *Database vendors*

Canadian Centre for Occupational Health and Safety
250 Main Street East
Hamilton, Ontario
CANADA L8N 1H6
(416) 572-2981
Website: http://www.ccohs.ca/

DIALOG Information Services, Inc.
3460 Hillview Avenue
Palo Alto, CA 94304
(415) 858-3785, (800) 334-2564
Website: http://www.dialog.com/

National Library of Medicine
8600 Rockville Pike
Bethesda, MD 20894
(888) FIND-NLM
(301) 594-5983 (local and international calls)
Website: http://www.nlm.nih.gov

SilverPlatter Information, Ltd.
100 River Ridge Drive
Norwood, MA 02062-5026
(617) 769-2599, (800) 334-0064
Website: http://www.silverplatter.com/

STN, Inc.
5113 Leesburg Pike
Falls Church, VA 22041
(800) 321-1969 or (703) 379-9700
Fax: (703) 824-0699
E-Mail: ask@stn.com
Website: http://www.stn.com/

TABLE A-2. *Selected toxicology databases*

Database name	Database	Format	Contents	Vendors
BIOSIS	Bibliographic	Online CD-ROM World Wide Web	Comprehensive coverage of biologic sciences	DIALOG, Ovid Silver Platter
Chemical Abstracts	Bibliographic	Online World Wide Web	Indexes world's chemical literature	DIALOG, Ovid
EMBASE	Bibliographic	Online CD-ROM	Contains references to all fields of medicine	DIALOG, Ovid, STN Silver Platter
MEDLINE	Bibliographic	Online CD-ROM World Wide Web	Medical and toxicologic literature	DIALOG, NLM, Ovid Silver Platter
NIOSHTIC	Bibliographic	Online CD-ROM	Literature in the field of occupational safety and health	CCOHS, DIALOG, STN CCOHS, Silver Platter
TOXLINE	Bibliographic	Online CD-ROM World Wide Web	Provides references to human and animal toxicity studies	DIALOG, NLM, Ovid, STN Silver Platter
TOXNET	Factual	Online, CD-ROM	Integrated system of peer-reviewed databases of toxicity and biomedical effects summaries	
Hazardous Substances Databank (HSDB)	Factual	Online, CD-ROM	Focuses on the toxicology of potentially hazardous chemicals	NLM (free access) http://toxnet.nlm.nih.gov/ Micromedex/Tomes Plus, Silver Platter
Chemical Carcinogenesis Research Information System (CCRIS)	Factual	Online, World Wide Web	Contains chemical records with carcinogenicity, mutagenicity, tumor promotion, and tumor inhibition test results	NLM NLM (free access) http://toxnet.nlm.nih.gov/
Registry of Toxic Effects of Chemical Substances (RTECS)	Factual	Online, CD-ROM	Toxicologic information compiled and updated by NIOSH	CCOHS Chemical Information System Data-Star DIALOG Micromedex/Tomes Plus

base can be obtained through Silver Platter, the CCOHS, Dialog Information Services, or STN International (Columbus, OH).

Other sources of data exist in various formats on the Internet. Government agencies throughout the world provide study results, abstracts, and technical reports for current and past projects in formats that usually include text, PDF, or HTML (e.g., the U.S. National Toxicology Program: http://ntp-server.niehs.nih.gov/). Similarly, profes-

sional societies and academic institutions provide public information with links to other resources on the Internet. Of particular usefulness are public health libraries with regularly updated links to toxicologic information, such as that at the University of California, Berkeley: http://www.lib.berkeley.edu/PUBL/Internet.html.

Retrieval of on-line information may be problematic owing to multiple synonyms for chemical substances; inexact matching for chemical names results in retrieval of many documents dealing with derivatives of the desired chemical or other chemical agents (10). Use of Chemical Abstracts Service (CAS) numbers when searching toxicologic databases may provide a higher degree of specificity but omits a significant number of references from which the CAS numbers are missing, miscoded, or not systematically used. Similarly, misspelled references to chemicals are difficult to obtain through traditional search strategies.

Some databases include descriptors that may represent a controlled vocabulary (e.g., Medical Subject Headings [MeSH], which are used by NLM databases) or uncontrolled keywords provided by article authors. Free-text searches are available for some databases, such as BIOSIS.

There are several general limitations on the information available in on-line databases, including the following:

1. *Time scope:* Entry of information into most on-line databases did not begin until the 1970s. Earlier literature may be missed unless more traditional searches using printed literature indices (e.g., Index Medicus) are conducted. Some bibliographic databases have entered historically important material; for example, NIOSHTIC includes selected references published since the 1890s.
2. *Content scope:* Selection of journals referenced may limit the applicability to a particular question. Most databases include only English-language journals. A few databases include German, French, Swedish, or other language references.
3. *Timeliness of data:* Most databases are updated with a time lag of up to 6 months from date of publication to inclusion in the database.

Selection of databases to search depends on several factors, including method of access (on-line, CD, or other), cost (per hit, per hour, or without extra charge), scope, ease of use and familiarity with search strategies,

and timeliness of data (11). Some have advocated use of multiple databases to increase the likelihood of completeness (e.g., to 80% with the combination of BIOSIS and MED-LINE, or to 90% if three or more databases are searched), but costs for commercial databases may prohibit such a strategy (12).

Future trends in toxicologic information access include Internet access without added use charges, use of more intuitive and simplified user interfaces, incorporation of systems to enhance symptom-related and other natural language searches, integration of multiple databases, and indicators of degree of peer review in databases (13).

STEP 3. INTERPRET THE DATA

Rarely does the available information about a chemical allow a clear and simple evaluation of the risk posed by that chemical. Generally you will need to bring a critical eye, using many of the principles of epidemiology discussed in Chapter 6, to evaluate a developing set of differing and perhaps conflicting data. Be aware that the abstract of a research article may not accurately summarize the full content of the article: it may represent the authors' own interpretation of the data, reflecting personal biases or preconceptions. When any publication forms a substantial part of the basis for your evaluation, you should review its methods and results and draw your own conclusions. Look at statistical confidence limits and risk ratios or odds ratios; never let the simplistic reporting of results as "positive" or "negative," based on an arbitrary statistical significance level, dominate your evaluation (see Chapter 6).

Be alert to the many different kinds of potential epidemiologic biases and confounding variables. Such factors are described in Chapter 6 and include the following:

1. *Selection bias.*
2. *Participation bias* can occur if exposed subjects and controls differentially choose to participate in a study on the basis of some factor associated with the outcome.

3. *Misclassification bias.*
4. *Imperfect ascertainment* of health outcomes can bias a study; nondifferential error produces bias toward the null, and a systematic error could produce bias in either direction.
5. *Confounding.*
6. *Statistical power.* Most studies do not have the power to detect relative risks below 2.0, but such risks may be of great public health significance.
7. The *low-exposure* problem is related to the exposure-misclassification and inadequate-power issues. In some epidemiologic studies, the "exposed" group actually has only a very modest exposure to the study agent; in essence, most members of this group are misclassified. Such a study is virtually guaranteed to obtain a "negative" result. Even a very large study with great statistical power and narrow confidence limits does not have the power to detect the very small but real risk that may accompany a very small exposure. This error has led to unfounded rejection of positive animal studies in favor of the "gold standard" of epidemiologic studies that appear negative but are actually consistent with the positive animal studies (14).
8. *High-exposure* problem. Conversely, experimental studies must address the question of excessive exposure. Animal studies usually have very limited statistical power. To produce a detectable effect, the animals usually are given large exposures. It is widely held that these high exposures can exceed acutely toxic levels and produce effects that would not occur with realistic human exposure levels.
9. *Control for smoking.* Epidemiologic studies are often criticized for failure to account for cigarette smoking. Usually the possibility is raised that differential smoking between exposed and unexposed groups could be responsible for an observed increase in tobacco-related health effects. However, it has been shown that differential smoking rates are unlikely to

explain more than a 1.5 relative risk, even for an outcome as overwhelmingly tobacco-related as lung cancer (15). The reality is that smoking greatly interferes with the ability to detect tobacco-related health effects caused by other agents. For example, relative risks for lung cancer among workers exposed to fibrous glass or other synthetic mineral fibers have typically been around 1.2 to 1.3, too low to attribute confidently to the fibers. But if all lung cancers induced by smoking could be subtracted out of both the exposed and unexposed groups, the relative risk could be about 3.0, a much more convincing indication of a real toxicity (16).

One mark of a high-quality modern study is the calculation of a benchmark dose rather than reliance on a simple "No Observed Adverse Effect" Level (NOAEL). A *benchmark dose* is a dose calculated to produce a prespecified level of effect, typically a 5% incidence (or a 10% modification in a continuous-variable effect, such as birthweight). The benchmark dose amount is often similar to the NOAEL, but it has the advantages of being independent of the choice of experimental exposure levels, using all of the available statistical data, and providing a quantified evaluation of risk across a broad range of realistic exposure levels.

Studies of developmental toxicity require evaluation of certain additional considerations. Perhaps chief among them is a concept sometimes called the *adult:developmental ratio.* For many chemicals, perhaps most, the lowest dose capable of causing developmental toxicity is very close to the lowest dose capable of causing toxicity to some other system. It has been argued that fetal defects seen in the presence of substantial maternal toxicity can be dismissed as nothing more than maternally mediated effects bearing no significance with regard to any lower exposure. However, it has been shown that even severe maternal toxicity with a significant mortality rate does not necessarily produce fetal defects (17,18); therefore, effects on the offspring should be considered independently of maternal effects.

In developmental toxicity testing, a broad range of end points should be examined. In some cases, an effect on one end point can obscure an effect on a different end point. For example, fetal defects can lead to early or late fetal loss, with an effect on either preimplantation or postimplantation fetal losses and little effect on the incidence of congenital defects in liveborn offspring.

The considerations described here will not only help you to evaluate the quality and validity of each study you must weigh but may also allow you to reconcile studies that appear, at first glance—and even in their authors' opinions, to have reached conflicting results.

REFERENCES

1. Gosselin RE, Smith RP, Hodge HC, Braddock JE. Clinical toxicology of commercial production. 8th ed. Williams & Wilkins, Baltimore, 1984.
2. U.S. Environmental Protection Agency. Catalog of Pesticide Chemical Names and Their Synonyms, 2nd ed. Washington DC: EPA, 1990.
3. Meister RT, Sine C, eds. Farm chemicals handbook '99. Meister Publishing Co., Willoughby OH, 1999.
4. Hayes WJ Jr., Laws ER Jr., eds. Handbook of pesticide toxicology. Academic Press Inc./Harcourt Brace Jovanovich, San Diego, 1991—three volumes.
5. Register of American Manufacturers. Thomas Publishing Co. New York, 1994.
6. National Institute for Occupational Safety & Health, CDCP, PHS, U.S. DHHS. NIOSH pocket guide to chemical hazards. NIOSH Publication no. 97–140, 1997.
7. National Institute for Occupational Safety & Health. CDC, PHS, U.S. DHHS. NIOSH Publications no. 81–123, 1981; 88–118, 1988; 89–104, 1989; 92–110; 1992; 95–121, 1995.
8. Voigt K, Bruggeman R. Toxicology databases in the metadatabank of on-line databases. Toxicology 1995;100:225–240.
9. Evangelisti M, Bolognesi C, Rabboni R, Ugolini D. Chemical hazard evaluation: the use of factual health and safety databanks. Sci Total Environ 1994;153:21 1–217.
10. Ludl H, Schope K, Mangelsdorf I. Searching for information on chemical substances in selected biomedical bibliographical databases. Chemosphere 1995;31:2611–2628.
11. Ludl H, Schope K, Mangelsdorf I. Searching for information on toxicological data of chemical substances in selected bibliographic databases: Selection of essential databases for toxicological researches. Chemosphere 1996;32:867–880.
12. Gehanno JF, Paris C, Thirion B, Caillard JF. Assessment of bibliographic databases' performance in information retrieval for occupational and environmental toxicology. Occup Environ Med 1998; 55:562–566.
13. Liverman CT, Ingalls CE, Fulco CE. Toxicology and environmental health information resources for health professionals: a report by the Institute of Medicine. Bull Med Libr Assoc 1997;84:432–425.
14. Hertz-Picciotto I, Neutra RR. Resolving discrepancies among studies: the influence of dose on effect size. Epidemiology 1994;5:156–163.
15. Siemiatycki J, Wacholder S, Dewar R, Cardis E, Greenwood C, Richardson L. Degree of confounding bias related to smoking, ethnic group and socioeconomic status in estimates of the associations between occupation and cancer. J Occup Med 1988; 30:617–625.
16. Levin LI, Silverman DT, Hartge P, Fears TR, Hoover RN. Smoking patterns by occupation and duration of employment. Am J Ind Med 1990;17: 711–715.
17. Chernoff N, Setzer RW, Miller DB, Rosen MB, Rogers JM. Effects of chemically induced maternal toxicity on prenatal development in the rat. Teratology 1990;42:651–655.
18. Kavlock RJ, Logsdon T, Gray JA. Fetal development in the rat following disruption of maternal renal function during pregnancy. Teratology 1993: 48;247–258.

Appendix B: Other Sources of Information

Marianne Parker Brown and Terrence A. Valen

GENERAL RESOURCES

Emergency Services

Telephone numbers are listed for technical, medical, regulatory, or reporting information in the event of an exposure-related problem or medical emergency.

Centers for Disease Control and Prevention

CDC Emergency Response
  770-488-7100
  This emergency response hotline addresses issues from chemical, biologic, and radiologic public health hazards to natural disasters. The hotline accesses all resources at CDC to handle or redirect concerns.

Chemical Transportation Emergency Center

CHEMTREC
  800-424-9300
  This 24-hour hotline relays information to manufacturers about a spill involving their products or employees. Staff can rapidly determine for the caller the contents of a spilled container and suggest appropriate action for containment.

M. P. Brown and T. A. Valen: UCLA-Labor Occupational Safety and Health (LOSH) Program, University of California—Los Angeles, Los Angeles, California 90095-1478.

For additional information or questions:
CHEMTREC
1300 Wilson Boulevard
Arlington, VA 22209
703-741-5524
E-mail: cwcchemica@aol.com
Web site: http://www.cwc-chemical.com/chemtrec.htm

National Pesticide Telecommunications Network

National Pesticide Telecommunications Network
  800-858-7378
  The network provides information on pesticide products, safety and health concerns, and disposal procedures. Available 7 days a week, from 6:30 a.m. to 4:30 p.m., Pacific Time.
  For additional information or questions:
  Oregon State University
  333 Weniger Hall
  Corvallis, OR 97331-6502
  E-mail: nptn@ace.orst.edu
  Web site: http://ace.orst.edu/info/nptn/

National Response Center

National Response Center Hotline
  800-424-8802
  This is the first federal point of contact for reporting of, and guidance on, oil and hazardous spills. The 24-hour hotline relays information on regulatory and response agencies and immediately contacts appro-

priate federal on-scene coordinators to respond to the spill.

For additional information or questions:
E-mail: info@nrc450.comdt.uscg.mil
Web site: http://www.nrc.uscg.mil

Occupational Safety and Health Administration

OSHA Emergency Hotline
800-321-OSHA (800-321-6742)
This is a toll-free number for occupational safety and health emergencies that relate to a fatality or imminent threat to life.

Poison Control Centers

Each state or region has a 24-hour poison control center offering expert emergency medical information and referrals. Centers are listed on the inside cover of local phone directories. They can also be reached by dialing 911 and asking for the Poison Control Center or by calling the National Response Center Hotline, listed earlier.

Guides, Catalogs, and Directories

National Institute for Occupational Safety and Health

NIOSH Publications
Mail Stop C-13
4676 Columbia Parkway
Cincinnati, OH 45226-1998
800-35-NIOSH (800-356-4674)
E-mail: pubstaft@cdc.gov
Web site: http://www.cdc.gov/niosh/pubs.html

NIOSH Publications Catalog
This is a faxed list of all NIOSH publications, including criteria documents, manuals, reports, health hazard evaluations, and databases. The toll-free number also provides ordering information for all available publications.

Occupational Safety and Health Administration

OSHA Publications

U.S. Department of Labor
Attn: OSHA Publications
Post Office Box 37535
Washington, DC 20013-7535
202-693-1888
Fax: 202-693-2498
Web site: http://www.osha-slc.gov/Osh Doc/Additional.html

OSHA Publications
This is a list of available OSHA publications and ordering information.

OSHA Computerized Information System (OCIS) on CD-ROM

U.S. Government Printing Office
Superintendent of Documents
Post Office Box 371954
Pittsburgh, PA 15250-7940
202-512-1800
To order on-line:
Web site: http://www.access.gpo.gov/su_docs/sale/order001.html

OSHA Regulations, Documents, and Technical Information on CD-ROM
This CD-ROM publication (stock no. 729-013-00000-5) contains agency documents, technical information, and training materials files. The new format is an electronic copy of selected information from the www.osha.gov and www.osha-slc.gov Web sites; it is available for $43 a year (4 issues).

Films and Other Audiovisuals

American Federation of Labor and Congress of Industrial Organizations

AFL-CIO Support Services Department
815 16th Street, NW
2nd Floor, Room 209
Washington, DC 20006
202-637-5041

Films and Video Tapes for Labor

Lists a number of films and video tapes on occupational health and safety, including training on regulations, workers' rights, and union strategies. For rental only.

Bureau of National Affairs Communications, Inc.

BNA Communications, Inc.
9439 Key West Avenue
Rockville, MD 20850-3396
800-233-6067
E-mail: bnac@bna.com
Web site: http://www.bna.com/bnac/sfprod/
sftymain.html

BNA Communications Safety Catalog

BNA Communications produces and distributes safety, environmental, and regulatory compliance video tapes, including a right-to-know package and a hazardous waste management series for purchase or rental.

International Film Bureau

International Film Bureau
332 South Michigan Avenue
Chicago, IL 60604
312-427-4545

The IFB has video tapes for safety and health training in industrial hygiene, personal protection and operational safety, hazardous materials handling, and other areas. Some are available in Spanish.

National Audiovisual Center, National Technical Information Service

Technology Administration
U.S. Department of Commerce
Springfield, VA 22161
703-605-6000
800-553-NTIS (800-553-6847)
E-mail: info@ntis.fedworld.gov
http://www.ntis.gov/prs/pr996.htm

Occupational Safety and Health Audiovisual Training Programs

These programs were all produced by safety experts at OSHA, NIOSH, EPA, and other federal agencies. The catalog contains training materials on federal programs and regulations, workplace safety, workplace hazards, work-related pathologies, hazardous materials, indoor air quality, transport safety, emergency response, and storms.

National Clearinghouse for Worker Safety and Health Training for Hazardous Materials, Waste Operations and Emergency Response

National Clearinghouse
5107 Benton
Bethesda, MD 20814
301-571-4226
E-mail: chouse@dgs.dgsys.com
Web site: http://www.niehs.nih.gov/wetp/
clear.htm

Ask for catalog of information on hazardous waste worker training programs. The National Clearinghouse also maintains a library of curricula developed and used by organizations involved in the training of hazardous waste workers and emergency responders.

National Information Center for Educational Media

National Information Center for Educational Media
Post Office Box 8640
Albuquerque, NM 87198-8640
800-926-8328
E-mail: nicemnet@nicem.com
Web site: http://www.nicem.com

NICEM Database

Exhaustive compilation of educational media categorized by subject and title. Available at most research libraries. NICEM Net Reference is available for on-line subscription. On-line version of the NICEM database contains more than 440,000 bibliographic records of nonprint educational materials, searchable by title, date, age level, subject area, and media types. The database is also available in CD-ROM and print form.

National Institute for Occupational Safety and Health

NIOSH-TV, NIOSH Publications

NIOSH-TV (C-12)
  4676 Columbia Parkway
  Cincinnati, OH 45226
  Web site: http://www.cdc.gov/niosh/tapes. html
  Also available from:
  NIOSH Publications
  4676 Columbia Parkway, Mail Stop C-13
  Cincinnati, OH 45226-1998
  800-35-NIOSH (1-800-356-4674)
  E-mail: pubstaff@cdc.gov
  Web site: http://www.cdc.gov/niosh/pubs. html
  Fifty-eight NIOSH-produced films available for either purchase or loan.

Resource Center of the Environmental and Occupational Health Sciences Institute

Public Education and Risk Communication Division
  EOHSI
  170 Frelinghuysen Road
  Piscataway, NJ 08854
  732-445-0110

Occupational Health Resource Guide
  From the University of Medicine and Dentistry of New Jersey (UMDNJ) and New Jersey Department of Health, 1989. Valuable guide to occupational health-related books, journals, pamphlets, audiovisual materials, databases, organizations, vendors, and more. Resources on hazard identification and control, laws and regulations, occupational diseases, specific industries and processes, and more. Can be obtained for $15.

ORGANIZATIONS

Federal Government

Hazardous Materials Information Center

Office of Hazardous Materials
  Standards Office
  U.S. Department of Transportation
  400 7th Street, SW
  Washington, DC 20590-0001
  800-467-4922; in Washington, DC, only, call 202-366-4488
  E-mail: infocntr@rspa.dot.gov
  Web site: http://hazmat.dot.gov/infocent. htm
  Provides information on, and interpretation of, hazardous materials transportation regulations.

U.S. Department of Health and Human Services

DHHS Web site: http://www.os.dhhs.gov

Centers for Disease Control and Prevention

CDC Web site: http://www.cdc.gov

National Institute for Occupational Safety and Health (NIOSH)

Main Office

National Institute for Occupational Safety and Health
  Centers for Disease Control and Prevention
  Department of Health and Human Services
  200 Independence Avenue, SW, 317B
  Washington, DC 20201
  202-401-3747

Directory of NIOSH Offices
  Web site: http://www.cdc.gov/niosh/offices.html

NIOSH Technical Information Service

NIOSH Technical Information Service
  800-35-NIOSH (800-356-4674)
  E-mail: pubstaft@cdc.gov
  Web site: http://www.cdc.gov/niosh/inquiry.
  html
Provides information and technical assistance on many aspects of occupational safety and health, including ordering information for NIOSH publications. NIOSH Technical Information Specialists are available to assist the public with questions about NIOSH activities.

NIOSH Research Laboratories

NIOSH maintains research laboratories in Cincinnati, Ohio (Alice Hamilton Laboratory, Robert A. Taft Laboratory); Morgantown, West Virginia; Spokane, Washington; and Pittsburgh, Pennsylvania (Pittsburgh Research Center).

NIOSH Field Offices

The following field offices cover three regions of the United States and perform functions such as health hazard evaluations.

New England Field Office (CT, MA, ME, NH, RI, and VT)
  Post Office Box 87040
  South Dartmouth, MA 02748-0701

Atlanta Field Office (AL, FL, GA, KY, MS, NC, SC, and TN)
  1600 Clifton Road, NE
  Atlanta, GA 30333

Denver Field Office (CO, MT, ND, SD, UT, and WY)
  Denver Federal Center
  Building 53, Room 14051
  Post Office Box 252261
  Denver, CO 80225-0226

Canadian Centre for Occupational Health and Safety

Canadian Centre for Occupational Health and Safety
  250 Main Street East
  Hamilton, Ontario
  Canada, L8N 1H6
  905-572-4400
  E-mail: inquiries@ccohs.ca
  Web site: http://www.ccohs.ca/Resources
CCOHS is Canada's national Center for occupational health and safety information. Its Inquiries Service provides information that enables people to make informed decisions about specific workplace issues and to take action to prevent occupational injuries and diseases.

Occupational Safety and Health Administration

Main Office

Occupational Safety and Health Administration
  200 Constitution Avenue, NW
  Washington, DC 20210
  Web site: http://www.osha.gov

Salt Lake City Technical Links
  Web site: http://www.osha-slc.gov/SLTC

OSHA Directorate of Technical Support

OSHA Directorate of Technical Support
  Frances Perkins Building, Room # N3653
  200 Constitution Avenue
  Washington, DC 20210
  202-693-2300
  Web site: http://www.osha-slc.gov/SLTC/
  dts/index.html
This central number for federal OSHA provides rapid technical and occupational medical support. In the event of a significant emergency, staff will notify the OSHA Salt Lake Technical Center National Response Team (801-487-0521). Also provides phone numbers for regional technical support offices for further assistance.

OSHA Regional Offices

Region I (CT, MA, ME, NH, RI, and VT)
Boston, MA 02203

Region II (NJ, NY, Puerto Rico, and Virgin Islands)
New York, NY 10014

Region III (DC, DE, MD, PA, VA, and WV)
Philadelphia, PA 19104

Region IV (AL, FL, GA, KY, MS, NC, SC, and TN)
Atlanta, GA 30303

Region V (IL, IN, MI, MN, OH, and WI)
Chicago, IL 60604

Region VI (AR, LA, NM, OK, and TX)
Dallas, TX 75202

Region VII (IA, KS, MO, and NE)
Kansas City, MO 64105

Region VIII (CO, MT, SD, ND, UT, and WY)
Denver, CO 80202-5716

Region IX (AZ, CA, HI, NV, American Samoa, Guam, and Trust Territories of the Pacific)
San Francisco, CA 94105

Region X (AK, ID OR, and WA)
Seattle, WA 98101-3212

State Agencies*

States with OSHA-approved plans are listed here. Unless indicated, plans cover both public and private workers.

ALASKA
Division of Labor Standards and Safety
Occupational Safety and Health
3301 Eagle Street, Suite 305
Post Office Box 107022
Anchorage, AK 995107022
907-269-4940
E-mail: timothy_bundy@labor.state.ak.us

* Connecticut and New York have public sector programs only.

ARIZONA
Arizona Division of Occupational Safety and Health (ADOSH)
800 West Washington
Phoenix, AZ 85007
602-542-5795
Web site: http://www.primenet.com/~azcompl/

CALIFORNIA
Department of Industrial Relations
CAL/OSHA
45 Freemont Street, Suite 1200
San Francisco, CA 94105
415-972-8500
Web site: http://www.dir.ca.gov/DIR/OS&H/DOSH/dosh1.html

CONNECTICUT*
Connecticut Department of Labor
Division of Occupational Safety and Health (CONN-OSHA)
38 Wolcott Hill Road
Wethersfield, CT 06109
860-566-4550
E-mail: director.connosha@po.state.ct.us
Web site: http://www.ctdol.state.ct.us/osha/osha.htm

HAWAII
Hawaii Occupational Safety and Health Division (HIOSH)
Consultation and Training Branch
Division of Occupational Safety and Health
Department of Labor and Industrial Relations
830 Punchbowl Street, Room 423
Honolulu, Hl 96813
808-5483150
Web site: http://www.aloha.net/~edpso/index.html

INDIANA
Indiana Department of Labor (IOSHA)
Bureau of Safety, Education, and Training
402 West Washington Street, Room W195
Indianapolis, IN 46204

317-232-2693 (Enforcement and Compliance)
Web site: http://www.ai.org/labor

IOWA
Occupational Safety and Health Administration (IOSHA)
1000 East Grand Avenue
Des Moines, IA 50319
515-281-3606

KENTUCKY
Kentucky Labor Cabinet
Division of Occupational Safety and Health Compliance
1047 U.S. Highway 127 South, Suite 4
Frankfort, KY 40601
502-564-3070
Web site: http://www.state.ky.us/agencies/labor/kyosh.htm

MARYLAND
Maryland Occupational Safety and Health
1100 North Eutaw Street, Room 613
Baltimore, MD 21201
410-767-2215
E-mail: dllr@maryland-e-baltimore.osha.gov
Web site: http://www.dllr.state.md.us/labor/mosh.html

MICHIGAN
Department of Consumer and Industry Services
Bureau of Safety and Regulation
7150 Harris Drive
Post Office Box 30643
Lansing, Ml 48909-8143
517-322-1814
E-mail: bsrinfo@cis.state.mi.us
Web site: http://www.commerce.state.mi.us/bsr

MINNESOTA
Minnesota-OSHA
Occupational Safety and Health Division
Department of Labor and Industry
443 Lafayette Road

St Paul, MN 55155-4307
651-296-2116 (Compliance)
Web site: http://www.doli.state.mn.us/mnosha.html

NEVADA
OSH Enforcement
1301 North Green Valley Parkway
Las Vegas, NV 89104
702-486-9020
Web site: http://www.state.nv.us/b&i/ir/

NEW MEXICO
New Mexico Environment Department
Occupational Health and Safety Bureau
525 Camino de Los Marquez, Suite 3
Post Office Box 26110
Santa Fe, NM 87502
505-827-4230
Web site: http://www.nmenv.state.nm.us/env_prot.html#OHSB

NEW YORK*
Division of Safety and Health
State Office Campus
Building 12, Room 130
Albany, NY 12240
518-457-2238

New York Public Sector Consultation Program
New York State Department of Labor
Building #12
State Building Campus
Albany, NY 12240
518-457-1263

E-mail: james.rush@ny-ce-albany.osha.gov
Web site: http://www.labor.state.ny.us/safety/saf_hlth.htm

NORTH CAROLINA
North Carolina Department of Labor
Division of Occupational Safety and Health
4 West Edenton Street

Raleigh, NC 27601
919-662-4575
E-mail: bandrews@mail.dol.state.nc.us
Web site: http://www.dol.state.nc.us/DOL/
osh.htm

OREGON
Oregon Department of Consumer and
Business Services
Occupational Safety and Health Division
(OROSHA)
350 Winter Street NE, Room 430
Salem, OR 97310
503-378-3272
Web site: http://www.cbs.state.or.us/
external/osha

PUERTO RICO
Occupational Safety and Health Office
505 Munoz Rivera Avenue
Hato Rey, PR 00918
787-754-2171

SOUTH CAROLINA
South Carolina Department of Labor
Licensing and Regulation
3600 Forest Drive
Post Office Box 11329
Columbia, SC 292111329
803-734-9594
Web site: http://www.llr.sc.edu/OCSAFE.
HTM

TENNESSEE
Tennessee Department of Labor
Tennessee OSHA
710 James Robertson Parkway, 3rd Floor
Nashville, TN 37243-0659
615-741-2793
E-mail: cynthia.mayberry@tn-c-nashville.
osha.gov
Web site: http://www.state.tn.us/labor/

UTAH
Utah OSHA
Industrial Commission
160 East 300 South
Post Office Box 146650

Salt Lake City, UT 841146650
801-5306901
Web site: http://www.labor.state.ut.us/
uosha.htm

VERMONT
Vermont Department of Labor and
Industry
Division of Occupational Safety and
Health (Vermont OSHA)
National Life Building, Drawer 20
Montpelier, VT 05620-3401
802-828-2765
Web site: http://www.state.vt.us/labind/
vosha.htm

VIRGIN ISLANDS
Virgin Islands Department of Labor
Division of Occupational Safety and
Health
3012 Golden Rock
Christiansted
St. Croix, Virgin Islands 00820
340-772-1315

VIRGINIA
Virginia Department of Labor and
Industry
Occupational Safety and Health
Training and Consultation
13 South Street
Richmond, VA 23219
804-786-8707
Web site: http://www/dli.state.va.us/home.
htm

WASHINGTON
Washington Department of Labor and
Industries
Safety and Health Division (WISHA)
Post Office Box 44600
Olympia, WA 98504
800-423-7233
Web site: http://www.wa.gov/lni/wisha

WYOMING
Wyoming Department of Employment
Workers' Safety and Compensation
Division
22 West 25th Street

Herschler Building, 2 East
Cheyenne, WY 82002
307-777-7786
E-mail: wwalm@missc.state.wy.us
Web site: http://wydoe.state.wy.us/wscd/
  osha/

Other Federal Agencies

Agency for Toxic Substances and Disease Registry

Agency for Toxic Substances and Disease
Registry
1600 Clifton Road, NE
Atlanta, GA 30333
800-447-1544
E-mail: ATSDRIC@cdc.gov
Web site: http://atsdr1.atsdr.cdc.gov:8080/
  atsdrhome.html
The ATSDR is a U.S. Public Health Service agency that implements the health-related sections of the "Superfund" Act and its
amendments and the Resource Conservation
and Recovery Act (RCRA). It is involved
in the areas of emergency response, health
assessments, health effects research, literature inventory and dissemination, exposure
and disease registries, toxicologic profiles,
health professional training, and worker
health. The agency maintains a list of toxic
waste sites that are closed to the public.

U.S. Department of Labor

Bureau of Labor Statistics

Web site: http://www.bls.gov/oshhome.
  htm
The Office of Safety, Health and Working
Conditions at the BLS has two departments
covering fatal and nonfatal (injuries and illnesses) workplace statistics. National data
are published approximately 8 months after
the reference year in a news release, and
more detailed data are published later in a
bulletin. State-specific data on workplace fatalities may be requested from the state agencies participating with BLS in the census and

its surveys. Staff members in both departments are available for assistance.

Fatality Data

Census of Fatal Occupational Injuries
  Division of Program Analysis and Control
  Office of Safety, Health and Working
    Conditions
  U.S. Department of Labor
  2 Massachusetts Avenue, NE
  Washington, DC 20212
  202-606-6175
  E-mail: cfoistaff@bls.gov
  Researchers may apply to BLS for access
to the CFOI research file.

Summary Data

Survey of Occupational Injuries and Illnesses
  Division of Safety and Health Statistics
  Office of Safety, Health and Working
    Conditions
  U.S. Department of Labor
  2 Massachusetts Avenue, NE
  Washington, DC 20212
  202-606-6179
  E-mail: oshstaff@bls.gov

Environmental Protection Agency

EPA Headquarters Information Resources Center

Headquarters Information Resources
Center
  Environmental Protection Agency
  401 M Street, SW (mailstop 3404)
  Washington, DC 20460
  202-260-5922
  E-mail: PUBLIC-ACCESS@epamail.epa.
    gov
  Web site: http://www.epa.gov/natlibra/
    hqirc/services.htm
The IRC provides access to EPA information for United States and international requests and has a range of information services consisting of environmental and related
subjects of interest to EPA staff and the gen-

eral public, including on-line searching of commercial databases. The focus of the IRC collection is on environmental regulations, policy, planning, and administration. The IRC also maintains a large collection of EPA documents on microfiche and in hard copy.

Mine Safety and Health Administration

Mine Safety and Health Administration
  U.S. Department of Labor
  Ballston Towers #3
  4015 Wilson Boulevard
  Arlington, VA 22203
  703-235-1452
  E-mail: asmsha@msha.gov
  Web site: http://www.msha.gov
  MSHA regulates health and safety in the mining industry.

National Institute of Environmental Health Sciences

National Institute of Environmental Health Sciences
  Post Office Box 12233
  Research Triangle Park, NC 27709
  919-541-3345 (Office of Communications, Public Affairs)
  Web site: http://www.niehs.nih.gov
  NIEHS is the principal federal agency for biomedical research on the effects of chemical, physical, and biologic environmental agents on human health and well-being. It supports and conducts basic research focused on the interaction between humans and potentially hazardous environmental agents. NIEHS administers hazardous waste training and research grants programs.

National Toxicology Program

National Toxicology Program
  Post Office Box 12233
  M.D. B2-04
  Research Triangle Park, NC 27709
  919-541-0530

Web site: http://ntp-server.niehs.nih.gov/default-text.html
NTP was established in 1978 to develop scientific information needed to determine the toxic effects of chemicals and to develop better, faster, and less expensive test methods.

U.S. Nuclear Regulatory Commission

U.S. Nuclear Regulatory Commission
  One White Flint North
  11555 Rockville Pike
  Rockville, MD 20852-2738
  301-415-7000
  Web site: http://www.nrc.gov/nrc.html
  The NRC regulates the commercial use of nuclear materials and issues licenses for such use. It ensures adequate protection of public health and safety, the common defense and security, and the environment in the use of nuclear materials in the United States.

Professional Organizations in the United States

American Association of Occupational Health Nurses

American Association of Occupational Health Nurses
  2120 Brandywine Road, Suite 100
  Atlanta, GA 30341-4146
  770-455-7757
  E-mail: aaohn@aaohn.org
  Web site: http://www.aaohn.org
  AAOHN is a professional organization of registered nurses engaged in occupational health and safety and occupational health nursing. Major activities are formulating and developing principles and standards of occupational health nursing practice; promoting, by means of publications, conferences, continuing education courses, and symposia, educational programs designed specifically for the occupational health nurse; and impressing on managers, physicians, and others the importance of integrating occupational

health and safety services into employee activities.

American College of Occupational and Environmental Medicine

American College of Occupational and Environmental Medicine
1114 North Arlington Heights Road
Arlington Heights, IL 60004-4770
847-818-1800
Web site: http://www.acoem.org

ACOEM is the largest organization of occupational and environmental physicians dedicated to promoting and protecting the health of workers through preventive services, clinical care, research, and educational programs. It provides courses and conferences, publishes newsletters and the *Journal of Occupational and Environmental Medicine,* and provides numerous other services.

American Conference of Governmental Industrial Hygienists

American Conference of Governmental Industrial Hygienists
1330 Kemper Meadow Drive, Suite 600
Cincinnati, OH 45240
513-742-2020
Web site: http://www.acgih.org

ACGIH is an organization of industrial hygiene and occupational health and safety professionals who are engaged in occupational health and safety services, consultation, enforcement, research, or education. It publishes information for occupational health and safety professionals to assist them in providing health and safety services for workers. Publications focus on specific areas such as industrial hygiene, health and safety, environment, laboratory/quality control, toxicology, medical, hazardous materials, hazardous waste, controls, physical agents, ergonomics, safety, computer, and professional development.

American Industrial Hygiene Association

American Industrial Hygiene Association
2700 Prosperity Avenue, Suite 250
Fairfax, VA 22031-4319
703-849-8888
E-mail: infonet@aiha.org
Web site: http://www.aiha.org

AIHA is a professional organization of industrial hygienists and allied specialists. It publishes the *AIHA Journal* and extensive literature on all phases of industrial hygiene and promotes information on career opportunities in industrial hygiene.

American Medical Student Association

American Medical Student Association
Environmental and Occupational Health Interest Group
1902 Association Drive
Reston, VA 20191
703-620-6600
E-mail: amsa@www.amsa.org
Web site: http://views.vcu.edu/amsa/namsa/eoh

AMSA develops and publicizes educational materials and facilitates student participation in occupational and environmental health projects.

American National Standards Institute

American National Standards Institute
11 West 42nd Street
New York, NY 10036
212-642-4900
Web site: http://www.ansi.org

ANSI coordinates the voluntary development of national standards. It serves as a clearinghouse and information center for national and international safety standards. ANSI is the official U.S. member of the International Organization for Standardization (ISO) and the International Electro-Technical Commission (IEC), via the U.S. National Committee.

American Public Health Association

American Public Health Association
  Occupational Health and Safety Section
  800 I Street, NW
  Washington, DC 20001-3710
  202-777-APHA
  E-mail: comments@apha.org
  Web site: http://www.apha.org
APHA presents numerous sessions on occupational health and other topics at its annual meeting each fall, develops public policy
statements on occupational health and safety
issues, and publishes a newsletter. It develops links between occupational health and
public health.

American Society for Safety Engineers

American Society for Safety Engineers
  Customer Service
  1800 East Oakton St.
  Des Plaines, IL 60018
  847-699-2929
  E-mail: customerservice@asse.org
  Web site: http://www.asse.org
ASSE publishes *Professional Safety;* supports safety professionals in accident, injury,
and illness prevention; and offers certification and training, technical publications, and
an annual convention.

Human Factors and Ergonomics Society

Human Factors and Ergonomics Society
  Post Office Box 1369
  Santa Monica, CA 90406
  310-394-1811
  E-mail: hfes@compuserve.com
  http://hfes.org/HFES.html
This organization of ergonomics and human factors professionals holds an annual
meeting, and its technical groups also present
regular meetings. It publishes *Human Factors, Ergonomics in Design,* and the proceedings of its annual meeting. Its Industrial Ergonomics Technical Group is of special
interest.

National Safety Council

National Safety Council
  1121 Spring Lake Drive
  Itasca, IL 60143-3201
  630-285-1121
  Web site: http://www.nsc.org/index.htm
NSC is a not-for-profit, nongovernmental,
international public service organization that
offers occupational safety and health training
programs and literature. The Council
Library is available for safety and health
research, and experts in safety, industrial
hygiene, and occupational health offer consultations.

Society for Occupational and Environmental Health

Society for Occupational and Environmental Health
  6728 Old McLean Village Drive
  McLean, VA 22101-3906
  703-556-9222
  E-mail: soeh@soeh.org
  Web site: http://www.soeh.org
SOEH seeks to improve the health quality
of the workplace by holding open forums
that focus public attention on scientific, social and regulatory problems.

Occupational Health Clinics

Association of Occupational and Environmental Clinics
  1010 Vermont Avenue, NW, Suite 513
  Washington, DC 20005
  202-347-4976
  E-mail: aoec@aoec.org
  Web site: http://occ-env-med.mc.duke.
    edu/oem/aoec.html
AOEC is an association of clinics organized to aid in identifying, reporting, and
preventing occupational and environmental
health hazards nationwide; to increase communication among such clinics concerning
issues of patient care; and to provide data
for occupational and environmental research projects.

Committees on Occupational Safety and Health (COSH Groups)

COSH groups are grassroots, voluntary advocacy organizations in the United States consisting of health professionals, legal professionals, and labor representatives who work to improve health and safety in the workplace through education and policy change.

COSH Groups

ALASKA
  Alaska Health Project
  90 Anchorage, AK 99501
  907-276-2864

CALIFORNIA
  Worksafe/Fran Schrieberg
  San Francisco, CA 94109
  415-433-5077

  LACOSH (Los Angeles COSH)
  Los Angeles, CA 90019
  213-931-9000

  SA-COSH (Sacramento COSH)
  Sacramento, CA 95820
  916-442-4390

  SCCOSH (Santa Clara COSH)
  San Jose, CA 95112
  408-998-4050

CONNECTICUT
  ConnectiCOSH (Connecticut)
  Hartford, CT 06106
  860-549-1877

ILLINOIS
  CACOSH (Chicago area)
  Chicago, IL 60612-7260
  312-996-2747

MAINE
  Maine Labor Group on Health
  Augusta, ME 04330
  207-622-7823

MARYLAND
  Alice Hamilton Occupational Health Center
  Silver Spring, MD 20910-3354
  301-565-4590

MASSACHUSETTS
  MassCOSH
  Boston, MA 02130
  617-524-6686

  Western MassCOSH
  Springfield, MA 01103
  413-731-0760

MICHIGAN
  SEMCOSH (Southeast Michigan)
  Detroit, Ml 48216
  313-961-3345

MINNESOTA
  MnCOSH
  Minneapolis, MN 55430
  612-572-6997

NEW HAMPSHIRE
  NHCOSH
  Pembroke, NH 03275
  603-226-0516

NEW YORK
  ALCOSH (Alleghany COSH)
  Jamestown, NY 14701
  716-448-0720

  CNYCOSH (Central New York)
  Syracuse, NY 13204
  315-471-6187

  ENYCOSH (Eastern New York)
  Schenectady, NY 12305
  518-372-4308

NYCOSH (NYC Metro)
New York, NY 10001
212-627-3900

ROCOSH (Rochester COSH)
Rochester, NY 14607
716-244-0420

WNYCOSH (Western NY)
Buffalo, NY 14214
716-833-5416

NORTH CAROLINA
NCCOSH
Durham, NC 27715
919-286-9249

OREGON
ICWUPortland
Portland, OR 07223
503-244-8429

PENNSYLVANIA
PhilaPOSH
Philadelphia, PA 19104
215-386-7000

RHODE ISLAND
RICOSH
Providence, RI 02903
401-751-2015

TENNESSEE
TNCOSH
Maryville, TN 37801
615-983-7864

TEXAS
TexCOSH
Beaumont, TX 77706
409-898-1427

WASHINGTON
WASHCOSH
Seattle, WA 98121
206-762-8337

WISCONSIN
WisCOSH (Wisconsin COSH)
Milwaukee, Wl 53230
414-933-2338

CANADA
Ontario
WOSH (Windsor OSH)
Windsor, Ontario N9A 4N1
519-973-4800

COSH-Related Groups

CALIFORNIA
Labor Occupational Health Program
(LOHP)
Berkeley, CA 94720-5120
510-642-5507

UCLA-Labor Occupational Safety and
Health (LOSH) Program
Los Angeles, CA 90095-1478
310-794-5964

DISTRICT OF COLUMBIA
Workers Institute for Occupational Safety
and Health
Washington, DC 20036
202-887-1980

LOUISIANA
Labor Studies Program/LA Watch
New Orleans, LA 70118
504-861-5830

MASSACHUSETTS
Massachusetts Coalition on New Office
Technology (CNOT)
Boston, MA 02215
617-247-6827

MICHIGAN
Michigan Right-to-Act Campaign
Ann Arbor, MI 48104
734-663-2404

NEW JERSEY
New Jersey Work Environment Council
Moorestown, NJ 08057
609-866-9405

NEW YORK
Midstate Central Labor Coalition
Ithaca, NY 14850
607-277-5670

OHIO
Greater Cincinnati Occupational Health
Center
Cincinnati, OH 45220
513-569-0561

WEST VIRGINIA
Institute for Labor Studies and Research
West Virginia University
Morgantown, WV 26506
304-293-3323

CANADA
Windsor Occupational Health Information Service
Windsor, Ontario N9A 4N1
519-254-5157

ENGLAND
London Hazards Centre
London, NW5 3NQ, England
(0171) 267-3387

WHIN—Workers Health International
Newsletter and HAZARDS Magazine
Sheffield, S1 4YL, England
(+44 14) 276-56-95

University Labor Education Programs

*University and College Labor
Education Association*

University of California-Los Angeles
Center for Labor Research and Education
Attn: Kent Wong
Institute of Industrial Relations
School of Public Policy and Social Re-
search

6350 B Public Policy Building, Box 951478
Los Angeles, CA 90095-1478
The UCLEA is an organization of more
than 60 academic-based labor education programs in the United States and Canada. Expertise in health and safety varies among the
programs, but they can be good sources of
background information on general labor issues and can help in the design of worker
education programs.

**University Professional
Training Programs**

*Southern California Center for
Occupational and
Environmental Health*

Southern California Center for Occupational
and Environmental Health
UCLA School of Public Health
Box 951772
Los Angeles, CA 90095-1772
310-206-6920
Fax: 310-206-9903
E-mail: coeh@ucla.edu
Web site: http://www.ph.ucla.edu/coeh
Located on the campus of UCLA, this
COEH conducts occupational and environ-
mental training and research in the areas of
industrial hygiene, medicine, toxicology, epi-
demiology, ergonomics, health education,
and policy.

*UC Irvine Center for Occupational and
Environmental Health*

UC Irvine Center for Occupational and En-
vironmental Health
College of Medicine
19722 MacArthur Boulevard
Irvine, CA 92715
714-824-8690
Fax: 714-824-2345
E-mail: dbbaker@uci.edu
Located on the campus of UC Irvine, this
COEH conducts occupational training and
research in the areas of industrial hygiene,
medicine, toxicology and epidemiology.

Educational Resource Centers

NIOSH funds 15 university programs that train occupational health professionals, conduct research, and provide services in this area. A number of other programs have been awarded NIOSH training funds to support graduate studies in the various disciplines (occupational medicine, occupational nursing, industrial hygiene, occupational ergonomics, and occupational safety). A complete list of these programs can be obtained from NIOSH.

ALABAMA
  Deep South Center for Occupational Health and Safety
  University of Alabama at Birmingham
  School of Public Health
  Birmingham, AL 35294
  Web site: http://www.uab.edu/dsc

CALIFORNIA—NORTHERN
  Northern California Center for Occupational and Environmental Health
  University of California, Berkeley, Richmond Field Station
  Richmond, CA 94804

CALIFORNIA—SOUTHERN
  Southern California Education and Research Center
  University of Southern California
  Los Angeles, CA 90033

CINCINNATI
  NIOSH Education and Research Center
  University of Cincinnati
  Cincinnati, OH 452670056
  Web site: http://www.uc.edu/www/envhealth/conted

FLORIDA
  Florida Educational Resource Center
  University of South Florida
  College of Public Health
  Department of Environmental and Occupational Health
  Tampa, FL 33612-3805

HARVARD
  Harvard Education and Research Center
  Center for Continuing Professional Education
  Harvard School of Public Health
  Boston, MA 02115-6023
  Web site: http://www.sph.harvard.edu/ccpe/ccpe.html

ILLINOIS
  Great Lakes Center for Occupational and Environmental Safety and Health
  University of Illinois at Chicago
  School of Public Health
  Chicago, IL 60612-7260

JOHNS HOPKINS
  Johns Hopkins Education and Research Center
  Johns Hopkins University
  School of Hygiene and Public Health
  Baltimore, MD 21205

MICHIGAN
  Michigan Education and Research Center
  University of Michigan
  Center for Occupational Health and Safety Engineering
  Ann Arbor, MI 48109-2117
  Web site: http://www.engin.umich.edu/dept/ioe/COHSE/

MINNESOTA
  Minnesota Education and Research Center
  Midwest Center for Occupational Safety and Health
  St. Paul, MN 55101
  Web site: http://www.umn.edu/mcohs

NEW YORK/NEW JERSEY
  New York/New Jersey Education and Research Center
  EOHSI Centers for Education and Training
  Piscataway, NJ 08854-3923
  Web site: http://www.eohsi.rutgers.edu/cet

NORTH CAROLINA

North Carolina Education and Research Center
University of North Carolina at Chapel Hill
Chapel Hill, NC 27514

TEXAS

Southwest Center for Occupational and Environmental Health
The University of Texas Health Science Center at Houston
School of Public Health
Houston, TX 77030
Web site: http://www.sph.uth.tmc.edu/www/ctr/swcoeh/swcoeh01.htm

UTAH

University of Utah Education and Research Center
Rocky Mountain Center for Occupational and Environmental Health
University of Utah
Salt Lake City, UT 84112
Web site: http://www.rmcoeh.utah.edu

WASHINGTON

Northwest Center for Occupational Health and Safety
Department of Environmental Health
University of Washington
Seattle, WA 98105
Web site: http://cucme.sphcm.washington.edu/DEHWEB/index.html

Agricultural Research Centers

High Plains Intermountain Center for Agricultural Health and Safety
133 Environmental Health Bldg
Colorado State University
Fort Collins, CO 80523-1681
800-622-8673
E-mail: hicahs@lamar.colostate.edu
States served: CO, MO, WY, ND, SD, UT

Pacific Northwest Agricultural Safety and Health Center
Department of Environmental Health
Box 357234
University of Washington
Seattle, WA 98195-7234
800-330-0827
E-mail: pnash@u.washington.edu
States served: WA, ID, OR, AL

Northeast Center for Agricultural Medicine and Health
One Atwell Road
Cooperstown, NY 13326
607-547-6023
E-mail: nycamh@lakenet.org
States served: NY, MA, RI, CT, VT, DE, ME, NH, NJ, MD, PA, VA, WV

Southeast Center for Agricultural Health and Injury Prevention
Department of Preventive Medicine
University of Kentucky
1141 Red Mile Road, Suite 102
Lexington, KY 40504-9842
606-323-6836
E-mail: lcharles@pop.uky.edu
States served: KY, GA, NC, SC, TN

Great Plains Center for Agricultural Health
Institute for Rural and Environmental Health
100 Oakdale Campus, IREH #103
The University of Iowa
Iowa City, IA 52242-5000
319-335-4887
E-mail: risto-rautiainen@uiowa.edu
States served: IA, NE, MS, KS

Midwest Center for Agricultural Disease and Injury Research, Education and Prevention
1000 North Oak Avenue
Marshfield, WI 54449-5790
800-662-6900
E-mail: youngn@mfldclin.edu
States served: WI, OH, IL, MI, MN, IN

UC Agricultural Health and Safety Center at Davis
  University of California at Davis
  Davis, CA 95616
  530-752-5253
  E-mail: ewwood@ucdavis.edu
  States served: CA, NV, AZ, HA

Deep-South Center for Agricultural Disease and Injury Research, Education and Prevention
  University of South Florida
  College of Public Health
  13201 Bruce B. Downs Boulevard
  Tampa FL 33612-3805
  813-974-6625
  E-mail: yhammad@com1.med.usf.edu
  States served: FL, MS, AL

Southwest Center for Agricultural Health, Injury, Prevention and Education
  University of Texas Health Center at Tyler
  Post Office Box 2003
  Highway 271 at Highway 155
  Tyler, TX 75710
  903-877-5896
  E-mail: gilmore@uthct.edu
  States served: TX, OK, NM, LA, AR

Industry-Sponsored Advisory and Research Groups

American Petroleum Institute

American Petroleum Institute
  1220 L Street, NW
  Washington, DC 20005
  202-682-8000
  Web site: http://www.api.org
The institute conducts research programs on occupational health and industrial hygiene aspects of all phases of the petroleum industry. These projects may result in manuals, guides, or research reports, which are available to interested parties.

American Welding Society

American Welding Society
  550 NW LeJeune Road
  Miami, FL 33126
  800-443-9353
  Web site: http://www.aws.org
The society offers seminars and home-study courses designed specifically to relate welding and cutting operations to plant environmental safety and health.

Chemical Industry Institute of Toxicology

Chemical Industry Institute of Toxicology
  6 Davis Drive
  Post Office Box 12137
  Research Triangle Park, NC 27709-2137
  919-558-1200
  E-mail: CIITinfo@ciit.org
  Web site: http://www.ciit.org
CIIT conducts basic toxicologic research to provide an improved scientific basis for understanding and assessing the potential adverse effects of chemicals, pharmaceuticals, and consumer products on human health. It acquires, interprets, and disseminates technical information and test data. CIIT trains toxicologists and scientists in related fields.

Chemical Manufacturers Association

Chemical Manufacturers Association
  13 Wilson Boulevard
  Arlington, VA 22209
  703-741-5122
  http://www.cmahq.com/
The association publishes monthly newsletter, *ChemEcology*. It has a team that deals with occupational safety and health matters. The association has approximately 200 member companies. It sponsors projects and occasionally publishes educational materials.

The Chlorine Institute

The Chlorine Institute
2001 L Street, NW, Suite 506
Washington, DC 20036
202-775-2790
Web site: http://cl2.com
The institute promotes safety in the manufacturing and handling of chlorine products. It provides educational materials to more than 200 member companies and to the public.

Industrial Health Foundation

Industrial Health Foundation
34 Penn Circle West
Pittsburgh, PA 15206-3612
412-363-6600
E-mail: ihfincorp@aol.com
Web site: http://members.aol.com/
ihfincorp/ihf.htm
The foundation conducts engineering plant visits to review industrial hygiene and health and safety aspects, publishes extensive abstracts, provides toxicology consultation, conducts training courses in occupational health and safety, and performs analytical testing of environmental and biologic samples.

International Lead Zinc Research Organization

International Lead Zinc Research Organization
2525 Meridian Parkway
Post Office Box 12036
Research Triangle Park, NC 27709-2036
919-361-4647
E-mail: rputnam@ilzro.org
Web site: http://www.ilzro.org
The organization conducts occupational and environmental health research on such subjects as pediatric lead absorption, occupational lead and cadmium exposure, and biologic interactions. It publishes an annual lead/cadmium research digest.

Plastics Education Foundation

Plastics Education Foundation
14 Fairfield Drive
Brookfield, CT 06804
203-775-0471
Web site: http://www.4spe.org/PEF.
HTML
This foundation, within the Society for Plastics Engineers, provides audiovisual materials, including films and video tapes, on occupational health and safety in the plastics and manufacturing industries.

Semiconductor Industry Association

Semiconductor Industry Association
181 Metro Drive, Suite 450
San Jose, CA 95110
408-436-6600
Web site: http://www.semichips.org/
index.htm
The mission of the SIA is to provide leadership for U.S. chip manufacturers on the critical issues of trade, technology, environmental protection, and worker safety and health.

Labor-Oriented Advisory and Research Groups

American Labor Education Center

American Labor Education Center
2000 P Street NW, Suite 300
Washington, DC 20036
202-828-5170
E-mail: amlabor@aol.com
ALEC provides education and educational materials to workers about health and safety.

Center for Safety in the Arts

Center for Safety in the Arts
Mailbox 310
2123 Broadway
New York, NY 10023
E-mail: csa@artswire.org

Web site: http://artsnet.heinz.cmu.edu:70/
0/csa/resource/csainfo

CSA provides information on hazards in the visual arts, the performing arts, children's and school arts programs, and museums, as well as general health and safety information and laws relevant to the arts. It produces fact sheets, video tapes, a newsletter (*Art Hazards News*), educational programs, and consultations on hazards in the arts.

Center to Protect Workers' Rights

Center to Protect Workers' Rights
  111 Massachusetts Avenue, NW, 5th Floor
  Washington, DC 20001
  202-962-8490
  E-mail: cpwr@cpwr.com
  Web site: http://www.cpwr.com/

The nonprofit research arm of the Building and Construction Trades Department, AFL-CIO, CPWR focuses on construction worker safety and health. Activities include labor-management cooperation, mainly in safety planning and training; medical surveillance of occupational disease, lost-time injuries, and deaths; interventions to reduce the hazards and exposures; analysis of government and industry data to provide benchmarks; and the application of ergonomic and other new technologies.

Environmental Defense Fund

Environmental Defense Fund
  257 Park Avenue South
  New York, NY 10010
  212-505-2100
  E-mail: Contact-EDF@edf.org
  Web site: http://www.edf.org

EDF addresses a broad range of regional, national, and international environmental issues. It provides environmental information by mail and electronically. EDF participates in environmental education projects; maintains a Member Action Network to influence national environmental policy; and works with grassroots groups at the local and regional levels in the United States and abroad.

It publishes a bimonthly newsletter, annual report, books and reports, news releases, committee testimony, environmental fact sheets, and brochures.

Farmworker Health and Safety Institute

Farmworker Health and Safety Institute
  4 South Delsea Drive South
  Glassboro, NJ 08028
  609-881-2507

FHSI is a consortium of the George Washington University School of Medicine, CATA (Farmworkers Support Committee), the Farmworker Association of Central Florida, and the Farmworker Justice Fund. It trains clinicians and farmworkers about farmworker health and safety issues, including pesticides, workplace injuries, and women's health, using "popular education" methods. FHSI develops training materials in English and Spanish at appropriate literacy levels.

Public Citizen

Public Citizen
  Health Research Group
  1600 20th Street, NW
  Washington, DC 20009
  202-588-1000
  Web site: http://www.citizen.org/HRG/

This a public interest group that studies many subjects, including occupational health hazards, and disseminates information to the working public. It produces numerous publications and other informational resources that are of value to health sciences students, physicians, and other health professionals. In certain circumstances, it can also investigate specific hazardous situations.

Highlander Research and Education Center

Highlander Research and Education Center
  1959 Highlander Way
  New Market, TN 37820
  423-933-3443

E-mail: hrec@igc.apc.org

This major labor education center, among other projects, conducts educational conferences on occupational safety and health for labor activists, health science students, and others.

The Labor Institute

The Labor Institute
853 Broadway, Suite 2014
New York, NY 10003
212-674-3322
E-mail: laborinst@aol.com

The institute specializes in developing economic analyses of the problems workers face and presenting them in easy-to-understand language from the workers' point of view. It has produced workbooks and video tapes on Hazardous Waste Training, Hazards of the Modern Office, Electromagnetic Fields, Sexual Harassment, and Multiple Chemical Sensitivities.

Migrant Legal Action Program

Migrant Legal Action Program
2001 S Street, NW, Suite 310
Washington, DC 20009-1125
202-462-7744

MLAP is a Legal Services Corporation–funded national support center. It represents migrant and seasonal farmworkers in numerous areas, including health and safety, housing, health care, and child labor.

Occupational Health Foundation

Occupational Health Foundation
815 16th Street, NW, Suite 312
Washington, DC 20006-4104
202-842-7840

This is a labor-supported organization that provides technical, scientific, and educational services in occupational health and safety to unions. It responds to requests for information.

9to5 National Association of Working Women

9to5 National Association of Working Women
1245 East Colfax Avenue, Suite 211
Denver, CO 80218
303-866-0925 or 800-522-0925
Web site: http://www.quikpage.com/9/
9to5naww/

This national association works to win rights and respect for female workers. It is engaged in a specific effort to address health and safety problems of office workers with research and education. 9to5's Job Survival Hotline provides callers with free and confidential information, resources, and referrals in English and Spanish concerning sexual harassment, job discrimination, work and family issues, and other workplace problems.

International Organizations

International Commission on Occupational Health

International Commission on Occupational Health
c/o Secretary General, ICOH
Department of Community, Occupational and Family Medicine
Faculty of Medicine MD3
National University of Singapore
Lower Kent Ridge Road
Singapore 119260
Web site: http://www.icoh.org.sg/
ICOHE1.htm

ICOH is an international scientific society that fosters scientific progress, knowledge, and development of all aspects of occupational health on an international basis. It holds a triennial international congress, maintains scientific committees, and collects and disseminates information on occupational health.

International Labor Organization

International Labor Organization
  Occupational Safety and Health Branch
    Secretariat
  International Labor Office
  4 route des Morillons
  1211 Geneva 22
  Switzerland
  Web site: http://www.ilo.org/public/
    english/90travai/sechyg
  Washington branch:
  1828 L Street, NW, Suite 801
  Washington, DC 20036
  Web site: http://www.ilo.org/publns (for
    ILO publications)

ILO focuses on the prevention of occupational injuries and diseases; promotion of safety, health, and well-being in all occupations; and identification and elimination of problems of working environments. Involvement in occupational safety and health is articulated in its International Programme for the Improvement of Working Conditions and Environment (PIACT). Modes of action include adoption of international conventions and recommendations, establishment of model codes of practices, convening of tripartite meetings, collection and dissemination of information (including publication of the *Encyclopedia of Occupational Health and Safety*), and technical cooperation.

World Health Organization

World Health Organization
  Avenue Appia 20
  1211 Geneva 27
  Switzerland
  (2D41 22) 791-21-11
  E-mail: info@who.ch
  Web site: http://www.who.ch

WHO offers a wide range of services in occupational health and safety. It publishes documents and sponsors a variety of educational programs.

Labor Organizations

The following list of labor organizations is not all-inclusive but includes some of those with the most active health and safety departments.

American Federation of Labor and Congress of Industrial Organizations

AFL-CIO
  815 16th Street, NW
  Washington, DC 20006
  202-637-5206
  Web site: http://www.aflcio.org
  Health and Safety Office:
  202-637-5366

This is a confederation of labor unions that serves as their advocate in health and safety matters. The Building and Construction Trades Department and the Food and Allied Services Trades Department work extensively on safety and health issues for unions in their respective sectors.

Building and Construction Trades Department
  1155 15th Street, NW, 4th Floor
  Washington, DC 20005
  202-347-1461
  E-mail: HB4@compuserve.com
  Web site: http://www.buildingtrades.org/

Food and Allied Services Trades Department
  815 16th Street, NW
  Washington, DC 20006
  202-737-7200
  E-mail: fast3@igc.org
  Web site: http://www.fastaflcio.org/

AFL-CIO Affiliates

Some of the affiliated unions with the most active health and safety departments are listed here.

American Federation of Government Employees (AFGE)
  80 F Street, NW
  Washington, DC 20001

202-639-6406
E-mail: paystobe@afge.org
Web site: http://www.afge.org

American Federation of State, County and
Municipal Employees (AFSCME)
1625 L Street, NW
Washington, DC 20036
202-429-1233
Web site: http://www.afscme.org

Communications Workers of America
(CWA)
501 3rd Street, NW
Washington, DC 20001-2797
202-434-1160
E-mail: cwaweb@earthlink.net
Web site: http://www.cwa-union.org

International Chemical Workers Union
(ICWU)/ United Food and Commercial
Workers International Union (UFCW)
1655 West Market Street
Akron, OH 44313
330-867-2444
E-mail: 104525.706@compuserve.com

International Union of Automobile, Aerospace and Agricultural Implement Workers (UAW)
8000 East Jefferson Avenue
Detroit, MI 48214
313-926-5000
Web site: http://www.uaw.org

Paper, Allied-Industrial, Chemical & Energy
Workers Union (PACE)
3340 Perimeter Hill Drive
Nashville, TN 37211
615-834-8590
Web site: http://www.upiu.org

Service Employees International Union
(SEIU)
1313 L Street, NW
Washington DC, 20005
202-898-3385
E-mail: webworker@seiu.org
Web site: http://www.seiu.org

United Mine Workers of America (UMWA)
900 15th Street, NW
Washington, DC 20005
202-842-7300
Web site: http://www.ACCESS.digex.net/
~miner/

Union of Needletrades, Industrial and Textile Employees (UNITE!)
275 7th Avenue, 6th Floor
New York, NY 10001
212-691-1691
E-mail: webmaster@uniteunion.org
Web site: http://www.uniteunion.org

Additional Resources on the World Wide Web

Duke University Occupational and Environmental Medicine
Web site: http://occ-env-med.mc.duke.
edu/oem
The Occup-Env-Med Mail-list provides a
forum for announcements, dissemination of
text files, and academic discussion. The forum is designed to allow presentation of clinical vignettes, synopses of new regulatory issues, and reports of interesting items from
publications elsewhere (both the medical
and the nonmedical journals). For more information on how to subscribe to the Duke
University occupational and environmental
e-mail list serve, visit the website: http://
gilligan.mc.duke.edu/oem/occ-env-.htm.

Occupational Health: Recognizing and Preventing Work-Related Disease and Injury, 4th ed.
Edited by Barry S. Levy and David H. Wegman, Lippincott Williams & Wilkins, Philadelphia © 2000.

Subject Index